ENVIRONMENTAL GEOLOGY
OF
URBAN AREAS

GEOTEXT 3

ENVIRONMENTAL GEOLOGY
OF
URBAN AREAS

Edited by

Nicholas Eyles

Environmental Earth Sciences
University of Toronto, Scarborough Campus
Scarborough, Ontario M1C 1A4

Geological Association of Canada

1997

Canadian Cataloguing in Publication Data

Main entry under title:
    Environmental geology of urban areas

(GEOtext, ISSN 1208–2260 ; 3)
Includes bibliographic references.
ISBN 0–919216–58–7

    1. Environmental geology. 2. Metropolitan areas--
Environmental aspects--Canada. 3. Urban ecology--Canada.
I. Eyles, Nicholas II. Geological Association of Canada
III. Series.

QE39.5.U7E48 1996   551.1′09173′2   C96–900933–X

PUBLISHER:

Geological Association of Canada
c/o Department of Earth Sciences
Memorial University of Newfoundland
St. John's, Newfoundland
A1B 3X5 Canada

The GEOLOGICAL ASSOCIATION OF CANADA (GAC) is Canada's national society for the geosciences. It was established in 1947 to advance geology and its understanding among both professionals and the general public. The GAC membership of 2500 includes representatives of all geological disciplines from across Canada and many parts of the world employed by government, industry and academia. The Association was incorporated under the Canada Corporations Act in January 1974.

There are twelve specialist divisions: Canadian Geomorphology Research Group, Canadian Sedimentology Research Group, Environmental Earth Sciences, Geographic Information Systems, Geophysics, Marine Geosciences, Mineral Deposits, Paleontology, Precambrian, Remote Sensing, Structural Geology and Tectonics, and Volcanology and Igneous Petrology. Regional sections have been set up in Victoria, Vancouver, Edmonton, Winnipeg and St. John's.

Activities of the Geological Association of Canada include the organization and sponsorship of conferences (Annual General Meeting, NUNA Research Conferences, Field Conferences), seminars, short courses, field trips, lecture tours, and student and professional awards and grants. The Association publishes the quarterly journal *Geoscience Canada*, the quarterly newsmagazine *GEOLOG*, the *Special Paper* series of research volumes, the *Geoscience Canada Reprint Series* of reference books, the new *GEOtext* series of textbooks, short course notes, and other miscellany including guidebooks and division newsletters.

GAC is associated with the Atlantic Geoscience Society (AGS), the Canadian Geophysical Union (CGU), the Canadian Quaternary Association (CANQUA), the Canadian Society of Petroleum Geologists (CSPG) and the Toronto Geological Discussion Group (TGDG). GAC is also affiliated with the Mineralogical Association of Canada (MAC) and the Canadian Geoscience Council (CGC).

For information contact: Geological Association of Canada, c/o Department of Earth Sciences, Memorial University of Newfoundland, St. John's, Newfoundland A1B 3X5 Canada.

L'ASSOCIATION GÉOLOGIQUE DU CANADA (AGC) est la société professionelle canadienne dans le domaine des géosciences. Elle a été fondée en 1947 pour la promotion de la géologie et l'avancement des connaissances géologiques, autant chez les professionnels que pour le grand public. Ses 2500 membres proviennent de toutes les disciplines géologiques du pays et d'ailleurs dans le monde, et ils oeuvrent aussi bien dans le secteur gouvernemental, universitaire que dans le secteur des sociétés privées de l'industrie. Elle a été constituée officiellement en janvier 1974 selon la Loi concernant les corporations canadiennes.

Elle compte douze divisions selon les spécialités qui suivent : Groupe canadien de recherche en géomorphologie, Group canadien de recherche en sédimentologie, sciences géo-environnementles, systèmes d'information géographique, géophysique, géosciences marines, gîtes minéraux, paléontologie, géologie du Précambrien, télédétection, géologie structurale et tectonique, et volcanologie et pétrologie ignée. Des sections régionales ont été mises sur pied à Victoria, à Vancouver, à Edmonton, à Winnipeg et à St-John's.

L'Association géologique du Canada organise et parraine la tenue de congrès (réunion générale annuelle, conférences scientifiques NUNA, ateliers-congrès sur le terrain), des séminaires, des cours intensifs, des excursions, des séries de conférences, et offre des bourses et des prix au étudiants comme aux praticiens méritants. L'Association publie une revue professionnelle trimestrielle *Geoscience Canada*, un bulletin d'information trimestriel *GEOLOG*, la collection de monographies scientifiques *Publications spéciales*, la collection d'ouvrages de référence *Rééditions de Geoscience Canada*, la nouvelle collection de manuels scolaires *GEOtext*, des notes de cours intensifs et diverses autres publications telles des livrets-guides et des bulletins de liaison pour ses divisions.

L'AGC est associée à la Société géoscientifique de l'Atlantique (SGA), à l'Union canadienne de Géophysique (UCG), l'Association canadienne pour l'étude du Quaternaire (CANQUA), la Société canadienne des géologues du pétrole (SCGP), et le Groupe de discussion géologique de Toronto (GDGT). L'AGC est également affiliée à l'Association minéralogique du Canada (AMC) et au Conseil géoscientifique canadien.

Pour de plus amples information, adressez-vous à : Association géologique du Canada, c/o Department of Earth Sciences, Memorial University of Newfoundland, St. John's, Newfoundland A1B 3X5 Canada.

Typesetting: Editorial Office [Sudbury, ON] of the Geological Association of Canada, St. John's, NF
Printing: Litho Acme Quebec Inc. /Transcontinental Printing Inc.   PRINTED IN CANADA

CONTENTS

PREFACE

This volume was conceived at a conference held at the Ontario Science Centre, Toronto, in February, 1994. The meeting was sponsored by the Geological Association of Canada (GAC) through its Canadian Sedimentological Research Group (CSRG). The Ontario Science Centre has a well-deserved reputation for scientific outreach to the public initiated under its first director, the late and great geologist, J. Tuzo Wilson. It was, therefore, a very appropriate locale for a meeting of academics, agency workers, consultants and the public sharing a common interest in the environmental geology of urban areas. Indeed, it was the first-ever GAC-sponsored meeting to which the public were invited as active participants and presenters. Dr. Emlyn H. Koster, Director of the Ontario Science Centre (also a geologist) placed the outstanding facilities of the Centre at our disposal, Vic Tyrer, geologist and Program Coordinator at the Science Centre, gave crucial logistical help and Dr. Darrel Long, President of the CSRG, provided encouragement and, most importantly, secured funds to allow publication of this book.

There are 39 papers in this volume representing the efforts of 67 contributors from the consulting sector, academia and government. Most of the papers are review articles that bring together original data previously found only in the grey literature of unpublished reports and documents and never previously assembled in one source. It is hoped that the volume will illustrate what environmental geologists do for a living and stimulate students to choose a career in this new and rapidly expanding field.

I thank the contributors for their timely submission of manuscripts and their willingness to readily accept technical and editorial suggestions, and the external reviewers for their comments. This volume would not have been possible without the skilled assistance of Judith James who completed the mammoth task of copy editing the entire volume, thereby ensuring a consistent format and style. Leslie King provided typesetting, and Monica Easton gave valuable editorial advice and assistance and provided that critical link with the printer during the final stages of the project. Completion of this volume is due in no small measure to Mike Doughty, who computer draughted many of the figures, and to the advice, encouragement and support of Carolyn Eyles.

Nick Eyles
Toronto, Ontario

1. Environmental Geology of Urban Areas

Nicholas Eyles

Environmental Earth Sciences, University of Toronto, Scarborough Campus
1265 Military Trail, Scarborough, Ontario M1C 1A4

INTRODUCTION

This volume is primarily intended to be a source of readings for senior undergraduate and graduate students. Many of the papers are state-of-the art summaries and will also be essential reading for professional environmental geologists already working in urban areas. The future well-being of urban society is reliant on an informed and proactive citizenry and it is hoped that the lay reader, urban planner, and business person can also learn something of the environmental geological problems associated with urban areas. For this reason, an extensive glossary of commonly used terms is provided. This book will be an aid to the teaching of academic courses and also be of practical use. Above all, we hope it will inform, interest and stimulate the reader.

ENVIRONMENTAL GEOLOGY

Geology has a long record as an academic discipline, and geologists play a key role in society. However, in recent years the focus of geological applications has broadened. Environmental geology has emerged as a distinct field, and traditional discipline boundaries have dissolved. There have been major changes in employer needs and job markets.

Environmental geology can be formally defined as the application of geological science to practical issues relating to human activity (Chan *et al.,* 1987; Dunlap *et al.,* 1992; National Research Council, 1993, 1994; Montgomery, 1994; Innes Lumsden, 1994; Figs. 1, 2). The central characteristic that distinguishes environmental geologists is their strongly interdisciplinary approach, drawing on such areas as chemistry, physics, biology and, increasingly, economics, urban planning, and the law. This is reflected in changing curricula in universities and in the integration of geology with other disciplines formerly taught as separate entities. A training rooted only in geology is no longer sufficient and there is now increasing emphasis on wider systems approaches to education and training (what has been called "earth system science"; National Research Council, 1993).

URBAN GROWTH
AND ENVIRONMENTAL DEGRADATION:
A FOCUS FOR ENVIRONMENTAL GEOLOGY

On a global scale, the combined forces of urbanization and a rapidly expanding chemical industry have been responsible for great improvements in the quality of life over the last 150 years. Ironically, these same forces now threaten the global environment. Urban areas are widely characterized by intense and often unregulated industrial activity, rapid and poorly planned growth, the fragmentation of natural habitats, and the degradation of surface and ground waters by a wide range of chemical contaminants.

More than 70% of the world's population now lives in urban areas and there are more than 20 so-called "supercities," each containing more than 10% of their respective national populations. Examples include Mexico City (31% of national population), Buenos Aires (42%), Cairo (36%), and Sao Paulo (17%; Turner *et al.,* 1990; United Nations Population Fund, 1991). Whereas the growth of cities is the engine of the world economy and generates enormous social benefits by concentrating human creativity, most global environmental problems of the late 20th century can be linked directly to the massive transfers of resources and waste products required by large cities. Terms such as "urbanization," "economic imbalance," and "environmentally unsustainable" are becoming synonomous (Tabibzadeh *et al.,* 1989; Stren *et al.,* 1992; Socolow *et al.,* 1994; Drakakis-Smith, 1995). In 1950, there were 83 cities with populations greater than 1 million; today, there are more than 280 cities and this total will double by 2015. The ten largest supercities are listed in Figure 3.

Whereas geologists have a long history of working in cities alongside geotechnical engineers (*e.g.,* Legget, 1973; Leveson, 1980), the artificial, built landscapes of urban areas have, by and large, been seen seen as "poor places to do geology" (Walton, 1982). Traditional approaches to "urban geology" have focussed largely on the engineering behaviour of the

---

Geology: The study of planet Earth

Environment: The sum of all the features and conditions surrounding an organism that may influence it

Environmental Geology: Since planet Earth provides the fundamental physical environment for organisms, then all of geology can be regarded as environmental geology. However, the term is increasingly restricted to:

- Understanding geologic processes as they relate directly to human activities

- Identifying constraints imposed on human activities by geologic processes, availability of resources, and disposal of wastes

Figure 1 *Definition of environmental geology (after Montgomery, 1994).*

various geological materials below cities in terms of their stability for roads, foundations, *etc.* Provision of adequate construction materials has also been a major concern. Despite, however, a wealth of subsurface geological and hydrogeological data in urban areas, there has been little effort in Canada to systematically collect and consolidate such information by government agencies; mining and other resource-based projects in the far north have traditionally taken precedence over the environmental needs of the immediate south. This situation is rapidly changing in the face of heightened public concern with urban environmental and health issues and is reflected in new planning legislation and land use regulations. In turn, environmental geological investigations centred on urban areas encompass a very broad range of issues.

The principal environmental geological concerns in urban areas includes the provision of adequate drinking water, waste disposal, soil and landscape degradation, and the increasing vulnerability of densely populated urban areas to geological hazards and environmental disasters. Environmental geologists working on such problems are, in the main, employed by the environmental-consulting sector. There is increasing interest by private companies and financial institutions (particularly insurance and banking) concerned with the safety of investments arising from litigation over contaminated properties or geological hazards, and the need for clients to recognize and comply with environmental regulations.

Concern with the sustainability and viability of urban environments is promoting collaboration between environmental geologists, planners, legislators and workers in public health. This interdisciplinary approach has spawned several new journals, such as *Environmental Geology* (Springer), *Environmental & Engineering Geoscience* (Geological Society of America and Association of Engineering Geologists), *Environmental Modelling and Assessment* (Baltzer Science Publishers), *Journal of Environmental Planning and Management* (Carfax Pub-

lishing), and *Environment and Urbanization* (International Institute for Environment and Development). In addition to the established journals serving specific areas such as ground water, planning and public health, these new journals provide a forum for environmental geological studies of urban areas and avenues for collaboration between scientists and policy makers. 1991 saw the formation of the International Working Group on Urban Geology, with its primary focus the improvement of communication between environmental geologists and city planners (*Urban Geology News*, 1995).

Growing recognition of the importance of environmental geology, and the need for an increased body of professionals, is reflected in new programs in universities and colleges and the availability of textbooks (*e.g.*, Keller, 1992; Montgomery, 1994; Pipkin, 1994; Murck *et al.*, 1996). Their use to a Canadian audience is strictly limited however, because most of their contents focus on American experience emphasizing large-scale clean-up projects funded by the United States Environmental Protection Agency Superfund. Published case examples of the wide range of everyday urban environmental geology investigations being undertaken in Canadian urban communities are still relatively few and found mostly in the "grey literature" of government and private-sector reports.

OBJECTIVES OF THIS VOLUME

The principal objective of this volume is to provide case examples of environmental geological investigations in Canadian urban areas. Despite a picture postcard image of vast tracts of wilderness, the country is, in fact, heavily urbanized. Statistics Canada (1994) defines an urban area as a community with a total population of more than 1000 persons and having a density of at least 400 persons per square kilometre. In 1871, 19% of the nation's population lived in an urban area. This figure had risen to 76% a century later. The latest available (1991) figure is 76.6%, which is close to the propor-

---

Scope of Environmental Geology

1. Understanding and managing earth processes

2. Providing sufficient resources, *e.g.,* water, minerals and fuels; coping with hazards *e.g.,* volcanic eruptions, earthquakes, landslides, floods, shoreline erosion, subsidence, *etc.*

3. Avoiding excessive disturbance and pollution of geological environments, *e.g.,* soil erosion, ground and surface water contamination, improper mining and waste disposal practices, *etc.*

4. Anticipating and adjusting to environmental and global changes

5. Ensuring long-term viability of society as it relates to public heath/safety and geologic processes

6. Providing reliable technical information for environmental decision making

7. Ensuring a sufficient number of well-qualified professionals

8. Effective collaboration with other sciences

9. Public education

Figure 2   *Scope of environmental geology.*

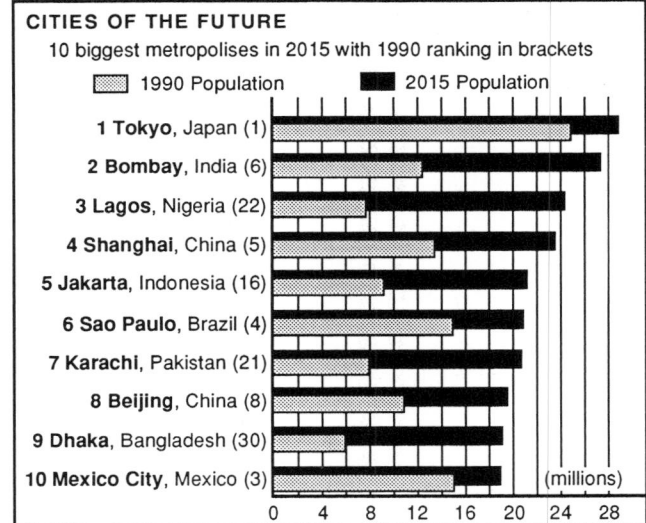

CITIES OF THE FUTURE
10 biggest metropolises in 2015 with 1990 ranking in brackets

▨ 1990 Population   ■ 2015 Population

1 **Tokyo**, Japan (1)
2 **Bombay**, India (6)
3 **Lagos**, Nigeria (22)
4 **Shanghai**, China (5)
5 **Jakarta**, Indonesia (16)
6 **Sao Paulo**, Brazil (4)
7 **Karachi**, Pakistan (21)
8 **Beijing**, China (8)
9 **Dhaka**, Bangladesh (30)
10 **Mexico City**, Mexico (3)

(millions)
0   4   8   12   16   20   24   28

Figure 3   *Expected ten largest supercities as of 2015, with 1990 ranking in parentheses. Emerging supercities of the near future include the Indian cities of Bangalore, Bombay, Delhi and Hyderabad; Karachi and Lahore in Pakistan; Dhaka in Bangladesh; Istanbul in Turkey; and Kinshasa and Lagos in Africa: all with current growth rates of more than 3% per annum. By 2015, it is expected that there will be 4.1 billion city dwellers worldwide; 3.2 billion in the developing world. An urban future is inevitable and urban areas will be increasingly the focus of geological investigations and training. (Data from the United Nations Population Fund, 1991).*

tion of the world's population currently identified as urban (74%). Illustrative figures for other countries are 74.1% for the United States, 76.9% for Japan, 85.5% for Australia, 21.4% for China, 83.9% for Sweden, and 72.6% for Mexico.

In Ontario and Quebec, the most populous provinces in Canada, the present-day urban population accounts for 81.8% and 77.6%, respectively, of the provincial totals (Statistics Canada, 1994). The provinces of British Columbia and Alberta also have relatively high proportions of urban residents (80.4% and 79.8%, respectively) compared with the global average. Only New Brunswick, Prince Edward Island and the Northwest Territories have fewer urban residents compared to those living in rural areas. Close to one-third of the national population of Canada (28,118,000 as of 1991), is now concentrated in three emerging supercities, Vancouver, Toronto and Montreal, all clustered along the United States border.

Canadian developments parallel the global transfer of population from rural to urban areas. But, contrary to the global trend for the development of sick cities, Canada has a well-deserved image of having one of the best-housed populations in the world. Our cities are safe and vital, and there is considerable interest in the Canadian experience of regulating, preventing and dealing with urban environmental geological problems. The environmental sector is a major employer of university and college graduates; in 1994 some 55% of geology graduates across North America found employment in the environmental sector. In the province of Ontario alone, the sector employs more than 38,000 people, working for 2200 companies, and generates $3.5 billion annual revenue for the province, much of which comes from the export of environmental expertise and technologies outside Canada (Canadian Environmental Directory, 1994-95). For comparison, the provincial mining industry employs 25,000 people and generates some $2.5 billion in revenues.

THEMES
This book is composed of 39 papers grouped into ten broad themes with considerable overlap between each theme. Most of the papers focus on environmental geology problems in the heavily urbanized core of central Canada, in Ontario. These are intended to provide important case examples for comparison with other urban communities across the nation.

Environmental Geology of Urban Areas
The first theme establishes the general scope of environmental geology of urban areas. Nick Eyles describes the environmental geological setting of the Greater Toronto Area, the largest urban complex in Canada, containing 16% of the national population. The contamination of ground and surface waters from a wide range of point and non-point sources, the remediation of contaminated industrial sites, and the legacy left by improper disposal of a wide range and large volume of municipal, industrial and nuclear wastes are the principal issues and are common to most urban areas in southern Canada.

Urbanization north of 60° latitude presents a rather different set of challenges. Because of cold conditions that persist year round, more than 50% of Canada is underlain by some form of permanently frozen ground (permafrost), and this has historically impeded northern development. About 50,000 Canadians live on permafrost. The major problems associated with urban development in permafrost terrains are reviewed by Hugh French and relate to the ease with which permafrost is made unstable when disturbed by construction. Because of unique ground-water conditions in permafrost

areas, the provision of water to urban communities is a major challenge and a constraint to future growth.

Urban Ground Waters: Sources of Contamination
Ground-water contamination in urban settings is the second and major theme of the volume. Most urban centres are faced with an increasing demand for responsible environmental policies regarding the safety of drinking-water supplies. Ken Howard identifies the principal sources of contaminants in urban areas and, in a subsequent paper with Steve Livingstone, shows how to conduct a quantitative assessment of the potential impact of such sources. Stanley Feenstra highlights the widespread problem of contamination by chlorinated solvents emanating from the dry-cleaning industry. In addition, there are approximately 200,000 underground fuel storage tanks in Canada, of which between 20,000 and 40,000 are estimated to leak. Ground-water contamination by hydrocarbons from such tanks is reviewed by Jean Pierre François and Hugh Molyneux; this widespread problem is also touched upon by several other papers in later sections of the book. Ken Howard and Janet Haynes identify the serious environmental impacts of the widespread Canadian practice of using salt as a road de-icing chemical in winter.

Urban Ground Waters:
Resource Evaluation and Protection
Resolving the sources of contamination of urban ground waters goes hand in hand with ground-water resource evaluation and protection. In mid-continent North America, about 110 million people live within the area directly affected by Pleistocene glaciations. Aquifers within complex successions of glacial sediments are a significant source of municipal drinking water for many communities. In southern Ontario, the largest aquifer system within the Toronto-centred region occurs within the Oak Ridges Moraine, which is threatened by urban development. An overview paper by Ken Howard and his colleagues Nick Eyles, Phil Smart, Joe Boyce, Rick Gerber, Sean Salvatori and Mike Doughty describes the formation and geology of the moraine and gives the broader context for a detailed paper by Rick Gerber and Ken Howard on quantitative evaluation of the moraine's ground-water resources. The need for proactive legislation to prevent contamination of ground-water resources by restricting land uses in the vicinity of municipal wells (wellhead protection) is an emerging issue in municipal planning, nation wide. Steve Livingstone, Thomas Franz and Nilson Guiguer review the types of information that are required to carry out wellhead protection and demonstrate the application of two-dimensional and three-dimensional computer modelling.

Surface-water Contamination in Urban Areas
Ground water is only one element of the urban hydrological cycle and the fourth theme of the volume is that of surface-water contamination. This is a problem associated with all urban areas and arises from the flushing of contaminants deposited on urban streets and other built surfaces into nearby waterbodies. Miron Berezowsky reviews the history of using constructed wetlands to assist in remediation of surface-water quality and provides examples of their use in controlling urban storm runoff and mine drainage waters. Given the historic dominance of a resource-based economy in Canada, the management and treatment of waste waters derived from either acidic or basic mine tailings is a major concern in many urban communities. Andrée Bolduc, Marc Laflèche and Lucie Talbot describe the geochemical impacts

on surface waters caused by basic mine drainage at Mont-auban, Québec. Nelson Belzile, Douglas Goldsack, Ste-phanie Maki and Andrew McDonald describe the Sudbury urban area of Ontario where highly acidic surface waters are the legacy of a long history of nickel mining and smelting and where some 35,000 tonnes of sulphide-rich tailings continue to be dumped every day.

Urban Waterfronts

Many environmental geologists are particularly concerned with urban waterfronts with their long history of industrial use and contamination from urban-impacted surface waters, sewage and contaminated ground water. John Coakley and Alena Mudroch describe case examples of contaminated harbours in the Great Lakes and St. Lawrence regions and argue that assessment and design of waste-water manage-ment systems is highly dependant upon knowledge of sedi-ment transport and deposition. Hamilton Harbour, at the western end of Lake Ontario, has been identified as one of over 40 severely affected areas in the Great Lakes basin. Ken Versteeg, Bill Morris and Norm Rukavina identify and map the impact of the iron and steel industry on sediment quality in the harbour, using magnetic properties of contaminated sedi-ment. Gordon Fader and Dale Buckley review a long history of waste-water management problems in Halifax Harbour, Nova Scotia where bottom sediments affected by sewage now contain the highest levels of metal contamination anywhere recorded in Canada. These workers similarly conclude that knowledge of geological processes is fundamental to re-mediation and management of contaminated waterfronts and harbours.

Rising sea and lake levels result in significant erosion problems along urban shorelines, requiring costly, ongoing remediation. However, the construction of hardened, artificial shorelines has to be balanced with the competing need to preserve areas of natural habitat and public recreation. These issues are illustrated by Rob Nairn and Nigel Cowie using the heavily urbanized Scarborough waterfront area of Lake On-tario, near Toronto. Bruce Hart and Vaughn Barrie describe the offshore environmental geology of the Fraser Delta in the Vancouver area of British Columbia, paying particular atten-tion to the hazard posed by submarine landsliding caused by earthquakes, together with debris and wastes left by previous industrial activity.

Urban Waste Management

Problems of urban waste management are common to all urban areas worldwide. Nick Eyles and Joe Boyce review geological constraints on municipal waste management prac-tices in southern Ontario that, hitherto, have focussed on disposing of wastes in the ground (landfills). An historical perspective by Richard Anderson reminds us that waste management is not a new problem in the Toronto area, and further, that the number and location of historic dumps is not at all well known. Modern landfills are highly engineered structures designed to protect underlying ground waters from contaminants that are leached from the waste pile; a guide to detailed siting and design requirements is presented by Kerry Rowe and Mike Fraser. Detailed studies of the geology of waste sites are essential; Laurence Andriashek, David Thompson and Reed Jackson show how an understanding of geological complexity arising from glacially deformed sub-strates below the municipal landfill at Edmonton, Alberta, is necessary to understand the site's hydrogeology. Leachate migration away from landfills sited in former quarries along

the Niagara Escarpment, near the city of Hamilton, Ontario, is documented by Jean Birks and Carolyn Eyles. Landfills also produce voluminous amounts of methane gas; the isotopic and geochemical characteristics of landfill methane are es-tablished by Steve Desrocher and Barbara Sherwood Lollar and compared with background methane sourced from gla-cial sediments and bedrock. All of the above papers provide case studies that can assist investigations elsewhere.

Contaminated Substrates in Urban Areas

Many urban areas are characterized by contaminated sub-strates arising from historic landfilling of low-lying areas, to create new land for development, and unregulated industrial activity. Kim Bolton and Les Evans identify the sources and geochemistry of the principal inorganic and organic contami-nants found in such sites. Many contaminated sites occur in downtown, inner-city locations that are now being rezoned and developed for residential and mixed-use purposes. Monica Campbell and her co-authors, Joan Campbell, Ste-phen McKenna, Scott MacRitchie and Miriam Diamond, show how development applications in contaminated areas in the city of Toronto are now being scrutinized from a public health perspective before being given approval to proceed. The intention is to identify the most appropriate clean-up technol-ogy that will not only protect the public during site remediation but ensure the health of future users.

Urban Geological Hazards

The effects of geological hazards are magnified in densely populated urban areas. The natural release of radioactive radon gas in urban areas is a geologic hazard that has only recently been recognized and documented in the United States and Europe. Imshun Je argues that systematic data collection regarding background radon levels in Canadian urban communities is urgently required. This hazard has been compounded in many urban communities by the historic unregulated dumping of low-level radioactive wastes. Alex Mohajer points to a heightened risk of damaging earthquakes in the mid-continent Toronto area, hitherto regarded as geo-logically stable and risk free. There, mid-continent earth-quakes (intraplate earthquakes) are produced by the reac-tivation of old geological structures deeply buried within the interior of the North American plate. John Clague provides an overview of the certain hazard facing the Greater Vancouver area as a consequence of large magnitude plate margin earthquakes produced where the North American tectonic plate is overriding the Pacific plate.

Site Investigation, Remediation and Data Management

Identification, assessment and remediation of many urban environmental geological problems is fundamentally depend-ent on site investigation, remediation and data management techniques. Identification of the subsurface geology of urban sites, where there has usually been a long history of site use and disturbance, is not straightforward. The traditional tech-nique of drilling boreholes is expensive, provides data for a few (unrepresentative?) points only and can accelerate the migration of contaminants by physically disturbing the under-lying substrate. In contrast, walk-over geophysical surveying techniques are rapid, non-invasive, inexpensive and, as a result, are being increasingly used to build a picture of what lies just below the surface of urban areas. John Scaife demon-strates the use of several of the more commonly used geo-physical methods of site survey and assessment in urban areas. Joe Boyce and Berkant Koseoglu show how seismic reflection profiling provides an image of deeper geological strata below waste sites. Significant progress has been made in adapting geophysical borehole logging techniques in use in the oil industry to ground-water resource investigations; these downhole techniques are demonstrated in the Kitchener-Waterloo region of Ontario by George Schneider, David Noble, Michael Lockhard and John Greenhouse.

Iqbal Noor shows how soil gas surveys are employed with the purpose of distinguishing background, naturally occurring hydrocarbons from petroleum products leaking from storage tanks. Once characterized, the clean-up of contaminated urban sites requires the selection of an appropriate remedial technology from a very wide range of available techniques, many of which are new and untested. The wide array of possible remediation techniques is comprehensively reviewed by Paul Beck, paying particular attention to the advantages and disadvantages of each. Bioremediation, using bacteria, is a relatively new technology that is finding increasing application in urban areas. Paul Hubley, Andrew Panko and Doug Boocock describe the bioremediation of sites affected by petroleum contamination.

"Data-rich but information-poor" describes an all too common situation where environmental geological data are scattered across different sources from where they cannot readily be retrieved and used. Much public money is often wasted recollecting or reformatting the same data. The field of computer-based Geoscientific Information Systems (GSIS) is rapidly evolving as a means of storing and presenting pictures of complex three-dimensional geological and environmental data. Nick Eyles, Mike Doughty and Derek Mack-Mumford describe the characteristics of the more commonly available systems and provide examples of their use in urban areas. Frank Kenny shows how satellite imagery can be used to map geology and land use on the rapidly changing urban fringe.

Environmental Assessment Legislation

Today's environmental geologist needs to be part biologist, part chemist, part physicist, part computer technician, and increasingly, part lawyer. For the practising environmental geologist, environmental assessment legislation provides the framework for nearly all their activities, ranging from the collection of field samples, analytical techniques, analysis of data, to the reporting and dissemination of results and final decision making. These activities are commonly conducted under intense public, scientific and legal scrutiny. Existing regulations are complex and often not user-friendly to non-lawyers. Carolyn Eyles reviews the current federal and provincial environmental regulatory framework in Canada, emphasizing the system of legislation in place in Ontario. That province has the largest urban population and the most stringently implemented environmental legislation anywhere

in the country.

Finally, a geologic time scale, a glossary of commonly used terms, and units of measurement are provided as appendices.

REFERENCES

Canadian Environmental Directory, 1994-95, Canadian Almanac & Directory Publishing Ltd., Toronto, ON, 2281 p.

Chan, M.W.H., Hoare, R.W.M., Holmes, P.R., Law, R.J.S. and Reed, S.B., 1987, eds., Pollution in the Urban Environment: Elsevier, London, UK, 699 p.

Drakakis-Smith, D., 1995, Third world cities: Sustainable urban development: Urban Studies, v. 32, p. 459-678.

Dunlap, R., Gallup, G., Jr. and Gallup, A., 1992, The health of the planet survey, Preliminary report: The George G.H. Gallup International Institute, Princeton, NJ, 45 p.

Innes Lumsden, G., 1994, ed., Geology and the Environment in Western Europe: Clarendon Press, Oxford, UK, 325 p.

Keller, E.A., 1992, Environmental Geology: Macmillan Publishing, New York, 521 p.

Legget, R.F., 1973, Cities and Geology: McGraw-Hill, New York, 624p.

Leveson, D., 1980, Geology and the Urban Environment: Oxford University Press, Oxford, UK, 386 p.

Montgomery, C., 1994, Environmental Geology, 3rd Edition: W.C. Brown Publishing, Dubuque, IA, 336 p.

Murck, B.W., Skinner, B.J. and Porter, S.C., 1996, Environmental Geology: John Wiley and Sons Inc., New York, 535 p.

National Research Council, 1993, Solid Earth Sciences and Society: National Academy Press, Washington, DC, 346 p.

National Research Council, 1994, Groundwater Cleanup Alternatives: National Academy Press, Washington, DC, 346 p.

Pipkin, B.W., 1994, Geology and the Environment: West Publishing, Minneapolis, MN, 476 p.

Socolow, R., Andrews, C., Berkhout, F. and Thomas, V., eds., 1994, Industrial Ecology and Global Change: Cambridge University Press, Cambridge, UK, 500 p.

Statistics Canada, 1994, Human Activity and the Environment: Ottawa, ON, 300 p.

Stren, R., White, R. and Whitney, J., 1992, Sustainable Cities: Urbanization and the Environment in International Perspective: Westview Press, Oxford, UK, 365 p.

Tabibzadeh, I., Rossi-Espagnet, A. and Maxwell, R., 1989, Spotlight on the Cities: Improving Urban Health in Developing Countries: World Health Organization, Geneva, Switzerland, 174 p.

United Nations Population Fund, 1991, Population, Resources and the Environment: The Critical Challenges: New York, 154 p.

Turner, B.L., II, Clark, W.C., Kates, R.W., Richards, J.F., Mathew, J.T. and Meyer, W.B., eds., 1990, The Earth as Transformed by Human Action: Global and Regional Changes in the Biosphere over the Past 300 Years: Cambridge University Press, Cambridge, UK, 454 p.

Walton, M., 1982, Engineering geology of the Twin Cities area, Minnesota, *in* Legget, R.F., ed., Geology under Cities: Reviews in Engineering Geology, v. 5, p. 125-131.

2. Environmental Geology of a Supercity: The Greater Toronto Area

Nicholas Eyles

Environmental Earth Sciences, University of Toronto, Scarborough Campus
1265 Military Trail, Scarborough, Ontario M1C 1A4

SUMMARY

This paper provides a case study of the environmental geology of Canada's largest and most densely populated urban area. The Greater Toronto Area is an emerging "supercity" extending over some 7200 km². Since 1945, the area has experienced a tidal wave of urbanization. The need for a more environmentally sustainable approach to urban development (now legislated under the reformed Planning Act of Ontario) has drawn attention to major gaps in environmental geological data available to municipal planners.

Major data gaps occur with respect to geological processes, resources and waste management. Particular problems centre on earthquake risk, ongoing neotectonic activity and resulting structures in rock and sediment, flooding and soil loss in urbanized watersheds, shoreline erosion, the impact of future climate change on lake levels and municipal water supplies, identification of aquifers and regional ground-water regimes, contamination of ground and surface waters from a wide range of urban and rural non-point and point sources, the environmental impact of sand and gravel mining both on land and offshore in Lake Ontario, rehabilitation of waste disposal sites, public health issues arising from residential developments close to such sites, management of contaminated storm-water runoff, the disposal of municipal, hazardous industrial and nuclear wastes and the remediation of contaminated substrates resulting from historic landfilling along lakeshore areas and industrial activity.

THE GREATER TORONTO AREA

The Greater Toronto Area (GTA) of southern Ontario (Fig. 1) is the largest Canadian metropolis and an emerging supercity with 4,324,997 residents. This is equivalent to 16% of the national population, an increase of 6% since 1970. The GTA consists of five rapidly growing Regional Municipalities (Peel, Halton, York, Durham and Metropolitan Toronto) that extend over some 7200 km² (< 1% of the area of the province). It is the largest Canadian urban complex in the Great Lakes basin, extending north-south from Lake Simcoe to Lake Ontario and west-east from the Niagara Escarpment almost to Rice Lake (Figs. 1, 2, 3). The GTA forms part of a wider urban conglomeration, extending from the Canada-United States border at Niagara Falls to Kingston, commonly referred to as the Golden Horseshoe (Fig. 2).

A west-east trending belt of glacial morainic uplands, the Oak Ridges Moraine (ORM), rises up to 400 m above the level of Lake Ontario (75 metres above sea level; m asl), and extends over some 18% of the GTA. The moraine forms a regional surface drainage divide and is a major ground-water resource not yet protected by legislation (Chapter 9). Legislation is urgently needed since a population increase of at least 50% is expected within the GTA over the next 30 years and most of this growth will be directed toward the southern flanks of the ORM, particularly in the communities of Newmarket and Richmond Hill (Fig. 3). The eastern sector of the GTA, in Durham Region, is also rapidly urbanizing. The largest current urban developments in the GTA are Seaton (90,000 population, 2750 ha) in North Pickering, and Springdale (74,000; 2500 ha) in Brampton (Fig. 3).

Purpose and Organization of this Paper

The paper commences with a brief historical overview of the growth of the GTA and the current state of the urban environment across the region. The remainder of the paper is divided into four sections. The first describes what is known of the geologic evolution of the area, focussing on key geologic events and the principal environmental geologic issues associated with each. The second part examines regional management of ground and surface waters. The management of municipal waste, hazardous waste and nuclear waste is examined in the third part. Lastly, the relationship between new initiatives in urban planning and the discipline of environmental geology, together with the different approaches followed by GTA municipalities is outlined. The paper concludes with a brief summary of the principal environmental geologic issues and some recommendations.

HISTORICAL GROWTH AND ENVIRONMENTAL IMPACTS

Early European Colonization: 1615 to 1840

The first Europeans arrived in what is now southern Ontario in 1615, following the access route from Lake Ontario to the upper Great Lakes. This route followed the Rouge and Holland rivers to Lake Simcoe and thence, to Georgian Bay (the Passage de Toronto). By 1700, southern Ontario, then part of Upper Canada, was largely depopulated (by disease) and under the influence of the Ojibway Nation. This resulted in accelerated European colonization because, unlike earlier Iroquois and Huron nations, the Ojibway placed no importance on settled agriculture and were easily dispossessed. Southern Ontario came under British authority by the Treaty of Paris (1763) and Loyalist refugees from the United States began arriving in large numbers after 1783. The Ojibway ceded virtually the entire northern shoreline of Lake Ontario to the Crown as a result of the Gun Shot Treaty in 1783. The treaty was not validated in law until 1923 when 1328 Indians received $375 each and gave up "all right, title claim or demand" to an area that included almost all the city of Toronto east to Trenton (Johnson, 1973). By 1800, a rectilinear road system had been etched onto the landscape regardless of topography or drainage; Yonge Street, the principal north-

south access route to the interior, was built in 1798.

City Growth: 1840 to 1953

Toronto, designated a city in 1834, is the historic core of the GTA and outgrew its early competitors, Hamilton and Kingston, because of its access to rich farmland, railways and ability to attract immigrants. Extensive clearance of forest covers and watershed denudation began in earnest shortly after 1840. The building of hundreds of grist mills and ponds

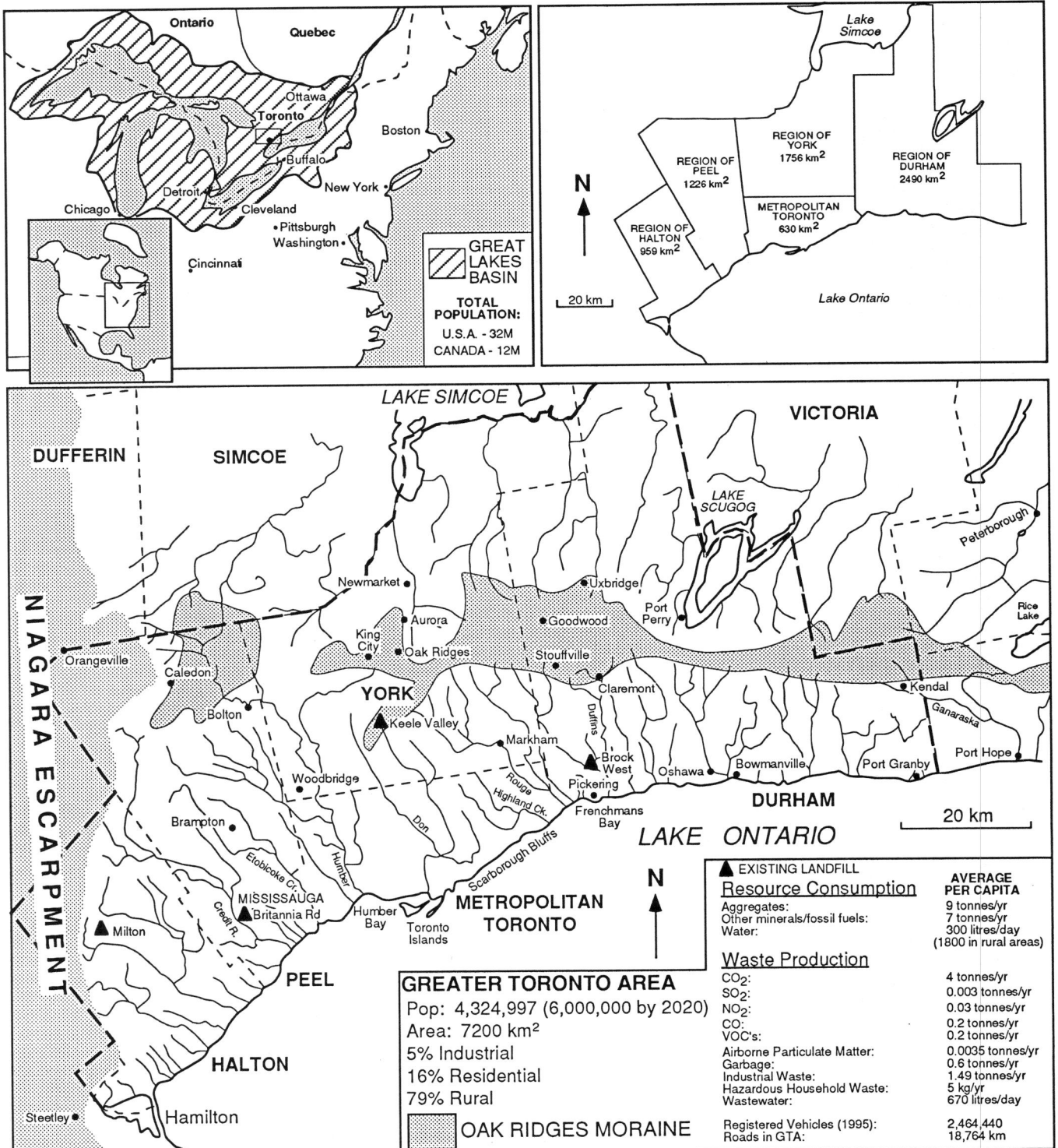

Figure 1   *Location of Greater Toronto Area (GTA), principal rivers, place names used in text, and vital statistics. Waste production figures do not include emissions from airports, nuclear plants and other sources (see Fig. 39). Data from Environment Canada.*

and the clogging of waterways with sawdust and logs destroyed the Atlantic salmon fishery and prompted regulation of navigation and riparian rights under the Public Works Act of 1859.

Between 1851 and 1954, the areal extent of forest cover in the Credit River drainage basin (Fig. 1) was reduced from 67% to 16.3%. Between 1837 and 1937, 320 km of streams (some 87% of the total) became intermittent and 17% of wells dried up in King Township (Fig. 3) because of reduced infiltration (Mayall, 1939; Richardson, 1974). Soil erosion resulted in rapid

Figure 2 *False-colour satellite image showing western Lake Ontario region. Urban areas are shown in blue and rural areas in red. The Greater Toronto Area lies in the centre of the image which covers an area 240 km long and 150 km wide.*

sedimentation along the lower reaches of rivers and damaging floods (Fig. 4). Loss of soil to regional rivers is recorded in sediment cores taken from Lake Ontario by the first appearance of *Ambrosia* (ragweed) pollen (McAndrews and Power, 1973). By the 1930s, the effects of deforestation on regional climate and ground-water supplies, particularly within the ORM, were recognized as a major problem worthy of federal and provincial attention (Richardson, 1944; Whitney, 1994; Chapter 9).

Disastrous flooding and loss of life caused by Hurricane Hazel in October 1954 (see below) showed that urban residents were not immune to geologic processes and demonstrated the need to incorporate an understanding of such processes into urban planning. Conservation authorities (*e.g.*, Metropolitan Toronto and Region Conservation Authority; MTRCA) were established and held responsible for watersheds and flood plains. The creation of Metropolitan Toronto (the City of Toronto and the new urban boroughs of Scarborough, York and Etobicoke), in 1953, recognized the

reality of urban sprawl while complete dependence on automobiles was fostered by a program of highway construction (the Superhighway Plan of 1943) modelled on Los Angeles. Canada's first four-lane, controlled-access highway (the Queen Elizabeth Highway) was opened in 1939, the nation's first subway (Yonge line) in 1954, Highway 401 in 1956 and the first elevated urban highway in 1966 (Gardiner Expressway).

The Urban Explosion after 1953

The pattern of urban development until 1953 was for rural settlements to expand and to develop their own local municipal government (*e.g.*, Weston in 1881; Mimico, 1911; North York, 1913; and East York, 1924). Most defaulted on debt repayments in the Depression and essential municipal services (drinking water, sewage) were postponed, such that they were subsequently unprepared for growth that occurred after 1939. Particular problems emerged in North York which had no direct access to Lake Ontario for water and sewage disposal. A key event was the formation of Metropolitan

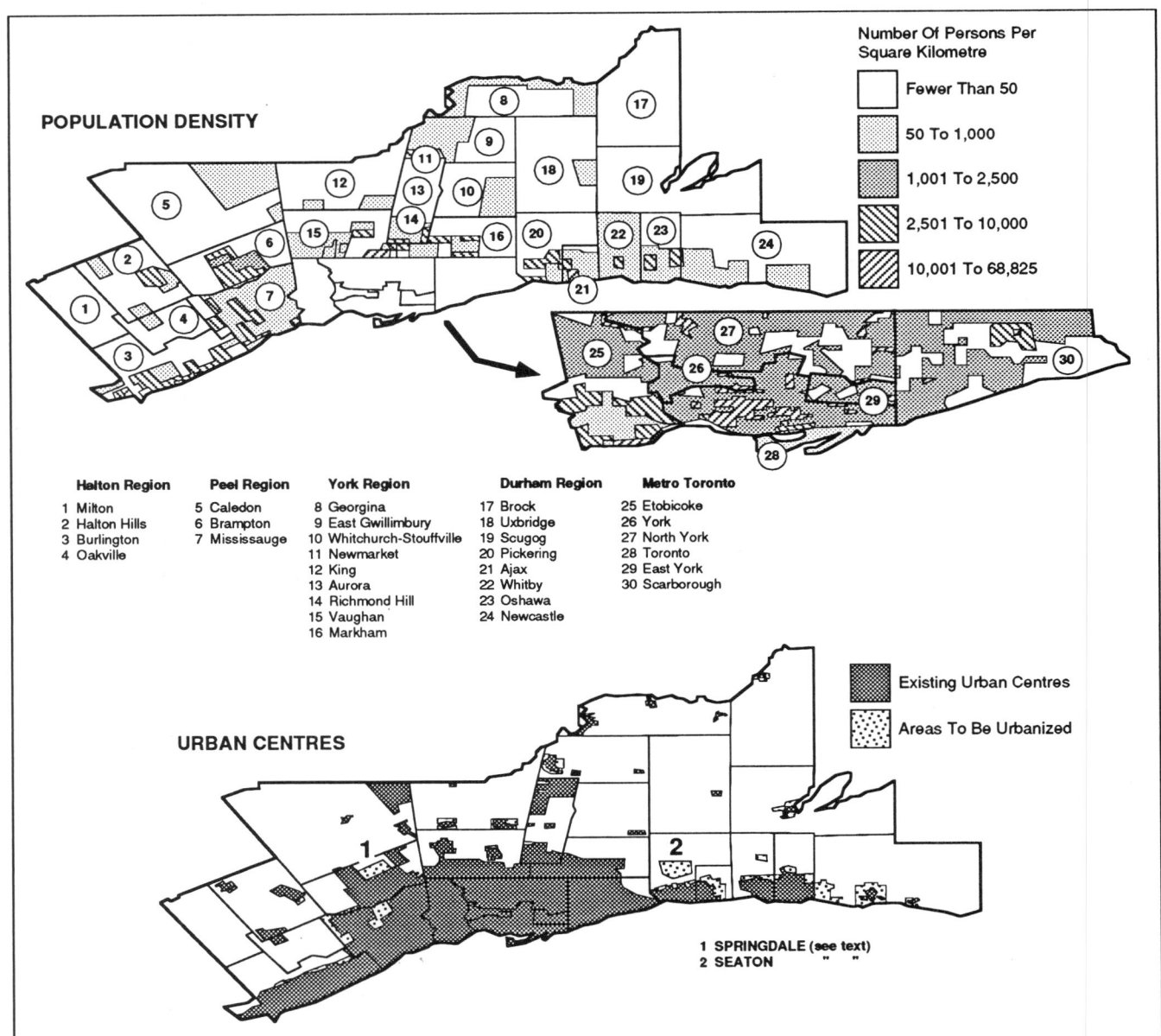

Figure 3 *Extent and density of urban area across GTA. See text for details.*

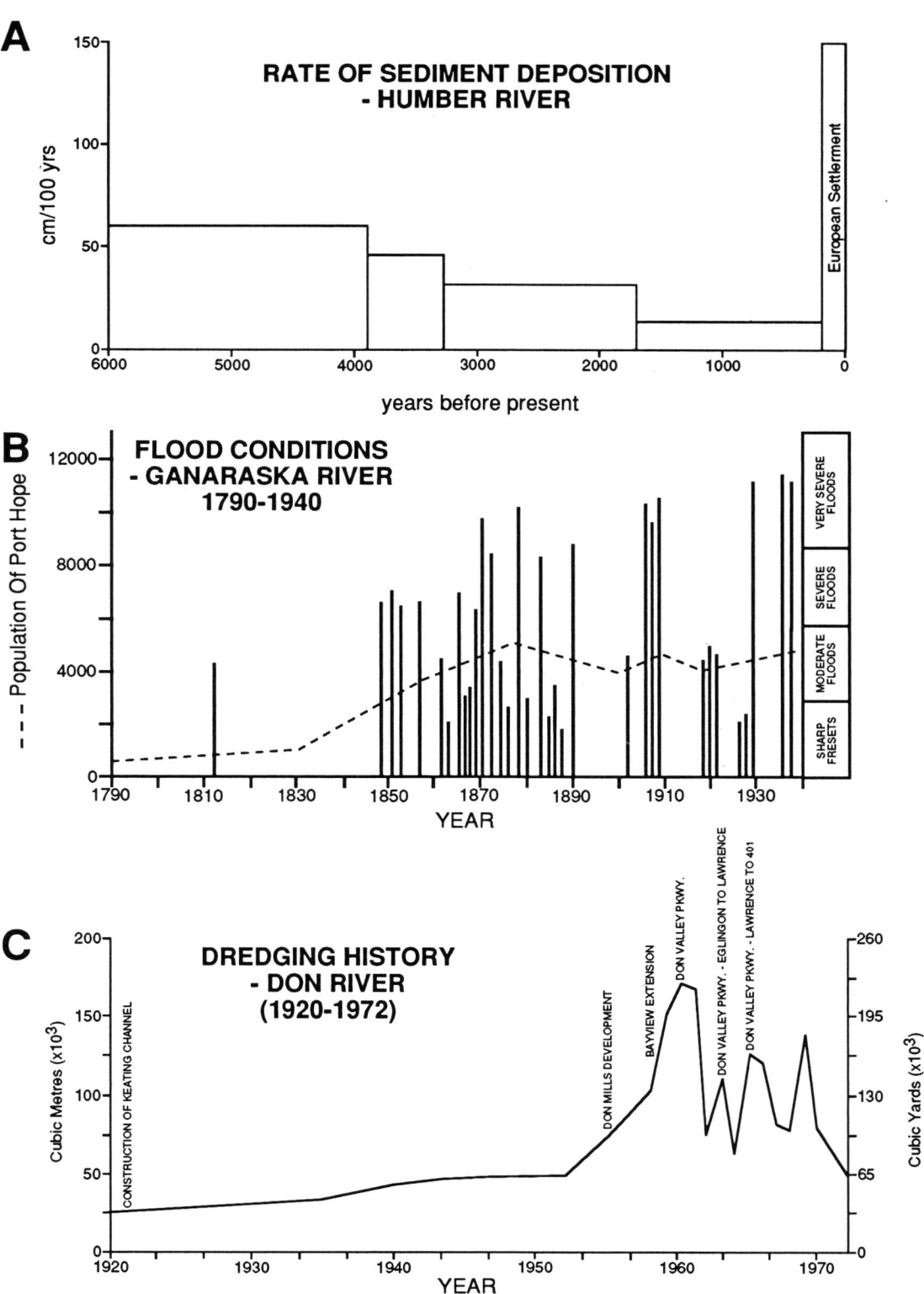

Figure 4 *Impact of urbanization on river dynamics:* (**A**) *Long-term decline in rate of sedimentation along Humber River, likely due to rising level of Lake Ontario and decreased river gradients (Fig. 20; data from Weninger and McAndrews, 1989). Note abrupt increase in rates as a result of European settlement.* (**B**) *Impact of European settlement and forest clearance between 1790 and 1940 along Ganaraska River (Fig. 1). All GTA rivers have experienced a similar history; in 1861, there were 50 watermills along the lower reaches of the Rouge River (Riley and Mohr, 1994).* (**C**) *Volume of sediment dredged from Keating Channel on Don River, 1920 to 1972, as an index of extent of urbanization and resulting soil disturbance in the watershed (see Figs. 26, 32, 36, 37; data courtesy of Toronto Harbour Commission). Don Mills was a large-scale (5000 ha) development on land held by business magnate E.P. Taylor. It initiated tract development by large corporations that has typified suburban growth in the GTA since the mid-fifties. Dredging of the Keating Channel was halted in 1972 because of concerns with toxicity of contaminated sediment, but recommenced in 1987 because of increased flood risk resulting from constriction of the channel (Fig. 30B).*

Toronto in 1953. The new government assumed responsibility for sewage and drinking water. Essentially, ground water was written off as an urban resource and replaced by water taken from Lake Ontario. Equalized assessment and taxation provided the necessary capital base for investment in infrastructure and this literally paved the way for accelerated urban expansion. Between 1951 and 1961, the population of Metropolitan Toronto increased by 50% (to 1.6 million) largely as a result of immigration from Europe. This rate of increase was only exceeded, in the whole of North America, by Houston and Denver. In 1961, over 55% of the total population had changed their principal residence at least once during the preceeding five-year period (Murdie, 1969) and promoted rapid urbanization of outlying areas ("a tidal wave of metropolitan expansion"; Blumenfeld, 1954). Between 1971 and 1991, the GTA experienced the largest absolute increase in population in Canada (1,265,000) while several GTA communities (e.g., Oshawa) posted the highest relative population increases in Canada during the same time (Statistics Canada, 1994).

Because of easy access to municipal drinking water and sewage disposal using Lake Ontario, post-war urbanization has taken place with little or no regard for the natural environment. Currently, about 21% of the GTA is urbanized (5% industrial, 16% residential) with the remainder being rural. Until very recently, rural land was considered as awaiting better and more productive uses, a throwback to early colonial times when forested property was regarded as waste land. The total area of rural land in active agricultural cultivation has decreased from 64% to 40% during the last 20 years (Riley and Mohr, 1994). More than 28,000 ha of the highest quality agricultural land were lost between 1981 and 1986. At the same time, however, the productivity of remaining agricultural land has sharply increased as a result of intensified tillage. Unfortunately, as a result of new farming practices, southern Ontario farms now have the highest average rate of soil erosion by water anywhere in Canada (4.4 t·ha^{-1}; Ketcheson and Webber, 1978; Statistics Canada, 1994). Ground-water contamination from fertilizers and pesticides is widespread (Rudolph and Goss, 1992, 1993).

Less than 50% of the GTA's former wetlands survive; lakeshore marshlands have been especially hard hit with a loss of 90% by area in Metropolitan Toronto (Varga, 1980; Whillans, 1982; Snell, 1987; Crombie, 1990). Fragmentation and loss of biohabitat is particularly unfortunate because the GTA extends across a very large part of Canada's most southerly and most species-rich terrestrial region (the Carolinian zone) which accounts for less than 0.25% of the national area (see Riley and Mohr, 1994). Cultural destruction has also occurred; some 5000 archeological sites were destroyed between 1951 and 1971. Predictive statistical models are now used to assess the likelihood of finding sites in rapidly urbanizing areas, using environmental variables such as proximity to waterbodies, slope, soil type, etc. (Young, 1994).

The Present-Day Urban Shadow

Most environmental problems in the GTA stem directly from high levels of per capita consumption (Fig. 1). As a result, the needs of the urban population dictate land uses in the outlying rural areas (the so-called urban shadow effect). This effect includes demands made by transportation, housing, energy needs, recreation, water supply and the export of wastes and pollutants. This shadow is particularly evident with issues such as waste management, where a variety of wastes are disposed of in landfills, the mining of aggregate (i.e., sand and

gravel) which blights large tracts of rural land, water supply and sewage disposal. The economic costs of urban sprawl and the need for more compact development, including re-use of existing industrial lands, have been highlighted by the Greater Toronto Area Task Force (1996).

In 1988, the GTA produced 44 million tonnes (t) of aggregate (about 60% of GTA requirements), most of which was

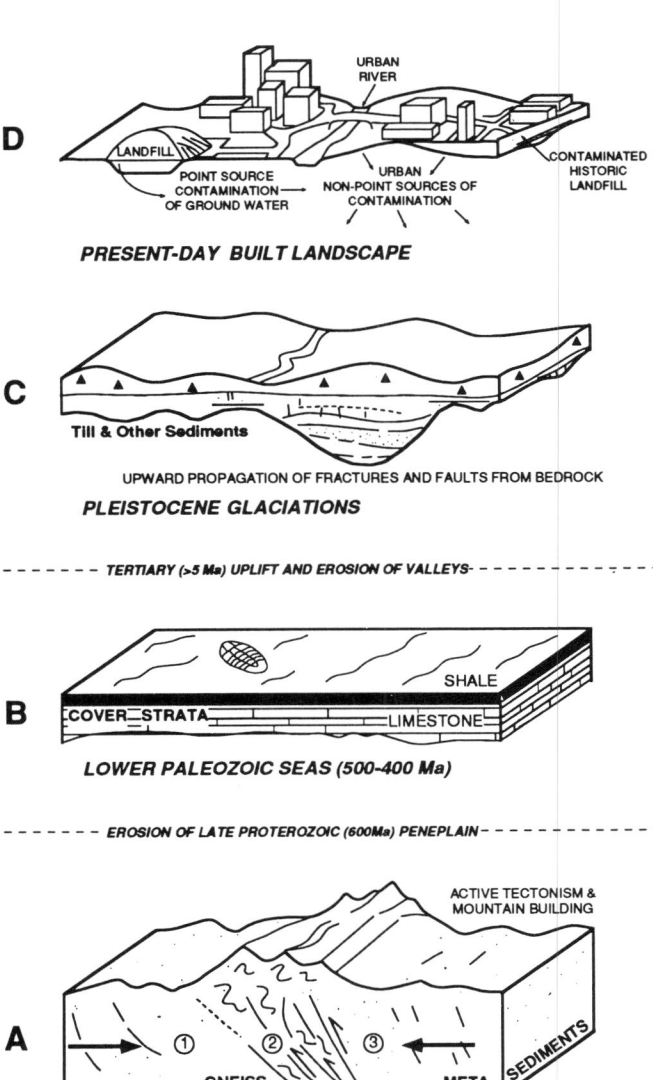

Figure 5 *Simplified four-layer model of the geological evolution of the GTA. (A) Mid-Proterozoic tectonism and collision of Central Gneiss Belt (1) and Central Metasedimentary Belt (3) to form Central Metasedimentary Belt Boundary Zone (2) (Figs. 7, 8, 9). (B) Deposition of lower Paleozoic cover strata (shales, carbonates) deposited in shallow seas that invaded peneplained basement (Figs. 6, 11). These deposits were uplifted and dissected by valleys during the last 5 million years. Valleys follow trend of underlying basement structures such as terrane boundaries and faults (Figs. 9, 10, 12). (C) Deposition of Pleistocene glacial sediments under and along the margins of continental ice sheets. Thickest deposits, and major aquifers, occur where sediments infill buried bedrock valleys (Fig. 12). Fractures and faults have been propagated upward into sediments from bedrock. (D) Waste material used to infill topography (termed historic landfilling; e.g., Fig. 27) and construction of waste dumps in former sand and gravel quarries (landfill sites; e.g., Fig. 43). Urban ground waters are contaminated by many non-point sources (e.g., road de-icing chemicals) and point sources (e.g., landfill sites; Chapters 4, 5).*

mined from large pits in the Oak Ridges Moraine. Unfortunately, this area is the principal ground-water recharge zone for GTA aquifers. Excessive usage of municipally supplied water (from Lake Ontario) in the urbanized part of the GTA (see inset on Fig. 1) is a result of water rates being among the lowest in the Canada; the average annual bill of $250 per household is less than 65% of the real cost. The provision of municipal water has not only drawn attention away from the need to protect ground-water and surface-water resources, but also results in the discharge of very large volumes of poor-quality effluent from sewage treatment plants back to Lake Ontario. Regional per capita sewage flow to the lake is about 670 L per day. The GTA routinely exports large volumes of municipal, industrial and nuclear wastes to surrounding com-

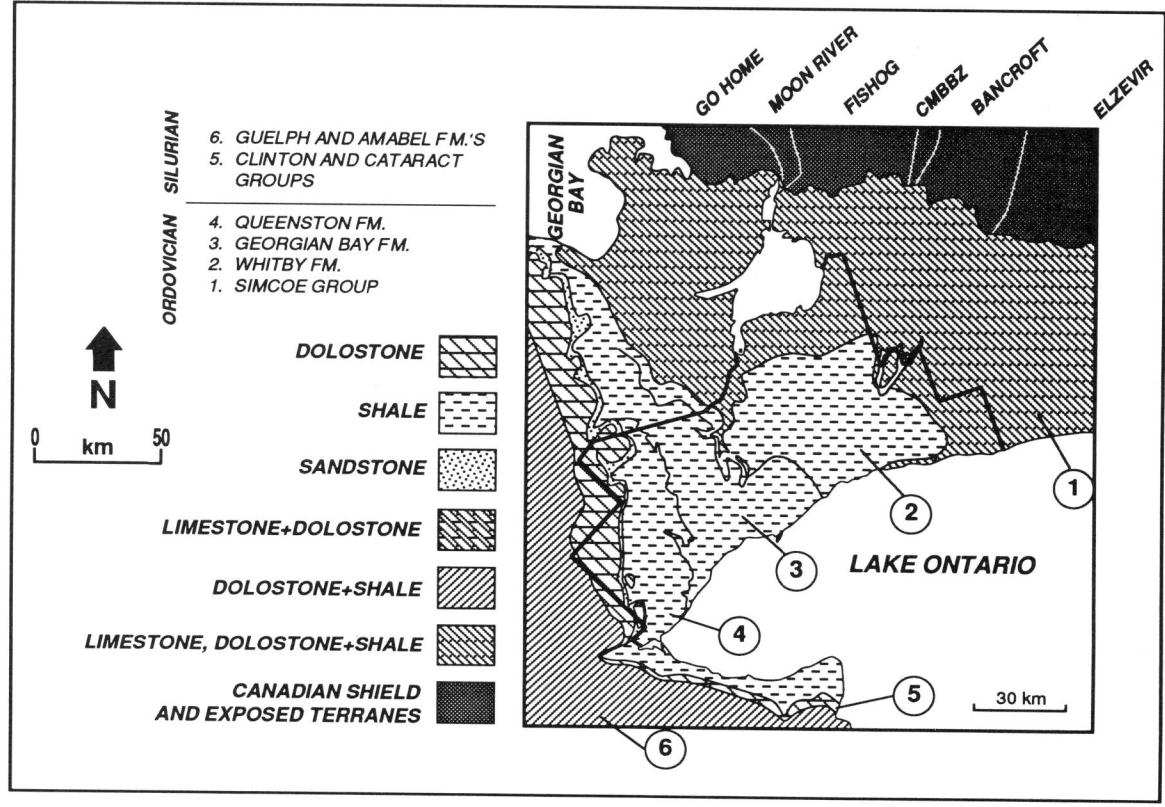

Figure 6 *(top) Distribution of Paleozoic cover strata in southern Ontario. Names at top of figure identify terranes shown in Figure 7. Rocks dip southwestward (about 6 m·km⁻¹) such that successively younger strata are exposed at surface to the west. GTA outlined. (bottom) Pop-up near Port Colborne, formed by upward buckling of near-surface Paleozoic limestone in response to relatively high horizontal compressive stresses in bedrock.*

munities (see below).

The GTA, in turn, lies within the shadows cast by other large urban centres. Airborne sulphate, dioxin, PCBs and pesticides are all imported into central Ontario by transcontinental transport from the United States and Mexico (*e.g.*, Macdonald and Metcalfe, 1991; MacDonald *et al.*, 1991; Hoff, 1994; Hoff *et al.*, 1992; Cohen *et al.*, 1995; Thurston *et al.*, 1994). Other contaminants are imported into Lake Ontario from the Niagara River, which contributes approximately 70% of the toxic contaminant loading to the lake (*e.g.*, Duran and Oliver, 1983; United States Environmental Protection Agency, 1993; International Joint Commission, 1994; Environment Canada, 1992, 1993, 1995a).

PART I: GEOLOGICAL SETTING OF THE GREATER TORONTO AREA

Understanding and management of many urban environmental problems require a thorough grasp of the geological history of cities. Surface environments, even in densely urbanized areas, are influenced in many different ways by geologic structures that are now deeply buried below younger materials. These materials include a wide variety of waste materials used over the last 100 years to fill existing topography and create new land.

The geology of the GTA can be visualized as being composed of four layers, which from bottom (oldest) to top (youngest) consist of:

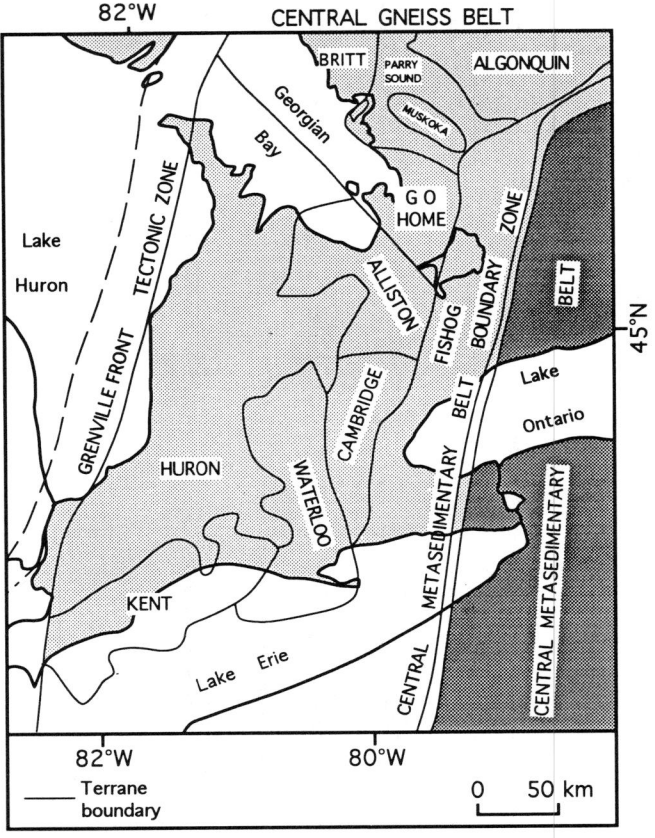

Figure 7 *(right)* Proterozoic (basement) terranes in southern Ontario (after Easton, 1992).

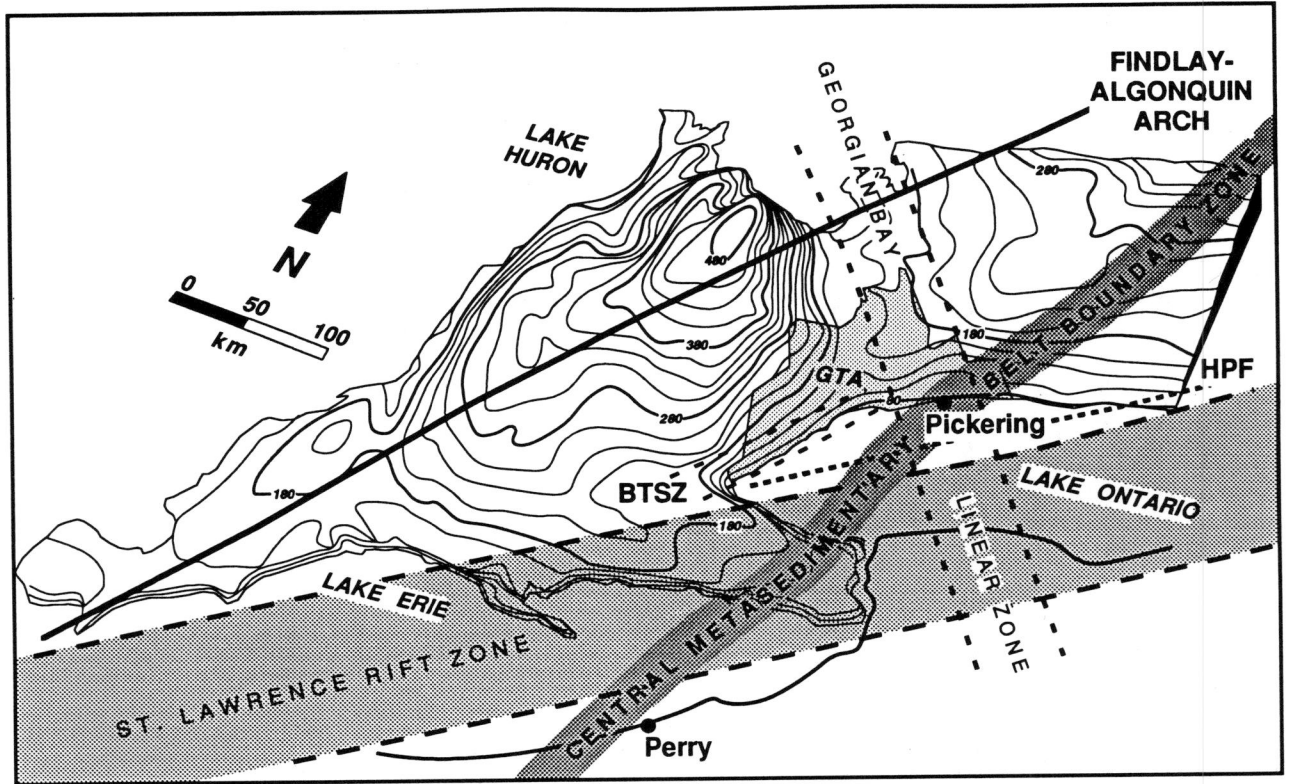

Figure 8 *Oblique view of regional bedrock topography of southern Ontario (m asl) showing Laurentian Channel below GTA and principal structural trends in basement. Abbreviations: **GBLZ**, Georgian Bay Linear Zone; **BTSZ**, Burlington-Hamilton Seismic Zone; **HPF**, Hamilton-Presqu'ile Fault; **CMBBZ**, Central Metasedimentary Belt Boundary Zone; **SLRS**, St. Lawrence Rift System. The BTSZ is located above the western margin of the Fishog terrane (Fig. 7). Lacking hard evidence to the contrary, all structures must be considered potentially capable of generating destructive earthquakes. Note position of Pickering and Perry nuclear power station above CMBBZ; an earthquake closed the American station in 1986 (see Seeber and Armbruster, 1993; Chapter 29).*

(1) complexly structured Proterozoic basement rocks between 1300 and 1100 million years old (1300 Ma and 1100 Ma; see Appendices A and B). These strata extend to the base of the continental crust some 50 km below the ground surface (Easton, 1992; Fig. 5A);

(2) gently dipping lower Paleozoic sedimentary cover strata, up to 500 m thick, deposited between 500 and 450 million years ago(500 to 400 Ma; Johnson *et al.*, 1992; Fig. 5B);

(3) Late Pleistocene glacial sediments younger than 100 thousand years (100 ka) in age (Eyles and Williams, 1992), but as much as 200 m thick (Fig. 5C); and

(4) historic landfill which is the result of the practice over the last 100 years of using waste soil and other materials, often highly contaminated, to infill shorelines and areas of irregular topography (Fig. 5D).

Each of the geological layers identified above is described below, paying particular regard to environmental geological issues associated with each layer.

PROTEROZOIC BASEMENT

The term basement refers to complexly structured mid-Proterozoic strata, mostly gneisses and other highly altered (metamorphosed) sedimentary rocks, that are about 1200 million years old. These rocks are the extension of the Canadian Shield southward under younger Paleozoic sedimentary strata (Fig. 6). It is important to understand the nature of the deeply buried basement structures because they dictate the distribution of earthquakes and, by promoting fracturing of overlying strata, create pathways for the movement of subsurface water, oil and gas, and other contaminants (see below). The plate tectonic paradigm that began in the 1970s transformed the ability of geologists to understand the global workings of planet Earth and the evolution of the continents and oceans. This model is now being applied locally to explain the geologic history of basement rocks in southern Ontario.

Mapping of the exposed shield, and drilling to the south, have demonstrated that the basement is composed of a complex collage of terranes (or micro-tectonic plates) from tens to hundreds of km² in extent (*e.g.*, Fishog, Go Home, Alliston terranes, *etc.*; Fig. 7). These record the growth of the ancestral North American continent by the collision and amalgamation of smaller micro-continents (*e.g.*, the United Plates of America; Hoffman, 1989) just as the larger tectonic plates of India and Eurasia are colliding at the present time. Basement below the GTA lies within the Grenville Province of the Canadian Shield (Easton, 1992) and is subdivided into the Central Gneiss Belt (CGB), dated at >1350 Ma, and a younger Central Metasedimentary Belt (CMB) about 1100 Ma in age (Easton, 1992; Figs. 7, 8). The boundary zone between these two geological provinces (called the Central Metasedimentary Belt Boundary Zone; CMBBZ) is of particular interest to environmental geologists as it passes directly below the GTA.

The Central Metasedimentary Belt Boundary Zone

Mapping of the exposed shield, together with geophysical data, show a pronounced northeast-trending grain within the basement of southern Ontario. The most conspicuous element within this grain is the Central Metasedimentary Belt Boundary Zone (CMBBZ). This structure contains rocks that are more highly magnetized than surrounding strata and can be mapped by a prominent aeromagnetic anomaly, termed an aeromagnetic lineament (Mohajer, 1993; Wallach and Heginbottom, 1993; Chapter 29). This anomaly extends from Ohio (*e.g.*, the Akron magnetic lineament; Seeber and Armbruster, 1993) to western Quebec, and passes below the eastern half of the GTA (Fig. 8). The CMBBZ structure is a major crustal weld, or suture, that records Himalayan-style collision (obduction) along a mid-Proterozoic continental margin of North America (then called Laurentia; Hoffman, 1989). In this process, the CMB was thrust over the CGB to produce a complex boundary zone, at least 20 km wide (Fig. 9). The zone consists of closely spaced thrust faults that trend (strike) southwest/northeast and dip to the southeast (Fig. 9B).

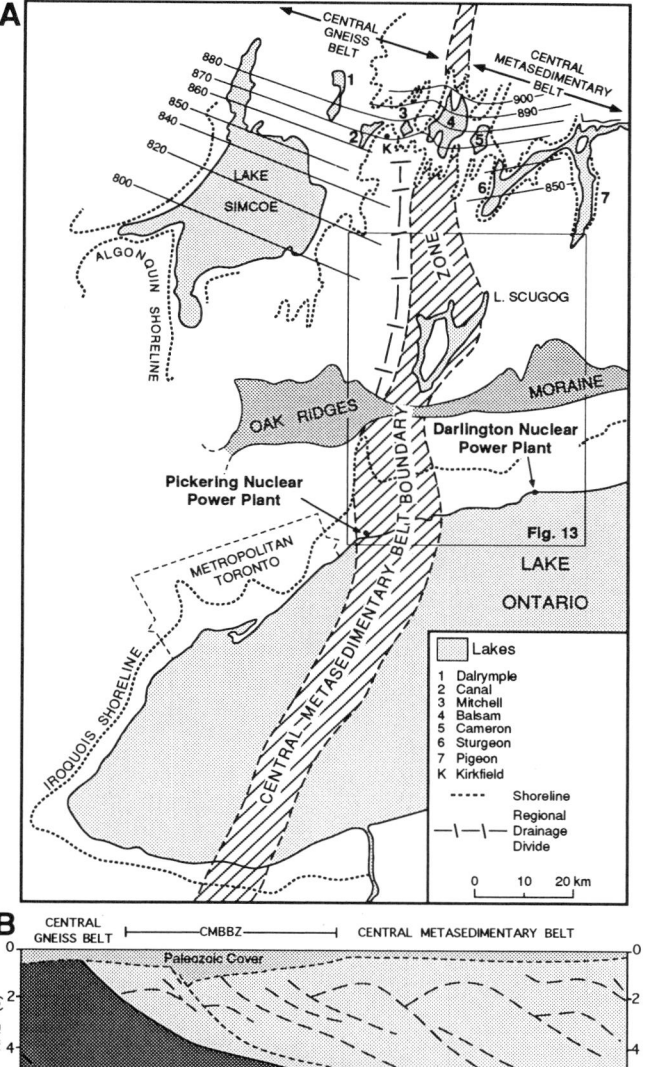

Figure 9 (**A**) *Trend of CMBBZ across Lake Ontario and eastern GTA showing associated geomorphic features. A pronounced embayment in the shoreline of Glacial Lake Iroquois, north of Pickering nuclear power plant, occurs along the western margin of the CMBBZ. Note the distinct thinning of the Oak Ridges Moraine above the structure. Contours (m asl) show elevation of Glacial Lake Algonquin shoreline and identify northeastward tilt of shoreline in response to more rapid post-glacial uplift in that direction (Fig. 20). Note the distinct warping in the elevation contours where they cross the CMBBZ near Kirkfield, suggesting differential movement either side of the structure. Differential movement is also suggested by a pronounced change in the slope of the upwarped shoreline of Glacial Lake Iroquois (Fig. 20B).* (**B**) *Schematic west-east cross-section of thrust faults within CMBBZ structure identified by seismic refraction data (after Milkereit et al., 1992). Reactivation of faults and fracturing of overlying Paleozoic cover and Pleistocene sediment cover is suggested by many data (e.g., Fig. 10).*

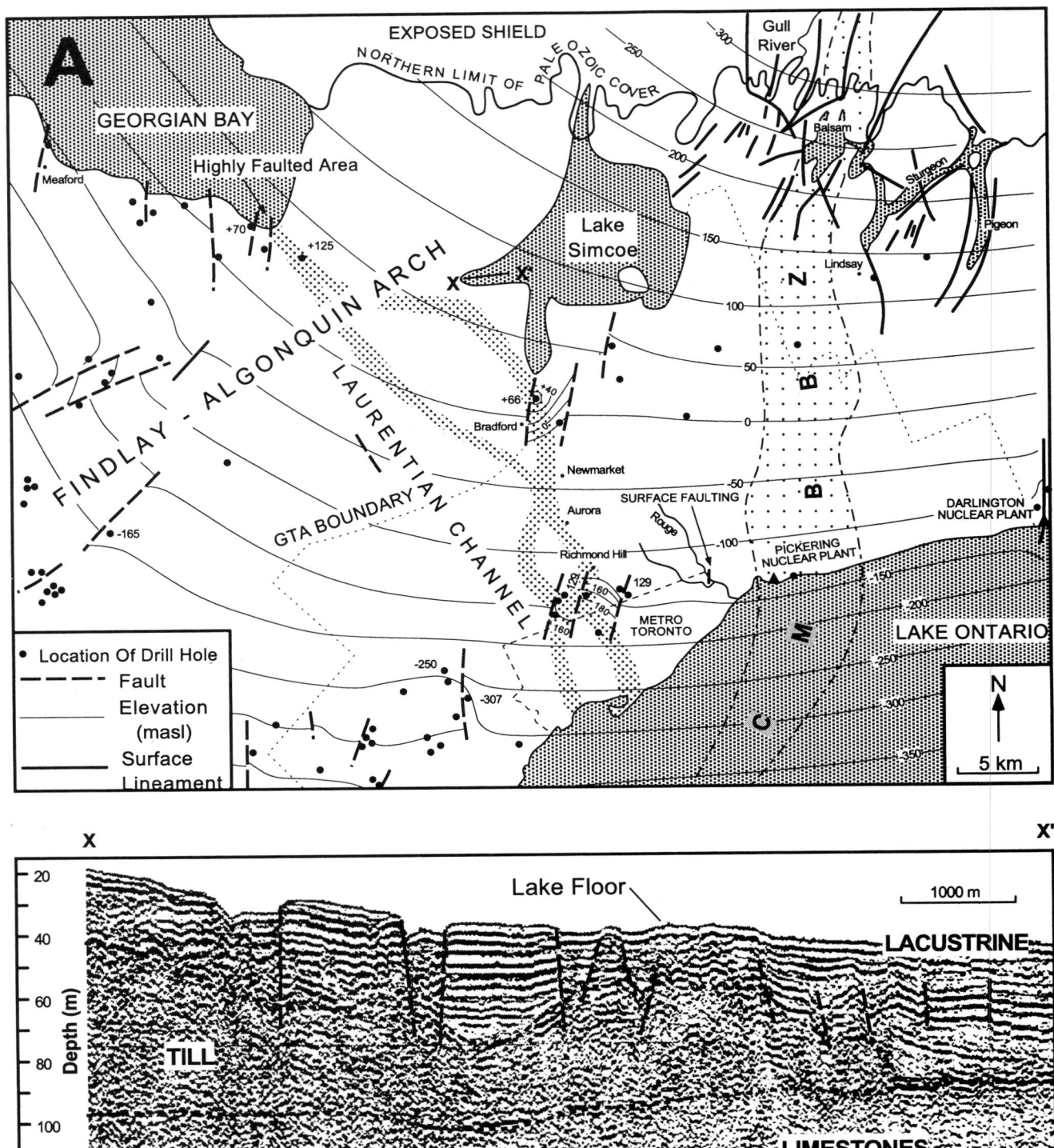

Figure 10 (**A**) *Topography of late Proterozoic peneplain (elevation in metres above or below sea level) below the GTA and surrounding area (after Bailey Geological Services, 1984). Upward bulging of this surface along the Findlay-Algonquin Arch (Fig. 8) is clearly identified. Dashed lines are faults, but note sparcity of data; the likely number of faults is much greater (see Fig. 12). Recent structural mapping of the exposed shield surface has identified faults whose abundance and extent of offset were previously unknown (e.g., Easton and Ford, 1991, p. 101). Section x-x' is a seismic reflection profile collected from western Lake Simcoe showing faulting of horizontally layered lacustrine sediments and till. Data are too poor to resolve structures in underlying bedrock, but a possible relationship with faults in the underlying Proterozoic peneplain was suggested by Todd and Lewis (1993). Faults are exposed in outcrop along the Rouge Valley (Mohajer et al., 1992). Surface lineaments north of GTA taken from Sanford (1993) and Rutty and Cruden (1993); many lineaments and lakes are the topographic expression of faults. Faults near Bradford lie along the contact of the Alliston and Fishog terranes (Fig. 7) and have influenced the erosion and location of the Laurentian Channel (Fig. 12). Much better data are needed in the vicinity of the Central Metasedimentary Belt Boundary Zone to assess extent of faulting in vicinity of nuclear power stations.*
(**B**) (**photograph on facing page**) *Flat-lying deltaic sediments of the Scarborough Formation (Figs. 17, 18) showing (at head level of figure) deformed bed recording liquefaction and slumping on a delta front, possibly in response to ground motion during an earthquake. Holes are the result of ground-water seepage and birds.*

Related Structures

Several other basement structures cross the GTA and the immediate area. A prominent structure is the St. Lawrence Rift System (SLRS; Fig. 8; Kumarapeli and Saull, 1966; Adams and Basham, 1991) which follows the axis of lakes Erie and Ontario and which may be associated with the Hamilton-Presqu'ile fault (HPF; Fig. 8) as identified by Thurston (1991). Other structures include the Georgian Bay Linear Zone (GBLZ; Wallach, 1990), and the western boundary of the Fishog Terrane (Fig. 7) which is associated with the Burlington-Toronto Seismic Zone (BTSZ; Fig. 8; Mohajer, 1993;

Chapter 29). Some of these structures, such as the SLRS and HPF, record fracturing (rifting) of the Earth's crust during the breakup of a large Late Proterozoic supercontinent (*Rodinia*), of which Ontario was then part, sometime between 600 Ma and 550 Ma. The larger structures such as the SLRS have influenced the location of the St. Lawrence River and possibly lakes Ontario and Erie (see Adams and Basham, 1991). Basement structures have not completely healed and have experienced repeated reactivation; the timing of these episodes and the potential for damaging earthquakes is much discussed (see below).

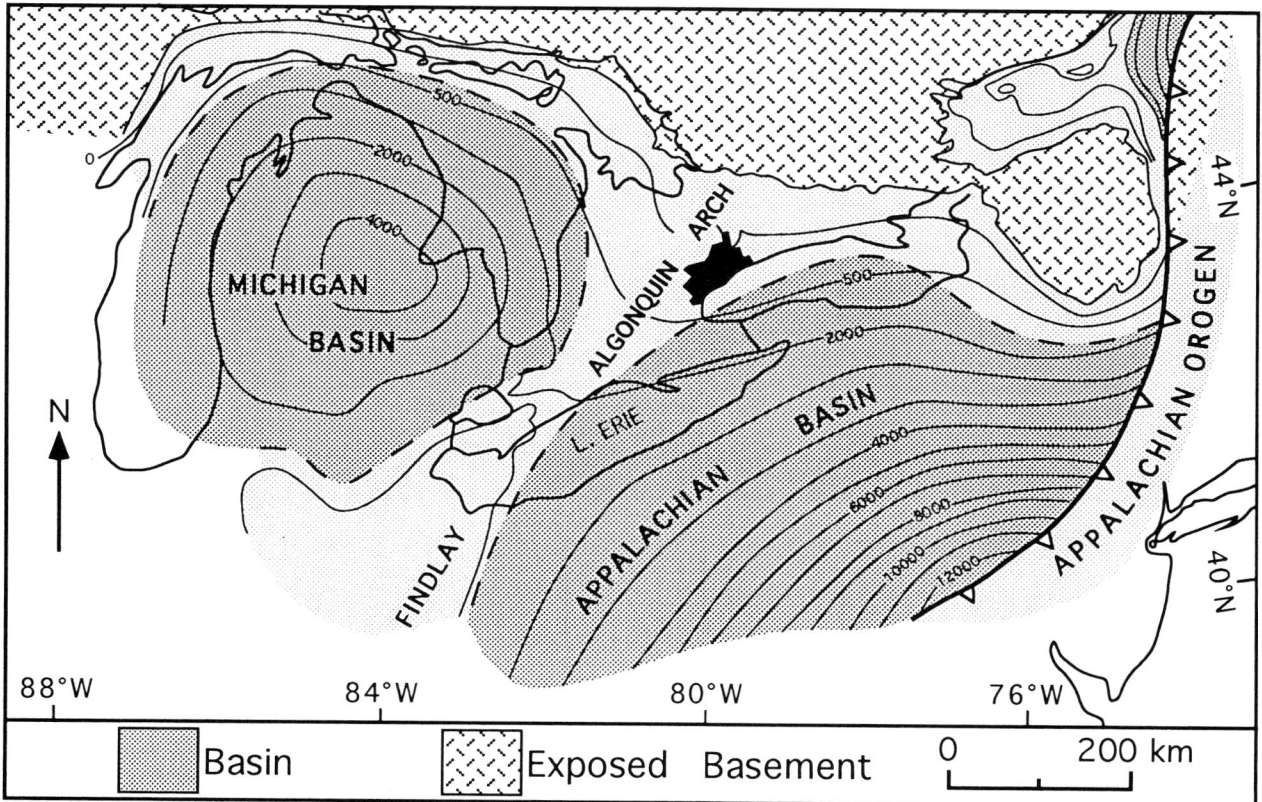

Figure 11 *Relationship of Findlay-Algonquin Arch to Michigan and Appalachian sedimentary basins. Contours are of thickness of Paleozoic sedimentary rock within basins (after Johnson et al., 1992). Arch results from upward bulging of basement in response to crustal load of adjacent sedimentary basins. GTA shown in black.*

The Findlay-Algonquin Arch:
The Backbone of Southern Ontario

No extensive sedimentary record of the time interval between 1200 Ma and about 500 Ma survives in southern Ontario (Fig. 5). From 1200 Ma to about 800 Ma, deep crustal rocks of the Proterozoic basement underwent cooling, uplift and regional erosion of several tens of kilometres; marine and volcanogenic sediments were deposited in local basins on the shield surface between 800 Ma and 500 Ma (see Johnson *et al.*, 1992; Easton, 1992). During this time, any high-standing basement topography was eroded to a low-lying surface devoid of much relief (peneplain; Fig. 10). Regionally, this surface is bulged upward along the Findlay-Algonquin Arch which forms a northeast-trending spine (Figs. 8, 10). The GTA lies on the south slope of the arch, such that the buried Late Proterozoic peneplain slopes southward from elevations of about 100 m above sea level near Lake Simcoe to below 300 m below sea level under Lake Ontario (Fig. 10A).

The Findlay-Algonquin Arch is overlain by a relatively thin cover of younger lower Paleozoic strata compared to the much thicker Paleozoic infills contained within the Michigan Basin to the west and the Appalachian Basin to the southeast (Fig. 11). The arch is partly a response to upward bulging of the basement necessary to accomodate the very thick sedimentary rock loads of the adjacent basins (see Sanford *et al.*, 1985; Sanford and Smith, 1987). The basins were initiated during major tectonic events along the eastern margin of North America (*e.g.*, Taconic Orogeny; Quinlan and Beaumont, 1984; Appendix A).

Repeated upward bulging of the Findlay-Algonquin Arch has likely been accomplished by reactivation of faults within the brittle basement rocks. Mapping of the Proterozoic peneplain where it is exposed north of the GTA, identifies prominent northeast-trending faults that affect both Proterozoic and overlying Paleozoic strata. The many *en echelon* lakes (such as Pigeon Lake, Balsam Lake, Chemung Lake, *etc.*; Fig. 10) record erosion of faulted and fractured Paleozoic strata. Subsurface mapping of the peneplain surface to the south also identifies northeast-trending faults passing below the GTA (Bailey Geological Sevices Ltd., 1984; Fig. 10A), which have the same trend as faults within the CMBBZ (Figs. 9B) suggesting a genetic relationship.

LOWER PALEOZOIC COVER ROCKS

The gently dipping basement peneplain in southern Ontario is covered by a layer-cake succession of Paleozoic sedimentary strata (limestones, dolostones, shales; Fig. 6). To the northwest of the Findlay-Algonquin Arch, the succession thickens and dips gently (< 2°) southward into the Michigan Basin (Fig. 11). Southeast of the arch, the succession thickens eastward into the Appalachian Basin. The northern margin of Paleozoic cover rocks runs approximately west-east some 50 km north of the GTA (Figs. 6, 11); to the north lies the exposed Proterozoic basement (shield).

Johnson *et al.* (1992) showed that the vertical succession of Paleozoic cover strata of southern Ontario can be subdivided into a number of depositional sequences separated by regional erosion surfaces (unconformities). These record regional uplift of the Findlay-Algonquin Arch, resulting in exposure, cutting of unconformities alternating with subsidence, and renewed sedimentation. The GTA is underlain by shales of the upper Ordovician Whitby (now Blue Mountain), Georgian Bay and Queenston formations (Fig. 6; Churcher *et al.*, 1991; Johnson *et al.*, 1992; Lehmann *et al.*, 1995). These overlie older limestones of the Simcoe Group exposed to the

east (Fig. 6). The change from limestone to shale records regional subsidence at about 440 Ma in response to the load created by thickened crust being generated by the Taconic Orogeny to the east (Brett and Brookfield, 1984; Lehmann *et al.*, 1995). Shales are exposed along river valleys close to Lake Ontario (*e.g.*, Kerr and Eyles, 1991; Rogojina, 1993) and their presence below the GTA has significance with regard to radon hazards (Chapter 28). Westward, the Niagara Escarpment marks the strike of Silurian dolostones that are resistant to erosion (Figs. 1, 6).

Neotectonic Stresses

There is a growing awareness of ongoing structural activity (neotectonics; Stewart and Hancock, 1994) in mid-continent North America. This area has traditionally, and incorrectly, been regarded as geologically stable (see below).

The GTA is slowly drifting southwestward as the North Atlantic Ocean continues to open and North America is pushed westward at about 3 cm·a^{-1}. The drift of the crust over the deeper rocks of the Earth's mantle creates stresses in overlying continental rocks and results in compressive forces acting horizontally in an ENE-WSW direction (Zoback, 1992). At the same time, the ground surface across the GTA is slowly rising in elevation (about 20 cm·a^{-1}) following the removal of the heavy load of the last ice sheet (termed glacioisostatic rebound; see below). This uplift also sets up horizontal and vertical stresses in near-surface rock strata (Neuzil and Pollock, 1983; James and Morgan, 1990). Near-surface, horizontal compressive stresses as high as to 14.5 MPa are reported by Palmer and Lo (1978).

Neotectonic forces reactivate deeper geologic structures, such as the numerous Proterozoic terrane boundaries and faults discussed above, resulting in earthquakes and the propagation of faults and fractures upward to the surface (Sanford *et al.*, 1985; Eyles *et al.*, 1993). In addition, the same stresses are sufficiently strong to create upward buckling (pop-ups; Fig. 6) and fracturing (jointing) of the Paleozoic bedrock surface (Wallach *et al.*, 1993; Scheidegger, 1980, 1993). Joints are present in relatively young (< 18 ka) Pleistocene sediments (Daniels, 1990; Wills *et al.*, 1992; see below) showing that such sediments are also being stressed and fractured in response to current stresses. Bedrock joints in southern Ontario show strongly preferred regional trends at 54°, 104° and 164° (Eyles and Scheidegger, 1995). Joint-forming mechanisms are discussed by Lorenz *et al.* (1991), Engelder (1994) and Scheidegger (1995).

In the oil-producing areas of southwestern Ontario, Sanford *et al.* (1985) showed that reactivation of the Findlay-Algonquin Arch and fracturing of overlying strata has controlled the migration pathways of oil and gas and the geometry of hydrocarbon reservoirs within Paleozoic rocks. Gases, such as methane and radioactive radon, escape to the surface along such structures (Sanford *et al.*, 1985; McCarthy and Reimer, 1986; Raven, *et al.*, 1992; Kappel and Tepper, 1992; Wallach and Heginbottom, 1993; Sherwood Lollar *et al.*, 1994; Noor *et al.*, 1992; Chapters 28, 34). In southwestern Ontario, oil has escaped to the surface and is found in young Pleistocene sediments resting on bedrock (Weaver *et al.*, 1995), explaining the early (mid-nineteenth century) exploitation of this resource.

Earthquake Risk

The word Toronto is inherited from the 17th century Huron Indian word meaning "the meeting place by the lake". It is an apt and prophetic title for an urban area now known to lie directly above the intersection of many geological structures

(Fig. 8). Such intersections (known as tectonic knots) are loci for earthquakes (Talwani, 1988), and there has been intense discussion of earthquake risk within the densely populated GTA, which contains nuclear generating stations at Pickering and Darlington (Fig. 1).

Traditional geologic notions held that frequent and powerful earthquake activity is limited to the active margins of the Earth's tectonic plates (*e.g.*, British Columbia, California) and that plate interiors such as mid-continent North America are stable. This attitude is epitomized by the Hare Commission Report of 1977 on the storage of nuclear waste below the Canadian Shield which concluded that the Shield has been stable for hundreds of millions of years and will continue to be stable for further millions of years. There is now increasing recognition of the hazard posed by infrequent, but large magnitude, intraplate earthquakes to eastern North American cities, including the GTA. Such earthquakes are the result of episodic reactivation of buried basement structures (Park and Jaroszewski, 1994). Johnston *et al.* (1994) overcame the lack of an historical earthquake record in southern Ontario by substituting "space for time" and reviewed the incidence of intraplate seismic activity worldwide. They determined a distinct correlation between large intraplate earthquakes ($M > 4.5$) with zones of broken and rifted crust such as occurs under the GTA and concluded that Pickering Nuclear Generating Station was "particularly vulnerable to earthquakes" (v. 1, p. 3).

Prompted to respond to public concern arising from reports of surface faulting along the Rouge River in the eastern part of Metropolitan Toronto (Fig. 10; Mohajer *et al.*, 1992, 1995; Hibbert *et al.*, 1994), the Geological Survey of Canada has stated that an earthquake of $M = 7$ can be expected anywhere in the western Lake Ontario region (Adams *et al.*, 1993a, b). Wallach (1995) concluded on the basis of available geologic information that the region encompassing the nuclear power plants may be subject to a greater seismic hazard than previously suspected (or considered when the plants were designed). The effects of earthquakes on the stability of downtown buildings and other infrastructure underlain by historic fill is also a cause for concern (see below).

Destructive, large-magnitude intraplate earthquakes have happened recently in Australia and India, areas regarded previously as stable (see Crone *et al.*, 1992). A problem is that the recurrence interval of such earthquakes is much longer than the historical record of seismic activity that has occurred since European colonization. Also, in central Canada, the bedrock surface is covered by thick glacial sediments. As a result, the pitfalls of using historical records as an index of seismic hazard in mid-continent are now being realized. Short episodes of intraplate earthquake activity are separated by long periods of inactivity and each new event redefines existing seismic zonation and hazard maps.

The role of reactivated basement structures.   In the absence of data to the contrary, it is safe to assume that the many faults and basement lineaments recognized close to the GTA (*e.g.*, Figs. 7, 8, 9, 10A) are potentially active. According to the Geological Survey of Canada, as many as 40 large-magnitude earthquakes ($M > 6$) might have occurred in Ontario during the 10,000 to 14,000 years since deglaciation, with a number of these having broken the surface to produce scarps (Adams, 1995). Recurring small-scale tectonic activity along basement structures is suggested by episodic low-magnitude seismic activity in the GTA, particularly along the Burlington-Toronto Seismic Zone (Fig. 8; Mohajer, 1993; Chapter 29).

The CMBBZ is blanketed by Paleozoic and Pleistocene cover strata. Despite this, there is an excellent correspondence between the location of this structure (Fig. 9A), the position of channels cut on the Paleozoic bedrock surface (Figs. 12, 13), and the form of modern surface topography in the area between lakes Ontario and Scucog (Fig. 14). Near Pickering, the shoreline of an ice-dammed lake (Glacial Lake Iroquois; see below), formed about 12.5 ka, shows a pronounced northward embayment directly above the western margin of the CMBBZ (Fig. 9A) suggesting a structural control. In the same area, a topographic lineament crosses thick Pleistocene sediments of the Oak Ridges Moraine (Fig. 14). The Iroquois shoreline is tilted up to the east, as a result of differential post-glacial uplift, and, where the shoreline crosses the CMBBZ, it shows a pronounced steepening (see below). Further north, near Kirkfield, shorelines of another glacial lake (Glacial Lake Algonquin), also cut about 12,000 years ago (see below), are offset where they cross the CMBBZ (Johnson, 1916; Fig. 9A). In this area, small-scale faults have been mapped by Sanford (1993) and Wallach (1995). Elevated concentrations of hydrocarbon gases occur in soils above the CMBBZ, suggesting upward outgassing along the structure (Chapter 34).

In summary, a wide range of topographic, seismic and geologic data strongly suggest reactivation of the CMBBZ (Wallach and Heginbottom, 1993; Seeber and Armbruster, 1993) and the potential for damaging earthquakes; there is a critical need for detailed studies of this structure and the many others that cross the GTA (Figs. 8, 9, 10, 13). The GTA contains a thick sedimentary record of the last 100,000 years (see below). Systematic examination of such strata for earthquake-related structures is needed (Fig. 10B).

Neotectonics and Waste Management

Lack of understanding of neotectonic stresses in bedrock below the GTA has been brought to attention by recent proposals to construct landfill waste sites in abandoned bedrock quarries along the Niagara Escarpment.

Case study: Steetley (South Quarry).   A recent proposal by Redland Quarries (formerly Steetley Quarries) called for the landfilling of 26 million tonnes of municipal and non-hazardous industrial waste (see Chapter 39). The site is an old quarry in fractured Guelph Formation dolostones at Steetley (South Quarry) along the Niagara Escarpment (Fig. 1). An engineered leachate collection system, consisting of a polyethylene liner and drainage control layers, was to be used to control the escape of leachate from the base of the waste pile (*e.g.*, Chapters 20, 23). After a lengthy public hearing, the undertaking was rejected (see Environmental Assessment Board & The Joint Board, 1995), in part because of the risk of bedrock pop-ups occurring directly under the liner and thereby rupturing it. Initial engineering plans called for closely spaced trenches to be excavated into the top of bedrock so as to allow expansion of the rock mass and release of stress. These were shown, during the course of the public hearing, to likely promote the ingress of chloride-rich ground waters into the landfill from deeper shales. The presence of saline water would complicate attempts to monitor and model chloride migration arising from the landfill operation (see Chapter 24 for discussion of this problem). The absence of any natural protection should the synthetic liner fail, the difficulty of modelling ground-water flows in fractured bedrock, and the likelihood of karst (subsurface caves) in the vicinity of the site were also cited. The Joint Board specifically identified a poor understanding of the stress field in bedrock as a reason for their decision. These considerations are likely to figure promi-

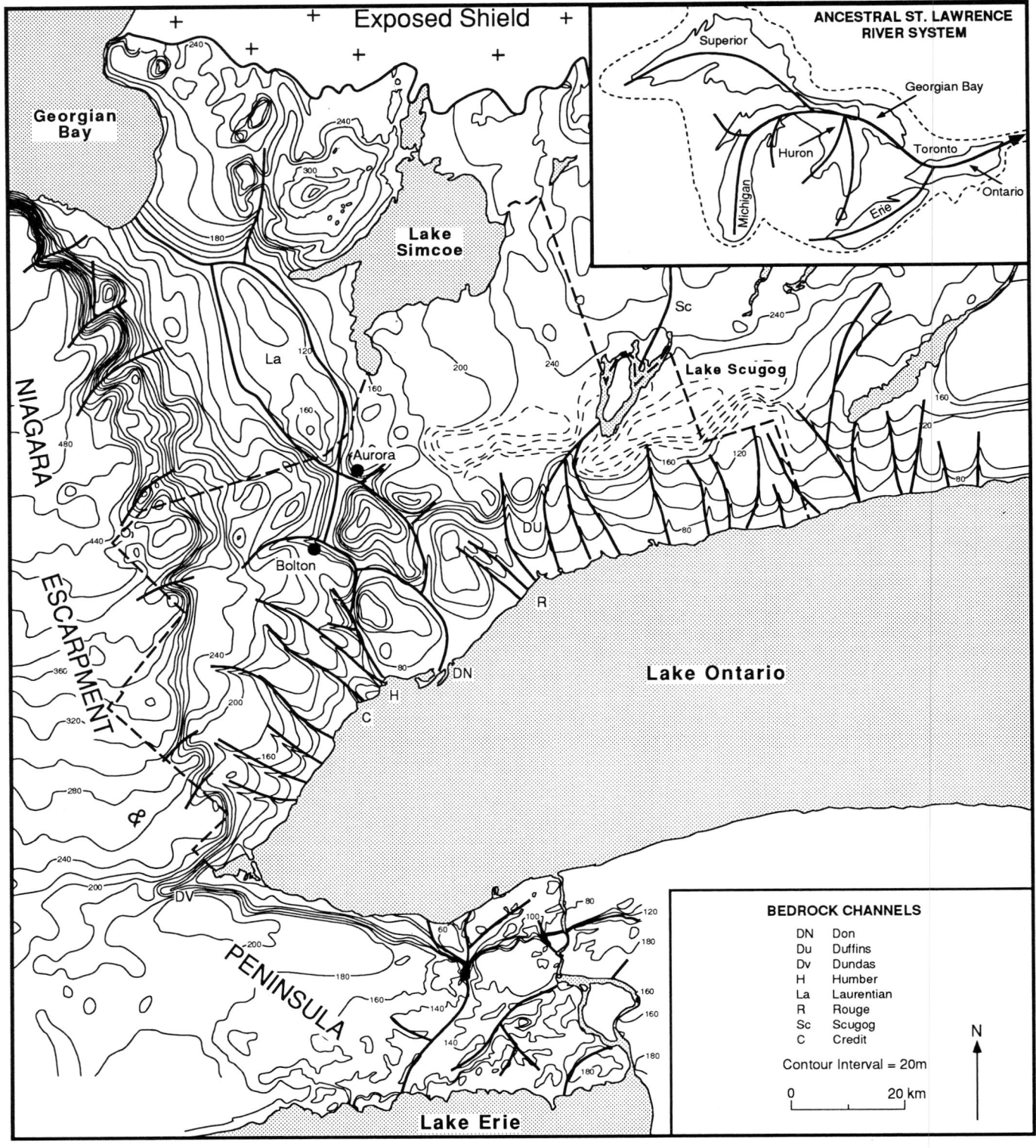

Figure 12   *Bedrock topography and buried bedrock channels (black lines) of southern Ontario (after Eyles et al., 1993). GTA (dashed line) lies at southern end of a broad bedrock channel network (Laurentian Channel) connecting Georgian Bay and the bedrock basin now occupied by Lake Ontario. Thick Pleistocene glacial sediments infill the bedrock topography; cross-sections near Bolton and Aurora are shown in Figs. 18A and 41, respectively; a section across the Niagara Peninsula is depicted in Fig. 18C. Inset shows Spencer's (1890) reconstruction of the Laurentian River.*

*Lake Scugog was enlarged by damming in the 1830s, but occupies a low in the bedrock surface eroded above the CMBBZ (Fig. 9). Elongate finger lakes northeast of GTA (e.g., Fig. 2) are structurally controlled by underlying basement structures. The Lake Simcoe basin can be essentially regarded as an extension of Georgian Bay; both features are structurally controlled by the Georgian Bay Linear Zone (**GBLZ**; Fig. 8). In view of the likely structural control on the location of bedrock channels, bedrock topography can probably be used to predict the presence of faults.*

nently in any future attempts to locate waste sites in bedrock quarries along the Niagara Escarpment (*e.g.*, Taro Aggregates Ltd., 1995; Chapter 39). Other examples illustrating the importance of understanding neotectonic structures (joints) in bedrock and Pleistocene sediments are provided later in this paper with regard to disposal of hazardous industrial waste.

Buried Bedrock Topography and Pre-glacial Drainage

There is no sedimentary record of the time period between the Silurian and the Pleistocene in the GTA. Any sedimentary record from the long time interval between the Silurian and the Pleistocene, if it was present, has been eroded. Late Jurassic igneous rocks occur as intrusive dykes just beyond

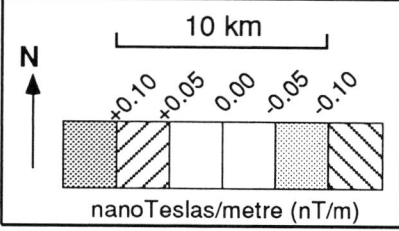

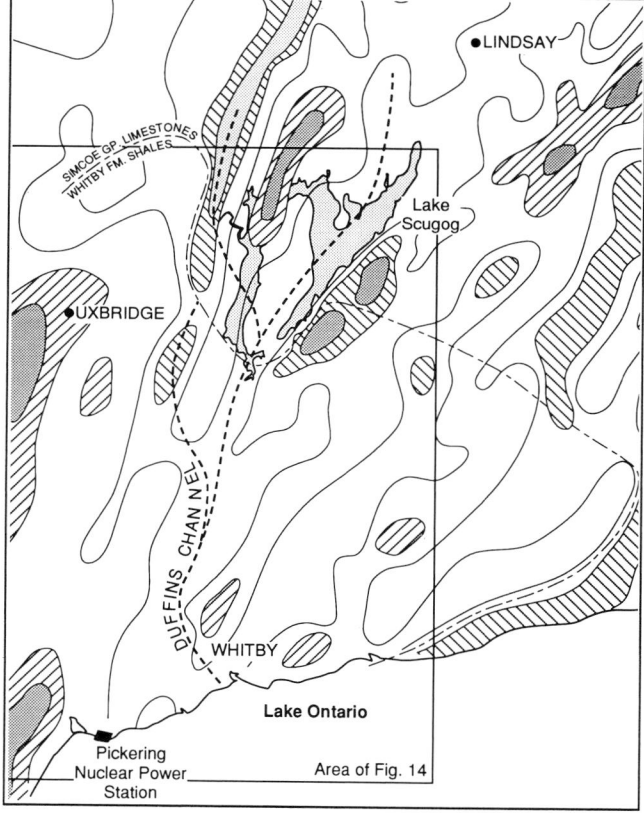

Figure 13 *Simplified contour map showing aeromagnetic gradient (in nanoteslas per metre: nT·m⁻¹) across eastern GTA near Pickering nuclear plant. Contours outline boundaries of geologic units of contrasting magnetic properties and likely identify northeast-trending thrust faults within the Central Metasedimentary Boundary Zone (**CMBBZ**; Figs. 7, 9). These structures lie below a cover of Paleozoic strata several hundred metres thick, but have controlled the position of the Duffins bedrock channel cut on top of cover strata (see Fig. 12) and the contact between the Simcoe Group limestones and Whitby Formation shales (Fig. 6). Note how Lake Scugog is nestled between basement aeromagnetic lineaments suggesting a close structural control. In addition, the topography of the modern day ground surface, clearly reflects the position of the underlying Proterozoic structures (Fig. 14).*

the eastern border of the GTA in Belleville. These record the opening of the North Atlantic Ocean at about 175 Ma (Barnett *et al.*, 1984) when southern Ontario was subject to uplift, extension and erosion (Miller and Duddy, 1989).

The upper surface of Paleozoic strata, now blanketed with Pleistocene glacial deposits, forms a distinct paleotopography. This surface is in places deeply weathered (to depths of 10 m) and is cut by numerous channels. These channels identify a former mid-continent river system that either drained southwestward through the Erie Basin to the Mississippi River (Grabau, 1901), or more likely, east to the Atlantic (Fig. 12). The age of this system is unknown, but it has been modified, and in places over-deepened below the natural grade of the channel, by glacial erosion. The orientation of bedrock channel segments, bedrock joints and basement structures is, in many areas, virtually coincident, suggesting a structural pre-design to the evolution of the channel system (Eyles and Scheidegger, 1995). The buried bedrock topography has subsequently been infilled with Pleistocene sediments that record major climatic changes in recent Earth history.

PLEISTOCENE SEDIMENTS

The last 2.5 Ma of Earth's history has been characterized by cycles of cold glacial episodes, that allowed the growth of continental ice sheets, and warmer periods (interglacials) akin to the present day. Glacial-interglacial cycles are controlled predominantly by so-called Milankovitch variations in Earth's orbit around the sun (see Hays *et al.*, 1976; Imbrie and Imbrie, 1976; Berger and Loutre, 1992; Eyles, 1993; Jouzel *et al.*, 1993). The last glaciation in North America (named the Wisconsin Glaciation) started in Canada at about 100 ka and finished about 10 ka (Fig. 15). As a result of erosion and transport below and on the margins of the Canadian (Laurentide) ice sheet, glacial sediment, as much as 200 m thick, smothers the bedrock topography of the GTA and gives rise to much of the present-day topography (Figs. 16, 17, 18, 19).

In addition to erosion and deposition of sediments, glaciations also result in significant vertical crustal movements which are created by the thickening and thinning of ice sheets. The Earth's crust bends below the heavy load of a continental ice sheet and forces underlying asthenosphere rocks to flow away from the load to form a peripheral bulge well beyond the ice sheet margin. At equilibrium, the amount of material displaced is equal to the mass of the ice sheet, such that, at any one point, the crust is depressed (glacio-isostatic depression) by an amount equal to ice thickness multiplied by the ratio of the density of ice to asthenosphere rocks (about 0.30). Given ice thicknesses over the GTA of about 1 km during the last glaciation, the region would have been depressed by no more than about 300 m. Because of the high viscosity of the asthenosphere, full crustal depression takes more than 20 ka and, as this is usually longer than the lifetime of any ice sheet, full depression is seldom attained (see Morner, 1971; Oerlemans and van der Veen, 1984). In turn, crustal rebound occurs when the ice sheet thins and retreats; full recovery takes a similar length of time and is not yet complete in the Great Lakes area following retreat of the last ice sheet. The environmental geologic effects of ongoing glacio-isostatic movements can be clearly recognized in the GTA in the form of changing lake levels (Figs. 4A, 20, 21A; see below). The horizontal and vertical stresses that affect bedrock during rebound have already been discussed (see above).

Pleistocene Geology of the Greater Toronto Area

The distribution of different Pleistocene sediment types at

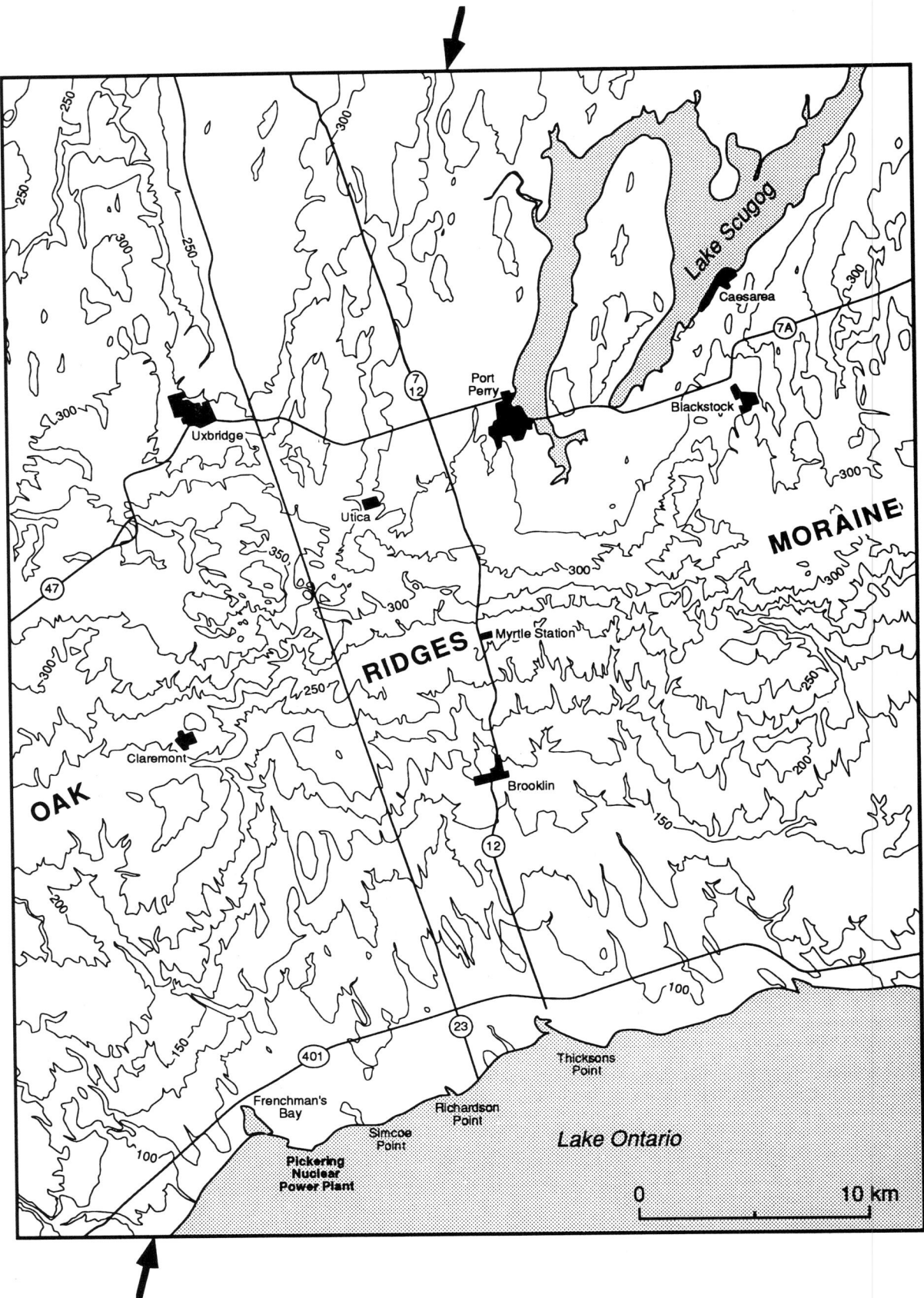

Figure 14 *Elevation contours (metres above sea level) of modern ground surface north of Pickering nuclear plant (see Figs. 9, 13 for location) showing a surface lineament (between arrows) that crosses a belt of thick Pleistocene sediment cover (the Oak Ridges Moraine; Fig. 8) and which lies directly above the western margin of the CMBBZ as defined by magnetic properties (Fig. 13). This relationship, together with warping of lake shorelines to the north (Fig. 9A), is strong evidence for geologically recent movement along the CMBBZ. The Oak Ridges Moraine is no older than 13.3 ka.*

surface across the GTA is well known as a result of a long history of geological mapping (*e.g.*, Miller, 1913; Coleman, 1932; Chapman and Putnam, 1984; Sharpe, 1980; Barnett *et al.*, 1991; Fig. 16). Coleman's Pleistocene map of Toronto was especially influential because it prompted early examination of the topographic and geologic controls on the growth of Toronto (*e.g.*, Taylor, 1936; Deacon, 1944; Kerr and Spelt, 1965). Statistical techniques developed by geologists for mapping spatial variations in rock types (*e.g.*, Krumbein, 1955) were also employed to investigate population structure and incomes across the city (Murdie, 1969). Given the stratigraphic and sedimentologic complexity of glacial sediments, existing maps of surficial geology of the GTA convey both an unrealistic sense of homogeneity and little information of subsurface conditions. Much remains to be learnt of the three-dimensional subsurface geometry and variability of the major Pleistocene stratigraphic units and their role as aquitards and aquifer units controlling regional and local groundwater flow (Chapters 9, 10).

As with European experience, much of the early knowledge of the geology and topography of the region was gained during the construction of canals (Angus, 1988). Subsequent advances in understanding of subsurface conditions have resulted from infrastructure projects such as subway construction (*e.g.*, Watt 1954; Legget and Schriever, 1960; Karrow, 1969; Lajtai, 1969; Sharpe, 1980; Eyles, 1987), regional sewer construction (*e.g.*, Marshall Macklin Monaghan Ltd., 1974), canal realignments (Owen, 1969, 1972), evaluation of regional ground-water and aggregate resources (Watt, 1968;

Hewitt and Yundt, 1971; Sibul *et al.*, 1977; Fligg, 1983) and the recent searches for waste disposal facilities (Cooper *et al.*, 1989; M.M. Dillon Ltd., 1994; Golder and Associates Ltd., 1994; Boyce *et al.*, 1995; Figs. 17, 18). Records of water wells on file with the Ontario Ministry of Environment and Energy allow bedrock topographic and drift thickness maps to be produced for most areas. In general, however, the subsurface geological and hydrogeological database for the GTA is disjointed with no central electronic clearing house where such information can be readily stored and retrieved (Chapter 37).

Pleistocene Strata older than the Last Glaciation (>25 ka)

Sediments older than the last (Wisconsin) glaciation are rarely preserved across the GTA. The oldest known Pleistocene deposit is the York Till (*ca.* 135 ka) which was deposited during the penultimate (Illinoian) glaciation (Figs. 15, 17) and rests directly on bedrock (Fig. 18A, B). It is exposed in a railway cut at Woodbridge and in a former brick pit (the Don Valley Brickyard) close to downtown Toronto (White, 1975; Karrow, 1989; Eyles and Williams, 1992) and is commonly encountered during downtown construction activity (Trow and Bradstock, 1972). Overlying lacustrine sediments of the Don Beds (Figs. 17, 18B) are richly fossiliferous and record the warm climate of the last (Sangamon) interglacial (Fig. 15). These deposits are only exposed in the Don Valley Brickyard quarry and sporadically along the Rouge River (Coleman, 1932; Eyles and Clark, 1988a; Karrow, 1989; Eyles and Williams, 1992; Williams and Eyles, 1995). The peak warmth of the last interglacial was at about 125 ka. The age of the Don

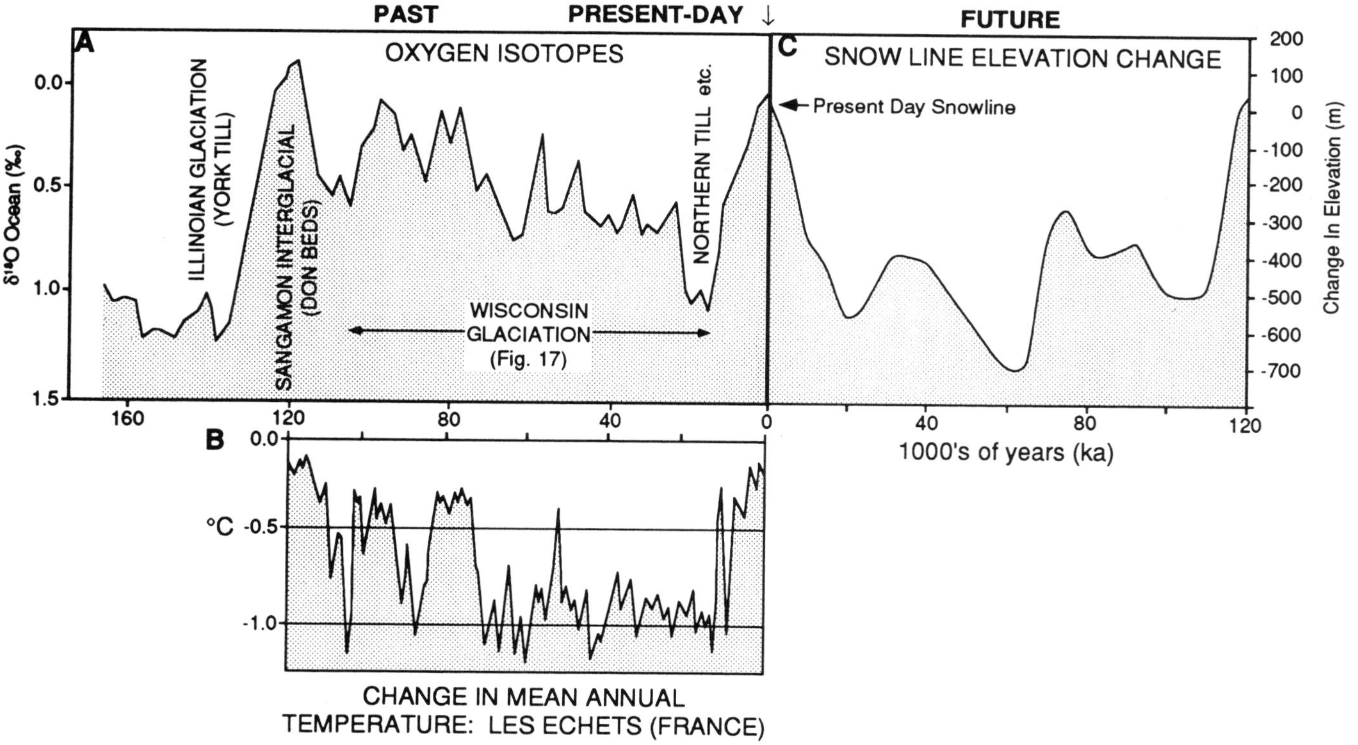

Figure 15 *Reconstruction of GTA climate over last 140,000 years and over next 120,000 years in absence of greenhouse gas global warming.* (**A**) *Oxygen isotope variation from analysis of deep ocean sediments deposited during last glaciation from which volume of glacier ice on planet Earth is derived (after Jouzel* et al., *1993). The Laurentide Ice Sheet in Canada was the largest glacier on the planet so these data give a crude indication of extent of the ice sheet over Canada and climate trends. Principal geologic units in GTA are indicated (Fig. 17).* (**B**) *Long-term record of climate from Les Echets, France derived from investigation of pollen in sediment; note similarity with trends shown by oceanic isotope data (from Boulton and Payne, 1993). Temperature trends across GTA were probably similar, but no continuous sedimentary record of this time span survives in southern Ontario.* (**C**) *Results of modelling of future change in snow line elevation in northwest Europe using Milankovitch astronomical variables; data are a crude proxy for expected temperature trends in GTA. The most severe phase of the next glaciation may occur between 50 and 70 ka in the future when the GTA will be below ice as much as 1000 m thick.*

Beds is not precisely known, but is thought to be about 80 ka, so much of the last interglacial may have gone unrecorded in the Toronto area.

A key Pleistocene event in the GTA was deposition of the Scarborough Formation. This is a thick (> 40 m) and regionally extensive fluvio-deltaic deposit which records an abrupt increase in the size and depth of an ancestral Lake Ontario during the early stages of the last glaciation (80-60 ka?; Fig. 17). This lake stood some 50 m above the modern level of Lake Ontario (Kelly and Martini, 1986) and was fed by large braided melt-water rivers draining from the ice sheet. On the northern margins of the GTA, the Scarborough Formation can be identified in the subsurface as elongate bodies of fluvial sands and gravels infilling over-deepened bedrock lows along the Laurentian Channel (Figs. 12, 18A). Below Toronto, the Scarborough Formation was deposited as extensive lobes on

a delta, and deltaic sediments are well exposed along Scarborough Bluffs (Fig. 18B). There, they comprise a well-defined coarsening-upward succession (silts to sands), up to 45 m thick.

The Scarborough Formation is the principal aquifer system for many urban communities sited above the Laurentian Channel (e.g., the Bolton and Aurora aquifers; Fig. 18A). Methane gas (CH_4), derived partly from peat within the Scarborough Formation, was formerly used for heating purposes (see Coleman, 1932; Chapter 25), a resource that could still be used.

The Scarborough Formation is draped by the Sunnybrook Diamict (Figs. 17, 18B), which was deposited during the early phases of the last glaciation, possibly at about 40 ka. The term diamict is used in recognition that the origin of the unit is not yet fully understood. The Sunnybrook has the appearance of a massive or weakly laminated pebbly silty-clay, varyingly interpreted either as a deformation till resulting from glacial over-

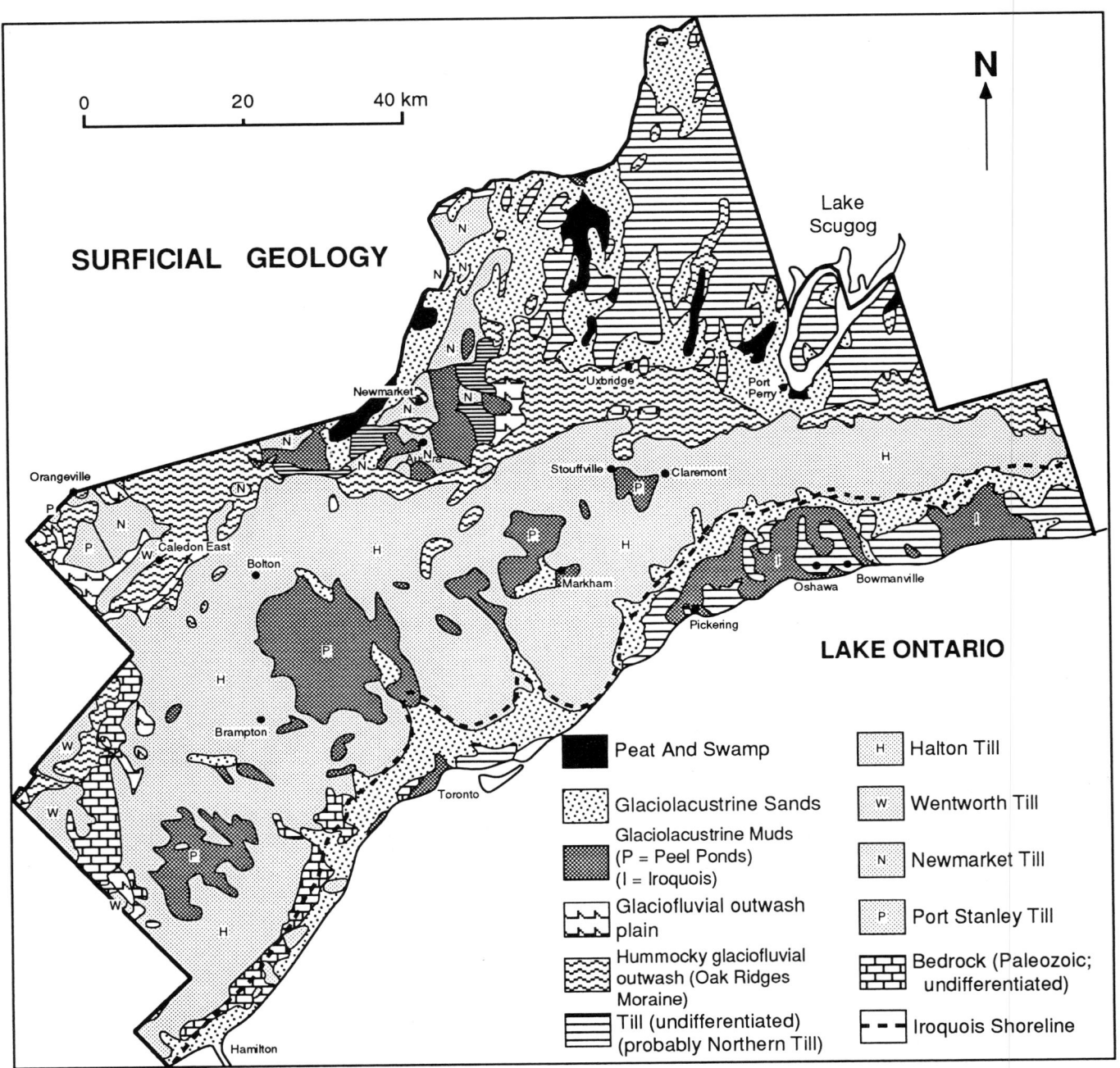

Figure 16   *Surficial Pleistocene geology of the GTA (after Barnett et al., 1991 and Ontario Geological Survey, 1992). For stratigraphic designations (e.g., Northern till, etc.), see Figure 17.*

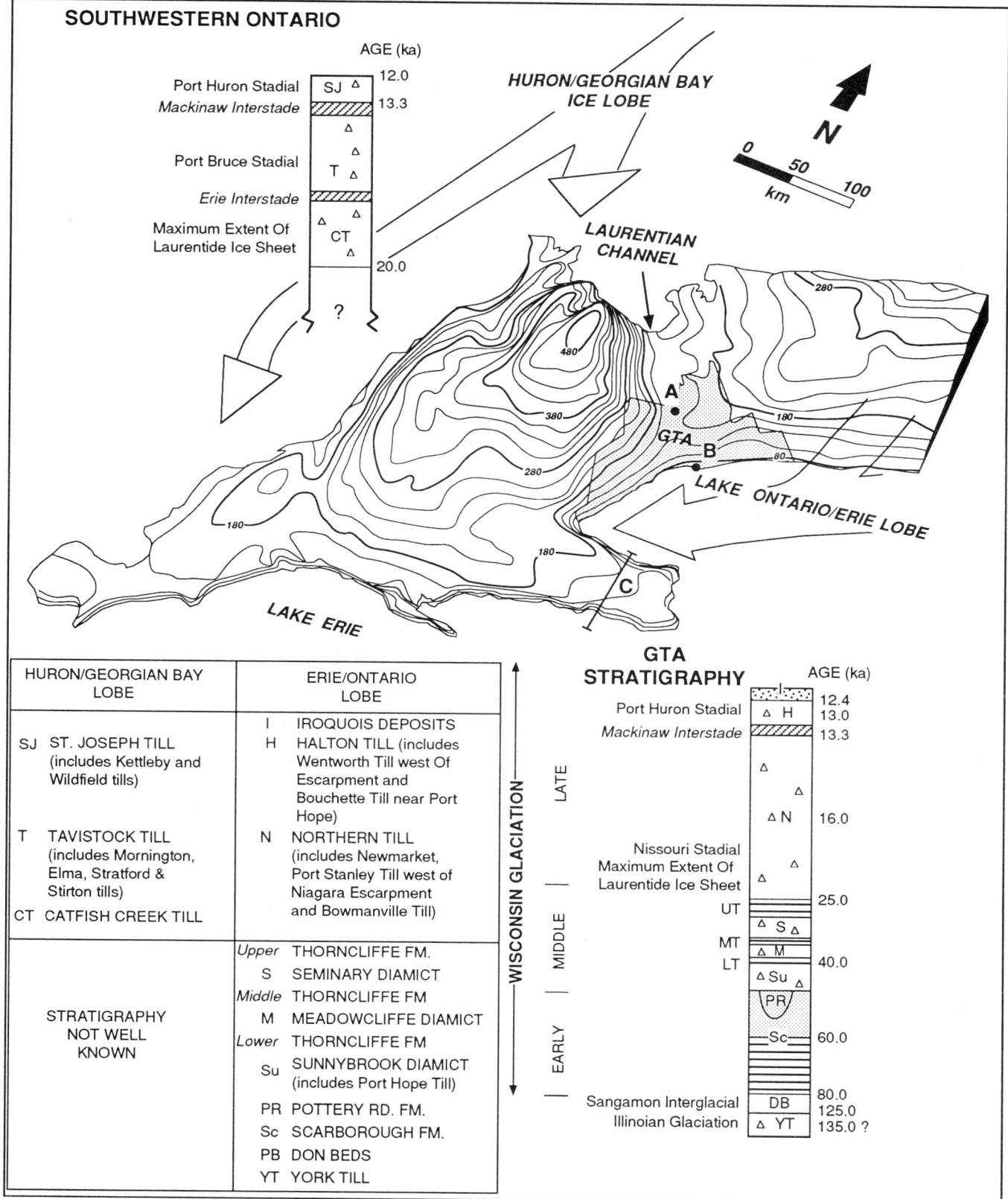

Figure 17 *Principal late Pleistocene strata in southern Ontario superimposed on oblique view of bedrock topography. GTA shaded. Bedrock surface shows topographic high developed over Findlay-Algonquin Arch (Fig. 8) that separates bedrock basins now occupied by lakes Huron, Erie and Ontario. These basins controlled the flow of two lobes of the Laurentide Ice Sheet during ice retreat following the late Wisconsin glacial maximum. Southwestern Ontario was affected by the Huron-Georgian Bay lobe. The GTA area and the eastern portion of Lake Erie basin was influenced by the Erie-Ontario lobe.*

Note that late Wisconsin stratigraphy of GTA contains last interglacial (Sangamon) and penultimate glacial (Illinoian) deposits (see Eyles and Williams, 1992; data from Karrow, 1989; Berger and Eyles, 1994; and Boyce et al., 1995). Location of cross-sections on Figure 18 shown at A, B and C.

riding of lake clays (Hicock and Dreimanis, 1989), or as a glaciolacustrine mud containing ice-rafted debris (Eyles and Westgate, 1987; Schwarcz and Eyles, 1991). It is well exposed along Scarborough Bluffs (Fig. 18B), can be traced eastward as far as Port Hope (Port Hope Till; Brookfield *et al.*, 1982; Fig.

17) and has been identified inland at Woodbridge (Fig. 1) as a pebble-free mud (White, 1975). The Sunnybrook is relatively impermeable and forms an extensive aquiclude that confines a regional aquifer within the underlying Scarborough Formation.

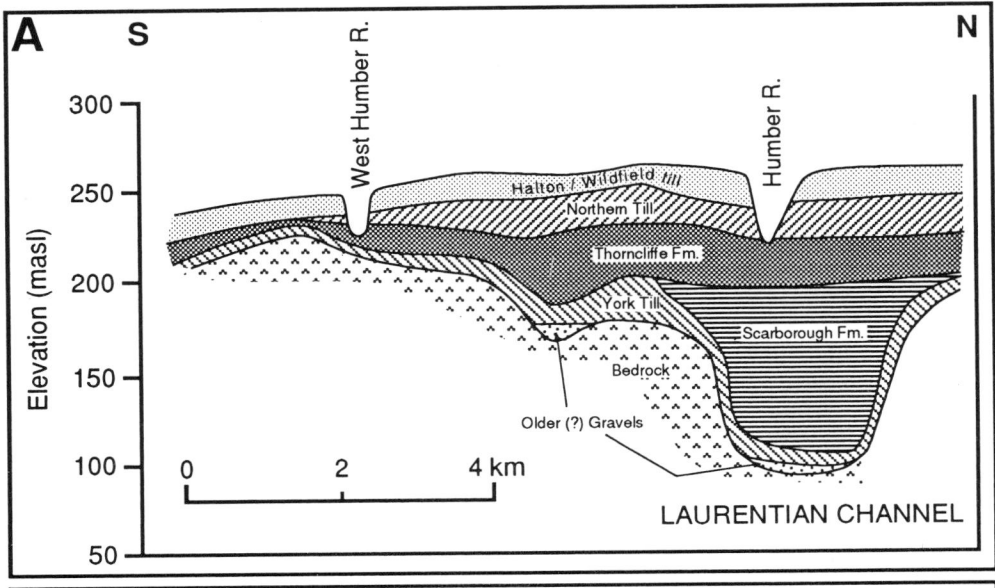

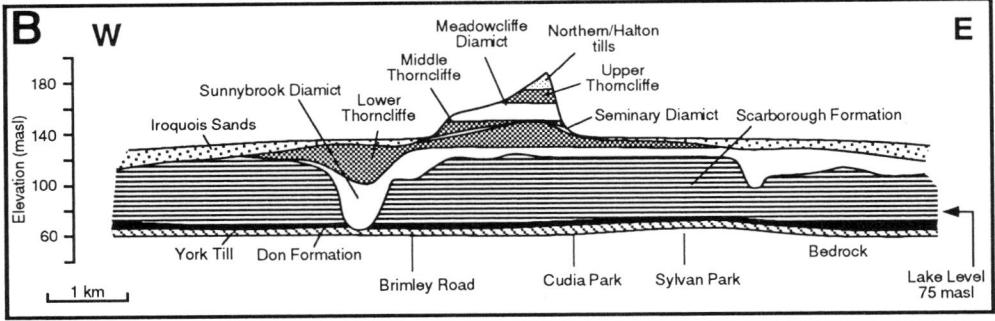

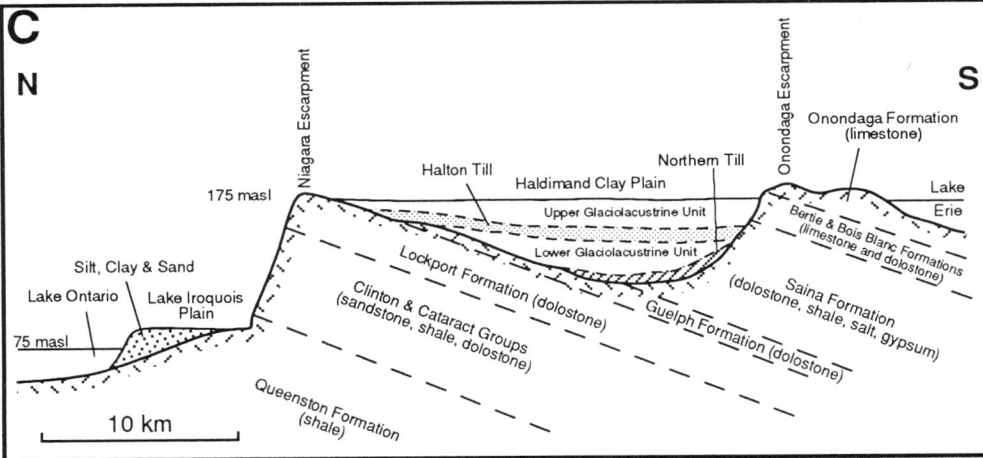

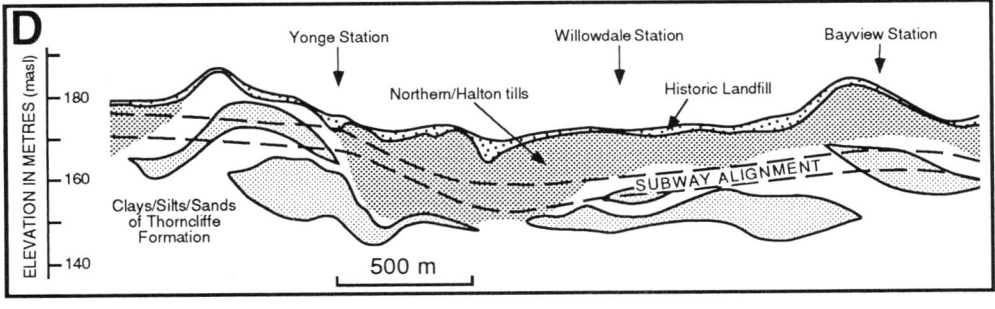

Figure 18 *Geologic cross-sections showing sequence of Pleistocene glacial sediments in GTA. For locations, see Figure 17.*
(A) *Section through the Laurentian Channel near Bolton (Fig. 12).*
(B) *Section along the Lake Ontario shoreline in the city of Scarborough where the Laurentian Channel broadens along the southern margin of the GTA. Note the changing geometry of the Scarborough Formation in A and B. At Bolton, the Scarborough Formation consists of sands and gravels, but, at Scarborough, it is composed of silts and sands.*
(C) *North-south section through the Niagara Peninsula (modified from Golder and Associates, 1994; Gartner Lee, 1987a; and work of author).*
(D) *Late Wisconsin tills encountered along planned west-east Sheppard Avenue subway. Section runs through the drumlin ridges and associated Thorncliffe Formation strata deformed by subglacial ice push. Note extensive cover of historic landfill up to 5 m thick (modified from Golder and Associates Ltd., 1995).*

The glaciolacustrine Thorncliffe Formation comprises thick silts and sands and, locally near Scarborough, pebbly muds (*e.g.*, Seminary, Meadowcliffe Diamict units; Fig. 17, 18B). The formation was deposited by glacial melt waters entering a deep, ice-dammed ancestral Lake Ontario (Eyles and Clark, 1988b). Silty sands of the Thorncliffe Formation infill the Laurentian Channel and host the largest aquifer complex in southern Ontario. The aquifer extends over 2500 km² north of the Oak Ridges Moraine (*e.g.*, Alliston Aquifer; Aravena and Wassenaar, 1993), and can be traced southward under the moraine to Lake Ontario (Chapter 9). Below the rapidly urbanizing south flank of the Oak Ridges Moraine, it comprises many interlinked aquifers under a cover of late Wisconsin tills (Haefeli, 1972; Sibul *et al.*, 1977; Howard and Beck, 1988; M.M. Dillon Ltd., 1994; Golder and Associates,

1994). The deeper parts of the Alliston Aquifer along the Laurentian Channel contain old waters (*ca.* 18 ka) recharged during the last glaciation (Aravena *et al.*, 1995). Locally, Thorncliffe sediments contain high concentrations of methane gas (> 1000 µmol·L⁻¹), produced by bacterial degradation of peat, that is isotopically distinct from thermogenic methane derived from underlying Paleozoic shales (Barker and Pollock, 1987). The Thorncliffe Formation can be traced eastward into the Bowmanville and Port Hope area where it has been mapped as Clarke Deposits (Brookfield *et al.*, 1982; Fig. 17).

Deposits of the Late Wisconsin Glaciation (25-13 ka)
During the maximum extent of the Laurentide Ice Sheet (Nissouri Stadial; about 22 to 16 ka; Karrow, 1989; Boyce *et al.*, 1995; Figs. 15A, 17), ice filled the Great Lakes basin and

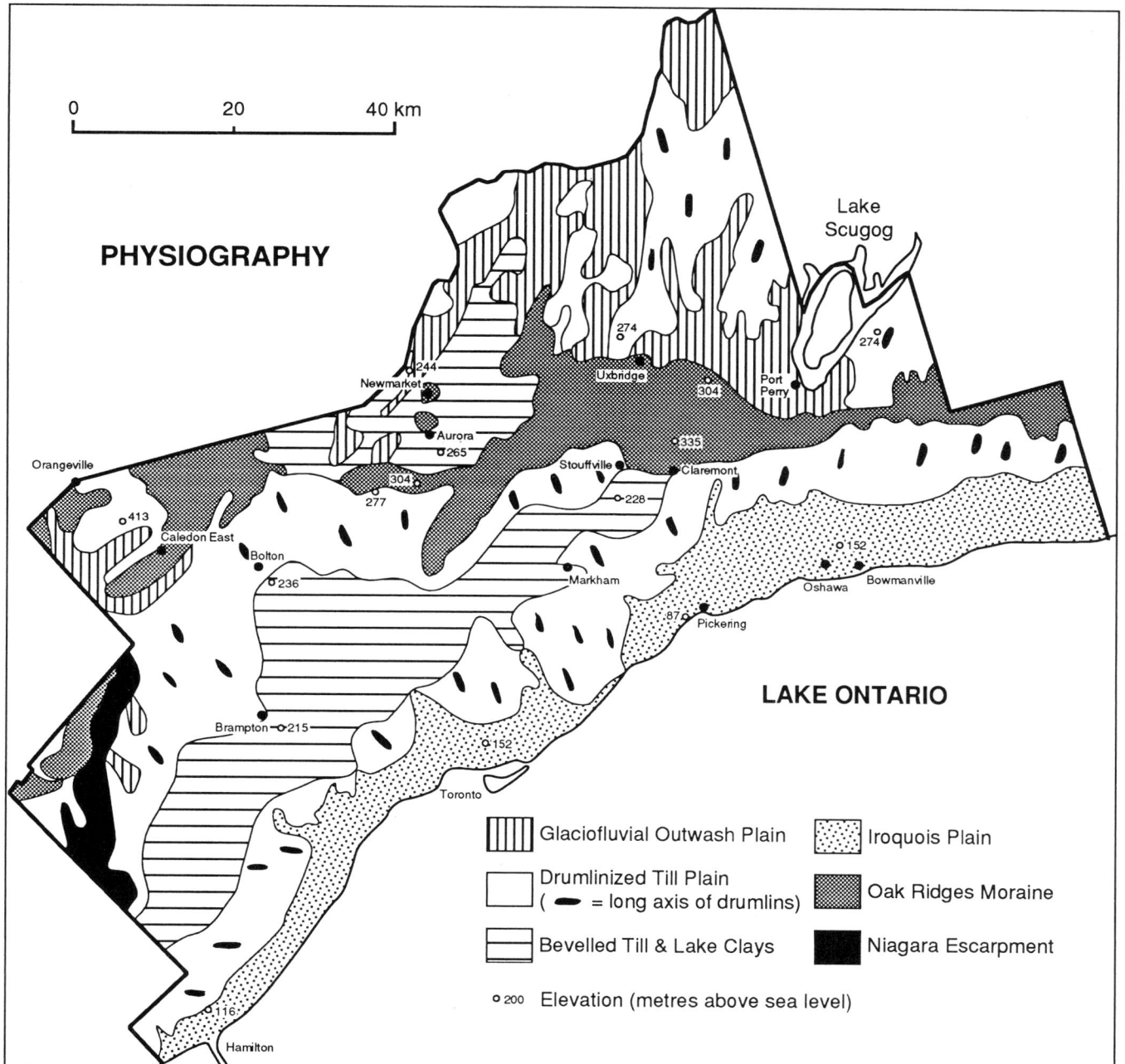

Figure 19 *Physiography of the GTA (after Chapman and Putnam, 1984). Undulating, hummocky topography of the Oak Ridges Moraine extends over 18% of GTA area; elsewhere, the low relief till plain and lake clays give rise to high-quality farmland, but, unfortunately, offer few topographic constraints to rapid urbanization (Fig. 37).*

covered the GTA. The Northern till is a dense and heavily over-consolidated diamict, locally as thick as 60 m, and can be traced as a stratigraphic marker across most of the GTA. The Northern till is correlative with the Newmarket Till to the north of the ORM (Gwyn, 1972) and the Catfish Creek Till which occurs in southwestern Ontario west of the Niagara Escarpment (Karrow, 1989; Fig. 17). These tills record deposition below the Laurentide Ice Sheet when it was at its greatest extent in mid-continent (Fig. 13). Thereafter, as the Laurentide Ice Sheet thinned and retreated from mid-continent, southwestern Ontario was affected by the Huron-Georgian Bay ice lobe, and the GTA by ice of the Ontario-Erie lobe (Fig. 17).

Because of its dense and supposedly impermeable character, the Northern till has been regarded as an aquitard preventing the infiltration of leachate to underlying aquifers and as a suitable substrate for landfills for waste disposal. However, recent work shows that the till contains extensive, sheet-like beds of sand and gravel recording melt waters flowing below the ice base. Such beds, with their high permeability, provide so-called hydraulic windows through which surface water and contaminants can penetrate through the till to underlying aquifers (Boyce et al., 1995; Chapters 9, 10, 20).

Mackinaw Interstadial sands, silts and gravels overlie the Northern till and record a cold, ice-free interval at about 13.3 ka, when the margin of the Laurentide Ice Sheet temporarily withdrew from the GTA (Fig. 17). Large areas of terrain were exposed in southern Ontario for colonization by plants and animals (e.g., muskox; Morris et al., 1993). A severely cold climate at that time is recorded by ice wedge casts that record sediment falling in open contraction cracks in the deeply frozen top of the Northern till. Deposits of the Mackinaw Stadial comprise several local aquifers in the GTA because they are overlain by the Halton Till.

The Halton Till records a subsequent short-lived re-advance (surge) of the Ontario-Erie ice lobe during the Port Huron stadial at about 13 ka (Fig. 17). This ice lobe moved northwestward to abut the Niagara Escarpment and dammed a high-level lake (Lake Whittlesey; Morris et al., 1993) in the Erie basin. Ice flow directions are recorded by the long axis of the many drumlins on the Halton Till (Fig. 19). These likely record deformation and moulding of the substrate below the ice (e.g., Boulton and Hindmarsh, 1989; Boyce and Eyles, 1991; Muller and Pair, 1992; Fig. 18D). A separate Wentworth Till is mapped to the west of the Niagara Escarpment (Fig. 17), but this simply represents the coarse-grained outer margin of the Halton Till sheet where it incorporated dolostone bedrock of the escarpment (White, 1975). At this time, the Oak Ridges Moraine was deposited between the northwestward flowing Ontario lobe and southward ice of the Huron-Georgian Bay lobe. The moraine is composed of ice-contact sands and silts deposited within a large interlobate lake that overflowed westward over the Niagara Escarpment into the Erie Basin. It is the principal recharge area for the many aquifers of the GTA (Chapters 9, 10).

Late Glacial Record (13-11.4 ka)

Downslope retreat of the Ontario ice lobe from the ORM resulted in the ponding of small ice-frontal lakes, the erosional bevelling of the Halton Till surface and deposition of a thin, discontinuous cover of lacustrine clays (e.g., the Peel Ponds; Figs. 16, 19). Eastward retreat of ice to the vicinity of Kingston resulted in damming of a high-level ancestral phase of Lake Ontario. The most prominent lake phase was that of Glacial Lake Iroquois, which stood about 35 m higher than the present-day level of Lake Ontario (Coleman, 1932, 1936; Figs. 9,

20). This lake formed between 12.5 and 11.7 ka, when retreating ice blocked the St. Lawrence Valley and drainage to the Atlantic was via the Mohawk River in New York State. The prominent Iroquois shoreline bluff, in places up to 30 m high, and associated gravel bar deposits mark the southern extent of thick late Wisconsin till across the GTA (Figs. 16, 19). At sites below the Iroquois bluff, the cover of late Wisconsin tills was thinned or completely removed by the waters of Lake Iroquois which left a blanket of beach sands and clays. Because of the steep Iroquois shoreline bluff encircling the city, Toronto earned a reputation in the early 19th century for being difficult to get into and out of by overland trail (Walker, 1965). The shoreline has been tilted up to the northeast (0.5 m·km⁻¹) by more rapid post-glacial rebound in that direction (Fig. 20). Iroquois beach deposits have been extensively

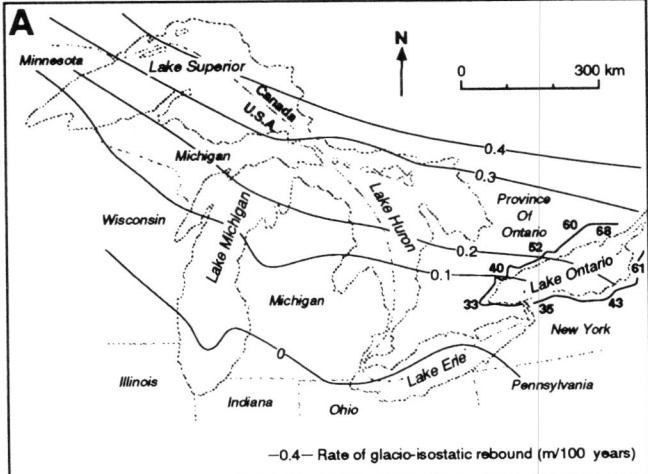

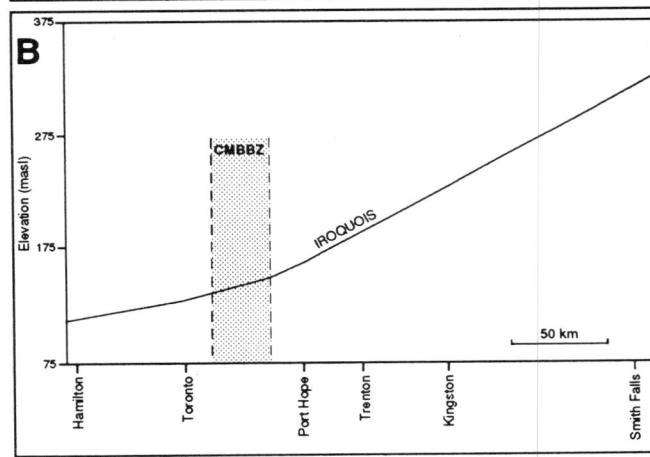

Figure 20 (**A**) *Rate of upward movement of the ground surface (cm per century) across the Great Lakes basin in response to postglacial unloading from load of last ice sheet (glacio-isostatic rebound). Topographic features such as the Lake Iroquois and Algonquin shorelines (that were originally horizontal) have been tilted up to the northeast; small bold numbers on diagram show elevation of Lake Iroquois shoreline **above modern level** of Lake Ontario (see also Figs. 9, 16, 43). The outlet of Lake Ontario near Kingston is rising faster than areas to the west resulting in rising water levels along the GTA shoreline (about 23 cm per century; Fig. 21) and flooding of river mouths (Figs. 4A, 22). (**B**) Elevation in metres above sea level of upwarped glacial Lake Iroquois shoreline from Hamilton in the west to eastern end of Lake Ontario basin. Note steepening of gradient where shoreline passes across the Central Metasedimentary Belt Boundary Zone (CMBBZ; Figs. 7, 9) suggesting reactivation of this structure in the form of a hinge. See also Figure 9A.*

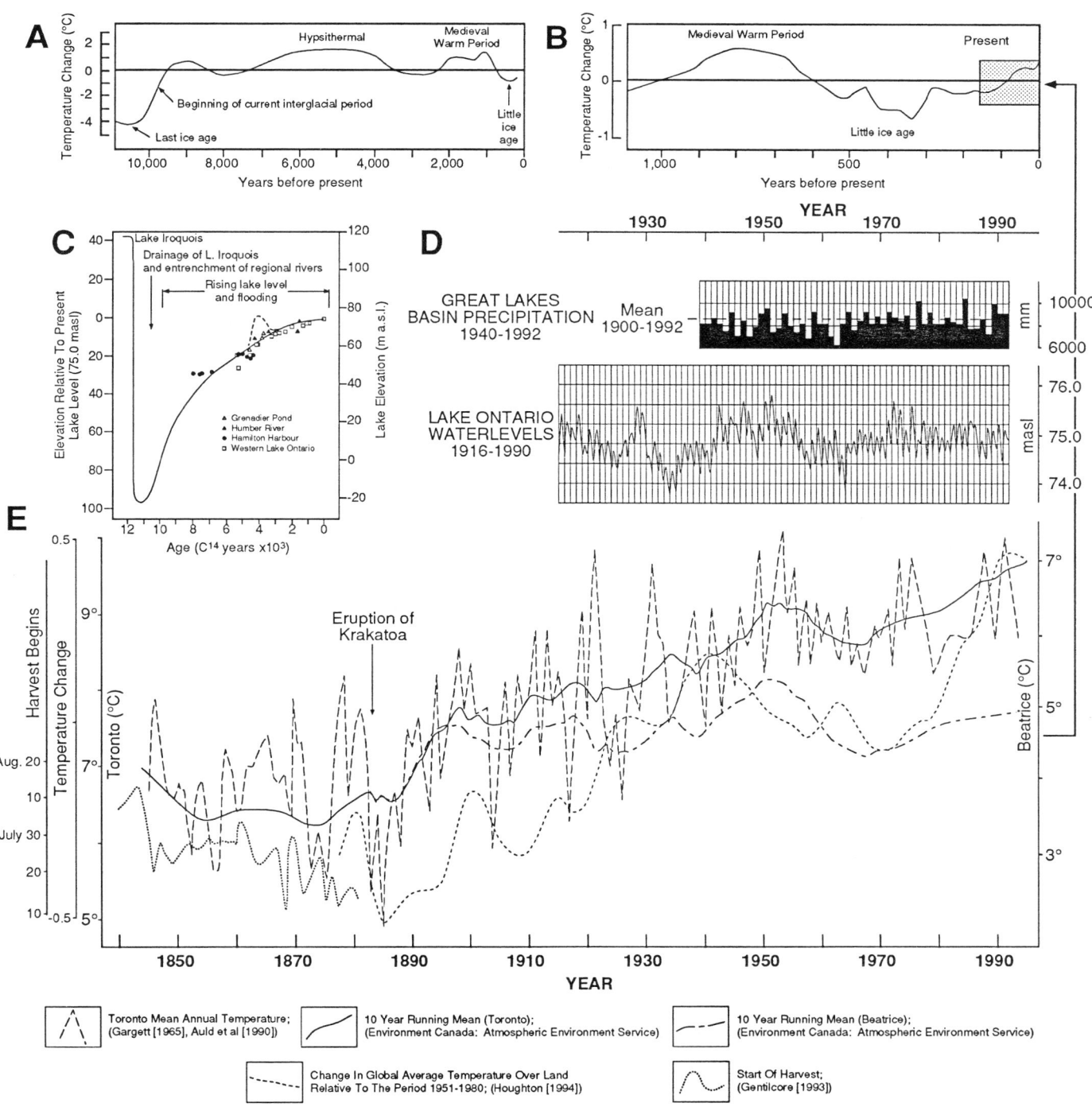

Figure 21 *Changing climate and level of Lake Ontario. (**A**) Climate over the last 10,000 years. (**B**) Over the last 1000 years. (**C**) Long-term changes in elevation of Lake Ontario after drainage of ice-dammed Glacial Lake Iroquois; data from the Humber River (Weninger and McAndrews, 1989), Grenadier Pond (McCarthy and McAndrews, 1988), western Lake Ontario and Hamilton harbour (Coakley and Karrow, 1994). Dashed line is postulated Nipissing Flood of Anderson and Lewis (1985) when lake level is argued to have stood 2 m higher than at present. The rate of crustal rebound at the outlet of Lake Ontario has been decreasing exponentially, resulting in decreasing rates of lake level rise over the last 10,000 years. Present rate of shoreline transgression is 23 cm per century (Fig. 16). (**D**) Short-term change in elevation of Lake Ontario water levels, 1916 to 1992 (after Environment Canada, 1994). (**E**) Effect of urbanization on Toronto climate. The growth of the urban area has resulted in a significant heat-island effect compared with outlying rural areas such as Beatrice, 150 km due north of Toronto. Note increased divergence of 10-year running mean annual temperatures for Toronto and Beatrice as the urban population of Toronto has grown. Overall temperature trends at both locations parallel global temperature trends after the termination of the Little Ice Age (ca. 1200 to ca. 1880 AD), but note pronounced global cooling effect caused by the eruption of Krakatoa in 1883. Walter Riddell kept a diary of farm life from 1841 to 1880 near Bowmanville in the eastern part of the GTA (Fig. 1); start of harvesting and length of growing season decreases as a result of cooling trend over this time interval (data from Gentilcore, 1993).*

worked for gravel, and little of the shoreline survives un-modified by mining or urbanization. Unfortunately, aban-doned gravel quarries have been rehabilitated by being filled with a wide variety of industrial and municipal wastes with no provision for ground-water protection (see below).

Shorelines of another late-glacial lake cross the extreme northern margin of the GTA, recording Glacial Lake Algon-quin, one of the largest lakes dammed in the upper Great Lakes region during the retreat of the last ice sheet, some time between 12.5 and 10.5 ka (Finamore, 1985; Fig. 9). This lake was contemporaneous with Glacial Lake Iroquois in the On-tario basin and, like the Iroquois shoreline deposits, Algon-quin beaches have undergone differential uplift as a result of greater postglacial glacio-isostatic rebound toward the north-east. The main Algonquin shoreline is warped where it crosses the CMBBZ (Johnson, 1916) suggesting recent neo-tectonic reactivation of that structure (see above).

Post-glacial Changes in Lake Level and Climate (11.4 ka to present)

Lake level.   With final ice retreat from the St. Lawrence valley, Glacial Lake Iroquois was drained, and lake levels in the Ontario basin fell to as much as 100 m below the modern level at about 11.4 ka (Coakley and Karrow, 1994; Fig. 21C). Rivers draining to Lake Ontario were rejuvenated and under-went entrenchment such that they now occupy narrow, steep-sided valleys (e.g., Humber, Don, Rouge). Major transporta-tion routes now follow these valleys and the stability of valley sideslopes is a recurring problem (e.g., Quigley et al., 1971).

In response to differential glacio-isostatic uplift toward the northeast (Fig. 20), the outlet of Lake Ontario east of Kingston is currently rising faster than the western end of the basin. As a result, the present-day level of Lake Ontario near Toronto (about 75 m asl) is rising at a rate of over 20 cm per century

(Kite, 1972; Tushingham, 1992) and has risen more than 100 m since 11.4 ka (Fig. 21C). With rising lake levels, the mouths of GTA river valleys are being flooded to create large wetlands and lagoons (e.g., Grenadier Pond, Frenchmans Bay; McCar-thy and McAndrews, 1988). Some have experienced rapid urbanization (Fig. 22; see also Warwick, 1980). Another con-sequence of the post-glacial rise in lake level has been the rapid erosion of coastal cliffs cut across unconsolidated fine-grained Pleistocene glacial sediments (termed cohesive shorelines). The management of cohesive shorelines in the face of encroaching urbanization is a problem requiring ex-tensive engineering, considerable cost and a careful balance between residential and industrial needs, recreation and pre-servation of natural habitat (Chapter 18).

The mean elevation of Lake Ontario is 74.6 m asl with a seasonal range of plus or minus 100 cm. Rodionov (1994) identified that seasonal lake level variations are a response to the amount of precipitation falling during the preceding winter. The same worker also showed that longer term (approx-imately decadal) fluctuations in lake level, of about 150 cm (Fig. 21D), are synchronous throughout the Great Lake basins, and are related to changing frequency of cyclones originating from Alberta (dry air masses; lake levels de-crease) and the Gulf of Mexico (moist air; lake levels rise). The outflow of Lake Ontario has been regulated since 1960 by a dam at Iroquois Falls near Cornwall designed to alleviate downstream flooding along the St. Lawrence River. Satellite radar altimetry has proved useful in monitoring the level of Lake Ontario, essentially treating the basin as a large-scale rain-water gauge (Birkett, 1994).

Climate.   There have been significant changes in climate across the GTA during the last 10 ka. The early post-glacial (approx. 12-8 ka) was characterized by a cold, dry climate and was followed by the Hypsithermal (about 7.4 ka to 4 ka; Fig.

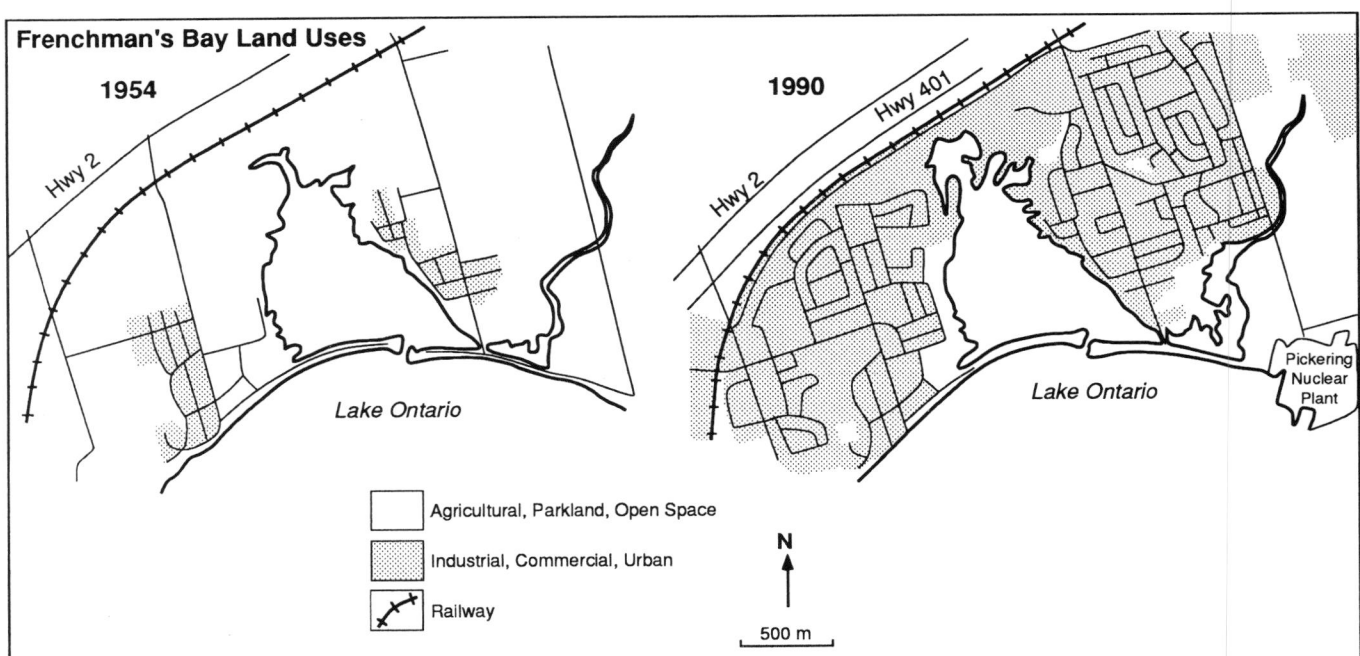

Figure 22   *Urbanization of coastal lagoons and loss of wetlands along the shoreline of Lake Ontario. Frenchman's Bay is a large lagoon formed by rising lake levels (Fig. 1). Urbanization around the bay, and within the drainage catchment that feeds into it, is reflected in increased input of contaminated storm water, accelerated deposition of fine sediment, deteriorating water quality and loss of aquatic habitat (Lemay and Mulamoottil, 1981; Stephenson, 1990; Nelson, 1991). The Metropolitan Toronto and Region Conservation Authority are actively acquiring land along the waterfront for improved shoreline management and habitat restoration (Metropolitan Toronto Region Conservation Authority, 1991; Waterfront Regeneration Trust, 1994).*

21A) when most of mid-continent North America experienced warm dry conditions (Edwards and Fritz, 1988). Mean July temperatures at this time were about 1°C higher than at present and annual precipitation about 10% less (Yu and McAndrews, 1994) resulting in decreased ground-water recharge and a decrease in the size and depth of lakes (e.g., Hendry et al., 1986). Cooler and wetter conditions returned after 4 ka and were coincident with expansion of glaciers in western North America and Europe. Short-lived wetter phases, at about 3 ka and about 2.1 ka, resulted in small (< 1 m), but abrupt, increases in the level of Lake Ontario (McCarthy and McAndrews, 1988).

The GTA region experienced so-called Little Ice Age cooling (a 2°C decrease in mean annual temperature) after A.D. 1200 when the multi-century medieval warm period came to a close (see Briffa et al., 1995; Fig. 21B). This cool phase was just coming to an end when large-scale European colonization started around A.D. 1840 (Beltrami and Mareschal, 1992; Campbell and McAndrews, 1993), but regional climate did not begin to warm until about 1890 (Fig. 21E). Since 1890, there has been a statistically significant warming of 0.7°C in mean annual temperature in the Great Lakes region (Environment Canada, 1992) with the period 1980-1989 being the warmest on record.

Urban growth in the GTA has resulted in a pronounced "heat island" effect where mean annual temperatures have been raised about 2°C compared with outlying rural communities (e.g., Beatrice, Fig. 21E). The present GTA climate is characterized by a mean annual temperature of 8.9°C and annual precipitation of between 700 to 900 mm, depending on location (Environment Canada, 1990; Chapter 10). Precipitation is greater over the urban area as a result of the presence of dust particles which act as condensation nucleii.

Future Climate (0 to +70 ka)

The direction and magnitude of future climate changes is of considerable importance to an urban area such as the GTA which is heavily dependent on lake water for municipal drinking supplies and disposal of sewage. In response to a doubling of carbon dioxide in the Earth's atmosphere as a result of industrialization (the $2\times CO_2$ climate scenario) it has been predicted that global temperatures may increase by 1.5°C to 4.5°C over the next few decades (Cohen, 1986; Smit, 1987; Crowley, 1990; Hartmann, 1991; Smith, 1991; Houghton, 1994). There is, however, considerable uncertainty regarding the direction and magnitude of climate changes in the Lake Ontario basin. Mean annual temperature in the GTA are expected to increase by about 4.5°C with average winter temperatures elevated by as much as 5°C (Sanderson, 1987; Fig. 23).

Smith (1991) used three different General Circulation Models (GCMs) to identify possible Great Lake water levels under the $2\times CO_2$ climate scenario; all models show increased annual precipitation but higher evaporative losses, resulting in a decrease in recharge to ground water and rivers and a fall of lake levels and reduced outflow from the upper Great Lakes. Discharge of the Niagara River (about 8000 m³·s⁻¹; DeCooke and Witherspoon, 1981) may be reduced by as much as 6% (Sanderson, 1987). With future urban growth and climate warming, withdrawals from ground-water and surface-water supplies will increase, thereby further decreasing flow to Lake Ontario. Any decrease in regional river flow will exacerbate existing surface water quality problems. Schindler et al. (1990) identified significant limnological effects, such as increased water temperature, a reduction in wetland area, greatly lengthened water renewal times in lakes, and in-

creased chemical and nutrient concentrations resulting from climate warming and reduced runoff. Detailed assessments of hydrologic and hydrogeologic impacts of climate warming have yet to be completed for the GTA.

In the absence of run-away global warming due to greenhouse gases, a new chapter could be added to the glacial geological record of the GTA when the Laurentide Ice Sheet begins to grow some 6000 years from now in Quebec and Labrador (Fig. 15C). Future climate change can be computed from Milankovitch astronomical variations (see above). Such estimates are of considerable applied significance with regard to determining the depth of glacial erosion likely to be experienced at sites on the Canadian Shield where long-lived high-level nuclear waste is to be stored underground (e.g., Boulton and Payne, 1993; Atomic Energy Canada Limited, 1994). There is common agreement that major cold phases will occur between 15 and 35 ka in the future; the GTA could be deeply buried below ice between 54 and 70 ka hence (Fig. 15C).

Sand and Gravel Mining

Pleistocene glacial and late-glacial sediments represent the largest source of sand and gravel (aggregate) for the construction industry. In 1988, at the peak of recent urban growth, the GTA demand for aggregate was about 75 million tonnes. Because of the high-bulk–low-value nature of aggregate, inexpensive supplies must be found within 50 km of the market. The bulk of the GTAs need is produced from the glacial sands and gravels of the Oak Ridges Moraine (Chapter 9) with much of the remainder from bedrock quarries along the Niagara Escarpment. Supplies from the Lake Iroquois shoreline are all but exhausted. There is renewed interest in mining sand from the floor of Lake Ontario adjacent to Toronto.

About half (53%) of the total aggregate supply to the GTA is used for road construction and maintenance; on average, 1 km of two-lane highway requires 15,400 t of aggregate, a six-lane highway 51,800 t per km. Nearly 97% of all aggregate, regardless of source, is moved by road. Because of its high-bulk, low-value characteristics, less than 40% of all aggregate is transported more than 50 km; current delivery costs for a tonne of aggregate are $2.00 for a haul of 10 km and $5.50 for 50 km. The cost of rail transport is prohibitive because of the rehandling involved and lack of lines.

Government Policies

Provincial regulation of the aggregate industry began in 1971 with the Pits and Quarries Control Act (PQCA) which was strengthened by the Aggregate Resources Act (1990). Until 1990, municipalities were hamstrung in their ability to develop long-term official plans because existing regulations prevented them from developing any land underlain by significant aggregate deposits, a constraint which greatly hindered the development of comprehensive long-term municipal planning in the GTA. Ontario government policy on aggregate resources is designed to promote the finding of supplies close to markets and, by so doing, promotes land use conflicts. Policy has consistently given the needs of the aggregate industry preference over those of the public and environment (see Estrin and Swaigen, 1993 for a history of regulation and litigation). Other problems within the aggregate industry have been a lack of rehabilitation of old pits in operation prior to 1990, and a lack of adequate enforcement of current regulations by the Ministry of Natural Resources (Estrin and Swaigen, 1993).

Aggregate mining in the Oak Ridges Moraine. That portion of the ORM that falls within the GTA (Fig. 1), termed ORMGTA, has about 95 active licensed aggregate operations centred mostly in the Town of Whitchurch-Stouffville and the Township of Uxbridge; average life expectancy of any one operation is about 16 years.

Licensed commercial pits account for 60% of total production from the ORMGTA with the remainder from temporary (wayside), non-commercial, municipally or provincially owned pits opened for a specific project. Licensees currently pay 8 cents per tonne per year of aggregate produced into a Rehabilitation Security Deposit System. Unfortunately, existing pits are located in environmentally sensitive areas; the ORM is the principal ground-water recharge area for the GTA, but the impact of mining on ground water is not well studied. Many former quarries have been rehabilitated by being filled with municipal and industrial wastes which create long-term environmental impacts on ground waters. Destruction of original drainage and topography, the loss of habitat, dust, noise, road traffic and impacts on ground and surface water are commonly cited as problems associated with aggregate mining. The drawdown of local water tables as pits are dewatered, the plugging of porosity by the discharge of silt into settling ponds, and the reduction of infiltration due to ground compaction after rehabilitation are other deleterious effects of aggregate operations in the ORM (Intera-Kenting, 1990; Oak Ridges Moraine Aggregate Resources Study, 1994).

More than 50% of the GTA's known aggregate resources are located in the ORM and it is the view of the aggregate industry that environmental protection must be balanced by the need to access aggregate deposits. The industry argues that serious losses of potential resource areas to pre-emptive land uses has already occurred.

Lake dredging.   Recent concerns with the impact of aggregate mining in the Oak Ridges Moraine have rekindled interest in lake dredging. Sand has been dredged from Lake Ontario since 1916 and is currently taken from the mouth of the Niagara River where 1989 production was 95,000 t. The practice of stone-hooking, *i.e.*, removal of boulders from the lake floor, was common in the past, resulting in destruction of habitat and spawning areas for cold-water fish such as lake trout (Bailey, 1973; Sly, 1991; Strus, 1994). In 1990, Bedrock Resources Inc. submitted an application for a permit to dredge sand from a 1400-ha block extending 7 km from Ashbridge's Bay to east of Scarborough Bluffs, lying between 1.5 and 2 km offshore (Fig. 24). Permission was sought to suction dredge a surficial unit of fine to coarse sand, up to 5 m thick, with a total volume of 28 million m³ (54 million tonnes). This is a relict sandbody, originally formed by wave action in shallow water and now preserved offshore as a result of transgression of the shoreline in response to post-glacial lake level rise (Fig. 21A). The deposit lies on a flat shelf overlooking a drowned shorebluff (the Toronto scarp; Fig. 24) cut during the low stage of Lake Ontario immediately following the drainage of Glacial Lake Iroquois (Fig. 21A; Lewis and Sly, 1971).

At intended production rates (about 1 million tonnes a year), the offshore sand reserve could be expected to remain productive for 60 years. The sand contains small amounts of silt and clay and dredging will result in the release of turbid plumes of suspended sediment together with any resuspended pollutants. A study by B.A.R. Environmental Inc. (1992) identified elevated nitrogen levels in the sand thought to be derived from the outfall of the Main Sewage Treatment Plant at Ashbridge's Bay. This and other contaminants, such as arsenic and cadmium, could be resuspended by dredging and there are concerns that water quality could be adversely

affected in the vicinity of the main municipal drinking-water intakes for the GTA (Fig. 24). The proposed dredge area lies within 400 m of the intake pipes for the R.C. Harris filtration plant which has the capacity to supply 45% of the municipal drinking water for Metropolitan Toronto. Quantitative hydraulic modelling suggests that dredging will have no impact on shore processes or erosion rates along Scarborough Bluffs (Atria Engineering Hydraulics Inc., 1994), but these results are controversial (*The Toronto Star*, September 9, 1995). It should be noted that massive hydraulic dredging of offshore sand along the western bluffs between 1910 and 1914 resulted in accelerated shoreline erosion (Chapter 18). Sand was used to infill Ashbridge's Bay (see below).

Bedrock sources of aggregate.   Rising seaway transportation costs are the limiting factor on freighter-borne, crushed-rock aggregate shipped from Manitoulin Island in Lake Huron. Underground mining of Ordovician limestones (Gull River and Bobcaygeon formations of the Simcoe Group; Fig. 6) below Lake Ontario, close to Metropolitan Toronto, is technically feasible, but precluded in a regional market by trucking costs. Within the GTA, potential bedrock sources of aggregate are defined as those areas having less than 8 m-thick cover of glacial sediment; about 80% of potential bedrock aggregate resources occur in Halton Region adjacent to the Niagara Escarpment. Silurian dolostones of the escarpment (Fig. 6) are used for the manufacture of the highest quality concrete such as that used for the construction of the CN Tower, for Skydome and for highway overpasses.

Bedrock mining has severely impacted the natural integrity of the Niagara Escarpment and, in 1973, the Government of Ontario made a commitment to protect the escarpment. Urbanization and quarrying operations along the escarpment are now regulated by the Niagara Escarpment Plan (Ontario Ministry of Environment and Energy, 1994c) and the Niagara Escarpment Planning and Development Act. The escarpment was designated a World Biosphere Reserve by UNESCO in 1990.

LANDFILL

The term landfill is potentially confusing as it is used with two meanings. It is widely employed to refer to the widespread practice of using surplus soil, coal ash, construction rubble and municipal wastes to infill pre-existing topography to

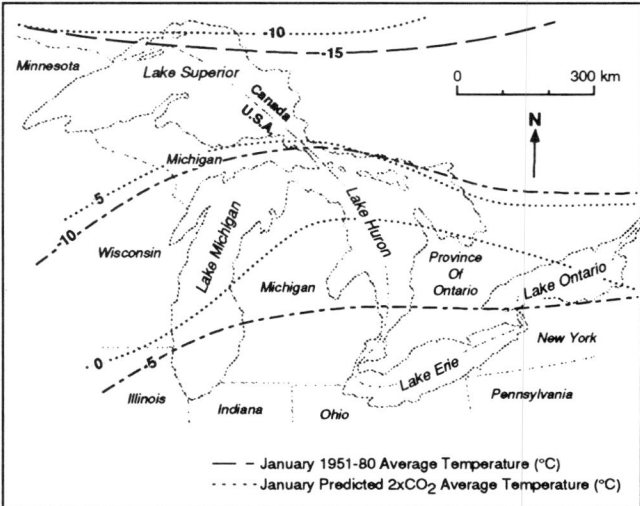

Figure 23 *Modelled effect of greenhouse warming for the Great Lakes basin assuming doubling of CO_2 content in atmosphere. A 5°C warming is modelled for average January temperatures in the GTA (after Sanderson, 1987).*

create new land (termed herein historic landfilling). In addition, the term landfill is employed to refer to disposal of hazardous and non-hazardous wastes in municipal waste dumps (termed herein landfill sites). The following section examines the environmental impacts arising from historic landfilling; impacts from municipal landfill sites are examined later (see also Chapters 20-25).

Historic landfilling has been widespread in many cities (*e.g.*, San Francisco, 1969; British Columbia Ministry of the Environment, 1991; Walsh and Lafleur, 1995; Meyer *et al.*, 1995). Across the GTA, major topographic changes have resulted from the dumping of contaminated fill along ravines and the lakeshore, especially along the lower Don Valley and Toronto waterfront (Figs. 25, 26, 27). Site assessments to identify landfilled areas and the nature of the fill have now become a routine consideration in real estate transactions. Environmental problems arising from the presence of contaminated landfill are of potential liability to buyers, sellers, developers and financiers, and need to be clearly identified

prior to the transfer of property (Brinke and Parkin, 1991; see below). There are also important public health issues arising from the presence of historic, contaminated fill in downtown residential communities and a need to decommission former industrial sites (Chapter 27). A further cause for concern is the stability of landfilled substrates underlain by high water tables. Such materials are prone to liquefaction in the event of earthquakes (*e.g.*, Chameau *et al.*, 1991).

Historic Landfilling in the City of Toronto

In the city of Toronto, 18 km², or about 20% of the total city area (similar to that of New York City), is designated as landfill reflecting the practice of infilling watercourses and heavily polluted coastal embayments (Figs. 26, 27, 28, 29). Landfill was used after 1850 to create new land for industry and railways and to reclaim coastal marshes polluted by sewage. More recently, the practice has been driven by the need of the development industry for inexpensive disposal sites located close to the downtown area, for waste soil material from

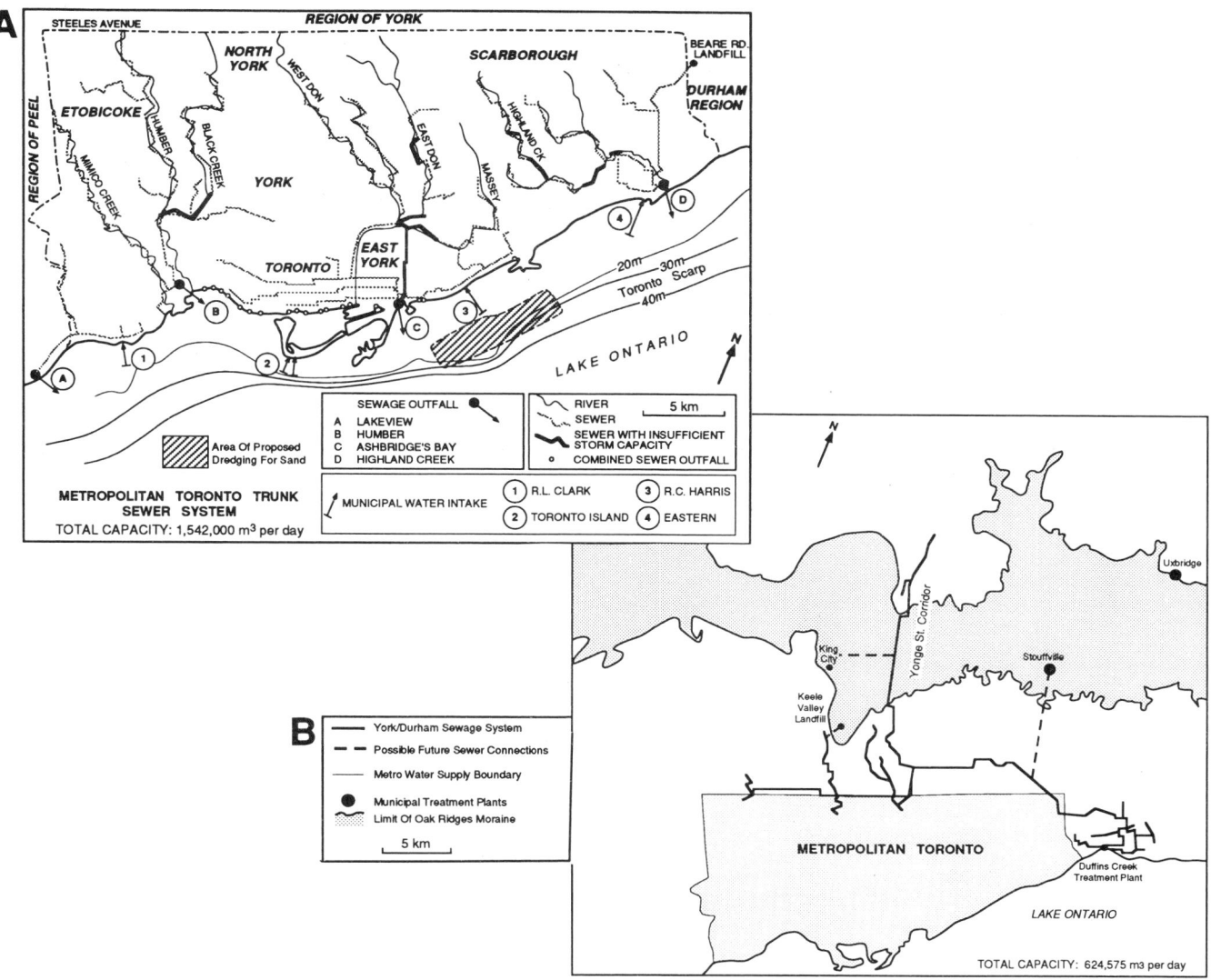

Figure 24 (**A**) *Extent of Metropolitan Toronto Trunk Sewer System (**MTTSS**) with treatment plants and outfall pipes. Note combined sewers (**CS**; Fig. 31) with insufficient storm-water capacity and associated outfalls along rivers and Lake Ontario shoreline (not all outfalls shown). Note proximity of municipal drinking-water intakes to offshore area where a licence application has been submitted to dredge sand from the lake floor. (**B**) Extent of York-Durham Sewer System (**YDSS**) and the area serviced within the Oak Ridges Moraine and the Yonge Street corridor, which drains to Duffins Creek Sewage Treatment Plant. Extension of this system northward, together with the provision of municipal (lake) drinking water, would promote rapid urbanization of the ecologically sensitive moraine area (Chapter 9). In late 1996, extension of the YDSS to Stouffville and other rural communities on the moraine was announced by the provincial government.*

excavations such as the subway and ash from thermal power stations (see Munson, 1991).

In 1993, the Medical Officer of Health for the City of Toronto in a State of the Environment Baseline Report concluded that as much as 25% of the city is "contaminated and unfit for residential use" (City of Toronto Public Health Department, 1993). A quarter of the city's schools are built on landfill. These estimates are based on historical land use data only, not field investigations, and so are likely to be underestimates. The areas that are the most contaminated include former industrial areas, waste disposal sites, lead reduction zones, salt storage areas, coal gasification sites, lakefill areas and former ravines, now infilled. Problems are often only brought to light by planned land use changes from industrial to residential or where slope stability problems develop (Fig. 25).

Current Policy for Management of Contaminated Fill
In Ontario, sites that are being decommissioned or subject to projected land use change must meet Ontario Ministry of the Environment Decommissioning Guidelines (1989); proposed changes to the guidelines are contained in Ontario Ministry of Environment and Energy (1994a; OMOEE). Space does not permit a detailed discussion and this issue is dealt with further in Chapters 26 and 27. Current decommissioning guidelines have established clean-up criteria for contaminants and soil quality parameters for land uses ranging from (1) old urban, new urban and rural land residential, to (2) parkland, to (3) agricultural to (4) commercial/industrial/transportation uses. Three basic approaches to site clean-up are allowed by the OMOEE: on-site isolation, treatment either on or off site, and/or disposal.

To date, only on-site treatment is covered by the proposed guidelines (Ontario Ministry of Environment and Energy, 1994a). These are stringent and allow for either full-depth clean-ups to the full vertical extent of subsurface contamination, or stratified clean-ups extending to 1.5 m depth with less

stringent criteria for deeper soils; any remaining contaminants are then registered on the legal title to the property. In contrast, those soils that need to be removed from the property for off-site treatment or disposal are covered by the Proposed Policy For Management Of Excess Soil, Rock and Like Materials (Ontario Ministry of Environment and Energy, 1992). This policy recognizes four categories of material ranging from inert fill (no restrictions on use), urban residential fill, urban industrial fill and controlled fill (to be landfilled at a licensed site).

On-site treatment options include soil dilution (mixing with non-contaminated soils to achieve appropriate clean-up criteria), isolation and capping by clean fill, buildings or roadways, regeneration using vegetative covers; processing and recycling of non-soil components, removal and disposal in a landfill, or *in situ* treatments involving destruction, separation and immobilization techniques (Eslinger *et al.*, 1994; Beak and Raven Beck Environmental, 1994; Chapter 35).

Case Studies

The Port Industrial District
The most severely contaminated area within the city of Toronto is the 500-ha Port Industrial District (PID: Fig. 29). This area is estimated to contain over 2 million tonnes of historic landfill contaminated by coal ash, arsenic, heavy metals, oils and greases (Munson, 1990, 1991; Persaud *et al.*, 1991; Koester and Hites, 1992). The PID occupies a former embayment in the Lake Ontario shoreline (Ashbridge's Bay). This was landfilled after 1912 as a means of remediating a marshland heavily polluted by sewage and offal from hundreds of slaughterhouses. In 1900, Toronto's population (181,000) was pro-

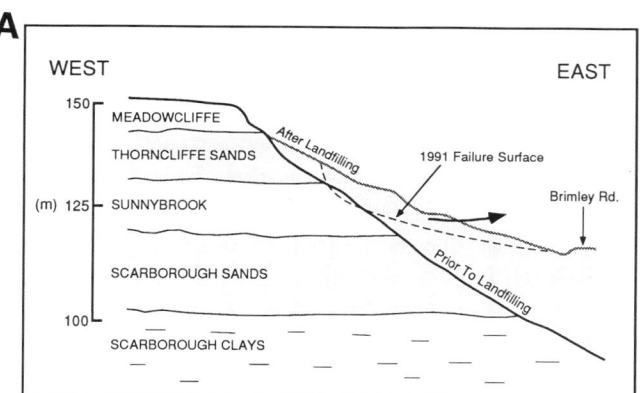

Figure 25 *In mid-April 1991, a massive landslide took place along Brimley Road, along the Lake Ontario shoreline in Scarborough (Fig. 18B). The slide involved the downslope collapse of municipal garbage dumped to a thickness of 25 m in Brimley Road Ravine between 1960 and 1968. Dumping was intended to solve a coastal erosion and garbage disposal problem (e.g., Chapter 18). Construction of a road through the ravine in the 1970s to service Bluffers Park Marina (in foreground) resulted in oversteepened sideslopes composed of garbage. These became saturated and unstable during heavy rain in March and April 1991; previously installed drainage measures were found to be completely inoperative (Golder and Associates, 1991). The resulting slide released significant volumes of contaminated leachate. The total cost of the clean-up to the tax payer was $2.1 million. Horizontal scale same as vertical. View of photograph is to the north.*

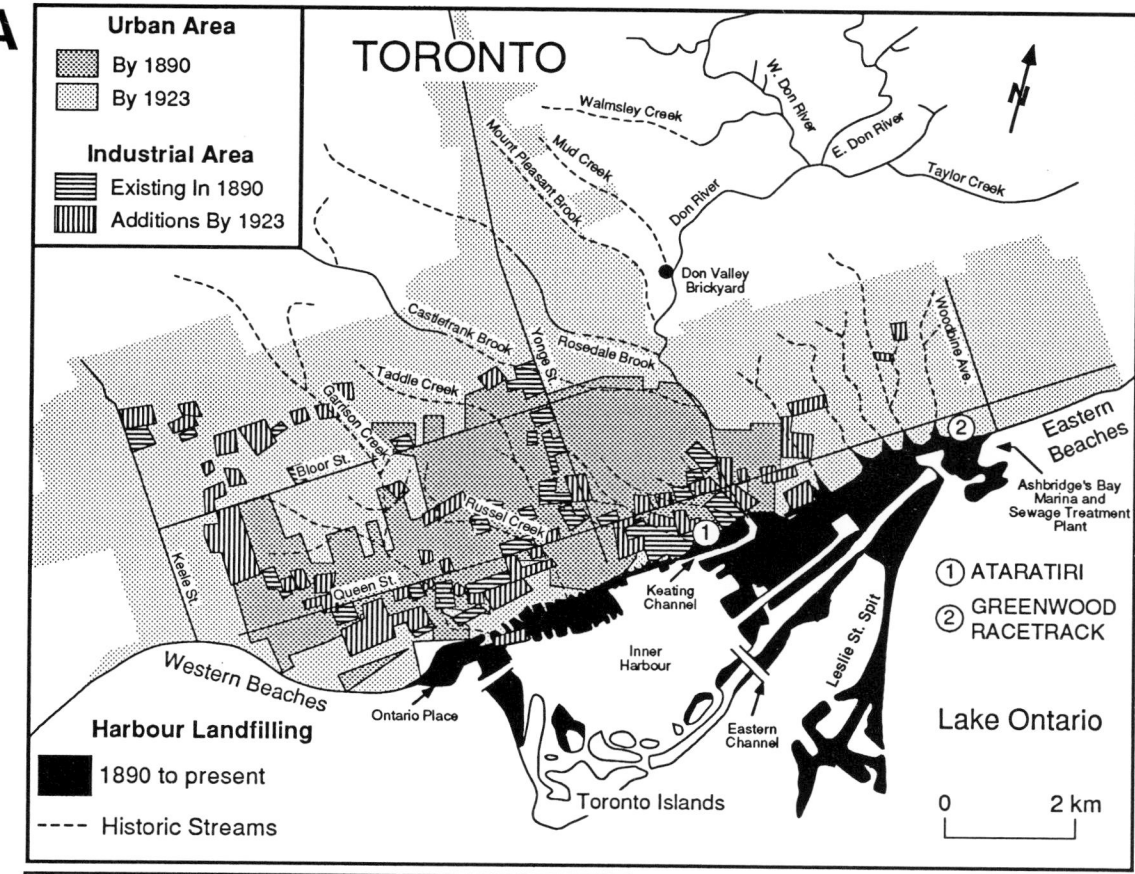

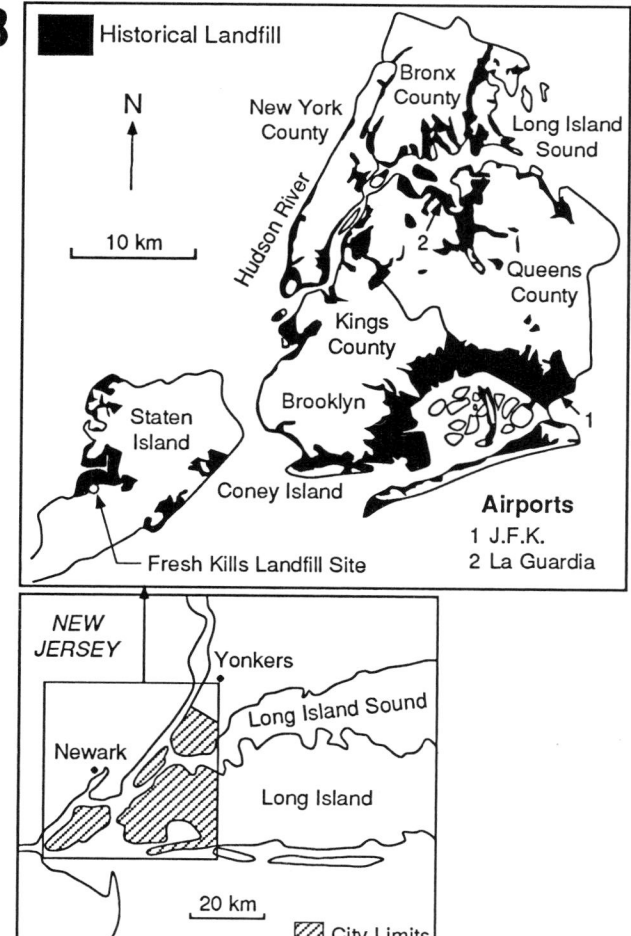

Figure 26 (**A**) *Growth of the city of Toronto between 1890 and 1923 (when present-day city limits were essentially established), showing loss of natural creeks and extent of lakeshore landfilling to present day. Extent of historic land-filling inland is not shown. Sediment dredged from Keating Channel is heavily contaminated with urban pollutants and is dredged and dis-posed of in containment ponds within the Tommy Thompson Aquatic Park along Leslie Street spit (Figs. 4C, 29). (**B**) Extent of historic landfilling (since 1844) in New York City (after Walsh and Lafleur, 1995). Landfill typically consists of a wide variety of municipal wastes and over 50% of ash residue from coal and has been used to infill tidal wetlands over some 20% (185 km²) of the city area. Leachate is now migrating to adjacent waterbodies. Note major airports sited on land-fill. The largest active municipal landfill in the world is at Fresh Kills on Staten Island, extend-ing over 1500 ha.*

A

B

C

D

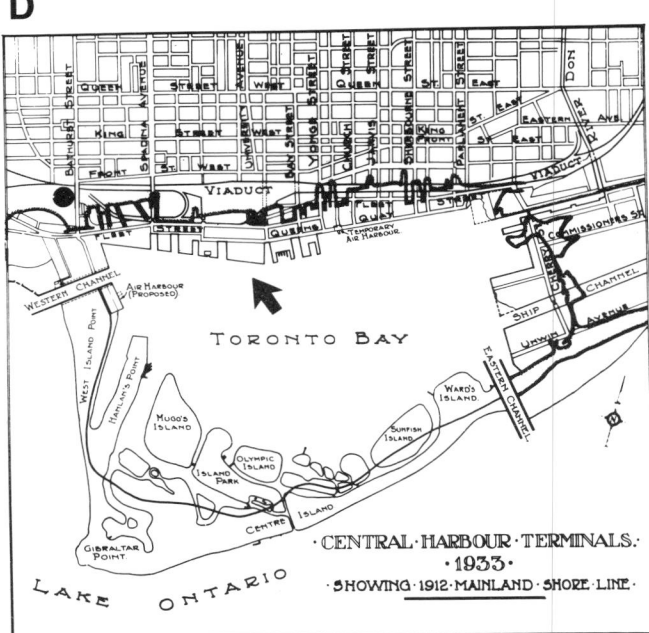

Figure 27 *Successive views of Lake Ontario shoreline looking westward from Toronto Harbour Commission building (**A**) in 1913 and (**B**) in 1921, showing extent of lakeshore landfilling. Toronto Harbour Commission (**THC**) Archive photographs PC1/1/2840 and 1/1/9417. The Toronto Harbour Commission was established in 1912 to upgrade and extend harbour facilities. (**C**) Aerial photograph (1921) of harbourhead wall consisting of sheetpile between Spadina Avenue and York Street*

ducing almost 80,000 m³ of liquid sewage and animal waste every day. Some 1000 m³ of solid matter was being deposited in Ashbridge's Bay every year. The Atlantic salmon fishery was destroyed and there were serious problems caused by flooding (by ice jams) and epidemics of water-borne diseases such as typhoid and cholera; more than 600 people died of outbreaks in 1832. Installation of sewer drains in the 1830s alleviated cholera outbreaks and the opening of the first modern sewage treatment plant at Ashbridge's Bay (1911) effectively ended the incidence of typhus (Brace, 1995). In order to literally cover up the problem of pollution from sewage, engineers undertook between 1912 and 1928, what was then the largest lake-filling project anywhere in North America. The former marsh was infilled, the mouth of the Don River was moved northwest, and the artificial Keating Channel completed in 1922 at right angles to the newly canalized river (Fig. 26). This destroyed the original hydraulic gradient of the river necessitating dredging and removal of contaminated sediment that accumulates in the channel (see below).

The Depression of the 1930s killed plans to make the PID a model for a mixed residential, industrial and manufacturing area, and instead, the area was used for the low-cost bulk storage of commodities such as road salt and fuel oils and

*with Front Street in background. Area within wall awaits landfilling with municipal wastes (Fig. 28) and sediment pumped from lake floor. THC archive photograph PC1/1/5869. (**D**) 1933 map of Toronto waterfront showing shoreline of 1912. Location of photograph C is shown by arrow. Front Street is the original natural shoreline; black circle marks site of Fort York the original British military base of 1793, on the shore of Lake Ontario. THC archive photograph PC1/1/10185. See Figure 26 for scale.*

Figure 28 *Municipal garbage and street sweepings being used as landfill along the Toronto waterfront in 1923. Photograph PA-84921/ Public Archives Canada.*

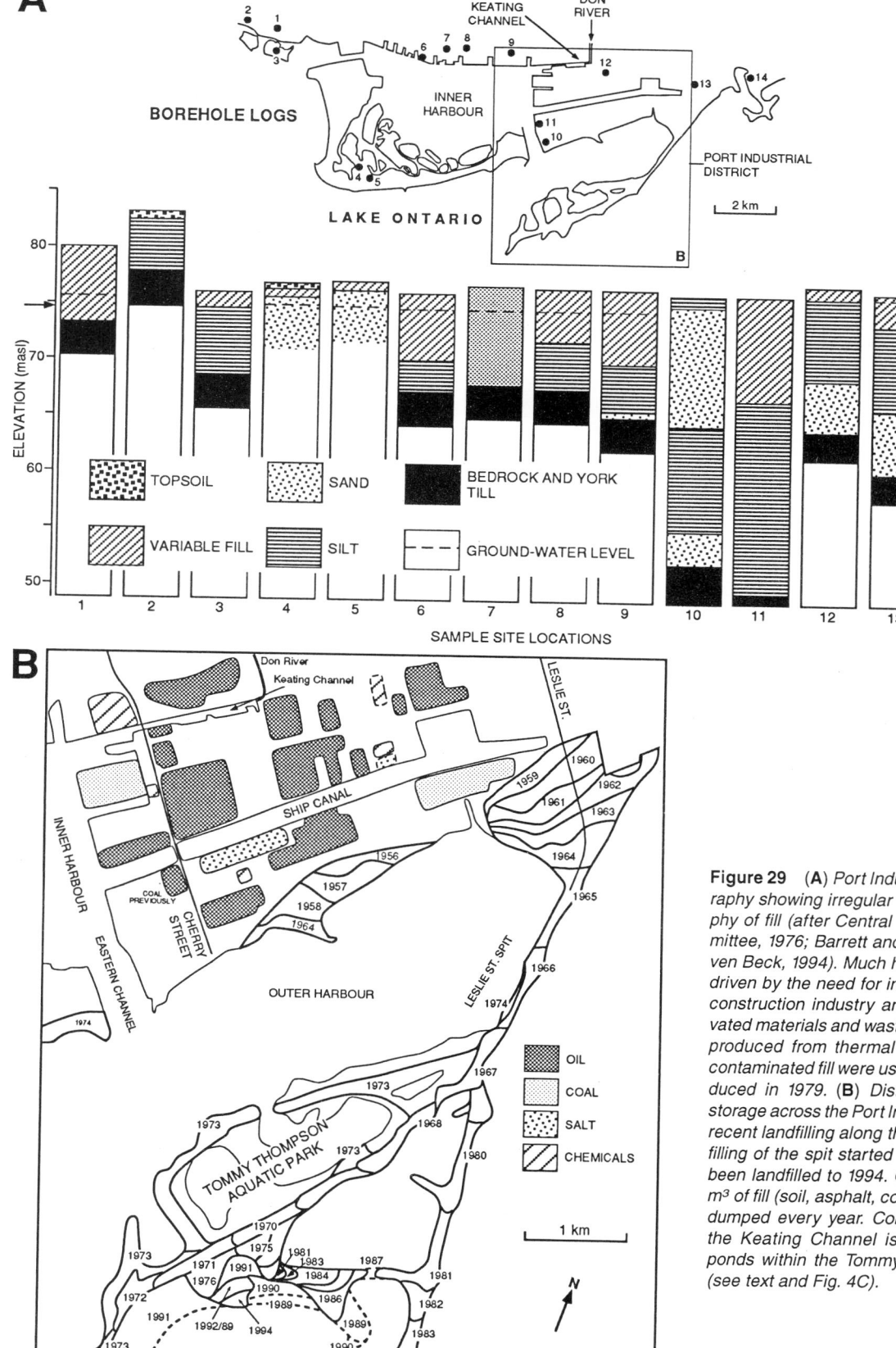

Figure 29 (**A**) *Port Industrial District (**PID**) stratigraphy showing irregular bedrock surface, stratigraphy of fill (after Central Waterfront Planning Committee, 1976; Barrett and Kidd, 1991; Beak and Raven Beck, 1994). Much historic landfilling has been driven by the need for inexpensive disposal by the construction industry and public utilities for excavated materials and waste; large volumes of fly ash, produced from thermal generating stations, and contaminated fill were used until controls were introduced in 1979.* (**B**) *Distribution of bulk chemical storage across the Port Industrial District and rate of recent landfilling along the Leslie Street spit. Landfilling of the spit started in 1956; about 140 ha has been landfilled to 1994. On average, some 84,000 m³ of fill (soil, asphalt, concrete and brick rubble) is dumped every year. Contaminated dredgate from the Keating Channel is dumped in containment ponds within the Tommy Thompson Aquatic Park (see text and Fig. 4C).*

refining and recycling operations. This has resulted in widespread contamination of underlying substrates and ground water (Barrett and Kidd, 1991). The geology of the PID (Fig. 29) consists, from bottom to top, of an irregular surface of Georgian Bay shales, cut by the buried Don bedrock channel (see Fig. 12), overlain by a thin, and intermittent, York Till, peat-rich silts and sands of unknown age (possibly Don Beds; Fig. 17). These deposits are overlain by up to 10 m of landfill consisting of street sweepings, municipal wastes, fly ash, scrap metal, dredged lake sediments and construction rubble and surplus soil. Buried basements of demolished buildings, sewers and abandoned fuel distribution pipelines are also present.

Contaminants found at elevated levels in fill and ground waters below the PID are free-phase products (non-aqueous phase liquids; NAPLs), inorganics (such as lead, cadmium, chromium, zinc, nickel, *etc.*), volatile organic compounds (VOCs such as benzene, xylene, trichloroethene), polycyclic aromatic hydrocarbons (PAHs such as pyrene, fluorene, anthracene and napthalene), and polychlorinated biphenyls (PCBs). Free-phase products include petroleum hydrocarbons that are less dense than water (LNAPLs) and compounds that are more dense than water (DNAPLs) such as chlorinated solvents, tars, creosotes and PCBs.

Present remediation plans for the PID call for on-site treatment in preference to removal and off-site disposal in landfills where space is at a premium; the bulk of surficial soil could be treated at a large-capacity central processing facility. Cost estimates range from $100 to $200 per tonne for remediation to a level sufficient for industrial use and up to $500 per tonne for residential usage. In addition, ground water (and surface runoff) in this and other landfilled sites is heavily contaminated, and requires collection, treatment and monitoring systems. Much of the Port Industrial District and surrounding area is prone to flooding by the Don River (Fig. 30) requiring additional clean fill and protective measures. The water table is also rising as a result of the gradual increase (20 cm per century) in lake levels at the western end of Lake Ontario (Fig. 21A; Kite, 1972).

Ataratiri

Ataratiri (a 17th-century Huron word meaning village) was a large-scale development project initiated in July of 1988 to deal with the severe shortage of affordable housing in the city of Toronto (Figs. 26, 30). Ataratiri was designed as a joint proposal between the city and the province to convert a largely abandoned, 32.5-ha industrial area prone to flooding, into a community of 7000 homes (City of Toronto Housing Department, 1991). As such, it was one of the largest downtown redevelopment projects being undertaken by any city in North America. The site has a long history of heavy industrial use since the turn of the century. The underlying stratigraphy of the site consists of historic landfill (up to 3 m thick) composed of fly ash, rubble, wood, street sweepings and hydraulic fill from the lake floor (Central Waterfront Planning Committee, 1976) heavily contaminated with metals and polycyclic aromatic hydrocarbons, resting on till and shale bedrock. The landfill material across most of the area fails to meet industrial/commercial quality guidelines, and the original quality of the material is questionable and may have contributed to the degraded condition of the site (Trow, Dames and Moore, 1991a, b). Particular problems are remediation of contaminated ground water and storm-water runoff entering Lake Ontario and the Don River, the risk of flooding, and the removal of DNAPL trapped within buried concrete foundations and fractured shale bedrock (MacViro Consultants Inc., 1991).

Ataratiri was cancelled in September 1992 because of the magnitude of costs required to satisfy residential clean-up guidelines and insufficient understanding of potential health risks posed by contaminants (City of Toronto Housing Department, 1992; Chapter 27). Toxicological assessments for Ataratiri have considered human receptor pathways, such as soil ingestion, inhalation of contaminated dust and skin contact with soil, under alternative land use scenarios; as expected, young children in residential areas with gardens are at greatest risk (Beak and Raven Beck Environmental, 1994; Chapter 27). Discussions of acceptable contaminant levels at former industrial sites such as Ataritiri have had the effect of focussing attention on existing soil contamination found in the older residential neighbourhoods of downtown and suburban Toronto, particularly with regard to lead levels (Ontario Ministry of Environment and Energy, 1993).

Greenwood Racetrack

Greenwood Racetrack occupies a 37-ha site that straddles the landfilled 1870 shoreline of Ashbridge's Bay (Figs. 26, 31). The area is landfilled with more than 1 m of dredged silt and sand and nineteenth-century household waste overlying lacustrine peat deposits. In June of 1994, following environmental assessments of the property as required for decommissioning purposes, a dispute arose after the site had been sold for residential development. The quality of the fill below the racetrack was found to exceed provincial guidelines for hydrocarbons, iron, arsenic, molybdenum, selenium and barium. The site is also underlain by a high water table and potentially explosive levels of methane gas derived from marsh peat buried by historic landfilling. The quality of ground water is also poor, especially with regard to chloride attributable to the use of road salt in parking areas. On the basis of the cost of removing the peat (about $6 million) and site remediation, the buyer cancelled the purchase agreement. Subsequent discussion has focussed on whether the initial site assessment had fully characterized the true extent and variability of subsurface contamination prior to estimates being made of remediation costs (City of Toronto policy now requires 8 boreholes for every 1000 m^2). Present proposals call for partial excavation of fill materials and restoration of the 1870 shoreline as open space and wetland (Barry Lyon Ltd., 1994).

Discussion

Examples such as the Port Industrial District, Ataratiri and Greenwood Racetrack illustrate the difficulties (*i.e.*, cost) of rehabilitating land in old industrial areas of inner cities. Costs may prohibit the level of clean-up required for residential development. This emphasizes the need to carefully regulate current industrial activities in urban areas to prevent land of great potential value becoming worthless. Remediation costs are a huge disincentive to inner-city redevelopment and promote consumption of pristine green-field sites in outlying rural areas. The GTA is currently experiencing a severe shortage of space available for cemeteries; mixed parkland/cemetery use has been suggested for the more contaminated sites. Hydrogeologic studies of the impact of cemeteries on ground waters are reported by Soo Chan *et al.* (1992). A brown-field policy for downtown sites, where redevelopment for residential uses is scaled back and continuing industrial/commercial activity is fostered, is also being considered. A particular difficulty is created by the wide range of available soil remediation techniques (see Chapters 27, 35), but a lack of detailed and objective assessments of the advantages and disadvantages of each technique.

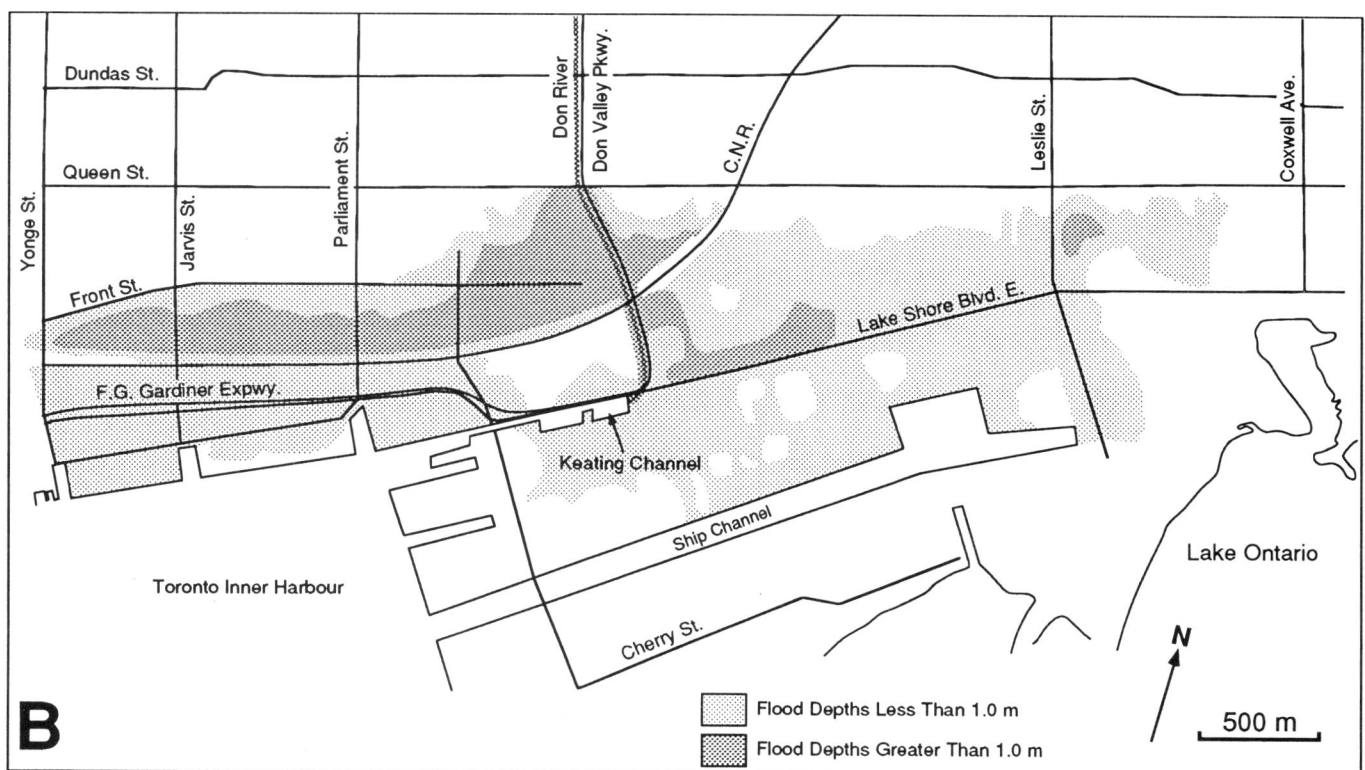

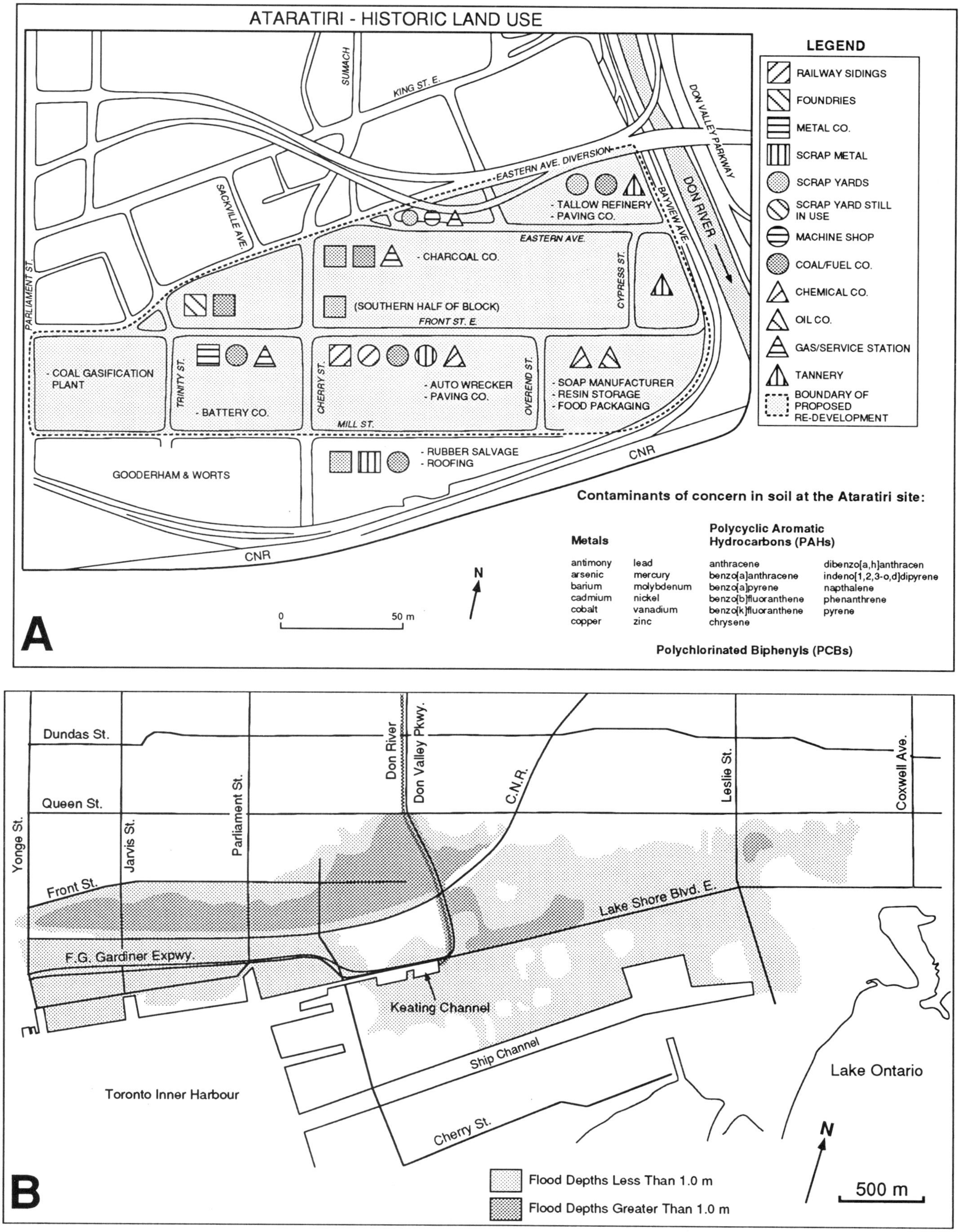

Figure 30 (**A**) *Ataratiri; history of industrial land use across proposed residential development site and principal contaminants. The project was to be the largest inner-city affordable housing development in North America, but was abandoned in 1992 due to the excessive cost of soil and ground-water remediation.* (**B**) *Extent of flood risk from the Don River across Ataritiri. Flooding is the result of right angle bend in the Don River at the Keating Channel.*

Regardless of future land use policy for contaminated sites, a major challenge is that of adequately characterizing subsurface geologic conditions and the extent of contamination at sites with complex fill histories and stratigraphies. Relatively inexpensive and rapid geophysical techniques are being increasingly employed; drilling is still required for ground-truthing and sampling, but it may, by physically disturbing the substrate, promote the migration of contaminants (Riggs, 1993; Chapter 31).

PART II: REGIONAL WATER MANAGEMENT

The intent of this section is to briefly review the impact of urbanization on surface and ground waters of the GTA. It is emphasized that this review is necessarily incomplete because fundamental data are as yet unavailable.

Urbanized Rivers

The spatial character of urbanization varies across the GTA from intense blanket urbanization of the Don River watershed (80% urbanized) to nodal patterns typical of the Rouge basin (18%). Regardless of these differences and because of the need to manage urban storm water and other waste waters, surface rivers have become essentially a sub-component of the regional waste-water treatment system.

The effects of European settlement and rapid urbanization have almost completely masked any differences in regional flow regimes attributable to watershed soils and geology. Large-scale forest clearance by European settlers after 1840 resulted in rapid soil erosion and enhanced spring runoff, flooding and loss of navigation along many creeks (Fig. 4). The practice of mining sand and gravel (aggregate) directly from river beds has also been prevalent (*e.g.*, Varga *et al.*, 1991; Eyles and Boyce, 1991; Fig. 32).

Urbanization and Flooding

Severe, damaging floods have occurred in the Metropolitan Toronto area on average every 1.3 years since 1804 (Metropolitan Toronto and Region Conservation Authority, 1980). Despite the reduced risk of flooding from spring ice jams, because thermal pollution has elevated water temperatures in some cases by as much as 5°C, the year-round risk of flooding has increased as a result of reduced infiltration through urbanized substrates covered in concrete and asphalt. The peak lag time, which is the time elapsed between the precipitation event and the peak flood discharge of an urban river, decreases with increased urbanization of the drainage catchment; the peak discharge is also greatly increased (Fig. 33A, B). Erosion of stream banks and loss of aquatic habitat are well-known effects. Comprehensive reviews of urban hydrology can be found in Wanielista (1990), Chapman (1992), Field *et al.* (1993) and Debo and Reese (1995).

Case Study

Highland Creek

The effects of urbanization on hydrology are clearly illustrated with regard to the Highland Creek catchment (Fig. 1) where a severe storm on August 27-28, 1976, caused damage to nearby properties comparable to that caused by Hurricane Hazel in the same watershed. Hurricane Hazel (October 14-16, 1954) produced 280 mm of precipitation, affected the entire GTA and killed 84 people (Kennedy, 1979). It is the standard regional storm with which all other storms are now compared. The 1976 storm, which produced over 60 to 80 mm of rainfall, was not as severe as Hurricane Hazel, but the watershed had since become 60% urbanized, resulting in a peak discharge equal to that experienced in the Humber Creek drainage basin during the 1954 hurricane (Fig. 33C). Highland Creek was able to contain the storm overflow during the 1976 storm, but damage to public property was considerable. This storm has a return period of as little as 15 years and costly containment and erosion control measures are now required along the lower reaches of GTA rivers.

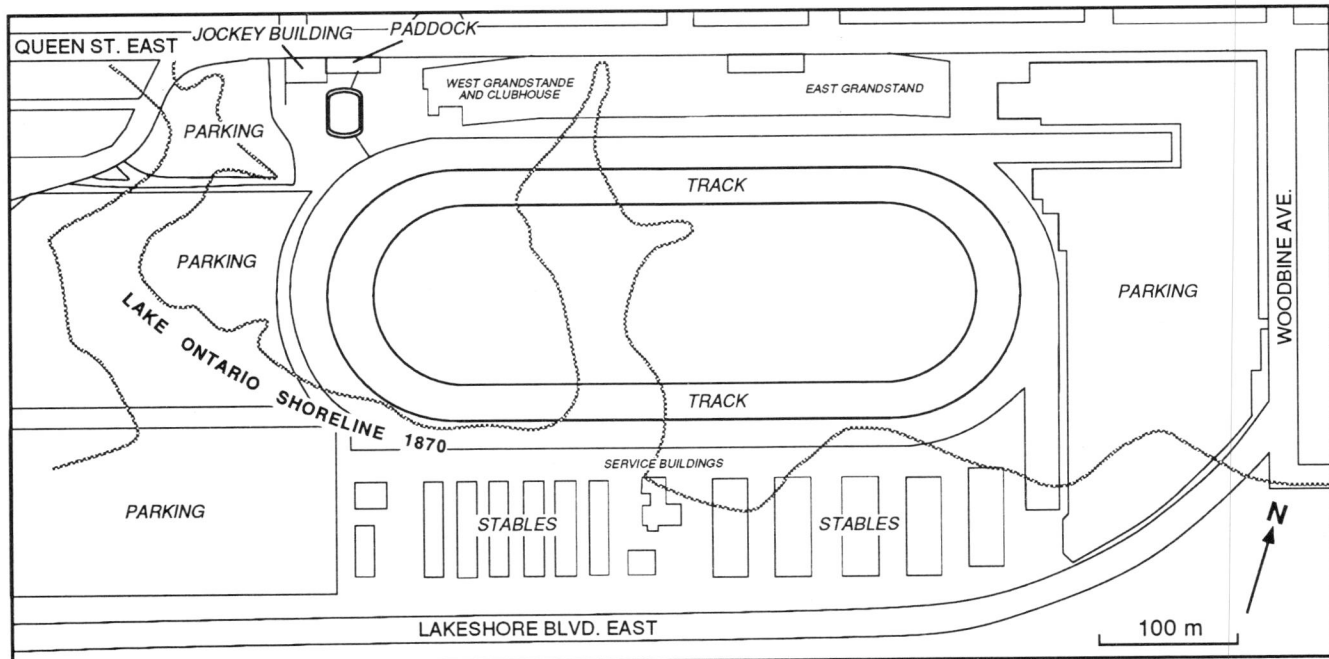

Figure 31 *Greenwood Racetrack (for location, see Fig. 26A). The site lies on landfill used to reclaim coastal marshes along the shoreline of Lake Ontario. Methane released from buried peat, and the presence of contaminated fill, prevented redevelopment of the site to residential use. Present plans call for excavation of the fill and restoration of the original shoreline.*

Storm-water Management Practices

Current storm-water management practices in the GTA are reviewed in guidelines published by the Ontario Ministry of the Environment and Energy (1994a). Concern has shifted from simple floodpeak shaving where flood water is retained in detention ponds, to that of the removal of contaminants found in urban storm waters and the design of preventative remediation techniques in source areas (Marsalek, 1990).

Surface-Water Quality

Urban runoff is contaminated by pollutants from many sources, principally from street runoff and particularly from areas where vehicles are parked or serviced (Fig. 34; Bannerman *et al.*, 1993; Schueler, 1994). Pollutant loadings are greatest during the initial first flush of runoff when most contaminants are captured (see Chang, 1994); an effect which is compounded by flushing from the base of snowpacks along roadsides or in snow dumps, especially during end-of-season melt (see Oberts, 1994).

Typical contaminants found in urban storm-water runoff are listed in Figure 34B. Pitt *et al.* (1994) showed that high counts of pathogens such as *Pseudomonas aeruginosa, Shigella* and enteroviruses are typical for areas where dogs are walked or birds congregate and can persist for up to 5 years in contaminated soil. Nitrate levels are high, resulting from municipal and household fertilizing of lawns. Under Canadian conditions, cool temperatures significantly reduce denitrification, allowing nitrate to stay in solution and reach the water table. Particularly mobile pesticides include 2,4-D, acenaphthylene, alachor, atrazine, cyanazine, dacthal, diazinon, dicamba and malathion. Organic compounds such as phthalate esters, phenolic compounds and polycyclic aromatic hydrocarbons (PAHs) such as benzo(a)anthracene, chrysene, anthracene and benzo(b)fluoroanthenene (Fig. 34B) are common near industrial sites or where vehicles are parked. The most common metals are aluminum, arsenic, cadmium, chromium, mercury, nickel, iron, lead, copper and zinc. Chapter 8 reviews the particular problem of sodium and chloride applied on roads during winter.

The use of constructed wetlands to remediate urban storm water is reviewed in Chapter 12; urban storm-water runoff is commonly contaminated to levels that exceed leachates from mine tailings (*e.g.*, Chapters 13, 14). Experience with pervious pipes, porous pavement and other techniques that allow urban storm waters to infiltrate into underlying strata (Fig. 35), is not encouraging. These systems require detailed geological information as they cannot be employed in areas of high water table or poorly permeable soils, are easily clogged with sediments, and difficult to maintain (Ontario Ministry of Environment and Energy, 1994b). Problems emerge in sandy soils where runoff rapidly infiltrates to the underlying ground water and there is minimal opportunity for biological activity. Because of this, porous pipes, porous pavement and dry wells have been found to promote contamination of ground

A

B

C

Figure 32 **(A)** *Mining of the flood plain of the Don River for sand and gravel in 1954; the effect was a dramatic increase in downstream sedimentation and the need for frequent dredging of the Keating Channel (Fig. 4C). Toronto Harbour Commission photograph PC 14/4572. Such practice was widespread across the GTA, particularly where rivers cross the former shoreline deposits of Glacial Lake Iroquois (e.g., Varga et al., 1991).* **(B)** *Steep-cut bank along Don River in the vicinity of O'Connor Drive (see Fig. 43) in 1949. Trees for scale.* **(C)** *Same view, 20 years later, after construction of the Don Valley Parkway; note embankment composed of municipal garbage against former cut bank. Garbage has been widely employed to infill topography across the GTA (e.g., Fig. 25). Photographs* **(B)** *and* **(C)** *from Lockwood Survey Corporation.*

water. Grass filter strips and grassed swales are effective in trapping suspended sediment in urban runoff and allow sorption of pollutants to soil particles where biodegradation may occur. Detention ponds incorporating constructed wetlands are a common and effective management practice (see Chapter 12 and Pitt *et al.*, 1994).

Several studies have been completed under the Great Lakes Action Plan Cleanup Fund to determine the annual loads of selected toxic contaminants from urban sources into Lake Ontario. Pollutants are commonly bound to suspended sediment leading to many cases of chronic toxicity to alongshore aquatic organisms (*e.g.*, Free and Mulamoottil, 1983; Salomons, 1985; Rowney *et al.*, 1986; Bodo, 1989; Marsalek and Ng, 1989; Servizi and Martens, 1991; Warren and Zimmerman, 1994a). These studies have also demonstrated that storm-water runoff, derived from a multitude of urban non-point sources, is a major contributor to the destruction of aquatic habitats and degradation of water quality in Lake Ontario (Ontario Ministry of Environment, 1991).

Sediment Loads

Urbanization results in greatly increased sediment loads to rivers (Fig. 4). The largest single source of suspended sediment in urban watercourses is site disturbance and vegetation clearance during construction activity. Soil erosion from uncontrolled construction sites can remove as much as 450 t of soil per ha every year (Schueler, 1987). Local geologic factors will dictate the grain-size characteristics of eroded sediment; much of the material can be trapped by retaining such waters in constructed ponds and leaving the sediment to settle out. OMOEE (1994b) guidelines require the capture of first-flush runoff in storm-water ponds to allow settling of sediment and contaminants; ponds must contain the runoff from a 25 mm storm and release it over 24 hours. Other techniques (*e.g.*, oil/grit separators) suffer from sediment resuspension and very poor trapping of sediment (Ontario Ministry of Environment and Energy, 1994b). Metropolitan Toronto and Region Remedial Action Plan (1990) concluded that current regulatory efforts are inadequate to control the entry of urban sediment to rivers and provide poor protection of aquatic resources. Current practice favours at-site prevention of sediment release rather than downstream removal in settling ponds since the latter involves the need to remove large volumes of sediment, some of which will be the result of natural erosion processes. Current guidelines for new developments are set out in Ontario Ministry of the Environment and Energy (1991). With regard to construction sites in general, there is little understanding of the potential effects of soil wash off following clear-cutting of surface vegetation on surface-water quality, and the effects of soil compaction during construction on subsequent recharge through grassed areas.

Case Study

The Don River

The Don River (Figs. 1, 4, 36) is one of Canada's most degraded urban rivers (Metropolitan Toronto and Region Conservation Authority, 1991, 1992, 1994). The present-day characteristics of this heavily urbanized watershed can be used as a predictor of changes that will occur in other rapidly urbanizing basins in the absence of any preventative planning. In 1950, only 15% of the watershed was urbanized; it is now 80% urbanized and is the most intensely urbanized watershed in the GTA, with a population of over 800,000. By the year 2021, the watershed is expected to become 91%

urbanized (Metropolitan Toronto and Region Conservation Authority, 1994). Many tributaries have been paved over (Steedman, 1986; Figs. 26, 37). Several areas are prone to severe flood damage and flood-water levels are projected to rise more than 0.5 m on the Don River with future urbanization (Marshall Monaghan Ltd., 1992a).

Water quality. The Don River is one of the largest sources of lead contamination to nearshore waters of the Toronto waterfront (Fig. 38). Don River water exceeds the Provincial Water Quality Objectives for phosphorus, fecal coliform, chlorides (from road salt), and heavy metals such as copper, zinc, and lead during wet weather. Recent work has focussed on metal partitioning between supended particulate and dissolved phases and the effects of high chloride levels (Warren and Zimmerman, 1994a, b) resulting from the application of over 6000 t of road salt to the Don Valley Parkway each winter. Occasional exceedances of pesticides and industrial organic chemicals, including DDT (banned 1988), still persist in Don River water. The major source of these contaminants is effluent from the North Toronto Sewage Treatment Plant (NTSTP; Fig. 36) and combined sewer overflows (Fig. 35; see below) along Massey/Taylor Creek and the Lower Don (Fig. 24A). Unfortunately, the NTSTP contributes 20% of the baseflow of the lower Don River and is essential to maintaining adequate flow in the lower Don. Throughout the Don watershed, an unknown number of old landfill sites are leaking into ground

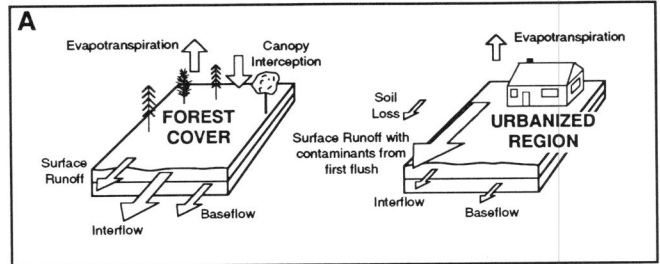

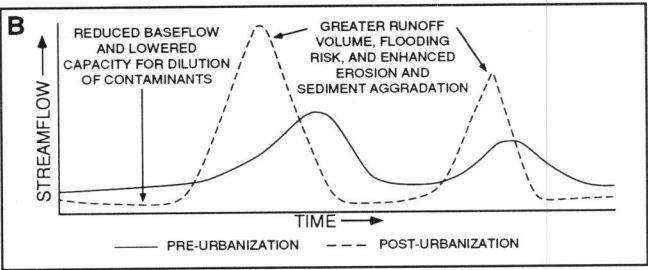

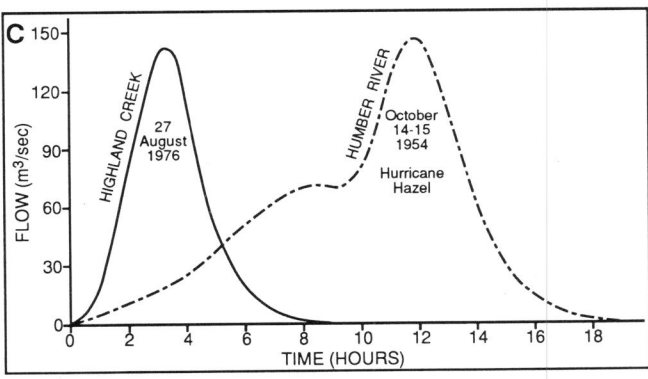

Figure 33   (A, B) *Effects of urbanization in watersheds.* **(C)** *Flow of urbanized Highland Creek during moderate storm compared to response of then undeveloped Humber River during the severe storm of Hurricane Hazel (data courtesy of W. Tovell).*

water and streams (Chapters 5, 20; see below).

Contaminated sediment.   By canalizing the Don River and placing a 90° bend in its channel where it enters the inner harbour, the original hydraulic gradient of the lower river was destroyed. Urbanization of the Don River catchment has given rise to repeated flooding, the need for frequent dredging

of the Keating Channel (Metropolitan Toronto and Region Conservation Authority, 1992b) and the disposal of contaminated fluvial sediment. Increased soil runoff resulting from historic gravel mining of the river bed and channel straightening, combined with successive phases of accelerated development in the Don River watershed, are clearly reflected in

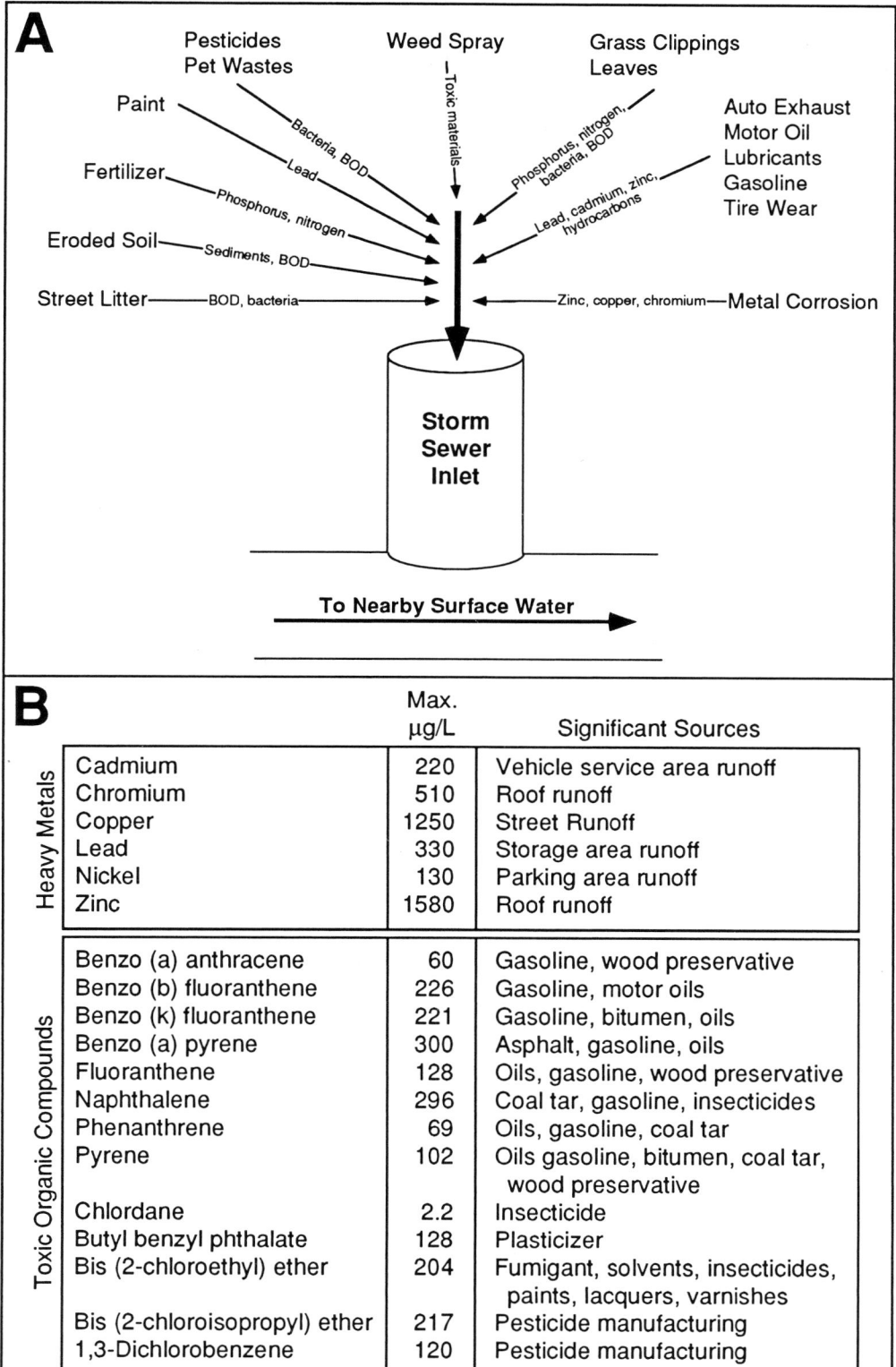

Figure 34   (**A**) *Sources of contamination in storm-water runoff from urban areas.* (**B**) *Common contaminants in urban storm water with maximum reported concentrations (after Pitt et al., 1994).*

the volume of contaminated dredgate needed to be removed from the channel (Fig. 4C). Dredgate was dumped in deep water of Lake Ontario. In 1972, the United States and Canada established maximum contaminant levels for open-water dumping and agreed to halt open-water dumping (OWD) of contaminated sediment (Persaud et al., 1985, 1991; Boyd, 1988). Dredging of the Keating Channel was stopped. Sedimentation within the channel quickly resulted in impeded navigation within the inner harbour and the increased incidence and severity of flooding of areas such as the proposed Ataratiri redevelopment site (Fig. 30B).

In 1987, dredging of the Keating Channel recommenced and dredgate was dumped in contained disposal cells within Tommy Thompson Park along the Leslie Street spit (Fig. 29B). This was done with a bottom-dumping hopper scow able to dump contaminated sediment in water less than 1 m deep. Under Ontario Ministry of Environment policy for Management of Excess Soil, Rock and Like Materials (1992) Don sediments are classified as controlled, and can only be disposed off in a secure landfill facility (see above). The Keating

Channel Environmental Monitoring Program Study Area was established to assess the quality of dredgate and any leakage of contaminants from the containment cells into the outer harbour (Metropolitan Toronto and Region Conservation Authority, 1992b).

Dredgate from the Keating Channel typically consists of fine sand (~45%), silt (~40%) and some clay (~15%). In 1989, over 90% of all samples exceeded OWD guidelines for lead, cadmium and zinc, and over 50% exceeded guidelines for loss on ignition oil and grease levels, while more than 25% of dredgate samples failed for total phosphorus and nitrogen. Organic compounds such as p,p'-DDE, DDT metabolites, chlordane and dieldrin, together with total PCBs, are frequently detected. Investigations of lake bottom sediment and benthic invertebrates such as clams, adjacent to the disposal cells at Tommy Thompson Park, shows that there is no leakage of resuspended sediment and contaminants to the surrounding lake (Boyd, 1986; Metropolitan Toronto and Region Conservation Authority, 1992b). No impacts have been detected to date in fish populations. Plans call for increasing

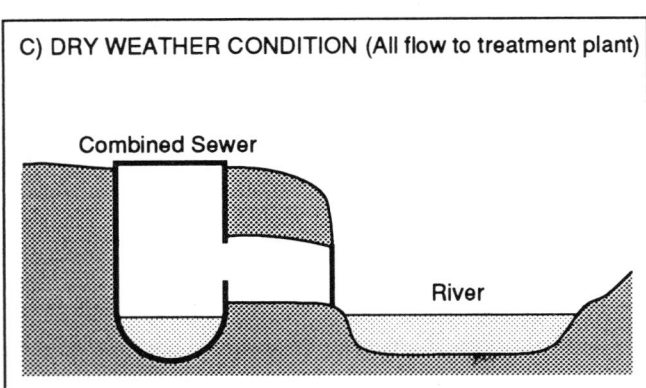

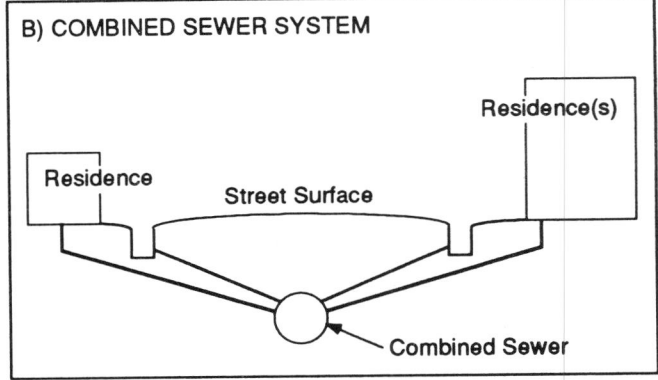

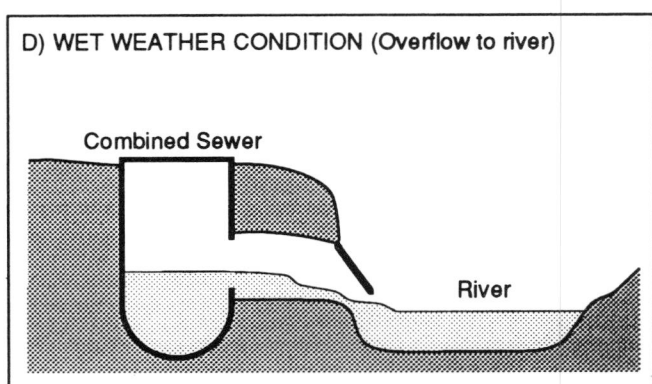

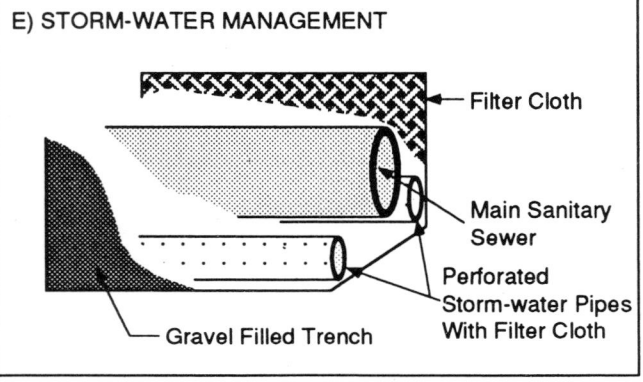

Figure 35 Distinction between (A) separate and (B) combined sewers, showing performance of sewer under (C) dry-weather and (D) wet-weather conditions. Sewage and street runoff are discharged directly into rivers during storms. (E) Design of perforated storm-water pipe to exfiltrate water into underlying soils. This is impractical in areas of impermeable substrates because clogging is a major problem in addition to contamination of ground water by urban contaminants (see text; after Metropolitan Toronto and Region Remedial Action Plan, 1994).

the longitudinal gradient of the Don River and the flushing of sediment into a 125 ha marsh at the mouth of the Keating Channel. This would re-establish what was the largest migratory bird assembly point on Lake Ontario prior to landfilling of Ashbridge's Bay (Crombie, 1994).

Sanitary Waste Water
Sanitary waste water produced within the GTA is handled by three different systems: (1) the Metropolitan Toronto Trunk Sewer System, (2) the York-Durham Sewer System, and (3) septic tanks and small municipal treatment plants (Fig. 24). The sewage system provides a good example of how large volumes of waste water are exported across the GTA to lakeshore sewage treatment plants; the environmental cost of treatment is then borne by distant communities.

Metropolitan Toronto Trunk Sewer System
The Metropolitan Toronto Trunk Sewer System (MTTSS) is about 350 km in length, and its separate sewersheds follow natural watersheds, thereby allowing the slope of the natural terrain to convey flows to treatment plants by gravity. Trunk sewers are up to 3 m in diameter and as much as 30 m deep. There are four sewersheds: Highland Creek which outfalls at the Highland Creek Treatment Plant (TP), the Main Treatment Plant Sewershed outfalling at Ashbridge's Bay TP (ABTP), and that of the Humber; in addition, the Region of Peel operates the Lakeview Treatment Plant in the city of Mississauga which serves the Etobicoke Creek Sewershed (Fig. 24).
Combined sewers.   Combined sewers (CS) handle both storm waters and sanitary sewage. These overflow into surface watercourses during storms and are a major source of

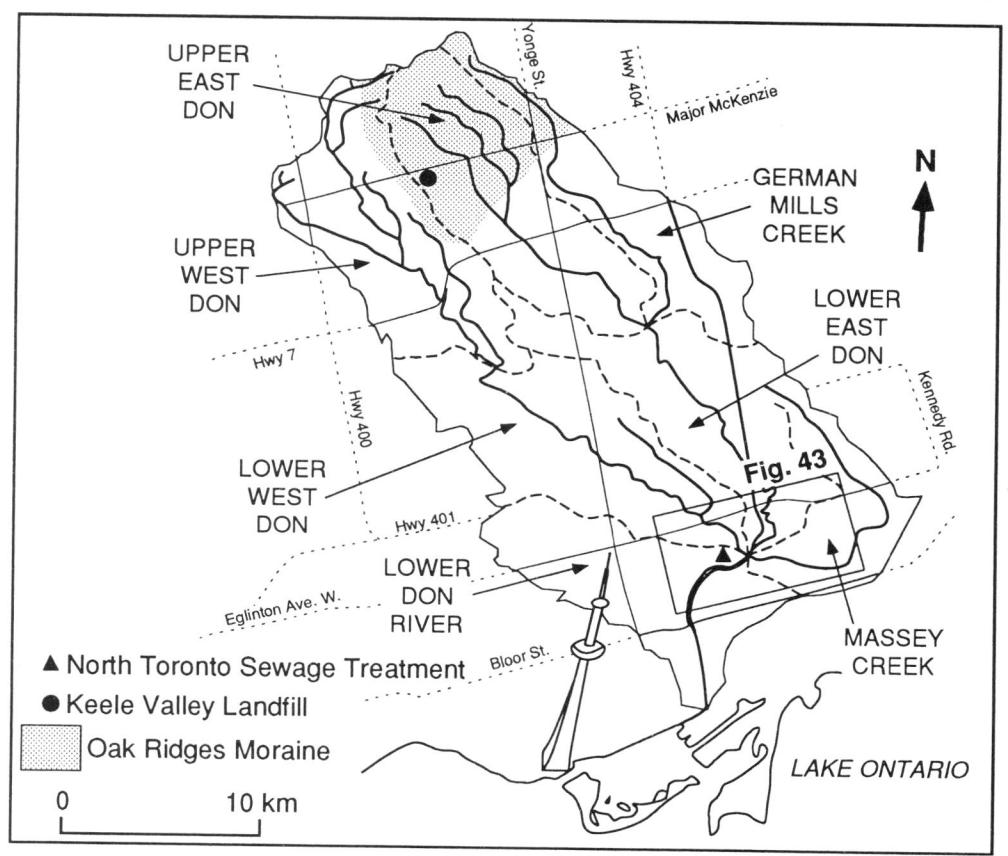

Figure 36 *The Don River drainage basin; one of Canada's most degraded urban rivers (see Fig. 1). Currently, 80% of the 360 km² area is urbanized. In 1992, Metropolitan Toronto and Region Conservation Authority (MTRCA) created a Don Watershed Taskforce to generate an action plan for remediation (MTRCA, 1992a, 1994). The location of numerous closed landfill sites within the drainage basin is shown in Figure 43A. Remediation cannot deal with a wide range of contaminants already present in ground waters and which will discharge to surface waters, including Lake Ontario, over the next few hundred years (see Chapters 4, 5). This frustrates recent policies that attempt to achieve "zero-discharge" to the Great Lakes by phasing out the manufacture and use of certain chemicals. Those chemicals are already in urban ground waters.*

Figure 37 *Blanket urbanization in the City of Scarborough (Fig. 3); the intersection of Lawrence Avenue east and McCowan Road in 1954 (**left**) and 1969 (**right**). Photographs from Lockwood Survey Corporation.*

pollution (Figs. 24, 34, 35). Combined sewers were constructed in many areas because only a single pipe was required and they were, therefore, inexpensive. Their real cost, measured in terms of environmental impacts, is very considerable. The city of Toronto has 25 CS waterfront outfalls that discharge directly into Lake Ontario; Metro Toronto has 21 waterfront outfalls. There are 22 combined sewer outfalls on the Don River; 5 on Black Creek. As much as 15% of the Lower Don watershed is drained by combined sewers; during wet weather, Massey Creek (Fig. 36) contributes over 80% of total pollutant loadings to the Don.

Work by D'Andrea *et al.* (1993) at CS outfalls in the city of Scarborough identified many organic compounds, identified by OMOEE as candidate substances for bans or phase-outs, at higher concentration than in the effluent of sewage treatment plants (see also Snodgrass, 1993). In Metro Toronto alone, over 386,000 m³ of surface storage and 718,000 m³ of deep-tunnel storage are used at a initial construction cost of

$500 million (1993 dollars) in order to detain and release runoff at a reduced flow rate. However, parts of the MTTSS are operating close to capacity and there is increasing pressure to use any excess capacity for new urban development rather than treating delayed flows from combined sewers. An example of the use of a constructed wetland to alleviate contamination near a combined sewer outfall on Lake Ontario is given in Chapter 12.

Water entering the Toronto waterfront from the Humber and Don rivers at any time of the year, irrespective of weather conditions, exceeds Provincial Water Quality Guidelines (PWQG; 100 *Escherischia coli* organisms per 100 mL of water; Pulton, 1986). Combined sewer overflows result in closing of the Western Beaches, Islands and Eastern Beaches (see Fig. 26) when bacteria counts in swimming areas exceed recommended PWQG; beaches are now permanently posted to warn the public of elevated bacteria counts for 48 hours after storms. In 1990, the City of Toronto constructed a detention

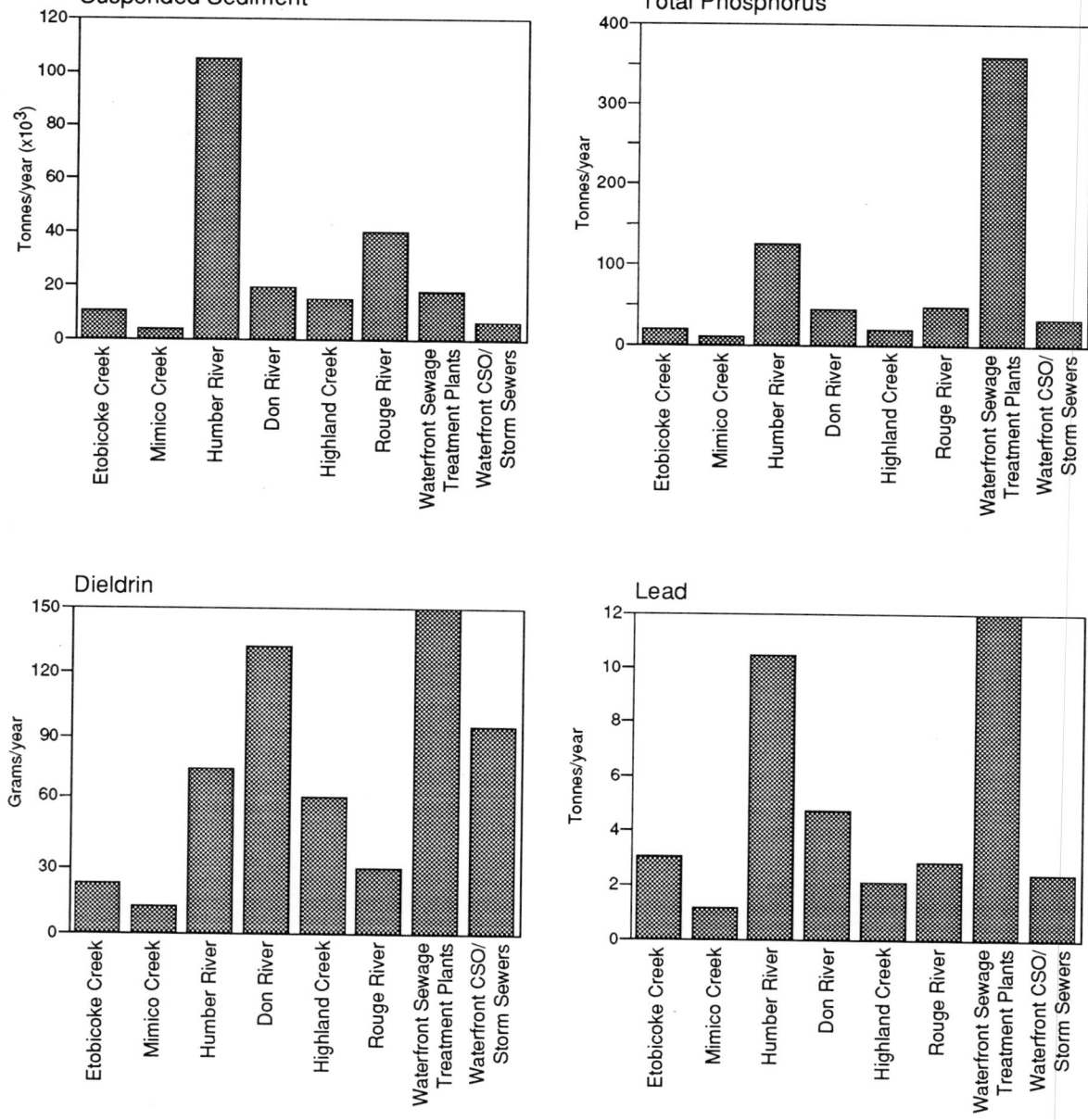

Figure 38 *Contaminant loading, and sources, along the Toronto waterfront area of Lake Ontario. The biggest sources are heavily urbanized rivers (Don, Humber) and sewage treatment plants. Data from Toronto Waterfront Regeneration Trust.*

tank to collect discharges from several combined sewers to alleviate conditions in the western portion of the Eastern Beaches. In 1991, the Western Beaches were placarded for almost the entire season, primarily due to excess storm-water runoff from combined sewers. A proposal to construct a $57 million bored-bedrock tunnel, designed to capture storm-water/sewage overflow in the Western Beaches and later transport it for treatment to ABTP, was dropped by the province in late 1994. The reason for cancellation was simply the perception of exporting west-end sewage across town to unwilling east-enders. The tunnel would have produced 368 t of sewage sludge each year, amounting to an increase of only 0.5% in the volume treated at ABTP (*The Toronto Star*, November 12, 1994).

Contaminant Load to Lake Ontario.   Lake Ontario is the eleventh largest fresh-water lake in the world, contains 1.64 trillion litres of water, and is 311 km long, 85 km at its widest and 243 m at its deepest point (Environment Canada, 1995a). Being the outlet for the entire Great Lake system, it faces the greatest environmental threat of all; 70% of the toxic contaminant load to the lake is derived from the Niagara River. It takes 6 years for incoming water to move through the lake. Urban contaminants entering ground water and surface waters ultimately find their way to Lake Ontario. The International Joint Commission has designated the Toronto waterfront and adjacent downtown area as an Area of Concern (AOC); one of 43 across the Great Lakes basin (see Zarull and Mudroch, 1993; Chapter 15). Remedial Action Plans (RAPs) designed to restore polluted waterways and waterfronts are underway in all these areas (Hartig and Zarull, 1991, 1992).

Estimates of total watershed and sewage treatment plant contributions to pollution loads entering Toronto's inner harbour area are shown in Figure 38. The greatest loadings for many pollutants in nearshore waters along the Toronto waterfront are derived from sewage treatment plants (Marsalek and Schroeter, 1988: Marsalek and Ng, 1989; Schroeter, 1993; see below). Data from 1990 with regard to effluent released from Ashbridge's Bay Treatment Plant show acute toxic effects for ammonia and chlorine within 500 m of the outfall; levels of phosphorus still exceed background levels $(0.015 \text{ mg·L}^{-1})$ at a distance of 3 km; PCB concentrations also exceed PWQG. In terms of BOD (a measure of oxygen-consuming wastes in water) surface lake waters around the treatment plant are grossly polluted; values consistently exceed 16 mg·L^{-1}. In general, pollutant loadings of trihalomethanes, total PCBs and chlorinated benzenes increase shoreward in surface waters of Lake Ontario and suggest many different sources (ground-water discharge, storm-water runoff, routine illegal discharges, accidental spills, *etc.*). Illegal household sanitary connections to storm sewers are a continuing problem in the city of Toronto and legislation requiring inspection of sewer connections as a condition of sale is under discussion (Metropolitan Toronto and Region Remedial Action Plan, 1990, 1991). As a result of past historic landfilling and the industrial legacy along the waterfront, the contribution to pollutant loadings of surface rivers and Lake Ontario derived from ground water is also of concern (see below). There are, however, few quantitative estimates available (Chapter 5).

Progress in identifying contaminant releases within the Lake Ontario basin has recently been made with information systematically gathered by the new National Pollutant Release Inventory (NPRI; Environment Canada, 1995b). This is the first of its kind in the world and requires any industrial facility releasing more than 10 t·a^{-1} of any one of 178 listed pollutants to identify the quantity released in air, water and on

land, or transferred off site, each year. A total of 738 facilities in Ontario made reports in 1993. In that year, 13,763,698 kg of industrial pollutants were released by industry across the GTA (Fig. 39), but the NPRI does not take into account pollutant released from nuclear power stations (see below). The biggest polluters are in Durham Region (General Motors, Oshawa and Co-Steel Lasco in Whitby; NPRI; Environment Canada, 1995b). A comprehensive contaminant audit for the GTA, involving water (ground water, surface water), air and land-based components should be possible in the near future. Quantitative data are critically important to assessment of the long-term state of health of Lake Ontario and the surrounding urban communities.

Impact on municipal drinking water.   Municipal drinking water in Metropolitan Toronto is taken from Lake Ontario (Fig. 24). A study of the quality of Toronto's drinking water (City of Toronto, 1990) found six contaminants with concentrations above acceptable levels (lead, chloroform, aluminum, alpha-hexachlorocyclohexane, bis(2-ethylhexyl)pthalate and tetra-chloroethylene). A further seven were identified at concentrations with a low margin of safety (chromium, trichloroethylene, barium, lindane, cyanide, bromodichloromethane and dibromochloromethane). The report concluded that "prudent public health policy would suggest undertaking measures to reduce the levels" (City of Toronto, 1990, p. 38). Other concerns are discussed by Moorhead *et al.* (1995), Morris *et al.* (1992) and MacKenzie *et al.* (1994).

The York-Durham Sewer System, Municipal Treatment Plants and Septic Systems

Outside of Metro Toronto, waste water is handled by the York-Durham Sewer System (YDSS) which discharges to the Duffins Creek STP (Fig. 24), by various municipal treatment plants (MTPs) serving small rural communities (*e.g.*, Stouffville, Uxbridge; Fig. 24), and many thousands of septic tanks in outlying rural areas. The extent and capacity of the YDSS dictates where urbanization can occur and where it cannot. Planned extensions of the existing system, northward along the Yonge Street corridor onto the Oak Ridges Moraine (Figs. 1, 2, 24), will allow wider suburbanization and threaten the integrity of one of the largest aquifer systems in southern Ontario (Chapter 9). The Stouffville MTP is at capacity and connection to the YDSS is planned for both these communities.

King City (Fig. 1) has a population of over 5000 people and is the largest community in the GTA totally reliant on septic systems (Proctor & Redfern Ltd., 1993a). Septic tanks are effective in removing about 50% of total suspended solids, nitrate, and phosphorus, but performance is dependent upon soil type; ground-water contamination from septic tanks is discussed by Robertson and Cherry (1992), OMOEE (1994d) and in Chapters 4 and 5.

Urban-Rural Impacts on Ground Water in the Greater Toronto Area

The discussion above has focussed on the quality of surface river and lake waters impacted by urbanization; these impacts are reasonably well known and mitigative work is starting to be put into place (*e.g.*, Marsalek and Ng, 1989; Schroeter, 1993; see above and Chapter 12). In contrast, there are limited data regarding the regional dynamics of ground-water flow and associated contaminant fluxes across the GTA. Quantitative estimates of ground-water flux to Lake Ontario have been made for a few GTA watersheds (*e.g.*, Haefeli, 1972; Singer, 1974; Ostry, 1979a, b; Chapter 5) and for the Lake Ontario basin as a whole by DeCooke and Witherspoon (1981); studies elsewhere in the Great Lakes basin show that traditionally computed ground-water flows (baseflow) can be considerable underestimates (Sellinger, 1995).

Urban Areas

There are very few data regarding the regional magnitude of contaminant loadings introduced to, and transported by, urban ground waters. A recent study of a representative urban catchment (600 km²) in the GTA by Livingstone (1993) identified the total ground-water contaminant loadings from point sources (such as open and closed landfills, underground storage tanks, snow dumps, coal tar sites and septic systems) and non-point sources (road de-icing chemicals and agrochemicals). The major non-point and point sources (road salt, landfills, respectively) accounted for over 95% of the total contaminant load. The potential volume of ground water that could be degraded to exceed Provincial Drinking Water Guidelines by such a load was estimated to be almost equivalent to the existing volume of water in Lake Ontario (Chapter 5). These numbers ignore contaminant attenuation as a result of the very large volumes of water stored in aquifers, but they at least suggest that, as currently managed, ground-water resources are not adequately safeguarded in urban areas under existing land uses. In reality, because municipal drinking water is derived from Lake Ontario, urban ground waters across a large part of the GTA have been written off as a resource. Ground waters and contaminants discharge to surface water courses and, ultimately, to Lake Ontario.

Rural Areas

The picture emerging from rural areas of the GTA reveals widespread contamination of ground water by nitrate, pesticides and bacteria. Rudolph and Goss (1992, 1993) reported results of a ground-water quality survey in rural areas of southern Ontario, and found that 37% of all wells tested (1300) showed levels of target contaminants well above Provincial Drinking Water Guidelines; 8% had detectable levels of pesticides. Comparable results are widely reported across mid-continent North America as a result of modern farming techniques (*e.g.*, Murphy, 1992; Sievers and Fulhage, 1992; Gish *et al.*, 1992; Rudoph and Goss, 1992 and references therein) and are likely representative of the rural areas of the GTA.

Discussion

The data reviewed above with regard to current ground-water conditions across the urban and rural parts of the GTA, allow the case to be made that current land stewardship is unsustainable in terms of water quality. Systematic mapping of aquifers, the identification of hydrogeologic constraints on development, water resource management, proactive planning and legislation for ground-water protection are urgently required (National Research Council, 1994; Chapter 11). A recent international report on progress in protecting urban ground-water resources in Canada and the United States concluded that "specific regulation is conspicuously absent" in Ontario (International Joint Commission, 1995, p. 103). The report identified, further, that municipalities dependent on ground water have received little assistance from the Ontario Ministry of the Environment and Energy in identifying and protecting recharge areas and devising wellhead protection programs (see example of Town of Caledon, below). In contrast, municipalities using Lake Ontario or other surface water have benefitted from government assistance programs. A provincial ground-water strategy is urgently needed.

PART III: REGIONAL WASTE MANAGEMENT

Municipal Waste

Dumping of municipal wastes in large landfills has been the preferred management option for many North American cities (see Chapter 20). The largest municipal landfill in the world is at Fresh Kills, Staten Island, New York, U.S.A. (Fig. 26) which accepts 13,000 t of waste every day; the largest in Canada is Keele Valley in the GTA (Figs. 1, 24, 36) which, in 1990, handled 6000 t of waste every day. Over 80% of solid non-hazardous (municipal) waste created in the GTA (about 4 million tonnes annually) is landfilled in four existing landfills (Milton, Brock West, Brittania Road and Keele Valley; Fig. 1); two of these (BW and BR) are near to capacity and scheduled to close by 1997. The search for new sites initiated one of the most detailed geologic and hydrogeologic investigations ever completed in the GTA (Chapter 20).

The aim here is to present case studies of existing and

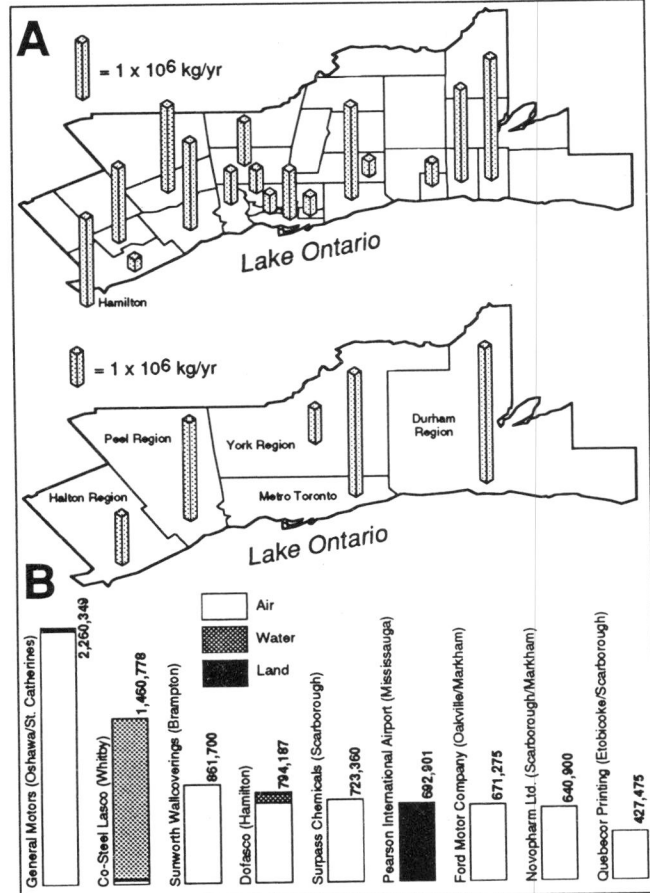

Figure 39 (**A**) *Releases, in kg, of industrial pollutants by GTA municipalities and regions for 1993 (based on NPRI, see Environment Canada, 1995b). See Figure 3 for municipalities. Data includes pollutants released into air, water and on land, but not those released from aircraft, by the marine sector, vehicles, railways or fuel distributors or nuclear industry. Figures are likely considerable underestimates of the total industrial contaminant load released across the Greater Toronto Area.* (**B**) *Ten largest industrial emitters in the GTA in 1995 (data from NPRI, see Environment Canada, 1995b) excluding nuclear generating plants and airports (data in kg). Significant airborne pollutant emissions to the surrounding GTA from Toronto International Airport together with the impact on nearby watercourses (Etobicoke and Mimico creeks) of stormwater contaminated with aircraft and runway de-icing compounds (glycol) are identified in Transport Canada (1991).*

closed municipal landfill sites on different geologic substrates across the GTA to show the nature of environmental impacts arising from the off-site migration of contaminated water (leachate). The various geological settings of landfills across the GTA can be divided into permeable substrates, such as offered by sands and gravels of the Oak Ridges Moraine and sands along the Iroquois shoreline deposits, and relatively impermeable substrates offered by tills.

Nearly 1200 abandoned landfill sites have so far been documented in southern Ontario, but detailed hydrogeologic investigations have been carried out at only a handful of sites (*e.g.*, Cherry, 1983; Barker *et al.*, 1987; Chapter 24). As of May 1995, only 13 closed landfills have been investigated (Environment Canada, 1995c). Many landfills are now surrounded by densely populated residential areas and some lie

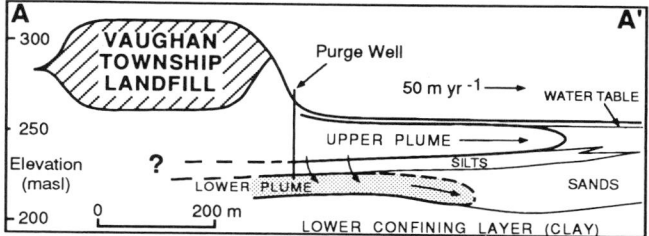

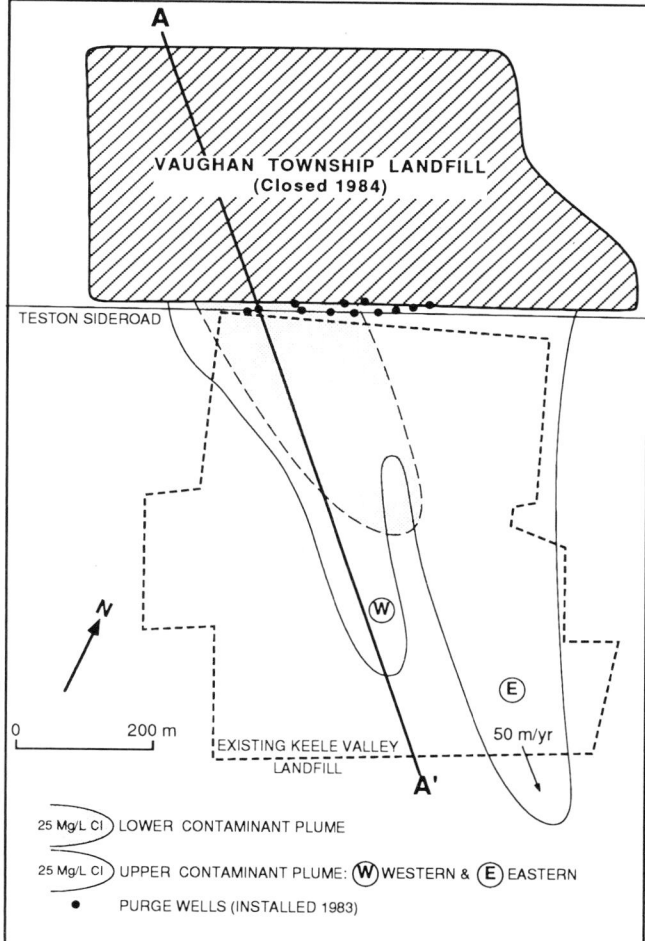

Figure 40 *Hydrostratigraphic section (top) and map (bottom) of multi-level contaminant plume associated with Vaughan Township landfill (closed 1984). The plume extends below existing Keele Valley landfill, the largest in Canada sited on a spur of the Oak Ridges Moraine (Figs. 1, 36; modified from Technos, 1983). The plumes follow channels of coarser sediment within glaciolacustrine silts.*

near eroding coastlines (see below and Coakley, 1990). The geologic setting of most is known to be completely inappropriate for either containment or attenuation of landfill-derived leachate and methane gas (Howard *et al.*, 1996).

Landfills on the Oak Ridges Moraine
About 60 landfills have so far been identified in former gravel pits located within the ORM. These landfills are unlined and, because they rest directly on highly permeable silt, sands and gravels, are a source of contaminants entering underlying aquifers.

Keele Valley Landfill–Vaughan Township Dump
Keele Valley sanitary landfill site is the third largest active landfill in North America and lies adjacent to the Vaughan Township dump which closed in 1984. These two landfills lie within a spur of the ORM (Figs, 1, 24, 36, 40). A major reason for the establishment of landfill sites in the area was to facilitate the rehabilitation of disused sand and gravel pits. Aggregate extraction, beginning in 1908, had left large pits, some of which were in excess of 30 m deep with steep side slopes scarred by erosion. In 1980, an OMOE landfill permit was issued to Superior Sand and Gravel Ltd., to operate the Keele Valley site. In 1983, the site was sold for $38 million to Metropolitan Toronto to accept waste previously taken to sites such as Beare Road which closed in 1985 (see below). The landfill site currently contains 20 million m³ of waste and is licensed to accept another 10 million m³. Of a total area of 376 ha, 100 ha are used for landfilling.

The local geology consists of Halton Till resting on thick silts and sands deposited during the Mackinaw Interstadial (Fig. 17). Two distinct aquifers occur within the Mackinaw deposits and are separated by an intermediate silt-rich confining layer. The Halton Till cap was removed during aggregate extraction and as a result, the landfills rest directly on the upper sand aquifer. Keele Valley Landfill utilizes a compacted-till liner (1.2 m thick) of low permeability (1×10^{-8} cm·s⁻¹ or less; Richards and Thompson, 1989; King *et al.*, 1993) designed to slow the escape of contaminated water (leachate) into the underlying aquifer. Accelerated compaction of the waste pile (and therefore increased capacity) is achieved by recirculating leachate through the waste pile, thereby increasing moisture content to 50%.

The 30-ha Town of Vaughan dump lies north of the present Keele Valley landfill and operated until 1984 (Fig. 40). A chloride contaminant plume, defined by chloride concentrations greater than 25 mg·L⁻¹, is migrating away from the site to the southeast. The southern edge of the plume lies 1750 m downstream of the Vaughan landfill, is about 500 m wide at Teston Sideroad and is advancing at about 50 m per year (Metropolitan Toronto Works Department, 1994). The plume consists of both shallow and deep components (upper and lower plumes; Fig. 40), following the two aquifers and separated by a silt-rich confining layer. The deeper plume is the result of contaminant migration through the confining layer to the lower aquifer. Downhole gamma-logging and conductivity data presented by Dewaele *et al.* (1992) suggest that the upper plume has separated into eastern and western lobes. This can be attributed to differential flow within buried channels in the upper aquifer. Purge wells installed along Teston Sideroad are designed to slow the advance of the plumes, and leachate is discharged to the YDSS (Fig. 24).

Aurora Landfill
The Waste Management of Canada Sanitary Landfill Site No. 1 (the Aurora landfill; 26 ha) lies close to the crest of the ORM,

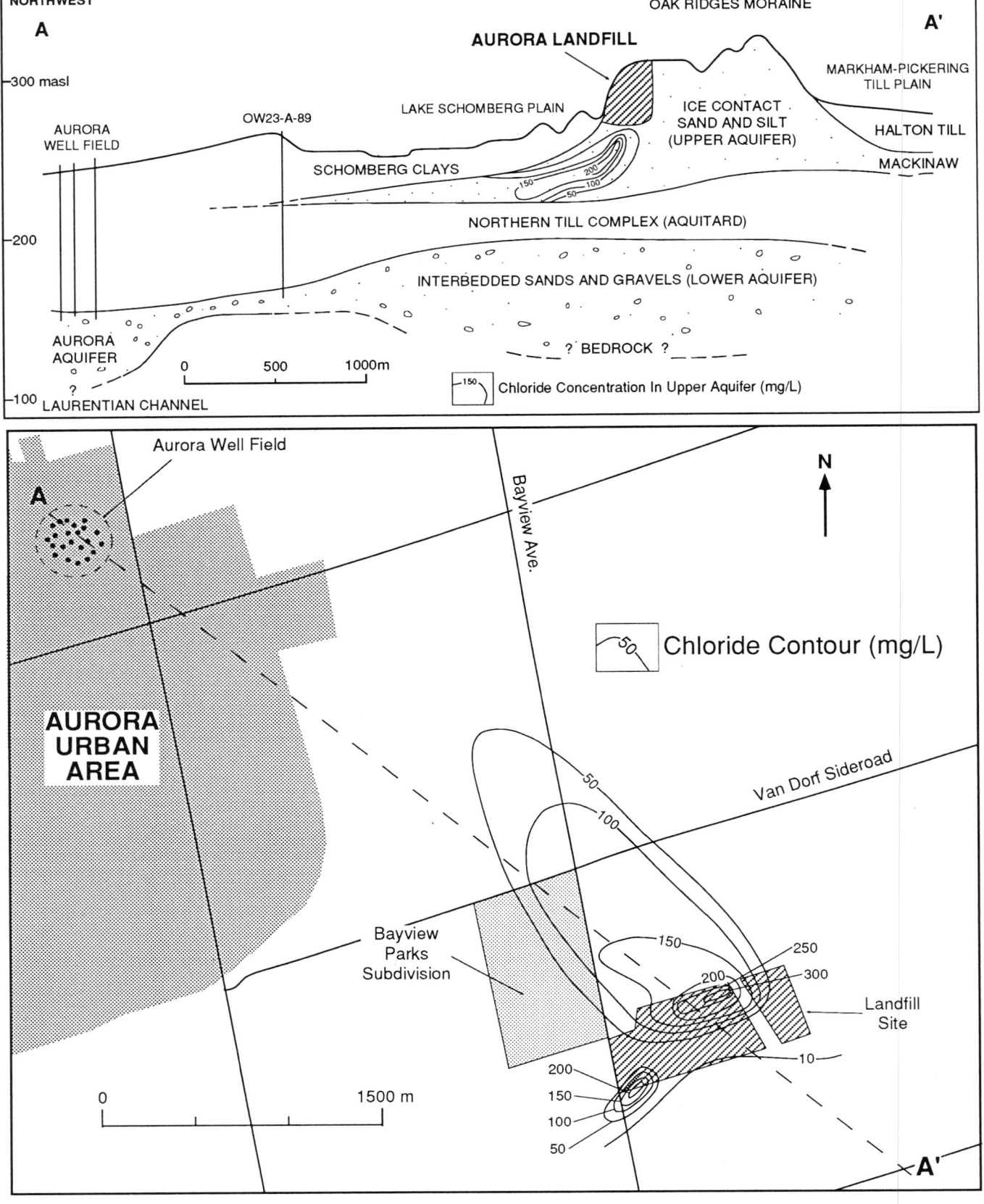

Figure 41 *Aurora landfill (1938 to mid-1980s): (top) hydrostratigraphic cross-section (A–A') and (bottom) distribution of subsurface contaminant plume. Note location of new residential subdivision. Phenols are reported in well OW23-A-89. Data from Conestoga Rovers and Associates (1992). The Aurora Aquifer sits within a deep bedrock valley (the Laurentian Channel; Fig. 12) and is composed of sands and gravels of the Scarborough Formation (Fig. 18A).*

some 1.5 km southeast of the township of Aurora (Figs. 1, 41). The site lies in an abandoned aggregate quarry and opened as a liquid and solid waste disposal facility in 1938, primarily for liquid tannery waste from Collis Leather Company in Aurora (until 1976). The site continued to accept municipal waste until the mid-1980s.

The Aurora landfill lies close to the axis of the Laurentian Channel (Fig. 12) where drift thickness approaches 150 m (*e.g.*, Fig. 18A). Below the site, an aquitard (Northern till, sand and clay) of restricted thickness separates an upper aquifer formed by ice-contact lacustrine sands and silts of the ORM, from a gravelly lower aquifer (the Aurora Aquifer, *i.e.*, Scarborough Formation; Figs. 18A, 41). The upper aquifer thins northward below thick (50 m) silty-clay deposits and discharges into a tributary of the Holland River. Linear groundwater velocities in the upper aquifer are as much as 30 m·a^{-1} and a well-defined leachate plume, defined by chloride concentrations greater than 100 mg·L^{-1} and phenol concentrations as high as 0.0656 mg·L^{-1}, extends almost 2 km northwest of the site directly below a new residential development (Bayview Parks subdivision; Fig. 41). Water quality in the lower aquifer was unimpaired as of 1991, but significant detections of phenols were recorded in a nearby monitoring well (OW23A-89; Fig. 41) and volatile organic compounds in a private well 2 km northwest of the landfill (Conestoga-Rovers and Associates, 1992). The town of Aurora obtains its municipal water supply from the lower aquifer immediately downflow of the contaminant plume.

Stouffville Landfill

Stouffville landfill (York Sanitation No. 4; Figs. 1, 42) began to accept domestic waste in 1962. Liquid industrial waste was also dumped in 6 lagoons at the site until 1970, when they were filled with solid and domestic waste and capped with till to reduce infiltration from rainfall and snow melt; the last lagoon was capped in 1976. No accurate information on waste types received at the landfill was collected, but subsequent studies have identified the presence of PCBs, benzenes, chlorinated phenols, various pthalates and esters, a wide range of petroleum hydrocarbons and volatile solvents and the pesticides D-BHC, endrin and 4,4-DDT. Concentrations of these contaminants vary widely across the site reflecting a complex filling history. Plans for expansion of the site were refused in 1981 in the face of vigorous local concerns with drinking-water quality and possible adverse health impacts; the site was, however, allowed to operate until 1985.

The geology of the site consists of a thick sand aquifer of Mackinaw Interstadial age (Fig. 17), separating an underlying Northern till from Halton Till and associated silts (upper fine-grained unit on Fig. 42). The chloride-rich contaminant plume is travelling southwestward within the aquifer at about 100-450 m·a^{-1}. Municipal drinking-water wells for Stouffville use the same aquifer (Proulx and Farvolden, 1989). The eastern margin of the contaminant plume is separated from the capture area of the wells by less than 200 m. With increased demand for water as the township grows, interception of the plume is possible. The population of Stouffville is expected to double over the next 20 years as a result of the extension of the York-Durham Sewer System (YDSS; Fig. 24) to outlying rural areas.

Landfills Along the Lake Iroquois Shoreline

Abandoned landfills, forming hills along the former shoreline of Glacial Lake Iroquois (Fig. 43), are prominent topographic features of the Toronto urban landscape. There is a long history of using wastes to rehabilate abandoned quarries, but there is little public appreciation of the origin of these hills, nor of the threat to ground and surface waters that they represent.

Former beach deposits of Lake Iroquois form a continuous belt across the southern part of the GTA and have been mined extensively for aggregate; more than 80 aggregate operations are known in Metro alone (see Ontario Ministry of Natural Resources, 1992), but many go unrecorded (Chapter 21). Many are in areas that have since become heavily urbanized (*e.g.*, the Don Valley; Fig. 43A).

Brock West, Brock North, Brock South Landfills

These landfills all occupy former aggregate pits in Iroquois beach deposits (Fig. 43B) where late Wisconsin till has been removed or thinned by shoreline erosion in Lake Iroquois, such that beach deposits rest directly on the underlying aquifer of the Thorncliffe Formation (Fig. 17). Brock West landfill (65 ha; Fig. 44) has been in operation since 1975 and contains 17 million tonnes of waste, in places up to 60 m thick. The fill includes liquid industrial wastes. Data regarding the waste types received by the landfill in 1991, after industrial waste was banned, are shown in Table 1. The site is located less than 500 m from a residential subdivision (Bertell, 1995) and other residential development is planned in the immediate vicinity of the site.

A bentonite clay liner was installed in those areas of the landfill underlain by sand or less than 3 m of till (about 90% of the site). A 1991 water budget shows that 114,000 m^3 of water (some 25% of annual precipitation) infiltrates the landfill each year in addition to 15,300 m^3 of sewage sludge from Duffins Creek treatment plant (Fig. 24B). Currently, more than 70% of this water is estimated to be stored within the refuse which, because of an unexpectedly high permeability, has not yet reached field capacity. Leachate production will increase substantially, perhaps by a factor of 8, when field capacity is reached (Dames and Moore, 1991, 1992 and studies cited therein). Currently, leachate (with a mean chloride concentration of 3300 mg·L^{-1} and a maximum value of 4470 mg·L^{-1} in 1991) is mounded within the landfill. The mound is some 20 m thick and results from insufficient underdrainage through a malfunctioning (clogged?) sub-drain system. Because of mounding, leachate is able to spill over the liner and is migrating off site in several directions (Fig. 44). In 1991, about 30% of fluid entering the landfill penetrated the liner (about 36,500 m^3·a^{-1}, of which 7,300 m^3 is captured by sub-drains).

The primary contaminant plume at Brock West, with a core chloride strength of 1000 mg·L^{-1} immediately below the landfill, follows the prevailing direction of ground-water flow to the south within a bedrock valley that joins West Duffins Creek (Fig. 44). As of 1991, this plume had advanced 450 m from the landfill boundary within Iroquois deposits and underlying fractured bedrock (Dames and Moore, 1991, 1992; Figs. 17, 44). Though the Sunnybrook Diamict is partially present across the site, it is of insufficient thickness to retard contaminant migration. The site of EE11, the former preferred long-term site for Durham Region waste disposal proposed by IWA Ltd., lies immediately to the north. The site, cancelled in July 1995, was to have been fully engineered with a synthetic liner system designed to prevent the problems that have developed at Brock West.

Brock North and South.   Brock North (72 ha) and South (102 ha; Fig. 43B) were identified and approved by OMOE as so-called temporary sites for landfilling during a previous shortfall in landfill capacity in 1976. Unfortunately, even temporary sites have long-term environmental impacts. In 1978, a

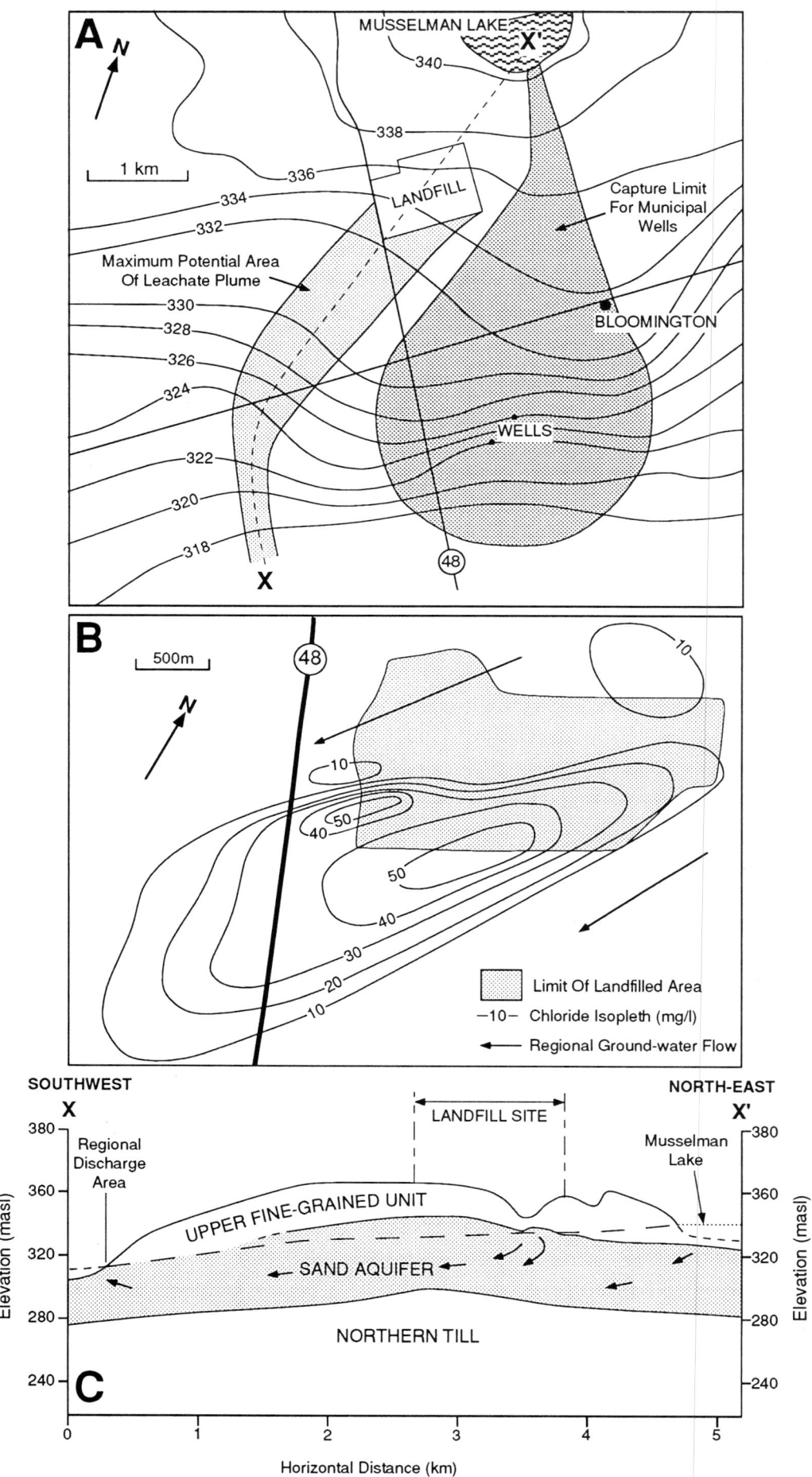

Figure 42 *Stouffville landfill (in use 1962 to 1985). For location, see Figures 1 and 16.*
(**A**) *Location of landfill, potentiometric surface (in m asl) with dimensions of contaminant plume and zone of influence of Whitchurch-Stouffville Township wells; enhanced use of the wells will expand capture limit and risk interception of plume.*
(**B**) *Plume dimensions as reported by Charlesworth (1987).*
(**C**) *Hydrostratigraphic cross-section X–X¹. Sand aquifer is that used by township wells and may be Mackinaw Interstadial deposit; upper fine-grained unit is in part, Halton Till (Figs. 16, 17).*

small area of the Brock North site (4.9 ha) was filled with 142,000 t of garbage. A perimeter tile system was installed in 1981 to drain leachate to an underground tank which is pumped daily (75,000 L) and trucked to Duffins Creek Sewage Treatment Plant. Removal of the entire waste pile to the Brock West site has been discussed. Brock South has a smaller landfilled area. Planned expansion of the sites was driven by the now-discredited assumption that leachate is attenuated by moving through underlying sediments (see next section). Golder and Associates (1987) described the sites as being completely inappropriate as long-term landfills because of the reduced thickness of the Northern till over much of the area and the direct physical contact of waste with

the underlying Thorncliffe Formation (Greenwood Aquifer; Sibul *et al.*, 1977).

Landfills on Till

Landfilling practices on till have undergone a profound change in the last 10 years. Formerly, till sites were regarded as providing a high degree of protection against contamination of underlying aquifers on the basis of the supposed low permeability of till and natural attenuation of leachate contaminants as ground waters slowly migrate through such materials (see Ontario Waste Management Corporation below). As late as 1977, the definition of a sanitary landfill referred only to the requirement for a daily soil cover; condi-

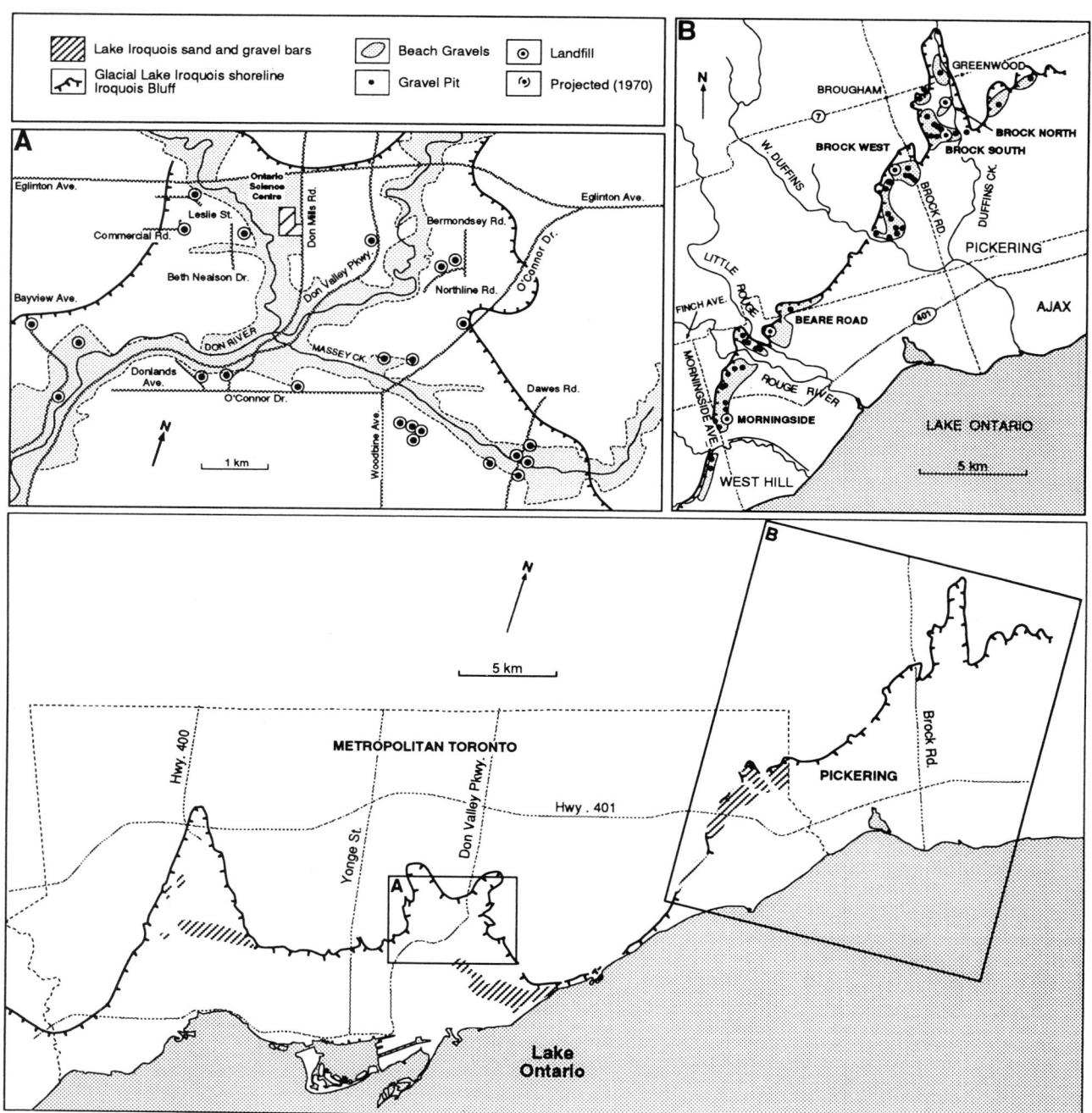

Figure 43 *Shoreline and beach deposits of former Glacial Lake Iroquois, with (**A, B**) active and abandoned landfills sited in former gravel pits. Hydrostratigraphy and subsurface contaminant plume at Brock West and Beare Road landfills are shown in Figures 43 and 45, respectively. Area of Don Valley shown in A is now heavily urbanized with landfills immediately adjacent to, and within, residential developments. In many cases, garbage was used to infill the ravine topography (Chapter 21; Fig. 25). A good example of landfills is that near O'Connor Drive (see Fig. 32A).*

tions at the base of such sites were disregarded, a state of affairs best summed up as "out of sight and out of mind". A conception widely held in the late 1970s was to regard as anomalous any evidence of off-site ground-water contamination in the vicinity of a landfill sited on till. The modern position recognizes that tills are not homogenous, low-permeability materials, but are fractured, contain sand and gravel beds and, locally, host active ground-water systems (Boyce et al., 1995) necessitating engineered landfill designs to contain leachate (Chapters 20, 23).

Beare Road Landfill

Beare Road sanitary landfill (1967-1983; Fig. 45) received a total of 9,617,698 t of domestic, commercial, medical and industrial wastes (including 180 million litres of liquid waste to 1978), but no detailed data are available as to waste types or volumes. Beare Road landfill occupies an abandoned gravel pit along the Iroquois shoreline (Fig. 43B) and is underlain by Halton and Northern tills. A leachate collector system was installed within the waste pile only, because of the assumption that the underlying till was an excellent attenuating medium

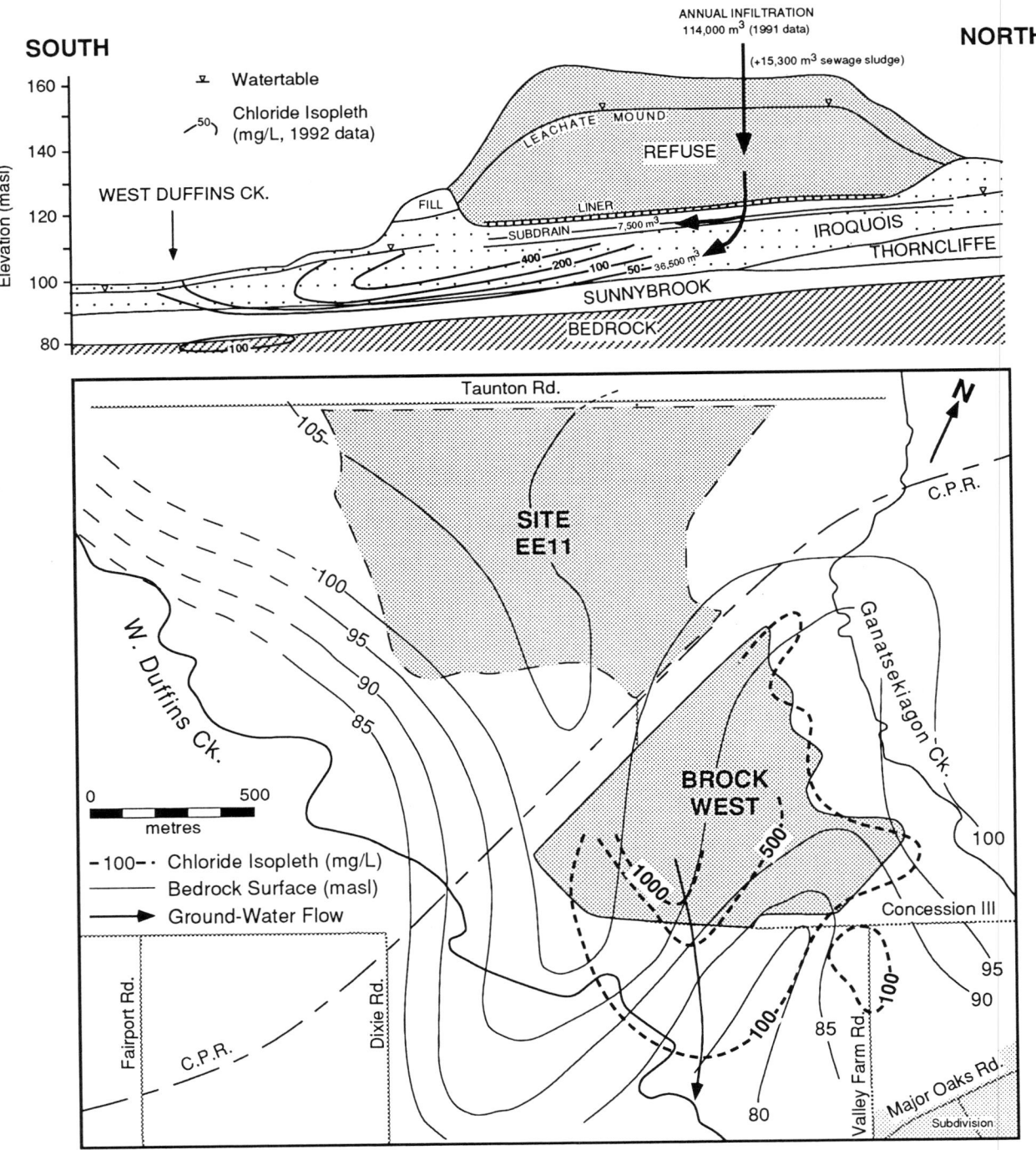

Figure 44 *Hydrostratigraphy and contaminant plumes at the active Brock West landfill (in use since 1975) showing water budget for site (after Dames and Moore, Canada, 1991, 1992 and work of author). The site occupies an abandoned gravel pit along the shoreline of Glacial Lake Iroquois (Fig. 43B) and is only partially lined; subdrains capture only a small proportion of leachate leaving base of landfill. Note leachate mound within landfill; contaminant plume follows bedrock channel below Ganatsekiagon Creek. The preferred site for the new Durham long-term landfill lies to the north (site EE11; Chapter 20). Health impacts among nearby residents were identified by Bertell (1995).*

(Hydrology Consultants Ltd, 1979; Johnson Sustronk Weinstein and Associates, 1980). However, waste lies in direct contact with Iroquois sands and gravels along the southern perimeter of the site (Fig. 45). The presence, and hydrogeologic significance, of sand and gravel beds within the till stratigraphy also went unrecognized. In 1979, ground-water leachate contamination was identified within both bedrock and the overlying Thorncliffe Formation beyond the southwestern corner of the landfill (Hydrology Consultants Ltd., 1979), moving southwestward at about 3 m·a^{-1}. Problems arose with the functioning of the leachate collector system as a result of pipe breakage caused by compaction and settlement of the fill.

Leachate from Beare Road is collected and piped, without any pretreatment, to Highland Creek sewage treatment plant (Fig. 24) under an "overstrength agreement" under By-Law 153-89 of Metropolitan Toronto which limits designated hazardous wastes that can be released into a sewer (*cf.* Britannia Road landfill; see below). Gartner Lee Ltd (1990, 1993a) estimated that the quantity of leachate produced annually is about 114,000,000 L. Leachate mounding is occurring within the waste pile, indicating the potential for off-site leachate migration in all directions. Extensive leachate springs and associated vegetation kills are present on the steep side slopes of the landfill (Fig. 45).

A Certificate of Approval for Beare Road landfill was only issued by the Ontario Ministry of Environment in 1982, some fifteen years after the site began to receive waste and one year before it closed. Conditions attached to the certificate state that "the overall hydrogeologic setting should be investigated and the leachate plume defined in the context of the impact at the site boundary", but to date, the work has not been completed. Based on reported ground-water velocities of 10 m·a^{-1}, the leading edge of the plume may be as much as 350 m beyond the landfill property (Fig. 45). Beare Road has a rudimentary system for flaring methane gas; most escapes to the atmosphere (Chapter 25), but an energy-from-methane project is in hand. New residential subdivisions abut the landfill to the south (Marshall Macklin Monaghan Ltd., 1992b).

Table 1   Waste stream entering Brock West landfill in 1991 (from Bertell, 1995). In 1988, the site accepted a total of 1,041,515 t of waste including 112,000 t of sewage sludge and 440,000 t of mixed industrial waste of unknown provenance. As with nearly all landfills (except those constructed very recently), a complete inventory of wastes in the landfill is not available.

WASTE TYPE	TONNAGE	PERCENT
Domestic solid waste	117,707	23.6
Mixed industrial waste	96,345	19.3
Mixed commercial waste	97,165	19.5
Demolition and construction	21,049	4.2
Misc. municipal solid waste	39,668	8.0
Incinerator ash	6,288	1.3
Catch basin/street sweepings	2,505	0.5
Sewage sludge	16,733	3.4
Clean fill	5	--
Lumber, logs and brush	1,609	0.3
Hospital waste	8,380	1.7
Animal carcasses	76	--
Asbestos	1,735	0.3
Contaminated fill	89,645	18.0
Total tonnage	**498,908**	**100.0**

Britannia Road Landfill

Britannia Road landfill (Fig. 46; 61 ha) is one of four currently active landfills in the GTA (Milton, Keele Valley and Brock West; Fig. 1). The site opened in 1980 to handle municipal wastes for Brampton and Mississauga, the two largest cities in Peel Region. It was expected to close in 1992, but, because of the expected shortfall in landfill capacity in the GTA (the garbage gap; Chapter 20), was extended by ministerial decree until 1997, when the site is expected to contain nearly 9 million tonnes of waste. The site will eventually be used as a golf course and, on the expectation of closure in 1992, is now surrounded by planned urban residential development. Local residents refer to the landfill as "the hole in the donut".

Britannia Road landfill lies on the Halton Till plain and, as a child of its time, was conceived as an attenuation site where a portion of leachate was expected to migrate downward into the till; as a consequence, no liner was installed (Gartner Lee Ltd., 1977). A leachate collection system system, consisting of 4" PVC pipes under each of 8 cells, feeding to a perimeter collector trench, was installed to prevent leachate mounding within the cells and surface breakouts such as occur on the sideslopes of Beare Road landfill (Fig. 45). The perimeter trench extends around the site with the exception of the northwest corner. Leachate is collected on site, pretreated using a Rotating Biological Contactor (RBC), and pumped to Creditview sanitary sewer and, eventually, the Clarkson Treatment Plant. The performance of the RBC is not known and pretreatment is designed to meet the Region of Peel Sewer Use By-Law; the annual volume treated averages about 15,000 m^3.

Britannia Road landfill lies directly above the Credit bedrock channel (Figs. 12, 46). Halton Till, up to 30 m thick, rests on sands and gravels underlain by weathered Georgian Bay shales. A permeable sand unit lies along the floor of the bedrock channel and forms a lower aquifer unit. The hydraulic conductivity of the weathered and fractured upper part of the till (about 5 m deep) is between 10^{-7} and 10^{-9} m·s^{-1}; that of the underlying unweathered till is in the 10^{-10} to 5 × 10^{-10} m·s^{-1} range. The total volume of leachate flowing into the Halton Till, and eventually into the lower aquifer, is estimated at between 3.6 and 17.5 L per minute, equivalent to between 0.3 and 2% of total precipitation in the area (Fig. 46). Leachate breakthrough into the aquifer is expected to occur within the next 10 years, and Reasonable Use Policy guidelines of the Ontario Ministry of Environment and Energy (see Chapters 20, 23) that specify allowable water quality deterioration at the site boundary will be exceeded (Gartner Lee Ltd., 1992). By then, the landfill site will be surrounded by residential development. It has been recommended that a contaminant attenuation zone (CAZ) be established around the site (Fig. 46) and legal title to all ground-water rights be obtained by the province within this area. As yet, there is no recognizable impact of leachate on surface waters; indeed, water quality in nearby Carolyn Creek is more noticeably affected by road salt contamination. An interesting experiment could be conducted at Britannia Road by comparing the impact of landfill (point source) on the underlying aquifer to that from encroaching residential development (non-point source).

Town of Cobourg Landfill

This site has been described as "a dinosaur and a relic of a bygone era" by a Hearings Board which rejected a proposal to extend the site for further use (Environmental Assessment Board, 1989). Waste disposal began in 1957 in a pit (7.2 ha; Fig. 47) excavated into a drumlin. The site stratigraphy consists of channeled sands (probably Mackinaw Interstadial deposits; Fig. 17) forming an aquifer confined above and below by late Wisconsin tills. No detailed records were kept of waste entering the pit, and the presence of hazardous wastes is suspected. An impermeable daily cover was not adequately installed, resulting in considerable infiltration of rain water and snow melt into the waste, and associated leachate mounding. A leachate collection system is less than 50% effective, and about 15,000 m³ of leachate enters the aquifer each year. By 1989, the leachate plume extended 350 m from the filled area,

ultimately discharging to a nearby creek (Fig. 47). The leachate plume is moving southwestward at a velocity of 30 m·a⁻¹. This site illustrates very clearly the hydrogeologic role of interstadial deposits and interbeds in controlling contaminant migration through till deposits.

Landfills and Public Health
Public health is an emerging issue in many urban communities as a result of the close proximity of residential developments to many closed and active landfills that historically

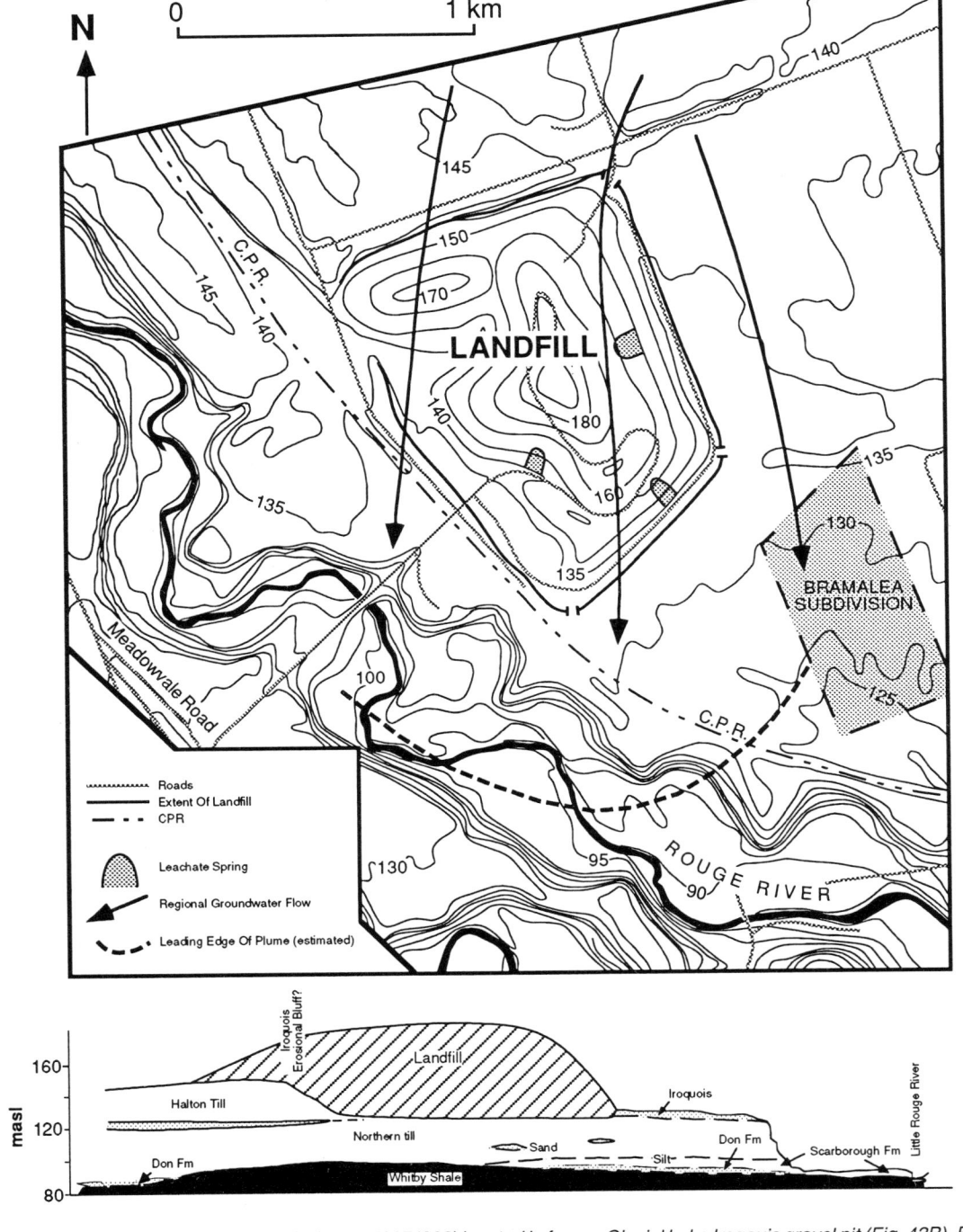

Figure 45 *Map and geology of Beare Road landfill (in use 1967-1983) located in former Glacial Lake Iroquois gravel pit (Fig. 43B). Exact extent of leachate plume has never been determined (estimated extent shown here), but construction of nearby residential subdivisions is proceeding. Leachate mounding and emergence of leachate springs on sideslopes is active and poses potential for off-site leachate migration. Plume follows interbeds in Northern till and has reached bedrock (after Hydrology Consultants, 1979; Gartner Lee Ltd., 1990; and work of author). Isotopic and geochemical characteristics of landfill methane from the Beare Road site are described in Chapter 25. Geologic section is oriented north-south through central part of site.*

received hazardous industrial wastes (*e.g.*, Figs. 41, 44, 45, 46; Deloraine *et al.*, 1995). Unfortunately, epidemiological studies of community health are few. In general, urban popu-

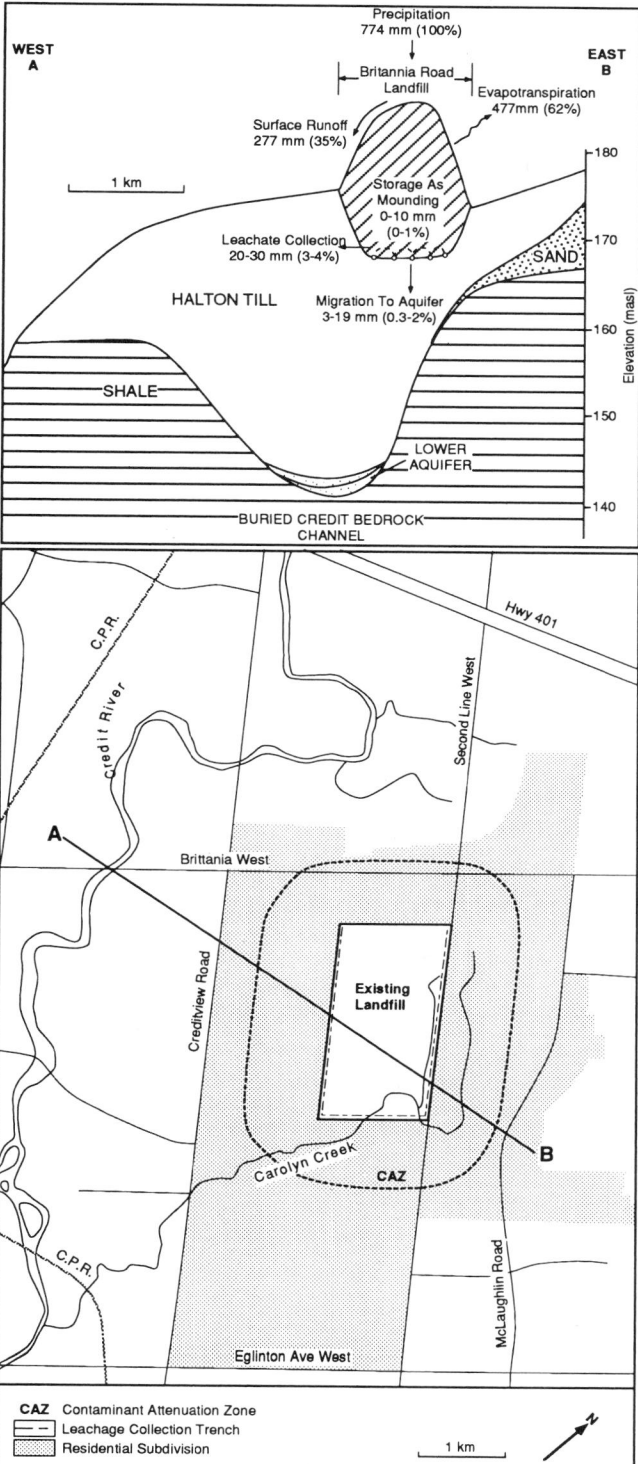

Figure 46 *(top) Geology and water budget for Britannia Road landfill (1980 to 1997; Fig. 1; after Gartner Lee Ltd., 1992). The site was expected to close in 1992, but was extended without any review to meet demand until new long-term sites are available (see Chapter 20). Residential development is planned around site. Note that leachate collection system (bottom) does not extend around entire site. The contaminant attenuation zone (CAZ) is designed to allow acceptable dilution of contaminant plume in lower aquifer along buried Credit bedrock channel (see Fig. 12); all ground-water rights in this zone will be ceded to province.*

lations are routinely exposed to contaminants, but accurate measures of exposure and quantitative environmental data, such as the type and distribution of contaminants and their pathways to humans are not yet available (see Taylor *et al.*, 1993, for a review). Nonetheless, the medical effects of exposure to contaminants that are present in large amounts in older landfills are well known (Buffler *et al.*, 1985; Table 2).

Bourns *et al.* (1986) were able to identify severe health effects among residents and former workers living in the vicinity of the Upper Ottawa Street landfill, near the city of Hamilton, which were attributable to long-term exposure to landfill-derived contaminants (see also Hertzman *et al.*, 1987). Britannia Road landfill (Fig. 46) releases about 2000 t per year of gaseous non-methane organic compounds (NMOC) such as vinyl chloride which is a known carcinogen (see Chapter 20). At the Stouffville landfill (Fig. 42), an epidemiological study found no evidence of any short-term adverse impacts on health among the community (Charlesworth, 1987; Goss, Gilroy and Associates Ltd., 1987), but these results are disputed by local residents. A more recent study of community health in the vicinity of the Brock West landfill (Fig. 44) demonstrated a clear impact on community health (Bertell, 1995). The report identified an increase in children with respiratory illness and skin problems in an area, downslope from the landfill, that has been plagued by chemical odours and air with detectable benzene, carbon tetrachloride, t-1,2 dichloroethene and trichloroethene (Bertell, 1995). The number of children with asthma was also above regional norms as established by a national study (Dales *et al.*, 1994). Most Toronto area landfills form hills of waste that are higher in elevation than the surrounding urban communities built on flat till plains or lake floors. At night, air that has been cooled by radiative heat loss from the landfill mound moves down into surrounding communities. Most complaints of odours are made by residents in the early morning and late evening, at times of marked temperature inversions when airborne contaminants are trapped close to the ground surface.

Direct links between environmental contamination from landfills and public health in surrounding urban areas are difficult to prove conclusively. However, it is now recognized

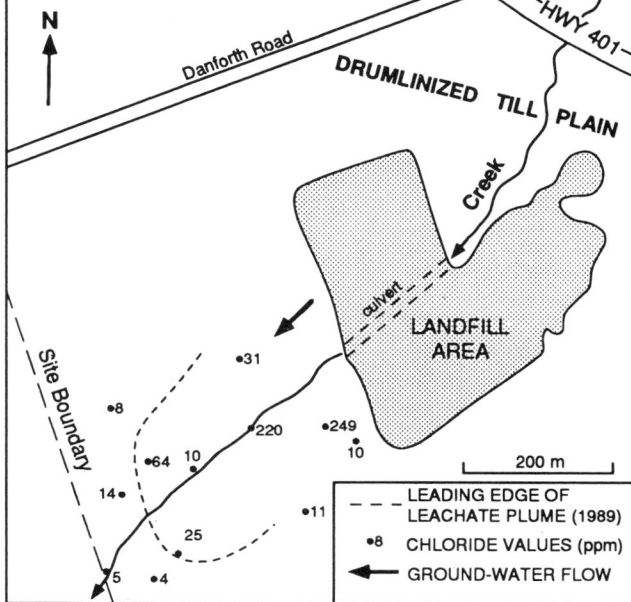

Figure 47 *Town of Cobourg landfill: contaminant plume is migrating off site through sand beds within Northern till.*

that responses to environmental contaminants in humans (cancers, *etc.*) have a complex etiology that is not simply dependent on direct physical exposure, but also perceived exposure (Taylor *et al.*, 1991). In other words, if a person believes he or she has been exposed, this can trigger a psychosocial, but no less real, physical response (see Edelstein, 1988; Taylor *et al.*, 1991; Flint and Vena, 1991; Kraus *et al.*, 1991; National Wildlife Federation, 1991; Hovinga *et al.*, 1993). Clearly, a major role for environmental geologists is one of compiling comprehensive environmental audits of urban areas, including past and present land use mapping and quantitative assessment of associated contaminants. Such data are urgently required for studies of community health and future planning of residential communities adjacent to landfill sites.

Hazardous Waste

Space does not permit a wide-ranging review of the geological aspects of hazardous waste management and the interested reader is referred to recent work by Arnoud *et al.* (1993), Grasso (1993) and Testa (1994). This section will briefly review the magnitude of the problem in southern Ontario and provide a case example (the Ontario Waste Management Corporation) illustrating the type of geologic information that needs to be collected.

Hazardous Waste Volumes

The volume of hazardous waste production in Canada is not well constrained (Statistics Canada, 1994). Total production for 1991 is estimated to have been about 3.2 million tonnes which can be expressed as 5,770 kg of hazardous wastes for every million dollars of Gross Domestic Product (GDP). In contrast, in the same year, the economy of Japan produced only 226 kg of hazardous waste per million dollars of GDP, reflecting the much higher proportion of primary industry in Canada (World Resources Institute, 1993; Statistics Canada, 1994). The equivalent figure for the United States is 44,186 kg per million dollars GDP.

Southern Ontario is the industrial heartland of Canada and produces 68% of the country's hazardous waste; this is equivalent to about 8,300 kg per million dollars of Ontario's GDP, which is almost double the national average. Despite the large amounts of hazardous waste produced, there is no waste facility in southern Ontario able to handle hazardous solid or liquid industrial waste. In this regard, it is noted that household waste is now a significance source of untreated hazardous waste entering municipal landfills in the GTA.

Existing Hazardous Waste Treatment and Disposal Facilities

The only industrial waste disposal facility in the province of Ontario is the privately owned Laidlaw incinerator at Corunna, near Sarnia in southwestern Ontario. This is a liquid injection facility and is not able to treat solids, sludges, PCBs, pathological wastes, halogenated solvents or compressed gases, for which the only disposal options are storage or export to the United States (with the exception of PCBs). Exports of hazardous wastes to the United States rose over 300% from 1986 to 1991 (42,000 t to 163,000 t). Imports totalled 126,069 t in 1991. There is also a hazardous waste landfill at Corunna (Laidlaw, formerly Tricil), but a land-ban regulation is anticipated which would greatly restrict (as in the United States) the types of wastes that can be disposed of in landfills.

The Laidlaw hazardous waste landfill site is located in the Lambton clay plain (Chapman and Putnam, 1984), underlain by up to 50 m of low-permeability glaciolacustrine clays. The

---

Table 2   Known effects of contaminants found in landfill leachates (from Buffler *et al.*, 1985). Chemistry of leachate in discussed further in Chapters 5 and 24.

Organic Compounds

Acetone
- Skin irritant

Benzene
- Skin irritant
- Central nervous system (CNS) depressant
- Bone marrow depression
- Teratogen

Carbon tetrachloride
- Hepatitc cirrhosis

Phenol
- Skin irritant
- Asthma

Polychorinated biphenyls
- Chloracne
- Skin cancer (uncertain)
- Hepatitis
- Sub-acute hepatic necrosis with possible cirrhosis
- Embryotoxicity
- Spontaneous abortion/fetal death

TCDD
- Chloracne
- Hepatitis
- Spontaneous abortion/fetal death suspected

1,1,2,2-tetrachloroethane
- Hepatitis
- Sub-acute hepatic necrosis with possible cirrhosis

Toluene
- Sub-acute hepatitic necrosis with possible cirrhosis
- CNS depression

Trichloroethylene
- Skin irritant
- CNS depression

Vinyl chloride
- CNS cancer
- Sub-acute hepatitc necrosis with possible cirrhosis
- Liver cancer
- Lung and respiratory cancer

Xylene
- CNS depression

Inorganic Compounds

Arsenic
- Skin irritant and contact allergen
- Skin cancer
- Neurasthenia, irritability and other mild CNS symptoms
- Acute hepatitic necrosis with possible cirrhosis

Asbestos
- Lung and respiratory tract cancer

Cadmium
- Pulmonary edema
- Hypertension

Chromium
- Skin irritant and contact allergen
- Lung and respiratory cancer
- Asthma

Sodium or potassium cyanide
- Skin irritant

Isocyanates
- Asthma

Toluene diisocyanate
- Hypersensitivity pneumonitis

Lead
- Emotional instability
- Neurasthenia, irritability and other mild CNS symptoms
- Embryotoxicity
- Spontaneous abortion/fetal death

Mercury salts
- Skin irritant and contact allergen
- Neurasthenia, irritability and other mild CNS symptoms

Nickel salts and oxides
- Skin irritant and contact allergen

Nickel sulphate
- Asthma

role of fractures and diffusion in controlling contaminant migration through glaciolacustrine clays has been the focus of a large number of hydrogeological studies (*e.g.*, Desaulniers *et al.*, 1981; D'Astous *et al.*, 1989; Ruland *et al.*, 1991; McKay and Cherry, 1992; McKay *et al.*, 1993; Parker *et al.*, 1994; Weaver *et al.*, 1995). Whereas solute transport in glaciolacustrine clays is very slow (mm to cm per year) and due almost entirely to diffusion, the bulk of subsurface contaminant transport is facilitated by fractures. This finding is very appropriate to reviewing attempts to locate a hazardous waste treatment and landfill facility close to the GTA.

Ontario Waste Management Corporation

In November 1980, the Ontario Waste Management Corporation (OWMC) was established (as a provincial Crown Corporation) to construct and manage a provincial hazardous industrial waste facility. Since 70% of provincial hazardous waste is generated in the GTA, the ensuing site selection process focussed on finding a candidate site, located on extensive areas of thick glaciolacustrine silts and clays having geologic predictability, close to the GTA. In 1981, a preferred site for an integrated treatment and disposal facility was identified at South Cayuga near the Grand River. The presence of thick glaciolacustrine clays below the site was instrumental in its selection, following the then accepted practice of reliance on the supposed natural containment afforded by such sediments (see above). In late 1981, the site was rejected as unsuitable because of flood hazard. After a wide-ranging search of prospective sites in Ontario (see Ontario Waste Management Corporation, 1988a, b), OWMC proposed in September 1985, to operate an incineration (30,000 t·a^{-1}), solidification, waste treatment (120,000 t·a^{-1}) and landfill facility on the Haldimand Clay Plain at West Lincoln, in the Regional Municipality of Niagara (site LF-9C; Fig. 48).

OWMC Site LF-9C

Geology. The geology of site LF-9C consists of heavily fractured, dolostone bedrock of the Guelph and Lockport formations overlain by a succession of tills and glaciolacustrine silty clays preserved between the Niagara and Onondaga escarpments (Feenstra, 1981; Figs. 18C, 49A). The Pleistocene stratigraphy below the site is between 30 and 38 m thick and consists of a lower coarse-grained till (probably Northern till; Fig. 17) resting on fractured bedrock. The till is overlain by Mackinaw Interstadial deposits (lower glaciolacustrine unit; Figs. 17, 18C) that, in turn, are overlain by a unit mapped as the Halton Till and an upper glaciolacustrine unit (*e.g.*, Peel Pond deposits; Fig. 16). The Halton Till, in fact, consists of laminated clayey silts that, in places, are extensively deformed as a result of subglacial deformation and incorporation as the Halton ice margin advanced into the eastern Lake Erie basin at about 13.3 ka (Fig. 17).

Despite the considerable sedimentological and stratigraphic variability below site LF-9C, the geology was described as possessing lateral and textural uniformity by Gartner Lee Ltd. (1987a). Fractures are abundant in the upper 3-4 m of the upper glaciolacustrine unit, and deeper fractures (to 12 m) occur in the Halton Till complex. Recent data suggests deep fractures have propagated up from bedrock (Daniels, 1990; Wills *et al.*, 1992; Eyles and Scheidegger, 1995) and provide potential pathways for the movement of leachate. Offsets of stratigraphic contacts by as much as 2-3 m were also identified on seismic reflection profiles of the site (Slaine *et al.*, 1990), suggesting faults are present. Earthquakes occur in the area (residents report sulphurous odours from

waterwells immediately thereafter), and a geotechnical assessment of the excavated side slopes of the proposed landfill concluded that "short-term earthquake-induced localized failures might occur" (Gartner Lee Ltd., 1987b, p. 30).

Landfill design. In addition to industrial hazardous waste such as PCBs, the proposed facility at site LF-9C was also designed to accept contaminated sediment dredged from Hamilton Harbour and the St. Lawrence River. The site was termed an integrated site because of the need to safely accept, store and incinerate hazardous waste and to landfill the incinerator residue. Incineration of PCBs produces chloride-rich ashes which originally were to be immobilized using Portland cement and disposed within an unlined (non-engineered) shallow-entombed landfill some 15 m deep (Ontario Waste Management Corporation, 1988a; Fig. 49B). Protection of underlying ground-water resources was argued to be assured by the natural attenuation properties of the Halton Till and underlying clay strata (Gartner Lee Ltd., 1987c).

Environmental suitability. Public hearings before a Joint Board (under the Environmental Assessment Act and other relevant acts together known as the Consolidated Hearings Act; Chapter 39) were convened to determine the environmental suitability of site LF-9C in November 1989. These concluded almost four years later in September 1993. The decision of the Joint Board was handed down in November 1994 (see Environmental Assessment Board and the Joint Board, 1994) and the site was rejected, some ten years after the initial identification of the preferred site. This decision was made on the basis of several considerations, but principally on the inadequacy of the landfill design and its ability to protect underlying ground waters.

Not until after site LF-9C had been chosen as the preferred site was it demonstrated that reliance on natural attenuation properties of the underlying till and clays was misplaced. Leachate from the landfill site would contaminate an extensive (1200 ha) aquifer in the underlying Guelph Formation dolostone used for local municipal water supplies (Figs. 18C, 49A). After this finding, OWMC pursued an alternative proposal to recover high-grade salt from the ashes and to dispose of the residue either by deep well injection or storage in a salt mine. The costs of the new chloride management plan were very significant over the lifetime of the facility and, if known earlier, would have profoundly changed the relative ranking of prospective sites that had earlier been rejected during the initial province-wide search (Fig. 48). Use of an engineered landfill design, rather than reliance on natural protection of thick clays, would have greatly expanded the possible number of potential candidate sites during the initial assessment. Timar (1988, p. 6) correctly identified that "the principal hazard is the contamination of the bedrock aquifer" and recommended employment of an engineered landfill designed to collect and control leachate migration. Six years later, the issue of inadequate mitigative action to control leachate came back to haunt the hearings, and eventually to kill the undertaking after an expenditure of $140 million of public funds.

Discussion

The story of the OWMC hazardous waste facility siting process provides a striking illustration of the concept of iteration (the re-visiting of earlier decisions in the light of new data). Iteration is virtually impossible with very large environmental geological undertakings that take many years to complete. As with the analogy of the rudderless supertanker heading for the rocks, the inertia in the system is simply too great to change course when new data emerge that question earlier decisions. The failure to iterate the OWMC site selection procedure, in the face of new, and what in hindsight were unambiguous, geological and hydrogeological concerns with site LF-9C (*e.g.*, Timar, 1988), is the principal reason that the OWMC proposal was denied under the Environmental Assessment Act.

Hydrogeological data from a nearby site at Smithville played an important role in the decision to kill site LF-9C. Between 1978 and 1986, Chemical Waste Management Ltd. operated a transfer and surface storage facility which received 434,000 L of liquid wastes (mostly polychlorinated biphenyl wastes). By 1987, contaminants spilled from drums

during handling had reached underlying bedrock (Wills *et al.*, 1992), and showed that rapid subsurface contaminant migration occurs through clays as a result of fracturing (Fig. 49). This finding has major implications for the more than 200 existing hazardous waste sites that dot the Canadian and American border near the Niagara River (*e.g.*, Kappel and Tepper, 1992; Kolmer, 1992). These are underlain by the same Pleistocene stratigraphy and fractured bedrock. At the notorious Love Canal site, where Hooker Chemical Co. placed drums of liquid waste in an old canal, a clay cap has been installed, designed to prevent surface escape of volatiles. In addition, perimeter drains were implemented to prevent off-site migration through nearby residential sewer systems (Fig.

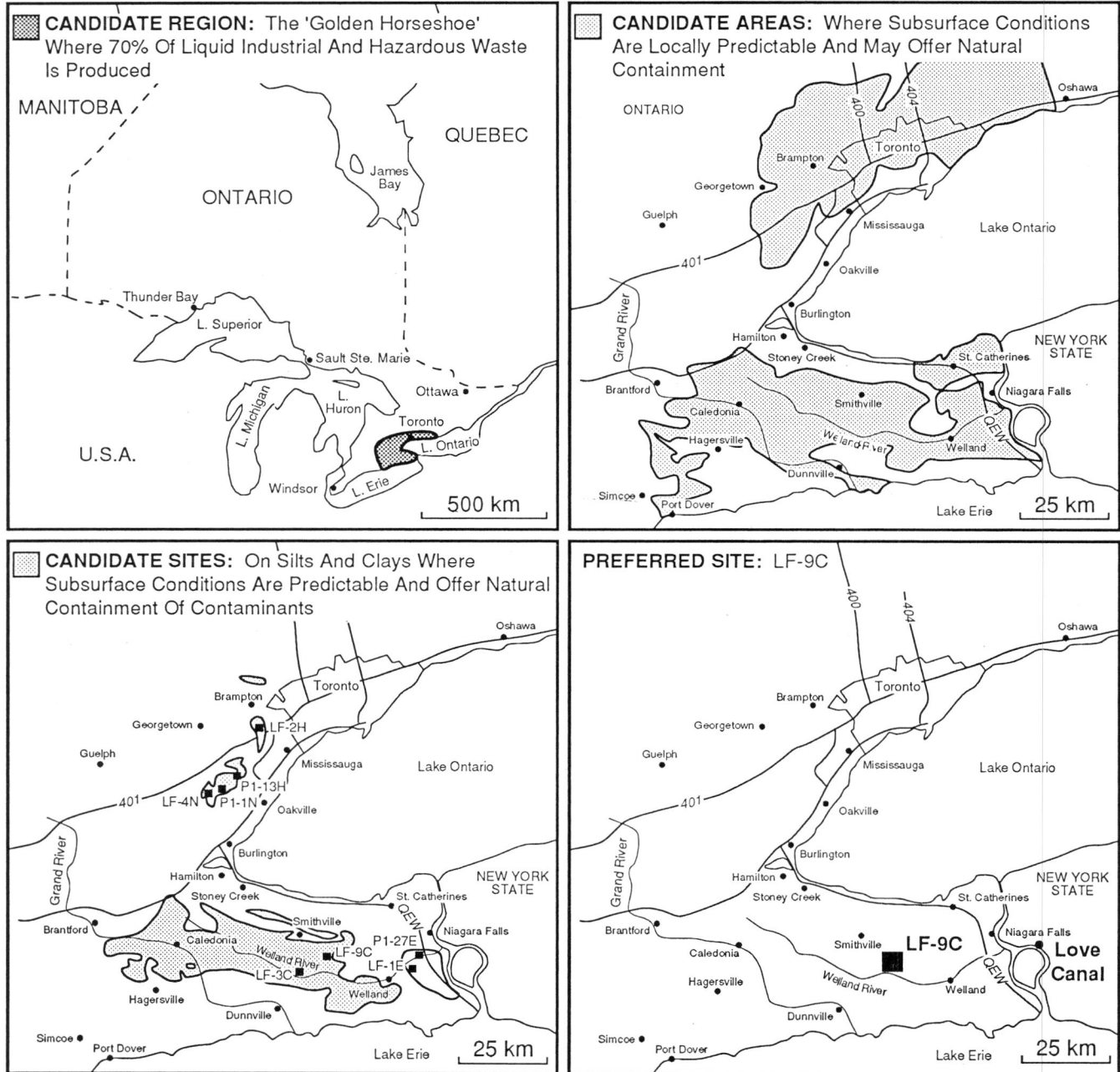

Figure 48 *Diagrammatic representation of selection process used by Ontario Waste Management Corporation to identify a hazardous waste treatment (incinerator) and disposal (landfill) site within the GTA and adjacent region. The site search focussed on the need for natural containment of landfill-derived leachate within thick, and supposedly geologically predictable, silt and clay deposits. Other sediments, requiring complex engineered systems to contain contaminants, were avoided (see text). For other examples of site searches for municipal waste, see Chapter 20.*

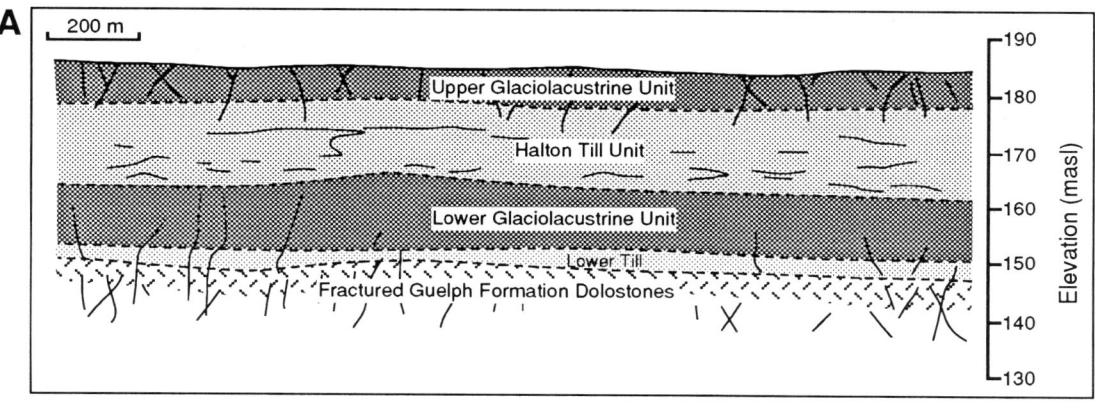

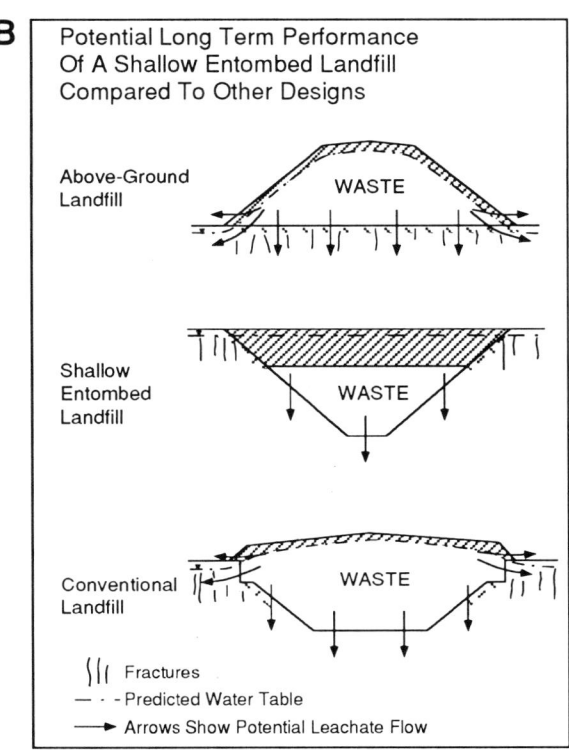

Figure 49 (**A**) *Geology of the OWMC hazardous waste site LF-9C showing fractured Guelph Formation dolostones (Fig. 6) overlain by a lower till (most likely Northern till; Figs. 17, 18C) and glaciolacustrine silts and clays. Halton Till records complex interbedding and incorporation of glaciolacustrine deposits into till as ice overrode glacial lake deposits. Fractures penetrate stratigraphy from the surface downward and from bedrock upward (see text).* (**B**) *Design of shallow-entombed landfill proposed by OWMC compared with other designs; note intended burial of incinerator waste (chloride-rich ash) below high-permeability zone of surface fracturing. Design ignores deep fractures. The site and landfill design was rejected in 1994 because of inability of site sediments to afford natural containment of leachate and prevent excessive contamination of underlying bedrock aquifer (modified from Gartner Lee, 1987a, c). The same concerns with fractures apply to Love Canal hazardous waste site (Fig. 50) and many others in the Niagara area.* (**C**) *Typical structure of glaciolacustrine clays of southern Ontario showing distribution and frequency of fractures (modified from Ruland et al., 1991). A similar structure has been identified in glaciomarine clays in eastern Ontario (O'Shaughnessy and Garga, 1994). Deeper fractures penetrate to bedrock and result from upward propagation of bedrock fracture sets (Eyles and Scheidegger, 1995).*

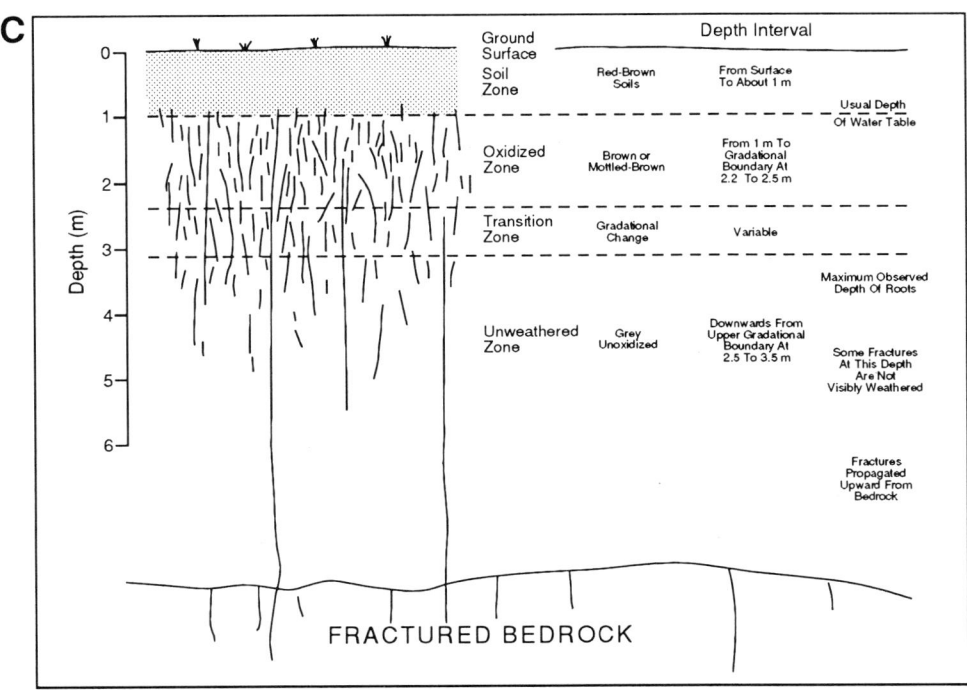

50). However, remedial installation of an engineered liner system, to prevent the much more substantial loss of contaminant through the fractured substrate, is not possible unless the entire waste pile and contaminated substrate is removed.

Waste as an International Commodity

Hazardous (and non-hazardous) waste has changed from being an expensive nuisance requiring public funding and the attention of government agencies to manage, to a scarce, and much sought after, international commodity. Waste management has become resource management. As a direct result of United States Environmental Protection Agency directives in the late 1980s, stating that hazardous materials could not be safely landfilled, many states in the United States built their own incineration facilities. There are now 21 sites in the United States. Consequently, there is a huge overcapacity in the industry which has resulted in a sharp fall in prices and intense competition for waste. The Swan Hills treatment facility in Alberta, consisting of state-of-the-art rotary kilns, intends to attract Ontario and Quebec waste, but it is cheaper for eastern Canadians to export to the United States market. In recent years, there have also been significant technological advances in at-source reduction and treatment of hazardous wastes, reducing the need for transport to central facilities. The placing of highly engineered facilities in willing-host communities is the most likely management option. Detailed understanding of the geologic history and stratigraphy of such sites is an essential component of contingency measures designed to safeguard ground waters in the case of failure of engineered systems.

Nuclear Waste

The GTA is one of the most "nuclearized" jurisdictions in the world (Kock, 1994); each stage of the nuclear fuel chain from uranium processing, nuclear reactor operation and radioactive waste management takes place within or close to the border of the GTA. A large sector of the population is routinely exposed to pollution emanating from nuclear reactors.

Nuclear Reactors in the Greater Toronto Area

The four "A" reactors at Pickering (Figs. 1, 22) are the oldest commercial reactors in Ontario and commenced operation between 1971 and 1973. The "B" reactors were added between 1982 and 1985. These reactors have projected life spans of 40 years. High-level radioactive waste (spent fuel bundles) is stored on-site within water-filled concrete tanks; Ontario Hydro is also experimenting with dry storage of 6200 t of used fuel bundles in concrete integrated canisters, which are designed to last 50 years and eventually be transported intact to a permanent high-level storage facility somewhere in the Canadian Shield (Atomic Energy of Canada Ltd., 1994). The Pickering nuclear reactors have a very high population density in the immediate vicinity, and are seriously underdesigned compared to American plants with regard to their ability to withstand earthquakes (Chapter 29). In addition, there is no budgetary provision made by Ontario Hydro for costs of decommissioning the reactors (Kock, 1994).

The national nuclear power industry is regulated by the Atomic Energy Control Board of Canada (AECB) formed in 1946. The AECB issues operating licences for nuclear power plants for periods of up to two years and it is their mandate to assess *all* hazards that could affect safe operation. In 1987, the public advocacy group Energy Probe challenged the legal standing of the *Nuclear Liability Act* (1970) which circumscribed the extent of financial liability in the case of a nuclear accident based on the then current understanding of risk. Energy Probe showed that the real risk, and thus extent of liability, had been grossly underestimated for the densely urbanized GTA. This challenge was dismissed, after appeal, in 1994 (Anon., 1994). Later that year, when the operating licence for Pickering Nuclear Generating Station (NGS) was due for renewal, public groups demanded that the case for

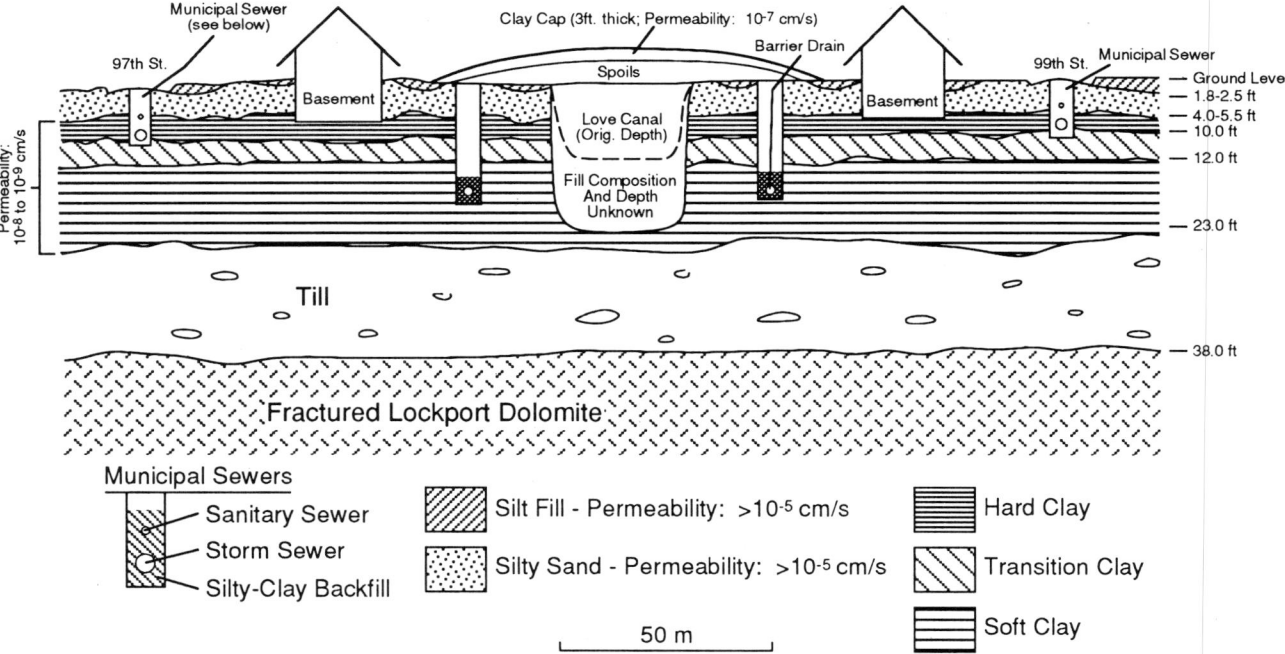

Figure 50   *Stratigraphy of Love Canal site, New York State (see Figure 48 for location). The clay cap is to prevent surface release of volatiles; no mitigative engineering is in place to prevent subsurface migration of DNAPLs through supposed low-permeability clays into the Lockport Dolostone, but sediments are likely fractured (see Figs. 18C, 49C). Former canal contains 20,000 tonnes of chemical wastes, extends over 10 km and is one of nearly 250 former landfills in the Buffalo-Niagara urban area.*

renewal before the AECB be aired in a full public hearing under the Federal Environmental Assessment and Review Process Guidelines Order which determines those projects that are required to undergo a full environmental assessment (see Chapter 39). This, too, was dismissed. Although unsuccessful, these recent legal challenges to the AECB have brought to light much documentation of inadequate operating and safety procedures at the Pickering NGS.

With privatization of the electricity generating industry now in sight, with the need to retrofit many ageing systems, and the enormous "hidden" costs of producing power from nuclear stations, it is likely that the Pickering NGS, at least, will be mothballed. This issue of the safety of the plant in the event of an earthquake (see above and Chapter 29), including the fate of nuclear wastes stored on site and ongoing tritium emissions (see below), will remain as a pre-eminent environmental concern in the GTA. Nigge (1995) has shown clearly how prevailing political pressures affected the assessment of new geological data showing a heightened risk of earthquakes in the vicinity of the Pickering NGS. In August 1996, the Director-General of the Atomic Energy Control Board, responsible for safe operation and licensing of nuclear generating stations in Canada, publicly acknowledged that "had the AECB known then what it knows now" regarding the geological setting of the Pickering NGS, it would not have been built.

The nuclear reactor at Darlington (Fig. 1) was scheduled to enter service in the mid-1980s, but did not become fully operative until 1993 after substantial cost overruns. The site includes a Tritium Recovery Facility designed to decontaminate heavy water by extracting pure tritium gas and reduce environmental contamination (see below). In addition to the commercial reactors at Pickering and Darlington, two university campuses (Toronto, and McMaster in Hamilton) and the Royal Military College in Kingston have nuclear reactors for research purposes; McMaster produces iodine-125 for the commercial market, but the reactor is to be decommissioned.

Uranium Processing and Nuclear Waste Management

Yellowcake (uranium oxide powder; U_3O_8) is used for nuclear-power generation in southern Ontario, and is imported to the GTA from the Rio Algom Stanleigh mine near Elliot Lake until June 1996, and thereafter from Saskatchewan where ore grades are higher. It is received at the Canadian Mining and Energy Corporation (CAMECO) facility at Blind River, on the northern shore of Georgian Bay, where yellowcake is processed to uranium trioxide (UO_3); the operating licence allows for the release of 44 t of airborne uranium dust per year.

Refined uranium is shipped to the Eldorado-CAMECO facility at the mouth of the Ganaraska River in Port Hope (Fig. 1), where about 20% is converted into uranium dioxide powder (UO_2), for use in CANDU reactors, and the remainder converted to uranium hexafluoride (UF_6) for export. Ammonium nitrate is produced as a by-product of the UO_2 conversion process and sold locally for fertilizer. In 1991, 16.6 kg of uranium was discharged into the sanitary sewer system in Port Hope through runoff and ground-water collection systems and laboratory drainage. Airborne dust emissions in 1991 were 162 kg (Atomic Energy Control Board of Canada, 1992). Coolant water from the conversion process is released directly into Port Hope Harbour; uranium and radium measurements in the community consistently exceed background levels (Ontario Ministry of Environment and Energy, 1991; Chapter 28). Uranium dioxide powder is compressed into ceramic pellets in Port Hope and at the Canadian General Electric plant in Toronto, with final fuel bundle assembly in

Peterborough. In 1991, airborne uranium emissions from the plant in Toronto were 12 g, but 1780 g were released to the Metro Toronto Trunk Sewer System.

From 1932 to 1948, radioactive waste was stockpiled in and around the Port Hope area; much was used as historic landfill for building foundations and filling ravines. Over 220,000 m^3 of waste has so far been identified. Port Hope Harbour is one of over 40 Areas of Concern identified by the International Joint Commission; the biological impacts of the disposal of historic radioactive wastes are discussed in Chapter 28. In addition to waste material disposed of in the harbour and among the surrounding urban community, radioactive solid wastes have also been landfilled close to an eroding cliff line at Port Granby (1955-1988), 16 km west of Port Hope (Figs. 1, 51), and inland at the Welcome site (1948-1955), 2 km west of the town. Wastes include limed raffinate (calcium sulphate with 0.5% unrecovered uranium and lesser radium and thorium), ammonium nitrate, magnesium fluoride (a radioactive slag), and calcium fluoride, an alkaline filter cake produced during the conversion from UO_3 to UF_6.

Port Granby Nuclear Waste Management Site

At Port Granby (Fig. 1), some 200,000 m^3 of wastes were dumped in coastal ravines and storage pits over a total area of some 10 ha (MacLaren, 1977). Pits were excavated to a depth of 6 m in Glacial Lake Iroquois beach sands (Fig. 51) and backfilled without any consideration of the need for a liner to prevent the outward migration of radionuclides to Lake Ontario. Bobba and Joshi (1988) showed that about 2.5×10^5 m^3 of ground water flowed annually from the site into Lake Ontario and transported about 2.5×10^7 becquerels (Bq) of ^{226}Ra and about 25 kg of uranium. About 90% of ground water leaving the site, together with an estimated 75% of surface runoff, is now intercepted. This water is treated with ferric chloride, which precipitates radium and arsenic in a sedimentation lagoon, and discharged into Lake Ontario. Coastal bluffs are eroding at rates of up to 1 m a year, necessitating coastal protection measures (Golder and Associates Ltd., 1990).

Low-level nuclear wastes from Port Granby and Port Hope, and other sites in the city of Scarborough (see below), are to be moved to a permanent site. In 1986, a Siting Task Force on Low-Level Radioactive Waste was established to seek a willing-host community in a geologically acceptable area. In the interim, low-level waste continues to accumulate in the Port Hope area, pending selection of a permanent low-level waste dump, either at Deep River, adjacent to Chalk River Nuclear Laboratories along the Ottawa River, or in bedrock caverns at Port Hope (*The Toronto Star*, July 16, 1995). In a July 1995 referendum, citizens of Port Hope rejected the latter option in the light of well-justified concerns that the hydrogeological setting of such a repository was poorly understood and that the site could eventually be used to store high-level nuclear wastes. In September 1995, residents of Deep River, 160 km northwest of Ottawa voted in favour of an underground repository for more than 880,000 m^3 of low-level nuclear waste to be removed from various urban areas in southern Ontario.

Radon

Radon (^{222}Rn) is a naturally occurring gaseous daughter product of uranium (^{238}U) that has a half-life of 3.8 days and breaks down into two radioactive polonium daughters (^{218}Po, ^{214}Po) that can cause lung cancer (Otton, 1992). Radon enters buildings *via* basements and ground water (Gundersen and Wanty, 1991; Lubin, 1992; Stone, 1993) The GTA has a relatively high radon potential since it is underlain by uranium-

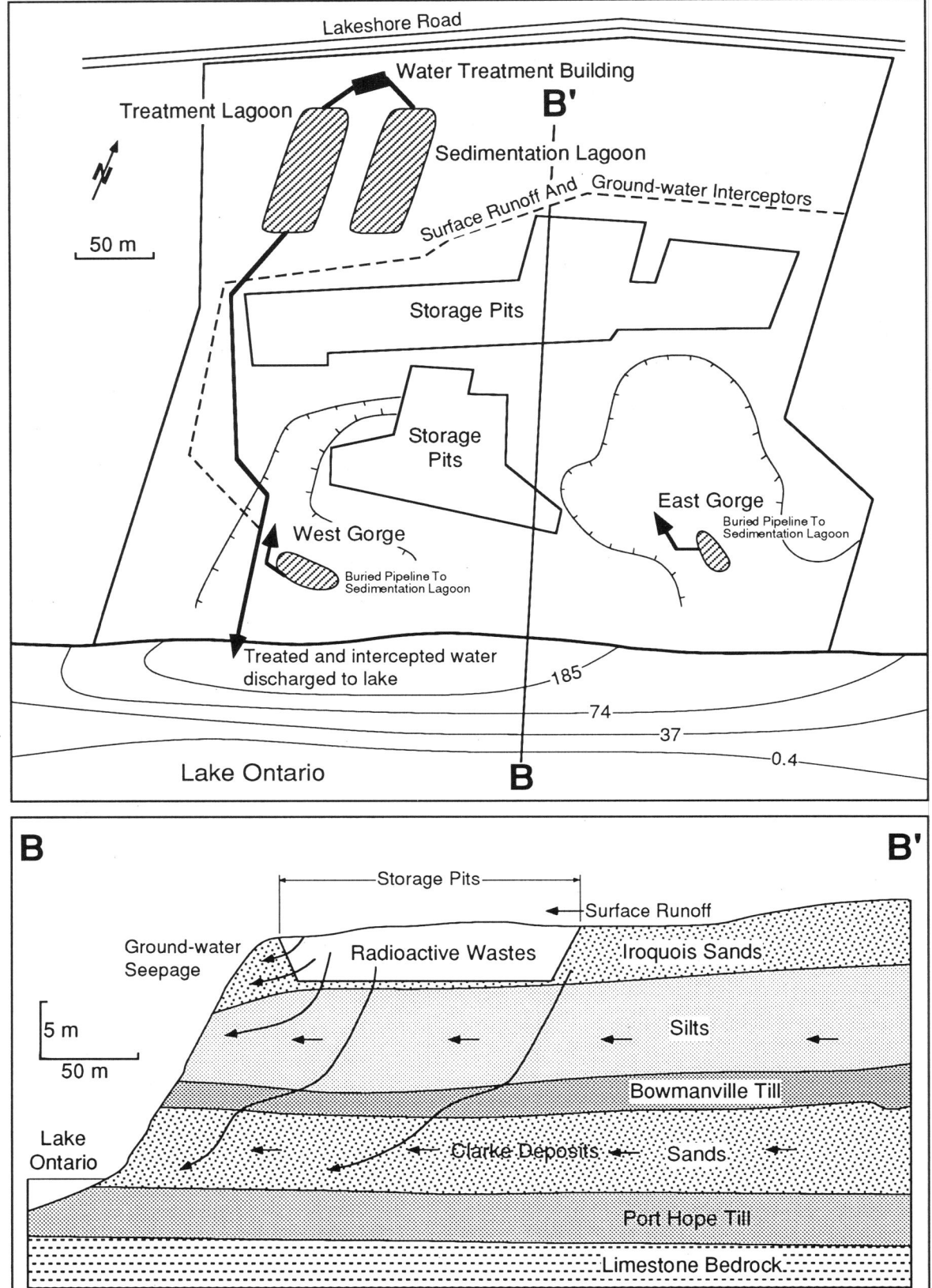

Figure 51 *(top) Low-level radioactive waste management facility along shoreline of Lake Ontario at Port Granby (Fig. 1). The site extends over 10 ha along the eroding shoreline and was in use between 1955 and 1988. In addition to storage pits, radioactive waste was used to fill two deep ravines (west and east gorges). Contours in lake water define the plume of ^{226}Ra in mBq·L^{-1} (after Bobba and Joshi, 1988). (**bottom**) Geologic cross-section through site showing ground-water flow (arrows). Bobba and Joshi (1988) argued that radionuclides would reach the Clarke Deposits by 1993 and eventually penetrate through the clay-rich Port Hope Till to bedrock as a result of diffusion. For stratigraphy, see Fig. 17.*

bearing shales and, at depth, by basement rocks such as granites and gneisses (Fig. 6), but little is known of actual radon levels and variability across the region. Existing data suggest the importance of bedrock fractures and faults in localizing where radon is released to surface environments. Other sources of radon include landfill sites where radioactive wastes were dumped from wartime factories (Acres International Ltd, 1993; Chapter 28).

Tritium

Tritium is a radioactive form of hydrogen, and is produced in large quantities by nuclear facilities in southern Ontario as an unintentional by-product of reactor operations by the conversion of deuterium in heavy water. Tritium is a known human carcinogen. Tritiated water escapes into the environment as steam; the release of pure tritium gas from the Tritium Recovery Facility at Darlington is a potential problem, as is the escape of fission products such as Iodine-131 (^{131}I) from fuel bundles into coolant. Tritium levels in ambient air in the vicinity of the nuclear reactors are routinely monitored; levels surrounding the Pickering reactors average about 40 times the provincial average, and range up to as much as 200 times (Atomic Energy Control Board, 1991). The annual average at Pickering is about 20 times that for Oshawa, some 20 km to the east (Fig. 1). Present background levels for water in the Great Lakes are between 5 and 10 Bq·L^{-1}; the water intake plant at Ajax, 6 km east of Pickering, averages 4 times the background level with peak levels correlating with episodic tritium spills at Darlington and Pickering (Kock, 1994).

In August 1992, more than 2000 L of heavy water containing tritium was released accidentally into Lake Ontario from Pickering. The resulting tritium plume travelled past the municipal drinking-water intakes for Toronto (21 km west), Scarborough (10 km west), Whitby (13 km east) and Ajax; a peak level of 1,600 Bq·L^{-1} was measured at the Ajax intake on August 7 (Kock, 1994). This spill focussed attention on the Region of Durham's plan to construct a new municipal drinking-water treatment facility on the coast at Ajax. Calls by the public for a full Environmental Assessment (EA) were rejected, but the new plant will have to comply with all provincial drinking-water standards; real-time monitoring of tritium is proposed and the independent Advisory Committee on Environmental Standards (ACES) recommended (1994) an immediate cut in the provincial drinking-water standard for tritium from 40,000 Bq·L^{-1} to 100 Bq·L^{-1} and ultimately to 20 Bq·L^{-1} in 5 years. In early 1995, The Ontario Ministry of the Environment and Energy set the drinking-water objective at 7000 Bq·L^{-1}, thereby ignoring the recommendations of ACES.

Health Impacts

The International Joint Commission has recently added radionuclides with half-lives greater than 6 months to their strategy of virtual elimination of persistent toxic substances in the Great Lakes. In contrast, the Ontario Municipal-Industrial Strategy for Pollution Abatement (MISA), which is designed to develop additional legislation to reduce industrial wastewater discharges at source, has exempted radionuclides because they fall under the exclusive federal jurisdiction of the Atomic Energy Control Board of Canada (Chapter 39).

Space does not permit a full review of health impacts arising from the activities of the nuclear industry in the GTA. The topic is briefly introduced here because the effects of routine exposure to radioactive pollution, the determination of allowable limits, and acceptable health effects are the subject of much debate (Clarke *et al.*, 1991; Straume, 1993;

Kock, 1994). A sievert (Sv) is a measure of biological damage arising from radiation. The background radiation dose for people living at remote locations in Ontario is between 2 and 3 mSv. Ontario Hydro estimates the total dose from environmental exposure in terms of a critical group (persons living at the site boundary of a nuclear power station) and the collective population, consisting of the larger population around the power station. Critical group doses at Pickering are generally less than 1% of the maximum allowable limit of 5 mSv. The collective population dose is the total dose received by all members of the community living within a certain distance; the total collective dose around Pickering in 1992 was 1.37 person-sieverts (reflecting the high population in Pickering) and 0.07 person-sieverts at Darlington. The question arises as to the magnitude of background radiation exposure from natural geological materials across the GTA, but data are few (Chapter 28).

PART IV: ENVIRONMENTAL GEOLOGY AND URBAN PLANNING

It is now widely accepted that future urban development across the GTA must embody the concept of sustainability, where land or resources are used supposedly without any loss or reduction in the health or integrity of local and regional ecosystems (Crombie, 1994). However, definition of sustainability is not well constrained scientifically, and there is a clear need to identify long-term environmental impacts of various land use options, using quantitative models that incorporate scientifically credible geological and hydrogeological data. Significant movement in this direction is represented by new legislation incorporated into the amended Planning Act of Ontario, which recognizes ecosystems and watersheds as fundamental units in planning and the need to identify environmental impacts throughout such systems. Another intention of the new legislation is to devolve responsibility for environmentally sensitive planning from the province to local municipal government and citizens. Such changes have major implications and these are identified in the following sections, paying particular regard to the role of environmental geology.

Ontario Planning Act and Environmental Protection Policies

Urban development is based, essentially, on municipal Official Plans which identify present land uses and guide allowable changes over a timeframe which, in practice, is usually about two decades. The Official Plans of most GTA municipalities were last formally revised in their entirety in the 1970s; most are now under revision and draft documents are available for public discussion.

Present urban planning across the GTA is based on decisions made up to 20 years ago when environmental data were scarcely considered in Official Plans. To date, planning can be categorized as business-as-usual, where urban development proceeds in stages and the resulting effects are loosely monitored (if at all). Crombie (1994) showed that the mere existence of an Official Plan does not ensure a long-term framework for rational land use changes in most urban communities in the GTA. Most plans are reactive in that they simply respond in *ad hoc* fashion to individual development applications and revisions to the plan are permitted regardless of any pre-existing designation as agricultural land or open space. Moreover, there is a tendency to develop to the maximum permitted capacity allowed by services such as trunk sewer systems. The impact of urbanization on ground-water resources is only briefly recognized in many Official Plans, because critical data have not been available and are too expensive to collect on a meaningful scale.

Planning Act Reform

In recognition of the inadequacies of existing planning approaches to safeguard the environment, in 1993, the Ontario Commission on Planning and Development Reform outlined proposals designed to ensure environmental protection by requiring long-term strategic municipal planning. The document introduced terms such as ecosystems, landscape systems and natural functions of different landscapes, such as ground-water recharge and discharge, that henceforth were to be identified in revised Official Plans (see Walker, 1994; Riley and Mohr, 1994; Crombie, 1994). Recognition was given to the need to identify the watershed as a fundamental unit for planning and to identify hydrologic and hydrogeologic resources in need of protection (Ontario Ministry of the Environment and Energy, 1994e,f).

Formal amendment to the Planning Act of Ontario, designed to promote sustainable economic development in a healthy natural environment, received Royal Assent in December 1994 and was proclaimed in March 1995. Comprehensive policy statements and detailed implementation guidelines were released in May 1995 (Ontario Ministry of Municipal Affairs, 1995). A full review of these guidelines is beyond the scope of this paper, but general principles are established for protection of water resources, habitat, and significant landscapes, including the identification of sensitive areas in need of protection. Guidelines are also established for identification of natural hazards such as created by flooding and erosion. Proponents bringing forward development proposals to local councils are now required to conduct a detailed environmental impact study incorporating comments from the public. Municipalities are required to carry out environmental audits using the guidelines as a basis and use such data to update Official Plans.

The largest environmental-planning study currently underway in the GTA is that dealing with the Oak Ridges Moraine, and this is discussed separately elsewhere in this volume (Chapter 9). The most extensive urban developments are located on the Halton Till plain in Peel (City of Brampton) and Durham (Seaton; Township of Pickering) regions (Fig. 3). These case studies are reviewed below to illustrate the current state of practice in using environmental geological information in the urban development process. The implications for future planning and the discipline of environmental geology presented by the revised Planning Act are then identified.

City of Brampton

Figure 52 shows the large area of industrial and residential development on the northern and eastern margins of the city of Brampton (see Fig. 3). The largest residential community under construction in this area is Springdale with a projected population of 74,000. This development is based on planning decisions made more than a decade ago, when environmental inputs to the planning process were minimal. To redress this situation, the City of Brampton (1992) set out an environment-first policy framework to guide future urban expansion. The policy represents a move away from simple designation of features to be preserved to an ecosystem-oriented planning framework. Urban expansion areas were identified outside the present urban limits (Fig. 52), and were subjected to a detailed comparative ranking based on environmental criteria such as woodlot area and sensitivity, watercourse and fishery resources,

ground-water resources, wetlands and significant species. This listing exercise cannot be described as an integrated ecosystem approach, but is, at least, a start. The specific issue of the impact of urbanization on ground-water resources is only briefly recognized in the policy framework because critical data are not available and are too expensive to collect, yet the issue is fundamental to the health of surface ecosystems.

In general, the need to conserve ground- and surface-water resources is made redundant in the face of development strategies based on full municipal drinking-water and sewage services and delivery of contaminated storm water to watercourses. In general, there is little recognition of the impact of urbanization on ground water. There is over-reliance on the supposed impermeability of surface tills and their ability to protect underlying aquifers from urban-sourced contaminants, in the face of increasing evidence to the contrary (see above; Boyce et al., 1995). In reality, ground water is written off as a resource in those areas that employ full municipal services; the much vaunted sustainable urban development is a contradiction in terms. This situation obtains over much of the GTA (e.g., Fig. 24) with the exception of new ground-water conservation measures in hand in Halton Region (Holysh, 1995).

Township of Pickering

The Township of Pickering, in the eastern part of the GTA (Figs. 1, 3), is presently conducting a public review of its draft Official Plan (Township of Pickering, 1995). The plan is examined here briefly because it illustrates a common problem where inadequate environmental information or expertise is available, giving rise to plans that are flawed by technical data gaps.

More than 75% of Pickering Township is rural and includes part of the Oak Ridges Moraine and land formerly expropriated for an international airport by the federal government (Airport Lands; Stewart, 1979; Fig. 53). The regulation of land use to protect ground-water resources in these undeveloped areas is a key issue among residents (see Chapter 9). The urban portion of the township lies in south Pickering and consists of intensive urban development sprawled across the Halton Till plain and Iroquois lake floor (Fig. 16).

The draft Official Plan issued by the Township of Pickering presents a natural-systems plan that identifies core areas, corridors (e.g., river valleys) and linkages (linking core areas and corridors; Township of Pickering, 1994; Fig. 54). This follows a widely used system for defining natural heritage areas (see Riley and Mohr, 1994 for a detailed review). However, the definition and purpose of the core areas, as identified in Pickering, is not clear; some include wetlands and forests and so presumably are for habitat protection. Others include sand and gravel quarries and have been designated for their resource value. Hydrogeological data are not used in the natural systems plan despite the dependence of a large rural area of the township on well water and septic systems. The draft Pickering Official Plan does set out criteria for developers for completing a brief environmental report for proposed development within set (and quite arbitrary) distances of the designated areas (e.g., within 120 m of a core area). The scope of such reports, however, is narrow when compared to the breadth of the environmental impact study requirements identified by the new Planning Act Guidelines (see above).

The draft Pickering Official Plan is straight-jacketed by large-scale urban development projects already in hand. The proposed new city of Seaton (90,000 people; 2,750 ha; Figs. 3, 53) will blanket one of the largest core areas identified in the township (Fig. 54); the new Highway 407, expected to be completed by 2005, will also sever several core areas and corridors. A large tract of the township is also reserved for a future regional airport (Fig. 53). These developments, already in the pipeline, represent "trojan horses" and defeat any realistic identification and protection of areas deemed environmentally sensitive in the draft Official Plan.

The Township of Pickering draft Official Plan is further flawed by poor use of terminology. The plan is heavily prefaced with notions such as ecological carrying capacity, healthy settings, global connectivity, systems, sustainable community development, *etc.* There is also reference to hydrological cycles, nitrogen cycles, oxygen-carbon cycles, but the purpose is not evident because neither the plan or supporting documentation contains any indication as to how they are defined scientifically or to be measured and applied to the urban planning process. These terms have become modern-day mantras, routinely applied without any clear understanding, or definition, of the precise meaning of these concepts (see comments by Jacobs and Sadler, 1992; Jakimchuk, 1992; Maddox, 1995).

As a consequence of the amended Planning Act, the province of Ontario has delegated environmental planning to local government, but planning departments seldom have the expertise to collect, marshall and interpret the type of detailed environmental geological data that is now required under the new guidelines. A fundamental flaw in the current Township of Pickering draft Official Plan, for example, is that it fails to recognize and incorporate a wide range of geological and hydrogeological assessments carried out since 1970, including highly detailed studies in the vicinity of the many landfills in the township. A wealth of local information, collected at considerable public expense and summarized in many reports (*e.g.*, Hewitt, 1969; Morton, 1973; Gwyn, 1976; Hydrology Consultants Ltd., 1976; Sibul *et al.*, 1977; Ostry, 1979a;

Golder and Associates, 1987; Dames and Moore, 1991, 1992; Nelson, 1991; Geomatics International, 1991; Gartner Lee Ltd., 1990, 1993a; Procter and Redfern, 1993b; HBT AGRA, 1994; M.M. Dillon Ltd., 1994), provides the foundation for a strengthened scientific approach to planning.

Town of Caledon

Arguably, the most environmentally rigorous approach to urban development currently in practice anywhere in the GTA has been adopted by the Town of Caledon. The township lies adjacent to the Niagara Escarpment (Fig. 1), and is wholly dependent on ground water for municipal drinking water (*e.g.*, Bolton Aquifer; Fig. 18A). A substantial planning document (By-law 94-101) established environmental and open-space policies designed to "identify, maintain and protect the physical features, ecological forms, functions and linkages which sustain ecosystem integrity" (Town of Caledon, 1994, p. 5). Each of these terms, and many others, is clearly defined in a glossary. Components such as natural core areas, natural corridors and natural linkages (see above) form the heart of the town's ecosystem and developers are required to carry out detailed work as part of an Environmental Impact Study and Management Plan (EIS & MP) to provide site-specific detail. Studies have to include a minimum footage of drilled holes, including one or more deep holes, and carry a requirement to correlate local geological data with the regional stratigraphy. Areas of ground-water recharge and discharge, identified from broader scale studies, are subject to detailed work and excluded from development as appropriate. A precedent is provided by the 1:50,000 scale DRASTIC mapping by the United States National Water Well Association which provides a standardized system for evaluating and mapping ground-water pollution potential (see Aller *et al.*, 1987). Previous efforts in southern Ontario to map the sensitivity of ground water to contamination were, unfortunately, limited to

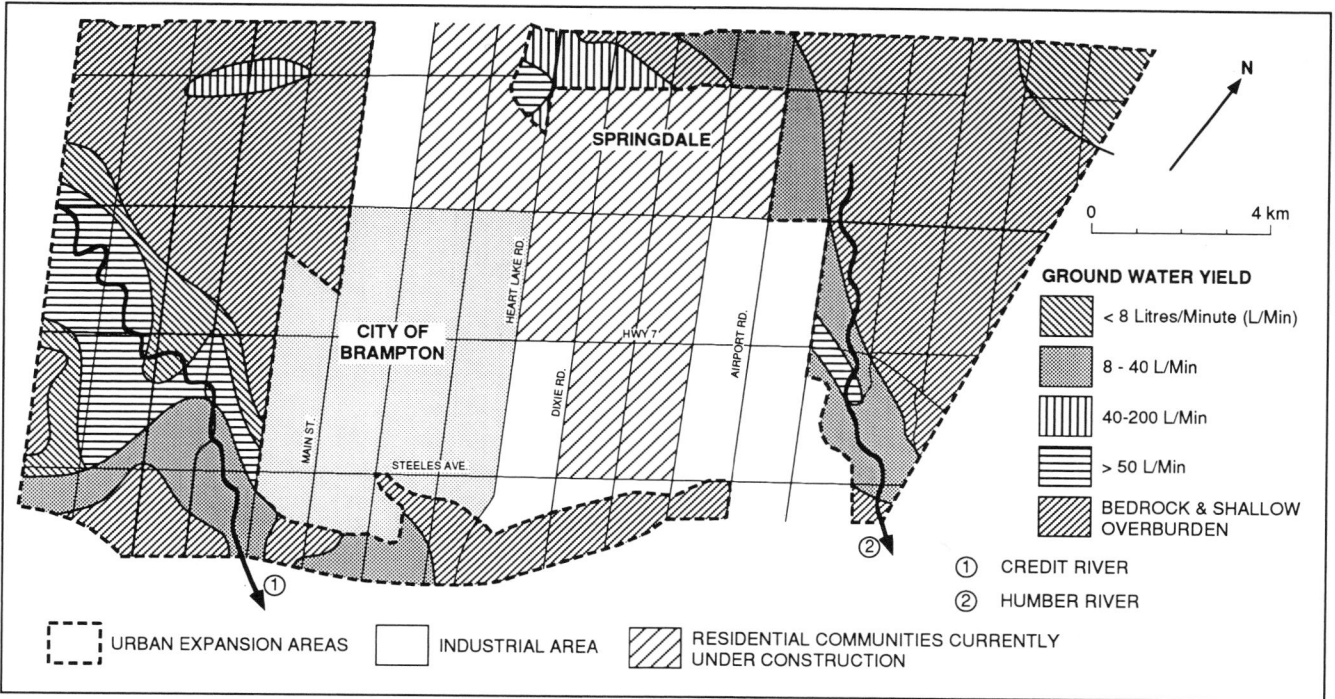

Figure 52 *Environmental geology constraints on urban development; ecologically based municipal planning in the city of Brampton (Fig. 3). The city has established urban expansion areas which have been ranked according to different environmental criteria such as ground-water resources (see text). Existing residential developments, planned with little or no regard for the environment, include Springdale (Fig. 3). Other examples of environmentally sensitive municipal plans are provided by Gore and Storrie (1992; Town of Markham) and Gartner Lee (1993b; City of Vaughan).*

Ministry of Environment (1980). Programs of wellhead protection (see Chapter 11) are being implemented in Caledon along with assessments of water quality and quantity in order to identify the impact of new water-taking uses such as municipal water supply production wells, golf course irrigation wells, commercial water suppliers and aggregate quarries.

The planning process followed by the Town of Caledon ensures that sufficient pre-development geological and hydrogeological work is done to model long-term impacts and thereby prevent costly future remediation work. It ensures land uses are compatible with environmental protection. Futhermore, the costs of environmental investigations are borne by the developer and, being passed on to the consumer, are simply part of the cost of doing development.

DISCUSSION

The principal environmental geological issues in the Greater Toronto Area of southern Ontario can be grouped under the headings of geological processes, geological resources, and wastes. Technical data gaps exist in all these areas (Fig. 55).

Geological Processes

The disastrous effects of Hurricane Hazel in 1954 provided a much needed wake-up call with regard to geological processes and their impact on the urban population of the GTA. Since then, other issues have emerged which require better understanding of the operation of basic geological processes. One such problem is that of earthquake risk. Basic information needs to be collected to identify regional basement

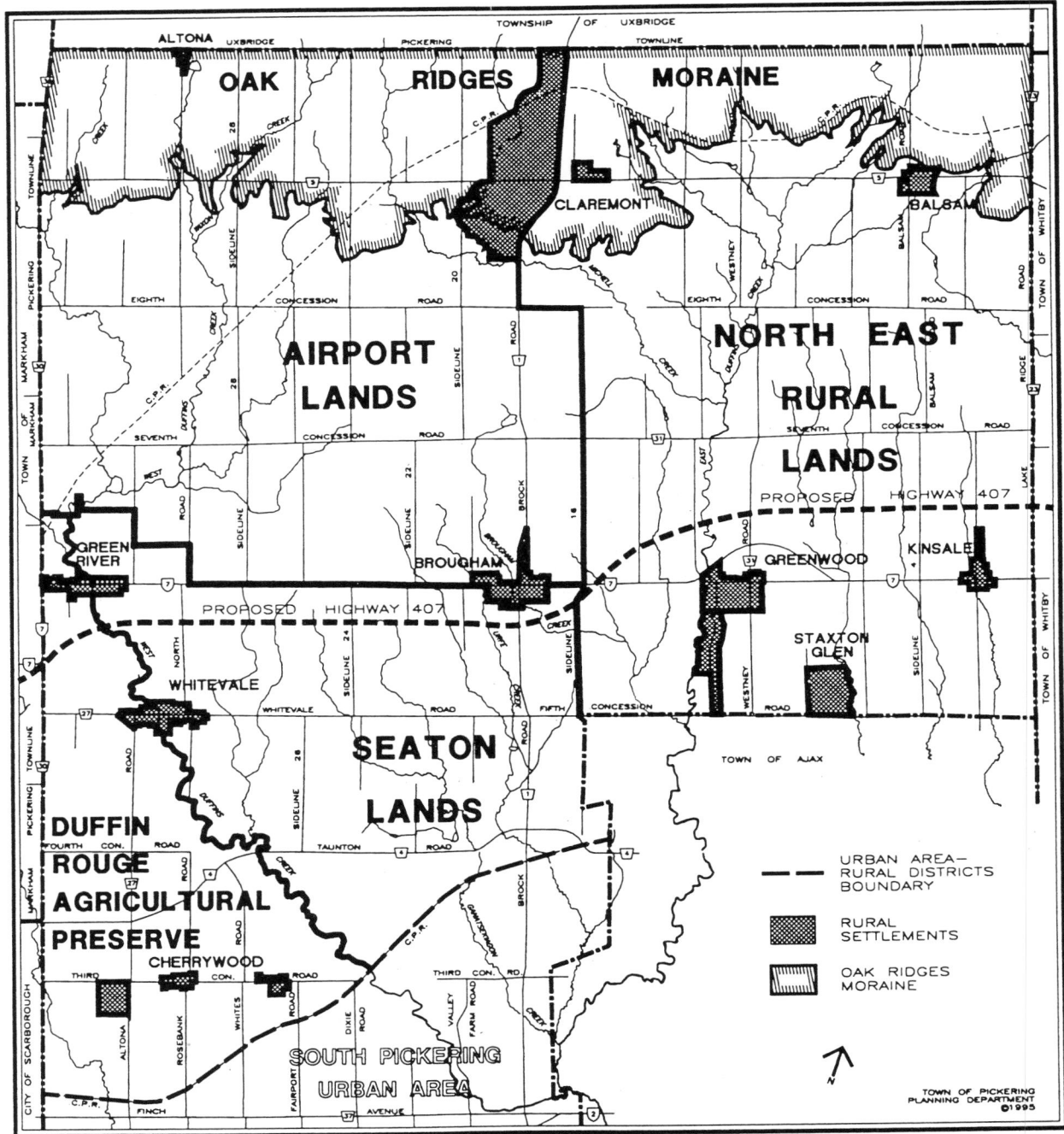

Figure 53   *Pickering Township (see Fig. 3) showing area reserved for future airport, for the urban community of Seaton (pop'n. 90,000); surplus airport land has been set aside for 20 years as an agricultural preserve (from Township of Pickering, 1995). Same scale as Fig. 54.*

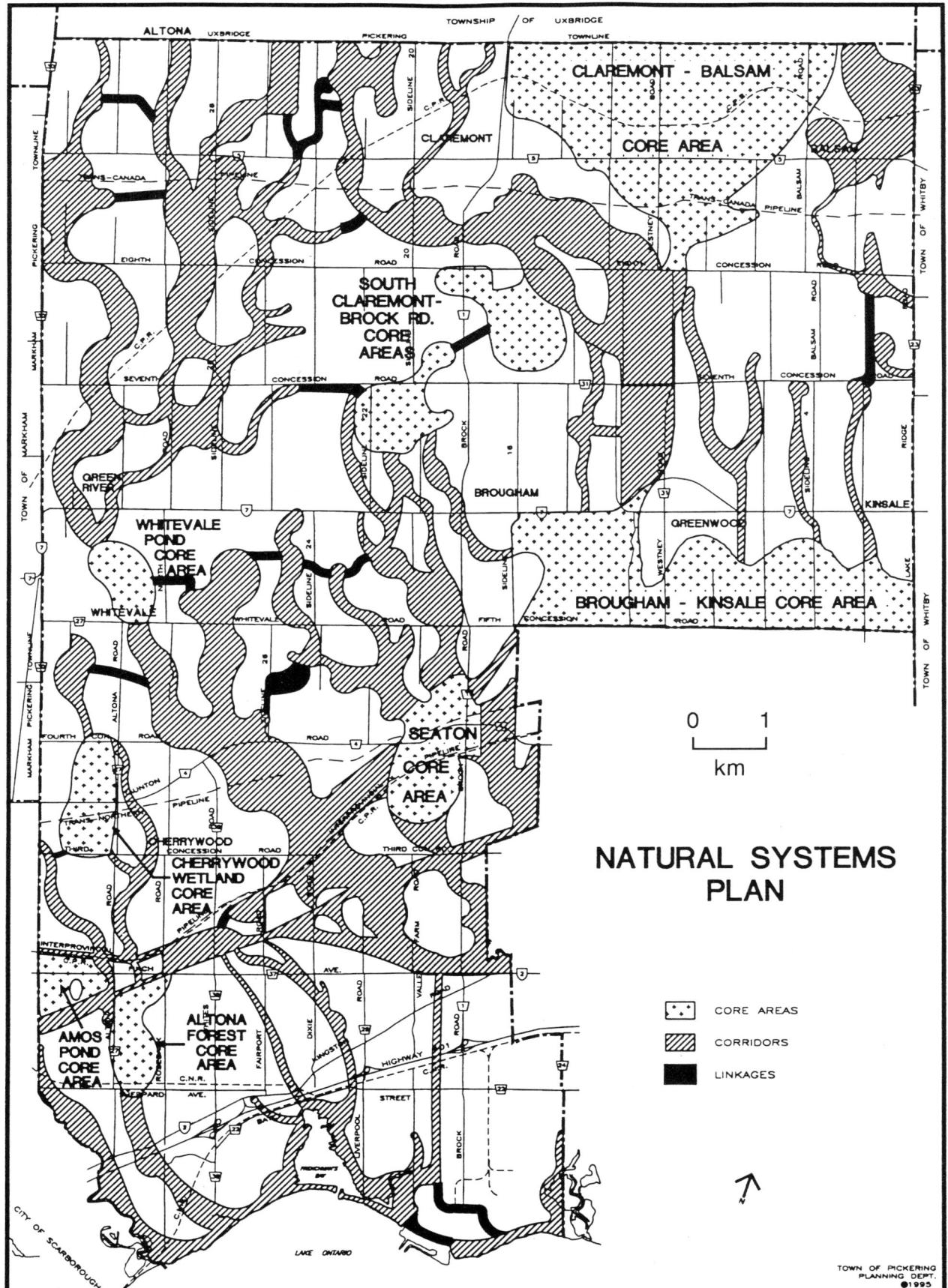

Figure 54 *Natural systems plan as identified in Pickering Township showing designated core areas, corridors and linkages. Definition is based on limited environmental data; substantial preservation of core areas has already been pre-empted because of the large tracts of land already set aside for a regional airport and the major urban community of Seaton. From Township of Pickering (1995). See Chapter 38 for an example of how remote sensing techniques have been employed in mapping natural systems in north Pickering.*

WORK REQUIRED

- Contaminant audits and identification of sources and discharge points
- Inventories of landfill sites
- Mapping of historic landfilling
- Mapping of groundwater pollution potential
- Public health / risk assessment incorporating environmental geological data
- 3-D characterization of Pleistocene sediments and associated regional / local ground water regimes (hydrostratigraphy) especially with regard to stratigraphic variability in tills
- Regional groundwater flow modeling
- Identification of pre-historic record of earthquakes in Pleistocene sediments
- Regional mapping of fractures in sediments
- Local mapping of buried bedrock topography and sediment thickness
- Mapping of fractures / faults in bedrock
- Regional / local ground-water flow modelling in fractured bedrock such as limestones
- Role of fractures in localizing radon release from shales in urban environments
- Subsurface mapping of basement structures and identification of seismically active structures

GEOLOGY

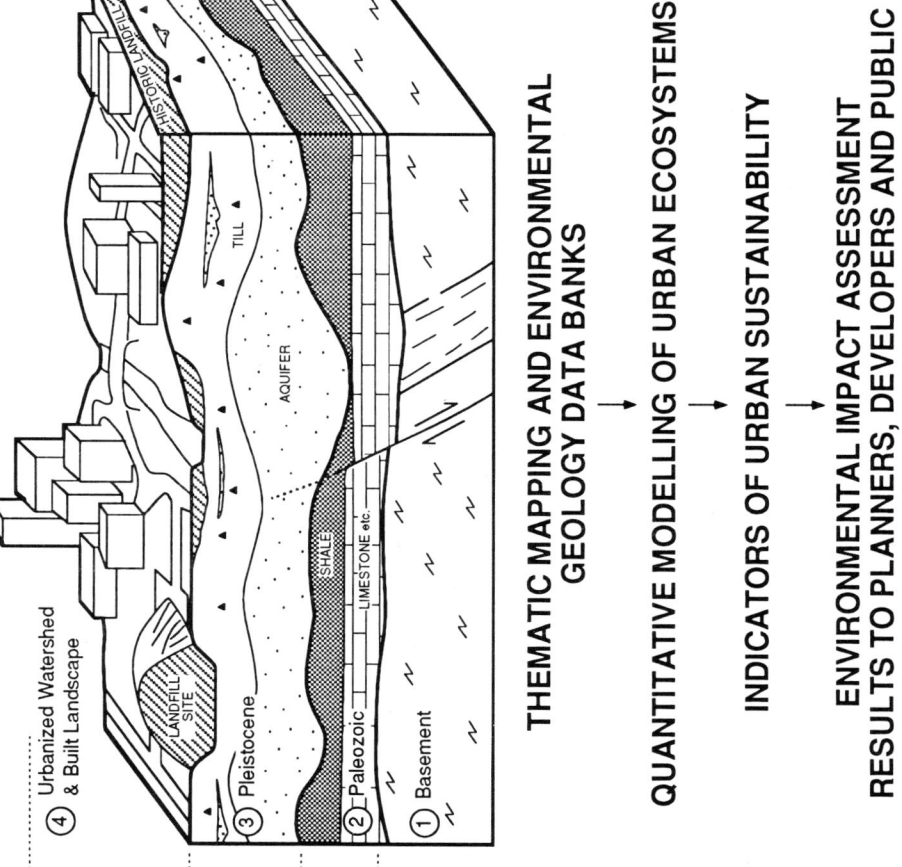

④ Urbanized Watershed & Built Landscape

③ Pleistocene

② Paleozoic

① Basement

HISTORIC LANDFILL

TILL

AQUIFER

LANDFILL SITE

SHALE

LIMESTONE etc.

THEMATIC MAPPING AND ENVIRONMENTAL GEOLOGY DATA BANKS

QUANTITATIVE MODELLING OF URBAN ECOSYSTEMS

INDICATORS OF URBAN SUSTAINABILITY

ENVIRONMENTAL IMPACT ASSESSMENT

RESULTS TO PLANNERS, DEVELOPERS AND PUBLIC

DATA GAPS

- Sources, types and total masses of contaminants moving to urban rivers, lakes, ground waters and humans from point and non-point sources
- Impact of major users (municipal, industrial) on groundwater quality / quantity
- Extent & stratigraphy of historic landfilling
- Stability of landfill during earthquakes
- Number, content and impact of landfill sites
- Flood hazard, shoreline erosion, slope stability
- Greenhouse gas-related climate changes and hydrologic and hydrogeological response
- Origin and extent of fractures in tills and clays
- Geometry / extent of aquifers / aquitards
- Impacts of aggregate mining
- Neotectonic stress field and ongoing deformation
- Long term record of earthquake activity and faulting
- Structure of basement

Figure 55 *Diagrammatic summary of significant data gaps with regard to the environmental geology of the Greater Toronto Area. These needs are common to most large urban areas in glaciated terrains where bedrock is covered by thick and complexly stratified Pleistocene glacial deposits.*

structures not only for understanding seismic risk to urban infrastructure, such as nuclear reactors, but for assessment of radon hazard, which appears to be localized along basement structures and fractures (Chapters 28, 29).

A high priority should be given to identification of landslide susceptibility and liquefaction potential of different substrates including areas of historic landfilling, in the event of moderate- to large-magnitude earthquakes. Assessment of the recurrence interval and magnitudes of past earthquake activity is dependent on study of the sedimentary record of seismic events preserved in Pleistocene sediments outcropping along lakeshore and ravine bluffs. Much of this record is being lost as critical outcrops are destroyed by urbanization. In particular, better understanding is needed of neotectonic stress fields in bedrock and sediments and the formation of bedrock joints and their propagation into Pleistocene sediments.

The interactions of future climate change and urbanization on river flood frequency, regional ground-water flows, water table levels, lake levels and shoreline erosion need to be identified and predictive models developed.

Geological Resources
Resources include ground and surface waters, sediments and rock. Quantitative understanding of regional water resources is non-existent. There is an immediate requirement to map subsurface Pleistocene deposits and associated ground-water regimes (hydrostratigraphy) in the GTA. Such work should include systematic mapping of ground-water recharge and discharge areas, aquifers and aquicludes, water quality, and quantitative flow modelling from recharge areas to Lake Ontario. Emphasis should be placed on identifying where contaminants are released from point and non-point sources in urban environments and their migration pathways. Quantitative flow modelling will allow forecasting of environmental impacts arising from different land use options and future climate changes.

The impacts of large-scale bedrock, sand and gravel mining on ground-water regimes in areas such as the Oak Ridges Moraine, Niagara Escarpment and offshore across the floor of Lake Ontario need to be identified and modelled.

Wastes
Large volumes of waste have been spread about the landscape in the form of municipal and industrial dumps and historic landfilling. Areas of historic landfill need to be mapped along with documentation of fill types, chemistry, stratigraphy and ground-water impacts. Landfill sites need to be inventoried, their contents identified, and impacts on surrounding environments defined. Priority sites should be listed for remediation when economic conditions allow. Impacts on public health among communities adjacent to municipal, hazardous and nuclear waste sites should be assessed. Urban development in the vicinity of such sites should be halted until impact studies have been completed by proponents.

Waste materials are emitted to air, land and water by a wide range of urban activities. Better quantification of pollutant loads resulting from each activity is central to assessments of public health and ecosystem health. The latest report of the International Joint Commission (1995) provides a very useful review of environmental epidemiological studies of what is increasingly termed "environmental burden of illness" in the Great Lakes basin. Many data regarding the geographic incidence of diseases have been collected (*e.g.*, Johnson *et al.*, 1992; Mills and Semenciw, 1992; Gillman *et al.*, 1992), but contaminant sources, quantities, pathways to re-

ceptors, and the effects of background geological factors are poorly understood. The International Joint Commission (1995) makes clear that existing data collected by many disciplines, including environmental geology, are fragmented, incomplete and unintegrated.

Final Comment
Filling the data gaps identified above is part of a larger effort to create credible scientific assessments (report cards) of the state of health and environmental sensitivity of urban areas such as the Greater Toronto Area of southern Ontario. Are current and projected urban land uses environmentally sustainable? What are the long-term cumulative environmental effects and impacts of changes in land use? These questions cannot be addressed until regional data bases are established that will allow quantitative modelling of land use changes and conflicts. Despite a political willingness for ecosystem planning, many policy statements are strictly qualitative in nature. At present, no single political jurisdiction within the GTA possesses sufficient environmental geological data on which to base quantitative, scientifically credible assessments of either the current state of the environment or the impacts of future urban land uses.

ACKNOWLEDGEMENTS
I thank the Natural Sciences and Engineering Research Council of Canada for their generous support, the many agencies, organizations, individuals and consulting companies for access to data, and the following for discussions; Paul Beck, Hugo Bolton, Gary Bowen, Jean Blundell, Larry Conrad, Mike D'Andrea, Adele Freeman, Ed Freeman, Bill Hogg, Helen Juhola, Irene Kock, Dave Neufeld, Alex Mohajer, Tom Mohr, Michael Moir, Ralph Moulton, Kathleen Murray, Derek Mack Mumford, Tim Myles, Bob Ostry, Sabina Nagpal, Jonathan P'ng, John Riley, Ed Sado, Sean Salvatori, Bryan Smith, Bill Snodgrass, Walter Tovell, Vic Tyrer, Joe Wallach, Dave Winfield and John Westgate. I am especially grateful to Mike Doughty for technical assistance and to Emmanuelle Arnaud, Carolyn Eyles, Tom Mohr, Kathleen Murray and David Steele for formal reviews of the manuscript. The views presented herein remain those of the author.

REFERENCES

Acres International Ltd., 1993, Revised environmental screening of the Malvern remedial project – Draft for public review: Report M93-05, 21 p.

Adams, J., 1995, The tectonic stress field in eastern Canada: its use in seismic source interpretation: Atomic Energy Control Board, Workshop on Seismic Hazard Assessment in Southern Ontario, Ottawa, 19-21 June.

Adams, J. and Basham, P.W., 1991, The seismicity and seismotectonics of eastern Canada, *in* Slemmon, D.B., ed., Neotectonics of North America: Geological Society of America, p. 261-276.

Adams, J., Dredge, L., Fenton, M., Grant, D.R. and Shilts, W.W., 1993a, Late Quaternary faulting in the Rouge River Valley, southern Ontario: seismotectonic or glaciotectonic?: Geological Survey of Canada, Open File Report 2652, 60 p.

Adams, J., Dredge, L., Fenton, M., Grant, D.R. and Shilts, W.W., 1993b, Neotectonic faulting in Metropolitan Toronto: Implications for earthquake hazard assessment in the Lake Ontario region: Comment and Reply: Geology, v. 21, p. 863-864.

Aller, L., Bennett, T., Lehr, J.H., Petty, R.J. and Hackett, J., 1987, DRASTIC: A standardized system for evaluating groundwater pollution potential using hydrogeologic settings: National Water Well Association, Dublin, OH, 66 p.

Anderson, T. and Lewis, C.F.M., 1985, Postglacial water level history of the Lake Ontario Basin, *in* Karrow, P.F. and Calkin, P.E., eds., Quaternary Evolution of the Great Lake Basins: Geological Association of Canada, Special Paper 30, p. 231-258.

Angus, J.T., 1988, A Respectable Ditch: A History of the Trent-Severn Waterway 1833-1920: McGill-Queen's University Press, Montreal, PQ, 455 p.

Anon., 1994, Decision at Pickering: Canadian Nuclear Society Bulletin, v. 15, p. 13-16.

Aravena, R. and Wassenaar, L. I., 1993, Dissolved organic carbon and methane in a regional confined aquifer, southern Ontario, Canada: Carbon isotope evidence for associated subsurface sources: Applied Geochemistry, v. 8, p. 483-493.

Aravena, R., Wassenaar, L.I. and Plummer, L.N., 1995, Estimating ^{14}C groundwater ages in a methanogenic aquifer: Water Resources Research, v. 31, p. 2307-2318.

Arnoud, M., Barres, M. and Come, B., 1993, eds., Geology and Confinement of Toxic Wastes: A.A. Balkema, Rotterdam, 610 p.

Atomic Energy of Canada Limited, 1994, Summary of the environmental impact statement on the concept for disposal of Canada's nuclear fuel waste: AECL-10721, COG-93-11, 470 p.

Atomic Energy Control Board of Canada, 1991, Tritium releases from Pickering Nuclear Generating Station and birth defects and infant mortality, 1973-1988: Research Report INFO-0401, 125 p.

Atomic Energy Control Board of Canada, 1992, Renewal of Cameco, Port Hope, fuel facility operating licence: AECB BMD 92-181, 68 p.

Atria Engineering Hydraulics Inc., 1994, Lake Ontario aggregate mining project shoreline processes study: submitted to B.A.R. Environmental Inc., 27 p.

B.A.R. Environmental Inc., 1992, Environmental review Bedrock Resources, Inc. permit application for aggregate dredging in Lake Ontario: 67 p.

Bailey, B.D., 1973, A history of the Toronto waterfront from Etobicoke to Pickering: Metropolitan Toronto and Region Conservation Authority, 61 p.

Bailey Geological Services Ltd., 1984, Evaluation of the conventional and potential oil and gas reserves of the Cambrian of Ontario: Ontario Geological Survey, Open File Report 5499, 63 p.

Bannerman, R.D., Owens, R., Dodds, R. and Hornewer, N., 1993, Sources of pollutants in Wisconsin storm water: Water Science and Technology, v. 28, p. 241-259.

Barker, J.F., Cherry, J.A., Reinhard, M., Pankow, J.F. and Zapico, M., 1987, The occurrence and mobility of hazardous organic chemicals in groundwater at several Ontario landfills: University of Waterloo, Lottery Trust Fund Project 118PL, Final Report, 116 p.

Barker, J.F. and Pollock, S.J., 1987, The geochemistry and origin of natural gases in southern Ontario: Bulletin of Canadian Petroleum Geology, v. 32, p. 313-326.

Barnett, P.J., Cowan, W.R. and Henry, A.P., 1991, Quaternary Geology of Ontario: southern sheet: Ontario Geological Survey, Map 2556, scale 1:1,000,000.

Barnett, R.L., Arima, M., Blackwell, J.D. and Winder, C.G., 1984, The Picton and Varty Lake ultramafic dykes; Jurassic magmatism in the St. Lawrence Platform near Belleville, Ontario: Canadian Journal of Earth Sciences, v. 21, p. 1460-1472.

Barrett, S. and Kidd, J., 1991, Pathways: Towards an ecosystem approach: East Bayfront and Port Industrial District: Royal Commission on the Future of the Toronto Waterfront, 165 p.

Barry Lyon Ltd., 1994, Financial analysis and feasibility study, Greenwood Racetrack, Toronto, Final report: 22 p.

Beak and Raven Beck Environmental Ltd., 1994, Lower Don Lands site characterization and remedial options study: Report to Waterfront Regeneration Trust, 150 p.

Beltrami, H. and Mareschal, J.C., 1992, Ground temperature histories for central and eastern Canada from geothermal measurements: Little Ice Age signature: Geophysical Research Letters, v. 19, p. 689-692.

Berger, A. and Loutre, M.F., 1992, Astronomical solutions for paleoclimate studies over the last 3 million years: Earth and Planetary Science Letters, v. 111, p. 369-82.

Berger, G. and Eyles, N. 1994, Thermoluminescence chronology of the Toronto area Quaternary sediments and implications for the extent of the mid-continent ice sheets: Geology, v. 23, p. 31-34.

Bertell, R., 1995, Health Profile of area children, Brock West Landfill, Pickering: International Institute of Concern for Public Health, Toronto, 87 p.

Birkett, C.M., 1994, Radar altimetry: A new concept in monitoring lake level changes: EOS, v. 75, p. 273-275.

Blumenfeld, H., 1954, The tidal wave of metropolitan expansion: Journal of the American Institute of Planners, v. XX, p. 3-14.

Bodo, B.A., 1989, Heavy metals in water and suspended particulates from an urban basin impacting Lake Ontario: Science of the Total Environment, v. 88, p. 329-344.

Bobba, A.G. and Joshi, S.R., 1988, Groundwater transport of radium-226 and uranium from Port Granby waste management site to Lake Ontario: Nuclear and Chemical Waste Management, v. 8, p. 199-209.

Boulton, G.S. and Hindmarsh, R.C.A., 1989, Sediment deformation beneath glaciers: Rheology and geological consequences: Journal of Geophysical Research, v. 92, p. 9059-9082.

Boulton, G.S. and Payne, A., 1993, Simulation of the European ice sheet through the last glacial cycle and prediction of future glaciation: Swedish Nuclear Fuel and Waste Management Co., Stockholm, Technical Report 93-14, 137 p.

Bourns, A.N., McCalla, D.R., McCallion, A.L. and Osbaldeston, J.B., 1986, Upper Ottawa Street Landfill site study: Final report to Ontario Ministry of Health, Toronto, ON, 89 p.

Boyce, J. and Eyles, N., 1991, Drumlins carved by deforming till streams below the Laurentide Ice Sheet: Geology, v. 19, p. 787-790.

Boyce, J., Eyles, N. and Pugin, A., 1995, Seismic, outcrop and borehole geometry of Late Wisconsin tills near Toronto, Canada: Canadian Journal of Earth Sciences, v. 32, p. 1331-1349.

Boyd, D., 1986, Effects of dredging and lakefilling in the Toronto waterfront during 1984: Ontario Ministry of the Environment, 26 p.

Boyd, D., 1988, The effect of contaminants associated with suspended sediment on water quality in the Toronto waterfront during 1985: Ontario Ministry of Environment, Water Resources Branch, 46 p.

Brace, C., 1995, Public works in the Canadian city; the provision of sewers in Toronto 1870-1913: Urban History Review, v. XXIII, p. 33-43.

Brett, C.E. and Brookfield, M.E., 1984, Morphology, faunas and genesis of Ordovician hardgrounds from southern Ontario, Canada: Palaeogeography, Palaeoclimatology, Palaeoecology, v. 46, p. 233-290.

Briffa, K.R., Jones, P.D., Sheingruber, F.H., Shiyatov, S.G. and Cook, E.R., 1995, Unusual twentieth century summer warmth in a 1,000 year temperature record from Siberia: Nature, v. 375, p. 156-159.

Brinke, M.L. and Parkin, W.P., 1991, Site Auditing: Environmental Assessment of Property: Specialty Technical Publishers, North Vancouver, BC, 850 p.

British Columbia Ministry of the Environment, 1991, Pacific Place soil remediation plan, Parcel 2: 106 p.

Brookfield, M.E., Gwyn, Q.H.J. and Martini, I.P., 1982, Quaternary sequences along the north shore of Lake Ontario, Oshawa, Port Hope: Canadian Journal of Earth Sciences, v. 19, p. 1836-1850.

Buffler, P.A., Crane, M. and Key, M.M., 1985, Possibilities of detecting health effects by studies of populations exposed to chemicals from waste disposal sites: Environmental Health Perspectives, v. 62, p. 423-456.

Campbell, I.D. and McAndrews, J.H., 1993, Forest disequilibrium caused by rapid Little Ice Age cooling: Nature, v. 366, p. 336-338.

Central Waterfront Planning Committee, 1976, Information base study: The environment: 260 p.

Chameau, J.L., Clough, G.W., Reyna, F. and Frost, J.D., 1991, Lique-faction response of San Francisco bayshore fills: Seismological Society of America, Bulletin, v. 81, p. 1998-2018.

Chang, G.J., 1994, First flush of stormwater pollutants investigated in Texas: Watershed Protection Techniques, v. 1, p. 88-90.

Chapman, D., 1992, ed., Water Quality Assessments: A Guide to the Use of Biota, Sediments and Water in Environmental Monitoring: Chapman and Hall, London, UK, 585 p.

Chapman, L.J. and Putnam, D.F., 1984, Physiography of Southern Ontario, Third Edition: Ontario Geological Survey, 270 p. with maps. [First Edition printed, 1951 by Ontario Research Foundation; Second Edition printed 1976 by University of Toronto Press, Toronto, ON]

Charlesworth, D.L., 1987, Development of a risk model for the Whitchurch-Stouffville health study: Hydrogeological report by MacLaren Plansearch Inc. (Lavalin) to Goss, Gilroy and Associates Ltd., 37 p.

Cherry, J.A., 1983, ed., Migration of contaminants in groundwater at a landfill: a case study: Journal of Hydrology, v. 63, p. 1-192.

Churcher, P.L., Johnson. M.D., Telford, P.G. and Barker, J.F., 1991, Stratigraphy and oil shale resource potential of the upper Ordovician Collingwood member, Lindsay Formation, southwestern Ontario: Ontario Geological Survey, Open File Report 5817, 98 p.

City of Brampton, 1992, Official Plan review: Environmental component, urban boundary Decision: 120 p.

City of Toronto, Housing Department, 1991, Ataratiri draft environmental evaluation study report: Toronto, ON, 51 p.

City of Toronto, Housing Department, 1992, Ataratiri, proposed cancellation: wind-down work plan: 3 p.

City of Toronto, Public Health Department, 1990, The quality of drinking water in Toronto: Summary of findings: 39 p.

City of Toronto, Public Health Department, 1993, State of the environment baseline report: 63 p.

Clarke, E.A., McLaughlin, J. and Anderson, T.W., 1991, Childhood leukemia around Canadian nuclear facilities, Phase II: Final report to the Atomic Energy Control Board of Canada, Ottawa, ON, 263 p.

Coakley, J.P., 1990, Contamination hazard from waste disposal sites near receding Great Lakes shorelines: Water Pollution Research Journal of Canada, v. 24, p. 81-100.

Coakley, J.P. and Karrow, P., 1994, Reconstruction of the post-Iroquois shoreline evolution in western Lake Ontario: Canadian Journal of Earth Sciences, v. 31, p. 1618-1629.

Cohen, M. and 7 others, 1995, Quantitative estimation of the entry of dioxins, furans and hexachlorobenzene into the Great Lakes from airborne and waterborne sources: Centre for the Biology of Natural Systems, Queens College, City University of New York, Flushing, NY, 102 p.

Cohen, S.J., 1986, Impacts of CO_2-induced climatic change on water resources in the Great Lakes basin: Climatic Change, v. 8, p. 135-153.

Coleman, A.P., 1932, The Pleistocene of the Toronto region: Ontario Department of Mines, v. XLI, p. 1-55.

Coleman, A.P. 1936, Lake Iroquois: Ontario Department of Mines, Annual Report, v. 45, pt. 7, p. 1-36.

Conestoga-Rovers & Associates, 1992, Hydrogeologic investigation Program, Aurora landfill: Canadian Landfill Division, 0181-80, 210 p.

Cooper, A.J., Funk, G.H. and Anderson, E.G., 1989, Using Quaternary stratigraphy to help locate a hazardous waste treatment site, *in* De Mulder, F.J. and Hageman, B.P., eds., Applied Quaternary Research: A.A. Balkema, Rotterdam, p. 1-14.

Crombie, D., 1990, Watershed: Royal Commission on the Future of the Toronto Waterfront, Interim report: Queen's Printer for Ontario, 207 p.

Crombie, D., 1994, Regeneration: Toronto's Waterfront and the Sustainable City: Royal Commission on the Future of the Toronto Waterfront, Final Report: Queen's Printer for Ontario, 506 p.

Crone, A.J., Machette, M.N. and Bowman, J.R., 1992, Geologic investigations of the 1988 Tennant Creek, Australia, earthquakes – Implications for paleoseismicity in stable continental regions: United States Geological Survey, Bulletin 2032-A, 52 p.

Crowley, T.E., II, 1990, Laurentian Great Lakes – Double CO_2 climate change hydrological impacts: Journal of Climatic Change, v. 17, p. 27-47.

Dales, R.E., Raizenne, M., El-Saadany, S., Brook, J. and Burnett, R., 1994, Prevalence of childhood asthma across Canada: International Journal of Epidemiology, v. 23, p. 775-781.

Dames and Moore, Canada, 1991 & 1992, Annual monitoring reports Brock West sanitary landfill site: prepared for Metropolitan Toronto Department of Works, 3 vols.

Daniels, S., 1990, Joint Orientations in south-central Ontario: unpublished M.Sc. thesis, McMaster University, Hamilton, ON, 76 p.

Davies, K., 1991, Towards ecosystem-based planning; a perspective on cumulative environmental effects: Royal Commission on the Future of the Toronto Waterfront, Working Paper 8, 106 p.

D'Andrea, M., Maunder, D.E. and Snodgrass, W.J., 1993, Characterization of stormwater and combined sewer overflows in Metropolitan Toronto: Stormwater Management Conference, Ontario Ministry of the Environment, 20 p.

D'Astous, A.Y., Ruland, W.W., Bruce, J.R.G., Cherry J.A. and Gillham, R., 1989, Fracture effects in the shallow groundwater zone in weathered Sarnia area clay: Canadian Geotechnical Journal, v. 26, p. 43-56.

Deacon, N.A., 1944, Geographic factors and land use in Toronto: Canadian Geographical Journal, v. XXIX, p. 80-99.

Debo, T.N. and Reese, A.J., 1995, Municipal Stormwater Management: Lewis Publishers, Boca Raton, FL, 756 p.

DeCooke, B.G. and Witherspoon, D.F., 1981, Terrestrial water balance, *in* Aubert, E.J. and Richards, T.L., eds., IFYGL: The International Field Year for the Great Lakes: National Oceanic and Atmospheric Administration, Ann Arbor, MI, p. 199-219.

Deloraine, A., Zmirou, D., Tillier, C. and Boucharlat, A., 1995, Case control assessment of the short-term health effects of an industrial toxic waste landfill: Environmental Research, v. 68, p. 124-133.

Desaulniers, D.E., Cherry, J.A. and Fritz, P., 1981, Origin, age and movement of porewater in argillaceous Quaternary deposits at four sites in southwestern Ontario: Journal of Hydrology, v. 50, p. 231-257.

Dewaele, P., Pehme, P. and Edelenbos, M., 1992, Applications of borehole geophysics to geochemical plume interpretation: International Association of Hydrogeologists, Hamilton meeting, Proceedings, p. 263-278.

Dillon, M.M., Ltd., 1994, Detailed assessment of the proposed site EE11 for Durham Region landfill site search: Interim Waste Authority Ltd., 546 p.

Duran, R.W. and Oliver, B.G., 1983, History of Lake Ontario contamination from the Niagara River by sediment radiodating and chlorinated hydrocarbon analysis: Journal of Great Lakes Research, v. 9, p.160-168.

Easton, R.M., 1992, The Grenville Province and the Proterozoic history of central and southern Ontario: Geology of Ontario: Ontario Geological Survey, Special Volume 4, Part 2, p. 714-904.

Easton, R.M. and Ford, F.D., 1991, Geology of the Mazinaw area: Ontario Geological Survey, Miscellaneous Paper 157, p. 95-106.

Edelstein, M.R., 1988, Contaminated Communities: The Social and Psychological Impacts of Residential Toxic Exposure: Westview Press, Boulder, CO, 217 p.

Edwards, T.W.D. and Fritz, P., 1988, Stable isotope paleoclimate records for southern Ontario: Comparison of results from marl and wood: Canadian Journal of Earth Sciences, v. 25, p. 1397-1406.

Engelder, T.A., 1994, Brittle crack propagation, *in* Hancock, P.L., ed., Continental Deformation: Pergamon Press, Oxford, UK, p. 43-52.

Environment Canada, 1990, The Climate of Metropolitan Toronto: Climatological Studies 41, The Climate of Canadian Cities, No. 5, 139 p.

Environment Canada, 1992, The State of Canada's Climate: Temperature Change in Canada 1895-1991: State of Environment Report No. 92-2, 36 p.

Environment Canada, 1993, Air Quality Trends in Canadian Cities 1979-1992: 5 p.

Environment Canada, 1994, Great Lakes water levels. Fact sheet: 3 p.

Environment Canada, 1995a, State of the Great Lakes: EN40-11/35-- 1995, 56 p.

Environment Canada, 1995b, Summary report of the National Pollutant Release Inventory for 1993: Ministry of Supply and Services Canada, EN40-495-1/1995E, 119 p.

Environment Canada, 1995c, First progress report under the 1994 Canada-Ontario agreement respecting the Great Lakes ecosystem: 34 p.

Environment Canada, 1995d, Trends in Contaminant Levels in the Niagara River: State of the Environment Fact Sheet No. 93-2, 12 p.

Environmental Assessment Board, 1989, Town of Cobourg: Decision EP-89-01, 60 p.

Environmental Assessment Board & The Joint Board (Consolidated Hearings Act), 1994, Ontario Waste Management Corporation: Decision CH-87-02, 12 chapters.

Environmental Assessment Board & The Joint Board (Consolidated Hearings Act), 1995, Steetley-South Quarry landfill site decision and reasons for decision: 215 p.

Eslinger, E., Oko, U., Smith, J.A. and Holliday, G.H., 1994, Introduction to Environmental Hydrogeology: Society for Sedimentary Geology, Tulsa, OK, Short Course 32, 136 p.

Estrin, D. and Swaigen, J., 1993, Environment on Trial: A Guide to Ontario Environmental Law and Policy: Canadian Institute for Environmental Law and Policy, Emond Montgomery Publications Ltd., Toronto, ON, 909 p.

Eyles, N., 1987, Late Pleistocene depositional systems of Metropolitan Toronto and their engineering and glacial geological significance: Canadian Journal of Earth Sciences, v. 24, p. 1009-1021.

Eyles, N., 1993, Earth's glacial record and its tectonic setting: Earth Science Reviews, v. 35, p. 1-248.

Eyles, N. and Boyce, J., 1991, Earth science survey of the Rouge Valley Park: Ontario Ministry of Natural Resources, Central Region, Open File Report 9113, 82 p.

Eyles, N., Boyce, J. and Mohajer, A., 1993, Bedrock topography in the western Lake Ontario region: Evidence of reactivated basement structures?: Géographie Physique et Quaternaire, v. 47, p. 269-283.

Eyles, N. and Clark, B., 1988a, Last interglacial sediments of the Don Valley brickyard and their paleoenvironmental significance: Canadian Journal of Earth Sciences, v. 25, p. 1108-1122.

Eyles, N. and Clark, B., 1988b, Storm-influenced deltas and ice-scouring in a Late Pleistocene glacial lake: Geological Society of America, Bulletin, v. 100, p. 793-809.

Eyles, N. and Scheidegger, A.E., 1995, Environmental significance of bedrock jointing in southern Ontario: Environmental Geology, v. 26, p. 269-277.

Eyles, N. and Westgate, J.A., 1987, Restricted regional extent of the Laurentide Ice Sheet in the Great Lake basin during early Wisconsin glaciation: Geology, v. 15, p. 537-540.

Eyles, N. and Williams, N.E., 1992, The sedimentary and biological record of the last interglacial-glacial transition at Toronto, Canada, in Clark, P.U. and Lea, P.D., eds., The Last Interglacial-Glacial Transition in North America: Geological Society of America, Special Paper 270, p. 119-137.

Feenstra, B.H., 1981, Quaternary geology and industrial minerals of the Niagara-Welland Area: Ontario Geological Survey, Open File Report 5361, 66 p.

Field, R., O'Shea, M.L. and Chin, K.K., 1993, eds., Integrated Stormwater Management: Lewis Publishers, Boca Raton, FL, 383 p.

Finamore, P., 1985, Glacial Lake Algonquin and the Fenelon Falls outlet, in Karrow, P.F. and Calkin, P.E., eds., Quaternary Evolution of the Great Lakes: Geological Association of Canada, Special Paper 30, p. 125-132.

Fligg, K., 1983, Geophysical well log correlations between Barrie and the Oak Ridges Moraine: Water Resources Branch, Ontario Ministry of the Environment, Map 2273.

Flint, R.W. and Vena, J., 1991, Human Health Risk from Chemical Exposure: The Great Lakes Ecosystem: Lewis Publishers, Boca Raton, FL, 314 p.

Free, B.M. and Mulamoottil, G.G., 1983, The limnology of Lake Wabukayne, a storm water impoundment: Water Research Bulletin, v. 19, p. 821-827.

Gargett, A., 1965, Long-term fluctuations in the Toronto temperature and precipitation record: Department of Transport, Meteorological Branch, Circular 4199, 21 p.

Gartner Lee Ltd., 1977, Hydrogeological Study final report: Central Britannia Road landfill for Region of Peel: Project No. 78-68, 89 p.

Gartner Lee Ltd., 1987a, Ontario Waste Management Corporation; Site assessment Phase 4B: Geology, Hydrogeology and geotechnics-baseline conditions, v. 1, 112 p.

Gartner Lee Ltd., 1987b, Ontario Waste Management Corporation; Site assessment Phase 4B: Geotechnical input to facilities design: 67 p.

Gartner Lee Ltd., 1987c, Ontario Waste Management Corporation; Site assessment Phase 4B: Potential Impacts on groundwater from the landfill: 125 p.

Gartner Lee Ltd., 1990, Beare Road site closure report volume 1, a review of monitoring data at Beare Road landfill: Report GLL90-128, 89 p.

Gartner Lee Ltd., 1992, Britannia Landfill expansion study, v. 2B, 2C: 210 p.

Gartner Lee Ltd., 1993a, Beare Landfill 1992 Monitoring report: Report GLL 91-204, 87 p.

Gartner Lee Ltd., 1993b, Subwatershed Study (part ii). A functional ecosystem approach: City of Vaughan Planning Department, 110 p.

Gentilcore, R.L., 1993, ed., Historical Atlas of Canada, v. III; The Land Transformed 1800-1891: University of Toronto Press, Toronto, ON, 184 p.

Geomatics International, 1991, Seaton Lands Environmental hazard study – geology and groundwater: prepared for Ontario Ministry of Housing, 97 p.

Gilman, A.P., Mao, Y., Burnett, R. and Semenciw, R., 1992, Linking administrative databases relating to cancer incidence and environmental contaminants in the Great Lakes Basin: Chronic Diseases in Canada, v. 13, p. S15-22.

Gish, T.J., Isensee, A.R., Nash, R.G. and Helling, C.S., 1992, Impact of pesticides on shallow groundwater quality: American Society of Agricultural Engineers, Transactions, v. 34, p. 1745-1753.

Golder and Associates Ltd., 1987, Review of Hydrogeology and assessment of existing designs, Brock North and South landfills: Report to Metropolitan Toronto, 71 p.

Golder and Associates Ltd., 1990, Field investigation location of waste storage pits, Port Granby Waste management facility: Report to Cameco Fuel Services, Port Hope, 121 p.

Golder and Associates Ltd., 1991, Investigation and Phase I remediation of the Brimley Road landslide for Scarborough Works and Environment Department: Report 911-1336, 44 p.

Golder and Associates Ltd., 1994, Detailed Assessment of the proposed site C-34B, Peel Region, Appendix C – Geology/hydrogeology: Interim Waste Authority Ltd., 135 p.

Golder and Associates Ltd., 1995, Geotechnical design report for twin tunnels, Sheppard Avenue East route: Report S-G3R3-D-3, 120 p.

Gore and Storrie Ltd., 1992, Town of Markham Natural features study. Phase I: Town of Markham Planning Department, 52 p.

Goss, Gilroy and Associates Ltd., 1987, A Community health study of the residents of Whitchurch-Stouffville: 119 p.

Government of Canada, 1995, The Great Lakes – An Environmental Atlas and Resource Book, 3rd Edition, Ottawa, ON.

Grabau, A.W., 1901, Guide to the geology and paleontology of Niagara Falls and vicinity: New York State Museum, Bulletin 45, p.1-284.

Grasso, D., 1993, Hazardous Waste Site Remediation: Lewis Publishers, Boca Raton, FL, 520 p.

Greater Toronto Area Task Force, 1996, Letter of Transmittal and Recommendations: Ontario Ministry of Municipal Affairs and Housing, 58 p.

Gundersen, L.C.S. and Wanty, R.B., 1991, Field studies of radon in rocks, soil and water: United States Geological Survey, Bulletin 1971, 334 p.

Gwyn, Q.H.J., 1972, Quaternary geology of the Alliston-Newmarket area, southern Ontario: Ontario Department of Mines, Miscellaneous Paper 53, p. 144-147.

Gwyn, Q.H.J., 1976, Quaternary geology resources of the western part of the municipality of Durham, southern Ontario: Ontario Division of Mines, Open File Report 5161, map with marginal notes.

Haefeli, C.J., 1972, Groundwater Inflow into Lake Ontario from the Canadian Side: Department of Energy, Mines and Resources, Inland Waters Branch, Scientific Series 9, 101 p.

Hartig, J.H. and Zarull, M.A., 1991, Methods of restoring degraded areas in the Great Lakes: Review of Environmental Contamination and Toxicology, v. 117, p. 127-154.

Hartig, J.H. and Zarull, M.A., 1992, eds., Under RAPs, Towards Grassroots Ecological Democracy in the Great Lakes Basin: University of Michigan Press, Ann Arbor, MI, 192 p.

Hartmann, H.C., 1991, Climate change impacts on Laurentian Great Lakes: Journal of Climatic Change, v. 17, p. 49-67

Hays, J. D., Imbrie, J. and Shackleton, N., 1976, Variations in the earth's orbit: pacemaker of the ice ages: Science, v. 194, p. 1121-1132.

HBT AGRA Ltd, 1994, Seaton Lands as a natural ecosystem: Ontario Ministry of Housing, TC 36232, 139 p.

Hendry, M.J., Cherry, J.A. and Wallick, E.I., 1986, Origin and distribution of sulfate in a fractured till in southern Alberta: Water Resources Research, v. 22, p. 45-61.

Hertzman, C., Hayes, M., Singer, J. and Highland, J., 1987, The Upper Ottawa landfill site health study: Environmental Health Perspectives, v. 75, p.173-195.

Hewitt, D.F., 1969, Industrial Mineral resources of the Markham-Newmarket area: Ontario Department of Mines, Industrial Mineral Report 24, 41 p.

Hewitt, D.F. and Yundt, S.E., 1971, Mineral resources of the Toronto-centred region: Ontario Department of Mines and Northern Affairs, Industrial Mineral Report 38, 67 p.

Hibbert, J., Mohajer, A.A. and Eyles, N., 1994, Faulting in unconsolidated sediments and bedrock east of Toronto, Phase 2: Atomic Energy Control Board, Report 2.263.2, 50 p.

Hicock, S.R. and Dreimanis, A., 1989, Sunnybrook drift indicates a grounded, early Wisconsin glacier in the Lake Ontario basin: Geology, v. 17, p. 169-172.

Hoff, R.M., 1994, An error budget for the determination of the atmospheric mass loading of toxic chemicals in the Great Lakes: Journal of Great Lakes Research, v. 20, p. 229-239.

Hoff, R.M., Muir, D.C.G. and Grift, N.P., 1992, Annual cycle of polychlorinated biphenyls and organohalogen pesticides in air in southern Ontario. 2. Atmospheric transport and sources: Environmental Science and Technology, v. 26, p. 276-283.

Hoffman, P.F., 1989, Precambrian geology and tectonic history of North America, in Bally, A.W. and Palmer, A.R., eds., The Geology of North America: An Overview: Geological Society of America, p. 447-512.

Holysh, S., 1995, Halton Region Aquifer management plan. Phase 1 report, Background hydrogeology: Regional Municipality of Halton, 90 p.

Houghton, J., 1994, Global Warming: The Complete Briefing: Lion Publishing, Oxford, UK, 192 p.

Hovinga, M.E., Sowers, M. and Humphrey, H.E.B., 1993, Environmental exposure and lifestyle predictions of lead, cadmium, PCB, and DDT levels in Great Lakes fish levels: Archives of Environmental Health, v. 48, p. 98-104.

Howard, K.W.F. and Beck, P., 1988, Hydrochemical interpretation of groundwater flow systems in Quaternary sediments of southern Ontario: Canadian Journal of Earth Sciences, v. 23, p. 938-947.

Howard, K.W.F., Eyles, N. and Livingstone, S., 1996, Municipal landfilling practice and its impact on groundwater resources in and around urban Toronto, Canada: Hydrogeology Journal, v. 4, p. 64-79.

Hydrology Consultants Ltd., 1976, Detailed terrain constraints survey of the urban area of the North Pickering Planning area: Ministry of Housing, North Pickering Development Corporation, 126 p.

Hydrology Consultants Ltd., 1979, Hydrogeological Investigation and proposed remedial works for control of leachate at the Beare Road Sanitary landfill: File 8742, 110 p.

Imbrie, J. and Imbrie, K.P., 1976, Ice Ages: Solving the Mystery: McMillan Press, New York, 224 p.

Intera-Kenting, 1990, The hydrogeological significance of the Oak Ridges Moraine: prepared for the Greater Toronto Greenlands Strategy, 82 p.

International Joint Commission, 1994, Canada-United States Air Quality Agreement: Progress Report EN40-388, 64 p.

International Joint Commission, 1995, Priorities and progress, 1993-5, under the Great Lakes Water Quality Agreement: 184 p.

Jacobs, P. and Sadler, B., eds., 1992, Sustainable development and environmental assessment: Perspectives on planning for a common future: Canadian Environmental Assessment Resarch Council, Ottawa, ON, 184 p.

Jakimchuk, R.D., 1992, The role of environmental assessment in sustainable development, in Jacobs, P. and Sadler, B., eds, Sustainable Development and Environmental Assessment: Perspectives on Planning for a Common Future: Canadian Environmental Assessment Resarch Council, Ottawa, ON, p. 81-92.

James, T.S. and Morgan, W.J., 1990, Horizontal motions due to postglacial rebound: Geophysical Research Letters, v. 17, p. 957-960.

Johnson, K., Rouleau, J. and Stewart, C., 1992, Atlas I: Birth Defects Atlas of Ontario: 1978-1988: Laboratory Centre for Disease Control, Health Canada, Ottawa, ON, 21 p.

Johnson, L.E., 1973, History of the County of Ontario, 1615-1875: The Corporation of the County of Ontario, Whitby, ON, 386 p.

Johnson, M.D., Armstrong, D.K., Sanford, B.V., Telford, P.G. and Rutka, M.A., 1992, Paleozoic and Mesozoic geology of Ontario: Geology of Ontario, Ontario Geological Survey, Special Volume 4, Part 2, p. 907-1010.

Johnson, W.A., 1916, The Trent Valley outlet of Lake Algonquin and the deformation of the Algonquin water plane in the Lake Simcoe district, Ontario: Geological Survey of Canada, Museum Bulletin 23, 27 p.

Johnson Sustronk Weinstein and Associates Ltd., 1980, Beare Road Landfill: Report 80-06, 72 p.

Johnston, A.C., et al., 1994, The Earthquakes of Stable Continental Regions: Assessment of Large Earthquake Potential: Electric Power Research Institute (EPRI), Report TR-102261, Palo Alto, CA, 335 p.

Jouzel, J. and 16 others, 1993, Extending the Vostok ice-core record of paleoclimate to the penultimate glacial period: Nature, v. 364, p. 407-412.

Kappel, W.M. and Tepper, D., 1992, An overview of the recent United States Geological Survey study of the hydrogeology of the Niagara Falls area of New York: International Association of Hydrogeologists, Hamilton Meeting, Proceedings, p. 609-622.

Karrow, P.F., 1969, Stratigraphic studies in the Toronto Pleistocene: Geological Association of Canada, Proceedings, v. 20, p. 4-16.

Karrow, P.F., 1989, Quaternary geology of the Great Lakes sub-region: Quaternary Geology of Canada and Greenland, Geological Survey of Canada, v. 1, p. 326-350 [a Geology of Canada/GSA Geology of North America volume].

Kelly, R.I. and Martini, I.P., 1986, Pleistocene glaciolacustrine deltaic deposits of the Scarborough Formation, Ontario, Canada: Sedimentary Geology, v. 47, p. 27-52.

Kennedy, B., 1979, Hurricane Hazel: Macmillan Co. of Canada, Toronto, ON, 176 p.

Kettles, I.M. and Rodriques, C.G., 1993, Evaluation of Glacial Lake Iroquois shoreline data from south-central and eastern Ontario: Geological Survey of Canada, Paper 93-1E, p. 271-274.

Kerr, D. and Spelt, J., 1965, The changing face of Toronto: Department of Mines and Technical Surveys, Geographical Branch, 81 p.

Kerr, M. and Eyles, N., 1991, Storm-deposited sandstones (tempestites) and related ichnofossils of the Late Ordovician Georgian Bay Formation, southern Ontario, Canada: Canadian Journal of Earth Sciences, v. 28, p. 266-282.

Ketcheson, J.W. and Webber, L.R., 1978, Effects of soil erosion on yield of corn: Canadian Journal of Soil Science, v. 58, p. 459-463.

King, K.S., Quigley, R.M., Fernandez, F., Reades, D.W. and Bacopoulos, A., 1993, Hydraulic conductivity and diffusion monitoring of the Keele Valley Landfill liner, Maple, Ontario: Canadian Geotechnical Journal, v. 30, p. 124-134.

Kite, G.W., 1972, An engineering study of crustal movement around the Great Lakes: Department of the Environment, Inland Waters Branch, Technical Bulletin 63, 56 p.

Kock, I., 1994, Nuclear hazard report: Durham Nuclear Awareness Project, Oshawa, ON, 63 p.

Koester, C.J. and Hites, R.A., 1992, Photodegradation of polychlorinated dioxins and dibenzofurans adsorbed on fly ash [abstract]: Environmental Science and Technology, v. 26, p. 502.

Kolmer, J.R., 1992, Quantitative analysis of DNAPL migration at Love Canal, Niagara Falls, New York: International Association of Hydrogeologists, Hamilton meeting, Proceedings, p. 516-517.

Kraus, N., Malmfors, T. and Slovic, P., 1991, Intuitive toxicology: expert and lay opinion of chemical risks: Risk Analysis, v. 12, p. 215-232.

Krumbein, W.C., 1955, Statistical analysis of facies maps: Journal of Geology, v. LXIII, p. 453-70.

Kumarapeli, P.S. and Saull, V.A., 1966, The St. Lawrence valley system: A North American equivalent of the East African rift system: Canadian Journal of Earth Sciences, v. 3, p. 639-658.

Lajtai, E.Z., 1969, Stratigraphy of the University subway, Toronto, Canada: Geological Association of Canada, Proceedings, v. 20, p. 17-23.

Lehmann, D., Brett, C.E., Cole, R. and Baird, G., 1995, Distal sedimentation in a peripheral foreland basin: Ordovician black shales and associated flysch of the western Taconic foreland, New York State and Ontario: Geological Society of America, Bulletin, v. 107, p. 708-724.

Legget, R.F. and Schriever, W.R., 1960, Site investigation for Canada's first underground railway: Civil Engineering and Public Works Review, v. 55, p. 73-77.

Lemay, M. and Mulamoottil, G., 1981, A limnological survey of eight waterfront marshes: Urban Ecology, v. 1, p. 55-67.

Lewis, C.F.M. and Sly, P.G., 1971, Seismic profiling of the Toronto waterfront area of Lake Ontario: International Association of Great Lakes Research, 14th Conference on Great Lakes Research, p. 303-354.

Livingstone, S.J., 1993, Assessment of the impact on Lake Ontario of groundwater contaminant mass loading from a representative region of the Greater Toronto Area: unpublished M.Sc thesis, Scarborough Campus, University of Toronto, Toronto, ON, 177 p.

Lorenz, J.C., Teufel, L.W. and Warpinski, N.R., 1991, Regional fractures I: A mechanism for the formation of regional fractures at depth in flat lying reservoirs: American Association of Petroleum Geologists, Bulletin, v. 75, p. 1714-1737.

Lubin, J.H., 1992, Lung cancer and exposure to residential radon: American Journal of Epidemiology, v. 140, p. 323-32.

Macdonald, C.R. and Metcalfe, C.D., 1991. Concentration and distribution of PCB congeners in isolated Ontario lakes contaminated by atmospheric deposition: Canadian Journal of Fisheries and Aquatic Science: v. 48, p. 371-381.

MacDonald, N.W., Burton, A.J., Jurgensen, M.P., McLughlin, J.W. and Mroz, G., 1991, Variation in forest soil properties along a Great Lakes air pollution gradient: Soil Science Society of America, v. 55, p. 1709-1715.

MacKenzie, W.R., and 10 others, 1994, A massive outbreak in Milwaukee of cryptosporidium infection transmitted through the public water supply: New England Journal of Medicine, v. 331, p. 161-167.

MacLaren, J.F., Ltd., 1977, Environmental Impact assessment of the Port Granby project: Prepared for Eldorado Nuclear Limited, 3 volumes.

MacViro Consultants Inc., 1991, Ataratiri PCB waste management strategy report: Report 3122-10, 66 p.

Maddox, J., 1995, Sustainable development unsustainable: Nature, v. 374, p. 305.

Marsalek, J., 1990, Stormwater management technology: Recent developments and experience: Urban Water Infrastructure, v. 217, p. 239-246.

Marsalek, J. and Ng, J., 1989, Evaluation of pollution loadings from urban non-point sources: methodology and applications: Journal of Great Lakes Research, v. 15, p. 444-451.

Marsalek, J. and Schroeter, H., 1988, Annual loadings of toxic contaminants in urban runoff from the Canadian Great Lake basin: Water Pollution Research Journal of Canada, v. 23, p. 360-378.

Marshall Macklin Monaghan Ltd., 1974, Site investigation for route selection, York/Pickering Trunk Sewer: 40 p.

Marshall Macklin Monaghan Ltd., 1992a, Don River Hydrology and hydraulics update: prepared for Metropolitan Toronto and Region Conservation Authority, 31 p.

Marshall Macklin Monaghan Ltd., 1992b, Hydrogeology of the Bramalea Residential 7, 8, & 9 site: Draft plan of subdivision 18T-91019.

Mayall, K.M., 1939, The natural resources of King Township, Ontario: Royal Canadian Institute, Transactions, v. XXII, p. 217-258.

McAndrews, J.A.H. and Power, J.H., 1973, Palynology of the Great Lakes: The surficial sediments of Lake Ontario: Canadian Journal of Earth Sciences, v. 10, p. 772-792.

McCarthy, F.M.G. and McAndrews, J.A.H., 1988, Water levels in Lake Ontario 4230-2000 years BP: evidence from Grenadier Pond, Toronto, Canada: Journal of Paleolimnology, v. 1, p. 99-113.

McCarthy, J.H., Jr. and Reimer, G.M., 1986, Advances in soil gas geochemical exploration for natural resources: some current examples and practices: Journal of Geophysical Research, v. 91, p. 12,327-12,338.

McKay, L.D. and Cherry, J.A., 1992, Groundwater research in clay-rich glacial tills in southwestern Ontario: International Association of Hydrogeologists, Hamilton, Ontario, meeting, Proceedings, p. 477-500.

McKay, L.D., Cherry, J.A. and Gillham, R.W., 1993, Field experiments in a fractured clay till 1. Hydraulic conductivity and fracture aperture: Water Resources Research, v. 29, p. 1149-1162.

Metropolitan Toronto and Region Conservation Authority (MTRCA), 1980, A history of flooding in the Metropolitan Toronto and Region watersheds, 6 p. and tables.

Metropolitan Toronto and Region Conservation Authority, 1991, Lake Ontario Waterfront Regeneration Project 1991-4: 82 p.

Metropolitan Toronto and Region Conservation Authority, 1992a, Don River watershed: state of the ecosystem: 162 p.

Metropolitan Toronto and Region Conservation Authority, 1992b, Keating Channel environmental monitoring program 1989-1990 report: 106 p.

Metropolitan Toronto and Region Conservation Authority, 1994, Forty steps to a new Don: Report of the Don Watershed Task Force, 10 p.

Metropolitan Toronto & Region Remedial Action Plan (MTRRAP), 1990, Strategies for restoring our waters: City of Toronto, 86 p.

Metropolitan Toronto & Region Remedial Action Plan (MTRRAP), 1991, Draft discussion paper: Review of the Public Advisory Committee, 36 p.

Metropolitan Toronto & Region Remedial Action Plan (MTRRAP), 1994, Clean waters, clear choices: Recommendations for action: City of Toronto, 108 p.

Metropolitan Toronto Works Department, 1994, Operation of Keele Valley landfill: Annual Report, 53 p.

Meyer, P.B., Williams, R.H. and Yount, K.R., 1995, eds., Contaminated Land: Reclamation, Redevelopment and Re-Use in the United States and the European Union: Edward Elgar Publishing, Brookfield, VT, 176 p.

Milkereit, B., Forsyth, D.A., Green, A.G., Davidson, A., Hanmer, S., Hutchinson, D.R., Hinze, W.J. and Mereu, R.F., 1992, Seismic images of a Grenvillian terrane boundary: Geology, v. 20, p. 1027-1030.

Miller, D.S. and Duddy, I.R., 1989, Early Cretaceous uplift and erosion of the northern Appalachian Basin, New York, based on apatite fission track analysis: Earth and Planetary Science Letters, v. 93, p. 35-49.

Miller, W.G., 1913, Map of Toronto and vicinity: Ontario Department of Lands, Forests and Mines, 1:63,360.

Mills, C. and Semenciw, R., 1992, Atlas II: Cancer incidence In the Great Lakes region of Ontario: 1984-1988: Laboratory Centre for Disease Control, Health Canada, Ottawa, ON, 46 p.

Mohajer, A., 1993, Seismicity and seismotectonics of the western Lake Ontario region: Géographie physique et Quaternaire, v. 47, p. 353-362.

Mohajer, A., Boyce, J.I. and Eyles, N., 1995, Shallow seismic reflection investigation of bedrock and Quaternary strata in the Rouge River valley, Toronto: Atomic Energy Control Board, Report INFO-0557, 26 p.

Mohajer, A., Eyles, N. and Rogojina, C., 1992, Neotectonic faulting in Metropolitan Toronto: Implications for earthquake hazard assessment in the Lake Ontario Region: Geology, v. 20, p. 1003-1006.

Moorhead, W.P.R., Guasparini, R., Donovan, C.A., Mathias, R.G., Cottle, R., and Bayhalen, G., 1990, Giardiasis outbreak from a chlorinated water supply: Canadian Journal of Public Health, v. 81, p. 358-362.

Morner, N-A., 1971, Eustatic changes during the last 20,000 years and a method of separating the isostatic and eustatic factors in an uplifted area: Palaeogeography, Palaeoclimatology, Palaeoecology, v. 9, p. 153-181.

Morris, R.D., Audet, A., Angelillo, I.F., Chalmers, T.C. and Mosteller, F., 1992, Chlorination, chlorination by-products and cancer: American Journal of Public Health, v. 82, p. 955-963.

Morris, T.F., McAndrews, J.H. and Seymour, K.L., 1993, Glacial Lake Arkona-Whittlesey transition near Leamington, Ontario: Geology, plant and muskox fossils: Canadian Journal of Earth Sciences, v. 30, p. 2436-2447.

Morton, J., Ltd., 1973, Interim report on ground classification related to foundation characteristics: North Pickering Community Development Project, 86 p.

Muller, E. and Pair, D.L., 1992, Evidence for large-scale subglacial meltwater flood events in southern Ontario and northern New York State: Geology, v. 20, p. 90-91.

Munson, W.E, 1990, Soil contamination and port redevelopment in Toronto: Royal Commission on the Future of the Toronto Waterfront: Working paper of the Canadian Waterfront Resource Centre, no. 3, 13 p.

Munson, W.E., 1991, The disposal of coal ash at Toronto's outer harbour: Working paper of the Canadian Waterfront Resource Centre, no. 7, 37 p.

Murdie, R.A., 1969, Factorial ecology of Metropolitan Toronto 1951 to 1961: University of Chicago, Department of Geography, Research Paper 116, 212 p.

Murphy, E., 1992, Nitrate in drinking water wells in Burlington and Mercer counties, New Jersey: Journal of Soil and Water Conservation, v. 47, p. 183-187.

National Research Council, 1993, Solid Earth Sciences and Society: National Academy Press, Washington, DC, 346 p.

National Research Council, 1994, Groundwater Cleanup Alternatives: National Academy Press, Washington, DC, 315 p.

National Wildlife Federation and the Canadian Institute for Environmental Law and Policy, 1991, A prescription for a healthy Great Lakes: National Wildlife Federation, Washington, DC, 102 p.

Nelson, J.G., 1991, Urbanization, conservation and sustainable development: The case of Frenchman's Bay, Toronto, Ontario: Heritage Resources Centre, University of Waterloo, Waterloo, ON, Technical Paper 5, 103 p.

Neuzil, C.E. and Pollock, D.W., 1983, Erosional unloading and fluid pressures in hydraulically "tight" rocks: Journal of Geology, v. 91, p. 179-193.

Nigge, K.M., 1995, Seismotectonic Boundary Work: A Case Study of Seismic Hazard Assessment in the Regulation of Nuclear Energy in Canada: Unpub. M.Env.Sc. thesis, York University, Toronto, ON, 156 p.

Noor, I., Novakowski, K.S. and Egden, J., 1992, Soil gas surveys as a tool for delineating pathways of natural hydrocarbon loading in southern Ontario: Implications for shallow groundwater: International Association of Hydrogeologists, Hamilton, Ontario meeting, Proceedings, p. 709-723.

Oak Ridges Moraine Aggregate Resources Study, 1994, Oak Ridges Moraine: Technical Working Committee, Background Study 10, 51 p.

Oberts, G.L., 1994, Influence of snowmelt dynamics on stormwater runoff quality: Watershed Protection Techniques, v. 1, p. 55-61.

Oerlemans, J. and van der Veen, C.J., 1984, Ice sheets and Climate: D. Reidel Publishing Co., Dordrecht, The Netherlands, 245 p.

Ontario Commission on Planning and Development Reform, 1993, New Planning for Ontario: Toronto, ON, 207 p.

Ontario Geological Survey, 1992, Geology of Ontario: Special Volume 4, 2 volumes (1525 p.) and 34 maps.

Ontario Ministry of Environment, 1980, Hydrogeologic environments and susceptibility of groundwater to contamination: Map S-100, scale 1:1,000,000.

Ontario Ministry of Environment, 1989, Guidelines for decommissioning and cleanup of sites in Ontario: 89 p.

Ontario Ministry of Environment and Energy, 1991, Interim stormwater quality control guidelines for new development: 67 p.

Ontario Ministry of Environment and Energy, 1992, Proposed policy for management of excesss soil, rock and like materials: 153 p.

Ontario Ministry of Environment and Energy, 1993, Rationale for the development of soil, drinking water and air quality criteria for lead: Hazardous Contaminants Branch, 16 p.

Ontario Ministry of Environment and Energy, 1994a, Proposed guidelines for the clean-up of contaminated sites in Ontario: 57 p.

Ontario Ministry of Environment and Energy, 1994b, Stormwater management practices planning and design manual: 260 p.

Ontario Ministry of Environment and Energy, 1994c, Niagara Escarpment Plan: 80 p.

Ontario Ministry of Environment and Energy, 1994d, An introduction to communal Sewage Systems: 27 p.

Ontario Ministry of Environment and Energy, 1994e, Water management on a watershed basis; Implementing an ecosystem approach: 32 p.

Ontario Ministry of Environment and Energy, 1994f, Subwatershed planning: 38 p.

Ontario Ministry of Municipal Affairs, 1995, Amendments to the Planning Act, Comprehensive Policy Statements and Implementation Guidelines: 245 p.

Ontario Ministry of Natural Resources, 1992, From pits to playgrounds: Aggregate extraction and pit rehabilitation in Toronto: 23 p.

Ontario Ministry of Transportation, 1990, Highway 407 – Route planning and environmental assessment study: Technical Paper T11547/53425, 30 p.

Ontario Waste Management Corporation, 1988a, Alternative Methods; Site Selection: v. III, 256 p.

Ontario Waste Management Corporation, 1988b, Executive summary of environmental assessment for a waste management system: 43 p.

O'Shaughnessy, V. and Garga, V.K., 1994, The hydrogeological and contaminant transport properties of fractured Champlain Sea clay in eastern Ontario. Part I. Hydrogeological properties: Canadian Geotechnical Journal, v. 31, p. 885-901.

Ostry, R.C., 1979a, The hydrogeology of the IFYGL Duffins Creek study area: Ministry of the Environment, Water Resources Report 5c, 87 p.

Ostry, R.C., 1979b, The hydrogeology of the IFYGL Forty Mile and Oakville Creeks study areas: Ministry of the Environment, Water Resources Report 5b, 101 p.

Otton, J.K., 1992, The geology of radon: United States Geological Survey, Reston, VA, 29 p.

Owen, E.B., 1969, Stratigraphy and engineering description of the soils exposed on a section of the Welland Canal by-pass project, Ontario, Canada: Geological Survey of Canada, Paper 69-31, 21 p.

Owen, E.B., 1972, Geology and engineering description of the soils of the Welland-Port Colborne area: Geological Survey of Canada, Paper 71-49, 32 p.

Park, R.G. and Jaroszewski, W., 1994, Craton tectonics, stress and seismicity, in Hancock, P.L., ed., Continental Deformation: Pergamon Press, Oxford, UK, p. 200-222.

Parker, B.L., Gillham, R.W. nd Cherry, J.A., 1994, Diffusive disappearance of immiscible phase organic liquids in fractured geologic media: Ground Water, v. 32, p. 805-820.

Palmer, J.H.L. and Lo, K.Y., 1978, In situ stress measurements in some near surface rock formations: Thorold, Ontario: Canadian Geotechnical Journal, v. 13, p. 1-7.

Persaud, D., Lomas, T., Boyd, D. and Mathai, S.K., 1985, Historical development and quality of the Toronto waterfront sediments, Part 1: Ontario Ministry of the Environment, 51 p.

Persaud, D.R., Jaagumagi, A. and Hayton, A., 1991, Provincial sediment quality guidelines: Water Resources Branch, Ontario Ministry of the Environment, 21 p.

Pitt, R., Clark, S. and Palmer, K., 1994, Potential groundwater contamination from intentional and nonintentional stormwater infiltration: United States Environmental Protection Agency, Project Summary EPA/600/SR-94/051, 7 p.

Proctor and Redfern, Ltd., 1993a, Water supply and sewage treatment systems in the Oak Ridges Moraine area: Background Report 12, Oak Ridges Moraine Planning Study, 110 p.

Procter and Redfern, Ltd., 1993b, Pickering surplus lands: Federal Environmental and Review Process, Initial Assessment. Report EO: 93209, 46 p.

Proulx, I. and Farvolden, R.N., 1989, Analysis of the contaminant plume in the Oak Ridges aquifer: Ontario Ministry of the Environment. R.A.C. Project 261, 134 p.

Pulton, D.J., 1986, Toronto Waterfront general water quality, 1976-1983: Ontario Ministry of the Environment, 26 p.

Quigley, R.M., Matich, M.A.J., Horvath, R.G. and Hawson, H.H., 1971, Swelling clays in two slope failures at Toronto: Canadian Geotechnical Journal, v. 8, p. 417-424.

Quinlan, G.M. and Beaumont, C., 1984, Appalachian thrusting, lithospheric flexure and the Paleozoic stratigraphy of the eastern interior of North America: Canadian Journal of Earth Sciences, v. 21, p. 973-996.

Raven, K.G., Novakowski, K.S., Yager, R.M. and Haystee, R.C., 1992, Supernormal fluid pressures in sedimentary rocks of southern Ontario, western New York State: Canadian Geotechnical Journal, v. 29, p. 80-93.

Regional Municipality of Halton, 1995, Halton aquifer management plan, Phase 1: Background hydrogeology: 92 p.

Richards, P.A.L. and Thompson, C.D., 1989, Permeability protocol for the compacted clay liner at Metropolitan Toronto's Keele Valley landfill: Canadian Journal of Civil Engineering, v. 16, p. 552-559.

Richardson, A.H., 1944, The Ganaraska Watershed: Province of Ontario Printer, 248 p.

Richardson, A.H., 1974, Conservation by the People: The History of the Conservation Movement in Ontario to 1970: University of Toronto Press, Toronto, ON, 154 p.

Riggs, C.O., 1993, Soil exploration at contaminated sites, in Daniel, D.E., ed., Geotechnical Practice for Waste Disposal: Chapman and Hall, London, UK, p. 358-378.

Riley, J. and Mohr, P., 1994, The natural heritage of southern Ontario; A review of conservation and restoration ecology for land-use and landscape planning: Ontario Ministry of Natural Resources, Technical Report TR-001, 78 p.

Robertson, W.D. and Cherry, J. A., 1992, Persistence of nitrate in three septic system plumes on unconfined sands in Ontario: International Association of Hydrogeologists, Hamilton meeting, Proceedings, p. 541-555.

Rodionov, S.N., 1994, Association between winter precipitation and water level fluctuations in the Great Lakes and atmospheric circulation patterns: Journal of Climate, v. 7, p. 1693-1706.

Rogojina, C., 1993, Neotectonic bedrock joints and pop-ups in the Metropolitan Toronto area, Toronto: unpublished M.Sc. thesis, University of Toronto, Toronto, ON, 46 p.

Rowney, A.C., Droste, R.L. and MacRae, C.R., 1986, Sediment and ecosystem characteristics of a detention lake receiving urban runoff: Water Pollution Research Journal of Canada, v. 21, p. 460-473.

Rudolph, D. and Goss, M., 1992, eds., Ontario farm groundwater quality survey – Summer 1992: Agriculture Canada, 162 p.

Rudolph, D. and Goss, M., 1993, eds., Ontario farm groundwater quality survey-Winter 1991/2: Agriculture Canada, 151 p.

Ruland, W.W., Cherry, J.A. and Feenstra, S., 1991, The depth of fractures and active groundwater flow in a clayey till plain in southwestern Ontario: Ground Water, v. 29, p. 405-417.

Rutty, A.L. and Cruden, A.R., 1993, Pop-up structures and the fracture pattern in the Balsam Lake area, southern Ontario: Géographie physique et Quaternaire, v. 47, p. 379-388.

Salomons, W., 1985, Sediments and water quality: Environmental Technology Letters, v. 6, p. 315-326.

San Francisco Bay Conservation and Development Commission, 1969, San Francisco Bay Plan Supplement, 201 p.

Sanderson, M., 1987, Implications of climatic change for navigation and power generation in the Great Lakes: Environment Canada, Atmospheric Environment Service, Climate Change Digest, 20 p.

Sanford, B.V., 1993, Stratigraphic and structural framework of upper Middle Ordovician rocks in the Head Lake-Burleigh Falls area of south-central Ontario: Géographie physique et Quaternaire, v. 47, p. 253-268.

Sanford, B.V. and Smith, G.W., 1987, Paleozoic geology of the Hudson platform, in Beamont, C. and Tankard, A.J., eds., Basin Forming Mechanisms: Canadian Society of Petroleum Geologists, Memoir 12, p. 483-505.

Sanford, B.V., Thompson, F.J. and McFall, G., 1985, Plate tectonics – A possible controlling mechanism in the development of hydrocarbon traps in southwestern Ontario: Bulletin of Canadian Petroleum Geology, v. 33, p. 52-71.

Scheidegger, A.E., 1980, The orientation of valley trends in Ontario: Zeitschrift für Geomorphologie, v. 24, p. 19-30.

Scheidegger, A.E., 1993, Joints as neotectonic plate signatures: Tectonophysics, v. 219, p. 235-239.

Scheidegger, A.E., 1995, Geojoints and geostresses, in Rossmanith, H.R., ed., Second International Conference on the Mechanics of Jointed and Faulted Rock: A.A. Balkema, Rotterdam, p. 3-35.

Schindler, D.W. and 9 others, 1990, Effects of climatic warming on lakes of the central boreal forest: Science, v. 250, p. 967-970.

Schueler, T.P., 1987, Controlling urban runoff; a practical manual for planning and designing urban best management practices: Metropolitan Washington Council of Governments, Washington, DC, 157 p.

Schueler, T.P., 1994, Hydrocarbon hotspots in the urban landscape: Can they be controlled?: Watershed Protection Techniques, v. 1, p. 3-5.

Schroeter, H.O., 1993, Loadings of toxic contaminants from urban nonpoint sources to the Great Lakes from Ontario communities: Stormwater Management and Combined Sewers: Ontario Ministry of Environment and Energy, Proceedings, p. 69-88.

Schwarcz, H. and Eyles, N., 1991, Laurentide Ice Sheet extent inferred from isotopic composition (O,C) of ostracods at Toronto, Canada: Quaternary Research, v. 35, p. 305-320.

Seeber, L. and Armbruster, J.G., 1993, Natural and induced seismicity in the Lake Erie-Lake Ontario region: Reactivation of ancient faults with little neotectonic displacement: Géographie Physique et Quaternaire, v. 47, p. 363-378.

Sellinger, C.E., 1995, Groundwater flux into a portion of eastern Lake Michigan: Journal of Great Lakes Research, v. 21, p. 53-63.

Servizi, J.A. and Martens, D.W., 1991, Effect of temperature, season and fish size on acute lethality of suspended sediments to Coho salmon: Canadian Journal of Fisheries and Aquatic Science, v. 48, p. 493-497.

Sharpe, D.R., 1980, Quaternary geology of Toronto and surrounding area: Ontario Geological Survey, Preliminary Map P2204.

Sherwood Lollar, B., Weise, S.M., Frape, S.K. and Barker, J., 1994, Geochemical and isotopic trends in natural gas in southwest Ontario – implications for the origin of hydrocarbon gas and helium: Bulletin of Canadian Petroleum Geology, v. 42, p. 283-292.

Sibul, U., Wang, K.T. and Vallery, D., 1977, Groundwater resources of the Duffins Creek, Rouge River drainage basins: Ontario Ministry of the Environment, Water Resources Report 8, 109 p.

Sievers, D.M. and Fulhage, C.D., 1992, Survey of rural wells in Missouri for pesticides and nitrate: Groundwater Monitoring Review, v. 12, p. 142-150.

Singer, S.N., 1974, A hydrogeological study along the north shore of Lake Ontario in the Bowmanville-Newcastle area: Ministry of the Environment, Water Resources Report 5d, 76 p.

Slaine, D.D., Pehme, P. E., Hunter, J.A., Pullan, S.E. and Greenhouse, J.P., 1990, Mapping overburden stratigraphy at a proposed hazardous waste facility using shallow seismic reflection methods, in Ward, S.H., ed., Geotechnical and Environmental Geophysics: Society of Exploration Geophysics, v. 2, p. 273-280.

Sly, P.G., 1991, The effects of land use and cultural development on the Lake Ontario ecosystem since 1750: Hydrobiologia, v. 213, p. 1-75.

Smit, B., 1987, Implications of climatic change for agriculture in Ontario: Environment Canada, Atmospheric Environment Service, Climate Change Digest 87-02, Downsview, ON, 91 p.

Smith, J.B., 1991, The potential impacts of climate change on the Great Lakes: American Meteorological Society, Bulletin, v. 72, p. 21-28.

Snell, E.A., 1987, Wetland distribution and conversion in southern Ontario: Environment Canada, Inland Waters and Lands Directorate, Working Paper 48, 53 p.

Snodgrass, W.J., 1993, Dry water discharges to the Metropolitan Toronto waterfront: Queen's Printer for Ontario, Toronto, ON, 42 p.

Soo Chan, G., Scafe, M. and Emami, S., 1992, Cemeteries and groundwater: An examination of the potential contamination of groundwater by preservatives containing formaldehyde: Ontario Ministry of Environment and Energy, Report PIBS 1813, 14 p.

Spencer, J.W., 1890, Origin of the Great Lakes of America: American Geologist, v. 7, p. 86-97.

Statistics Canada, 1994, Human Activity and the Environment: Ottawa, 300 p.

Steedman, R.J., 1986, Historical streams of Toronto: Toronto Field Naturalist, v. 382, p. 15-18.

Stephenson, T.D., 1990, Fish reproduction utilization of coastal marshes of Lake Ontario near Toronto: Journal of Great Lakes Research, v. 16, p. 71-81.

Stewart, W., 1979, Paper Juggernaut: McClelland and Stewart, Toronto, ON, 207 p.

Stewart, I.S. and Hancock, P.L., 1994, Neotectonics, in Hancock, P.L., ed., Continental Deformation: Pergamon Press, Oxford, UK, p. 370-410.

Stone, R., 1993, EPA analysis of radon in water is hard to swallow: Science, v. 261, p. 1514-1516.

Straume, T., 1993, Tritium risk assessment: Health Physics, v. 65, p. 671-682.

Strus, R.H., 1994, Metro Toronto waterfront fish communities: Summary and assessment 1989-1993: Ministry of Natural Resources, Greater Toronto Area District, Maple, 49 p.

Talwani, P., 1988, The intersection model for intraplate earthquakes: Seismological Research Letters, v. 59, p. 305-310.

Taro Aggregates Ltd., 1995, Proposed East Quarry landfill environmental assessment.: Executive Summary and Volumes I and II.

Taylor, G., 1936, Topographic controls in the Toronto region: Canadian Journal of Economics, v. II, p. 493-511.

Taylor, S.M., and 7 others, 1991, Psychosocial impacts in populations exposed to solid waste facilities: Social Science and Medicine, v. 33, p. 441-447.

Taylor, S.M., Streiner, D., Eyles, J. and Elliot, S.J., 1993, Psychosocial effects in populations exposed to contaminants in the Great Lakes Basin: Institute of Environment and Health, McMaster University, Hamilton, ON, 62 p.

Technos Inc., 1983, Delineation of a contaminant plume at the proposed Keele Valley landfill: WMI Waste Management of Canada Ltd., Report 82-152, 67 p.

Testa, S.M., 1994, Geological Aspects of Hazardous Waste Management: Lewis Publishers, Boca Raton, FL, 536 p.

Thurston, G.D., and 8 others, 1994, The nature and origin of acid summer haze air pollution in Metropolitan Toronto: Environmental Research, v. 65, p. 254-270.

Thurston, P.C., 1991, Structural framework of Ontario, in Eastern Seismicity Source Zones for the 1995 Seismic Hazard Maps: Geological Survey of Canada, Open File Report 2437, 89 p.

Timar, G.S., 1988, Review of site assessment report (geology, hydrogeology and geotechnics): Ontario Waste Management Corporation, 9 p.

Todd, B.J. and Lewis, C.F.M., 1993, A reconnaissance geophysical survey of the Kawartha Lakes and Lake Simcoe, Ontario: Géographie physique et Quaternaire, v. 47, p. 313-324.

The Toronto Star, November 12, 1994, Sewage tunnel plans blocked by province: Toronto, ON.

The Toronto Star, July 16, 1995, Hope turns to fear: Toronto, ON.

Town of Caledon, 1994, Amendment No. 124 to the Official Plan: Caledon, ON, 40 p.

Township of Pickering, 1994, Backgrounder No. 2, Addressing the issues: Pickering Planning Department, 165 p.

Township of Pickering, 1995, Draft Official Plan, Pickering Planning Department, 198 p.

Transport Canada, 1991, Lester B. Pearson International Airport Environmental Impact Statement: Report TP 10675E, 31 p.

Trow Hydrology Consultants Ltd., 1986, Halton Region landfill technical report site D, Milton. Hydrogeologic Report, 3 volumes.

Trow, W. and Bradstock, J., 1972, Instrumented foundations for two 43 storey buildings on till, Metropolitan Toronto: Canadian Geotechnical Journal, v. 9, p. 290-303.

Trow, Dames and Moore, 1991a, Ataratiri soil management plan: prepared for City of Toronto Housing Department, 98 p.

Trow, Dames and Moore, 1991b, Ataratiri groundwater management plan: prepared for City of Toronto Housing Department, 83 p.

Tushingham, A.M., 1992, Postglacial uplift predictions and historical water levels of the Great Lakes: Journal of Great Lakes Research, v. 18, p. 440-445.

United States Environmental Protection Agency, 1993, Reduction of toxics loadings to the Niagara River from hazardous waste sites in the United States: A progress report: Report USEPA/NYS DEC, 38 p.

Varga, S., 1980, Progress on the preservation of Metro's natural areas, in Barrett, S.W. and Riley, J.L., eds., Protection of Natural Areas in Ontario: York University, Toronto, ON, Faculty of Environmental Studies, Working Paper 3, p. 144-156.

Varga, S., Jalava, J. and Riley, J.L., 1991, Ecological Survey of the Rouge Valley Park: Ontario Ministry of Natural Resources, Central Region, Open File Report, 282 p.

Walker, F.N., 1965, Sketches of Old Toronto: Longmans, Toronto, ON, 350 p.

Walker, G., 1994, Planning the future of rural Ontario: Structure planning in the Greater Toronto Area: The Great Lakes Geographer, v. 1, p. 53-66.

Wallach, J.L., 1990, Newly discovered geological features and their potential impacts on Darlington and Pickering: Atomic Energy Control Board, Technical Report INFO-0342, 20 p.

Wallach, J.L., 1995, Characteristics of the Niagara-Pickering and Georgian Bay linear zones and their implications for a large magnitude earthquake in the vicinity of the Darlington and Pickering Nuclear Power Plants, near Toronto: Atomic Energy Control Board, Workshop on Seismic Hazard Assessment in southern Ontario, Ottawa, ON, 19-21 June, 8 p.

Wallach, J.L. and Heginbottom, J.A., 1993, eds., Neotectonics of the Great Lakes Area: Géographie physique et Quaternaire, v. 47, no.3, 398 p.

Wallach, J.L., Mohajer, A.A., Bowlby, J.R., Pearce, M. and McKay, D.A., 1993, Pop-ups as geological indicators of earthquake-prone areas in intraplate eastern North America: Quaternary Proceedings, v. 3, p. 67-83.

Walsh, D.C. and Lafleur, R.G., 1995, Landfills in New York City 1844-1994: Ground Water, v. 33, p. 556-560.

Wanielista, M.P., 1990, Hydrology and Water Quality Control: John Wiley & Sons Inc., New York, 565 p.

Warren, L.A. and Zimmerman, A.P., 1994a, Suspended particulate grain size dynamics and their implications for trace metal sorption in the Don River: Aquatic Science, v. 56, p. 348-362.

Warren, L.A. and Zimmerman, A.P., 1994b, The influence of temperature and NaCl on cadmium, copper and zinc partitioning among suspended particulate and dissolved phases in an urban river: Water Research, v. 28, p. 1921-1931.

Warwick, W.F., 1980, Palaeolimnology of the Bay of Quinte, Lake Ontario: 2800 years of cultural influence: Department of Fisheries and Oceans, Bulletin 206, 118 p.

Waterfront Regeneration Trust, 1994, Ecosystem approach to shoreline management: Toronto, Ontario: 175 p.

Watt, A.K., 1954, Correlation of the Pleistocene geology as seen in the subway with that of the Toronto region: Geological Association of Canada, Proceedings, v. 6, p. 69-81.

Watt, A.K., 1968, Pleistocene geology and groundwater resources, Township of Etobicoke: Ontario Department of Mines, Geological Report 59, 50 p.

Weaver, T.R., Frape, S.K. and Cherry, J.A., 1995, Recent cross-formational fluid flow and mixing in the shallow Michigan Basin: Geological Society of America, Bulletin, v. 107, p. 697-707.

Weninger, J.M. and McAndrews, J.H., 1989, Late Holocene aggradation in the lower Humber River valley, Toronto, Ontario: Canadian Journal of Earth Sciences, v. 26, p. 1842-1849.

Whillans, T.H., 1982, Changes in marsh area along the Canadian shore of Lake Ontario: Journal of Great Lakes Research, v. 8, p. 570-577.

White, O.L., 1975, Quaternary geology of the Bolton Area, southern Ontario: Ontario Division of Mines, Report 117, 119 p.

Whitney, G.G., 1994, From Coastal Wilderness to Fruited Plain: A History of Environmental Change in Temperate North America From 1500 to the Present: Cambridge University Press, Cambridge, MA, 451 p.

Williams, N.D. and Eyles, N., 1995, Sedimentary and paleoclimatic controls on caddisfly (insecta:Trichoptera) assemblages during the last interglacial-to-glacial transition in southern Ontario: Quaternary Research, v. 43, p. 90-105.

Wills, J., Howell, L., McKay, L., Parker, B. and Walter, A., 1992, Smithville C.W.M.L. site: Characterization of overburden fractures and implications for DNAPL transport: International Association of Hydrogeologists, Hamilton meeting, Proceedings, p. 501-515.

World Resources Institute, 1993, The 1993 Information Please Almanac: Houghton Mifflin, New York, 315 p.

Young, P.M., 1994, A biophysical model for prehistoric archaeological sites in southern Ontario: Ontario Ministry of Transportation, Research and Development Branch, Report MAT-94-02, 73 p.

Yu, Z. and McAndrews, J.H., 1994, Holocene water levels at Rice Lake, Ontario, Canada: sediment, pollen and plant macrofossil evidence: The Holocene, v. 4, p. 141-152.

Zarull, M.A. and Mudroch, A., 1993, Remediation of contaminated sediments in the Laurentian Great Lakes: Reviews of Environmental Contamination and Toxicology, v. 132, p. 93-115.

Zoback, M.L., 1992, Stress field constraints on intraplate seismicity in eastern North America: Journal of Geophysical Research, v. 97, p. 11,761-11,782.

3. Living on Ice: Problems of Urban Development in Canada's North

Hugh M. French

Department of Geography and Geology, and Ottawa Carleton Geoscience Centre
University of Ottawa, P.O. Box 450, Station A, Ottawa, Ontario K1N 6N5

SUMMARY
Canada is a cold country. It is only along the maritime lowland fringes of the Pacific coast that snow and sub-freezing temperatures are rare. By contrast, those areas along the southern borders, where most of the Canadian population resides, are seasonally cold. In these regions, seasonal agriculture is possible, plant and animal productivity is high, and the constraints of cold can be temporarily forgotten during the summer months. Elsewhere, over the vast majority of the Canadian landmass, and certainly north of 60°N, the problems created by coldness persist throughout the year. Although there are few urban settlements in excess of 5000 people, these constraints dominate urban and socio-economic activities. The purpose of this chapter is to illustrate the nature of such constraints, paying particular attention to the character of perennially frozen substrates (permafrost), terrain disturbances caused by various types of construction activity, and problems associated with ground and surface waters.

INTRODUCTION
The severity of Canada's northern climate is best illustrated by the comparison of freezing degree-days (FDD) and growing degree-days (GDD) for some of Canada's largest cities with similar data for some of the more important northern settlements (Table 1). The northern localities all experience more than 3000 FDDs and fewer than 1000 GDDs. The one exception is Whitehorse, Yukon, because of its mountain location.

An important consequence of the long period of winter cold and the relatively short period of summer thaw in northern Canada is the formation of a layer of frozen ground that does not completely thaw during the summer. This perennially frozen ground is termed permafrost, a word first coined by Muller (1943) of the United States Army Corps of Engineers following his experiences in building the Alaska Highway during the Second World War. Referring specifically to Alaska, Muller (1943) wrote:

> "The destructive action of permafrost phenomenon has materially impeded the colonization and development of extensive and potentially rich areas in the north. Roads, railroads, bridges, houses and factories have suffered deformation, at times beyond repair, because the condition of permafrost ground was not examined beforehand, and because the behaviour of frozen ground was little if at all understood." (p. 12)

Such a comment applies equally to Canada, where approximately one-half of the country (5.7 million km²) is underlain by permafrost of one sort or another (Fig. 1). The big difference between the 1940s and today, however, is that there is a better understanding of the problems associated with permafrost terrain. In this context, the aim here is to summarize the geotechnical and engineering challenges presented by permafrost, and to outline the impact which permafrost has upon settlement and economic development in northern Canada.

Permafrost
Traditionally, permafrost is defined on the basis of temperature; that is, ground (*i.e.*, soil and/or rock) that remains at or below 0°C. Therefore, to differentiate between the temperature and to state conditions (*i.e.*, frozen or unfrozen) of permafrost, the terms cryotic and noncryotic have been proposed. These terms refer solely to the temperature of the material independent of its water/ice content (Associate Committee on Geotechnical Research, 1988). Perennially cryotic ground is, therefore, synonymous with permafrost, and permafrost may be unfrozen, partially frozen, and frozen, depending upon the state of the ice/water content.

Several other terms are used to describe the stratigraphy of permafrost. The permafrost table is the upper surface of the permafrost, and the ground above the permafrost table is called the supra-permafrost layer (Fig. 2). The active layer is that part of the supra-permafrost layer that freezes in the winter and thaws during the summer, *i.e.*, it is seasonally frozen ground. Although seasonal frost usually penetrates to the permafrost table in most areas, in some areas an unfrozen zone exists between the bottom of seasonal frost and the permafrost table. This unfrozen zone is called a talik. Unfrozen zones within and below the permafrost are also termed taliks.

Extent and Thickness of Permafrost in Canada
In Canada, the broad outlines of permafrost distribution are relatively well known and a number of detailed permafrost maps are now available (Heginbottom and Radforth, 1992; Natural Resources Canada, 1995). Some of the more important urban centres located in permafrost terrain include Yellowknife, Inuvik, Churchill, and Resolute (Fig. 1).

Permafrost is usually classified as being either continuous or discontinous in nature. In areas of continuous permafrost, frozen ground is present at all localities except for localized thawed zones (taliks) existing beneath lakes and river channels. In discontinuous permafrost, bodies of frozen ground are separated by areas of unfrozen ground. At the southern latitudinal limit of this zone, permafrost becomes restricted to isolated islands, typically occurring beneath peaty organic sediments. At the local level, variations in permafrost conditions are determined by a variety of terrain and other factors. Of widespread importance are the effects of relief, and the nature of the physical properties of soil and rock. More complex are controls exerted by vegetation, snowcover, waterbodies, drainage and fire.

The thickness to which permafrost develops is determined by a balance between the internal heat gain with depth and the heat loss from the surface. Heat flow from the Earth's interior normally results in a temperature increase of approximately 1°C per 30-60 m increase in depth. This is known as the geothermal gradient. Thus, the lower limit of permafrost occurs at that depth at which the temperature increase due to the geothermal gradient just offsets the amount by which the freezing point exceeds the mean surface temperature. If there is a change in the climatic conditions at the ground surface, the thickness of the permafrost will change appropriately. For example, an increase in mean surface temperature will result in a decrease in permafrost thickness, while a decrease in surface temperature will give the reverse.

Ice Within Permafrost

The amount of ground ice present within permafrost can vary from negligible, as in certain igneous and metamorphic rocks, to considerable, as in the case of unconsolidated, fine-grained Quaternary-age sediments. In general, it is highest near the permafrost table and decreases with depth (Pollard and French, 1980). Table 2 lists typical ground ice volumes that exist in the upper 5 m of permafrost at three localities in the western Canadian Arctic. The total volumetric ice content

usually varies between 35% and 60%, of which by far the majority (66-80%) forms either distinct ice masses (segregated ice) or fills pores (pore ice). Segregated ice may grow to form large, massive icy bodies (termed ground ice), commonly found as downward tapering ice wedges. The presence of pore ice gives rise to icy sediments. In regions underlain by

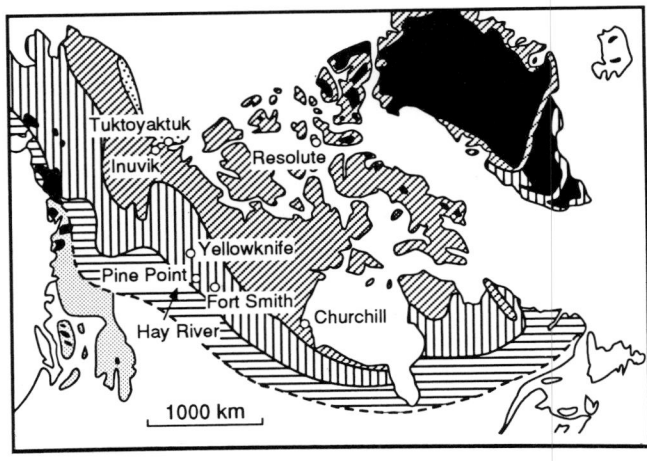

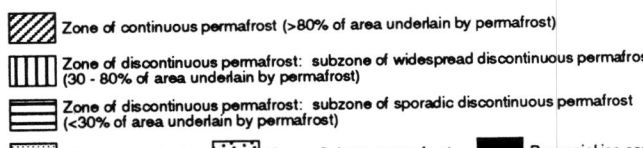

Figure 1 *Distribution of permafrost in North America (from Associate Committee on Geotechnical Research, 1988) and location of the larger communities in northern Canada.*

Table 1   Climatic statistics for selected Canadian stations in northern and non-northern locations (data from Hare and Thomas, 1974; and Wahl et al., 1987).

Station	Latitude (N)	Longitude (E)	Mean Annual Temperature (°C)	GDD	Frost-free Period (days)	Snow Cover (days)	FDD	Annual Total Precipitation (mm)
A "Non-northern"								
Vancouver	49°11'	123°10'	9.8	2019	212	7	45	1068
Edmonton	53°34'	113°31'	2.8	1516	127	117	1501	477
Saskatoon	52°10'	106°41'	1.6	1618	110	128	1977	353
Winnipeg	40°54'	97°14'	2.3	1791	118	122	1903	535
Toronto	43°40'	79°24'	8.9	2434	192	59	438	790
Ottawa	45°19'	75°40'	5.8	2069	137	117	1040	851
Montreal	45°30'	73°35'	7.2	2301	183	117	813	999
Quebec City	46°48'	71°23'	4.4	1729	132	140	1153	1089
Halifax	44°38'	63°30'	6.8	1708	183	63	422	1381
B "Northern"								
Churchill, MN	58°45'	94°04'	−7.3	688	81	213	3791	397
Baker Lake, NWT	64°18'	96°00'	−12.3	326	61	240	5206	213
Resolute, NWT	74°43'	94°59'	−16.4	36	9	283	6238	136
Whitehorse, YT	60°43'	135°04'	−0.8	897	87	167	1970	260
Yellowknife, NWT	62°28'	114°27'	−5.6	955	108	190	3614	250
Dawson, YT	64°04'	139°24'	−4.7	1014	91	n/a	8135	325
Iqualuit, NWT	63°45'	68°34'	−5.2	159	59	237	3903	415
Aklavik, NWT	68°14'	135°00'	−8.9	859	77	234	4474	236

fine-grained, unconsolidated surficial materials (*e.g.*, till, lacustrine clay), ground ice can be an extremely important component of permafrost. These fine-grained and ice-rich permafrost sediments are regarded as thaw-sensitive because of the large-scale disturbance to sediments that results from the thawing of such ice. Exensive subsidence of the ground surface results. An extreme case of high-ice-content permafrost occurs in the Mackenzie Delta area, in the vicinity of the settlement of Tuktoyaktuk. There, massive icy bodies and icy sediments, several tens of metres thick, occur under the townsite. If these ice masses and sediments were to thaw, the entire townsite would disappear beneath sea level.

PERMAFROST AND URBAN DEVELOPMENT

As stressed by Muller (1943), many important geotechnical and engineering problems result from the occurrence of permafrost. For the most part they relate to the water and/or ice content of the permafrost. They are summarized below.

Frost Heave

Pure water freezes at 0°C and, in doing so, expands by approximately 9% of its volume. The most obvious result of soil freezing is the volume increase that results and the associated deformation of host sediment and rock. The resulting frost heave has considerable practical significance since it causes the displacement of buildings, foundations and road surfaces. For the soil to heave, the ice must first overcome the resistance to its expansion caused by the strength of the overlying frozen soil. This usually occurs only when segregated ice lenses form. Frost heave results in significant damage to structures and foundations (Ferrians *et al.*, 1969). The annual cost of rectifying seasonal frost damage to roads, utility foundations, and buildings in areas of permafrost and deep seasonal frost is considerable.

Subsidence

As outlined above, ground ice is a major component of permafrost, particularly in unconsolidated sediments. Numerous case studies have been documented in Canada where, following the thaw of permafrost, ground subsidence

Figure 2   *Exposure showing permafrost table (arrowed), underlying frozen silts (permafrost) and overlying supra-permafrost layer that freezes in winter and thaws in summer. Figure for scale.*

has resulted (Mackay, 1970). Thaw consolidation may also occur as thawed sediments compact and settle under their own weight; the high pore-water pressures that are generated may also favour soil instability and slumping. The various processes associated with permafrost degradation are termed thermokarst. A related problem is that the physical properties of icy sediments, in which soil particles are cemented together by pore ice, may be considerably different to those of the same material in an unfrozen state. For example, in unconsolidated and/or soft sediments there is often a significant loss of bearing strength upon thawing. Beneath heated buildings, therefore, it is often essential to maintain the frozen state of the underlying material in order to support the structure.

Ground Water

The hydrological and ground-water characteristics of permafrost terrain are different from those of non-permafrost terrain (Sloan and van Everdingen, 1988). For example, the presence of both perennially and seasonally frozen ground prevents the infiltration of water into the ground or, at best, confines it to the active layer. At the same time, subsurface flow is restricted to unfrozen zones or taliks. A high degree of mineralization in subsurface permafrost is often typical, caused by the restricted circulation imposed by permafrost and the concentration of dissolved solids in the taliks. Thus, frozen ground eliminates many shallow-depth aquifers, reduces the volume of unconsolidated deposits or bedrock in which water is stored, influences the quality of ground-water supply, and necessitates that wells be drilled deeper than in non-permafrost regions.

Thermokarst and Site Disturbance

The thawing of permafrost, and the heaving and subsidence caused by frost action, result in severe damage to roads, bridges and other structures. In Alaska, following the realization of these effects in the 1940s, a determined effort was made by federal and state agencies to improve construction practices and to document permafrost problems (Ferrians *et al.*, 1969; Péwé, 1966; Péwé and Paige, 1963). In Canada, where large-scale development projects in permafrost regions occurred slightly later, it was possible to benefit from Alaskan experience.

A general problem related to urban development and economic activity in and around Canada's northern communities relates to artificial disturbances to the forest tundra and tundra vegetation (Fig. 3). Where the underlying permafrost is ice-rich, the thawing of the near-surface soils results in a thickening of the active layer. Distinctive hummocky terrain and thermokarst mounds can develop (French, 1975, 1987).

Table 2   Ground ice volumes in the upper 5 m of permafrost in three localities in the western Canadian Arctic (from French, 1993).

Area	Pore/ Segregated Ice (%)		Wedge Ice (%)		Total Ice (%)
King Point, Yukon (100)	43.5	(79.2)	11.4	(20.8)	54.9
Richards Island, NWT (100)	28.3	(79.3)	7.5	(21.0)	35.7
SW Banks Island, NWT (100)	38.7	(66.2)	19.8	(33.8)	58.5

Buildings and other structures can be undermined or eroded, requiring costly renovation and maintenance. In some instances, structures have to be abandoned due to excessive terrain disturbance and thermokarst activity. In an attempt to minimize such disturbances, especially in terrain underlain by thaw-sensitive permafrost, the Federal Government, through the Territorial Arctic Land Use Act and Regulations, has imposed strict regulations associated with the movement of heavy machinery, the mining of aggregate sources, the timing of exploratory drilling and seismic activity, and the unnecessary disturbance of vegetation. All aspects of oil exploration activity are closely regulated (French, 1980, 1984) and the right-of-way associated with the recently completed Norman Wells pipeline is monitored regularly to ensure that terrain stability is being maintained (Burgess and Harry, 1990). Recent diamond exploration activities in the Northwest Territories are also subject to these land use regulations.

It must be emphasized that even very small disturbances to the surface may be sufficient to induce thermokarst activity. Mackay (1970), for example, describes how a sled dog in the Mackenzie Delta was tied to a stake with a 1.5 m long chain. In the ten days of tether, the animal trampled and destroyed the tundra vegetation of that area. Within two years, the site had subsided like a pie dish by a depth of 18-23 cm, and the active-layer thickness had increased by more than 10 cm within the depression.

Several case studies of man-induced disturbance provide insight into the nature and speed at which such terrain becomes and remains unstable. In many cases, stabilization only begins 10 years to 15 years after the initial disturbance and is not complete until 30 or more years have passed. Thus, man-induced disturbances to permafrost terrain have the potential to permanently scar the landscape for decades, and the minimization of terrain disturbance is an important consideration in the municipal planning of all northern communities.

With respect to minimizing the effects of construction on permafrost, a number of approaches are available, depending on site conditions and fiscal limitations. If the site is under-

lain by hard consolidated bedrock, as is the case for some regions of the Canadian Shield, ground ice is usually non-existent and permafrost problems can be largely ignored. In most areas, however, this simple approach is not feasible, since an overburden of unconsolidated silty or organic sediments is usually present. In the majority of cases, therefore, construction techniques are employed which aim to maintain the thermal equilibrium of the permafrost and avoid the onset of thermokarst.

The most common technique is the use of a pad or some sort of fill which is placed on the surface (Fig. 4). This compensates for the increase in thaw which results from the warmth of the structure. By using a pad of appropriate thickness the thermal regime of the underlying permafrost is unaltered. It is possible, given the thermal conductivity of the materials involved and the mean air and ground temperatures at the site, to calculate the thickness of fill required. Too little fill, plus the increased conductivity of the compacted active layer beneath the fill, will result in thawing of the permafrost and in subsidence (Morgenstern and Nixon, 1971; Fig. 4A). On the other hand, too much fill will provide too much insulation, and the permafrost source will aggrade on account of the reduced amplitude of the seasonal temperature fluctuation (Figs. 4B, C). In northern Canada and Alaska, gravel is the most common aggregate used, since it is reasonably widely available and is not as susceptible to frost heave as more fine-grained sediments.

In instances where the structure concerned is capable of supplying significant quantities of heat to the underlying permafrost, as is the case of a heated building or a warm oil pipeline, additional measures are frequently adopted. Usually the structure is mounted on piles which are inserted into the permafrost. An air space left between the ground surface and the structure enables the free circulation of cold air which dissipates the heat emanating from the structure. Other techniques used include the insertion of open-ended culverts into the pad, the placing of insulating matting immediately beneath the pad and, if the nature of the structure justifies it,

A

B

Figure 3 *Examples of man-induced terrain disturbances in northern Canada. (**A**) Themokarst mounds and standing waterbodies developed in disturbed terrrain adjacent to the SOBC Blackstone D-77 exploratory well site, interior Yukon (65° 40′ N; 137° 15′ W). The well was drilled in the summer of 1962. Thaw depths in the disturbed terrain now commonly exceed 90 cm; those in undisturbed terrain are less than 50 cm (photograph taken July 1979). (**B**) Gully erosion along a vehicle track made in the summer of 1970 near the site of the Drake Point blow-out (76° 26′ N; 108° 55′ W), Sabine Peninsula, Melville Island, NWT. The terrain is underlain by ice-rich shale (photograph taken August 1976).*

the insertion of refrigeration units or Cryo-Anchors (Hayley, 1982) around the pad or through the pilings.

The following sections provide illustrations of the various problems and solutions in dealing with permafrost terrain, using examples from a number of Canadian urban communities.

Inuvik

The construction of the town of Inuvik in the Mackenzie Delta in the early 1960s was an example of the careful manner in which large-scale construction projects need to be undertaken in permafrost regions. A major factor governing the location of the town was the presence of a large body of fluvioglacial gravel a few kilometres to the south that could be used to construct a gravel pad below the entire townsite. Today, the gravel deposit has been exhausted and the future growth of the community is dependent upon the exploitation of more distant aggregate sources with their associated higher costs of haulage.

The provision of municipal services such as water supply and sewage disposal are particularly difficult in permafrost regions. Pipes to carry these services cannot be laid below ground beneath the depth of seasonal frost, as is the case in non-permafrost regions, since the heat from the pipes will promote thawing of the surrounding permafrost and subsequent subsidence and fracture of the pipe. At Inuvik, the provision of these utilities has been achieved through the use of utilidors: continuously insulated aluminium boxes that run

above ground on supports and link each building to a central system (Fig. 5A, B). The cost of such utilidor systems is high, involving a high degree of town planning and constant maintenance, and can only be justified in the larger settlements.

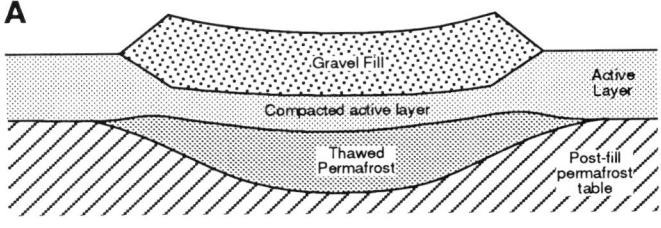

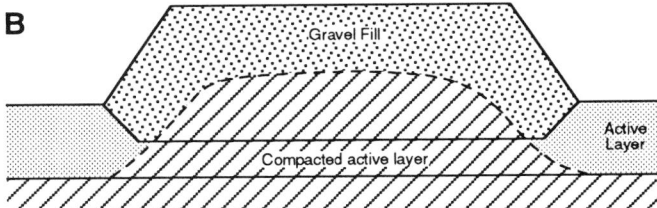

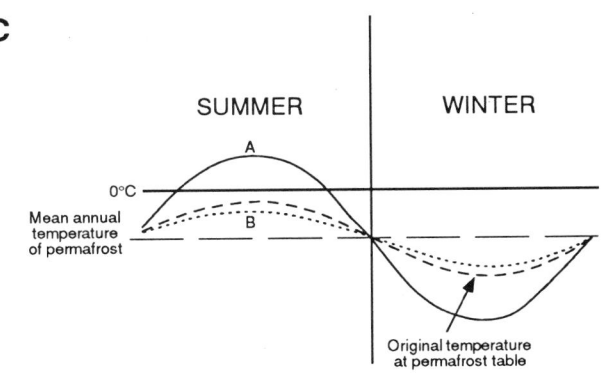

Figure 4 *Diagram illustrating the effects of a gravel fill upon the thermal regime and thickness of the active layer. (**A**) Too little fill; (**B**) Too much fill; (**C**) Effects of cases (A) and (B) upon the thermal regime; too little fill increases the amplitude of seasonal temperature fluctuation at the permafrost table and promotes the development of subsidence (thermokarst; from Ferrians et al., 1969).*

Figure 5 *Inuvik, NWT, buildings are placed upon wooden piles. Services such as water and sewage are effected by a utilidor system which links each building to a central plant. (**A**) Typical utilidor topography. (**B**) Roads pass over utilidors on bridges.*

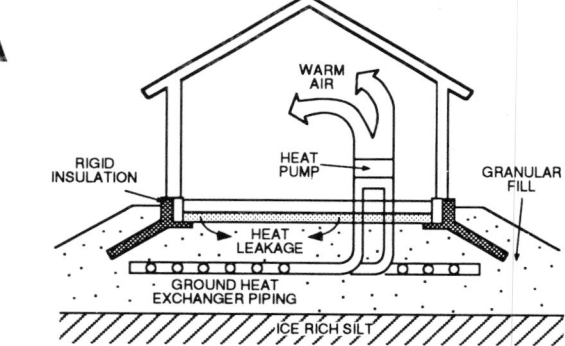

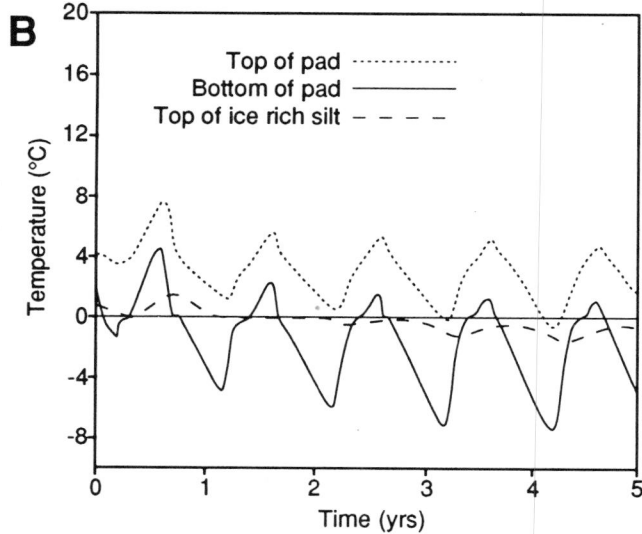

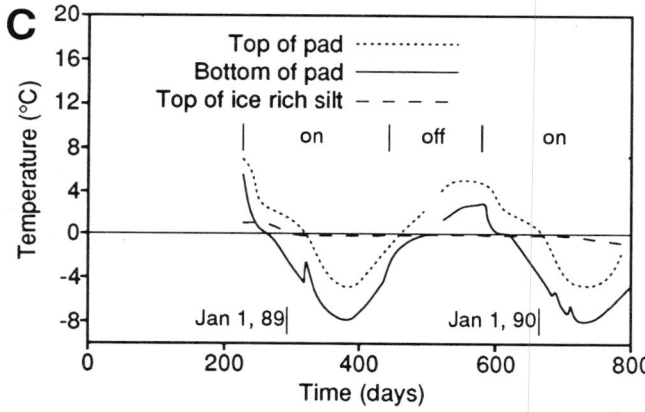

Figure 7 *(above)* Ross River, Yukon. (**A**) *Design of insulated building using heat pump chilled foundations. (**B**) Calculated temperatures. (**C**) Measured temperatures (from Baker and Goodrich, 1990).*

Figure 6 *(left)* Dawson City, Yukon. (**A**) *Abandoned building illustrating severe settlement due to permafrost degradation. (**B**) Heritage building restored by Parks Canada and placed upon non-frost susceptible granular infill following removal of silty permafrost sediments. Note the adjacent modern fire hydrant requiring deep burial of municipal services. (**C**) The installation of modern municipal services in 1980 was by means of trenches excavated to a minimum depth of 2.0 m and backfilled with coarse non-frost susceptible aggregate.*

Dawson City

In Dawson City, Yukon, a slightly different approach has been taken by Parks Canada and the Yukon government in restoration and maintenance of historic buildings, and in the provision of utilities to the townsite. The city of Dawson is located on a restricted area of the flood plain of the Yukon River, underlain by 2-4 m of silt over alluvial gravel. Permafrost at temperatures between $-3°C$ and $-1°C$, is present to a depth of about 20 m below the city. The occurrence of segregated ice lenses and ice wedges means that the soils are thaw sensitive and subject to settlement if disturbed. The earliest buildings in Dawson were log structures or frame buildings placed on large squared timbers laid at or near the surface. Virtually all old buildings still remaining have settled differentially, necessitating periodic jacking and levelling with additional cribbing, and/or eventual abandonment (Fig. 6A). Similar deformations in old structures placed directly upon ice-rich permafrost have been described from Alaska and Siberia (Péwé, 1983, p. 11-62). Since the early 1960s all new buildings in Dawson have been constructed on wooden piles or gravel pads. In restoring some of the historic buildings, Parks Canada has tried to maintain the original levels of the buildings with respect to the streets, ruling out the emplacement of thick gravel pads or the use of piles. Instead, the silty ice-rich material has been excavated and replaced by thaw-stable granular material to a depth of 5-7 m, and the historic buildings have been replaced in their original positions, supported by adjustable jacks (Fig. 6B).

The provision of municipal services in Dawson has also been upgraded at considerable cost. The town uses water from infiltration wells situated near the bank of the Klondike River. Prior to 1980, the city water distribution and sewage systems were those that had been constructed in 1904. They consisted of woodstave pipes laid in gravel in the active layer: all were shallower than 1.2 m. In winter, the water was heated by electricity to $+5.5°C$ and enough flow was maintained to prevent freezing by bleeding into each house. At the end of the circulation system, the water temperature was about $1.1°C$. Needless to say, these water and sewage systems required frequent repairs due to seasonal frost heave, settlement of the pipes through thaw, and frost deterioration of the pipes.

During the winter and spring of 1979-1980, a new system of underground services was installed in trenches that were excavated to a minimum depth of 2.0 m and backfilled with coarse (frost-stable) gravel fill (Fig. 6C). Similar costly procedures are now regarded as inevitable in many of the smaller northern communities in Canada if reliable and up-to-date utilities are to be provided.

Ross River and Old Crow

The biggest disadvantage of pile foundations is their cost. This is especially the case for very small buildings, such as individual houses. Equally, in certain communities where good-quality non-frost-susceptible aggregate is scarce, the cost of gravel pad construction is also high. As a result, additional new technologies are being tested. For example, in 1987, the Institute for Research in Construction, National Research Council of Canada, joined with the Yukon government in constructing two 350 m^2 multi-purpose municipal buildings in the communities of Ross River and Old Crow, on heat-pump-chilled foundations (Goodrich and Plunkett, 1990). The aim here is to prevent the thaw of permafrost beneath the buildings. Ross River has a mean annual ground temperature greater than $-0.5°C$ and is typical of relatively warm permafrost in the discontinuous zone. Old Crow has a mean annual ground temperature of $-5°C$ to $-7°C$ and is more typical of cold permafrost in the continuous zone. The design of the heat-pump-chilled foundation is illustrated in Fig. 7A. The heat exchangers were placed in a sand layer within the granular fill used to level the site. Heat flowing down through the floor is captured by the heat exchangers and pumped back into the building. Thus, while the house is being heated, the ground is being chilled. Comparisons of predicted and measured temperatures at the top and bottom of the gravel pad and at the top of the ice-rich silt layer (Fig. 7 B,C), for the first two years of operation at Ross River, suggest that the system is working well and that permafrost is being maintained beneath the structure.

Bridge Construction in the North

Frost heaving of the seasonally frozen, supra-permafrost zone is a major engineering problem encountered in permafrost regions, and is particularly damaging to urban infrastructure. Not only can differential heave cause structural damage to buildings, but equally important, frost heaving affects the use of piles for the support of structures. While in warmer climates the chief problem of piles is to obtain sufficient bearing strength, in permafrost regions the problem is to keep the pilings in the ground since frost action tends to heave them upward. Since heaving becomes progressively greater as the active layer freezes, it follows that the thicker the active layer, the greater is the upward heaving force. In the discontinuous permafrost zone, where the active layer may exceed 2 m in thickness, frost heaving of piles assumes critical importance. In parts of Alaska, for example, old bridge structures illustrate the effects of differential frost heave (see Péwé, 1983). In these regions, it is not uncommon for a thawed zone to exist beneath the river channel. Thus, piles inserted in the stream bed itself experience little or no frost heave, and piles inserted within permafrost on either side of the river are also unaffected. However, the piles adjacent to the river bank experience repeated heave since they are located in the zone of seasonal freezing. As a result, arching of both ends of the bridge may occur.

In order to prevent these problems in Canada, alternative structures involving minimal pile support have been consid-

Figure 8 *The Eagle River bridge, Dempster Highway, Yukon. This is a single-span structure with minimal pile support in the river.*

ered. A case in point is the recent construction of the Eagle River bridge on the Dempster Highway, northern Yukon (Fig. 8). The bridge consists of a single 100-m-long steel span with the footings on the north side placed in ice-rich permafrost. Drilling prior to construction indicated that permafrost was present on the north bank to a depth of ≈90 m. However, a deep near-isothermal talik existed beneath the main river channel, while near the proposed south bridge abutment, relatively warm (~0.4°C) permafrost was present to depths of 8-9 m.

In order to maintain the delicate permafrost conditions and to provide structural integrity, 15 steel piles were inserted at each abutment (Fig. 9). Conventional engineering and freeze

analysis indicated the optimum depth of emplacement of each pile was ≈5 m. However, because of the very warm permafrost at the south abutment, the piles were driven to a depth of about 30 m. On the north abutment, where the permafrost was colder, the piles were installed in holes augered to a depth of 12 m. These were then backfilled with a sand slurry to promote freezing. A further complexity was that construction had to be carried out during the winter (1976–1977), in order to minimize surface terrain damage. Subsequent monitoring has indicated that the piles have experienced minimal heave, the thermal regime of the permafrost has been maintained, and the bridge structure is performing

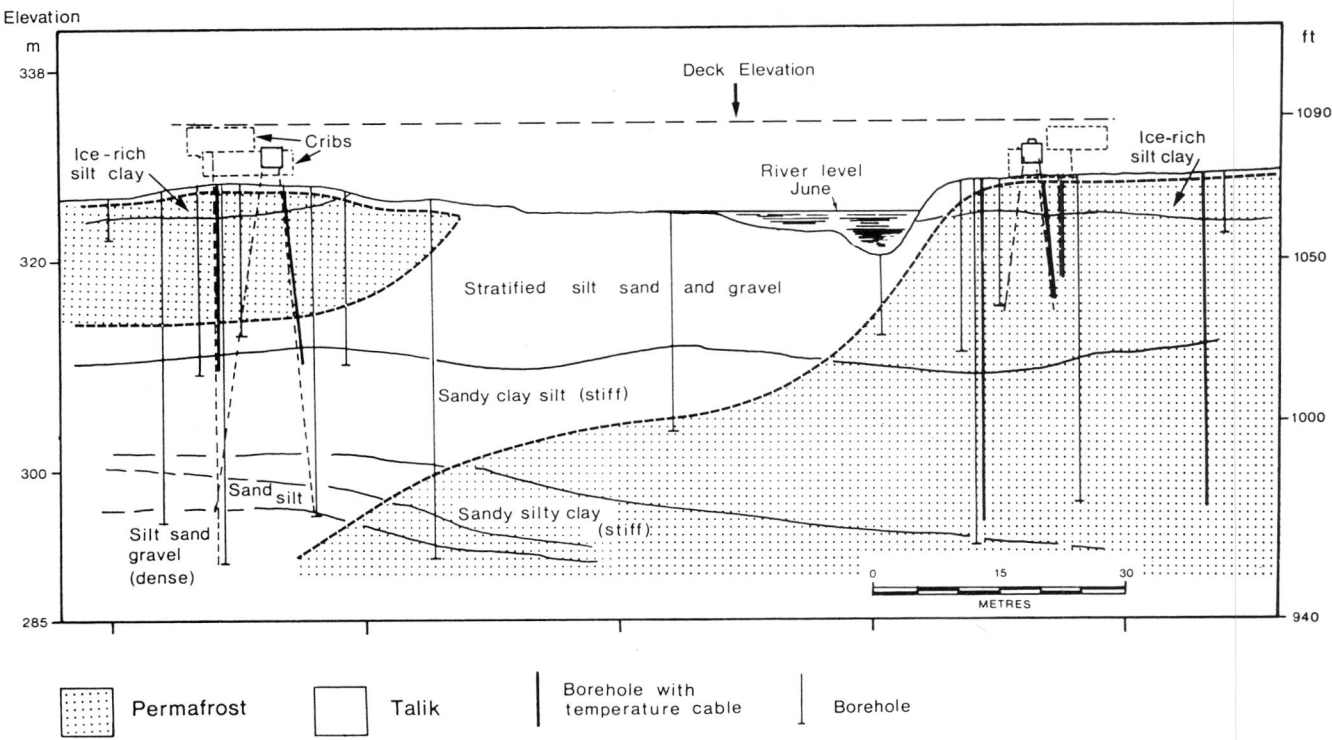

Figure 9 *Design plan and subsurface profile at the Eagle River bridge site, with stratigraphic and permafrost information determined by drilling (from Johnston, 1980).*

Figure 10 *Trans-Alaska oil pipeline using vertical support members (VSMs) with cooling devices (fins) designed to prevent heat transfer to ice-rich, thaw-sensitive permafrost.*

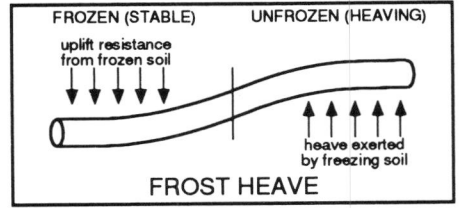

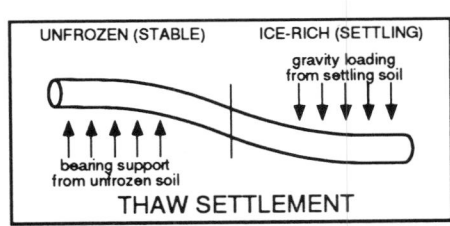

Figure 11 *Conceptual illustration of the freezing and thawing effects of a pipeline crossing from unfrozen to frozen terrain, or vice versa, in the discontinuous permafrost zone in Canada (from Nixon, 1990).*

satisfactorily (Johnston, 1980).

A major consideration in recent road building in northern Canada has been the design of river-crossing facilities that avoid the expense associated with bridges. Large diameter culverts have been employed along the Dempster Highway, with their diameter based upon a flood discharge with a recurrence interval of 50 years. Since the higher and sustained water velocities that occur within culverts may block fish migration in the upstream direction, culverts are installed such that the mean water velocity does not usually exceed 0.9 m·s⁻¹ except during the mean annual flood and during the 50 year design flood. To further minimize culvert utilization, roads such as the Dempster Highway follow upland interfluve locations wherever possible. They are also installed with their bases at or below the natural stream bed elevation to discourage the formation of upstream ponds or downstream plunge pools.

In the case of larger river crossings, such as the Peel and Mackenzie on the Dempster Highway, it has been found that summer ferries and winter ice crossings are the best alternative to expensive and difficult bridge designs.

Oil and Gas Pipelines

The construction of warm oil pipelines through permafrost terrain further illustrates the complexity of frost heave and related permafrost problems. For example, the construction of the Trans-Alaska Pipeline System (TAPS) from Prudhoe Bay on the North Slope to Valdez on the Pacific Coast between 1974 and 1977 used many procedures designed to minimize permafrost problems (Metz *et al.*, 1982; Heuer *et al.*, 1982). Approximately half of the route was elevated on verti-

cal support members (VSMs), many with cooling devices (heat tubes) to prevent heat transfer to ice-rich (*i.e.*, thaw-sensitive) permafrost (Fig. 10).

In Canada, the recently completed small-diameter Norman Wells pipeline did not have to address the problems of thaw subsidence to the same extent as the Alaska line, since it operates at or close to the prevailing ground temperature. Many of the terrain problems associated with this pipeline lie in the stability of wood-chip covered embankments along the right of way, and in the crossing of streams (Burgess and Harry, 1990).

The proposed construction of buried chilled-gas pipelines presents more complex problems that are, as yet, not completely resolved. Here, the problem is one of prolonged frost heave adjacent to the pipe with the possibility of eventual rupture (Fig. 11). This might occur in the discontinuous permafrost zone wherever the pipe crosses unfrozen ground and where there would be relatively unlimited moisture migration towards the cold pipe. Equally, when the pipe passes from unfrozen (stable) to ice-rich (unstable) terrain, or *vice versa*, thaw settlement and/or frost heave may result depending upon the situation.

In order to understand these problems, several natural-scale experiments are currently in progress at Calgary, Caen (France) and Fairbanks (Alaska), which aim to study the behaviour of soil around a refrigerated pipeline (Fig. 12). The Calgary frost heave test facility has been in operation since 1974 and circulates air at −10°C in a 1.2 m diameter pipe buried to represent a number of possible gas pipeline modes (Carlson *et al.*, 1982). Within a couple of years of operation, a frost bulb had formed around the pipe, and in the deep burial mode, the pipe had heaved more than 60 cm, while frost depths had penetrated to 3 m below the pipe. At Caen, a non-insulated pipe, 2.7 m in diameter, was buried in initially unfrozen soil with a lateral transition from a frost-susceptible silt to a non-frost-susceptible sand, thereby simulating a major soil type boundary common to permafrost terrain. During a first freezing experiment run in 1982-1983, with a pipe temperature of −2°C

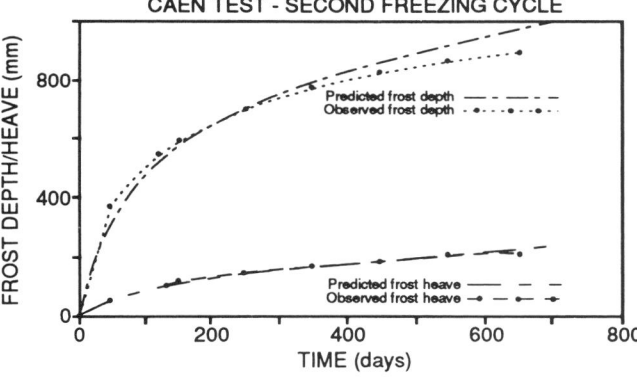

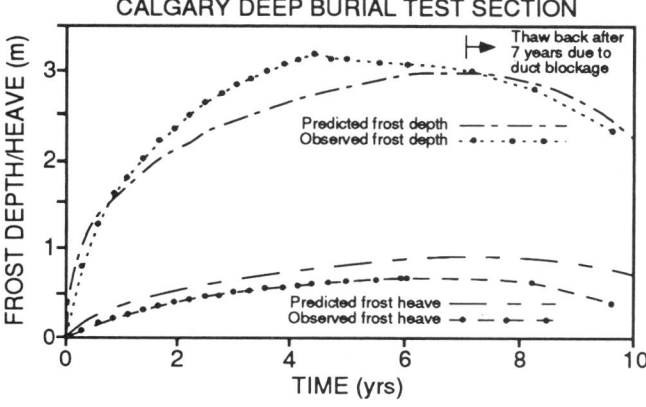

Figure 12 *Results of the Calgary and Caen frost heave test facilities. Observed frost heave and frost depth are compared to predicted values (from Nixon, 1990).*

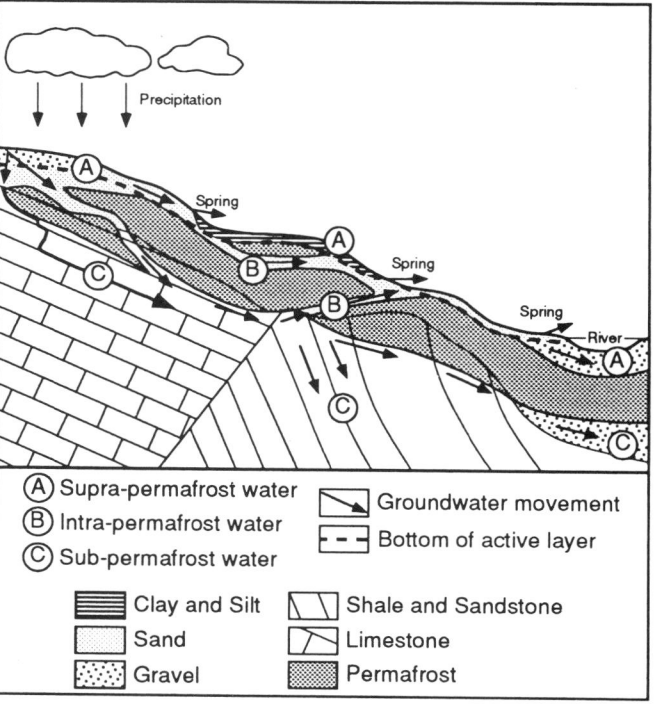

Figure 13 *Generalized model for the occurrence of ground water in permafrost regions.*

and a chamber temperature of $-7°C$, the pipe heaved 11 cm on the 16-m-long section, and frost penetrated 45 cm beneath the pipe in the sand and 30 cm beneath the pipe in the silt (Burgess, 1985). Results from the Fairbanks test facility are not yet published; preliminary data indicate that in the first 166 days of operation, the pipe heaved at least 10 cm at the critical permafrost–non-permafrost boundary. One can afford to be reasonably optimistic that solutions to these problems will be found. The observed magnitudes of frost heave and the frost penetration depths at the experimental sites relate well with values obtained *via* numerical simulation. They suggest that the amounts of heave and settlement that will be experienced by northern pipelines are predictable to engineering levels of accuracy using existing methods (Nixon, 1990).

GROUND AND SURFACE WATER PROBLEMS

The ground-water system of permafrost regions is unlike that of non-permafrost regions since permafrost acts as an impermeable layer. Under these conditions, the movement of ground water is restricted to taliks (Fig. 13). These may be of three types. First, a supra-permafrost talik may exist immediately above the permafrost table, but below the depth of seasonal frost. In the continuous permafrost zone, supra-permafrost taliks are rare. In the discontinuous permafrost zone, however, the depth of seasonal frost frequently fails to reach the top of the permafrost since the latter is often relict and unrelated to present climatic conditions. In these areas, supra-permafrost taliks are widespread and may be several metres or more thick. Second, intra-permafrost taliks are thawed zones confined within the permafrost and, third, sub-permafrost taliks refer to thawed zones beneath the permafrost.

Given these hydrological characteristics, a difficult problem in many permafrost regions is the provision of drinking water to settlements. Since supra-permafrost water is subject to contamination and usually of small volume, and intra-permafrost water is often highly mineralized and difficult to locate, the mapping of sub-permafrost water is vital.

In the discontinous permafrost zone, opportunities exist for ground-water recharge, and in parts of central Alaska and the Mackenzie Valley, North West Territories, extensive alluvial deposits provide an abundant source of ground water. In Fairbanks, houses rely on numerous small-diameter private wells (see Péwé, 1983). In parts of Siberia, the occurrence of perennial springs fed by sub-permafrost water assumes special importance since these may be the sole source of water available over large areas. However, in many areas of northern Canada (the Canadian Shield) the permafrost is several hundred metres deep and perennial springs are absent. In these situations, drilling is either not possible, since the well would freeze, or too costly. As a result, surface waterbodies, particularly those that do not freeze to their bottoms in winter, must be used, and great care taken to prevent contamination.

The supply of water is a severe limitation to any large-scale permanent settlement in much of the continuous permafrost zone. For example, the water supply problem of Sachs Harbour, a small Inuit community of approximately 250 people on southwest Banks Island, is typical of many situations. There, the water supply is derived from a lake approximately 3 km from the townsite. It is trucked, every 3-4 days, by water tanker, to individual homes that have indoor storage containers. Contamination is a problem, and the size of the lake, one of the few deep enough not to freeze to its bottom during winter, limits growth of the community.

A different group of hydrological problems relates to the formation of icings. These are sheet-like masses of ice that form at the surface in winter where water issues from the ground, usually from a supra-permafrost talik. Icings are of great practical concern as regards highway and railway construction and, in fact, are a distinct hazard to construction activity. These problems are most common in the discontinuous permafrost zone. Although sub- and intra-permafrost waters may be involved, the most frequently occurring icings are those associated with supra-permafrost water. A common occurrence is where a roadcut or other man-made excavation intersects with the supra-permafrost ground-water table. Seepage occurs and a sheet of ice forms, often over many tens of square metres in extent. In North America, icings were first encountered on a large scale during the building of the Alaskan Highway in the 1940s, and they occur widely in Alaska (see Péwé, 1983). Unless precautions are taken, icings can occur on most northern highways that traverse sloping terrain. Counter measures to reduce icing problems include the avoidance of roadcuts wherever possible, the installation of high-arch culverts to divert water from the source of the icing, and the provision of large drainage ditches adjacent to the road. Icings may also block culverts placed beneath road embankments and, by diverting melt water, initiate washouts in the spring thaw period. The costs of icing control and/or remedial measures can be considerable: van Everdingen (1982) provides a conservative estimate of $20,000 for ice control at a single locality on the Alaska Highway, Yukon, for the 1979-1980 winter.

DISCUSSION

Permafrost, with its particular terrain, ground ice and hydrological conditions, exerts a dominant influence over urban activities in northern Canada, and poses unique environmental geological problems. Although the settlements in Canada's permafrost terrain are not numerous or large, the urban geology of such settlements, and the economic activity upon which they often rely, cannot be adequately understood without detailed appreciation of the peculiarities of frozen ground.

ACKNOWLEDGEMENTS

The author wishes to thank the many research agencies that have supported his work in Canada's northlands.

REFERENCES

Associate Committee on Geotechnical Research, 1988, Glossary of permafrost and related ground-ice terms: National Research Council, Permafrost sub-committee, Technical Memorandum 142, 156 p.

Baker, T.H.W. and Goodrich, L.E., 1990, Heat pump chilled foundations for buildings on permafrost: Geotechnical News, v. 8, p. 26-28.

Burgess, M., 1985, Permafrost: Large-scale research at Calgary and Caen: Geoscience, v. 2, p. 19-22.

Burgess, M.M. and Harry, D.G., 1990, Norman Wells pipeline permafrost and terrain monitoring; geothermal and geomorphic observations: Canadian Geotechnical Journal, v. 27, p. 233-244.

Carlson, L.E., Ellwood, J.R., Nixon, J.F. and Slusarchuk, W.A., 1982, Field test results of operating a chilled buried pipeline in unfrozen ground, in Fourth Canadian Permafrost Conference: National Research Council of Canada, Ottawa, ON, p. 475-480.

Ferrians, O., Kachadoorian, R. and Green, G.W., 1969, Permafrost and related engineering problems in Alaska: United States Geological Survey, Professional Paper 678, 37 p.

French, H.M., 1975, Man-induced thermokarst, Sachs Harbour airstrip, Banks Island, NWT: Canadian Journal of Earth Sciences, v. 12, p. 132-144.

French, H.M., 1980, Terrain, land use and waste drilling fluid disposal problems, Arctic Canada: Arctic, v. 33, p. 794-806.

French, H.M., 1984, Some terrain and land use problems associated with exploratory wellsites, northern interior Yukon, *in* Olsen, J., ed., Northern Ecology and Resources: University of Alberta Press, Edmonton, AB, p. 365-385.

French, H.M., 1987, Permafrost and ground ice, *in* Gregory, K.J. and Walling, D., eds., Man and Environment Activity: J. Wiley and Sons Ltd., Chichester, UK, p. 237-269.

French, H.M., 1993, Cold-climate processes and landforms, *in* French, H.M. and Slaymaker, O., eds., Canada's Cold Environments: McGill-Queen's University Press, Montreal, PQ, p. 143-167.

Goodrich, L.E. and Plunkett, J.C., 1990, Performance of heat pump chilled foundations, in 4th Canadian Permafrost Conference, Université Laval: Nordicana, v. 54, p. 409-418.

Hare, F.K. and Thomas, M.K., 1974, Climate Canada: Wiley Publishers of Canada Limited, Toronto, ON, 256 p.

Hayley, D.W., 1982, Application of heat pipes to design of shallow foundations on permafrost, *in* Fourth Canadian Permafrost Conference: National Research Council of Canada, Ottawa, ON, p. 535-544.

Heginbottom, J.A. and Radburn, L.K., 1992, Permafrost and ground ice conditions of Northwestern Canada: Geological Survey of Canada, Map 1691A, scale 1:1,000,000.

Heuer, C.E., Krzewinski, T.G. and Metz, M.C., 1982, Special thermal design to prevent thaw settlement and liquefaction, *in* Fourth Canadian Permafrost Conference: National Research Council of Canada, Ottawa, ON, p. 507-522.

Johnston, G.H., 1980, Permafrost and the Eagle River bridge, Yukon Territory, Canada, *in* Permafrost Engineering Workshop: National Research Council of Canada, Technical Memorandum 130, p. 12-28.

Mackay, J.R., 1970, Disturbances to the tundra and forest tundra environment of the Western Arctic: Canadian Geotechnical Journal, v. 7, p. 430-432.

Metz, M.C., Krzewinski, T.G. and Clarke, E.S., 1982, The Trans-Alaska Pipeline workpad: an evaluation of present conditions, *in* Fourth Canadian Permafrost Conference: National Research Council of Canada, Ottawa, ON, p. 523-534.

Morgenstern, N.R. and Nixon, J.F., 1971, One-dimensional consolidation of thawing soils: Canadian Geotechnical Journal, v. 8, p. 558-565.

Muller, S.W., 1943, Permafrost or permanently frozen ground and related engineering problems: United States Army, Office Chief of Engineers, Strategic Engineering Study Special Report 562, p. 231 [1945, second printing with corrections; reprinted in 1947, J.W. Edwards, Inc., Ann Arbor, MI].

Natural Resources Canada, 1995, Canada Permafrost, The National Atlas of Canada, 5th Edition, Map MCR 4177, scale 1:1,500,000, Canada Map Office, Ottawa, ON.

Nixon, J.F., 1990, Northern pipelines in permafrost terrain: Geotechnical News, v. 8, p. 25-26.

Péwé, T.L., 1966, Permafrost and its effect on life in the north, *in* Hansen, H.P., ed., Arctic Biology, 2nd edition: Oregon State University Press, Corvallis, OR, p. 27-66.

Péwé, T.L., 1983, Geologic hazards of the Fairbanks area, Alaska: Division of Geology and Geophysical Surveys, Alaska, Special Report 15, 109 p.

Péwé, T.L. and Paige, R.A., 1963, Frost heaving of piles with an example from the Fairbanks area, Alaska: United States Geological Survey, Report 1111-I, p. 333-407.

Pollard, W.H. and French, H.M., 1980, A first approximation of the volume of ground ice, Richards Island, Pleistocene Mackenzie Delta: Canadian Geotechnical Journal, v. 17, p. 509-516.

Sloan, C.E. and van Everdingen, R.O., 1988, Region 28, Permafrost region, *in* Back, W., Seaber, P.R. and Rosenshein, J.S., eds., Hydrogeology: Geological Society of America, The Geology of North America, v. O-2, p. 263-270.

van Everdingen, R.O., 1982, Management of groundwater discharge for the solution of icing problems in the Yukon, *in* Fourth Canadian Permafrost Conference: National Research Council of Canada, Ottawa, ON, p. 212-228.

Wahl, H.E., Fraser, D.B., Harry, R.C. and Maxwell, J.B., 1987, Climate of Yukon: Atmospheric Environment Service, Environmental Canada, Climatological Studies 40, 323 p.

4. Impacts of Urban Development on Ground Water

Ken W.F. Howard

Environmental Earth Sciences, University of Toronto, Scarborough Campus
1265 Military Trail, Scarborough, Ontario M1C 1A4

SUMMARY

On a global scale, ground water represents the largest and most important source of fresh potable water. Urban development stresses water resources by increasing demand, affects rates and distribution of aquifer recharge, and introduces contaminants that can seriously degrade drinking-water quality.

INTRODUCTION

Interest in the relationship between urbanization and water was initiated during the 1950s and 1960s when the post-war construction boom began to produce a range of hydrological problems. For example, construction of houses, commercial buildings, parking lots, and paved roads and streets increases the impervious cover in a watershed, reduces direct infiltration, and stimulates the hydraulic efficiency of flow through artificial channels, gutters and storm-water collection systems (Chow *et al.*, 1988). The overall effect is to boost the volume and velocity of surface-water runoff and produce larger peak flood discharges (Wanielista, 1979; Chapter 2). The result is urban flooding, channel erosion and uncontrolled deposition of sediment. Brown (1990) describes some of the more serious flood events seen in Canada. In 1950, floods in Winnipeg damaged 10,500 homes and caused the evacuation of 100,000 residents. In 1954, Hurricane Hazel dumped as much as 280 mm of rain on Toronto in a 48-hour period (Den Otter, 1988) killing 84 persons.

The science of urban hydrology is now well established and several texts have appeared on the subject (see Lazaro, 1979 and Hall, 1984, for example). In general, the key processes are well understood, and methods of calculation and hydrological prediction are well advanced. These range from the well-entrenched Rational Method (Pilgrim, 1986; Linsley, 1986) to advanced computer simulation models such as the Storm Water Management Model (SWMM; Huber *et al.*, 1977; Huber and Dickinson, 1988) and the Stormwater Management and Design Aid (SMADA; Wanielista, 1990). A compendium of recent advances in the modelling of hydrological systems is provided by Bowles and O'Connell (1991). To a large extent, predictive skills and developments in the design and construction of engineered structures such as barriers, floodways and storm detention ponds now afford a high degree of flood protection to most major cities in the developed world. Nevertheless, as seen by the "floods of the century" that devastated large low-lying areas of mainland Europe in January, 1995, even the most advanced engineering technologies will not deter the severest of storms.

While urban hydrology and the demand for flood control remain high on the urban agenda, other environmental effects of urbanization have now come to the forefront. In particular, serious questions have been raised as to the effects of urbanization on underground water, *i.e.*, urban hydrogeology has become an issue of environmental concern. On a global scale, ground water represents the world's largest and most important source of fresh potable water. Yet, in many countries both the quantity and quality of this resource are being threatened by urban growth and associated industrial development. Urban development not only imposes stress on the resource through increasing demand (*e.g.*, Shahin, 1990), but often reduces the amount of infiltration that naturally replenishes underlying aquifer systems. Urban development may also introduce contaminants that can seriously and irreversibly degrade drinking-water quality.

Unfortunately, scientific research into urban hydrogeology has been limited. Most research studies have been initiated in response to a particular ground-water quality or quantity problem; the result is that remediation measures tend to vastly outnumber protection measures when urban hydrogeology is concerned. Nevertheless, research into urban ground water has accelerated in recent years and the importance of subsurface problems is becoming well recognized. Ground-water issues formed a major component of Urban Water '88, a recent UNESCO symposium on hydrological processes and water management in urban areas. More recently, the International Association of Hydrogeologists (IAH) has established a Commission on Ground Water in Urban Areas, whose endeavours currently include an Internet database and a series of workshops related to ground-water protection and chemical processes.

As an environmental science, urban hydrogeology remains very much in its infancy. This paper reviews many of the key issues, and examines, on a global scale, the effects of urbanization on either the quality or quantity of ground-water resources.

IMPACTS ON GROUND-WATER QUANTITY

Direct and Indirect Aquifer Recharge

In natural systems, most ground-water recharge is due to the direct and indirect infiltration of rainfall and snow melt (Fig. 1). Direct recharge refers to water that infiltrates into the soil immediately following its arrival as incident precipitation, and which passes directly to the underlying aquifer. Indirect recharge normally occurs when water enters the deep subsurface after moving across the land surface as runoff. It most commonly occurs in stream channels or beneath shallow temporary ponds that form following intense periods of rain. Mechanisms of direct and indirect recharge can be complex, with the timing and volume of the recharge dependent on such factors as soil type, soil moisture, soil permeability, ground cover, ground surface slope, drainage conditions, depth to water table, free water evaporation, and the intensity and volume of precipitation. Urbanization can affect these parameters and influence recharge, either in a subtle way by altering the microclimate, or more overtly by sealing large areas of the ground surface with impermeable materials and increasing surface runoff.

According to Oke (1982), the effect of urbanization on microclimate is well documented, but poorly understood. Urban heat islands are generally warmer, more cloudy, and wetter than surrounding areas (Landsberg, 1981). They also tend to be less windy and less humid. Overall, the effect is to increase rainfall volumes and intensities (Changnon *et al.*, 1977); however, since the effects on evapotranspiration are extremely difficult to quantify (Van de Ven, 1990), the net effect on direct recharge rates is difficult to estimate.

The hydrogeological effects of introducing large areas of impermeable cover can be substantial. Impermeable cover includes roofs, sealed roads, parking lots and pedestrian pathways. In essence, it may also include sports fields and parks where underground drains have been installed to intercept any water that is in excess of basic evapotranspiration requirements. Low-density residential development generally creates an impermeable seal over 20% of the land area; in intensively urbanized areas, this value can easily reach 60% or more. For a typical urban environment where 50% of the land area is impermeable, direct recharge will be reduced by a

comparable amount, *i.e.*, in a temperate climate it will be halved, from about 250 mm to 125 mm per year (Fig. 2). Such a reduction can be critical in areas where urbanization occurs in the recharge area of a major aquifer. It will be less important where underlying strata exhibit a low permeability, and recharging water is forced to move laterally as interflow until it eventually discharges to drains and surface streams.

The depletion in direct recharge can often be offset by an increase in indirect recharge. Large areas of impermeable cover will cause a significant rise in the surface runoff and a proportion of this water will infiltrate naturally, depending largely on surface topography, permeability of the soil and underlying rocks, local head conditions and the hydraulic efficiency of the storm-water collection systems. In some cases, indirect recharge can be encouraged artificially by the use of soakaways and recharge basins that drain to the subsurface with minimal evaporation. On Long Island, New York, for example, it has been argued that storm-water recharge basins fully offset losses that result from urbanization (Seaburn and Aronson, 1974), even though the spatial and temporal distribution of the recharge is radically altered (Ku *et al.*, 1992). Similar results have been reported in Bermuda (Thomson and Foster, 1986), South Africa (Wright and Parsons, 1994) and Australia (Martin and Gerges, 1994). In some areas, the process of artificial recharge can be taken one stage further with the use of injection wells to recharge storm water into underlying aquifers. As explained by Dillon *et al.* (1994), in their excellent review of injection well methodology, the approach is particularly appropriate where aquifers are confined by overlying strata of low permeability. It should be noted, however, that while any attempt to conserve water by artificial recharge is often commendable in principle, it is not without risk. Particular problems include the introduction of

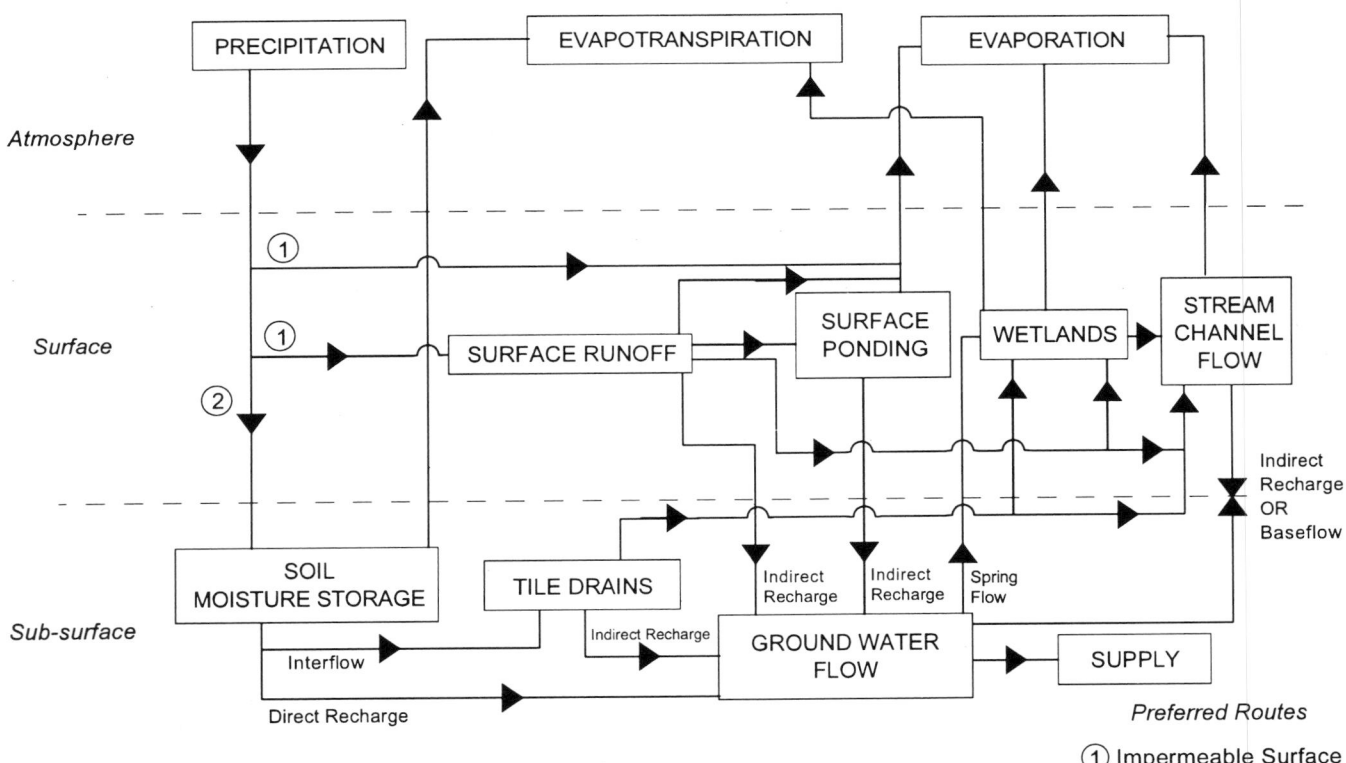

Figure 1   *Mechanisms of ground-water recharge.*

urban contaminants (discussed below), and clogging of infiltration lagoons, recharge wells, and even the aquifer, as a result of sedimentation and chemical and biological processes.

Aquifer Replenishment

In most natural situations, aquifer replenishment depends entirely on direct and indirect aquifer recharge. As discussed, urban development may seriously deplete direct recharge, and the deficit is rarely offset by increased indirect recharge even when artificial recharge is practised. However, this does not automatically mean that urban development causes a net loss in aquifer replenishment. As shown by Lerner (1990a, b, c) and Foster (1990), urbanization can radically alter the entire water balance of an area, thereby introducing sources of aquifer recharge that are entirely new to the region. Potential recharge sources include (1) septic tanks, (2) leaking sewers, (3) leaking water mains, (4) over-irrigation of domestic and municipal gardens, (5) infiltration of effluent (deliberate or otherwise), and (6) infiltration of storm runoff (deliberate or otherwise).

The contribution of these sources to aquifer recharge can be extremely difficult to quantify; however, leaking water mains are believed to be one of the major contributors. As discussed by Lerner (1986, 1990a; Cabrera, 1995), all water supply networks leak. Well-maintained systems may lose only 10% of supply; at the other extreme, losses as high as 50% have been reported. Hueb (1986) reports an average of 17% for 18 Latin America cities, a rate which, according to Foster (1990), effectively doubles the natural rate of aquifer replenishment. Of course, high rates of recharge due to supply

network losses mean very little in cities where the water supply is derived entirely from ground water. They simply represent an inefficiency of delivery which, if rectified, would not lead to any net change in the ground-water budget; less water would be required and less would be pumped. Instead, the effects of leakage tend to be most crucial in cities where large quantities of water are imported for water supply. Typically, for example, water imports for a large city may amount to an equivalent depth of between 300 and 5000 mm per annum (mm·a^{-1}), the upper estimate applying to densely populated centres such as Hong Kong. Clearly, network losses of 25% would make a substantial contribution to the underlying aquifer. In temperate regions, this leakage may merely offset the loss of direct recharge caused by extensive impermeable cover; in arid and semi-arid areas, leakage can be the primary source of ground-water recharge.

In some parts of the world, urbanization has reduced total recharge and/or affected water quality of ground water to the extent that wells have been abandoned in favour of imported surface-water supplies. The long-term effect of importing water and rejecting previously utilized ground-water reserves has been to cause a dramatic recovery in ground-water levels, the results of which include the re-establishment of springs, water-logging of low-lying residential areas and an upward flushing of salts and contaminants previously held in the shallow unsaturated zone. Lerner (1994) describes the effects of rising water levels in Brighton, Birmingham, London and Liverpool, UK. Similar problems have been encountered in Brisbane, Australia (Cox and Hillier, 1994).

Sewage can also contribute significantly to aquifer replenishment. In urban areas serviced by septic systems, it can be assumed that all water received by the systems eventually replenishes the aquifer. In Bermuda, for example, septic discharge has been shown to account for over 35% of the total annual aquifer replenishment. In Buenos Aires, Argentina, over 50% of the urban area is served by septic tanks, the gross recharge from which is estimated to be 3000 mm·a^{-1}, a six-fold increase over recharge in uninhabited areas (Foster, 1990). In other cities where sewage and waste water are less capably managed, rates of aquifer replenishment are virtually impossible to estimate. In Mexico City, for example, it is not known how much of the 13 million kg of sewage the city produces each day eventually enters the underlying aquifer. In some parts of the city, sewage is discharged to open canals and downward leakage of this water is believed to explain the local presence of elevated concentrations of nitrate. Concerns have also been raised for wells located closely adjacent to the Chalco Canal, one of the main passageways for waste water leaving the city. A particular dilemma facing the Mexican authorities is that while leakage of waste and sewage water may degrade ground-water quality, local ground-water resources are so seriously overexploited that ground-water recharge is at a premium.

In cities serviced by sewers, it must be recognized that, like water mains, sewers also leak (Hornef, 1985; Seyfried, 1984). Unfortunately, leakage rates are especially difficult to estimate. Most published work concerns the leakage of ground water into sewers that have been constructed beneath the water table. Comparable flows might be expected to occur in the reverse direction when sewers are constructed in the unsaturated zone; however, most studies of sewer pipe leakage have concerned water quality rather than water quantity. In the Permo-Triassic aquifer underlying Liverpool, UK, a water balance study conducted by the University of Birmingham indicated that leakage from the largely antiqu-

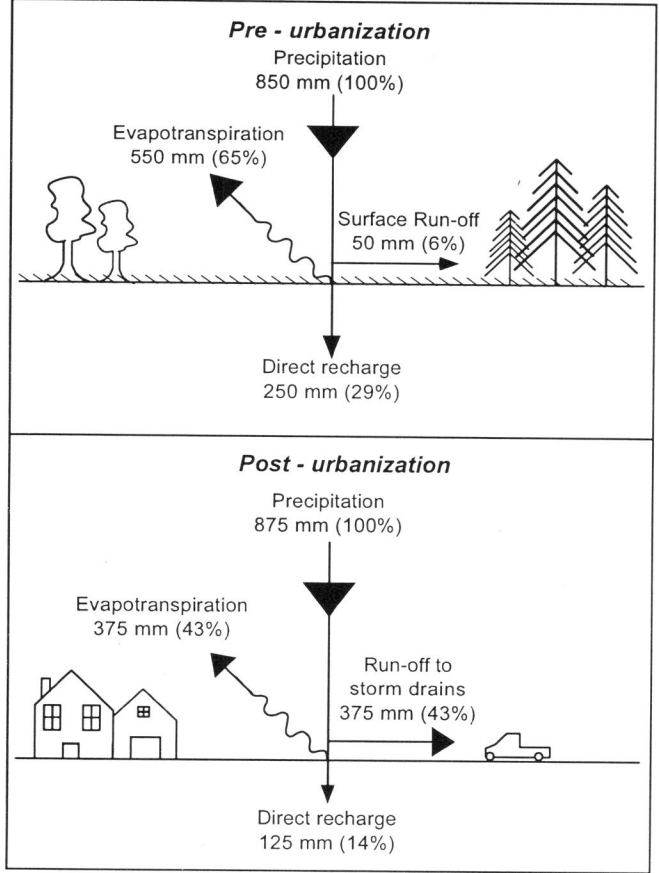

Figure 2   *Recharge pathways before and after urbanization.*

ated storm sewer system was comparable in volume to water mains leakage. Lerner (1986) suggests, however, that since sewer pipes are normally unpressurized, leakage from sewer pipes should normally be quite small.

As a general rule, in cities where sewage is not exported, as much as 90% of the total water imports may eventually recharge the local ground-water system (Lerner, 1990a). Clearly, the age and degree of maintenance of the water and sewage distribution systems are critical in determining the degree to which recharge is supplemented. While leakage from water supply and sewage systems appears to be a major source of recharge in such cities as Liverpool, UK, (University of Birmingham, 1984), Lima, Peru (Lerner, 1986), and Doha, Qatar (Lerner, 1990b), it is not clear that supplementary recharge is similarly significant in younger cities.

IMPACTS ON GROUND-WATER QUALITY

A primary concern of urban development is the introduction of contaminants that can seriously and irreversibly degrade drinking-water quality. As documented by Lerner (1990b, c) potential threats to water quality in urban areas include waste disposal (landfilling), industrial premises (where leaks and spills of chemicals are common), chemical stockpiles (raw materials and wastes), spillages during road and rail transport of chemicals, sewers and septic tanks, lawn/parkland fertilizers and pesticides, road de-icing chemicals, storage tanks, vehicular oils and greases, wet and dry deposition from smoke stacks, and oil and chemical pipelines (Chapter 5).

Timing of Impact

In general, the impacts of urban pollutant sources on surface-water quality are well known and well documented. Models such as RUNQUAL (Roesner et al., 1977), HSPF (Barnwell and Kittle, 1984) and AQUALM (Philips et al., 1992) can be very useful for minimizing impacts at the urban planning stage.

Usually, the impacts of urban pollutants on ground-water (Fig. 3) are considerably more serious. In the first case, many urban pollutants are either stored underground (e.g., gasoline) or originate underground (e.g., landfill leachate), and thus pose an immediate and direct threat to the quality of subsurface water (i.e., ground water). Furthermore, the movement of pollutants in the subsurface is generally slow and impacts are frequently not recognized until several decades following pollutant release. For example, chemically conservative contaminants such as chloride and nitrate move at or close to the velocity of the ground water in which they are dissolved. Typically, conservative contaminants are carried to the water table at rates of <1 m per year; within the aquifer, the contaminants are transported by advecting ground water at velocities that rarely exceed between 10 and 100 m per year. It follows that by the time ground-water pollution appears at a wellpoint or spring, perhaps several kilometres away, the plume of contamination is so well established that it is often impossible to treat and restore the ground water to its original quality. Non-conservative contaminants such as metals and many organic chemicals are delayed due to chemical reaction with the aquifer matrix and may move at velocities many orders of magnitude more slowly. In some cases, slow-moving chemically persistent chemicals may not emerge for tens of thousands of years. When surface water is found to be contaminated, it is often possible to trace and seal the source, thereby minimizing impacts before serious and irreversible damage is done. When ground water is found to be contaminated, it is not unusual to find that the source responsible

disappeared many years ago and that damage has become serious and extensive. Clearly, prevention of ground-water contamination is considerably more effective than any possible cure.

Organic *versus* Inorganic Chemicals — Water Quality Standards

There are now numerous documented cases of ground water being contaminated by products of urban development. In some instances, they represent a legacy of urban activity dating back several generations. Organic contamination is often considered to be the most serious since many commonly used organic chemicals are carcinogenic, mutagenic or teratogenic; many others are acutely toxic and can severely damage the human nervous and respiratory systems. In many cases, only very small amounts of the chemical will severely affect water potability (Table 1). As often cited (for example, Freeze and Cherry, 1979), a million litres of water will be rendered undrinkable by the organic constituents contained in just one litre of gasoline.

From another perspective, most organic chemicals are introduced to ground water as point sources (i.e., they emanate from a relatively discrete location, such as a storage tank), and many are poorly soluble or are subject to chemical sorption, biodegradation, and volatilization, all of which can severely constrain their range of influence (see Lyman et al., 1992, for example). Thus, while some organic chemicals will always be a serious threat to ground-water quality, particularly those that are relatively mobile, chemically persistent, or are associated with regionally distributed sources such as agricultural pesticides, others will never move very far from source and will constitute a problem only on a very local scale, i.e., in respect of individual wells and springs. The incidence of organic contamination of ground water would appear to be far more frequent in developed than in lesser developed nations, but, in all likelihood, this difference may largely reflect the high cost of performing organic analysis and the relative availability of analytical facilities. Somasundaram et al. (1993), for example, describe severe inorganic contamination of ground waters underlying Madras, India, but, in the absence of reliable data, can only speculate that the ground waters are just as seriously contaminated by organic chemical species.

Compared to the more common organic pollutants, common inorganic chemicals such as chloride, sodium, sulphate and nitrate are much less harmful to humans. Standards for drinking-water quality are shown in Table 1 and show that, where human health is at issue, concentration guidelines for the more common inorganic chemicals tend to be far less rigorous than those established for most organics. In addition, common inorganic chemicals tend to be more soluble and mobile than organics, and are more frequently associated with non-point contaminant sources, i.e., distributed and line sources of contamination such as acid precipitation, agriculture and de-icing salts applied to roads and highways. As a general rule, point sources can cause severe pollution of ground water on a very localized scale, whereas distributed and line sources of contamination generally cause widespread contamination of water at relatively low levels. Most pragmatists would argue that once drinking-water quality guidelines have been exceeded, mildly contaminated water is just as serious a concern as water that has been contaminated to a much higher degree.

Most cases of inorganic contamination of ground water in urban areas involve the ions chloride, nitrate and sulphate. Of

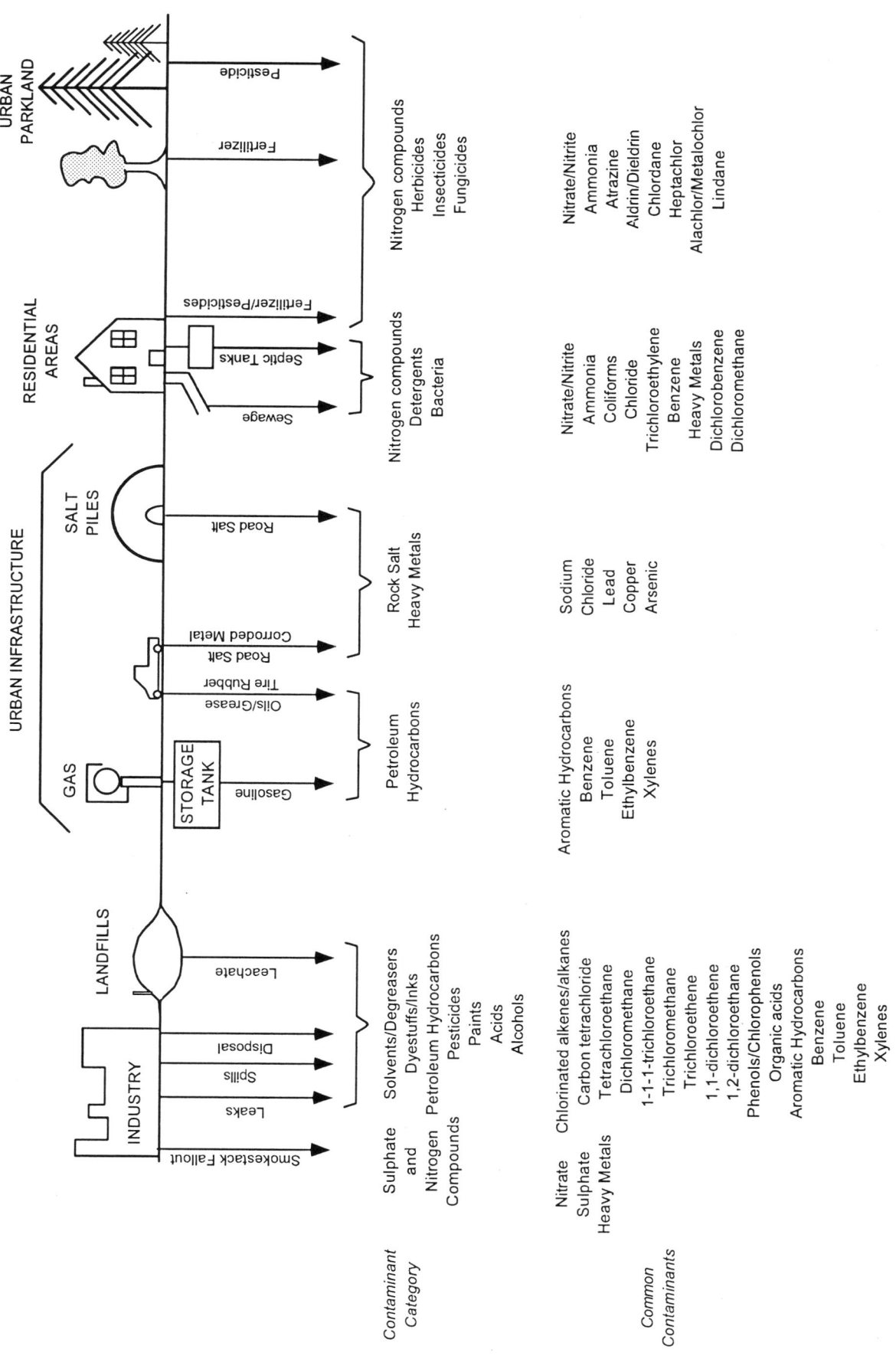

Figure 3 *Subsurface pathways for urban pollutants.*

these, only nitrate and, to a lesser extent, sulphate, are seriously implicated from a health standpoint. However, chloride can impart an unpleasant taste to the water, and its presence is often an indication that other more deleterious chemicals will be found in the water. Other inorganics of health-related concern include barium, lead, chromium, zinc, cyanide, arsenic, cadmium and mercury. Elevated iron and manganese are also frequently documented; however, these metals are not a health concern, and water quality standards, where established, are related to aesthetic issues such as staining of laundry and plumbing fixtures. Similarly, elevated boron is not generally regarded as a health issue; however, its almost ubiquitous use in industry and the home, together with its high mobility in ground water, give it the distinction of being an excellent indicator of anthropogenic pollution. Phosphate is another widely used chemical in the urban-industrial environment and is often found in surface waters that drain urban areas. However, phosphate is rarely a concern in ground waters since the neutral to alkaline pH conditions normally encountered tend to limit its mobility.

Urban Contaminant Sources

Industrial Activity

Some of the most serious cases of ground-water contamination are reported in urban centres with a long history of industrial activity (e.g., metal processing including electroplating, electronics, textiles, laundries, vehicle construction and maintenance, printing, brewing, pharmaceuticals, food processing, tanneries, etc.). Industry stores, uses and produces a wide range of organic and inorganic chemicals, and it is inevitable that at least some of this material will eventually enter the subsurface, with the potential to degrade ground-water water quality. Many of the more toxic chemicals, such as PCBs (polychlorinated biphenyls) and the larger molecule PAHs (polycyclic aromatic hydrocarbons) have a very limited solubility in water and are readily adsorbed on to clay and organic matter in the soils and sediments. As a result, these materials rarely move very far from source and human exposure due to ground-water pathways is rare. The most severe threat comes from toxic organic chemicals that are sufficiently soluble, mobile and persistent in water to migrate some distance from their source and enter wells, surface streams and lakes. In industrial environments, the chlorinated hydrocarbon solvents (CHS) represent one such group (Chapter 6).

In Europe, widespread contamination by CHS has been reported in heavily industrialized cities such as Milan, Italy (Cavallaro et al., 1986) and in the Midlands of England (Rivett et al., 1989, 1990; Nazari et al., 1993; Burston et al., 1993). CHS contamination is also common in North America with typical case studies described in New Jersey by Roux and Althoff (1980), in Indiana by Cookson and Leszcynski (1990) and in Nebraska by Kalinski et al. (1994). In Australia, Benker et al.

(1994) report extensive CHS contamination underlying a residential area in Perth. Chlorinated hydrocarbon solvents belong to a group of liquids known as DNAPLs (dense non-aqueous phase liquids). DNAPLs are only partially soluble in water and tend to form relatively discrete plumes that sink very rapidly toward the bottom of the aquifer. Here, they are gradually assimilated into the surrounding ground water to produce an aqueous or dissolved phase which moves readily in accord with the prevailing ground-water flow regime. These issues are discussed at length in Chapters 5, 6, 7 and 35. Trichloroethylene (TCE), tetrachloroethylene (also known as perchloroethylene or PCE), 1,1,1–trichloroethane (TCA), carbon tetrachloride (CTC), and chloroform (trichloromethane or TCM) are usually the most problematical DNAPLs. Most are released into the aquifer as point sources due to inappropriate or inadequate handling, storage or disposal by industrial users. However, significant quantities are also contained in domestic landfills (Howard and Livingstone, 1994) and can be

Table 1	World water quality guidelines for drinking water.			
Parameter	Canada	United States EPA	World Health Organization	Council of European Communities
INORGANIC				
Aluminum	—	—	0.2*	0.05
Arsenic	0.05	0.050	0.05	0.05
Barium	1.0	2.0	—	0.1
Boron	5.0	—	—	1
Cadmium	0.005	0.005	0.005	0.005
Chloride	250*	—	250*	—
Chromium	0.05	0.1	0.05	0.05
Copper	1.0*	1.3	1.0*	0.1
Cyanide	0.2	0.2	0.1	0.05
Fluoride	1.5	4.0	1.5	0.7
Lead	0.05	0.015	0.05	0.05
Mercury	0.001	0.002	0.001	0.001
Nickel	—	0.0	—	0.05
Nitrate (as N)	10.0	10	10	—
Sodium	200*	—	200	20
Silver	—	0.10*	—	0.01
Sulphate	500	400–500	400*	250
ORGANIC				
Alachlor	—	0.002	—	—
Aldicarb	0.009	0.003	—	—
Atrazene	0.06	0.003	—	—
Benzene	0.005	0.005	0.010	—
Carbon tetrachloride	0.005	0.005	0.003	—
1,2–dichlorobenzene	0.2	—	—	—
1,1–dichloroethane	—	—	0.000	—
1,2–dichloroethane	0.005	0.005	0.01	—
1,2–dichloroethylene	—	0.007	—	—
Ethylbenzene	0.002	0.7	—	—
Tetrachloroethylene	—	0.5	0.01	—
Toluene	0.024	1.0	—	—
Trichloroethylene (TCE)	0.05	0.005	0.030	—
1,1,1–trichloroethane	—	0.2	—	—
Xylene	0.3*	10.0	—	—

Note: all units expressed as $mg \cdot L^{-1}$; * denotes aesthetic objective

released to the environment as a component of landfill leachate (see below). In Birmingham, UK, Rivett et al. (1989; 1990) detected CHS in 78% of 59 supply boreholes tested. Forty percent of the boreholes contained TCE in excess of the 30 $\mu g \cdot L^{-1}$ World Health Organization guideline value. With respect to the 1 $\mu g \cdot L^{-1}$ guideline value set for all organochlorine compounds in the European Communities drinking-water directive (CEC, 1980), TCE, TCA, TCM and PCE showed exceedance in 62%, 22%, 17% and 9% of the boreholes, respectively. For the Coventry area, Nazari et al. (1993) and Burston et al. (1993) similarly reported TCE as the most ubiquitous with TCA ranking second. Comparable results were reported for aquifers beneath Milan by Cavallaro et al. (1986). Other DNAPLs commonly found in ground waters affected by industrial activity include the solvents methylene chloride (dichloromethane), 1,1–dichloroethane, 1,2–dichloroethane, chlorobenzene, and 1,2–dichloroethylene (Fusillo et al., 1985).

Inorganic contamination of ground water is frequently reported in industrialized areas with heavy metals, cyanide and boron the most common offenders. In Madras, India, Somasundaram et al. (1993) associate high ground-water concentrations of arsenic, mercury, lead and cadmium with industrial activity and inadequate waste disposal facilities. A lack of sewers in industrial areas has also been blamed for heavy metal contamination of ground waters in parts of South America (Foster, 1990). In Odessa, Texas, severe contamination of ground water by hexavalent chromium (locally as high as 72 $mg \cdot L^{-1}$) has been associated with the direct discharge into the soil of waste water and rinse water from chromium-plating operations (Henderson, 1994).

In Birmingham (Ford and Tellam, 1994) and Coventry (Nazari et al., 1993), two of the largest and oldest industrial conurbations in Britain, inorganic contamination is not regarded as a serious problem, with only a handful of wells displaying trace metal concentrations at the $mg \cdot L^{-1}$ level. Where elevated concentrations of copper, zinc, chromium, nickel and cadmium occur, large metal industry sites appear to be responsible. A particular concern in the Birmingham area is that many of the metals exhibit a mobility far greater than would be expected in an aquifer known for its near-neutral pH and high sorption capacity. Ford and Tellam (1994) speculate that the enhanced mobility can be explained by high metal supply rates, colloidal transport and/or complexation conditions and, in the case of chromium, its anionic form.

Underground Storage Tanks

Throughout urban communities and along major highways, large volumes of gasoline are stored in underground storage tanks (USTs; Chapter 7). It is believed that as many as 2.5 million such tanks exist in the United States (Fetter, 1993). Of these, the United States Environmental Protection Agency (US EPA) has estimated that as many as 35% eventually lose product in some way, usually due to spills or leaks associated with pressure delivery pipes. From a water quality standpoint, the most serious threat comes from four relatively soluble aromatic hydrocarbons: benzene, toluene, ethylbenzene and xylene, often referred to collectively as BTEX. BTEX chemicals are less dense than water and are only partially water soluble. They belong to a group of organic liquids known as LNAPLs (light non-aqueous phase liquids) which move less readily through the vadose zone than their heavier (DNAPL) counterparts and will tend to float on the water table or on top of the capillary fringe (Chapters 6, 7, 35).

BTEX leaking from underground storage tanks is one of the most serious potential threats to ground-water quality in a typical urban community (Chapter 5). Fortunately, the extent and severity of the problem are frequently moderated by the processes of volatilization, sorption and biodegradation. Volatilization will be most effective when the contaminants are in contact with air in soil pores in the vadose zone above the water table. However, its importance will decrease considerably when the contaminants enter the aqueous phase at depth in the aquifer. Sorption can virtually immobilize hydrocarbons that have low water solubilities and low vapour pressures, but will have little influence on the relatively soluble and volatile BTEX other than to slow its migration. Biodegradation (Davis et al., 1994) appears to be the key process if serious water quality degradation is to be avoided. Barker et al. (1987) have shown that BTEX will biodegrade if aerobic conditions are maintained, and Chiang et al. (1989) have shown that the process can be effective provided dissolved oxygen concentrations exceed 0.9 $mg \cdot L^{-1}$. Hadley and Armstrong (1991) suggest that biodegradation must account for the unexpected absence of benzene in ground waters underlying parts of California.

Septic Systems

Septic systems are used worldwide for the disposal of waste water. In the United States, for example, it has been estimated that septic systems are used by approximately one-third of the population. North American septic tanks may constitute 20 million potential point sources of ground-water contamination (Wilhelm et al., 1994). In principle, septic systems operate by discharging waste water into a tank, where solids will settle as a sludge and begin a process of anaerobic decomposition. The remaining liquid effluent is directed into a leaching bed, a system of buried open-jointed tiles or perforated pipes arranged in parallel rows; from here the liquid drains vertically through the unsaturated zone to the water table. Theoretically, aerobic oxidation of organic carbon and ammonia in the leaching bed produces carbon dioxide and nitrate. In turn, the nitrate is reduced to nitrogen gas as the waste water moves into the saturated zone beneath the drain field. If the system works efficiently, natural processes should deplete contaminants in the effluent to harmless levels within a short distance of the leaching bed.

Geochemical conditions in the soils and sediments commonly fail to provide the degree of natural attenuation necessary to adequately treat the liquid effluent. As described by Wilhelm et al. (1994), if adequate oxygen is not available in the leaching bed, aerobic digestion is incomplete and the accumulation of organic matter can cause the system to fail. Also, the reduction of nitrate to nitrogen gas is frequently inhibited by the lack of labile carbon in the natural setting. In Canada, contaminant plumes over 100 m in length have been observed in sand aquifers (Robertson et al., 1991). Typically, these plumes showed depressed levels of pH and dissolved oxygen, and elevated concentrations of chloride, nitrate, sodium, calcium, potassium, alkalinity and dissolved organic carbon. While the exact nature of the dissolved organic carbon was not identified in the Canadian study, the use of septic tank cleaning fluids containing trichloroethylene, benzene and methylene chloride has been blamed for the contamination of ground water by these chemicals on Long Island, New York (Eckhardt and Oaksford, 1988). In a study of septic systems at two sites in Australia, Hoxley and Dudding (1994) additionally report contamination of ground waters by fecal bacteria. In one case, Escherischia coli was detected at distances greater than 500 m from the suspected source.

Leaking Sewers

Contamination by leaking sewers is a common problem in heavily urbanized areas. In the Federal Republic of Germany (Eiswirth and Hotzl, 1994), it is estimated several hundred million m³ of waste water leaks from partly damaged sewage systems each year. The range of contaminants is wide and varied, but includes sulphate, chloride and nitrogen compounds, fecal pathogens, heavy metals and numerous hydrocarbons including BTEX. In Cairo, Shahin (1990) reports that sulphate derived from leaking sewers is locally causing severe damage to the concrete foundations of buildings.

Impacts from sewage wastes are even more serious in many of the less-developed nations where drainage ditches, unlined open canals and rivers are frequently used to convey a large variety of urban wastes. In Brazil, the Paraiba do Sul passes through the industrial towns of Barra Mansa and Volta Redonda, where a population of over 250,000 contributes 14,200 kg of BOD (biochemical oxygen demand) and 1790 kg of nitrogen each day (Foster, 1990; Hydroscience Inc., 1977). Obviously, even small amounts of leakage to the subsurface can generate severe ground-water quality problems. Contamination by fecal coliforms and nitrogen compounds such as nitrate and ammonium are an obvious concern and many cases have been reported (Lerner, 1994). Foster (1990), however, suggests that an even greater threat may be posed by synthetic organic chemicals contained in the sewage waste, including dichlorobenzene disinfectants and deodorants (Baxter, 1982).

Fertilizers and Pesticides

Contamination of ground waters by nitrogen fertilizers has become a common problem in urban areas (e.g., Morton et al., 1988). In most cases, the fertilizer is applied by urban residents to maintain and enhance the growth of lawns. However, indiscriminate and excessive use, together with intermittent heavy watering, can frequently lead to heavy losses by leaching. In Long Island, New York, Flipse et al. (1984) showed that fertilizer use could account for over 70% of the nitrate detected in ground waters beneath a sewered housing development in Suffolk County. In Perth, Australia, Sharma et al. (1994) installed suction lysimeters beneath fertilized urban lawns and showed that between 16 and 47% of incident water passed below the root zone carrying nutrients with flow-weighted nitrate-N concentrations ranging up to 5.37 mg·L⁻¹.

Pesticides, including insecticides, herbicides and fungicides have also been used widely in urban areas to control insects and weeds (e.g., Loague, 1994). Most incidents of ground-water contamination by pesticides appears to be confined to rural areas (Crowe and Mutch, 1994; Ricketts and Liebscher, 1994) where the practice of crop spraying is extensive. While few data are available, it is suspected that most cases of pesticide contamination in urban areas relate to spills during production, storage and transport.

Landfills

Contamination of ground waters by landfills is a concern throughout the world. In the United States, for example, Peterson (1983) reported the existence of almost 13,000 landfills including nearly 2400 open (and generally unregulated) dumps; at least 1200 are known in southern Ontario (Chapter 20).

The primary concern associated with landfills is the production of leachate, a contaminant soup which can leave the site and pollute both ground- and surface-water resources (Chapters 2, 23, 24). Leachate is produced when moisture enters the landfill and gradually dissolves the waste materials. Analyses of leachate from municipal sites all over North America show similarities that must reflect the common refuse types in municipal waste (Barker et al., 1989). Inorganic parameters are normally dominant and typically range up to 50,000 mg·L⁻¹. The primary inorganic contaminants are chloride, calcium, magnesium, sodium, potassium, iron, bicarbonate, sulphate, ammonia, nitrite/nitrate, and various forms of phosphorous; other inorganics often include nickel, manganese, zinc, copper, lead, chromium, arsenic, cyanide, mercury, strontium, molybdenum, aluminum, boron and fluoride. Leachate may also contain significant concentrations of organic acids and synthetic organic compounds such as components of petroleum, paints, household chemicals, solvents, cleaners, glues, inks and pesticides. Characteristic organics usually include volatile organic hydrocarbons and halocarbons such as benzene, ethylbenzene, toluene, xylene, dichloromethane, tetrachloroethylene, trichloroethylene, phenolics, cresol, phthalates, chlorobenzenes and other substituted benzenes. In Wisconsin, analyses of total organic carbon from municipal solid-waste landfills typically range between 400 and 6000 mg·L⁻¹ (Fetter, 1993).

Closed, unlined landfills are the primary source of potential ground-water contamination in a typical urban area (Chapter 5). In most cases, such landfills are small enough to behave as point sources of contamination, the outcome being severe degradation of ground-water quality in close proximity to the site with very limited impacts on a regional scale.

Road Salt Chemicals

Each year, sodium chloride is applied extensively to urban roads and highways throughout many snow belt regions of Europe, Canada and the United States. These chemicals are well known for damaging vegetation, weakening concrete engineering structures and corroding vehicles (Scott and Wylie, 1980). The salts are also readily mobilized by rain and melt waters, and enter rivers, lakes and shallow ground-water systems where they can cause serious degradation of water quality. Over 50% of the salt applied each year in southern Ontario eventually reaches the water table. If present rates of salt application continue, it is expected that sodium and chloride concentrations in ground waters from the region will increase gradually over the next few hundred years, until mean steady-state concentrations, in excess of 250 and 400 mg·L⁻¹, respectively, are approached. These values represent a three-fold increase over concentrations presently observed in baseflow entering urban streams and well exceed guidelines for drinking-water quality (see Chapter 8). Significantly, Pilon and Howard (1987) report chloride concentrations as high as 13,000 mg·L⁻¹ in shallow subsurface waters adjacent to a southern Ontario highway, and a survey of springs issuing from shallow sites along the shoreline of Lake Ontario by Eyles and Howard (1988) reveals average chloride concentrations close to 400 mg·L⁻¹.

Impacts on ground water are documented by numerous other workers including Huling and Hollocker (1972), Diment et al. (1973), Field et al. (1974), and Locat and Gélinas (1989). For example, in Burlington, Massachusetts (Toler and Pollock, 1974), the use of sodium and calcium chloride on state highways has been held responsible for chloride concentrations in local ground waters exceeding the drinking-water standard of 250 mg·L⁻¹. Similar degrees of water quality deterioration have been reported in Illinois (Wulkowicz and Saleem, 1974), and Wisconsin (Eisen and Anderson, 1980). In the eastern United States, concern for the impact of road de-icing chemicals on water quality and vegetation has led to the ban of these chemicals on environmentally sensitive sections of some state highways.

Urban Storm Water

Artificial recharge of ground water using urban storm-water runoff is an attractive means of augmenting ground-water supply. Unfortunately, the quality of urban runoff is often extremely poor (Lazaro, 1979); thus, considerable care is required if the quality of ground water is not to be compromised for the sake of increasing ground-water reserves (German, 1987). In Auckland, New Zealand, where storm water artificially supplies 80% of recharge to the shallow fractured aquifer, real and potential impacts on ground-water quality have generated considerable debate. As discussed by Smaill (1994), the practice of artificial recharge will likely continue because the cost of obtaining alternative fresh-water supplies is prohibitively expensive.

Urban storm water has a high sediment load, in part attributable to the hydraulic efficiency of urban channels and the high erosive capability of rapidly flowing water. Urban storm water will also carry pollutants from automobile traffic, litter, domestic refuse, dustfall and spills, lawn chemicals, animal fecal matter and, in some parts of the world, large quantities of road de-icing chemicals (Marsalek and Ng, 1987; Marsalek, 1990; Chapter 2). Various types of storm-water detention or retention ponds can be effective in removing suspended sediments, and the use of grass swales or constructed wetlands can, in addition, reduce dissolved loads, either through chemical transformation in the water column or by plant uptake (Chapter 12). However, while these types of natural treatment can often handle suspended sediment, some organics, heavy metals and nutrients, they are largely incapable of removing many of the more mobile contaminants, notably the inorganic major ions such as chloride. Furthermore, such facilities require regular maintenance, and in the case of wetlands, may simply accumulate contaminants, especially metals, which can subsequently threaten wetland wildlife.

DISCUSSION

Ground water represents the world's largest and most important source of fresh potable water, but is increasingly threatened by human activity. Unfortunately, urban ground-water systems are poorly understood and an obstacle to the management of urban watersheds. Urban development can effect considerable change to the water balance, altering not only the volumes of ground and surface water, but the distribution of water movement in space and time. Furthermore, the water balance is highly susceptible to climate change and to variations in the efficiency and maintenance of water and sewage distribution systems. A particular problem is that, in the vast majority of cases, the mechanisms, rates and distribution of aquifer recharge under natural pre-urban conditions are complex and seldom well known. On a global scale, it is the *contamination* of ground and surface waters that represents the greatest risk associated with urban development. A fundamental problem is the sheer volume of point and non-point source contaminants that urban development automatically generates: underground storage tanks containing gasoline fuels, sewer pipes, road de-icing salts, fertilizers, *etc.* Contaminants stored, transported and released at the surface commonly pollute surface water run-off, and the effects are

soon manifest in rivers and streams. Where serious degradation occurs, resolution of the problem can be equally as rapid. By comparison, contaminants stored and released directly into the sub-surface represent a far more serious concern. Contaminants transported by ground water can move extremely slowly, and impacts on the quality of the ground water may take many years to materialize; moreover, any attempt to reverse or ameliorate such impacts can take even longer to effect.

REFERENCES

Barker, J.F., Cherry, J.A., Reindard, M., Pankow, J.F. and Zapico, M., 1989, Final report: The occurrence and mobility of hazardous organic chemicals in groundwater at several Ontario landfills: R.A.C. Project No. 118 PL for Ontario Ministry of Environment, Toronto, 120 p.

Barker, J.F., Patrick, G.C. and Major, D., 1987, Natural attenuation of aromatic hydrocarbons in a shallow sand aquifer: Ground Water Monitoring Review, v. 7, p. 64-71.

Barnwell, T.O., Jr. and Kittle, J.L., 1984, Hydrologic Simulation Program – Fortran: development, maintenance and applications: Third International Conference on Stormwater Drainage, Goteburg, Sweden, Proceedings, p. 493-502.

Baxter, K.M., 1982, The effects of the disposal of sewage effluents on groundwater quality in the United Kingdom: Ground Water, v. 17, p. 429-437.

Benker, E., Davis, G.B., Appleyard, S., Barry, D.A. and Power, T.R., 1994, Groundwater contamination by trichloroethene (TCE) in a residential area of Perth: distribution, mobility and implications for management: 25th Congress of the International Association of Hydrogeologists, Adelaide, Australia, 21-25 November, Preprints of Papers NCPN No. 94/14, p. 261-266.

Bowles, D.S. and O'Connell, P.E., 1991, Recent Advances in the Modelling of Hydrologic Systems: NATO ASI Series: C. Mathematical and Physical Sciences, Kluwer Publishing, Rotterdam, v. 345, 667 p.

Brown, R.K., 1990, Urban water resources management, *in* Schilling, K.E. and Porter, E., eds., Urban Water Infrastructure: NATO ASI Series E, Applied Sciences, v. 180, Kluwer Publishing, Rotterdam, p. 71-81.

Burston, M.W., Nazari, M.M., Bishop, P.K. and Lerner, D.N., 1993, Pollution of groundwater in the Coventry region (UK) by chlorinated hydrocarbon solvents: Journal Hydrology, v. 149, p. 137-161.

Cabrera, E., 1995, ed., Improving Efficiency and Reliability in Water Distribution Systems: Kluwer Publishing, Norwell, MA, 428 p.

Cavallaro, A., Corradi, C., De Felice G. and Grassi, P.,1986, Underground water pollution in Milan and the Province by industrial chlorinated organic compounds, *in* Solbe, J.F. de L.G., ed., Effects of Land Use on Fresh Waters: Ellis Horwood, Chichester, UK, p. 68-84.

Changnon, S.A., Huff, F.A., Schickedanz, P.T. and Vogel, J.L., 1977, Summary of METROMEX, vol. 1: Weather anomalies and impacts: Illinois State Water Survey Bulletin 62, 260 p.

Chiang, C.Y., Salanitro, J.P., Chai, E.Y., Colthart, J.D. and Klein C.L., 1989, Aerobic biodegradation of benzene, toluene, and xylene in a sandy aquifer – data analysis and computer modeling: Ground Water, v. 27, p. 823-834.

Chow, V.T., Maidment, D.R. and Mays, L.W., 1988, Applied Hydrology: McGraw-Hill Book Co., New York, 572 p.

Cookson, J.T., Jr. and Leszcynski, J.E., 1990, Restoration of a contaminated drinking water aquifer: Fourth National Outdoor Action Conference on Aquifer Restoration, Ground Water Monitoring and Geophysical Methods, Dublin, OH, National Water Well Association, p. 669-681.

Council for the European Communities (CEC), 1980, Directive relating to the quality of water intended for human consumption, 15th July, 1980, (80/778/EEC): Official Journal European Communities, L229, p. 11-29.

Cox, M. and Hillier, J., 1994, Impacts on groundwater resources by urban expansion; the Brisbane case: 25th Congress of the International Association of Hydrogeologists, Adelaide, Australia, 21-25 November, Preprints of Papers NCPN No. 94/14, p. 267-270.

Crowe, A.S. and Mutch, A.S. , 1994, An expert systems approach or assessing the potential for pesticide contamination of ground water: Ground Water, v. 32, p. 487-498.

Davis, J.W., Klier, N.J. and Carpenter, C.L., 1994, Natural biological attenuation of benzene in ground water beneath a manufacturing facility: Ground Water, v. 32, p. 215-226.

Den Otter, A.A., 1988, Building Canada, Irrigation and Flood Control: University of Toronto Press, Toronto, ON, p. 162-165.

Dillon, P.J., Hickinbotham, M.R. and Pavelic, P., 1994, Review of international experience in injecting water into aquifers for storage and reuse: 25th Congress of the International Association of Hydrogeologists, Adelaide, Australia, 21-25 November, Preprints of Papers NCPN No. 94/14, p. 13-19.

Diment, W.H., Bubeck, R.C. and Deck, B.L., 1973, Some effects of de-icing salts in Irondequoit Bay and its drainage basin: Highway Research Record, v. 425, p. 23-35.

Eckhardt, D.A. and Oaksford, E.T. 1988, Relation of land use to ground-water quality in the upper glacial aquifer, Long Island, New York, in National Water Summary 1986 – Hydrologic Events and Ground Water Quality: United States Geological Survey, Water Supply Paper 2325, p. 115-121.

Eisen, C. and Anderson, M.P., 1980, The effects of urbanization on groundwater quality, Milwaukee, Wisconsin, USA, in Jackson, R.E., ed., Aquifer Contamination and Protection: UNESCO Press, Paris, SRH 30, p. 378-390.

Eiswirth, M. and Hotzl, H., 1994, Groundwater contamination by leaking sewerage systems: 25th Congress of the International Association of Hydrogeologists, Adelaide, Australia, 21-25 November, Preprints of Papers NCPN No. 94/10, 111-114.

Eyles, N. and Howard, K.W.F. 1988, Urban landsliding caused by heavy rain; geochemical identification of recharge waters along Scarborough Bluffs, Toronto, Ontario: Canadian Geotechnical Journal, v. 25, p. 455-466.

Fetter, C.W., 1993, Contaminant Hydrogeology: Macmillan Publishing Co., New York, 458 p.

Field, R., Stuzeski, E.J., Masters, H.E. and Tafuri, A.N., 1974, Water pollution and associated effects from street salting: American Society of Civil Engineering, Environmental Engineering Division, Journal, v. 100, p. 459-477.

Flipse, W.J., Jr., Katz, B.G., Lindner, J.B. and Markel, R., 1984, Sources of nitrate in ground water in a sewered housing development, central Long Island, New York: Ground Water, v. 22, p. 418-426.

Ford, M. and Tellam, J.H., 1994, Source, type and extent of inorganic contamination within the Birmingham aquifer system, UK: Journal Hydrology, v. 156, p. 101-135

Foster, S.S.D., 1990, Impacts of urbanization on groundwater, in Hydrologic Processes and Water Management in Urban Areas, Duisberg Symposium, April, 1988, International Association of Hydrological Sciences (IAHS), Publication 198, p. 187-207.

Freeze, R.A. and Cherry, J.A., 1979, Groundwater: Prentice-Hall Inc., Englewood Cliffs, NJ, 604 p.

Fusillo, T.V., Hochreiter, J.J., Jr. and Lord, D.G., 1985, Distribution of volatile organic compounds in a New Jersey coastal plain aquifer system: Ground Water, v. 23, p. 354-360.

German, E.R., 1987, Quantity and quality of drainage-well inflow in Orlando, Florida, in Gujer, W. and Krejci, V., eds., Topics in Urban Storm Water Quality, Planning and Management: IV Conference on Urban Storm Water Drainage, Lausanne, August 31–September 4, 1987, Proceedings, p. 177-178.

Hadley P.W. and Armstrong, R., 1991, Where's the benzene? – Examining California ground-water quality surveys: Ground Water, v. 29, p. 35-40.

Hall, M.J., 1984, Urban Hydrology: Elsevier Science Publishing Co., New York, 299 p.

Henderson, T., 1994, Geochemical reduction of hexavalent chromium in the Trinity Sand Aquifer: Ground Water, v. 32, p. 477-486.

Hornef, H., 1985, Leaky sewers – problems and responsibilities: Korrespondenz Abwasser, v. 32, p. 816-818

Howard, K.W.F., Boyce, J.I., Livingstone, S. and Salvatori, S.L., 1993, Road salt impacts on groundwater quality – the worst is yet to come!: Geological Society of America, GSA Today, v. 3, p. 301-321.

Howard, K.W.F. and Livingstone, S.J., 1994, Impact of urban development on surface water as a result of shallow groundwater flow – a case study from the Great Lakes Basin of North America: 25th Congress of the International Association of Hydrogeologists, Adelaide, Australia, 21-25 November, Preprints of Papers NCPN No. 94/10, p. 133-138.

Hoxley, G. and Dudding, M. 1994. Groundwater contamination by septic tank effluent: two case studies in Victoria, Australia: 25th Congress of the International Association of Hydrogeologists, Adelaide, Australia, 21-25 November, Preprints of Papers NCPN No. 94/10, p. 145-152.

Huber, W.C. and Dickinson, R.E., 1988, Storm water management model, Version 4: User's Manual: United States Environmental Protection Agency, Athens, GA, 12 p.

Huber, W.C., Heaney, J.P., Medina, W.A., Peltz, W.A., Sheikhj, H. and Smith, G.F., 1977, Storm water management model users' manual, Version II: Environmental Protection Technology Series, Municipal Environmental Research Laboratory, United States Environmental Protection Agency, EPA-670/2-75-017, 12 p.

Hueb, J.A., 1986, El programa de control de perdidas como estregis para el desarrollo de instituciones de agua potable y saneamiento: OPS-CEPIS Hojas de Divulgacion Tecnica 34 (Lima, Peru), 24 p.

Huling, E.E. and Hollocker T.C., 1972, Groundwater contamination by road salt; steady-state concentrations in east central Massachusetts: Science, v. 176, p. 288-290.

Hydroscience Inc., 1977, The Paraiba do Sul river water quality study: WHO-UNDP BRA-73/003, Technical Report 6, 14 p.

Kalinski, R.J., Kelly, W.E., Bogardi, I., Ehrman, R.L.,and Yamamoto, P.D., 1994, Correlation between DRASTIC vulnerabilities and incidents of VOC contamination of municipal wells in Nebraska: Ground Water, v. 32, p. 31-34.

Ku, H.F.H., Hagelin, N.W. and Buxton, H.T., 1992, Effects of urban storm-runoff control on ground-water recharge in Nassau County, New York: Ground Water, v. 30, p. 507-514.

Landsberg, H.E., 1981, The Urban Climate: Academic Press, New York, 275 p.

Lazaro, T.R., 1979, Urban Hydrology: Ann Arbor Science Publishers, Ann Arbor, MI, 249 p.

Lerner, D.N., 1986, Leaking pipes recharge ground water: Ground Water, v. 24, p. 654-662.

Lerner, D.N., 1990a, Recharge due to urbanization, in Lerner, D.N., Issar, A.S. and Simmers, I., eds., Groundwater Recharge: A Guide Book for Estimation of Natural Recharge: International Association of Hydrogeologists, International Contributions of Hydrogeology, Hannover, Heise, v. 8, p. 201-214.

Lerner, D.N., 1990b, Groundwater recharge in urban areas: Atmospheric Environment, v. 24, p. 29-33.

Lerner, D.N., 1990c, Groundwater recharge in urban areas, in Hydrologic Processes and Water Management in Urban Areas: Duisberg Symposium, April, 1988: International Association of Hydrological Sciences, Publication 198, p. 59-65.

Lerner, D.N., 1994, Urban groundwater issues in the U.K.: 25th Congress of the International Association of Hydrogeologists, Adelaide, Australia, 21-25 November, Preprints of Papers NCPN No. 94/14, p. 289-293.

Linsley, R.K., 1986, Flood estimates: how good are they?: Water Resources Research, v. 22, supplement, p. 159S-164S.

Loague, K., 1994, Regional scale ground-water vulnerability estimates: impact of reducing data uncertainties for assessments in Hawaii: Ground Water, v. 32, p. 605-616.

Locat, J. and Gélinas, P., 1989, Infiltration of de-icing salts in aquifers: the Trois-Rivières-Ouest case, Quebec, Canada: Canadian Journal of Earth Sciences, v. 26, p. 2186-2193.

Lyman, W.J., Reidy, P.J. and Levy, B., 1992, Mobility and Degradation of Organic Contaminants in Subsurface Environments: C.K. Smoley, Chelsea, MD, 395 p.

Marsalek, J., 1990, Stormwater management technology: recent developments and experience, *in* Schilling, K.E. and Porter, E., eds., Urban Water Infrastructure: NATO ASI Series E, Applied Sciences, Kluwer Publishing, Dordrecht, Holland, v. 180, p. 217-240.

Marsalek, J. and Ng, H.Y.F., 1987, Contaminants in urban runoff in the upper Great lakes connecting channels area: National Water Research Institute, Burlington, ON, June, 1987, Contribution 87-112, 54 p.

Martin, R.R. and Gerges, N.Z., 1994, Replenishment of an urban dune system: 25th Congress of the International Association of Hydrogeologists, Adelaide, Australia, 21-25 November, Preprints of Papers NCPN No. 94/14, p. 21-25.

Morton, T.G., Gold, A.J. and Sullivan, W.M., 1988, Influence of over watering and fertilization on nitrogen losses from home lawns: Journal Environment, v. 17, p. 124-130.

Nazari, M.M., Burston, M.W., Bishop, P.K. and Lerner, D.N., 1993, Urban ground-water pollution – A case study from Coventry, United Kingdom: Ground Water, v. 31, p. 417-424.

Oke, T.R., 1982, The energetic basis for the urban heat island: Royal Meterological Society, Quarterly Journal, v. 108, p. 1-24.

Peterson, N.M., 1983, 1983 survey of landfills: Waste Age, March, 1983, p. 37-40.

Philips, B.C., Lawrence, A.I. and Bogiatzis, T., 1992, An integrated water quality and streamflow model suite: International Symposium on Urban Stormwater Management, Sydney, Australian National Conference Publishing, v. 92, p. 402-407.

Pilgrim, D.H., 1986, Bridging the gap between flood research and design practice: Water Resources Research, v. 22, supplement, p. 165S-176S.

Pilon, P.E. and Howard, K.W.F., 1987, Contamination of subsurface waters by road de-icing salts: Water Pollution Research Journal of Canada, v. 22, p. 157-171.

Ricketts, B.D. and Liebscher, H., 1994, The geological framework of groundwater in the Greater Vancouver area: Geological Survey of Canada, Bulletin 481, p. 287-298.

Rivett, M.O., Lerner, D.N., Lloyd, J.W. and Clark, L., 1989, Organic contamination of the Birmingham aquifer: Water Research Centre, Marlow, UK, Report PRS 2064-M, 27 p.

Rivett, M.O., Lerner, D.N., Lloyd, J.W. and Clark, L., 1990, Organic contamination of the Birmingham aquifer: Journal of Hydrology, v. 113, p. 307-323.

Robertson, W.D., Cherry, J.A. and Sudicky, E.A., 1991, Ground-water contamination from two small septic systems on sand aquifers: Ground Water, v. 29, p. 82-92.

Roesner, L.A., Giguere, P.R. and Davis, L.C., 1977, Users' manual for the Storm Runoff Quality Model RUNQUAL, prepared for Southeast Michigan Council of Governments, Detroit, MI.

Roux, P.H. and Althoff, W.F., 1980, Investigation of organic contamination of ground water in South Brunswick Township, New Jersey: Ground Water, v. 18, p. 464-471.

Seaburn, G.E. and Aronson, D.A., 1974, Influence of recharge basins on the hydrology of Nassau and Suffolk Counties, Long Island, New York: United States Geological Survey, Water Supply Paper 2031, 66 p.

Scott, W.S. and Wylie, N.P., 1980, The environmental effects of snow dumping: A literature review: Journal of Environmental Management, v. 10, p. 219-240.

Seyfried, C.S., 1984, Municipal sewage treatment: Umwelt, v. 5, p. 413-415.

Shahin, M., 1990, Impacts of urbanization of the Greater Cairo area on the groundwater in the underlying aquifer, *in* Hydrologic Processes and Water Management in Urban Areas: Proceedings of the Duisberg Symposium, April, 1988, International Association of Hydrological Sciences (IAHS) Publication 198, p. 243-249.

Sharma, M.L., Herne, D. and Byrne, J.D., 1994, Nutrient discharge beneath urban lawns to a sandy aquifer: 25th Congress of the International Association of Hydrogeologists, Adelaide, Australia, 21-25 November, Preprints of Papers NCPN No. 94/14, p. 309-315.

Smaill, A., 1994, Water resource management conflicts in a shallow fractured rock aquifer underlying Auckland City, New Zealand: 25th Congress of the International Association of Hydrogeologists, Adelaide, Australia, 21-25 November, Preprints of Papers NCPN No. 94/14, p. 325-328.

Somasundaram, M.V., Ravindran, G. and Tellam, J.H., 1993, Groundwater pollution of the Madras urban aquifer: Ground Water, v. 31, p. 4-11.

Thomson, J.A.M. and Foster, S.S.D., 1986, Effects of urbanization on groundwater in limestone islands: an analysis of the Bermuda case: Institute of Water Engineers and Scientists, Journal, v. 40, p. 527-540.

Toler, L.G. and Pollock, S.J., 1974, Retention of chloride in the unsaturated zone: United States Geological Survey, Journal Research, v. 2, p. 119-123.

Van de Ven, F.H.M., 1990, Water balances of urban areas, *in* Hydrologic Processes and Water Management in Urban Areas: Proceedings of the Duisberg Symposium, April, 1988, International Association of Hydrological Sciences (IAHS) Publication 198, p. 21-33.

Wanielista, M.P., 1979, Stormwater Management, Quantity and Quality: Ann Arbor Science, MI, 383 p.

Wanielista, M.P., 1990, Hydrology and Water Quality Control: John Wiley & Sons Inc., New York, NY, 565 p.

Wilhelm, S.R., Schiff, S.L. and Cherry, J.A., 1994, Biogeochemical evolution of domestic waste water in septic systems: 1 Conceptual model: Ground Water, v. 32, p. 905-916.

Wright, A. and Parsons, R., 1994, Artificial recharge of urban wastewater, the key component in the development of an industrial town on the arid west coast of South Africa: 25th Congress of the International Association of Hydrogeologists, Adelaide, Australia, 21-25 November, Preprints of Papers NCPN No. 94/14, p. 39-41.

Wulkowicz, G.M. and Saleem, Z.A., 1974, Chloride balance of an urban basin in the Chicago area: Water Resources Research, v. 10, p. 974-982.

5. Contaminant Source Audits and Ground-water Quality Assessment

Ken W.F. Howard

Environmental Earth Sciences, University of Toronto, Scarborough Campus
1265 Military Trail, Scarborough, Ontario M1C 1A4

Stephen Livingstone

Beatty Franz & Associates Ltd., 315 Zephyr Avenue, Ottawa, Ontario K2B 5Z7

SUMMARY

Urbanization and related industrial activities release a wide range of chemical contaminants that can seriously degrade the quality of both surface and ground waters. A critical need is to identify contaminant sources as a starting point for quantitative modelling of contaminant pathways. This paper shows how a contaminant source audit can be conducted in an urban area and used to estimate potential chemical loadings on the ground-water system. Audits allow identification of particular contaminant sources and chemicals that represent the greatest threat to ground-water resources.

INTRODUCTION

One of the most serious impacts of urbanization concerns the potential for human assimilation of urban-sourced toxic chemicals. Human health is at risk when there is a chemical source, a human receptor and a pathway for chemical transmission. Transmission usually takes place either by air (intake of chemical vapour, gases or airborne particulate matter during respiration), or through water. Where chemicals are transmitted by water, human ingestion can take place either directly, through consumption of liquids, or indirectly, through intake of food (*e.g.*, fish or vegetables). All humans are at risk, but the degree can only be evaluated with a detailed knowledge of chemical sources and the environmental pathways. Combinations of these two variables are numerous and the analysis can be complex. This is especially so where transmission of chemicals takes place in the subsurface by ground-water flow. Ground water tends to move very slowly when compared to surface water; consequently, contaminants often move undetected for many years. By the time contaminated ground-water emerges as well water or surface water in streams and lakes, the contamination is often extensive and irreversible.

SUBSURFACE FLOW PATHS

In most cases, urban contaminants that are released to the subsurface will migrate gradually to the water table where they are assimilated in ground water and slowly transported to pumping wells, wetlands and surface water bodies such as streams and lakes. Impact analysis generally involves a two-stage process: (1) identifying and quantifying the source contaminants, and (2) estimating transport times for these contaminants using a knowledge of ground-water flow paths.

Transport times for the contaminants will depend strongly on local hydrogeological conditions. Rates of advective transport for chemically conservative contaminants will be determined by aquifer recharge rates, basin dimensions and aquifer properties including effective porosity, dispersivity and hydraulic conductivity. Some contaminants, notably organics, will be slowed by chemical reaction with the aquifer materials; others will undergo volatilization and various forms of biotransformation. In such cases, critical controlling factors will include pH, redox potential and the presence of certain types of bacteria, as well as the clay and organic content of the host aquifer materials. In most cases, contaminant transport times can be estimated using a calibrated ground-water flow model and a knowledge of likely physical, biological and chemical reaction processes. Techniques for modelling contaminant processes are well developed; for further detail, the reader is referred to Domenico and Schwartz (1990), and Anderson and Woessner (1991).

While modelling techniques are well advanced, the task of modelling the fate of an essentially infinite number of chemical sources in an urban environment soon becomes an impossible challenge. Clearly, some contaminant sources are likely to be more hazardous than others. Similarly, individual chemicals within each source may constitute a greater potential problem than other chemicals. In some cases, the sources may no longer exist, and the contaminants have already been released to the aquifer system. Obviously, there is a strong need for evaluation and rationalization of the contaminant sources before modelling is attempted.

The objective of this paper is to focus on the sources of ground-water contamination in a typical urban environment.

This is an essential starting point for any contaminant impact investigation. The primary intentions are as follows: (1) to demonstrate how a contaminant source audit can be used to estimate potential chemical loadings on the ground-water system; (2) to show how an understanding of contaminant release mechanisms can be used to assess the risk for ground-water contamination; and (3) to identify urban contaminant sources and individual chemicals that represent the greatest threat to surface- and ground-water resources.

STUDY AREA

Since the end of World War II, the Greater Toronto Area (GTA) has experienced rapid urbanization, marked by a dramatic shift in land use and in the use, storage and disposal of a wide range of chemicals. Point sources include numerous active and abandoned landfill sites, underground storage tanks, snow dumps and septic systems. Seasonal applications of road de-icing agents and agricultural fertilizers and pesticides represent significant seasonal non-point sources.

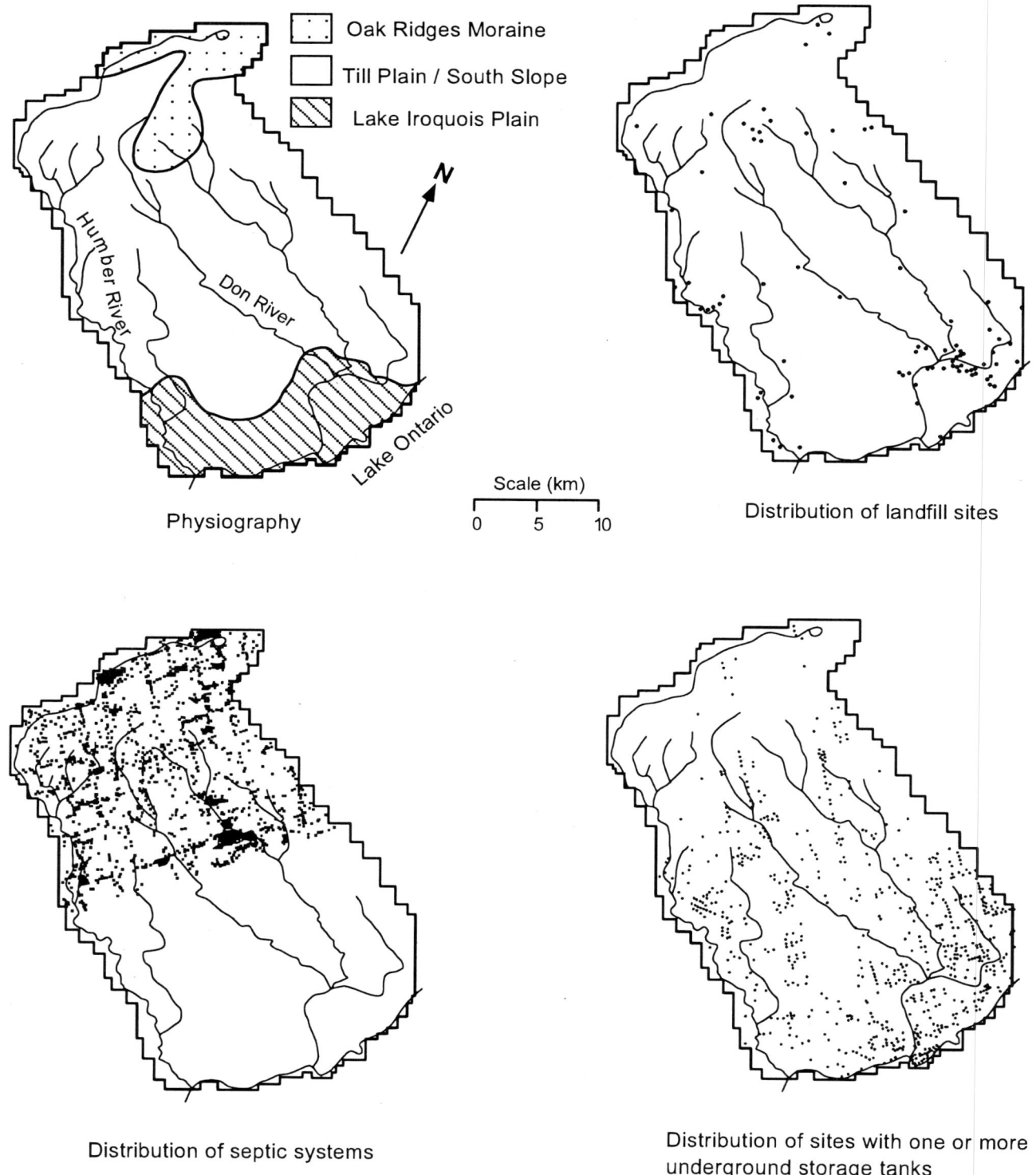

Figure 1 *(this page and facing page) Study area and location of contaminant sources (from Livingstone, 1993).*

The contaminant source audit was performed for a sub-region of the GTA (GTASR; Fig. 1), an area of some 700 km² or approximately 10% of the GTA. The GTASR extends from the downtown core on the northern shoreline of Lake Ontario to the Oak Ridges Moraine (Chapter 9). The area is representative of a typical expanding urban community of the Great Lakes basin. Under the Canada-United States Great Lakes Water Quality Agreement of 1972, the Pollution from Land Use Activities Reference Group, an organization of the International Joint Commission (IJC), was established to study contaminant loadings in the Great Lakes basin. Research activities commissioned by this group focussed mainly on atmospheric, urban runoff and surface-water pathways, but gave little regard for the transport of chemicals *via* ground water. For over a decade, in-depth research on ground-water contaminant loadings was typically limited to the study of specific contaminant sources in pilot watersheds (*e.g.*, Veska, 1977; Chan, 1978; Gillham *et al.*, 1978). As amended by Protocol in 1987, Annex 16 of the Great Lakes Water Quality Agreement of 1978 now recognizes ground-water transport of contaminants to be critical to the long-term health and viability of the Great Lakes basin.

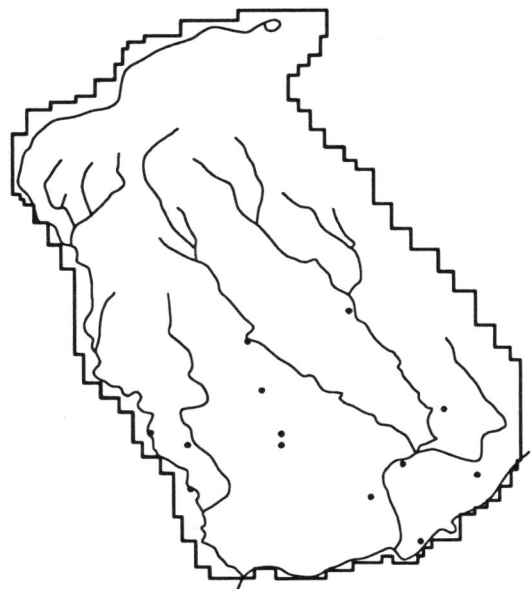

Distribution of snow dumps

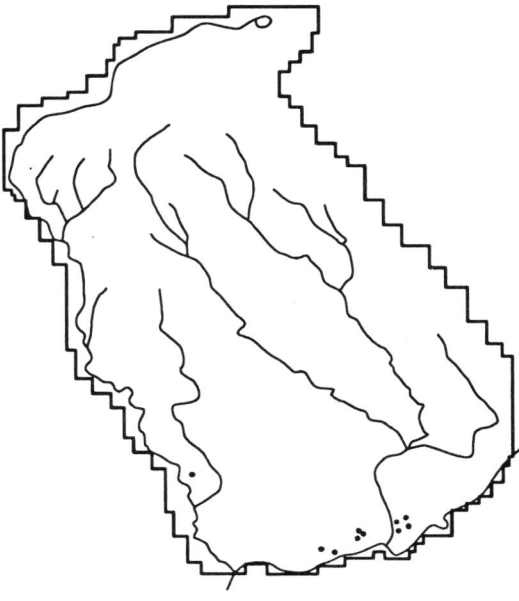

Distribution of former industrial and municipal sites producing or using coal tar

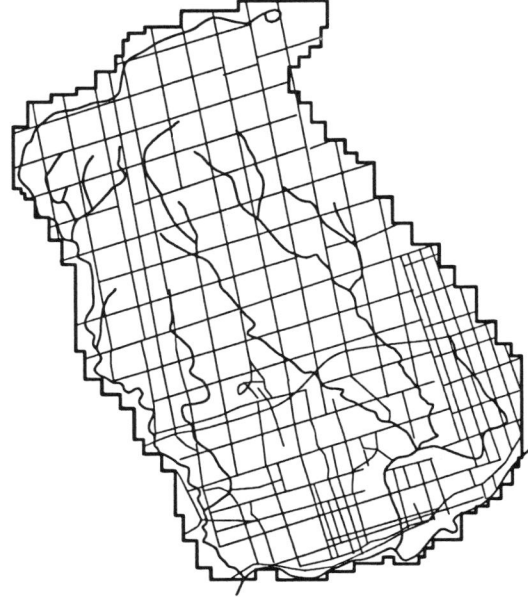

Distribution of primary and secondary roads receiving deicing agents

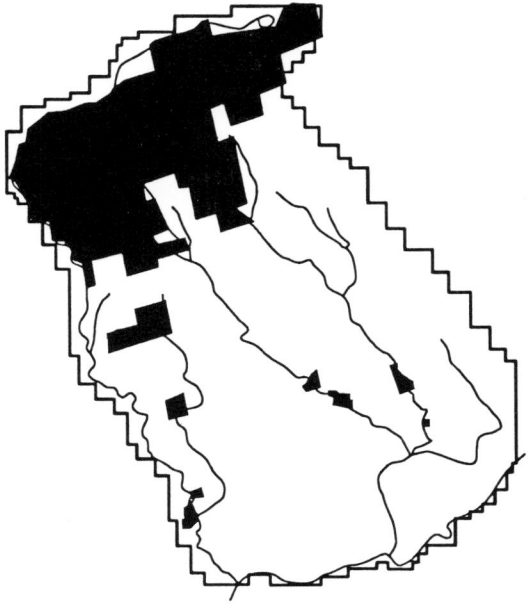

Distribution of areas receiving fertilizer and pesticide applications

CONTAMINANT SOURCES AND RELEASE CHARACTERISTICS

An audit was conducted with the intention of quantifying, for each contaminant source category (Table 1), the nature and quantity of a wide range of contaminating chemicals. Both historical and existing sources were considered; these included sources that are managed using supposedly environmentally acceptable practices. Point sources considered include abandoned and active landfill sites, underground storage tanks containing hydrocarbon fuels, snow dumps, domestic septic systems and sites known to have used or produced coal tar. Non-point sources are represented by road de-icing agents and agricultural fertilizer and pesticide applications.

Landfill Sites

Files available from the Ontario Ministry of the Environment and Energy (OMOEE) document 82 landfill sites within the GTASR. Only one is currently receiving significant quantities of municipal waste. The majority are older, closed landfills which operated mainly throughout the 1940s to 1970s and served as receptors for a wide range of refuse types, including domestic solid wastes, liquid industrial wastes, incinerator ashes, construction and commercial debris and, in some cases, hazardous wastes. Detailed information regarding volumes and exact types of waste accepted is not available since records were rarely kept at this time. It is rumoured that hazardous industrial wastes have been dumped in some of the older sites, but this is not easily verified.

Most of the landfills were sited to meet the ancillary demand for land reclamation; thus, many were located in eroding ravines and quarries. Moreover, many of the older sites were designed to leak, using the natural sediments as an attenuation zone. It can be assumed that the landfills are not unlike the 1200 others distributed across southern Ontario (Chapter 20). The vast majority of the sites are less than 10 ha in size with waste extending to a depth of <25 m below ground surface. Statistical analysis of dimensional data for landfills within and around the GTASR suggests that an area of 5 ha and a waste depth of 12 m are appropriate estimates for sites where these data are lacking. Assuming an average landfill density of 600 kg·m^{-3}, the total mass of waste landfilled in the GTASR is estimated to be 4.6×10^{10} kg.

The primary concern associated with landfills is that contaminants can be mobilized by infiltrating waters and released to the subsurface environment as leachate (Chapters 2, 4, 20, 23, 24). Currently, only two sites in the GTASR have leachate collection systems, the remainder discharge leachate directly into the subsurface.

The amount and chemical composition of the leachate produced by a landfill can be extremely variable. According to Farquhar (1989), the major cations (calcium, magnesium, sodium, and potassium) and the major anions (chloride, sulphate and carbonate) appear first in the leachate, reach a peak concentration, then slowly reduce with time. In turn, readily biodegradable organics appear, reach peak concentrations, then similarly diminish. Poorly biodegradable, low-solubility contaminants are the last to enter the leachate and last to leave the landfill. Though it is acknowledged that there is no typical leachate, there is evidence that many chemical constituents occur within relatively consistent ranges of concentration at most landfills (Table 2).

Domestic Septic Systems

In septic systems, raw domestic sewage is first directed to a septic tank where solids are segregated from the liquid effluent. Anaerobic bacterial action breaks down the solids into liquids and gases and the residual sludge is stored. Meanwhile, the liquid effluent passes to a leaching bed, and is released to the subsurface environment where natural attenuation and biodegradation normally completes the treatment process. The typical design life of a septic system is 10 to 15 years.

It is estimated that approximately 3000-5000 domestic septic systems exist in the GTASR. Most are 15 to 20 years old and are located in the north of the study area, in the Regional Municipality of York, and are centred around small hamlets and developments. Few were designed to cope with modern conveniences such as automatic washers, dishwashers and oversized bathtubs. As a consequence, complaints of faulty systems are common throughout the study area. Septic system failure can cause serious ground-water contamination (Robertson et al., 1991). In Vaughan Township of the GTASR, Chan (1978) has reported a 24-year-old septic system with a plume extending more than 50 m from the tile bed. Contaminant ground-water plumes developed by septic systems are usually characterized by elevated concentrations of chloride, nitrogen species, sodium, calcium and potassium (Chan, 1978; Robertson et al., 1991; Yates, 1985). Testing for volatile organic compounds (VOCs) by Anderson et al. (1989), Sherman and Anderson (1991) and DeWalle et al. (1985) has detected contaminants such as toluene, 1,4–dichlorobenzene, chloroethane, chloromethane, dichloromethane, and trichloromethane.

Underground Storage Tanks

According to provincial legislation, the locations and characteristics of underground storage tanks (USTs) containing petroleum hydrocarbons must be registered with the Ontario Ministry of Commercial and Consumer Relations (OMCCR) Fuel Safety Branch. Within the GTASR, one or more USTs can be found at over 700 industrial and retail sites. Detailed information is available for over 500 of these sites; just over 200 are classified as status-unknown or have incomplete data. Based on the data available, it is estimated that each site houses an average of three USTs, suggesting that approximately 2100 tanks exist throughout the GTASR. About 69% of these contain gasoline, while the remainder store diesel fuels. Tank sizes range from 800 L to over 400,000 L. However, approximately 50% are in the 20,000 to 30,000 L range with the majority (97%) ranging between 5000 to 50,000 L. Information on nearly 1200 tanks show that 62% are constructed of steel and 38% are fibreglass. Steel tanks are generally associated with installation dates before 1985 and are now typically protected from tank corrosion by internal cathodes.

Fuel tanks are stored underground to reduce the risk of fire and explosion. A drawback of underground storage, however, is that most hydrocarbon leaks go unrecognized and cause serious impact on shallow ground water (Chapters 7, 35).

Table 1	Contaminant source categories.
POINT SOURCES	**NON-POINT SOURCES**
Landfills	
Underground Storage Tanks	Road De-Icing Salts
Snow Dumps	Agricultural Fertilizers
Septic Systems	Pesticide Applications
Coal Tar Sites	

Gasoline fuels typically contain a mixture of hydrocarbon compounds generally having molecular weights below 150. Benzene, toluene, ethylbenzene, and xylenes (collectively referred to as BTEX) are the characteristic contaminants and represent approximately 25% of the total product. Diesel fuels typically contain a mixture of hydrocarbons having a molecular weight above 150 and contain BTEX components at concentrations of approximately 0.1-0.5% by weight.

Based on field observations and data from approximately 175,000 tanks nationwide, the United States Environmental Protection Agency (US EPA; 1987) estimated that 35% of all USTs leak. Environment Canada (1987) suggests a more conservative 5-10%. Direct leakage through the tank wall is usually confined to steel tanks over ten years old and normally

Table 2   Organic and inorganic constituents of Ontario landfills.

Compound/Ion	Range (mg·L^{-1})			Mean (mg·L^{-1})	Sample Size
Organic					
Benzene	<0.0001	–	0.149	0.0065	18
Chlorobenzene	<0.0002	–	0.046	0.00044	12
Ethylbenzene	<0.0002	–	0.404	0.005	14
2–butanone (MEK)	<0.0067	–	2.7	0.218	6
Butyric acid	14	–	229.2	56.7	2
1,1,1–trichloroethane	<0.0002	–	1.388	0.0005	14
1,1,2,2–tetrachloroethane	<0.0003			<0.0003	14
1,1–dichloroethane	<0.0002	–	0.1205	0.0013	13
1,2–dichloroethane	0.0002	–	2.976	0.0014	14
Ether, extractable	9	–	160	22.5	28
1,1–dichloroethene	<0.005	–	4.216	0.0017	16
Tetrachloroethene	<0.0005	–	0.36	0.004	16
Trichloroethene	<0.0002	–	0.94	0.0017	16
Cresol	<0.01	–	0.168	0.026	3
Dichloromethane	<0.0004	–	0.39	0.0019	15
Phenol	0.016	–	472	1.292	119
Diethyl phthalate	<0.002	–	0.021	0.006	3
Dimethyl phthalate	<0.002	–	<0.6	0.037	6
1,2–dichlorobenzene	<0.0002	–	<0.002	0.0003	11
1,3–dichlorobenzene	<0.0002	–	0.0084	0.0004	11
1,4–dichlorobenzene	<0.0002	–	0.024	0.0019	12
Styrene	<0.0015	–	0.0809	0.0081	5
Toluene	<0.0005	–	0.95	0.009	15
Trichlorofluoromethane	<0.001	–	<0.085	0.0027	14
Trihalomethanes (THM)	<0.0002	–	0.16	0.00075	12
Xylene (total)	<0.0002	–	5.245	0.0041	22
Inorganic					
Aluminum	ND	–	14	0.482	42
Chloride	0.4	–	12000	269.8	159
Calcium	2.8	–	4580	579.8	113
Magnesium	0.14	–	1750	265.8	104
Sodium	13.8	–	16000	935.7	118
Potassium	1.5	–	1680	259.1	74
Iron	0.01	–	1300	58.05	131
Bicarbonate	2050	–	18910	8367	8
Ammonia-N	7.6	–	1820	174.5	95
Total Kjeldahl Nitrogen (TKN)	3.5	–	2550	154	92
Nitrate-N	0.004	–	137	2.49	76
Nitrite-N	<0.001	–	66	0.077	47
Zinc	ND	–	16	1.416	92
Copper	0.007	–	7	0.045	89
Lead	0.001	–	2.1	0.068	96
Chromium	0.006	–	2.5	0.09	101
Sulphate	<1	–	4200	139.4	93
Cyanide	<0.001	–	1.2	0.136	36
Mercury	<0.00003	–	0.002	0.00016	19
Molybdenum	<0.005	–	2	0.057	37
Nickel	0.025	–	2.9	0.279	84
Boron	0.8	–	52	10.43	41
Arsenic	<0.001	–	0.32	0.0062	44
Manganese	0.03	–	793	3.54	105

accounts for just 15-20% of the total product loss. When a spill occurs, the average release is 1400-1600 L. In comparison, detailed audits have indicated that most product losses (80-85%) can be attributed to problems with spills, overflows and leaks in pressure delivery pipes. When releases of this nature occur, product losses commonly fall in the 1400 to 14,000 L range. Both steel and fibreglass tanks are similarly affected.

Snow Dumps

In a typical winter, the GTASR receives approximately 140 cm of snow, including two or three major storms with accumulations of 15 cm or more. Snow on major roadways is generally ploughed to the side and eventually removed and dumped. Johnston (1984) reported that in a typical winter 500,000 tonnes (t) of snow are removed from Toronto roads. Until 1972, approximately 20% of the snow removed was dumped into the harbour; this practice was eventually stopped due to environmental concerns. In 1969, Toronto had 13 land disposal sites; since about 1975, this has been reduced to eight.

Snow ploughed into roadside banks accumulates domestic inert particulate matter, and a wide range of contaminants such as domestic garbage, road de-icing chemicals (e.g., chloride and sodium), heavy metals (e.g., lead, arsenic, copper), oils and organic matter (Johnston, 1984). These chemicals will accompany the snow to the snow dump, and can accumulate in soils and ground water beneath the site when the snow eventually melts. Analytical data for soils beneath the snow dump site are provided in Table 3 (City of Toronto Public Works, 1985). Similar types of contamination were reported in shallow ground water by Pilon and Howard (1987).

Road De-icing Agents

In a normal winter, approximately 3200 kilometres of Metropolitan Toronto roads and highways receive approximately 65,000 t of sodium chloride (NaCl; Howard and Beck, 1993). Several thousand tons of salt are also applied to parking lots and residential driveways. In the past, it was commonly assumed that the majority of the road salt applied during the winter season is washed directly into sewers, streams and lakes by spring rains and snow melt. Recent salt balance studies in eastern Metropolitan Toronto have shown, however, that as little as 45% of the salt applied each year is flushed from the system (Chapter 8). The remainder migrates to the water table where it causes widespread degradation of ground-water quality. If these findings are extended to the GTASR, calculations suggest that contaminant loading to the ground-water system during the past 30 years amounts to over 900,000 t of chloride and nearly 600,000 t of sodium.

Former Sites Producing or Using Coal Tar

In the early 1900s, municipal coal gasification plants turned coal into a gas used for heating and lighting. Coal and coal tar wastes were also used for a variety of secondary industries, including the manufacture of creosote, varnish and abrasives. Usually waste products were buried or stored on the plant property. Within the GTASR, ten sites are known to have used or produced coal tar (Intera Technologies, 1988): three municipal coal gasification plants, four coal tar distillation plants, two plants producing roofing felt and tarred paper products, and one private gas-manufacturing plant. Many of the sites operated throughout the early 1900s, but most closed between 1940 and the early 1960s. Recent studies have found coal tar, buried wastes and contaminated ground water in evidence at six of the ten sites. Particular chemicals of concern include polycyclic aromatic hydrocarbons (PAHs),

phenolics, and light aromatic hydrocarbons (i.e., benzene, toluene, ethylbenzene and xylenes).

Agricultural Chemicals

Data on land use and the application of agricultural chemicals are available from the Ontario Ministry of Agriculture and Food. Approximately 159 km^2 of the GTASR can be classified as agricultural, but specific uses have varied according to consumer demand and the economic returns associated with various types of crop and livestock. During the 1950s and 1960s, mixed crop types, notably corn, vegetables and mixed grains, were in favour with high yields being maintained by the heavy use of fertilizer and pesticides. In more recent times, farmers seem to have reverted to a more mixed agricultural base, and are reducing chemical use in an attempt to lower costs. More seasonal crop rotations are also being implemented to take advantage of residual effects of fertilizers and to fix nitrogen in the soils.

Fertilizer materials used to supplement nitrogen include traditional manure applications and nitrogenous fertilizers (e.g., ammonium nitrate, urea and ammonium sulphate). Phosphate is supplemented by the addition of materials such as diammonium phosphate, monoammonium phosphate and single or triple superphosphates. Potash fertilizers include muriate of potash (KCl), sulphate of potash (K$_2$SO$_4$) or potassium nitrate (KNO$_3$).

Leaching of fertilizers can lead to serious ground-water contamination (e.g., Gillham et al., 1978; Gormley and Spalding, 1979; Hill, 1982; Saffigna et al., 1977). Nitrate-nitrogen is the most frequent offender. Nitrogen mass balance calculations have shown that between 25 and 50% of applied nitrogen will normally be leached into the ground water (Gillham et al., 1978; National Research Council, 1978; Lee, 1966; Sonzogni and Lee, 1974). Studies undertaken by Hill (1982), in an agricultural area just north of the GTASR, revealed that 68 of 164 ground-water samples contained nitrate-N in excess of the 10 mg·L^{-1} water quality guideline.

The use of pesticides also represents a potential source of ground-water contamination (Cohen et al., 1986). Pesticide is a general term that covers all chemicals used for pest or fungi control, i.e., insecticides, herbicides, and fungicides. Insecticides, such as diazinon, are seldom used on crops in the study area and are applied on an as-needed basis with crop rotation preferred as a means of insect control. Herbicides such as atrazine and metolachlor are applied frequently in the spring for weed control. However, fungicides applied to fruits and vegetables are normally used very sparingly.

While some applied pesticides tend to volatilize, photodegrade or degrade microbially, others are slow to decompose, are highly soluble in water and have a high potential to impact on ground-water quality. In Ontario, the Ministry of Environment has funded, over the years, a pesticide-monitoring program to identify pesticide ground-water impacts in

Table 3	Representative inorganic contaminants in soils at a snow dump.

Parameter	Average Concentration (mg·kg^{-1})	Range (mg·kg^{-1})
Chloride	805.4	36–2697
Copper	23.2	9–34
Lead	21.1	11–57
Arsenic	1.2	<1–2

Table 4   Chemicals characteristically associated with various contaminant sources.

Parameters	Landfills	Septic Systems	Underground Storage Tanks	Snow Dumps	Former Coal Tar Plants	Road De-icing Chemicals	Agriculture
Inorganic							
Aluminum	•						
Chloride	•	•		•		•	•
Calcium	•	•					
Magnesium	•	•					
Sodium	•	•		•		•	
Potassium	•	•					
Iron	•	•		•			•
Bicarbonate	•						
Ammonia-N	•	•					
Total Kjeldahl Nitrogen	•	•					•
Nitrate-N	•	•					•
Nitrite-N	•	•					•
Zinc	•						•
Copper	•			•			
Lead	•			•			
Chromium	•						
Sulphate	•	•					
Cyanide	•						
Mercury	•						
Molybdenum	•						
Nickel	•						
Boron	•						
Arsenic	•			•			
Manganese	•						
Organic							
Benzene	•		•		•		
Chlorobenzene	•						
Ethylbenzene	•		•		•		
2-Butanone (MEK)	•						
Butyric acid	•						
1,1,1–trichloroethane	•						
1,1,2,2–tetrachloroethane	•						
1,1–dichloroethane	•						
1,2–dichloroethane	•						
Ether, extractable	•						
1,1–dichloroethene	•						
Tetrachloroethene	•						
Trichloroethene	•						
Cresol	•						
Dichloromethane	•	•					
Phenol	•						
Diethyl phthalate	•						
Dimethyl phthalate	•						
1,2–dichlorobenzene	•						
1,3–dichlorobenzene	•						
1,4–dichlorobenzene	•	•					
Styrene	•						
Toluene	•	•	•		•		
Trichlorofluoromethane	•						
Trihalomethanes (THM)	•						
Xylene (total)	•		•		•		
Chloroethane							
Chloromethane					•		
Naphthalene					•		
Benzo[a]pyrene					•		
Acenaphthlene					•		
Acenaphthene					•		
Fluorene					•		
Fluoranthene					•		
Anthracene					•		
Pyrene					•		
Chrysene					•		
Phenanthracene					•		
Perylene					•		
Benzo[a]anthracene					•		
Atrazine							•

rural domestic and municipal drinking-water wells. The results reveal that the herbicides, atrazine and d-ethyl atrazine, occur in fifty percent of the wells sampled (OMOE, 1987). One well in the central region of Ontario (*i.e.*, the GTA) showed persistent levels of atrazine at a depth of approximately 40 m. Moreover, pesticide residues were frequently found in significant concentrations in rivers draining heavily agriculturalized watersheds. The evidence of herbicide impacts to ground water is not unexpected since recent figures (Ontario Ministry of Agriculture and Food, 1988) suggest that herbicides represent 70% of all pesticides applied to Ontario agricultural lands (mostly field crops). The herbicides used in the largest quantities were metolachlor (1709 t) and atrazine (1041 t). Atrazine has been linked to skin irritation, lung and respiratory effects, blood cell disorders, embrotoxicity and mutagenicity (Blair *et al.*, 1985).

SELECTING CHEMICALS FOR MASS-LOADING CALCULATIONS: THE SCREENING PROCESS

Each of the seven contaminant sources considered here contains numerous inorganic and organic chemicals (Table 4). In order to reduce the number of chemicals to a manageable size, a screening process was developed to identify chemicals of greatest concern. In this process, each of the chemicals listed in Table 4 was examined in light of the following:

1.   Has the chemical been classified as persistent and undesirable in the environment according to the International Joint Commission Critical Contaminants list?

2.   Have Canadian or United States government agencies established drinking-water quality guidelines for the chemical?

3.   Is the chemical present in highly significant quantities or can it be regarded as an indicator parameter for a particular source category?

4.   Has the chemical been identified as significant in two or more source categories?

Chemicals which satisfied question 1 were automatically selected for mass-loading calculations. Chemicals which could be shown to fully satisfy conditions 2, 3 and 4 were also selected provided the water quality criteria were established primarily for health, rather than aesthetic, purposes. Contaminants satisfying only conditions 2 and 3 were chosen if they were organic. Inorganic chemicals satisfying the same conditions were selected if they were deemed to be especially hazardous (*e.g.*, mercury and cyanide) or present in extremely high quantities. Table 5 lists the chemicals finally selected for the mass-loading estimate together with the more commonly quoted guidelines for drinking-water quality. It should be noted that not all of these guidelines are currently recognized by the Province of Ontario.

CHEMICAL MASS-LOADING ESTIMATES

Chemical mass loading will depend on the total mass of chemical within each contaminant source and the loading history, which describes the rate at which the contaminant is released to the subsurface. Loading generally occurs in one of four ways: (1) a spill or pulse load (*e.g.*, contaminant releases from underground storage tanks); (2) continuous loading with variable concentration from a decaying finite mass source (*e.g.*, landfill sites, coal tar sites); (3) a pulse load which is seasonally variable (*e.g.*, road salting, agricultural applications, snow dumps); or (4) a continuous source loading with constant concentration (*e.g.*, septic systems).

Chemical mass-loading estimates were made for contaminants known to have been stored or used in the GTASR up to the present time. For most chemicals, this would mean a time frame of about 30 to 50 years. No attempt was made to predict how policy change, *e.g.*, changes in landfill design or road salting procedures, might influence loadings in the future. Also, while it is acknowledged that many contaminants are susceptible to some decline in source strength caused by biological and chemical transformations, these were not included. In this regard, estimates will tend to err on the side of conservatism. In many cases, estimates were constrained by a limited database or an inadequate knowledge of contaminant release behaviour. In general, therefore, two estimates were made to represent best and worst case scenarios.

Landfill Sites

Landfill sites contain a finite mass of contaminants which declines as contaminants are released in the form of leachate. Landfill leachate will normally take several years to achieve its peak concentration (Farquhar, 1989); thereafter the loss of contaminant mass will result in a reduction of leachate strength with time. Most of the readily dissolvable mass will be depleted within 50 years; low-solubility contaminants may, however, persist considerably longer (Chapter 24).

The vast majority of landfill sites in the study area have not been monitored to determine their impacts on ground-water quality or contaminant mass loadings. Reliance must therefore be placed on data from existing landfills. Hughes *et al.* (1971) and Reitzel (1990) suggest that chloride represents between 0.1% and 0.2% of the mass of the landfill waste. However, few data are available for other chemical species, particularly organics. Rowe (1990) has proposed that an upper-limit mass estimate can be derived by assuming a constant ratio exists between the mass percent of a particular contaminant and its peak concentration in the leachate. Rowe (pers. comm.) also suggests that the peak concentration for chloride in leachate from Ontario landfills is best represented by a value of 1600 mg·L^{-1}. Estimates of potential contaminant loading shown in Table 6 are based on Rowe's approach. They assume that chloride represents 0.2% of

---

Table 5   Contaminants selected for mass-loading calculations with water quality guidelines.

Parameter	Water Quality Guidelines (mg·L^{-1})
Inorganic	
Chloride	250
Sodium	200
Nitrate-N	10
Copper	1
Lead	0.01
Chromium	0.05
Cyanide	0.2
Mercury	0.001
Arsenic	0.05
Organic	
Benzene	0.005
Ethylbenzene	0.0024
Tetrachloroethene	0.5
Trichloroethene	0.05
Dichloromethane	0.05
Phenol	0.002
1,4–dichlorobenzene	0.001
Toluene	0.024
Xylene	0.3
Naphthalene	—
Atrazine	0.06

the total landfill mass and that the peak chloride concentration is 1600 mg·L⁻¹. Peak concentrations for other solutes are taken from Table 2. Also shown in Table 6 are more moderate estimates of mass using mean rather than peak concentrations.

Domestic Septic Systems

Within the GTASR, most of the septic systems have operated for approximately 20 years with an average effluent discharge of approximately 1000 L per day per dwelling unit (OMOE, 1992). Between 3000 and 5000 septic systems exist in the GTASR and these numbers were used to provide the mass-

loading estimates shown in Table 7. As shown, chloride is the dominant inorganic contaminant with a maximum load approaching 3600 t, followed by nitrate-N (maximum 3100 t). Toluene is the primary organic contaminant (7.8 t) followed by dichloromethane and 1,4–dichlorobenzene. It should be noted that values for NO_3-N assume that all nitrogen species in the effluent are converted to nitrate-N, a behaviour observed in field studies by Robertson et al. (1991).

Underground Storage Tanks

Releases from underground storage tanks and distribution

Table 6   Contaminant mass loadings for GTASR landfill sites (total waste mass = 4.6×10^{10} kg).

Parameter	Peak concentration in leachate (mg·L⁻¹)	Total mass estimated from peak concentration (tonnes)	Mean concentration in leachate (mg·L⁻¹)	Total mass estimated from mean concentration (tonnes)
Chloride	1600	92,000	270	92,000
Sodium	16,000	920,000	935.7	318,800
Cyanide	1.2	69	0.136	46.3
Mercury	0.002	0.1	0.000164	0.1
Lead	2.1	120.8	0.068	23.2
Nitrate-N	137	7877.5	2.49	848
Arsenic	0.32	18.4	0.0062	2.1
Chromium	2.5	143.8	0.9	307
Copper	7	402.5	0.045	15.3
Phenol	472	27,140.0	1.292	440
Benzene	0.149	8.6	0.0065	2.2
Ethylbenzene	0.404	23.2	0.005	1.7
Toluene	0.95	54.6	0.009	3.1
Xylene	5.245	301.6	0.0041	1.4
Tetrachloroethylene	0.36	20.7	0.004	1.4
Dichloromethane	0.39	22.4	0.0019	0.6
1,4–dichlorobenzene	0.024	1.4	0.0019	0.6
Trichloroethene	0.94	54.1	0.0017	0.6
Naphthalene	0.04	2.3	0.02	6.8

Table 7   Mass loading from septic systems.

Mass Contaminant Loading: Based on Peak Concentrations and 5000 Septic Systems

Parameter	Peak concentration (mg·L⁻¹)	Contaminant Mass per System each year (kg)	Total Mass Loading in GTASR (tonnes per year)	Total Mass Loading over 20 years (tonnes)
Chloride	98	0.017542	179.0	3580
Nitrate-N	84	0.012852	153.0	3070
Toluene	0.213	0.078	0.4	7.8
Dichloromethane	0.007	0.0026	0.01	0.26
1,4–dichlorobenzene	0.0033	0.0012	0.01	0.20

****

Mass Contaminant Loading: Based on Low Concentrations and 3000 Septic Systems

Parameter	Low Concentration (mg·L⁻¹)	Contaminant Mass per System each year (kg)	Total Mass Loading in GTASR (tonnes per year)	Total Mass Loading over 20 years (tonnes)
Chloride	35	0.0013405	38.3	767
Nitrate-N	40	0.001752	43.8	876
Toluene	0.007	0.0026	0.008	0.15
Dichloromethane	0.0019	0.0007	0.002	0.04
1,4-dichlorobenzene	0.0021	0.0008	0.002	0.05

lines occur as discrete leaks or spills. Given that most of the tanks within the GTASR have been installed since 1970, loading estimates are based on potential for leakage over a 20-year period. Maximum and minimum estimates of contaminant loading are shown in Table 8. The estimates assume that tank failure produces an average product leak of 1500 L while distribution line failure releases an average of 7500 L. It was further assumed that benzene, toluene, ethylbenzene, and xylenes represent 3%, 10%, 2% and 11%, respectively (typical for unleaded gas), of the total product released. The broad range of total mass loadings shown in Table 8 reflects the sensitivity of the assumptions used, particularly the percentage of tank failures. Given the uncertainties, the results provided represent little more than order of magnitude approximations.

Snow Dumps

Within the GTASR, snow dumps have been releasing contaminants to the subsurface for the last 20 to 30 years. Mass-loading estimates shown in Table 9 assume that approximately 5% (water equivalent) of all snow disposed at snow dumps recharges the ground water, with the remainder discharging into surface waterbodies or storm sewers as runoff. As expected, chloride and sodium are the dominant contaminants; however, loadings of copper, lead and arsenic can also be significant.

Former Sites Producing Or Using Coal Tar

Very approximate estimates of contaminant loading from coal tar sites in the GTASR are shown in Table 10. A review of historical site maps suggests that heaviest contamination is typically constrained to an area 25 m by 25 m and extends to a depth of about 1 m. If a saturated porosity of 30% can be assumed, approximately 187,500 L of water will be severely contaminated at source. Actual loadings will depend on the concentration of contaminants in the water and the regularity with which the water is replaced. In Table 10, upper mass limits are based on the maximum solubility of the contaminants, but assume that contamination is limited to only 187,500 L of water per site. While the former assumption is likely to err severely on the side of conservatism, any bias in the estimate will be compensated at least partially by the latter which, depending on recharge conditions and soil permeability, could be underestimated by a factor of at least ten.

Road De-icing Chemicals

Mass-loading estimates for NaCl road de-icing chemicals are shown in Table 11. The estimates assume that salt is applied only to urbanized regions of the GTASR (460 km^2) and that the application rate has averaged 200 g·m^{-2} per year for a period of 30 years. 55% of the salt enters the ground water while the remainder is flushed from the area by snow melt and spring rains.

Table 8   **Mass loading from underground storage tanks.**

Low Estimate Based on 5% Leakage Rate †

	Mass of Product (L)	Benzene (tonnes)	Toluene (tonnes)	Ethylbenzene (tonnes)	Xylenes (tonnes)
Leakage from Tank	24,000	0.7	2.4	0.5	2.6
Leakage from Delivery Pipes	675,000	20.3	67.5	13.5	74.3
Totals		21.0	69.9	14.0	76.9

†   Criteria: 1)   5% of 2124 tanks leak = 106
              2)   15% of all leaks are from tank storage: Average leak = 1500 L
              3)   85% of all leaks are from delivery pipes: Average leak = 7500 L

Peak Estimate Based on 35% Leakage Rate ††

	Mass of Product (L)	Benzene (tonnes)	Toluene (tonnes)	Ethylbenzene (tonnes)	Xylenes (tonnes)
Leakage from Tank	169,500	5.1	17.0	3.4	18.6
Leakage from Delivery Pipes	4,785,000	143.5	478.5	95.7	526.4
Totals		148.6	495.5	99.1	545.0

††   Criteria: 1)   35% of 2124 tanks leak = 750
               2)   15% of all leaks are from tank storage: Average leak = 1500 L
               3)   85% of all leaks are from delivery pipes: Average leak = 7500 L

Table 9   **Mass loading from snow dumps (estimated total snow volume ≃ 2.7 × 10^5 m^3).**

	Average Peak Concentration (mg·L^{-1})	Peak Mass Loading in GTASR (tonnes)	Minimum Concentration (mg·L^{-1})	Minimum Loading in GTASR (tonnes)
Chloride	805	218	36	9.7
Sodium	536	145	26	6.9
Copper	23.2	6.3	9	2.4
Lead	21.1	5.7	11	3.0
Arsenic	1.2	0.32	1	0.27

Table 10   **Mass loading from former coal tar sites.**

	Low Mass Loading Estimate Based on Peak Concentrations from Ataritiri Site		*Peak Mass Loading Estimate Based on Chemical Solubility*	
Parameter	Peak concentrations observed at Ataritiri site (mg·L^{-1})	Low mass loading estimate for all 6 GTASR sites (tonnes)	Maximum concentration based on chemical solubility (mg·L^{-1})	Peak mass loading estimate for all 6 GTASR sites (tonnes)
Benzene	8.4	0.0016	1780	2.00
Toluene	5.1	0.0010	500	0.56
Ethylbenzene	1.9	0.0004	150	0.17
Xylenes	3.3	0.0006	150	0.17
Naphthalene	6.7	0.0013	31.7	0.04

Note: Assumes contamination at each site saturates an area 25 m × 25 m to a depth of 1 m (n=30%).

Table 11   **Road salt mass loading.**

Parameter	Tonnes applied annually to GTASR	Tonnes released annually to ground water	Total mass loading over 30 years (tonnes)
Chloride	56,000	30,800	924,000
Sodium	36,000	19,800	594,000

Agricultural Chemicals

Agricultural chemicals have been applied during the past 30 years to approximately 159 km^2 of the GTASR. Estimates of contaminant loading from agricultural activities are shown on Table 12 and assume that between 25 and 50% of the applied chemicals are leached into the ground water. Fertilizer application rates were estimated based on the rates recommended for corn by the Ontario Ministry of Agriculture and Food (1992a, b). Application rates for atrazine are typically in the range 1 to 1.5 kg·ha^{-1} and the upper value was taken to err on the side of conservatism.

COMPARISON OF CONTAMINANT MASS LOADINGS

The relative importance of the various contaminant sources in terms of the total mass impact is shown in Figure 2. To date, the total mass loading from all sources may have exceeded 2,600,000 t, equivalent to 4 kg of chemical for every square metre of the GTASR study area. As indicated, road de-icing chemicals are the primary contributor, with an estimated loading of over 1.5 million tonnes or 58% of the total. Landfills are close behind, contributing perhaps 1 million tonnes or

40% of the total. Other sources pale in comparison with agricultural chemicals, accounting for just 1% of the total, and the remaining sources accounting for another 1%.

POTENTIAL IMPACTS ON WATER QUALITY

To provide effective comparisons of potential impacts, the mass of each chemical released by the contaminant sources can be divided by the corresponding drinking-water criterion. This procedure essentially determines the volume of water that would be contaminated to the drinking-water standard by the available mass. Volumes are shown in Table 13. To put these values in context, the volume of ground water immediately underlying the GTASR study area is approximately 10^{13} L and the volume of ground-water recharge in the GTASR each year is 10^{11} L. With regard to Lake Ontario, it is not clear how much of its 1.64 × 10^{15} L (Schertzer, 1982) are likely to be impacted by the influx of contaminants. Neilson and Stevens (1986) used a variety of multivariant statistical analysis to divide Lake Ontario into six major water quality zones (Fig. 3). The zonation observed correlates with the general pattern of nearshore counter-clockwise flow (Simons and Schertzer, 1987). Contaminants from the study area will enter Zone 4 which, accounting for nearshore depth, represents just 5% of the volume of Lake Ontario or approximately 8.0 × 10^{13} L. It is this zone that supplies municipal drinking water for Metropolitan Toronto, and into which large volumes of sewage are discharged (Chapter 2).

Given inadequacies in the data set and the many broad assumptions made during the analysis, the volumes of contaminated water shown in Table 13 must be treated with a considerable degree of caution. Nevertheless, the results do reveal several features of interest; they also highlight areas

Table 12   **Mass loading to agricultural land.**

		Peak Estimate: Assumes 50% Leaching Rate		*Low Estimate: Assumes 25% Leaching Rate*	
Parameter	Application rates (kg·ha^{-1})	Mass loading (tonnes per year)	Mass loading over 30 years (tonnes)	Mass loading (tonnes per year)	Mass loading over 30 years (tonnes)
Nitrate-N	100	797	23,910	399	11,955
Chloride	14	112	3347	56	1674
Atrazine	1.5	12	359	6	179

requiring closer re-examination. Key observations are as follows:
1.   Contaminant sources contributing organic contaminants to the subsurface appear to have the greatest impact potential. This is largely a reflection of their stringent water quality criteria. Unfortunately, since the source sizes and biochemi-

cal behaviour of the organic contaminants are often least well known, estimates of impact potential for the organic contaminants are generally less reliable than those for the inorganic contaminants.
2.   Landfill leachate represents the greatest threat to

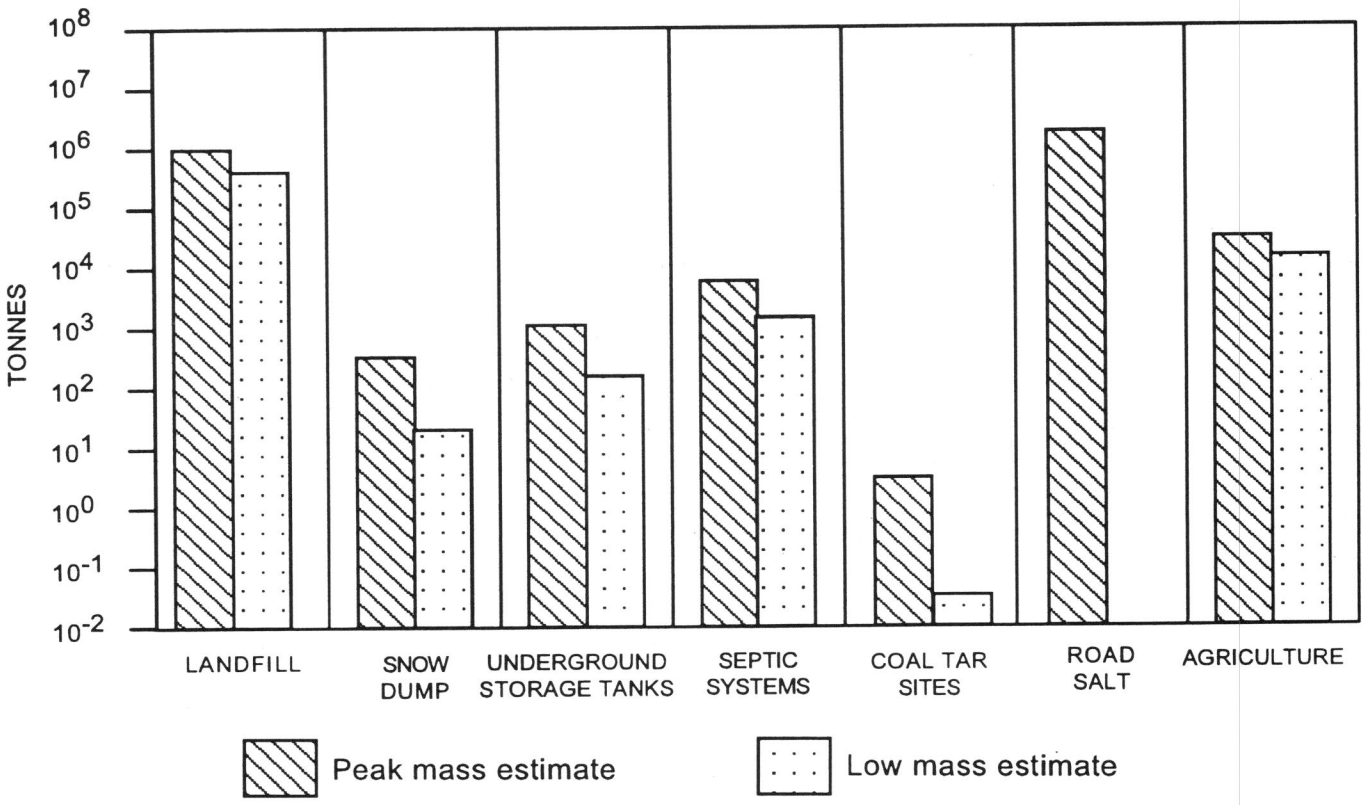

Figure 2   *A comparison of total mass impacts for the various contaminant sources.*

Table 13	Impact potential: volume of water (in litres) contaminated to the drinking water standard.						
Parameter	Landfills	Septic Systems	Underground Storage Tanks	Snow Dumps	Former Coal Tar Plants	Road De-icing Chemicals	Agriculture
Inorganic							
Chloride	3.7×10^{11}	1.4×10^{10}		8.7×10^{8}		3.7×10^{12}	9.2×10^{10}
Sodium	4.6×10^{12}			7.3×10^{8}		3.0×10^{12}	
Nitrate-N	7.9×10^{11}	3.1×10^{11}					3.3×10^{11}
Copper	4.0×10^{11}			6.3×10^{9}			
Lead	1.2×10^{13}			5.7×10^{11}			
Chromium	2.9×10^{12}						
Cyanide	3.5×10^{11}						
Mercury	1.0×10^{11}						
Arsenic	3.7×10^{11}			6.3×10^{9}			
Organic							
Benzene	1.7×10^{12}		3.0×10^{13}		4.0×10^{11}		
Ethylbenzene	9.7×10^{12}		4.2×10^{13}		7.1×10^{10}		
Tetrachloroethene	4.1×10^{10}						
Trichloroethene	1.1×10^{12}						
Dichloromethane	4.5×10^{11}	5.2×10^{9}					
Phenol	1.4×10^{15}						
1,4–dichlorobenzene	1.4×10^{12}	2.0×10^{11}					
Toluene	2.3×10^{12}	3.3×10^{11}	2.1×10^{13}		2.1×10^{13}		
Xylene (total)	1.0×10^{12}		1.8×10^{12}		5.7×10^{8}		
Naphthalene	No quality standard						
Atrazine							6.0×10^{12}

ground-water quality. Many organics, and phenols in particular, are capable of contaminating very significant volumes of water. It should be emphasized, however, that the risk assessment depends heavily on the estimates of contaminant mass at source, and these, in turn, were estimated indirectly using a knowledge of the contaminant's peak concentration as observed in Ontario leachates (Table 2). The use of mean concentrations to estimate contaminant masses (Table 4) would, in many cases, have reduced estimated volumes of water contaminated by up to two orders of magnitude. The present landfill database is woefully inadequate for reliable impact assessment, and there is a serious need to improve both knowledge and understanding of chemicals stored in, or emanating from, GTASR landfills.

3.   Underground storage tanks containing benzene, toluene, ethylbenzene and xylenes constitute a serious potential threat to ground water in an urban environment. Volatilization and biodegradation will obviously alleviate the risk in many situations and a full appreciation of these processes and the environmental conditions under which they operate will be critical if our present understanding of risk potential is to be refined.

4.   Road de-icing agents and applications of agricultural chemicals during the past 30 years probably rank third and fourth in terms of their present impact potential . However, these sources must be taken very seriously given that there is a strong likelihood that use of these chemicals will continue for many years to come. During the past thirty years, the amount of water in storage beneath the GTASR (approximately 10^{13} L) has been sufficient to dilute the influx of road salt and agricultural chemicals to relatively safe levels. However, continued use of these chemicals will cause the quality of the ground water to deteriorate until its quality eventually matches that of the contaminated recharge water. In this regard, it should be noted that annual recharge in the GTASR (10^{11} L) is less than the volumes of water that would be contaminated over the same period by chloride in road salt (3.7×10^{12} L) or by atrazine from agricultural chemicals (6 ×

10^{12} L). Given that the residence time for ground water in the region averages approximately 100 years, the evidence suggests that the water quality guidelines for chloride will be exceeded well within this time frame. In the case of atrazine, the degree and rate of degradation will be somewhat lessened by the ameliorating effects of biodegradation and soil-water reaction.

5.   Septic systems, snow dump sites and coal tar sites represent the least threatening sources on a regional scale. However, it should be noted that, as with all the sources considered, serious degradation may be anticipated on a very local scale. Lead derived from snow dump sites is particularly worthy of close re-examination. Similarly, toluenes from former coal tar sites deserve close attention. Significantly, in the GTASR at least, septic systems may have contributed almost as much nitrate to the ground water as agricultural fertilizer.

DISCUSSION

A contaminant source audit represents an important starting point in any contaminant transport study. It provides a means of identifying contaminant sources and individual chemicals worthy of detailed consideration. It also highlights inequities and inadequacies in the chemical dataset. In the case study described, the GTASR represents a typical urban area in the Great Lakes basin. Landfills and underground storage tanks represent the most serious threat to ground-water quality. However, closer accounting of existing landfill contents is required if reliable impacts are to be made. In the future, changes in landfilling practice and storage tank design should reduce or even eliminate additional impacts to the system from these sources. However, continued deployment of road de-icing chemicals and agricultural chemicals may cause serious degradation in the longer term.

ACKNOWLEDGEMENTS

The authors gratefully acknowledge numerous University of

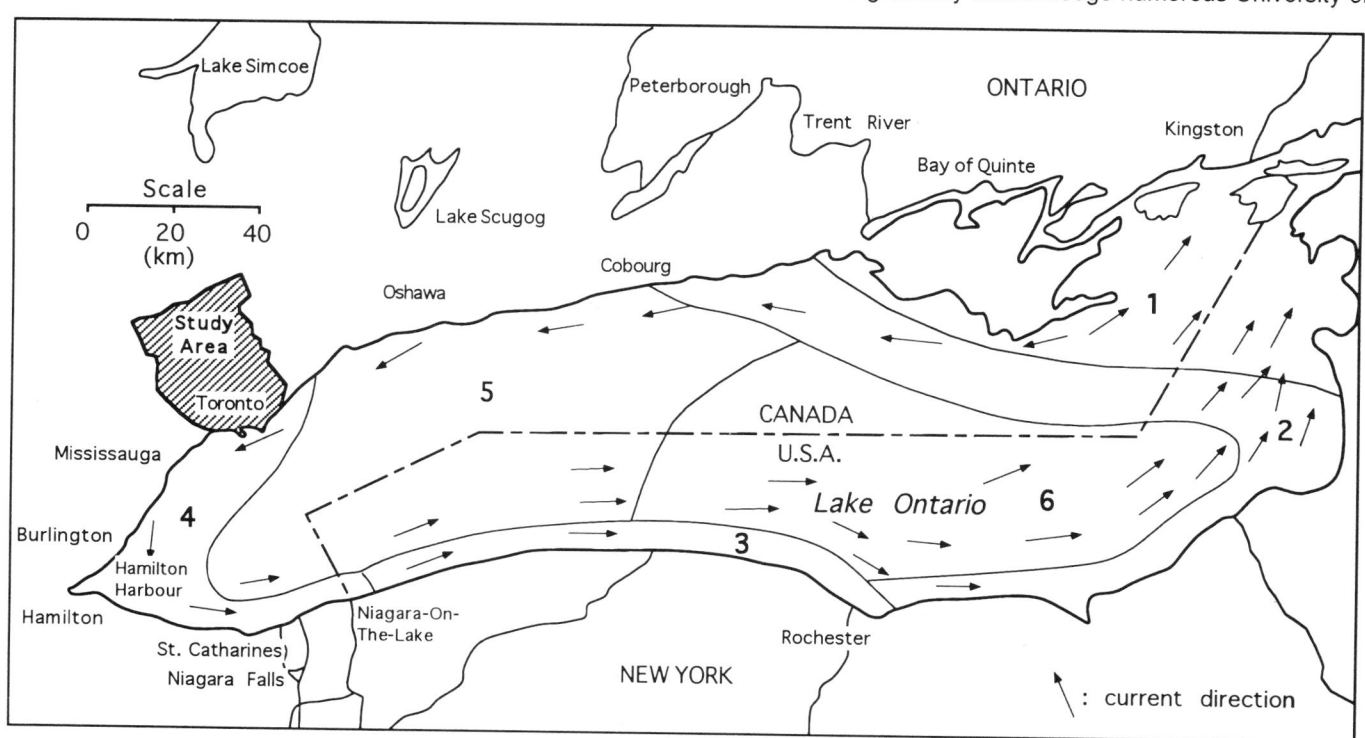

Figure 3   *Water quality zonation in Lake Ontario (after Neilson and Stevens, 1986).*

Toronto students and colleagues who assisted in the collection and interpretation of the data. Particular thanks go to Sean Salvatori, Philip Smart, Richard Gerber and Janet Haynes. The work was supported by research grants to the senior author from the Great Lakes University Research Fund (sponsored by Environment Canada and the Natural Sciences and Engineering Research Council) and the Ontario Ministry of the Environment. The views presented here are those of the authors and are not necessarily endorsed by the funding agencies.

REFERENCES

Anderson, D.L., Lewis, A.L. and K.M. Sherman, 1989, Unsaturated zone monitoring below subsurface wastewater infiltration systems serving individual homes in Florida, in Focus Conference on Eastern Regional Groundwater Issues, Kitchener, ON, October 17-19, 1989: National Water Well Association, 216 p.

Anderson, M.P. and Woessner, W.W., 1991, Applied Groundwater Modelling, Simulation of Flow and Advective Transport: Academic Press Inc., Harcourt Brace Jovanovich, San Diego, CA, 381 p.

Blair, A., Malker, H., Cantor, K.P., Burmeister, L. and Wiklund, K., 1985, Cancer among farmers: A review: Scandinavian Journal of Work and Environmental Health, v. 11, p. 397-407.

Chan, H.T., 1978, Contamination of the Great Lakes by private wastes: Report prepared by the International Joint Commission Pollution from Land Use Activities Reference Group, Task Group C (Canadian Section) Parts 1 and 2, Activity 3, Windsor, ON, PLUARG 78-080.

City of Toronto, Department of Public Works, 1985, Analysis of snow samples: Unwin Avenue & Pottery Road snow disposal areas: 56 p.

Cohen, S.Z., Eiden, C. and Lorber, M.N., 1986, Monitoring groundwater for pesticides, in Garner, W.Y., Honeycutt, R.C. and Nigg, H.N., eds., Evaluation of Pesticides in Groundwater: American Chemical Society, Washington, DC, ACS Symposium Series No. 315, p. 170-196.

DeWalle, F.B., Kalman, D.A., Normam, D., Sung, J. and Plens, G., 1985, Determination of toxic chemicals in effluent from household septic tanks: United States Environmental Protection Agency, Office of Research and Development, EPA 600/2-85/050, Cincinnati, OH, 25 p.

Domenico, P.A. and Schwartz, F.W., 1990, Physical and Chemical Hydrogeology: John Wiley and Sons, New York, 824 p.

Environment Canada, 1987, Leaking UST Newsletter, Sept-Nov, 1987, v. 1. [ISSN 0832-7580]

Farquhar, G.F., 1989, Landfill leachate migration in soil: Canadian Geotechnical Journal, v. 6, p. 51-87.

Gillham, R.W., Cherry, J.A., Hendry, M.J., Sklash, M.G., Frind, E.O., Blackport, R.J. and Pucovsky, M.G., 1978, Studies of the agricultural contribution to nitrate enrichment of groundwater and subsequent loading to surface water, Final report: International Joint Commission, Pollution from Land Use Activities, Task Group C, Activity 1 Project 14, PLUARG 78-064, Windsor, ON, 203 p.

Gormley, J.R. and Spalding, R.F., 1979, Sources and concentrations of nitrate-nitrogen in ground water of the central Platte region, Nebraska: Ground Water, v. 17, p. 291-300.

Hill, A.R., 1982, Nitrate distribution in the ground water of the Alliston Region of Ontario, Canada: Ground Water, v. 20, p. 696-702.

Howard, K.W.F. and Beck, P.J., 1993, Hydrogeochemical implications of groundwater contamination by road de-icing chemicals: Journal of Contaminant Hydrology, v. 12, p. 245-268.

Hughes, G.M., Landon R.A. and Farvolden, R.N., 1971, Hydrogeology of solid waste disposal sites in northeastern Illinois: United States Environmental Protection Agency, Report SW-12d, 152 p.

Intera Technologies Ltd., 1988, Inventory of Coal gasification waste sites in Ontario, v. I and II, 129 p.

Johnston, T., 1984, Snow disposal in Metropolitan Toronto, in Minimizing the environmental impact of the disposal of snow from urban areas: Proceedings of Workshop June 11-12, 1984, Report EPS 2/UP/1, October 1985.

Lee, G.F., 1966, Report on the nutrient sources subcommittee of Lake Mendota problems: International Joint Commission, Madison, WI, 76 p.

Livingstone, S.J., 1993, Assessment of the impact on Lake Ontario of ground-water contaminant mass loading from a representative region of the GTA: unpublished M.Sc. thesis, University of Toronto, Toronto, ON, 186 p.

National Research Council, 1978, Nitrates: An Environmental Assessment: Environmental Studies Board, Commission on Natural Resources, Washington, DC, United States Environmental Protection Agency, Contract No. 68-01-3253, 488 p.

Neilson, M.A. and Stevens, R.J.J., 1986, Determination of water quality zonation in Lake Ontario using multivariate techniques, in El-Shaarawi, A.H. and Kwiatkowski, R.E., eds., Statistical Aspects of Water Quality Monitoring: Developments in Water Science, v. 47: Elsevier Science Publishers, Amsterdam, p. 99-116.

Ontario Ministry of Agriculture and Food, 1988, Survey of pesticide use in Ontario: Economics Information Report 89-08, 40 p.

Ontario Ministry of Agriculture and Food, 1992a, 1991-1992 Field crop recommendations: Publication 296, 95 p.

Ontario Ministry of Agriculture and Food, 1992b, Guide to weed control: Publication 75, 208 p.

Ontario Ministry of the Environment, 1987, Pesticides in Ontario drinking water, 1986: 56 p. [ISBN 0-7729-2915-7]

Ontario Ministry of the Environment, 1992, Environment Servicing Development: Conference, February 25-26, 1992, Kingston, ON, 104 p.

Pilon, P.E. and K.W.F. Howard, 1987, Contamination of subsurface waters by road de-icing salts: Water Pollution Research Journal of Canada, v. 22, p. 157-171.

Reitzel, A., 1990, The temporal characterization of municipal solid waste leachate: unpublished M.App.Sc. thesis, Department of Civil Engineering, University of Waterloo, Waterloo, ON, 121 p.

Robertson, W.D., Cherry, J.A. and Sudicky, E.A., 1991, Groundwater contamination from two small septic systems on sand aquifers: Ground Water, v. 29, p. 82-92.

Rowe, K.R., 1990, Clayey Barriers for mitigation of contaminant impact: Evaluation and design, Volume 1: University of Western Ontario, London, ON, 66 p.

Saffigna, P.G., Keeney, D.R. and Tanner, C.B., 1977, Nitrogen, chloride and water balance with irrigated Russet Burbank potatoes in a sandy soil: Agronomy Journal, v. 69, p. 251-257.

Schertzer, W.M., 1982, Physical limnology of the Great Lakes basin: Investigation and Modeling Section, Aquatic Physics and System Division, Canada Centre for Inland Waters, 121 p.

Sherman K.M. and Anderson, D.L., 1991, An evaluation of volatile organic compounds and conventional parameters from on-site disposal systems in Florida, in On-Site Wastewater Treatment: Sixth National Symposium on Individual and Small Community Sewage Systems, December 16-17, 1991, Chicago, IL, v. 6, p. 67-72.

Simons, T.J. and W.M. Schertzer, 1987, Stratification, currents and upwelling in Lake Ontario, Summer 1982: Canadian Journal of Fisheries and Aquatic Sciences, v. 44, p. 2047-2058.

Sonzogni, W.C. and Lee, G.F., 1974, Nutrient sources for Lake Mendota, Wisconsin: Academy of Science, Transactions, v. 62, p. 133-164.

United States Environmental Protection Agency, 1987, Causes of Release from UST Systems: EPA Contract 68-01-7053, 52 p.

Veska, E., 1977, Geochemistry and hydrogeology of agricultural watershed no. 10 and their influence on the chemical composition of water and sediments: International Joint Commission on Pollution from Land Use Activities, Task Group C (Canadian Section), Activity 1, Project 23, PLUARG 77-075, Windsor, ON, 119 p.

Yates, M.V., 1985, Septic tank density and groundwater contamination: Groundwater, v. 23, p. 586-591.

6. Chlorinated Solvents in Urban Ground Waters

Stanley Feenstra

Applied Groundwater Research Ltd., 207–2550 Argentia Rd., Mississauga, Ontario L5N 5R1

SUMMARY

Chlorinated solvents have been and are currently used for many applications in both small and large industries, and have contaminated many individual and community wells in North America. The most insidious potential sources of solvent contamination in urban areas, such as dry cleaners and small manufacturing industries, are generally not subjected to stringent controls and monitoring. With the presence of poorly regulated potential sources, and because of the lack of routine water quality testing and the high odour thresholds for chlorinated solvents, the users of ground water from individual wells in urban areas experience the greatest risk of exposure to ground water contaminated by chlorinated solvents.

INTRODUCTION

Chlorinated solvents are important contaminants in ground water used for drinking-water supply because they are considered to represent significant risks to human health at very low concentrations. As a result, drinking-water standards specified by environmental regulatory agencies are in the range of parts per billion, or micrograms per litre ($\mu g \cdot L^{-1}$). Chlorinated solvents such as 1,1,1–trichloroethane (TCA), dichloromethane (DCM), perchloroethylene (PCE) and trichloroethylene (TCE) are used extensively by small and large industries for metal degreasing and dry cleaning, in paints and adhesives, and for the manufacture of electronics and instruments. These chlorinated solvents have been in widespread use since the early 1900s and are produced currently in quantities of many millions of kilograms per year (see Table 1). A survey of ground-water monitoring data from 183 hazardous waste disposal sites in the United States found that the chlorinated solvents TCE, DCM, and PCE were among the five most frequently identified ground-water contaminants (Plumb and Pitchford, 1985). A history of discovery of chlorinated solvents in ground water is described by Pankow *et al.* (1995).

PROPERTIES OF DENSE NON-AQUEOUS PHASE LIQUID SOLVENTS

Chlorinated solvents, in their pure form, are liquids which are immiscible with water and denser than water. In North America, such chemicals are generally referred to as dense non-aqueous phase liquids (DNAPLs). The liquid densities of these chlorinated solvents range from 1.33 $g \cdot cm^{-3}$ for dichloromethane to 1.63 $g \cdot cm^{-3}$ for perchloroethylene. The mixture of substantial quantities of chlorinated organic substances with other organic substances can also result in immiscible liquids with the properties of a DNAPL. For example, a mixture of petroleum hydrocarbons (density of 0.9 $g \cdot cm^{-3}$) and TCE (density of 1.47 $g \cdot cm^{-3}$) would be a DNAPL if the TCE concentration in the mixture exceeded about 25%. The occurrence of chlorinated solvents in DNAPLs and the migration of DNAPL through the subsurface contribute significantly to the common and widespread contamination of ground water by chlorinated solvents.

PROPERTIES OF AQUEOUS-PHASE CHLORINATED SOLVENTS

The properties of aqueous-phase or dissolved chlorinated solvents also contribute to their importance as ground-water contaminants. Although the solubility of some chlorinated solvents is sufficiently low that they are immiscible in water, their solubilities may be very high compared to their respective drinking-water standards (see Table 2). The ratios of solubility to drinking-water standard for the chlorinated solvents, trichloroethylene and dichloromethane, are 280,000 and 4,000,000, respectively. This means that these chemicals have the potential for being released to the ground water at concentrations that are many orders of magnitude higher than drinking-water standards. The ratios of solubility to drinking-water standard are generally much lower for other com-

Table 1   Production quantities of selected chlorinated solvents in the United States in 1990 (data from United States International Trade Commission, 1991).

Solvent	Quantity (millions of kg)
Dichloromethane (DCM)	210
Perchloroethylene (PCE)	170
1,1,1–trichloroethane (TCA)	365
Trichloroethylene (TCE)	50

Table 2   Solubility of chlorinated solvents compared to drinking water standards. US EPA MCL is the current or proposed maximum concentration limit for drinking water specified by the United States Environmental Protection Agency.

Solvent	Solubility ($mg \cdot L^{-1}$)	US EPA MCL ($mg \cdot L^{-1}$)	Solubility: MCL Ratio
Dichloromethane	20,000	0.005	4,000,000
Perchloroethylene	240	0.005	48,000
1,1,1–trichloroethane	1250	0.2	6300
Trichloroethylene	1400	0.005	280,000

mon ground-water contaminants such as monocyclic and polycyclic aromatic hydrocarbons.

In addition to their high potential for ground-water contamination, chlorinated solvents have high odour thresholds and relatively innocuous odours so that when contamination does exist, concentrations far in excess of drinking-water standards go unnoticed by ground-water users (Table 3). Because of this, ground-water users could be unknowingly exposed to ground water contaminated by chlorinated solvents for many years. In contrast, the odour thresholds for petroleum hydrocarbons found in fuel products (gasoline, diesel, *etc.*) are sufficiently low, and the odours sufficiently objectionable, to identify and preclude long-term use of ground water that exceeds drinking-water standards for these compounds.

Some degree of attenuation (*i.e.*, reduction in concentration) will occur during migration of dissolved organic substances in the ground water as a result of sorption on geologic solids. Sorption of dissolved organic compounds is considered to occur principally due to partitioning of the dissolved organic substances from the ground water into solid organic

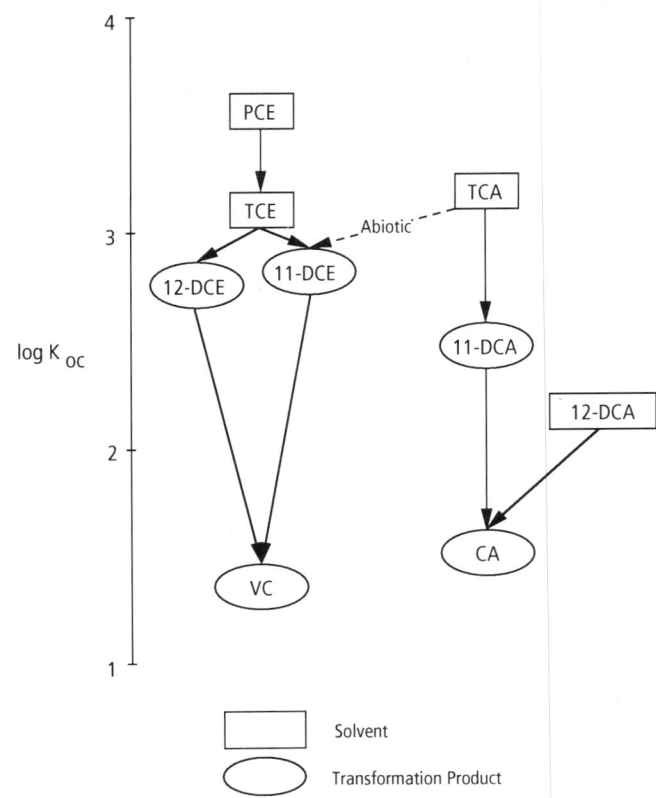

Figure 1 *Sequence of biotic and abiotic transformations of selected chlorinated solvents (from Vogel et al., 1987). Abbreviations: **PCE**, perchloroethylene; **TCE**, trichloroethylene; **TCA**, 1,1,1–trichloroethane; **12-DCE**, 1,2–dichloroethylene; **11-DCE**, 1,1–dichloroethylene; **VC**, vinyl chloride; **11-DCA**, 1,1–dichloroethane; **12-DCA**, 1,2–dichloroethane; **CA**, chloroethane. K_{oc} is the organic carbon-water partition coefficient (see text).*

Table 3   Odour thresholds of chlorinated solvents compared to drinking water standards. US EPA MCL is the current or proposed maximum concentration limit for drinking water specified by the United States Environmental Protection Agency.

Solvent	Odour Threshold (mg·L⁻¹)	US EPA MCL (mg·L⁻¹)	Odour Threshold: MCL Ratio
Dichloromethane	0.79	0.005	160
Perchloroethylene	0.25	0.005	50
1,1,1–trichloroethane	6.7	0.2	34
Trichloroethylene	0.05	0.005	10

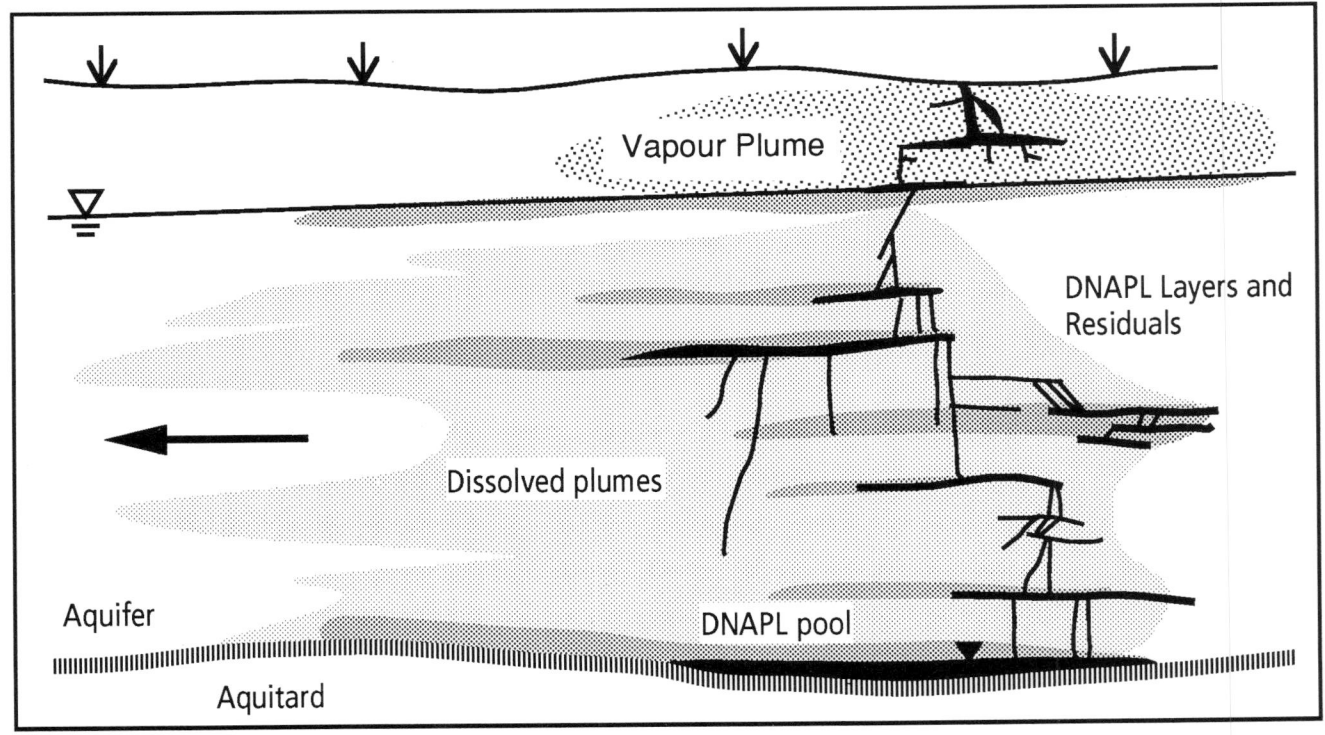

Figure 2   *Conceptual model for ground-water contamination by DNAPLs in a sand aquifer. Arrow shows direction of ground-water flow.*

carbon in the geologic medium (Karickhoff, 1984). The affinity of an organic chemical for sorption on geologic media is generally expressed by K_{oc}, the organic carbon-water partition coefficient for the chemical. The higher the K_{oc} value, the greater the degree of sorption and the less mobile the contaminant will be in ground water. Chlorinated solvents have relatively low K_{oc} values and would be expected to be relatively mobile in ground water.

The migration of dissolved organic substances in the ground water can also be attenuated by degradation processes. However, chlorinated solvents commonly exhibit a substantial resistance to biotic degradation compared to non-chlorinated organic contaminants. Monocyclic aromatic hydrocarbons such as benzene and toluene are generally degraded at a significant rate to harmless by-products through aerobic degradation mediated by native soil bacteria. Conditions in the vadose zone and shallow ground water are commonly amenable to such degradation processes. In con-

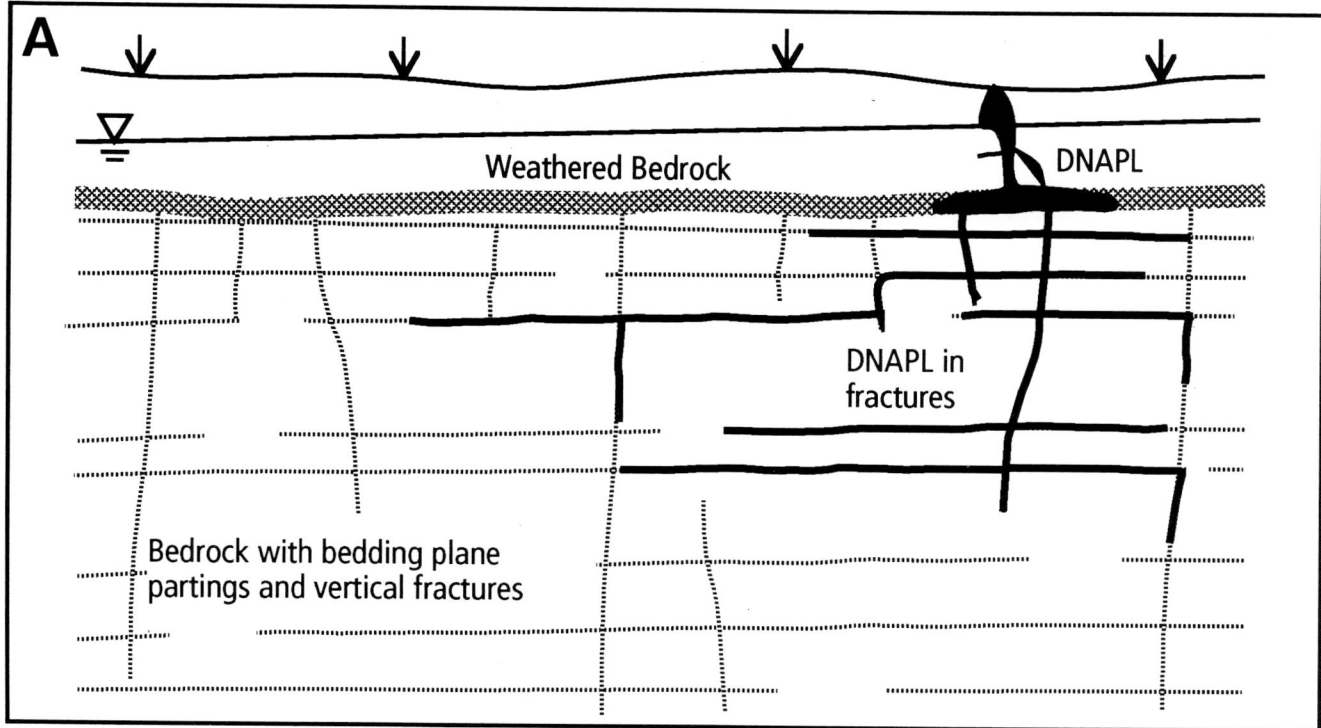

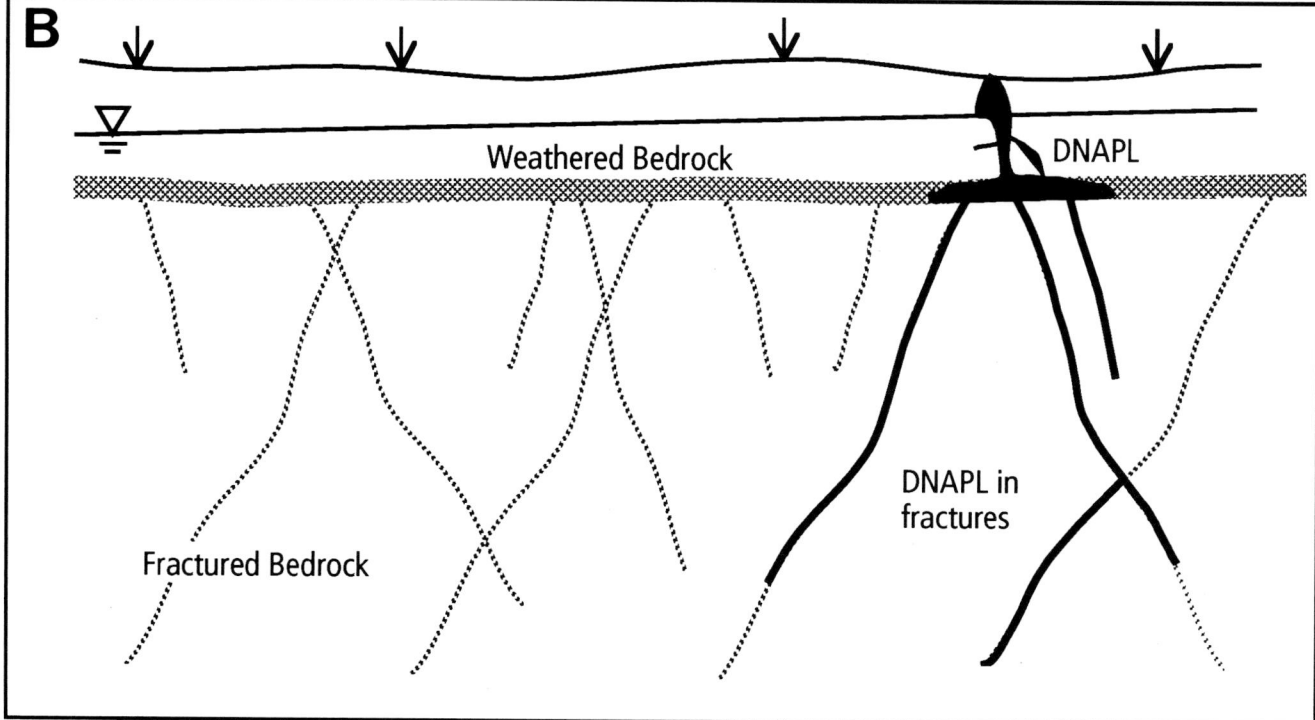

Figure 3 *Conceptual models for contamination by DNAPLs in fractured strata: (**A**) horizontally bedded sedimentary rock; (**B**) crystalline rock.*

trast, there is little evidence that chlorinated solvents undergo aerobic degradation at a significant rate. Anaerobic degradation of chlorinated organic substances does occur in the subsurface, but commonly as sequential dechlorination reactions which lead to transformation products that can be more mobile than the parent compound (Fig. 1). Transformation products, such as vinyl chloride from the anaerobic degradation of PCE and TCE, may have drinking-water standards that are even lower than the parent compounds.

CONCEPTUAL MODELS FOR SUBSURFACE BEHAVIOUR OF SOLVENTS

The physical and chemical properties of chlorinated solvents define the behaviour of these chemicals in the subsurface, and distinguish them from all other ground-water contaminants. The potential for significant ground-water contamination is greatest when chlorinated solvents are released to the subsurface as DNAPLs. A conceptual model for ground-water contamination in a sandy aquifer, resulting from the release of a DNAPL, such as a chlorinated solvent, is shown in Figure 2. This conceptual model was developed during the 1980s as a result of laboratory studies, computer modelling, controlled field experiments, and observations at many chemical spill and waste disposal sites in North America (Feenstra et al., 1996).

This conceptual model is likely applicable to a great many commercial and industrial sites where such chemicals have been produced or used, and to most liquid waste disposal facilities. The dissolution of DNAPL in the ground-water zone represents a source of ground-water contamination that is significantly different from the leaching of solid wastes or contaminated soils in the vadose zone, or dissolution of non-aqueous phase liquids less dense than water (such as petroleum fuels) situated in the vadose zone and zone of water table fluctuation. When DNAPL is released into the subsurface environment, its low solubility and high density allows the DNAPL to penetrate downward through the vadose zone and below the water table into the ground-water zone as a separate liquid phase. The ability of DNAPL to penetrate into the ground-water zone differs from that of petroleum fuels which will not penetrate significantly below the water table because of their lower density.

During its migration through the subsurface, a portion of the DNAPL remains trapped as residual in the soil pores. In these areas of DNAPL residual, DNAPL exists as immobile, disconnected and partially connected blobs and filaments of immiscible liquid. In stratified aquifers, DNAPL may also accumulate in zones, commonly layers or pools, where its further movement is limited by capillary forces. DNAPL migration and distribution is strongly influenced by structure and variability of the underlying geology. DNAPL will migrate along the most permeable pathways, and even subtle variation in permeability, on a scale of millimetres, can result in a highly variable spatial distribution of the DNAPL (Kueper et al., 1989, 1993; Poulsen and Kueper, 1992). In the ground-water zone, the migration of the DNAPL will be controlled principally by the geologic structure and DNAPL fluid pressures. In many cases, ground-water flow may have no significant influence on the direction or extent of DNAPL migration.

DNAPL can also penetrate into fractured media such as weathered silt and clay and fractured bedrock aquifers (see Figs. 3A, B). Kueper and McWhorter (1991) illustrated that DNAPL can penetrate fractures having a very small opening size (aperture) of a few tens of micrometres or less. Because

of the low porosity in a fractured medium, a given volume of DNAPL will spread through a much larger area and depth in a fractured medium compared to a porous medium.

Ground water which comes into contact with these residual zones or DNAPL accumulations will slowly dissolve the DNAPL, and create a plume of dissolved contaminants in the ground water. Because chlorinated solvents are not strongly sorbed or rapidly degraded in ground water, the dissolved plumes produced by DNAPL source zones may be very large. The vertical and spatial distribution of the dissolved plume will be determined by the depth and location of the DNAPL source zone. Because of the ability of DNAPL to migrate to significant depths below the water table, and to migrate uninfluenced by the direction of ground-water flow, the location and pattern of the resultant dissolved plumes may be highly variable. In particular, the dissolved plume resulting from a DNAPL source zone below the water table may be much deeper than expected from a source of contamination in the vadose zone. Variability in the nature of DNAPL occurrence (i.e., as residual and pools) and spatial variability in the distribution of DNAPL will frequently result in dissolved concentrations in monitoring wells that are much lower than the solubility of the contaminants (Feenstra and Guiguer, 1995; Feenstra and Cherry, 1995).

DNAPL which resides in the vadose zone may create plumes of chemical vapours which may also contribute to ground-water contamination. Many chlorinated solvents have relatively high vapor pressure and may create vapour plumes. Vapour plumes migrate outward from the source zone by diffusion (Hughes et al., 1992). In some cases, vapour concentrations can be sufficiently high that the contaminated soil air is significantly more dense than the surrounding soil air and result in advective sinking of vapours (Mendoza and Frind, 1990a, b). Ground-water contamination develops when water infiltrates through the DNAPL zone above the water table and through the vapour plume, and carries dissolved contaminants to the water table. The dissolved plumes from such vadose zone source may have very high concentrations, but are very thin in vertical extent and occur close to the water table.

Because of their low solubility, the zones of DNAPL residual and DNAPL accumulation in the ground-water zone can contribute dissolved contaminants to the ground water for centuries or longer. Thus, plumes may continue to grow until they encounter water supply wells or ground-water discharge

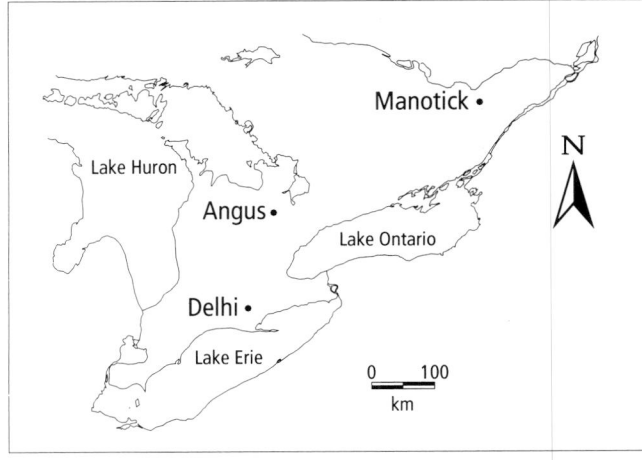

Figure 4 *Locations of the Angus, Manotick and Delhi, Ontario sites discussed in text.*

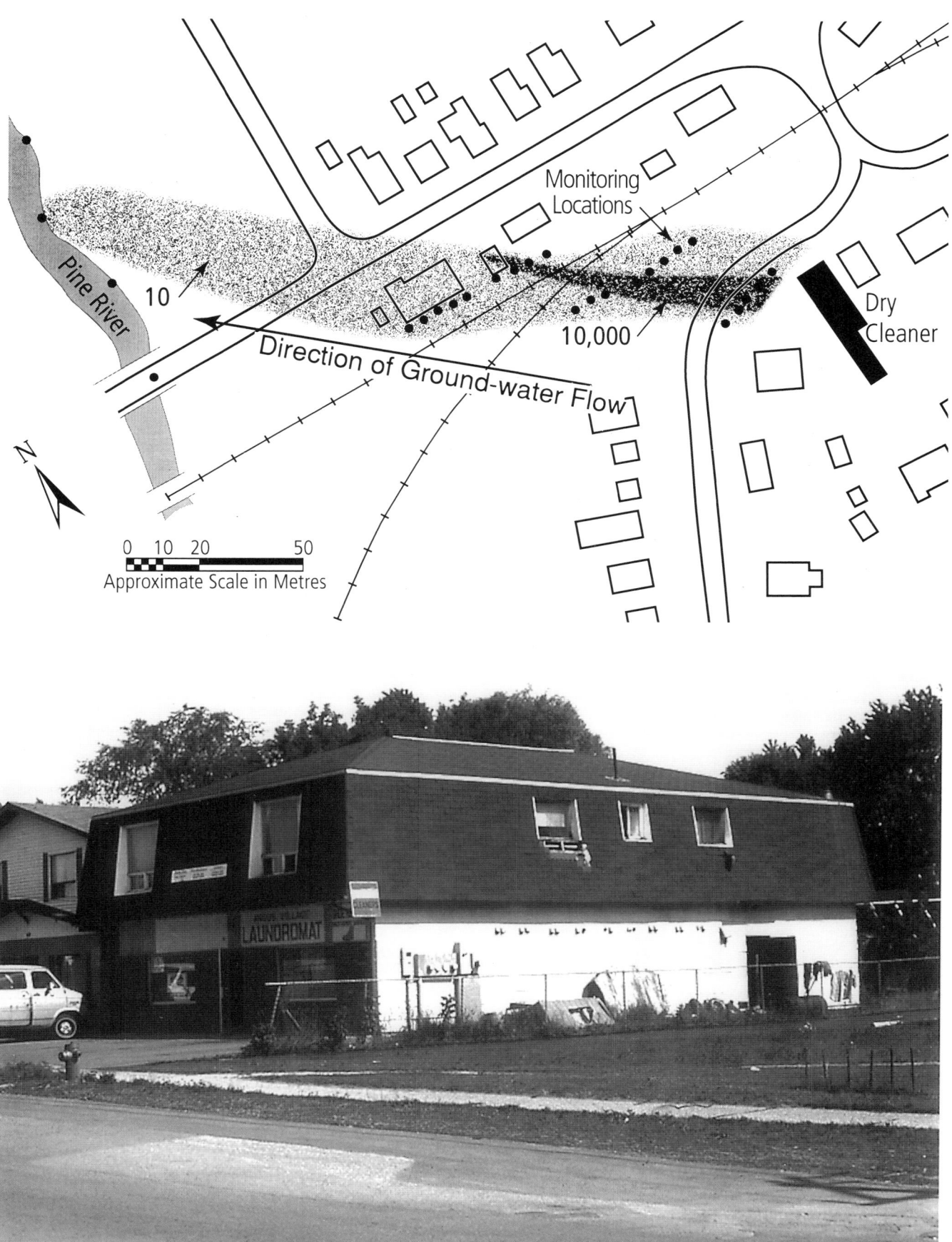

Figure 5 *Configuration of the PCE (perchloroethylene) plume from a retail dry cleaner (photograph) in Angus, Ontario. Concentrations in* $\mu g \cdot L^{-1}$ *as measured in 1992-1993 (modified from Pitkin et al., 1994).*

areas such as rivers or lakes. DNAPL in the vadose zone, comprised of high-vapour pressure compound solvents, may persist only for much shorter times before they are exhausted by vapourization.

GROUND-WATER CONTAMINATION IN URBAN AREAS

Three examples illustrating some of the key features of solvent contamination in urban areas are described below The examples include retail dry cleaner sites in Angus and Manotick, Ontario, and sources unknown in Delhi, Ontario. The location of these sites is shown in Figure 4.

Retail Dry Cleaner, Angus, Ontario

Angus is a small village located about 80 km north of Toronto. Many residents in the village obtained water from individual wells in a shallow sandy aquifer. In 1992, a disagreeable taste was noted in a well by a tenant who had just occupied a rental unit. There had been no complaints by prior tenants. Test conducted by the Ontario Ministry of the Environment and Energy indicated a PCE concentration of 27,000 $\mu g \cdot L^{-1}$, far in excess of the World Health Organization drinking-water guideline of 10 $\mu g \cdot L^{-1}$. Further sampling revealed several other contaminated wells and, subsequently, these residences were provided with piped water from another area of the village. In 1993, researchers from the Waterloo Centre for Groundwater Research, University of Waterloo, conducted a detailed delineation of the PCE plume (Pitkin et al., 1994). The PCE plume originates from a small retail dry cleaner and extends a distance of about 250 m to the Pine River (see Fig. 5). The ground-water velocity is about 10 to 20 cm·a⁻¹.

Based on recollections of the former and present owners of the dry cleaners, the PCE contamination may be due to a spill of several hundred litres that occurred in the 1970s. If this is the case, it is possible that previous tenants and residents may have used PCE-contaminated ground water for as many as 10 to 20 years.

The maximum PCE concentration measured in the plume was 44,000 $\mu g \cdot L^{-1}$ (about 20% of solubility) and suggests that the plume originated from a DNAPL solvent source zone (Feenstra and Cherry, 1995). The PCE plume is long and narrow, and caused contamination of only a few wells in the area down-gradient from the dry cleaner. A narrow plume is consistent with a small source area and with weak transverse dispersion of the dissolved plume. These effects were demonstrated in detailed field experiments in a similar sandy aquifer at the neaby C.F.B. Borden test site (Rivett et al., 1994). There would appear to be little lateral spreading of the plume caused by pumping of surrounding wells. It would be expected that individual residential wells drawing only low pumping rates (~5 L·min⁻¹ or less) would not create large zones of influence and spreading of the plume.

Retail Dry Cleaner, Manotick, Ontario

Manotick is a small village situated about 20 km south of Ottawa. Residences and businesses in the community obtained water from individual wells in a fractured limestone/dolostone aquifer at depths of 15 m to 30 m. In response to complaints of disagreeable taste and odour in late 1991, the testing of four private wells in the downtown area of the village by the Ontario Ministry of the Environment and Energy indicated elevated concentrations of PCE. Further testing of surrounding wells indicated PCE concentrations as high as 66,000 $\mu g \cdot L^{-1}$ and a total of 62 nearby wells in which PCE concentrations exceeded the World Health Organization drinking-water guideline of 10 $\mu g \cdot L^{-1}$ (see Fig. 6). Subsequent

investigation revealed that, in 1984, two businesses had replaced their wells with deeper wells in response to disagreeable taste and odour. It was also revealed that several residences and businesses relied for many years on bottled water for drinking because of disagreeable taste and odour. These well owners continued to use well water for other household and commercial purposes. In response to these findings, the residences and businesses were supplied initially with bottled water or point-of-use treatment systems. By 1993, the area was serviced with municipal piped water.

Detailed hydrogeological studies conducted in 1993-1994 (Raven Beck Environmental Ltd., 1994) confirmed that the source of PCE contamination in this area was a former retail dry cleaner and that the PCE plume was about 400 m in length and 200 m in width. The general direction of ground-water flow is toward the Rideau River.

The dry cleaner ceased operation about 1985. At that time, it was discovered that a concrete holding tank that had been used for storage of waste PCE was badly deteriorated. It is believed that this tank was the source of releases to the subsurface. The maximum PCE concentration in the plume was 66,000 $mg \cdot L^{-1}$ (about 28% of solubility) and suggests that DNAPL solvent is likely present in the bedrock aquifer and is the source of the dissolved PCE plume (Feenstra and Cherry, 1995).

The PCE plume is longest in the direction of ground-water flow, but also extends a substantial distance laterally and up-gradient from the source area. As a result, the Manotick plume is much wider than that at Angus. Spreading of the Manotick plume may be the result of the migration of DNAPL PCE in various directions in the fractured bedrock, or the effect of ground-water pumping from the surrounding water wells. In a fractured bedrock aquifer, individual residential wells drawing only low pumping rates (~5 L·min⁻¹ or less) could create relatively large zones of influence and spreading of the dissolved plume in various directions.

Sources Unknown, Delhi, Ontario

The Town of Delhi is situated about 150 km west of Toronto. The town draws about 900,000 L per day or 20% of its municipal water supply from a pond fed by ground water from a shallow sandy aquifer. In 1989, TCE was detected in the supply at a concentration of 6 $\mu g \cdot L^{-1}$ during sampling performed by the Ontario Ministry of the Environment and Energy Drinking Water Surveillance Program. Subsequent sampling revealed gradually increasing concentrations, reaching 40 $\mu g \cdot L^{-1}$ in 1993 before the use of the pond ceased in late 1993. Although this level was below the Ontario guideline of 50 $\mu g \cdot L^{-1}$, hydrogeological investigations were initiated to define the nature and extent of the problem. A summary of the investigations performed at the site is given by Terraqua Investigations Limited (1994).

Studies around the spring pond determined that the pond intercepts a TCE plume in the ground water having concentrations as high as 520 $\mu g \cdot L^{-1}$. TCE concentrations are as high as 750 $\mu g \cdot L^{-1}$ about 240 m up-gradient of the spring pond (Fig. 7). A large former tobacco-processing plant is situated 240 m up-gradient of the spring pond. No detailed studies have yet been conducted around this plant to assess its potential contribution to the TCE plume. It is believed that machine shop activities at the plant used TCE for degreasing.

Studies conducted around two manufacturing sites about 1100 m away from the spring pond revealed low TCE concentrations of about 10 $\mu g \cdot L^{-1}$ or less over a significant area. However, the maximum TCE concentration found at the two sites was only 200 µg·L⁻¹. Given the magnitude of the TCE concentrations found to date and the distance of these industrial sites from the spring pond, it is not clear that these sites have contributed to the TCE contamination at the spring pond.

Important Features of Contamination in Urban Areas

Chlorinated solvents can enter the subsurface environment from a wide variety of sources. At commercial and industrial sites, contamination can originate from leakage of chemical products or wastes from underground or above-ground storage tanks, drum storage areas, distribution pipelines, and on-

loading or off-loading areas. At sites where intentional disposal activities have occurred, such as landfills, settling ponds, lagoons and septic tile fields, contamination can originate from the disposal of immiscible liquid organic wastes, aqueous wastes or leaching from waste sludges or solids.

Because chlorinated solvents have been and are currently used for many applications in both small and large industry, typically there is a multitude of potential sources of contamination to aquifers in urban areas. Such sources range from small retail dry cleaners and machine shops that may use only several drums of solvent per month, to large indus-

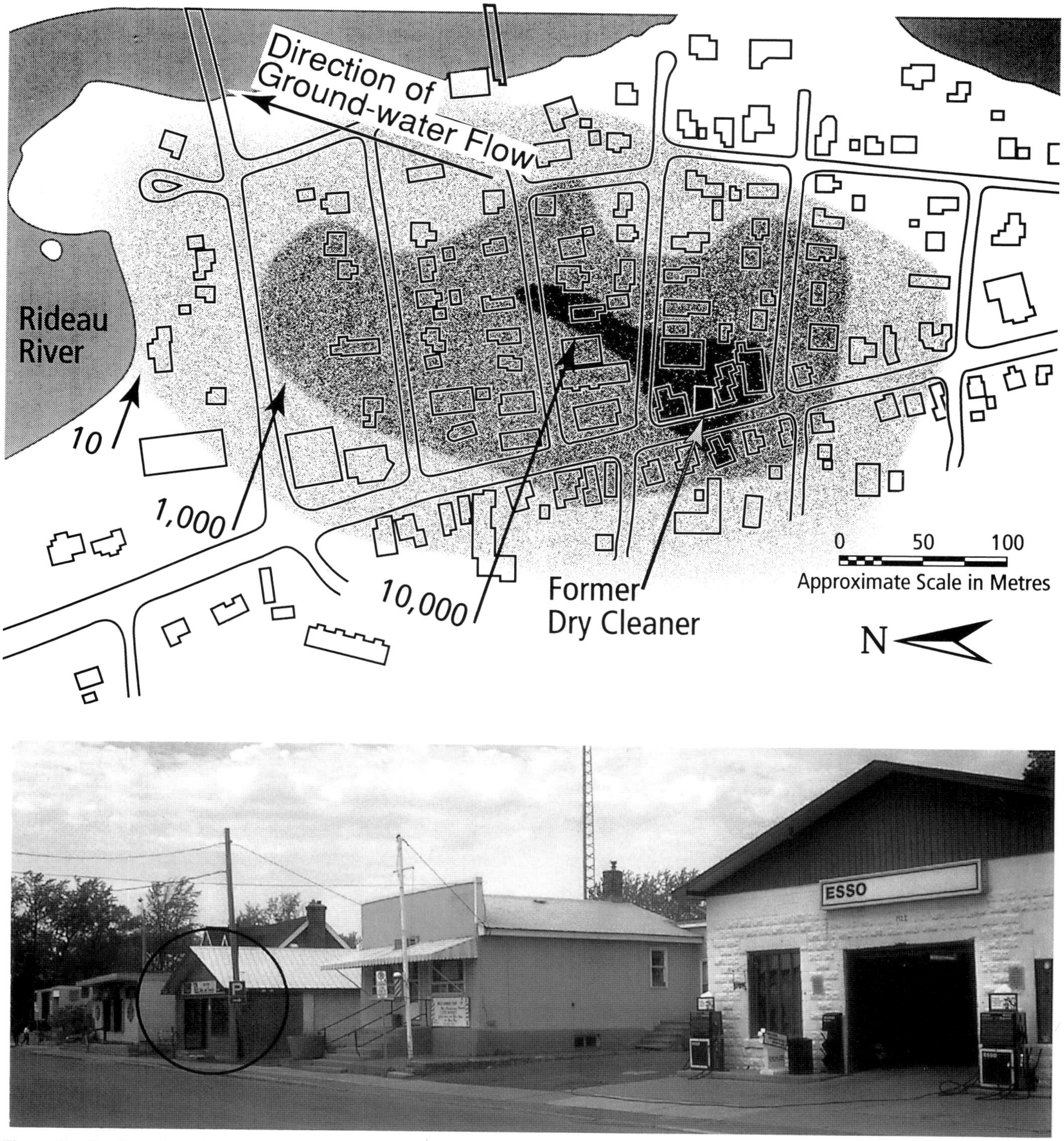

Figure 6   *Configuration of the PCE (perchloroethylene) plume from a former dry cleaner in Manotick, Ontario (photograph). Concentrations are maximum values measured in each well during 1992-93. Concentrations in µg·L⁻¹ (modified from Raven Beck Environmental Ltd., 1994).*

trial complexes where solvents may be used or produced in quantities of thousands of drums per year. Disposal sites may range from small areas of past sporadic dumping or septic systems on industrial lands, to large landfills.

Users of ground water in urban areas are relatively well protected from contamination that may originate from large industrial sites or modern landfills. Most environmental regulatory agencies in North America require identification and ongoing monitoring of such sites, with governments and large corporations generally accepting responsibility for these efforts. However, many old and abandoned industrial sites and landfills within urban areas go unidentifed or unmonitored (Chapters 2, 20, 21).

In addition, other significant potential sources of solvent contamination, such as dry cleaners and small manufacturing industries, are subjected to far less stringent controls and monitoring than larger facilities. It is precisely these types of potential sources that are most likely to be situated within urban areas. This is illustrated in the Angus, Manotick and Delhi examples. These potential sources are numerous in urban areas. In 1986, an estimated 1300 dry cleaners were in operation in Ontario and used over 4 million kilograms of PCE annually (Beak Consultants Limited, 1990).

Users of ground water provided by municipalities and private water companies are relatively well protected from contamination as a result of regulatory requirements for frequent monitoring of water quality. Most environmental and health regulatory agencies now specify analysis of common chlorinated solvents for drinking-water testing programs. The provincial program for drinking-water testing identified TCE in Delhi before concentrations reached unacceptable levels. In contrast, users of ground water supplied by individual wells would seldom, if ever, have their water tested for chlorinated solvents. The cost of analysis for chlorinated solvents is typically $100 to $200 per sample and would likely be viewed as excessive for most homeowners.

Because of the lack of routine water quality testing, the presence of poorly regulated potential sources situated within urban areas, and the high odour thresholds for chlorinated solvents, the users of ground water from individual wells in urban areas experience the greatest risk of exposure to ground water contaminated by chlorinated solvents. This is clearly illustrated in the Angus and Manotick examples. In these cases, ground-water users were exposed to highly contaminated water for many years. Depending on the hydrogeological conditions, even small sources of solvents can cause very severe ground-water contamination. In the case of Manotick, a single retail dry cleaner caused the contamination of 62 water supply wells.

TREATMENT OF CONTAMINATED GROUND WATER
When urban aquifers become contaminated by chlorinated solvents, it is common that use of the aquifer ceases and alternate sources of water are developed. For individual households, alternate supplies may be bottled or trucked water, or water piped from other areas. For community supplies, alternate sources may be surface-water resources or ground water extracted from other areas.

Ground water contaminated by chlorinated solvents can be readily treated. Conventional water treatment technologies such as air stripping and carbon absorption are usually very effective in reducing concentrations of chlorinated solvents to far below present drinking-water standards. Treatment systems available range from point-of-use systems intended for individual households treating a few litres per minute, to large systems capable of treating thousands of litres per minute (see Chapter 35).

Unfortunately, there is often great resistance by both private and community water users to utilizing ground water that has been treated, despite the fact that the development of alternate water supplies is considerably more costly than the treatment of existing ground-water supplies. This is probably

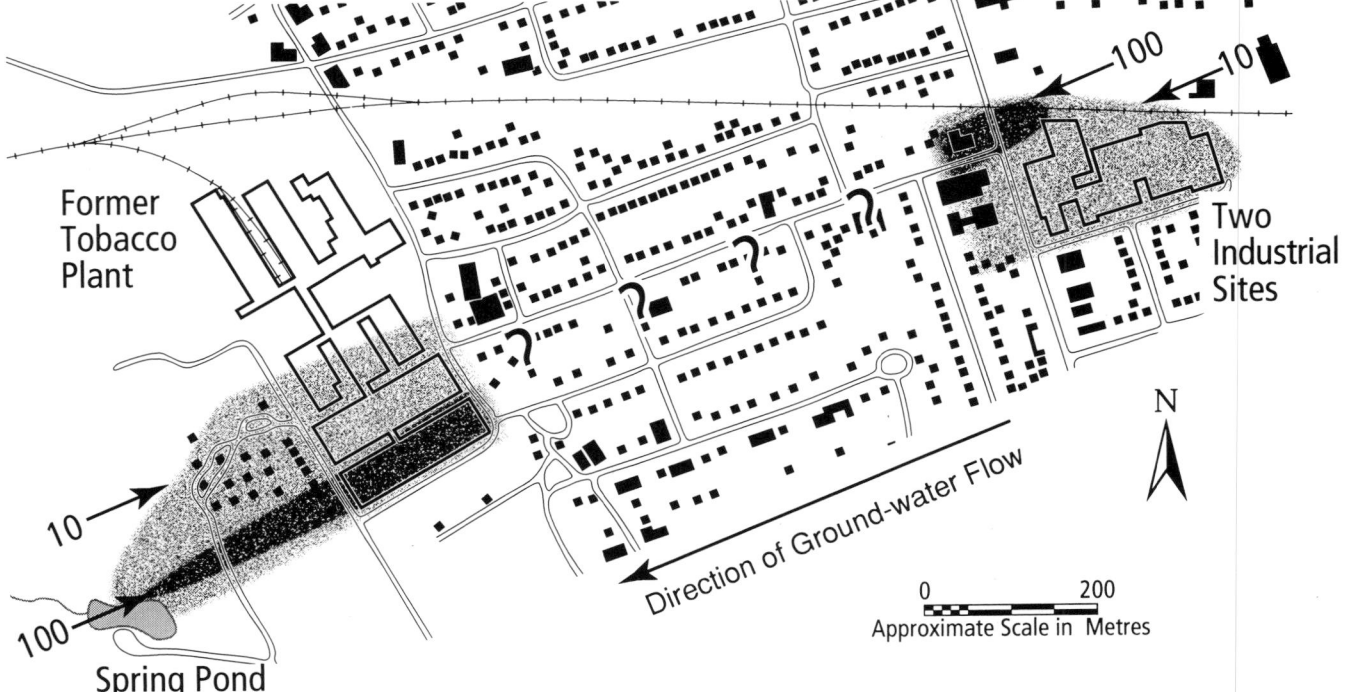

Figure 7 *Configuration of the TCE (trichloroethylene) plumes in Dehi, Ontario. Concentrations are maximum values measured at each location during 1992-93; concentrations in µg·L⁻¹ (modified from Terraqua Investigations Limited, 1994).*

based on the unrealistic expectation that ground water is a resource that should be pure and uncontaminated, even in urban areas. This is also paradoxical because few people would reconsider using surface water from urban areas that is typically subject to sophisticated pretreatment.

Because of the reluctance to treat and use ground water that has been contaminated by chlorinated solvents, and because of society's general desire to eliminate environmental contamination of all types, there have been literally billions of dollars spent unsuccessfully in North America to attempt to restore contaminated aquifers to their original condition.

PROSPECTS FOR AQUIFER RESTORATION

Goals of Aquifer Restoration

It is the zone of DNAPL residual and DNAPL accumulation below the water table which is the principal source of chlorinated organic contaminant plumes in ground water. During the 1980s, the prevailing conceptual model for ground-water contamination was that plumes were caused by leaching of contaminants from waste materials and from contaminated soil above the water table. During this time, remedial measures were implemented at many such sites with the goal of aquifer restoration. The goal of restoration is to permit the use of ground water for drinking water without the use of engineered systems to control contaminant migration. Restoration of the ground water requires removal of both the source of the contamination and the dissolved plume. In the case of the previous conceptual model, the source of contamination consisted of the waste materials and contaminated soil in the vadose zone. Restoration required excavation of the waste and soil to eliminate the source and ground-water pump-and-treat to eliminate the plume.

However, it is now evident that this previous conceptual model does not apply to sites where DNAPL solvents are present below the water table. In this case, restoration of an aquifer becomes a daunting task because restoration can only be achieved if the principal source of contamination, the DNAPL below the water table, is completely removed. This has yet to be accomplished successfully at any site in North America.

Reality of Source Zone Restoration

For most chlorinated solvents, complete removal or restoration of DNAPL zones will require soil clean-up targets of about $1 \ mg \cdot kg^{-1}$ or lower. For most DNAPL source zones, reductions in soil concentrations to these levels will require mass removal efficiencies of 99.9% or greater. The magnitude of this challenge has generally gone unrecognized.

Excavation is rarely a technically feasible option for complete removal of DNAPL from below the water table, especially in urban areas. Variability in DNAPL distribution, depth of DNAPL penetration, and migration of DNAPL beneath buildings and into bedrock formations can make definition of the DNAPL zone extremely difficult. Excavation may also release vapours into the atmosphere and cause unacceptable effects when performed in urban areas. Excavation below the water table will also require dewatering of the DNAPL zone, which may cause remobilization of the DNAPL into zones previously uncontaminated by DNAPL.

Ground-water pump-and-treat is rarely an effective measure for complete removal of DNAPL from below the water table. Although it is possible to increase ground-water flow through DNAPL source zones and accelerate the rate of mass removal by dissolution to some degree, the low solubility of the DNAPL and kinetic limitations on the rate of dissolution from DNAPL accumulations (*i.e.*, layers and pools) will generally prevent complete removal of DNAPL source zones in any practical time period of ten to twenty years, or less.

Many technologies are currently in development which may be applicable for the removal of DNAPL from below the water table. These technologies and concepts for their application are described in Cherry *et al.* (1995). Experimental technologies include: DNAPL pumping; water flooding; steam flooding; mobilization using surfactants; solubilization using surfactants or co-solvents; radio frequency (RF) heating; *in situ* air injection; dewatering followed by soil vapour extraction; microbial degradation; and chemical degradation.

Technologies such as DNAPL pumping and water flooding will always leave a substantial proportion of the DNAPL in the subsurface. Other technologies such as mobilization by surfactants or dewatering followed by soil vapour extraction have the potential to remobilize DNAPL and expand the size of the DNAPL source zone. The effectiveness of virtually all *in situ* remedial technologies will be impaired by geological variability. Although these technologies will accomplish varying degrees of removal of the DNAPL chemical mass, the technical necessity of achieving greater than 99.9% mass removal remains as an insurmountable obstacle to complete aquifer restoration. Of the many thousands of sites in North America found to be contaminated by chlorinated solvents, very few have been fully restored because of difficulties with restoration of DNAPL source zones.

Necessity for Containment

In most cases of ground-water contamination, the reduction or elimination of environmental risk does not require full aquifer restoration. The discharge of contaminated ground water to surface water, or the further expansion of dissolved ground-water plumes, can be reduced or eliminated by containment of the dissolved plumes. Ground-water pump-and-treat can be considered a proven technology for the containment of dissolved plumes. Low-permeability barriers such as slurry walls, grout curtains and sheet pile walls may also be considered to be proven technologies to provide plume containment, although the size and complexity of some dissolved plumes may prohibit the use of such technologies. Experimental technologies in development for containment of dissolved plumes include permeable reaction walls which induce chemical or microbial degradation of the contaminants as the plumes pass through the walls (Chapter 35). In particular, chlorinated solvents may be amenable to treatment by metal-catalyzed dechlorination reactions (Gillham and O Hannesin, 1994).

Remedial measures which provide plume containment must be maintained for as long as the DNAPL source zone persists. In some cases, effective containment of the source zone, together with ground-water pump-and-treat of the dissolved plume, can lead to restoration of that part of the aquifer outside the source zone in a practical period of time.

DISCUSSION

Full restoration of contaminated aquifers to drinking-water use requires that both the dissolved plumes and source zones be removed from the subsurface. This is not possible when the principal source of chlorinated organic substances is DNAPL located below the water table, because there are presently no remedial technologies which have been proven capable of complete clean-up of such DNAPL zones. Containment of dissolved plumes to prevent further migration is possible using proven remedial technologies such as ground-water pump-and-treat. Experience from sites indicates that partial restoration of plumes of dissolved chlorinated organic substances may also be possible using ground-water pump-and-treat because many chlorinated solvents are weakly sorbed on geologic materials. However, partial restoration of plume zones requires ongoing containment of the DNAPL source zone. As shown at some sites, failure to provide containment of the source zone can result in regrowth of the dissolved plumes.

REFERENCES

Beak Consultants Limited, 1990, Hazardous waste management practices in the Ontario dry-cleaning industry: Report to the Waste Managment Branch of the Ontario Ministry of the Environment and Energy, December 1990, 86 p.

Cherry, J.A., Feenstra S. and Mackay, D.M., 1996, Concepts for remediation of DNAPL sites, in Pankow, J.F. and Cherry, J.A., eds., Dense Organic Solvents and Other DNAPLs in Groundwater Systems: History, Behavior and Remediation: Waterloo Press, Portland, OR, p. 475-506.

Feenstra, S. and Cherry, J.A., 1995, Diagnosis and assessment of DNAPL sites, in Pankow, J.F. and Cherry, J.A., eds., Dense Organic Solvents and Other DNAPLs in Groundwater Systems: History, Behavior and Remediation: Waterloo Press, Portland, OR, p. 395-473.

Feenstra, S., Cherry, J.A. and Parker, B.L., 1995, Conceptual models for dense chlorinated solvents in groundwater, in Pankow, J.F. and Cherry, J.A., eds., Dense Organic Solvents and Other DNAPLs in Groundwater Systems: History, Behavior and Remediation: Waterloo Press, Portland, OR, p. 53-88.

Feenstra, S. and Guiguer, N., 1995, Dissolution of dense non-aqueous phase liquids (DNAPL) in the subsurface, in Pankow, J.F. and Cherry, J.A., eds., Dense Organic Solvents and Other DNAPLs in Groundwater Systems: History, Behavior and Remediation: Waterloo Press, Portland, OR, p. 203-232.

Gillham, R.W. and O'Hannesin, S.F., 1994, Enhanced degradation of halogenated aliphatics by zero-valent iron: Ground Water, v. 32, p. 958-967.

Hughes, B.M., Gillham, R.W. and Mendoza, C.A., 1992, Transport of trichloroethylene vapours in the unsaturated zone: A field experiment, in Weyer, K.U., ed., Conference on Subsurface Contamination by Immiscible Fluids, Calgary, Alberta, April 18-20, 1990, International Association of Hydrogeologists: A.A. Balkema, Rotterdam, p. 81-88.

Karickhoff, S., 1984, Organic pollutant sorption in aquatic systems: Journal of Hydraulic Engineering, v. 110, p. 707-735.

Kueper, B.H., Abbot, W. and Farquhar, G.J., 1989, Experimental observations of multiphase flow in heterogeneous porous media: Journal of Contaminant Hydrology, v. 5, p. 83-96.

Kueper, B.H., Redman, D., Starr, R.C., Reitsma, S. and Mah, M., 1993, A field experiment to study the behaviour of tetrachloroethylene below the water table: Spatial distribution of residual and pooled DNAPL: Ground Water, v. 31, p. 756-766.

Kueper, B.H. and McWhorter, D.B., 1991, The behavior of dense, non-aqueous phase liquids in fractured clay and rock: Ground Water, v. 29, p. 716-728.

Mendoza, C.A. and Frind, E.O., 1990a, Advective-dispersive transport of dense organic vapours in the unsaturated zone, 1. Model development: Water Resources Research, v. 26, p. 379-387.

Mendoza, C.A. and Frind, E.O., 1990b, Advective-dispersive transport of dense organic vapours in the unsaturated zone, 2. Sensitivity analysis: Water Resources Research, v. 26, p. 388-398.

Pankow, J.F., Feenstra, S., Cherry, J.A. and Ryan, M.C., 1995, Dense chlorinated solvents in groundwater: Background and history of the problem, in Pankow, J.F. and Cherry, J.A., eds., Dense Organic Solvents and other DNAPLs in Groundwater Systems: History, Behavior and Remediation: Waterloo Press, Portland, OR, p. 1-52.

Pitkin, S., Ingleton, R.A. and Cherry, J.A., 1994, Use of a drive point sampling device for detailed characterization of a PCE plume in a sand aquifer at a dry cleaning facility [abstract]: Outdoor Action Conference, Minneapolis, MN, May 19-21, 1994, National Ground Water Association, p. 16

Plumb, R.H., Jr. and Pitchford, A.M., 1985, Volatile organic scans: Implications for ground water monitoring: Petroleum Hydrocarbons and Organic Chemicals in Groundwater: Houston, TX, November 13-15, 1985, National Water Well Association, p. 207-222.

Poulsen, M. and Kueper, B.H., 1992, A field experiment to study the behaviour of tetrachloroethylene in unsaturated porous media: Environmental Science and Technology, v. 26, p. 889-895.

Raven Beck Environmental Ltd., 1994, Soils and hydrogeological investigation of PCE and Petroleum contamination: Village of Manotick: report to the Ontario Ministry of Environment and Energy, Ottawa District Office, 140 p.

Rivett, M.O., Feenstra, S. and Cherry, J.A., 1994, Transport of a dissolved-phase plume from a residual solvent source in a sand aquifer: Journal of Hydrology, v. 15, p. 27-41.

Terraqua Investigations Limited, 1994, Phase 1 study of the TCE (Trichloroethylene) contamination of the spring water supply in Delhi: Report to the Ontario Ministry of Environment and Energy, Hamilton Regional Office, 67 p.

United States International Trade Commission, 1991, Synthetic Organic chemicals: United States Production and Sales, USITC Publication 2470, Washington, DC, 82 p.

Vogel, T.M., Criddle, C.S. and McCarty, P.L., 1987, Transformations of halogenated aliphatic compounds: Environmental Science and Technology, v. 21, p. 722-736.

7. Hydrocarbon Fuels in Urban Ground Waters and Soils

Jean-Pierre François
Hugh Molyneux

Environmental Management Services, 15 Cameron Avenue, North York, Ontario M2N 1C9

SUMMARY

Hydrocarbon fuels and oils (light non-aqueous phase liquids) leaking from above-ground and underground storage tanks represent one of the greatest threats to the quality of urban soil and ground water because they are so widely stored in urban areas. Fuel and oil leakage from automobile service stations, industrial facilities and distribution terminals has lead to a degradation of the overall quality of the soil and ground-water resources in large sectors of the urban environment. In the more serious cases, the impacts can result in immediate and severe health effects. The following sections examine the effects of leakage to the subsurface, as well as methodologies to investigate, remediate and clean up hydrocarbon spills.

INTRODUCTION

There are about 200,000 underground storage tanks (USTs) in service in Canada (Rush and Metzger, 1991). Most of these tanks were installed in the 1950s and 1960s and are now well beyond their 20-year life expectancy. It has been variously estimated that between 20,000 and 40,000 of the in-service tanks leak (Canadian Institute for Environmental Law and Policy, 1995). On a national scale, the impact comes clearly into focus when it is considered that approximately 38% of all municipalities in Canada rely on ground water as a potable source (Environment Canada, 1987). Health effects of persistent exposure to hydrocarbon fuels such as gasoline are reviewed by National Institutes of Health (1993) and Knox (1994).

It is widely held that oil floats on water, but, in reality, hydrocarbons will dissolve in ground water to varying degrees. For example, the solubility of benzene, a common compound in gasoline and one which is carcinogenic, is 1808 $mg \cdot L^{-1}$ (Table 1). This value greatly exceeds the drinking-water standard for this compound set by many regulatory agencies in North America (usually about 0.7 ppb). Ground-water contamination by hydrocarbons is a widespread problem and examples of subsurface contaminant plumes from storage facilities are depicted in Chapters 34, 35, 36 and 37. This chapter emphasizes ground-water and soil comtamination by hydrocarbon fuels. However, there is increasing concern with highly soluble additives (such as methyl tertiary butyl ether: MTBE) routinely added as an octane enhancer in unleaded gasoline. These are more mobile in ground water than most hydrocarbons and their presence in wells is often the first sign of contamination (Bass and Riley, 1995).

PHYSICAL AND CHEMICAL CHARACTERISTICS OF FUELS AND LUBRICANTS

Hydrocarbon compounds are the family of substances composed primarily of carbon and hydrogen atoms and are the simplest of the organic chemicals where carbon and hydrogen atoms are bound covalently. Carbon can form single, double and triple bonds, such that a variety of compounds,

Table 1 Typical properties of organic contaminants commonly found in urban ground waters (from Kehew, 1995). Properties may vary according to temperature and pH.

Compound	PROPERTY			
	Density $(g \cdot mL^{-1})$	Solubility $(mg \cdot L^{-1})$	Henry's constant (atm)	$Log_{10}K_{ow}$
Trichloroethene	1.4	1100	550	2.29
Tetrachloroethene	1.63	200	1100	2.88
Chloroform	1.49	8200	170	1.95
Benzene	0.876	1808	240	2.01
Toluene	0.876	535	308	2.69
1,2–dichlorobenzene	1.305	145	90	3.38
Phenol	1.07	93,000	0.04	1.49
1,1–dichloroethene	1.013	250	1400	0.73
1,1,1–trichloroethane	1.435	480	860	2.49
Vinyl chloride	gas	1100	35,500	0.60
Methyl ethyl ketone	0.805	260,000	1.5	0.26
Acetone	0.79	1,000,000	1.0	–0.24
Ethylene dibromide	2.18	3400	26	1.80

each with very different characteristics, are formed. The basic families most commonly encountered in fuels and lubricants are alkanes, alkenes, alkynes and aromatic hydrocarbons. The basic arrangement of the molecules for each family is shown on Figure 1. Generally, the alkanes are characterized by carbon atoms with single covalent bonds; alkenes by carbon atoms with double bonds; alkynes by carbon atoms with triple bonds, and aromatic hydrocarbons by those whose structure is characterized by the presence of a benzene ring. For the sake of simplicity, environmental geologists commonly focus on the number of carbon atoms to evaluate the general physical properties of a compound. For example,

ALKANES- Single Covalent Bond Hydrocarbon Compounds

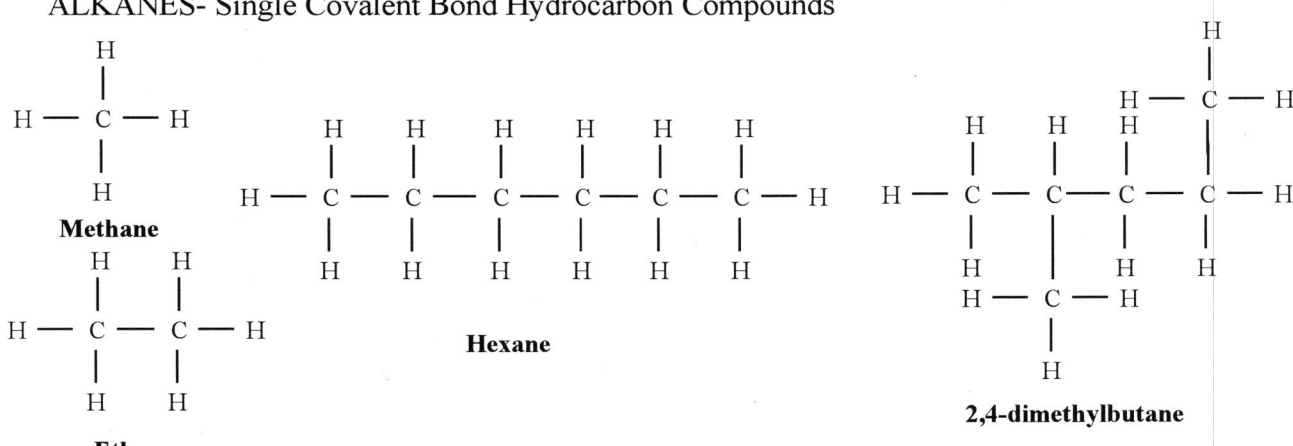

ALKENES- Double Covalent Bond Hydrocarbon Compounds

ALKYNES- Triple Covalent Bond Hydrocarbon Compounds

AROMATIC HYDROCARBONS- Benzene Ring Hydrocarbon Compounds

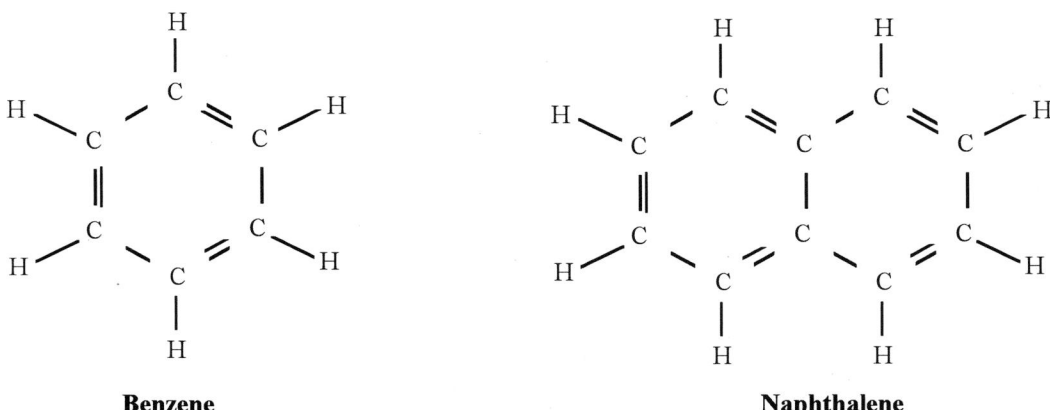

Figure 1 *Chemical structure for alkanes, alkenes, alkynes and aromatic hydrocarbons.*

methane (CH_4) is a C_1 compound, ethane (C_2H_6) is a C_2 compound and hexane (C_6H_{14}) and benzene (C_6H_6) are C_6 compounds (see also Chapters 25, 34).

Boiling Points and Viscosity

As a generalization, the boiling point of a hydrocarbon compound increases with its C_{NUMBER}. It is important to remem-ber that hydrocarbon fuels are fractionated from crude oil by distillation. Compounds such as methane and ethane are gases at room temperature; gasoline (C_4-C_{11} compounds) is a volatile liquid which distills between 30°C and 210°C; diesel oil (C_{10}-C_{19}) is essentially a non-volatile liquid which distills between 200°C and 400°C, and waxes (C_{20+}) are solid. The range of $C_{NUMBERS}$ for common hydrocarbon fuels and lubri-

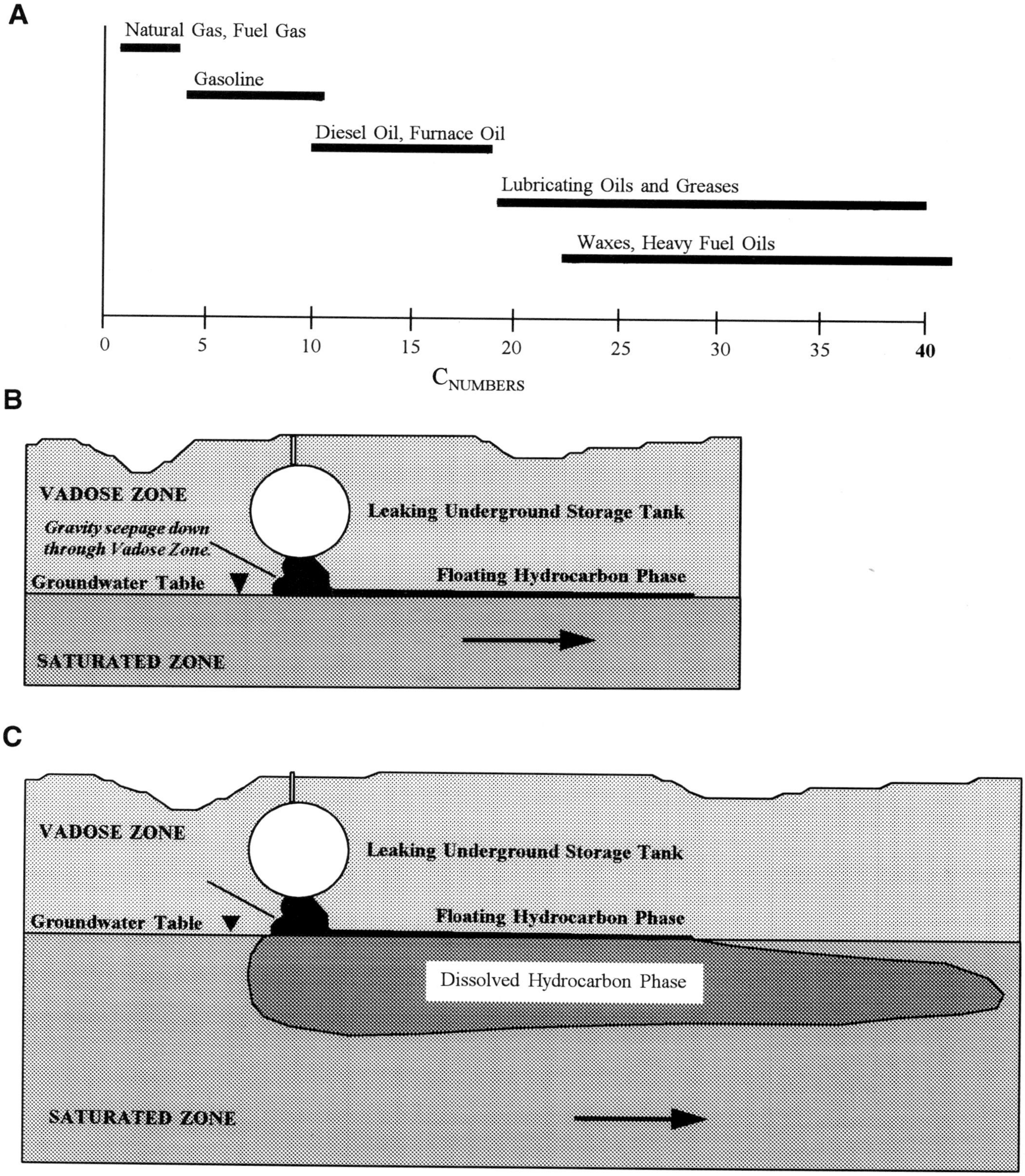

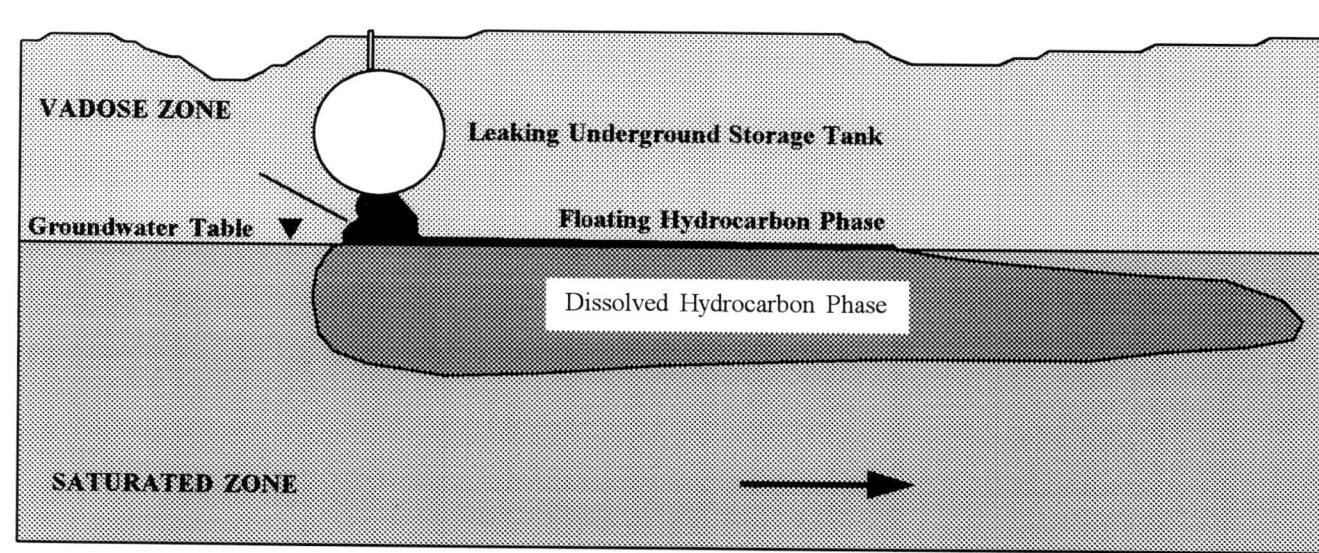

Figure 2   (**A**) $C_{NUMBERS}$ for common hydrocarbon compounds. (**B**) Floating non-aqueous phase hydrocarbon fuel known as free product. (**C**) Dissolved phase hydrocarbon fuel moving as a plume within the subsurface.

cants is presented in Figure 2A. The characteristics of commonly found organic contaminants are shown in Table 1. The relative viscosity of a fuel or lubricant at room temperature has a profound effect on the mobility of that compound once it has spilled onto the ground surface or leaked from an underground tank. Gasoline tends to migrate and be distributed over a far wider area than diesel or heavy oils over any one period of time.

Solubility

Another important attribute related to the C_{NUMBER} is the aqueous solubility of a compound. As a general statement, the solubility of a hydrocarbon compound tends to decrease with increasing C_{NUMBER}. Thus, the dissolved plumes associated with gasoline spills are more extensive than for diesel or heavier oils (assuming all other factors are equal).

Density

For the most part, the constituent compounds that make up hydrocarbon fuels and lubricants have a specific gravity of less than 1 $g \cdot cm^{-3}$. Consequently, when fuels and lubricants are spilled onto or below the ground-water table, the non-aqueous phase floats on top of the table, as shown on Figure 2B, as a pool of free product. This will be the source for a plume of dissolved hydrocarbons (Fig. 2C). These compounds are referred to as light non-aqueous phase liquids (LNAPLs).

Most of the commonly encountered industrial and commercial solvents are halogenated hydrocarbon compounds (*i.e.*, hydrocarbon compounds that have chlorine, fluorine or bromine atoms as part of their structure). In contrast to LNAPLs, these compounds generally have a specific gravity of more than 1 $g \cdot cm^{-3}$, and sink down through the water table when spilled. These compounds are referred to as dense non-aqueous phase liquids (DNAPLs) and examples are napthalene, dichlorobenzene and 1,1,1–trichloroethane. The behaviour and characteristics of DNAPLs are addressed in Chapters 6 and 35.

HYDROGEOLOGICAL INFLUENCES ON LIGHT NON-AQUEOUS PHASE LIQUID CONTAMINANT TRANSPORT

Permeability

The principal control on the extent of migration of leaking fuels and lubricants such as LNAPLs is the primary and secondary permeability of subsurface geological units. The primary permeability of a geological unit may loosely be defined as how it is able to transmit ground-water flow. Sands and gravels have high primary permeabilities, and ground-water movement through such units is often on the order of tens of centimetres to metres per day. Clay silt tills have low primary permeabilities, and ground-water movement is more commonly on the order of a few millimetres per day. Secondary permeability is an important factor to consider in low permeability units such as clays, tills or in bedrock, and is essentially the effective permeability of a unit based on ground-water flow through connecting fractures or voids. It is prudent never to discount the possibility that considerable migration may have occurred through a unit with low primary permeability as a result of fracture flow. The various origins of fractures are discussed in Chapter 2.

Contaminant Migration Through the Unsaturated Zone

Migration of contaminants through the unsaturated (vadose) zone is by gravity flow (Fig. 2B). The principal factors affecting the rate of downward migration of the contaminant are the permeability of the soils and the viscosity of the liquid. Clearly, the lower the permeability of the soils and the more viscous the contaminant, the slower the rate of downward migration. Another important influence on the rate of migration is the sorption capacity of the soils. This is the degree to which the soil particles act to absorb and retain the hydrocarbon liquid. As a general statement, soils which are high in organic content tend to have correspondingly higher sorption capacities.

Contaminant Migration Through the Saturated Zone

Dissolved hydrocarbons are subject to diffusion, advection and dispersion and, in addition, are adsorbed onto organic particles and clay minerals; in combination, these processes act to retard the movement of hydrocarbons in ground water. Organic compounds are also subject to bacterial degradation. The two most important transport mechanisms that govern the movement of dissolved hydrocarbons are advection and dispersion. Advective movement is governed by the hydraulic conductivity (or permeability) of the aquifer, the hydraulic gradient (slope of the water table) and the porosity of the aquifer. Dispersive movement is generally a much smaller component of overall flow, and is the result of the combined action of mechanical mixing and random movement of the molecules (molecular diffusion).

The mobility of dissolved-phase chemicals is dependent on a number of physical and chemical factors. These include the solubility of the compound, the density and the Henry's Law constant which indicates the tendency of a compound to volatize from the dissolved phase. For any organic compound that is dissolved in ground water, the higher the Henry's Law constant for that particular compound, the greater ease at which it will volatize and migrate as a vapour. Benzene is more likely to move as a vapour than phenol (Table 1). Other factors include the biodegradation rate (the rate at which compounds are broken down by bacteria in soil), and the octonal-water partition coefficient which describes the degree to which a compound is soluble in water, and is adsorbed onto the soil or other receptors (*e.g.*, organic particles), and the degree to which it will be concentrated in successively higher trophic levels of food chains (bioconcentration). This coefficient is expressed as a quantity K_{ow}. Compounds with low values of K_{ow} readily dissolve in water and are less likely to be adsorbed in the soil. Compounds having higher values of K_{ow} (commonly expressed as a logarithmic value; Table 1) have a greater tendency to form non-aqueous phases and are described as hydrophobic. A detailed treatment of these factors is beyond the scope of the present paper and can be found in Fetter (1993) and various papers in Daniel (1993). Useful overviews of hydrocarbon contamination and remediation are given by Abdul (1988), Eslinger *et al.* (1994) and Kehew (1995).

INVESTIGATION OF HYDROCARBON FUEL CONTAMINATED SITES

The investigation of sites contaminated by hydrocarbon fuel is generally undertaken at two levels: (1) preliminary investigation; and (2) if necessary, a more detailed program to delineate the extent of impacts and identify remedial options.

It is very important to have a detailed understanding of the site setting and the nature of the contamination, and to develop a preliminary idea of the extent of impact. The site setting is described in terms of the geology (*i.e.*, the permeability of the native geological units, their thickness and spatial distribution, *etc.*), the hydrogeology (depth to ground water, ground-water flow direction, the community dependence on ground water as a potable source) and location of the closest sinks (rivers, lakes and ground-water supply wells). Many of these aspects are reviewed elsewhere in this volume.

The nature of contamination is of critical importance. There are three principle states in which LNAPL hydrocarbon fuel contamination can exist (Fig. 2). These are (1) as vapour in the unsaturated (vadose) zone lying above the ground-water table; (2) as dissolved phase in ground water; and (3) as light liquid hydrocarbon present within unsaturated soil in the vadose zone or found as free product floating on the water table.

The vapour phase is the most immediate concern because of the potential for explosions if the vapour collects in basements or sewers. It is worth noting that the contact between the water table and free-product LNAPL is not smooth, but irregular as a result of the LNAPL resting on the capillary fringe where capillary forces pull water up from the water table. The layer of LNAPL will, in turn, have its own capillary fringe (Fig. 3). In many cases, it may be difficult to resolve the

true thickness of the LNAPL layer from measurements conducted in wells. In a well, the capillary fringe above the water table is no longer maintained, and LNAPL will tend to flow into the well and displace water. As a result, the oil layer measured in a well may be unrepresentative of the true LNAPL thickness (Fig. 3). This problem is discussed at length by Fetter (1993). An interface probe is used to measure the thickness of oil penetrated by a well (Fig. 4).

The dissolved phase is a public health concern in terms of ingestion where ground water is used for municipal drinking-water supplies. LNAPL can continue to dissolve and contaminate ground water for many decades. Any preliminary investigation should provide a general picture of the size of the affected area in terms of vapour impacts, dissolved phase and LNAPL distribution. It is worth emphasizing that hydrocarbons also occur naturally at many sites and need to be discriminated from those arising from the leakage of storage tanks (see Chapter 34).

Field Procedures

Soil Contamination

Boreholes are the primary means of investigating soil conditions at a site contaminated by hydrocarbons because of the need to collect samples for analysis. Boreholes are advanced at predetermined locations at a site using a rotary drill rig equipped with augers (Fig. 5A). Soil samples are retrieved using a split-spoon sampler, which is hammered into the soil in advance of the auger to obtain an undisturbed sample (Fig. 5B). The soil sample is carefully removed from the split spoon and divided into two sub-samples. One fraction of the sample is tightly packed into a 250-mL clean glass jar for later laboratory analysis. It is important that the soil sample be packed to the brim of the jar to leave as little headspace as possible. This maintains the hydrocarbon concentrations in the soil and minimizes volatilization into the headspace (Fig. 5C).

The second fraction is placed into a sealable plastic bag or a clean 500-mL glass jar for field testing. There are a number of field test methods that can be used to obtain a semi-quantifiable

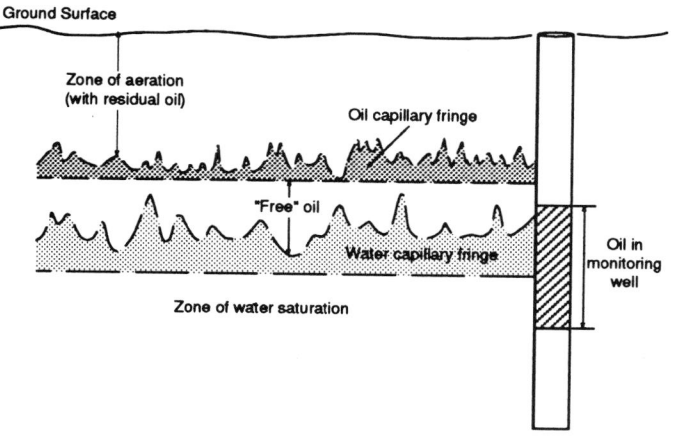

Figure 3 *Contrast in thickness of free-product oil measured away from, and inside, a well. Note irregular lower surface of oil created by the water capillary fringe. Thickness of oil from centimetres to metres.*

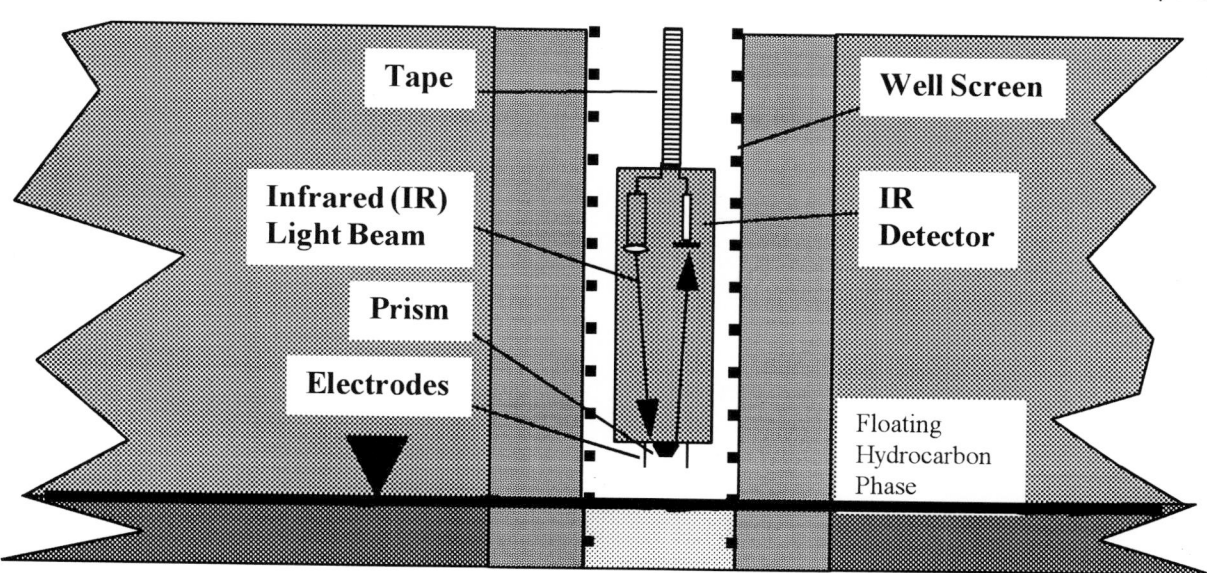

Figure 4 *Interface probe. The interface probe is used to measure the thickness of a floating hydrocarbon layer in a well (Fig. 3). A beam of infrared (IR) light is directed to a prism at the base of the probe. The electrodes at the tip of the probe indicate first contact with the hydrocarbon layer (HCL) and a constant beep is returned to surface via wires embedded in the tape. IR light is absorbed by the hydrocarbon compounds when the prism is immersed in the HCL. As soon as the prism breaks through the HCL to the water below, the beam is returned to the IR detector and an intermittent beep is signaled at surface.*

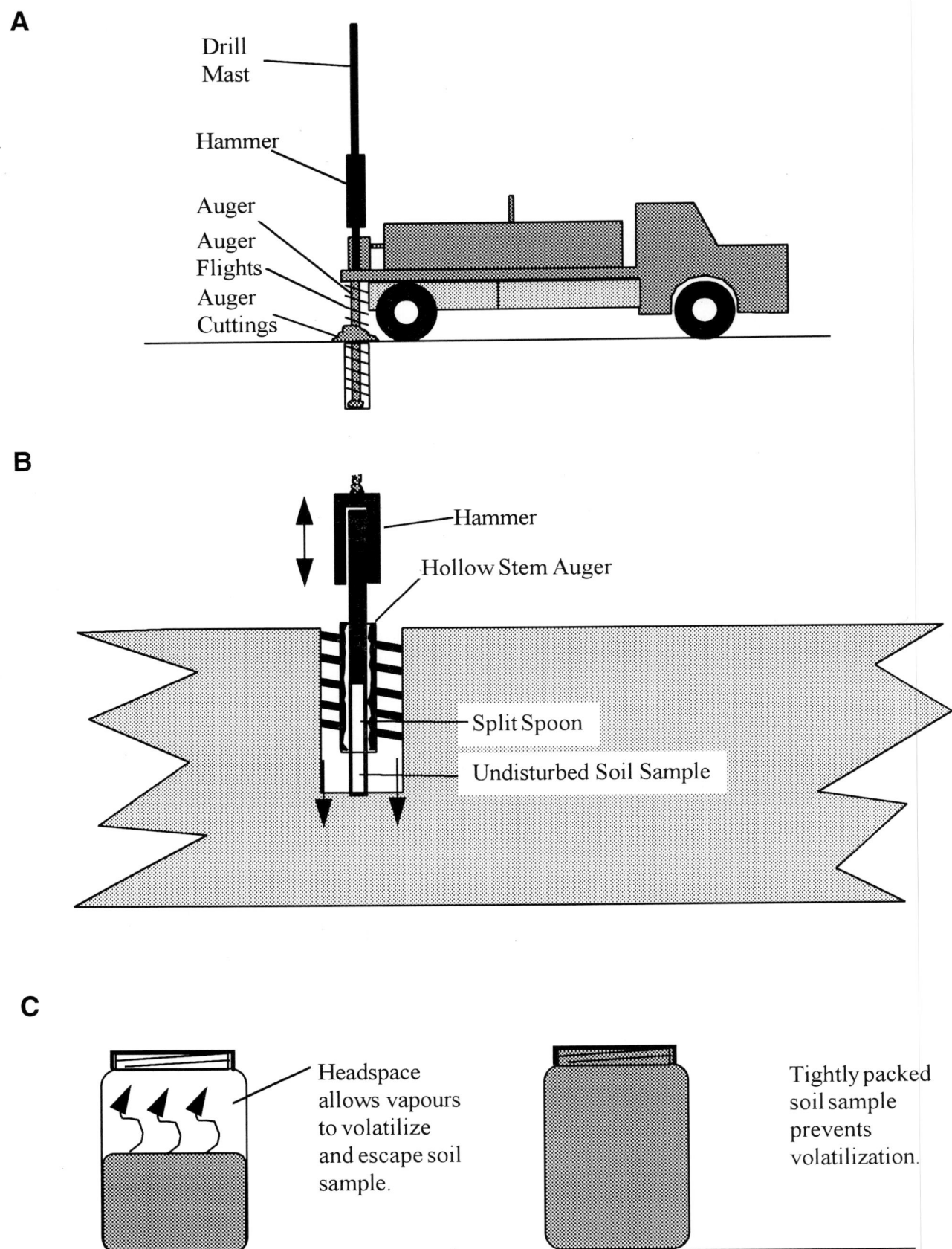

Figure 5 (**A**) *Truck-mounted rotary auger drilling rig.* (**B**) *Split-spoon sampler is driven into soil through the inside of the hollow stem auger. The split-spoon sampler is hammered in ahead of the auger to retrieve an undisturbed soil sample.* (**C**) *Headspace loss in hydrocarbon soil samples.*

(indirect measure) concentration of hydrocarbon compounds in the soil. The field screening device shown on Figure 6 is a photoionization detector, which relies on measurements of the concentration of the volatile phase of the hydrocarbon contaminant. Most field screening methods are only representative for volatile fuels such as gasoline, and are proportionally less reliable as the C_{NUMBER} of the hydrocarbon compound increases (and the volatility decreases). Generally, only the soil samples with the highest field readings are sent to the laboratory for chemical testing.

Monitoring Wells

Once the borehole has been advanced, a monitoring well can be installed, as shown on Figure 7. The monitoring well is comprised of a screen, which is a slotted section of polyvinyl chloride (PVC) pipe, and a riser which is not slotted. The well is set in silica sand, which acts as a filter and prevents fine soil particles from clogging the screen. A bentonite clay seal is installed at the top of the well (Fig. 8) to prevent direct infiltration of rain water into the well. It is important to install the screen so that it straddles the water table. This ensures that any floating LNAPL present on top of the water

table enters the well, where it can be measured (Fig. 3) by using an instrument called an interface probe which records the electrical resistance of the various liquid phases present in the soil.

Extreme care must be taken to properly seal the well to ensure contamination is not carried down to lower aquifers (Figs. 7, 8). It should be noted that contamination of lower aquifers tends to be less of a concern when drilling to investigate LNAPLs, as opposed to DNAPLs which can sink down to the bottom of monitoring wells and start a plume in a lower, previously uncontaminated aquifer (Chapters 6, 35).

Water samples are usually collected from each well using a dedicated sampling system to avoid cross contamination between wells (Fig. 9). A bailer can be used, but it must be carefully washed between monitoring wells. The collected ground-water samples are stored in a cooler on ice and should be delivered to the laboratory as quickly as is possible.

Cross Contamination of
Soil and Ground-water Samples

Chemical testing conducted at commercial laboratories routinely measures compound concentrations at the part-per-million level

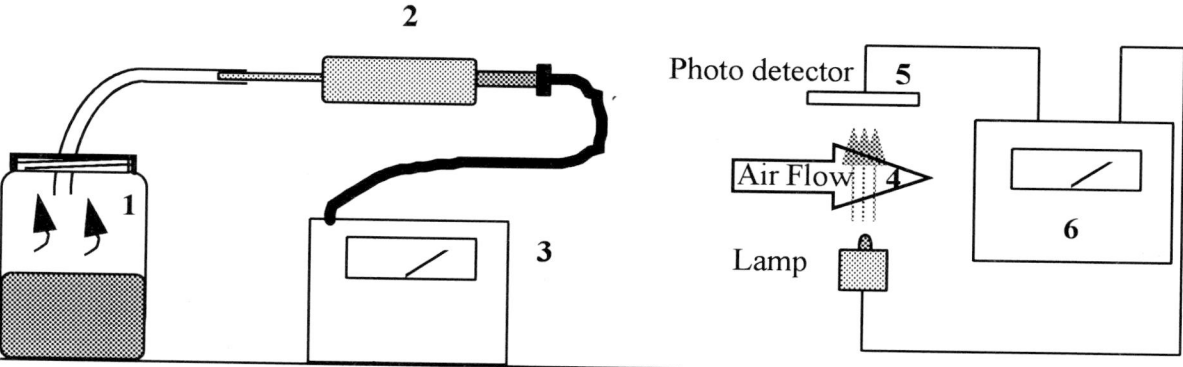

Figure 6   (**A**) *Photoionization detector. (1) Vapours are drawn from the sample jar into the photoionization detector (2). A signal is sent to the meter (3) that is proportional to the concentration of ionizable hydrocarbon compounds present in the air stream. The greater the concentration of ionizable hydrocarbon vapours (4), the less light reaches the photo detector (5) and the greater the reading on the meter (6).*

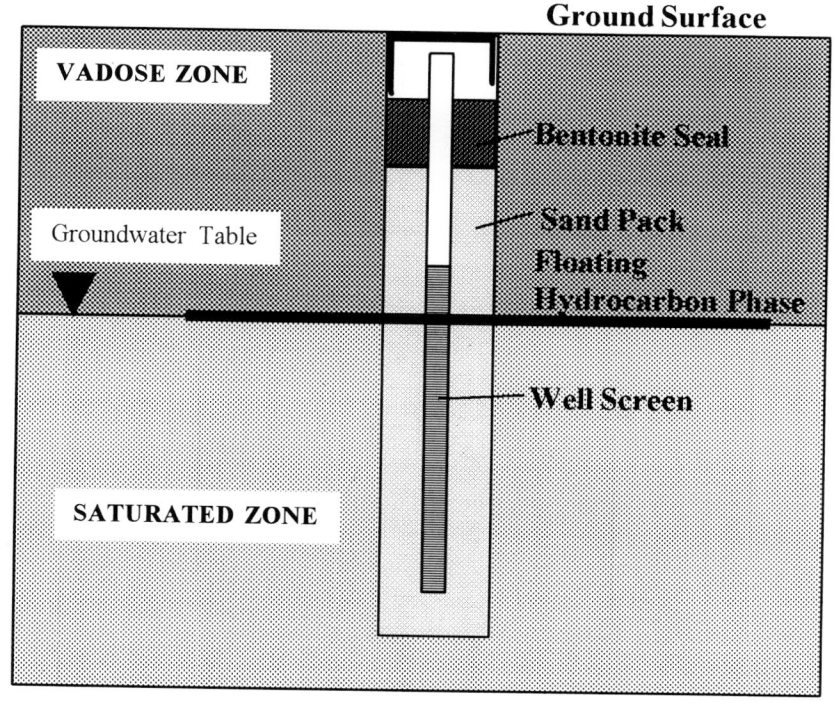

Figure 7   *Typical monitoring well installation at a site underlain by LNAPL.*

Figure 8 *Cased monitoring well set through contaminated zone. (1) A large-diameter (25 cm) borehole is augured through the upper sand layer, a small distance into the clay layer, using a hollow-stem auger. A bentonite clay slurry is poured down the inside of the auger and a steel tube called a well casing (2) is inserted inside the auger. The well casing is cemented into place and the auger is withdrawn. A small-diameter hollow-stem auger drills through the bentonite inside the casing, through the clay layer and into the lower sand aquifer. At no time is the second, smaller diameter auger in contact with the contaminated zone, precluding the possibility of dragging down contamination. Also, the casing is set into the low-permeability clay precluding downward migration of contaminate.*

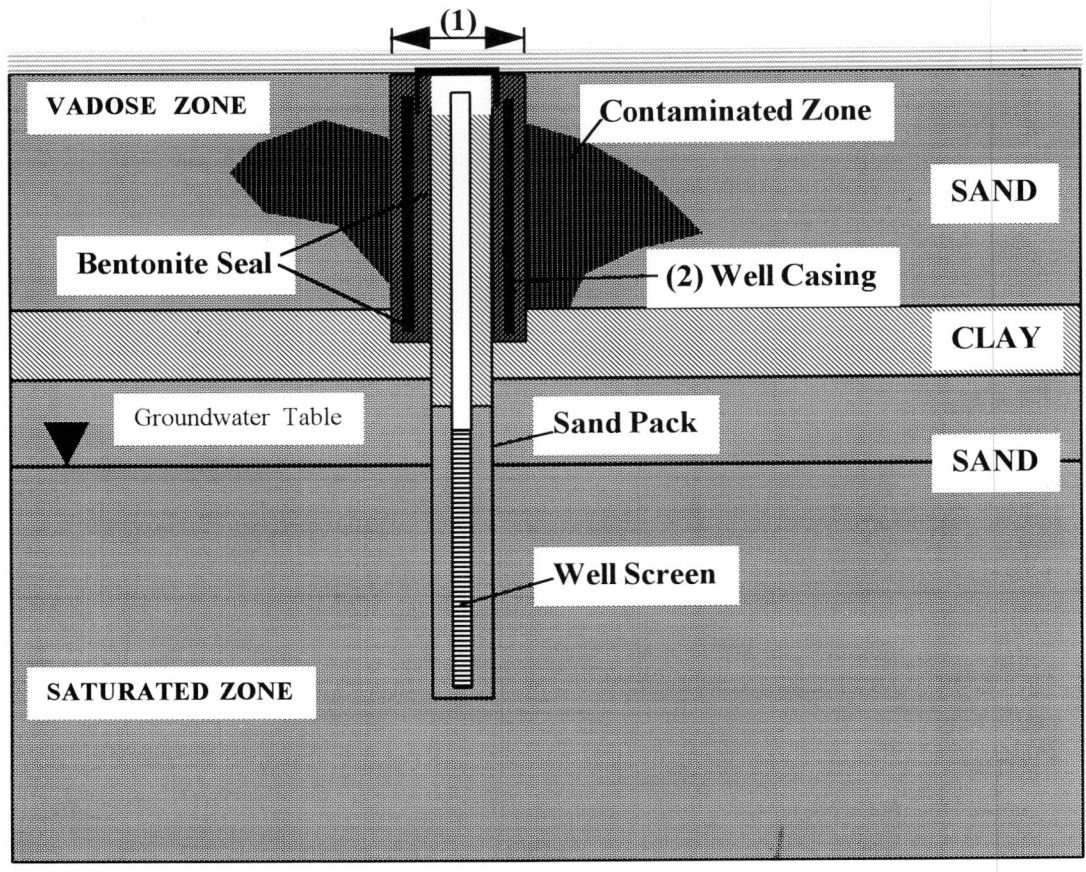

Figure 9 *Well-sampling equipment. As a bailer is lowered into the well, water entering the base of the bailer pushes up a ball (1). As the bailer is raised the ball falls back (2), plugging the orifice at the base and trapping the water inside the bailer. The same principal applies to the dedicated sampling system. A bailer-like foot valve screws onto the base of the tubing. As the tubing is pumped up and down from surface, the foot valve allows water in (down stroke), traps the water (up stroke) and then allows more water in (repeat down stroke). By continuing the pumping action, water is brought to surface through the tube.*

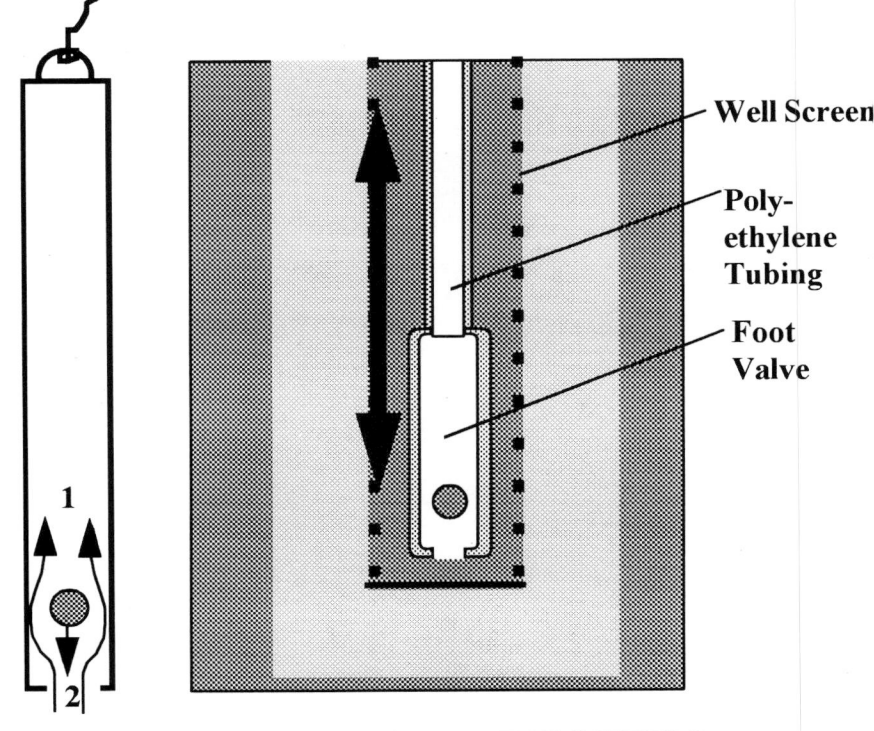

BAILER   **DEDICATED SAMPLING SYSTEM**

(soil) or part-per-billion level (ground water). This is primarily because of the stringent regulations for the clean-up of hydrocarbon fuels (particularly gasoline). It is therefore critical that strict cleaning and handling protocols be adhered to to prevent contamination from one sample being transferred to another, producing a false high reading. Also, improper handling can result in loss of hydrocarbon from a sample, resulting in a false low reading (Fig. 5C).

Augers should be cleaned between drilling each borehole, using a pressure steam cleaner to prevent contaminated soil from being carried from one borehole location to the next. The split-spoon sampling device is cleaned between each sample through a three-stage cycle, consisting of a wash in a bucket of warm water charged with phosphate-free detergent, a solvent rinse in acetone and a final rinse in distilled water. Samples and sampling equipment are handled using disposable surgical gloves, which are changed between samples. Proper labeling of

samples is key to preventing confusion over where a sample was collected from. A unique sample identifier (usually comprising the borehole or monitoring well number and sequential sample number) is used; the date of sample collection, supervising scientist or engineer, *etc.* are noted on the label. A chain of custody form is filled out and accompanies the samples to record who had access to the samples at all times.

SOIL AND GROUND-WATER REMEDIATION TECHNIQUES FOR SITES CONTAMINATED BY HYDROCARBON FUELS

A comprehensive review of soil and ground-water remediation techniques is presented in Chapter 35; this section focusses on the remediation of sites impacted by hydrocarbon fuels. Many remedial technologies are in their infancy. New technologies are constantly being developed, and present-day technologies improved, to meet the demand for cost effective and timely re-

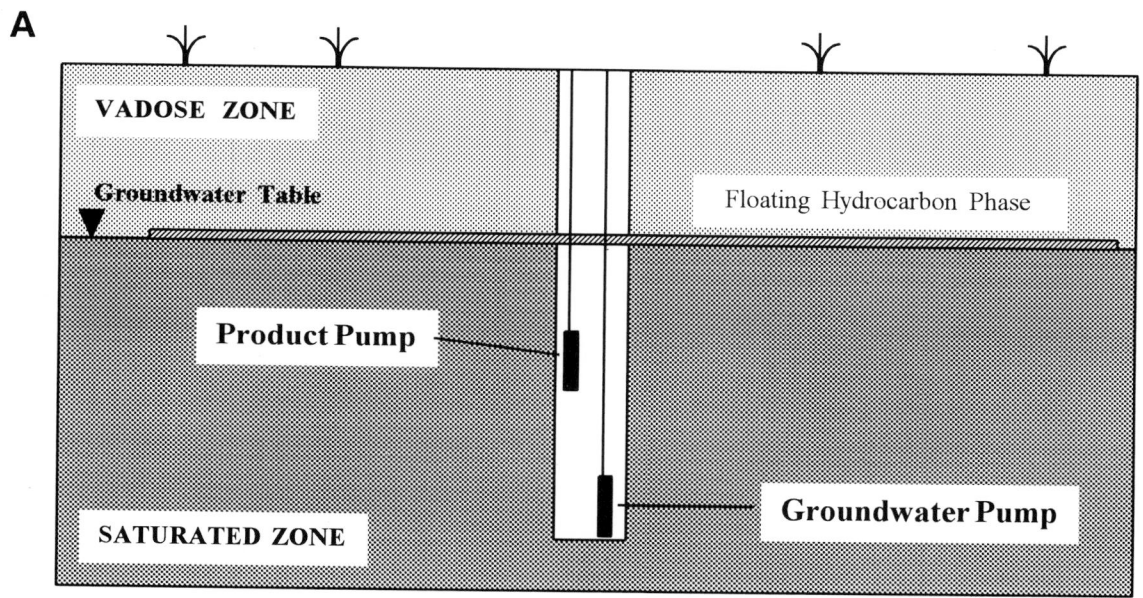

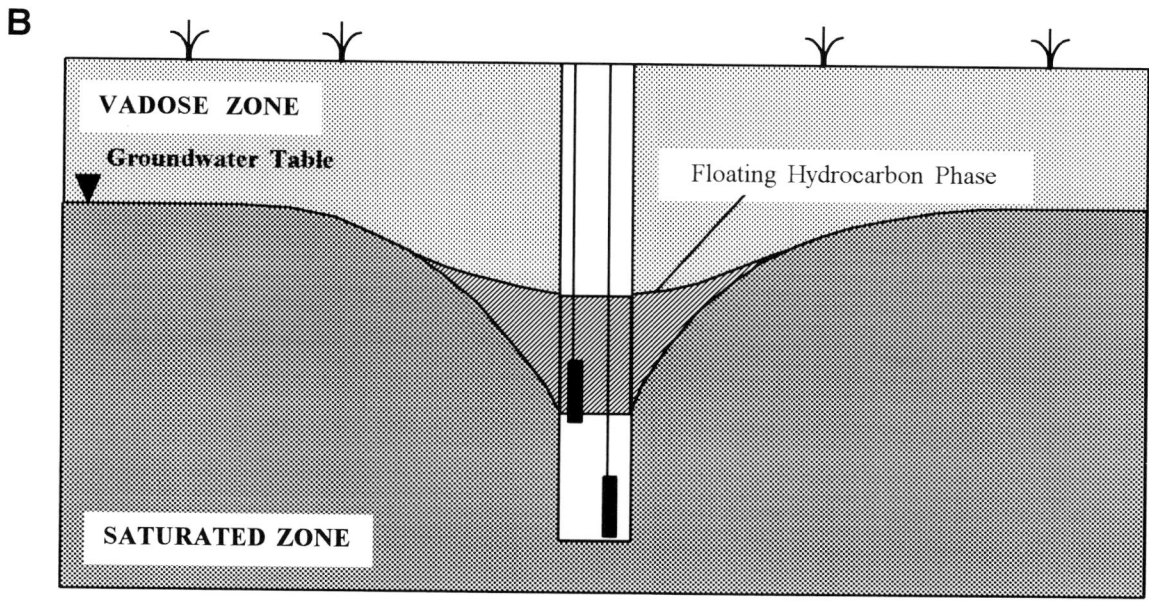

Figure 10   (**A**) *Removal of LNAPL by ground-water depression/product pumping.* (**B**) *Ground-water pumping creates a cone of depression in the water table, allowing liquid hydrocarbon to migrate towards the centre of the cone where it is removed by the pump (Fig. 11A).*

mediation. Along with low cost and short duration, techniques must be applicable in a large number of geological and climatic settings. Some techniques work well in sands and gravels, but tend to be less effective and efficient in remediating tills and clays. In general, water management is crucial during all aspects of a remediation and can be very expensive. Significant cost overruns can occur in the absence of a detailed understanding of hydrogeological conditions.

Time constraints with regards to remediating the site are an important factor. Remediation may have to be rapidly completed due to seasonal weather conditions or in cases of impending property transfers. For this reason, long-duration remedial techniques must be ruled out. Some remedial techniques require a large area for placement of equipment and soil. Therefore, the site setting and layout need to be carefully examined to assess the viability of a particular technique. Weather conditions can also play a significant role. Some techniques are very inefficient at low temperatures and are not suited for use during winter. In general, remediations conducted in winter or during rainy seasons are less efficient and more costly.

After determining the nature and extent of contamination at a site and which remedial technologies are applicable, a remedial plan can be formulated (see Chapter 36 for an example). It is essential to first remove the primary source of contamination, i.e., any leaking tanks and/or LNAPL in the subsurface. After this is completed any contaminated soil is remediated. After removal of LNAPL and the clean-up of impacted soil, contaminated ground water can then be treated.

REMOVAL OF LIGHT NON-AQUEOUS PHASE LIQUIDS FROM GROUND WATER
Floating LNAPLs are commonly removed using such techniques as ground-water depression–product collection, surface separation and skimming.

Ground-water Depression–Product Collection
This has been employed for many years as a pump-and-treat method to contain and collect floating LNAPL. This involves creating a cone of depression by removing water from a well or sump with a submersible pump set several metres below the water table. Any LNAPL floating on the water table moves to the center of the cone due to the induced hydraulic gradients (Fig. 10). As the product collects at the centre of the cone it will increase in thickness due to the dimensional constraints of the cone. It is then collected by a second pump set in the centre of the cone at the oil-water interface. The pump used to collect the LNAPL at the interface is usually set in a canister, as shown in Figure 11A, which reduces the amount of water removed with the LNAPL. Because it is very costly to treat or dispose of the water recovered during the process of removal, it is important to limit the amount of water brought to the surface. This technique works well in sands and gravels, but is less effective in clay-rich soils. The main drawback to using this technique is the potential for smearing of the LNAPL across uncontaminated soil as the water table is lowered to form a cone of depression.

Surface Separation
This is a very simple technique and consists of a pump

with canister set in a recovery well at or just below the water table. Both water and LNAPL are pumped to the surface where they are separated in an oil-water separator. This technique is less cost effective if large volumes of water are brought to the surface requiring treatment or disposal, but it does reduce the amount of product smeared across uncontaminated soil.

Skimming
This technique involves the skimming of product off the surface of the water table with the use of a special pump (Fig. 11B). The pump will successfully reduce the volume of LNAPL in the subsurface without the expense of treating ground water or causing additional soil contamination.

SOIL REMEDIATION
After removal of LNAPL, soil can be remediated in a timely and efficient manner using either an *in situ* or *ex situ* technique (Chapters 35, 36). The type of technique selected is usually

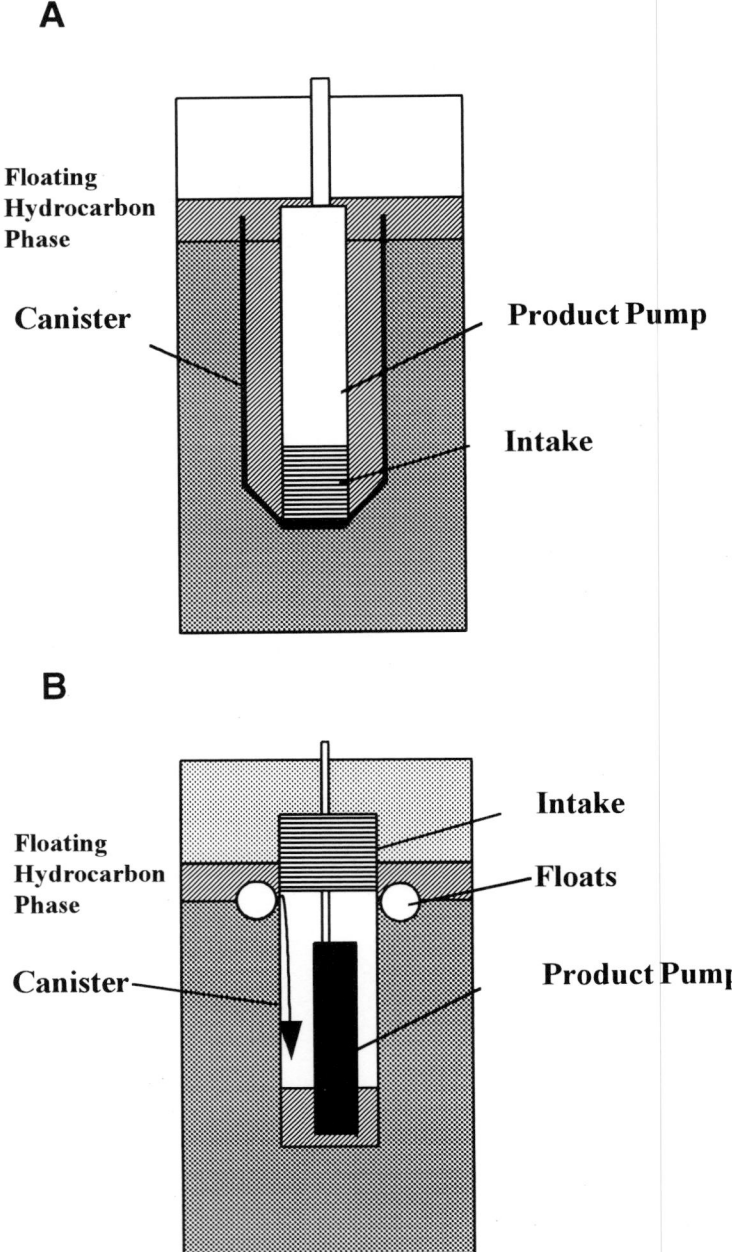

Figure 11 (**A**) *Product removal pump with canister.*
(**B**) *Schematic diagram showing a skimmer pump system.*

dependent on cost, site geology, site hydrogeology and time constraints. *In situ* techniques, where the soil is treated in place, though generally less expensive than *ex situ* alternatives involving excavation, are not very effective in clay-rich soils and require a detailed understanding of site hydrogeology.

In Situ Techniques

The most commonly applied *in situ* soil remediation techniques include vapour extraction, air sparging and bioremediation.

Vapour Extraction

This technique has been used for many years to remove hydrocarbon compounds as a vapour phase from void spaces in unsaturated soil (Silka and Jordan, 1993, and references therein). A network of wells and/or trenches is used to remediate soils lying above the ground-water table (Fig. 12). Pipes installed within the wells or trenches are perforated, and the gap (annulus) between the walls of the wells or trenches and the piping is packed with gravel. This is used to reduce the ambient vapour pressure in the subsurface, resulting in transport of contaminants to the surface for treatment. At the ground surface, the pipes are connected to a vacuum which induces the flow of air through subsurface soils. The technique also aids in the volatilization of liquid or residual organic compounds that have sorbed onto soil particles.

To clean contaminated soils below the water table, either the water table can be lowered *via* pumping or air sparging can be employed. Extensive dewatering of the contaminated area to create unsaturated conditions can be extremely costly because of the large volumes of water involved. Air sparging is used to avoid this problem.

Air Sparging

Air sparging simply involves the pumping of air or oxygen into the subsurface soil below the ground-water table in order to enhance the mobility of contaminants (see Marley *et al.*, 1992; Chapter 35). The principal advantage of the system is that it obviates the need to pump water to the surface. Air bubbles move horizontally and vertically through the soil and create air-filled regimes within the saturated soil. Hydrocarbon compounds exposed to air evaporate into a gaseous phase, and are transported by the air stream into the vadose zone. Low-molecular-weight hydrocarbons are more easily stripped from air because of their greater volatility (see Henry's Law constant, above). Once in the vadose zone, the hydrocarbon compounds are collected and treated either by vapour combustion units, where the vapours are incinerated; by catalytic oxidation units, where vapours are heated and passed over a catalytic bed where hydrocarbon compounds are stripped from the air; or, by carbon beds, where vapours are passed through granular activated carbon (GAC) filters. One or more of the systems may be used to ensure complete and cost-effective clean-up; GAC units may be used as polishers of ground water treated initially by air stripping. In general, the cost of GAC systems is much greater than the other treatment methods because of the need to replace carbon on a regular basis and dispose of carbon saturated with contaminants.

Bioremediation

This was developed by the waste-water treatment industry as an alternative to pump-and-treat methods of remediating ground water, largely because of the failure to remove the actual source of contamination within soils (see Brubaker, 1993; Chapter 35). Bioremediation is based on the ability of micro-organisms (bacteria) to degrade and destroy specific hydrocarbon compounds. There are basically two methods to bioremediate soils. The first (biostimulation) encourages the growth of indigenous bacteria capable of destroying hydrocarbon compounds by injection of oxygen and/or nutrients (N, P, *etc.*) into the subsurface soils *via*

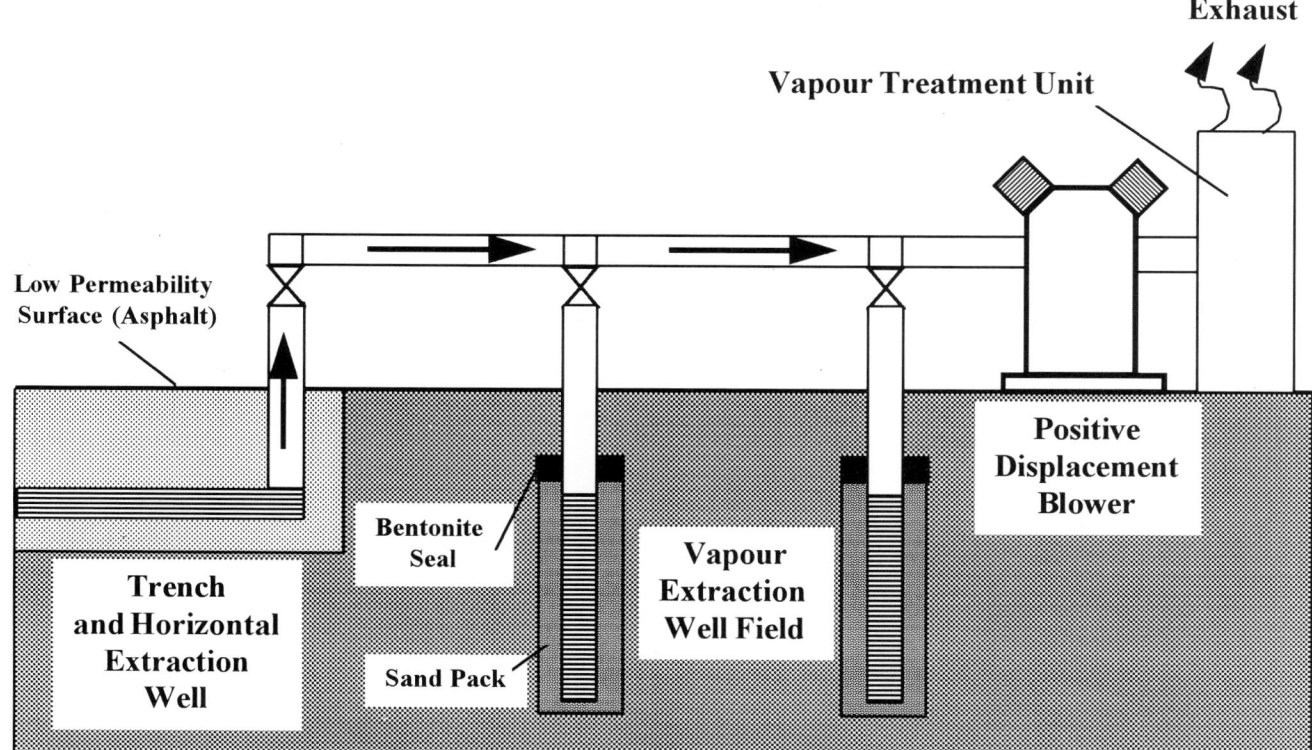

Figure 12   *Schematic diagram showing a typical trench and well field vapour extraction system.*

wells and/or trenches. A second approach (bioaugmentation) involves the injection of specialized strains of bacteria, which are isolated and concentrated from soils originating from the contaminated site. This concentrated bacterial strain or mixture of strains is called a microbial inoculum. Inoculum, oxygen and nutrients are inserted into contaminated soils *via* wells and or trenches.

Any bioremedial method to be implemented requires a biofeasibility study to be conducted, which includes the collection of contaminated soil samples from the site, a nutrient analysis, plate counts for total bacteria, plate counts for contaminant degrading bacteria, interpretation of data for nutrient augmentation requirements, and calculation of quantity and type of nutrients to be added. As in soil venting, bioremediation occurs primarily above the water table. Lack of oxygen below the water table inhibits proliferation of the aerobic bacteria which are best suited to degrade hydrocarbons. As a result, it may be necessary to lower the water table to complete the remediation of contaminated soils. As discussed previously, an alternative to lowering the water table would be air sparging which creates oxygen-rich conditions below the water table.

Ex Situ Techniques

The presence of clay-rich tills in many parts of Canada limits the use of *in situ* techniques. The alternative is an *ex situ* process (where soil is excavated) which is relatively more expensive, but generally less time consuming. Some of the more common *ex situ* techniques include disposal of contaminated soils at a landfill, land farming, biopiles, low thermal temperature desorption, and soil washing.

Disposal of Contaminated Soil At a Landfill

This is possibly the most common technique applied in the remediation of soils; advantages and disadvantages are discussed in Chapter 35.

Land Farming

In this process, contaminated soil is excavated and spread on a prepared engineered surface designed to collect runoff and prevent percolation of contaminated water into underlying clean soils. Hydrocarbon compounds are allowed to volatilize into the atmosphere. Soils are turned daily to aid in the desorption of contaminants from soil particles and subsequent volatilization of the contaminants into the atmosphere. To enhance this technique, bioremediation can also be conducted. The soil to be remediated can also be bio-inoculated, or proliferation of indigenous bacteria can be encouraged by the introduction of nutrients to the soils. In many municipalities, land farming is not permitted unless there is a bioremediation component involving the use of biopiles (see below). It needs to be scientifically proven that the majority of hydrocarbon contamination is truly being biodegraded, as opposed to simply volatilizing into the atmosphere and being exported to adjacent sites and communities.

Biopiles

As the name implies this technique involves the construction of a pile (also called an engineered bioremediation cell) designed to bioremediate soils. A case study is provided in Chapter 36.

Low Thermal Temperature Desorption

The technique of low thermal temperature desorption (LTTD) is derived from the thermal destruction units used to destroy

hazardous substances such as PCBs at temperatures of 1500°C or more. Disaggregated soil is fed through a furnace where the hydrocarbon contaminants desorb from the soil and are subsequently oxidized. Most thermal units are mobile and can treat the soil on site. They do, however, require a significant amount of space. Soil from small sites can be transported to a central unit, treated and returned to the site as backfill material. Chapter 35 shows a schematic of a LTTD unit.

Soil Washing

In this process, impacted soil is washed with surfactants (essentially detergents) to release the hydrocarbons. Contaminated material is first crushed and then transferred to washing tanks. Surfactants are added to the tanks and the material agitated with the use of water jets. The water jets cause a shearing action, effectively scrubbing the soil particles. The slurry is pumped through manifolds from the bottom of the washing tanks onto shaker screens. Larger clean particles are retained by the screen and transferred to a clean soil pile. Smaller particles and liquids pass through the screen into hydrocyclones, where a flocculating agent is added which agglomerates small micron-size particles. Oil, water and surfactant are skimmed from the surface of the last washing tank and passed through an oil-water separator.

GROUND-WATER REMEDIATION

After removal of LNAPL and contaminated soil, the remediation of contaminated ground water can proceed. Pump-and-treat is the most widely used method for remediating hydrocarbon-contaminated ground water, but has limitations in the absence of efforts being made to remove or isolate the source of contamination. A full review of ground-water remediation techniques, including pump-and-treat, is provided in Chapter 35.

CASE STUDIES

Example 1

A chemical company with many on-site storage tanks initiated an environmental site assessment (ESA) as part of a site closure plan when intending to sell the property for redevelopment. Studies revealed that xylene had impacted both the soil and ground water in the vicinity of two solvent storage tanks

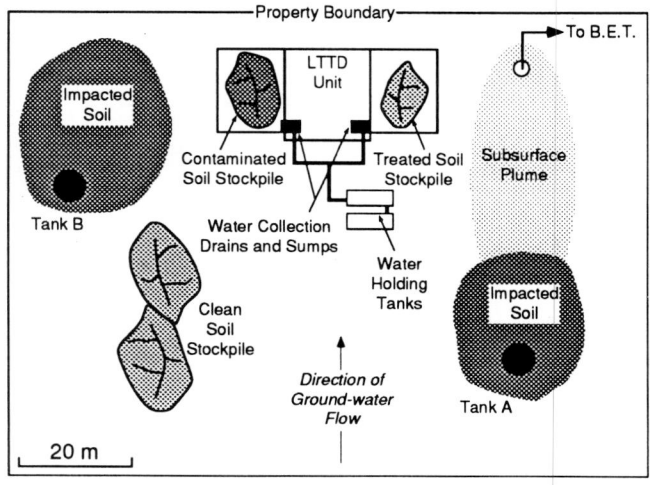

Figure 13 *Site layout for case study 1.*

(Fig. 13). In addition, a long, narrow plume of contaminated ground water was discovered within a subsurface sandbody and found to extend from one storage tank (Tank A; Fig. 14) to the property boundary. Investigations revealed that strata below the site consisted of a surficial layer of historic landfill, up to 2 m thick, consisting of sand and gravel, resting on a 18-m-thick till unit resting on shale bedrock. The till contained an intraformational sandbody, at least 4 m thick, which extended off site, and it was in this unit that the xylene contaminated plume of ground water was present. The presence of sand layers in till is discussed in Chapters 2, 20 and 32. Near-surface till was fractured and weathered, allowing contaminants to move down to the sand layer. The hydraulic conductivity of the weathered till was estimated (using slug tests) to be in the range of 10^{-7} to 10^{-9} metres per second ($m \cdot s^{-1}$), and that of the sand was determined to be in the range of 10^{-4} to 10^{-6} $m \cdot s^{-1}$. Drilling showed that about 3000 cubic metres (m^3) of till had been contaminated in the vicinity of Tank A and an additional 5100 cubic m^3 in the vicinity of Tank B.

The first remediation tasks were to remove the tanks and to prevent any off-site migration of xylene-contaminated ground water. A recovery well was installed in close proximity to the property boundary and screened across the sand. Water was pumped from the well at a rate of 10 $L \cdot min^{-1}$ to produce a cone of depression (Fig. 10B) which captured LNAPL within a 10-m radius of the well. With this control in place, the removal of LNAPL in the vicinity of the tanks could be started. A trench sump network was constructed to drain LNAPL from pockets of fill and transport it to a central sump, from where water and LNAPL were pumped and separated using an on-site Biological Effluent Treatment facility (BET). Treated water was discharged to the sanitary sewer.

While LNAPL was being removed, a technology search for remediating soils was initiated. Because of the low permeability of the till, the use of *in situ* techniques was ruled out. The client preferred to have the contaminant destroyed rather than have contaminated soil trucked to a landfill site and incur possible future liability for the dumping of contaminated soil. Low thermal temperature desorption was selected as the preferred technology to remediate on-site soils. The site was large enough to use a LTTD facility, which, together with stockpiles of contaminated till, was set up on an impermeable surface with hydraulic controls to collect and direct rain water (Fig. 13). Treated till was sampled daily and, if deemed analytically clean, it was moved to a clean soil stock pile. The walls and floor of the excavation were checked daily to determine when uncontaminated soil had been reached. After verifying the entire excavation was free of contaminated soil, treated till was used to backfill the excavations.

The remediation took four months to complete and a total of 20,000 t of till was excavated, treated and returned to the ground as backfill. All water collected from the excavation, together with rain water, was pumped to holding tanks for testing. If the test showed elevated levels of xylene, the water was drained to the BET for treatment. Air was monitored throughout the project at the property boundary downwind from the LTTD. At no time during the project was air quality impeded at the sampling points with respect to particulate and xylene vapors. Upon completion of the project, excavated areas were sloped for drainage and seeded with grass. A verification report was completed and filed with the provincial environmental agency and attached to the property title.

Example 2

An environmental site assessment conducted at a city works yard revealed the presence of LNAPL and contaminated soil in the vicinity of a gasoline UST. The tank was pressure tested and found to be leaking. The city immediately removed and replaced the tank. All soil excavated during tank removal was transported to a landfill for disposal. Subsurface strata at the site consisted of 1 m of fill on top of interbedded fine sand and silt. The hydraulic conductivity of the silt and sand was determined to be in the range of 10^{-8} $m \cdot s^{-1}$ and the water table was encountered at 1.5 to 2 m below ground surface.

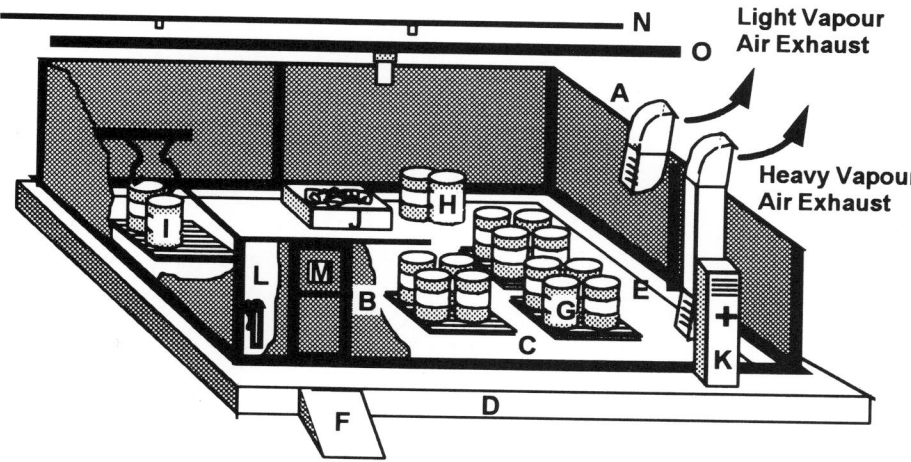

Figure 14   *Idealized drum storage area.*

A - Chain Link Fence
B - Padlocked Door
C - Epoxy Coated Floor
D - Concrete Floor (no drains)
E - Sealing Compound in Edges
F - Ramp Over Curb
G - Drums - Non Flammable Waste
H - Spare Drums for Clean-Ups

I - Drums
J - Bin Holding Clean-Up Equipment
K - Locker for Garments and First Aid Kit
L - Fire Extinguisher
M - Emergency Procedures
N - Halon Automated Fire Suppression System
O - Explosion - Proof Lighting

To remove LNAPL in the subsurface, a dual pump system was installed within a 1-m-diameter, 7-m-deep culvert, dug 2 m from the location of the UST. LNAPL and water collected by the product pump (Fig. 10A) were directed to an oil-water separator at the surface. Recovered LNAPL was drained to a storage tank and water discharged to the city's sanitary sewer. After six months of operation, all recoverable LNAPL had been removed from the subsurface. The vapour extraction technique was selected to remediate contaminated soil. Perforated pipes were installed within a network of trenches and connected to a positive displacement pump at the surface, which set up a vacuum throughout the pipe system and surrounding sand and silt. Recovered vapours were directed through carbon filters and treated air was then exhausted to the atmosphere.

MANAGEMENT OF HYDROCARBON COMPOUNDS AND EQUIPMENT

Prevention of ground-water and soil contamination is more cost effective than remediation and preventative measures are now in routine use.

Drum Storage Areas

The basic elements of a hydrocarbon compound storage facility are shown on Figure 14. Drum storage areas should be conveniently located, preferably in close proximity to where the wastes are generated.

Above-ground Storage Tanks Holding Bulk Hydrocarbon Compounds

Generally, it is far preferable to store hydrocarbon com-

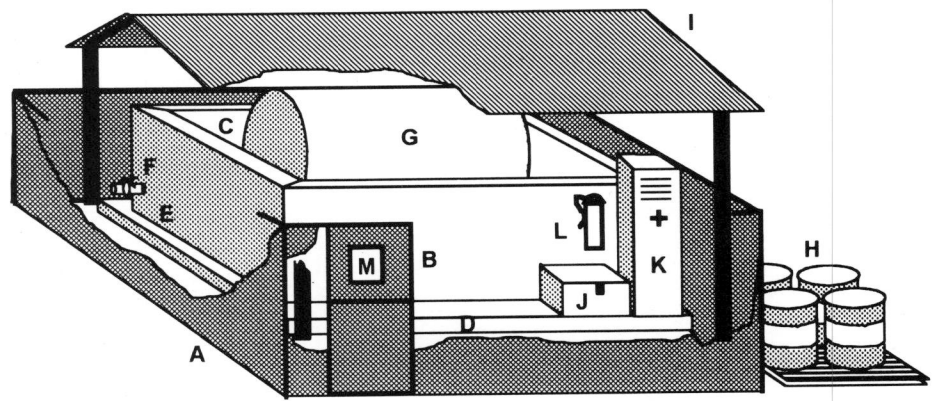

Figure 15 *Idealized above-ground storage tank containment area.*

A - Chain Link Fence
B - Padlocked Door
C - Epoxy Coated Walls and Floor
D - Concrete Floor (no drains)
E - Sealing Compound in Edges
F - Lockable Tap
G - Above Ground Tank

H - Spare Drums for Clean-Ups
I - Roof Structure
J - Bin Holding Clean-Up Equipment
K - Locker for Garments and First Aid Kit
L - Fire Extinguisher
M - Emergency Procedures

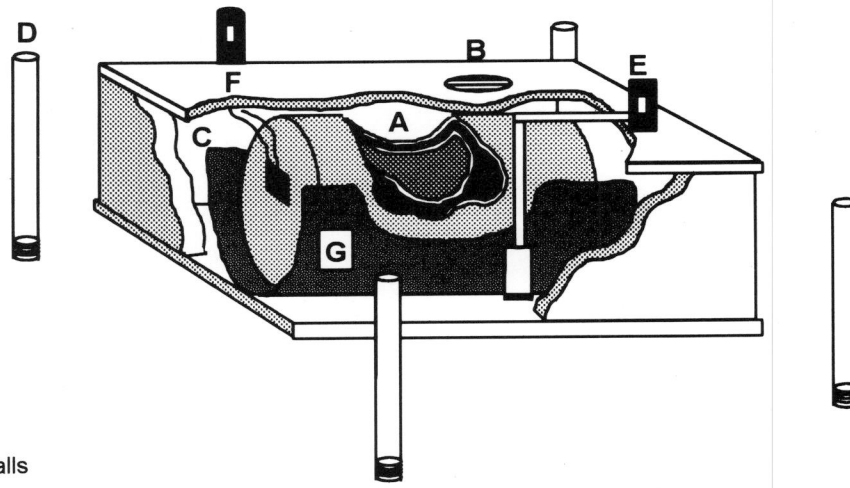

A - Double Walled Tank
B - Leak Detector System Between Walls
C - Epoxy Coated Vault
D - Monitoring Wells
E - External Leak Detector System
F - Cathodic Protection
G - Liner

Figure 16 *Idealized underground storage tank containment and monitoring system.*

pounds in above-ground storage tanks (AGSTs) rather than in underground storage tanks (USTs). This is because AGSTs are entirely visible and can easily be inspected to verify condition and integrity. Leaks from AGSTs are immediately evident, and can, therefore, be acted on before any significant spreading of contamination takes place. Steel USTs are subject to the corrosive effects of the subsurface environment, and require specialized corrosion protection. In general AGSTs should be constructed within impermeable secondary-spill protection structures capable of holding at least 110% of the volume of the tank. An example of a modern AGST storage structure is presented as Figure 15.

Underground Storage Tanks
Holding Bulk Hydrocarbon Compounds

A significant advantage of holding bulk hydrocarbon compounds in underground storage tanks (USTs) is that USTs are inherently much safer from accidental impact or, in the case of flammable liquids, fire or explosion. However, from an environmental protection standpoint, USTs are difficult to manage. Leakage often goes unnoticed until significant environmental damage has occurred. Containment and monitoring are the keys to limiting impacts from hydrocarbon compound USTs. Figure 16 shows the various available features for UST protection. The double-walled tank serves to provide two levels of protection. If the interior wall of the tank is breached, a detector positioned in the annular space between the interior and exterior wall sounds an alarm in response to detecting hydrocarbon liquids. Similarly, if the exterior wall leaks, ground water entering the tank triggers a second alarm. It is equally important to install double-walled piping where possible, to offer the similar benefits of dual protection. A liner can be installed around the tank to provide a low-permeability barrier. The liner can be constructed of clay, although hydrocarbon-resistant geotextiles offer a greater level of protection and are easier to install. Generally, liners are not considered to provide a high level of protection and are not commonly used, except as vapour barriers. If the UST is constructed with steel, a cathodic protection device is required to prevent corrosion of the tank. USTs can corrode and leak in as little as five to ten years, depending on the chemistry of the host soils surrounding the tank.

Over the past 15 years, the petroleum industry has moved to the installation of fibreglass tanks, which are inert and do not corrode. Monitoring wells offer a remote means to alert of a possible leak. However, the wells must be sampled and tested on a regular basis if they are to serve as an effective warning tool. Wells should be set into the permeable backfill around the tank, where leaks to the ground water can be quickly detected by a leak detector system which avoids the expense of monthly chemical testing of ground water.

DISCUSSION

Fuel storage tanks represent one of the greatest threats to the quality of urban soil and ground water because they are so numerous in urban communities and prone to leakage. The cost to clean up individual sites ranges from tens of thousands to millions of dollars, depending on the extent of impact and stringency of clean-up. The total remediation cost for dealing with the problem of leaking fuel storage tanks in Canada has been estimated to be many tens of billions of dollars, *i.e.*, about the same order of magnitude as the present federal government deficit (Rush and Metzger, 1991).

REFERENCES

Abdul, S.A., 1988, Migration of petroleum products through sandy hydrogeological systems: Ground Water Monitoring Review, v. 8, p. 73-81.

Bass, D.H. and Riley, B., 1995, Highly soluble MTBE responds to pumping: International Ground Water Technology, v. 1, p. 13-17.

Brubaker, G.R., 1993, *In situ* bioremediation of groundwater, *in* Daniel, D.E., ed., Geotechnical Practice in Waste Disposal: Chapman and Hall, London, UK, p. 551-584.

Canadian Institute for Environmental Law and Policy (CIELAP), 1995, Toxic Time Bombs: the Regulation of Canada's Underground Storage Tanks: Emond Montgomery Publications, Toronto, 200 p.

Daniel, D.E., 1993, Geotechnical Practice for Waste Disposal: Chapman and Hall, London, UK, 683 p.

Environment Canada, 1987, Leaking underground storage tanks: Environment Canada Fact Sheet, EN 40-203/4-1987, 6 p.

Eslimger, E., Oko, U., Smith, J.A. and Holliday, G.H., 1994, Introduction to Environmental Hydrogeology: Society for Sedimentary Geology, Short Course 32, Tulsa, OK, 198 p.

Fetter, C.W., 1993, Contaminant Hydrogeology: Macmillan, New York, 458 p.

Kehew, A.E., 1995, Geology for Engineers and Environmental Scientists: Prentice Hall, NJ, 574 p.

Knox, E.G., 1994, Leukaemia clusters in childhood; geographical analysis in Britain: Journal of Epidemiology and Community Health, v. 48, p. 369-376.

Marley, M.C., Hazebrouck, D.J. and Walsh, M.T., 1992, The application of *in situ* air sparging as innovative soils and groundwater remediation technology: Ground Water Monitoring Review, Spring, p. 137-145.

National Institutes of Health, 1993, Health Effects of Gasoline: Environmental Health Perspectives Supplements, v. 101, supplement 6, 212 p.

Rush, R. and Metzger K., 1991, Leaking storage tank costs could rival our federal deficit: Environmental Science and Engineering, v. 9, p.38-41.

Silka, L.R. and Jordan, L.R., 1993, Vapor analysis/extraction, *in* Daniel, D.E., ed., Geotechnical Practice in Waste Disposal: Chapman and Hall, London, UK, p. 379-429.

8. Contamination of Urban Ground Water by Road De-icing Chemicals

Ken W.F. Howard

Environmental Earth Sciences, University of Toronto, Scarborough Campus
Scarborough, Ontario M1C 1A4

Janet Haynes

David L. Charlesworth and Associates Inc., Suite 110, 77 Mowatt Avenue, Toronto, Ontario M6K 3E3

SUMMARY

Every year, roads and highways in Metropolitan Toronto receive more than 100,000 tonnes (t) of NaCl road de-icing chemicals. While much of this salt is flushed from the region every winter season by overland flow, a proportion will enter the sub-surface and eventually discharge to urban streams as baseflow. To determine annual retention rates of de-icing salts in an urban watershed, a chloride mass balance has been applied to the Highland Creek basin, a typical urban catchment in eastern Metropolitan Toronto. The catchment has an area of 104 km² and ground-water recharge is estimated to be 162 mm per year. Chloride input to the catchment was determined from municipal records. These show that the catchment receives approximately 10,000 t of chloride annually, predominantly in the form of NaCl de-icing chemicals which are applied to roads, highways and parking lots during the winter months. Chloride output was estimated from stream flow data.

The balance reveals that only 45% of the salt applied to the catchment is being removed annually and that the remainder is entering temporary storage in shallow sub-surface waters. If present rates of salt application are maintained, it is predicted that *average* steady-state chloride concentrations in ground waters discharging as springs in the basin will reach an unacceptable 426 ± 50 mg·L⁻¹ possibly within a 20-year time frame. The value of 426 mg·L⁻¹ represents a three-fold increase over present average baseflow concentrations, and is nearly twice the drinking water quality objective of 250 mg·L⁻¹ maximum acceptable concentration.

INTRODUCTION

During the past 40 years, residents living in the snowbelt regions of Europe, Canada and the United States have come to expect bare-pavement driving conditions throughout the winter. As a result, millions of tonnes of de-icing agents, usually in the form of sodium chloride (NaCl), have been applied to urban roads and highways. The general assumptions have been that the majority of the applied salt is flushed from the basin every season by rain and snow-melt, and environmental impacts are minimal. There is increasing evidence, however, that a significant proportion of the salt may be retained in the basin, entering the shallow sub-surface and migrating gradually to the watertable (*e.g.*, Diment *et al.*, 1973; Eisen and Anderson, 1980; Pilon and Howard, 1987). In such cases, serious degradation of ground-water quality could be anticipated (Howard and Beck, 1993; Howard *et al.*, 1993). The scenario shows an ominous resemblance to the situation in the United Kingdom and other European countries (Howard, 1985) where nitrate, innocently applied as fertilizer during post-war years, migrated through the unsaturated zone at a rate of ~1 m per year and eventually caused extensive contamination of major aquifers. In the case of NaCl de-icing chemicals, the most common concern is an increase in salinity to levels which would make the water unsuitable for human consumption and some industrial applications. There is also evidence, however, that sodium, the counter ion to chloride in most road de-icing salts, may have serious health implications. The ion has been strongly linked with the development of hypertension, a condition affecting perhaps 20% of the United States population (Moses, 1980; Craun, 1984; Tuthill and Calabrese, 1979). Elevated sodium intake has also been associated indirectly with hypernatraemia (World Health Organization, 1984). It is generally recommended that sodium concentrations in drinking water should not exceed 20 mg·L⁻¹ for patients with hypertension or congestive heart failure.

Concern for contamination by deicing salts is particularly acute in the Toronto region, where >100,000 t of NaCl are applied to urban roads and highways every year. Within Metropolitan Toronto, major streams regularly contain several hundred mg·L⁻¹ chloride; in winter months, values greater than 1000 mg·L⁻¹ are common. Recent records suggest that values of this order have shown little change in recent years and, while these levels are unacceptably high, there has been little stream quality evidence to indicate that the situation will become any worse.

It was Paine (1979) who first suggested that road de-icing

chemicals may be accumulating in the sub-surface and could one day re-appear to cause serious damage to urban water-ways. Paine performed a relatively coarse chloride mass balance on the Don River watershed (Fig. 1) and suggested that as little as 50% of the applied chloride was being removed from the basin annually and that the remainder was being stored in the shallow sub-surface. Follow-up studies by Pilon and Howard (1987), Eyles and Howard (1988), and Taylor et al. (1991) confirmed that concentrations of chloride as high as 14,000 mg·L⁻¹ were accumulating in shallow ground waters beneath Metropolitan Toronto, raising concern that the waters would ultimately enter local streams where serious and uncontrollable contamination would occur.

Recent and current studies at the University of Toronto have been concerned with the impact of urban development on ground-water quality and have focussed especially on the movement and behaviour of shallow ground waters containing elevated concentrations of chloride. An overall goal of the current work is to develop a series of numerical models that will permit the movement of de-icing salts in a catchment to be simulated and thereby allow the impact of alternative salting strategies to be evaluated. The success of these models depends, in turn, on an understanding of the nature of salt behaviour in a typical urban catchment, and, in particular, on an accurate knowledge of the rate at which salt is retained on an annual basis.

The retention rate of de-icing salts in a watershed is most reliably determined using a catchment mass balance approach. In this approach, salt input, represented by the mass of salt applied to the catchment during a specified time frame, is budgeted against salt output in the form of salt loads in the exiting stream. The net difference represents the mass of salt that is stored (retained) within the catchment.

In practice, the rates of salt retention will vary with time. At early times, the amount of salt entering the sub-surface will far exceed the amount leaving in baseflow. As time proceeds, salt concentrations in baseflow will increase and, on an annual basis, losses will start to approach the amounts entering the sub-surface. Steady state will be reached when the net rate of salt retention equals zero and the inflow of salt matches outflow. While a simple salt balance for a specific drainage system will allow the rate of salt retention to be determined for any one time period, corrections for baseflow contributions to the drainage system must be made if the annual rate of contribution of salt to the sub-surface is to be estimated.

The accuracy of the balance is determined by the quality, frequency and time frame of the data used. Previous studies,

conducted in differing types of catchment, have produced figures for the amount of salt (chloride) retained annually, ranging from 19% to 65% (Wulkowicz and Saleem, 1974; Scott, 1980; Paine, 1979). Diment et al., (1973) showed that the amount retained can fluctuate each year, finding that unusually high summer rainfall in the second year of a two-year study resulted in much higher (20%) salt removal than seen in the first year. A criticism of much of the earlier work, however, is that the balances were conducted over a relatively short time (often one year or less) and/or used daily stream flow data. The latter is of particular concern in urban catchments, where stream flow rates and chemical concentrations can vary by three orders of magnitude within a matter of hours. In the present study, conducted on Highland Creek from March 1989 to April 1991, this shortcoming has been surmounted by installation of a data logger to collect the necessary data at 15-minute intervals.

STUDY AREA

The chloride mass balance study was performed on Highland Creek basin, one of 14 major sub-catchments in the Metropolitan Toronto and Region watershed (Fig. 1). Annual precipitation in this catchment is approximately 850 mm and Thornthwaite evapotranspiration is estimated to be 600 mm. The catchment has a total area of 104 km² and an Environment Canada gauging station is located 5 km upstream of its discharge into Lake Ontario. A review of storm sewer drainage areas and topographic data suggests that the catchment area upstream of the weir has an area of approximately 82 km². This area was used for the mass balance calculations.

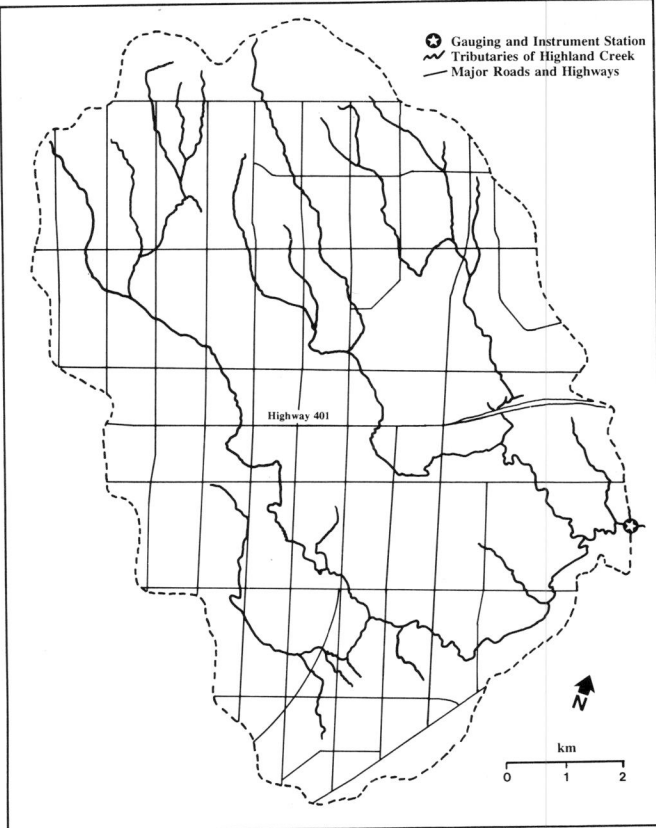

Figure 2 *The Highland Creek basin showing distribution of major salted roads and highways. The basin is highly urbanized; minor salted roads (not shown) comprise 75% of the total length of salted roads and highways in the basin.*

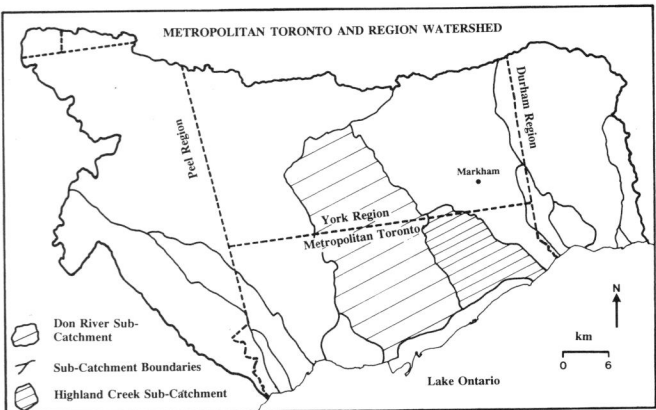

Figure 1 *Sub-catchments of the Metropolitan Toronto and Region Watershed.*

The basin is almost entirely urbanized (Fig. 2), with recreational open space along the main Creek valley and some remaining undeveloped land in the extreme northeast. The surface sediments are predominantly silty sand till (Karrow, 1967), but recent flood plain deposits occur extensively along the valley floor. The most important aquifers include the Scarborough Sands and the Lake Iroquois terrace deposits (Ontario Ministry of Natural Resources, 1980). Locally these are separated by the Sunnybrook diamict. Total aquifer thickness is estimated to be 30 m and the water table depth varies up to 20 m. Groundwater flow velocities are thought to be in the range of 10-100 m per year.

The basin is crossed by Highway 401 (12 lanes wide), and by a grid of two- and four-lane arterial roads about 1.5-2 km apart. These arteries and numerous secondary roads are regularly salted throughout the winter season by five agencies: the Ontario Ministry of Transportation (MTO), Metropolitan Toronto, City of Scarborough, Town of Markham, and the Regional Municipality of York. The basin receives ~17,000 t of NaCl road de-icing chemicals each year. This represents approximately 200 g of NaCl for every square metre of the catchment.

Approximately two years of average daily stream flow data for Highland Creek are shown in Figure 3. Baseflow contributions vary between 0.3 $m^3 \cdot s^{-1}$ and 1.2 $m^3 \cdot s^{-1}$, peaking during the spring, and average 0.42 $m^3 \cdot s^{-1}$ over the year. This represents an average annual recharge to the 82 km^2 catchment of 162 mm. If all of the applied salt were to enter the sub-surface *via* recharge, average steady-state sodium and chloride concentrations in ground water would approach 500 $mg \cdot L^{-1}$ and 800 $mg \cdot L^{-1}$, respectively.

INSTRUMENTATION

Because no reasonably priced, reliable and rugged chloride sensor was available, electrical conductivity, which is directly related to chloride concentration, was selected as a surrogate for monitoring in the field. Work began in July of 1988 when a YSI conductivity cell was installed at the Environment Canada weir to take advantage of the flow data which were already being collected at 15-minute intervals. Within days, the meter was damaged by a nearby lightning strike and the probe was buried in the stream bed. A second attempt to obtain data was made in March 1989, when an IC Controls temperature-compensated conductivity probe, model CC01, was installed near mid-stream at the weir. This probe is designed specifically for industrial applications, with both the sensor and sensor electronics sealed in PVC. The probe was connected by cable to a conductivity meter which was housed in the Environment Canada monitoring station, and provided a continuous output of electrical voltage as a measure of the conductivity. This output was recorded every 15 minutes by a Lakewood LE7110 data logger, which is capable of storing up to six months of data. When possible, data were retrieved from the data logger monthly using a laptop computer.

The probe sensors were cleaned every 3-4 weeks with a very fine abrasive. Algal growth that accumulates during the summer on the protective mesh around the probe tip was removed manually and appears to have had no effect on the probe function.

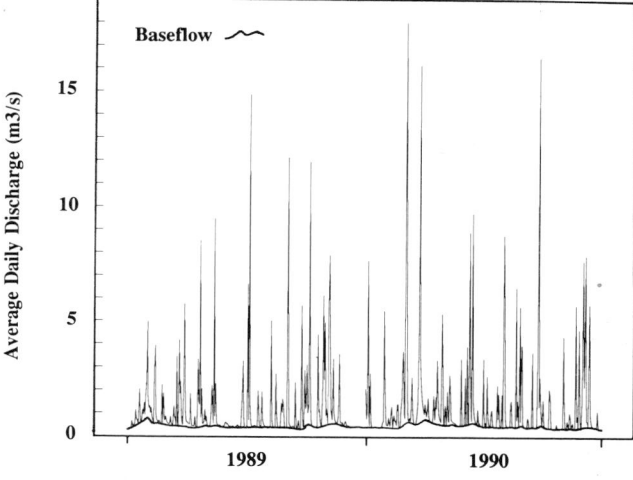

Figure 3 *Highland Creek stream hydrograph for the period 1 January 1989 to 30 November 1990.*

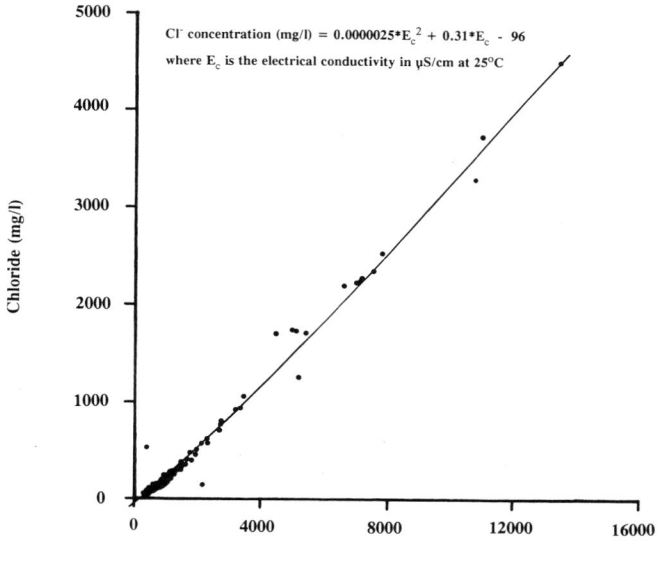

$$Cl^- \text{ concentration (mg/l)} = 0.0000025 * E_c^2 + 0.31 * E_c - 96$$

where E_c is the electrical conductivity in $\mu S/cm$ at 25°C

Figure 4 *Relationship between chloride and electrical conductivity used for site calibration.*

Table 1	Annual rates of NaCl application to the Highland Creek catchment.				
	SODIUM CHLORIDE				**CHLORIDE**
Salting Season	**Applied to Roads/ Highways** (t)	**Applied to Parking Lots** (t)	**Domestic Use** (t)	**Basin Total** (t)	**Basin Total** (t)
1988-89	15,031	2104	150	17,285	10,486
1989-90	16,095	2253	161	18,510	11,228
1990-91	13,149	1841	131	15,122	9,173
Annual Average	14,758	2066	147	16,972	10,295

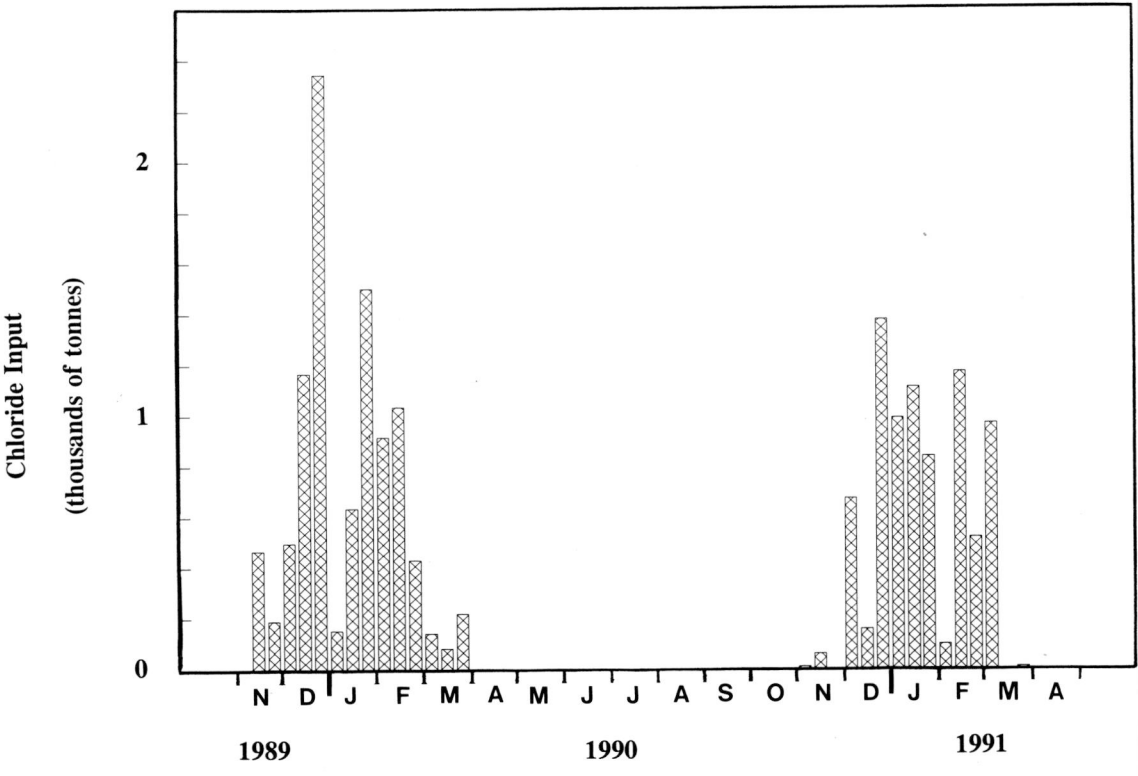

Figure 5   *Temporal distribution of chloride input for the period 1 November 1989 to 30 April 1991. Data are plotted for periods ending on the tenth, twentieth and last day of each month.*

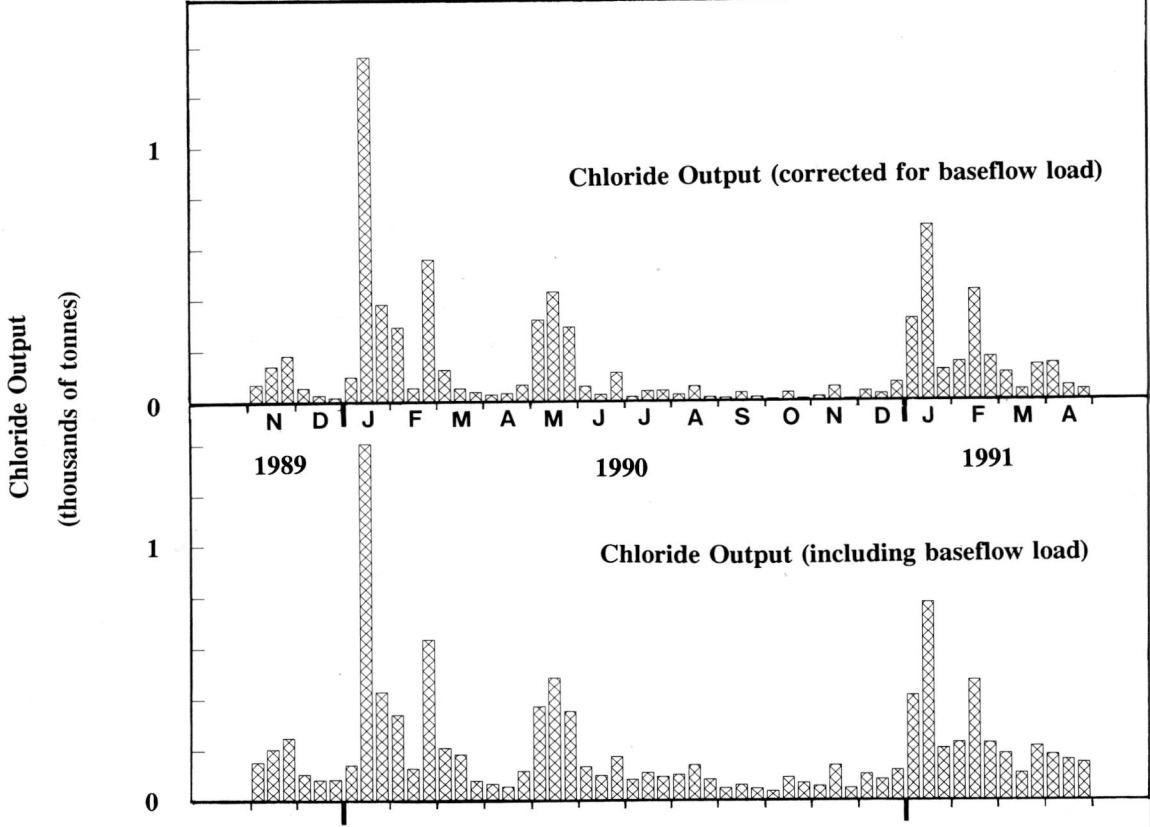

Figure 6   *Temporal distribution of chloride output in Highland Creek for the period 1 November 1989 to 30 April 1991. Data are plotted for periods ending on the tenth, twentieth and last day of each month.*

Calibration

The probe was initially calibrated and tested for drift in the laboratory using NaCl solutions of known conductivity. After installation in the creek, however, a consistent difference was observed between conductivity values calculated from the *in situ* conductivity readings and values measured from stream samples. This discrepancy is largely the result of monitoring in a flowing stream *versus* calibrating in static solutions. A small correction factor is now applied to the field measurements.

To permit conversion of measured electrical conductivity values to chloride concentration, samples of water were collected from the creek on a regular basis and analyzed for both parameters. As shown in Figure 4, the data correlate well. The best line fit is represented by a second order quadratic of the form:

$$\text{Cl}^- \text{ concentration (mg·L}^{-1}) = 0.0000025*E_c^2 + 0.31*E_c - 96 \qquad (1)$$

where E_c is the electrical conductivity in $\mu S \cdot cm^{-1}$ at 25°C.

Table 2   Chloride mass balance components for the period 1 December 1988 to 31 October 1991.

Month Period	Input (t)	Baseflow load (t)	Output¹ (t)	Input (t)	Baseflow load (t)	Output¹ (t)	Input (t)	Baseflow load (t)	Output¹ (t)
	1988								
Dec. 1-10	228.62	–	–						
11-20	1802.65	–	–						
21-31	973.62	–	–						
	1989			**1990**			**1991**		
Jan. 1-10	1761.81	–	–	173.83	47.25	96.54*	1136.75	94.05	319.84
11-20	526.17	–	–	728.48	47.25	1359.52*	1276.50	94.50	684.27
21-31	412.93	–	–	1725.53	47.25	381.74*	962.56	90.00	117.50
Feb. 1-10	1031.09	–	–	1052.91	47.25	291.94*	111.95	81.00	147.93
11-20	239.59	–	–	1189.35	76.32	51.33	1347.16	43.20	429.98
21-28	1453.80	–	–	494.25	75.60	558.13	597.11	61.20	166.43
March 1-10	735.89	–	–	160.74	87.08	123.51	1111.99	81.90	103.40*
11-20	1024.05	112.50	810.70	95.32	135.00	49.88	0.67	70.20	37.09*
21-31	164.70	121.50	588.53	251.77	44.55	34.93	9.00	79.56	134.65*
Apr. 1-10	130.88	108.00	112.21	0.00	40.50	24.70	0.00	40.50	140.76*
11-20	0.00	99.00	109.54	0.00	26.64	28.93	0.00	108.00	51.73
21-30	0.00	45.00	29.60	0.00	54.00	62.38	0.00	112.50	36.11
May 1-10	0.00	61.20	235.56	0.00	51.75	319.30	–	–	–
11-20	0.00	76.50	201.86	0.00	56.25	427.58	–	–	–
21-31	0.00	45.90	187.65	0.00	63.00	289.23	–	–	–
June 1-10	0.00	45.00	104.50*	0.00	79.20	56.15	–	–	–
11-20	0.00	36.00	92.94*	0.00	75.26	23.45	–	–	–
21-30	0.00	34.20	47.23	0.00	64.80	110.19	–	–	–
July 1-10	0.00	45.36	49.14	0.00	69.84	14.00	–	–	–
11-20	0.00	48.60	18.70	0.00	73.80	36.68	–	–	–
21-31	0.00	54.00	9.17*	0.00	55.44	38.98	–	–	–
Aug. 1-10	0.00	63.00	353.27*	0.00	79.70	22.13	–	–	–
11-20	0.00	54.72	174.85	0.00	85.32	55.15	–	–	–
21-31	0.00	51.98	87.81	0.00	72.00	11.96	–	–	–
Sept. 1-10	0.00	51.84	60.61	0.00	38.93	9.07	–	–	–
11-20	0.00	70.20	14.62	0.00	32.63	27.76	–	–	–
21-30	0.00	36.99	63.30	0.00	35.10	11.56	–	–	–
Oct. 1-10	0.00	47.25	66.64	0.00	31.86	3.87	–	–	–
11-20	0.00	27.00	2.01	0.00	60.75	30.18	–	–	–
21-31	0.00	16.88	1.14	0.00	64.08	5.49	–	–	–
Nov. 1-10	0.00	86.76	67.49	10.89	43.88	12.98			
11-20	534.28	67.50	138.83*	70.76	88.20	50.47			
21-30	217.00	67.50	182.03*	0.00	44.10	3.51*			
Dec. 1-10	571.04	54.00	52.50	773.43	70.56	34.43*			
11-20	1340.71	60.75	23.50	181.72	61.74	22.42*			
21-31	2,693.18	70.20	14.60*	1582.66	52.92	67.85*			

Notes:
¹ Output has been corrected for baseflow load.
– Not determined.
* Some data infilling required.

CHLORIDE BALANCE CALCULATION

The chloride balance study was carried out from December 1988 to April 1991, inclusive, and included three winter seasons' salting. It was performed by estimating the total chloride input to the catchment over a pre-determined period and subtracting the amount of this chloride leaving the watershed during the same period by overland flow. The difference represents the mass of chloride retained, at least temporarily, in surface waters, soils and subsurface waters. In the study, the amount of chloride leaving the basin through stream sediment load was assumed to be negligible. Corrections were made, however, to account for the mass of chloride leaving the basin as baseflow, since very little of this chloride would have originated from the previous salting season.

Chloride Input Data

Road salt is the major source of chloride entering the basin. It is applied as pure salt (NaCl) or as a salt/sand mixture. Daily salt application was determined from the yard records of the five agencies applying salt. Where salt/sanding routes straddled the catchment boundary, the total salt applied was apportioned to the catchments on the assumption that rates of application remained consistent along the routes. Total salt application in the study area during the winters of 1988-89, 1989-90 and 1990-91 is shown in Table 1. The annual distribution of chloride application from 1 November 1989 to 30 April 1991 is shown in Figure 5. It is estimated that these figures are accurate to ±5%.

Also included in Table 1 are estimates of other sources of chloride input. These include the amount of salt applied to parking lots, for which it was assumed that rates of application for shopping centre lots were the same as those for major roads, and for other lots, one-half this rate (Scott, 1980). Salt applied to parking lots represented about 14% of the total salt applied. Application by private home-owners was estimated by multiplying the approximate number of single-family residences by an average annual rate of 4 kg per household. This represented <1% of the total salt application.

Other sources of chloride are relatively insignificant. Chloride concentration in precipitation around Metropolitan Toronto during 1986 averaged 0.2 mg·L^{-1} (Ontario Ministry of the Environment, 1988), which converts to ~0.1% of the total chloride input. A calcium chloride de-icing additive is used occasionally in the area, but contributes <0.1% of the total chloride input. There are no known sanitary-storm sewer interconnections in the basin (Scarborough Works; pers. comm.). Depending on their chemical formulation, fertilizers may contain a significant amount of chloride; however, this potential contribution is not thought to be significant and has not been quantified in this study.

Chloride Output Data

Chloride discharge or "loading" in a stream can be calculated by:

$$Cl = Cm * Q \qquad (2)$$

where Cm is mass concentration and Q is stream discharge. This calculation was performed for 15-minute intervals using a Lotus 123 spreadsheet. Totals were output three times per month, at the end of the tenth, twentieth and last day of each month. Estimates for the period from 11 March 1989 to 30 April 1991 are shown on Table 2; data for the period from 1 November 1989 to 30 April 1991 are plotted in Figure 6. Some values in Table 2 are marked with an asterisk to indicate that some data were lost during the time period: data extrapolation was required in those cases.

Also shown in Table 2 are estimates of the mass of chloride entering the stream as a result of baseflow. These were determined by multiplying the average baseflow for the time interval of interest by the estimated chloride concentration. Normally, this would be the chloride concentration in the stsream at a time, usually following several days of dry weather, when stream flows are maintained entirely by baseflow. Most of the chloride contained in the baseflow is unlikely to have originated during the previous winter's salting, and instead represents chloride that has accumulated in the basin over its history of salting. As such, the baseflow contributions must be subtracted from the total chloride load to calculate the amount of chloride retained in the basin as a result of the previous winter's salting. A summary of the results is shown in Table 3.

Table 3	Chloride balance summary for the period 1 November 1988 to 31 October 1991.						
	Salting Season	Total Input (t)	Total Output (t)	Baseflow Load (t)	Corrected Output[1] (t)	Salt Output (as % of salt applied during salt season)	Total %
1988-89	Winter (1 Nov.–30 April)	10,486	2137[2]	486[2]	1651[2]	>15[2]	>34
	Summer (1 May–31 Oct.)	NIL	2889	867	2022	19	
1989-90	Winter (1 Nov.–30 April)	11,228	4562	1135	3427	31	45
	Summer (1 May–31 Oct.)	NIL	2699	1089	1609	14	
1990-91	Winter (1 Nov.–30 April)	9173	3651	1318	2333	26	>26
	Summer (1 May–31 Oct.)	NIL	–	–	–	–	
Notes	[1] Output has been corrected for baseflow load. [2] Data only available for March and April. – Not determined.						

RESULTS AND CONCLUDING DISCUSSION

Urbanized catchments of the Metropolitan Toronto and Region watershed have, for several decades, received more than one hundred thousand tonnes of de-icing salts annually. For much of this time, it has been assumed that most of this salt is flushed from the catchments each season by overland flow and that sub-surface impacts are minimal. However, recent evidence of elevated chloride in ground water beneath the catchments suggests that a significant proportion of the applied salt may be retained in the basin each season and may, therefore, be responsible for the observed degradation in ground-water quality.

The salt balance performed on the Highland Creek basin of Metropolitan Toronto has generated a large volume of reliable data. Most of the data gaps are due to the failure of flow logging instruments; these gaps were usually filled by correlation with data from neighbouring catchments. In total, a full chloride balance was completed for 26 months, extending over three salting seasons.

The results (Table 3) are consistent over the period of study. Each year, during the winter period (1 November to 30 April), approximately 10,000 t of chloride was applied to the Highland Creek catchment. Only 45% of this was removed by surface run-off before the following winter, when a new salting season began. Most of the chloride removed was flushed from the catchment during the winter in which it was applied. In the 1989-90 salting season, for example, 3427 t of chloride left the catchment by overland flow before the end of April, representing 31% of the total chloride applied. A further 1609 t (or 14% of the total) were removed by summer rain between 1 April and 30 October. Data are incomplete for the salting seasons 1988-89 and 1990-91, but the available results are comparable to those for 1989-90. For example, in 1990-91, 26% of the total chloride applied left the basin during the winter months; in 1988-89, 19% of the total chloride applied was removed during the summer.

If only 45% of the salt applied to the catchment is being removed annually, then the remainder is being stored, presumably in underlying ground waters. The rate of accumulation will depend on the rate of ground-water movement in the basin. While the total mass of chloride entering the sub-surface is greater than the mass of chloride leaving as baseflow to the stream, chloride will accumulate in the ground water, and ground-water chloride concentrations will increase. For the period of study, total input of chloride far exceeds output (including baseflow load) and significant accumulation occurs. Eventually, chloride in the ground water will reach a level at which annual baseflow losses will match the amount of chloride entering the sub-surface. At this stage, steady-state will be reached and no further deterioration of ground-water or stream-water quality will occur. Numerical modelling will allow the rates of change of ground-water and stream-water quality to be predicted with more certainty. Assuming an annual recharge of 162 mm, an aquifer thickness of 30 m, a specific yield of 20%, and ground-water flow velocities of 100 m per year, deployment of analytical transient solutions for the transport of contaminants from continuous line sources (Domenico and Schwartz, 1990) indicates that steady state concentrations could be achieved within 60 years of initial salt application, or approximately 20 years from the present.

When steady state is reached, inflow and outflow of chloride will be in balance and the mass of chloride stored within the basin will remain unchanged. The stored chloride may not be evenly distributed within the sub-surface and water quality stratification will occur depending on the local ground-water flow regime. If conditions in the catchment remain the same and present rates of salt application are maintained, however, a simple division of the total chloride entering the sub-surface by the annual recharge reveals that *average* chloride concentrations in ground waters discharging as baseflow in the basin will reach 426 ± 50 mg·L^{-1}. This value represents a three-fold increase over present average baseflow concentrations, and is nearly twice the drinking water quality objective of 250 mg·L^{-1} maximum acceptable concentration.

ACKNOWLEDGEMENTS

The authors gratefully acknowledge numerous University of Toronto students and colleagues who provided technical and research assistance. Notable contributions were made by Mark Hughes and J.P Francois, who helped install the monitoring station and performed calibrations, and by Sean Salvatori, Connie Romano and Julia Montgomery, who processed much of the data. Special thanks go to Environment Canada for providing much of the flow data, and to the Ontario Ministry of Transport and the local municipalities for their co-operation in providing the necessary salt application data.

REFERENCES

Craun, G.F., 1984, Health aspects of groundwater pollution, in Bitton, G. and Gerba, C.P., eds., Groundwater Pollution Microbiology: John Wiley, New York, p. 135-179.

Diment, W.H., Bubeck, R.C. and Deck, B.L., 1973, Some effects of de-icing salts on Irondequoit Bay and its drainage basin: Highway Research Record 425, p. 23-25.

Domenico, P.A. and Schwartz, F.W., 1990, Physical and Chemical Hydrogeology: John Wiley and Sons, New York, 824 p.

Eisen, C. and Anderson, M.P., 1980, The effects of urbanization on groundwater quality, Milwaukee, Wisconsin, U.S.A., in Jackson, R.E., ed., Aquifer Contamination and Protection: UNESCO Press, Paris, SRH no. 30, p. 378-390.

Eyles, N. and Howard, K.W.F., 1988, A hydrochemical study of urban landslides caused by heavy rain: Scarborough Bluffs, Ontario, Canada: Canadian Geotechnical Journal, v. 25, p. 455-466.

Howard, K.W.F., 1985, An approach to recognizing low level contamination of groundwaters in shallow sedimentary aquifers: Water Pollution Research Journal of Canada, v. 20, p. 1-11.

Howard, K.W.F. and Beck, P., 1993, Hydrogeochemical implications of groundwater contamination by road de-icing chemicals: Journal of Contaminant Hydrology, v. 12, p. 245-268.

Howard, K.W.F., Boyce, J.I., Livingstone, S. and Salvatori, S.L., 1993, Road salt impacts on groundwater quality – the worst is yet to come!: Geological Society of America, GSA Today, v. 3, p. 301-321.

Karrow, P.F., 1967, Pleistocene Geology of the Scarborough Area: Ontario Department of Mines, Geological Report 46, 107 p.

Moses, C., 1980, Sodium in Medicine and Health, A Monograph: Reese Press Inc., Baltimore, MD, 122 p.

Ontario Ministry of the Environment, 1988, Acidic precipitation in Ontario study, cumulative precipitation chemistry listings 1986: Ministry of the Environment, Report ARB-034-88.

Ontario Ministry of Natural Resources, 1980, Quaternary Geology of Toronto and Surrounding Area: Preliminary Map P. 2204.

Paine, R.L., 1979, Chlorides in the Don River Watershed Resulting from Road De-Icing Salt: University of Toronto, Institute for Environmental Studies, Snow and Ice Control Working Group, Working Paper SIC-3, 23 p.

Pilon, P.E. and Howard, K.W.F., 1987, Contamination of subsurface waters by road de-icing salts: Water Pollution Research Journal of Canada, v. 22, p. 157-171.

Scott, W.S., 1980, Road salt movement into two Toronto streams: ASCE Journal of Environmental Engineering, v. 106, p. 547-560.

Taylor, L.C., Howard, K.W.F. and Chambers, L.K., 1991, The use of spring-dwelling ostracodes as bio-monitors for inorganic contamination of groundwaters along an urban-rural transect in southern Ontario: Geological Association of Canada–Mineralogical Association of Canada, Program with Abstracts, v. 16, p. A123.

Tuthill, R.W. and Calabrese, E.J., 1979, Elevated sodium levels in the public drinking water as a contributor to elevated blood pressure levels in the community: Arch. Environmental Health, v. 34, p. 197-203.

World Health Organization, 1984, Guidelines for Drinking Water Quality, Vol. 1. Recommendations: World Health Organization, Geneva, 130 p.

Wulkowicz, G.M. and Saleem, Z.A., 1974, Chloride balance of an urban basin in the Chicago area: Water Resources Research, v. 10, p. 974-982.

9. The Oak Ridges Moraine of Southern Ontario: A Ground-water Resource at Risk

Ken W.F. Howard, Nicholas Eyles, Philip J. Smart, Joseph I. Boyce, Richard E. Gerber, Sean L. Salvatori and Mike Doughty

Environmental Earth Sciences, University of Toronto, Scarborough Campus
1265 Military Trail, Scarborough, Ontario M1C 1A4

SUMMARY

Protection of ground-water resources is an emerging theme in many urban areas. In south-central Ontario, the Oak Ridges Moraine is a prominent glacial moraine complex on the rapidly urbanizing northern margin of the Greater Toronto Area and constitutes a major regional aquifer system. Given the pre-eminent economic importance of the region, the ground-water resource can be argued to have provincial significance, and there is much current debate regarding the impact of urbanization and other anthropogenic activities on ground-water quality and supply. Particular issues are the degree of acceptable future development and the definition of the most environmentally and hydrogeologically sensitive areas. This chapter reviews what is currently understood of the geology and hydrogeology of the Oak Ridges Moraine, and provides an overview of quantitative modelling studies.

INTRODUCTION

Throughout the Great Lakes basin, the quantity and quality of ground waters have become serious issues for concern. In 1983, the Great Lakes Advisory Board of the International Joint Commission recognized major deficiencies in knowledge about the nature and extent of ground water within the basin. Concerns were also expressed that "existing estimates of ground-water flow based on general geology were inadequate".

In south-central Ontario, the Oak Ridges Moraine (ORM; ~1400 km²) is a prominent physiographic feature now being threatened by rapid and poorly planned urbanization on the northern fringe of the Greater Toronto Area (GTA). The moraine forms a west-east trending belt of undulating, kettled topography, between 5 km and 20 km wide, that extends for more than 140 km eastward from the Niagara Escarpment (Fig. 1A). The ORM extends across three regions (Peel, York, Durham) of the GTA and 14 townships (Fig. 1B), and accounts for more than 18% of the total GTA area. The moraine forms the surface-water drainage divide between Georgian Bay and Lake Ontario and provides baseflow to more than 30 major streams. The ORM is the focus of much current debate regarding the impact of urbanization and other anthropogenic activities on ground-water quality and supply, together with the degree of acceptable future development and definition of the most environmentally sensitive areas. The ORM aquifer is recognized as being of provincial importance given its strategic location adjacent to the rapidly growing GTA.

HISTORICAL OVERVIEW

The recent history of the ORM provides a classic illustration of unsustainable land use practices. The moraine is of considerable significance in the economic and social development of the province of Ontario; this history is briefly reviewed below and in Figure 2.

European Settlement

Permanent European settlement of the ORM started in 1783 when the British Government sought homes for Loyalists driven from the newly independent United States of America. Many settlers were part of religious groups such as Quakers, Amish, Hutterites and Mennonites from Pennsylvania (the plain folk), who were attracted by an exemption from military service issued by Lieutenant-Governor John Graves Simcoe in 1792. Quakers settled the townships of Uxbridge and Newmarket; Mennonites settled Altona. A growing influx of American settlers, who had dubious loyalty to the Crown, threatened the political security of the fledgling province of Upper Canada, and large-scale European migration was promoted after the War of 1812-1814. This coincided with the end of the Napoleonic Wars in Europe and the arrival of war veterans entitled to free lots of wild land; other arrivals to the ORM (described as "bold sweeping hills" by Catharine Parr Traill in 1836) included displaced Scottish Highlanders and craftsmen made redundant by mechanization in the factories of Britain. Men with military experience were favoured so as to enable the mustering of local militias. Settlement of the ORM was seen as a first line of defence for York (Toronto) with Yonge Street, the main north-south route across the ORM, primarily a military trail. Unfortunately, much land was obtained by Crown grants, not for immediate settlement, but for future sale for profit after it had been cleared of forest. Unbridled land speculation, inadequately taxed uncultivated land, the prevalent practice of squatting, and government corruption were leading causes of the Rebellion of 1837 when William Lyon Mackenzie and his Reformers marched south along Yonge Street to the skirmish at Montgomery's Farm.

Habitat Destruction and Loss of Sustainability

Municipal reforms of Lord Durham, following the Rebellion of 1837, ushered in several decades of prosperity across the ORM (or Pickering Sandhills as it was widely known), accelerated by the introduction of steam-driven saw mills (1850), plank roads, and railways (the Northern Railway arrived in Aurora in 1853). The 1854 Reciprocity Treaty between Canada and the United States removed the duty from Canadian wheat and lumber and the so-called White Pine Clause that had previously reserved the cutting of pine for masts destined for the Royal Navy. The thick forest soils of the ORM were recognized for their agricultural potential, and wholesale forest clearance began. To avoid labour-intensive tree cutting, ring barking was used, whereby the trees were simply stripped of bark and allowed to rot. Because labour was expensive, logging and stumping bees were organized to haul logs into enormous piles for burning to supply potash fertilizer; the burning of 10 acres of forest typically produced about 5 barrels (2500 lbs) of potash (Johnson, 1973). Contemporary journals describe widespread devastation to the habitats of animals and indigenous people; Johnson (1973) remarks that the area was not so much settled as overrun. An excellent account of the harsh backwoods life at this time is provided by Susanna

Moodie in *Roughing It in the Bush*, published in 1852. The largest man-made impact on the landscape at this time was the construction of a dam on the Scugog River at Lindsay and the creation of Lake Scugog in the late 1830s. A canal was also planned across the ORM, from Lake Simcoe to Lake Scucog across the Dividing Ridges to Whitby, but never built (Keefer, 1863).

Widespread clearance of the forest cover across the ORM resulted in environmental degradation and, ultimately, loss of sustainability in the latter half of the 19th century. The widespread practice of slash and burn resulted in rapid soil erosion, and created extensive tracts of derelict land of little agricultural value. Exposed soils, devoid of a thick cover of leaf litter, became frozen in winter, reducing recharge to underlying aquifers, creating rapid spring runoff and flooding along many creeks (*e.g.*, the Ganaraska watershed; Fig. 3). Large volumes of sediment accumulated along the lower reaches of rivers and in Lake Ontario (see Chapter 2); springs

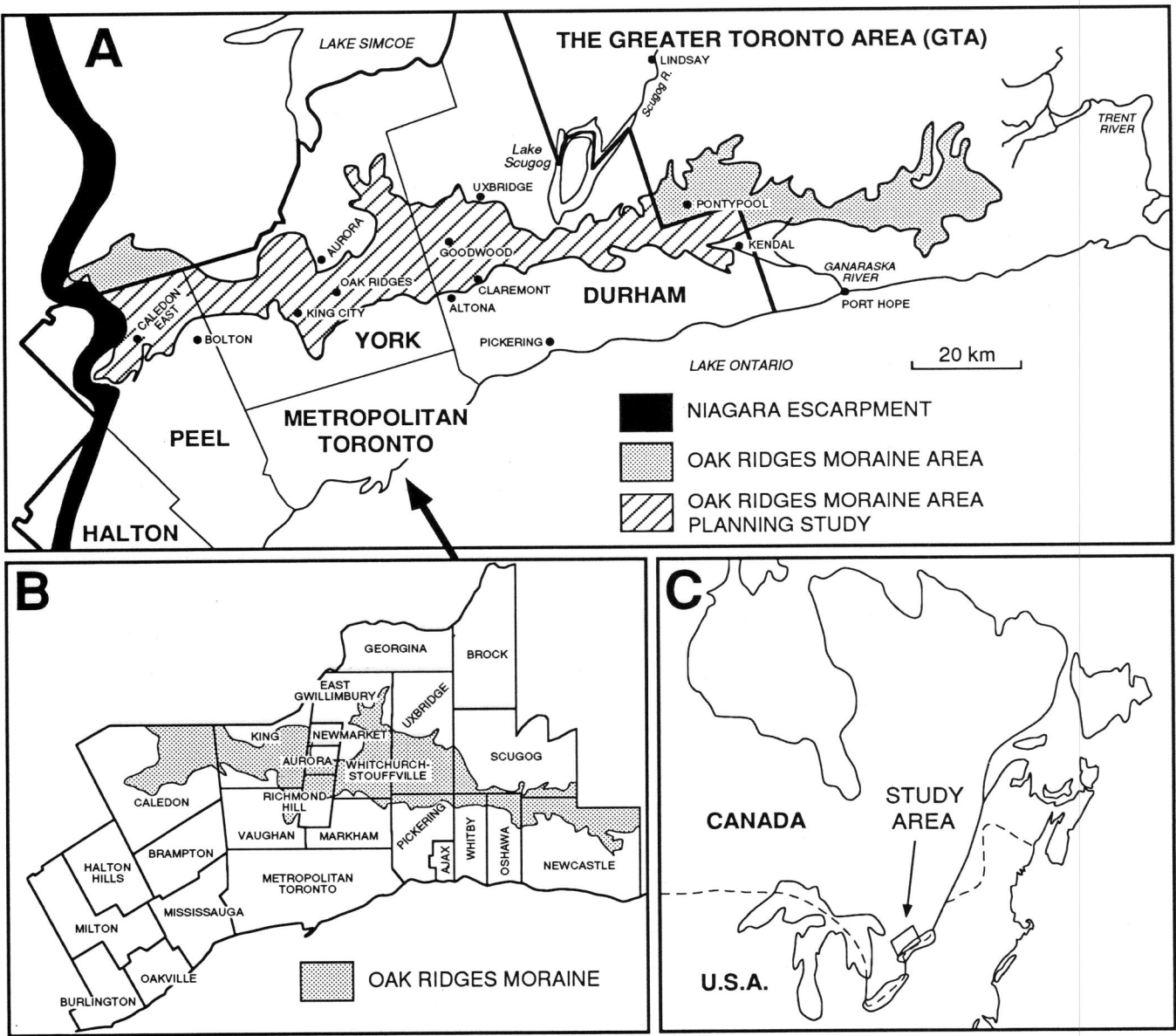

Figure 1 *Oak Ridges Moraine area showing (**A**) the five regions composing the Greater Toronto Area and (**B**) townships. The regional location is shown in (**C**). The area shown under Oak Ridges Moraine Area Planning Study area is the focus of the ORM Technical Working Committee, having the objective of formulating legislation to protect the natural resources of the area.*

dried up, requiring wells to be dug at great expense. Sustained agriculture became impossible on nutrient-starved sandy soils, and formerly cultivated areas were converted to grazing, resulting in intensification of soil loss from overgrazing, trampling and the widespread generation of sand dunes and blowouts.

The resident population of the ORM area peaked in 1861 (Fig. 2) and began to decline thereafter as attempts at agriculture proved futile; many small communities disappeared (Caddy, 1861). The McKinley Tariff, introduced by the United States in 1890, virtually shut off the lucrative United States market for squared timber and other commodities such as wheat. This coincided with the appearance of the destructive wheat midge and resulted in widespread economic depression and loss of population across the ORM (Johnson, 1973). Increased reliance on farm machinery, access to the rich soils of the Prairies facilitated by the completion of the Canadian Pacific Railroad in 1888, and the rapid growth of the city of Toronto further drained the ORM of its rural population (Johnson, 1973; Brown, 1978; Darroch and Soltow, 1994; Fig. 2).

Early Conservation Measures

The lack of sustainability clearly evident in the ORM in the late 1930s, and comparison with American conservation efforts in the Tennessee Valley, prompted the first provincial steps in environmental conservation in Ontario; these were considered of national significance and regarded as blueprints for postwar work in other areas of the country (Coventry, 1945). The connection between deforestation, downstream flooding, and declining ground-water levels was realized and conservation projects were initiated in the Ganaraska watershed to curtail repeated flooding of Port Hope. Carman (1941, p. 12) reviewed the then state of the environment across the ORM and concluded that "past and present generations were wholly occupied with the immediate problem of making a living, and in so doing have mined their natural resources of forest, soil and water"; he urged the planting of protective forests. The Reforestation Act of 1921 had earlier empowered municipalities to buy and reforest barren land.

Carman (1941) first recognized the importance of the numerous kettle lakes that dot the ORM for retaining surface-

BIGSBY (1824)	"BOLD LINE OF HEIGHTS .. BREAKING INTO CONFUSED RIDGES AND HUMMOCKS"
1840	EUROPEAN SETTLEMENT & CLEARANCES
1861	PEAK OF POPULATION
TAYLOR (1913)	RECOGNITION OF INTERLOBATE ORIGIN BETWEEN NORTHERN AND SOUTHERN ICE MASSES
1920	INCREASED RUN-OFF FLOODING AND LOSS OF SUSTAINABILITY - POPULATION DECREASE REDUCED RECHARGE, LOWERED GROUNDWATER TABLE
CHAPMAN & PUTNAM (1943)	DETAILED PHYSIOGRAPHIC MAPPING
(1944)	DEPARTMENT OF PLANNING & DEVELOPMENT
1946	CONSERVATION AUTHORITIES ACT
CALEY, etc. 1940's	FIRST GROUND WATER STUDIES & FARM PONDS POLICY
1954	HURRICANE HAZEL: INCREASED ROLE FOR CONSERVATION AUTHORITIES
HEWITT, KARROW 1960's	SURFICIAL MAPPING: EVALUATION OF AGGREGATE RESOURCES
SINGER, SIBUL, etc. 1970's	GROUND WATER RESOURCES OF SELECTED WATERSHEDS
DUCKWORTH (1979)	SUBSURFACE STRATIGRAPHY
1980's	POPULATION RETURNS TO 1860 LEVELS
MTRCA (1989)	GREENSPACE STRATEGY
INTERA-KENTING (1990), CROMBIE COMMISSION (1990)	'ECOLOGICALLY-BASED WATERSHED PLANNING'
OAK RIDGES MORAINE TECHNICAL WORKING COMMITTEE (1994a,b)	OAK RIDGES MORAINE AREA STRATEGY FOR THE GREATER TORONTO AREA

Figure 2 *Important milestones in the post-European settlement history and study of the Oak Ridges Moraine.*

water runoff and promoting ground-water recharge. Carman recognized that this type of topography trapped surface water that would normally run off to streams. His work prompted a wider appreciation of the need to store spring runoff, and by 1950 as many as 5000 farm ponds had been constructed to augment ground-water supplies (Richardson, 1944, 1974). The first major initiative in conservation was made in 1944 when George Drew, Premier of Ontario, established the Department of Planning and Development (Fig. 2) under the direction of geologist George Langford. The 1946 Conservation Authorities Act established a framework for protecting watersheds by provincial and municipal governments. The London Conference of 1944 had earlier set the agenda for this effort when R.F. Legget, another geologist, had emphasized that ground-water supplies were central to all renewable resources, and identified the need for an inventory of existing hydrogeological conditions. A number of ground-water supply papers were produced under the auspices of the Ontario Water Resources Commission (e.g., Hainstock et al., 1948, 1952). However, because of the $100 million destruction caused by Hurricane Hazel in 1954, when 300 million tonnes of rain fell on Toronto, emphasis shifted to surface-water hydrology and flood protection. Only recently has attention been refocussed on the hydrogeological significance of the Oak Ridges Moraine after a hiatus of 50 years. A long history of rural depopulation has now come to an end and the population at the present time (>200,000) exceeds the previous peak reached in 1861 (Fig. 2).

URBANIZATION AND THE OAK RIDGES MORAINE

The present-day population of the GTA is 4.2 million and is expected to increase to 6.5 million by 2021, with most of this growth taking place on the northern outskirts of Metropolitan Toronto along the southern flanks of the ORM (Fig. 1). As a result of rapid urbanization along the Yonge Street corridor, extending north to Aurora and Richmond Hill, developmental pressures on the Oak Ridges Moraine have greatly intensified. Poorly planned residential, recreational and industrial activity along the moraine is recognized as a direct threat to the area's ecological integrity (Crombie, 1990; Intera-Kenting, 1990; Kanter, 1990). The major concerns that need to be addressed include the effects of land use changes on aquifer recharge, present and potential impacts from landfills, the effects of extensive aggregate extraction and rural urbanization requiring septic systems on well-water supplies, large-scale withdrawal of ground water by golf courses, farms, fish hatcheries, and batch (ready-mix) aggregate operations, and the effects of widespread application of agro-chemicals and road de-icing agents. The ORM has more than 100 active licensed aggregate operations, centred mostly in the town of Whitchurch-Stouffville and near Pontypool; average life expectancy of any one operation is about 16 years (Oak Ridges Moraine Technical Working Committee, 1994a). Licensees pay 8 cents per tonne per year of aggregate produced into a Rehabilitation Security Deposit System; a total of 486 ha has been rehabilitated to date. More than 1378 ha are classified as disturbed, i.e., awaiting rehabilitation. Some 27 of the 95

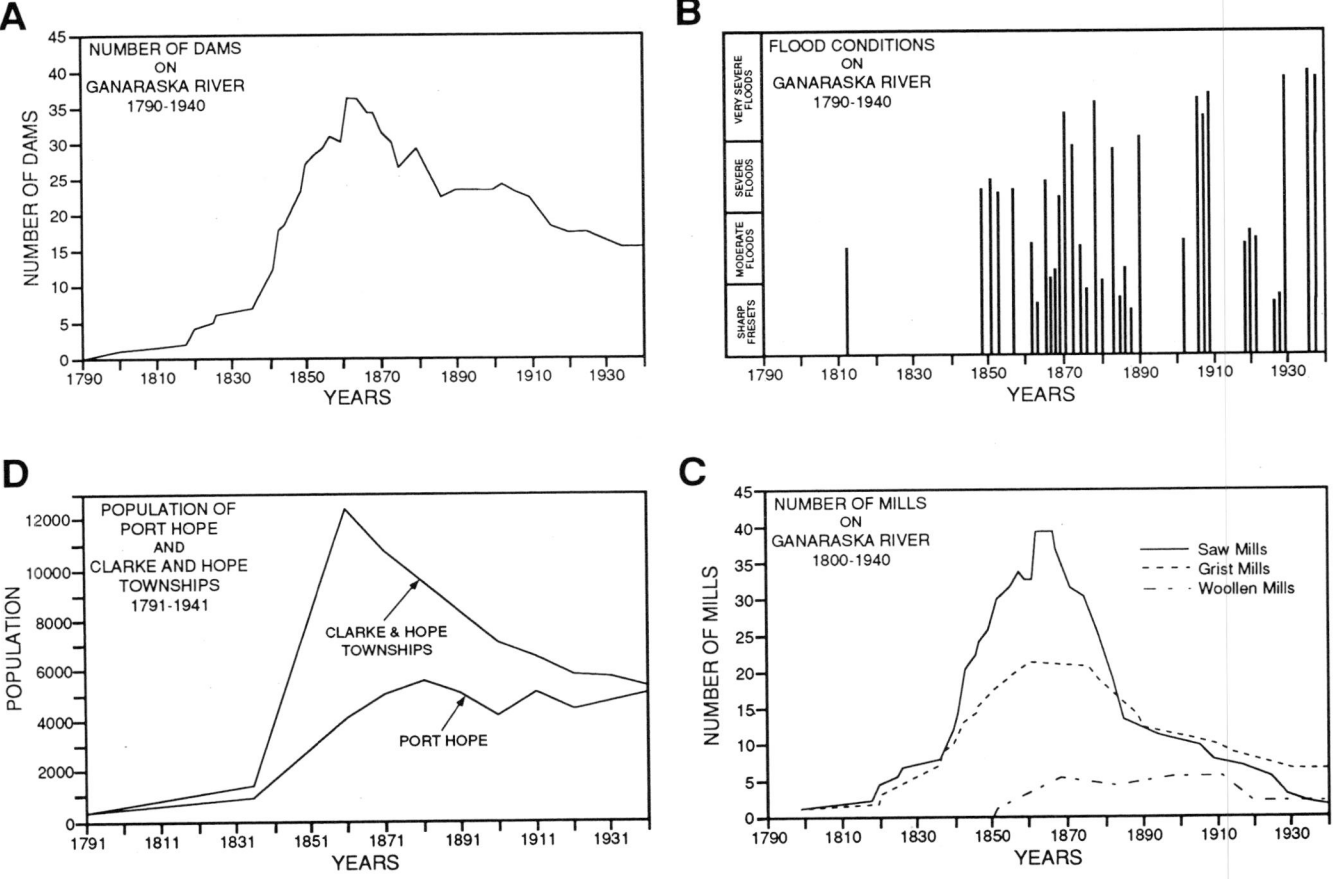

Figure 3 *Nineteenth-century environmental changes resulting from forest clearance across the Oak Ridges Moraine area; an example from the Ganaraska watershed (Fig. 1) (**A**) number of dams; (**B**) incidence of flooding; (**C**) population; (**D**) number of mills. Data from Richardson (1974). Note peak of population in 1860 and greatly increased incidence of flooding after 1850 when large-scale forest clearance began.*

active operators are authorized to excavate below the water table. This activity, together with large ground-water abstractions by batch plants producing ready-mix concrete for immediate delivery, are sources for concern. The key to evaluating the impact of aggregate extraction and many other activities is an understanding of the geological and hydrogeological complexity of the moraine.

Existing Water Usage

Within the ORM, about 60,000 people are dependent on about 30,000 private wells as listed by the Ontario Ministry of Environment and Energy, while the remaining population is dependent on municipal water (Lake Ontario and municipal ground-water production wells). To date, there has been no shortfall in supply within the ORM, but many areas are experiencing water quality problems related to septic tank operation and road salt. In the more heavily developed western half of the ORM, along the Yonge Street corridor, municipal water (from Lake Ontario) is supplied to about 70,000 people in Richmond Hill, and the remainder are supplied from municipal wells that draw on the Yonge Street aquifer and individual wells. In the east, municipal systems will shortly extend as far north as Brooklin, some 16 km north of Lake Ontario. Total municipal well production across the ORM is in the order of about 22 million $m^3 \cdot a^{-1}$, with greater than 50% of this total pumped from the Yonge Street aquifer alone. Other large users include farms, golf courses (about 60 in the ORM), and aggregate wash plants. As yet, a comprehensive water budget encompassing all users across the ORM has not been completed.

The provision of adequate sewage treatment systems is currently a severe restraint on development outside the Yonge Street corridor. Clearly, the further extension of municipal facilities, and the accompanying high-density urban development that follows, will require close regulation (see below). The York-Durham Sewage System (YDSS) services 140,000 residents of Richmond Hill, Aurora, Oak Ridges and Newmarket (Chapter 2). Municipal communal systems service the communities of Stouffville and Uxbridge; the treatment plant for Stouffville (pop'n: 18,500) discharges into a tributary of Duffins Creek, that for Uxbridge into the Nonquon River and, ultimately, Lake Scucog. King City (pop'n: 5000) is noteworthy as being the largest community in Ontario totally dependent on septic tanks, most of which are elderly and on small lots. Ironically, the shortfall in adequate sewage treatment facilities has had the beneficial effect of being a major impediment to urban growth. Extension of the YDSS has been suggested, but it is certain that this would promote rapid urbanization of the ORM. The provision of adequate facilities to handle sewage (whether private single dwelling, communal or municipal), and identification of the relative impact of each system on ground- and surface-water quality, are among the more pressing problems to be addressed, and will strongly influence the level and style of future development.

HYDROGEOLOGICAL SIGNIFICANCE OF THE OAK RIDGES MORAINE

Previous Work

Haefeli (1970) was the first to address the nature of large-scale ground-water flow in the region between Lake Simcoe and Lake Ontario and the role of the ORM in recharging the system. The first detailed evaluation of the ORM is that of Sibul *et al.* (1977) and Turner (1977) working for the Ontario Ministry of Environment. Based on the evaluation of many hundreds of water well records, Turner noted that yields of

ground water were generally excellent and that the potential for the development of future high-capacity wells was good in most areas of the moraine. Sibul *et al.* (1977) recognized the role of the ORM in supplying baseflow to the head-water streams that source the tributaries of over 30 major rivers and creeks within the GTA.

While the hydrogeological importance of the ORM is well established in a strategic sense (Kanter, 1990), a comprehensive functional understanding is impeded by its hydrostratigraphic complexity. This difficulty was first recognized by Turner (1977) and Sibul *et al.* (1977), who were unable to establish the hydrogeological relationships between the Oak Ridges Aquifer Complex (term introduced by Turner, 1977) and other related aquifers identified locally at King City, Woodbridge, Nobleton, Richmond Hill, Aurora, Newmarket, Mount Albert, Uxbridge and Port Perry. Understanding the subsurface relationship between these aquifers is hampered by poor-quality stratigraphic information contained in many well logs and the virtual absence of deep boreholes. The vast majority of wells penetrate to a depth of less than 75 m, which is less than one-third of the total thickness of glacial sediments in the ORM. In contrast to detailed mapping of surface sediments (*e.g.*, Gwyn, 1972, 1976a, b; White, 1975; Gwyn and Cowan, 1978; Duckworth, 1979; Sharpe and Finley, 1993), little is known of the deeper stratigraphy and sedimentology of the ORM. Provisional geophysical studies of the deep moraine stratigraphy and regional correlations of the most prominent glacial stratigraphic units were presented by Fligg and Rodriques (1983) and Eyles *et al.* (1985).

The work of Sibul *et al.* (1977) drew attention to the importance of hydrostratigraphic detail, particularly with respect to surface tills which locally drape the moraine and its flanks. The authors correctly recognized that the distribution and hydrogeological characteristics of surficial units influence the recharge characteristics of the aquifer and its susceptibility to contamination. More recently, the work of Howard and Beck (1986), Kaye (1986), and Proulx and Farvolden (1989) has resolved a number of specific issues concerning the hydrogeological role of the ORM. Howard and Beck (1986) were able to establish the hydraulic relationship between the ORM and 13 other aquifer systems in the Duffins Creek-Rouge River drainage basins. Kaye (1986) and Proulx and Farvolden (1989) were able to explain site-specific hydrogeological anomalies related to recharge and contaminant movement, anomalies that were notably complicated by the hydrostratigraphic complexity of the moraine and the local occurrence of perched aquifers isolated from underlying aquifer systems.

In 1989, the Metropolitan Toronto Region Conservation Authority (MTRCA) released its Greenspace Strategy, which addressed the protection of the hydrology of the ORM. In this document, the ground-water component of the hydrological cycle was identified as being "poorly understood and an obstacle to the development of land use controls and management practices". A key requirement identified by the MTRCA was the development of ground-water management objectives that the authority could recommend to head-water municipalities. Also in 1989, the Greater Toronto Greenlands Strategy was initiated by the provincial government in an effort to protect the natural landscape. Background papers prepared for this document highlighted the hydrogeological significance of the ORM and recommended a comprehensive hydrogeological investigation of the entire ORM area. Similar sentiments were echoed in the "Interim Report of the Royal Commission on the Future of the Toronto Waterfront" (Crom-

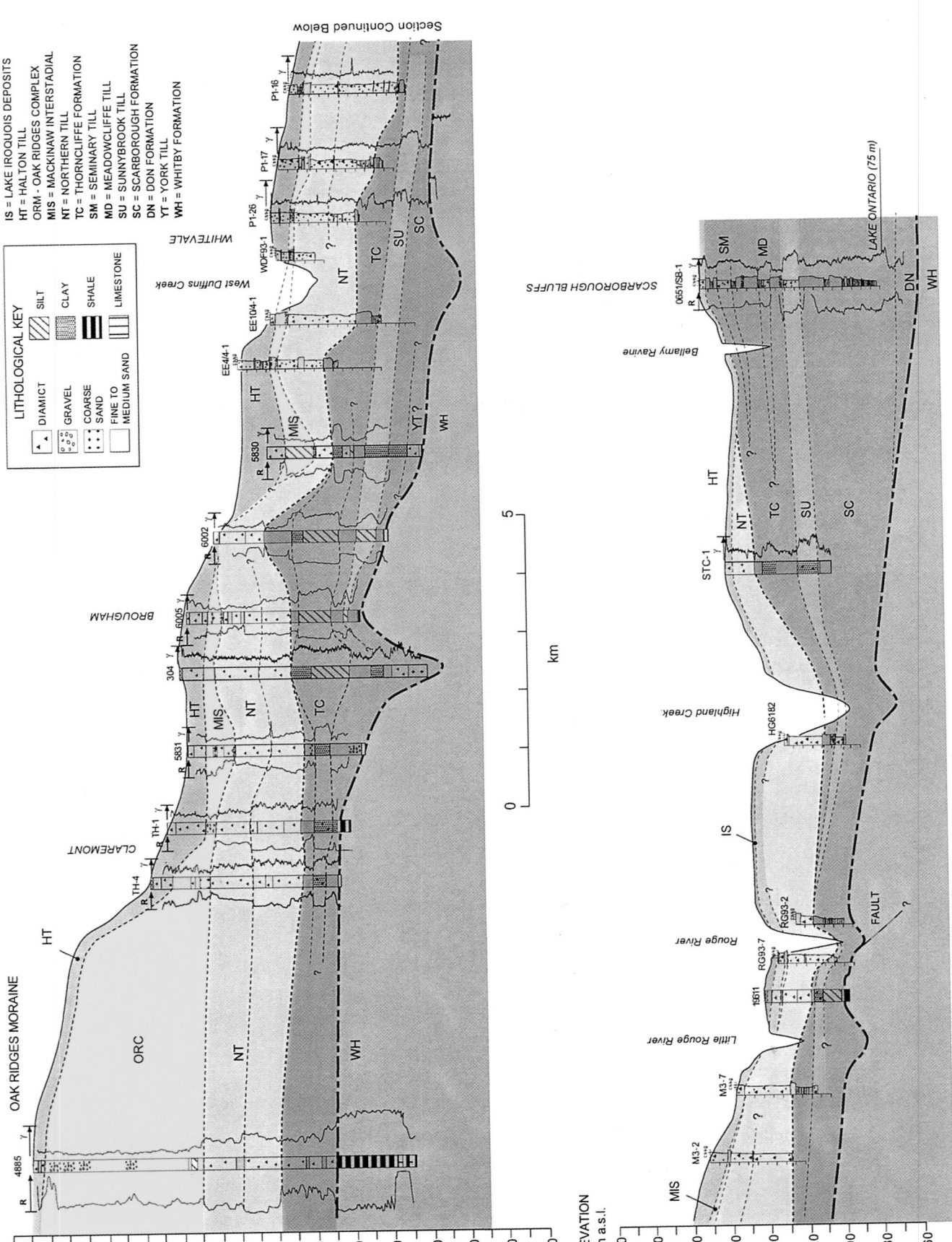

Figure 4 Compilation of subsurface data from outcrop, downhole geophysical and drill core data from the crest of the ORM to Lake Ontario showing inferred distribution of principal Late Pleistocene stratigraphic units. For a highly simplified version, see Figure 5. For location of section, see Figure 10. Abbreviation: ORC, Oak Ridges (Aquifer) Complex.

bie, 1990). The report argued that an adequate understanding of the waterfront region, its shoreline, rivers, wetlands and associated biota could only be established through study of the entire Toronto watershed area. The ORM was identified as "unique, varied and complex, under great stress and in need of comprehensive studies of its geology and hydrogeology" (Crombie, 1990).

The province of Ontario established the Oak Ridges Moraine Technical Working Committee (ORMTWC) in June 1991 to develop long-term strategies and legislation that would not only protect the biological and cultural integrity of the ORM, but would ensure abundant clean water, the maintenance of baseflow in head-water streams, storage in kettle lakes, and sustainable use of water by residents on and adjacent to the moraine (Oak Ridges Moraine Technical Working Committee, 1994a, b).

GEOLOGICAL FRAMEWORK OF THE OAK RIDGES MORAINE

The geology of the ORM was discussed as early as 1863 by William Logan, but the first surficial geology map of the ORM was produced for the Ontario Research Foundation in 1951 by L.J. Chapman and D.F. Putnam. The deep stratigraphy of the ORM has only recently been explored using high-resolution seismic reflection surveys, drilling, geophysical borehole logging, coring, outcrop description, and queries of computerized water well records (*e.g.*, Fig. 4). The regional geology of the study area consists of thick Late Pleistocene glacial sediments (Fig. 5) resting on southwesterly dipping Lower Paleozoic strata which, in turn, cover a complexly structured basement of mid-Proterozoic age (Chapter 2).

Location of the Oak Ridges Moraine

Funk (1977) suggested that the west-east alignment of the ORM reflected the existence of an underlying ancestral bedrock drainage divide; recent data confirm this. The position of the ORM is controlled by the topography of the underlying Paleozoic bedrock surface, which shows a prominent west-east trending bedrock high (informally named the Pontypool ridge by Eyles *et al.*, 1993; Fig. 6). A narrow, northeast-trending bedrock valley punctures the ridge, extending northward to Lake Simcoe, as a result of selective erosion along fractured Paleozoic strata lying above a thrust zone in underlying mid-Proterozoic basement strata (the Central Metasedimentary Belt Boundary Zone; see Chapter 2). To the west, the ORM rests on a thick Late Pleistocene stratigraphy, selectively preserved in a major bedrock low (the Laurentian

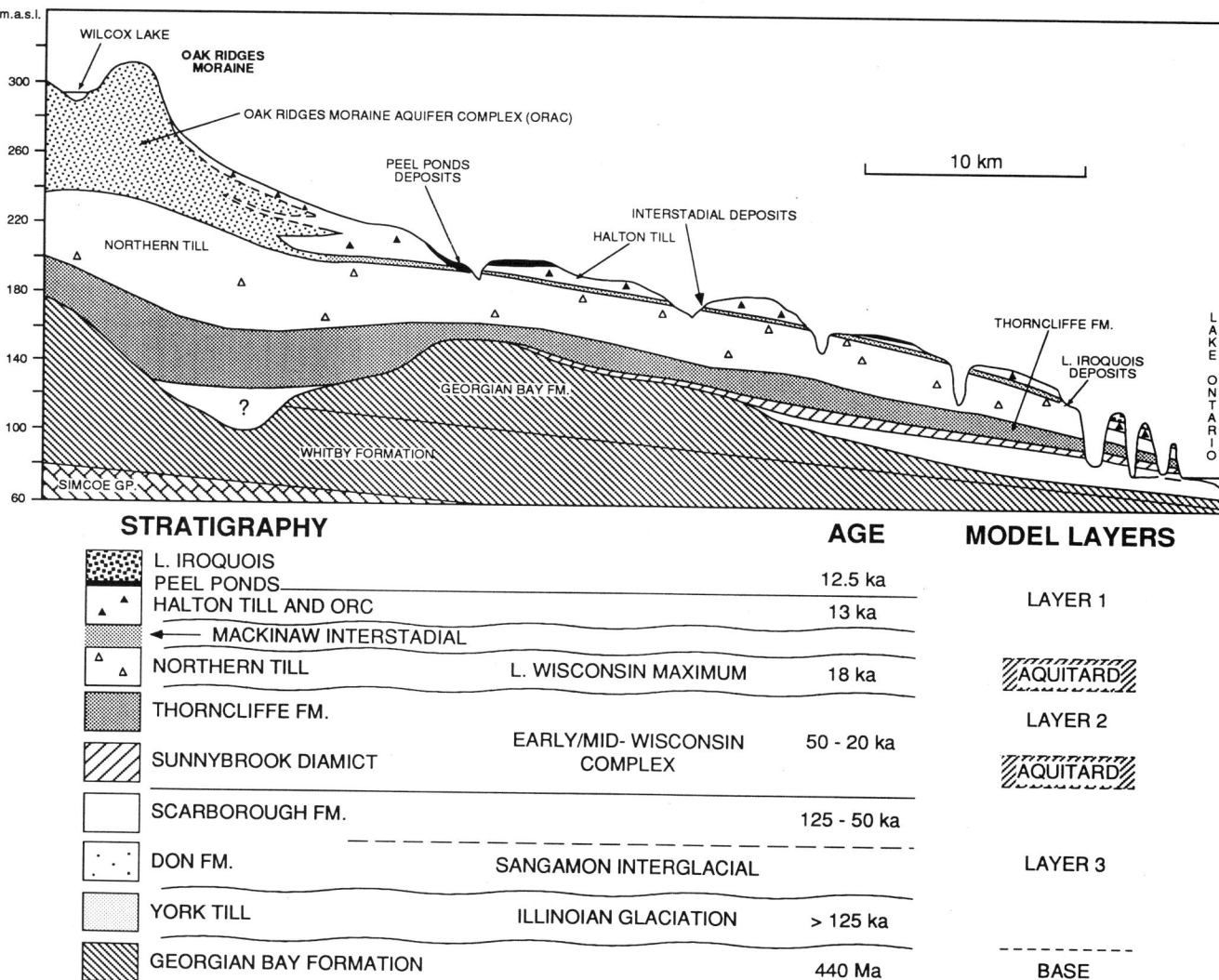

Figure 5 *Highly schematic north-south cross-section through the ORM south to Lake Ontario and component stratigraphy and hydrogeological behaviour. Layers 1, 2 and 3 identify modelled layers discussed in text (see Fig. 11).*

Channel), that connects Georgian Bay and the Lake Ontario basin and which formerly carried the drainage from the upper Great Lakes to the St. Lawrence River (Spencer 1881, 1889; Fig. 6). The Laurentian Channel has played a fundamental role in allowing the preservation of a thick Late Pleistocene stratigraphy in southern Ontario that contains significant aquifers.

Glacial Stratigraphy of the Moraine

The deeper Late Pleistocene glacial strata infilling the Laurentian Channel and related bedrock channels below the ORM comprise strata from the Illinoian glaciation (York Till; >100 ka), the last (Sangamon) interglacial (Don Formation, *ca.* 80 ka) and the early and middle phases of the last (Wisconsin) glaciation (Scarborough Formation, Sunnybrook Diamict and Thorncliffe Formation; 60-45 ka; Fig. 5; Chapter 2). Deltaic sands of the Thorncliffe Formation are a particularly distinct geophysical marker unit throughout the length of the Laurentian Channel (*e.g.*, Fligg, 1983) forming a regional aquifer system north, south and below the ORM (*e.g.*, Aravena and Wassenaar, 1993).

Late Wisconsin Events

During the main phase of the last (Late Wisconsin) glaciation (*ca.* 18 ka), the Laurentide Ice Sheet covered all of southern Ontario and extended well into New York State and Ohio

(Karrow, 1989; Fig. 7A). This event resulted in deposition of the Northern till. The Northern till can be traced below the ORM and is equivalent to the Newmarket Till, mapped north of the ORM by Gwyn (1972), and the Bowmanville Till, mapped to the east by Brookfield *et al.* (1982; see also Singer, 1973, 1974; Gwyn, 1976a, b). The Northern till (up to 60 m thick) contains sheet-like sand and gravel interbeds that play a critical role in providing potential ground-water flow paths through otherwise impermeable till (Boyce *et al.*, 1995; Chapters 20, 32). Investigations using environmental isotopes (M.M. Dillon Ltd., 1990) and water balance investigations (Chapter 10), suggest that a significant proportion of recharge to the confined ORM aquifer complex originates along the flanks of the ORM in areas mapped as till. Significant recharge to the deeper aquifers occurs outside the boundary of the ORM as demarcated by hummocky surface topography. This has important implications for protection of the hydrogeological integrity of the ORM and surrounding rivers and definition of the precise areas to be protected (see below).

Formation of the Oak Ridges Moraine. Despite its thickness and wide geographic extent, the ORM records only a brief episode of glacial sedimentation at the close of the last glaciation. The ORM is classified as an interlobate moraine (Chapman and Putnam, 1984) and records sedimentation between two lobes of the Laurentide Ice Sheet during the Port Huron Stadial, sometime after 13 ka (Fig. 7B; Terasmae and

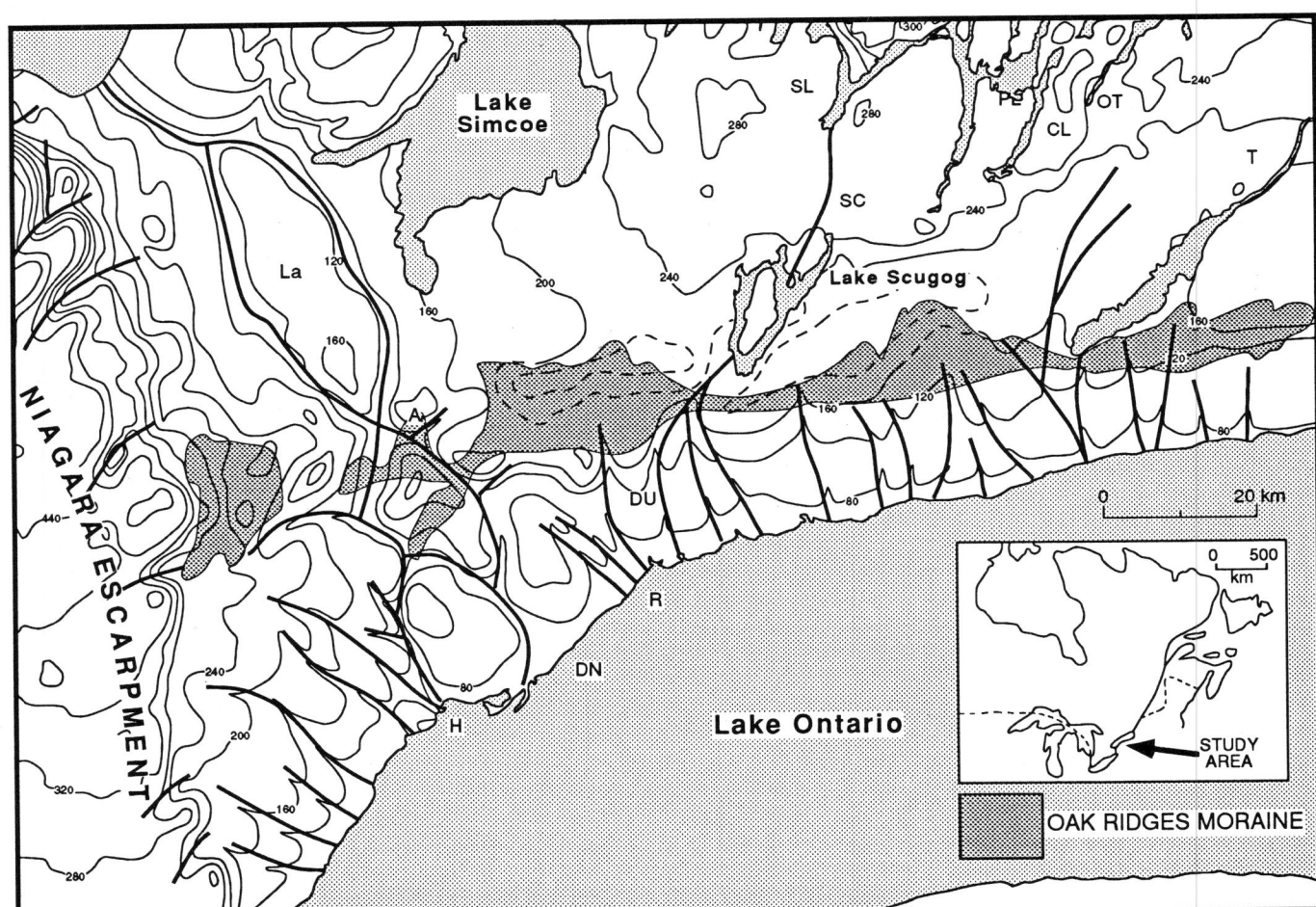

Figure 6 *The bedrock surface of south-central Ontario (after Eyles et al., 1993). Dashed lines on bedrock contours indicate bedrock high (Pontypool ridge) which controls location of ORM. The bedrock surface shows several major channel systems; Dn, Don channel; Du, Duffins Creek channel; H, Humber channel; La, Laurentian channel; OT, Otanabee channel; R, Rouge River channels; Sc, Scugog channel; T, Trent channel. Lakes named are; BL, Balsam Lake; CL, Chemung Lake; PL, Pigeon Lake; SL, Sturgeon Lake. Lake Scugog is man-made (see text).*

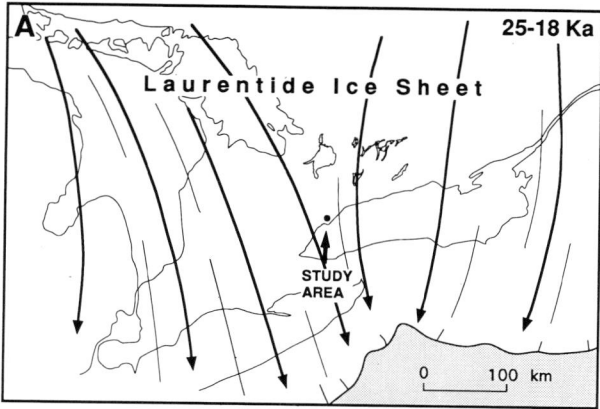

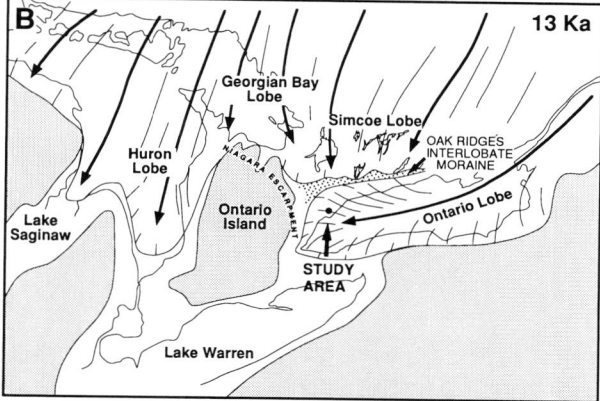

Figure 7 *Ice flow directions during (A) maximum extent of the late Wisconsin Laurentide Ice Sheet and deposition of Northern till (Figs. 4, 5, 8) and (B) during the Port Huron Stadial when the Halton Till and the ORM were deposited (after Boyce et al., 1995).*

Matthews, 1980). The moraine is, in fact, composed of glaciolacustrine fan-delta and outwash deposits that accumulated in an interlobate lake (Fig. 8), with water depths controlled by the elevation of overflow channels across the Niagara Escarpment in the west (*e.g.*, Duckworth, 1979; Chapman, 1985). This lake was dammed between the northwestward flowing Ontario lobe, which deposited the Halton Till south of the ORM, and the Simcoe Lobe to the north (Figs. 7, 8; Hewitt, 1969; Barnett *et al.*, 1991). Large volumes of dead ice were trapped within the lake and were buried by sediment; their later melt resulted in the many hundreds of kettle lakes that dot the moraine and impart the typical hummocky surface topography. The outer margins of the moraine show many steep slopes and deep gullies as a result of late-glacial meltwater erosion as the ice margins receded. The Halton Till is the most widespread late Wisconsin glacial unit at surface on the southern flank of the ORM (Fig. 7B). Regionally extensive sands and gravels lying between the base of the Halton Till and the Northern till were deposited during a brief ice-free episode in southern Ontario (Mackinaw Interstade; 13.3 ka; Fig. 5). The ORM area became ice free about 12 ka.

RECHARGE AND THE REGIONAL WATER BALANCE OF THE OAK RIDGES MORAINE

Development of regional ground-water flow models depends on a quantitative understanding of aquifer recharge, discharge and the regional water balance; these components of the regional ground-water system of the ORM have been studied on the Beaverton River, Pefferlaw Brook, Black River, East Holland River, Rouge River, and Duffins Creek watersheds (Fig. 9). Several independent approaches have been used to perform the analysis, including separation of stream and spring flow measurements and estimates of direct re-

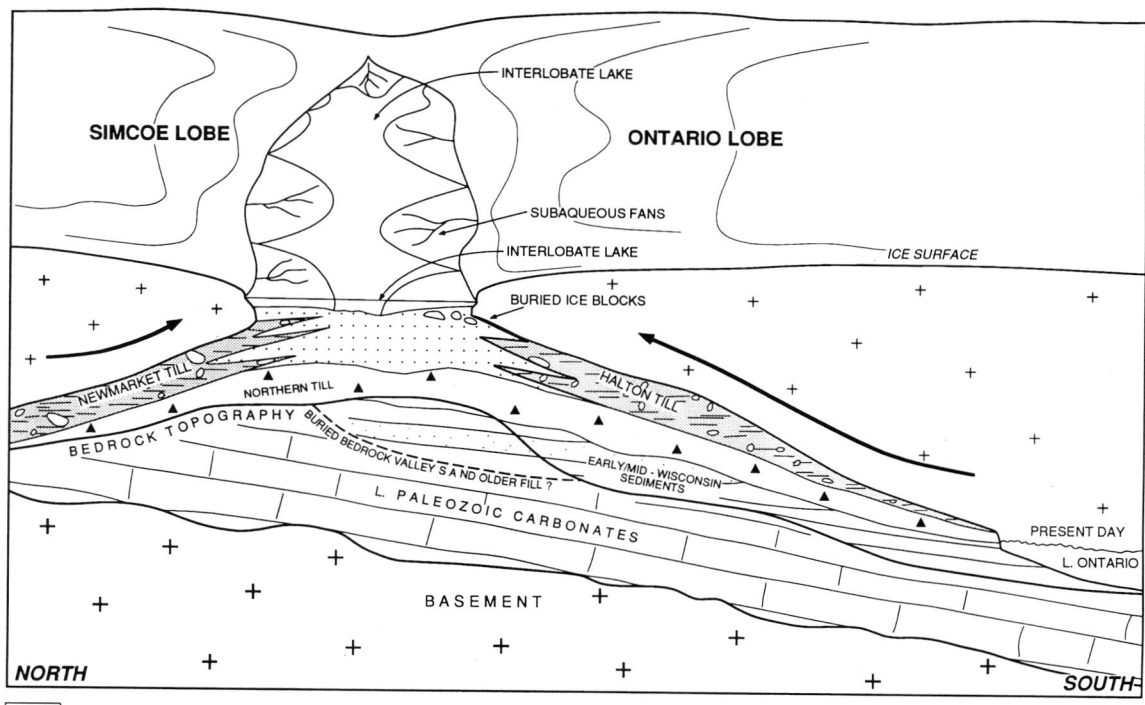

Figure 8 *Eastward-looking, highly schematic view of formation of the interlobate Oak Ridges Moraine as a series of overlapping glaciolacustrine fans deposited in a large interlobate lake between Lake Simcoe and Lake Ontario ice lobes (see Fig. 7). Burial of stranded ice blocks below sediment results in the typical hummocky, kettled topography of much of the moraine. Lake level was controlled by the Niagara Escarpment in the west (Fig. 7B). The Northern till comprises an extensive aquitard below the ridge (Figs. 5, 11).*

charge using soil moisture balance techniques. These techniques are widely used in regional assessments of ground-water regimes, and results are described in detail in Chapter 10.

HYDROGEOLOGICAL MODELLING OF THE OAK RIDGES MORAINE

Ultimately, quantitative ground-water flow models can be used to evaluate the potential effects of changing land use, climate and competing resource management practices in the ORM watershed, and to understand the sensitivity and sustainability of the ground-water resource. At present, hydrogeological models are being used primarily to test specific hypotheses concerning the nature and extent of deep ground-water flow and the influence of surface tills on aquifer recharge.

Ground-water flow models are being developed as a two-stage process. During the first stage, modelling work has focussed on FLOWPATH, a finite-difference steady-state flow model that can simulate ground-water flow and perform particle tracking in heterogeneous, anisotropic aquifers of two dimensions (Franz and Guiguer, 1990). FLOWPATH has been used to model a sub-domain of the regional study area (Fig. 10), thereby providing important data regarding ground-water flow paths, flow rates, and contaminant transport. As part of the second stage of the modelling work, a multi-dimensional transient ground-water flow model is under development that incorporates FLOWPATH calibrations, but has been extended in three dimensions to cover the entire ORM area. The transient model is based on MODFLOW, a United States Geological Survey (USGS) finite-difference model coded by McDonald and Harbaugh (1984). MODFLOW can be used to simulate steady-state, transient, confined and unconfined ground-water flow in three dimensions or as quasi three-dimensional flow. In addition, MODFLOW allows for spatial and temporal variation of boundary conditions and recharge, and specification of any number of injection or withdrawal

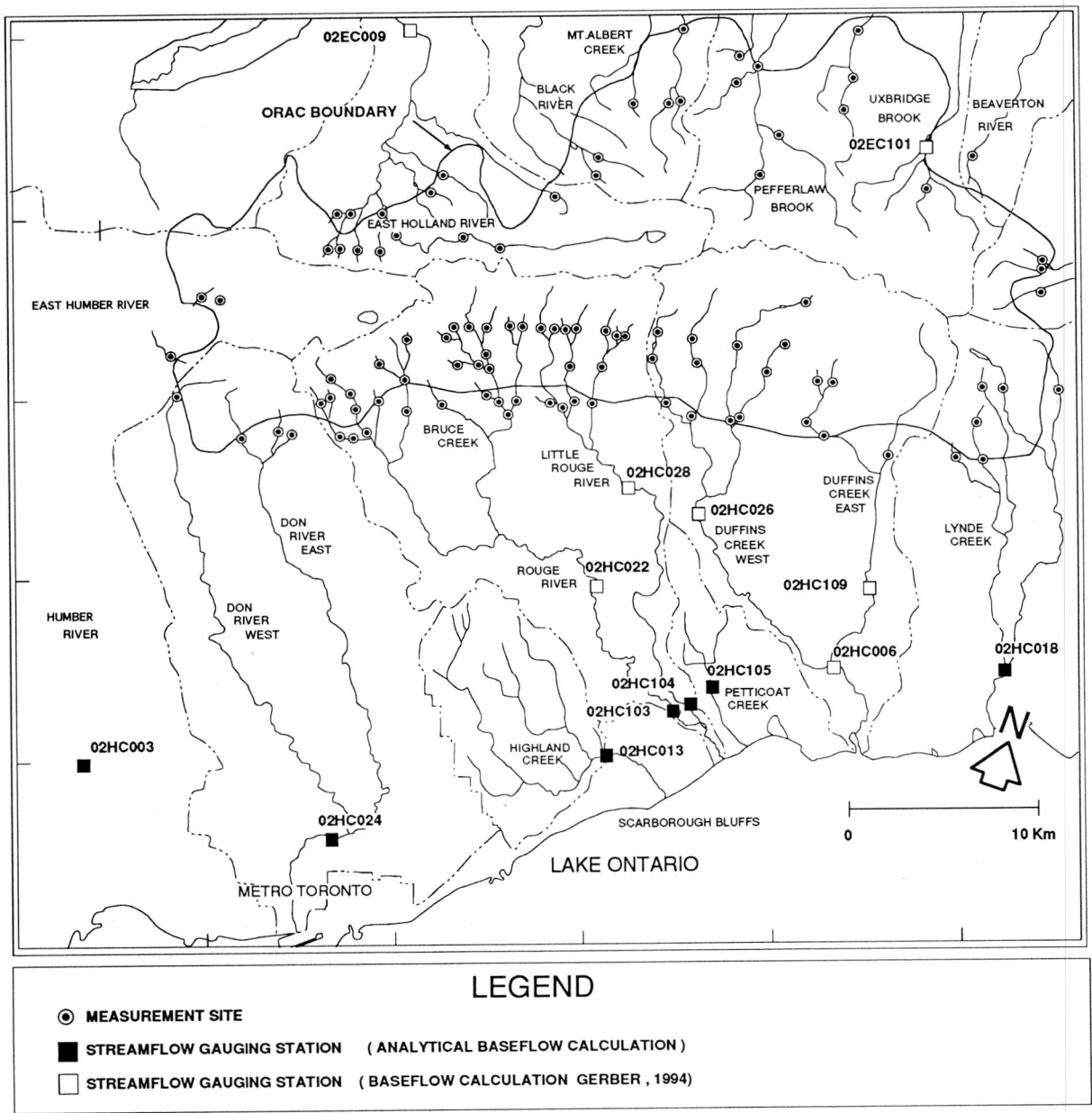

Figure 9 *Streamflow measurement sites and gauging stations.*

wells. MODFLOW has been extensively used and tested and is widely accepted as a verified model.

The governing equation on which MODFLOW is based is given by Equation 1. This equation is a mathematical representation of ground-water flow, which describes three-dimensional ground-water flow of constant density through porous media.

$$\frac{\partial}{\partial x}\left(K_{xx}\frac{\partial h}{\partial x}\right) + \frac{\partial}{\partial y}\left(K_{yy}\frac{\partial h}{\partial y}\right) + \frac{\partial}{\partial z}\left(K_{zz}\frac{\partial h}{\partial z}\right) - W = S_s\frac{\partial h}{\partial t} \qquad (1)$$

where K_{xx}, K_{yy}, K_{zz}, are values of hydraulic conductivity along the x, y and z co-ordinate axes, which are assumed to be aligned with the principal axes of hydraulic conductivity; h is the potentiometric head; W is the volumetric flux per unit volume and represents sources and/or sinks of water; S_s is the specific storage of the porous material; and t is time (Table 1). Hydraulic conductivity (K) is a measure of the ease with which water passes through a medium. The storage coefficient refers to the amount of water that an aquifer releases from, or takes into, storage. Equation 1 describes ground-water flow under non-equilibrium conditions in a heterogenous and anisotropic medium. The solution to this equation yields

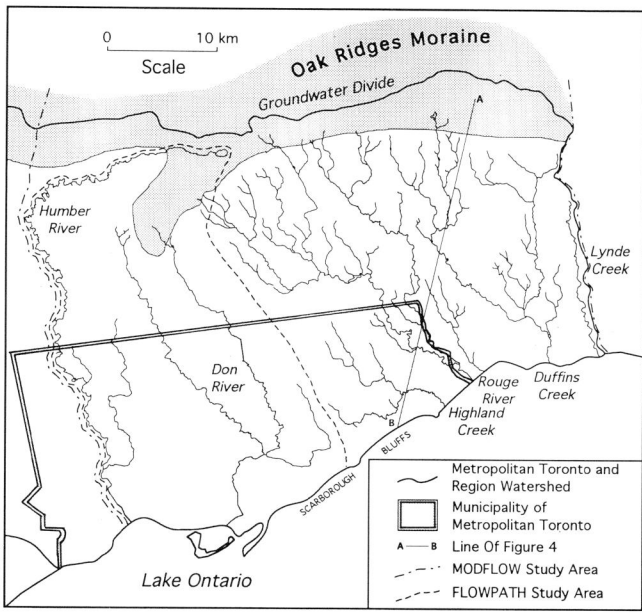

Figure 10 *MODFLOW and FLOWPATH study areas (from Smart, 1994). The line of section for Figure 4 is also shown.*

Table 1   Input parameters for transient ground-water flow modelling (MODFLOW). All model units are in metres *(L)* and days *(T)*; recharge has been converted to mm·a⁻¹ to maintain consistency with recharge estimates in the study area (Gerber and Howard, 1994).

PARAMETER	DIMENSIONS	RANGE
Hydraulic Conductivity (K_x, K_y)	L/T	0.1-10
Transmissivity (T_x, T_y)	L^2/T	1.0–4000
Specific Yield (Sy, unconfined aquifer)	dimensionless	0.25–0.06
Storativity (S, confined aquifer)	dimensionless	0.005–0.00001
Layer Thickness Unconfined (bottom elevation) Confined (top and bottom elevation)	L	10–80
Recharge	L/T	0–450
Leakance	T^{-1}	1.0×10^{-1}–1.0×10^{-8}

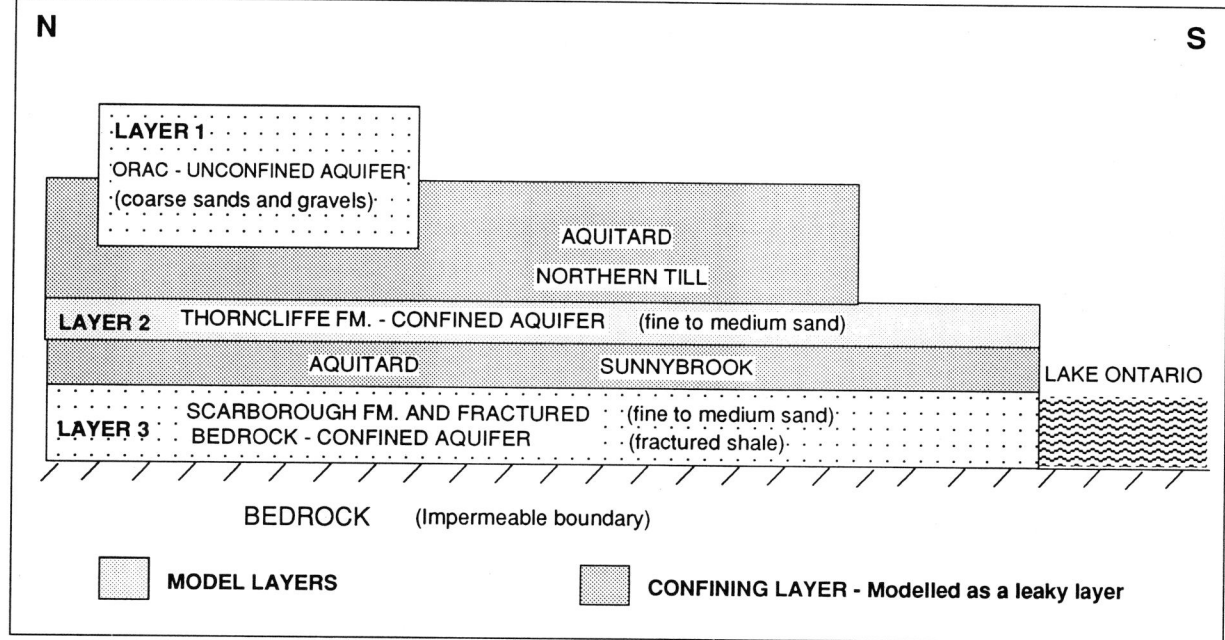

Figure 11 *Design of quasi three-dimensional ground-water flow model based on north-south stratigraphy shown in Figure 5 (from Smart, 1994). The figure identifies model layers only and does not attempt to portray thickness or extent (see Fig. 4).*

values of hydraulic head (water levels) at specific points and time.

MODFLOW employs the finite difference method to solve Equation 1; detailed derivation of this method is given by Rushton and Redshaw (1979). Essentially, the finite difference method involves approximating the partial derivatives ∂x, ∂y and ∂z by discretizing the problem domain into a large number of smaller subdomains, which are termed blocks or cells, and approximating the partial derivative ∂t using discrete time steps (Mercer and Faust, 1992). Each finite difference block contains a node, and the collection of blocks produces a rectangular grid. This grid is superimposed over the study area, and each block is assigned aquifer parameters appropriate to the field situation. To fully define the flow regime, internal sources, sinks and boundary conditions must also be specified.

For the purposes of MODFLOW, the regional ground-water flow regime of the ORM is defined as being bounded by Lake Ontario in the south, Lynde Creek in the east, the East Humber River to the west, and the furthest extent of the ORM to the north (Fig. 10). Aquifers in this region have been divided into a three-layer system consisting of an upper sand and gravel complex of the moraine proper (Oak Ridges Aquifer Complex; 200-325 m asl), an intermediate aquifer (equivalent to the Thorncliffe Formation; 140-215 m asl), and a lower aquifer defined between bedrock and 140 m asl (Scarborough Formation; Figs. 5, 11). Model transmissivities were estimated on the basis of aquifer lithology and thickness; aquifer recharge and the regional water budget data were obtained using various methods described above (Table 1). Recharge zones are shown in Figure 12. A steady-state version of the model has been calibrated by comparing potentiometric heads and streamflows generated by the model with comparable field data.

Potentiometric Surface

The potentiometric surface of the Oak Ridges Aquifer Complex (ORAC; Fig. 12A) is a subdued replica of the surface topography of the moraine, and ranges from 320 m asl in elevation at the centre of the complex to 220 m asl along the flanks. Water levels are highest in the east, where the drainage basin divide separates the Rouge River Basin and the Duffins Creek Basin to the south from drainage basins to the north, and far northwest, where a drainage basin divide separates the Humber River basin from the Holland River basin. These high water levels are roughly coincident with the crest of the ORM, and form the regional ground-water divide. Local highs up to 290 m asl occur at the northern extent of the

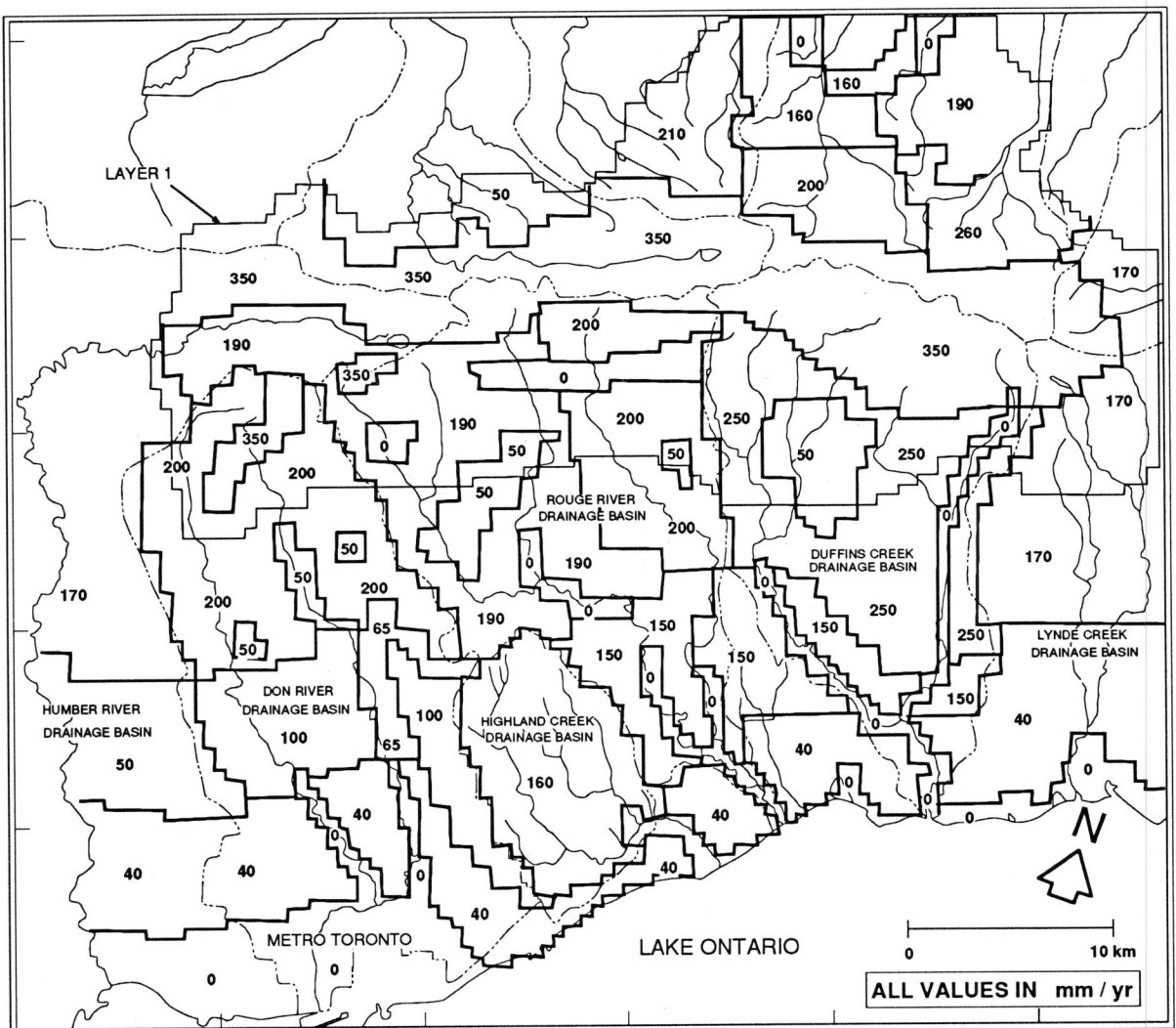

Figure 12 *Recharge zones and values used in ground-water flow model.*

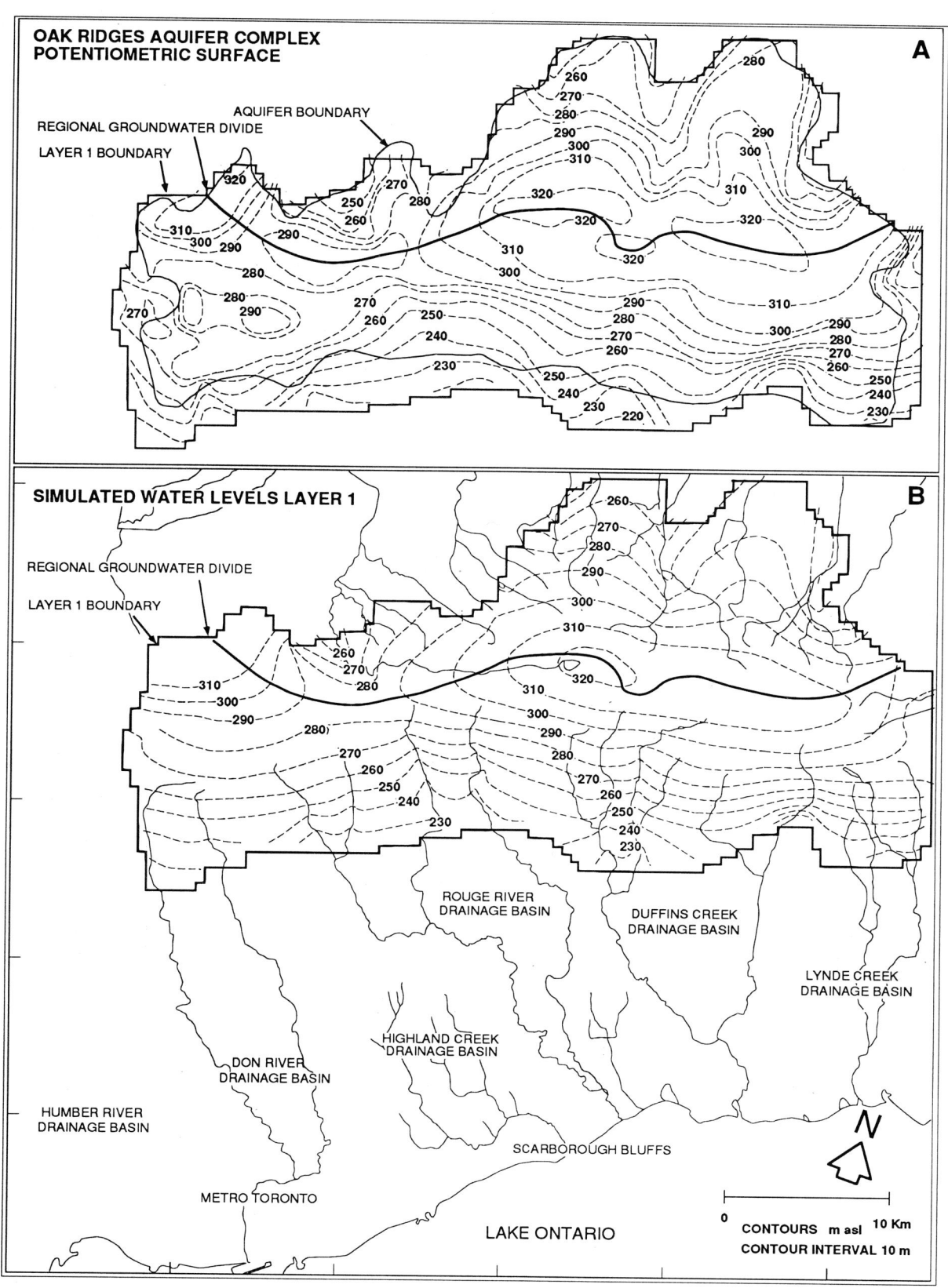

Figure 13 (**A**) *Potentiometric surface for the Oak Ridges Aquifer Complex (layer 1; Fig. 11);* (**B**) *modelled surface. See text for details.*

Don River drainage basin, and are due to perched aquifer systems situated above the regional ORAC water table.

The regional horizontal hydraulic gradient averages 0.012. Horizontal gradients range from 0.004 near the centre of the complex to 0.025 at discharge points such as East Duffins Creek and the East Holland River. Ground-water contours deflect around tributaries of major streams originating on the moraine, indicating a significant contribution of ground water to these streams. Along the northern and southern boundaries of the ORAC, regional horizontal gradients increase to 0.017. Ground-water contours are grouped closer together in these areas, due to a decrease in transmissivity resulting from aquifer deposits interfingering with the less permeable Halton and Northern tills that flank the ORM (e.g., Fig. 4).

Simulated Water Levels in the Oak Ridges Aquifer Complex

The steady-state model was calibrated, in part, by comparing model heads (Fig. 13B) with the observed potentiometric surface (Fig. 13A). A good correlation is obtained. Model heads rise from 230 m asl along the southern boundary of layer 1 to 320 m asl at the centre. A ground-water divide runs generally east-west across the moraine, and this is also well represented by the model data. In general, regional hydraulic gradients are well reproduced, but increased gradients observed near the northern and southern boundaries of the complex are not well represented. The highest gradients are observed at discharge points such as East Duffins Creek and the East Holland River, and most contours deflect around streams in sympathy with the field data. Since no effort was made to simulate the perched aquifer systems, local water level highs in the north Don River basin are not reproduced by the model.

The quality of the simulation can also be established by a statistical analysis of the field heads and model heads. The difference between these values is known as the residual head. In general, a good calibration is achieved if the mean of the residual is close to zero and the standard deviation is less than 10% of the overall range in head for the model layer. The mean error for the ORAC model layer (layer 1) is –0.5 m, and the standard deviation is 8 m, which are both within acceptable limits. Model heads are plotted against field heads in Figure 14; a perfect calibration is achieved when points fall along the 1:1 line. It can be seen from the figure that most data points fall close to the straight line, with a few anomalous highs. Most of these anomalies are caused by perched water tables or by high gradients along the northern and southern

boundaries of the ORAC, which are not well simulated by the model.

Simulated Flow in the Oak Ridges Aquifer Complex

A fundamental element of the calibration process has been to determine whether the model generates the correct ground-water flux. Commonly, this is achieved by comparing model outflows with baseflow observed in receiving streams (Table 2). The influence of receiving streams on regional flow is shown by the ground-water flowlines on Figure 15. Under steady-state ground-water flow conditions, the total ground-water inflow to the model is balanced by ground-water outflow. To estimate outflow, more than 100 stream and spring flow measurements were taken around the boundary of the ORM during a summer dry spell. Flow measurements were subject to error depending on the size, cross-sectional shape, and flow conditions of the stream. However, comparisons to permanent stream gauge stations in the Holland River, Black River, and Mt. Albert Creek basins suggest that the estimates are within ±30%. In basins containing only a few head-water tributaries, errors of this magnitude are small in comparison to total ground-water fluxes. However, measurement errors become much more significant in basins such as the Rouge River and Duffins Creek, which are fed by many tributaries. For this reason, while many streamflow measurements were taken in the Rouge and Duffins basins, greater reliance is placed on gauging station data for estimating outflows in

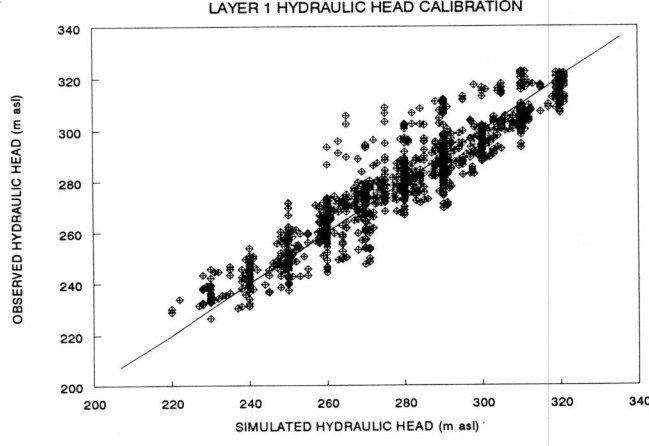

Figure 14   *Observed hydraulic head plotted against simulated head. See text for details.*

Table 2	Ground-water inflows and outflows; all values in m³·d⁻¹. Measured and calculated outflows in *italics*.					
	INFLOWS			**OUTFLOWS**		
	RECHARGE	LEAKAGE	RIVER AND DRAIN BOUNDARY	LEAKAGE	GENERAL HEAD BOUNDARY	CONSTANT HEAD BOUNDARY
LAYER 1	555,000	255,000	300,000 *267,000*			
LAYER 2	283,800	300,000	350,400	158,000	75,400 *75,600*	
LAYER 3	85,200		211,000		18,000 *16,000*	14,200 *7,500*
	Total Simulated Baseflow 816,400			**Total Calculated Baseflow** *770,600*		

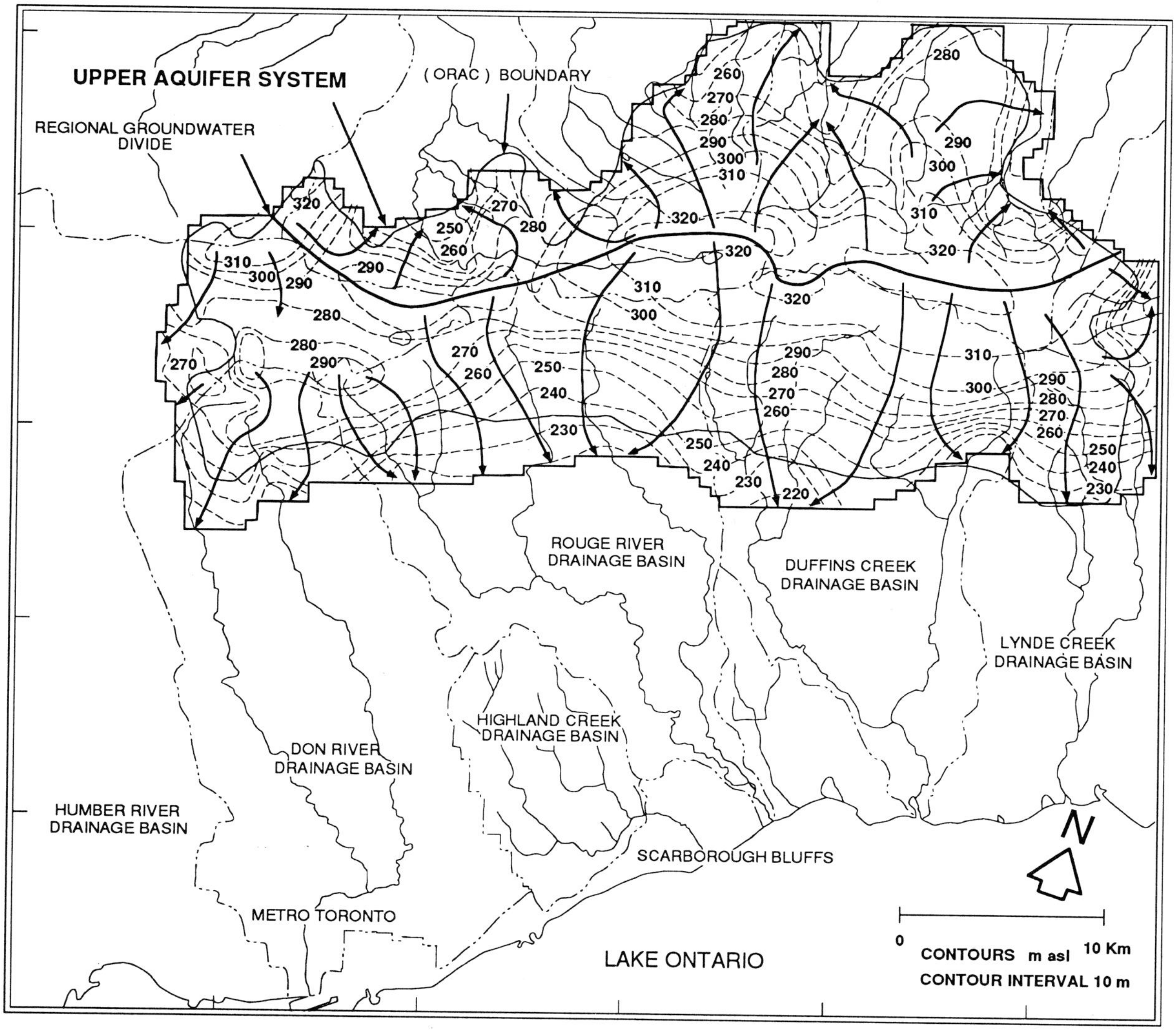

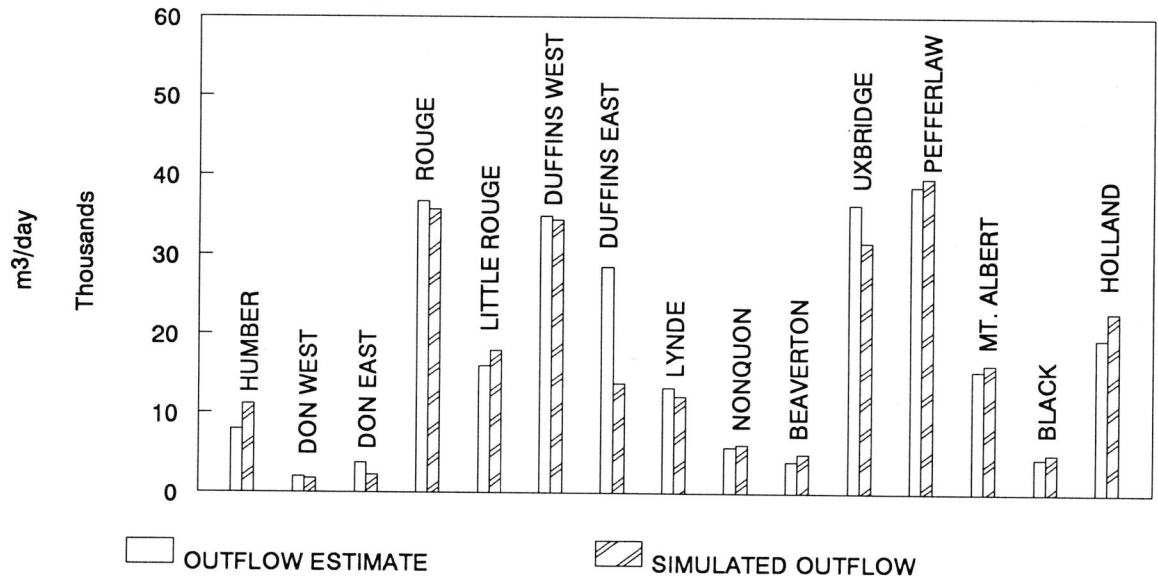

Figure 15   *Regional flowlines for the Oak Ridges Aquifer Complex (layer 1; Figs. 5, 11).*

LAYER 1 STREAM AND SPRING OUTFLOW CALIBRATION

Figure 16   *A comparison of model outflows with field-based estimates.*

these basins.

Model outflows are compared with field outflow estimates in Figure 16. A good correlation is generally observed, with the exception of East Duffins Creek, where it is suspected that additional inflows are being received from neighbouring basins to the east, possibly as a result of the buried bedrock channels shown in Figure 6. Under steady-state conditions, total ground-water outflows will match the incoming recharge. Based on recharge calculations, baseflow studies, and model behaviour, total recharge to the model layer 1 is 5.55×10^5 $m^3 \cdot d^{-1}$ and model outflow to streams, springs and marshes is 2.55×10^5 $m^3 \cdot d^{-1}$. The remaining 3.00×10^5 $m^3 \cdot d^{-1}$ represents the amount of water which must infiltrate or leak *via* the Northern till to deeper aquifers contained within bedrock channels.

Model Leakance

The model is referred to as a quasi three-dimensional ground-water flow model because intervening aquitard layers are not explicitly represented. A quasi three-dimensional model uses a leakage term, referred to as leakance, to simulate the effects of an aquitard layer.

Model leakance is used to simulate the behaviour of an aquitard, *i.e.*, a layer that has a lower hydraulic conductivity than overlying or underlying aquifers. Leakance and "leak-

age" are related, but differ in the following manner: leakance is equal to the hydraulic conductivity of an aquitard divided by the thickness of the aquitard, whereas leakage refers to the rate at which water is transmitted across the aquitard (as a volume per unit time).

Leakance was used to simulate the influence of the Northern till. This unit has been widely used as a substrata for landfills, but problems are emerging with higher-than-expected ground-water flows through the till (see Chapters 2, 10, 20, 32). Leakance values were estimated using hydraulic conductivity values typical of till deposits and using an arbitrary aquitard thickness of 10 m. Hydraulic conductivity values typical of till deposits range from 1×10^{-1} $m \cdot d^{-1}$ (1×10^{-4} $cm \cdot s^{-1}$) to 1×10^{-6} $m \cdot d^{-1}$ (1×10^{-9} $cm \cdot s^{-1}$; Freeze and Cherry, 1979). Based on these hydraulic conductivity estimates and an aquitard thickness of 10 m, the leakance values used during the calibration of the model ranged from $1\times10^{-1} \cdot d^{-1}$ to $1\times10^{-8} \cdot d^{-1}$ (Table 1).

In order to calibrate the model, two leakance zones, which are within an order of magnitude, were used: one with a value of $2.5\times10^{-5} \cdot d^{-1}$ and the other with a value of $1\times10^{-4} \cdot d^{-1}$. In general, the lower leakance value was applied to the central portion of layer 1, and the higher leakance value was applied to the perimeter of layer 1. These leakance values convert to a hydraulic conductivity of 2.5×10^{-4} $m \cdot d^{-1}$ (2.8×10^{-7} $cm \cdot s^{-1}$) for

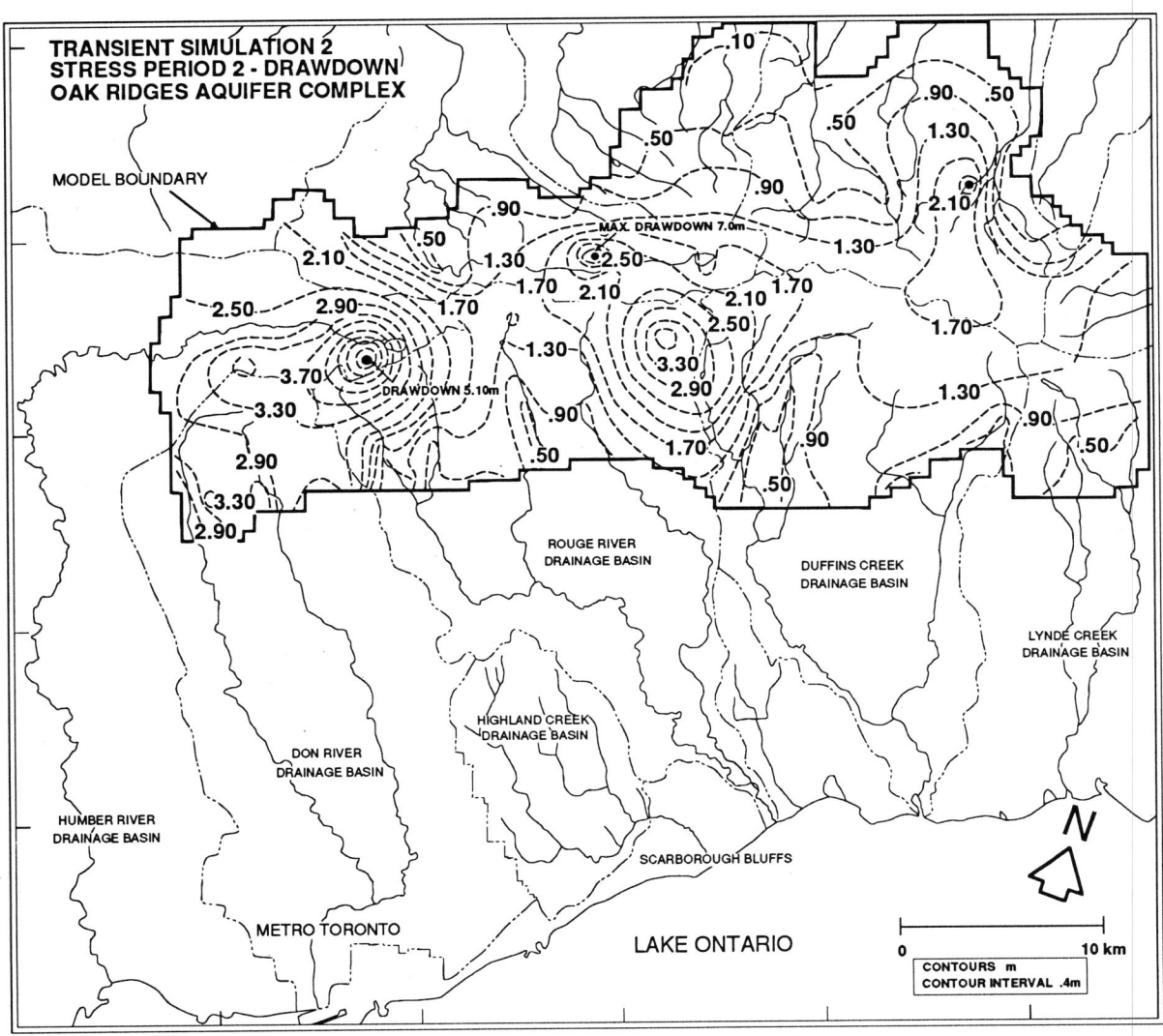

Figure 17 *Drawdown in the Oak Ridges Aquifer Complex resulting from increased municipal well abstraction.*

the central portion and 1×10^{-3} m·d^{-1} (1×10^{-6} cm·s^{-1}) for the edges of layer 1, based on a 10-m aquitard thickness. This range is in good agreement with the known hydraulic conductivity of the Northern till underlying the ORAC, and allows for reliable estimates of the amount of recharge received by aquifers underlying the ORAC as vertical leakage to be made. This deep infiltration, spread over the 849 km^2 area of the ORAC, is equal to an average infiltration rate of 129 mm·a^{-1}.

A relatively high infiltration rate (156 mm·a^{-1}) in the Holland River basin contrasts with a low infiltration rate (41 mm·a^{-1}) obtained for the Pefferlaw drainage basin. The high downward movement of ground water in the Holland River basin is facilitated by a high recharge to the ORAC and the under-drainage resulting from steep vertical gradients between the aquifer complex and deep aquifers below the ground-water divide area. Low vertical infiltration in the Pefferlaw basin is probably the result of lower vertical gradients between the ORAC and deep aquifers, and from upward gradients from deep aquifers discharging to the Pefferlaw River. These data emphasize the need for a detailed understanding of the subsurface geology across the entire ORM area in order to more closely constrain the modelling exercise.

MODEL PREDICTION FOR EXPANDED URBANIZATION
Predictive transient simulations were used to estimate the drawdown and aerial cone of depression that may result if urbanization expands onto the ORAC. The transient simulation presented here was carried out to study the effects of simultaneously increasing municipal water abstractions to the maximum capacity of the wells and decreasing recharge by 5%. In some situations, recharge losses due to urbanization can be as much as 50%, which, for example, would decrease recharge in a temperate climate from 300 mm·a^{-1} to 150 mm·a^{-1} (Smart, 1994).

Recharge losses can be supplemented with the use of soak aways and infiltration ponds; however, recent studies in the Oak Ridges Moraine area indicate that, even with these protective measures, recharge will be depleted by at least 5% (Gore and Storie, 1993). For the purposes of this simulation, recharge was decreased for layer 1 by 5%, representing moderate urbanization with population densities approximately equivalent to the Aurora-Newmarket area. The model was allowed to reach a new steady state in response to the stress. In practice, it may take tens of years to reach the predicted drawdowns. The decline in water levels and the cone of depression surrounding the municipal wells are shown in Figure 17. Water levels are predicted to decline up to a maximum of 5 m, which occurs at the community of Oak Ridges, and 1 m to 3 m in outlying areas. A second simulation in which recharge to layer 1 was decreased by 10% and municipal water abstractions were raised to twice the existing maximum capacity of the wells, caused a widespread decline of water levels. A maximum decline of 19 m was observed in the community of Oak Ridges; in several other areas, water level depressions approached 11 m. As anticipated, both simulations caused a significant reduction in the discharge of ground water to head-water streams originating along the flanks of the moraine. It should be noted that the model has not been fully calibrated in transient mode due to a lack of long-term data. However, predicted drawdowns are consistent with observed declines in the Aurora-Newmarket area (International Water Consultants Ltd., 1991).

The model also allows estimates to be made of travel times for pollutants that enter the ORAC from septic systems and landfills and from other land uses. Smart (1994) showed that conservative inorganic contaminants (*i.e.*, those such as chloride or nitrate) which undergo no adsorption by binding to soil particles, and which travel at the average linear flow velocity of ground water released close to the ground-water divide, take more than 800 years to discharge to streams on the south slope of the ORM.

DISCUSSION
Work completed to date demonstrates the existence of a regionally extensive hydrostratigraphy within the Oak Ridges Moraine, comprising three principal aquifer systems. These include an upper sand and gravel complex (Oak Ridges Aquifer Complex (ORAC; 200-325 m asl), an intermediate aquifer (equivalent to the Thorncliffe Formation; 140-215 m asl), and a lower aquifer defined between bedrock and 140 m asl (Scarborough Formation). Water balance and modelling studies suggest the deeper aquifers systems are recharged by the ORAC, close to the crest of the ridge, but also by recharge through the Halton and Northern tills along the moraine flanks. This finding has very important implications for managing the extent and density of urban development outside the topographically defined limits of the moraine proper.

Ground-water Management
The central issue with regard to ground-water resource management is precisely what area should be included within the ORM and be protected from further development. Interim definition of the ORM, as currently proposed by the Ontario Ministry of Natural Resources (OMNR), is arbitrarily based on the 275 m asl contour because it encloses most of the hummocky, ice-contact deposits that make up the central geographical core of the moraine. Implicit in this definition is the assumption that this geographical zone includes the bulk of the Oak Ridges Aquifer Complex (ORAC), and that restricted urban development in this area will safeguard deeper aquifers. This assumption is difficult to sustain because there is good evidence that the ORM aquifer receives recharge through flanking till deposits both to the north, *via* deep bedrock channels that underdrain the moraine, and along the south slope from Richmond Hill east to Pickering. This area contains the head waters of the Don, Rouge and Duffins rivers, where the most intense developmental pressures are likely to be felt in response to proximity to the planned west-east 407 highway from Markham to Whitby and the proposed development of Seaton, a planned community of 90,000 people, in North Pickering (see Seaton Advisory Committee, 1994; Chapter 2).

An important goal must be to develop the regional ground-water flow model into fully operational transient mode, thereby permitting the impact of various land use changes to be evaluated. The reliability of such a model will depend heavily on the ability, in the interim, to refine hydrogeological knowledge in several key areas. While the hydrogeological behaviour of the ORAC is generally well established, and geophysical data are beginning to provide insight into the deeper stratigraphy, further hydrochemical and potentiometric data are required for the deeper aquifer units. In addition, while the till and clay deposits flanking the moraine are capable of transmitting significant quantities of water, further work is required to determine the nature of transmission, the velocities of flow, and the regional variability of these sediments. This knowledge is critical both for understanding contaminant impact potential and for quantifying the rates at which deeper aquifers are replenished *via* overlying aquitards. There remains a significant risk that the rate of urban development along and adjacent to the moraine will outpace the ability to make reliable predictions of potential impacts.

ACKNOWLEDGEMENTS

Research is supported by research grants from the Great Lakes University Research Fund, Environment Canada, the Natural Sciences and Engineering Research Council, and the Ontario Ministry of the Environment and Energy. The authors gratefully acknowledge many University of Toronto students and colleagues who assisted in the collection and interpretation of data, particularly Steve Livingstone, Mike Bromley and Kit Soo. We also thank members of the Oak Ridges Moraine Technical Working Committee, under the Chairmanship of Ron Christie. The views presented here are those of the authors and are not necessarily endorsed by the funding agencies named above.

REFERENCES

Aravena, R. and Wassenaar, L.I., 1993, Dissolved organic carbon and methane in a regional confined aquifer, southern Ontario, Canada: Carbon isotope evidence for associated subsurface sources: Applied Geochemistry, v. 8, p. 483-493.

Barnett, P.J., Cowan, W.R. and Henry, A.P., 1991, Quaternary Geology of Ontario, Southern Sheet: Ontario Geological Survey, Map 2556, Scale 1:1,000,000.

Bigsby, J.J., 1824, Notes on the geogaphy and geology of Lake Huron: Geological Society, Transactions, 2nd series, v. 1, p. 6-61.

Boyce, J.I., Eyles, N. and Pugin, A., 1995, Seismic reflection, borehole and outcrop geometry of Late Wisconsin tills at a proposed landfill near Toronto: Canadian Journal of Earth Sciences, v. 32, p. 1331-1349.

Brookfield, M.E., Gwyn, Q.H.J. and Martini I.P., 1982, Quaternary sequences along the north shore of Lake Ontario: Oshawa-Port Hope: Canadian Journal of Earth Sciences, v. 19, p. 1836-1850.

Brown, R., 1978, Ghost Towns of Ontario v. 1. Southern Ontario: Stagecoach Publishing Co., Langley, BC, 200 p.

Caddy, E.C., 1861, Map of the counties of Northumberland, Durham, Peterborough and Victoria: lithographed by J. Ellis, Toronto, ON, print 198×134 cm.

Carman, R.S., 1941, The glacial pot hole area, Durham County, Ontario: The Forestry Chronicle, September, p. 110-120.

Chapman, L.J., 1985, On the origin of the Oak Ridges Moraine, southern Ontario: Canadian Journal of Earth Sciences, v. 22, p. 300-303.

Chapman, L.J. and Putnam, D.F., 1943, The moraines of southern Ontario: Royal Society of Canada, Transactions, v. 37, p. 33-41.

Chapman, L.J. and Putnam, D.F., 1951, Physiography of Southern Ontario: Ontario Research Foundation, Map 1:250,000.

Chapman, L.J. and Putnam, D.F., 1984, The Physiography of Southern Ontario, Third Edition: Ontario Geological Survey, Special Volume 2, 270 p.

Coventry, A.P., 1945, The Need for River Valley Management in Southern Ontario: Department of Planning and Development, Ontario, 57 p.

Crombie, D., 1990, Watershed: Royal Commission on the Future of the Toronto Waterfront, 207 p.

Darroch, G. and Soltow, L., 1994, Property and Inequality in Victorian Ontario: Structural Patterns and Cultural Communities in the 1871 Census: University of Toronto Press, Toronto, ON, 68 p.

Dillon, M.M., Ltd., 1990, Regional Municipality of Durham Contingency Landfill Site Assessment, Technical report, v. B - Hydrology, 225 p.

Duckworth, P.B., 1979, The late depositional history of the western end of the Oak Ridges Moraine, Ontario: Canadian Journal of Earth Sciences, v. 16, p. 1094-1107.

Eyles, N., Boyce, J.I. and Mohajer, A.A., 1993, The bedrock surface of the western Lake Ontario region: Evidence of reactivated basement structures?: Géographie Physique et Quaternaire, v. 47, p. 269-283.

Eyles, N., Clark, B.M., Kaye, B.G., Howard K.W.F. and Eyles, C.H., 1985, The application of basin analysis techniques to glaciated terrains: An example from the Lake Ontario basin, Canada: Geoscience Canada, v. 12, p. 22-32.

Fligg, K., 1983, Geophysical Well Log Correlations between Barrie and the Oak Ridges Moraine: Water Resources Branch, Ontario Ministry of the Environment, Map 2273.

Franz, T. and Guiguer, N., 1990, FLOWPATH, Two-dimensional Horizontal Aquifer Simulation Model: Waterloo Hydrogeologic Software, Waterloo, ON, 74 p.

Freeze, R.A. and Cherry, J.A., 1979, Groundwater: Prentice-Hall, Englewood Cliffs, NJ, 804 p.

Funk, G., 1977, Geology and Water Resources for the Bowmanville, Soper and Wilmot Creeks IHD Representa-tive Drainage Basin: Ontario Ministry of the Environment, Water Resources Report 9a, 113 p.

Gerber, R.E., 1994, Recharge Analysis for the Central Portion of the Oak Ridges Moraine: unpublished M.Sc. thesis, University of Toronto, Toronto, ON, 172 p.

Gore and Storie, 1993, Phase II Environmental Studies Draft Summary Report OPA 71: Town of Richmond Hill, ON, 82 p.

Gwyn, Q.H.J., 1972, Quaternary geology of the Alliston-Newmarket Area, Southern Ontario: Ontario Department of Mines, Miscellaneous Paper 53, p. 144-147.

Gwyn, Q.H.J., 1976a, Quaternary Geology Resources of the Western Part of the Municipality of Durham, Southern Ontario: Ontario Department of Mines, Open File Report 516, Map with Marginal Notes.

Gwyn, Q.H.J., 1976b, Quaternary Geology Resources of the Central and Eastern Part of the Municipality of Durham, Southern Ontario: Ontario Department of Mines, Open File Report 5176, Map with Marginal Notes.

Gwyn, Q.H.J. and Cowan, W.R., 1978, The origin of the Oak Ridges and Orangeville moraines of southern Ontario: Canadian Geographer, v. 22, p. 36-47.

Haefeli, C.J., 1970, Regional Groundwater Flow between Lake Simcoe and Lake Ontario: Department of Energy, Mines and Resources, Inland Waters Branch, Technical Bulletin 23, 52 p.

Hainstock, H.N., Owen, E.B. and Caley, J.F., 1948, Groundwater Resources of Scarborough Township, York County, Ontario: Geological Survey of Canada, Water Supply Paper 29, 62 p.

Hainstock, H.N., Owen, E.B. and Caley, J.F., 1952, Groundwater Resources of Whitchurch Township, York County, Ontario: Geological Survey of Canada, Water Supply Paper 320, 51 p.

Hewitt, D.F., 1969, Industrial Mineral Resources of the Markham-Newmarket area: Ontario Department of Mines, Industrial Mineral Report 24, p. 41.

Howard, K.W.F. and Beck, P., 1986, Hydrochemical interpretation of groundwater flow systems in Quaternary sediments of southern Ontario: Canadian Journal of Earth Sciences, v. 23, p. 938-947.

Intera-Kenting, 1990, The Hydrogeological Significance of the Oak Ridges Moraine: prepared for the Greater Toronto Greenlands Strategy, 82 p.

International Water Consultants Ltd., 1991, Regional Municipality of York Aquifer Performance Assessment, Oak Ridges, Aurora, Newmarket, East Gwillimbury and Bradford: prepared for the Regional Municipality of York, 19 p.

Johnson, L.A., 1973, History of the County of Ontario, 1615-1875: The Corporation of the County of Ontario, 386 p.

Kanter, R., 1990, Space for All: Options for a Greater Toronto Area Greenlands Strategy: Queen's Printer, Toronto, ON, 163 p.

Karrow, P.F., 1989, Quaternary geology of the Great Lakes subregion, Chapter 4, , in Fulton, R.J., ed., Quaternary Geology of Canada and Greenland: Geological Survey of Canada, Geology of Canada, v. 1, p. 326-350. [also Geological Society of America, The Geology of North America, v. K-1]

Kaye, B.G., 1986, Recharge characteristics of the Oak Ridges Complex: the tole of Musselman Lake: unpublished M.Sc. thesis, University of Waterloo, Waterloo, ON, 187 p.

Keefer, T. C., 1863, Plan of Survey for Georgian Bay Canal through the Dividing Ridges of the County of Ontario: W.C. Chewett & Co., Toronto, ON, print 72×100 cm.

Mercer, J.W. and Faust, C.R., 1992, Ground-Water Modelling, 3rd edition: National Groundwater Association, 60 p.

Metropolitan Toronto and Region Conservation Area (MTRCA), 1989, Greenspace strategy for the Great Toronto Area, 62 p.

McDonald, M.G. and Harbaugh, A., 1984, A Modular Three-dimensional Finite-difference Groundwater Flow Model: United States Geological Survey, Open File Report 83-875, 528 p.

Oak Ridges Moraine Technical Working Committee, 1994a, Oak Ridges Moraine Area, Strategy for the Greater Toronto Area: An Ecological Approach to the Protection and Management of the Oak Ridges Moraine, 106 p.

Oak Ridges Moraine Technical Working Committee, 1994b, Oak Ridges Moraine Aggregate Resources Study, Background Study 10, 51 p.

Proulx, I. and Farvolden, R.N., 1989, Analysis of the Contaminant Plume in the Oak Ridges Aquifer: Ontario Ministry of the Environment, R.A.C. Project 261, 134 p.

Richardson, A.H., 1944, The Ganaraska Watershed: Province of Ontario Printer, 248 p.

Richardson, A.H., 1974, Conservation by the People: The History of the Conservation Movement in Ontario to 1970: University of Toronto Press, Toronto, ON, 154 p.

Rushton, K.R. and Redshaw S.C., 1979, Seepage and Groundwater Flow Numerical Analysis By Analog And Digital Methods: John Wiley and Sons, New York, 339 p.

Seaton Advisory Committee, 1994, Phase 3 Design Brief: Ontario Ministry of Housing, 53 p.

Sharpe, D.R. and Finley, W.D., 1993, Surficial Geology Greater Toronto Area: Markham: Geological Survey of Canada, Map Sheet 30M/14.

Sibul, U., Wang, K.T. and Vallery, D., 1977, Groundwater Resources of the Duffins Creek-Rouge River Drainage Basins: Ontario Ministry of the Environment, Water Resources Report 8, 109 p.

Singer, S., 1973, Surficial geology along the north shore of Lake Ontario in the Bowmanville-Newcastle Area: 16th Conference on Great Lakes Research, Proceedings, p. 441-453.

Singer, S., 1974, A hydrological study along the north shore of Lake Ontario in the Bowmanville-Newcastle area: Ontario Ministry of the Environment, Water Resources Report 5d, 72 p.

Smart, P., 1994, A Water Balance Numerical Groundwater Flow Model Analysis of the Oak Ridges Aquifer Complex, south-central Ontario: unpublished M.Sc. Thesis, University of Toronto, Toronto, ON, 155 p.

Spencer, J.W., 1881, Discovery of the preglacial outlet of the basin of Lake Erie into that of Lake Ontario; with notes on the origin of our Great Lakes: American Philosophical Society, Proceedings, v. 19, p. 300-337.

Spencer, J.W., 1889, Upon the Origin of Alpine and Italian Lakes; With an Introduction and Notes Upon the American Lakes: Humboldt Publishing Co., New York, 148 p.

Taylor, F.B., 1913, The moraine system of southwestern Ontario: Canadian Institute, Transactions, v. 10, p. 57-68.

Terasmae, J. and Matthews, H.L., 1980, Late Wisconsin white spruce (Picea glauca (Moench) Voss) at Brampton, Ontario: Canadian Journal of Earth Sciences, v. 17, p. 1087-1095.

Traill, C.P., 1836, The Backwoods of Canada: Charles Knight, London, UK [republished 1989, McClelland and Stewart, Toronto, ON, 243 p.]

Turner, M.E., 1977, Oak Ridges Moraine, Major Aquifers in Ontario Series, Ontario Ministry of the Environment, Hydrogeological Map 782.

White, O.L., 1975, Quaternary Geology of the Bolton Area, Southern Ontario: Ontario Department of Mines, Geological Report 117, 119 p.

10. Ground-water Recharge to the Oak Ridges Moraine

Richard E. Gerber
Ken W.F. Howard

Environmental Earth Sciences, University of Toronto, Scarborough Campus
1265 Military Trail, Scarborough, Ontario M1C 1A4

SUMMARY

In order to properly regulate and protect the quality and quantity of ground-water resources, an adequate understanding of the geology and hydrogeology of the regional flow system is necessary. Critical to this understanding is an estimate of ground-water recharge to the water table aquifer, and a knowledge of the degree to which this water subsequently replenishes underlying aquifers. This paper reviews two methods used to estimate recharge (soil moisture balance and stream baseflow separation) to the Oak Ridges Moraine area of south-central Ontario. Both methods are in close agreement, within 13% of the average annual precipitation.

INTRODUCTION

Ground-water recharge refers to the range of processes by which aquifers receive replenishment. Quantification of recharge rates is necessary in order to evaluate and manage ground-water resources, particularly in regard to forecasting the consequences of aquifer utilization (Walton, 1970). A knowledge of recharge rates and mechanisms is also essential if detrimental effects of land use change (*e.g.*, urbanization) are to be minimized. The primary sources of recharge to aquifers may include precipitation (direct recharge), river losses, inter-aquifer flow, irrigation losses and leakage from urban utilities such as water pipes, sewers and storm-water collection ponds. Precipitation and inter-aquifer flow are usually the primary source of ground-water recharge in south-central Ontario.

In general, direct recharge is believed to be the most important recharge mechanism and occurs when a proportion of precipitation enters the soil profile and passes directly to the shallow water table. Indirect recharge is normally less significant, and occurs when precipitation runs off across the soil surface and enters the subsurface along stream channels or beneath shallow temporary ponds. Inter-aquifer flows are internal transfers of water that take place as a function of local ground-water head conditions and the hydraulic conductivity of aquifer and aquitard materials.

Considerable variation in direct recharge occurs both spatially and temporally, and is dependant on a number of factors. These include quantity and intensity of precipitation, permeability of soils, depth to water table, soil type and moisture conditions within the unsaturated zone, topography, vegetation type and density, temperature and other meteorological factors (Walton, 1970; Lerner *et al.*, 1990). As such, no single easy and accurate method can be used to directly measure the rate and amount of recharge entering the ground-water system (Hensel, 1992).

A number of recharge estimation methods exist; and some of the more common methods are summarized in Figure 1. These estimation methods are separated into point and regional measurements based on the scale (spatial) of the measurement technique. Each method also has inherent time scales of influence. Point methods are generally considered to have a high degree of accuracy, but normally include only the direct recharge component. Regional methods are prone to larger error but determine the total recharge from all possible sources. Point measurements generally cannot be used to provide regional rates, unless the basin is homogenous or point estimates are made for each combination of geological terrain and land use found within the regional study area. A detailed discussion of various recharge estimation methods can be found in Lerner *et al.* (1990) and Hensel (1992).

Estimates of recharge by different techniques can yield

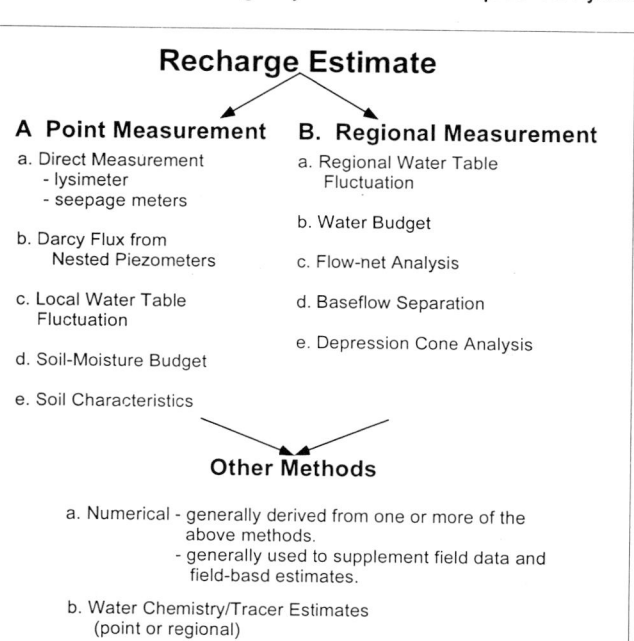

Figure 1 *Summary of common recharge estimation methods (compiled from Hensel, 1992; and Lerner et al., 1990).*

substantially different values. To attempt to reduce the resultant uncertainty, estimates should be compared to other independent recharge estimation methods whenever possible, and interpreted in light of the hydrogeological characteristics of the area being studied (Lerner *et al.*, 1990). Rushton and Ward (1979) and Johansson (1988) discuss difficulties of estimating recharge using conventional methods. A particular disadvantage of water balance methods is that recharge is often the residual between very large numbers (Lerner *et al.*, 1990).

In this study, we present ground-water recharge estimates for the central Oak Ridges Moraine area. A more detailed treatment is included in Gerber (1994). As discussed in Chapter 9, the Oak Ridges Moraine (ORM) of south-central Ontario is increasingly affected by urbanization that threatens the

quality and quantity of local ground-water resources. The study area includes parts of seven surficial drainage basins: Beaverton River, Uxbridge Brook, Pefferlaw Brook, Black River, Holland River, Rouge River and Duffins Creek (Fig. 2; see also Chapter 38). Included within this area are many municipalities, such as Stouffville, Aurora, Uxbridge and Newmarket, which obtain water supplies from ground water. A regional hydrogeological understanding is presently lacking in the study area as well as other areas of Ontario (Leech, 1994).

To date, estimates of recharge have been made on a drainage basin scale for the Rouge River and Duffins Creek drainage basins (Sibul *et al.*, 1977; Ostry, 1979) and the Holland and Black River drainage basins (Vallery *et al.*, 1982). Similar investigations have been conducted on Bowmanville,

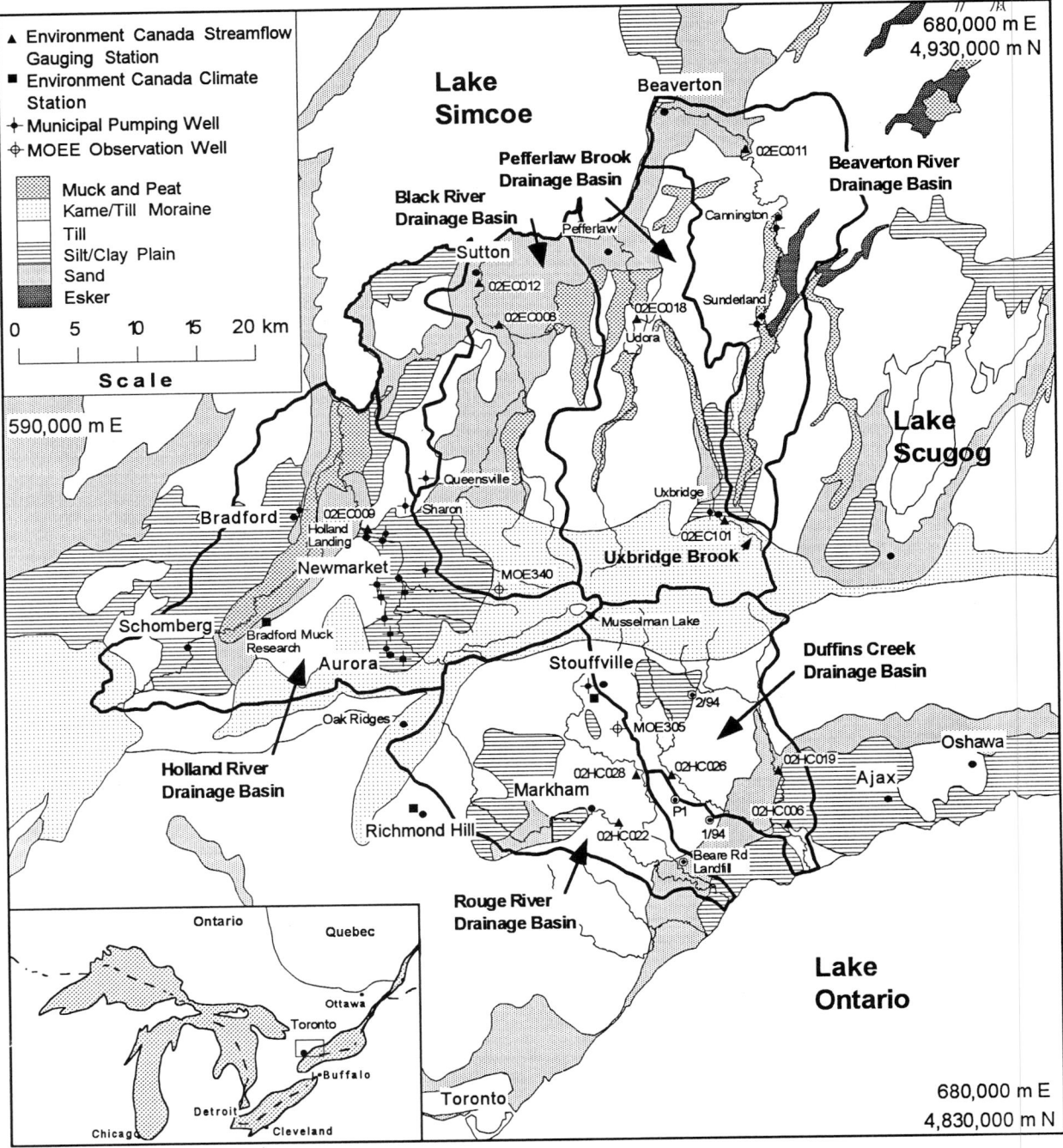

Figure 2   *Study area and surficial geology from Chapman and Putnam (1972) and Sibul* et al. *(1977).*

Wilmot and Soper creeks, (Singer, 1974; Funk, 1977; Singer, 1981), east of the study area. These studies utilized a range of methods to determine recharge rates.

The purpose of the present study has been to provide estimates of recharge on a regional scale using independent techniques. The approach combines soil moisture budget estimates (point estimates) with streamflow hydrograph separation (baseflow separation-regional estimates) using daily data (streamflow, precipitation, temperature).

Climate

The study area occurs within three climatic regions: the Simcoe and Kawartha lakes region north of the ORM, the southern slope of the ORM, and the Lake Ontario shore

(Brown *et al.*, 1980). Annual average mean daily temperatures range from 5.6° to 6.7°C in the Simcoe and Kawartha lakes region of the northern part of the study area to 6.7 to 7.8°C along the Lake Ontario shore. In the Simcoe and Kawartha lakes region, the mean daily temperature for January (coldest month) is from −8.9° to −7.8°C. The mean daily temperature for July (warmest month) is 20°C. For the Lake Ontario shore, mean daily temperatures for January and July are −6.7° to −4.4°C and 20° to 21.1°C, respectively (Brown *et al.*, 1980).

The mean annual precipitation for southern Ontario is 813 mm. Mean annual snowfall for the Great Lakes region is approximately 203 mm (water equivalent, 1931-1960; Phillips and McCulloch, 1972; Brown *et al.*, 1980). Precipitation over southern Ontario shows little seasonal variation (1931-1960)

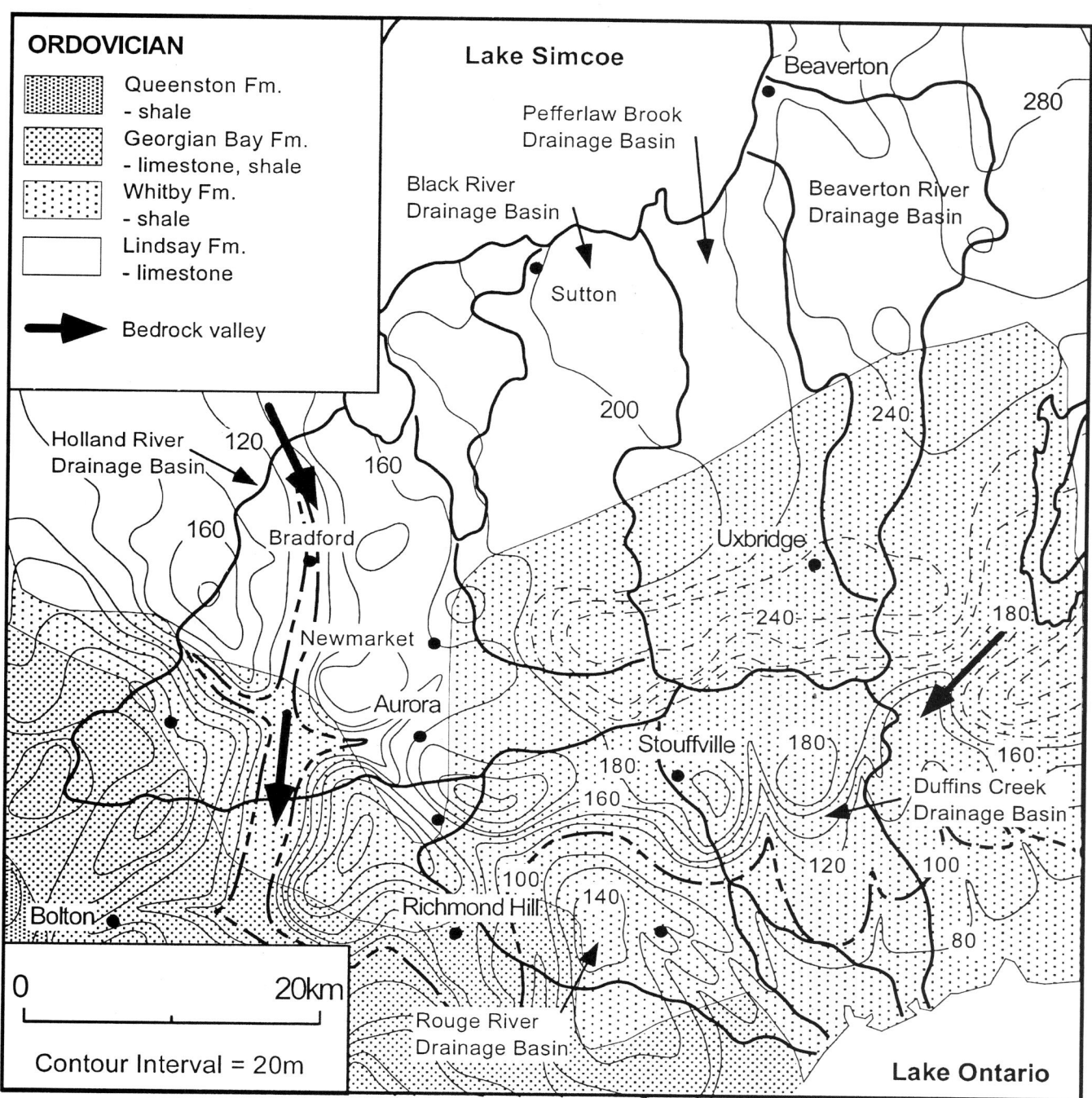

Figure 3 *Bedrock geology and topography (after Eyles* et al., *1993; and Hewitt, 1972).*

with mean precipitation in the growing season (May to September) ranging from 380 mm along the moraine to 356 mm along the Lake Ontario shore (Brown *et al.*, 1980).

The mean annual potential evapotranspiration (PET; calculated by the Thornthwaite method, see below) ranges from 584 mm along the moraine to 610 mm along the shore of Lake Ontario (Brown *et al.*, 1980; 1931-1960). This compares to estimates of 559 to 584 mm (1935-1964) by Phillips and McCulloch (1972) for the Great Lakes region, also using the Thornthwaite method. The mean annual actual evapotranspiration (AET) for the region including the study area is 533 to 559 mm·a^{-1}, reflecting seasonal periods of soil moisture limitations (Phillips and McCulloch, 1972; Brown *et al.*, 1980). During periods of soil moisture depletion, estimates of water deficiencies (PET-AET) range from 51 mm along the moraine to 76 mm along Lake Ontario. Estimates of water surplus during periods of soil moisture capacity range from 279 mm along the moraine highlands to 330 mm along the north and south flanks of the moraine. This water surplus represents that available as surface runoff and/or ground-water recharge (Brown *et al.*, 1980).

Bedrock Geology

The study area is underlain by sedimentary rocks of Middle and Upper Ordovician age which dip to the southwest (Fig. 2). The oldest rocks occur at the east end of the study area, and consist of Middle Ordovician limestone of the Lindsay Formation. Rock types become younger to the west, and consist of Upper Ordovician shales of the Whitby Formation, grey limestone with interbedded shale of the Georgian Bay Formation and red shale of the Queenston Formation (Hewitt, 1972; Liberty, 1969). Figure 3 shows Paleozoic geology for the study.

Bedrock valleys occur beneath the ORM (Fig. 3), having been cut by a drainage system that predates the last glacial advance. The major depression known as the Laurentian Channel provided drainage from Georgian Bay to Lake Ontario prior to development of the Great Lakes (Chapter 2).

Quaternary Geology

The surficial Quaternary geology/physiography for the study area is shown on Figure 2. Drift thickness ranges from less than 20 m along the Lake Simcoe shore to greater than 100 m in the Newmarket-Richmond Hill area and less than 25 m along the Lake Ontario shore (Haefeli, 1970). Briefly, the sand

and gravel sequence of the ORM is believed to be underlain by a late Wisconsin till (Gwyn, 1976a, b; Gwyn and Cowan, 1978; Turner, 1978) termed the Newmarket Till by Gwyn (1972) and Gwyn and DiLabio (1973) and the Northern till by Boyce *et al.* (1995). This till overlies fine-grained, predominantly deltaic and glaciolacustrine material (early to mid-Wisconsin Scarborough and Thorncliffe Formations) deposited in bedrock valleys beneath the central and western portions of the ORM. At some locations, the Northern/Newmarket till may be absent (Sharpe *et al.*, 1994). The older deltaic and glaciolacustrine deposits (Thorncliffe Formation) can be traced along the length of the Laurentian Channel from the Lake Ontario shoreline to the Barrie area to the north of the ORM (Fligg, 1983; Chapter 2).

Hydrogeology

Paleozoic shale bedrock is not a good aquifer from a quality and quantity perspective (Sibul *et al.*, 1977; Ostry, 1979; Vallery *et al.*, 1982), but adequate yields are obtained from the Middle Ordovician limestone (Lindsay Formation) in parts of the Pefferlaw Brook and Beaverton River watersheds (Haefeli, 1972).

Hydrogeologically, the overburden sequence is complex. Haefeli (1970) concluded that regional, extensive, well-defined aquifers do not exist in the study area, with aquifer units occurring as permeable lenses of variable thickness within otherwise very low permeability Quaternary deposits. Many overburden aquifers have been mapped from available water well records for part of the study area. Turner mapped the Alliston Aquifer Complex (1977) and the Oak Ridges Moraine Aquifer Complex (1978). Sibul *et al.* (1977) mapped fourteen different aquifers (including the Oak Ridges Aquifer Complex) within the Rouge River-Duffins Creek watersheds while Vallery *et al.* (1982) mapped seven aquifers (including the Alliston and Oak Ridges Moraine aquifer complexes) within the Holland River and Black River watersheds. Most of these aquifers can be classified as either shallow or deep overburden aquifers. Deep aquifers would be those occurring below the late Wisconsin till (Northern/Newmarket) within the Thorncliffe and Scarborough Formations. The deep aquifers provide municipal water supplies for Bradford, Newmarket, Aurora, Oak Ridges, Stouffville and Uxbridge.

Shallow aquifer systems are considered to occur above the Northern/Newmarket till within the Oak Ridges Moraine and associated glaciolacustrine deposits (see Chapter 9).

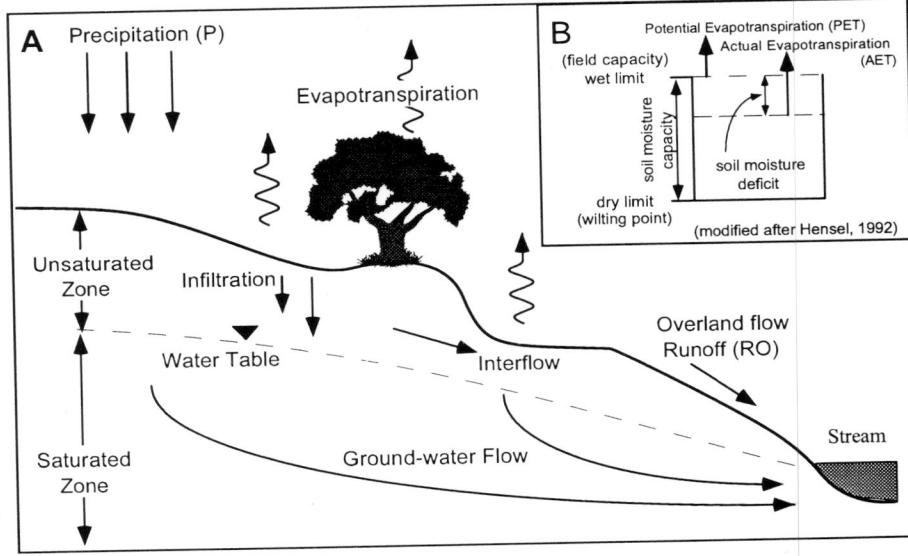

Figure 4 (**A**) *Basin hydrologic cycle.*
(**B**) *Summary of soil moisture terminology for processes within the unsaturated zone.*

The major shallow aquifer in the study area is the Oak Ridges Aquifer Complex, which is generally unconfined; however, local occurrences of less permeable units such as silt, clay and till result in localized perched and confined ground-water conditions. Ground-water flow within the Oak Ridges Aquifer Complex is generally downward and away from the divide along the crest of the moraine. This water discharges to head-water streams along the moraine edge and also into lakes (*e.g.*, Musselman Lake) along the moraine crest (Kaye, 1986; Intera Kenting Ltd., 1990). The complex recharges the deeper aquifers beneath and along the flanks of the moraine.

METHODS FOR CALCULATING RECHARGE

Recharge can normally be calculated in terms of the water balance components constituting the basin hydrological cycle shown in Figure 4. Precipitation reaching the land surface is distributed in numerous ways. When the surface is impermeable, precipitation runs off directly toward surface depressions and streams or evaporates back into the atmosphere. When precipitation falls on permeable soils, however, the runoff component is very small, and most of the precipitation enters the soil profile where it becomes subjected to free water evaporation, or if vegetation is present, transpiration. The combination of these two processes is termed evapotranspiration. The potential evapotranspiration (PET) is the amount of water that would evaporate and transpire if water was available to the plants and soils in unlimited supply. Since this is not always the case, the term actual evapotranspiration (AET) is used such that AET ≤ PET.

Water which remains after evapotranspiration has the potential to increase the soil moisture content of the soil. In theory, the soil moisture content cannot exceed its maximum, or field capacity, also known as wet limit, and any excess will drain from the soil as infiltration. Infiltrating water will eventually recharge the water table unless field drains or highly permeable layers in the unsaturated zone intercept the water and direct it horizontally as interflow to nearby streams. Figure 5 shows the typical variation of each component of the water balance, over a year, for the study area. Most of the ground-water recharge occurs during the spring, with smaller events during the fall when evapotranspiration rates are low.

Mathematically, the water balance can be expressed as follows:

$$P = RO + AET + I + B + A \pm \Delta s \pm \Delta g \qquad (1)$$

where P = precipitation, RO = surface runoff,
AET = actual evapotranspiration, I = interflow, B = baseflow, A = supply/abstraction, Δs = change in soil moisture storage, and Δg = change in ground-water storage. In turn,

$$\text{Stream Flow Discharge (SFD)} = I + B + RO \qquad (2)$$
$$\text{Infiltration (Inf)} = P - AET - RO - \Delta s \qquad (3)$$
and
$$\text{Aquifer Recharge (R)} = P - AET - RO - \Delta s - I \qquad (4)$$

Soil moisture storage may vary considerably on a daily basis, but the net change (Δs) over an annual cycle will be negligible compared to other water balance components. Similarly, ground-water storage may fluctuate on a monthly or annual basis, but Δg will approach zero (steady state) over an extended period of time provided other water balance components remain essentially constant. If Δs and Δg equal zero,

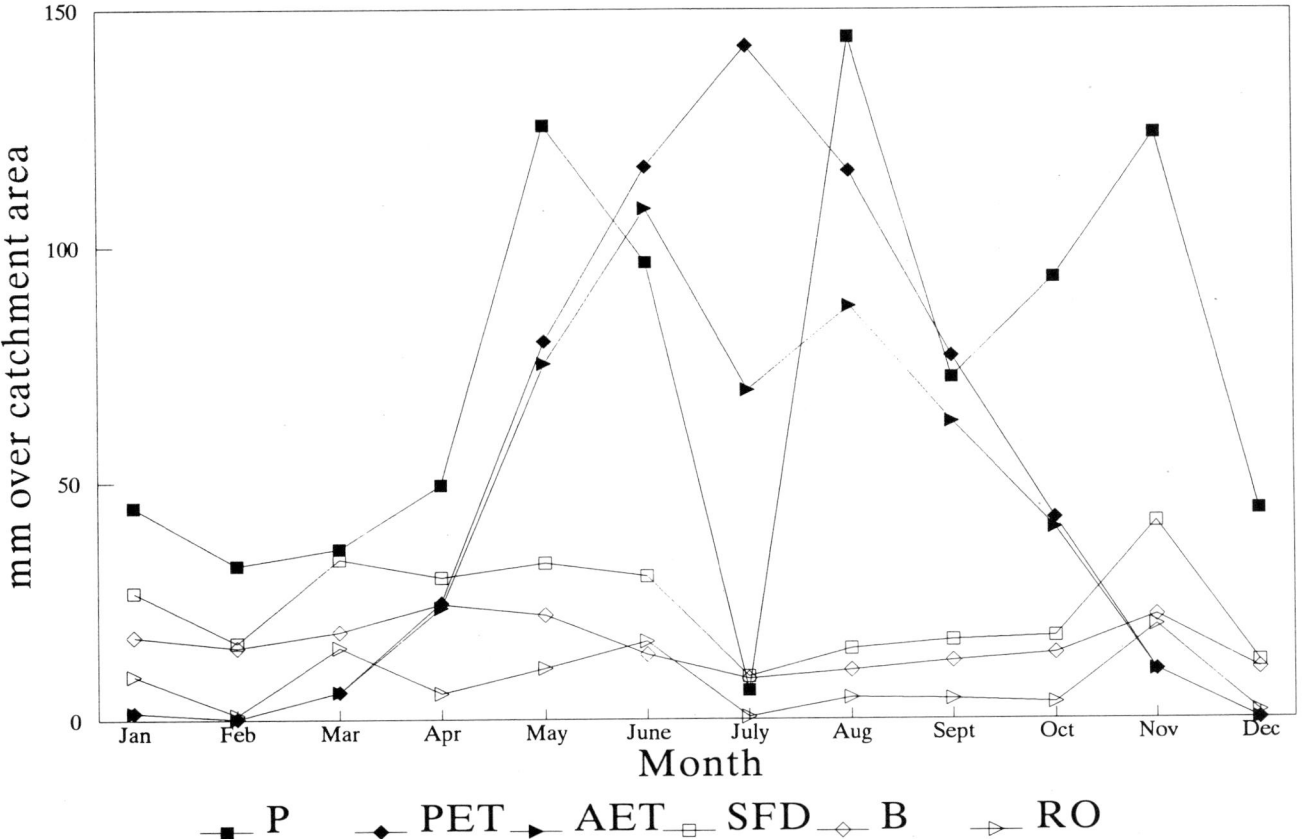

Figure 5 *Monthly variation of water balance components over a one-year period (1989). Climate data for Stouffville WPCP climate station. Streamflow data from Duffins Creek at Pickering gauging station (02HC006). PET and AET from daily calculations, B and RO from FIM (see text for methodology and explanation of terms).*

then substitution of Equation 4 into Equation 1 reveals that

$$\text{Aquifer Recharge (R)} = B + A \tag{5}$$

Similarly, substitution of Equation 2 into Equation 1 shows that

$$\text{Stream Flow Discharge (SFD)} = P - AET - A \tag{6}$$

If ground-water pumping is small, (*i.e.*, A→0), then annual recharge can be equated to baseflow,

$$R = B \tag{7}$$

and stream flow discharge will be equal to the difference between precipitation and actual evapotranspiration, *i.e.*,

$$SFD = P - AET \tag{8}$$

In the present study, recharge has been calculated using two independent approaches: (1) a baseflow separation method using Equation 7, and (2) a soil moisture balance technique based on Equation 4.

Baseflow Separation

Baseflow separation is a technique by which the streamflow hydrograph is separated into its primary components, base-flow and surface runoff. In permeable basins in temperate regions, ground-water discharge (baseflow) may account for 70 to 80% of the stream's annual discharge (United States Environmental Protection Agency, 1990; US EPA). In general, baseflow contributions to streamflow vary gradually in re-sponse to long-term changes in the ground-water flow sys-tem. In contrast, runoff is provided by storm and snow melt events, and these lead to sharp short peaks on the otherwise smooth streamflow hydrograph (Freeze and Cherry, 1979). Removing the run-off component will yield the quantity of ground-water discharge over time. Assuming there is no net change in ground-water storage over the long term (several years), then the amount of ground-water discharge to streams will equate to aquifer recharge.

In some cases, sewage treatment plant (STP) discharges provide considerable contributions to streamflow and these lead to an overestimate of recharge when they are mistakenly included as baseflow. Also, ground-water flow between sur-face drainage basins can lead to an overestimate of recharge. More often, however, baseflow separation leads to an under-estimate of recharge because (1) many gauging stations fail to intercept all the available ground water, the remaining underflow discharging to the river further downstream; (2) some recharge may leave the local catchment by leaking to deeper, semi-confined aquifers; and (3) abstractions from above the gauging station may be discharged at STPs down-stream from the gauging station.

Streamflow Hydrograph Separation Methods

With any streamflow hydrograph separation technique, the ac-curacy of the separation is always questionable, and different methods can yield different results for the same hydrograph (Singer, 1981; US EPA, 1990). As a general rule, separation techniques will lead to a range of baseflow values, which, at the very least, will allow comparisons between basins to be made.

The baseflow component of streams represents with-drawal of ground water from aquifer storage, and is termed a ground-water recession, which normally follows an exponen-tial decay law. The recession is determined by separating a stream hydrograph into its component parts: runoff (overland flow), interflow and baseflow (ground-water flow, see Fig. 4). Many field studies conclude that interflow is rare in space and

time, especially in humid and temperate vegetated basins. It has also been noted by Freeze (1972) that interflow is only significant on convex hillslopes that feed deeply incised chan-nels with very highly permeable soils. For this study, it was considered practical to separate the streamflow hydrograph into two components, runoff and baseflow. Any interflow which does occur would be included in the runoff or baseflow estimate, depending on the model for unsaturated zone flow.

In the upper reaches of a watershed, surface runoff contri-butions to streamflow aid in the build-up of flood waves. In the

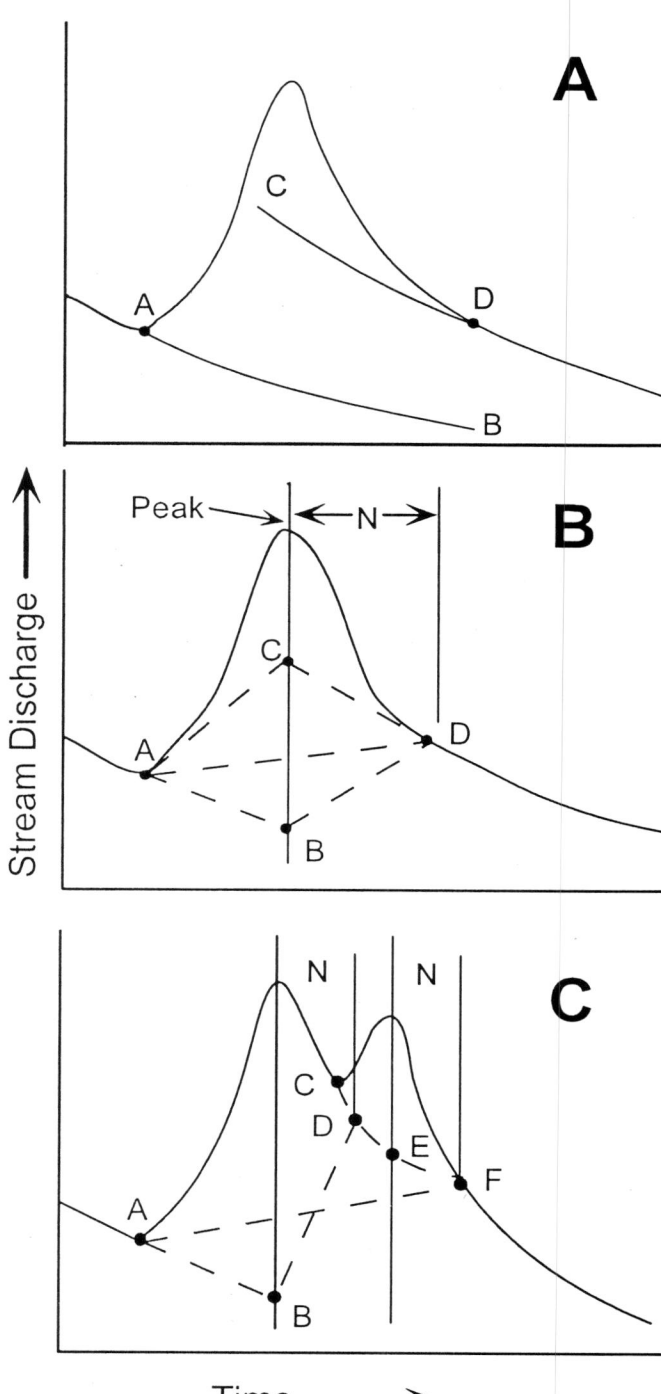

Figure 6 (**A, B**) *Basic streamflow hydrograph separation showing envelope of baseflow-runoff separation.* (**C**) *Multiple peak storm events separation (after Pettyjohn and Henning, 1979).*

lower reaches, bank storage often moderates the flood wave. Bank storage is the temporary storage of water in surficial deposits bordering a stream channel. Flood waves may induce flow into stream banks and, as the flood stage declines, the flow is reversed out of the stream banks (Freeze and Cherry, 1979). Bank storage is generally not considered in regional ground-water recharge analyses, because the water only remains in the ground (stream banks) for a short duration, on the order of hours or days (Pettyjohn and Henning, 1979). This assumption is accepted for this analysis.

The basic procedure for hydrograph separation used in this study is shown on Figure 6 (Pettyjohn and Henning,

1979). The onset of surface runoff is represented by point A on Figure 6A. The original recession can be extended to point B. The area below AB represents the baseflow that would have occurred if there had been no surface runoff or precipitation event. Point D represents the end of the surface runoff event. This point may be difficult to determine visually so the location of D may be determined by the empirical equation

$$N = A^{0.2} \qquad (9)$$

where N = number of days after peak when surface runoff ceases, and A = basin area in square miles. From this point onward, the total flow can be considered to be the baseflow component derived from ground-water discharge to the stream until the next storm event. The recession starting at time N can be described by the exponential decay law

$$Q = Q_o e^{-kt} \quad \textbf{\textit{or}} \quad k = (1/t)\ln(Q/Q_o) \qquad (10)$$

where Q_o = discharge at time N, Q = discharge at any time, k = recession constant, t = time since recession began. In general, permeable regions have flat recessions (k approximately 1) and less permeable regions have steep recessions (k approaching 0). Hall (1968) provides a historical summary of baseflow recession analysis including a list of equations used in streamflow hydrograph separation. The depletion curve (Figure 6A) can be extended back to C matching the recession limb from D. ABCD represents an envelope of the limits within which a line may be drawn to separate the streamflow hydrograph into runoff and baseflow. Three methods based on this technique and used for this study are described below.

Method 1 involves extending a recession curve back from D to a point C on a line drawn vertically through the peak of the hydrograph (see Fig. 6B). Point A, the start of surface-water runoff, is attached to point C. The area under ACD represents baseflow, while the area above represents surface-water runoff. This method tends to be more valid where baseflow is relatively large and reaches the stream relatively quickly. Such a stream would have a deeply incised valley where the stream channel is below the water table and bank storage is minimal. Using isotopic analysis of stream waters (^{18}O, ^{2}H and ^{3}H), Sklash and Farvolden (1979, 1982) found that ground-water discharge dominates most run-off events except for the most intense rain storms or melting days. This phenomenon was attributed to a rapid rise in hydraulic head in discharge areas where the capillary fringe was instantaneously converted into phreatic water. During peak runoff periods, the ground-water component of streamflow can be as high as half to two-thirds (Freeze and Cherry, 1979).

Method 2 simply involves extending a line from the start of surface runoff (point A) to the end of surface runoff (point D). The third method involves joining points ABD with the baseflow component representing the area under the line. This method would be more applicable to streams with a significant component of bank storage (see above).

Runoff events commonly occur at closely spaced intervals creating complex hydrographs where events are superimposed (Figure 6C). An example of a separation technique for complex hydrographs involves extending the recession curve preceding the first runoff event from A to B, with point B occurring on a vertical line extending from the peak of the first runoff event. The recession limb from the first runoff event is extended N days from the first peak, and the line BD is formed. The first recession trend is further continued through points C and D to point E on a vertical line extended through the peak of the second runoff event. From point D, the line is extended to E and then by N days to point F, the end of the surface-water

A

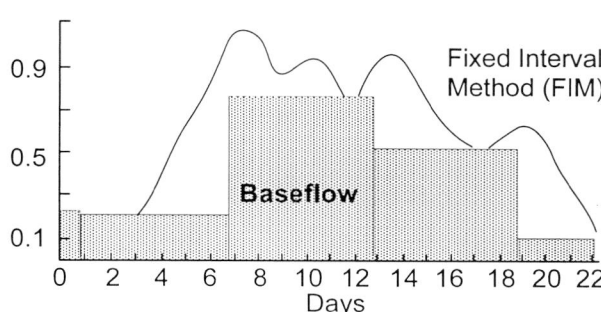

B

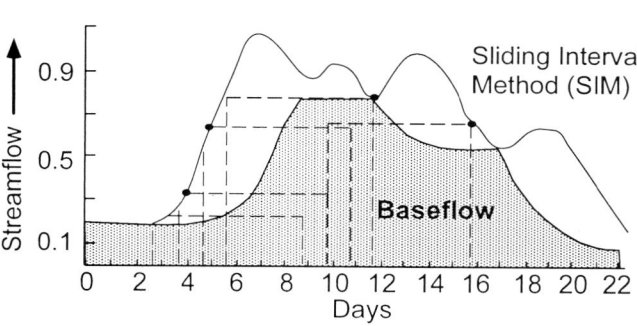

C

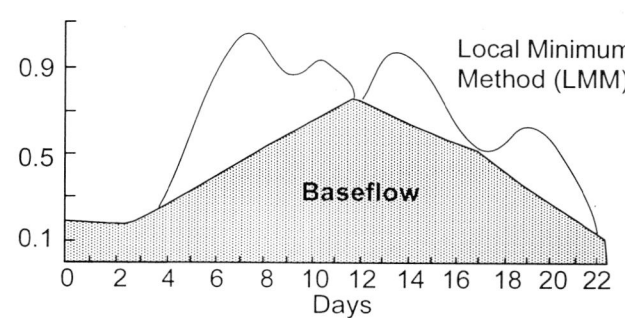

Figure 7 *Streamflow hydrograph separation subroutines used for this study:* (**A**) *fixed interval method;* (**B**) *sliding interval method; and* (**C**) *local minimum method (after Pettyjohn and Henning, 1979).* **N** *is equal to 3.*

Table 1   Study area baseflow estimate comparison.

Drainage Sub-basin	Gauging Station	Time Period	Total Streamflow Range (mm a⁻¹)	Total Streamflow Avg. (mm a⁻¹)	Baseflow FIM Range (%)	Baseflow FIM Avg. (mm a⁻¹)	(%)	Baseflow LMM Range (%)	Baseflow LMM Avg. (mm a⁻¹)	(%)	Q60 (mm a⁻¹)	(%)	Other (mm a⁻¹)	(%)	Source
Beaverton R.	02EC011	1980-89	214-448	323	68-77	237	73	47-72	207	64	126	39			
Uxbridge Bk.	02EC101	1980, 1982-84	460-527	499	84-91	441	88	83-91	435	87	473	95			
Pefferlaw Bk.	02EC018†	1980-86, 1988,89	242-380	323	67-83	245	76	56-79	215	67	205	63			
Black R.	02EC012/8	1980-89, 1965-68, 1970-76	176-351	267 222	64-81	191	72	47-74	160	60	132 86	50 39			Vallery *et al.* (1982)
Holland R.	02EC009	1980-89, 1966-76	155-340	241 223	55-68	147	61	47-64	137	57	117 103	48 46			Vallery *et al.* (1982) (unadjusted)
Rouge R.	02HC022	1980-89, 1962-70, 1961-67	160-438	289 227 159	47-60	153	53	42-54	138	48	110 62	38 27	72	46	Sibul *et al.* (1977) Haefeli (1972)
L. Rouge Ck.	02HC028	1980-89, 1964-70	245-517	357 262	55-71	222	62	48-58	188	53	126 85	35 32			Sibul *et al.* (1977)
W. Duffins Ck.	02HC026	1980-87, 1964-70	321-512	408 202	55-72	258	63	53-71	246	60	205 74	50 37			Sibul *et al.* (1977)
E. Duffins Ck.	02HC019	1980-89, 1962-70	346-570	461 385	60-71	303	66	57-70	295	64	268 233	58 61			Sibul *et al.* (1977)
Duffins Ck.	02HC006	1982-89, 1946-70, 1961-67	264-539	397 300 224	57-68	255	64	57-68	249	63	211 150	53 50	141	63	Sibul *et al.* (1977) Haefeli (1972)

Notes: % = percent of average annual streamflow
mm a⁻¹ : millimetres per annum (year) values represent estimates applied to entire drainage area, *i.e.*, including discharge areas
Q60 = daily streamflow value equalled or exceeded 60% of the time
SIM estimates are generally within 2% of FIM estimates and are not shown
† Station number changed from 02EC103 to 02EC018 during 1987

runoff events. Another method would be to extend a straight line from point A to point F.

Baseflow separation subroutines. To perform the base-flow separation, three subroutines (FIM, SIM and LMM) were used (Pettyjohn and Henning, 1979). These are illustrated in Figure 7. The Fixed Interval Method (FIM) separates the streamflow hydrograph by moving a bar of 2N width upward from a base line until a part of the bar intersects the hydrograph at one point. The bar is moved over 365/2N intervals. All elements of the line are then assigned that minimum discharge value (Fig. 7A). The Sliding Interval Method (SIM) involves sliding a bar 2N wide upward from the baseline to a point where it intersects the hydrograph. The discharge value of the hydrograph is assigned to the median element of the interval. The bar then slides over to the next day. The values of the elements on both sides of the point being considered for one-half of the interval minus one are scanned, and the element with lowest value is assigned as the ground-water

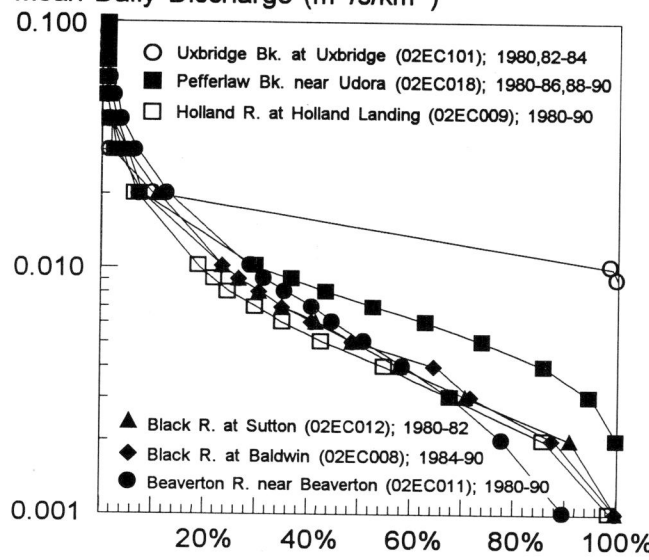

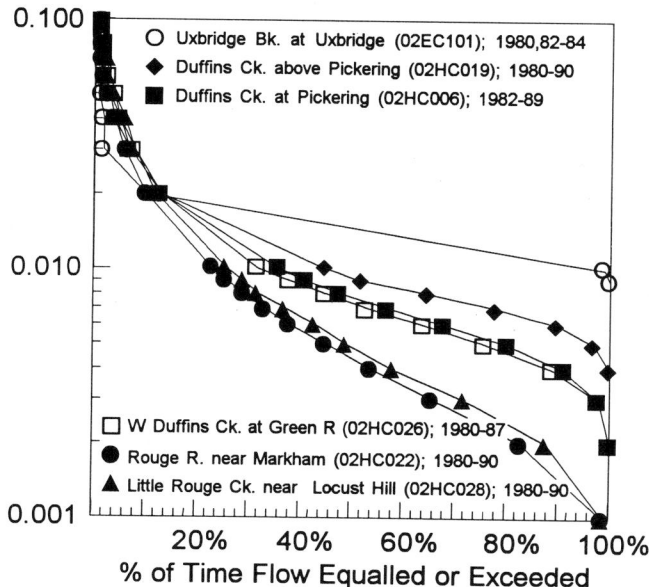

Figure 8   *Flow duration curves.*

component of the point being considered (Fig. 7B). The Local Minimum Method (LMM) scans the discharge array element by element to determine if it is the lowest element in the interval (1/2 the interval, 2N, minus one on either side of the element). The lowest element in the interval is a local minimum and is connected by lines to adjacent local minima. This method produces the most conservative (lowest) estimates of baseflow (Fig. 7C).

The three subroutines were applied to streamflow hydrographs prepared from daily mean-discharge data provided in digital format by the Water Survey of Canada and summarized in Water Survey of Canada (1992). Although it is acknowledged that considerable error can occur in stream-flow records, particularly with regard to the accuracy of winter streamflow data (Rosenberg and Pentland, 1983), no editing or correction of the data was undertaken for this study. Environment Canada streamflow gauging station locations used in this study are shown on Figure 2. Results are summarized in Table 1, which also includes Q_{60} baseflow estimates. Q_{60} estimates are according to the method of Bloyd (1975), where ground-water discharge or baseflow is assumed to equal the daily streamflow value equalled or exceeded 60% of the time. Q_{60} estimates for this study are from flow duration curves shown on Figure 8 and are provided for comparison to estimates of Sibul *et al.* (1977) for the Rouge River-Duffins Creek basin and Vallery *et al.* (1982) for the Holland River-Black River basin.

Once the baseflow component or quantity has been extracted from the streamflow hydrograph, this quantity is applied to the drainage area represented by the gauging station to give a regional estimate of the catchment recharge rate. Note that these estimates do not include underflow or take into account possible effects of STP discharges and flow regulation devices which are known to affect streamflow. Also, abstractions from the system, such as municipal water wells, will affect the calculated recharge rate, making STP discharge type and location important. It should finally be recognized that recharge estimates apply to the drainage area represented by the gauging station as a whole, but will vary from point to point within the drainage area as a result of the changing nature of surficial sediments. For example, areas of the ORM (sand and gravel) are expected to have higher recharge rates than areas of till and glaciolacustrine deposits.

Soil Moisture Budget

The soil moisture budget technique provides a means of calculating actual evapotranspiration as a function of potential evapotranspiration and local soil moisture conditions. The difference, or residual, between incoming precipitation and actual evapotranspiration will represent the amount of water contributing to recharge (R) and runoff (RO). The residual can be compared to estimates of R and RO from baseflow separation discussed above to provide estimates of underflow and inter-aquifer ground-water flow between neighbouring catchments.

The soil moisture balance approach is described by Howard and Lloyd (1979), who discuss the time period for the balance and demonstrate the sensitivity of the estimates to errors in Penman evaporation input data. The authors recommend using daily recharge balances whenever possible, followed by ten-daily, with monthly recharge balances being the least preferred. In the present study, actual evapotranspiration (AET) and ground-water recharge were calculated on a monthly and daily basis. Monthly balances remove or smooth

Table 2   Monthly and daily water balance summary.

Drainage Sub-basin	Gauging Station	Time Period	Area (km²)	P	PET Monthly	(%)	PET Daily	(%)	AET Monthly	(%)	AET Daily	(%)	R Monthly	(%)	R Daily	(%)	Streamflow Total	(%)	Baseflow	(%)	Runoff	(%)	Daily R – Baseflow	(%)	Municipal Ground-water Pumping	(%)	
Beaverton R.	02EC011	1982-89	282	908	593	65	619	68	538	59	493	54	271	30	315	35	323	36	224	25	99	11	91	10	1.4	0	Cannington, Sutherland
Uxbridge Bk.	02EC101	1982-84	24.3	926	587	63	613	66	520	56	471	51	353	38	403	43	490	53	438	47	52	6	-35	-4	20.3	2	
Pefferlaw Bk.	02EC018†	1982-89*	332	923	590	64	616	67	533	58	485	53	304	33	353	38	317	34	231	25	86	9	122	13		0	
Black R.	02EC012/8	1980-89	324/274	803	582	72	610	76	532	66	495	62	179	22	216	27	267	33	176	22	91	11	40	5	0.5	0	Mt. Albert
E. Holland R.	02EC009	1980-89	181	803	582	72	610	76	528	66	492	61	176	22	212	26	241	30	142	18	99	12	70	9	49.0	6	Aurora Newmarket, Sharon, Holland Landing, Queensville (38 mm·a⁻¹, 1980 to 65 mm·a⁻¹, 1989)
Rouge R.	02HC022	1980-89	186	908	603	66	629	69	548	60	511	56	217	24	254	28	289	32	146	16	143	16	108	12		0	
L. Rouge Ck.	02HC028	1982-89	77.7	908	593	65	619	68	542	60	498	55	211	23	253	28	371	41	216	24	155	17	37	4	16.0	2	Stouffville
W.Duffins Ck.	02HC026	1982-87	98.1	932	591	63	619	66	547	59	507	54	231	25	270	29	422	45	267	29	155	17	3	0		0	
E.Duffins Ck.	02HC019	1982-89	93.5	908	593	65	619	68	538	59	492	54	209	23	254	28	464	51	303	33	161	18	-49	-5		0	
Duffins Ck.	02HC006	1982-89	249	908	593	65	619	68	540	59	496	55	224	25	267	29	397	44	252	28	145	16	15	2		0	
Average				893	591	66%	617	69%	537	60%	494	55%	238	27%	280	31%	358	40%	240	27%	119	13%	40	4%	8.7	1%	

Notes:

P = Precipitation
PET = Potential Evapotranspiration
AET = Actual Evapotranspiration
Runoff = Storm and Snow Melt Runoff
R = Recharge is equal to P – AET – RO
All values represent annual average for period analyzed in mm·a⁻¹ over the catchment area. % represents percentage of precipitation
* Not including 1987
Baseflow = Average of FIM and LMM method estimates for period analyzed
Municipal Groundwater Pumping data provided by the regions of York and Durham
† Station number changed from 02EC103 to 02EC018 during 1987

the effects of recharge from precipitation storm events, particularly when soil moisture is limited during the summer period; they also tend to remove the effects of rainfall intensity which controls effective infiltration (Dunne *et al.*, 1991). Consequently, they were used primarily as an initial indicator of recharge rate. Daily balances are expected to more accurately define soil moisture surplus, particularly during summer storm events. It should be noted, however, that all estimates are limited by the accuracy of the available input data, and all estimates need to be validated by other methods whenever possible.

Potential Evapotranspiration

Evapotranspiration is generally considered to be the most difficult hydrological process to quantify, given the complexity of the mechanisms involved and the variability of transpiration rates for different soil and vegetative covers. Methods to calculate PET include those of Blaney and Criddle, and Thornthwaite, which are based on empirical correlations between evapotranspiration and climate; and Penman, and VanBavel which are energy budget approaches considered to have better physical foundations, but require more meteorological data (Lerner *et al.*, 1990).

For the purposes of the present study, PET was calculated according to the method of Thornthwaite (Thornthwaite and Mather, 1957; Dunne and Leopold, 1978). This method is an empirical formula which uses air temperature as an index of the energy available for potential evapotranspiration. The Thornthwaite formula was chosen because of the availability of daily temperature data (for the study period 1980-1989) and the lack of coverage of more detailed climatological data necessary for the Penman equation. The Thornthwaite equation is as follows:

$$PET = 1.6[10T_a/I]^a \qquad (11)$$

where PET = potential evapotranspiration (cm per month),
T_a = mean monthly air temperature (°C),
I = annual heat index = $\sum[T_{ai}/5]^{1.5}$, where i = 1 to 12 for each month, and
$a = 0.49 + 0.0179I - 0.0000771I^2 + 0.000000675I^3$.
The value of PET calculated according to Equation 11 is for a standard month of 360 hours of daylight, and must be adjusted appropriately as the number of sunshine hours vary according to latitude. PET was calculated for stations at Bradford, Richmond Hill and Stouffville, and corrected for a latitude of 44°N (Thornthwaite and Mather, 1957). The corrected values of PET calculated on a monthly and daily basis are included in Table 2.

Soil Moisture and Actual Evapotranspiration

When available water for potential evapotranspiration is limited (P<PET), actual evapotranspiration may occur at a reduced rate, depending largely on the ability of vegetation to draw water out of the soil column. Soil moisture can take one of three forms (Singer, 1981): (1) hygroscopic water absorbed onto soil particles and held with considerable force, (2) water held by surface tension in capillary spaces which moves under capillary forces and is available to plants, or (3) gravitational water.

The amount of moisture which can be retained in a soil and made available to plants (soil moisture capacity) falls between two limits: an upper limit termed the wet limit or field capacity, and a lower limit termed the dry limit or wilting point. The wet limit (field capacity) is the amount of moisture remaining after the gravitational water has been allowed to drain away (close to percentage of volumetric moisture content at 0.33 bar). The dry limit (wilting point) is the percentage of volumetric moisture content at 15 bar where plants wilt and fail in a dark humid atmosphere. The soil moisture capacity (SMC) is the amount of water in the soil between the wet and dry limits or the amount of water available for evapotranspiration (Fig. 4B). Soils in the study area are at the wet limit in winter, during the spring snow melt and in late fall. It is during this period that much of the ground-water recharge occurs. Soils approach the dry limit during the summer months, where much of the precipitation is returned to the atmosphere as evapotranspiration or used to satisfy soil moisture deficits.

Soils maps of York County and Ontario County (now York and Durham regions) are available from the Ontario Soil Survey (Hoffman and Richards, 1955; Olding *et al.*, 1956). The soils are grouped into series based on similar profiles and parent materials. For this study, soil moisture capacities were designated according to parent materials (surficial geological map units). The surficial geology coverage for each gauging station drainage area was determined from the physiographic mapping of Chapman and Putnam (1972), with modifications to Peel Pond deposit coverage in the Rouge River and Duffins Creek basins from Sibul *et al.* (1977). This mapping was used because of the lack of consistent surficial geological mapping for the entire study area at the time of the study.

Soil moisture capacities (SMC) were assigned to areas of surficial geology according to Table 3. These soil moisture capacities are comparable to the measured values used by Singer (1981) for a ground-water resource study of the Bowmanville, Wilmot and Soper Creeks (east of the study area). Estimates of AET for this study were based on soil type only, with soil types simplified to represent parent overburden

Table 3	Soil moisture capacities used for this study.		
Surficial Geologic/Physiographic Unit	**Soil Type**	**Soil Moisture Capacity** (mm)	
		This Study	**Singer (1981)**
Oak Ridges Moraine	sand and sandy loam	100	
Newmarket Till	sandy loam	125	107
Halton Till	loam and silty loam	150	157
Lake Iroquois Glaciolacustrine	silty clay, silty sand	200	
	clay loam and clay silty loam		183
	clay and silty clay		198

Note: Soil moisture capacities used by Singer (1981) for study of Bowmanville, Soper and Wilmot creeks to the east of the Study Area.

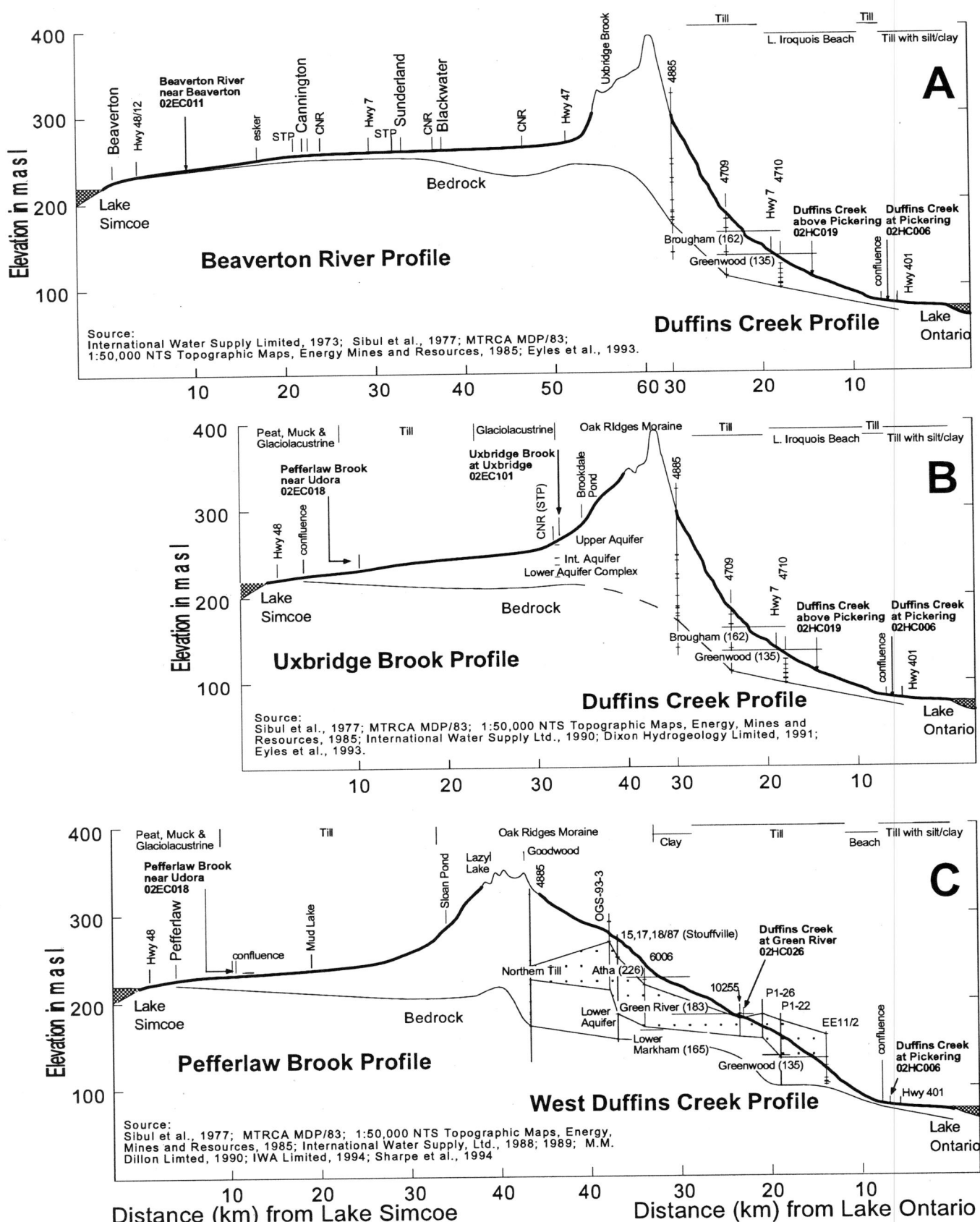

Figure 9 *River profiles of Beaverton River, Uxbridge Brook, Pefferlaw Brook, Duffins Creek and West Duffins Creek. Numbers in parentheses represent elevation (in metres above sea level; m asl) of top of aquifer complex.*

materials. Land use or vegetative cover were not directly included in the calculations.

There are many proposed methods which attempt to relate PET and AET. Some authors (Veihmeyer, 1964 *in* Dunne and Leopold, 1978) suggest that evapotranspiration will proceed at the potential rate until the permanent wilting point is reached. Others (Thornthwaite and Mather, 1957) suggest that the ratio of AET to PET decreases as a linear function of the amount of available water. The AET rate is expected to fall within these two extremes, depending on soil type. For example, the Penman (1949) root constant concept is considered a compromise between the above two models (Alley, 1984). For the present study, the Thornthwaite relationship was utilized; this is expected to represent a minimum estimate of AET as described above. Thornthwaite and Mather (1957) provide values for the water retained in different textured soils with different vegetative covers which have been subjected to an accumulated potential water loss (APWL) depending on the initial SMC of the soil. The APWL represents the sum of consecutive negative values of P–PET representing potential moisture loss calculated on a daily or monthly basis. The AET estimate will then be derived by adding the change in soil moisture (water evapotranspired from soil moisture storage) to the precipitation value. For months or days with positive P–PET values, evapotranspiration occurs at the potential rate and the remaining precipitation (P) is added to the soil moisture storage to replenish any existing deficit.

Storm and Snow Melt Runoff

There remains some debate as to whether storm and snow melt runoff (RO) should be removed from P before or after calculating AET. Dunne and Leopold (1978) and Ward (1976) recognize that storm runoff can occur during periods of soil moisture deficit (dry, summer months), and argue that RO should be removed before the soil is considered, *i.e.*, the effective moisture available for soil moisture balance and evapotranspiration demands would be equal to P–RO. As part of the present study, RO was determined from stream hydrograph separation and subtracted from P prior to calculating AET on a monthly basis for six of the gauging station drainage areas (Beaverton River, Uxbridge Brook, Pefferlaw Brook, Black River, Holland River and the Rouge River). The procedure resulted in a maximum reduction in calculated AET (and corresponding increase in R) of 11 mm. Subtracting RO from P before AET on a daily basis is a little more complex, because a single storm or melt event can lead to RO, on that day and subsequent days, which can overlap with subsequent storm events. It is suggested that the procedure is likely to have a negligible effect for the study area because when RO is low (summer), AET is high, and when RO is high (spring and fall), AET is low. For this study, daily calculations of AET assume that the effective available moisture for evapotranspiration is represented by P, not P–RO. For the purposes of calculating recharge, RO is subtracted after AET requirements are met. The water content retained in soil was calculated through the time step for corresponding APWL values for different soils. Recharge was not allowed to occur until soil moisture deficits had been satisfied.

It should also be noted that the estimates of recharge do not consider snow melt. The actual storm and snow melt runoff values are considered to be included within the RO values obtained from streamflow data. As a result, some of the recharge listed for January and February may not be realized until March and April during the peak snow melt period. The results of estimated PET and AET rates for the study area are included on Table 2 and generally fall within the range of published yearly values.

Municipal Ground-water Pumping and Sewage Treatment Plant Discharge

Municipal water supplies from ground water are obtained mainly within the central portion of the study area; municipalities at the northern and southern ends of the study area obtain drinking water from Lake Simcoe and Lake Ontario, respectively. Sites of municipal pumping wells are shown on Figure 2. Quantities pumped for each location were obtained from records on file at the regions of York (west study area) and Durham (east study area).

Municipal ground-water withdrawals are important to the water balance because they represent a quantity of recharge which, if not returned to the system, can have a significant effect on aquifer storage and/or baseflow to a stream. Ground-water withdrawals in the study area are generally discharged through sewage treatment plants (STP) to streams, lagoons or large lakes. Private residence ground-water withdrawal and sewage discharge to septic tanks were not incorporated into this water balance, as it was assumed that the net withdrawal is negligible.

RESULTS AND DISCUSSION

Table 1 provides a summary of baseflow estimates from this study, and also compares to estimates from other studies. Table 2 summarizes the water balance results for each of the drainage areas shown on Figure 2. As discussed below, the water balance characteristics of each catchment are strongly influenced by local geology and physiography.

Soil Moisture Balance *versus* Baseflow Estimates

Comparing the estimates of recharge obtained by the two methods, soil moisture balance and baseflow separation, may allow for the degree of inter-catchment aquifer flow and underflow at the gauging station to be estimated. Table 2 includes the differences between the two recharge estimation methods.

Soil Moisture Balance >> Baseflow

Four catchment areas (Beaverton River near Beaverton, Pefferlaw Brook near Udora, Holland River at Holland Landing and Rouge River near Markham) have daily soil moisture balance recharge estimates greater than average baseflow estimates by at least 9% of average total annual precipitation. This is interpreted as significant underflow at the streamflow gauging station. The river profiles for Beaverton River and Pefferlaw Brook are shown on Figure 9. It would appear that streamflow data at these gauging stations should include ground-water discharge from all aquifers within the catchment areas, including bedrock. The underflow is expected to occur within limestone bedrock which is a significant aquifer (Haefeli, 1970). The thickness of the bedrock aquifer depends on the depth of fracturing and thickness of permeable beds within the limestone.

For the Holland River at Holland Landing catchment area, soil moisture balance recharge estimates are greater than baseflow estimates by an average of 70 mm·a^{-1} (9% of average annual precipitation). A river profile for this catchment is shown in Figure 10. Streamflow data at the gauging station are believed to primarily include ground-water discharge from upper and intermediate aquifers; however, underflow can be expected within deep aquifers (Alliston Aquifer Complex) based on ground-water flow directions reported by

Vallery *et al.* (1982) and International Water Consultants Ltd. (1991).

The water balance for the Holland River at Holland River catchment area is complicated by significant municipal ground-water pumping in Aurora, Newmarket, Holland Landing, Sharon and Queensville. A trend of note is the decrease in streamflow and baseflow since 1985, despite significant precipitation years during this period. This trend may be attributed to re-routing sewage treatment plant (STP) discharge at Aurora and Newmarket, starting in 1985, from the Holland River to the York-Durham sewer, which discharges to Duffins Creek; and to the increase in municipal ground-water pumping. Vallery *et al.* (1982) estimate the STP contribution to streamflow to be 35 mm·a⁻¹ or 15% of average annual streamflow of 229 mm·a⁻¹ (6.9 ft³·s⁻¹; 1970-1976). For the period 1980 to 1989, STP discharge is at least 55 mm·a⁻¹ (Ontario Ministry of the Environment, 1991). Most of this STP discharge water is expected to have originated as ground water from municipal pumping wells abstracting from deep aquifers shown on Figure 10.

In work related to the York Region water supply, Gartner Lee Ltd. (1986, 1988) estimate recharge to deep aquifers to be 30 mm·a⁻¹ per km² over the aquifer area, based on ¹⁴C dating of deep ground waters (2900 years) travelling through 100 m of overburden. International Water Consultants (1991) estimate recharge to deep aquifers through surficial materials at:

Oak Ridges Moraine = 47 mm·a⁻¹

till = 26 mm·a⁻¹

lacustrine = 11 mm·a⁻¹

Peat and Muck = 0, discharge area.

The same report suggests that total recharge rates range from 64 to 138 mm·a⁻¹ (recharge through the soil zone) depending on the surficial deposit. York Region municipal ground-water withdrawals for this period average 49 mm·a⁻¹, increasing from 38 mm in 1980 to 65 mm in 1989. For years with low precipitation, the ground-water withdrawals are similar to the underflow or deep percolation estimates (70 mm·a⁻¹; Table 2). It is not known how much of the underflow is destined to re-emerge as baseflow in adjoining catchments, however, the comparably high rates of abstraction suggest there is considerable risk of drawing on deep aquifer storage. International Water Consultants Ltd. (1991) and Kuehl *et al.* (1992) show declining water levels throughout the 1980s for monitoring wells in deep aquifers forming part of the York Region supply. It should be noted that most of the monitoring wells are situated near pumping wells. In the longer term it is anticipated that further declines in potentiometric levels within the deep aquifers will locally increase vertically downward gradients, and thereby induce additional recharge to the deep aquifers. While it is possible that steady-state conditions will be achieved in the deep aquifers, further declines in baseflow discharge from the shallow system can be expected. Concern about ground-water withdrawal exceeding recharge from deep aquifers in the ORM area has been expressed by Intera Kenting Ltd. (1990); however, Kuehl *et al.* (1992) suggest that additional ground-water withdrawals from the deep aquifer can be sustained with modifications to the existing extraction system.

Significant underflow is also suggested for the Rouge River catchment near Markham. A profile of the Rouge River is shown in Figure 10. Baseflow estimates at the gauging station near Markham (02HC022) are believed to represent ground-water discharges from the upper and lower Unionville, upper and lower Victoria and Oak Ridges Aquifer Complex aquifers. Recharge estimates from baseflow separation

methods at this station are unlikely to include underflow within theLower Unionville or deeper aquifers (*i.e.*, bedrock). Recharge estimates from this study are considerably larger (double) than those of Sibul *et al.* (1977), who suggested that recharge rates for the Rouge and Duffins drainage basins average 140 mm·a⁻¹.

Soil Moisture Balance > Baseflow

Recharge estimates from both methods are within 5% of average annual precipitation for four catchment areas: Black River near Udora, Little Rouge Creek near Locust Hill, West Duffins Creek at Green River and Duffins Creek at Pickering, suggesting less than 40 mm·a⁻¹ underflow at these gauging stations. The profile of Black River is shown in Figure 10. Streamflow data at the gauging stations (02EC012, 02EC008) are believed to include ground-water discharge from upper and intermediate aquifers (Holt, Algonquin and Oak Ridges Aquifer Complex); however, there remains opportunity for underflow at each of the two stations *via* intermediate aquifers (Algonquin Aquifer Complex) and deep aquifers (Mt. Albert Aquifer). Vallery *et al.* (1982) and International Water Consultants Ltd. (1991) report flow in intermediate aquifers toward the rivers with flow in the deep aquifers also deflected toward the river (upward gradients occur along the lower flat sections of the river). Thus, recharge estimates from baseflow separation methods at these stations on the Black River can be considered to represent minimum values, with deep percolation (underflow) to intermediate and deep aquifers not completely accounted for.

As shown by the profile of Little Rouge Creek in Figure 10, streamflow data at the gauging station near Locust Hill (02HC028) are believed to represent discharge from the Oak Ridges Aquifer Complex and parts of the upper and lower Markham aquifers. Ground water entering the deeper aquifers will leave the catchment as underflow (37 mm·a⁻¹). A large part of the drainage area represented by this gauging station is documented as having upward gradients and flowing wells (International Water Supply, Ltd., 1988, 1989; Sibul *et al.*, 1977). It is expected, therefore, that most of the recharge to the deep aquifers would be provided outside the basin, by deep recharge from the Oak Ridges Moraine or inter-aquifer flow from the West Duffins Creek drainage basin to the east, via the Lower Markham Aquifer as mapped by Sibul *et al.* (1977).

Municipal ground-water abstractions at Stouffville average 16 mm·a⁻¹ over the Little Rouge Creek catchment for the eight-year period 1982 to 1989. STP discharges from Stouffville are discharged to Stouffville Creek, which is a tributary of Reesor Creek which is, in turn, a tributary of West Duffins Creek north of Green River. This discharge is equal to 13 mm·a⁻¹ over the West Duffins Creek at Green River catchment area (98.1 km²; Ontario Ministry of the Environment, 1991) which would be realized as baseflow. Removing this value from the baseflow estimate leads to an underflow estimate of 17 mm·a⁻¹ for the West Duffins Creek at Green River catchment within the Lower Markham and Greenwood aquifers (Fig. 9).

Figure 9 illustrates profiles of the west and east branches of Duffins Creek and the location of the gauging station at Duffins Creek at Pickering (02HC006). As indicated, underflow is estimated to be very small (15 mm·a⁻¹; Table 2), with the gauging station intercepting ground-water flow from all aquifers in the basin. Ostry (1979) and Haefeli (1972) estimate ground-water inflow to Lake Ontario to be 0.092 ft³·s⁻¹ per mile of shoreline (51,000 (m³·a⁻¹)·km⁻¹). This is equivalent to

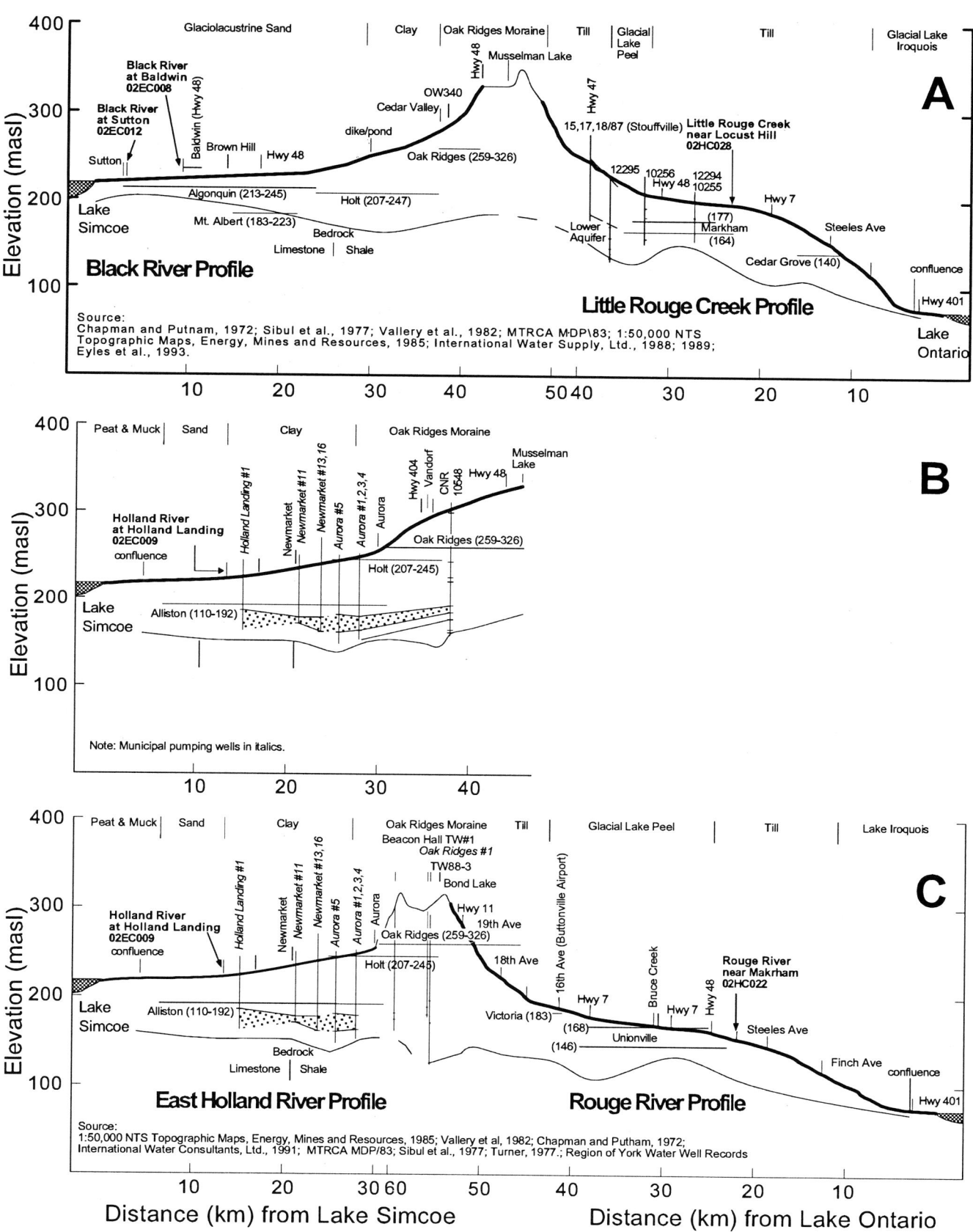

Figure 10   *River profiles of Black River, East Holland River, Little Rouge Creek and Rouge River. Numbers in parentheses represent elevation (in metres above sea level; m asl) of top of aquifer complex or range of elevation of aquifer complex.*

just 3 mm·a⁻¹ recharge over the 30.5 km² area between this gauging station and the Lake Ontario shoreline.

Baseflow > *Soil Moisture Balance*

Two catchment areas (Uxbridge Brook at Uxbridge, 02EC101, and the east branch of Duffins Creek above Pickering, 02HC019) have baseflow recharge estimates greater than recharge estimates from the soil moisture balance method. Figure 9 shows a profile of Uxbridge Brook including Pefferlaw Brook below the confluence. Recharge estimates based on streamflow data at gauging station 02EC101 are believed to represent partial discharge from the upper and intermediate aquifers as defined from Uxbridge municipal ground-water supply information (Fig. 9). Upward gradients exist between the lower and upper/intermediate aquifers which are sometimes reversed by municipal ground-water pumping (International Water Supply, Ltd., 1990). Recharge calculations span the period 1982-1984, which are the only years to contain complete daily streamflow records during the period 1980 to 1990.

The baseflow estimate of recharge is greater than the soil moisture balance estimate by 35 mm·a⁻¹ over the catchment area. This could be indicative of many factors: (1) ground-water flow is entering into this drainage area from neighbouring catchments; (2) stream flow is representative of a much larger drainage area than documented; (3) deeper aquifers recharged over a larger area are discharging to the Creek (upward gradients noted at Uxbridge; International Water Supply, Ltd., 1990); (4) SMC values are lower than the 100 mm assumed (*i.e.*, lower AET); (5) soil moisture balance techniques do not reflect recharge mechanisms for the hummocky terrain comprising much of the ORM. For a large portion of the ORM, RO to defined streams is negligible. Water that does undergo short term RO will collect in low areas and enter the aquifer as indirect recharge.

In work related to the Uxbridge municipal ground-water supply, International Water Supply, Ltd. (1990) estimates recharge to the lower aquifer complex (Fig. 9) at 64 to 96 mm·a⁻¹. Since hydraulic gradients between the upper/intermediate and lower aquifer are vertically upward, this recharge must originate from outside the surface drainage area. Uxbridge municipal ground-water supply abstractions average 20 mm·a⁻¹ which is 20% to 33% of deep recharge estimates. This abstraction, 20 mm·a⁻¹, when added to the estimate of ground-water flow into the basin, 35 mm·a⁻¹ (Table 2), compares to the deep recharge estimates, 64 to 96 mm·a⁻¹, described above.

Figure 9 illustrates a profile of the east branch of Duffins Creek. Streamflow data at the gauging station north of Pickering (02HC019) are believed to represent ground-water discharge from the Oak Ridges Aquifer Complex, Brougham and Greenwood aquifers. Baseflow recharge is estimated to be 49 mm·a⁻¹ greater than soil moisture balance estimates, which suggests that this drainage area is receiving ground-water flow from a neighbouring surface drainage basin, possibly within the Brougham and Greenwood aquifers. Alternatively, the baseflow estimate values could be indicative of increased recharge through areas mapped as till.

Soil Moisture Balance Estimates and Water Level Fluctuation

In an attempt to validate the water balance, soil moisture surplus estimates (potential water available for recharge and runoff) was compared to water table response. Figure 11 shows these data, plotted on a daily basis for the years 1984 and 1985 using OMOE monitoring well 340 at Cedar Valley. This well (Fig. 2) is considered representative of the water table within sand and gravel deposits of the ORM.

Well hydrographs for 1984 are compared with soil moisture surplus estimates calculated using Bradford (Fig. 11A) and

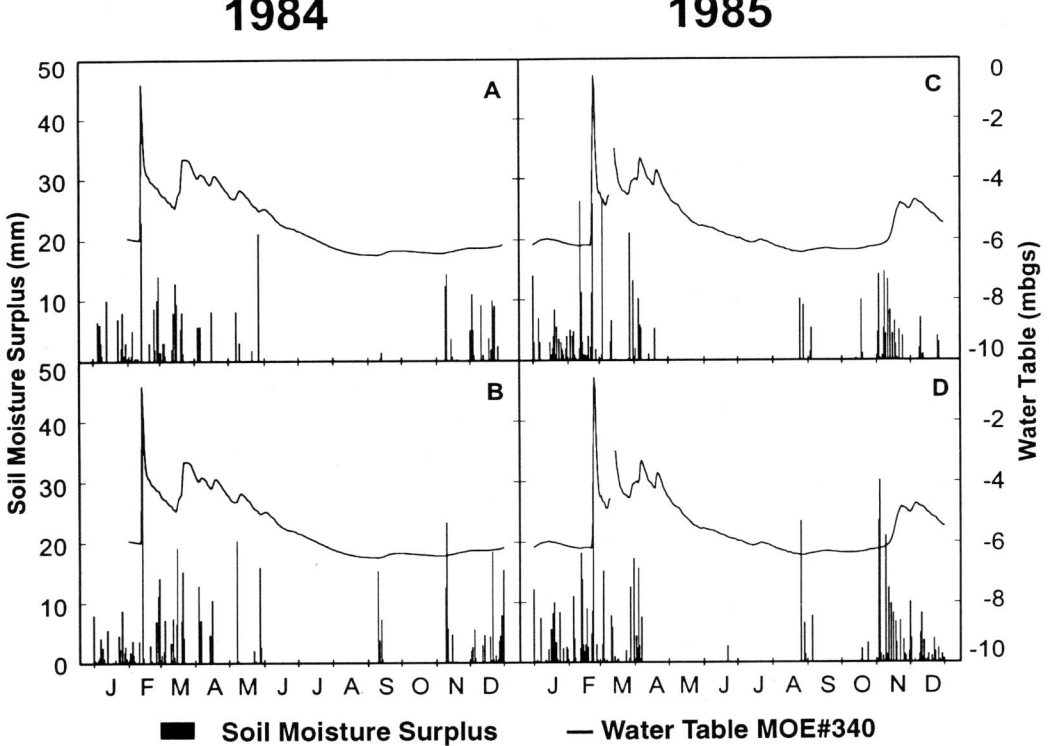

1984   **1985**

Figure 11 *Predicted soil moisture surplus* versus *water table response within sand and gravel (SMC equals 100 mm) of the Oak Ridges Moraine (MOE 340).* **(A)** and **(C)** *Bradford climate station data;* **(B)** and **(D)** *Richmond Hill climate station data.*

Richmond Hill (Fig. 11B) climate station data. The well hydrographs show six peaks between February and June, each of which correlates with large surpluses of soil moisture. Significantly, surpluses generated during the months November to February do not generate a water table response. Much of this surplus may contribute to surface runoff over frozen ground and some may be stored as snowpack or within frozen waterlogged soil and released as recharge when warming occurs in March.

Water table hydrographs for 1985 (Figs. 11C and 11D) show that the decline in water levels in mid-April corresponds to the beginning of the soil moisture deficit predicted by the daily soil moisture balance using both Bradford (Fig. 11C) and Richmond Hill (Fig. 11D) climate data. The water table rise of early November is indicative of the replenishment of the soil moisture deficit as predicted by soil moisture balance calculations. The four hydrograph peaks from late February to early April are predicted by daily water balance calculations for both climate stations; however, only calculations for the Bradford data correlate with the late April water table peak. Data from neither station can account for the slight water table rise in July. Estimates from both stations predict surplus in January and early February which is not reflected by water table response. Again, it is expected that much of this water contributes to runoff over frozen ground.

Figure 12 compares daily soil moisture surplus estimates (Stouffville WPCP climate station) with water table fluctuations (1983) for OMOE monitor 305 (Fig. 2) installed in sandy silt till. Figure 12A shows estimates using a soil moisture capacity of 150 mm; a soil moisture capacity of 100 mm is used in Figure 12B. During periods when the soil moisture content is at field capacity (January to May), the method predicts a soil moisture surplus which correspond to water table fluctuations. However, while the method also predicts fairly accu-

rately the onset of soil moisture deficit (early June), it fails to predict water table fluctuations in late August and September. Lowering the soil moisture capacity to 100 mm (Fig. 12B) increases recharge, but still does not predict water table fluctuations in August and September. A similar trend was found by Rushton and Ward (1979) in an analysis using data from a chalk aquifer in England. For their analysis, the total recharge predicted by the method was redistributed, whereby a certain percentage of precipitation was allowed to bypass soil moisture processes and infiltrate rapidly.

It should be noted that water table rise can occur disproportionately in response to rainfall. Novakowski and Gillham (1988) found the magnitude of a water table response was much greater than expected based on the specific yield of the soil materials studied. Such a response was attributed to the presence of a capillary fringe and an increase in gas phase pressure caused by an infiltrating wetting front. In fine-grained soils, the capillary fringe can extend for several metres above the water table. The capillary fringe has a small percentage of non-saturated porosity; thus, a small recharge event can produce a large water table rise (Trudell *et al.*, 1986).

Regional Trends

Water balance estimates for the study area catchments reveal some interesting trends. Figure 13 shows average yearly streamflow, recharge and baseflow estimates for the period 1980 to 1990. An evident pattern is the significant decrease in streamflow and baseflow and a slight drop in soil moisture balance recharge from east to west. Four factors can be considered: (1) till permeability, (2) land use, (3) drift thickness, and (4) secondary permeability. South of the ORM, the Halton Till becomes more clay-rich to the west; whereas to the north of the moraine, glaciolacustrine silt and clay deposits

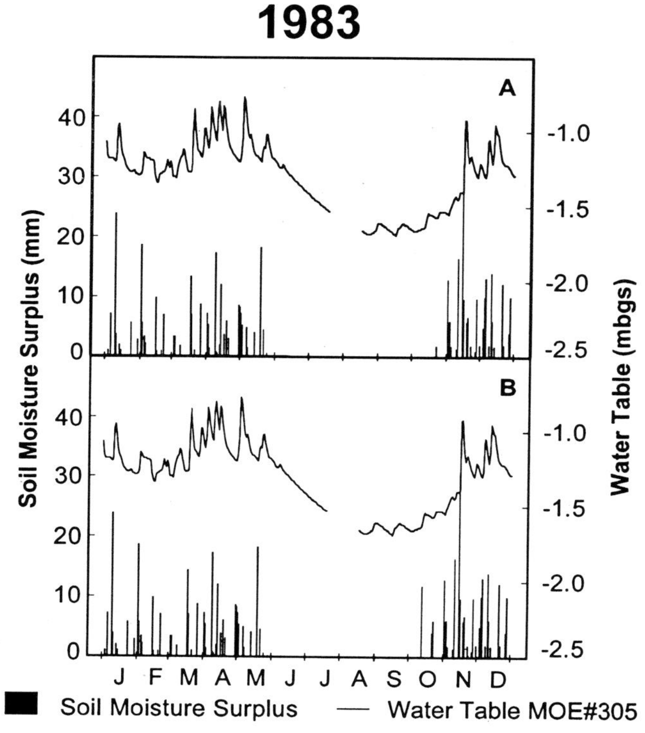

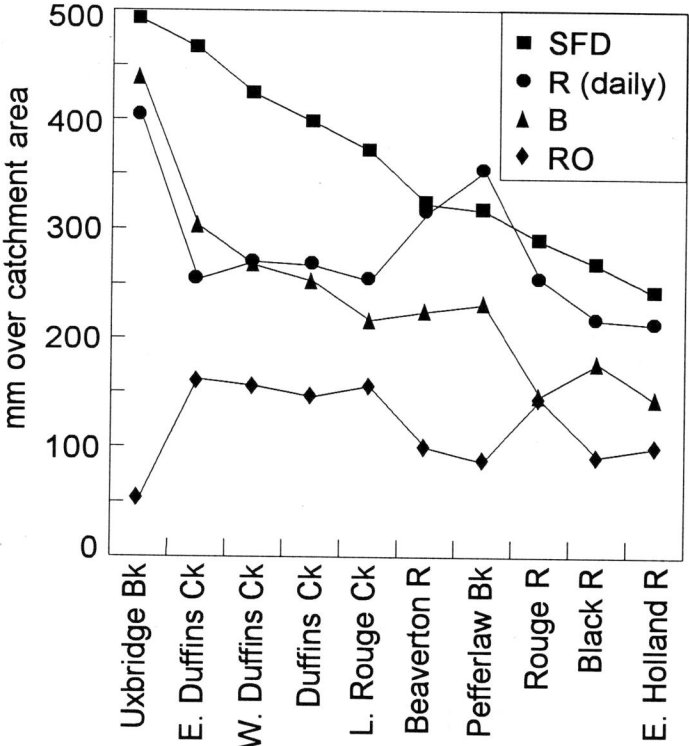

Figure 12 *Predicted soil moisture surplus (Stouffville climate station) versus water table response within Halton Till (MOE 305).* **(A)** *SMC equals 150 mm;* **(B)** *SMC equals 100 mm.*

Figure 13 *Water balance results — regional trends. See text for explanation.*

also become more prevalent to the west, particularly near Bradford, Schomberg, Newmarket and Aurora (Barnett et al., 1991). Land use to the west becomes more urbanized, which may also have a pronounced affect on ground-water recharge. Another factor for the apparent decrease in recharge rates to the west could be drift thickness. Drift thickness increases to the west, with the result that more underflow may be occurring within catchment areas in this direction, particularly the Black and East Holland river basins. Also, higher recharge rates to the east may reflect an increased permeability from secondary structures (sand/boulder layers, fractures) within confining till units, particularly within the Duffins Creek and eastern part of the Rouge River drainage basins (Gerber, 1994; Boyce et al., 1995; Chapters 20, 32). Isotope studies on pore waters extracted from till core samples of Halton and Northern till indicate tritiated or young ground waters (post-1953) to depths exceeding 40 m within the Northern till (Gerber and Howard, 1996, Sites 1/94 and 2/94; M.M. Dillon Limited, 1990; Site P1, see Fig. 2). Also studies by Kaye (1986) and Proulx and Farvolden (1989) indicate areas of increased permeability or windows through the Halton Till near Stouffville. Hydrology Consultants Ltd. (1981) indicate the presence of landfill leachate in an aquifer beneath 38 m of Northern till (Beare Road landfill) and invoke fractures as the transport pathway. While pathways indicated by young ground waters exist within the Halton Till and underlying Northern till, the magnitude of the ground-water flux through the till on a local and regional scale is still under investigation.

CONCLUSIONS

Methodology

For the study area, ground-water recharge mainly occurs during the spring snow melt and, to a lesser extent, during the fall. Although some recharge does occur during summer storm events, most of the summer precipitation is used to satisfy the soil moisture deficit, and its contribution to the yearly recharge quantity is relatively small. By comparison, the fall recharge is significant, and is not always accurately predicted using conventional soil moisture balance calculation techniques. In such cases, the soil moisture balance estimates appear to underestimate actual recharge. It is suggested that the technique utilized in this study could be improved by allowing recharge to occur during periods of soil moisture deficit during large summer storms and rain events during the fall when the soil moisture content is being replenished. In addition, there appears to be a need to incorporate the role of vegetation, and in particular, the reduction of AET caused by the fall harvest. Clearly, recharge predictions also need to be calibrated with site-specific water level measurements.

The Pettyjohn and Henning (1979) method of stream baseflow separation yields estimates of recharge greater than Q_{60} values by between 5 and 50%. Five of the ten drainage basin baseflow estimates from the two methods were within 15% relative to estimates by the method of Pettyjohn and Henning (Uxbridge Brook, Pefferlaw Brook, East Holland River, Duffins Creek above Pickering and Duffins Creek at Pickering). The Q_{60} estimates from this study are comparable to those of earlier studies by others (Sibul et al., 1977; Vallery et al., 1982) when taking into account total annual streamflow.

Table 2 summarizes recharge estimates calculated by soil moisture balance and baseflow separation techniques. Average recharge estimates from soil moisture balance methods

calculated daily (Thornthwaite method; 1980-1990) range from 26 to 43% of annual precipitation. Baseflow estimates of recharge (1980-1990) range from 16 to 47% of annual precipitation. Estimates of recharge from both methods are within 13% with respect to total annual precipitation for all ten catchment areas studied. In general, both daily and monthly soil moisture balance recharge estimates differ by <5% of total annual precipitation. Daily estimates are considered more accurate, however, in that they are believed to more closely approximate actual soil moisture processes.

Regional Trends

An apparent trend within the study area is the decrease in streamflow, baseflow and recharge from east to west. This is attributed to a combination of increasing drift thickness to the west and increased permeability of surficial geological materials to the east. Other trends of note include the apparent ground-water flow from the east into the Uxbridge Brook and Duffins Creek catchment areas, the large underflow within limestone in the Pefferlaw Brook and Beaverton River catchment areas, and underflow or deep recharge estimates approximating municipal ground-water pumping within the East Holland River catchment area.

Role of the Oak Ridges Moraine

While it is acknowledged that significant recharge occurs along the crest of the moraine, research suggests that significant recharge (direct and indirect) also occurs locally along the moraine flanks. This is illustrated by recharge estimates from the three gauging stations situated in the Duffins Creek drainage basin. While the areal coverage of ORM deposits ranges from 21% to 45%, average annual baseflow estimates are within 7% of average annual precipitation. Recharge along the moraine flanks can be explained by vertical leakage through areas mapped as till. Sensitivity analyses for this basin suggest that, even when the recharge through the ORM is set between 300 mm·a^{-1} to 400 mm·a^{-1}, recharge through areas mapped as till must range from 330 mm·a^{-1} to 190 mm·a^{-1} if the baseflow estimates are to be reproduced. It is also suggested that significant indirect recharge enters the eastern part of the Duffins Creek basin as ground-water flow from outside the surface water catchment area.

ACKNOWLEDGEMENTS

Andy Mellary (OMOEE), Phil Harrison and Bruce MacGregor (Region of York) and Ron Motum (Region of Durham) provided municipal ground-water usage details. Dan Sharma (OMOEE) provided observation well data. Herman Goertz (Water Survey of Canada) and Mike Webb (Canadian Climate Centre, Atmospheric Environment Service) provided streamflow and climate data, respectively. Sean Salvatori assisted with all phases of the study. Funding was provided by the Ontario Ministry of the Environment and Energy.

REFERENCES

Alley, W.M., 1984, On the treatment of evapotranspiration, soil moisture accounting, and aquifer recharge in monthly water balance models: Water Resources Research, v. 20, p. 1137-1149.

Barnett, P.J., Cowan, W.R. and Henry, A.P., 1991, Quaternary geology of Ontario, southern sheet: Ontario Geological Survey, Map 2556, scale 1:1,000,000.

Bloyd, R.M., Jr., 1975, Summary appraisals of the nation's ground-water resources, Upper Mississippi region: United States Geological Survey, Professional Paper 813-B, 36 p.

Boyce, J.I., Eyles, N. and Pugin, A., 1995 Seismic reflection, borehole and outcrop geometry of Late Wisconsin tills at a proposed landfill site near Toronto, Ontario: Canadian Journal of Earth Sciences, v. 32, p. 1331-1349.

Brown, D.M., McKay, G.A. and Chapman, L.J., 1980, The climate of southern Ontario: Environment Canada, Atmospheric Environment Service, Climatological Studies 5, 67 p.

Chapman, L.J. and Putnam, D.F., 1972, Physiography of the south central portion of southern Ontario: Ontario Department of Mines and Northern Affairs, Map 2226, Scale 1:253,440.

Dillon, M.M., Ltd., 1990, Regional Municipality of Durham contingency landfill site assessment technical support volume b, Technical report, hydrogeology: prepared for the Regional Municipality of Durham, September 1990, 93 p.

Dixon Hydrogeology Limited, 1991, Review of water works projects, Town of Uxbridge, Regional Municipality of Durham: prepared for the Regional Municipality of Durham, ON, 13 p.

Dunne, T. and Leopold, L.B., 1978, Water in Environmental Planning: W.H. Freeman and Company, San Francisco, 818 p.

Dunne, T., Zhang, W. and Aubry, B.F., 1991, Effects of rainfall, vegetation, and microtopography on infiltration and runoff: Water Resources Research, v. 27, p. 2271-2285.

Eyles, N.E., Boyce, J. and Mohajer, A.A., 1993, The bedrock surface of the western Lake Ontario region: Evidence of reactivated basement structures?: Géographie Physique et Quaternaire, v. 47, p. 269-283.

Freeze, R.A., 1972, Role of subsurface flow in generating surface runoff: 2. Upstream source areas: Water Resources Research, v. 8, p. 1272-1283.

Freeze, R.A. and Cherry, J.A., 1979, Groundwater: Prentice-Hall Inc., Englewood Cliffs, NJ, 604 p.

Fligg, K., 1983, Geophysical well log correlations between Barrie and the Oak Ridges Moraine: Water Resources Branch, Ontario Ministry of the Environment, Map 2273.

Funk, G., 1977, Geology and water resources of the Bowmanville, Soper and Wilmot Creeks IHD representative drainage basin: Ontario Ministry of the Environment, Water Resources Report 9a, 113 p.

Gartner Lee Limited, 1986, Part 1: Preliminary groundwater study for Newmarket, Aurora and the Town of East Gwillimbury: prepared for Regional Municipality of York, ON, 1986, 51 p.

Gartner Lee Limited, 1988, Groundwater supply investigation Oak Ridges, Town of Richmond Hill: prepared for the Regional Municipality of York, ON, 28 p.

Gerber, R.E., 1994, Recharge analysis for the central portion of the Oak Ridges Moraine: unpublished M.Sc. thesis, University of Toronto, Toronto, ON, 172 p.

Gerber, R.E. and Howard, K.W.F., 1996, Evidence for recent groundwater flow through Late Wisconsinan till near Toronto, Ontario: Canadian Geotechnical Journal, v. 33, p. 538-555.

Gwyn, Q.H.J., 1972, Quaternary geology of the Alliston-Newmarket area, southern Ontario: Ontario Department of Mines, Miscellaneous Paper 53, p. 144-147.

Gwyn, Q.H.J., 1976a, Quaternary geology and granular resources of the western part of the Regional Municipality of Durham (parts of Uxbridge, Pickering, Reach, and Whitby Townships), southern Ontario: Ontario Division of Mines, Open File Report 5161, 4 p.

Gwyn, Q.H.J., 1976b, Quaternary geology and granular resources of the eastern part of the Regional Municipality of Durham (parts of Reach, Whitby, Cartwright, Darlington, and Clarke Townships), southern Ontario: Ontario Division of Mines, Open File Report 5176, 6 p.

Gwyn, Q.H.J. and Cowan, W.R., 1978, The origin of the Oak Ridges and Orangeville Moraines of southern Ontario: Canadian Geographer, v. XXII, p. 345-352.

Gwyn, Q.H.J. and DiLabio, R.N.W., 1973, Quaternary geology of the Newmarket area, southern Ontario: Ontario Division of Mines, Preliminary Map P836, scale 1:50,000.

Haefeli, C.J., 1970, Regional Groundwater Flow Between Lake Simcoe and Lake Ontario: Department of Energy, Mines and Resources, [Canada], Inland Waters Branch, Technical Bulletin 23, 42 p.

Haefeli, C.J., 1972, Groundwater inflow into Lake Ontario from the Canadian side: Department of the Environment, [Canada], Inland Waters Branch, Scientific Series 9, 101 p.

Hall, F.R., 1968, Base-flow recessions – a review: Water Resources Research, v. 4, p. 973-983.

Hensel, B., 1992, Natural recharge of groundwater in Illinois: Illinois State Geological Survey, Environmental Geology 143, 33 p.

Hewitt, D.F., 1972, Palaeozoic geology of southern Ontario: Ontario Department of Mines, Geological Report 105, 18 p.

Hoffman, D.W. and Richards, N.R., 1955, Soil survey of York County, Ontario Soil Survey, Report 19, 104 p.

Howard, K.W.F. and Lloyd, J.W., 1979, The sensitivity of parameters in the Penman evaporation equations and direct recharge balance: Journal of Hydrology, v. 41, p. 329-344.

Hydrology Consultants Ltd., 1981, Deep groundwater flow system study, Beare Road sanitary landfill: prepared for Metropolitan Toronto Works Department, 17 p.

Intera Kenting Ltd., 1990, The hydrogeological significance of the Oak Ridges Moraine: prepared for The Greater Toronto Greenlands Strategy, 82 p.

Interim Waste Authority Limited, 1994, Detailed assessment of the proposed site Eell; for Durham Region landfill site search. Parts 1 and 3 of 4: prepared by M.M. Dillon Limited, October, Appendix C, C1-1 to C6-13.

International Water Consultants Ltd., 1991, Regional Municipality of York aquifer performance assessment, Oak Ridges, Aurora, Newmarket, East Gwillimbury and Bradford: prepared for the Regional Municipality of York.

International Water Supply, Ltd., 1988, Regional Municipality of York. Town of Whitchurch-Stouffville, Community of Stouffville groundwater investigation – 1987: prepared for the Regional Municipality of York, 5 p.

International Water Supply, Ltd., 1989, Regional Municipality of York, Stouffville Groundwater Investigation 1988/1989: prepared for the Regional Municipality of York, 5 p.

International Water Supply, Ltd., 1990, Regional Municipality of Durham, Uxbridge, Proposed New Municipal Well Groundwater Impact Study: prepared for the Regional Municipality of Durham, 4 p.

Johansson, P., 1988, Methods for the estimation of natural groundwater recharge directly from precipitation — comparative studies in sandy till, *in* Simmers, I., ed., Estimation of Natural Groundwater Recharge: D. Reidel Publishing Company, Dordrecht, Holland, p. 239-270.

Kaye, B.G., 1986, Recharge characteristics of the Oak Ridges Complex: the role of Musselman Lake: unpublished M.Sc. thesis, University of Waterloo, Waterloo, ON, 187 p.

Kuehl, G.A., Harris, J.A. and Turnbull, D.R., 1992, Assessment of aquifer performance and sustainable yield, a case study, *in* Modern Trends in Hydrogeology, Hamilton, ON, 1992 Conference of the Canadian National Chapter, International Association of Hydrogeologists, p. 407-421.

Leech, R.E.J., 1994, Ontario's water resources; the future of groundwater resources: Environmental Science and Engineering, June/July, p. 18-19.

Lerner, D.N., Issar, A. and Simmers, I., 1990, Groundwater Recharge, A Guide to Understanding and Estimating Natural Recharge: International Association of Hydrogeologists, v. 8, Verlag Heinz Heise, 510 p.

Liberty, B.A., 1969, Palaeozoic geology of the Lake Simcoe area, Ontario: Geological Survey Canada, Memoir 355, 201 p.

Metropolitan Toronto and Region Conservation Authority, 1983, MDP/83 Stream Profiles, maps.

Novakowski, K.S. and Gillham, R.W., 1988, Field investigations of the nature of water-table response to precipitation in shallow water-table environments: Journal of Hydrology, v. 97, p. 23-32.

Olding, A.B., Wicklund, R.E. and Richards, N.R., 1956, Soil survey of Ontario County: Ontario Soil Survey, Report 23, 60 p.

Ontario Ministry of the Environment, 1991, Report on the 1989 discharges from municipal sewage treatment plants in Ontario: prepared for the Water Resources Branch, 35 p.

Ostry, R.C., 1979, The Hydrogeology of the IFYGL Duffins Creek study area: Ministry of the Environment, Water Resources Branch, Water Resources Report 5c, Toronto, ON, 39 p.

Penman, H.L., 1949, The dependence of transpiration on weather and soil conditions; Journal of Soil Science, v. 1, p.74-89.

Pettyjohn, W.A. and Henning, R., 1979, Preliminary estimate of ground-water recharge rates, related streamflow and water quality in Ohio: State of Ohio Water Resources Centre, Ohio State University, Report 552, Contract A-051-Ohio, 323 p.

Phillips, D.W. and McCulloch, J.A.W., 1972, The climate of the Great Lakes basin: Environment Canada, Atmospheric Environment Service, Climatological Studies 20, 40 p.

Proulx, I. and Farvolden, R.N., 1989, Analysis of the contaminant plume in the Oak Ridges Aquifer; prepared for Environment Ontario, R.A.C. Project No. 261 RR, 134 p.

Rosenberg, H.B. and Pentland, R.L., 1983, Accuracy of winter streamflow records: Environment Canada, Inland Waters Directorate, 30 p.

Rushton, K.R. and Ward, C., 1979, The estimation of groundwater recharge: Journal of Hydrology, v. 41, p. 345-361.

Sharpe, D.R., 1980, Quaternary geology of Toronto and surrounding area: Ontario Geological Survey, Preliminary Map P.2204, Geological Series. Scale 1:100,000, Compiled 1980.

Sharpe, D.R., Barnett, P.J., Dyke, L.D., Howard, K.W.F., Hunter, G.T., Gerber, R.E., Paterson, J. and Pullan, S.E., 1994a, Quaternary geology and hydrogeology of the Oak Ridges Moraine area: Geological Association of Canada–Mineralogical Association of Canada, Joint Annual Meeting, Waterloo 1994, Field Trip A7, Guidebook, 32 p.

Sibul, U., Wang, K.T. and Vallery, D., 1977, Ground-water resources of the Duffins Creek-Rouge River drainage basins: Ontario Ministry of the Environment, Water Resources Report 8, 109 p.

Singer, S.N., 1974, A hydrogeological study along the north shore of Lake Ontario in the Bowmanville-Newcastle area: Ontario Ministry of the Environment, Water Resources Report 5d, 72 p.

Singer, S.N., 1981, Evaluation of the Groundwater Resources Applied to the Bowmanville, Soper and Wilmot Creeks IHD Representative Drainage Basin: Ontario Ministry of the Environment, Water Resources Report 9b, 153 p.

Sklash, M.G. and Farvolden, R.N., 1979, The role of groundwater in storm runoff: Journal of Hydrology, v. 43, p. 45-65.

Sklash, M.G. and Farvolden, R.N., 1982, The use of environmental isotopes in the study of high-runoff episodes in streams, in Perry, E.C., Jr. and Montgomery, C.W., eds., Isotope Studies of Hydrologic Processes: Northern Illinois University Press, DeKalb, IL, p. 65-73.

Thornthwaite, C.W. and Mather, J.R., 1957, Instructions and Tables for Computing Potential Evapotranspiration and the Water Balance: Drexel Institute of Technology, Laboratory of Climatology, Publications in Climatology, v. X, Centerton, NJ, 311 p.

Trudell, M.R., Howard, A.E. and Moran, S.R., 1986, Quantification of groundwater recharge in a till environment, East-Central Alberta: Third Canadian Hydrogeological Conference, April 20-23, 1986, International Association of Hydrogeologists, Proceedings, p. 147-158.

Turner, M.E., 1977, Alliston Aquifer Complex: Ontario Ministry of the Environment, Water Resources Branch, Hydrogeological Map 77-1.

Turner, M.E., 1978, Oak Ridges Aquifer Complex: Ontario Ministry of the Environment, Water Resources Branch, Hydrogeological Map 78-2.

United States Environmental Protection Agency, 1990, Handbook, Ground Water, Volume I: Groundwater and Contamination: EPA/625/6-90/016a, Centre for Environmental Research Information, Cincinnati, OH, p. 50-73.

Vallery, D.J., Wang, K.T. and Chin, V.I., 1982, Water resources of the Holland and Black River basins — Summary: Ontario Ministry of the Environment, Water Resources Report 15, 9 p.

Walton, W.C., 1970, Groundwater Resource Evaluation: McGraw-Hill Book Company, Toronto, ON, 664 p.

Ward, C., 1976, The calculation of recharge from meteorological data, with particular reference to the South Humberside chalk aquifer: unpublished M.Sc. thesis, Department of Civil Engineering, University of Birmingham, Birmingham, UK, 40 p.

Water Survey of Canada, 1992, Historical Streamflow Summary, Ontario, to 1990: Inland Waters Directorate [Canada], Water Resources Branch, 663 p.

11. Managing Ground-water Resources Using Wellhead Protection Programs

Stephen Livingstone

Beatty Franz & Associates Limited, 315 Zephyr Avenue, Ottawa, Ontario K2B 5Z7

Thomas Franz

*Beatty Franz & Associates Limited, 18 King Street East, Bolton, Ontario L7E 1E8 and
Waterloo Hydrogeologic Inc., 180 Columbia Street West, Waterloo, Ontario N2L 3L3*

Nilson Guiguer

Waterloo Hydrogeologic Inc., 180 Columbia Street West, Waterloo, Ontario N2L 3L3

SUMMARY

Significant efforts have been completed by the United States to legislate guidelines for the long-term protection of the recharge areas around water supply wells through wellhead protection programs. Currently, Canada does not have national wellhead protection guidelines, and it is the responsibility of proactive local governments, municipalities or regions to implement their own wellhead protection programs. This paper presents the wellhead protection terminology and methodologies currently used to define wellhead protection areas (WHPAs). Delineation methodologies and evaluation techniques for wellhead protection are presented using a hypothetical case study. Based on the case study, it is shown that three-dimensional numerical modelling provides more accurate WHPAs than analytical and two-dimensional numerical models. The risks of delineation errors are discussed and the potential risks of over- or under-protecting the WHPA are demonstrated.

INTRODUCTION

Throughout Canada, ground water is an invaluable natural resource that sustains the growth and development of communities, industries and agricultural activities. Ground-water surveys completed in the 1960s and the 1980s indicate that ground-water usage in Canada has grown from approximately 10% to about 26% of the population (Hess, 1986). In addition, about 40% of the municipalities in Canada use ground water for a significant portion of their domestic water supply. In Prince Edward Island, more than 99% of the population uses ground water for domestic use due to a lack of acceptable surface-water sources. Domestic use of ground water is most prevalent in Ontario and Quebec, whereas agricultural activities use the most ground water in the Prairies. Industries are sustained by ground-water use in British Columbia (Cherry, 1987).

The growth and development of communities, industries and agriculture has resulted in the storage and discharge of chemicals in the form of point and non-point sources (Chapters 4, 5). Point sources include landfill sites and lagoons, storage tanks, snow dumps, septic systems, and chemical discharges. Seasonal applications of road de-icing agents, agricultural fertilizers, and pesticides represent significant non-point sources. These contaminant sources pose a serious threat to the long-term health and potability of ground water as a drinking-water resource.

The management of ground-water resources and the significance of keeping ground-water supplies safe have been relatively recent concerns. As shown in Table 1, many communities in Ontario have had the costly task of finding alternative drinking-water sources due to ground-water contamination of their municipal water wells. Chemicals commonly detected in ground water include PCBs, volatile organic compounds (VOCs), nitrates and bacteria associated with faulty septic system designs. Many of these are known carcinogens (*e.g.,* benzene). In the United States, more than 200 different harmful chemicals have been detected in ground water (United States Environmental Protection Agency [US EPA], 1993). Monitoring and chemical testing of Canadian drinking-water supplies will likely show the same results with time and with the development of site-specific sampling programs.

Significant efforts have been completed by the United States to legislate guidelines for the long-term protection and risk assessment of the recharge areas around water supply wells through wellhead protection programs. Wellhead protection may be defined as managing a land area around a well or well field to prevent ground-water contamination and guarantee a high-quality water supply source (Cleary and Cleary, 1991; US EPA, 1993). Currently, Canada does not have national wellhead protection guidelines. It is the responsibility of proactive local governments, municipalities or regions to implement wellhead protection programs, as, for example, the Kitchener-Waterloo region of Ontario has. This region is the largest urban area in Canada depending entirely on ground water.

Five steps to implementing a wellhead protection program, as defined by the United States Environmental Protection Agency (1993) are (1) forming a community planning team, (2) delineating the wellhead protection area, (3) identifying and locating potential sources of contamination, (4) managing the wellhead protection area, and (5) planning for the future.

In this paper, the methodologies and evaluation techniques used to delineate the wellhead protection area (step 2) will be discussed with an emphasis on the use of computer modelling. A hypothetical case study is presented to illustrate the differences of various modelling techniques.

WELLHEAD PROTECTION TERMINOLOGY

The term wellhead protection can be misleading. It does not refer to the protection of the mechanical equipment or well construction materials used at the individual wells. Rather, wellhead protection is concerned with the land area or zones around a pumping well which supplies ground water to the well. Delineation of the wellhead protection area represents the definition of the geographical limits most critical to the protection of the wellfield from unexpected contaminant releases to the aquifer recharge area of the aquifer. These zones around the wells are referred to as wellhead protection areas (WHPA).

Four terms — cone of depression, zone of influence (ZOI), zone of contribution (ZOC), and zone of transport (ZOT) — are commonly used to assess and describe the risk of contaminants captured by a pumping well (Fig. 1).

When a well is pumped, drawdown of the water table (unconfined aquifer) or potentiometric surface (confined aquifer) occurs. This drawdown in water levels by pumping is called the cone of depression. The size and shape of the cone of depression is related to the well pumping rate and time period, the physical parameters of the aquifer material (i.e., hydraulic conductivity, porosity), the aquifer boundary condi-

---

Table 1   Municipal ground-water supplies in Ontario; contamination incidents and/or concerns.

MUNICIPALITY	CONTAMINANT	RESOLUTION/ CURRENT SITUATION
WEST CENTRAL REGION		
Smithville	PCBs	• Municipal wells out of operation • Extended water from Lake Ontario
Elmira	NDMA	• One well field out of operation • Extended supply from St. Jacobs
Kitchener/Waterloo	VOCs	• Contaminants in a number of well fields
Town of Delhi (spring supply)	VOCs-benzene	• Spring supply shut down • New municipal well drilled
Town of Simcoe	VOCs	• Contaminants at low levels • Monitoring municipal wells
Guelph	VOCs-TCE	• One well field shut down
Erin	VOCs	• One or two wells shut down • Two new wells drilled
SOUTHWEST REGION		
Ingersol	VOCs	• Contaminants detected in one well • Wells monitored
SOUTHEAST REGION		
Trenton	Nitrates Bacteria	• Well shut down
Frankford	Nitrates	• Well shut down
Manotick	VOCs	• Contamination of private wells • Surface water extended from Ottawa
MID-ONTARIO REGION		
Penetanguishene	VOCs	• Well field shut down
Orillia	VOCs	• Contamination of two wells
Barrie	VOCs	• Contamination of well fields
NORTHWEST REGION		
Manitouwadge	VOCs	• Contamination detected (low levels)

tions (*i.e.*, rivers, faults, recharge zones), and the hydro-stratigraphic setting (*i.e.*, unconfined, confined aquifers). In general, the cone of depression for unconfined aquifers is relatively small in comparison to confined aquifers. For confined aquifers, the cone of depression may extend outward for kilometres.

The zone of influence (ZOI) is synonymous with cone of depression. ZOI is defined as the distance from the well where

changes in the ground-water surface (water levels) can be measured or inferred as a result of pumping (US EPA, 1994). Theoretically, the ZOI in a homogeneous, isotropic porous aquifer will be circular. Most natural hydrogeological settings, however, are complex. Thus, in heterogeneous, anisotropic porous and fractured aquifers, the ZOI often has an irregular shape. Ideally, the ZOI is measured in the field by means of water level response in monitoring wells and is based on a

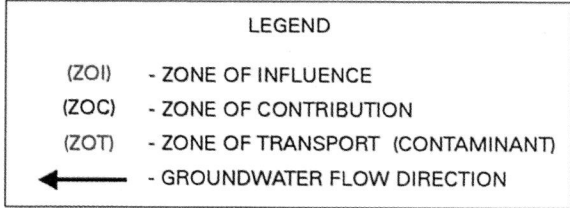

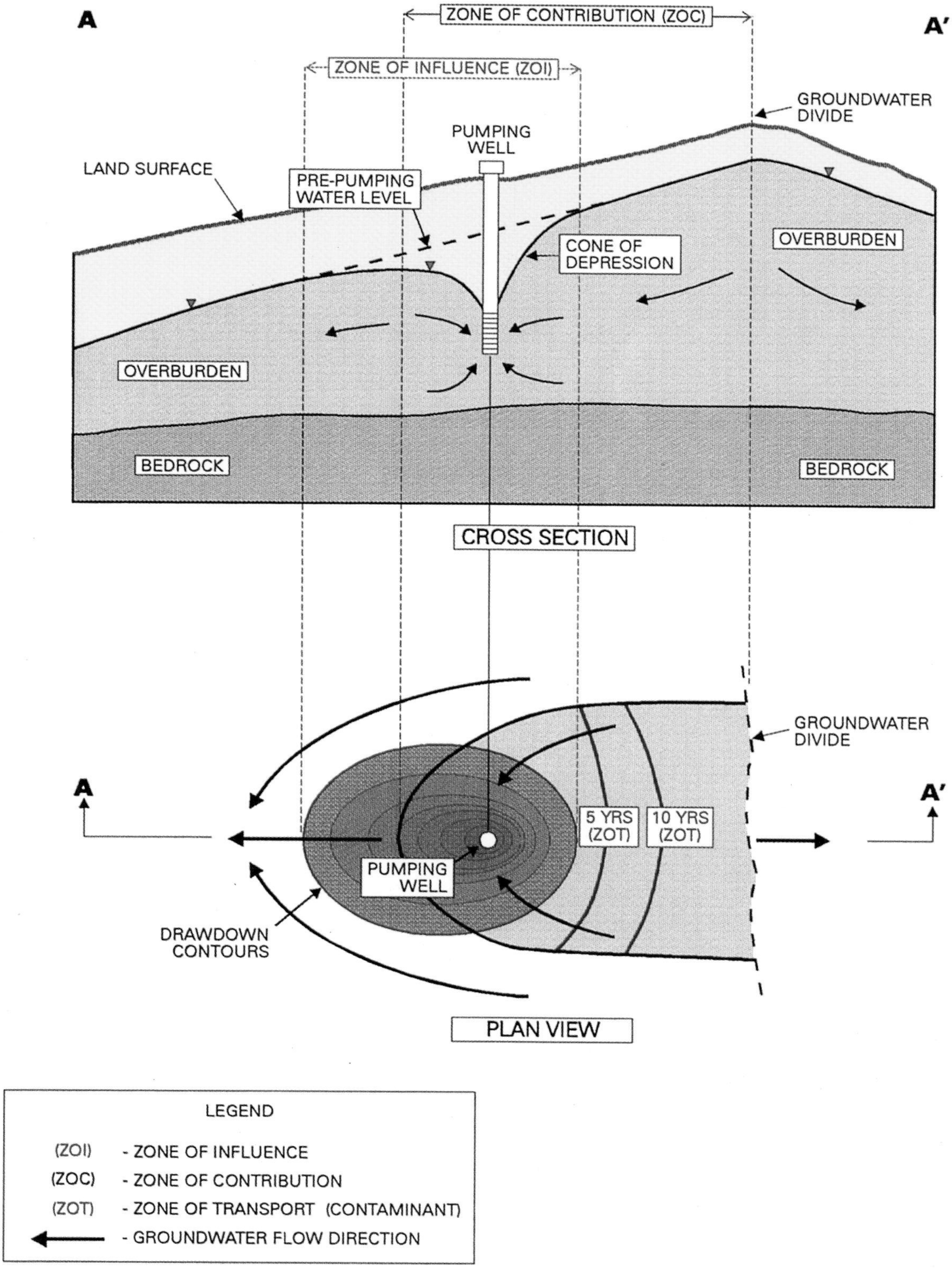

Figure 1 *Conceptual model of wellhead protection areas and terminology.*

comparison of pre-pumping and pumping conditions. However, the pumping wells may have been operating for many years and have modified the natural flow conditions. Thus, it may be difficult to determine the exact limits of the ZOI due to pumping of the municipal well or well field.

The zone of contribution or capture (ZOC) is defined as the area of the aquifer that recharges the well or well field (US EPA, 1994). Ground-water contaminants discharged into the ZOC area will be captured by the water well. Thus, there is a high risk of contamination to the ground water within the ZOC. As shown on Figure 1, the ZOC and the ZOI do not coincide. For the ZOC, ground water is removed from the pumping well from only a relatively small portion of the downstream area of the well, but it may extend as far as the ground-water divide on the up-gradient side of the well. By contrast, the down-gradient portion of the ground water within the ZOI is not drawn toward the pumping well but continues down gradient, while the ZOI does not extend to the up-gradient limit of the ZOC. As a result, the ZOI overprotects the down-gradient ground water and underprotects the up-gradient ground water. In our experience, however, if the ZOC is small, then the ZOC and ZOI will generally overlap.

The ZOC can be further delineated by the time of travel within the ZOC. The time of travel (TOT) is generally presented as isochrones (contours of equal travel time) that indicate the time required for a contaminant to reach a pumping well from a contaminant source within the ZOC. The time of travel depends on the ground-water flow velocity, the contaminant characteristics, and the properties and composition of the aquifer material. Figure 1 depicts the delineation of the travel times (TOT) for contaminants within the ZOC.

Another term used in wellhead protection plans is the zone of attenuation (ZOA). The ZOA is always smaller than the ZOC, because it takes into account that contaminants can be immobilized or attenuated in the subsurface to an acceptable concentration before reaching the pumping well. The chemical processes that reduce the contaminant concentrations along its flowpath include sorption, chemical precipitation and degradation. Such processes are discussed in several chapters in this volume (Chapters 4, 6, 35). Typically, the zone of attenuation is adjusted through the time of travel calculations. Retardation factors, which represent the combined effects of attenuation for the chemicals, are calculated; these shift the isochrones closer to the pumping well. This procedure results in a reduction of the WHPA for certain chemical parameters that may discharge in the ZOC. It is often impossible to predict what types of contaminants may be released within the ZOC, and, therefore, it is often impractical to apply attenuation as a means to reduce the size of the WHPA. This approach does apply, however, when a known potential source is located within the ZOC, and when the contaminants potentially released at that location are known to attenuate.

WELLHEAD PROTECTION DELINEATION METHODS

The US EPA has identified the following methods as practical procedures to delineate wellhead protection zones: (1) hydrogeological mapping, (2) fixed radius around a well, (3) calculated radius, (4) analytical modelling, and (5) numerical modelling.

Hydrogeological mapping (method 1) requires the identification of the zone of contribution (ZOC) based on the physical and chemical characteristics of the aquifer. This approach is sometimes used to develop a conceptual model, which is then simulated in an analytical or numerical computer model. The geometric methods (methods 2 and 3) use predetermined fixed radius and aquifer geometry without any

consideration for the natural ground-water system. The analytical method (method 4) uses simplifications and assumptions to describe the hydrogeological setting for the determination of the time of travel of contaminants and the drawdown calculations for wellhead protection. The fifth method involves computer modelling. Advanced numerical models are employed to simulate ground-water flow and contaminant transport in two or three dimensions (see Mercer and Faust, 1980; Konikow and Mercer, 1988; Bear et al., 1992; Anderson and Woessner, 1992).

Two-dimensional modelling can be completed in plan view or cross-section. The use of a two-dimensional model has some implications in terms of the size and shape of the predicted WPHA. The two-dimensional formulation neglects any vertical gradients and, therefore, usually results in a good representation of the regional scale model when flow predominantly occurs in the horizontal plan. However, in the immediate vicinity of wells and shallow-surface waterbodies, and in strongly variable topography, three-dimensional flow conditions with significant vertical components exist. A two-dimensional horizontal (plan view) model cannot resolve vertical layering of the aquifer sediments. Any vertical variations in hydraulic conductivity, porosity, etc., are averaged over the aquifer thickness.

Three-dimensional models can more fully represent hydrogeological conditions by incorporating multi-layers, partially screened water wells, and horizontal and vertical fluxes. The major advantage of using numerical methods is the greater flexibility and ability to incorporate varying hydrogeological and stress conditions into the computer model. Thus, the result is a more accurate description and simulation of the real physical conditions. Ground-water flow is typically calculated with the finite difference or finite element method, while contaminant transport can be simulated using particle-tracking techniques or finite-difference and finite-element methods.

These methodologies differ in their degree of complexity and accuracy. The US EPA (1987) concluded that numerical modelling achieves the most accurate and reliable WHPA delineation; however, numerical modelling can only achieve reliable results when the following conditions are met:
1.   There must be a minimum amount of knowledge of the ground-water flow system that is to be protected. The characterization of a municipal aquifer typically requires the input of professional hydrogeologists. The data required are (1) extent of aquifer (i.e., three-dimensional aquifer geometry), (2) stratigraphy of the subsurface near the well (possibly over several square kilometres, (3) pumping test data, (4) historic information on pumping schedules, (5) data on the interaction of ground water and surface water, (6) ground-water recharge and infiltration data, and (7) regional water level information.
2.   An appropriate model must be chosen for the calculation of ground-water flow and well capture zones. The model must be able to reflect the physical conditions in the ground-water system. An important consideration in the selection of a model is the choice of two-dimensional versus three-dimensional modelling. While three-dimensional modelling requires greater efforts, it may be the only way to achieve a realistic representation of the ground-water flow system.
3.   The model must be set up and calibrated properly by an experienced modeller. Non-calibrated models will not be any better than simplistic methods such as fixed radius or calculated radius.

Risks of Delineation Errors
Errors in the delineation of WHPAs can have two significantly

negative effects. For example, if an area is delineated greater than necessary (*i.e.*, when the predicted well capture zone is larger than the true capture zone), then development restrictions in the wellhead protection area will over protect. This could have a negative economic impact on the area, especially if the WHPA lies within a highly developed area, where restrictions on land use and/or monitoring requirements may be required from landowners who are, in fact, outside of the true catchment area of the well.

On the other hand, if a WHPA is underprotected (*i.e.*, when the delineated area is smaller than the true catchment area or if the delineated area is not in the right location), then environmental impacts can still occur and the ground water remains at risk. Thus, it is important to establish the degree of accuracy that may be required in the process of deciding on the WHPA delineation methodology. In this case, ground-water monitoring and chemical analysis programs would be developed which would not reflect the appropriate level of effort. Thus, receptors of the drinking water could be at risk from contaminated ground water.

WHPA Delineation Models

There are several computer models available for the delineation of WHPAs. Table 2 shows the most popular models and categorizes them according to their features and capabilities.

In this paper, we consider WHPA, FLOWPATH and VISUAL MODFLOW to highlight the differences between analytical, two-dimensional numerical, and three-dimensional numerical models. Of the models listed in Table 2, these models are probably the most popular. For example, WHPA was developed by the US EPA and, therefore, is in widespread use in the United States, while the proprietary FLOWPATH and VISUAL MODFLOW have gained acceptance as the most powerful numerical models on the market in the United States, Canada, and Europe (predominantly the United Kingdom).

Hypothetical Case Study

A hypothetical setting that exists near many municipal well fields was constructed to illustrate (1) the differences between analytical, numerical and two-dimensional *versus* three-dimensional modelling, and (2) potential problems associated with ground-water monitoring and chemical sampling programs.

In our example, a residential development is located east of the Beatty River (Fig. 2). All units are on septic systems and supplied water is taken from a municipal water well. South of the residential development, agricultural land (Palmer Farm)

is used for growing potatoes and cash crops. Land use on the west side of the river includes Parnham's Gas Bar, Rennie Paints and Dyes Inc., and Fraser and Brown Hardware. All industries and stores were established five to ten years ago. Currently, the municipality samples the water well twice per year for bacteria chemical analysis (fecal and total coliforms) to document the absence or presence of septic system effluent impacting the water supply.

The municipal well pumps at a rate of 250 $m^3 \cdot d^{-1}$ and it is a partially penetrating well with the well screen extending from 20 m to 30 m below ground surface. The surficial aquifer is made up of sand and gravel with a respectable transmissivity of 26 $m^2 \cdot d^{-1}$ (*i.e.*, a hydraulic conductivity of 10^{-3} cm·s^{-1} and a thickness of 30 m). The well is near a river that is situated within an area of alluvial gravel with a hydraulic conductivity of 10^{-5} cm·s^{-1} and a thickness of 10 m.

The ground-water flow conditions were simulated in this relatively simple hydrogeological setting using (1) the two-dimensional WHPA model, (2) the numerical two-dimensional model FLOWPATH, and (3) the fully three-dimensional numerical model VISUAL MODFLOW.

Two-Dimensional Analytical WHPA Model

WHPA is a modular semi-analytical ground-water flow model developed by the US EPA's Office of Groundwater Protection (currently, the Office of Groundwater and Drinking Water) primarily to assist state and local technical staff with WHPA delineation. The WHPA solves analytical equations for two-dimensional flow to a well under various hydrological conditions. The WHPA model contains independent modules such as MWCAP (Well Capture Zone) GPTRAC (General Particle Tracking) and MONTEC (Uncertainty Analysis).

The modelling of the case study involves rotating the model domain such that it best reflects regional ground-water flow, and such that the orientation of the river can be simulated. The WHPA resulting from this relatively quick simulation is shown in Figure 2. The figure suggests that all ground-water path lines captured by the municipal well originate on the east side of the river, and that the river contributes a substantial amount of water to the well. Based on this capture zone, it appears that ground-water protection efforts should be focussed on the area on the east side of the river. Industries on the west side of the river do not require any further attention. Sampling of the water well for bacteria alone is unsatisfactory considering that the major impacts on groundwater in the capture zone include septic system effluent and pesticides and fertilizers associated with agricultural land in the south.

Table 2	Commonly used models for wellhead protection area delineation.			
MODEL	**TYPE**	**DEVELOPER**		**COMMENTS**
FLOWPATH	numerical	Waterloo Hydrogeologic Inc.		2-D, flow and pathlines
VISUAL modflow	numerical	Waterloo Hydrogeologic Inc.; USGS		3-D, flow and pathlines
WHPA	analytical numerical	US EPA		2-D, analytical: flow and pathlines; numerical: pathlines only
GWPATH	numerical	Illinois State Water Survey		2-D, pathlines only
QUICKFLOW	semi-analytical	Geraghty & Miller Inc.		2-D, flow and pathlines

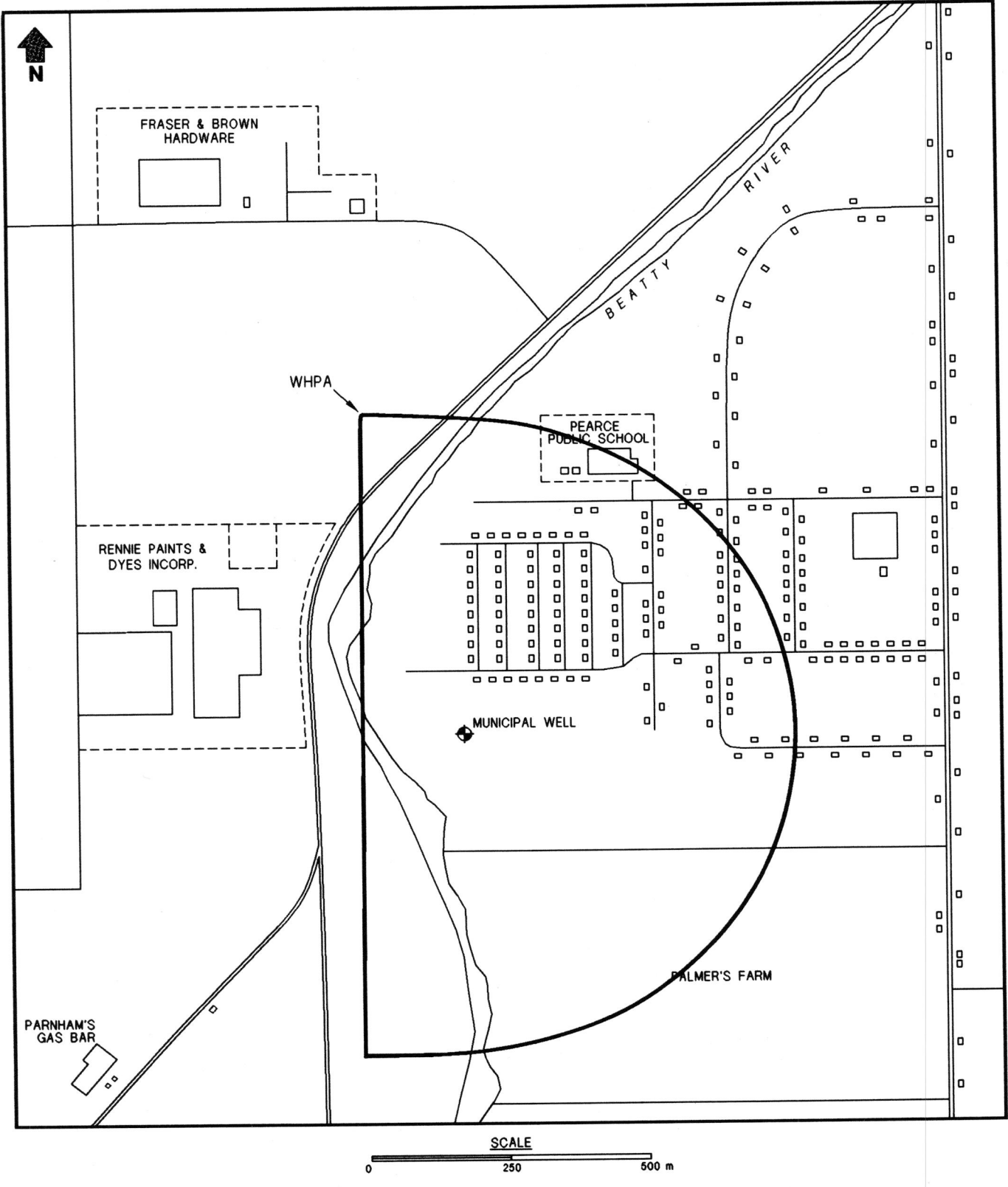

Figure 2 *Wellhead protection area for the community of Beatty, based on two-dimensional analytical WHPA model.*

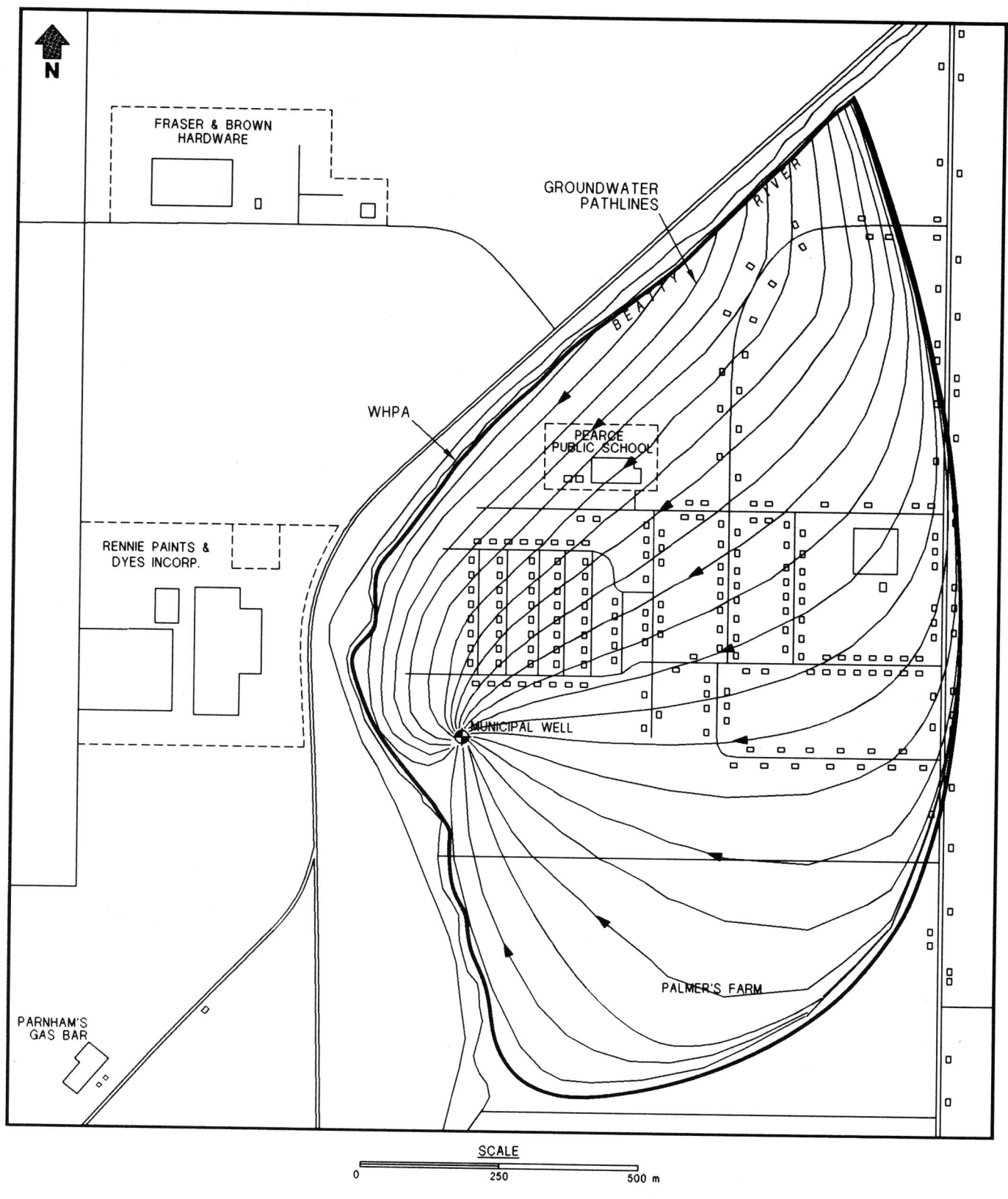

Figure 3 *Wellhead protection area based on two-dimensional numerical FLOWPATH model.*

Two-Dimensional Numerical FLOWPATH Model

Ground-water flow and the wellhead capture zone were as simulated using FLOWPATH Version 5.11 (Franz and Guiguer, 1989). FLOWPATH is a two-dimensional steady-state ground-water flow model based on the finite-difference method. The model can simulate horizontal ground-water flow in hetero-geneous, anisotrophic, confined/unconfined and leaky aqui-fers. It can handle withdrawal and injection of water at multiple wells, contaminant particle tracking and capture zones (WHPA).

Initially, FLOWPATH was calibrated to reflect the physical conditions described above. This was achieved by selecting appropriate boundary conditions along the edges of the model domain and by assigning a leakage factor to the river bed. The leakage factor assumes that a limited rate of ground-water flow is drawn from the river; this rate is simply based on Darcy's Law, and is in relation to the hydraulic conductivity of the river bed and the hydraulic-head difference between the water level in the river and the head in the underlying aquifer.

In contrast to the analytical solution discussed above, the FLOWPATH model results (Fig. 3) now indicate that the capture zone extends significantly to the north, following the Beatty River, and to the south within the agricultural land. Due to the added capability of the FLOWPATH model to simulate limited leakage from the river bed, the FLOWPATH model is more accurate than the WHPA modelling package in this situation. Based on this capture zone, it appears that ground-water protection efforts should be focussed on the area on the east side of the river and to the north and south. Industries on the west side of the river do not require any further attention. Sampling of the water well for bacteria alone would not be satisfactory, considering the major impact on ground-water in the capture zone would include both septic system effluent and agricultural chemicals, including pesticides and fertilizers.

Three-Dimensional Numerical VISUAL MODFLOW Model

As a final step in the comparison, a fully three-dimensional model is used to simulate the three-dimensional ground-water flow and contaminant migration using particle tracking. VISUAL MODFLOW Version 1.1 (Guiguer and Franz, 1995) is a fully integrated modelling platform which unites the USGS's MODFLOW and MODPATH in a graphical modelling environ-ment. Ground-water flow within the aquifer is simulated using a block-centred finite-difference approach. Layers can be simulated as confined, unconfined or a combination of both. Flow from external stresses, such as flow to wells, areal recharge, evapotranspiration, flow to drains and flow through riverbeds, can also be simulated.

Using MODFLOW, it is possible to explicitly represent the alluvial sand and gravel overlying the aquifer near the river. The results of this simulation reveal that a significant area on the west side of the river contributes to the water pumped at the well, and, therefore, should be protected (Fig. 4). This capture zone configuration shows that ground-water protec-tion efforts should be focussed on the area on the west and east side of the river. Industries on the west side of the river, specifically the Rennie Paints and Dyes Inc. plant, require further attention considering the capture zone now includes this property. It is important to note that both the analytical and numerical 2-D approaches failed to identify this area. Sampling of the water well for bacteria would not be satisfac-tory, considering the major impacts on ground-water in the capture zone would include septic system effluent and indus-

trial chemicals, including carcinogenic volatile organic com-pounds (VOCs). Thus, the present ground-water monitoring and sampling program is inadequate, and additional chemi-cals may be present in the ground water which pose a significant risk to human health.

DISCUSSION

Canada currently does not have mandatory national guide-lines or regulations for the implementation of wellhead protec-tion programs. As a result, communities in urbanized areas without wellhead protection programs will continue to be faced with the difficult and costly task of finding alternative drinking-water sources due to ground-water contamination of their municipal water wells. Many chemicals affecting the ground-water supplies are known carcinogens. Coupled with the lack of effective remediation technologies for removing the contaminants from the deep subsurface, contaminated aquifers will remain non-potable for potentially hundreds of years.

This paper has demonstrated that the choice of model may result in significant differences in WHPA delineation. Two-dimensional analytical, two-dimensional numerical, and three-dimensional numerical models have been used to simu-late a hypothetical, but realistic, hydrogeological setting where a municipal pumping well is located near a river. Due to the three-dimensional nature of the hypothetical setting, the three-dimensional numerical model was able to describe the ground-water flow pattern near the stream accurately, while the two-dimensional analytical and numerical models had to use approximations in order to idealize the hypothetical set-ting. The approximations involved the simulation of the shal-low river bed as a fully penetrating feature, and the partially penetrating well was simulated as a fully penetrating well. By definition, there are no vertical gradients in the two-dimen-sional model (i.e., the model is a plan view model) and, therefore, the two-dimensional models were unable to repre-sent the vertical flow components near the river. The three-dimensional model shows that a significant portion (approx-imately 40%) of the pumped water is withdrawn from an area on the opposite side of the stream; thus, this area should be protected as part of the WHPA. The two-dimensional models failed to identify this area on the opposite side of the river.

The use of numerical models, and specifically three-di-mensional models, is only possible if sufficient data are available to develop a valid conceptual model and to properly calibrate the analytical or numerical model. It is important to properly characterize the hydrogeological system in the field using monitoring wells and appropriate tests (e.g., pumping tests). The level of effort in characterizing and modelling a well or wellfield should be in proportion to the potential risks associated with over- or under-protecting the zone of capture. For example, near industrial facilities it is cost-effective in the long run to establish an accurate WHPA in order to (1) mini-mize risks of contamination and (2) minimize the required monitoring effort by focussing on the potential sources that really present a risk to the water supply.

The use of simple delineation methods, such as fixed radius or calculated radius, has not been discussed herein because the WHPAs illustrated in this paper have rather complex shapes, despite a relatively simple hydrogeological setting. Such simplistic methods are inappropriate unless they are applied in a highly conservative (i.e., over-protective) manner.

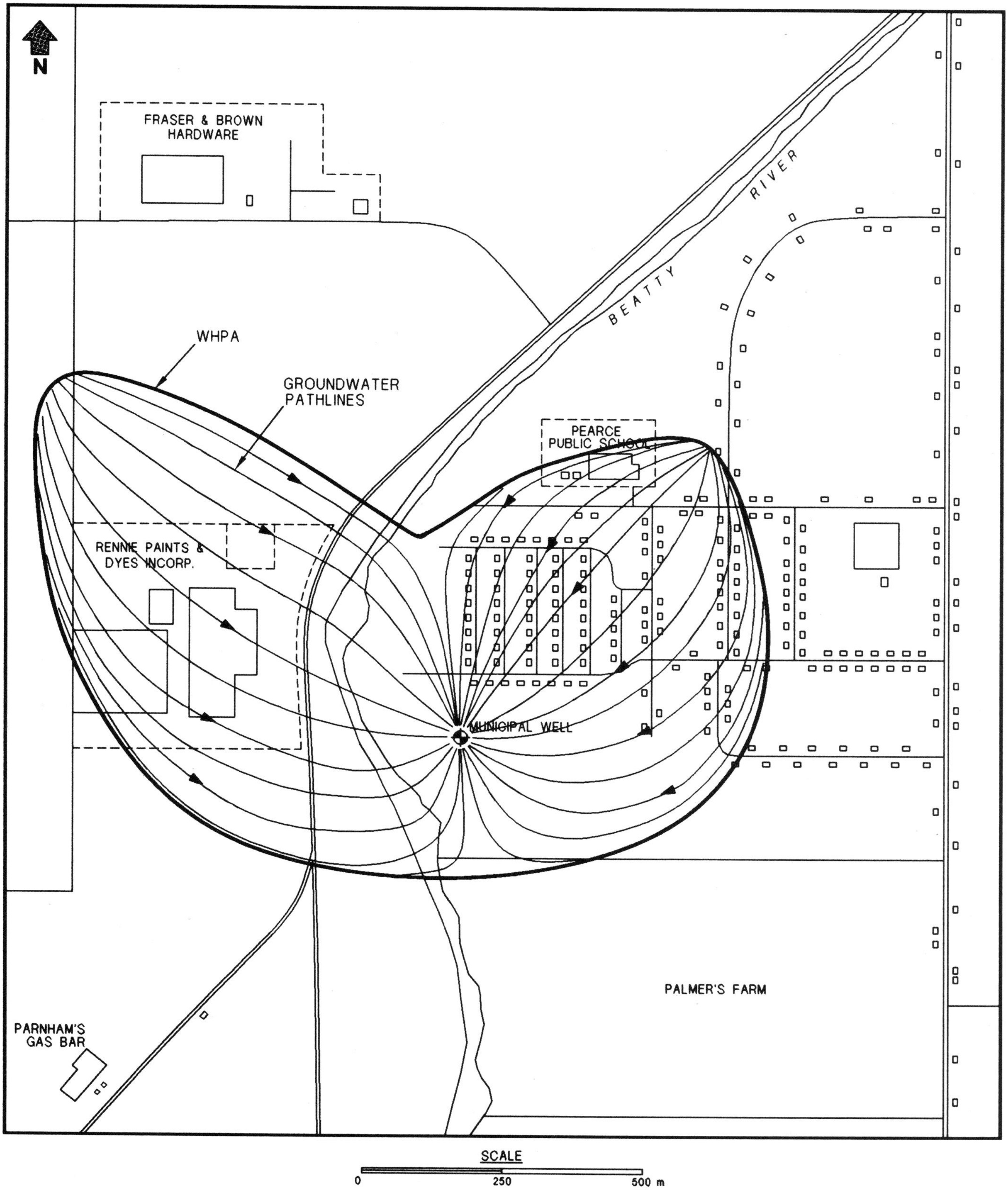

Figure 4 *Wellhead protection area based on three-dimensional numerical visual MODFLOW model.*

REFERENCES

Anderson, M.P. and Woessner, W.W., 1992, Applied Groundwater Modelling: Academic Press, New York, 256 p.

Bear, J., Beljin, M.S. and Ross, R.R., 1992, Fundamentals of groundwater modeling: EPA-540/S-92-005, 11 p.

Cherry, J.A., 1987, Groundwater occurrence and contamination in Canada, in Healy, M.C. and Wallace, R.R., eds., Canadian Aquatic Resources: Canadian Bulletin of Fisheries and Aquatic Science, v. 215. p. 387-426.

Cleary, T.C.B.F. and Cleary, R.W., 1991, Delineation of wellhead protection areas: theory and practice: Water Science and Technology, v., 24, p. 239-250.

Franz, T. and Guiguer, N., 1989-1995, FLOWPATH Version 5.11: Steady state two-dimensional horizontal aquifer simulation model: Waterloo Hydrogeologic Software.

Guiguer, N. and Franz, T., 1995, VISUAL MODFLOW, The integrated modelling environment for MODFLOW and MODPATH: Waterloo Hydrogeologic Software.

Hess, P.J. 1986, Groundwater use in Canada, 1981: Environment Canada, Inland Waters Directorate, Bulletin 140, 43 p.

Konikow, L.F. and Mercer, J., 1988, Groundwater flow and transport modeling: Journal of Hydrology, v. 100, p. 379-409.

Mercer J. and Faust, C.R., 1980, Groundwater modeling: applications: Ground Water, v. 18, p. 486-497.

United States Environmental Protection Agency, 1987, Guidelines for delineation of wellhead protection areas: Office of Groundwater Protection, Washington, DC, EPA 440/6-87/010, 98 p.

United States Environmental Protection Agency, 1993, Wellhead Protection: A Guide for Small Communities: Seminar Publication EPA/625/R-93-002, 406 p.

United States Environmental Protection Agency, 1994, Groundwater and Wellhead Protection: Office of Research and Development, Office of Water, Washington, DC, EPA/625/R-94/001, 115 p.

12. Constructed Wetlands for Remediation of Urban Waste Waters

Miron Berezowsky

Boojum Technologies Ltd., 468 Queen Street East, Toronto, Ontario M5A 1T7

SUMMARY

This paper provides a brief review of the characteristics of artificially constructed wetlands and their use in remediating water quality, principally in connection with municipal waste waters, urban storm-water runoff, and acid mine drainage. In the past 15 years, a heightened interest in the application of both natural and constructed wetland treatment for water quality improvement has resulted in the construction of more than 1500 wetland systems worldwide. This rapid growth appears to be part of a global movement that supports more resource conservation and greater reliance on natural ecological processes in preference to the energy- and chemical-intensive systems currently in use.

INTRODUCTION

Although not applicable to every situation, wetland treatment systems have been accepted as alternatives to many traditional and conventional engineering methods across a wide range of municipal, industrial and agricultural uses. In some cases, wetland systems do not replace, but merely complement or enhance, the performance of the mechanical and/or chemical systems already in place (Haven and Lothop, 1992; Anderson, 1993; Todd and Todd, 1994). The life support functions and values of natural wetlands have long been recognized. Their high primary biological productivity creates the potential to accumulate, transform and cycle organic material and nutrients, as well as providing a major sink for metals (Newman, 1993; Bastian and Hammer, 1993; Knight *et al.*, 1993). It is precisely for these material-processing capabilities that wetlands are exploited for a wide range of applications, namely in the treatment of municipal and domestic waste water, acid mine drainage, industrial waste water/effluent, agricultural waste water/effluent, urban storm-water runoff, compost leachate, sludge, pulp mill effluent, and landfill leachate. Although the first two applications — the treatment of municipal or domestic waste water and acid mine drainage — are by far the most abundant, there is a strong interest in incorporating wetlands as a key component in the management of urban storm-water runoff (Kadlec and Knight, 1995).

Historical Development

The earliest research specifically investigating the waste-water potential of wetlands was by Seidel at the Max Planck Institute in Germany in 1952. She began by examining the possible removal of phenols from waste water by *Scirpus lacustria*. Four years later, she used the same plant in experiments in dairy waste water and continued with her research with wetland plants and waste water until the mid-1970s (Bastian and Hammer, 1993). One of her students, Kickuth, studied the effectiveness of a natural reed marsh in treating municipal waste-water effluent. Upon the successful demonstration of the technical merits of this natural filtration system, more than 200 municipal and industrial waste treatment systems across Europe began using emergent macrophyte-based wetland systems, either as components of their conventional systems or as stand-alone systems. In the United Kingdom, the first emergent macrophyte treatment system, often referred to in the British literature as a rooted bed system, was set up in 1985, and by the summer of 1988, 81 beds had been constructed at 27 separate sites. More than 130 of these systems were constructed in Denmark between 1983 and 1988, while similar wetland systems are also operational in Belgium, Netherlands, Hungary and Sweden.

Developments in the United States and Canada

Throughout the 1970s, studies were carried out by numerous universities and government agencies. Research by the United States Environmental Protection Agency (US EPA), the United States Army Corp of Engineers, and the United States Department of Agriculture investigated wetlands as alternatives to chemical treatment systems. University research was focussed on two areas: municipal waste-water treatment and acid mine drainage (Wildeman, 1993), with the principal researchers being Kalbec and Kalbec at the University of Michigan, and Odum and Ewel at the University of Florida (Taylor, 1992). The first attempts at constructing wetlands for waste-water treatment were carried out at the Brookhaven National Laboratory in New York, followed by pilot scale studies at Santee and Arcata in California in the early 1980s. The earliest small-scale waste-water systems were constructed at Listowel, Ontario, Iselin, Pennsylvania, and Arcata, California, all of which became operational in the mid-1980s.

The application of wetlands for acid mine drainage (AMD) developed very quickly following studies of wetlands in the mid-1970s. Ironically, these studies were initiated to examine the degradation AMD had on wetlands in the Ohio and West Virginia coalfields, and not for evaluating wetlands as alternatives to the application of lime. The earliest pilot experiments appeared in 1978, with fully operational systems in place by 1982 (Taylor, 1992). By 1991, more than 200 wetland systems were treating municipal, industrial and agricultural waste waters and effluent in North America. In addition to Europe, systems are now operating in Australia, China, Egypt, Columbia, Brazil, India and South Africa (Cooper and Findlay, 1990). International conferences and symposia featuring wetlands have been held every few years since 1972. The proceedings of the last four major meetings resulted in the publication of more than 250 papers in four monographs (Hammer, 1989; Cooper and Findlay, 1990; Moshira, 1992; Mitsch, 1994). An excellent overview on the potential use of constructed wetlands for storm-water management in southern Ontario is presented by M.E. Taylor and Associates (1992). Numerous technical articles have appeared in a variety of journals, most notably *Water Environment Technology*.

NATURAL WETLANDS

Definition and Attributes
Wetlands is a generic term that spans the spectrum from mangrove and cypress swamps, through fresh-water and salt-water marshes, to bogs and fens (Hammer, 1991). Despite this variety, all wetlands share three common attributes. First, wetlands have soils, or substrates, that are saturated for long periods, or for much of the growing season. For that reason, they contain vegetation types with specialized structures that transport oxygen to their roots for respiration; this enables these plants to grow in an otherwise hostile environment. Roots of most terrestrial plants obtain oxygen for respiration from gases within soil pore spaces and if those spaces are filled with water lacking oxygen, the plant dies. Hydrophytic, or wetland plants have developed specialized physical structures called aerenchyma, best described as bundles of drinking straws, to transport atmospheric gases, including oxygen, through leaves and stems down to the roots to provide oxygen for respiration. Aerenchyma also transport respiratory by-products and other gases generated in the substrate back up the roots, stem and leaves for release to the atmosphere, thereby reducing potentially toxic accumulations in the region of the growing roots.

Second, inundation and aerobic conditions also cause specific changes in chemical substances found in most soils. Anoxic substrates with reducing environments cause many elements and compounds to occur in reduced states, creating characteristic colours, textures and compositions typical of hydric soils. Owing to the prevalence of iron in many of these soils, and its colour in reduced states, wetland soils often have a grey or greyish colour and a fine texture.

Third, wetlands are among the most productive eco-systems in the world. This is largely the result of abundant sources of water and nutrients, and the development of plants that have adapted to take full advantage of these optimum conditions. Table 1 shows ranges of primary productivity of selected species and systems under both tropical and temperate conditions. Agricultural productivity in both climatic zones is provided for comparison (Smith, 1992; Newman, 1993). The high primary productivity in wetlands results in a high microbial activity, which, in turn, leads to a high capacity to decompose organic matter and other substances.

Wetland Functions and Values
Wetlands have both functions and values, terms which, although often used interchangeably, are not synonymous. Function describes what a wetland does irrespective of any beneficial worth assigned by humans. It can be an objective process such as water purification, or an objective product such as mosquitoes produced per square metre per day. A value is a subjective interpretation of the relative worth of some wetland process or product such as wild rice, or of a recreational use, say duck hunting. Values can be negative, such as the cost of eradicating the mosquitoes, or positive, such as the flood storage capacity upstream from an urban area.

The functional values of wetlands most often cited are: (1) life support, which includes all types of microbial, invertebrate and vertebrate animals, and microscopic and macroscopic plants; (2) hydrologic modification, which includes flood storage and, conversely baseflow augmentation, ground-water recharge and discharge, altered precipitation and evaporation, and other physical influences on waters; (3) water quality changes, which include addition and/or removal of biological, chemical and sedimentary substances, changes in dissolved oxygen, pH and Eh, and other biological or chemical influences on waters; (4) erosion protection, which includes bank and shoreline stabilization, dissipation of wave energy, alterations in flow patterns, and current velocity; (5) open space and aesthetics, which include outdoor recreation, environmental education, research, scientific influences, and heritage preservation; and (6) geochemical storage, which includes carbon, sulphur, iron, manganese and other minerals.

It is obvious that wetlands are extremely multi-functional and, as will be shown below, wetlands constructed for one particular function or value will also provide other functions, and/or create other values (Hammer, 1991).

Wetland Hydrology
Hydrological factors, in particular water depth, modify or determine the structure and functioning of wetlands by controlling the composition of the plant community and, in turn, the animal community. Adaptations to inundation vary considerably, with fewer and fewer species able to survive under longer, deeper flooding. In general, those sites with short-term and/or shallow flooding will support a higher biodiversity of both plants and animals. Under prolonged inundation, many nutrients are immobilized under reducing conditions in the substrate, and are unavailable to plants. Periodic drying and oxidation returns these substances to the nutrient cycles, resulting in an explosive growth response. Changes in oxygen availability and concentration caused by inundation strongly influence decomposition rates (Hammer, 1991).

Table 1   Net primary productivity of selected plants and ecosystems.

VEGETATION TYPE	ANNUAL ORGANIC PRODUCTIVITY (tonnes dry weight/ha/year)	
Free-floating macrophytes *Eichhornia crassipes* (water hyacinth)	106–162	tropical
Emergent-rooted macrophytes *Typhi* sp. (cattails) *Phragmite* sp. (reeds)	18–97	temperate to tropical
Fully submergent macrophytes	12–20	tropical
	6–8	temperate
Reed swamps	64–87	tropical
	32–59	temperate
Phytoplankton	1–4	tropical
	7–18	temperate
Agriculture	24–36	tropical
	19–26	temperate

Wetland Soils

Wetland soils are the main medium for many chemical transformations, and serve as the principle reservoir for minerals and nutrients needed by other plants and a variety of other organisms. As previously mentioned, the principle differences with an upland soil are an abundance of water replacing air that typically fills soil pores or voids, and the isolation of the soil system from atmospheric oxygen. As a result, only a very thin (~m) boundary layer at the soil surface has adequate oxygen to maintain aerobic/oxidizing conditions, and almost everything below is anaerobic/reducing. Shortly after a soil is flooded, any oxygen present is consumed by microbial organisms and chemical oxidation. Diffusion of oxygen through water is many orders of magnitude slower than through well-drained soils, and so the lower layers quickly become, and remain, anaerobic. Many of the interrelated physical and chemical changes that occur are because of limited oxygen, rather than the direct effect of excess water.

Wetland soils are generally considered as hydric soils because they are saturated long enough to develop anaerobic conditions during the growing season to support hydrophytic vegetation. Hydric soils are subdivided into mineral soils with 12-20% organic matter, and organic soils with more than 12-20% organic matter. In well-developed wetlands the upper layers are often organic soils. Organic soils have a high percentage of pore spaces (80%), and, consequently, higher water-holding capacities than mineral soils. Organic soils also have a greater cation exchange capacity, and the major cations are different than in mineral soils. Metal cations (Ca^{2+}, Mg^{2+}, Na^+) dominate in mineral soils, while H^+ dominates in organic soils. Saturation and loss of oxygen generally cause wetland soils to have negative redox potentials; however, with fluctuations in water levels, the Eh can range from −300 to +300 mV. The pH of wetland soils varies from strongly acidic (3) to strongly alkaline (11), although most wetlands are neutral. Typical wetland soils have a pH of 7 and an Eh of −200 mV, in which case common substances occur in reduced form, nitrogen as N_2O, N_2, or NH_4^+, iron as Fe^{2+}, manganese as Mn^{2+}, carbon as CH_4, and sulphur as S^-. Decomposition rates under anaerobic conditions are 10% of aerobic decomposition rates and frequently much lower than carbon fixation or biomass production rates. It is these fluctuating conditions, characteristic of what is termed a pulsating ecosystem, that provide the conditions for a wide range of complex reactions, which in turn permit the wetland to be a sink for such a large variety of substances.

Wetland Vegetation

Throughout the literature, many terms are commonly applied to wetland plants. Some of these are phytoplankton, non-vascular aquatic plant, vascular aquatic plant, hydrophyte, aquatic microphyte, vascular hydrophyte, and, simply, aquatic plant. Plankton implies small and current borne, *i.e.*, suspended or floating, and with no rooted attachment. Non-vascular refers to small, simple plants that lack internal transport mechanisms. Macrophyte simply means larger than microscopic.

Common to all wetland plants is their ability to grow in an environment that is periodically but continuously inundated for more than five days during the growing season. Typically, this includes upland plants capable of surviving five days of flooding or saturated soils, as well as deep-water, rooted vascular plants in depths of 7-8 m in very clear waters. However, the vast majority of wetland plants are limited to water depths of less than 2 m.

Wetland plants are divided into free-floating and rooted forms, with the rooted group subdivided into submergent, emergent and floating-leaved types. Woody species can range from low-growing shrubs to towering cypress. As will be seen later, constructed wetlands are classified according to which wetland plant type is used, and descriptions of these plant types are found elsewhere in this paper. The five general wetland types common to southern and central Ontario, together with their pertinent features, are summarized in Table 2.

Problems in using natural wetlands for waste-water treatment. As identified above, natural wetlands are characterized by extreme variability in functional components, making it virtually impossible to predict responses to waste-water application and to translate results from one geographical location to another. Although significant improvement in the quality of the waste water is generally observed as a result of flow through natural wetlands, the extent of their treatment capacity is largely unknown. In addition, their performance may change over time as a consequence of changes in species composition and accumulation of pollutants. Therefore, the treatment capacity of natural wetlands is unpredict-

Table 2	Wetland types of southern and central Ontario (after Taylor, 1992).			
TYPE	WATER TABLE LEVEL	SURFACE WATER CHARACTERISTICS	SOIL	VEGETATION
BOG	High	Slow moving Acidic, pH <4.6 Mineral poor (low Ca, Mg)	Peaty Upper layer deficient in minerals Root zone low mineral levels	Sphagnum mosses Heath shrubs Low stunted trees
FEN	High	Very slow moving Alkaline High levels of Ca^{2+}, Mg^{2+}	Peaty Moderate mineral levels	Graminoid (grassy) fens Shrub fens Treed fens
SWAMP	Seasonally variable From complete submergence to below root zone, allowing surface layers to be aerated	Standing or gently moving Neutral to slightly acidic Moderate loadings of minerals	Peaty or mineral High nutrient level Good vegetation growth	Woody with deciduous or coniferous trees
MARSH	Daily or seasonally variable From high complete submergence to below root zone	Moving waters Neutral Well oxygenated	High organic content	Floating aquatics and emergents such as reeds, sedges or rushes (cattails)

able, and the design criteria for constructed wetlands cannot be extracted from results obtained in natural wetlands. There are still too few data from natural systems to allow performance predictions of the treatment capabilities of natural systems and the receiving ecosystems. Moreover, intentional or unintentional use of natural systems for remediating waste waters could lead to serious damage to the ecosystems, from which recovery could take hundreds or even thousands of years (Johns, 1995). Most workers and researchers recommend that natural systems be preserved for their other multifunctional values, and not deliberately used as waste-water treatment systems (Hammer, 1991; Brix, 1993).

CONSTRUCTED WETLANDS

Advantages of Constructed over Natural Wetlands in the Treatment of Waste Waters

Pollutant removal in all natural systems involves a combination of physical, chemical and biological processes. Sedimentation, precipitation, adsorption to soil particles, assimilation by the plant tissue, volatilization and microbial transformations are continuously taking place according to the schedule and needs of the ecosystem. These processes often occur at rates and on schedules that are unsuitable for treating large amounts of waste water. Constructed wetlands can be built with a much greater degree of control of substrate, vegetation types, and flow characteristics to enhance these natural processes. Other advantages are site selection, flexibility in sizing, and significantly, control over the hydraulic pathways and retention times (Brix, 1993).

Types of Constructed Wetlands and Pollution Removal Mechanisms

The dominant use of macrophyte- (wetland plant-based)

treatment systems is in the treatment of municipal and residential waste water, where four main types of systems are used. These systems are classified according to the life form of the dominating macrophyte. Systems for other purposes

Table 3	Removal mechanisms in macrophyte-based waste-water systems (from Brix, 1993).
CONSTITUENT	**REMOVAL MECHANISM**
Suspended Solids	• Sedimentation, filtration
Biochemical Oxygen Demand	• Microbial degradation: aerobic and anaerobic • Sedimentation: accumulation of organic matter/sludge on the sediment surface
Nitrogen	• Ammonification followed by microbial nitrification and denitrification • Plant uptake • Ammonia volatilization
Phosphorus	• Soil sorption: adsorption/precipitation reactions with Al, Fe, Ca and clay minerals in the soil • Plant uptake
Pathogens	• Sedimentation, filtration • Natural die-off • Ultra violet radiation • Excretion of antibiotics from roots of macrophytes
Trace Metals	• Adsorption and complexation with organic matter • Plant uptake • Microbial transformations

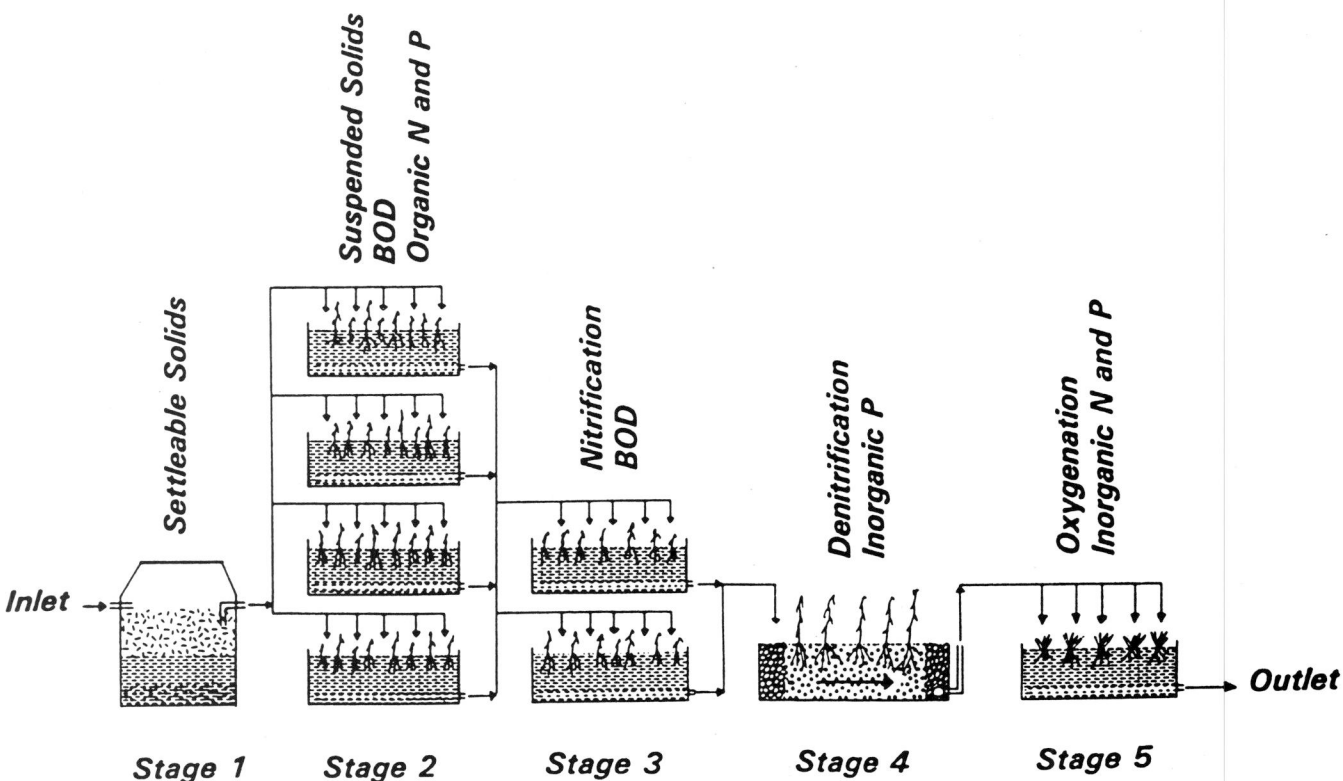

Figure 1 *Multi-stage constructed wetland showing where removal of waste-water components occurs within the system (after Taylor, 1992).*

are often no more than modified versions of these four systems: (1) free floating, (2) rooted emergent, (3) submergent, and (4) multi-stage systems consisting of a combination of (1) to (3) with the addition of other kinds of technology systems such as oxidation ponds and sand filtration systems.

Removal Mechanisms in Macrophyte-based Waste-water Treatment Systems

Regardless which type of system is employed, all have common removal mechanisms (Table 3). Figure 1 is a schematic showing a multi-stage macrophyte system, illustrating where the removal of the various waste-water constituents listed in Table 3 takes place. Typical systems can involve single cells, cells in series, in parallel, or both.

Free-floating Macrophyte Systems

Free-floating macrophytes are highly diverse in form and habit, ranging from large plants with aerial and/or floating leaves and well-developed submerged roots, such as the water hyacinth, *Eichhornia crassipes*, to minute surface-floating plants with few or no roots, such as various duckweeds, *Lemna*, *Spirodella*, *Wolffia*. Figure 2 is a schematic of a cell with free-floating macrophytes.

Water-hyacinth-based systems. The water hyacinth is one of the most prolific and productive plants in the world. Its rate of growth in the tropics and subtropics is so rapid that it is often regarded as a severe weed, blocking irrigation canals, clogging rivers and lakes, and generally making drainage difficult. In constructed wetlands, it is an ideal species for nutrient removal. When used in the tertiary stage of waste-water treatment (the nutrient removal stage) the water hyacinth removes nitrogen and phosphorus by uptake directly into the biomass. The biomass is harvested frequently to maintain maximum plant productivity and to remove the nitrogen and phosphorus. Some nitrogen may also be removed as a result of microbial denitrification.

Combined secondary and tertiary waste-water systems involve the removal of both biochemical oxygen demand (BOD) and nutrients. In these systems, the degradation of organic matter and the microbial transformations of nitrogen proceed simultaneously, so that the plant is only harvested for maintenance purposes. This system must also include free water surfaces to allow oxygen to be transferred into the water from the atmosphere by diffusion, and to provide areas where algal oxygen production can occur. Retention time is on the order of 5-15 days.

Most of the suspended solids are removed by sedimentation and subsequent degradation. Electrical charges associated with hyacinth roots are reported to react with opposite charges on the suspended solids, causing them to adhere to roots. They are slowly digested and assimilated by the plant and microorganisms. The extensive root system provides a huge surface area for attaching microorganisms, which, in turn, increases the potential for the decomposition of organic matter.

Water hyacinth systems are severely affected by frost. The growth rate is greatly reduced at air temperatures of less than 10°C, thus open-air applications are only possible in tropical and subtropical climates. In temperate climates, they can be used year-round in greenhouses, but only outdoors in the summer. For winter operations, pennywort, *Hydrocotyle umbellata*, a hardier plant with similarly high growth and uptake rates, can be substituted for water hyacinth (Brix, 1993).

Duckweed-based systems. Duckweeds have a much wider geographical range than water hyacinths and are able to grow at temperatures as low as 1° to 3°C. However, they lack an extensive root system, and, therefore, provide a smaller area for attached microbial growth. The main use of duckweeds is for recovering nutrients from secondary waste water.

A dense cover of duckweed inhibits sunlight penetration for oxygen production through phytoplankton photosynthe-

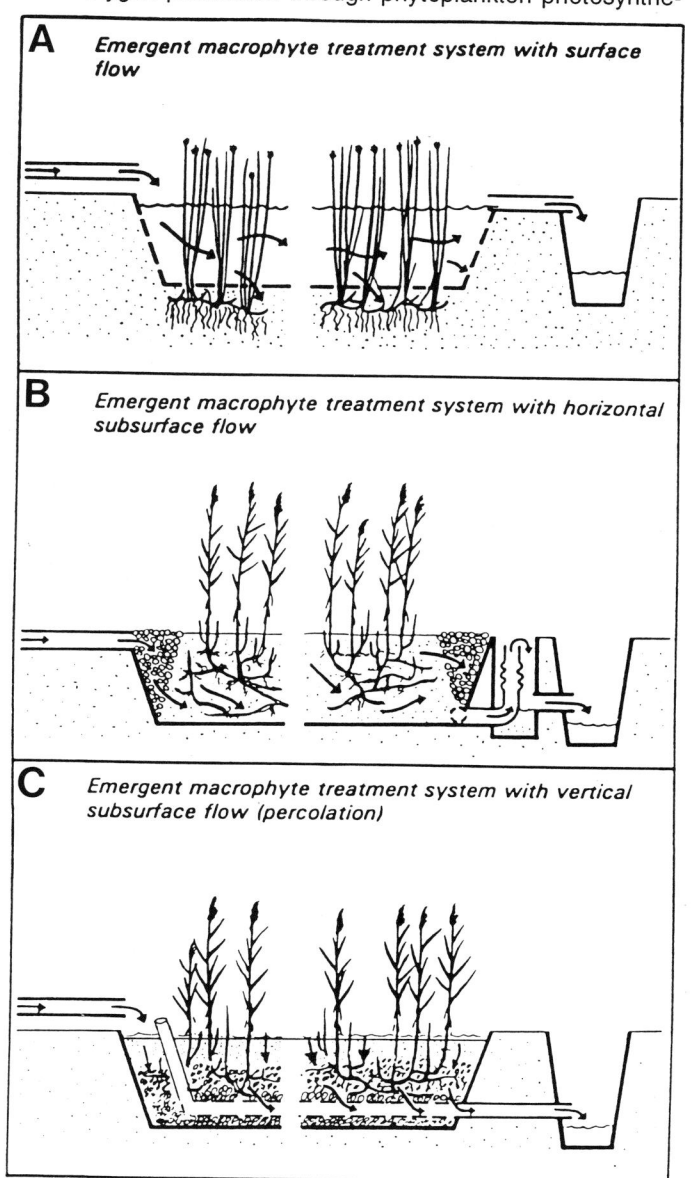

A *Emergent macrophyte treatment system with surface flow*

B *Emergent macrophyte treatment system with horizontal subsurface flow*

C *Emergent macrophyte treatment system with vertical subsurface flow (percolation)*

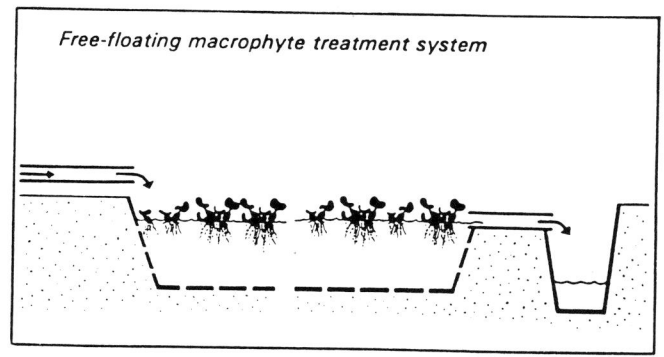

Free-floating macrophyte treatment system

Figure 2   *Typical free-floating macrophyte treatment system. Depth of system is usually less than 1.5 m (after Taylor, 1992).*

Figure 3   *Different flow systems associated with emergent macrophyte wetlands (after Taylor, 1992).*

sis, and also prevents the diffusion of oxygen into the water. As a result, the water quickly becomes anaerobic, thereby favouring denitrification. Duckweeds are easily harvested, and the nutritive value of the biomass contains twice as much protein, fat, nitrogen and phosphorus as an equivalent mass of water hyacinth. Retention times vary with the waste-water quality, the effluent quality desired, and climate, but are about 30 days in the summer, and several months in the winter. Winds can easily sweep the duckweeds into piles, so that barriers on the water surface are normally required.

Emergent Aquatic Macrophyte-based Systems
Rooted emergent aquatic macrophytes dominate most natural wetlands. They grow in water up to 1.5 m deep, producing aerial stems and leaves along with extensive root and rhizome systems. Typical species are (1) common reed (*Phragmites australis*), (2) cattail (*Typha latifolia*), and (3) bulrush (*Scirpus lacustris*). Three common system designs are shown in Figures 3A to 3C. All three are suitable for use in southern and central Ontario.

Emergent macrophyte-based systems with surface flow (Cattail-Bulrush). This system typically consists of ditches, 3-5 m wide and more than 100 m long, planted with bulrushes. The submerged portions of the stems and litter serve as the substrate for attached microbial growth. As will be described in more detail in the next section, most of the transformations take place in the soil and rhizosphere. The plant stems capture some of the suspended sediment, and aid in sedimentation by retarding the flow. A considerable amount of waste water can also drain out through the unsealed bottom (Fig. 3A). These systems have been used in Holland for almost 30 years.

Emergent microphyte-based systems with horizontal subsurface flow (Common Reed). This is the original system developed in Germany, and there are now several hundred of these systems in operation in Germany, Denmark and the United Kingdom. Typically, it consists of a bed of soil or gravel, planted with common reed, and underlain by an impermeable membrane to prevent seepage. As the waste water passes through the rhizosphere of the reeds, organic matter is decomposed microbiologically, nitrogen may be denitrified, and phosphorus and heavy metals fixed in the soil. In this system, the main function of the reeds is to supply oxgyen through aerenchyma to the heterotrophic micro-organisms in the rhizosphere (Fig. 3B). Uptake of nutrients in the plant tissue is negligible.

An evaluation of these systems shows that suspended sediments and BOD are generally removed effectively, with the effluent attaining advanced secondary treatment quality. Removal efficiencies for nitrogen and phosphorus are variable. Too-rapid runoff, thus preventing the waste water from coming in contact with the rhizosphere, is a common problem in all soil-based facilities, and the oxygen transport capacity of the reeds is often insufficient to ensure aerobic decomposition in the rhizosphere and for nitrification.

Emergent macrophyte-based systems with vertical subsurface flow. This system consists of several beds laid out in parallel. Percolating flow and intermittent loading increases soil oxygenation several-fold compared to horizontal subsurface flow systems. During the loading stage, air is forced out of the soil, while during the drying stage, atmospheric air is drawn into the soil pore spaces, replenishing the soil with oxygen (Fig. 3C). These alternating oxidation and reducing conditions in the substrate stimulate sequential nitrification-denitrification and promote phosphorus adsorptions. Treat-

ment performance of these systems is apparently very good with respect to suspended solids, BOD, ammonia and phosphorus (Brix, 1993).

Submergent Macrophyte-based Systems
Plants characteristic of these systems have their photosynthetic tissue entirely submerged, and assimilate nutrients from polluted waters (Fig. 4). As their growth is limited to well-oxygenated waters, they cannot be used for waste water with a high content of readily biodegradable organic matter, because the microbial decomposition of the organic matter will create anoxic conditions. These systems are mainly used for polishing secondary treated waste waters, and their prime area of application is as a final step in multistage systems. Some of the plant species under consideration are egeria, elodea, hornwort and hydrilla.

Multistage Macrophyte-based System
The numerous individual systems previously described may be combined with one another or with conventional treatment technologies. Two such systems are currently operational:
The marsh-pond-meadow system. This consists of (1) a bar screen and an aeration cell using a floating surface aerator, (2) a lateral-flow marsh planted with cattails in a sand medium, (3) a pond with aquatic macrophytes and herbivorous fish, (4) a meadow planted with red canary grass, and (5) a chlorination chamber. The removal efficiency is reported to be 77% for ammonia nitrogen and 82% for total phosphorus.
The Max-Planck-Institute Process. This design is used in France and was a model for a system implemented in Oaklands Park, United Kingdom. The system consists of four or five stages in cascade, each with several basins laid out in parallel and planted with emergent macrophytes in gravel. The flow pattern in the first two stages is vertical, while the final ones have horizontal flow. In the French system, removal of suspended solids and BOD is good, but with poor results for nitrogen and phosphorus, perhaps because of high loading rates. The United Kingdom system produces a nitrified effluent and 98% reduction of BOD and suspended solids (Brix, 1993).

EVALUATING WETLAND PERFORMANCE
The data presented in the previous sections should leave little doubt as to the potential of constructed wetlands in addressing at least some of the water problems prevalent in urban areas. However, as with any emerging technology, after the initial optimism comes the need to demonstrate an acceptable level of performance.

Knight *et al.* (1993) have catalogued information on wetland sites throughout North America and have released pre-

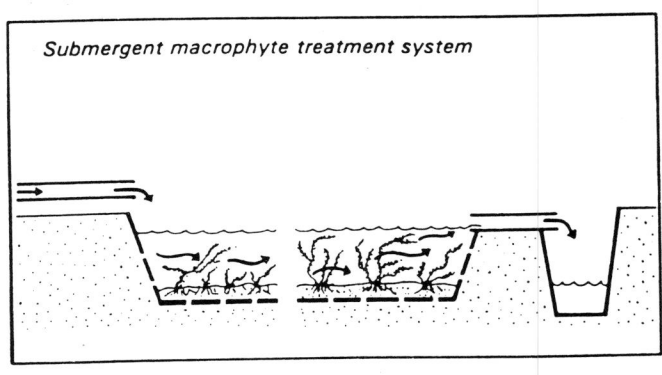

Figure 4 *Typical submergent macrophyte treatment system (after Taylor, 1992).*

liminary numbers (Table 4). Knight *et al.* (1993) list 127 systems at 96 sites; about 90% of the systems have operated less than three years. As a consequence, many data are preliminary and based on less than one year of operation. Only six systems at four sites have been operational for at least five years. Most systems are small, on pilot scale, often seasonal, treating less than 1000 $m^3 \cdot d^{-1}$. Only five systems averaged more than 10,000 $m^3 \cdot d^{-1}$. Some are treating primary effluent, some secondary, while many are polishers, or act as tertiary treatment systems. Some focus on one particular pollutant, while others are attempting to address the full complement. In general, a comparison between systems is difficult, if not impossible. A more realistic evaluation of performance can be determined by examining those systems with 5-15 years of operation statistics (Table 5).

Although none of these systems is capable of removing all of the pollutants at acceptable removal rates, what they do remove, they do extremely well, and so it is obvious why these systems are still operating after all these years. In addition to the group identified in Table 5, two systems not on the list of Knight *et al.* (1993) serve as excellent examples of the utility of constructed wetlands in remediating waste waters. Other examples are described by Jewell (1994).

American Crystal Sugar Company, Hillsboro Wetlands, North Dakota

This sugar beet operation releases 300-600 million litres of water into the Red River during the 210-day processing period (Haven and Lothop, 1992). Over the past 15 years, as all sugar refiners have done, the company has invested in many types of waste-water treatment systems that have high energy requirements and require skilled operators. A biological treatment process was being used, but the design capacity was often exceeded, resulting in a poor quality effluent. The end result was that years of discharging the nutrient-rich effluent into the stabilization pond, which, unfortunately, also served as the water supply reservoir, resulted in serious degradation of water quality. In 1989, the company began constructing a 64-ha wetland which, in addition to the seven cells, ponds and lagoons used for treatment, included nine nesting islands and 10 ha of grasslands for waterfowl habitat. In 1990, 30,000 cattails were planted in the first two cells and these quickly spread to other areas. In 1991, flow to the wetlands averaged 21.9 $L \cdot s^{-1}$, with an average BOD of 100 $mg \cdot L^{-1}$. Overall BOD removal efficiency was 85%, peaking at 93%, and well below the 25 $mg \cdot L^{-1}$ discharge limit. The 1992 flow was considerably increased through four cells, and the remaining three cells

---

Table 4   Performance data for constructed wetlands (from Knight *et al.*, 1993).

Waste-water source: 89 municipal (mun), 22 industrial (ind), 5 storm water, 11 unclassified
Treatment systems used: 69% free water type, 31% vegetation submergent type
Wetland area: 0.02 ha to 1093 ha
Vegetation types used: cattail, bulrush, pickerel weed, duck potato, duckweed, sedges, grasses (in decreasing abundance)

POLLUTANT	INFLOW ($mg \cdot L^{-1}$)	OUTFLOW ($mg \cdot L^{-1}$)	REMOVAL EFFICIENCY		PERMITTED LEVEL ($mg \cdot L^{-1}$)
Biochemical Oxygen Demand (BOD)	38.8	10.5	73%	59-90%	5-30
Total Suspended Solids (TSS)	49.1	15.4	69%	49-94%	10-30
Nitrate (NH_3-N)	7.5	4.2	44%	—	1-8 (mun) 50 (ind)
Total Nitrogen (TN)	14.0	5.0	64%	30-98%	—
Total Phosphorus (TP)	4.2	1.9	55%	20-90%	—
	AVERAGE	AVERAGE	AVERAGE	RANGE	

(The above data are based on n = from 28 to 58)

Table 5   Performance data for constructed wetlands in use for at least five years (from Knight *et al.*, 1993).

SYSTEM	FLOW ($m^3 \cdot d^{-1}$)	YEARS OF OPERATION	REMOVAL RATES (%)				
			BOD	TSS	NH_3-N	TN	TP
Reedy Creek	12,058	11	64	73	76	82	13
Bellaire	572	11	—	—	88	—	89
Houghton Lake	3374	15	—	—	93	97	—
Kinross	1350	15	64	94	98	—	—
Drummond	265	6	—	—	14	85	—

Notes: BOD = biochemical oxygen demand; TSS = total suspended solids; NH_3–N = nitrate; TN = total nitrogen; TP = total phosphorus

were completed later that year.

Total project cost for the wetlands was US $1.6 million, or about 20% of the cost of the biological treatment system. The wetlands have also become a popular haven for migrating waterfowl.

City of Orlando, Florida
(Easterly Wetlands Reclamation Project)
In the early 1980s, one of Orlando's waste-water facilities was operating near capacity with no opportunity to increase its existing waste load allocation, and the city sought alternative means of tertiary treatment and effluent disposal (Anderson,

1993). In 1984, a 12-ha water hyacinth treatment system was devised as an interim solution. This system was so successful at removing nitrogen and phosphorus, that the interim solution was treating 16 million litres per day. A 506-ha wetland, capable of treating 80 million litres per day of effluent, was then constructed, and full-scale operations began in 1987. Effluent is detained in the wetland for 30 days, and discharged into the environmentally sensitive St. John's River. The performance of the wetland is outstanding; treated effluent consistently has <1.0 mg·L^{-1} nitrogen and <0.1 mg·L^{-1} phosphorus. The wetland's success is also evident by the abundance of wildlife in the area, now designated as the Orlando Wilder-

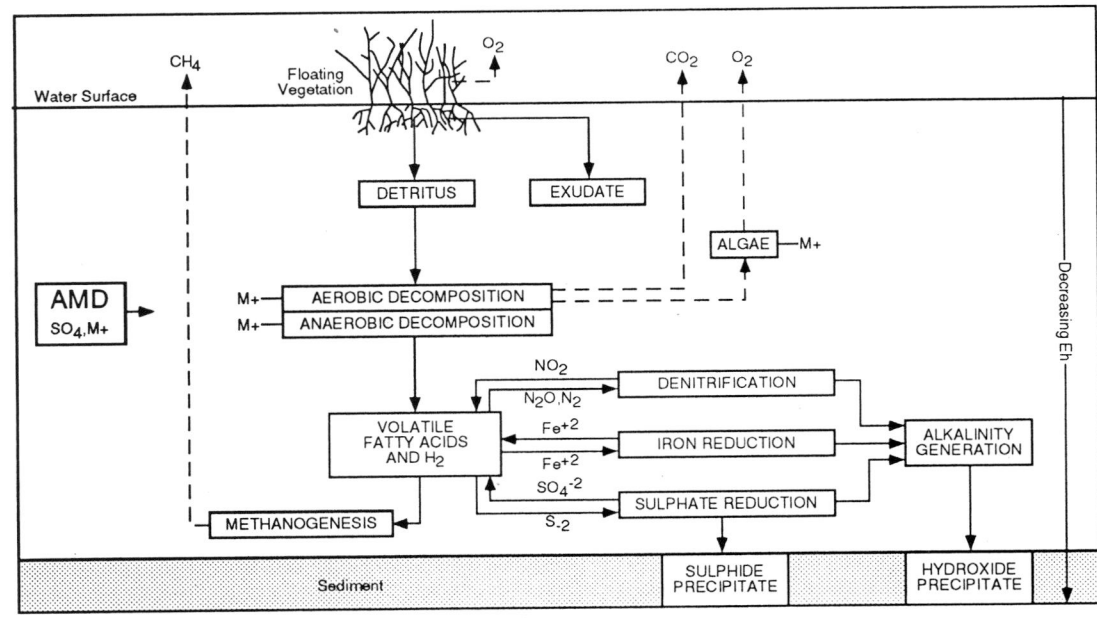

Figure 5 *Macrophyte-based wetland system for remediation of acid mine drainage (AMD) showing principal biogeochemical processes.*

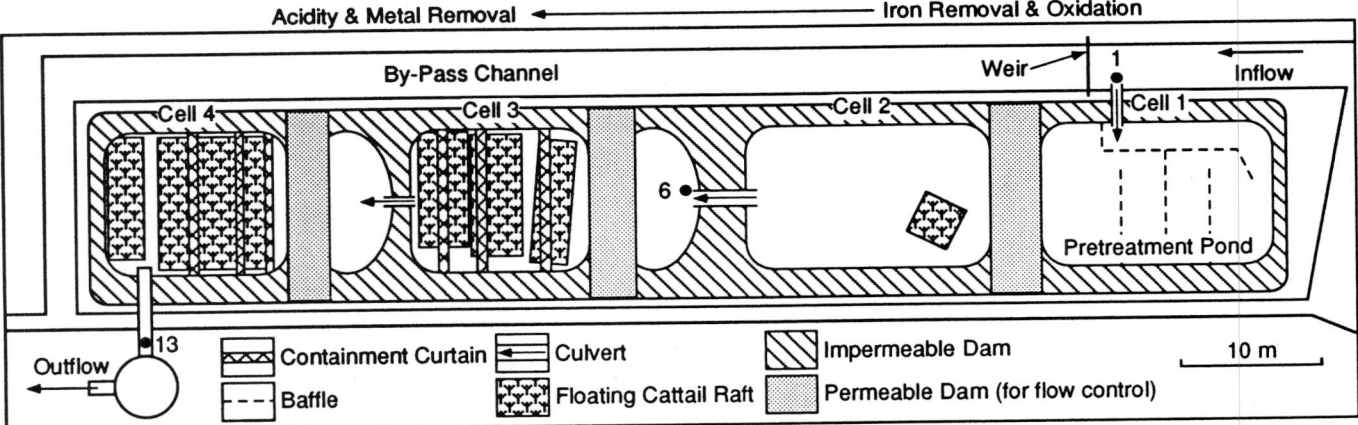

Figure 6 *Structure of a wetland used to remediate acid mine drainage at Sudbury, Ontario, with performance data.*

	July Flow = 1.125 L·min^{-1}			August Flow = 0.24 L·min^{-1}			October Flow - 0.1 L·min^{-1}		
	Stn 1	Stn 6	Stn 13	Stn 1	Stn 6	Stn 13	Stn 1	Stn 6	Stn 13
Temp [1]	17.2	15.5	21.5	15	20.9	19.3	8.4	8.2	8.1
pH	5.65	3.05	5.52	5.76	2.96	3.34	5.70	2.94	6.14
Eh [2]	284	687	305	260	688	653	289	680	234
Acidity [3]	1215	599	315.9	131	100	27.7	84.9	35.9	23.4
Al [3]	<1.6	29.2	<1.6	0.01	7.02	0.24	0.32	2.76	0.14
Cu [3]	<1.6	7.06	<1.6	0.01	1.9	0.05	0.03	0.4	0.001
Fe [3]	505	27.4	59	81.2	3.12	1.71	32	2.52	3.18
Ni [3]	60.4	70.6	14	8.71	13.2	1.75	3.66	4.2	0.46
S [3]	1408	948	855	292	237	190	122	86.7	60.1

Notes: [1] temperature in °C; [2] Eh in mV; [3] units are g per day

ness Park, which is somewhat remarkable, since five years previously the area was a cattle pasture.

DISCUSSION

These two case studies illustrate the current capabilities and strengths of constructed wetlands. They demonstrate that outstanding results can be achieved from full-scale systems that are low tech compared with conventional chemical-biological-physical treatment systems. Furthermore, constructed wetlands require considerably lower capital and operating costs. Additional benefits to the environment are obvious.

In both cases, wetlands were used for very well-defined problems, with the focus on only one or two contaminants. This is probably an accurate reflection of the current level of understanding of, and capabilities with, wetland technology. It can be concluded that, when operating under appropriate conditions, constructed wetlands are generally able to achieve high removal efficiencies for BOD, TSS and bacteria from municipal and some forms of industrial waste water (Bastian and Hammer, 1993). Success in ammonia conversion/removal by nutrification and denitrification can be highly variable, dependent on system design, retention times, oxygen supply, and other factors. Phosphorus removal rates tend to vary between projects, and may be effective only for limited periods unless large areas or special media are involved.

The application of constructed wetlands to more complex problems, such as urban storm-water runoff, is still very much problematic, and where they have been used, the focus appears to have been on one particular pollutant. At Lacamas Shores, Washington, for example, 95% of the phosphorus entering a recreational lake that had become eutrophic was from urban runoff. Although the problem with the phosphorus, initially the main concern, was resolved, nitrate concentrations continued to straddle the compliance level (Bautista and Geiger, 1993).

CURRENT AND PROPOSED APPLICATIONS IN ONTARIO

There has been considerable interest and research focussed on constructed wetlands, not only as an alternative for treating municipal waste water, but for addressing other urban environmental problems. Constructed wetlands have been proposed for storm-water management and wildlife habitat enhancement in recently developed regeneration plans for degraded urban watersheds (Taylor, 1992; Marshall Macklin Monaghan, 1992; M.M. Dillon, Ltd., 1993; Crombie, 1994). Research is ongoing for the treatment of a variety of waste waters, including those originating from mining wastes which continue to accumulate within urban centres such as Sudbury.

Municipal and Industrial Waste Water

Several experimental wetlands have been in operation since the mid- and late 1980s; however, scale-up has not yet occurred at any major urban sites. Municipal and industrial waste water is being treated and evaluated in five constructed wetlands in Listowel. Cells at Cobalt, Cochrane and, more recently, in Port Perry, are restricted to municipal waste water. In addition, small waste-water operations have been constructed at locations such as campgrounds, where demand is seasonal, and where the costs of installing a standard treatment system are prohibitive (Taylor, 1992). Monitoring and process modifications are ongoing in an attempt to improve cold weather performance to allow for scale-up.

Mine Waste Water–Acid Mine Drainage

Millions of tonnes of acid-generating waste rock and tailings are generated annually at numerous mine sites throughout Ontario, and these are added to the hundreds of millions of tonnes already present from more than 100 years of mining in the province. As a result, passive biological treatment systems have received much attention as an alternative to conventional chemical treatment of heavy metals and acidity that originate from the mining waste, and which contaminate both surface waters and ground water (Kalin et al., 1995).

A review of the performance data of emergent macrophyte systems operating at sites of former coal mines in the United States Appalachians indicates that the effectiveness of the biological systems has been highly variable, and generally only works well where the pH of the water is initially high, >4.5 (Kalin et al., 1995). In addition, the precipitation of metal

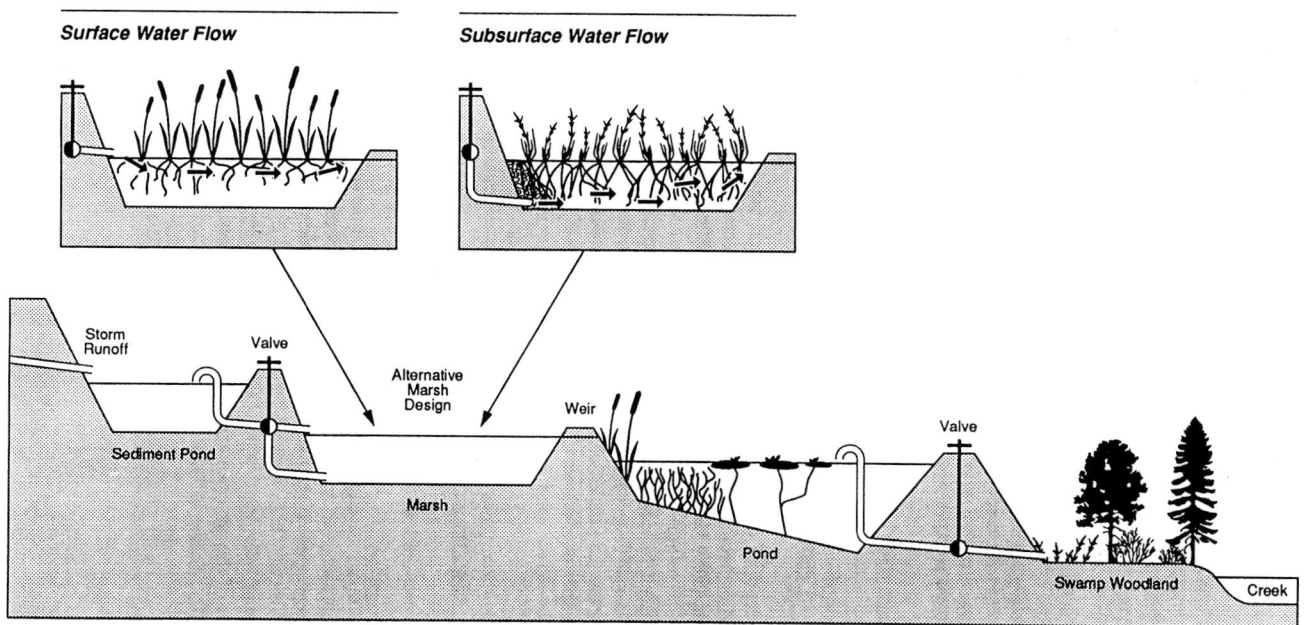

Figure 7 *Schematic of constructed wetland used for remediation of urban storm-water runoff.*

hydroxide alters the hydraulic conditions and leads to system failure. Ontario's base metal mines are high in sulphides and, therefore, the resulting waste water is lower in pH than can be effectively treated with aerobic wetlands. As a result, several alternative systems including compost wetlands, microbial reactor systems, and constructed wetland–sediment systems have been developed and are being tested (Kalin *et al.*, 1995).

A test cell of a constructed wetland–sediment system, referred to as ARUM (Acid Reduction Using Microbiology) has been treating acid mine waste waters from Copper Cliff tailings at Sudbury. In a departure from the processes described earlier in this paper, the constructed wetlands are floating cattail islands. The cattails provide a continuous supply of organic carbon (litter) for decomposition by anaerobic bacteria. In addition, the release of organic acids to the water column complexes metals and converts them to particulates that settle to the sediment. With a retention time of 131 days, a removal of 80-87% of the Ni, 77-98% of the Cu, and 10-20% of the S loadings, as well as 47-73% of the acidity, was achieved during the first year of operation at a flow rate of 1 to 2 L·m^{-1} (Figs. 5 and 6; Kalin *et al.*, 1995).

Storm-water Management

An extensive review and evaluation of constructed wetlands for storm-water management was recently completed by Taylor (1992). In contrast to the treatment of municipal or mining waste water, whose flow rates and compositions tend to be predictable and relatively constant, storm-water flows and compositions vary considerably, depending on the land uses in the catchment basin and the frequency and intensity of rain/snow events. In addition, constructed wetlands may not be effective in treating certain storm-water contaminants, such as road salt and some organic compounds. Pretreatment of sediment using sedimentation/siltation ponds may be necessary to prevent the sediment from choking the wetland soils and rendering them ineffective.

Considering the diverse nature of the contaminants found in urban storm water, the most effective solution appears to be a multi-stage system, such as the marsh-pond-meadow or Max-Planck-Institute process mentioned in an earlier section. A schematic of a multi-stage system with pretreatment sediment pond is illustrated in Figure 7 and the two pilot wetland storm-water projects currently under study/construction in the Toronto area (Emery Creek and Windego Creek-Grenadier Pond) are planned as multistage systems with pretreatment (Marshall Macklin Monaghan, 1992; M.M. Dillon, Ltd., 1993).

Combined Sewer Overflows

In many urban areas, surface water is affected by overflows from combined sewers, which carry both storm runoff and domestic sewage. Being of limited capacity, they overflow during storms and discharge into adjacent watercourses (see Chapter 2). Figure 8 shows a heavily urbanized portion of the city of Scarborough, Ontario, where dry weather flows are directed *via* pumping stations to the main sewage treatment plant. During significant rainfall events, flows in excess of the capacity of the pumping station are directed to an outfall in Lake Ontario. Flows from roof downspouts directly connected to the combined sewer are the primary cause of the combined sewer overflows which occur, on average, about 30 times a year. Some 60,000 kg of sediment, contaminated with metals and polycyclic aromatic hydrocarbons, are deposited in the lake; concentrations for many water quality parameters ex-

ceed Provincial Water Quality Objectives by more than one order of magnitude. Various alternatives for reducing pollutant loadings were evaluated by the city, and Figure 8 shows the design of a system, currently under construction, incorporating a constructed wetland and designed to reduce pollutant loadings to the lake through sedimentation of solids and pollutants with a series of cells (Dunkers Flow Balancing System). Initially, combined sewer overflows will discharge into cell 1, followed by cell 2 and cell 3 if the runoff event is significant. Runoff volumes exceeding the capacity of the first three cells (about 40,000 m^3) will discharge into the lake. After flows subside and settling of suspended particles has occurred (12 hours), water will be pumped from cells 2 and 3 into

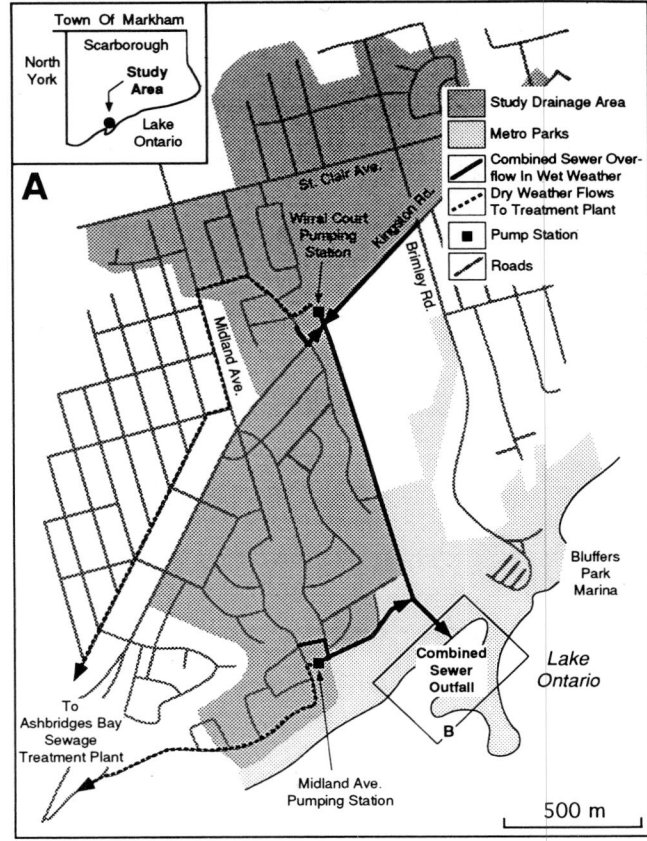

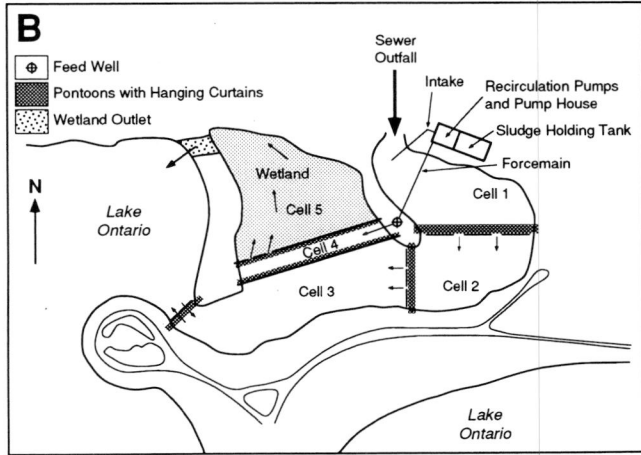

Figure 8 *Design of Dunkers Flow Balancing System at Scarborough, Ontario, designed to remediate overflows from a combined sewer in the adjacent urban area to Lake Ontario (after Aquafor Beech Ltd., 1994). See text for details and Chapter 18 for location.*

1 and thence to cells 4 and 5 *via* a forcemain (Fig. 8). Cell 5 consists of a constructed submergent-emergent wetland that will enhance pollutant removal by the uptake of nutrients. Treated water is then discharged to Lake Ontario. Sludge accumulating within the cells will be periodically removed and dumped on industrial lands. The capital cost of the system is $2.3 million, and annual operating costs are expected to be about $100,000 (Aquafor Beech Ltd., 1994). Mandatory disconnection of roof downspouts and the provision for infiltration of storm runoff will significantly reduce combined sewer overflow volumes.

Wildlife Habitat Enhancement

The construction of wetland areas appears to be a major component of many urban watershed restoration projects. The development of these wetlands should be viewed largely as restoration of natural-wetland areas that originally existed in these watersheds prior to urbanization. Restored natural-wetland areas should not be expected to take on the role of storm-water management. Where this has been allowed to occur, natural wetlands become choked with sediment and heavy metals, resulting in a severe drop in biodiversity. Thus, although increased biodiversity can be expected as an indirect or secondary benefit from wetlands constructed for treating waste water, secondary benefits such as improved stormwater quality from the restoration of natural wetlands are neither desirable nor should they be expected.

Landfill Leachate

In comparison to municipal waste water, landfill leachate is often a higher strength waste water with high levels of not only nutrients and BOD, but also heavy metals and an assortment of organic compounds (see Henry and Heinke, 1989; Chapters 4, 24). As a result, only a few constructed wetland systems are able to treat landfill leachate. The largest of these, in Escambia County, Florida, dilutes the incoming high-strength landfill leachate with rain water before passing it through a multi-stage system that includes constructed wetlands (Dohms, 1993). The landfill leachate forms a very small percentage of the overall volume. In Ontario, pilot scale tests of treating landfill leachate using constructed wetlands are being conducted at a few localities, principally at Storrington, near Kingston (D. Smith, pers. comm., 1995).

REFERENCES

Aquafor Beech Ltd., 1994, Environmental Study report – Brimley Road drainage area water quality enhancement strategy, Executive Summary: 18 p.

Anderson, P., 1993, Constructed wetlands are a sweet deal: Water Environment and Technology, v. 5, p. 56-59.

Bastian, R.K. and Hammer, D.A., 1993, The use of constructed wetlands for wastewater treatment and recycling, *in* Moshiri, G.A., ed., Constructed Wetlands for Water Quality Improvement: Lewis Publishers, Chelsea, MI, p. 59-68.

Bautista, M.F. and Geiger, N.S., 1993, Wetland for stormwater treatment: Water Environment and Technology, v. 5, p. 59-65.

Brix, H., 1993, Wastewater treatment in constructed wetlands: system design, removal processes, and treatment performance, *in* Moshiri, G.A., ed., Constructed Wetlands for Water Quality Improvement: Lewis Publishers, Chelsea, MI, p. 9-22.

Cooper, P.F. and Findlay, B.C., 1990, Constructed Wetlands in Water Pollution Control: Cambridge University Press, Cambridge, UK, 340 p.

Crombie, D., 1994, Regeneration: Toronto's Waterfront and the Sustainable City: Final Report: Royal Commission on the Future of the Toronto Waterfront, Queen's Printer, Ontario, 530 p.

Dillon, M.M., Ltd., 1993, Grenadier Pond Sedimentation Control Study in High Park: report to the City of Toronto, 69 p.

Dohms, P.H., 1993, Hydrology and groudwater monitoring, constructed wetlands system, Perido Landfill, Escambia County, Florida, *in* Moshiri, G.A., ed., Constructed Wetlands for Water Quality Improvement: Lewis Publishers, Chelsea, MI, p. 87-98.

Hammer, D.A., 1989, ed., Constructed Wetlands for Treatment: Municipal Industrial and Agricultural: Lewis Publications, Chelsea, MI, 856 p.

Hammer, D.A., 1991, Creating Freshwater Wetlands: Lewis Publishers, Chelsea, MI, 189 p.

Haven, R.C. and Lothop, T.L., 1992, A decade of accomplishment: Water Environment and Technology, v. 4, p. 40-45.

Henry, J.G. and Heinke, G.W., 1989, Environmental Science and Engineering: Prentice-Hall, Englewood Cliffs, NJ, 336 p.

Jewell, W.J., 1994, Resource recovery wastewater treatment: American Scientist, v. 82, p. 366-375.

Johns, C., 1995, Contamination of riparian wetlands from past copper mining and smelting in the headwaters region of the Clark Fork River, Montana: Journal of Geochemical Exploration, v. 52, p. 193-204.

Kadlec, R.H. and Knight, R.L., 1995, Treatment Wetlands – Theory and Implementation: Lewis Publishers, Boca Raton, FL, 848 p.

Kalin, M., Fyson, A. and Smith, M.P., 1995, Passive treatment processes for the mineral sector: Canadian Institute of Mining and Metallurgy, in press.

Knight, R.L., Ruble, R.W., Kadlec, R.H. and Reed, S., 1993, Wetlands for wastewater treatment: performance database, *in* Moshiri, G.A., ed., Constructed Wetlands for Water Quality Improvement: Lewis Publishers, Chelsea, MI, p. 35-58.

Marshall Macklin Monaghan Ltd., 1992, Emery Creek Environmental Assessment Interim Report: Municipality of Metropolitan Toronto, 76 p.

Mitsch, W.J., 1994, ed., Global Wetlands Old World and New: Elsevier, New York, 967 p.

Moshiri, G.A., 1993, ed., Constructed Wetlands for Water Quality Improvement: Lewis Publications, Chelsea, MI, 656 p.

Newman, E.I., 1993, Applied Ecology: Blackwell Scientific Publications, London, UK, 430 p.

Smith, R.L., 1992, Elements of Ecology: Harper Collins, New York, 230 p.

Taylor, M.E., & Associates, 1992, Constructed Wetlands for Stormwater Management: A Review: Ontario Ministry of Environment and Energy, PIBS 1907 E01, 52 p.

Todd, J. and Todd, N.J., 1994, From Eco-Cities to Living Machines: The Principles of Ecological Design: North Atlantic Books, Berkeley, CA, 99 p.

United States Environmental Protection Agency, 1993, Handbook for constructed wetlands receiving acid mine drainage: Emerging Technology Summary, SITE Program, EPA/540/SR93/523.

Wildeman, T., 1993, Wetland Design for Mining Operations: Bitech Publications, Richmond, BC, 108 p.

13. Basic Mine Drainage in the Montauban Area, Québec [1]

Andrée M. Bolduc

*Commission géologique du Canada, Centre géoscientifique de Québec
2700 rue Einstein, C.P. 7500, Sainte-Foy, Québec G1V 4C7*

Marc R. Laflèche

*INRS-Géoressources, Centre géoscientifique de Québec
2700 rue Einstein, C.P. 7500, Sainte-Foy, Québec G1V 4C7*

Lucie Talbot

Département de géologie, Université Laval, Cité Universitaire, Sainte-Foy, Québec G1K 7P4

SUMMARY
Many thousands of abandoned orphan and semi-orphan mine sites and associated tailings piles occur in Canada. While the problem of acid mine drainage waters is well studied, that of basic mine drainage is not. This paper provides a case study from Montauban-les-Mines, Quebec where mine tailings (mostly sphalerite, galena and metals rich in zinc, lead, gold and silver) are stored in a hilltop dump. Zinc, lead, cadmium, arsenic and other elements are released in the tailings by chemical reactions involving oxidation of sulphides in the residue. Metals are transported by ground and surface waters to the Batiscan River. Contaminants are transported on iron-manganese coatings on sediment particles and desorbed with changes in pH. Downstream, cadmium and zinc concentrations increase in the suspended and stream sediment, and in the dissolved component of stream water. Cadmium, in association with trace metals such as lead and zinc, is toxic to fish; uptake by vegetation may result in further bio-availability to wildlife and humans. The potential environmental impact of basic mine drainage is, as yet, poorly understood, and underscores the need for further evaluation.

INTRODUCTION
Canada's economy has traditionally been resource based. Mineral production has been an important economic activity, resulting in the opening of thousands of mines, small and large. Mines which were opened, operated, and later abandoned, have become a growing environmental concern. Old tailings from early mining activities have been left mostly unattended, resulting in what are now called orphan mine

sites. Partial responsibility can sometimes be attributed to a company, in which case the site is semi-orphan. Many former mine sites are located in urban areas, such as in Sudbury, Ontario (Peters, 1988; Ferrris *et al.*, 1991) and the Abitibi region of Québec (Firlotte *et al.*, 1992). These sites have been the subject of some, or substantial, remediation, because they are in urban settings (Marcotte, 1995). Studies of mine drainage and associated waters allow a better perception of the problem and its solutions.

Generally, mine drainage is acidic. The literature on acid mine drainage (AMD) is thus abundant (see Chapters 12, 14). Basic mine drainage (BMD), however, is not as common a problem and is poorly understood.

SITE DESCRIPTION

The Mine and Its Tailings
For over 80 years, polymetallic deposits were mined from carbonate-rich rocks at Montauban-les Mines, Québec. The minerals extracted were mainly sphalerite, galena, and various minerals rich in zinc, lead, gold and silver (Bernier *et al.*, 1987). Highly oxidized sandy tailings, as well as some early milling residue, are stored in a single tailings area of over 48,640 m². The total depth of the waste, at least 3.5 m, is unknown, but over 2.5 million tonnes of residue are present (R. Grenier, pers. comm., 1994). As with many other mine sites in Canada, tailings have been used in construction projects in the surrounding community. In this fashion, the problem of mine drainage is not limited to the immediate mine site. Some tailings were used to build segments of an access road to the Batiscan River, others were spread in the surroundings and are now partially vegetated. Two small settling

[1] Geological Survey of Canada Contribution No. 664994

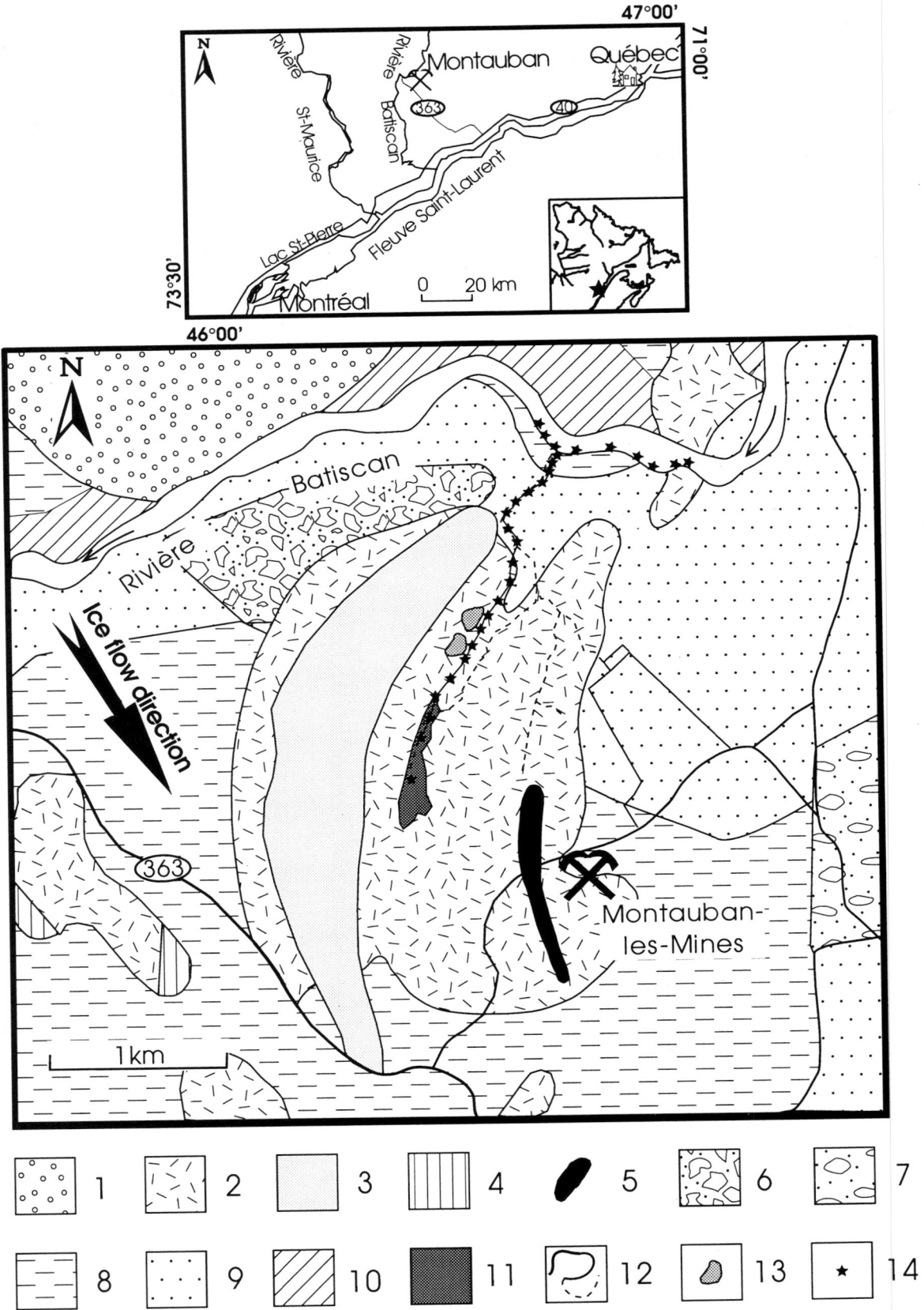

Figure 1 *Location and geology of the Montauban site; legend: 1, quartz diorite; 2, quartzo-feldspathic gneisses; 3, amphibolites and mafic gneisses; 4, quartzite; 5, mineralized zone; 6, till; 7, fluvioglacial sediments; 8, Quaternary marine sediments; 9, shallow-water marine sediments; 10, alluvial deposits; 11, tailings area; 12, major, minor, and dirt roads; 13, settling ponds; and 14, sampling locations.*

ponds used to decant the waste from the mill are located about 200 m downstream from the tailings impoundment (Fig. 1). The Montauban tailings park is a semi-orphan site with only limited monitoring because of the basic nature of its drainage. Basic mine drainage has generally been considered of less environmental concern than acid mine drainge.

The Surroundings: Stream and River

The tailings dump is located on top of a hill near the abandoned mine. Surface drainage from the tailings results in a small stream that is channelled through a drain to the Batiscan River (Fig. 2A, B). Eroded stream banks show that the stream runs through Quaternary glaciomarine sediments and an old tailings deposit, now partially vegetated. There is a fine reddish residue, rich in iron and manganese colloids (Figs. 2C, D), on sediment in the upper part of the stream. During wet periods, a sediment plume from the stream suggests volumetrically important discharge into the Batiscan River, although mixing occurs within a few tens of metres.

The sampling methodology and geochemical techniques used in this study are described in Lafléche *et al.* (1994). Two sampling campaigns, in late June and mid-October, under dry and wet conditions, were undertaken. To avoid contamination of the samples during sampling and analysis, ultra-pure water and chemicals were used throughout the procedures.

SEDIMENT GEOCHEMISTRY

The chemical compositions of stream sediments are highly variable and reflect the presence of both Quaternary sediments and mine tailings. This is shown by the very low sodium

and high calcium concentrations in the first 400-m length of the stream. In this section, sodium values are lower than 3,000 ppm, whereas, in Quaternary glaciomarine sediment, sodium reaches average concentrations of 10,000 ppm. In the same way, the sediments bordering the head of the stream display very high calcium concentrations (75,000 ppm) when compared to those of Quaternary sediments (average 30,000 ppm) that predominate in the lower part of the stream. At the head of the stream, pH is mildly basic (pH = 7.38), whereas in the lower part of the stream, pH is lower than neutrality (pH = 6.61), and in the Batiscan River the pH is weakly acid (pH = 5.18 ± 0.07).

WATER CHEMISTRY

The chemistry of the dissolved component of the stream is heterogeneous along its path toward the Batiscan River because of the mixing of the surface water with discharging ground waters. Mixing processes and changing redox conditions fractionate some elements particularly sensitive to sorption and precipitation processes. Waters overlying the tailings and in the first 200 m length of the stream are characterized by high SO_4^{2-} abundances (up to 1700 ppm), high conductivity (high total dissolved solids; average 2185 mhos·cm^{-1}) and relatively high alkalinity (approximately 105 ppm of equivalent $CaCO_3$). Despite their alkalinity, these waters are undersaturated with respect to carbonates.

After 200 m, ground waters coming from the mine tailings mix with the stream water, which becomes saturated in carbonates (*e.g.*, calcite, aragonite and dolomite) and sulphates (*e.g.*, barite; Lafléche *et al.*, 1994). These hybrid waters are

Figure 2   (**A**) *Surface drainage on the tailings at Montauban (GSC photo 95-074A),* (**B**) *Stream tailings area (GSC photo 95-074B),* (**C**) *Highly oxidized amorphous colloids in the bed of the stream (GSC photo 95-074C),* (**D**) *The Batiscan River (GSC photo 95-074D).*

characterized by a high pH (7.66) and high SO_4^{2-} contents (1103 ± 396 ppm), and are enriched in cations such as calcium (Ca: 281 ± 59 ppm) and magnesium (Mg: 93 ± 34 ppm). In this transition zone of the stream, which extends to 500 m downstream, iron and manganese concentrations drop rapidly. Field observation shows that this area is characterized by the precipitation of abundant Fe(Mn)–hydroxide that formed in a thin red colloidal deposit on the stream sediments. It is noteworthy that, in this section of the stream, silicon (a conservative element) and sodium, potassium and SO_4^{2-} abundances are relatively constant, which preclude dilution mechanisms to explain the systematic improverishment in dissolved iron and manganese. Trace metals such as lead and cadmium show a systematic impoverishment similar to that of iron and manganese in this section of the stream.

After approximately 900 m, stream water is undersaturated in carbonates, but saturated in iron-oxides and hydroxides (goethite, hematite, and amorphous $Fe(OH)_3$). Modification of the water chemistry in the lower part of the stream is further reflected by a decrease in the alkalinity (88 ± 3 ppm equiv. $CaCO_3$) and conductivity (110 ± 6 mhos·cm^{-1}) of the water. Such modifications probably reflect dilution processes involving relatively uncontaminated ground water coming from the Quaternary sediments in the watershed. Lowering of the stream-water pH by mixing with uncontaminated ground water agrees with the observation that the water which is in equilibrium with Quaternary sediments has a pH lower than 7.

In general, zinc concentrations in water tend to increase downstream, away from the tailings. The zinc concentrations are negligible in the first 900 m of the stream, and then abruptly rise to 4-5 mg·L^{-1} (Fig. 3). Cadmium concentrations

in the stream water (Fig. 4) are generally above maximum limits for human consumption (5 ppb; Environment Canada, 1993). As with zinc concentrations, cadmium concentrations increase steadily downstream. On the other hand, arsenic and lead concentrations in water decrease downstream from the tailings (e.g., Fig. 5), which appears to suggest that they are diluted or precipitated.

Water in the Batiscan River is highly undersaturated in carbonates and sulphates. These characteristics reflect the predominance of granitic and gneissic rocks and the absence of carbonates and evaporitic rocks in the Batiscan River watershed. Average pH in the river is 6.38 in the summer and the alkalinity and conductivity are very low (3.2 ± 0.5 ppm equiv. $CaCO_3$ and 22 ± 3 mhos·cm^{-1}, respectively). Early in the spring, the difference in pH is much higher between the stream and the river, and measured values for the Batiscan River range between 5.28 and 5.42. Such acidic conditions may strongly influence the speciation of trace metals and their toxicity in the aquatic system.

INTERPRETATION

Increasing concentrations of heavy metals such as zinc and cadmium in the stream water away from the tailings is unexpected. The concentration of trace metals in water is normally diluted in the lower reaches of streams and rivers by relatively uncontaminated ground water (e.g., Salomons and Förstner, 1984). A fraction of the metallic concentrations in the suspended and stream sediment is susceptible to release to the water if physico-chemical conditions change (e.g., Baudo et al., 1990). In particular, two processes may explain the observed enrichment in zinc and cadmium in the lower part of

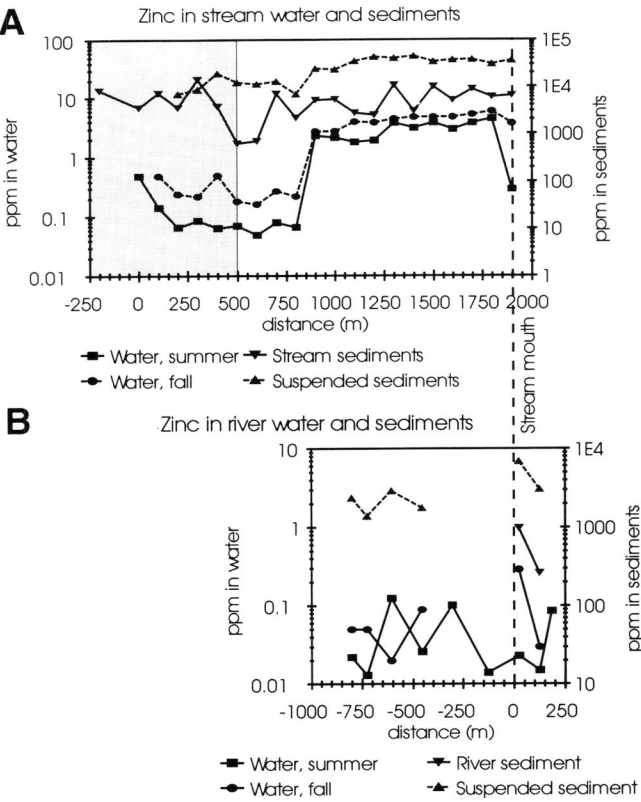

Figure 3 *Zinc concentrations in (A) stream water and sediments, and (B) river water and sediments. The shaded area on graph A represents that portion of the stream which flows over the tailings. The sample taken at –200 m is of dry tailings, upstream from the actual beginning of the stream.*

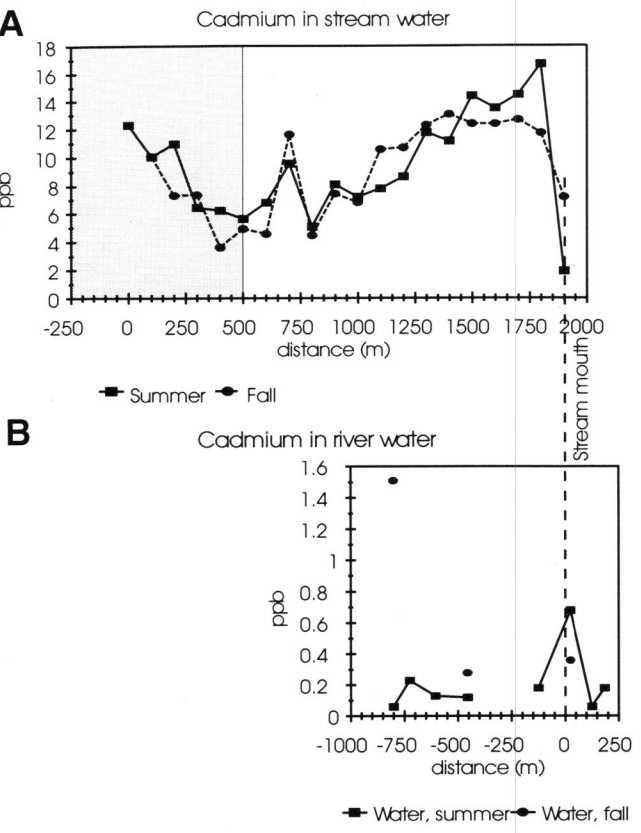

Figure 4 *Cadmium concentrations in (A) stream water, and (B) river water. The shaded area on graph A represents that portion of the stream which flows over the tailings.*

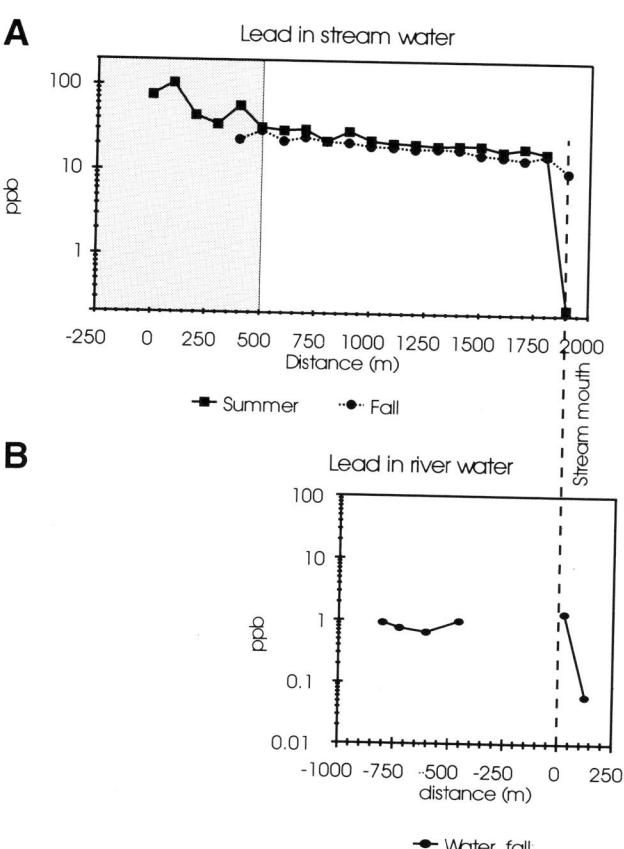

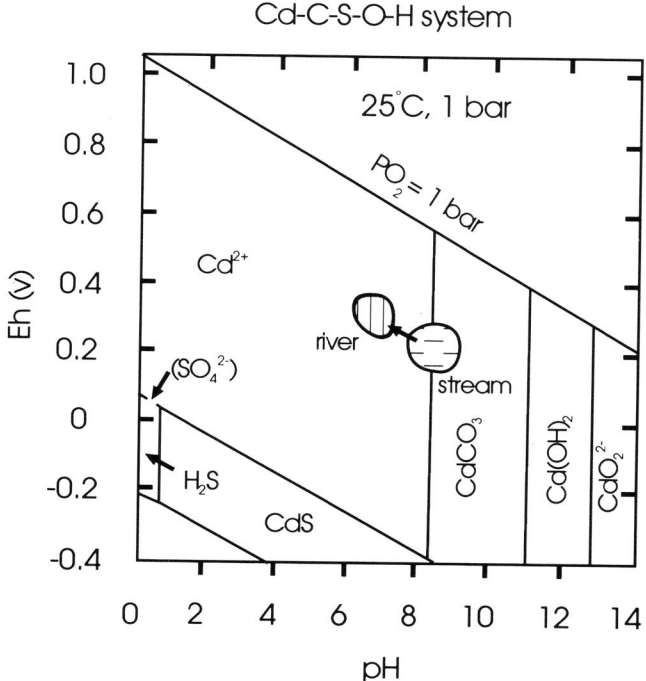

Figure 5 *Lead concentrations in (**A**) stream water, and (**B**) river water. The shaded area on graph A represents that portion of the stream which flows over the tailings.*

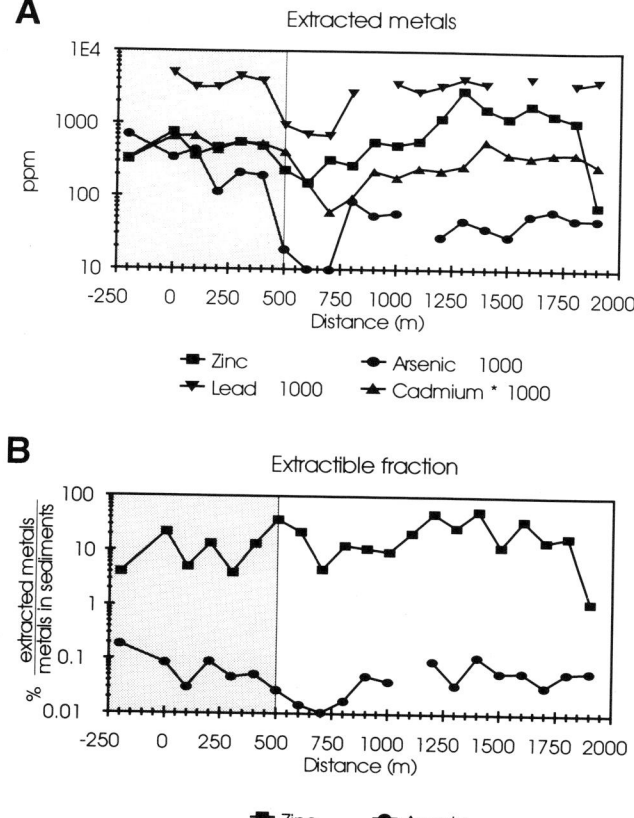

Figure 6 *Speciation of cadmium at varying conditions of pH and Eh. The stream and river fields for summer conditions are plotted, and show that cadmium may locally precipitate as CdCO₃ in this stream, but it will always be dissolved if the particles reach the Batiscan River. Note that the toxicity of Cd²⁺ is much higher than CdCO₃.*

the stream: (1) dissolution of mechanically added zinc-rich sediments or of iron-manganese colloids (chemical sediments), which are particularly abundant in the upper part of the stream, (2) desorption from iron and manganese hydroxide coatings on sediment particles, due to changes in pH. According to Papadopoulos and Rowell (1988), zinc speciation is mostly controlled by adsorption processes and precipitation of zinc is improbable in natural conditions. The speciation of cadmium, however, is mainly controlled by precipitation-dissolution processes due to changes in pH and Eh (Madrid and Diaz-Barientos, 1992). As shown on Figure 6, in the presence of carbonates, cadmium may locally precipitate as $CdCO_3$ or dissolve as Cd^{2+}, depending on pH-Eh conditions. While precipitation-dissolution may explain the sawtooth pattern in the cadmium concentration, we believe that because cadmium co-varies with zinc, adsorption-desorption best explains the high concentration of these metals in the stream water. Accordingly, desorption of zinc and cadmium from iron and manganese hydroxide coatings on sediment particles, due to changes in pH, is the more likely cause of the downstream increase in zinc concentrations.

To test the hypothesis that zinc and cadmium contamination of the stream water is related to water-sediment interactions, and that these trace metals are weakly bound to the sediments (extractable metals), subsamples of the stream sediments were exposed to a dilute acid solution (HCl, 0.3 M in ultra-pure water), and the resulting solutions analyzed. Figure 7 shows that zinc, cadmium and lead are released in large quantities, but not arsenic.

Figure 7 *HCl (0.3 M) extractible fraction of potential contaminants, arsenic, lead, cadmium, and zinc, after stream sediment has been exposed to an acid solution for 6 hours. (**A**) Concentration of metals found in the water after the experiment. (**B**) Extracted metals with respect to the initial concentration in the stream sediments.*

Table 1   Minimum and maximum concentrations of potential contaminants at Montauban-les-Mines. Bold numbers are tolerance limits published by Environment Canada (1993).

Element		Zinc (ppm)	Cadmium (ppm)	Lead (ppm)	Arsenic (ppm)
Ore	min	59	<1	18	<5
	max	>20,000	757	>20,000	719
Tailings	min	3450	—	—	127
	max	14,100	—	—	1450
Limit of tolerance level in sediments		**800**	**10**	**250**	**33**
Till	min	283	—	—	1
	max	98	—	—	13
Stream sediment	min	615	—	—	29
	max	10,700	—	—	499
Suspended sediment	min	13	—	—	0.09
	max	90	—	—	3.857
Maximum limit in water		**5**	**0.005**	**0.005**	**0.05**
Maximum limit for fish		**0.03**	**0.0002**	**0.002**	**0.05**
Water (summer)	min	0.049	0.0051	0.00024	0.0003
	max	4.45	0.0167	0.10609	0.0024
Water (fall)	min	0.16	0.0037	0.01018	0.0002
	max	5.9	0.0131	0.02964	0.0012

POTENTIAL CONTAMINANTS

Trace metals suspected to be present in significant quantities in the tailings or in the settling ponds include zinc, lead, arsenic, chromium, barium, copper, strontium, tungsten and cadmium (Jourdain, 1993; Table 1). Sulphates and phosphates are also present in high quantities. Phosphates can influence algal growth, and could contribute to eutrophication of the stream and Batiscan River. Lacking Canadian standards for clean sediments, the limits of tolerance for a few metals, as proposed by the Ontario Ministry of Environment (OMOE; Baudo *et al.*, 1990), have been used to evaluate the relative contamination of the sediments. In both the stream sediments and suspended sediment, zinc concentrations are variable and above the tolerance limit of 800 ppm proposed by the OMOE (Baudo *et al.*, 1990) for 19 of the 21 samples (Fig. 3). In the stream and suspended sediment arsenic occurs in concentrations well above the OMOE limit of 33 ppm (Baudo *et al.*, 1990) in the first few hundred metres of the stream (Fig. 8). The concentrations decrease significantly away from the tailings, but are still above the limit.

Of the 13 elements analyzed in water, zinc, cadmium, arsenic and lead occur in potentially toxic concentrations; cadmium and zinc concentrations exceed the standards for drinking water (Environment Canada, 1993), while lead and arsenic are well within acceptable limits. Other components, such as sulphates, total dissolved solids and iron, are just below the tolerance limit or cause only aesthetic concerns. Local municipal authorities have expressed concerns that the cyanide used during mineral processing may now be released into the local environment (L. Filteau, pers. comm., 1993). The recommended limit for arsenic in human consumption is 50 ppb, and does not appear to be a problem at the Montauban site.

CONTAMINANT SOURCES

There are five possible sources for the inorganic contami-

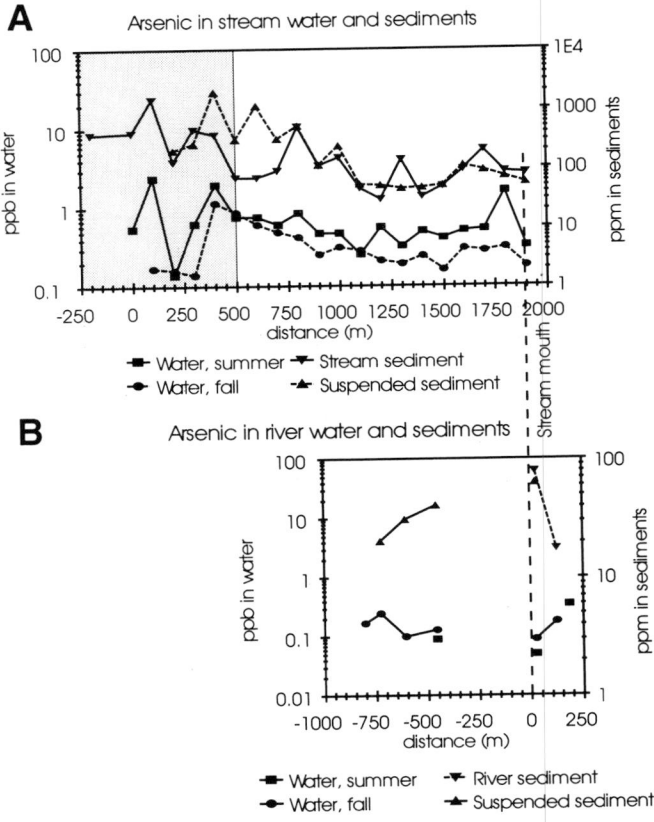

Figure 8 *Arsenic concentrations in (**A**) stream water and sediment, and (**B**) river water and sediment. The compositions of the river sediment and water are also plotted. The shaded area on graph A represents that portion of the stream which flows over the tailings. The sample taken at −200 m is of dry tailings, upstream from the actual beginning of the stream.*

nants at Montauban: (1) bedrock and associated mineral deposits, (2) Quaternary glaciomarine sediments, (3) ground water, (4) mine tailings, and (5) atmospheric contributions from precipitation and aerosol deposition. The bedrock is the ultimate source of metals, but the stream does not run over the mineralized portion of it, or even near the ore deposit (Fig. 1). Till rich in mineralized debris derived from the Montauban deposit probably occurs southeast of the site, since ice flow and sediment transport was in that direction (Bolduc, 1992). Ground water in bedrock, from two samples taken in the flooded ventilation shafts, appears essentially clean for most of the trace metals analyzed. Atmospheric contributions are likely negligible, because Montauban is located away from the main industrial sites such as Montréal and Trois-Riviéres, and snow analyses show that cadmium and zinc concentrations are less than 5 ppb and 0.02 ppm, respectively. This contribution can be considered as non-significant. Consequently, the tailings are most likely the direct source of the contaminants.

Rare earth elements (REE: from lanthanum [La] to lutetium [Lu] and scandium [Sc]) are highly insoluble, and can be used to measure quantitatively the importance of each component in the sediments. At Montauban, the REE signature of the stream sediment (light rare earth elements [LREE] from 30 to 70 times chondrite) suggests that it is derived from the local bedrock (Fig. 9A), while the suspended sediment deviates substantially from it, being systematically enriched in light rare earth elements (LREE from 180 to 400 times chondrite). Scandium reflects the presence of mafic mineral phases such as pyroxenes, while sodium comes essentially from plagioclases. Both of those minerals are major constituents of bedrock in the area. A plot of scandium *versus* sodium (Fig. 9B) shows that the stream sediment is a mixture between the tailings, local bedrock and till, and Quaternary marine silt, with increasingly larger local imput away from the tailings dump.

ENVIRONMENTAL IMPACT

Cadmium, along with mercury, is one of the most toxic elements (Salomons *et al.*, 1995). Unlike mercury, however, cadmium concentrations are not biomagnified with increasing position in the food chain of most species (Cossa and Lassus, 1989). Phytoplankton are particularly vulnerable, and concentrations as small as 1-10 ppb can decrease cellular growth of phytoplankton by more than 50% (Saunders and Cibik, 1985). The maximum metal concentrations acceptible for aquatic life are thus much lower than those set for human consumption (Environment Canada, 1993). Cadmium toxicity increases considerably from saline to soft waters and the larval stage of fish is particularly vulnerable; cadmium toxicity increases in various species when it is associated with metals such as zinc, lead, iron and copper. Fish exposed to zinc and cadmium, two known contaminants likely to reach the Batiscan River, have a mortality rate that is related to the additive effects of each contaminant (Wicklund Glynn *et al.*, 1992). Cadmium is thus a potentially important contaminant being introduced to the Batiscan River by drainage from the Montauban mine.

It is likely that the local vegetation growing on tailings is enriched in contaminants released by the oxidation into near-surface water (*e.g.*, Chambers and Sidle, 1991). These contaminants are then bio-available for wildlife that feeds on that vegetation, and so moves up the food chain to humans. Effects on local ground-water supplies are also presently unknown, and underscore the need for further study.

DISCUSSION

Site Restoration

Mining operations ceased at Montauban in 1992, and a local company has undertaken to restore the site by sealing mine shafts with concrete slabs, removing buildings and diverting the stream so that it will not run through the settling ponds. The Ministére de l'Environnement et de la Faune du Québec (MEFQ) monitors the restoration by visiting the site regularly, taking water samples in the flooded mine shafts and taking other samples as necessary. There is no trace of cyanide in ground water at the mine site, but there is likely direct recharge of surface precipitation and chemical leachate to

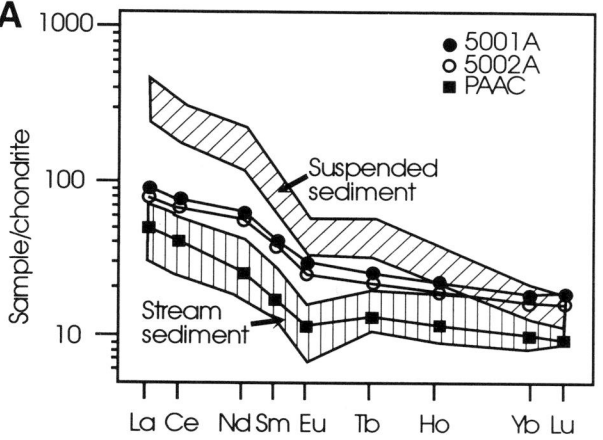

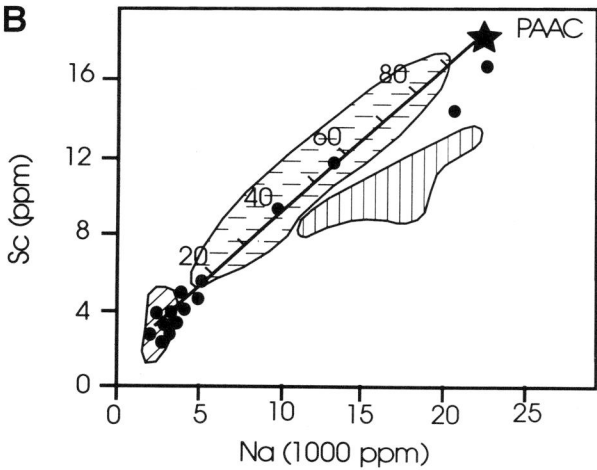

★ Post Archean Average Continental Crust

⬭ Mine tailings

⬯ Stream sediments

⬭ Batiscan River sediments

● Suspended sediment

Figure 9 *Geochemical composition of the stream and suspended sediments in relation to the post-Archean Average Continental Crust (PAAC) of Taylor and McLennan (1985). (A) Chondrite-normalized REE patterns of the stream and suspended sediment. Note the strong LREE enrichment of the suspended sediments. (B) Na-Sc diagram showing the tailings field, the PAAC, a calculated average binary mixing line, and the composition of the stream and suspended sediment. These diagrams show that the stream sediments are derived primarily from bedrock, with a small component of the sediment coming from the tailings, while the suspended sediments are primarily derived from tailings.*

underlying ground water outside the mine area. The volume of water entering the Batiscan River from the tailings area is small compared to the river, but the effects of enhanced flows during spring runoff occur at a critical period for fish reproduction. In terms of further remediation, a constructed wetland at the mouth of the stream could be used to decontaminate the stream discharge.

Wider Implications

The problem of metal contamination from basic (and acid) mine drainage waters is not limited to mine sites alone. Large volumes of tailings have been exported to surrounding urban communities for use as fill for road, rail and housing construction (see also the example of radioactive wastes, Chapter 28). Unfortunately, the magnitude of the wider contamination problem is unknown, and awaits detailed inventories of waste ore and disposal in urban areas. A key development is the generation of historical land use maps (*e.g.*, Chapter 27).

ACKNOWLEDGEMENTS

We thank Les Entrepreneurs Lefebvre of Saint-Séverin, Québec, for unlimited access to the Montauban property. Chemical analysis were performed by J. Bélanger and R. Gosselin of the INRS Géoressources laboratories. Funds for this project come from Geological Survey of Canada to Bolduc, and from the Natural Sciences and Engineering Council of Canada to Lafléche. Comments by T. Birkett greatly improved the manuscript.

REFERENCES

Baudo, R., Giesy, J.P. and Muntau, H., 1990, Sediments: Chemistry and Toxicity of In-place Pollutants: Lewis Publishers, Ann Arbor, MI, 405 p.

Bernier, L., Pouliot, G. and MacLean, W.H., 1987, Geology and metamorphism of the Montauban North Gold Zone: a metamorphosed polymetallic exhalative deposit, Grenville Province, Québec: Economic Geology, v. 82, p. 2076-2090.

Bolduc, A.M., 1992, Compilation cartographique et caractérisation des dépots de surface de la région de Shawinigan–Trois-Riviéres, Québec: Geological Survey of Canada, Paper 92-1D, p. 155-164.

Chambers, J.C. and Sidle, R., 1991, Fate of heavy metals in an abandoned lead-zinc tailings pond: I, Vegetation: Journal of Environmental Quality, v. 20, p. 745-751.

Cossa, D. and Lassus, P., 1989, Le cadmium en milieu marin, biogéochimie et écotoxicologue: IFREMER, Rapports scientifiques et techniques, no. 16, 111 p.

Environnement Canada, 1993, Recommandations pour la qualité des eaux au Canada : Document préparé par le groupe de travail sur les recommandations pour la qualité des eaux du Conseil canadien des ministéres des ressources et de l'environnement, 86 p.

Ferris, F.G., Beveridge, T.J., Fyfe, W.S., Mann, H., Tazaki, K., Kerrich, R. and Wiseman, M., 1991, Retardation of toxic heavy metal dispersion from nickel-copper mine tailings, Sudbury District, Ontario: role of acidophilic micro-organisms: Ontario Geological Survey, Open File Report 37, 15 p.

Firlotte, F.W., Gélinas, P., Knapp, R. and McMullen, J., 1992, Acid drainage treatment at la Mine Doyon: Evolution and future direction: Second International Conference on the Abatement of Acidic Drainage, Proceedings, v. 4, p. 119-140.

Jourdain, V., 1993, Géologie des amas sulfurés auriféres de la province de Grenville: unpublished Ph.D. thesis, Université du Québec á Chicoutimi, Chicoutimi, PQ, 153 p.

Lafléche, M.R., Bolduc, A., Camiré, G., Talbot, L. and Bélanger, J., 1994, Dispersion des métaux lourds dans l'eau et les sédiments d'un ruisseau s'écoulant du parc á résidus miniers de Montauban (comté de Portneuf, Québec): Geological Survey of Canada, Paper 94-1E, p. 233-241.

Madrid, L. and Diaz-Barrientos, E., 1992, Influence of carbonate on the reaction of heavy metals in soils: Journal of Soil Sciences, v. 43, p. 709-721.

Marcotte, R.R., 1995, The restoration of mining sites in Québec: Current status and future direction, *in* Hynes, T.P. and Blanchette, M.C., eds., Sudbury '95, Mining and the Environment, Conference proceedings, CANMET, Ottawa, ON, v. 1, p. 359-365.

Papadopoulos, P. and Rowell, D.L., 1988, The reaction of cadmium with calcium carbonate surfaces: Journal of Soil Sciences, v. 39, p. 23-36.

Peters, T.H., 1988, Mine tailings reclaimation: Inco Limited's experience with the reclaiming of sulphide tailings in the Sudbury area, Ontario, Canada, *in* Salomons, W. and Förstner, U., eds., Environmental Management of Solid Waste: Dredged Material and Mine Tailings: Delft Hydraulic Laboratory, Institute for Soil Fertilisation, Delft, The Netherlands, p. 152-165.

Salomons, W. and Förstner, V., 1984, Metals in the hydroxides: Springer-Verlag, Berlin-Heidelberg, 349 p.

Salomons, W., Förstner, U. and Mader, P., 1995, Heavy Metals – Problems and Solutions: Springer-Verlag, New York, 270 p.

Sanders, J.G. and Cibik, S.J., 1985, Reduction of growth rate and resting spore formation in a marine diatom to low levels of cadmium: Marine Environmental Research, v. 16, p. 165-180.

Taylor, S.R. and McLennan, S.M., 1985, The Continental Crust: its Composition and Evolution. An Examination of the Geological Record Preserved in Sedimentary Rocks: Blackwell Scientific Publishers, Oxford, UK, 307 p.

Wicklund Glynn A., Haux, C. and Hogstrand, C., 1992, Chronic toxicity and metabolism of Cd and Zn in juvenile minnows (*Phoximus phoximus*) exposed to a Cd and Zn mixture: Canadian Journal of Fisheries and Aquatic Science, v. 49, p. 2070-2079.

14. Acid Mine Drainage in the Sudbury Area, Ontario

Nelson Belzile
Douglas Goldsack

Department of Chemistry and Biochemistry, Centre in Mining and Mineral Exploration Research
Laurentian University, Sudbury, Ontario P3E 2C6

Stephanie Maki

Department of Chemistry and Biochemistry, Laurentian University, Sudbury, Ontario P3E 2C6

Andrew McDonald

Department of Earth Sciences, Laurentian University, Sudbury, Ontario P3E 2C6

SUMMARY
There is a long history of metal mining in Canada. Serious environmental problems are emerging in many urban and rural communities as a result of the large areas of waste rock and tailings that generate acidic waters and toxic metals for many hundreds of years after mine closure. Nationally, an area of about 150 km^2 is occupied by operating or abandoned acid-generating mine sites; remediation costs are substantial. This paper illustrates the problem of acid mine drainage and acid rock drainage associated with large-scale nickel mining and smelting near Sudbury, Ontario. INCO Limited has the largest currently operating site of acid mine drainage in the western world, and it is the purpose of this paper to identify the factors controlling the generation of acidic waters and techniques that are being used to control acid mine drainage.

INTRODUCTION
Acid mine drainage (AMD) and acid rock drainage (ARD) are two of the most serious environmental problems facing the mining industry and various government agencies in Canada. An estimated area of over 150 km^2 is occupied by operating or abandoned acid-generating sites across the country. The cost of stabilizing reactive wastes is highly site-specific, but at an average cost of at least $125,000 per hectare (ha), reclamation of non-ferrous metal mine sites can easily reach $3 billion over the next 20 years (Feasby and Sirois, 1991). In Ontario, more than 2000 abandoned mine sites have been identified, and many of them are potential sources of acidic drainage. Twenty of these abandoned sites, containing around 55,000,000 metric tonnes (t) of reactive sulphidic tailings, cover a surface area of 830 ha (Feasby and Sirois, 1991).

THE SUDBURY AREA
The main tailings site for INCO Limited's production in Sudbury covers an area of more than 2000 ha, with approximately 35,000 t of tailings being produced and stored every day (Heale, 1991). With approximately 500 million tonnes of tailings sitting in Copper Cliff, INCO Limited has the largest operating site of AMD in the western world (Fig. 1).

The principal tailings dump is located west of Sudbury, between the communities of Lively and Copper Cliff, and is divided into a number of impoundments (Fig. 1). Originally located in a series of bedrock lows, the dump site now stands topographically above the surrounding terrain, and acts as a ground-water recharge area. Precipitation is about 80 cm·a^{-1}, with about 22 cm being recharged. Ground water moves out radially from the dump area (McGregor et al., 1995).

In this paper, we will explain the chemical reactions involved in the generation of AMD, review the main factors affecting the oxidation of sulphidic mine wastes, and look at some of the possible solutions that have been devised to reduce the problem.

WHAT IS ACID MINE DRAINAGE?
The economic ores of base metals such as nickel, copper, zinc and the precious or radioactive metals such as gold and uranium are often intimately associated with sulphides such as pyrrhotite (Fe$_{1-x}$S) and pyrite (FeS$_2$; Fig. 2). When exposed to air and water, these sulphides can be oxidized with sulphide ions being converted to sulphate ions. Hydrogen ions, and therefore acid, are generated simultaneously in these reactions and this acid is the source of the AMD problem (see Reactions 1 and 2).

THE OXIDATION OF PYRITE AND PYRRHOTITE BY OXYGEN:

$$FeS_2 + 7/2O_2 + H_2O \rightarrow Fe^{2+} + 2SO_4^{2-} + 2H^+ \quad (1)$$

$$Fe_{1-x}S + (2-x/2)O_2 + xH_2O \rightarrow (1-x)Fe^{2+} + SO_4^{2-} + 2xH^+ \quad (2)$$

Another oxidant of both pyrite and pyrrhotite generated by the oxidation of Fe^{2+} in the presence of oxygen is the ferric ion Fe^{3+}. This reaction may occur by the following:

$$Fe^{2+} + 1/4O_2 + H^+ \rightarrow Fe^{3+} + 1/2H_2O \quad (3)$$

$$FeS_2 + 14Fe^{3+} + 8H_2O \rightarrow 15Fe^{2+} + 2SO_4^{2-} + 16H^+ \quad (4)$$

$$Fe_{1-x}S + (8-2x)Fe^{3+} + 4H_2O \rightarrow (9-3x)Fe^{2+} + SO_4^{2-} + 8H^+ \quad (5)$$

If Reaction 3 occurs under acidic conditions, a significant quantity of Fe^{3+} will remain in solution. This can react with pyrite or pyrrhotite (Reactions 4 and 5) and produce more acid. Note that Reactions (3), (4) and (5) constitute a cyclic process, so that low levels of ferric ion can catalyze the oxidation of the sulphides, thus speeding up reaction rates. If the pH is greater than 3.5, low concentrations of Fe^{3+} are maintained through the precipitation of ferric hydroxide and the generation of more acid, as shown in Reaction 6 (Nicholson, 1994):

$$Fe^{3+} + 3H_2O \leftrightarrow Fe(OH)_{3(s)} + 3H^+ \quad (6)$$

As the pH decreases with the production of hydrogen ions, ferric (Fe^{3+}) ions act as oxidizing agents in Reaction 4. Both methods of oxidation (by air and Fe^{3+}) are strongly catalyzed by bacteria such as *Thiobacillus ferrooxidans* (Backes *et al.*, 1986).

Acidic waters are thus generated and carry both toxic metals and high levels of dissolved salts, the latter resulting from the dissolution of other minerals and silicates (Table 1).

WHICH FACTORS AFFECT THE OXIDATION OF SULPHIDIC MINE WASTES AND THE PRODUCTION OF ACID MINE DRAINAGE?

Aside from bacteria, there are a host of other factors which can affect the rate and ease of oxidation of pyrite, pyrrhotite and other sulphide minerals such as marcasite (orthorhombic FeS_2), arsenopyrite (FeAsS), chalcopyrite ($CuFeS_2$), galena (PbS) or sphalerite (ZnS).

Crystal Structure and Morphology
The majority of tailings derived from sulphide-bearing ores

consist of quartz, various silicates, carbonates, and iron sulphides. In general, AMD is principally caused by the oxidation of pyrite (cubic FeS_2) and pyrrhotite ($Fe_{1-x}S$, monoclinic; Fe_7S_8, hexagonal; Fe_9S_{10}–$Fe_{11}S_{12}$, orthorhombic). There is a large difference in reactivity between pyrite and pyrrhotite, with the rate of pyrrhotite oxidation being 20 to 100 times higher than that of pyrite (Nicholson, 1994). Pyrite, however, produces more acid per mole than pyrrhotite because of its higher sulphur content. The difference in the composition and electronic structure implied by the crystal structure of the various forms of pyrrhotite also pre-supposes some difference in their reactivity and rate of oxidation.

Strains in the structure of the crystal, defects such as grain edges and crystals, solution pits, fluid inclusion pits and cleavages can increase the degree of chemical attack (Bierens de Haan, 1991). A difference in reactivity between pyritic samples containing mainly zinc and lead, compared to pyrite loaded with nickel and copper, has been observed (Lowson, 1982).

Partial pressure of oxygen
The rate of pyrite oxidation is dependent on the partial pressure of oxygen, and reaction orders from 0.5 to 1.0 have been proposed (Lowson, 1982). Hammack and Watzlaf (1990) showed that abiotic oxidation of fresh pyrite surfaces is independent of oxygen partial pressure above 10% (0-reaction order), and is proportional to oxygen partial pressure below 10% (first-order reaction). With bacteria present, the rate of pyrite oxidation is independent of the oxygen partial pressure down to 1%, with the reaction rate being proportional to oxygen partial pressures below that concentration. These results suggest that low oxygen levels should be maintained to prevent the oxidation of pyrite and pyrrhotite.

Surface area
In general, the oxidation rates reported in the literature are directly proportional to surface area (Nicholson *et al.*, 1988; Nicholson and Scharer, 1994). Thus, the smaller the size of the sulphide particle, the greater the surface area and the faster oxidation occurs. For very small particles (1 to 10 μm) sulphide dust explosions can occur due to the very rapid oxidation process.

Temperature
The oxidation rate of pyrite has been found to be particularly dependent on the temperature, with the rate doubling for each

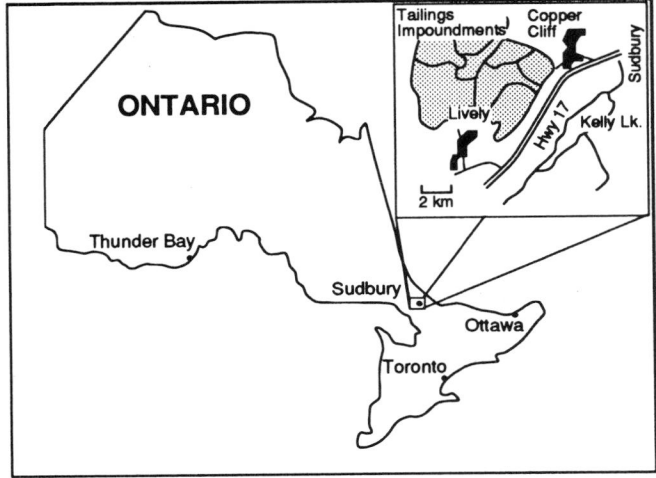

Figure 1 *Location of INCO Limited tailings area in Copper Cliff, Sudbury.*

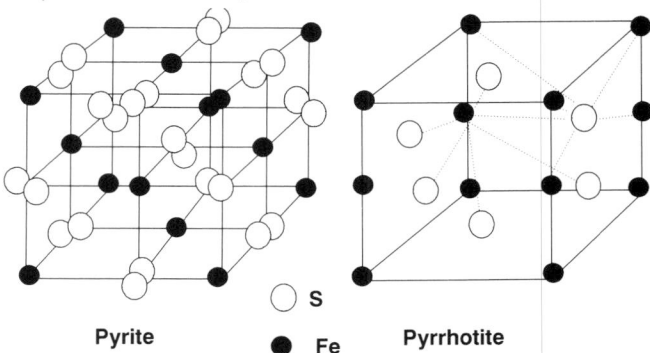

Figure 2 *Crystal structure of pyrite and pyrrhotite. The pyrite structure is cubic. The space lattices resemble those of sodium chloride (NaCl) with Fe^{2+} replacing Na^+ and S_2^{2-} replacing Cl^-. Pyrrhotite crystallizes in the hexagonal system with some of the positions normally occupied by Fe^{2+} vacant ($Fe_{1-x}S$). The vacancies (x) cause the mineral to be magnetic.*

ten degree increment in temperature (Caruccio *et al.*, 1988; Hutchison and Ellison, 1992). For pyrrhotite, trends are less clear, but still indicate that an increase in temperature will favour oxidation (Nicholson and Scharer, 1994).

pH, Redox Potential and Ferric Iron

The influence of pH on the oxidation rate of pyrite is closely related to the presence of ferric iron and to bacterial mediation (Nicholson *et al.*, 1988; Moses *et al.*, 1987). Most experiments conducted on the oxidation of pyrite by ferric iron have been carried out at pH less than 4 because of the stability of Fe^{3+} in acidic aqueous solutions. However, Moses *et al.* (1987) show that the oxidation is slightly increased as the pH is increased from 2 to 9. It has been suggested that pyrite is more readily oxidized by ferric iron than by molecular oxygen. At pH higher than 5, the precipitation of ferric iron and the role of a film of iron oxyhydroxide coating the grain of pyrite has to be considered. The oxidation of pyrrhotite does not appear to be strongly pH-dependent, but minimum rate values have been observed for pH of 3 to 4 (Nicholson and Scharer, 1994). It is expected that oxidation rates of both pyrite and pyrrhotite will correlate with redox potential (Eh), $Fe^{3+}:Fe^{2+}$ ratio or dissolved oxygen concentration, and will be consistent with the stability fields of the two minerals illustrated in Figure 3.

The Eh-pH diagram is a condensed illustration indicating the stability of chemical species and minerals under various conditions. The Eh scale refers to the redox potential (presence or absence of oxygen), and the pH scale refers to the acidity level. Figure 3 shows that pyrite and pyrrhotite are stable in a very limited area of Eh. This diagram shows why pyrite and the more reactive pyrrhotite are easily oxidized under acidic and oxidizing conditions.

CAN WE STOP OR PREVENT ACID MINE DRAINAGE?

The prevention or reduction of sulphidic wastes oxidation in tailings is necessary for the control of acid mine drainage waters. Monumental efforts have been expended in attempts to reduce or stop AMD, and many solutions have been proposed and presented in recent international conferences (*e.g.*, MEND, 1991; United States Department of the Interior, 1994). Over the past five years, over $18 million in research money, from provincial and federal governments as well as the mining industry, has been expended in an effort to understand and control AMD. Substantial expenditures have also

been made by other by other countries, such as United States, Australia, Sweden, *etc.*, since this is a worldwide problem.

The most important parameter to consider in the control of AMD is oxygen (see equations 1 to 3), and most of the solutions proposed to deal with the problem involve the creation of an air (oxygen) barrier which will minimize the penetration of oxygen into tailings or waste rocks.

Liming

A lime treatment (calcite, $CaCO_3$; portlandite, $Ca[OH]_2$; dolomite, $CaMg[O_3]_2$) can be used to neutralize acidity, control pH, and decrease oxidation of sulphidic tailings. This also prevents the infiltration of oxygen and controls the precipitation of ferric ions. The limitations of the technique are the difficulty in predicting the quantities required, the limited depth of incorporation of the lime material into the tailings site and the associated costs. Liming is used in combination with dry covers and revegetation as a remediation technique.

Dry and organic covers

Various types of dry and organic covers can be used on the surface of tailings to act as an oxygen barrier, thus reducing the oxidation of sulphidic mine wastes and the generation of AMD. The most common covers include clay or till, but soils, peat, hay, straw, sawdust or wood chips, sludge or compost can also be used. The application of cementitious materials is also used to solidify mine tailings and minimize water and air transfer to waste rock dumps. Synthetic geomembranes can play the same role. The main limiting factors in the use of dry covers are the costs associated with this technique on large surfaces and the necessity of long-term monitoring.

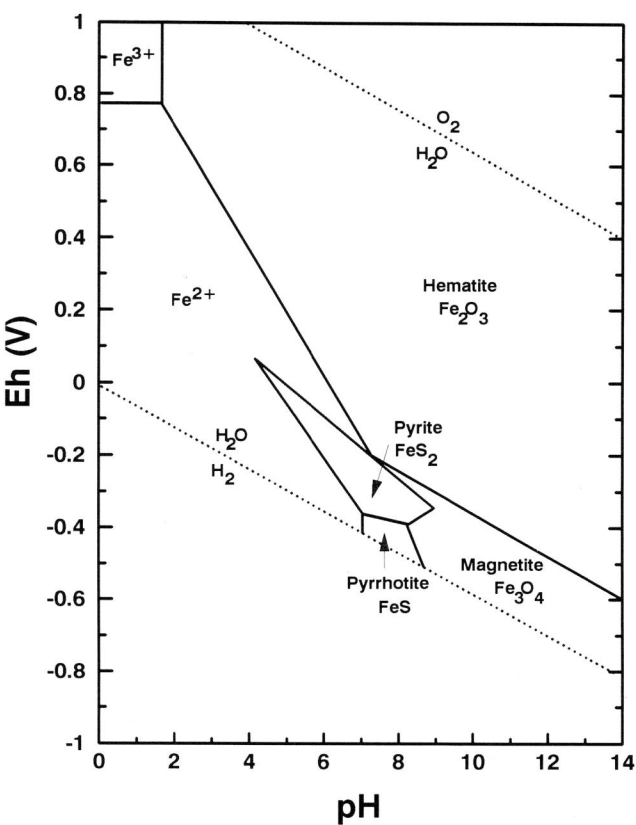

Figure 3 *Eh-pH diagram showing the stability contours for pyrite and pyrrhotite in water at 25° C when total S = 10⁻⁶ M (Brookins, 1988).*

Table 1   Characteristics of acid mine drainage (modified from Ritchie, 1994).

Products	Associated chemical species	Concentration range	Impact on the environment
Acidity	H_2SO_4	pH 2 to 4	mobilization of metal ions
Iron	Fe^{2+}, Fe^{3+}, iron oxides, iron sulfates	100–3000 mg·L⁻¹	turbidity and precipitation as pH increases
Toxic Metals & Elements	Cu, Ni, Zn, Pb, Cd, Hg, As, Se, Th, Ra, *etc.*	1–200 mg·L⁻¹	affect aquatic flora and fauna, water quality affected
Dissolved Salts	Ca^{2+}, Mg^{2+}, Na^+, K^+, Al^{3+}, SO_4^{2-}	100–30,000 mg·L⁻¹	water quality affected

Wet covers

The low solubility (typical concentration of 5-8 ppm in water *versus* 20% in air) and low diffusion rate of oxygen in water makes subaqueous disposal an effective way to prevent AMD in sulphidic mine wastes. This technique is particularly effective in lakes where anoxia can be maintained in the sediments by a continuous bacterial degradation of organic matter. The use of wetlands and cattails can maintain ideal conditions to stabilize certain tailing sites (Chapter 12). However, severe winter conditions in Canada limit the use of this technique because of the reduction in bacterial action at low temperatures. Underground mine workings and open pits can also be flooded to control acid generation, but underwater disposal is not always a viable option since it requires specific terrain and water conditions.

Passivation

In this technique, pyrite and pyrrhotite particles are coated with a substance which renders them impenetrable to oxidative attack. Various chemical products have been used for this purpose: phosphates, silicates, and organic reagents such as oxalic acid, acetyl acetone, humic acid, lignin, *etc*. The oxidation rate of pyrite can be decreased significantly by a coating treatment (Maki, 1995). The formation of a strong non-porous passive film on the pyrite surface could explain the inhibition of oxidation. Each coating agent has its own method of treatment and degree of success. Further studies are needed to examine the permanency of the treatment.

DISCUSSION

The rehabilitation of mine sites in urbanizing areas is a worldwide problem and centres on issues such as subsidence, toxic waters, and wastes. Mining has had a major impact on urban and rural environments in the Sudbury area over the past century. While surface roasting of nickel ores caused important local damage to soil prior to 1927, along with massive destruction of forest cover (for fuel), the major biological impacts have been tied to smelter emissions by way of acid precipitation and metal enrichment. Sustained efforts have been made by industry in recent years to reduce SO_2 and toxic metal emissions. This, in combination with major revegetation programs, has led to a significant improvement in the quality of air, water, soil, and biota in the Sudbury area (see Gunn, 1995). Problems associated with wind-blown fines from the tailings areas, which once plagued urban areas of Copper Cliff and Sudbury, have been largely eliminated by the establishment of a vegetation cover on the tailings sites. Despite these advances, the problem of AMD is not entirely solved, and large surfaces of exposed tailings and waste rock are still generating acid and toxic metals. Studies by Coggans (1992) suggest that, under the present hydrodynamic environment, products of sulphide oxidation will be discharged at elevated levels for the next 500 years.

REFERENCES

Backes, C.A., Pulford, I.D. and Duncan, H.J., 1986, Studies on the oxidation of pyrite in colliery spoil. I. The oxidation pathway and inhibition of the ferrous-ferric oxidation: Reclamation and Revegetation Research, v. 4, p. 279-291.

Bierens de Haan, S., 1991, A review of the rate of pyrite oxidation in aqueous systems at low temperature: Earth Science Reviews, v. 31, p. 1-10.

Brookins, D.G., 1988, Eh-pH Diagrams for Geochemistry: Springer-Verlag, Berlin, 121 p.

Caruccio, F.T., Hossner, L.R. and Geidel, G., 1988, Pyritic materials: Acid drainage, acidity, and liming, in Hossner, L.R., ed., Reclamation of Surface-mined Lands: CRC Press, Boca Raton, FL, v. 1, p. 159-189.

Coggans, C.J., 1992, Hydrogeology of the INCO Ltd. Copper Cliff, Ontario, mine tailings impoundments: unpublished M.Sc. thesis, University of Waterloo, Waterloo, ON, 159 p.

Feasby, D.G. and Sirois, L.L., 1991, The Mine Environmental Neutral Drainage (MEND) program, in 2nd International Conference on the Abatement of Acidic Drainage, Montréal, PQ: Proceedings, v. 1, p. 1-26.

Gunn, J.M., 1995, ed., Restoration and Recovery of an Industrial Region: Springer-Verlag, New York, 358 p.

Hammack, R.W. and Watzlaf, G.R., 1990, The effect of oxygen on pyrite oxidation, in Mining and Reclamation Conference, Charleston, West Virginia: University Publication Services, Morganton, WV, v. 1, p. 257-264.

Heale, E.L., 1991, Reclamation of tailings and stressed lands at the Sudbury, Ontario operations of INCO Limited, in Skousen, J., Sencindiver, J.C. and Samuel, D., eds., 2nd International Conference on the Abatement of Acidic Drainage, Montréal, PQ: Proceedings, v. 2, p. 529-541.

Hutchison, I.P. and Ellison, R.D., 1992, Acid Mine Drainage: Lewis Publishers, Boca Raton, FL, 686 p.

Lowson, R.T., 1982, Aqueous oxidation of pyrite by molecular oxygen: Chemical Reviews, v. 82, p. 461-493.

Maki, S., 1995, Passivation of pyrite as a means of decreasing pyritic oxidation: unpublished M.Sc. thesis, Laurentian University, Sudbury, ON, 72 p.

McGregor, R.M., Blowes, D.W. and Robertson, D.W., 1995, The application of chemical extractions to sulphide tailings at the Copper Cliff tailings area, Sudbury, Ontario, in Hynes, T.P. and Blanchette, M.C., eds., Sudbury '95: Mining and the Environment: CANMET, Ottawa, ON, Conference Proceedings, p. 1133-1142.

MEND, 1991, Proceedings of the 2nd International Conference on the Abatement of Acidic Drainage, Montréal, PQ: v. 1-4.

Moses, C.O., Nordstrom, D.K., Herman, J.S. and Mills, A.L., 1987, Aqueous pyrite oxidation by dissolved oxygen and by ferric iron: Geochimica et Cosmochimica Acta, v. 51, p. 1561-1572.

Nicholson, R.V., 1994, Iron-sulfide oxidation mechanisms: Laboratory studies, in Jambor, J.L. and Blowes, D.W., eds., Environmental Geochemistry of Sulfide Mine-Wastes: Mineralogical Association of Canada, Short Course Handbook Number 22, p. 163-183.

Nicholson, R.V., Gillham, R.W. and Reardon, E.J., 1988, Pyrite oxidation in carbonate-buffered solutions. 1. Experimental kinetics: Geochimica et Cosmochimica Acta, v. 52, p. 1077-1085.

Nicholson, R.V. and Scharer, J.M., 1994. Laboratory studies on pyrrhotite oxidation kinetics, in Alpers, C.N. and Blowes, D.W., eds., Environmental Geochemistry of Sulfide Oxidation: American Chemical Society, Symposium Series 550, p. 14-30.

Ritchie, A.I.M., 1994, The waste-rock environment, in Jambor, J.L. and Blowes, D.W., eds., Environmental Geochemistry of Sulfide Mine-Wastes: Mineralogical Association of Canada, Short Course Handbook Number 22, p. 133-161.

United States Department of the Interior, 1994, International Land Reclamation and Mine Drainage Conference and 3rd International Conference on the Abatement of Acidic Drainage: Bureau of Mines, Special Publication SP O6C-94, Pittsburg, PA, v. 1-4.

15. Contaminated Sediment in Urban Great Lakes Waterfronts

John P. Coakley
Alena Mudroch

Aquatic Ecosystem Restoration Branch, National Water Research Institute, Burlington, Ontario L7R 4A6

SUMMARY

Contaminated sediments in the Great Lakes are the legacy of decades of waste disposal into the aqueous environment by municipalities and industries. The capacity of the receiving waters to mix, dilute, neutralize, and export wastes has declined, with the result that there is now a stockpile of contaminants held within bottom sediments. This situation is presently creating a problem for water use management in a number of Great Lakes areas, 43 of which have been designated as Areas of Concern by the United States-Canada International Joint Commission.

This paper presents a review of contaminated sediment problems, and discusses important physical and chemical processes that influence the effect of contaminated sediments on the aquatic ecosystem. Processes discussed include those such as deposition-resuspension, mixing, and advection, as well as the more important chemical and biological transformations that take place in the sediments. Effective remediation depends ultimately on measures being taken to reduce or eliminate further contaminant inputs.

INTRODUCTION

Urban waterfront areas in the Great Lakes are usually located around sheltered coastal embayments that provide natural deep-water harbours. However, these conditions also promote deposition of fine-grained sediments, having a grain size diameter of less then 0.06 mm, thus falling within the silt and clay size fractions. These fine sediments are characterized by extremely large surface-to-volume ratios, and play a disproportionate role in the adsorption and transport of chemical contaminants. Excess deposition of fine particles results when natural processes that lead to sedimentation in a waterbody, such as watershed and shoreline erosion, and aqueous transport and deposition of soil material, are altered by urban activities. Fine sediment supply is aggravated by soil erosion caused by urbanization, primarily commercial and residential construction, and accelerated surface runoff from roadways and parking lots (Chapter 2). Fine particles are also supplied in great quantities by industrial discharges. As a result, fine materials comprise the bulk of sediments being presently deposited in urban waterfront areas in Canada.

Chemical species labelled as priority contaminants (Table 1) enter the aquatic environment through a variety of natural and anthropogenic pathways. The most common natural vectors are erosion and weathering of rocks and atmospheric transport of volcanic emissions. Anthropogenic sources include industrial, municipal, and agricultural discharges, and wastes from mineral processing or petroleum refining. In addition, there are a number of coastal areas that receive large volumes of wastes from mining and pulp-and-paper industries. Many contaminant elements and compounds enter the aquatic system in dissolved form, but most of the persistent contaminants that are of concern for human health are either hydrophobic (organic compounds having a low solubility in water), or are particle-reactive, *e.g.*, heavy metals that readily scavenged out of solution onto particle surfaces. The net result is that they tend to partition with the fine particulate phase. In the case of atmospheric inputs, contaminants enter the aquatic system as wet or dry deposition attached to fine particles or colloids. After entering the waterbody, these particles are carried by water currents into areas of reduced flow, where they settle and accumulate in bottom sediment deposits. Under energetic hydrodynamic conditions, contaminants in bottom sediments may be resuspended into the overlying waters, making them again accessible for uptake into the aquatic food chain. Consequently, bottom sediments can act both as sinks and as sources of contaminants in an aquatic environment. The United States-Canada International Joint Commission (IJC) and others have recognized contaminated sediments as representing one of the most serious problems in the sustainable management and remediation of Great Lakes urban waters, especially when navigation necessitates regular dredging (International Joint Commission Great Lakes Water Quality Board, 1988; Sérodes, 1978; Pollman and Danek, 1988).

In recent years, efforts have focussed primarily on the mobility of contaminants from the sediment into the water, bioavailability and biological impacts of sediment contaminants and interactive processes in contaminated sediments (Förstner and Patchineelam, 1980). Results of the research to date indicate that many sediment-associated contaminants can be remobilized within an aquatic ecosystem and have a negative effect on the health of the biota (see Calmano and Förstner, 1995). Further, the effects of sediment contaminants often are not related to high concentrations of the contaminants in the sediment, but rather to their chemical forms and their physico-chemical behaviour, particularly at the sediment-water interface (Zarull and Mudroch, 1993).

The visible impacts of contaminated sediments in urban waterfronts range from aesthetics (odours, gas emission, and

Table 1   Sources and effects of priority contaminants in urban waters (adapted from Sittig, 1991; and Pollman and Danek, 1988).

Priority Contaminants	Predominant Source	Input Point	Physiological and Environmental Effect
INORGANIC			
Mercury	Paper manufacturing, metal recovery, fossil fuel combustion, waste incineration, electronic and scientific instruments, chemical and pharmaceutical industry	Atmospheric deposition, aquatic biomagnification process, industrial discharges	Minimata disease (central nervous system), ulcerative gastroenteritis, tubular necrosis
Cadmium	Electroplating, Ni-Cd battery manufacture and disposal, paint and pigment process, fertilizers, pesticide catalysts, sewage sludge application	Urban runoff, industrial and sewage treatment plant (STP) outfalls	Chronic effect on lungs and kidneys, cancer and various dysfunctions
Zinc	Galvanizing of metals, zinc and phosphate fertilizers, paint and textile dyeing, combustion, rubber manufacture and tire wear	Industrial and STP outfalls, combined sewer overflows (CSO)	Gastrointestinal and central nervous system dysfunction
Lead	Paint manufacture and disposal, automobile exhaust, battery manufacture and disposal, metal plating	Atmospheric deposition, urban runoff, STP, CSO outfalls	Cumulative brain damage, hyperactivity, hypertension
Chromium	Leather tanning, electroplating, pigment and dye process, ferro-alloys, wood preservation, solid waste incineration	Atmospheric deposition, urban runoff, industrial discharges	Respiratory dysfunction, lung and nasal cancer, contact dermatitis and ulceration
Arsenic	Pesticide and fertilizers, refining and smelting, fossil fuel combustion, incineration of solid wastes, wood preservatives, ceramics, electrophotography	Atmospheric deposition, urban runoff, industrial discharges	Comprehensive poison for organisms, carcinogenic, mutagenic; extremely toxic to aquatic life
Selenium	Smelting and refining, electronic equipment manufacture, fertilizers, glass making and ceramics, photocopying, pigments, fossil fuel combustion	Atmospheric deposition, industrial discharges	Hair loss, nail and tooth pathology, liver effects
Copper	Metal working, waste incineration, fossil fuel and wood combustion, coal pile leaching, electroplating, insecticides and herbicides, antifouling paints, wood preservatives, water treatment, welding and soldering, wallpaper manufacture	Atmospheric deposition, industrial discharges, STP outfalls	Contact dermatitis, respiratory and renal dysfunction; toxic to aquatic life
Iron	Iron and steel manufacture, burning of coke and coal, landfill leachates	Steel plant discharges, STP outfalls	Toxic to aquatic life in high concentrations
ORGANIC			
Polychlorinated biphenyls (PCBs)	Electronic industry, plastics and adhesives manufacturing, plasticizer in paints, photocopiers, sealant	Urban landfill leachate, atmospheric deposition, spills	Central nervous system dysfunction, presumed carcinogenic and mutagenic in humans and fish
Polycyclic aromatic hydrocarbons (PAHs)	Metallurgical and coking plants, combustion processes of all kinds	Atmospheric deposition, urban runoff from roads and parking lots, STP and CSO outfalls	Presumed carcinogen, mutagen; promotes cancers in fish
Phenols	Coke oven by-products, refinery wastes		Presumed carcinogen
Oil and grease	Manufacturing losses and spills, crank-case oil disposal, tank leaks, water-cooled combustion engines	Landfill leachates, ground-water discharges and streams, accidental spills	
OTHERS			
Suspended solids	Refinery wastes, food processing, construction site runoff	Industrial discharges, urban creeks and runoff, CSO	Habitat loss, degraded aesthetics, pathogenic bacteria
Ammonia	Refinery wastes, sewage treatment, industrial cleaning products, municipal wastes, coal combustion, manure and fertilizer use	STP, CSO outfalls, urban creek runoffs	Eutrophication and loss of habitat, degraded aesthetics
Biochemical oxygen demand (BOD)	Organic waste disposal, food processing wastes, pulp and paper manufacture	STP, CSO outfalls, industrial discharges	Anoxia, loss of habitat, degraded aesthetics, pathogenic bacteria

fish kills), to loss of beneficial use, such as degraded sport fisheries (unsavoury fish with tumours and lesions, consumption advisories, low species diversity). Other less visible impacts are contaminated drinking water and closed or restricted bathing beaches. Arguably, the most deleterious effect is felt on aquatic biota which use such sediments as spawning sites, as nurseries for juvenile organisms, or as habitat. Bioassay testing of contaminated sediments has demonstrated serious effects of exposure to various contaminants, ranging from genetic defects to reproductive failure (Reynoldson *et al.*, 1991; Krantzberg, 1994).

The major urban centres on Canada's coastline where contaminated sediment problems have been most publicized are shown in Figure 1, and the nature of their contamination is summarized in Table 2. Several of these are discussed elsewhere in this volume (Chapters 16–19), and this paper will focus on those within the Great Lakes–St. Lawrence system. These areas have been affected by long-term waste disposal practices, and are now degraded to the point where industrial and port activities, navigation, and non-contact recreation are virtually the only uses. However, efforts to restore these areas have been initiated over the past decade in conjunction with mandated reductions of contaminant inputs. The costs of remedial efforts involving sediments are high, but nevertheless, these efforts are being actively pursued.

SEDIMENT CONTAMINATION IN GREAT LAKES URBAN WATERFRONTS

Scope Of The Problem
Urban waterfronts on fresh-water systems such as the Great Lakes (Fig. 1) have been affected more severely by sediment pollution than comparable marine environments. The Great Lakes basin, for example, is home to over 36 million inhabitants in Canada and the United States. In addition to being smaller in extent and lacking the strong flushing action of tides, the Great Lakes system serves as a supplier of industrial process and municipal water, and as a receiver of wastes from both. These waters are also intensely used for aquatic recreation.

In response to concerns about water quality in the Great Lakes, the IJC identified 43 Areas of Concern, defined as aquatic sites where failure to meet the objectives of the United States-Canada Great Lakes Water Quality Agreement has caused impairment of beneficial use or the ability to support aquatic life (International Joint Commission Great Lakes Water Quality Board, 1987). They include many harbours and embayments in the vicinity of urban areas on the lakeshore. Impairment of beneficial use is defined as a change in the chemical, physical, or biological integrity sufficient to cause any of the following: restriction on fish and wildlife consumption; tainting of fish and wildlife flavour; degradation of fish and wildlife populations; fish tumours and other deformities; bird or animal deformities or reproductive problems; degradation of benthos; restrictions on dredging activities; eutrophication or undesirable algae; restrictions on drinking-water consumption or taste and odour problems; beach closings; lowered aesthetic appeal; added cost to agriculture or industry; degradation of phytoplankton and zooplankton populations; or loss of fish and wildlife habitat (Hartig and Zarull, 1991).

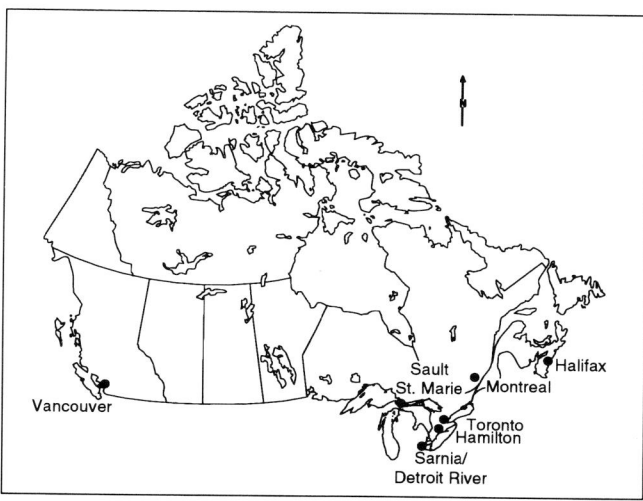

Figure 1 *Location map showing selected harbours and urban waters in Canada affected by contaminated sediments.*

Contaminant Sources And Levels
Of the 43 Areas of Concern identified for the Great Lakes, 42 have sediments that are contaminated with metals or persis-

Table 2 Examples of Canadian urban areas with sediment contamination problems. Data sources are given in the text.

Location	Site of Major Contamination	Principal Contaminants
Halifax Harbour	Bedford Basin	Pb, Ni, Cr
	Central Harbour	Pb, Cu, Cr, Cd, Zn, Hg
Toronto Waterfront	Inner Harbour	Solvent-extractable organics, nutrients, PCBs, P, Cr, Cu, Pb, Fe, Zn
	Keating Channel	Organics, P, metals, PCBs
	Humber Bay	Nutrients, metals, solvent-extractables, PCBs
Hamilton Harbour	Randle Reef	Coal tar, PAH, PCB
	Windermere Basin	PCBs, Hg, Zn, Pb, Cu
	Cootes Paradise	PCBs, Hg, Zn, Cu
Montreal area	Lac Saint-Louis	Hg, Cd, As, Pb, Zn
	Port of Montreal	PCB, hydrocarbons
Vancouver	Port Moody Arm	Cd, Cr, Cu, Pb, Hg, oil and grease, PAHs, PCBs

tent organic substances. Most of this contamination results from past discharge practices; however, there remain significant sources of these contaminants from point and non-point sources including the atmosphere (Zarull and Mudroch, 1993). The relatively large surface area of the five Laurentian Great Lakes (about 147,000 km²) indicates that atmospheric deposition is a major contributor of contaminants originating from urban areas. Atmospheric deposition is the principal source of many trace metals in lakes Superior and Huron, and accounts for around 50% of trace metals in lakes Erie and Ontario (Nriagu, 1986). However, in the specific cases discussed herein, contaminated sediments are confined to restricted, semi-enclosed bodies of water, and result from contaminant being discharged directly to coastal waters or into inflowing tributaries.

Table 2 summarizes the major contaminants found in sediments in various degraded coastal sites in Canada. It is difficult to compare levels quantitatively between the marine and fresh-water sites, but it is clear that both organic and inorganic contaminants are implicated. This paper emphasizes the dynamics of contaminated sediment deposition along urban waterfronts, a more detailed discussion of contaminant geochemistry is given in Chapter 26.

Metals in Sediment Profiles

In Great Lakes depositional basins, the concentrations of metals, nutrients and organic contaminants are greater in the surficial sediments (about 0 to 15 cm sediment depth) than those in the deeper sediments. An example of concentration profiles of lead in sediments in the depositional basins of Lake Ontario is shown in Figure 2. The concentration of lead below about 12-cm sediment depth is considered the background concentration, i.e., the natural concentration of lead in mate-

rials from which the bottom sediments were derived. The increase of lead concentrations toward the sediment surface reflects the anthropogenic input of lead into Lake Ontario. Automobile exhaust and atmospheric deposition are major contributors of lead to Lake Ontario sediments (Nriagu, 1986). Dating of Lake Ontario sediments, using ^{210}Pb and ^{137}Cs, indicated that the concentration peak of lead in Lake Ontario sediments, normally occurring between 3 and 4 cm sediment depth (Fig. 2), corresponds to sediments deposited around 1973 (Mudroch, 1993). The decrease of lead concentrations in the uppermost 3 cm of sediment appears to reflect the introduction of unleaded gasoline in Canada and the United States. Similar observations were reported for sediment in Lake Michigan, in precipitation in north-central Minnesota and in sediments in the Mississippi River delta (Christensen and Goetz, 1987; Eisenreich et al., 1986; Trefry et al., 1985). This trend indicates that the sediments containing the greatest concentrations of lead are being buried by settling particles containing lower concentrations of lead.

In contrast to sediments in offshore depositional areas, surficial sediments in the Great Lakes harbours can contain lower concentrations of some metals and organic contaminants than the deeper sediments. For example, the concentrations of mercury, zinc, copper and chromium in Lake Erie harbours is considerably lower in the harbour and nearshore than in the deeper sediments (Table 3). This may be due to activities in the harbours which cause mixing or overturning of the bottom sediments, such as dredging and disposal of sediments, shipping, etc. In more industrial Lake Ontario harbours, however, the concentrations of metals, nutrients and organic contaminants are generally greater than in sediments from the depositional areas of the lake (Table 3). This is due to the input of point sources at the harbour or from

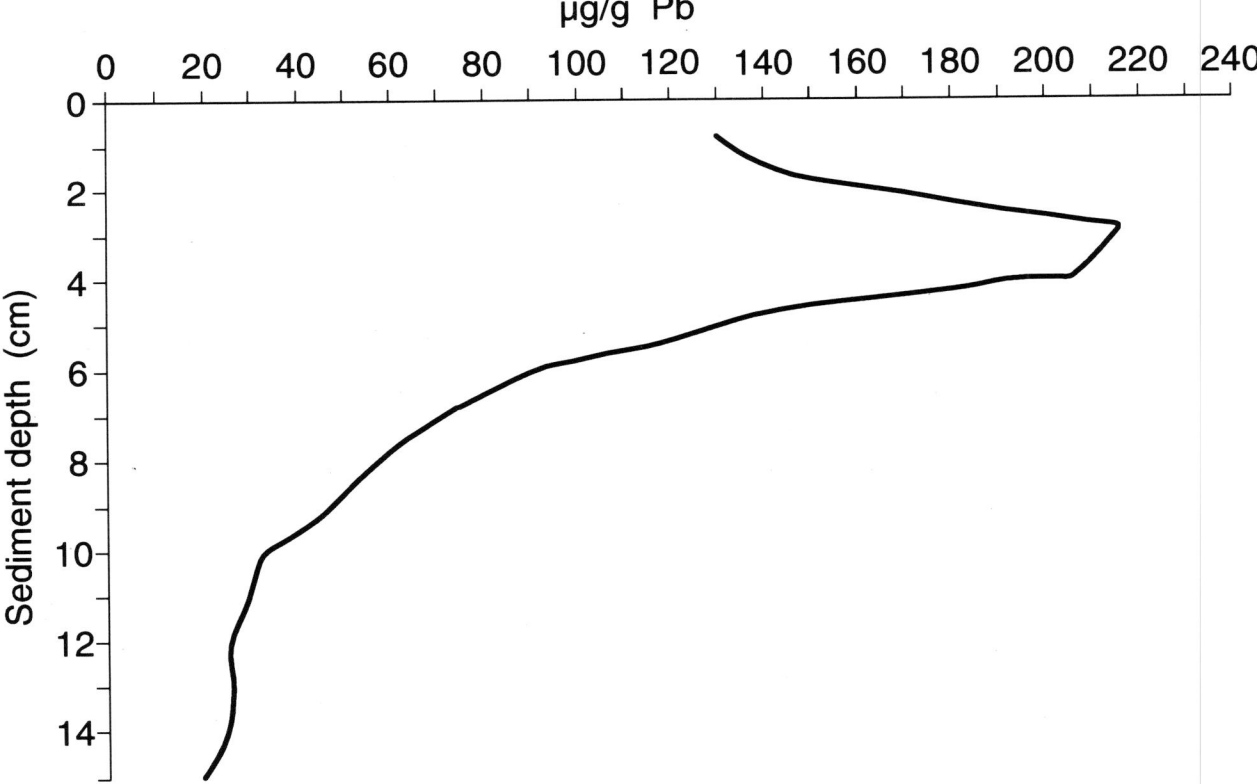

Figure 2 *Vertical profile of lead contamination in sediments from Lake Ontario showing peak concentrations (~3 cm) related to anthropogenic lead inputs (notably leaded gasoline) followed by decline after implementation of unleaded gasoline legislation.*

streams entering the harbour. The sediment in the harbour acts as a sink for some contaminants and nutrients collected by streams from rural or urbanized watersheds.

Recent monitoring of most of these areas shows that the remedial actions and abatement programs initiated over the past decade have begun to cause a reduction in contaminant levels in surface sediments. A similar reduction to that of lead mentioned above has occurred in lead and phenol levels in Hamilton Harbour and for mercury in the sediments of the Detroit River as a result of source reduction actions taken in the late 1970s and early 1980s.

Examples of Contaminated Sediment Problems in the Great Lakes–St. Lawrence System

Toronto Harbour. Metropolitan Toronto (Fig. 3) represents the largest Canadian urban complex on the Great Lakes, with a population of over 4 million and a waterfront extending over 20 km. Toronto Harbour has been designated as an Area of Concern because of the large volume of degraded sediments deposited over the past decades. The most seriously degraded areas are located near the Keating Channel (Fig. 3), which receives sediments from the heavily polluted Don River. The most important of the contaminants in the sediment are solvent-extractable organics, nutrients, polychlorinated biphenyls (PCBs), polycyclic aromatic hydrocarbons (PAHs), and oil/grease compounds, phosphorous, and heavy metals: chromium, copper, lead, iron and zinc (Persaud *et al.*, 1985; Chapter 2).

Hamilton Harbour. Hamilton Harbour (Fig. 4) is another Canadian Area of Concern severely impacted by industrial and municipal contaminant inputs (Chapter 16). The most degraded area, Randle Reef, immediately offshore from the Steel Company of Canada main loading docks, shows concentrations of PCBs, PAHs, and coal tars that are several hundred times the local background. Loadings of phenols from the nearby steel mills were problematic at one time, but, since the mid-1970s, have declined by almost 90%. The Windermere Basin of Hamilton Harbour (analogous with the Keating Channel area of Toronto Harbour) is heavily polluted by municipal (sewage) and industrial effluents. To the west of the harbour, sediments in Cootes Paradise also show contamination by a variety of metals and organics (Table 2). The major contaminants are PCB, PAH, and heavy metals such as mercury, derived from the Dundas Sewage Treatment Plant (Hamilton Harbour Remedial Action Plan, 1992).

Sault Ste. Marie. Sault Ste. Marie Harbour (Fig. 1) receives contaminants from industrial complexes producing steel, and pulp and paper. Concentrations of Pb, Zn and As in bottom sediments along the Canadian shore of the St. Mary's River at Sault Ste. Marie, Ontario, were as high as 1865, 4600 and 92 $\mu g \cdot g^{-1}$, respectively. These concentrations, among the highest observed in Great Lakes sediments, originated from the input of material from a slag dump site near a steel plant on the north shore of the river (Mudroch, 1991).

St. Clair and Detroit rivers. The area south of the city of Sarnia, on the St. Clair River, has been dubbed "chemical alley" because of the concentration of petroleum-refining and chemical-manufacturing plants located there (Fig. 5). This situation has led to a number of well-publicized chemical spills, the latest being in November 1985, when a large volume of perchloroethylene (used in dry cleaning) was accidentally discharged into the river. Because of its relatively high density and low solubility (Chapter 6), it was concentrated as puddles in the sandy sediments of the river bed. Although no loss of recreational or other uses has resulted from such events, there is always the threat to drinking-water intakes downstream and to fish resources exploited by native communities on Walpole Island in the St. Clair delta area. A similar situation exists downstream in the Detroit River segment (Fig. 5), where contaminated sediments downstream of the Detroit-Windsor metropolitan areas cause concern for fisheries. In the 1970s, mercury contamination from chemical manufacturers on the shores of the St. Clair River was responsible for the high concentrations of mercury in sediments and fish in the river and in the western basin of Lake Erie. More recently, toxic and highly persistent organic contaminants (PCBs and PAHs) have been accumulating in contaminated sediments in the area (Carter and Hites, 1992).

Montreal and environs. The St. Lawrence River in the vicinity of Montreal widens considerably to form riverine lakes, Lac St.-Louis and Lac des Deux Montagnes (Fig. 6). This widening, together with the deepening and isolation of the main channel by the St. Lawrence Seaway and an abundance of islands, causes a sharp reduction in the flow velocity of the river and encourages extensive macrophyte growth. All these factors cause increased deposition of fine sediments, which carry a full load of contaminants from the Great Lakes and from upstream and local industrial activities. Figure 6

Table 3	Concentrations of mercury (Hg), zinc (Zn), chromium (Cr) and copper (Cu) in surficial and deep sediments in Lake Erie offshore depositional areas and harbours ($\mu g \cdot g^{-1}$ dry weight; after Mudroch *et al.*, 1988).		
		Depositional Areas	Harbours
Hg	surficial	0.045–4.800	0.015–2.200
	deep	0.010–0.190	0.050–7.000
Zn	surficial	18–536	12–650
	deep	8–28	40–500
Cr	surficial	12–362	30–150
	deep	9–25	30–250
Cu	surficial	5–207	2–100
	deep	20–48	10–110

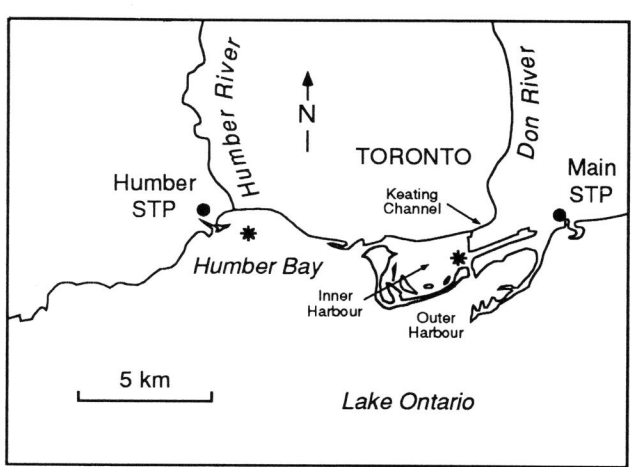

Figure 3 *Contaminated sediments in the Toronto Harbour and waterfront areas. Asterisks indicate sites of most serious contamination or hot spots; dots show location of sewage treatment plants (STPs; see also Chapter 2).*

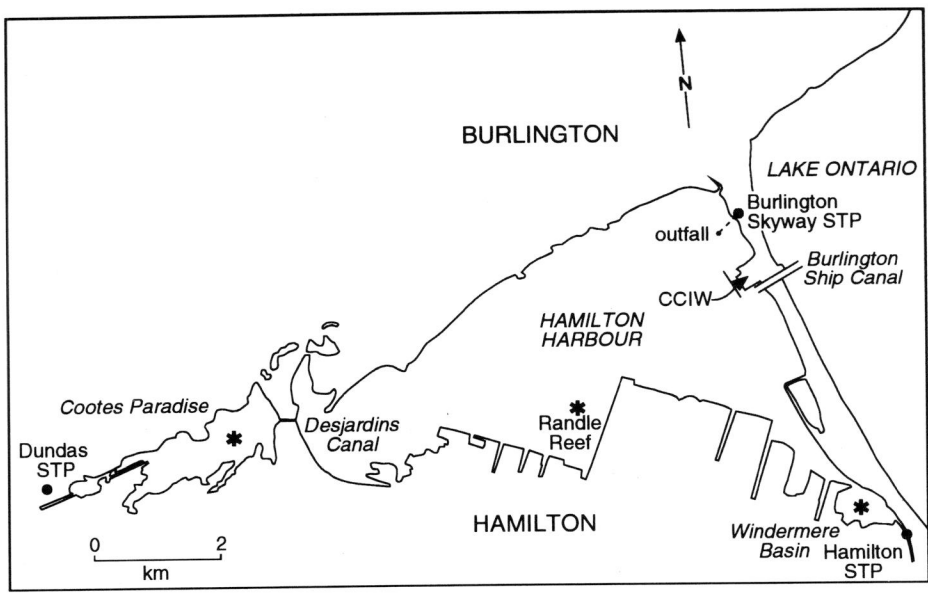

Figure 4 *Contaminated sediments in Hamilton Harbour. Symbols and abbreviations same as in Figure 3.*

shows the locations of the most contaminated areas, and Table 2 summarizes the nature of the contamination.

There is a continuous input of various contaminants and nutrients from many industrial and municipal discharges into the St. Lawrence River, as has been demonstrated in studies of surficial sediments in Lac St.-Louis (Rukavina *et al.*, 1990). Exceptions (mercury, zinc, copper and chromium) are attributed to inputs from local sources. Nevertheless, it was concluded that Lac St.-Louis acts as a conduit for suspended sediment rather than as an accumulation zone for deposited

sediment, such that contaminated sediment is eventually exported downstream by the St. Lawrence River. Montreal Harbour sediments are highly contaminated by metals, particularly cadmium, chromium, copper, lead, mercury, nickel and zinc (Olivier and Bérubé, 1993). However, recent monitoring of the sediment quality in the St. Lawrence River has shown an improvement in the quality of surface sediments; contamination by cadmium, chromium, copper, lead, mercury, nickel and zinc has decreased over time in areas where sources of contamination were controlled (Olivier and Bérubé, 1993).

NATURAL TRANSPORT PATHWAYS AND FATE OF CONTAMINATED FINE SEDIMENTS

Contaminants in sediments are gradually removed from the aquatic system by uptake and export by organisms, biochemical transformation, or by deep burial. Figure 7 summarizes, in cartoon form, the physical evolution of conservative contaminants, *i.e.*, those stable compounds whose spatial concentration is affected mainly by dilution away from their source. These comprise predominantly the organic, hydrophobic compounds (PAHs, PCBs, pesticides, mirex). However, many heavy metals tend to partition to the sediment

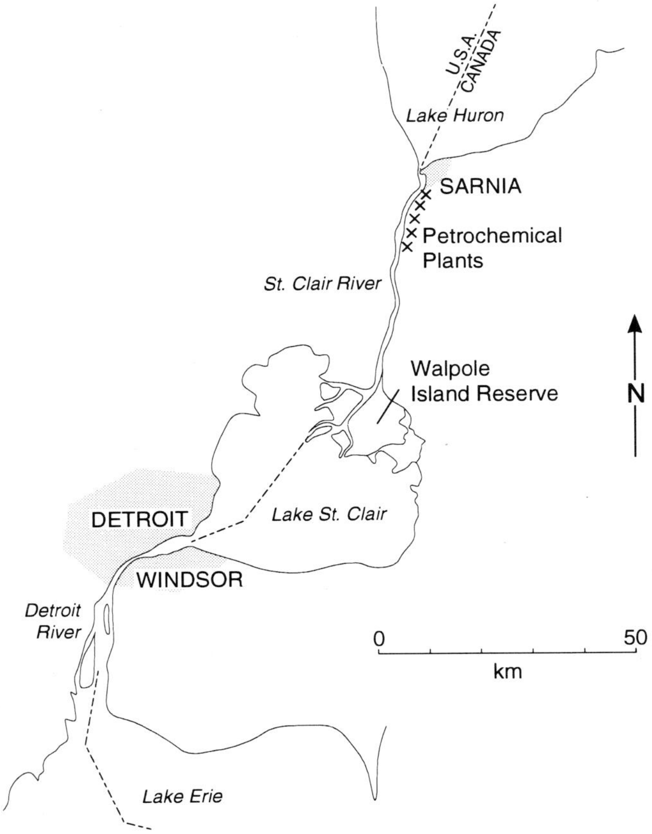

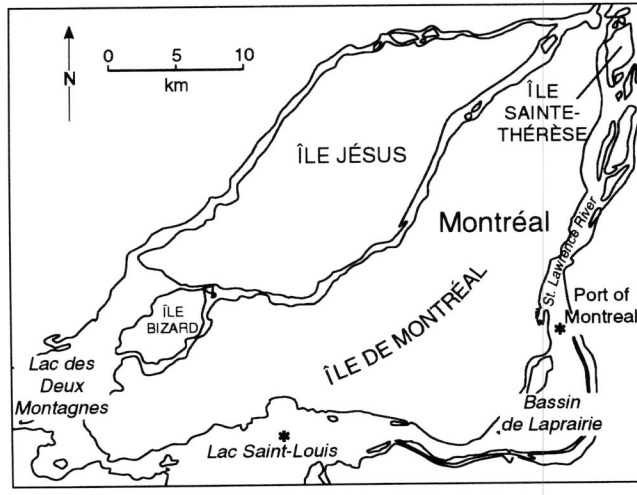

Figure 5 *The Sarnia area of the lower Great Lakes, showing the locations of contaminated sediment areas referred to in text. Asterisks show location of known hot-spots.*

Figure 6 *Contaminated sediments in the Montreal area. Asterisks show location of hot-spots.*

phase where they may persist until subjected to major changes in pH, Eh, or salinity (de Groot *et al.*, 1976).

In considering the fate of these contaminated sediments, it must be borne in mind that they may not stay in place on the bottom, but may be subject to several cycles of deposition and resuspension, thereby serving as contaminant sources for aquatic organisms. This is an important factor for water use managers and planners, because even after a contaminant source is shut down, deleterious effects still persist. In multi-use urban waterfront areas, where water intakes or bathing beaches must co-exist in proximity to contaminant source outfalls, it is critical to understand the local transport patterns and the efficiency of the hydrodynamic mixing processes involved. An important parameter in contaminant control and remediation, therefore, is the net pathway of these sediments through the aquatic environment, and the rate at which they are transported and buried.

The major process agents responsible for sediment transport are natural currents driven by either hydraulic gradients (in streams and rivers), or, in the case of open waters, by winds, density gradients, and wave action. However, in urban coastal areas where port facilities are important, processes such as propeller-wash and anchor dragging associated with ship passage, docking, dredging for navigational purposes, and shoreline construction projects are significant (see Chapter 17). Deterministic assessment of the dynamics of fine sediments (and their attached contaminants) has proved difficult as the parameters necessary for predicting wave transport and water circulation, such as wind speed and direction, are problematic to measure in restricted areas. For this reason, numerical models of fine sediment transport, based on bottom shear stress, resuspension, and advection have yet to overcome some major difficulties.

Modelling Of Fine Sediment Transport

One of the major difficulties in modelling the transport of contaminated fine sediment is in the parameterization of resuspension of these sediments. Unlike sand and silt, they are cohesive, *i.e.*, they are held together by electrostatic as well as gravitational forces. Added to these attractive forces are the effects of mucous and fibrous materials associated with bacteria and benthic infauna that tend to hold the deposited sediment in place. It is, therefore, very difficult to obtain consistent values for the critical shear stress, *i.e.*, the force on the bottom created by wave-induced and circulation currents, that is required to resuspend the sediment. A useful insight into the work carried out on modelling transport in such cases is obtained from Lick *et al.* (1994) and Tsai and Lick (1988). Another area where difficulty is encountered is in assessing the role of flocculation, *i.e.*, the agglomeration of electrically charged fine particles into irregular-shaped composite masses of variable sizes and densities. These parameters are crucial in modelling the transport of the advection and dispersion of these materials in the water column.

Despite the uncertainties in understanding critical transport processes, models have been developed to calculate the quantities of fine material transported in response to a given wave-induced (orbital) or unidirectional (fluvial) current regime (Lick *et al.*, 1994; Krishnappan, 1991). Without going into the details of these models, it should be noted that much effort is presently going into the inclusion of flocculation of fine particles as a determining factor in the transport process (Partheniades, 1986; Krishnappan, 1991).

Sediment Transport Tracing

A direct way of studying the dynamics of contaminated fine sediments in urban aquatic environments is by the use of

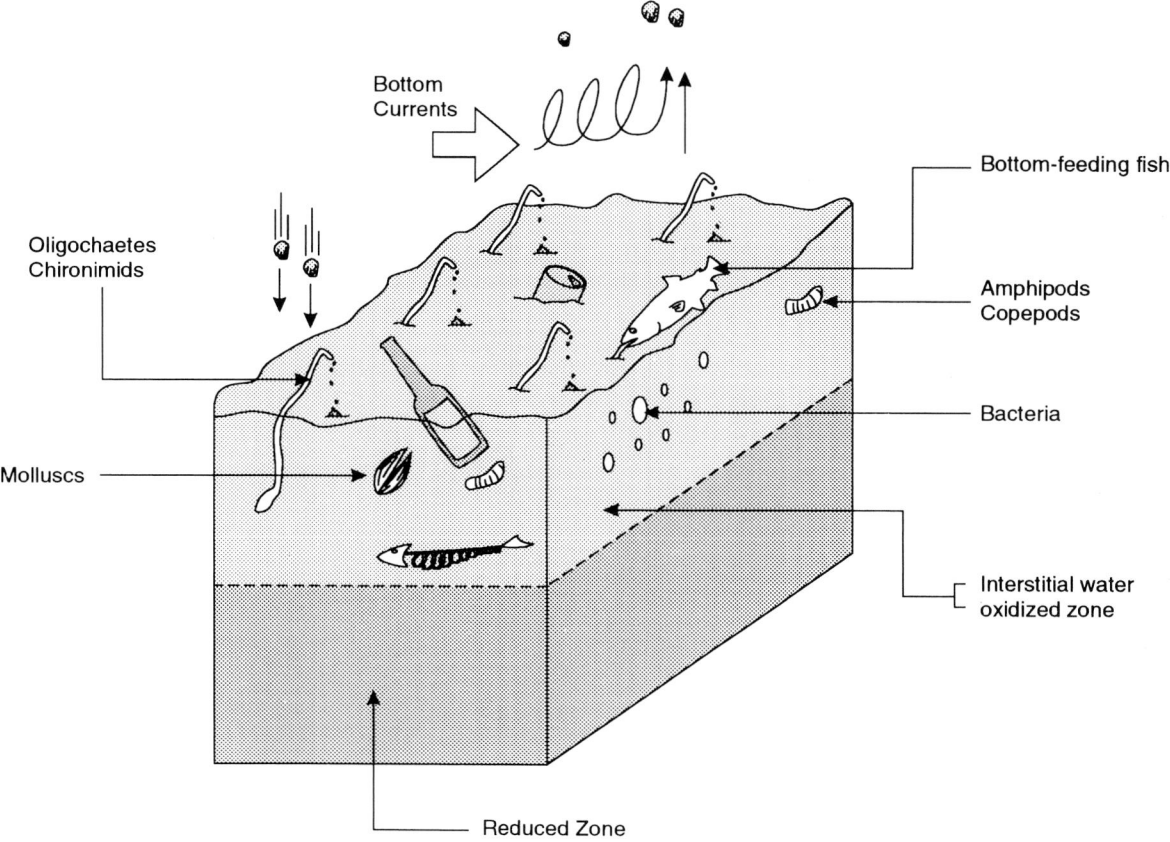

Figure 7 *Schematic illustration of important processes and effects in contaminated sediment deposits.*

tracers. Unlike the modelling approach discussed above, this approach is empirical, and provides a time-integrated view of sediment transport patterns at a specific site. The tracers found to be most useful consist of conservative chemical elements, compounds, or attributes of the sediments that may be linked to an identified point source. Such tracers may either be introduced artificially by the investigator, or supplied naturally. Natural tracer components also include contaminants discharged by industrial or other anthropogenic point sources that persist in the sediments at levels above background. The procedure for the use of these techniques is described in Coakley and Long (1988) and in references cited in the case studies below. Transport studies involving both artificial and natural tracers have been carried out at the following Areas of Concern in Lake Ontario: Humber Bay and Hamilton Harbour.

Humber Bay. In a study initiated in 1988, distribution patterns were examined for the bottom sediment concentrations of the naturally occurring elements thorium, cobalt, and scandium, as well as for a cesium-based artificial tracer (Fig. 8). Initial testing was also successfully carried out on organic contaminants (coprostanol, n-alkanes, vitamin-E acetate, PAHs) as sediment tracers (Coakley *et al.*, 1990; Coakley and Poulton, 1991). The long-term sediment transport trend inferred was southward along the west shore of Humber Bay, then curving southwestward parallel to the shore.

What is evident from the results of time-averaged studies of sediment dynamics is the variability of the transport patterns. Figure 8A shows secondary plumes in different directions from the main pattern traced by cobalt concentrations from the Humber River. The reasons for these divergences would be transport of the particles in different flow regions, due to grain size or density fractionation or to seasonal changes in discharge trajectories.

Hamilton Harbour. Spatial transport and dispersal of contaminated sediments were investigated using chemical markers for sewage effluents as tracers. A range of such markers, including the fecal sterol coprostanol, C- and N-isotope ratios, and physical sediment properties were used to trace the dispersal of the effluent from the Burlington Skyway sewage treatment plant (STP), on the northeast corner of the harbour (Bachtiar *et al.*, 1996). The plume for coprostanol (Fig. 8B) showed a dominant shore-parallel dispersion in both directions, in response to the prevailing SW-W wind regime.

A study of overall transport patterns in Hamilton Harbour (Coakley *et al.*, 1994) identified the importance elsewhere in the harbour of sediment focussing, the long-term transport of fine, contaminated sediments to the deepest parts of the basin. This process involves the periodic resuspension and steady displacement over time of fine sediments to deeper and deeper water, until resuspension eventually becomes non-existent. The lack of any indication of sediment focussing in the STP plume described above indicates the prevalence of wave-induced longshore drift in this area in response to prevailing westerly winds and waves. Other work in Hamilton Harbour using magnetic characteristics of sediment is described in Chapter 16.

Mixing, Dilution and Export

The rationale often given for the deliberate discharge of contaminants into adjacent waterbodies is that dynamic processes in these waters, be they flowing rivers, tidal streams, or wave-washed coasts, are able to mix the effluent with large volumes of clean water and sediment. Thus, the contaminants would tend to be diluted to acceptable concentrations. This is

true except for cases where the inputs exceed the diluting capacity of the waterbody. In such cases, the contaminated sediments are allowed to settle out and remain for considerable time in the basin, and undergo cycles of resuspension and deposition, as in the cases discussed above.

In most urban waterfronts, mixing and transport processes are sufficient to mobilize and flush out most of the finer contaminated sediments that remain in suspension, except in restricted arms and harbour boat-slips, where circulation and turbulence are diminished to levels insufficient to maintain particle suspension. Because these areas are often sites of effluent discharges (accidental or deliberate), the result over the years is the accumulation of contaminants into eventual hot-spots that later require remediation. In the Great Lakes, the absence of tides requires that all the physical mixing, resuspension, and export of contaminants and contaminated sediments be carried out by nearshore wave or circulation-related currents and river flow, respectively. In the case of Montreal, the St. Lawrence River is energetic enough to keep fine sediments in suspension in all areas except where the river widens (Lac Saint-Louis, west of Montreal) or where restricted arms of the river cause local flow reduction and stagnation (Rivière des Prairies; Fig. 6). The quasi-enclosed configuration of both Hamilton and Toronto harbours serves to reduce currents, and thus, deposition of contaminated sediments is the rule. Some contaminated sediments are resuspended and exported through the harbour inlets to the open lake, but this occurs mainly during storm events or during periods of intense ship traffic.

CHEMICAL DIAGENESIS, RELEASE OF CONTAMINANTS FROM SEDIMENTS AND UPTAKE BY ORGANISMS

Once a contaminant reaches the bottom sediments, it is commonly recycled, *via* biological and chemical agents, both within the sedimentary compartment and back into the water column. Freshly deposited particles consist of flocs lying on the sediment surface. Organic matter becomes decomposed by bacterial action, and the remaining material starts to consolidate. A layer of new particles buries what remains from the floc, which thus becomes incorporated on a more permanent basis into the bulk of the sediments. Thereafter, the sediments are affected by diagenetic processes, such as compaction, cementation, diffusion and mineral segregation, and by biochemical transformations. Diagenetic processes affect the bioavailability of major and trace elements, heavy metals, organic compounds and nutrients in the sediment.

Diagenesis

An example of a diagenetic process in contaminated sediment is provided by iron. Iron minerals are common in particles deposited on the bottom of lakes, rivers and oceans. Since iron exhibits two oxidation states, stability of the iron minerals is a function of the redox conditions in the sediments. Under oxic conditions, which usually occur at the sediment-water interface and within the surficial (3-5 cm thick) sediment layer, the presence of fine particles of iron-oxyhydroxides is common. These compounds can adsorb many metals, trace elements and nutrients. After burial, redox conditions change because of oxygen depletion in the fine-grained sediments. The oxidation state of iron changes, leading to dissolution of the iron-oxyhydroxides and release of adsorbed elements and compounds which, in turn, may become re-adsorbed on clay minerals or other fine particles. A portion of these elements and compounds may remain in dissolved or colloidal form in sediment interstitial water and equilibrate with their concentrations in the sediments.

Besides iron, the common elements in sediments which exhibit multiple valence states and are affected by oxidation and reduction are manganese, carbon, nitrogen, sulphur, hydrogen and oxygen. Changes in chemical form and mobilization of these species are largely affected by microbial

processes in the sediments. Diagenetic processes in sediment are described in detail by Berner (1971).

Organic contaminants entering the aquatic environment are preferentially associated with sediment organic matter. Less persistent organic contaminants become decomposed, but persistent organic contaminants, such as PCBs, reside in sediments for many years (Oliver and Durham, 1983; Charles

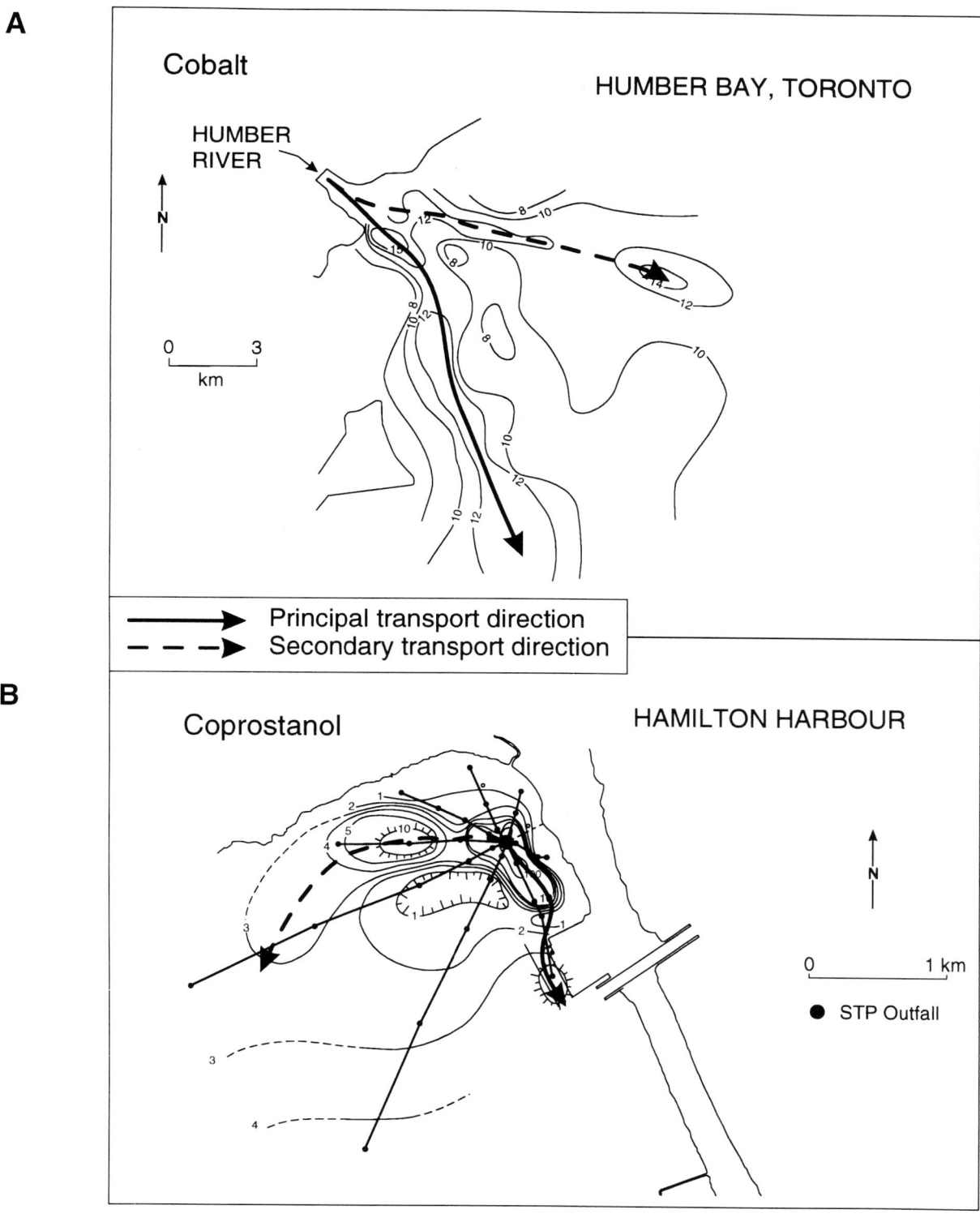

Figure 8   *Net (long-term) transport of contaminated sediment can be identified by the use of natural tracers.* (**A**) *Cobalt used as a tracer of sediment transport patterns at the mouth of the Humber River, Toronto waterfront (see Fig. 3 for location).* (**B**) *The fecal sterol, coprostanol, used as a tracer of fine sediments associated with the sewage plume for the Burlington STP, Hamilton Harbour. Note variability in inferred net transport directions (see Fig. 4 for location).*

and Hites, 1987). They can become adsorbed on various sediment particles after decomposition of the organic matter. Most of the organic contaminants are hydrophobic with low solubility in sediment interstitial water.

Release of Contaminants

Sediment-associated contaminants have the potential for release during changes in physiochemical and biological conditions at the sediment-water interface. For example, increased water salinity may lead to the release of some metals adsorbed on sediment particles. However, although this process is of major importance in the estuarine environment (de Groot *et al.*, 1976), it is not important in fresh water, such as the Great Lakes system, other than in urban rivers impacted by road salt (*e.g.*, Warren and Zimmerman, 1994). For a comprehensive treatment of transfer processes between the solid and aqueous phases in bottom sediment, the reader is referred to Förstner and Wittman (1981). A decrease in pH at the sediment-water interface can result in dissolution of carbonates and hydroxide minerals in the sediments, with subsequent release of contaminants associated with these two sediment components. Such a decrease in pH can occur in areas receiving intermittent deposition of SO_2 from industrial emissions and in waterbodies with low buffering capacity, such as waters with low concentrations of bicarbonate ions. While this is the case in the Canadian Shield portions of the Great Lakes watershed (Lake Superior), most of the other areas have abundant carbonate drainage areas for effective buffering.

Another way in which lowered pH can take place is through the decomposition of organic matter. Increased input of nutrients to a body of water from industrial and municipal wastes or agricultural runoff can lead to enhanced algal growth, and thus to increased deposition of organic matter in the bottom sediments. Carbon dioxide released during the decomposition of organic matter on the lake bottom can lead to the lowering of pH and to oxygen deficiency and changes in the redox potential at the sediment-water interface. These changes can be responsible for mobilization of various metals in the sediment, such as iron and manganese, and contaminants and nutrients associated with these elements.

Microbial activities in sediments can also be responsible for changes in redox potential and subsequent release of contaminants, especially metals. In addition, micro-organisms can convert inorganic metal and trace element compounds to organic compounds, such as the transformation of inorganic mercury to methyl mercury and other organo-mercuric compounds. These are all known to be both toxic and readily accessible to the aquatic food chain (Jackson, 1991).

Uptake By Organisms And Export

Organisms living in the sediments can take up contaminants from sediments either directly through their skin or as a result of passage through their digestive systems. As is schematically illustrated in Figure 7, benthic organisms such as oligochaete worms process large volumes of sediment and, after nutrients have been extracted, return the sediment to the surface as fecal pellets. Contaminants are retained in their tissues until after death, when such contaminants are released by decay of tissues and skeletal material. This recycling of sediment contaminants can also include their transfer through the food chain, such as sediment particles-to-bacteria-to-benthic invertebrates-to-bottom-feeding fish. The portion of contaminants which is incorporated into hard parts of the biota, such as shells and bones, is recycled more slowly

than that in the soft, faster decomposing parts (Warren, 1981). The quantity of many contaminants, particularly metals and trace elements, taken up by the biota often depends on the chemical form of the contaminant and its association with sediment particles. For example, lead bound to iron-oxides has been found to be less available than other forms of lead in the sediments (Luoma and Bryan, 1978).

Many fish species, as well as aquatic birds, spend a significant portion of their life cycle in the Great Lakes before migrating to distant areas. Examples are migratory waterfowl (ducks and geese) and fishes, such as eels, that spawn in the Great Lakes before spending their adult life in the Atlantic Ocean. These animals feed in coastal waters and streams of the lakes, thus taking up considerable amounts of persistent contaminants. After migration to other areas, they may die or be preyed upon by other species, thus passing on their contaminant burden to other environments. Eels have been cited as vectors for contamination of the St. Lawrence estuary beluga whale population, which feed on them after they exit the Great Lakes-St. Lawrence system (Comba *et al.*, 1993; Lum *et al.*, 1987). Lum *et al.* estimate that the amount of the organochlorine compound, mirex, transported out of Lake Ontario in the bodies of migrating eels was twice that exported on suspended sediment. Assuming that other persistent organics such as PCBs behave in a similar manner, this vector for removal of contaminants by migrating organisms may be very significant.

Not much has been published on the role of migrating waterfowl in contaminant export, but this vector has been considered in evaluating the presence of PCBs in breast milk of Inuits in the Canadian north. Although most of the organochlorine contamination in these remote areas has been attributed to atmospheric transport, it has been suggested that diet may be a factor. These Inuit communities consume migratory birds, such as Canada geese, that feed extensively in Great Lakes coastal waters and wetlands.

Benthic filter feeders, such as the exotic zebra mussel, have invaded the lower Great Lakes in vast quantities in recent years. In western Lake Erie, they have been credited with improving greatly the clarity of the water column by filtering out much of the suspended particulate material, along with the phytoplankton on which they feed. Studies of zebra mussel colonies suggest that the inorganic particulate material is later expelled onto the bottom as pseudo-feces which accumulate in the mussel colonies. This material is less readily resuspended than the original undigested particulates, and along with the dead mussel shells, may serve as stable reservoirs of contaminants (M.N. Charleton, National Water Research Institute, pers. comm., 1994). The net result is the removal and partial isolation of much of the contaminants entering the waterbody, and their eventual burial as these "reefs" accumulate.

ARTIFICIAL REMEDIATION
OF CONTAMINATED SEDIMENT

Until source reduction regulations now in force make their effect known, other techniques might have to be used in urgent cases. Many of the Areas of Concern in the Great Lakes are now considering engineered solutions to the contaminated sediment problem. These are discussed briefly.

Permanent Burial

For most of the persistent contaminants, such as the organic compounds and heavy metals mentioned earlier, the most ecologically desirable fate is burial below relatively clean sediment. This isolates them from the overlying water column, and thus, from resuspension and/or uptake into the aquatic food chain. Benthic organisms such as oligochaetes, chironomids, and burrowing molluscs are capable of mixing sediments down to depths of 20 cm or more, so any notion of permanent burial in sediments must assume burial below such depths.

In addition, permanent burial can only take place in areas (or at depths) where physical processes such as wave action are incapable of resuspending them. In lakes and enclosed bays, where strong residual currents are lacking, fine sediments are prone to steady displacement by repeated resuspension/depositional events into water depths below effective wave action. This condition has been mentioned earlier in connection with Hamilton Harbour and is active in offshore Lake Ontario. A prerequisite for elimination of the contaminated sediment problem through permanent natural burial is that subsequently deposited sediments be relatively clean. This solution thus requires that source reduction or elimination be implemented beforehand on all major discharge sites.

Dredging

To maintain safe navigation depths in harbours subject to sedimentation, bottom sediments need to be removed periodically by dredging. As discussed above much of these sediments are contaminated by a variety of harmful pollutants. This presents a problem for environmentally sensitive management of Great Lakes water resources. The two most important problems associated with these operations are (1) dredging the sediments without causing massive resuspen-

sion and enhanced uptake and dispersal of the contaminated material, and (2) disposal of the dredged material without polluting other areas of the aquatic ecosystem.

The first problem is an engineering one, and is readily addressed by the use of silt curtains, surrounding the dredge cutting-head, designed to contain any turbidity generated by dredging activity (Averett *et al.*, 1990; Raymond, 1984). The problem of disposal is the most difficult to address, primarily because of the large volumes involved and the fact that these sediments almost always exceed federal and provincial guidelines for open-water disposal. Disposal methods range from open-water dumping (for acceptable contaminant levels) to secure hazardous waste confinement. The mode and location of the dredged sediment disposal depends on the sediment quality, *i.e.*, the degree of contamination, expressed as concentrations of different organic compounds, metals, trace elements and nutrients, and availability of an acceptable site.

Environment Canada regulates the disposal of substances at sea through permits and inspections administered by Environment Canada pursuant to the provisions of Part VI of the Canadian Environmental Protection Act (CEPA), which became law in June 1988. As such, CEPA, Part VI (formerly the Ocean Dumping Control Act), implements domestically the provisions of the London Convention, an international convention on the Prevention of Marine Pollution by Dumping Wastes and Other Matter. The information on physical, chemical and biological characteristics of the material intended for dumping is required for the application for the Ocean Dumping Permit. Dredged material accounts for about 90% of the volume of material permitted for open-water disposal in the

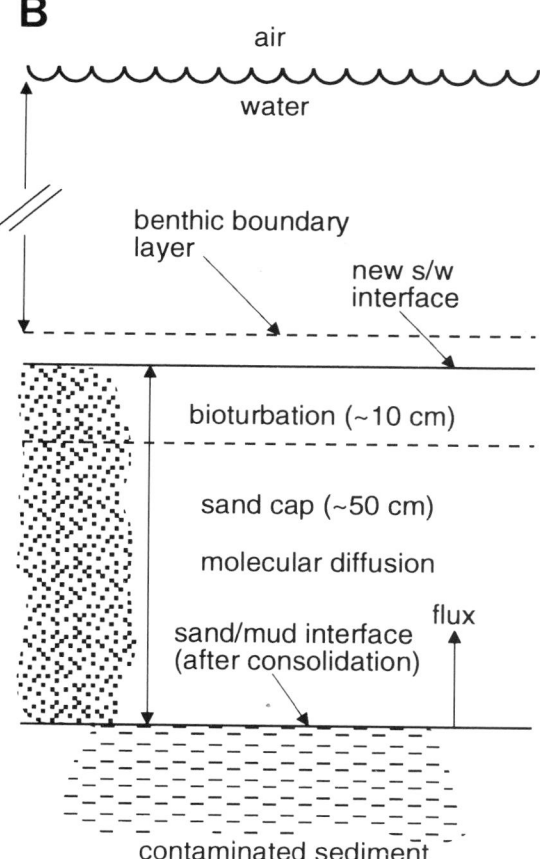

Figure 9   *Capping of contaminated sediments as a means of remedial action.* (**A**) *Uncapped contaminated sediment,* (**B**) *capped with clean sand. The sand cap is designed to be thick enough to isolate the contaminants from the local biosphere (after Zeman, 1994).*

Canadian marine environment.

On the Canadian side of the Great Lakes, disposal of dredged sediment is regulated by sediment guidelines developed by the Ontario Ministry of the Environment and Energy (OMOEE; Persaud et al., 1992). Sediments containing concentrations of contaminants greater than those in the OMOEE sediment guidelines need to be confined (Zarull and Mudroch, 1993); confined disposal facilities (CDFs) have been constructed along the Great Lakes shoreline for disposal of such material (e.g., Leslie Street spit along the Toronto waterfront; Chapter 2). However, there is a concern about contaminants re-entering the Great Lakes ecosystem from CDFs through uptake by cover vegetation, ingestion by soil invertebrates, and other biota utilizing the CDFs. Most of the over 30 CDFs constructed along the United States Great Lakes shoreline have become wildlife habitat. Appropriate management procedures are necessary to prevent movement of contaminants, establishment of undesirable animal species, and wildlife diseases.

In Situ Treatment

Many new technologies have been developed to clean up the contaminated sediments removed by dredging. These include, for example, thermal, chemical and biological treatments of the sediments (Zarull and Mudroch, 1993; Chapter 35). In situ treatment of contaminated sediments is an alternative for remediation of contaminated sediments. It has been considered in the clean-up of some of the Great Lakes Areas of Concern having a large area of contaminated sediments located in deep water where navigational dredging is not required. For example, subaqueous capping of highly contaminated sediments has been proposed as a method of isolation of the contaminants, particularly in cases where dredging could resuspend large quantities of contaminated sediment. The technique consists of placing by dredge a 50-cm layer of clean sediment, usually fine sand, over the contaminated sediment deposits (Fig. 9).

A demonstration project is presently underway in Hamilton Harbour (Zeman, 1994). In situ techniques of biochemical stabilization and/or neutralization of the contaminated sediments in urban watercourses have also been tested. An innovative system using a ship-mounted device for mechanical injection of a proprietary mixture of oxidants and nutrients into the sediments has also been tested in Hamilton Harbour (Murphy et al., 1994). This mixture was designed and tested in the laboratory to optimize bacterial decomposition of toxic organics such as PAHs and oil/grease complexes.

DISCUSSION

Contaminated sediments constitute a serious problem in lacustrine urban environments in Canada. They serve as stockpiles of contaminants that, even after source elimination, continue to supply contaminants to aquatic ecosystems. Understanding the eventual fate of contaminants in urban aquatic sediments is a complex task, largely because many of the key physical, biological and chemical processes that determine such a fate are poorly understood. Useful references on the topic are contained in Hites and Eisenreich (1987), Förstner and Whittman (1981), and Schmidtke (1988). The most desirable fate is permanent burial, whether by natural or artificial (engineered) means. In this way, the contaminants are placed below the level reached by burrowing organisms, and are thus removed from contact with the aquatic biosphere. Chemical transformation of contaminants into stable insoluble compounds, such as sulphides, phosphates, oxides, also plays a role. Until a sediment-bound contaminant reaches a deep-water sink, where conditions favour permanent burial, the most beneficial condition would be one of sufficient mixing with uncontaminated materials for a benign, or no-deleterious-effect, threshold to be reached. Such a condition is promoted by hydrodynamic flushing and resuspension, together with the reduction of additional contaminants through clean-up of sources.

ACKNOWLEDGEMENTS

The authors thank K. Lum and Stephane Lorain of the Centre Saint-Laurent, Environment Canada, Montreal for discussions, unpublished reports, and information on sediment contamination in the Montreal area. We are also grateful to M.N. Charlton and M.A. Zarull for comments on the manuscript and for suggesting relevant literature.

REFERENCES

Averett, D.E., Perry, B.D., Torrey, E.J. and Miller, J.A., 1990, Review of removal, containment and treatment technologies for remediation of contaminated sediment in the Great Lakes: United States Army Waterways Experimental Station, Vicksburg, MS, United States Army Corps of Engineers, Miscellaneous Paper EL-90-25, 67 p.

Bachtiar, T., Coakley, J.P. and Risk, M.J., 1996, Tracing sewage-contaminated sediments in Hamilton Harbour using selected geochemical indicators, in Bricker, S.B. and Valette-Silver, N.J., eds., Transport and Accumulation Processes of Contaminants: Science of the Total Environment, v. 179, p. 3-16.

Berner, R.A., 1971, Principles of Chemical Sedimentology: McGraw-Hill Book Company, New York, p. 86-135.

Calmano, W. and Förstner, V., 1995, Sediments and Toxic Substances: Springer-Verlag, New York, 740 p.

Carter, D.S. and Hites, R.A., 1992, Fate and transport of Detroit River derived pollutants throughout Lake Erie: Environmental Science and Technology, v. 26, p. 1333-1341.

Charles, M.J. and Hites, R.A., 1987, Sediments as archives of environmental pollution trends, in Hites, R.A. and Eisenreich, S.J., eds., Sources and Fates of Aquatic Pollutants: American Chemical Society, Washington, DC, p. 365-389.

Christensen, E.R. and Goetz, R.H., 1987, Historical fluxes of particle-bound pollutants from deconvolved sedimentary records: Environmental Science and Technology, v. 2, p. 1088-1096.

Coakley, J.P., Carey, J.H. and Eadie, B.J., 1990, Specific organic components as tracers of contaminated fine sediment dispersal in Lake Ontario near Toronto: Hydrobiologia, v. 235/236, p. 85-96.

Coakley, J.P. and Long, B.F.N., 1988. Tracing the movement of fine-grained sediment in aquatic systems: a literature review: Environment Canda, Inland Waters Branch, Scientific Series 174, 21 p.

Coakley, J.P. and Poulton, D.J., 1991, Tracers for fine sediment transport in Humber Bay, Lake Ontario: Journal of Great Lakes Research, v. 17, p. 289-303.

Coakley, J.P., Poulton, D.J. and Morris, W.A., 1994, Trace metals and selected elements in surficial sediments from Hamilton Harbour: National Water Research Institute, Contribution 94-43, 23 p.

Comba, M.E., Norstrom, R.J., Macdonald, C.R. and Kaiser, K.L.E., 1993, A Lake Ontario-Gulf of St. Lawrence dynamic mass budget for mirex: Environmental Science and Technology, v. 27, p. 2198-2206.

de Groot, A.J., Salomons, W. and Allersma, E., 1976, Processes affecting heavy metals in estuarine sediments, in Burton, J.D. and Liss, P.S., eds., Estuarine Chemistry: Academic Press, New York, p. 131-157.

Eisenreich, S.J., Metzer, N.A., Urban, N.R. and Robbins, J.A., 1986, Response of atmospheric lead to decreased use of lead in gasoline: Environmental Science and Technology, v. 20, p. 171-174.

Förstner, U. and Patchineelam, S, 1980, Chemical associations of heavy metals in polluted sediments from the lower Rhine River, *in* Kavannaugh, M. and Leckie, J., eds., Particulates in Water: American Chemical Society, Advances in Chemistry Series, v. 189, p. 177-193.

Förstner, U. and Wittman, G.T.W., 1981, Metal Pollution in the Aquatic Environment: Springer-Verlag, Berlin, 486 p.

Hamilton Harbour Remedial Action Plan, 1992, The Remedial Action Plan (RAP) for Hamilton Harbour. Stage II: Goals, Options and Recommendations: Hamilton Harbour RAP Office, Canada Centre for Inland Waters, 327 p.

Hartig, J.H. and Zarull, M.A., 1991, Methods of restoring degraded areas in the Great Lakes: Reviews of Environmental Contamination and Toxicology, v. 117, p. 127-154.

Hites, R.A. and Eisenreich, S.J., 1987, Sources and fates of aquatic pollutants: American Chemical Society, Advances in Chemistry Series 216, Washington, DC, 558 p.

International Joint Commission Great Lakes Water Quality Board, 1987, Report on Great Lakes Water Quality, Appendix A: 236 p.

International Joint Commission Great Lakes Water Quality Board, 1988, Procedures for the assessment of contaminated sediment problems in the Great Lakes: Report of the Sediment Subcommittee, 140 p.

Jackson, T.A., 1991, Effects of heavy metals and selenium on mercury methylation and other microbial activities in freshwater sediments, *in* Vernet, J.-P., ed., Heavy Metals in the Environment: Elsevier, Amsterdam, p. 191-217.

Krantzberg, G., 1994, Spatial and temporal variability in metal bioavailability and toxicity of sediment from Hamilton Harbour, Lake Ontario: Environmental Toxicology and Chemistry, v. 13, p. 1685-1698.

Krishnappan, B.G., 1991, Modelling of cohesive sediment transport: International Symposium on the Transport of Suspended Sediments and its Mathematical Modelling, Proceedings, International Association for Hydraulic Research, Florence, Italy, September 2, 1991, p. 433-448.

Lick, W., Lick, J. and Ziegler, C.K., 1994, The resuspension and transport of fine-grained sediments in Lake Erie: Journal of Great Lakes Resarch, v. 29, p. 599-612.

Lum, K.R., Kaiser, K.L.E. and Comba, M.E., 1987, Export of mirex from Lake Ontario to the St. Lawrence estuary: Science of the Total Environment, v. 67, p. 41-51.

Luoma, S.N. and Bryan, G.W., 1978, Factors controlling the availability of sediment-bound lead to the estuarine bivalve *Scrobicularia plana*: Journal of the Marine Biological Association of the United Kingdom, v. 58, p. 793-802.

Mudroch, A., 1991, Metal concentration in sediments (bottom and suspended) of St. Mary's River, Ontario/Michigan: Water Pollution Research Journal of Canada, v. 26, p. 119-143.

Mudroch, A., 1993, Lake Ontario sediments in monitoring pollution: Environmental Monitoring and Assessment, v. 28, p. 117-129.

Murphy, T.P., Moller, A. and Brouwer, H., 1994, In situ treatment of the Dofasco boat slip sediments: Canadian Institute of Mining, Metallurgy and Petroleum Geology, Annual Meeting, Toronto, ON, August 1994, Proceedings, p. 393-404.

Nriagu, J.O., 1986, Metal pollution of the Great Lakes and their carrying capacity, *in* Kullenberg, G., ed., The Role of Oceans as a Waste Disposal Option: Reidel Publishing Co., Dordrecht, The Netherlands, p. 441-468.

Oliver, B.G. and Durham, R.W., 1983, History of Lake Ontario contamination from the Niagara River by sediment radiodating and chlorinated hydrocarbon analysis: Journal of Great Lakes Research, v. 9, p. 169-168.

Olivier, L. and Bérubé, J., 1993, Sediment Quality and Dredging Activities in the St. Lawrence River: Environment Canada, Technology Development Branch, St. Lawrence Centre, EN 153-12/-1993E, 81 p.

Partheniades, E., 1986, The present state of knowledge and needs for future research on cohesive sediment dynamics: Third International Symposium on River Sedimentation, University of Mississippi, MS, Proceedings, v. III.

Persaud, D., Jaagumagi, R. and Hayton, A., 1992, Guidelines for the protection and management of aquatic sediment quality in Ontario: Water Resources Branch, Ontario Ministry of the Environment and Energy, Toronto, ON, 50 p.

Persaud, D., Lomas, T., Boyd, D. and Mathai, S., 1985, Historical development and quality of the Toronto waterfront sediments – Part I: Ontario Ministry of the Environment, Report ISBN 0-7729-0493-6, Toronto, ON, 62 p.

Pollman, C.D. and Danek, L.J., 1988, Contribution of urban activities to toxic contamination from large lakes, *in* Schmidtke, N.W., ed., Toxic Contamination in Large Lakes, Volume III – Sources, Fate and Controls of Toxic Contaminants: Lewis Publishers Inc., Chelsea, MI, p. 25-40.

Raymond, G.L., 1984, Techniques to reduce the sediment resuspenison caused by dredging: United States Army Engineer Waterways Experiment Station, Vicksburg, MS, United States Army Corps of Engineers, Miscellaneous Paper HL-84-3,. 121 p.

Reynoldson, T.B., Day, K.E., Thompson, S.P. and Ramsey, J.L., 1991, A sediment bioassay using the tubificid oligochaete worm *Tubifex tubifex*: Environmental Toxicology and Chemistry, v. 109, p. 1061-1072.

Rukavina, N.A., Mudroch, A. and Joshi, S.R., 1990, The geochemistry and sedimentology of the surficial sediments of Lac St.-Louis, St. Lawrence River: Science of the Total Environment, v. 97-9, p. 481-494.

Schmidtke, N.W., 1988, Toxic Contamination In Large Lakes. Volume III: Sources, Fate and Controls of Toxic Contaminants: Lewis Publishers, Chelsea, MI, 440 p.

Sérodes, J.-B., 1978, Qualité des sédiments de fond du fleuve Saint-Laurent entre Cornwall et Montmagny: Environment Canada, Inland Waters Directorate (Quebec region), Comité d'étude sur le fleuve Saint-Laurent, Technical Report 15, 467 p.

Sittig, M., 1991, Handbook For Toxic And Hazardous Chemicals And Carcinogens, Third Edition, Volumes I and II: Noyes Publications, Park Ridge, NJ, 1685 p.

Trefry, J.H., Metz, S., Trocine, R.P. and Nelsen, T.A., 1985, A decline in lead transport by the Mississippi River: Science, v. 230, p. 439-441.

Tsai, C. and Lick, W., 1988, Resuspension of sediments from Long Island Sound: Water Science and Technology, v. 29, p. 155-164.

Warren, L.J., 1981, Contamination of sediments by lead, zinc, and cadmium: a review: Environmental Pollution (Series B), v. 2, p. 401-436.

Warren, L.A. and Zimmerman, A.P., 1994, The influence of temperature, and NaCl on cadmium, copper and zinc partitioning among suspended particulate and dissolved phases in an urban river: Water Research, v. 28, p. 1921-1931.

Zarull, M.A. and Mudroch, A., 1993, Remediation of contaminated sediments in the Laurentian Great Lakes: Reviews of Environmental Contamination and Toxicology, v. 132, p. 93-115.

Zeman, A.J., 1994, Subaqueous capping of very soft contaminated sediments: Canadian Geotechnical Journal, v. 31, p. 570-577.

16. Mapping Contaminated Sediment in Hamilton Harbour, Ontario

J. Ken Versteeg
William A. Morris

Department of Geology, McMaster University, 1280 Main St. W., Hamilton, Ontario L8S 4M1

Norman A. Rukavina

Aquatic Ecosystem Restoration Branch, National Water Research Institute, Burlington, Ontario L7R 4A6

SUMMARY

Hamilton Harbour lies at the western end of Lake Ontario, and has a long history of industrial activity and urban development. Elevated levels of metals and organic compounds in bottom sediments led the International Joint Commission to designate Hamilton Harbour as an Area of Concern. Development of a remediation strategy requires definition of the distribution of the contaminants. Direct measurement of contamination by chemical analysis is prohibitively expensive. Previous studies on subsampled cores have shown that magnetic susceptibility closely tracks contaminant levels. This paper shows how non-destructive measurements of magnetic susceptibility (κ) on unopened cores can be used to map the distribution of post-industrial contaminated sediments and determine areas of recent sediment disturbance.

INTRODUCTION

Hamilton Harbour is a triangular-shaped embayment located at the western end of Lake Ontario (Fig. 1). During the past century, extensive urbanization and industrialization have resulted in the direct discharges of untreated sewage and industrial effluent into the harbour. Current daily discharge of water into the harbour is $2.6\text{-}3.8 \times 10^6$ m^3, of which 2-19% is runoff, 7-16% is from sewage treatment plants (STPs), and 72-87% is exchange with Lake Ontario. Two major iron and steel industries (STELCO and DOFASCO) withdraw and return 2×10^6 m$^3\cdot$d^{-1} which is used for contact cooling (Remedial Action Plan, 1989). The presence of heavy metals, toxic organics in fish, contaminated sediments, eutrophication and poor aesthetics led the International Joint Commission (IJC) to designate Hamilton Harbour an Area of Concern (International Joint Commission, 1985). Presently, a plan is being developed to study and restore the ecosystem of the harbour (Remedial Action Plan, 1989, 1991). Central to the development of this plan is an understanding of the distribution and volume of contaminated sediments.

The main limitation of previous attempts to estimate the extent of the contaminated sediment is that they are based on only a few cores. When one compares estimates of the thickness of contaminated sediments from isolated cores, it rapidly becomes apparent that there are large variations in thickness throughout the harbour and that the variation is dependent upon the sample location. While in theory, the solution would be to analyze more cores, in practice, this is often not feasible due to manpower requirements and consequent economic restraints. What is needed is a rapid, cost-efficient method that can identify the boundary between contaminated and non-contaminated sediments and be used to map the three-dimensional distribution of contaminated sediments.

Magnetic susceptibility is a measure of the ease of magnetization of a sample, and is related to the amount, size and composition of the magnetic minerals it contains (Thompson and Oldfield, 1986). A direct linkage between magnetic susceptibility and contamination in Hamilton Harbour sediments has been demonstrated in studies by Morris *et al.* (1994) and Versteeg *et al.* (1995). An order of magnitude increase in magnetic susceptibility in the upper 30-70 cm is related to a rise in magnetic mineral content and an increase in magnetic mineral grain size typical of industrial and urban sources. Confirmation that this magnetic mineral boundary represents the base of the post-industrial sediment is provided by concomitant increases in both polycyclic aromatic hydrocarbons (PAHs; Morris *et al.*, 1994) and metal content (Versteeg *et al.*, 1995). A detailed discussion of the geochemistry of contaminants found in Hamilton Harbour is given in Chapters 15 and 26.

In this paper, magnetic susceptibility was measured rapidly and non-destructively on 40 whole, unopened cores that were collected on a 500-m grid. The thickness of contaminated sediment was determined from these profiles, and contoured to produce a map of the distribution of contaminated sediments. Additional maps were produced by contouring magnetic susceptibility values averaged over 10-cm-thick sections, to map the distribution of contaminated sediments at various sediment depths.

Background

European settlement of the harbour region began in 1786 (Campbell, 1966), and has had a dramatic impact on the harbour ecosystem. In 1823, the Burlington Ship Canal between Lake Ontario and Hamilton Harbour was constructed, allowing the exchange of large volumes of water between the two waterbodies (Fig. 1). A similar canal, the Desjardins Canal, was constructed to join Cootes Paradise to the harbour in 1853, and the old fluvial channel filled in to make way for railroad construction. Since 1926, 25% of the open-water area in the harbour has been lost, as the marshes along the south shore were filled in to reclaim land for industrial use (Fig. 2). Steel production began in Hamilton around the turn of the century, and the south shore of the harbour is home to two large steel-manufacturing facilities, STELCO and DOFASCO. Historically, these industries discharged untreated process effluent containing contaminants such as metals, aromatic hydrocarbons and cyanide. In recent years, pollution control equipment, which greatly reduces or eliminates these contaminant loadings, has been installed, but an historical legacy of contaminated sediments remains.

Contaminated sediments have been investigated in Hamilton Harbour in two ways. The first involves assessing surficial sediments over a broad area to establish the distribution of contamination (Poulton, 1987; Poulton et al., 1988; Remedial Action Plan, 1989; Murphy et al., 1990); the second involves studying the historical record of pollution by analyzing sediment cores (Nriagu et al., 1983; Mayer and Johnson, 1993; Yang et al., 1993).

Studies that focus on the distribution of contaminants in surficial sediments point to enhanced levels of PAHs, PCBs and heavy metals in areas such as the sewage treatment outfalls and the steel mill outfalls (Remedial Action Plan, 1989). The main limitation of these studies is that they do not assess the depth of contamination and, therefore, have no way of computing volumes of contaminated material that may require remediation.

Studies of the depth variations of physical and chemical properties such as organic matter, bulk density, heavy metal and phosphorus concentration in cores (Mayer and Johnson, 1993) demonstrate the impact of settlement and industrialization on the sediment deposited in the harbour. Zinc, lead and cadmium concentrations in surficial sediments are 160, 28 and 19 times, respectively, those in the pre-colonial sediments (approximately 50 cm depth; Nriagu et al., 1983). PAH concentrations, associated with high-temperature combustion of fossil fuels, also have higher concentrations in the upper 70 cm relative to the deeper sediments (Morris et al., 1994). Yang et al. (1993) have reported microfossil changes which they interpret as being indicative of the eutrophication and pollution that accompanied the deforestation and settlement of the harbour's drainage basin. These methods, all of which show contamination of the upper sediments, require extensive sample preparation in order to acquire data. An additional limitation of existing studies is that they are often based on cores which are too short to penetrate the anthropogenic horizon.

A possible alternative approach to contaminated sediment mapping is through the use of magnetic properties, especially magnetic susceptibility. There is a demonstrated

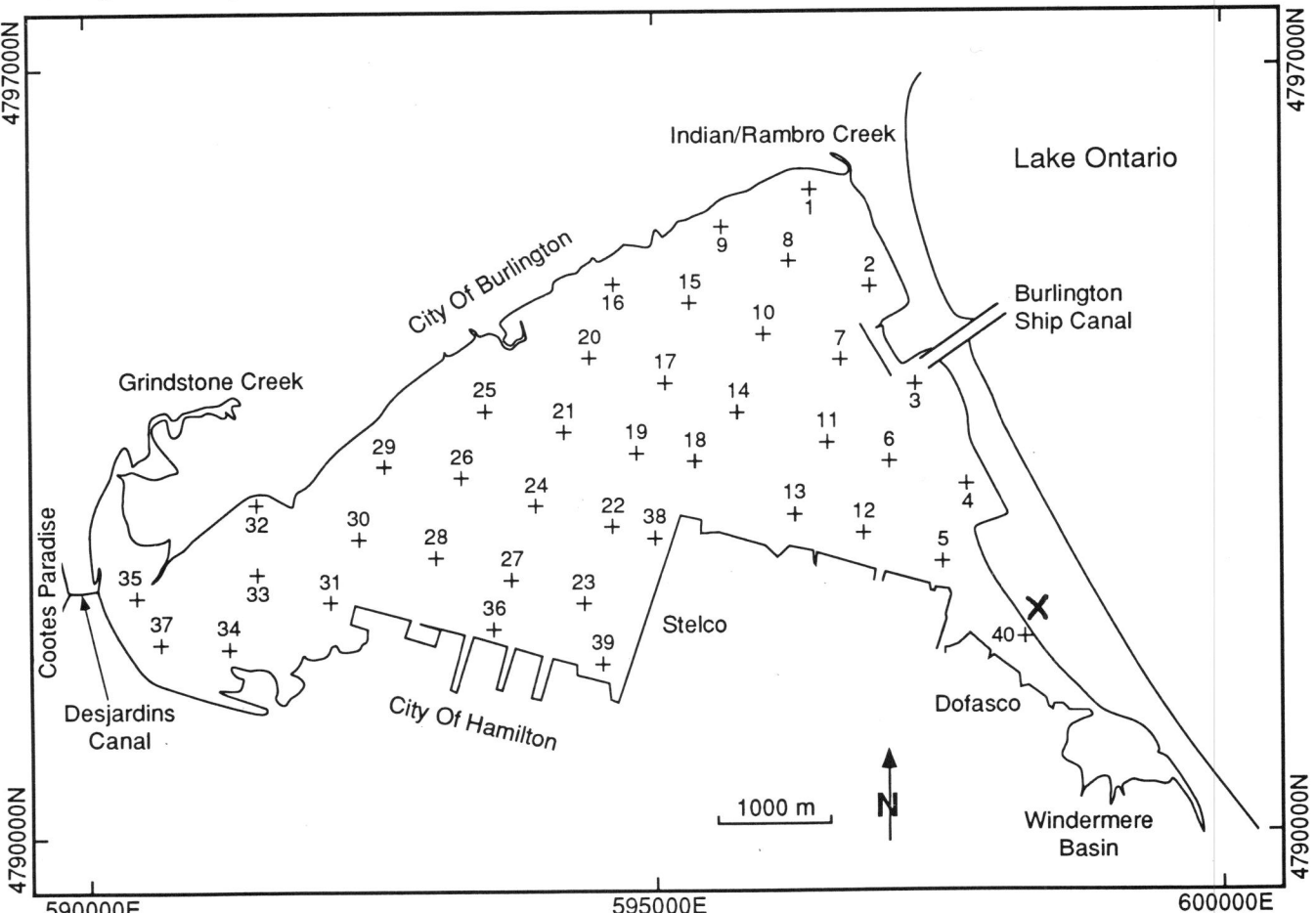

Figure 1 *Location map showing sites of cores taken from Hamilton Harbour. Figure 2 shows successive views from X over STELCO and DOFASCO property.*

A

B

Figure 2 *Photographs taken (**A**) in 1927 and (**B**) in 1985 looking west from site X in Figure 1 showing extent of lakefilling for STELCO steelworks (from Martin, 1988).*

relationship between increased magnetic susceptibility in sediments and contamination *via* industrial processes and urbanization. Coal fly ash, a by-product of the combustion of coal, contains spherules of magnetite (Locke and Bertine, 1986; Dekkers and Pietersen, 1992) which can be transported atmospherically. The enhanced susceptibility in the upper layers of cores from peat bogs has been related to increased atmospheric loading of magnetic particles related to increased coal burning associated with the Industrial Revolution (Thompson and Oldfield, 1986). Automobiles and urban construction materials can also lead to the formation of magnetic minerals (Beckwith *et al.*, 1984). While these iron-rich magnetic materials are not toxic themselves, they are often associated, either directly or indirectly, with true contaminants (*i.e.*, PAHs and heavy metals). Indirectly, contaminants that are produced by the same processes as the magnetic materials will follow the same pathways to sedimentation. Contaminants may adsorb onto the iron-rich particles (Stumm and Morgan, 1981).

In Hamilton Harbour, there are several possible sources that could result in elevated magnetic mineral production levels. The relative impact of each individual source may have changed with progressive urbanization, industrialization and environmental monitoring of the watershed. Original settlement of the area was first accomplished by burning much of the native forest (Weaver, 1982). This would certainly have been marked by an enhancement of fine-grained magnetite in the soils and, subsequently, the sediments, as demonstrated in other areas (Bloemendal *et al.*, 1979; Rummery *et al.*, 1979; Bloemendal, 1982; Rummery, 1983). As the local population grew, sewage and runoff were discharged directly into the harbour. This effluent contained magnetic iron-oxides from construction materials, coal fly-ash, and eventually automobiles. Since the turn of the century, the production of steel in Hamilton has released large quantities of iron into the harbour, through the atmospheric release of coal fly-ash and by direct discharge of process effluent high in iron into the harbour. These combined sources have produced a rise in the content of highly magnetic particles in recent sediments.

Measurements of magnetic susceptibility have several advantages over direct measures of contamination: they are rapid (10-15 seconds per measurement, *i.e.*, 10 minutes to analyze a 1-m long core, with measurements every 2 cm); they are non-destructive (measurements can be taken on unopened cores through plastic core tubes); and they are economical (a susceptibility meter is inexpensive).

METHODS

In July 1993, sediment cores were collected at 40 sites in Hamilton Harbour (Fig. 1), with a standard Benthos Corer, with 40 kg of weights and a 2-m, 7.5-cm ID plastic core, tube. The corer was lowered to within 2 m from the bottom, and then allowed to free-fall into the sediment. Recovery varied from 60 cm up to 140 cm long. Immediately upon collection, the cores

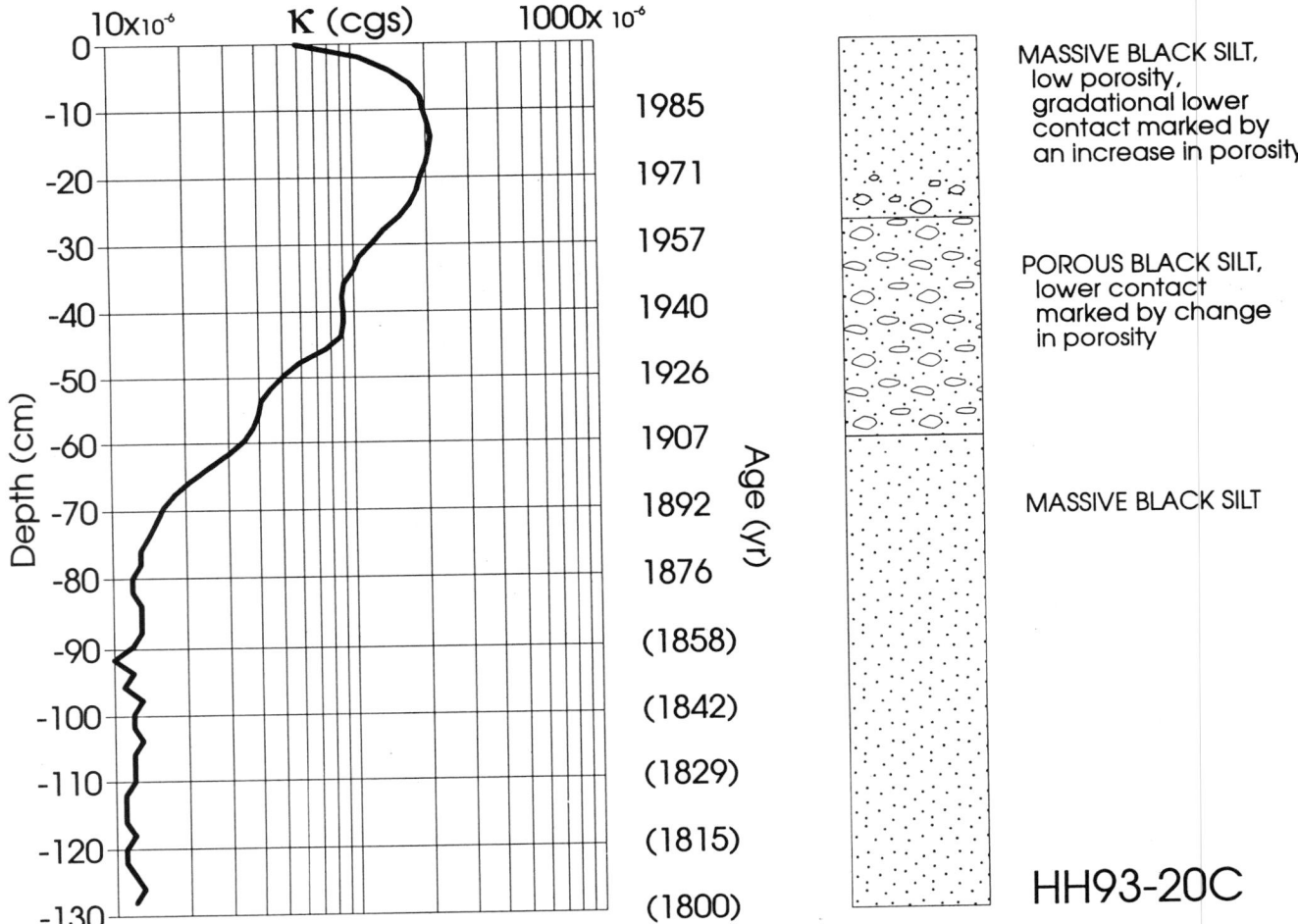

Figure 3   *Typical profile of magnetic susceptibility for Hamilton Harbour core taken at site 20 (Fig. 1). Measurement techniques are described in the text. Ages are ^{210}Pb ages. Ages in parentheses are extrapolated using calculated sedimentation rates. See Chapter 15 for typical profile of lead contamination in Lake Ontario sediments.*

were examined through the liner for colour and structural variations. The cores were then sealed and stored upright in their tubes, at 4°C.

Volume-specific magnetic susceptibility (κ) was measured on the whole core by passing it through a 100-mm ID coil attachment for a Bartington MS2 susceptibility meter. Measurements were made every 2 cm, from the sediment-water interface down. The instrument resolution is 1×10^{-6} cgs emu, (or $4 \, Pi \times 10^{-6}$ SI). Three repeat measurements made on one core (HH93-20a) show that the real precision is approximately ±5% of the instrument reading, although this is as great as ±20% for values less than 5×10^{-6} cgs. These errors are small in comparison to the variability of the data, which ranges over an order in magnitude.

RESULTS

Profiles

For each core, the profile of magnetic susceptibility was plotted *versus* depth (using a log-normal plot). Figure 3 shows a typical κ profile, along with lithologic descriptions. Many features of the κ profile are common to most cores. In general, low ($<2 \times 10^{-6}$ cgs) background values are followed by an order of magnitude rise in susceptibility at sediment depths ranging from 20-70 cm. Above the initial rise in κ, there is a plateau (45-35 cm in Fig. 3), followed by a second rise which peaks just below the surface, and then falls slightly. ^{210}Pb data (Turner, 1994) indicates that the date of the initial rise in κ at 70 cm is 1892, and that the second rise at 35 cm corresponds to a ^{210}Pb date of 1945.

Variations in the κ profiles can be used to identify correlatable horizons. Each horizon is characterized by a local peak or trough in the κ profile. As seen in Figure 4, features between adjacent cores define several magneto-stratigraphic units with a resolution of 6 cm (*i.e.*, 3 κ data points), which represent a combination of changes in magnetic mineral content and grain size. There is a reasonable correlation between these magneto-stratigraphic units and lithological units.

There are a number of cores from the south shore (*i.e.*, Fig 4B, cores 38 and 39) for which it is not possible to identify any obvious correlatable horizons. Lack of correlation indicates that these sediments have been disturbed. The distribution of these disturbed sediments, in a zone that includes the south-central and southeastern areas of the harbour, suggests that the disturbance may be related to shipping and dredging activity which is prominent in this area (Holmes and Whillans, 1984; Holmes, 1986). Other factors include the dumping of coarse-grained, highly magnetic sediment during land recla-

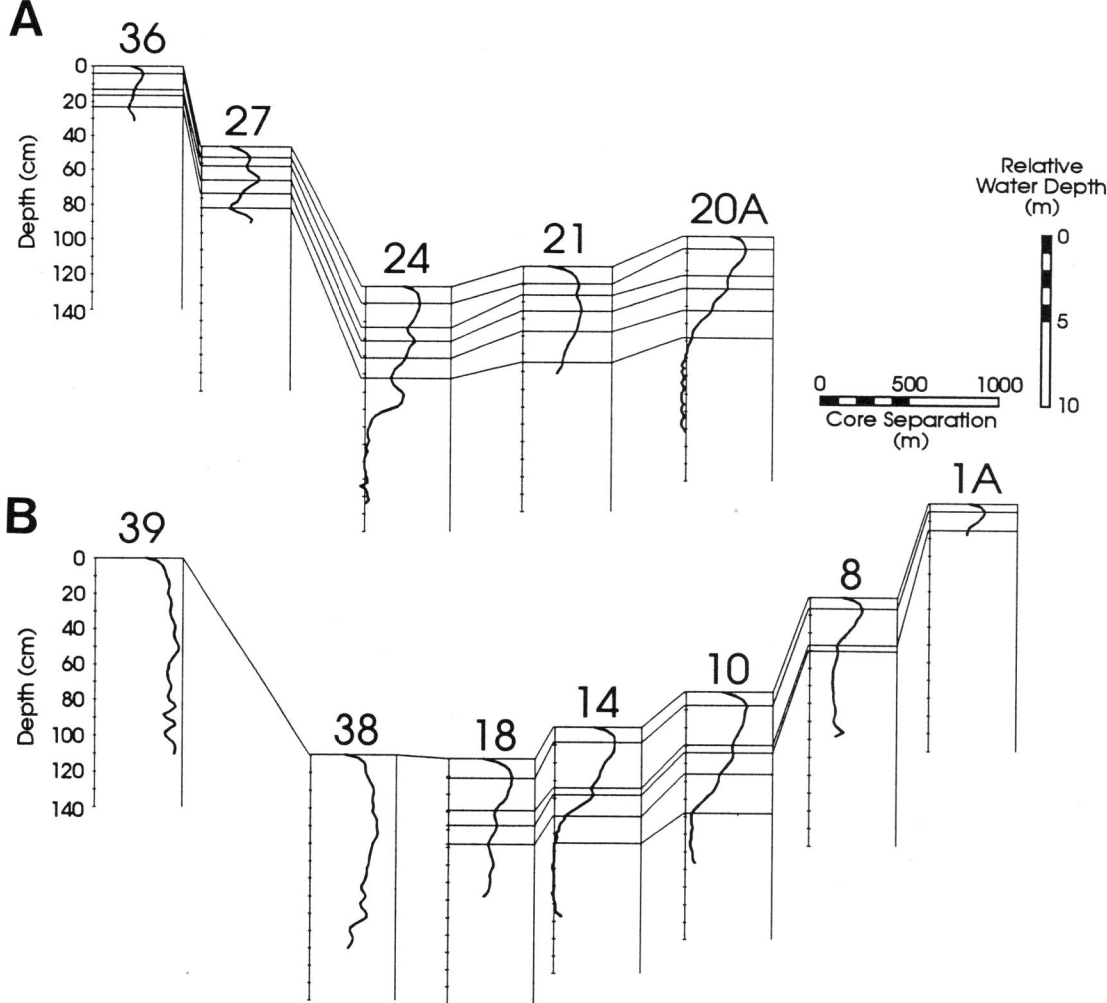

Figure 4 *Correlated cross-sections in Hamilton Harbour: (A) section next to Randle Reef, showing non-disturbed zone; (B) section through Randle Reef showing disturbed sediments along south shore at sites 38 and 39 (Fig. 1). For susceptibility scale, see Figure 3.*

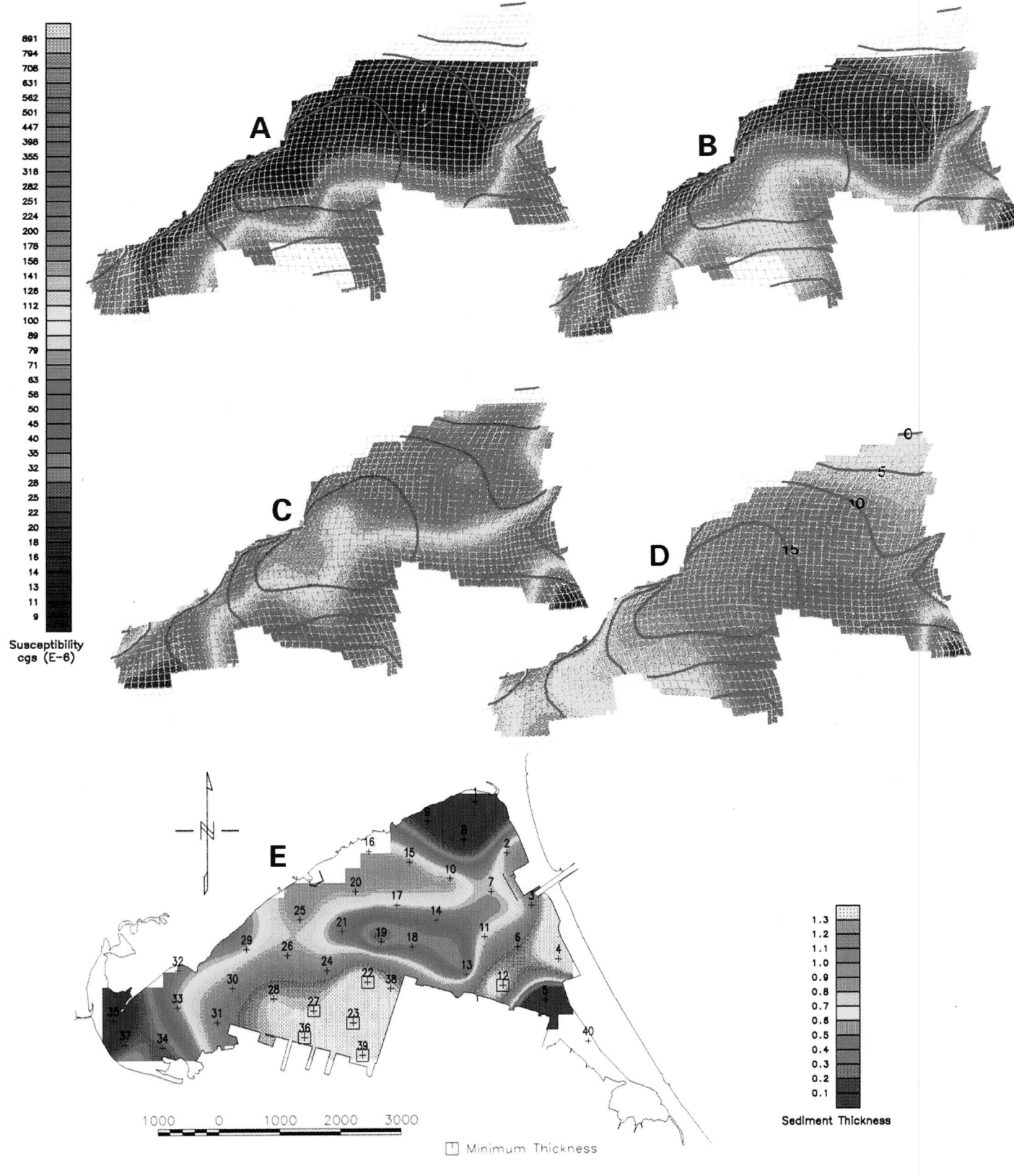

Figure 5   *Colour contour map of average susceptibility for 10-cm depth slices, draped onto the bathymetry of the harbour:* (**A**) *80-90 cm,* (**B**) *60-70 cm,* (**C**) *40-50 cm,* (**D**) *10-20 cm, bathymetry shown by red lines, depths in metres. Note northward expansion of comtamination in successively shallower depth slices.* (**E**) *Isopach map showing thickness of contaminated sediment in Hamilton Harbour. Thickness and horizontal scale in metres.*

mation (Ozanian, 1957) and the proximity to a source of highly magnetic material, *i.e.*, an industrial discharge or spills from iron ore ships.

Spatial Variations

For each core, the measured κ values were averaged for each successive 10 cm of sediment, *i.e.*, 0-10 cm, 10-20 cm, *etc.* A colour contour map of the magnetic susceptibility distribution across Hamilton Harbour was created for each slice between successive 10 cm isobaths below the sediment-water interface. The averaged κ values were gridded by the Geosoft RANGRID random-gridding algorithm, using a 250-m grid cell size with a 750-m blanking distance. The maps produced by this procedure provide a broad estimate of observed distributions within the harbour. Caution must be exercised in the interpretation of features along the shoreline, as the gridding process does not incorporate the boundary conditions necessarily imposed by the shoreline.

To emphasize the relationship between κ distribution and the shape of the depositional basin, the colour contour maps were draped over a three-dimensional projection of the bathymetry of the harbour, using the NETVIEW package by Geosoft. The bathymetry grid was interpolated from water depths measured at each station during core collection, and gridded using the same procedure as described above.

The depth-slices provide a picture of the variations in the distribution of κ at various sediment depths. Four of these are presented in Figure 5 representing a gradation from deep to surficial sediments. The deepest section (80-90 cm, Fig. 5A) shows a zone of high κ which is restricted to the south shore. At this sediment depth, the remainder of the harbour has low (*i.e.*, background) κ values and shows no other regional features. Moving up through the sedimentary column, this zone of high κ sediment appears to migrate northward (60-70 cm, Fig. 5B). A sharp boundary separating high and low κ is maintained until the 40-50 cm section (Fig. 5C). At this point, the values of κ begin to rise above background in the sediment over the remainder of the northern part of the harbour. With decreasing depth, the zonation of the harbour becomes gradually less distinct. In the near-surface sediments (10-20 cm, Fig. 5D), the previously distinct south shore zone is all but non-existent, having been succeeded by relatively uniform, high κ sediment covering most of the harbour.

There is a clear relationship between the shape of the boundary separating high and low κ, and the bathymetry of the harbour. As the boundary advances from south to north, it first progresses into deeper parts of the basin (Fig. 5B, 60-70 cm; Fig. 5C, 40-50 cm). The presence of a plume of high κ sediments into the deep basin is consistent with κ values being associated with the finest grain size fraction, which would be subject to depositional focussing into the deeper waters of the central basin.

A prominent feature brought out in Figure 5 is the effect on the distribution of κ of the influx of extraneous sediment introduced by tributaries into the harbour. The mouth of the Desjardins Canal, at the western end of the harbour, is marked by a persistent low κ value which is likely a result of the influx of fine-grained, silicate-rich sediments from Cootes Paradise, a wetland marsh drained by the canal. These sediments would be relatively non-magnetic, since the highly magnetic sediments are generally related to industrial sources, and Cootes Paradise drains a predominantly agricultural watershed. Another low κ zone in the southeastern harbour, at the mouth of Windermere Basin, in Figures 5B, C and D, is interpreted as being the result of dredging.

One must be careful in the interpretation of depositional pictures presented by these depth-slice images. It is tempting to treat them as being directly equivalent to time slices. Such a model would require uniform sedimentation rates at all points in the harbour, but currently available ^{210}Pb ages are inconclusive.

Contaminated Sediment Thickness

Previous chemical analyses have shown that there is a direct relationship between highly magnetic sediments and contaminants (Morris *et al.*, 1994; Versteeg *et al.*, 1995). It is possible to operationally define contaminated sediment as sediment having κ values above background. Since the rise of κ is so dramatic in most cores, there is little difficulty in picking the depth of this transition. Ambiguity does arise, however, when κ does not fall to a background value. Where this occurs (cores 12, 22, 23, 27, 36 and 39), we have assumed a minimum thickness of contaminated sediment, based on the profiles of adjacent cores. An isopach map of contaminated sediment thickness was derived using a gridding procedure similar to that used for the depth slices. In general, contaminant thickness appears to decrease with increasing water depth. The region of thinnest contaminated sediment corresponds to the deepest part of the harbour (Fig. 5E), which is in contrast to what is expected if the fine-grained contaminated sediments are being focussed in the central basin. It is possible that focussing is occuring, but sediment accumulation is miminized because of the small grain sizes (due to more efficient packing). Superimposed on this regional pattern are a number of localized features. The zone along the south shore, west of STELCO Pier, has the thickest zone of contaminated sediment, over 1.4 m. This zone includes Randle Reef, an area well known for high levels of contamination (Murphy *et al.*, 1990). Another thick zone of contaminated sediment occurs at the mouth of Windermere Basin.

The total volume of contaminated sediment, as estimated by this κ procedure, is over 12×10^6 m^3. Some caution should be applied in the use of this estimate of contaminated sediment thickness and volume. First of all, the thickness of contaminated sediment has been interpolated between cores which were spaced 500 m apart, and could, therefore, have missed localized variations. Greater confidence in the distribution could be obtained by analyzing more cores, with a finer sampling resolution. Second, it is possible that part of the inferred contamination could be related to colonial settlement activities (*i.e.*, forest clearance), and not solely to industrial operations. We propose that the κ maps be used as a reconnaissance tool, to identify the potentially contaminated zones, which can be verified through additional chemical analysis. This would significantly reduce the number of samples for chemical analysis, and since the κ analysis is non-destructive, the same samples can be used, eliminating the need to collect more cores.

DISCUSSION

By exploiting a previously demonstrated relationship between magnetic susceptibility and contaminants, the distribution of potentially contaminated sediment in Hamilton Harbour has been mapped. In comparison to conventional methods of contaminant detection, magnetic susceptibility measurements have the advantage of being completely non-destructive, rapid and inexpensive. Although susceptibility does not directly determine the presence of contamination, it is a rapid, inexpensive non-destructive tool which could identify potentially contaminated sediments. Such knowledge can substantially reduce the number of chemical analyses required to map the distribution of contaminated sediments in urban harbours.

ACKNOWLEDGEMENTS

The authors acknowledge the assistance of the Technical Operations staff at the National Water Research Institute in the core collection. Financial support for this project was provided by a Great Lakes University Research Fund grant and a Tri-council Hamilton Harbour Ecosystem research grant to Morris.

REFERENCES

Beckwith, P.R., Ellis, J.B., Revitt, D.M. and Oldfield, F., 1986, Heavy metal and magnetic relationships for urban source sediments: Physics of the Earth and Planetary Interiors, v. 42, p. 67-75.

Bloemendal, J., 1982, The quantification of rates of total sediment influx to Llyn Goddionduon, Gwynedd: unpublished Ph.D. thesis, University of Liverpool, UK, 206 p.

Bloemendal, J., Oldfield, F. and Thompson, R., 1979, Magnetic measurements used to assess sediment influx at Llyn Goddionduon: Nature, v. 280, p. 50-53.

Campbell, M.F., 1966, A Mountain and a City: The Story of Hamilton: McClelland and Stewart, Toronto, ON, 351 p.

Dekkers, M.J. and Pietersen, H.S., 1992, Magnetic properties of low-Ca fly ash: A rapid tool for Fe-assesment and a survey for potentially hazardous elements: Material Research Society, Symposium Proceedings, v. 245, p. 37-47.

Holmes, J.A., 1986, The impact of dredging and spoils disposal on Hamilton Harbour Fisheries: Implications for rehabilitation: Great Lakes Fisheries Research Branch, Canada Centre for Inland Waters, Canadian Technical Report of Fisheries and Aquatic Sciences 1498, 146 p.

Holmes, J.A. and Whillans, T.H., 1984, Historical review of Hamilton Harbour fisheries: Canadian Technical Report of Fisheries and Aquatic Sciences, n. 1257, 125 p.

International Joint Commission, 1985, Report on Great Lakes water quality: International Joint Commission, Great Lakes Water Quality Board, 68 p.

Locke, G. and Bertine, K.K., 1986, Magnetite in sediments as an indicator of coal combustion: Applied Geochemistry, v. 1, p. 345-356.

Martin, V., 1988, Changing Landscapes of Southern Ontario: Boston Mills Press, Erin, ON, 243 p.

Mayer, T. and Johnson, M.G., 1993, History of anthropogenic activities in Hamilton Harbour as determined from the sedimentary record: National Water Research Institute, Report 93-119, 14 p.

Morris, W.A., Versteeg, J.K., Marvin, C.H., McCarry, B.E. and Rukavina, N.A., 1994, Preliminary comparisons between magnetic susceptibility and polycyclic aromatic hydrocarbon content in sediments from Hamilton Harbour, western Lake Ontario: Science of the Total Environment, v. 152, p. 153-160.

Murphy, T.P., Brouwer, H., Fox, M.E., Nagy, E., McArdle, L. and Moller, A., 1990, Coal tar contamination near Randle Reef, Hamilton Harbour: National Water Research Institute, Report 90-17, 67 p.

Nriagu, J.O., Wong, H.K.T. and Snodgrass, W.J., 1983, Historical records of metal pollution in sediments of Toronto and Hamilton Harbours: Journal of Great Lakes Reearch, v. 9, p. 365-373.

Ozanian, S.M., 1957, A geographical study of the development of Hamilton Harbour: unpublished B.A. thesis, McMaster University, Hamilton, ON, 117 p.

Poulton, D.J., 1987, Trace contaminant status of Hamilton Harbour: Journal of Great Lakes Research, v. 13, p. 193-201.

Poulton, D.J., Simpson, K.J., Barton, D.R. and Lum, K.R., 1988, Trace metals and benthic invertebrates in sediments of nearshore Lake Ontario at Hamilton Harbour: Journal of Great Lakes Research, v. 14, p. 52-65.

Remedial Action Plan (RAP), 1989, Remedial Action Plan for Hamilton Harbour, Stage 1 report: Environmental conditions and problem definitions: Ontario Ministry of the Environment, Ontario Ministry of Natural Resources, Ontario Ministry of Agriculture and Food, Environment Canada, Fisheries and Oceans Canada, Royal Botanical Gardens, 153 p.

Remedial Action Plan (RAP), 1991, Remedial Action Plan for Hamilton Harbour, Draft Report: Ontario Ministry of the Environment, Ontario Ministry of Natural Resources, Ontario Ministry of Agriculture and Food, Environment Canada, Fisheries and Oceans Canada, Royal Botanical Gardens, 117 p.

Rummery, T.A., 1983, The use of magnetic measurements in interpreting the fire histories of lake drainage basins: Hydrobiologia, v. 103, p. 53-58.

Rummery, T.A., Bloemendal, J., Dearing, J. and Oldfield, F., 1979, The persistance of fire-induced magnetic oxides in soils and lake sediments: Annales de Géophysique, v. 2, p. 103-107.

Stumm, W. and Morgan, J.J., 1981, Aquatic Chemistry: An Introduction Emphasizing Chemical Equilibria In Natural Waters: Wiley and Sons, New York, 512 p.

Thompson, R. and Oldfield, F., 1986, Environmental Magnetism: Allen and Unwin, London, UK, 235 p.

Turner, L.J., 1994, ^{210}Pb dating of lacustrine sediments from Hamilton Harbour (Core 047, Station 37-A; Core 051, Station 22-D; and Core 052, Station 20-C), Lake Ontario: National Water Research Institute, Technical Note RAB-TN-93-42, 47 p.

Versteeg, J.K., Morris, W.A. and Rukavina, N.A., 1995, The utility of magnetic properties as a proxy for mapping contamination in Hamilton Harbour sediment: Journal of Great Lakes Research, v. 21, p. 71-83.

Weaver, J.C., 1982, The History of Canadian Cities. Hamilton, An Illustrated History: James Lorimer and Co., Toronto, ON, 356 p.

Yang, J.R., Duthie, H.C. and Delorme, L.D., 1993, Reconstruction of the recent environmental history of Hamilton Harbour (Lake Ontario, Canada) from analysis of siliceous microfossils: Journal of Great Lakes Research, v. 19, p. 55-71.

17. Environmental Geology of Halifax Harbour, Nova Scotia

Gordon B.J. Fader
Dale E. Buckley

Atlantic Geoscience Centre, Geological Survey of Canada, Bedford Institute of Oceanography
P.O. Box 1006, Dartmouth, Nova Scotia B2Y 4A2

SUMMARY

This paper demonstrates the importance of integrated studies of marine geology and geochemistry in the environmental management of an urbanized coastal inlet, using Halifax Harbour, Nova Scotia, as an example. The harbour receives 170 million litres of raw sewage per day; other sources of contamination include landfills, industrial activity, surface run-off and dredging. The level of contamination by metals in surficial sediments in Halifax Harbour is among the highest anywhere recorded in marine harbours, and is a result of sediment trapping and lack of flushing. Geological and oceanographic conditions strongly influence the present environmental quality of the harbour; assessment of environmental quality, and the design of waste-water management systems in urban harbours, are fundamentally dependant on detailed knowledge of sediment transport, deposition and erosion.

INTRODUCTION

Halifax Harbour is the largest port along the Atlantic seaboard of Canada, the east coast home of the Canadian Navy, and a major trans-shipment facility for large quantities of bulk and container materials (Fig. 1). The harbour is a significant lobster and finfish resource, and a major recreational facility for the surrounding communities. However, it also receives approximately 170 million litres of raw sewage per day from about 50 outfalls from the cities of Halifax and Dartmouth. Other sources of contamination are industrial activity related to shipbuilding and repair, surface drainage, old landfill sites, and dredge spoils and debris on the harbour bottom.

In 1977, the Metropolitan Area Planning Commission proposed that a single regional primary sewage treatment facility be constructed in the outer part of Halifax Harbour, with an outfall of treated effluent extending seaward into the approaches to Halifax Harbour. A federal-provincial agreement was signed in 1988 to develop this facility. In 1989, a report by the Nova Scotia Environmental Control Council identified major information gaps concerning the location of the plant and outfall, and the associated marine data base. As a result of this report, the Province of Nova Scotia commissioned the Halifax Harbour Task Force (Fournier, 1990), and together with a Federal Science Advisory Committee on Halifax Harbour (Nicholls, 1989), reviewed the proposed regional sewage treatment plan, focussing on the marine environment and related science and engineering issues. For further details of the assessment process itself, see Chapter 39.

The Atlantic Geoscience Centre, Geological Survey of Canada, participated as an active member in assessing the marine environmental conditions and quality of the harbour, and conducted considerable additional research on the marine geology and geochemistry. These data, together with oceanographic and other environmental information, were used by the Halifax Harbour Task Force to recommend an alternate location for the consolidated sewage outfall in the inner harbour, rather than the earlier proposal for discharge in the approaches to Halifax Harbour. This recommendation was supported by the subsequent Environmental Assessment Panel. This decision was considered unconventional to many accepted practices, where sewage outfalls are located as far at sea as possible. The philosophy adopted in the final choice was one of containment *versus* dispersion of wastes. Geological and geochemical data provided a basis for understanding the present distribution and transport pathways of sediments and contaminants, and a basis on which to assess various sewage management options.

The purpose of this chapter is to illustrate and summarize (1) the methodology used in the study of the harbour, (2) the importance and relative role of geological and geochemical studies, (3) the integration of the various disciplines, (4) the major scientific findings as they relate to waste-water management, and (5) recommendations for similar studies in other areas.

THE GEOLOGICAL FRAMEWORK OF HALIFAX HARBOUR

Data used in geological interpretations were mainly collected during surveys in 1988, 1989, and 1990 (OceanChem Group Ltd., 1988; Buckley *et al.*, 1989; Miller and Fader, 1989; Nichols, 1989; Miller *et al.*, 1990; Fader *et al.*, 1991; Fader and Miller, 1992). The data base consists of a dense regional grid of echo sounder data, 100 kHz and 500 kHz side-scan sonograms, mid- to high-resolution seismic reflection profiles, cores, grab samples, epibenthic dredge hauls, bottom photographs and remotely operated vehicle (ROV) video information. A variety of seismic reflection systems were tested in the harbour to provide a combination of very high resolution with maximum penetration. Modifications were made to existing systems, and the final, most appropriate configuration for the harbour used a Huntec boomer and a surface towed IKB Seistec with line and cone array, together with a Datasonics Bubblepulser and an Elac 12 kHz ship's echo sounder. Long piston cores and vibrocores of sediment were collected for subsurface stratigraphic information and geochemistry (Buckley, 1988; Fitzgerald *et al.*, 1989; Fader *et al.*, 1993).

Early reports on the formation of Halifax Harbour attributed its origin to the presence of suggested northwest-southeast structural lineaments or faults (Cameron, 1949). No evidence was found for structural offset and the presence of such interpreted faults in bedrock beneath the harbour. None of the major synclines and anticlines on either side of the harbour are horizontally displaced (Fig. 2). Aeromagnetic data also indicate continuity beneath the harbour.

An ancient Sackville River, and another drainage system originating from the northeast, flowed through Halifax Harbour. The second system existed in the location of a series of lakes in Dartmouth that presently empty into Dartmouth Cove. The term Ancient Sackville River, as defined by King (1970), is used for the entire system. Before Pleistocene glaciations, Bedford Basin likely did not exist. The Ancient Sackville River continued on its course across a gently dipping Cretaceous peneplain through the present area of Bedford Basin and out of the harbour. A major fractured bedrock anticline (Faribault, 1908) lies below the centre of Bedford Basin. With the onset of glaciation, the glaciers followed the course of the early fluvial drainage system, eroding the fractured rocks and excavating the present 70-m deep depression of Bedford Basin. The area now occupied by Halifax was left as a topographic high.

Widespread distribution of till beneath the harbour suggests that the area was occupied by glaciers several times during the Pleistocene. The last (late Wisconsin) glaciation covered the entire harbour area and extended across the

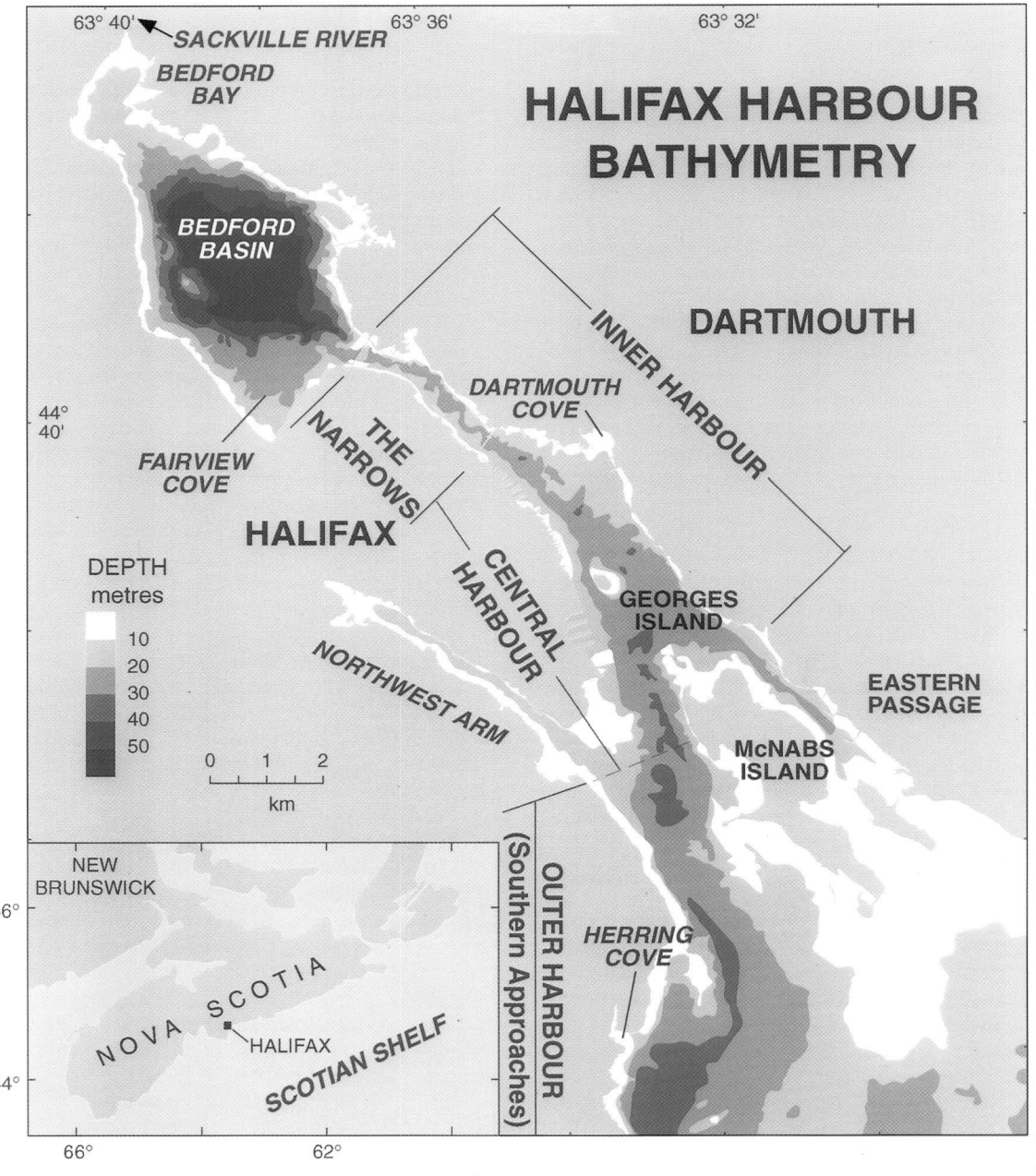

Figure 1 *Halifax Harbour including physiographic subdivisions named in text.*

Scotian Shelf to its edge (King and Fader, 1986).

Recent studies have identified a post-glacial low sea-level stand at 65-70 m water depth (Stea *et al.*, 1993; Fader *et al.*, 1993) at about 11.65 ka. Most of Halifax Harbour was subaerially exposed at this time, with the Ancient Sackville River draining through the harbour to Chebucto Head. A series of lakes existed in the inner harbour surrounding Georges Is-

land, in the Eastern Passage, the Northwest Arm, adjacent to western McNabs Island, and in several areas of the western part of the outer harbour (Fig. 3).

The post-glacial marine transgression of the harbour reworked glacial and lacustrine sediments into lag gravel and sandy deposits. As sea level rose, small lakes in the harbour were inundated, forming a series of estuarine embayments

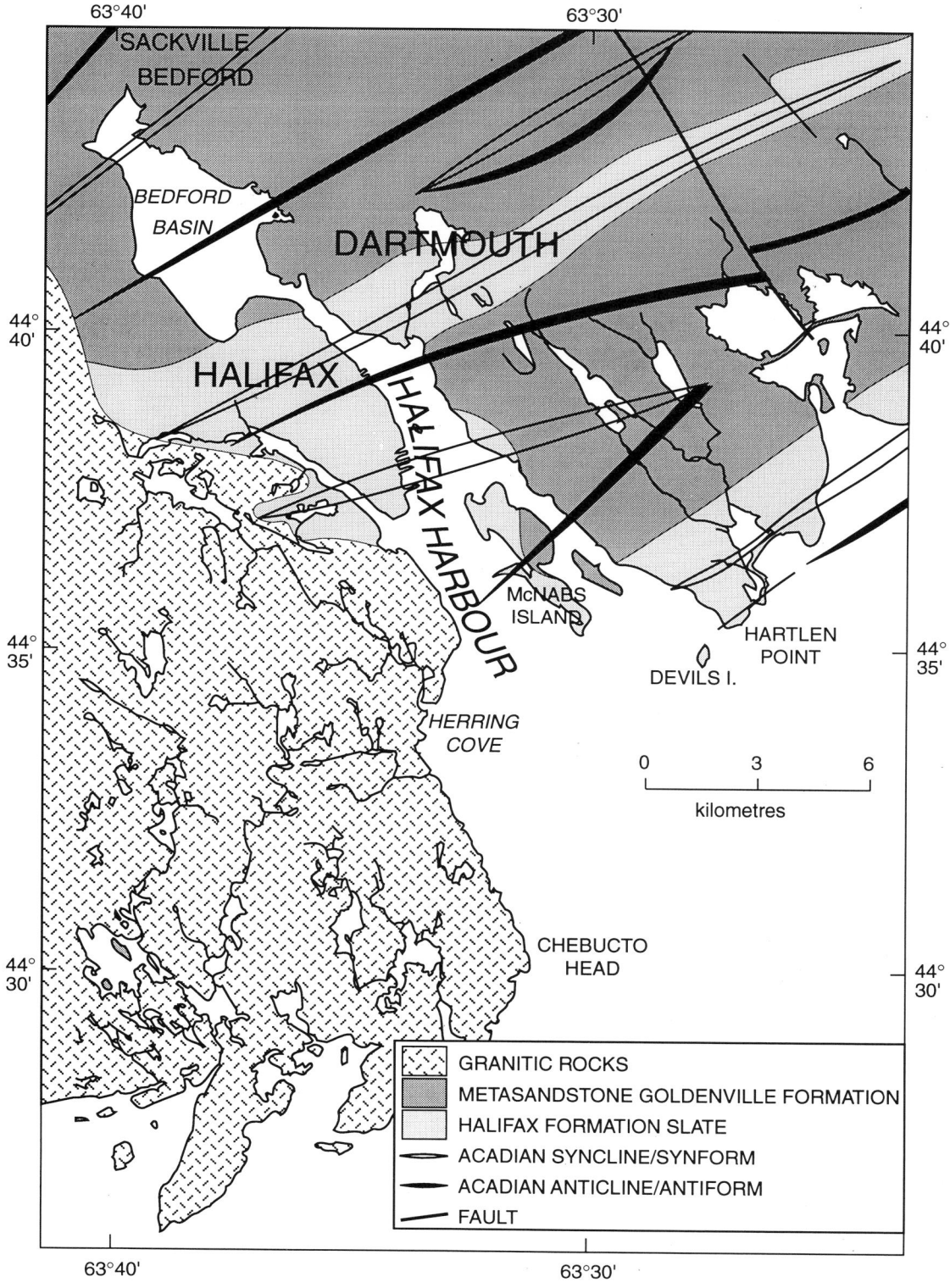

Figure 2 *Simplified bedrock geological map of the Halifax-Dartmouth area with the location of major structural elements (anticlines are solid slashes; synclines are open slashes; adapted from Keppie, 1982).*

(Edgecombe, 1994). The final breach of the Narrows sill, at 5.8 ka (Miller *et al.*, 1982), resulted in the modern configuration of the harbour. Many gravel areas of the inner harbour are relict surfaces formed during the Holocene marine transgression, and remain exposed on the seabed in areas where high current energy has prevented deposition of fine-grained Holocene mud. A sediment core sample from Bedford Basin (Buckley, 1988: Fitzgerald *et al.*, 1989) provided evidence of the transition from lake conditions, through brackish marine conditions, to restricted marine conditions over the past 6000 years (Fig. 4).

The geological framework of Halifax Harbour, which has evolved over millions of years, establishes the geomorphological control for hydrodynamic and sedimentary processes that determine the present nature of the harbour. Old fluvial, lacustrine and glacial deposits, buried under more recent sediments, can continue to influence the geochemical nature of recent sediments.

PHYSIOGRAPHIC DIVISIONS AND CHARACTERISTICS

We have divided Halifax Harbour into three divisions from south to north termed: outer Halifax Harbour; inner Halifax Harbour including the central harbour, Northwest Arm and Eastern Passage, and Bedford Basin including Bedford Bay (Fig. 1). The entire harbour has sometimes been referred to as Halifax inlet (Nicholls, 1989); however, we propose to use the term Halifax Harbour for the body of water extending north of a headland line from Hartlen Point to Chebucto Head (Fig. 5). Areas seaward, to the southeast of this line are part of the inner Scotian Shelf.

Bedford Basin

Bedford Basin is a bowl-shaped depression at the head of the harbour with a maximum depth of 71 m (Fig. 1). It is joined in the north with Bedford Bay, a shallow embayment into which the Sackville River drains. This river is the major drainage conduit for areas to the north and east of Halifax Harbour. The basin is mostly mud covered in areas deeper than 20 m, but the shallow areas consist of gravel in the cobble and boulder range with a few areas of outcropping bedrock.

Three separate bodies of Holocene mud up to 10 m thick occur in Bedford Basin, in Fairview Cove, Bedford Bay and the main central area of the basin (Fig. 5). They are separated from one another by bedrock-controlled sills, sometimes covered with till and/or gravel. These modern depocentres also

reveal a similar distribution of subsurface lacustrine sediments deposited in early post-glacial time. Bedrock crops out on the east side of the basin adjacent to Long Cove, Admiral Cove, Roach Cove, and Wrights Cove (Fig. 3). Other areas of hard seabed consist of transgressive sand and gravel deposits formed as lag gravel overlying till.

A parallel set of boulder berms, separated by 12-15 m horizontally, and 2 m in depth, surround the entire nearshore area of Bedford Basin at a present depth of 23 m. The boulder berms define the location of two paleo-shorelines of an ancient fresh-water lake which occupied the area of Bedford Basin in late glacial to early Holocene time (Fader *et al.*, 1994). Floating ice masses concentrated the boulders as push ridges.

Inner Halifax Harbour

A new technology for marine seabed mapping, multibeam bathymetry, was applied to the study of the inner (Courtney, 1993) and outer Halifax Harbour (Courtney and Fader, 1994; Fig. 6). These systems provide decimetre depth resolution and 100% coverage of seabed morphology, thus enabling correlation of morphologic features on conventional side-scan sonar or bottom profile records with greater accuracy (Courtney and Fader, 1994; Fader *et al.*, 1994).

The Narrows

The Narrows is a geographic junction between Bedford Basin and the central harbour, having a sill at a depth of approximately 20 m. The seabed of the Narrows is composed of gravel with boulders and some outcropping bedrock (Fader *et al.*, 1991), but with patches of a thin veneer of Holocene mud. A former channel of the Ancient Sackville River can be clearly identified (Courtney and Fader, 1994).

Central Harbour

This area of the harbour is adjacent to the downtown areas of the cities of Dartmouth and Halifax and is 20-30 m in depth.

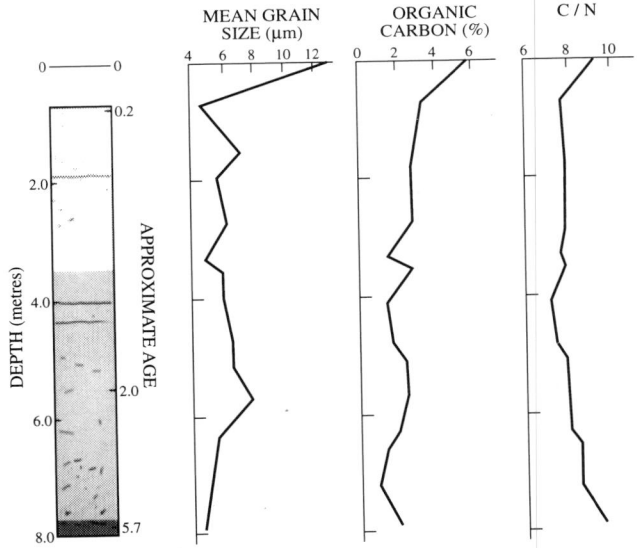

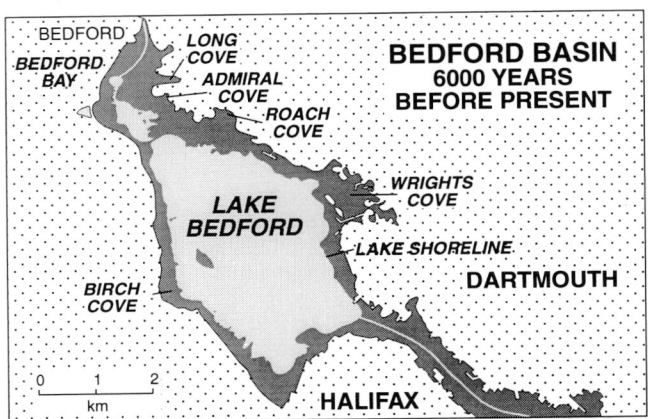

Figure 3 *Paleographic reconstruction of Bedford Basin just before the Holocene marine incursion. The location of the former shoreline is based on two basin-fringing boulder berms at a present depth of 23 m. See text for details.*

Figure 4 *Sediment core from Bedford Basin showing a sequence of sediments dating to 8 ka. The deepest sediments contain fresh-water brachiopods and marsh vegetation. The quantity of organic matter preserved increased during the past 5000 years as the basin became more marine; the carbon to nitrogen (C/N) ratio also reflects the transition from a terrestrial fresh-water or brackish environment, to a marine environment, up to about 200 years ago (0.2 ka). Recent changes probably reflect contamination.*

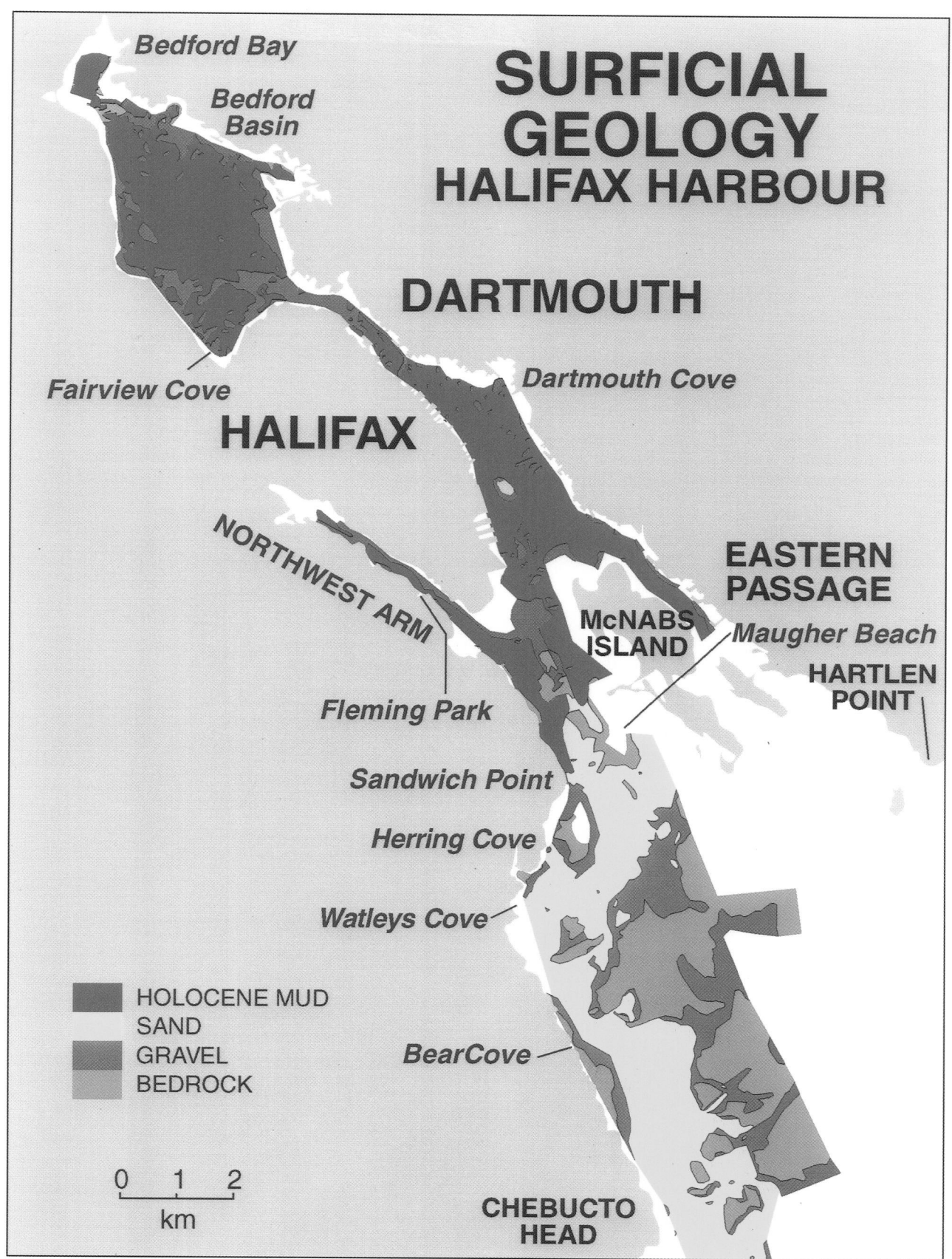

Figure 5 *Distribution of surficial sediment types in Halifax Harbour. Note the transition from inner harbour muddy sediments to outer harbour sands and gravels.*

The main body of Holocene mud begins in the southern Narrows and continues southward to the Maugher Beach area on McNabs Island (Fig. 5). Although the seabed appears generally flat and featureless, several drumlins protrude through the mud floor (Fig. 6). To the northeast of Georges Island is a small submerged drumlin with a dredged upper surface. The most northern mud deposit, up to 7 m thick (Fig. 7), formed in the lee of Georges Island under flood-tide-dominated flow. Seismic reflection data indicate the presence of a thick sequence of estuarine/lacustrine sediments overlain unconformably by a thin layer of sand and thick Holocene mud (Fig. 8). The largest continuous body of Holocene mud in the harbour is found southeast of Georges Island. It is more than 9 m thick and is charged with methane gas.

A large area of the seabed, between Ives Knoll and Halifax City, consists of gravel-covered till. Holocene mud is patchy or absent (Fig. 7) as a result of stronger currents in this constricted part of the harbour. Ives Knoll is an area of coarse sediment consisting of transgressive (lag) gravel overlying till. Ives Knoll has been the preferred location proposed by Halifax Harbour Cleanup Incorporated for the construction of an artificial island for a sewage treatment facility (see below and Chapter 39).

The Northwest Arm is a northern extension of the central harbour west of Halifax, and is mostly a mud-bottomed nar-row linear depression approximately 10 km in length (Fig. 5). Currents have been strong enough to prevent fine-grained sediment deposition, so that gravel and bedrock are exposed in the constricted area of the Arm. Eastern Passage is a southern extension of the central harbour to the east of McNabs Island; a small outlet connects Eastern Passage with the outer harbour.

Outer Halifax Harbour
To the north, in the inner harbour, the seabed is dominated by Holocene mud (Fig. 5). South of Sandwich Point, the outer harbour consists of exposed bedrock, gravel and sandy-silty sediment with little or no mud. The outer harbour is dominated by a deep sand-bottomed, western channel, 30-40 m in depth (see Fig. 1). On the eastern side of the outer harbour, the seabed is very shallow with large shoal areas of exposed bedrock and boulder gravel. The outer harbour is influenced by large storm-driven waves which move cobbles in 20 m water depth, thereby preventing the accumulation of fine sediment. The gravel is a nearshore equivalent of the Sable Island Sand and Gravel Formation (King, 1970; King and Fader, 1986) found on the inner shelf and bank areas of the Scotian Shelf.

The morphology of the outer harbour is dominated by a deep channel that hugs the western shoreline (see Fig. 1). The

Figure 6 *A multi-beam bathymetric image of a section of inner Halifax Harbour presented in a shadowgram format with artificial illumination from the northwest. Note the scour moat around Georges Island, the associated depositional mound and related sediment drifts recording principal current direction. See Figure 1 for location.*

channel continues for more than 30 km further seaward beyond Chebucto Head, cutting across the inner Scotian Shelf (Loncarevic *et al.*, 1994).

GAS-CHARGED SEDIMENTS

Both the Holocene mud and the older lacustrine/estuarine sediments of Halifax Harbour display large areas of gas-

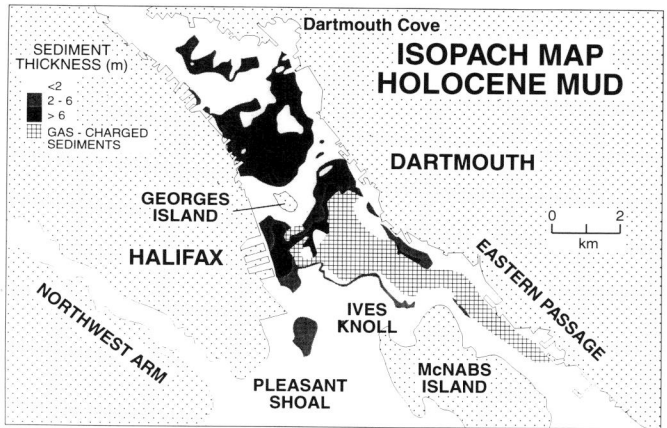

Figure 7 *Isopach map of Holocene mud in the inner harbour. Clearly defined depositional centres occur to the north and southeast of Georges Island. The thickness of the sediment southeast of Georges Island cannot be measured because of the presence of methane gas which prevents penetration of acoustic energy from the seismic reflection systems.*

charging, where acoustic energy from the high-resolution seismic reflection systems will not penetrate and resolve subsurface stratigraphic reflections (Fig. 9). It occurs at a variety of depths, ranging from virtually at the seabed in Bedford Basin to 15 m below the seabed in the outer harbour. Gas is represented on the seismic reflection profiles as zones of incoherent reflections accompanied by a lack of acoustic penetration and the presence of discontinuous high-intensity reflections (Knebel and Scanlon, 1985; Keen and Piper, 1976; and Fader, 1991). This effect is referred to as acoustic masking or acoustic blanking, and is attributed to the presence of interstitial gas within the sediments (Hovland and Judd, 1988; see also Chapter 19).

The gas most frequently associated with this characteristic is biogenic methane, which is formed during bacteriological decay of organic matter at shallow depth within anoxic unconsolidated sediments. The presence of shallow gas within sediments is often considered a nuisance during seismic reflection profiling, as the resolution of geological events beneath the area of gas is degraded. Lower frequency sound sources can better penetrate gas-charged sediments, but stratigraphic resolution is reduced. Reflection pull-downs, characterized by downward dipping reflections at the flanks of the gas zones, are frequently associated with acoustic masking. These arise from the reduction of the sound speed through the gassy sediments. The distribution of gas-charged sediment is shown in Figure 9.

Circular patches of high-intensity acoustic backscatter on side-scan sonograms occur on the muddy seabed of the

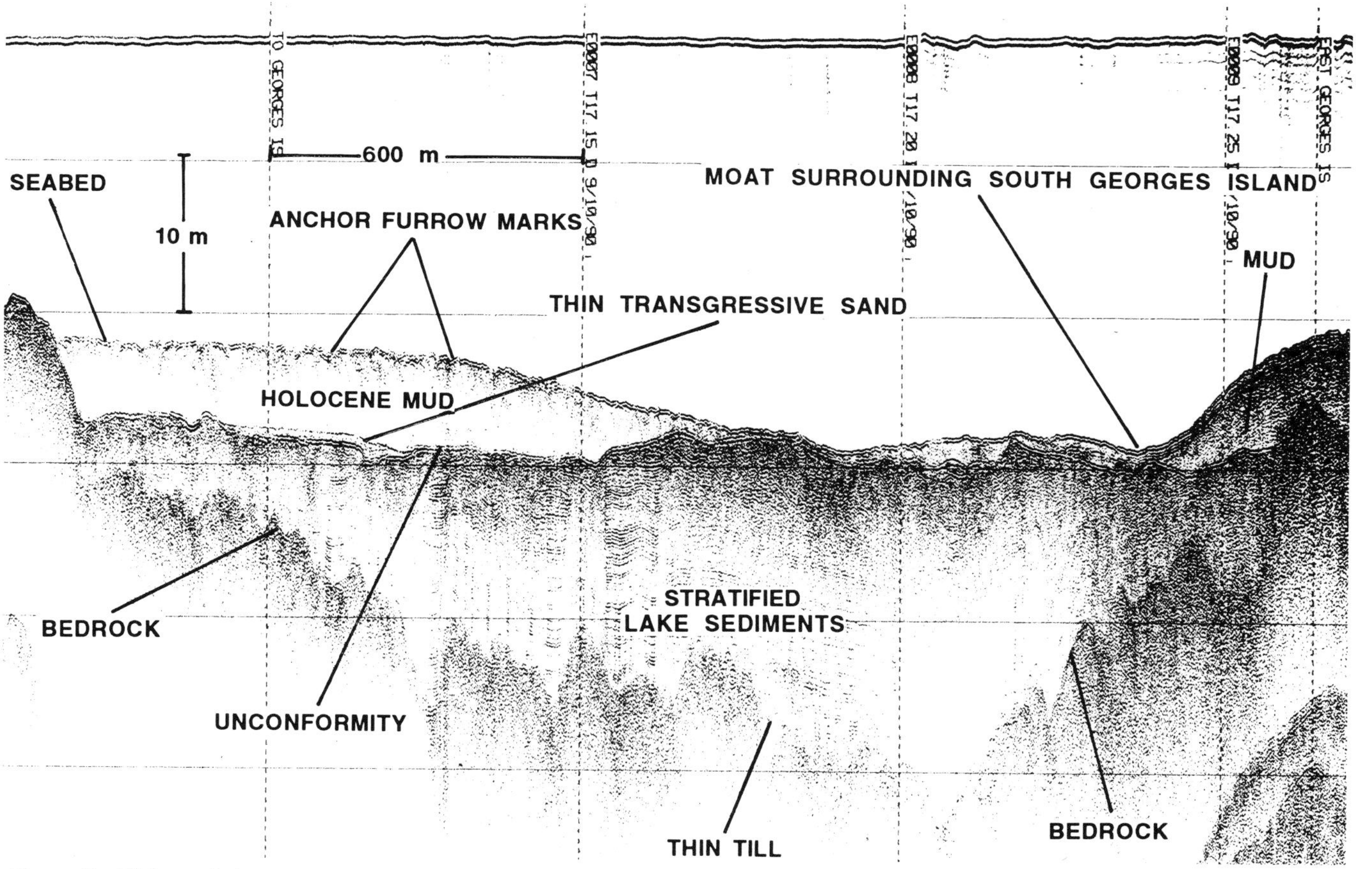

Figure 8 *High-resolution seismic reflection profile and interpretation from inner Halifax Harbour, to the east of Georges Island, showing the typical stratigraphic sequence. The basal sequence is a thin till overlying a rough bedrock surface. The basin infill of high-intensity continuous coherent reflections represent lacustrine/estuarine sediments. An unconformity on the surface of the basal sequence was formed during the Holocene marine transgression. Thin sands and gravels were deposited on this surface, overlain by Holocene mud as the harbour deepened. The roughness of the present seabed is mainly due to marks left from repeated ship anchoring (Fig. 12).*

northern flank of Bedford Basin, near the mouth of Bedford Bay (Fader *et al.*, 1994). They give the seabed a mottled character. The circular patches, up to 5 m in diameter and clustered, are interpreted to represent either localized effects of methane gas venting or the presence of concentrated benthic communities, also possibly associated with the venting of methane gas. The area occurs at the northern subsurface edge of gas-charged sediments in Bedford Basin (Fader *et al.*, 1991).

The largest zone of gas charging occurs in the Holocene mud in the inner harbour to the southeast of Georges Island, and continues to the south into Eastern Passage (Figs. 7 and 9). The upper gas-charged seismic reflector occurs at a consistent depth of between 2 m to 3 m. It is important to note that many anchor marks in the harbour (see below) penetrate the seabed to depths of between 1 and 3 m, suggesting that the gas-charged layer is broken through in many areas, possibly liberating some of the gas.

Pockmarks
Pockmarks are crater-like depressions on the sea floor first discovered by King and MacLean (1970) on the Scotian Shelf. They are interpreted as gas or fluid-escape features and can range to hundreds of metres in diameter and tens of metres in depth (Fader, 1991). They generally occur in thick muds, where gasses or fluids venting to the water erode the seabed. Fine-grained sediments are put into suspension by this pro-

cess, and ocean currents disperse the material away from the vent site.

Pockmarks exist in a dense distribution at the entrance to the Northwest Arm, and are approximately 3 m in diameter. Their depths into the seabed have not been measured. It is possible that these depressions could also be formed from current scour around objects. ROV surveys of these features (Fader and Miller, 1992) were inconclusive concerning the venting of methane gas as the process responsible for their formation.

HYDRODYNAMICS AND SEDIMENT DEPOSITIONAL FEATURES

Circulation Patterns in Halifax Habour
Oceanographic currents exchange water between the inner Scotian Shelf and Halifax Harbour. Early simplistic models suggested that the harbour was flushed regularly and sediments were removed by this process (ASA Consulting Ltd., 1986). Oceanographic data most recently collected, together with a knowledge of sediment and contaminant distributions on the sea floor (*e.g.*, Fader and Petrie, 1991), indicate that conditions are much more complex, and that most sediments discharged into the inner harbour remain in the inner harbour.

Halifax Harbour is an estuary, *i.e.*, a semi-enclosed body of water whose properties are influenced by fresh-water runoff from the land. The average annual fresh-water discharge into Halifax Harbour is 15.7 $m^3 \cdot s^{-1}$, of which 16% is through the sewer systems of the metropolitan areas. The largest river

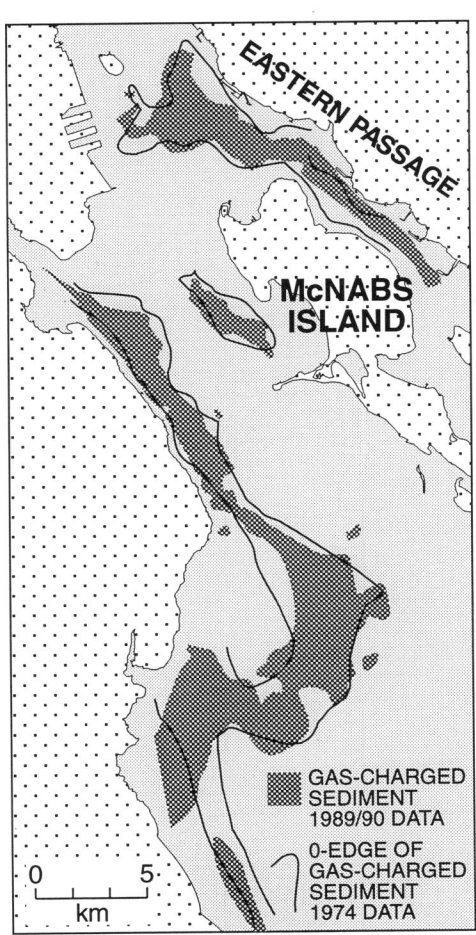

Figure 9 *Distribution of gas-charged sediments in a section of Halifax Harbour. Gas occurs at average depths of 2-3 m in mid-Holocene sediments in the inner harbour, and at depths up to 15 m in older lacustrine/estuarine sediments in the outer harbour that are buried beneath thick transgressive sands and gravels.*

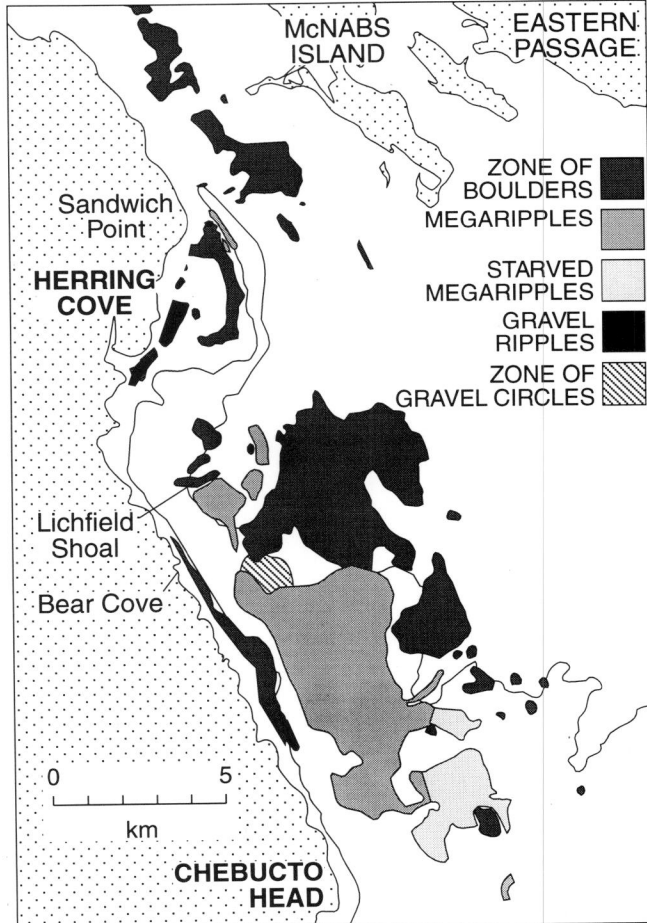

Figure 10 *Bedforms and zones of boulders in outer Halifax Harbour.*

discharge comes from the Sackville River which contributes 5.41 m³·s⁻¹ (Buckley and Winters, 1992). The near-surface watermass flows toward the ocean (286 m³·s⁻¹), becoming saltier as it moves through the harbour. Salt is supplied through mixing with a water mass from the adjacent continental shelf which moves into the harbour below the outgoing near-surface flow (Petrie and Yeats, 1990). In turn, this shelf water becomes less salty as it moves into the harbour because of mixing with the shallower, fresher water.

With regard to current velocity in the surface layer, the weakest outflow (0.2 cm·s⁻¹) is found in Bedford Basin and moves a parcel of water approximately 200 m in one day. The currents accelerate to their highest values of about 5 cm·s⁻¹ in the Narrows, slow to about 2 cm·s⁻¹ as the harbour widens adjacent to the downtown city areas, increase slightly with a

narrowing off Sandwich Point, and finally slow to about 1 cm·s⁻¹ before flowing out onto the shelf. The picture is much the same in the lower layer except in the opposite direction, *i.e.*, inflow instead of outflow.

In the harbour, currents change rapidly as a result of semi-diurnal tidal flows. Wind also can bring rapid, dramatic changes to the circulation by causing waterborne material to cross the harbour in about an hour (Lawrence, 1989), or by resuspending bottom sediments through wave action.

Sediment Transport

No direct measurement of sediment transport in Halifax Harbour has been undertaken. Such studies require the use of tracers together with subsequent monitoring programs; in this region, these studies have only been conducted on the adjacent Scotian Shelf (Amos, 1990). However, many other characteristics of seabed sediments can be used as qualitative indicators with respect to the responsible currents. These include distribution patterns of gravel, sand, silt and clay; bedforms in sand and gravel; scour or moats around seabed obstructions; the distribution of sewage banks; the distribution of geochemical anomalies relative to injection points; and the distribution of exposed bedrock. A complicating factor is that Halifax Harbour has been subjected to a series of varying geological processes that have produced relict seabed features that are not in equilibrium with present conditions.

Sedimentary features indicative of current patterns in the inner harbour are rare, in part either because bottom currents are weak in areas of sediment accumulation, or currents have been strong enough to sweep away most sediment (*e.g.*, the Narrows). South of Georges Island are sedimentary features, termed obstacle-induced sediment drifts, developed in the lee (north) side of bedrock highs (Courtney and Fader, 1994; Fig. 6). The only other features indicative of current patterns in the inner harbour are gravel ripples near the Dartmouth side of the harbour, resulting from wave action under strong westerly winds.

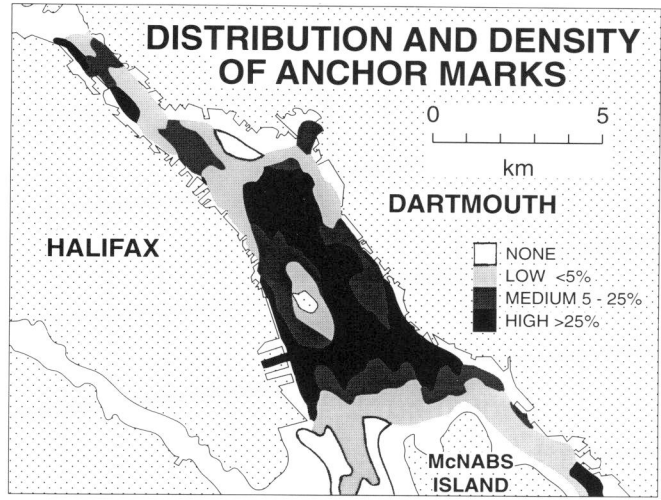

Figure 11 *Distribution and density of anchor marks in the inner harbour. Anchor marks include drag marks, pit marks, and chain marks (see text).*

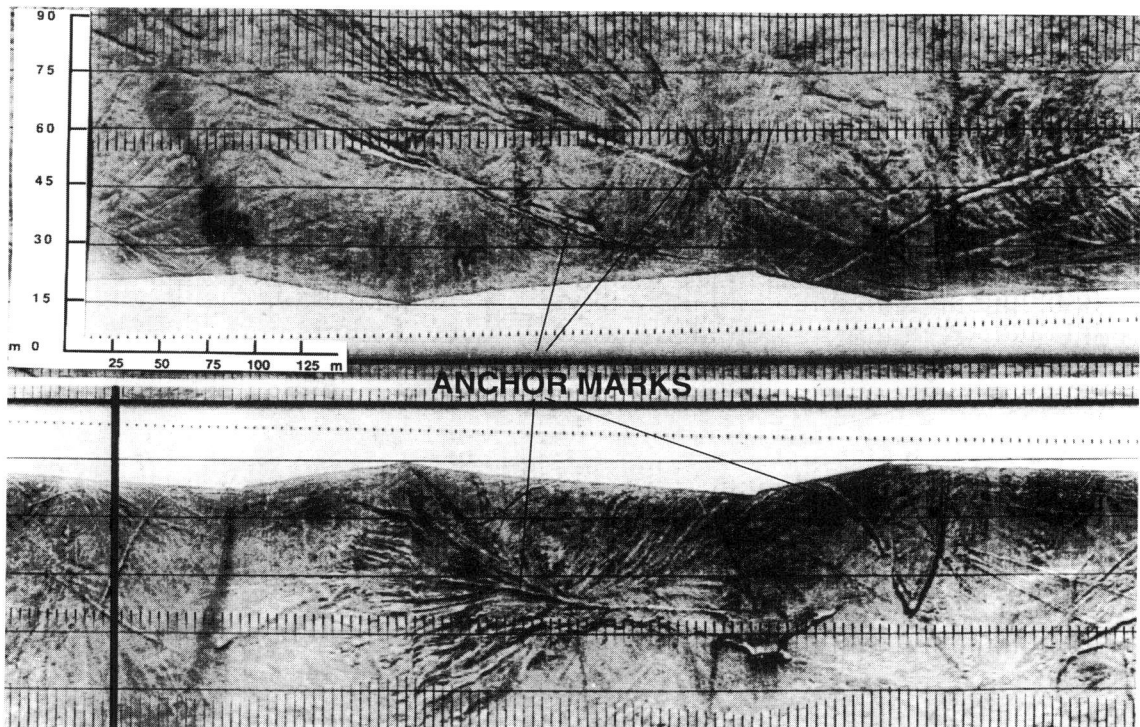

Figure 12 *Side-scan sonogram from the inner harbour showing typical distribution and shapes of anchor marks.*

Sedimentary Furrows

A series of linear erosional scours occurs in Holocene mud at the boundary between the inner and outer harbour adjacent to Sandwich Point (Fig. 10). These features parallel the adjacent shoreline and show higher acoustical reflectivity than the surrounding seabed, suggesting the presence of sand, debris or shells. They are up to 2 m in depth, and extend over 3.5 km. The furrows bifurcate, opening to the south, or seaward. Toward the north, they gradually shallow and merge with the flat mud seabed. Similar erosional features have been found in diverse sedimentary environments: such as the deep sea, lakes and estuaries (Dyer, 1970; Hollister et al., 1974; Flood, 1980, 1981). Furrows that exhibit tuning-fork junctions, where a single furrow opens into the direction of dominant current flow, can be used as current and sediment transport pathway indicators. Sedimentary furrows adjacent to Sandwich Point indicate that the currents that formed them move from the south to the north toward the inner harbour. Extensive data do not exist on the flow regimes necessary to create sedimentary furrows, but in estuaries where measurements have been made, and where the flow is tidal, currents range from 50 cm·s⁻¹ to much greater than 100 cm·s⁻¹. Dyer (1970) and Hollister et al. (1974) postulated that helical secondary flows operating close to the sea floor are responsible for formation of furrows; in addition, Flood (1981) suggested that abrasion by coarse sediment plays an important role in the initiation of furrows. Their presence indicates that the harbour is subjected to periodic, strong, directionally stable currents.

Sand Megaripples

The innermost bedforms in the harbour are subdued sand megaripples which occur in the deep narrow channel defined by the 30-m contour adjacent to Sandwich Point (Fig. 10). They are straight-crested, flow-transverse bedforms with a ripple-like profile. They are formed by currents with a near bedflow velocity of between 40-60 cm·s⁻¹ (Amos and King, 1984) moving from south to north. The major zone of megaripples occurs south of Lichfield Shoal, and continues south to Chebucto Head (Fig. 10).

Gravel Ripples

Large areas of gravel ripples are present in many areas of the outer harbour between 10 m and 20 m water depth (Fig. 10). These often flank the outcropping bedrock shoals in areas between the bedrock and sand megaripples. The ripples are characterized by a wave length of between 1 m and 2 m and

heights of less than 0.3 m, and are formed by oscillatory motions associated with large waves. Areas of gravel ripples and megaripples in the outer harbour indicate that fine-grained silt and clay-sized sediments are not deposited.

Gravel Circles

Approximately 90 gravel circles, each 25 m in diameter, occur in the outer harbour (Fig. 10). The circles are slight depressions, less than 0.2 m in depth from the surrounding sandy seabed, in about 30 m of water depth. Most occur in groups and many are aligned; in some cases, they are connected by narrow gravel deposits, giving them a beaded appearance on side-scan sonar imagery. In a few areas, the circles are singular isolated features. Small patches of sand occur in the centre of many of the circles. Initially, it was thought that they may represent dredge spoils dumped at the seabed of the outer harbour, but this is not a dumping ground for dredge spoils. Circles may have formed through the action of vortices which scoured the thin overlying sand in a circular pattern and exposed the underlying lag gravel. The steep bedrock topography of Bear Cove Shoal and the narrowing of the channel in this area may have generated a localized circulation in intense storms to form vortices, thereby attesting to periodic high-energy conditions at the seabed of the outer harbour.

ANTHROPOGENIC IMPRINTS AND ARTIFACTS

The impact of human activity on the character of Halifax Harbour has been profound (Figs. 6, 11, 12, 13, 14). The first significant anthropogenic impact on Halifax Harbour was the scuttling of a French fleet of 14 ships in Bedford Basin in 1746 (Raddall, 1965). Indeed, a very old shipwreck found in western Bedford Basin may be the remains of one of these vessels (Fader et al., 1994). After the founding of Halifax in 1749, the construction of fortifications and wharfs would have had some impact. With increasing urban growth in the 19th and 20th centuries, waste from domestic and industrial sources has significantly influenced water quality in the harbour. Dredging of the seafloor for aggregate material or for improved shipping access, dumping of spoils from construction projects, and construction of footings for bridges have all had an impact. Anthropogenic features on the seabed include anchor marks, shipwrecks, dredge spoils, borrow pits, cables, propeller scours, dredged areas, and unidentified debris at the seabed, on both the Holocene mud and the transgressive gravels.

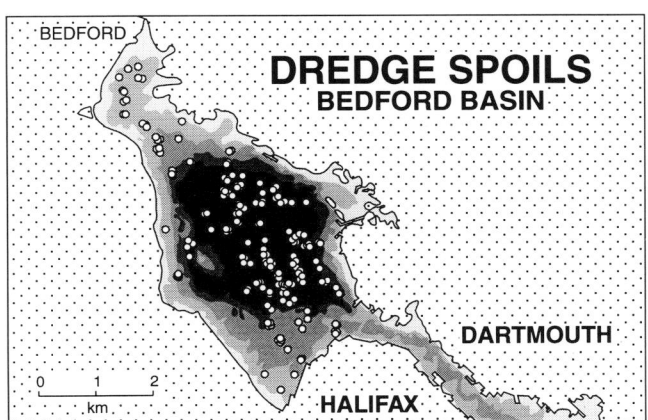

Figure 13 *Distribution of dredge spoils in Bedford Basin indicated by circles. Spoils are up to 45 m in diameter.*

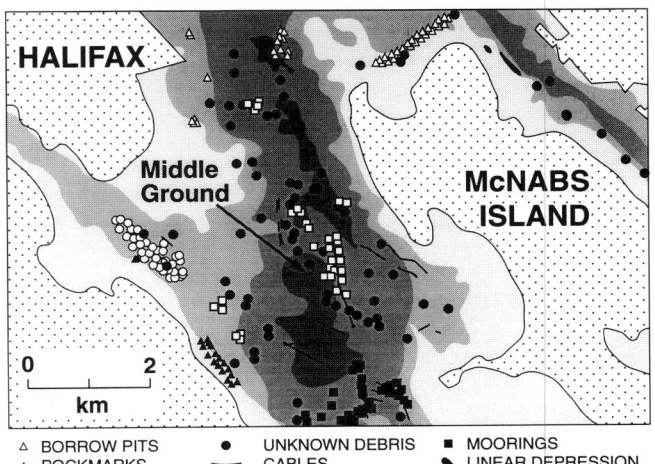

Figure 14 *Distribution of various anthropogenic and natural features in an area of inner Halifax Harbour.*

Anchor Marks

Marks on the seabed attributed to the process of ship anchoring are widespread, especially in Bedford Basin and the inner harbour (Figs. 11, 12). These features give the seabed a morphology similar to areas of offshore northern Canada, where iceberg and sea ice scours are common.

This process is a widespread and significant sediment turbator, and the term anchorturbation has been proposed for this process (Fader et al., 1991). Sediment cores must be collected away from areas where anchor marks are found, in order to sample an intact and meaningful stratigraphy. The interpretation of results from sediment samples must also be carefully evaluated in areas of high density of anchor marks, especially where the anchors penetrate several metres into the seabed and older sediments are exposed and overturned. The widespread distribution of anchoring throughout the inner harbour suggests that sediments and their associated contaminants are continually being exhumed, resuspended and redeposited.

Anchor marks have been classified into three types: anchor pits, anchor drag marks, and anchor chain marks (Fader et al., 1994). The initial impact of the anchor on the seabed often produces an amphitheatre-shaped, deep, circular depression (anchor pit). If the anchor has been dragged across the seabed in the process of setting the flukes or digging in and providing a hold, or in the process of being retrieved, long linear features termed anchor drag marks are formed. These can be kilometres in length and may also result from the dragging of anchors in response to the movement of ships under high winds. Anchor pit marks are often found at the beginning or termination of anchor drag marks.

Anchor chain marks form as the result of repeated touching down upon the seabed of anchor chains in response to winds and tides, as ships weathervane about their anchors. Such marks appear as a radial pattern of linear depressions, giving the seabed an imprint similar to a large feather. Some of these features can resemble other features termed plumose structures (Fader et al., 1994), which are interpreted as dewatering structures in fine-grained sediments, formed in response to seismic activity. However, in Halifax Harbour the features are interpreted as resulting from anchoring.

Anchor marks occur in a variety of sizes and shapes. The deepest ones are up to 2.5 m in depth and 5 m in width. The depth of some of these features may be increased by the release of methane gas from the seabed in areas where gas-charged sediments occur at shallow depth and anchors penetrate through to the gas-charged layer, facilitating its release. Anchor marks on gravelly hard seabeds are shallow, most often less than 1 m in depth. Some have been traced for over 2 km along the seabed of the harbour. Many of the large harbour docks have radiating patterns of anchor marks along them, suggesting that the deployment of anchors as a ship's speed and/or direction control mechanism is a common occurrence.

The densest distribution of anchor marks in the inner harbour, where greater than 25% of the seabed is disrupted, occurs to the north, east and southeast of Georges Island, in the area where present designated anchorages are located (Fig. 11). In Bedford Basin, more than 80% of the muddy seabed is criss-crossed with anchor marks. Several generations of anchor marks have been defined on the floor of Bedford Basin. Some appear fresher than others, having clearly defined berms and sharp contrasts in reflectivity, while others have gradational changes suggesting erosion and degradation of the original features. Because sedimentation since the founding of Halifax in 1749 is generally insufficient to

fill the anchor marks (about 41 cm; Buckley et al., 1995), it is possible that the entire population of anchor marks is still visible on the basin floor. The majority of the marks were likely made during the First and Second World Wars, when large convoys of cargo and warships assembled in Bedford Basin prior to embarking for Europe.

Borrow Pits

Borrow pits are seabed depressions formed by the removal of aggregate. The largest area of borrow pits in the harbour occurs north of McNabs Island, where there are many circular pits at the seabed, averaging 15 m in diameter and ranging up to 3 m in depth. A series of very large linear pits that were formed by dredging for aggregates occurs along the eastern side of McNabs Island. Some are over 200 m in length, 30 m in width and 3 m in depth.

Dredge Spoils

Dredge spoils are circular-shaped deposits of material discharged to the seabed, generally by barges. They range in diameter up to 40 m, and may exist as positive features several metres in height above the surrounding seabed, or as coarse debris in depressions several metres in depth. At the entrance to the Northwest Arm, dredge spoils have been dumped on Holocene mud, compressing and displacing the mud into features that resemble pockmarks on the side-scan sonar and seismic reflection records. In this area, the dredge spoils have resulted in the venting of gas from a zone directly beneath the spoil, as evidenced by the absence of gas-charged horizons on the seismic reflection profiles. Linear areas of high acoustic backscatter, connected with much larger circular features of the same acoustic signature, represent sporadic barge discharge of spoil while underway. These features appear concentrated near the Naval Dockyard in Halifax and are widespread in other areas of the inner harbour. Dredge spoils consist of a wide variety of materials such as construction debris, old wooden docks, gravel, boulders, muddy sediments, garbage and unidentified debris. Their presence on the harbour bottom makes it difficult to sample the naturally occurring sediments. Side-scan sonar data is essential for their recognition and accurate sampling. In Bedford Basin, dredge spoil covers approximately 5% of the basin floor (Fig. 13). The practice of dumping dredge spoils in the harbour is presently discontinued.

Shipwrecks

Shipwrecks normally exhibit unique characteristics on side-scan sonar data, primarily resulting from the hard nature of the material, the presence of large expanses of flat metal or wooden surfaces, high and unusual angles between structural elements, and the presence of diagnostic-shaped features such as railings, smokestacks, funnels, bows, anchors and chains, hatches and openings in the superstructure. Even when ships are badly broken, large fragments can be identified by the presence of unusual sonar characteristics, which are much different from natural geological signatures of sedimentsh and bedrock. The most famous shipping accident in Halifax Harbour was the explosion of the munitions ship Mont Blanc in 1917 that resulted in the destruction of much of the north end of the city of Halifax. Marine geological studies have been conducted to identify the exact location of the explosion and any evidence of a crater that may have been made by the explosion (Fader, 1994).

Approximately 30 shipwrecks or large pieces of shipwrecks have been located on the seabed of Halifax Harbour

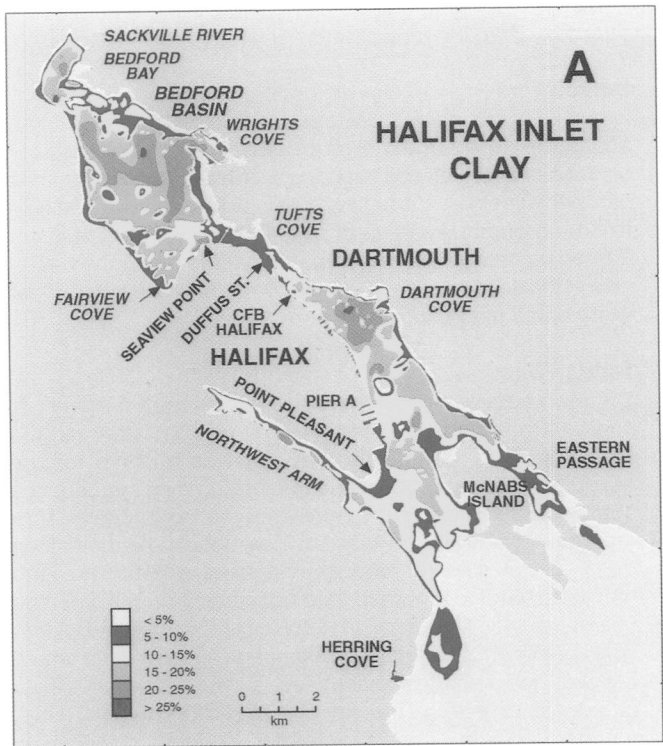

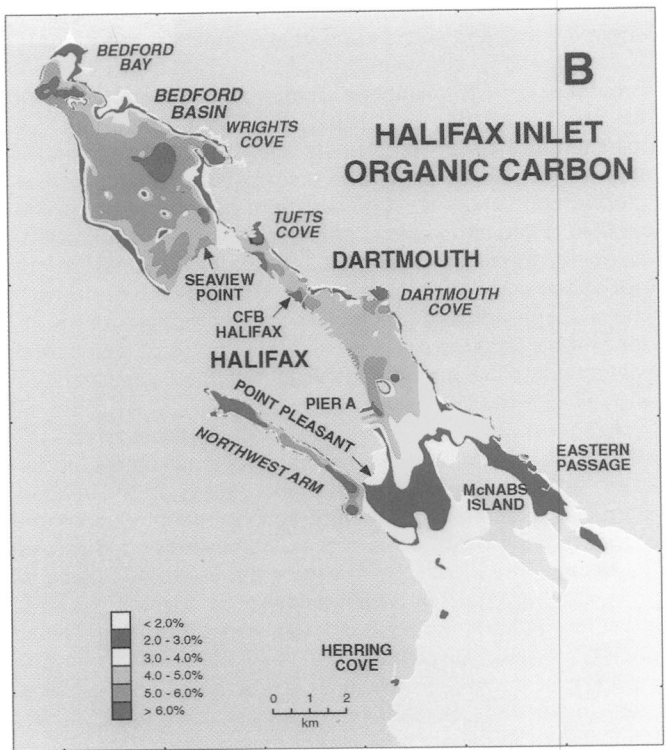

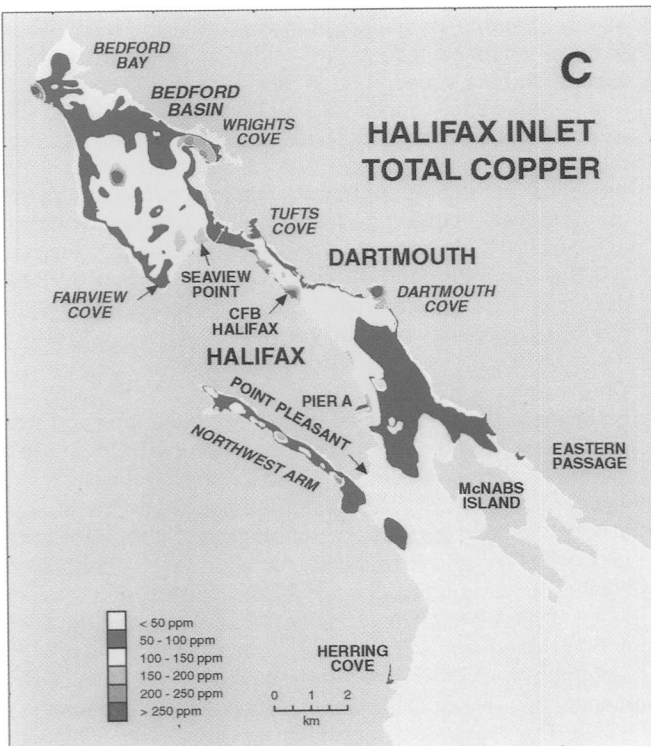

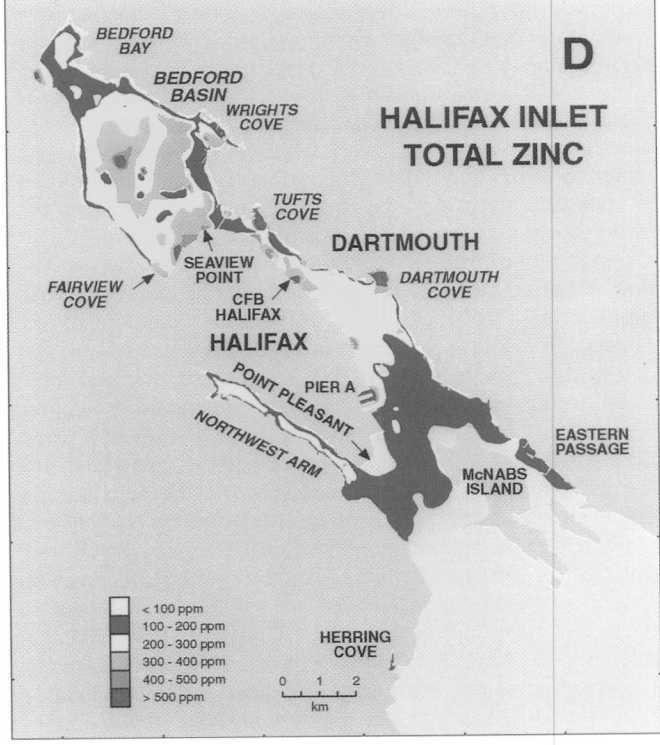

Figure 15   *Sedimentary and geochemical characteristics of surficial sediments (upper 2 cm) in Halifax Harbour.* (**A**) *Clay content (particles <4 μm diameter);* (**B**) *organic carbon derived from land and marine plants and animals and waste from sewage in sediment;* (**C**) *and* (**D**) *distribution of total copper and total zinc metal (after Buckley and Winters, 1992).*

(Fader *et al.*, 1994). Some are well-known wrecks, such as the ferry Governor Cornwallis, located on the southeast flank of Georges Island, while others remain unknown and require visual identification. Some of the vessels have recently been identified, such as the *Havana*, a schooner that sank in 1906 while attempting a salvage operation in the inner harbour. To the north of the *Havana*, off the south end of Halifax, lies another schooner which has been determined to be the *Gertrude de Costa*. It was involved in a collision in the harbour in 1951, and was never located until our recent survey. Seven seamen went down with the vessel after collision with an oil tanker. Both of these vessels lie in an area planned for the diffuser from the proposed sewage treatment plant and where considerable excavation of the seabed was planned for the placement of the diffuser. Of particular significance to this construction is not only the potential for disturbance of the wreck with human remains, but the presence of unexploded ordinance, discovered on the deck of the *Gertrude de Costa*, which may represent an engineering hazard.

One of the most difficult-to-interpret features found on the seabed of Halifax Harbour was a feature eventually recognized as the Trongate Depression (Fader *et al.*, 1991; Fig. 6), which is a linear depression 125 m in length, 3 m deep (deeper at one end), with flanking, asymmetrical berms of mud up to 2 m above the surrounding seabed. The feature was formed by the sinking of the Norwegian vessel, the *S.S. Trongate*, a seven thousand tonne merchant ship, which was purposely sunk in 1942 because of an on-board fire and an explosive cargo. The hull of the vessel was later salvaged, but an impression of the hull remained in the mud at the location of anchorage number 4, near the centre of the central harbour. ROV observations in the vicinity of the depression showed the presence of rolls of newsprint, wooden planks, boots and much debris at the location including unexploded ordinance such as primed 4-inch shells, .303-calibre rifle ammunition, and scattered cordite.

The most recently formed large anthropogenic features of the seabed of Halifax Harbour were formed by the container ship *Atlantic Conveyor*, which grounded on June 1991, while entering the Dartmouth side of the harbour in 5-m water depth. The grounding produced a 200-m-long gash on the seabed, cutting several metres through a gravel lag and into underlying till.

Dredged and Blasted Excavations

Many areas of the harbour bottom have been dredged to provide deeper draft for the large vessels which use Halifax Harbour. Some of these are long linear bucket-dragged zones, while others were deepened by vertical clam-shell dredging. The Dartmouth nearshore has been extensively dredged and used as a dump site for dredge spoils. The top of a drumlin in the inner harbour has been dredged with a clam shell bucket (Fig. 6).

Propeller Scours

The seabed appears to have been eroded to a maximum depth of 2 m directly adjacent to many of the docks along the harbour waterfront. The eroded areas appear on the sonograms as a series of scallop-shaped depressions. This unique morphology is interpreted to result from propeller wash associated with large vessels.

Remains of Narrows Bridges

The remains of the first two bridges that spanned the harbour, connecting Halifax and Dartmouth, lie in the Narrows, approx-

imately 500 m to the south of the existing A. Murray MacKay Bridge (Fader *et al.*, 1991). They were constructed in 1886 and 1892. The first bridge collapsed in a violent hurricane, on September 7, 1891, which destroyed many of the docks along the waterfront. A new piled bridge was unstable, and chains and granite blocks were attached for increased stability in 1892; geophysical data indicate that outcropping bedrock and gravel covers the majority of the seabed in the area and may have prevented the proper penetration of the bridge piles. The seabed is covered with cribwork, steel rail, and large timbers. Large carved rectangular granite blocks that may have been used to stabilize the second bridge also remain at the seabed.

Other Anthropogenic Features

Other anthropogenic features on the seabed of Halifax Harbour include large circular bermed pits, attributed to jack-up oil rig spud-can depressions (Fig. 6). These features are up to 3 m in depth and 15 m in diameter. Long linear parallel depressions record the grounding of the pontoons of a semi-submersible oil drilling rig (Fig. 14). Other scours have resulted from the presence of submarine nets that spanned the harbour during the Second World War (Fader *et al.*, 1994). Slight movements of the bottom of the nets in response to tides and currents eroded the parallel depressions.

CHEMICAL CONTAMINATION IN SEDIMENTS

The quality of water and sediments in a marine harbour reflects the anthropogenic influences of waste disposal, accidental and incidental losses of chemicals, and residues from domestic and industrial energy production.

Surficial Sediments

Systematic geochemical studies of the surficial sediments in Halifax began with the reports by Prouse and Hargrave (1987) and OceanChem Group Ltd. (1988). Samples were analyzed for total content of major and trace elements (silicon, aluminum, iron, calcium, magnesium, potassium, titanium, manganese, copper, zinc, nickel, lead, mercury and lithium), total and organic carbon, water content, and sediment textural characteristics (see Buckley *et al.*, 1989; Buckley and Hargrave, 1989).

Maps illustrating the distribution of some sedimentological and geochemical characteristics of harbour sediments (Fig. 15) indicate a great deal about the sources and dispersion of contaminants. Fine-grained sediments, indicated by higher percentages of clay, are confined almost entirely to the inner harbour, especially in the central harbour and in Bedford Basin. Similarly, sediments containing high content of organic carbon, indicative of high organic matter content, originate mainly from the major sewer outfall areas and are dispersed in the inner parts of the harbour, especially in the Northwest Arm and in Bedford Basin. Specific contaminant metals such as Cu and Zn are seen to be associated with point sources at Pier A, Dartmouth Cove, CFB Halifax, Tufts Cove, and in Bedford Bay. The indicated sources in the central harbour are at the location of major untreated sewer outfalls. The Bedford Bay anomaly is near the outfall from the sewage treatment plant for the town of Bedford, indicating the discharge of contaminant metals. The anomaly in the centre of Bedford Basin is associated with the dumping of contaminated dredge spoils from the central harbour (Lay *et al.*, 1993; Fader *et al.*, 1994; Buckley *et al.*, 1995). The anomalous zinc concentration near Seaview Point indicates that this metal is being leached from the site of the former Halifax City waste landfill.

In order to assist in the interpretation of a complex matrix of

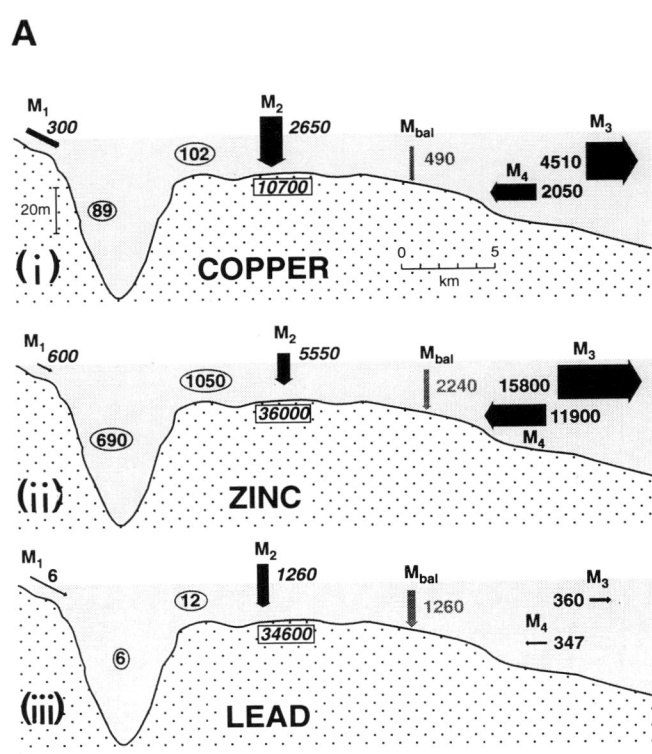

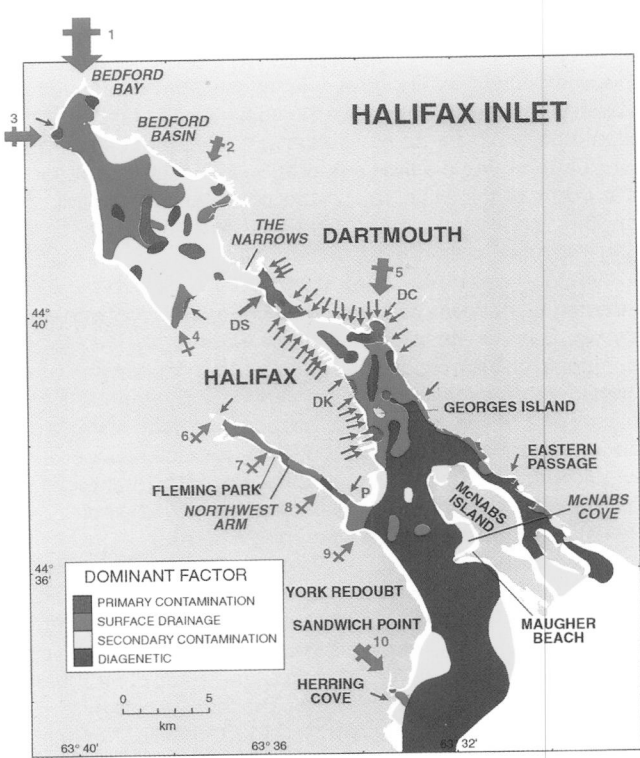

Figure 16 *(above)* Distribution of dominant factor type sediments, based on highest statistical factor loadings for individual samples. Sewage outfalls are identified by arrows with major designated outfalls being Duffus Street (**DS**), Duke Street (**DK**), Dartmouth Cove (**DC**), and Point Pleasant (**P**). Surface drainage and fluvial inputs are represented by numbered and scaled "t" vectors: **1**, Sackville River; **2**, Wrights Brook; **3**, Paper Mill Lake flume; **4**, Fairview Cove drainage; **5**, Banook Lake flume; **6**, Chocolate Lake brook; **7**, Frog Lake brook; **8**, Williams Lake brook; **9**, Purcells Lake brook; **10**, Powers Pond flume (after Buckley and Winters, 1992).

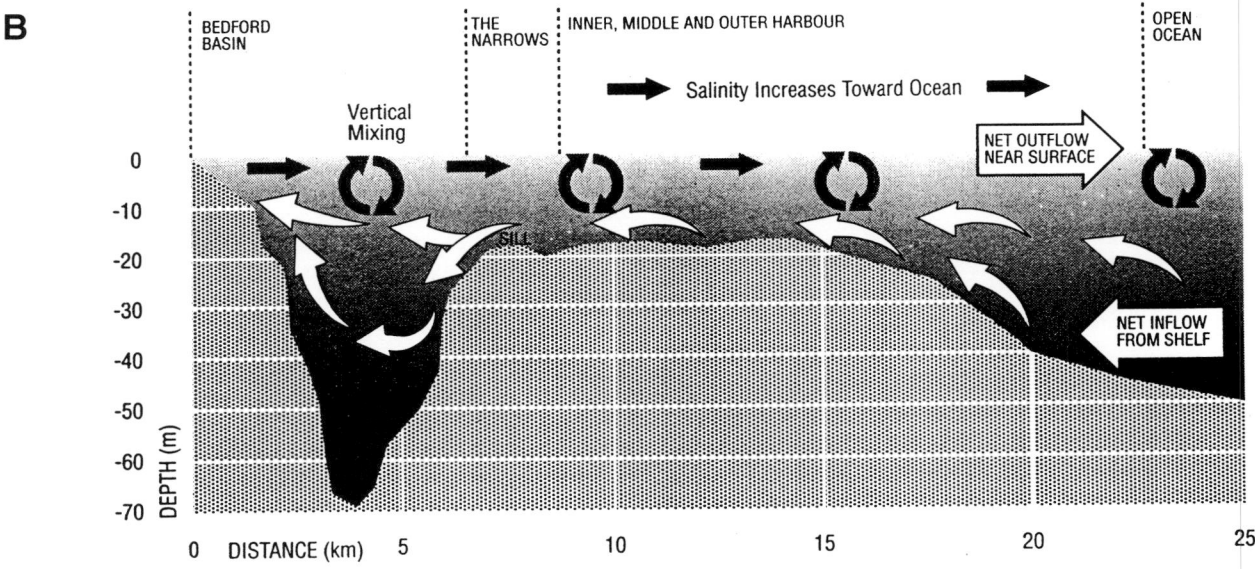

Figure 17 (**A**) Representation of longitudinal section through Halifax Harbour illustrating geochemical flux models for (i) copper (Cu), (ii) zinc (Zn) and (iii) lead (Pb). All flux quantities are in kilograms per year. Numbers in ovals in the water column represent the average metal mass (kg) contained in the upper and lower water layers. Numbers in boxes below the sediment surface represent the mass (kg) of potentially reactive metal deposited in sediments in the year 1990. Inputs of dissolved metals from surface drainage and fluvial sources are written beside the arrow designated as M1; dissolved inputs from the sewers are noted beside the vertical arrows designated as M2. Exchanges of dissolved metal between the harbour and the Approaches are designated by M3 and M4. Estimates of dissolved metal scavenged from the water column and added to the bottom sediments are indicated by Mbal, (after Buckley and Winters, 1992). (**B**) Estuarine circulation of Halifax Harbour. Dense ocean water flows in to the harbour at depth; outflow water is less dense because of the addition of urban runoff, sewage and fresh water from the Sackville River.

geochemical data, a factor analysis technique was employed to identify spatial variability across the harbour and source areas (Buckley and Winters, 1992; Winters and Buckley, 1992). Five factors were identified that could be used to characterize the surficial sediments. The first and most significant factor (41% of the variance in all geochemical variables), included four forms of zinc, two forms of copper, three forms of lead, nickel and chromium, as well as total cadmium, and organic carbon. These variables characterize contaminants that are associated with sewage discharges and contamination from industrial sites.

A second factor (8% of the total variance) included silicon, aluminum, magnesium, potassium, and iron, as well as organically bound iron and manganese. This factor also included the minor element lithium, which can be used as an indicator of clay minerals that are often concentrated in the finest sediment grain sizes (Buckley and Cranston, 1991). The second factor has been designated as representing sediments derived from surface drainage systems from urban areas, including the Sackville River and several other smaller systems.

The third factor group (6% of the total variance), included acid labile or reducible forms of zinc, lead and copper that are the product of some post-depositional secondary modification of metals originally associated with sewage or surface drainage sources.

A fourth factor group (5% of variance) associated several labile forms of manganese with lead and copper as well as titanium. These associations, and the location of samples with these characteristics, suggested that the labile metals were the product of diagenetic remobilization from subsurface layers into the surficial sediments, where the metals adsorbed to particle surfaces. A fifth, and least significant factor ($CaCO_3$ factor; 4% of variance), was one in which three forms of calcium were associated and is restricted to small areas of the outer harbour.

A map illustrating the distribution of factors 1 to 4 is shown in Figure 16, which readily identifies areas in the harbour that are dominated by a particular contamination process. It is evident that several areas in the central harbour adjacent to the many sewer outfalls are dominated by primary contamination (factor 1). Contamination from surface drainage (factor 2) from the Sackville River dominates the northwestern part of Bedford Basin and part of the central harbour. The presence of zinc, lead and copper contamination (factor 3) in the areas surrounding Seaview Point (site of old Halifax landfill) and adjacent to the industrial sites in the central harbour is an indication that secondary leaching of metal contaminants is taking place in these areas. The diagenetic factor (factor 4) indicates migration of metals from subsurface sediment layers, and dominates areas where subsurface methane has been found. This association may indicate that highly reduced sediments, in which organic matter is decomposing, may be a source for these metals.

Metal Budget

A simple box model was used by Buckley and Winters (1992) to estimate budgets of metals contributed to the harbour from urban drainage and the sewer outfalls, accumulated in surface sediments, and exchanged with the Atlantic Ocean. This budget is depicted for three metals in Figure 17. The budget for copper and zinc indicated that only about 10% of the dissolved metal enters the harbour from urban drainage systems, with 90% being contributed from the combined sewer outfalls. A small portion (17%) of the dissolved copper is sequestered to

the bottom sediments inside the harbour; however, about 36% of the dissolved zinc is sequestered to the bottom sediments. A much larger quantity of potentially labile particulate copper and zinc is deposited in the bottom sediments each year (10.3 tonnes [t] Cu; 33.7 t Zn). The net export of dissolved Cu from the harbour each year is about 2.5 t, whereas the net export of dissolved Zn is about 3.9 t. There is about an order of magnitude more dissolved zinc as compared with dissolved copper in the water column inside the harbour, with considerably more metal dissolved in the upper water layer as compared with the bottom layer. The budget model for lead contrasts with that of copper and zinc in that nearly all of the dissolved lead injected into the harbour from the sewers and surface drainage is sequestered to the bottom sediments. The annual potentially labile particulate lead accumulation in the sediments is much greater (27 times) than that from dissolved lead sequestering. There is no significant net exchange of dissolved lead with the Atlantic Ocean.

The level of contamination by metals in the surficial sediments of Halifax Harbour is among the highest recorded for marine harbours and estuaries in any developed country (Buckley and Winters, 1992). This may be surprising in view of the relatively small urban population and industrial base around Halifax Harbour as compared with many other harbours. The reason for the high level of contamination lies in the hydrodynamic characteristics of the harbour in comparison with other harbours and estuaries (Fig. 17B). Halifax Harbour is an estuary with a relatively small fresh-water inflow (17 ± 8 m³·s⁻¹) as compared with other systems with similar areas (e.g., Miramichi, 332 ± 264 m³·s⁻¹). The tidal to fresh-water volume ratio for Halifax Harbour is 373, as compared with the Miramichi estuary where this ratio is 73 (Buckley, 1994) and results in a much reduced flushing capacity. Moreover, the bathymetry of Halifax Harbour fosters the trapping of contaminants in the deep inner Bedford Basin. Partly as a result of these characteristics, about twice as much zinc and seven times as much lead is deposited annually in sediments of Halifax Harbour as compared with the Miramichi estuary (Buckley, 1995).

Subsurface Sediments

An objective of the geochemical study of Halifax Harbour was to determine the history of contamination as well as the present state of environmental quality. A series of core samples was collected between 1989 and 1991 (LeBlanc et al., 1991; Buckley et al., 1991, 1994; Fitzgerald et al., 1991). Sediment layers were dated by the use of ^{210}Pb and ^{137}Cs isotope analyses for determining high-resolution geochronology over the past 150 years. Hydrocarbon concentrations in the sediments increased 100-fold from about the year 1900 (depth 15-20 cm) to 1990 (Gearing et al., 1991). Aromatic hydrocarbons reach peak concentrations in sediments deposited around the 1950s, coinciding with the conversion from use of coal as a domestic and industrial fuel to the use of petroleum. Metal contamination profiles in this core also indicated peak concentrations for mercury, lead, zinc and some forms of copper occurred in the 1970s.

With detailed chemical analyses and dating of a number of cores located at various places throughout the harbour, it was possible to reconstruct a contamination history for the entire harbour (Buckley et al., 1995). Using the same factor analysis techniques as were applied in the study of the surficial sediments (Buckley and Winters, 1992; see above), four contamination factors were again identified, and represent the same processes as those that contributed to the contamination of the surficial sediments (see above). The value of this analysis in the study of the core samples was that it was possible to deduce when specific anthropogenic activities began to influence environmental quality of sediments.

Briefly, copper contamination began around 1900 in most of the harbour areas, but the extent of contamination was greatest in the central harbour (Fig. 18). Sediments in Bedford Basin have been dominated by contaminants sourced from urban drainage. Sediments from the central harbour were dominated by contamination derived from sewage and post-depositional diagenesis for most of the time since 1900. Sediments from the Northwest Arm show a transition from surface drainage dominance to that derived from sewage in most recent times. These historical trends reflect changing intensity of industrialization, urban growth influences, changes in the use of metals in paints and other chemicals, and in the nature of combustion fuels. Some changes can also be attributed to changes in the location and volume discharge of sewer outfalls.

Environmental Risks

The high concentration of contaminant metals and hydrocarbons in sediments in Halifax Harbour poses a risk to marine biological communities. Unfortunately, few studies of biological community distortions (decreased species diversity and/or biomass), biological uptake of contaminants, and toxicity reactions to contamination have been conducted in Halifax Harbour. Contamination of shellfish in Halifax Harbour by bacteria and other contaminants was considered a sufficient threat that this fishery was closed in the 1960s. Benthic ecological studies of a few areas in Bedford Basin and near Herring Cove revealed that overall the range of species was similar to other coastal areas in Atlantic Canada, but specific sites in Bedford Basin had low species numbers and low biomass (Hargrave et al., 1989). Analyses for specific toxic chemical uptake in lobsters found that concentrations of cadmium, copper, zinc, mercury and lead in the digestive glands and cooked meat were below levels judged to be hazardous to human health, although there was some elevation in some animals from certain areas of the harbour (Uthe et al., 1989). These investigators also found elevated levels of PCBs and PAHs in digestive glands, suggesting that they should be avoided in human consumption. In a more systematic biological survey, Tay et al. (1991) found significant biological damage to bottom-dwelling fish (flounder). All specimens had tissue deterioration commonly associated with severe contamination. They also demonstrated through experimental toxicity tests that the highly contaminated bottom sediments resulted in severe acute toxicity for bivalves. Cook (1995) demonstrated that surface sediments are marginally toxic, while subsurface sediments are considered to be toxic, based on Environment Canada guidelines.

The large inventory of contaminants in the surface and subsurface sediments of Halifax Harbour poses an environmental threat for many years in the future. Presently, most of the contaminated sediments are in a chemically reduced state because of the limited oxidation of the bottom sediments (Buckley et al., 1995). The high content of organic matter in these sediments prevents penetration of oxygen from the overlying water column. Even secondary oxidants such as sulfate are reduced at relatively shallow depths in the sediments. As a result of reducing conditions, some of the contaminants are sequestered in insoluble forms that prevent them from being remobilized back into the overlying water and becoming available to most of the biological community. However, if these sediments were exposed to oxidizing condi-

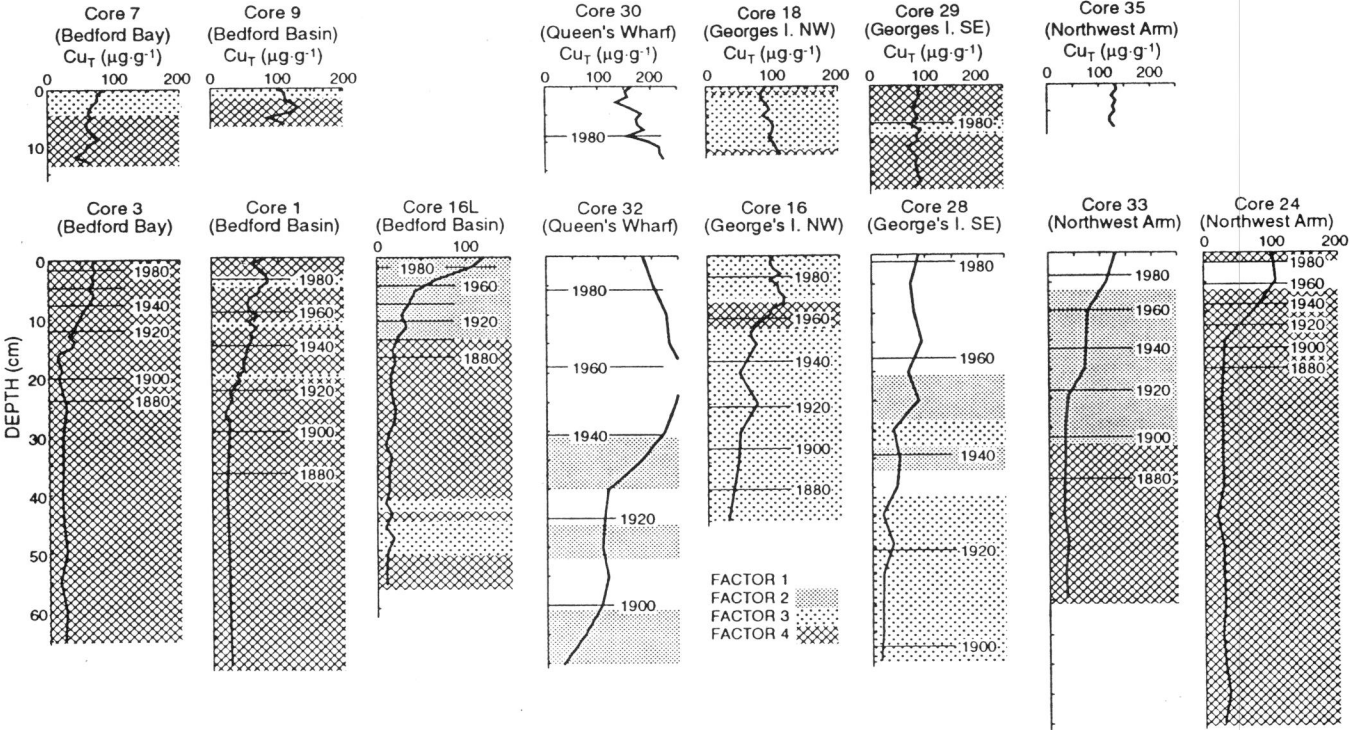

Figure 18 *Profiles of sediment cores illustrating the concentration of total copper with depth in dated sediments. Dominant geochemical factors representing dominant contamination processes are identified for depth zones in each core. Factor 1 is primary contamination from sewage; factor 2 is post-depositional, secondary contamination from urban drainage and storm runoff; factor 3 represents diagenetic alteration of metals from sewage and surface drainage; factor 4 is associated with diagenetic remobilization of metals from subsurface layers and is associated with areas of methane (see also Fig. 16).*

tions, such as may occur when the sediments are disturbed either by dragging of ships' anchors, or by dredging, then remobilization of contaminants is likely.

A WASTE-WATER TREATMENT FACILITY FOR HALIFAX HARBOUR

Many aspects of the geoscience investigations described above can be applied to the problem of developing a new waste-water treatment facility for the city of Halifax. In proposing and planning such facilities, environmental engineers and managers should be aware of information provided by the scientific investigations. On the basis of this knowledge, clearly stated objectives are required that can be understood and supported by the public. This was not done well in the case of Halifax Harbour. Establishment of Halifax Harbour Cleanup Incorporated (HHCI), to carry out design and environmental studies related to the waste-water facilities, created an expectation that the objective was to clean up the harbour. At best, the objective of the HHCI was to design a contamination abatement system by treating some of the raw sewage. It was never made clear to the public that the proposed sewage treatment facility would only address part of the public environmental concern for Halifax Harbour. These concerns can be classified as being: (1) aesthetics, (2) habitat restoration or preservation, and (3) public health. The proposed primary treatment facility would remove about 70% of the particulate matter, but less than 50% of contaminant metals. Such a facility would thus address mainly the concerns with aesthetics, in that particulates, especially the large floatables, could be removed during normal discharge conditions. The waste-water treatment facilities would do little to restore habitat conditions because these have been altered by many years of waste discharge in the harbour. Likewise, there would be little change in the human health risk because of continued discharge of contaminants, and because the large reservoir of contaminants in the harbour sediments would likely continue to release some contaminants.

Sewage Transport Paths

Problems regarding the siting of waste-water management facilities and outfalls often can be reduced essentially to predicting where sewage particulates, with their associated contaminants, will eventually be deposited. For Halifax Harbour, the majority of the evidence suggested that the harbour trapped sediments and only a small particulate fraction was transported seaward. There is a general tendency for the finer sedimentary particles on the bottom to move toward the head of the harbour, *i.e.*, toward Bedford Basin. Sewage particles, which enter the harbour waters in the surface layer, initially would be carried toward the shelf; however, as the sewage particles sink, they are caught up in the deeper inflow and move back up the harbour. Sewage-derived sediments are thus confined to the inner harbour, the Narrows and Bedford Basin. As a result, decisions regarding the location of effluent outfalls from waste-water treatment facilities in the sewage management system for Halifax Harbour were designed to enhance containment and deposition of effluent particles in the inner harbour (Halifax Harbour Task Force and Panel reports).

Given a containment philosophy for the choice of the outfall location, it became a rather easy task to locate the outfall in the inner harbour. Geotechnical characteristics required a suitable foundation for a marine outfall together with a location in an area of non-deposition so as not to bury the outfall. The preferred outfall location was chosen to the east of

Georges Island, at the Georges Island moat area in the inner harbour where sewage would be contained and deposited. This area is the site of the largest and thickest deposit of mud recording deposition and non-erosion (see above).

The decision to site a sewage outfall in the inner harbour is unconventional as outfalls are often chosen as far out at sea as possible, as is the case for Boston Harbour and the city of Victoria. However, long outfalls at sea simply export the problems to distal fishing and recreational zones. Earlier recommendations for a sewage treatment plant location at Sandwich Point and an adjacent outfall were based on little or no marine data. From new information, it is clear that the Sandwich Point location is one of the worst possible areas for a marine outfall. The large sedimentary furrows found directly off this area indicate that sewage discharge would periodically result in storm-driven mass transport directly up the harbour to the entrance of the Northwest Arm, an important recreational zone. It would also directly impact Pleasant Shoal, a large lobster-fishing ground.

Current Status of the Waste-water Management Plan

On March 31, 1995, funding agreements for the Halifax Harbour Cleanup Project between the federal government and the Nova Scotia provincial government expired and the project has been terminated. The Cleanup Corporation is being dismantled and no agency representative of all government stakeholders will exist after this time. In the first few months of 1995, there was a flurry of proposals to move a number of the highly visible and odorous sewage outfalls a few hundred metres out into the central harbour. These measures were being proposed simply to satisfy aesthetic concerns.

The final chapter in the clean-up of Halifax Harbour has not been written, but it appears that an opportunity to control waste from sewers has been missed. However, a large scientific data base has been assembled on which to base future environmental management decisions.

DISCUSSION

It is clear from research in Halifax Harbour that future studies in other similar areas should first undertake geophysical remote-sensing surveys (seismic reflection and side-scan sonar). These data sets can be interpreted to produce a wide variety of geological and thematic maps, in order to determine where appropriate and representative samples should be collected to ground truth the acoustic signatures, to define anomalous sediment distributions and seabed features including hazards; and to have an initial assessment of dynamic processes indicative of sediment transport. Similar observations are made in Chapters 15, 16, 18 and 19).

The most desirable primary tool for so-called first-look surveys is multi-beam bathymetry. Shadowgrams produced from these data can greatly enhance the interpretation of subtle dynamic features to further refine the selection of sample locations. Subsurface seismic reflection data are essential to measure sediment thickness, to outline subsurface stratigraphy, and to choose core locations. Critical relationships often exist between subsurface features and seabed characteristics which can be enhanced by the integration of seismic reflection and side-scan sonar data. The seabed can provide important clues as to subsurface processes that must be further investigated. Core locations should be chosen away from areas where significant erosion and sediment mixing occur, such as produced by ship anchoring and propeller scouring.

When working near urban centres, the anthropogenic

imprint can be overwhelming, making the interpretation and sampling of natural seabed materials and features very difficult. The widespread distribution of anchor marks limits areas where undisturbed sediments can be found and cored. Sample locations must be carefully selected taking this type of information into consideration. It is important to understand the historical aspects of waste discharge to harbours, and many archival agencies, museums, societies and special interest groups such as divers, industrial organizations and military associations can be valuable sources of information.

We lacked direct studies of sediment transport under conditions of waves and currents, and had to rely on other methods to assess the long-term transport pathways and locations of sediment sinks. Methods such as the use of grain size trends to define net sediment transport pathways are limited in areas like Halifax Harbour where relict sediments reflect glacial and post-glacial transgressive environments which are not in equilibrium with the present dynamics of the system. Geochemical anomaly distributions relative to input points greatly assisted in understanding transport directions. Other features, such as the location of sediment depocentres, scour moats, obstacle-induced sediment drifts, sedimentary furrows, sand bedforms, and regional distributions of hard seabeds, helped to infer the net direction of sediment transport.

ACKNOWLEDGEMENTS

We thank the captains and crews of the *C.S.S. Navicula*, *M.V. Frederick R. Creed*, *F.C.G. Smith*, *C.S.S. Hudson*, and the *C.S.S. Dawson* for their assistance in collection of samples and data. We thank Robert Miller for his assistance in field operations, map compilations and initial illustration designs. C. Schafer, C. Amos, D. Forbes, J. Syvitski and R. Cranston provided helpful advice in several geoscience interpretation aspects. We thank members of the Halifax Harbour Task Force and the Federal Scientific Advisory Committee for support and encouragement of the geoscience research program. Finally, we appreciate the reviews of an early draft version of the manuscript by Brian MacLean and Al Grant.

REFERENCES

Amos, C.L., 1990, Modern sedimentary processes, Chapter 11, *in* Keen, M.J. and Williams, G.L., eds., Geology of the Continental Margin of Eastern Canada: Geological Survey of Canada, Geology of Canada, v. 2, p. 609-673. [*also* Geological Society of America, The Geology of North America, v. I-1]

Amos, C.L. and King, E.L., 1984, Bedforms of the Canadian eastern seaboard: A comparison with global occurrences: Marine Geology, v. 57, p. 167-208.

ASA, 1986, The Halifax Inlet water quality study, phase 2: ASA Consulting Ltd., Dartmouth, NS, 136 p.

ASA, 1990, Residual circulation in Halifax inlet and its impact on water quality: Report for Nova Scotia Department of Environment, ASA Consulting Ltd., Dartmouth, NS, 245 p.

Buckley, D.E., 1988, Geochemistry, *in* Piper, D.J.W., ed., Cruise report HUDSON 88-010: Atlantic Geoscience Centre, Bedford Institute of Oceanography, Dartmouth, NS, p. 45-49.

Buckley, D.E., 1991, Deposition and diagenetic alteration of sediment in Emerald Basin, the Scotian Shelf: Continental Shelf Research, v. 11, p. 1099-1122.

Buckley, D.E., 1994, 25 years of environmental assessment of coastal estuarine systems: lessons for environmental quality management, *in* Wells, P.G. and Ricketts, P.J., eds., Coastal Zone Canada '94, "Cooperation in the Coastal Zone": Coastal Zone Canada Association, Bedford Institute of Oceanography, Dartmouth, NS, v. 3, p. 1304-1340.

Buckley, D.E., 1995, Sediments and environmental quality of the Miramichi Estuary: new perspectives, *in* Chadwick, E.M.P., ed., Water, Science, and the Public: The Miramichi Ecosystem: Fisheries and Aquatic Sciences, Canadian Special Publication, v. 123, p. 179-190.

Buckley, D.E. and Cranston, R.E., 1991, The use of grain size information in marine geochemistry, *in* Syvitski, J.P.M., ed., Principles, Methods, and Application of Particle Size Analysis: Cambridge University Press, New York, p. 311-331.

Buckley, D.E., Fitzgerald, R.A., Winters, G.V., LeBlanc, K.W.G. and Cranston, R.E., 1991, Geochemical data from analyses of sediments and pore waters obtained from cores collected in Halifax inlet, *M.V. Fredrick Creed* cruise '90 and *Hudson* cruise 89-039 ('90): Geological Survey of Canada, Open File Report 2410, 116 p.

Buckley, D.E. and Hargrave, B.T., 1989, Geochemical characteristics of surface sediments, *in* Nicholls, H.B., ed., Investigations of Marine Environmental Quality in Halifax Harbour: Fisheries and Aquatic Sciences, Canadian Technical Report 1693, p. 9-36.

Buckley, D.E., Hargrave, B.T. and Mudroch, P., 1989, Geochemical data from analyses of surface sediments obtained from Halifax inlet: Geological Survey of Canada, Open File Report 2042, v. 1 and v. 2, 24 p.

Buckley, D.E., MacKinnon, W.G., Cranston, R.E. and Christian, H.A., 1994, Problems with piston core sampling: mechanical and geochemical diagnosis: Marine Geology, v. 117, p. 95-106.

Buckley, D.E., Smith, J.N. and Winters, G.V., 1995, Accumulation of contaminated metals in marine sediments of Halifax Harbour, Nova Scotia: environmental factors and historical trends: Applied Geochemistry, v. 10, p. 175-195.

Buckley, D.E. and Winters, G.V., 1992, Geochemical characteristics of contaminated surficial sediments in Halifax Harbour: impact of waste discharge: Canadian Journal of Earth Sciences, v. 29, p. 2617-2639.

Cameron, H.L., 1949, Faulting in the vicinity of Halifax, Nova Scotia: Proceedings of the Nova Scotia Institute of Science, v. XXII, Part 3, p. 1-15.

Cook, N.H., 1995, Toxicity of Halifax Harbour sediments: an evaluation of the Microtox® solid phase bioassay: unpublished M. Env. Sc. thesis, Dalhousie University, Halifax, NS, 139 p.

Courtney, R., 1993, Halifax Harbour bathymetric morphology: Geological Survey of Canada, Open File Report 2637, colour map sheet.

Courtney, R. and Fader, G.B.J., 1994, A new understanding of the ocean floor through multibeam mapping: Science Review 1992 & '93, Bedford Institute of Oceanography, Halifax Fisheries Research Laboratory, St. Andrews Biological Station, Department of Fisheries and Oceans, Scotia-Fundy Region, p. 9-14.

Dyer, K.R., 1970, Linear erosional furrows in Southampton Water: Nature, v. 255, p. 56-58.

Edgecombe, R.B., 1994, Paleoenvironmental analysis of Halifax Harbour: sedimentology, paleoclimatology and Holocene sea level: unpublished B.Sc. Hons. thesis, Dalhousie University, Halifax, NS, 76 p.

Fader, G.B.J., 1991, Gas-related sedimentary features from the eastern Canadian continental shelf: Continental Shelf Research, v. 11, p. 1123-1153.

Fader, G.B.J., 1994, The marine geological setting and seabed impact of the 1917 explosion of the *Mont Blanc* in Halifax Harbour: Proceedings of the 75th Conference on the 1917 Explosion in Halifax Harbour, Special Publication of the Goresbrook Institute, St. Mary's University, Halifax, NS, p. 345-364.

Fader, G.B.J. and Miller, R.O., 1992, Cruise report *Navicula* 1990-010, Halifax Harbour, May 22-June 8, 1990: Geological Survey of Canada, Open File Report 2445, 26 p.

Fader, G.B.J., Miller, R.O. and Pecore, S.S., 1991, The marine geology of Halifax Harbour and adjacent areas: Geological Survey of Canada, Open File Report 2384, v. 1 and v. 2, 23 p.

Fader, G.B.J., Miller, R.O. and Pecore, S.S., 1994, Sample control, anchor marks, anthropogenic features and lacustrine sediments of Halifax Harbour: Geological Survey of Canada, Open File Report 2958, 32 p.

Fader, G.B.J., Miller, R.O., Stea, R.R. and Pecore, S.S., 1993, Cruise report 91-018, *C.S.S. Dawson*: operations on the inner Scotian Shelf: Geological Survey of Canada, Open File Report 2633, 30 p.

Fader, G.B.J. and Petrie, B., 1991, Halifax Harbour — How the currents affect sediment distributions, in Smith, T.E., ed., Science Review, 1988 and 1989: Halifax Research Laboratory and St, Andrews Biological Station, Scotia Fundy Region, Department of Fisheries and Oceans, Bedford Institute of Oceanography, p. 31-35.

Faribault, E.R., 1908, Province of Nova Scotia, Halifax County (City of Halifax Sheet No. 68) Department of Mines, Geological Survey Branch, Map.

Fitzgerald, R.A., Winters, G.V., Buckley, D.E. and LeBlanc, K.W.G., 1989, Geochemical data from analyses of sediments and pore water obtained from piston cores and box cores taken from Bedford Basin, LaHave Basin, Emerald Basin and the slope of the southern Scotian Shelf, *Hudson* cruise 88-010: Geological Survey of Canada, Open File Report 1984, 47 p.

Fitzgerald, R.A., Winters, G.V., LeBlanc, K.W.G., Buckley, D.E. and Cranston, R.E., 1991, Geochemical data from analyses of sediments and pore waters obtained from cores collected in Halifax inlet, *C.S.S. Navicula* cruise 90-010: Geological Survey of Canada, Open File Report 2449, 106 p.

Flood, R.D., 1980, Deep-sea sedimentary morphology: Modelling and interpretation of esho-sounding profiles: Marine Geology, v. 38, p. 77-92.

Flood, R.D., 1981, Distribution, morphology and origin of sedimentary furrows in cohesive sediments; Southampton Water: Sedimentology, v. 28, p. 511-529.

Flood, R.D., 1983, Classification of sedimentary furrows and a model of furrow initiation and evolution: Geological Society of America, Bulletin, v. 94, p. 630-639.

Fournier, R.O., 1990, Final report of the Halifax Harbour Task Force, submitted to the Minister of the Environment, The Honourable John Leefe: 84 p.

Gearing, J.N., Buckley, D.E. and Smith, J.N., 1991, Hydrocarbon and metal contents in a sediment core from Halifax Harbour: a chronology of contamination: Canadian Journal of Fisheries and Aquatic Sciences, v. 48, p. 2344-2354.

Geomarine Associates Ltd., 1975, Halifax Harbour bottom survey: Geological Survey of Canada, Open File Report 283, p. 1-100.

Hargrave B.T., Peer, D.L. and Wiele, H.F., 1989, Benthic biological observations, in Nicholls, H. B., ed., Investigations of Marine Environmental Quality in Halifax Harbour: Fisheries and Aquatic Sciences, Canadian Technical Report 1693, p. 37-45.

Hollister, C.D., Flood, R.D., Johnson, D.A., Lonsdale, P.F. and Southard, J.B., 1974, Abyssal furrows and hyperbolic echo traces on the Bahama Outer Ridge: Geology, v. 2, p. 395-400.

Hovland, M. and Judd, A.G., 1988, Seabed Pockmarks and Seepages: Graham and Trotman Inc., Sterling House, London, UK, 293 p.

Keen, M.J. and Piper, D.J.W., 1976, Kelp, methane, and an impenetrable reflector in a temperate bay: Canadian Journal of Earth Sciences, v. 13, p. 312-318.

Keppie, J.D., 1982, Tectonic map of the province of Nova Scotia, Department of Mines and Energy, Halifax, NS.

King, L.H., 1970, Surficial geology of the Halifax-Sable Island map area: Marine Science Paper 1, 16 p.

King, L.H. and Fader, G.B.J., 1986, Wisconsinan glaciation of the southeastern Canadian Continental Shelf: Geological Survey of Canada, Bulletin 363, 72 p.

King, L.H. and MacLean, B., 1970, Pockmarks on the Scotian Shelf: Geological Society of America, Bulletin, v. 81, p. 3141-3148.

Knebel, H.J. and Scanlon, K.M., 1985, Sedimentary framework of Penobscot Bay, Maine: Marine Geology, v. 65, p. 305-324.

Lawrence, D.J., 1989, Physical oceanography and modelling in Halifax Harbour: a review, in Nicholls, H.B., ed., Investigations of Marine Environmental Quality in Halifax Harbour: Fisheries and Aquatic Sciences, Canadian Technical Report 1693, p. 54-63.

Lay, G.F.T., Fader, G.B.J. and Buckley, D.E., 1993, Environmental implications from disturbance of contaminated marine dredge spoils in Halifax Harbour, Nova Scotia: 1993 Canadian Coastal Conference, Coastal Zone Engineering Program, National Research Council, Ottawa, ON, p. 253-269.

LeBlanc, K.W.G., Fitzgerald, R.A., Winters, G.V., Buckley, D.E. and Cranston, R.E., 1991, Geochemical data from analyses of sediments and pore waters obtained from cores collected in Halifax inlet: Geological Survey of Canada, Open File Report 2345, 116 p.

Loncarevic, B.D., Courtney, R.C., Fader, G.B.J., Giles, P.S., Piper, D.J.W., Costello, G., Hughes-Clark, J.E. and Stea, R.R., 1994, Sonography of a glaciated continental shelf: Geology, v. 22, p. 747-750.

Miller, A.A.L., Mudie, P.J. and Scott, D.B., 1982, Holocene history of Bedford Basin, Nova Scotia; foraminifera: Canadian Journal of Earth Sciences, v. 19, p. 2342-2367.

Miller, R.O. and Fader, G.B.J., 1989, Cruise report 88-018(A) Phase 1, *F.R.V. Navicula*, Halifax-Sambro, Nova Scotia, May 26-June 2, 1988: Geological Survey of Canada, Open File Report 2039, 22 p.

Miller R.O., Fader, G.B.J., Buckley, D.E., 1990, Cruise report 89-009, Phase A – Halifax Inlet, *F.R.V. Navicula*: Geological Survey of Canada, Open File Report 2242, 66 p.

Nicholls, H.B., 1989, ed., Investigations of marine environmental quality in Halifax Harbour: Fisheries and Aquatic Sciences, Canadian Technical Report 1693, 83 p.

OceanChem Group Ltd., 1988, Halifax Harbour sediment quality study, Phase 1: sample collection: Environment Canada, Dartmouth, NS, 26 p.

Petrie, B. and Yeats, P., 1990, Simple models of the circulation, dissolved metals, suspended solids and nutrients in Halifax Harbour: Water Pollution Research Journal of Canada, v. 25, p. 325-346.

Prouse, N.J. and Hargrave, B.T., 1987, Organic enrichment of sediments in Bedford Basin and Halifax Harbour: Fisheries and Aquatic Sciences, Canadian Technical Report 1571, 36 p.

Raddall, T.H., 1965, Halifax, Warden of the North: Doubleday, Garden City, NJ, 340 p.

Stea, R.R., Pecore, S.S. and Fader, G.B.J., 1993, Quaternary stratigraphy and placer gold potential of the inner Scotian Shelf, Nova Scotia: Department of Natural Resources, Mines and Energy Branches, Paper 93-2, 62 p.

Tay, K.L., Doe, K.G., Wade, S.J., Vaughan, J.D.A., Berigan, R.E. and Moore, M.J., 1991, Biological effects of contaminants in Halifax Harbour sediments: 17th Annual Aquatic Toxicity Workshop, Vancouver, BC, Fisheries and Aquatic Sciences, Canadian Technical Report 1774, p. 383-426.

Uthe, J.F., Chou, C.L., Prouse, N.J. and Musial, C.J., 1989, Heavy metal, polycyclic aromatic hydrocarbon, and polychlorinated biphenyl concentrations in American lobsters (*Homarus Americanus*) from Halifax Harbour, in Nicholls, H.B., ed., 1989, Investigations of marine environmental quality in Halifax Harbour: Fisheries and Aquatic Sciences, Canadian Technical Report 1693, Appendix A, p. 64-68.

Winters, G.V. and Buckley, D.E., 1992, Factor analyses as a method of evaluating sediment environmental quality in Halifax Harbour, Nova Scotia: Geological Survey of Canada, Paper 92-1D, p. 165-171.

Winters, G.V., Buckley, D.E., Fitzgerald, R.A. and LeBlanc, K.W.G., 1991, Inorganic geochemical data for surface sediments from Halifax Inlet: Geological Survey of Canada, Open File Report 2389, 64 p.

18. Management of the Scarborough Bluffs Shoreline of Lake Ontario

Robert B. Nairn

Baird & Associates, 221 Lakeshore Road E., Oakville, Ontario L6J 1H7

Nigel B. Cowie

Metropolitan Toronto and Region Conservation Authority, 5 Shoreham Drive, Downsview, Ontario M3N 1S4

SUMMARY

Along eroding shorelines, urban development ultimately results in pressures for the implementation of shoreline stabilization measures. All approaches to shoreline stabilization, whether they consist of soft approaches (such as beach replenishment) or hard approaches (such as seawall and armour stone revetment), alter the natural ecosystem of the coastal or shoreline zone. The impacts of development can alter the fish and wildlife communities in the coastal zone, change the physical nature of the coastline or shoreline, and modify water quality and quantity. Some of these changes are desirable and others are not; many are irreversible. This conflict of interest is particularly acute where intense development occurs along a shoreline or coastline which, in its natural state, is eroding rapidly and irreversibly. Scarborough Bluffs, along the northern shore of Lake Ontario, provides an example of an acute shoreline management problem.

INTRODUCTION

Scarborough Bluffs (Figs. 1, 2) is one of the most imposing sections of cliff shoreline anywhere on the Great Lakes. The rapid erosion of this shoreline has been of great interest and concern ever since the Toronto area was first settled by Europeans in the late eighteenth century. The bluffs extend for about 15 km northeast of the present boundary between the city of Toronto and the city of Scarborough near the R.C. Harris Filtration Plant. They reach a height of just over 100 m at the east end of Bluffers Park. A map of the existing conditions, indicating shore protection and recreational parks, is given in Figure 3.

The bluffs have been irreversibly altered by human development, and with this change, our understanding of the biophysical processes which have shaped and continue to shape this feature has also evolved. This integrated progression of human influences and understanding of the environment, sometimes in step and sometimes out of step, will be the focus of this paper.

The paper is divided into two sections. The first provides a description of the shoreline management efforts along Scar-

borough Bluffs and neighbouring shoreline for five periods since the initial European settlement of the area. The second section of the paper presents a description of the state of the art understanding of the biophysical processes for eroding bluff shorelines of this type.

EARLY INTEREST IN SHORELINE MANAGEMENT (1793 to 1900)

The location of Toronto was selected by Lord Simcoe owing to the excellent natural harbour provided by the presence of the Toronto Islands. The harbour as it was in the late eighteenth century is shown in Figure 4.

As early as 1840, there were concerns for the future function of the natural harbour. These concerns were initially related to the perceived threat of closure of the Western Gap, which was the only entrance to the harbour at that time. In 1852, an April storm opened a breach in the neck of the island near the present-day location of the Eastern Gap. The pres-

Figure 1 *Scarborough Bluffs looking west from East Point (January, 1989; courtesy MTRCA). See Figure 2 for location.*

ence of this gap was thought by some to threaten the future of the harbour. These problems had become so acute in 1854 that the Toronto Harbour Commission (THC) solicited reports on the preservation and future improvements to the harbour, with the offer of approximately 300 pounds sterling in awards for the best reports. The top four reports by Hind, Fleming, Tully and Richardson were published by the authority of the Toronto Harbour Commissioners (1854) in the *Canadian Journal*.

These four reports provide some of the first published insights into the nature of the geomorphology of both Scar-

borough Bluffs and the Toronto Islands and the interrelationship between these features. The understanding that the Toronto Islands feature (which is a classical compound recurved spit) owed its existence to sediments derived from erosion of Scarborough Bluffs may be attributed to an earlier paper by a civil engineer, Sandford Fleming (later Sir Sandford), read before the Canadian Institute in 1850 and published in the *Canadian Journal* in 1853 (Vol. II). An original illustration of Fleming's theory of how the ongoing erosion of Scarborough Heights has resulted in the formation of the Toronto Islands is given in Figure 5 (adopted from his *Canadi-*

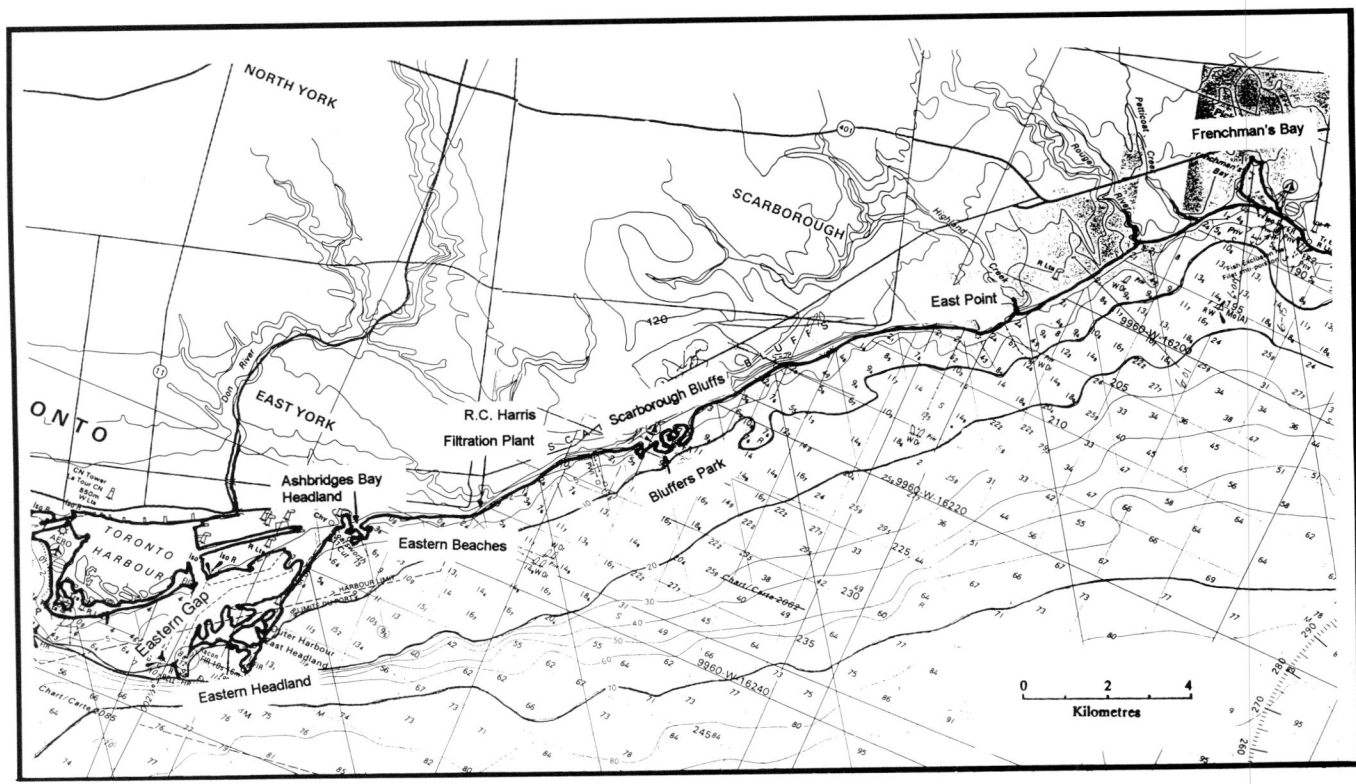

Figure 2   *Scarborough Bluffs shoreline. Bathymetry in fathoms; inland elevation in metres above sea level (m asl).*

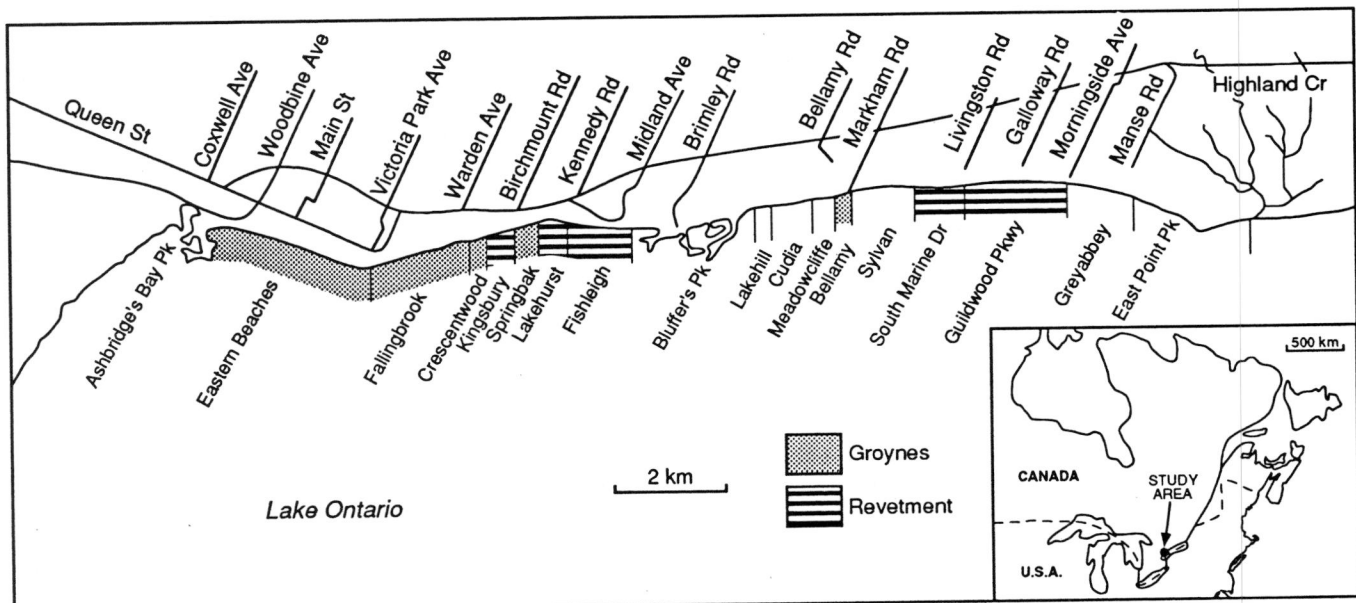

Figure 3   *Plan showing shoreline protection projects and parks along the bluffs (courtesy MTRCA).*

an Journal report in 1854). This theory was supported by Fleming's finding that the net direction of littoral drift or alongshore transport of sand was from east to west based on the observations of beach deposits on the east side of natural groynes (*e.g.*, stream mouths and fallen trees). Fleming correctly surmised that westward directed net transport of sand was a result of the longer lake fetches to the east, despite the fact that the prevailing winds are from the southwest. Fleming reports that farmers residing near the bluffs had measured recession rates of 10 to 12 feet (3.0-3.7 m) annually in the early 1850s. Fleming also recognized the important distinction between the erosion of clay shores, such as Scarborough Bluffs, and sandy shores. Where clay banks are eroding, the clay fraction of the eroded material is transported offshore into less energetic deep water, and is, therefore, lost from the littoral zone.

Henry Y. Hind, a professor of chemistry in the University of Trinity College, presented an interesting hypothesis that the clearing of trees for agricultural purposes resulted in considerable acceleration of recession rates along Scarborough Bluffs. Historically, the large number of trees that would fall to the base of the bluffs acted as a multitude of groynes, helping to sustain beaches along the base of the bluffs. The absence of the trees also increased surface runoff over the bluff face and consequent gullying. One of the earliest known human influences along the Scarborough Bluffs shoreline was the operation referred to as stone hooking (Snider, 1930). Over the latter part of the 1800s and the early 1900s, ships' crews hooked boulders and large slabs of bedrock from the shoreline and lakebed. The material was delivered to Toronto Harbour for use as building materials in roads and buildings. The net result of the accelerated erosion rates was the increased sedimentation of the western entrance to the harbour.

A permanent entrance at the Eastern Gap was eventually undertaken by the Toronto Harbour Commissioners (THC) in 1892. In recent investigations of the evolution of the Toronto Islands over the last 100 years, Baird & Associates (1994a, b) point to the construction of the Eastern Gap jetties, and the subsequent annual dredging of the created channel, as the most important anthropogenic factor leading to the massive erosion of the beaches along the south shore of the Toronto Islands.

URBAN DEVELOPMENT OF THE TORONTO HARBOUR AREA (1900 to 1945)

With the growth of Toronto as a city, neighbouring rural areas became attractive as locations for summer homes and cottages. Slowly, the agricultural land along the top of the bluffs was converted to country estates, summer cottages, country clubs and golf courses. The summer cottages, in particular, were located close to the bluff crest to take advantage of the dramatic views of Lake Ontario (see Bonis, 1968). Some of these dwellings were built in steep ravines along the bluffs, and all of them have since been removed by erosion. Further advancement in understanding of the bluffs and their evolution is related to the requirement for building supplies such as sand and gravel to support the urban development in the form of road and building construction. This was the impetus for the Ontario Department of Mines (1936) report by A.P. Coleman which provided the most comprehensive treatise of the time on both modern and Pleistocene processes.

An estimate of the sediment budget for the littoral cell between East Point and the Toronto Islands had been made earlier by Coleman (1913) based on the amount of sediment in the Toronto Islands feature and the annual rate of deposition next to the east jetty of the Eastern Gap (see Fig. 6) and corroborated by an estimate of the recession rates along Scarborough Bluffs. The average amount of sediment trapped next to the east jetty was 42,000 cubic yards per year

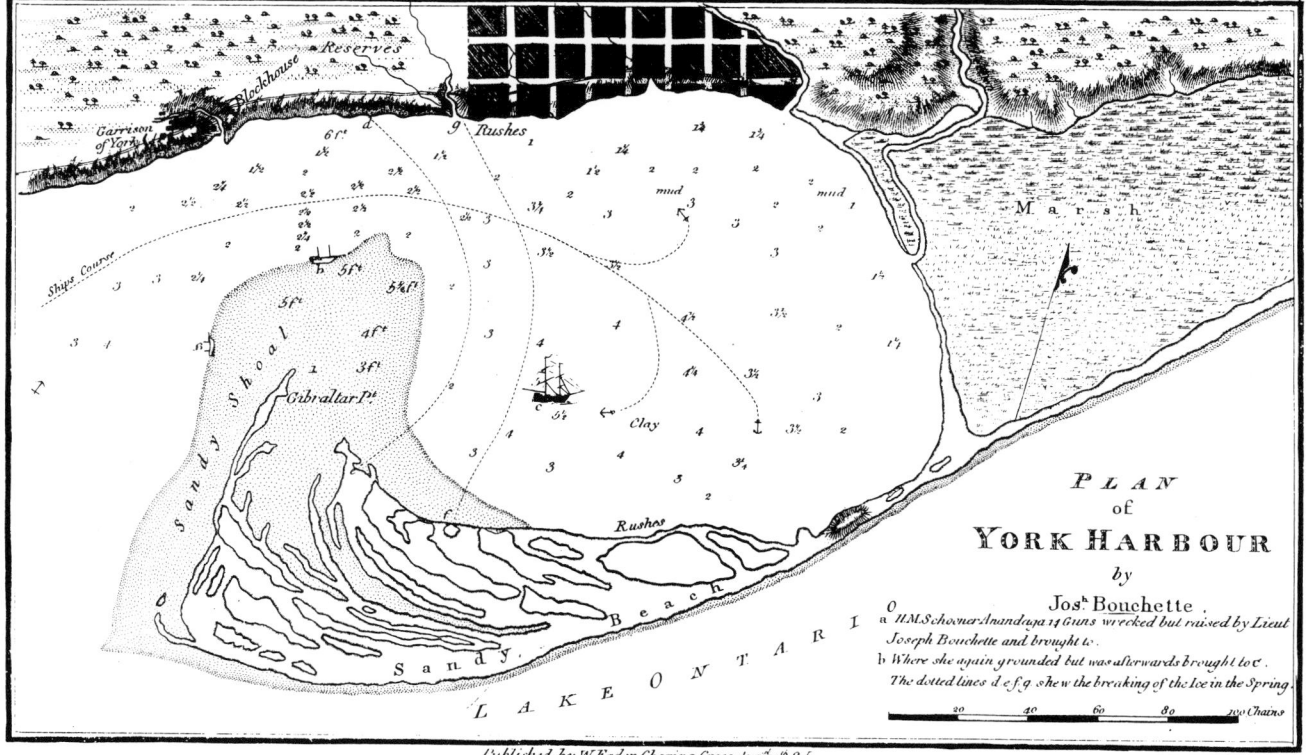

Figure 4   *Bouchette's map of York (Toronto) Harbour (1815). At that time, the only entrance was the Western Gap; the Eastern Gap (Fig. 2) was permanently opened in 1892. The large marsh area was subsequently named Ashbridge's Bay. The bay became heavily polluted in the late nineteenth century and was landfilled in the 1920s (see Chapter 2).*

(about 32,000 m³·a⁻¹) between 1892 and 1905 (see Fig. 6). The total volume of the islands deposit was determined by Coleman to be 337,000,000 cubic yards (about 260,000,000 m³) and, based on the calculated annual supply rate of 42,000 cubic yards (about 32,000 m³), the age of the islands deposit was estimated to be 8000 years. The age of the deposit could also be related to the distance that the bluffs had receded since the initiation of erosion at the present-day water level. The bluffs were estimated to be receding at an average rate of 1.6 feet per year (0.5 m·a⁻¹) based on a comparison of surveys between 1862 and 1912. Therefore, the 13,000 feet (3960 m) distance between the outer edge of the wave-cut shelf off-shore of the bluffs and the present shoreline would have been traversed by an eroding bluff face over a period of 8000 years. In the absence of any radiometric age-dating, this has proved to be a fairly accurate estimate of the age of the Toronto Islands.

Coleman (1936) showed that the rate at which the bluffs

recede was accelerated during years with high lake levels such as 1929 and 1930. Other high lake level periods occurred in 1788, 1838, 1853, 1863, 1887 and 1916-1919. More recently, high lake levels have occurred in 1952, 1973 and 1993. Later in this paper, the importance of low lake level years in the continual retreat of the bluffs will also be highlighted. Since the beginning of continuous and accurate lake level measurements at Toronto in 1916, the range between the record monthly high in June of 1952 and the record monthly low in December of 1934 is exactly 2 m. Figure 7 provides an indication of the fluctuation in monthly mean lake levels in Toronto Harbour between 1860 and 1990.

The origin of the Pleistocene glacial sequence exposed along the Scarborough Bluffs (Fig. 8) was also described by Coleman (1936).

The development of the Toronto Harbour area for industry, trade and recreation required large quantities of fill, much of which was taken by dredging of the lakebed. For example, the large Ashbridge's Bay wetland (Fig. 4) was entirely filled, starting in 1910, by the THC as part of their original Waterfront Plan. At the time, the stagnation of water in Ashbridge's Bay, no doubt influenced by the input of sewage from the city, was believed to represent a health hazard. The infilling of the wetland is now recognized to have resulted in the irreversible loss of an extremely important part of the ecosystem. It is documented in the "Report on the Select Committee of the Legislature on the Lakes Levels of the Great Lakes, 1953" that, in 1927, Scarborough Township claimed that the massive hydraulic sand dredging undertaken by the THC between 1910 and 1914 to fill the wetland resulted in the depletion of the sandbar offshore of the westerly end of the Scarborough Bluffs. It was alleged that this bar had once provided a natural barrier to wave attack at the base of the bluffs.

URBANIZATION OF THE SCARBOROUGH SHORELINE AND EARLY SHORE PROTECTION EFFORTS (1945 to 1970)

As the post-war population of Toronto increased in the baby boom years, remaining agricultural land, estates and golf clubs were rapidly transformed into residential subdivisions. To maximize the number of lots with a direct view of the lake, the roads of the new communities were built parallel to the bluff crest, examples include Guildwood Parkway, South Marine Drive, Sylvan Avenue and Fishleigh Drive (Figs. 9A, B). As we will see later, it is precisely these areas which have

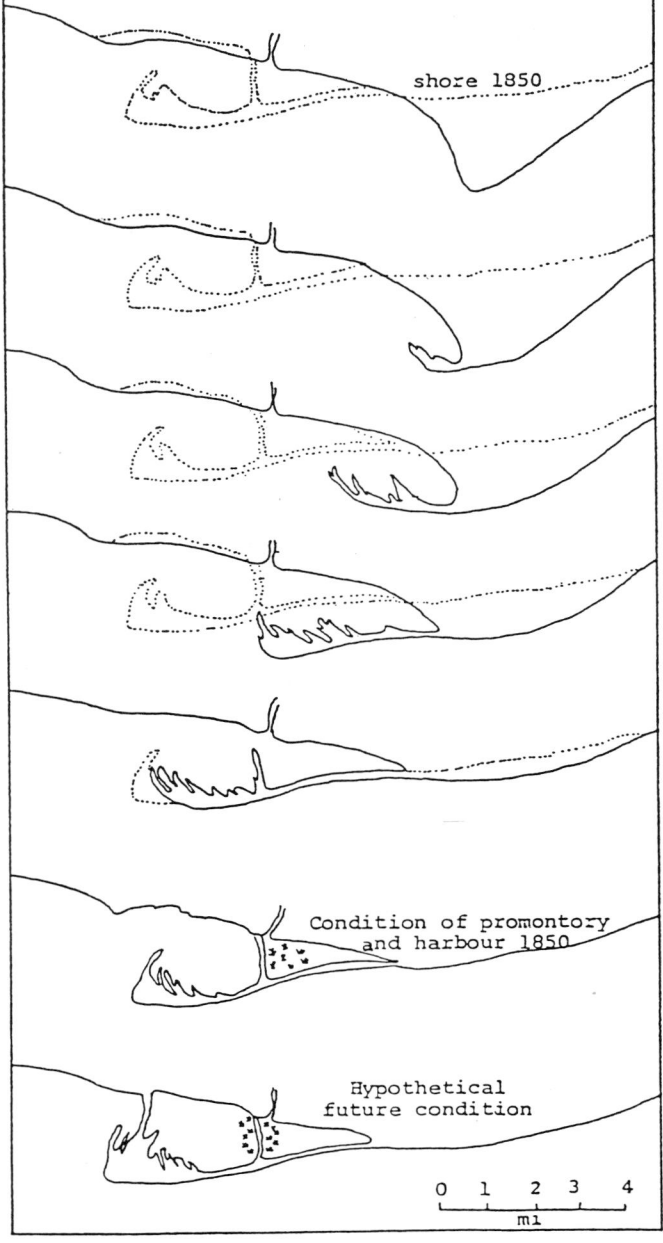

shore 1850

Condition of promontory and harbour 1850

Hypothetical future condition

0 1 2 3 4
ml

Figure 5 *Sir Sandford Fleming's promontory theory on the evolution of the Toronto Islands.*

Figure 6 *Sediment trapping against the eastern jetty of the Eastern Gap in the early 1950s (see Fig. 2). Sediment is moving to the west, i.e., to the left (e.g., Fig. 5).*

required the implementation of massive shore protection along the base of the bluffs in response to continuing bluff recession, as shown in Figure 9C for Fishleigh Drive. In the haste to meet the demand for new homes, careful planning was compromised, and the historic recession rates of the bluffs were disregarded.

Buttle and von Bulow (1986) provide a comprehensive discussion of recent and historic investigations of bluff crest recession rates for Scarborough Bluffs, and these are sum-

marized in Table 1. While these estimates indicate an annual average rate of about 0.4 m·a⁻¹, many landowners along the top of the bluff will attest to the fact that short-term recession rates can be much higher—up to several metres per year. The temporal and spatial variability in the bluff recession rates is most evident in the vicinity of gullies. Certain sections of the bluff are particularly prone to the development of large gullies due to patterns of ground-water flow and sediment liquefaction (see Eyles *et al.*, 1986). The manner in which the rapid

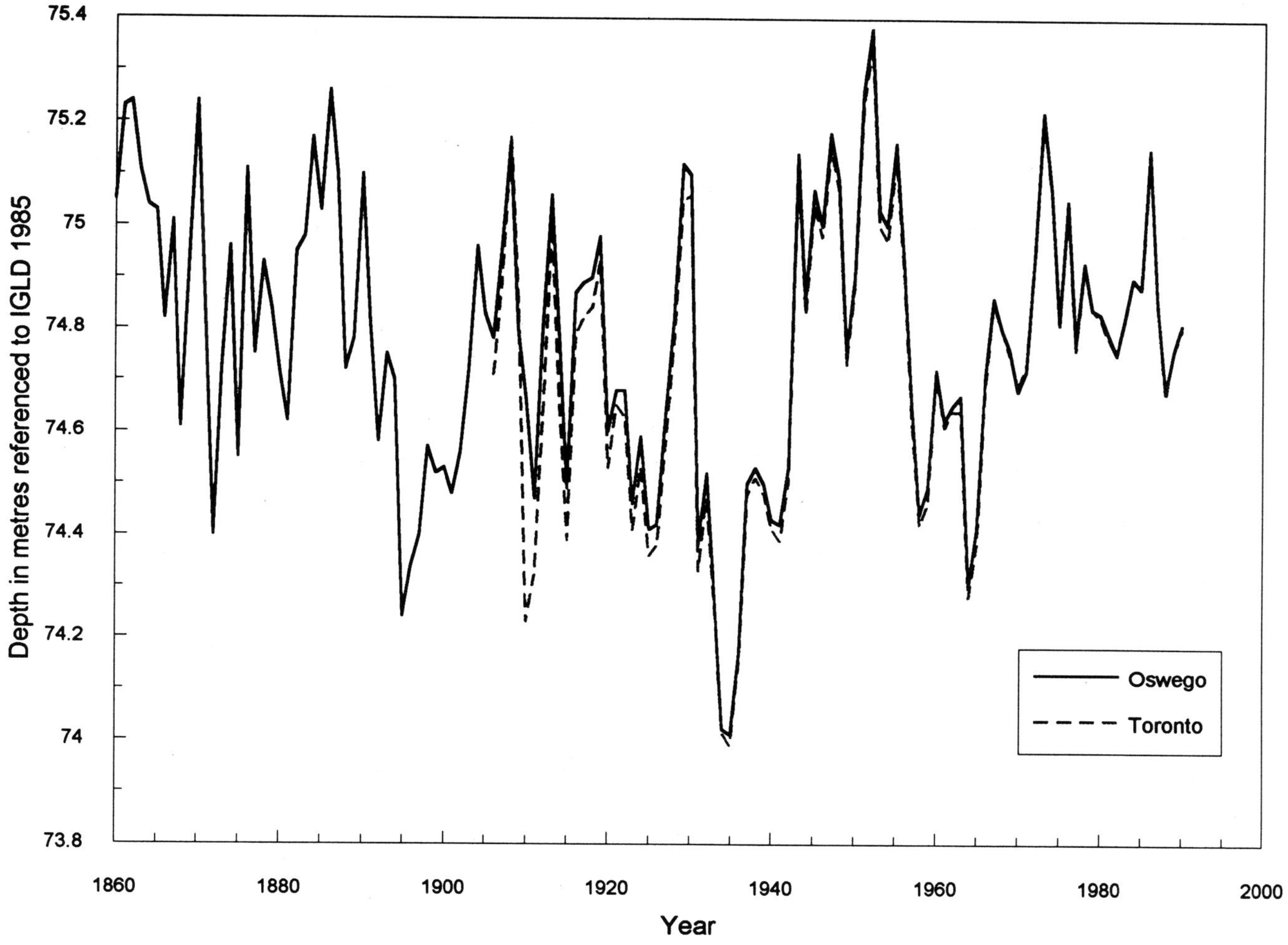

Figure 7 *Lake level records for Toronto Harbour, 1860-1990 in metres above sea level.. Data for Oswego, New York State, are shown for comparison. Record after 1916 is more accurate than those collected earlier. Abbreviation: **IGLD**, International Great Lakes Datum. Mean lake level is 74.6 m.*

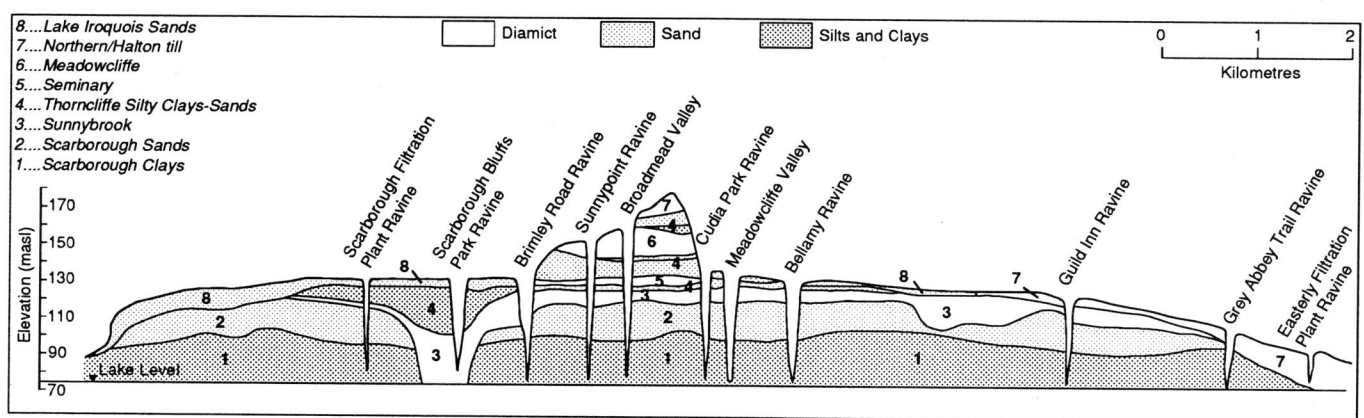

Figure 8 *Geologic cross-section of Scarborough Bluffs (after Coleman, 1936; and Eyles et al., 1985).*

growth of these gullies can threaten dwellings and roads close to the crest is illustrated in Figure 10.

Following the 1952 high lake level period, extensive measures were taken by property owners to protect the bluffs from wave action. Concrete groynes were constructed along the Toronto Hunt Club shoreline. Individual piecemeal attempts at shoreline protection by property owners included the top dumping of old cars, earth, rubble and garbage. The owner of the Guild Inn sank a series of steel barges and dump scows along the base of the bluffs. This was followed by the construction of an access road and filling along the base of the bluffs at this location in the late 1960s. This work has been relatively successful at stabilizing the toe of the bluffs. As a result, the bluffs along this section have stabilized naturally and have become vegetated. This self-stabilization approach, in conjunction with toe protection, became an important model for future shore protection efforts by the Metropolitan Toronto and Region Conservation Authority (MTRCA; Fig. 11).

LARGE-SCALE SHORE PROTECTION EFFORTS (1970 to 1990)

A return of high lake levels was experienced in 1973, and was associated with accelerated erosion damage along the Lake Ontario shoreline. By 1973, the development of subdivisions along Scarborough Bluffs was essentially complete, and the impact of the high lake levels affected many property owners. Some of the earlier attempts at individual shoreline protection were severely damaged or destroyed, and property owners began to realize that larger scale, properly designed and constructed measures were required. There was a concerted effort by the property owners to seek government funding to implement large-scale shoreline protection measures during the 1970s. By this time, the MTRCA was actively implementing the Metropolitan Toronto and Region Waterfront Plan (1967).

Further advancements were made during this time toward the understanding of the Pleistocene geology of the bluffs and the unique nature of the feature in this respect, as a record of the history of the last glacial period (see Karrow, 1967; Lewis and Sly, 1971; Eyles *et al.*, 1985; Fig. 8). This information was valuable to the assessment of the controls on slope stability of the bluffs (Geocon Inc., 1980, 1982; Eyles and Howard, 1988). Also, the uniqueness of the geological feature became a reason for preserving part of the bluffs in an eroding form (*i.e.*, instead of pursuing toe stabilization along the entire length of the bluffs, which would eventually eradicate any open exposures of the bluff face as the slopes became stabilized and overgrown with vegetation).

One of the key policies in the Waterfront Plan was the

A

B

C

Figure 9 (**A**) *Fishleigh Drive before (1944) and* (**B**) *after (1962) development (from Bonis, 1968).* (**C**) *Slope failures at Fishleigh Drive in 1987 (courtesy MTRCA). Urbanization has resulted in accelerated erosion.*

Table 1	Bluff recession rate investigations: Scarborough Bluffs (adapted from Buttle and von Bulow, 1986).	
Author	**Time Interval**	**Average Annual Recession Rate** (m·a⁻¹)
Coleman (1913)	8,000 years before 1913	0.49
Coleman (1932)	1860–1912	0.49
Langford (1952)	1791–1854–1942	0.37
Burt and Roosenboom (1952)	1922–1952	0.39
Bird and Armstrong (1970)	1931–1964	0.31
Bryan and Price (1980)	1952–1976	0.76
Geocon (1980)	1922–1980	0.41
Eyles *et al.* (1985)	8500 years before present	0.58

Shoreline Management Program. The plan required that the waterfront projects of MTRCA form an integrated management system for the entire shoreline, which will limit abnormal erosion at the land-water interface, will enable public access along the water's edge, and will be conducive to beach maintenance. The integrated management system would be based on a knowledge of natural lake processes (and the research required to further understand these processes) and co-operation with other levels of government.

Stabilization Projects

Two of the major constraints to undertaking stabilization projects along the Scarborough shoreline were property ownership and access for construction equipment and materials. Much of the shoreline was privately owned and agreements were required from benefiting property owners to transfer the riparian rights and ownership of the water lot and lower portion of the bluff to MTRCA. In 1979, MTRCA carried out a pilot stabilization project at Crescentwood Road. The project consisted of constructing 900 m of rock core revetment. Equipment was delivered to the site by barge, and materials were delivered by dumping over the edge of the bluff to fill in and arrest the erosion of gullies (Fig. 10). Through annual monitoring of the bluff recession rates, a noticeable general reduction in the rates has been observed due to the self-stabilization of the toe-protected slopes.

Following the successful completion of the Crescentwood project, property agreements were obtained for other sectors of the bluffs which allowed for construction of shore protection works to proceed along Kingsbury Crescent, Springbank Avenue and Lakehurst Crescent (Fig. 3). The works along Kingsbury Crescent comprised a short section of armour stone revetment below a recently filled gully. In 1980, at this location, the MTRCA had acquired three houses that had become endangered by an eroding gully at the corner of Kingsbury Crescent and Harding Boulevard and all were eventually demolished. Again, as at Crescentwood Road, this gully further threatened the street and services and, in response, the City of Scarborough filled the gully with rubble material. This site became another material delivery access point. Because of difficulties in obtaining all the required property agreements, the Kingsbury Crescent project, which started in 1980, is expected to be completed in 1995. An

offshore armour stone revetment has been constructed along the full extent of Kingsbury Crescent. In addition, MTRCA has acquired a total of five houses; four have been demolished and one has been relocated. MTRCA is currently completing slope stabilization works to protect two other houses in this sector. The Kingsbury sector has probably required more resources than any other along the length of the bluffs' shoreline.

In 1982, construction of a groyne and beach system commenced along the Springbank-Lakehurst sector. The bluffs along this sector were not as over-steepened as at Kingsbury and had more vegetation cover. The groyne-beach system, while being similar in cost to a revetment, provides a softer and more natural shoreline treatment (Fig. 12). The core of the groynes and beaches was constructed of broken concrete rubble with smaller sized concrete rubble placed to widen the beaches. The groynes were armoured with 3- to 5-tonne armour stone, with the heads protected by two layers of the armour stone.

Other erosion control projects completed in the 1980s included the Toronto Hunt Club (privately funded) and the Fallingbrook sector (Fig. 3). During the period 1984 to 1991, the South Marine Drive erosion control project was completed along a 1-km section of shoreline at a total cost of $2.5 million. Eyles and Howard (1988) showed how urban storm runoff had contributed to accelerated shoreline erosion along South Marine Drive as a result of enhanced ground-water discharges from the bluffs. An offshore armour stone revetment was constructed and the area behind was backfilled with imported clean earth fill (see Fig. 13). Partial slope stabilization work was also completed. An important component of this project was the upgrading of an existing access road to the shoreline on the Guild Inn property. This access road is integral to the long-term maintenance of all the shore protection works east of Bluffers Park, and is an important link in the continuation of the Waterfront Trail project of the Waterfront Regeneration Trust and its partners.

An important impact of the shore protection measures that were constructed along the base of the bluffs during this period is the ongoing reduction in sediment supply to the Eastern Beaches (*i.e.*, as erosion of the bluffs is halted). The Eastern Beaches Shoreline Management Study, commissioned by MTRCA in 1989 and completed by Sandwell (1991),

Figure 10 *Eroding gully endangering houses on Lakehurst Crescent (November 1983; courtesy MTRCA). This is typical of deep gullies affecting the western bluffs shoreline. For location, see Figure 3.*

Figure 11 *Early bluff toe stabilization efforts at the Guild Inn property (June 1979; courtesy MTRCA). For location, see Figure 3.*

addressed the possible implications of the reduced sediment supply, including a detailed sediment budget analysis. This report recommended that eventually some form of beach nourishment program would have to be implemented to recirculate sediment from where it is trapped at the Ashbridge's Bay headland to the updrift end of the beaches at the R.C. Harris Water Filtration Plant in order to address the problem of a reduced natural supply of sediment from bluff erosion.

With the rapid growth of residential communities came the requirement for public infrastructure to provide municipal drinking water from Lake Ontario (see Chapter 2). The R.C. Harris Water Filtration Plant was built at the westerly boundary of the bluffs. A massive concrete sea wall, constructed in 1933, provides protection to this facility from wave attack. An outfall pipe serves as a groyne extending into the littoral zone.

Further east, the Scarborough Water Treatment Plant was constructed at Fishleigh Drive. A pumping building was constructed on the shoreline at the base of Fishleigh ravine, and a section of vertical steel sheet pile wall was installed for protection in 1921. In 1955, the Highland Creek Water Pollution Control Plant was constructed on the west side of the mouth of Highland Creek. A short section of the natural beach shoreline was replaced with an armour stone revetment to protect a recent expansion to the plant.

At frequent intervals along the bluffs, storm sewer outfalls were constructed by tunnelling vertically down the bluffs and horizontally to a headwall structure at the toe of the bluffs as outlets for storm drainage. Typically, these concrete headwall structures were protected with short sections of vertical steel sheet pile walls. Remediation of contamination arising from a combined sewer outfall carrying domestic sewage and urban storm runoff is described in Chapter 12.

WATERFRONT PARKS (1970 to 1990)

A major component of the Metropolitan Toronto Waterfront Plan of 1967 was the proposed construction of two lakefill parks in the Scarborough area to address the need for a regional parks system to provide direct access to the waterfront. As early as 1967, Scarborough Township was filling into the lake at the base of Brimley Ravine. This work was taken over by the MTRCA in 1970, and over the next 15 years, massive volumes of earth and broken-concrete fill material were placed in the lake to create the 42-ha Bluffers Waterfront Park (see Fig. 14). This newly reclaimed land provided the

added benefit of stabilizing the toe of the bluffs over a 1.8-km section of the shoreline.

Further to the west along the Toronto shoreline, the Ashbridge's Bay Waterfront Park (Fig. 3) was created by lakefilling over the period from 1974 to 1977. Since its completion, the easterly headland of the park has trapped most, if not all, of the sand supplied by the erosion of the bluffs and transported along the Eastern Beaches shore from east to west. A second lakefill park is proposed at East Point (Fig. 3) at the east end of Scarborough Bluffs according to the Waterfront Plan. This project has not commenced and will be subject to further studies.

The potential impact of Bluffers Park on sediment supply to downdrift shorelines was the subject of research by Greenwood and McGillivray (1980) and McGillivray and Greenwood (1978). Both these investigations found that the Bluffers Park headland feature is probably a complete littoral barrier to alongshore transport. Sediment supplied from erosion of the bluffs east of Bluffers Park is trapped by the headland and prevented from continuing along the shore to the west, and particularly as a supply of sediment to the Eastern Beaches. As these investigations occurred after the decision to con-

Figure 12 *Groyne field along the Springbank-Lakehurst sector (October 1992; courtesy MTRCA). For location, see Figure 3.*

Figure 14 *Bluffers Park marina (July 1988; courtesy MTRCA).*

Figure 13 *South Marine Drive armour stone revetment (October 1987; courtesy MTRCA). For location, see Figure 3.*

struct Bluffers Park in the late 1960s, it is unclear how the implications of reducing the sediment supply to the Eastern Beaches were considered in the decision-making process. However, it should be noted that the important connection between natural maintenance of the Toronto beaches and the erosion of the bluffs had been made by Sandford Fleming as early as 1850.

An additional impact of the creation of these large waterfront parks through the placement of fill in the lake is the potential for contamination of the lake water. This issue has been addressed by many, and is summarized in the Regeneration report by the Royal Commission on the Future of the Toronto Waterfront (1991).

Significant efforts were made to improve the knowledge base of the coastal processes in the vicinity of Bluffers Park including detailed sounding surveys to define the local bathymetry and wave hindcasts to describe the wave conditions (*i.e.*, where the historic wave conditions are predicted on an hourly basis through a consideration of hourly wind records and fetches). In addition, Bluffers Park and the neighbouring nearshore zone have been the focus of ongoing monitoring of aquatic habitat as described by MacPherson and Piercy (1993).

AN ECOSYSTEM APPROACH
TO WATERFRONT MANAGEMENT (1990 to PRESENT)

The 1990s have thus far been characterized by two important trends that have influenced waterfront management, increased environmental awareness and reduced government spending. At the same time, the economic potential of waterfront development and resource utilization continues to fuel proposals for new waterfront projects.

Growing environmental awareness has been reflected in calls from MTRCA and the Waterfront Regeneration Trust (WRT) for ecosystem-based planning and decision making. The WRT was formed in 1992 to oversee and co-ordinate activities along the Greater Toronto Bioregion waterfront, following the recommendations of the Royal Commission on the Future of the Toronto Waterfront which were presented in

the Regeneration report (1991) by the commission. The WRT is presently completing a Lake Ontario Greenway Strategy to ensure that waterfront activities and development contribute to ecosystem health. One of the several recent workshops held by the WRT (1993) addressed the topic of an "Ecosystem Approach to Shoreline Treatment". The proceedings of this workshop provide extensive discussions of the issues and problems from the perspectives of the many municipal, provincial and federal agencies with a stake in shoreline management, as well as from the scientific, engineering and planning communities. A common finding of the workshop was the need for integrated, multi-disciplinary planning and decision making. Any decision on whether to and how to design shoreline treatment should give equal consideration to (a) protection from flood and erosion hazards, (b) biophysical impacts, and (c) socio-economic factors.

Also, the Ontario Ministry of Natural Resources has recently developed a Shoreline Policy which provides guidelines and standards to address flood- and erosion-related hazards and to prevent environmental impacts. The Shoreline Policy prescribes setbacks for development to protect development from flood and erosion hazards over a 100-year planning horizon. This policy has been incorporated into the Planning Act which was legislated by the government of Ontario in 1995 (see Chapter 2).

During this period, this new awareness has led to and/or is reflected by new attitudes toward bluff stabilization projects by MTRCA. The land inshore of the South Marine Drive revetment has been graded to create topographic relief and small ponds and wetlands fed by surficial drainage. The design for the recently initiated Sylvan Avenue shoreline regeneration project features many innovative aspects, which will effectively result in a more natural shoreline condition (including beaches, small shoals, mud flats and overwashed areas; Fig. 15) compared to the neighbouring formal and linear armour stone revetment approach of South Marine Drive. At Fishleigh Drive, MTRCA is presently reviewing the possibility of terminating the shoreline protection at its current

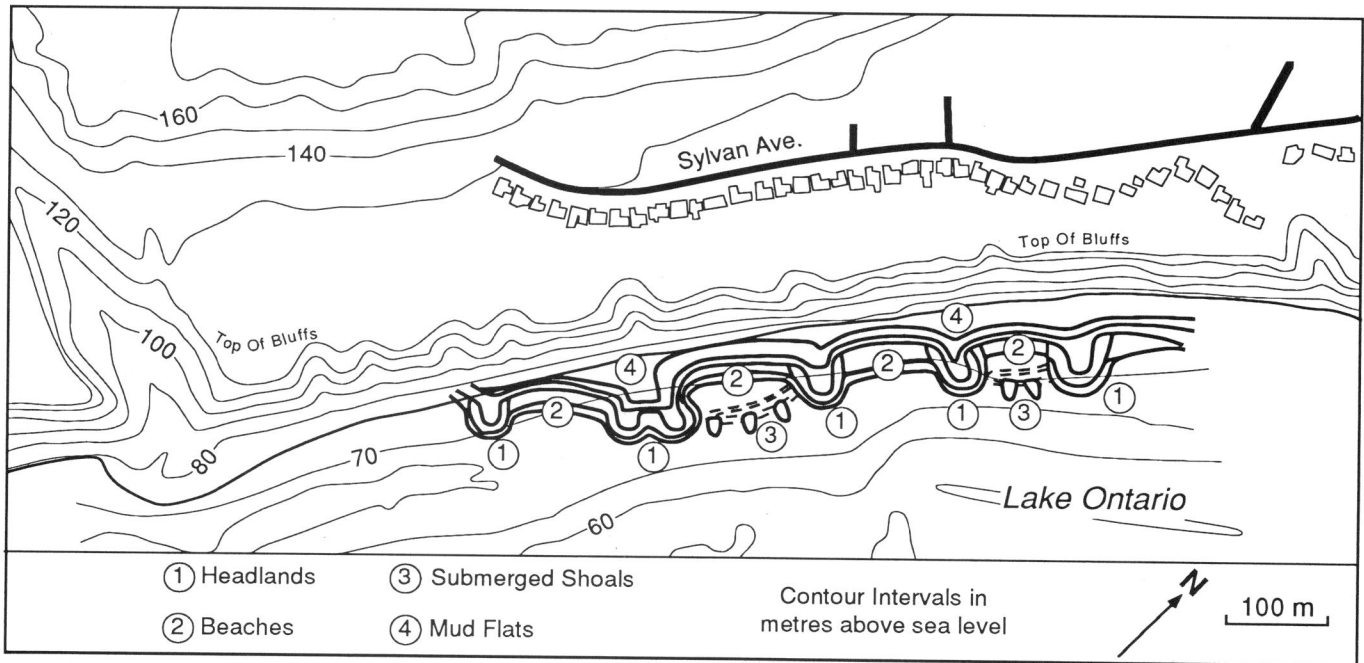

① Headlands   ③ Submerged Shoals
② Beaches   ④ Mud Flats

Contour Intervals in metres above sea level

100 m

Figure 15 *Proposed shoreline regeneration project at Sylvan Avenue (under construction; courtesy MTRCA). Contours in metres above sea level. Lake level is at 75 m asl.*

position instead of extending the revetment further toward Bluffers Park as called for in the original design. The consideration at Fishleigh Drive has been the preservation of the Needles Bluff feature which consists of dramatic near-vertical bluff faces with pinnacles (Fig. 16). It is recognized that further extension of the revetment toward Bluffers Park would lead to the eventual loss of the defining characteristics of the Needles feature.

Costs and Benefits of Shoreline Stabilization

Recent projects have highlighted the continuing problem of comparing the costs and benefits of bluff stabilization to that of the do-nothing (*i.e.*, property acquisition) alternative. When a direct comparison is made between the cost of acquiring properties and the cost of implementing bluff stabilization measures, the latter appear to be much less expensive (Metropolitan Toronto Region Conservation Authority, 1994). There are several important issues which are difficult to factor into this comparison: (a) the value of the exposed bluffs as a geological feature, (b) the implications and cost of impacts to biophysical processes, (c) the value of maintaining an existing community from a social perspective, (d) the value of the infrastructure (roads and services), and finally (e) a consideration of the timing of property acquisition (and the related deferral of acquisition costs).

Regarding the issue of the cost of a cost-benefit analysis to assess shore stabilization projects, the present Ontario Ministry of Natural Resources and MTRCA guidelines stipulate that the full purchase value of the houses must be considered for those homes which are at imminent risk. However, imminent risk is defined on the basis of homes falling lakeward of the stable bluff crest position. Because the bluff heights are in the range of 50 m, the stable crest position is located well inshore (over 75 m or more from the current crest position); therefore, in many cases, even if no toe stabilization were implemented, and bluff retreat continued at an average rate of about 0.5 m·a^{-1}, the imminent risk may translate to a requirement to abandon a home in 50 years' time or more. The interim value of these homes over this period is often not factored into the cost benefit analysis. MTRCA is not obligated to purchase homes that are deemed to be at imminent risk and does not have a mechanism for deferring the cost of the purchase of these homes to some future date (*i.e.*, there is no guarantee that future funding will be sufficient to make required purchases). Therefore, the true cost comparison between property acquisition and bluff stabilization is complicated by many factors, not the least of which is the funding policy and mandate under which MTRCA operates.

CONTEMPORARY UNDERSTANDING OF COHESIVE SHORE (ERODING BLUFF) PROCESSES

During the period between the mid-1980s and the present, the understanding of processes which control erosion on cohesive shores has rapidly developed. This leap in understanding was fueled by an investigation of erosion processes downdrift (east) of Port Burwell on the north shore of Lake Erie. A group of land owners filed a suit against the Government of Canada, claiming that the harbour jetties had accelerated bluff recession and loss of their land. The conventional coastal engineering wisdom held that a disruption to the alongshore transport of sand would always result in accelerated erosion downdrift of the obstruction. The Port Burwell jetties represent a very large obstruction, extending 1300 m offshore. In response, the government invested in a comprehensive study of the erosion processes to build a defence against the $30 million claim

(Philpot, 1986).

Based on a comparison of nearshore profiles dating back to 1896, one of the major findings of the Port Burwell investigation was that the controlling process of bluff recession is not restricted to wave action at the toe of the bluff (as suggested by Sunamura, 1984 for rocky coasts), but by the downcutting of the nearshore lake floor (see Philpott, 1986). Boyd (1992) cites many earlier references which also suggest that the near shore has a controlling influence on shoreline recession. The shoreward shift of the dynamic equilibrium profile implies that erosion or downcutting is proportional to the gradient of the nearshore profile and is, thus, greatest close to shore. Davidson-Arnott (1986) describes field measurements of downcutting for a till profile at a site near Grimsby on Lake Ontario. The results confirm the hypothesis on downcutting; the rates increase toward the shore in a manner related to the local bed slope, allowing for the preservation of the profile shape as it shifts shoreward with time. Kamphuis (1987) points out that cliff height does not exert much influence on the process (in fact, a distinct lack of correlation was noted) because erosional debris from a cohesive shore cliff is quickly swept away and winnowed offshore as suspended load or wash load.

Characteristics of Cohesive Shores

A shore is defined as cohesive when a cohesive substratum (such as glacial deposits, soft rock or other consolidated deposits) forms the erosion surface below near-shore sands (Fig. 17). Often the only visual distinction of a cohesive shore (above the sea or lake level) is the existence of a backshore bluff or cliff. Typically, the maturity of the vegetation on the bluff face provides some indication of the rate at which the shore is eroding. Wherever shoreline or bluff recession is a

Figure 16   *Needles Bluff (large arrow) west of Bluffers Park prior to construction of a revetment (April 1976; courtesy MTRCA). The small arrow shows the outfall of a combined sewer; remediation of contaminated overflows using a constructed wetland is described in Chapter 12.*

concern, there will be little in the way of vegetation on the bluff face. At the base of the bluff, there is often a sandy beach, and offshore, there may even be sand bars. In some cases, cohesive shores of relatively low profile will not feature an eroding bluff face; see, for example, the discussion by Riggs and Cleary (1993) of perched barrier beaches on the United States Atlantic coast.

Cohesive shores make up a large part of the Great Lakes shorelines with some continuous segments of over 100 km in length (see Boyd, 1981). Other examples of cohesive shores include a large proportion of the North Sea coast of England, sections of shoreline on each of the United States ocean coasts, some of the Black Sea coast, part of the Baltic coast, and possibly some sections of Arctic coasts (for the latter, see the discussion by Harper (1990) on the Canadian Beaufort Sea coast). Many other examples throughout the world, including erodible rocky coasts, are cited by Sunamura (1992).

Laboratory Experiments: The Role of Cohesionless Sediment

Cohesionless sediment, such as sand and gravel, overlying the cohesive profile, has an important interactive role. Sunamura (1976, 1992) describes the process of erosion observed in a laboratory model as a "feedback control mechanism". As the cliff was eroded, the sand introduced to the system acted as an abrasive agent on the cohesive foreshore, and increased the erosion rate, thus accelerating the cliff erosion, increasing sediment input and causing further accelerated erosion. However, there comes a point where the sediment begins to offer protection to the cliff and foreshore, and the erosion rates rapidly decrease thereafter. For shores consisting of glacial till (or similar cohesive or consolidated sediment), Kamphuis (1987, 1990) refers to laboratory experiments which show the fundamental importance of sand acting as an abrasive agent in the downcutting process. Other experiments investigating the role of sand in abrading the offshore profile along cohesive shorelines are reported by Bishop *et al.* (1992), Nairn (1992a), and Skafel and Bishop (1994).

The dual role of sand, sometimes accelerating erosion of the cohesive profile and sometimes protecting the profile, has important implications for the understanding of erosion processes on cohesive shores. For example, in the Port Burwell litigation mentioned above, it meant that the impoundment of

sand updrift of the Port Burwell jetties (and the deprivation of this sand from the downdrift shore) did not result in a significant acceleration of bluff recession rates downdrift of the harbour. Whether the cut off of sediment supply on a cohesive shore has an impact on downdrift bluff recession rates is determined by the pre-existing conditions, particularly the quantity of sand that historically covered the underlying cohesive profile. Nevertheless, for both the Port Burwell location and for the Scarborough Bluffs (see Baird & Associates, 1994c), the interruption of alongshore sediment transport will not result in a significant increase in bluff recession rates.

Numerical Modelling

The new understanding of the role played by sand in cohesive shore processes was applied to a series of nearshore profiles along the bluffs by Nairn (1992b), as part a study for the International Joint Commission (IJC) Water Levels Reference Board to evaluate erosion processes on the Great Lakes and the influence of fluctuating lake levels on erosion rates. The IJC used this information to evaluate the benefits of increased regulation of lake levels with respect to reduced erosion rates. Ultimately, the IJC opted not to consider increased regulation on the Great Lakes after considering many different factors.

Figure 17 shows the underlying cohesive profile and the sand cover at the same site, in 1952 and again in 1989, for a location offshore of Sylvan Avenue along Scarborough Bluffs. Focussing on the bluff, the bluff face has retreated approximately 30 m in this 37-year period, for an average long-term recession rate of 0.8 m·a^{-1}. However, the most important observation is that the nearshore lake bottom has been lowered through erosion processes despite the presence of a sand cover which exceeds 0.5 m in thickness. An equally important observation is that the underlying cohesive profile shape in 1952 is very similar to that of 1989; it has simply shifted shoreward by 30 m. There is not a cross-shore balance of erosion and deposition as might be expected from the application of the Bruun (1962) principle; all of the eroded material from the cohesive profile and the bluff is either winnowed offshore (for the clay and silt fractions) or transported alongshore (for the sand and gravel fractions).

The IJC study by Nairn (1992b) involved the application of a numerical model for the prediction of cohesive profile change over the 37-year period at Scarborough Bluffs under a variety of different lake level fluctuation scenarios. The model was developed to simulate the processes that occur on a sandy beach profile subjected to wave action (see Nairn, 1990; Southgate and Nairn, 1993; Nairn and Southgate, 1993). These processes include (1) wave action (including shoaling, breaking and decay), (2) hydrodynamics (including orbital velocities, undertow and longshore currents), (3) sediment transport (bed load, sheet flow and suspended load in the cross-shore and alongshore directions), and (4) profile adjustments due to gradients in the cross-shore transport pattern. Each of the processes is evaluated at approximately 200 finite-difference calculation points across the profile. This process-based numerical model, developed for sandy shores, has been adapted for application to cohesive shores.

Downcutting of cohesive sediment is a complicated process. It has been postulated by many researchers that shear stress at the bed created by orbital velocities, along with the presence of sand as an abrasive agent, may be responsible for irreversible downcutting of the cohesive surface. However, the shear stress at the bed due to orbital velocity decreases in an onshore direction through the surf zone due to the dimin-

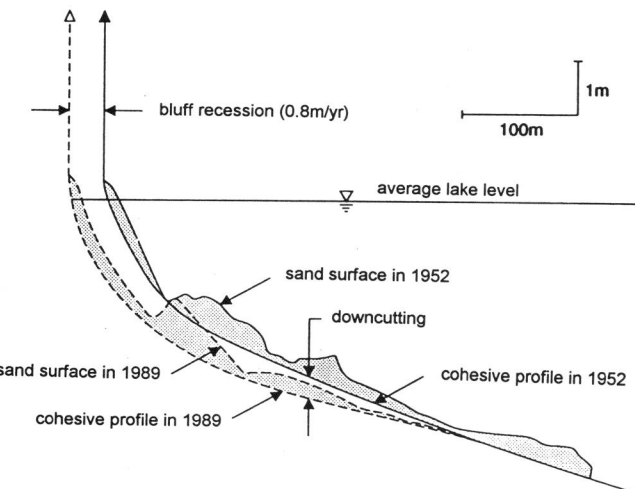

Figure 17 *Observed nearshore profile evolution offshore of Sylvan Avenue, 1952 to 1989 (from Nairn, 1992b). See Figures 3, 15 for location.*

ished wave heights in shallower water. This is at odds with the fact that downcutting must increase closer to shore in order for the profiles to maintain a similar shape as they recede shoreward (Fig. 17). Additional downcutting mechanisms in the surf zone include breaking-induced turbulence which is able to penetrate to the bed, as well as the shear stress attributable to the undertow velocity near the bed. Both the generation of turbulence and undertow may be quantified by the rate of wave energy decay. In the model used by Nairn (1992b), downcutting is attributed to two factors which describe the driving forces: shear stress at the bed caused by orbital velocities (for unbroken waves) and wave energy dissipation in the surf zone (for broken waves).

An example of the results from the numerical model investigation of erosion processes is given in Figure 18 for a profile at Sylvan Avenue. This figure shows the model prediction of profile and shoreline retreat for the 37-year period which is based on a calculation of lakebed erosion related to wave action (and the associated lake level) for every hour between 1952 and 1989. The influence of lake level fluctuations was determined by repeating the model test for the same input wave conditions with different lake levels for each hour of the 37-year period. The results indicated that changes to the range of lake level fluctuations did not have a measurable impact on the rate of bluff recession. Therefore, for this profile, erosion at low and average lake levels must be equally as important as erosion during periods of high lake levels. In contrast, bluff recession rates for profiles with a convex shape (*i.e.*, *versus* the concave shape of the example shown in Fig.

18) were found to be significantly influenced by the range of lake level fluctuations by Nairn (1992b). For the convex profiles, high lake levels have a greater impact because more wave energy is able to traverse the nearshore shelves associated with these profiles.

It must be recognized that numerical models of cohesive shoreline processes cannot, as yet, be used to make precise predictions of future conditions. Instead, the numerical model should be used as a tool to contribute to an improved understanding of a very complex system.

A BIOPHYSICAL MODEL OF COHESIVE SHORES AND SCARBOROUGH BLUFFS SHORELINE

As mentioned previously, the Waterfront Regeneration Trust is preparing a Lake Ontario Greenway Strategy to promote an ecosystem-based approach to shoreline management. A draft report prepared by the Shoreline Management Work Group of the Trust (1995) developed an ecosystem-based approach to shoreline classification. A descriptive model of the coastal zone ecosystem has been developed, and is based on the distinctive biophysical characteristics of different shore types. The Scarborough Bluffs segment of shoreline between the R.C. Harris Filtration Plant and the west side of East Point formed a specific shoreline unit. As has been explained, the primary characteristic of this unit is the underlying cohesive profile. This unit is distinguished from neighbouring cohesive shore units to the east by the shape of the nearshore profile. As shown in Figure 19A, the profile shape along the bluffs is distinctively concave. In contrast, for East

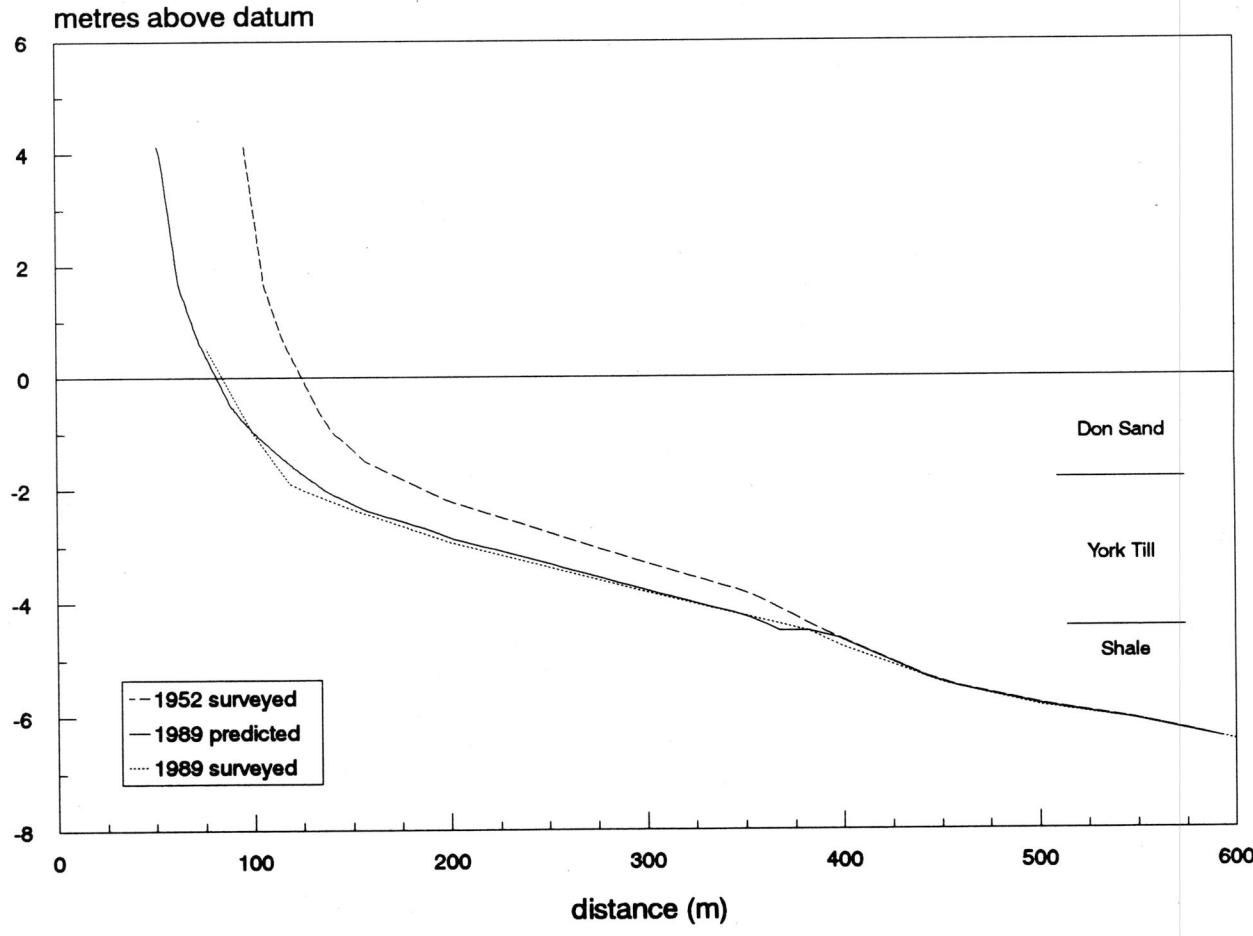

Figure 18 *Predicted* versus *observed profile change of lake floor offshore of Sylvan Avenue. Geologic units underlie Scarborough Clays (Fig. 8). Datum refers to lake level.*

Point and the shore further to the east the profile is convex in shape (see Fig. 19B). The convex shape is related to the formation of an erosion-resistant offshore shelf. The shelf is protected by cobble and boulder lag deposits derived from bluff and lake bed erosion in the past. Interestingly, this shelf develops at a consistent depth of about 2 m below chart datum throughout the Great Lakes.

The different nearshore profile shapes have important implications for shoreline management. The concave shores present a difficult problem for shore protection since the nearshore profile is always lowering (out to depths of well beyond 2 m). Therefore, shore protection structures will come

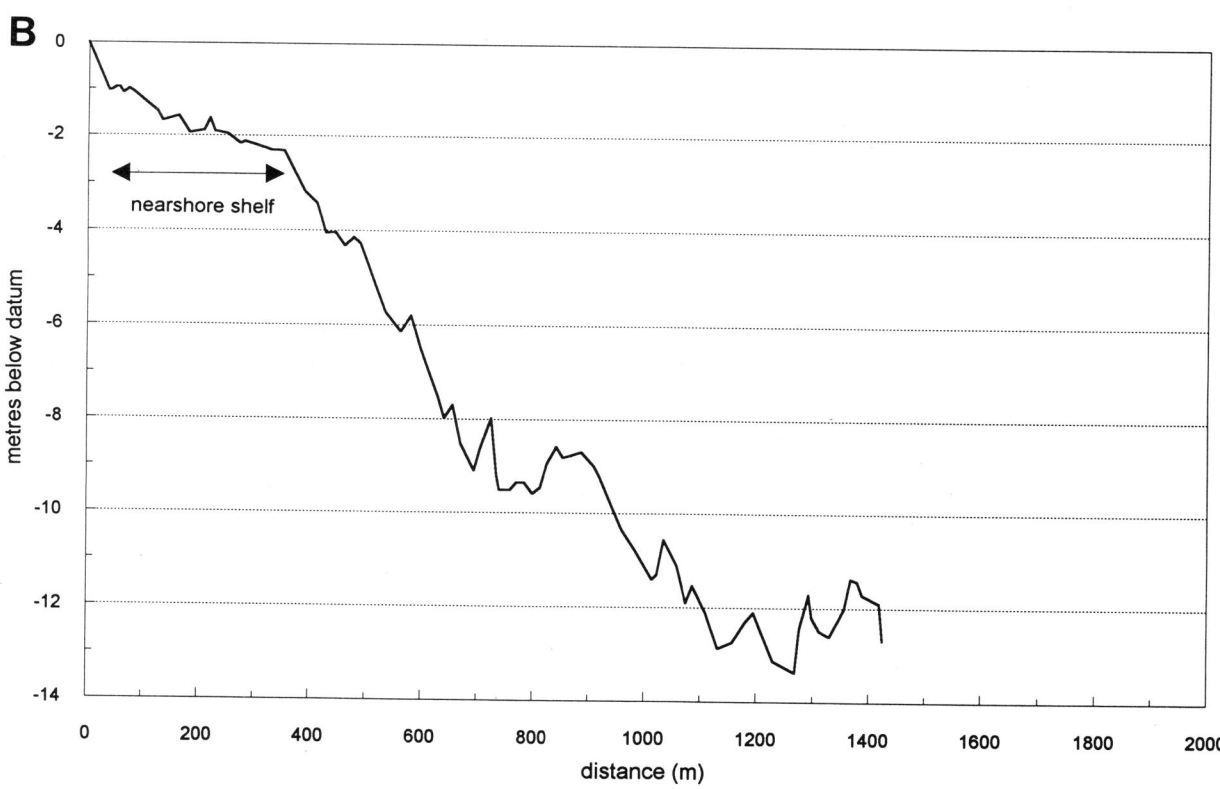

Figure 19   (**A**) *Example of a concave cohesive profile from Sylvan Avenue.* (**B**) *Example of a convex cohesive profile from shoreline east of Scarborough Bluffs. Datum refers to lake level.*

under increasing threat of toe erosion and larger waves (related to increasing depths) in the future. This has been recognized in recent designs such as the Sylvan Avenue project. In contrast, convex profiles can be protected more permanently by extending the shore protection to depths of 2 m below chart datum, where erosion no longer occurs due to the presence of the protective lag deposit.

Convex profiles provide more suitable aquatic habitat for cold-water species because of the presence of the exposed cobble/boulder substrate and because of a steep drop-off to deep water at the outer edge of the shelf. In contrast, concave profiles have a surficial substrate which consists of rapidly shifting sands over a relatively smooth exposed cohesive substratum which is generally a less hospitable environment for most species. The biological aspects of the descriptive model for this section of shoreline are discussed in greater detail in the final report of the Shoreline Management Work Group of the WRT (1995).

This type of integrated understanding of nearshore bio-physical processes must form the fundamental basis for all decision making and shoreline management planning in the future.

DISCUSSION

The history of human responses to the erosion of Scarborough Bluffs provides many classic examples of the conflicting issues and difficult problems associated with shoreline management. Through the benefit of hindsight, it can be demonstrated that poor decisions were often made as a result of either limitations to the knowledge base at the time, or to a complete disregard for the knowledge base. In other cases, the benefits of an action were found to outweigh negative impacts. However, in all of these cases, no consideration was given to the implications of cumulative impacts. The common thread in all of the decisions was a lack of foresight; they were short-term decisions made on a site-specific basis.

Future shoreline management actions must be based on the most current and complete knowledge base of biophysical processes and conditions. These actions must also consider future changes to these processes and conditions. Furthermore, some consideration must be given to the fact that this knowledge base remains imperfect and that predictions of future impacts will have some level of associated uncertainty. Socio-economic factors cannot be ignored and waterfronts will continue to be an important focal point for urban development. A changing attitude of the public toward the environment presents an opportunity to improve future shoreline management. One key to this improvement is the education of planners and the public to develop an appreciation for the complexity, sensitivity and importance of waterfront ecosystems. All management actions which have the potential to alter the waterfront ecosystem should involve an informed public in the decision-making process.

ACKNOWLEDGEMENTS

The senior author would like to acknowledge the important contributions made to the understanding of cohesive shorelines by Keith Philpott. The MTRCA have kindly provided many of the photographs for this paper. Thanks are due to Bill Baird and Pete Zuzek of Baird & Associates for their review of this pape and discussions.

REFERENCES

Baird & Associates, 1994a, Toronto Islands shoreline management study – Summary report: prepared for MTRCA, 68 p.

Baird & Associates, 1994b, Toronto Islands Nature school study – Final report: prepared for MTRCA, 91 p.

Baird & Associates, 1994c, Assessment of coastal processes at Sylvan Avenue: prepared for F.J. Reinders & Associates and MTRCA, 72 p.

Bird, S.J. and Armstrong, J.L., 1970, Scarborough Bluffs – A recessional study: International Association for Great Lakes Research, 13th Conference on Great Lakes Research, Proceedings, p. 187-197.

Bishop, C.T., Skafel, M.G. and Nairn, R.B., 1992, Cohesive profile erosion by waves: American Society of Coastal Engineers, 23rd International Conference on Coastal Engineering, p. 2976-2989.

Bonis, R.R., 1968, A History of Scarborough: Scarborough Public Library, Scarborough, ON, 327 p.

Boyd, G.L., 1981, Canada/Ontario Great Lakes monitoring programme final report: Dept. of Fisheries and Oceans Canada, Oceans and Aquatic Sciences, Manuscript Report 12, 200 p.

Boyd, G.L., 1992, A descriptive model of shoreline development showing nearshore control of coastal landform change: Late Wisconsinan to present, Lake Huron, Canada: unpublished Ph.D. thesis, Department of Geography, University of Waterloo, Waterloo, ON, 207 p.

Bruun, P., 1962, Sea-level rise as a cause of shore erosion: Journal of Waterways, v. 88, p. 117-130.

Bryan, R.B. and Price, A.G., 1980, Recession of Scarborough Bluffs, Ontario, Canada: Zeitschrift für Geomorphologie, Neues Folge Supplement Bund, v. 34, p. 48-62.

Burt, H.F. and Roosenboom, H., 1952, Scarborough Township bluffs lakeshore survey, Oct. 1952, 81 p.

Buttle, J.M. and von Bulow, P., 1986, Crest retreat along the Bluffers Park section of the Scarborough Bluffs: National Research Council of Canada, Symposium on Cohesive Shores, Ottawa, ON, p. 87-102.

Coleman, A.P., 1913, An estimate of post-glacial and inter-glacial time in North America: 12th International Geological Congress, Proceedings, p. 435-449.

Coleman, A.P., 1932, The Pleistocene of the Toronto region: Ontario Department of Mines, 41st Annual Report, v. XLI, Part VII, 54 p.

Coleman, A.P., 1936, Geology of the north shore of Lake Ontario: Ontario Department of Mines, Annual Report, v. XLV, Part VII, p. 37-74.

Davidson-Arnott, R.G.D., 1986, Rates of erosion of till in the nearshore zone: Earth Processes and Landforms, v. 11, p. 53-58.

Eyles, N., Buergin, R. and Hincenbergs, A., 1986, Sedimentological controls on piping structures and the development of scalloped slopes along an eroding shoreline, Scarborough Bluffs, Ontario: National Research Council of Canada, Symposium on Cohesive Shores, Ottawa, ON, p. 69-86.

Eyles, N., Eyles, C.H., Lau, K. and Clark, B., 1985, Applied sedimentology in an urban environment – The case of Scarborough Bluffs, Ontario: Canada's most intractable erosion problem: Geoscience Canada, v. 12, p. 91-104.

Eyles, N. and Howard, K.W.F., 1988, A hydrochemical study of urban landslides caused by heavy rain: Scarborough Bluffs, Ontario, Canada: Canadian Geotechnical Journal, v. 25, p. 455-466.

Fleming, S., 1853, Toronto Harbour – Its formation and preservation: Canadian Journal, v. II, December 1853, p. 1-15.

Geocon Inc., 1980, Erosion control study, Scarborough Bluffs – Stage 1. Report prepared for Metropolitan Toronto Region Conservation Authority: 291 p.

Geocon Inc., 1982, Erosion control study, Scarborough Bluffs – Stage 2. Report prepared for Metropolitan Toronto Region Conservation Authority: 235 p.

Greenwood, B. and McGillivray, D.G., 1980, Modelling the impact of large structures upon littoral transport in the central Toronto waterfront, Lake Ontario, Canada: Zeitschrift für Geomorphologie, Neues Folge Supplement Bund, v. 34, p. 97-110.

Harper, J.R., 1990, Morphology of the Canadian Beaufort Sea coast: Marine Geology, v. 91, p. 75-91.

Kamphuis, J.W., 1987, Recession rate of glacial till bluffs: Journal of Waterway, Port, Coastal and Ocean Engineering, v. 4, p. 60-73.

Kamphuis, J.W., 1990, Influence of sand or gravel on the erosion of cohesive sediment: Journal of Hydraulic Research, v. 28, p. 43-53.

Karrow, P.F., 1967, Pleistocene geology of the Scarborough area: Ontario Department of Mines, Geological Report 46, 108 p.

Langford, G.B., 1952, Report on lakeshore erosion, Lake Ontario from Niagara to Cobourg: Ontario Department of Planning and Development, Toronto, ON, 36 p.

Lewis, C.F.M. and Sly, P.G., 1971, Seismic profiling and geology of the Toronto waterfront area of Lake Ontario: International Association for Great Lakes Research, 14th Conference on Great Lakes Research, Proceedings, p. 250-270.

MacPherson, G. and Piercy, L., 1993, Shoreline treatment and fish and wildlife habitat: MTRCA Case Studies: Waterfront Regeneration Trust, Workshop on an Ecosystem Approach to Shoreline Treatment, Toronto, ON, p. 16-21.

McGillivray, D.G. and Greenwood, B., 1978, Physical Impact assessment and coastal structures, Toronto waterfront, Lake Ontario, Canada: National Research Council of Canada, Coastal Zone '78, Ottawa, ON, Proceedings, p. 2865-2884.

Metropolitan Toronto Planning Board, 1967, The Waterfront Plan for the Metropolitan Toronto Planning Area: 187 p.

Metro Toronto Region Conservation Authority, 1980, Shoreline Management Program – Watershed Plan: 12 p.

Metropolitan Toronto and Region Conservation Authority, 1994, Environmental study report for Sylvan Avenue shoreline regeneration report: prepared under the Class Environmental Assessment for Remedial Flood and Erosion Projects: 71 p.

Nairn, R.B., 1990, Prediction of cross-shore sediment transport and beach profile evolution: unpublished Ph.D. thesis, Imperial College of Science, Technology and Medicine, University of London, London, UK, 225 p.

Nairn, R.B., 1992a, Study of Toe scour and downcutting on cohesive shores protected with revetment: report prepared for Metropolitan Toronto Region Conservation Authority: 29 p.

Nairn, R.B., 1992b, Erosion Processes evaluation paper. Final report for International Joint Commission Great Lakes – St. Lawrence River Levels Reference Study Board: a report submitted to Water Planning and Management Branch, Environment Canada, Burlington, ON, 82 p.

Nairn, R.B. and Southgate, H.N., 1993, Deterministic profile modelling of nearshore processes. Part II. Sediment transport processes and beach profile development: Coastal Engineering, v. 19, p. 57-96.

Philpott, K.L., 1986, Coastal engineering aspects of the Port Burwell shore erosion damage litigation: National Research Council of Canada, Symposium on Cohesive Shores, Ottawa, ON, p. 309-338.

Riggs, S.R. and Cleary, W.J., 1993, Influence of inherited geologic framework upon barrier island morphology and shoreface dynamics, *in* List, J.H., ed., Large Scale Coastal Behaviour '93: United States Geological Survey, Open File Report 93-381, p. 173-176.

Report of the Select Committee of the Ontario Legislature on the Lake Levels of the Great Lakes, 1953, Queen's Printer, Toronto, ON: 87 p.

Royal Commission on the Future of the Toronto Waterfront, 1991, Regeneration – Toronto's Waterfront and the Sustainable City: Final Report: 530 p.

Sandwell Inc., 1991, Eastern Beaches Shoreline Management Study: report prepared for Metropolitan Toronto Region Conservation Authority: 82 p.

Skafel, M.G. and Bishop, C.T., 1994, Flume experiments on the erosion of till shores by waves: Coastal Engineering, v. 23, p. 329-348.

Snider, C.H.J., 1930, A series of articles titled "Schooner Days" that appeared in *The Toronto Evening Telegram* during the 1930s.

Southgate, H.N. and Nairn, R.B., 1993, Deterministic profile modelling of nearshore processes. Part I. Waves and currents: Coastal Engineering, v. 19, p. 27-56.

Sunamura, T., 1976, Feedback relationship in wave erosion of laboratory rocky coast: Journal of Geology, v. 84, p. 427-437.

Sunamura, T., 1984, Processes of sea cliff and platform erosion: Coastal Research Council, Handbook on Coastal Processes, p. 233-266.

Sunamura, T., 1992, Geomorphology of Rocky Coasts: John Wiley & Sons Ltd., London, UK, 302 p.

Toronto Harbour Commissioners, 1854, Reports on the Improvement and preservation of Toronto Harbour (including those of Hind, Fleming, Tully and Richardson): Supplement to the *Canadian Journal*.

Waterfront Regeneration Trust, 1993, Proceedings Workshop on an Ecosystem Approach to Shoreline Treatment: Waterfront Regeneration Trust, Toronto, ON, 86 p.

Waterfront Regeneration Trust, 1995, Final report of the Shoreline Management Work Group for the Lake Ontario Greenway Strategy. Draft: 54 p.

19. Environmental Geology of the Fraser Delta, Vancouver

Bruce S. Hart

Department of Geosciences, Pennsylvania State University, University Park, Pennsylvania 16802 USA

J. Vaughn Barrie

Pacific Geoscience Centre, Geological Survey of Canada, P.O. Box 6000, Sidney, British Columbia V8L 4B2

SUMMARY

Environmental geological and geophysical studies of the submarine portions of the Fraser delta at Vancouver, British Columbia (BC) reveal complex patterns of sedimentation, erosion and slope instability resulting from natural and industrial activity. Knowledge of the spatial and temporal variability of these processes is needed to help identify sea-floor hazards in areas of significant economic importance, and to monitor environmental contamination and changes to the physical environment due to industrial activity. Portions of the delta slope adjacent to the main fluvial distributary are prone to downslope mass movement in response to sediment loading, accentuated by the development of steep subaqueous slopes and interstitial (biogenic) gas. The presence of Canada's largest coal export facility, ferry terminals, and submarine high-voltage cables in close proximity to areas of mass movement are causes for concern. Triggering events for other large slope failure complexes (several tens of square kilometres in areal extent) include large earthquakes.

INTRODUCTION

Like many other large cities, Vancouver and its outlying urban areas (1994 population 1.7 million) lie in close proximity to the sea (Fig. 1). The inevitable consequence of this proximity is the exploitation of the sea floor and overlying water column for a variety of uses, which include sewage disposal, fishing, dredge spoil dumping, construction of jetties, navigation/transportation, and the laying of submarine cables (electrical transmission, telecommunications). A partial summary of some of the developments on the sea floor and adjacent tidal flats is listed in Table 1.

Unlike most other coastal cities in Canada (*e.g.*, Hamilton, Ontario, see Chapter 16; Halifax, Nova Scotia, see Chapter 17), the Vancouver urban area is adjacent to a rapidly prograding delta. The Fraser delta is a post-glacial Holocene feature, no older than 10,000 years, that is rapidly building out into the Strait of Georgia (Fig. 1). Submarine sedimentation rates of a few centimetres to a few decimetres per year have been measured from some portions of the delta, while other parts are known to be eroding (Mathews and Shepard, 1962; Hart *et al.*, 1993; Evoy *et al.*, 1993). High sedimentation rates such as

these are commonly associated with various forms of submarine slope instability (*e.g.*, Prior and Coleman, 1984). Furthermore, because this is a region of high seismic activity (see Chapter 30), the potential exists for earthquake-induced submarine slope failure. Large failures have the potential to damage not only sea-floor installations, but also structures on the adjacent tidal flats (Luternauer *et al.*, 1994). It has been

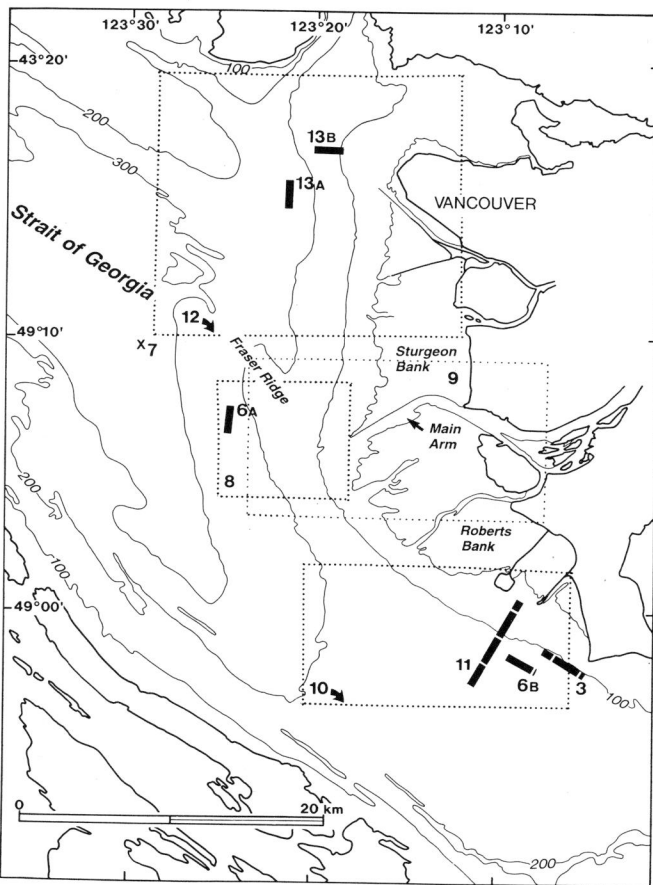

Figure 1 *Location map of Fraser delta showing bathymetry and location of subsequent figures.*

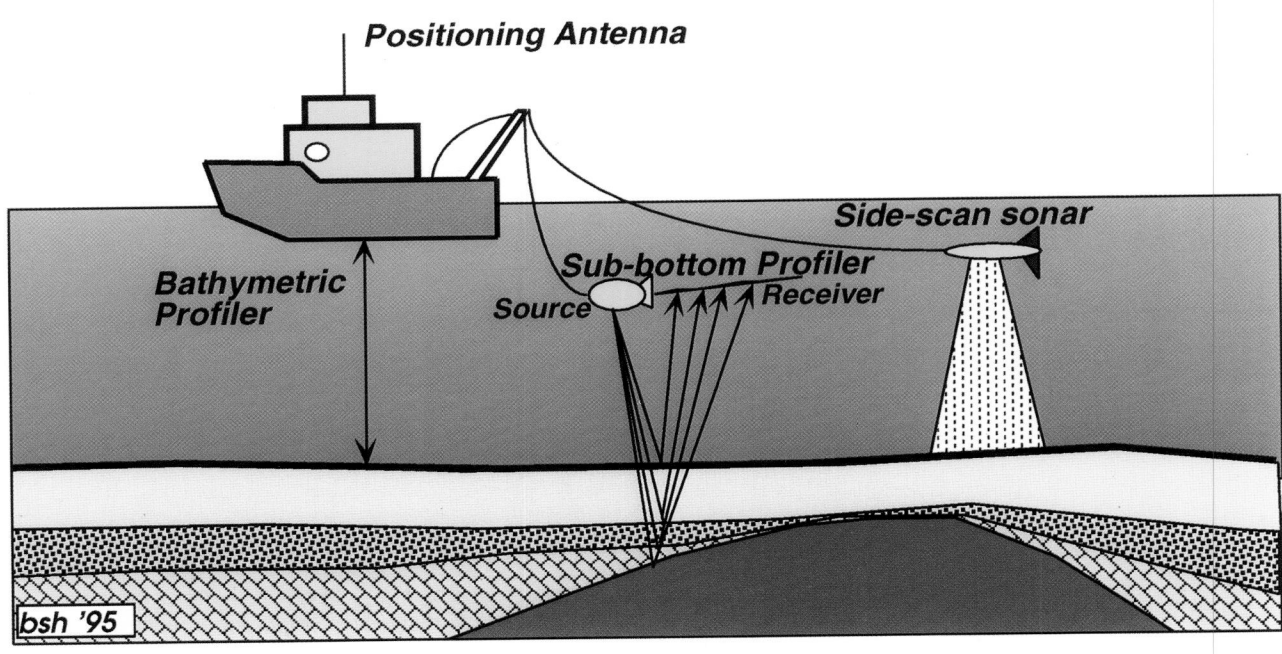

Figure 2   *Illustration of simultaneous collection of bathymetry (hull-mounted system), sub-bottom profiler (high-resolution seismic) and side-scan sonar data during marine geophysical survey. Ship's position fixed using satellite-based system such as GPS. Location of profiler and side-scan towfish with respect to antenna (setback) determined by acoustic positioning systems mounted to towfish (for detailed work) or (for reconnaissance work) using cable length out, visual estimate (surface tow systems) or post-survey calibration to bathymetry data.*

Table 1   Fraser Delta: selected sea-floor/tidal flat installations and activities.

WHAT	WHERE	HAZARDS/ISSUES

12 High Voltage Cables - BC Hydro
    Initially lain in mid-1950s
    Supply electricity to Vancouver Island
	Crossing southern Strait of
	    Georgia from Roberts Bank
	    to Galiano Island
		Trawling, burial by sediment
		    (reduces conductivity),
		    submarine failure

Fibre optics telecommunications cables
    Lain on seafloor in 1993
	Galiano Island to Sturgeon
	    Bank, Nanaimo Harbour
	    to Point Grey (English Bay)
		Trawling, illegal dumping,
		    submarine failure

Jetty/dyke construction to fix courses of distributaries
    and to protect low-lying delta plain
    Began in late 1800s
	Sturgeon and Roberts Banks	Habitat change, changes in
		    sedimentary dynamics

Sand Heads lighthouse
    Constructed mid-1960s
	Mouth of main Arm,
	    Fraser River
		River mouth failures

BC Ferries - Tsawwassen Terminal
    Constructed in 1959-60, expanded in 1991.
    5.9 million passengers and 2.0 million vehicles in 1992
	Edge of tidal flats,
	    Roberts Bank
		Delta slope failure, changes in
		    sedimentary dynamics

Iona Sewage outfall
    Treated sewage from greater Vancouver region
    Initially discharged onto tidal flats,
    submarine extension (down to 100 m depth) in 1988
	Sturgeon Bank,
	    near Middle Arm
		Discharge in proximity to trawling
		    grounds, delta slope failure

Point Grey dumpsite
    Dredge spoils and other land-derived wastes
    Established in 1968 by Department of Transportation
	Deep water (*ca.* 200 m)
	    offshore from Point Grey
		Illegal dumping outside limits
		    interferes with fishing

Roberts Bank "Superport"
    Originally constructed in 1969, expanded in 1981-82
    Canada's largest coal export facility
	Edge of tidal flats,
	    Roberts Bank
		Delta slope failure, changes in
		    sedimentary dynamics

A

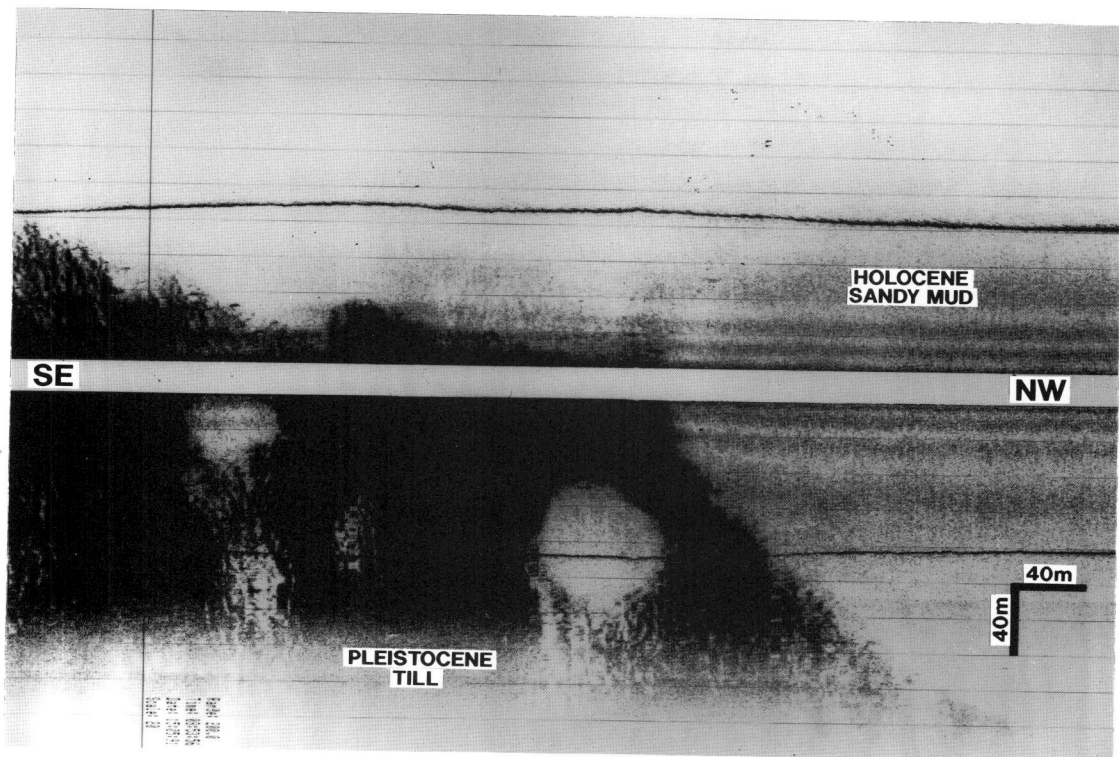

B

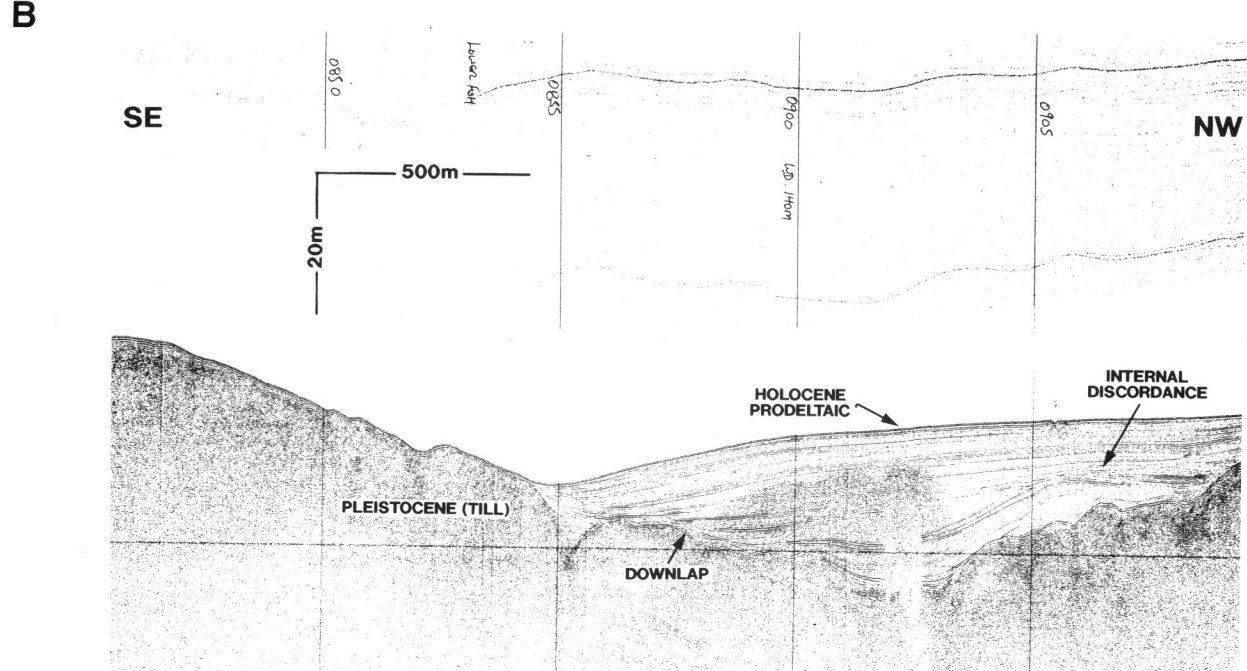

Figure 3 (**A**) *Side-scan sonar image (Klein 100 kHz) and* (**B**) *corresponding high-resolution seismic profile (Huntec Deep Tow) showing Pleistocene deposits overlain by Holocene prodeltaic sediments. Using only the sonograph, it would be difficult to determine whether the dark patches represented a surficial deposit of dense material (e.g., Fig. 13) or an outcrop exposure of an underlying unit. Note fine-scale stratigraphic detail visible on seismic profile. For location, see Figure 1.*

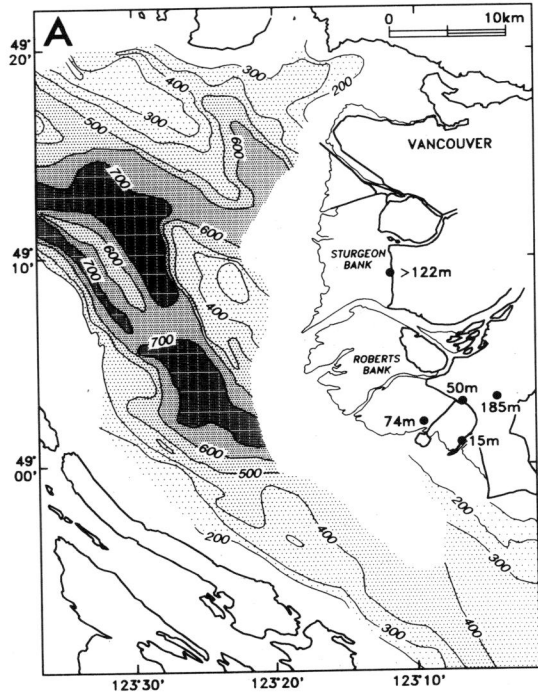

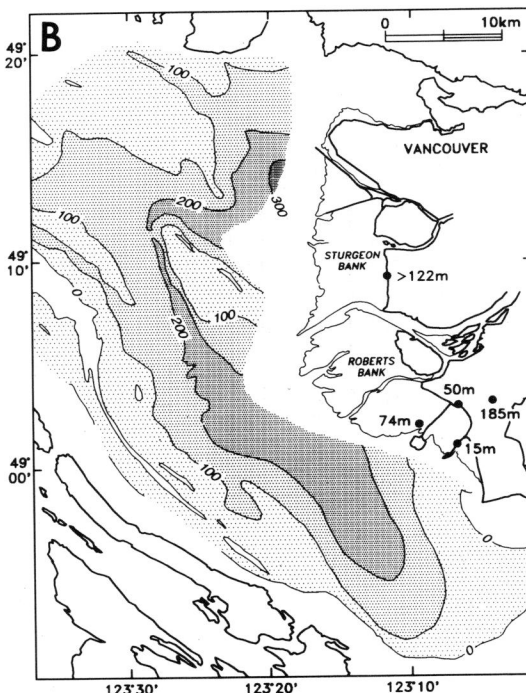

suggested that low-lying portions of the delta plain could also be affected if a large submarine failure were to induce a tsunami in the Strait of Georgia (Hamilton and Wigen, 1987).

This paper will synthesize the results of previous environmental geological and geophysical studies of the submarine portions of the Fraser delta (Fig. 1). The principal objective is to use the Fraser delta work as an example to examine the role of the environmental geologist in conducting environmental or slope stability studies.

MARINE GEOLOGICAL AND GEOPHYSICAL SURVEYING

Generally, the seabed and underlying strata cannot be directly observed, but must be examined indirectly, using geophysical methods (Fig. 2). Seismic, or sub-bottom, profilers are commonly used. There is a trade-off between the vertical resolution and the vertical penetration of seismic profiling systems: high-resolution systems such as the Huntec Deep-Tow Seismic system employed on the Fraser delta (see below) have a theoretical bed resolution of a few decimetres (based on dominant frequencies of 3-5 kHz), but typically penetrate only a few tens of metres, whereas single-channel airgun profiling systems (dominant frequencies of a few hundred Hz) may penetrate hundreds of metres, but have significantly less vertical resolution (metre scale). Bathymetric profilers operate on a similar princi-

Figure 4 *Example of (**A**) time structure map (two-way time, milliseconds; ms) from sea level to top of Pleistocene sediments or bedrock (where Pleistocene deposits are absent), and (**B**) isochron (two-way traveltime; thickness in milliseconds) map of Holocene deltaic and prodeltaic sediments, derived from seismic data. Isochron values can be converted to approximate thickness (metres) by dividing by 2 (to convert to one-way traveltime in ms) then multiplying by 1.5 m·ms⁻¹ ("average" velocity of sound in modern sediments). Note NW-SE trending ridges and troughs in structure map. Holocene sediments are absent in places, and are over 200 m thick offshore from Sturgeon Bank. Onshore points represent thicknesses of Holocene sediments (metres) measured from boreholes.*

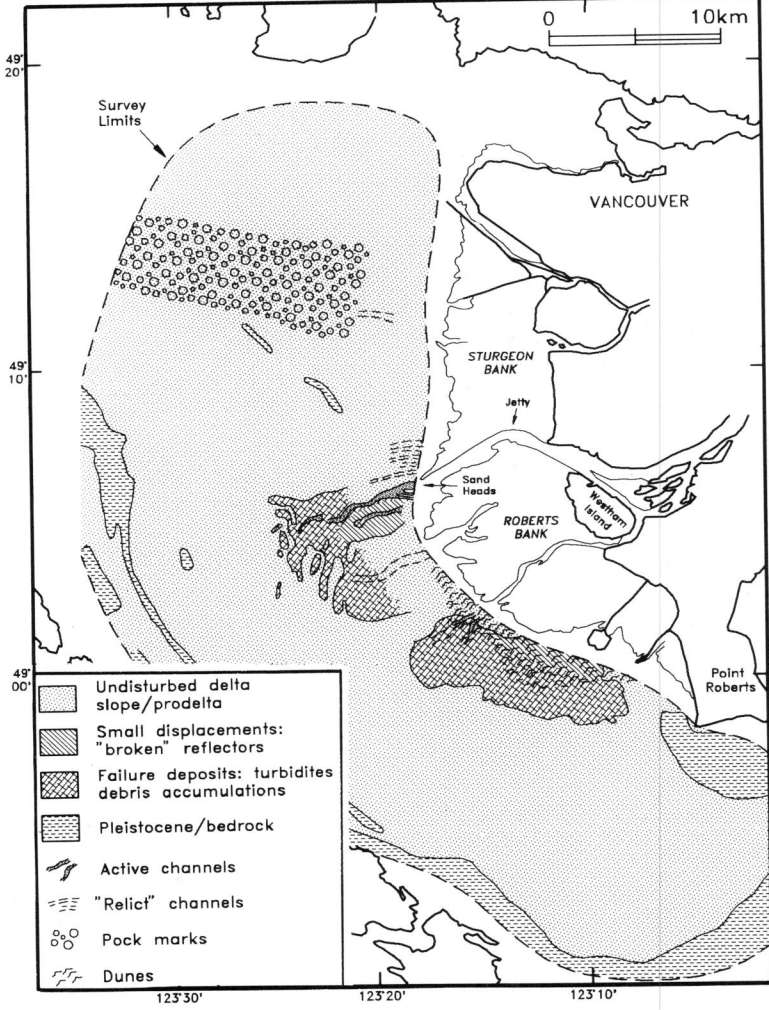

Figure 5 *Surficial geology map of Fraser delta slope and prodelta areas derived from seismic and side-scan sonar surveys. Failure complexes form prominent components of the system in the southern and central regions. The northern part of the delta slope is relatively unaffected by slope failures, although a field of pockmarks (gas escape features) is found at the base of the delta slope.*

pal to seismic systems, but use higher frequencies (typically 12 kHz or more) and sub-bottom penetration is usually negligible. Side-scan sonar systems use even higher frequencies (typically 100 kHz or 200 kHz) to produce images of the sea floor in a fashion somewhat analogous to airborne side-looking radar.

Distances for all acoustic-based systems are originally recorded in time units (seconds, or milliseconds), but these can be converted to depths (seismic, bathymetry) or distances (side-scan sonar) if the velocity of sound in the transporting medium (water, sediment) is known or can be assumed. Modern side-scan sonar systems can compensate for survey speed variations in real time, producing undistorted images of the sea floor. Ideally, seismic, side-scan sonar, and bathymetry data can be collected concurrently, so that the subsurface structure and surficial expression of sea-floor features can be determined to help facilitate interpretation (*e.g.*, Fig. 3A, B).

Much useful information about substrate properties can be obtained through seismic facies analysis incorporating information such as reflection character, amplitude and frequency (*e.g.*, Bouma *et al.*, 1983; see Chapters 31, 32). When combined with surface sediment sampling and coring, these marine geophysical methods allow mapping of surficial features and surficial geology and the determination of subsurface geology and stratigraphy. Experience has shown that the most beneficial approach to integrated marine geological and geophysical work involves conducting the geophysical work first, then using the coring or sampling to ground truth and sample acoustic facies, stratigraphic contacts, or other features of interest.

Several marine geological and geophysical surveys of the Fraser delta slope and adjacent Strait of Georgia have been conducted by the Geological Survey of Canada in the period 1982-1992. This work has resulted in the collection of (1) more than 1200 km of Huntec high-resolution seismic profiles along

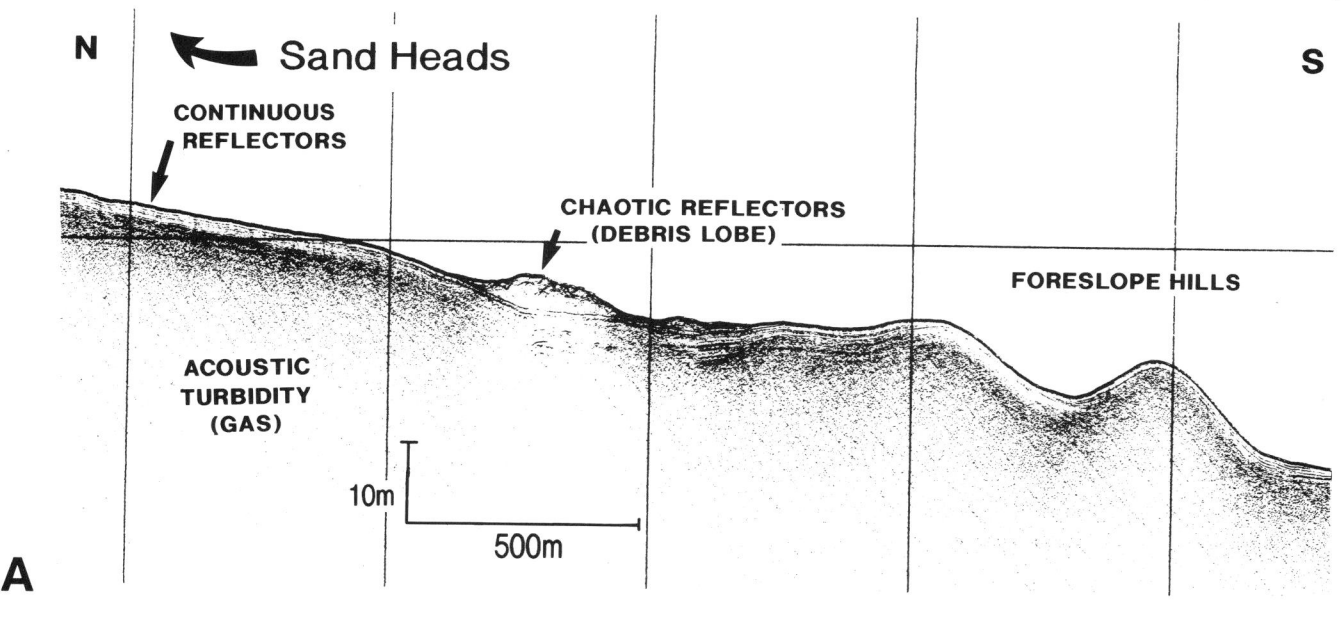

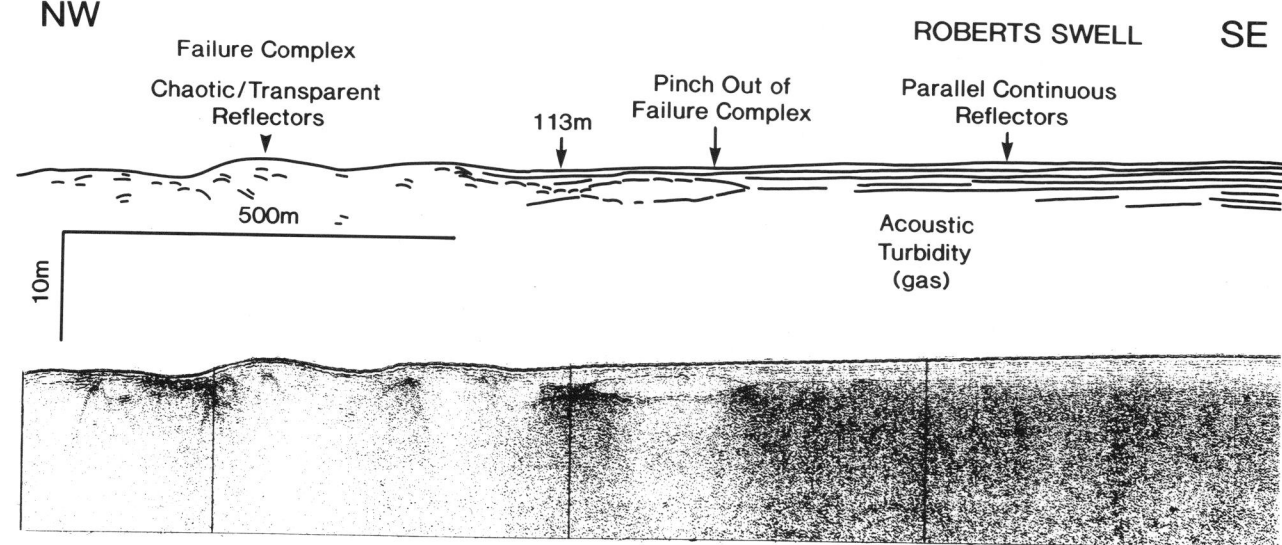

Figure 6 (**A**) *Huntec seismic profile from near base of delta slope seaward of Sand Heads. Parallel continuous reflectors with acoustic turbidity (fog-like seismic character which masks structure deeper than a few metres), characteristic of gassy, undisturbed muds and mounded chaotic reflectors of debris flow lobe. For location, see Figure 1.* (**B**) *Huntec profile, with interpretation from base of delta slope near Roberts Bank. Chaotic and transparent reflectors which attenuate seismic signal represent sandy failure deposits, and interfinger with gassy, undisturbed muds.*

with fully corrected 100 kHz side-scan sonar imagery of the sea floor, (2) more than 5000 km of airgun seismic profiles (of which 2500 km is near or on the delta slope), and (3) more than 50 vibrocores, gravity and piston cores.

STRATIGRAPHY AND MORPHOLOGIC FRAMEWORK

The stratigraphy of the submarine portions of the Fraser delta was described most recently by Hart *et al.* (1995). At the base of the section are thrusted Tertiary sedimentary rocks. Thrust sheets of these rocks form the Gulf Islands along the western margin of the Strait of Georgia and sea-floor ridges along the western part of the Strait. Stratified and unstratified deposits thought to be Pleistocene in age (based on coring, reflection character, stratigraphic position, and comparison with land-based studies) are present locally above bedrock. These deposits (tills, proglacial deposits, *etc.*) can be thick (many hundreds of metres in places). Glaciomarine sediments (clays with ice-rafted debris), deposited during glacial retreat, are found in the deep ice-scoured troughs and locally as a thin drape over ridges of bedrock or Pleistocene sediments. Holocene deltaic and prodeltaic sediments of the Fraser delta form the uppermost stratigraphic unit. These deposits are locally more than 200 m thick beneath the Strait of Georgia (Fig. 4), but are absent in places over ridges of Pleistocene or Tertiary deposits (*e.g.*, Figs. 3, 4). From coring and grab sampling (*e.g.*, Pharo and Barnes, 1976; Hart *et al.*, 1992a, b, c, 1995), delta slope and adjacent prodelta sediments are generally muddy, although sands are found locally (see below).

The Fraser delta is prograding into water that is locally more than 300 m deep. Three principal morphologic zones of the submarine delta can be identified. The *delta front* is the wave-influenced portion of the delta at the seaward limit of the tidal flat, the lower limit of which is approximately 10 m depth. The *delta slope* is the relatively steeply dipping (typically 2° to 3°, locally over 7°) portion of the delta below the delta front that grades offshore into the less steeply dipping (typically <1°) *prodelta zone*. The transition from the delta slope to prodelta occurs at depths that increase northward, from approximately 100 m offshore of Roberts Bank to more than 180 m offshore from Sturgeon Bank. This morphologic picture is broken where the deltaic sediments drape and partially bury antecedent relief (*e.g.*, Fraser Ridge; Fig. 1).

By integrating geophysical surveying results, sampling and coring, maps can be constructed that depict the thickness (in milliseconds) of Holocene deposits (Fig. 4) and the surficial geology of the delta slope and proximal prodelta areas (Fig. 5). Seismic profiling suggests that most of these areas are unaffected by sediment instability phenomena, and that failures are restricted to specific areas (see below). Parallel continuous seismic reflectors are characteristic of undisturbed sediments, whereas failure deposits are characterized by transparent, chaotic or truncated reflectors (Figs. 6 A, B). Unfortunately, interstitial methane gas (manifest as acoustic turbidity on seismic profiles) adversely affects both resolution and penetration of seismic profiling systems on much of the delta slope and prodelta (*e.g.*, Figs. 3, 6A, B; Hart and Hamilton, 1992), thereby limiting the ability to image much of the internal structure of these areas (see also Chapter 17). The gas is derived from bacterial degradation of buried organic matter in the sediments, and *in situ* gas saturations have been measured which exceed 6% (Hart *et al.*, 1993). By raising sediment pore pressures, the gas contributes to submarine slope instability (Prior and Coleman, 1984).

Areas of deposition and erosion can at times be dis-

tinguished using seismic profiles (*e.g.*, Hart *et al.*, 1992d) and can be identified by measuring activities of cesium-137 in cores. The presence of ^{137}Cs in a sediment qualitatively indicates recent deposition, since this man-made radioelement is known to be primarily a product of atmospheric nuclear weapons testing (there are no natural sources), which began in the mid-1950s. By measuring profiles of ^{137}Cs activity with depth in cores, it is sometimes possible to calculate sedimentation rates (Fig. 7). For cores of constant lithology, the relative variations in concentration of ^{137}Cs are determined by the fallout signal (which is well documented globally) and the rate of deposition. This technique has been applied to cores from the Fraser delta slope (Hart *et al.*, 1993; Evoy *et al.*, 1993), and it is known that sediment is currently being deposited on most of the delta slope at rates typically about 1-2 cm·a^{-1}, but locally more than 30 cm·a^{-1} near the mouth of the Main Arm. Sedimentation chronologies help establish contaminant-loading histories; Macdonald *et al.* (1991) have used this approach to examine heavy metal fluxes in the Strait of Georgia. The southern part of the delta slope (next to Roberts Bank) is swept by strong near-bottom tidal currents, and is a zone of non-deposition and erosion.

In the remainder of this section, three specific portions of the Fraser delta are examined in detail. These sites have been

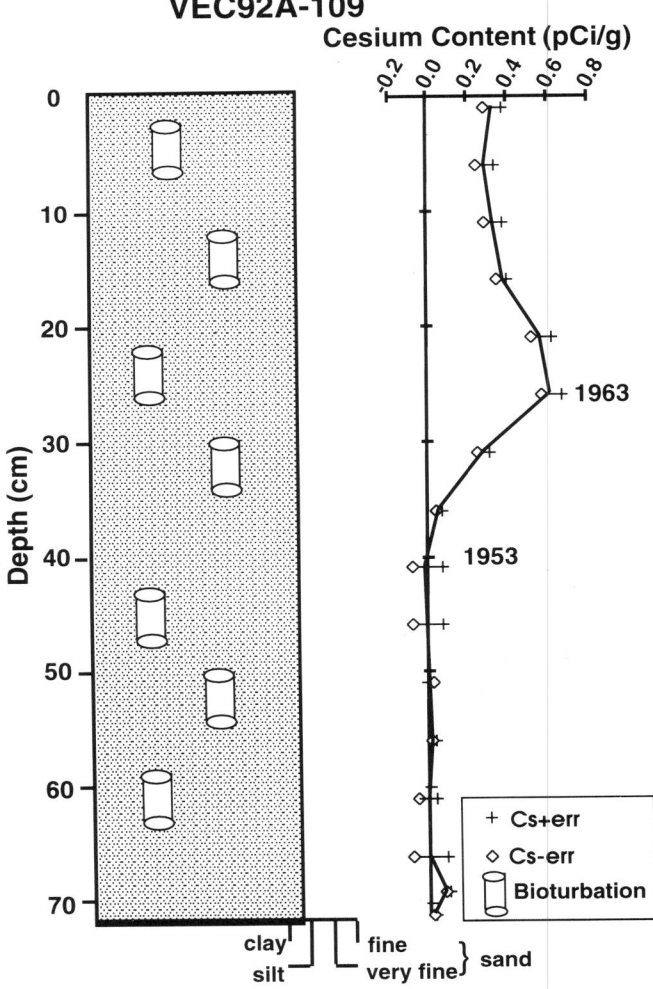

Figure 7 *Sample core showing profile of ^{137}Cs with depth. Error bars also shown. The peak in ^{137}Cs activity at 26 cm depth corresponds to 1963, suggesting an average sediment accumulation rate of 0.93 cm·a^{-1}. For location, see Figure 1.*

selected because they illustrate a range of specific environmental geological problems.

Sand Heads Failure Complex

A submarine channel and failure complex is found on the delta slope seaward of the main distributary at Sand Heads (Hart *et al.*, 1992c; Kostaschuk *et al.*, 1992; Figs. 5, 8). Several submarine channels are present on the sea floor here, and the largest (fed by tributary channels in its upper portions) has a depth of incision that decreases downslope, from more than 35 m on the upper delta slope to less than 3 m at the 200 m isobath. The width of this main channel is typically about 250 m. The channel has been cut by episodic density currents

derived from submarine failures at the river mouth such as those described by McKenna *et al.* (1992). Deposition at the river mouth reflects the interaction of the sediment-laden river discharge and the saline, tidally influenced waters of the strait (*e.g.*, Milliman, 1980).

The area north of the main submarine channel is characterized by undisturbed suspension deposits that blanket the slope and are infilling a series of relict (*i.e.*, inactive) channels on the upper slope. A smaller active submarine channel is found on the delta slope just south of the main channel. Shallow rotational slides composed of muddy sediments that are broken up into a series of rotated blocks, each many tens of square metres in areal extent, have developed south of the main

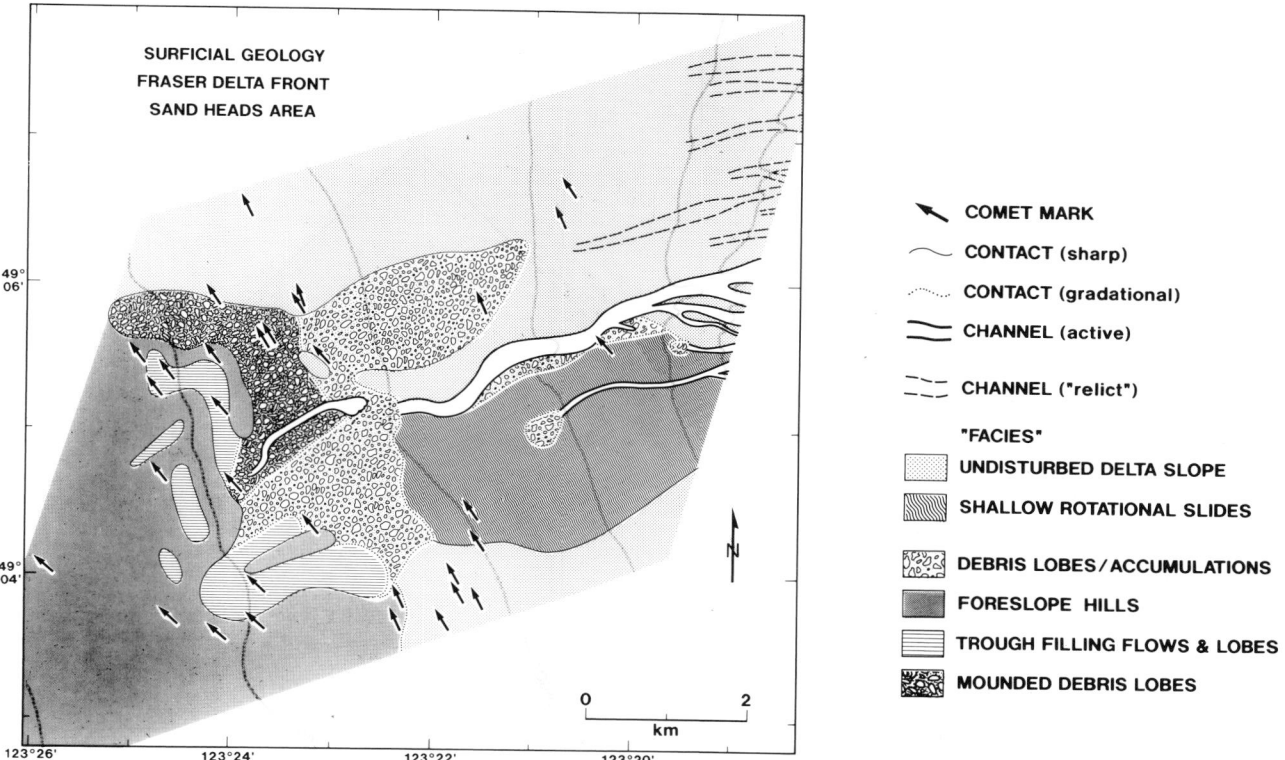

Figure 8 *Surficial geologic map of the sea floor seaward of the mouth of the Main Arm of the Fraser River (see Fig. 1), prepared using side-scan sonographs, high-resolution seismic profiles, bathymetry and cores. Rapid sediment deposition on relatively steep slopes has led to the development of various failure-related morphological elements. Comet marks are current-formed scour marks around sea-floor debris, and can be used to identify dominant current direction. These are equivalent to the obstacle-induced sediment drifts mentioned in Chapter 17.*

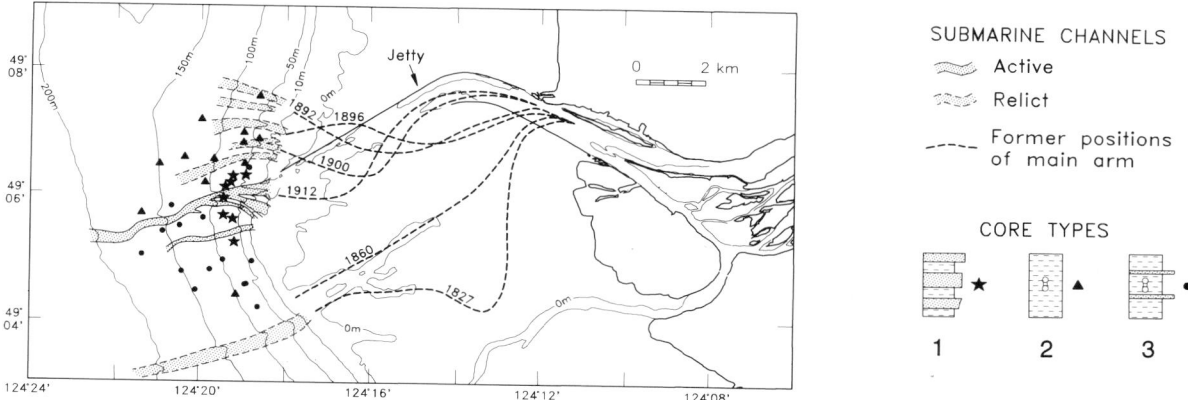

Figure 9 *Distribution of lithofacies from upper delta slope seaward of mouth of Main Arm of Fraser River at Sand Heads as derived from core data. Former positions of river mouth shown (from Clague et al., 1983). As discussed in text, stabilization of river course following jetty construction in early 1900s appears to have greatly influenced sedimentation patterns. See Figure 1 for location. 1 = sand beds; 2 = muds; 3 = interbedded sand and mud.*

channel in response to high sedimentation rates, steep sea-floor slopes (typically 6° to 7°) and elevated pore pressures.

Cores from the upper delta slope next to the Sand Heads submarine channel and its tributaries (Fig. 9) reveal the presence of centimetre- to decimetre-scale interbeds of sand (typically fine- to coarse-grained) and mud. Sand beds are generally absent from the delta slope more than about 250 m north of the main submarine channel, but are present on the delta slope south of the channel down to about 100 m water depth. The absence of sand beds to the north of the river mouth may be the result of jetty construction (Luternauer and Finn, 1983; Hart et al., 1992c).

The Main Arm is not constrained to the south, and it seems possible that during spring flood conditions sand could be supplied to the upper slope, leading to slope failures that could have generated the sand beds and laminae found in the mid-slope sediments. Morphological evidence presented by Hart et al. (1992c) from the upper delta slope south of the Main Arm (e.g., Fig. 8) indicates active failure on this portion of the delta, while north of the Main Arm, the upper delta slope is undisturbed by recent failures and sand laminae or beds are generally absent. These results suggest that sedimentation patterns at the river mouth are measurably affected by jetty construction.

Beds of fine- to medium-grained sand can also be found in density current (turbidite) deposits near the base of the delta slope at the base of the Sand Heads submarine channel (Hart et al., 1992c; Kostaschuk et al., 1992; Evoy et al., 1993; Fig. 5). Submarine channels are found incised in the sea floor off-shore from other past and present distributary mouths, and lobes of sandy failure deposits are found at their bases (Fig. 5). In the past, these channels acted as conduits for the downslope transport of sandy sediments. This association suggests that the failure complexes are part of the natural delta-building process. Since the Main Arm is no longer free to migrate, the sea floor seaward of the river mouth can be expected to continue to be a zone of ongoing slope instability.

Roberts Bank Failure Complex

The existence of a massive failure complex beneath much of the delta slope adjacent to Roberts Bank (Fig. 5) was first suggested by Hart et al. (1992b, d). Economically, this area is of major importance (Table 1; Fig. 10). Twelve of the high-voltage electrical cables that supply power from the mainland to Vancouver Island cross the tidal flats and the delta slope here. Canada's largest coal export facility (250 million tonnes loaded in the first 21 years of operations) was constructed at the seaward limit of the tidal flats (at the top of the delta slope), as was B.C. Ferry's Tsawwassen terminal, which handled nearly 6 million passengers and 2 million vehicles during 1992.

Subaqueous sand dunes cover an area of the upper delta slope 28 km² in breadth centred offshore from the coal port (Figs. 5, 10). Bedform orientations measured on the sono-graphs and current meter measurements indicate that the bedforms are generated by strong flood tide currents (maximum currents reaching 1 m·s⁻¹ during spring flood tides). In places, the dunes have buried electrical transmission cables laid in the mid-1950s, attesting to the mobility of the bedforms. Trawler marks and dredge spoil patches are locally found in proximity to the cables.

Seismic profiles reveal the existence of a failure complex which underlies at least 40 km² of the delta slope (Fig. 10). Discontinuous wavy reflectors, chaotic reflectors and transparent mounds are all interpreted as failure deposits (Figs. 6B, 11) which together form a volume of approximately 1×10^9 m³ (Hart and Olynyk, 1994). The lack of subsurface penetration evident on high-resolution seismic profiles suggests that

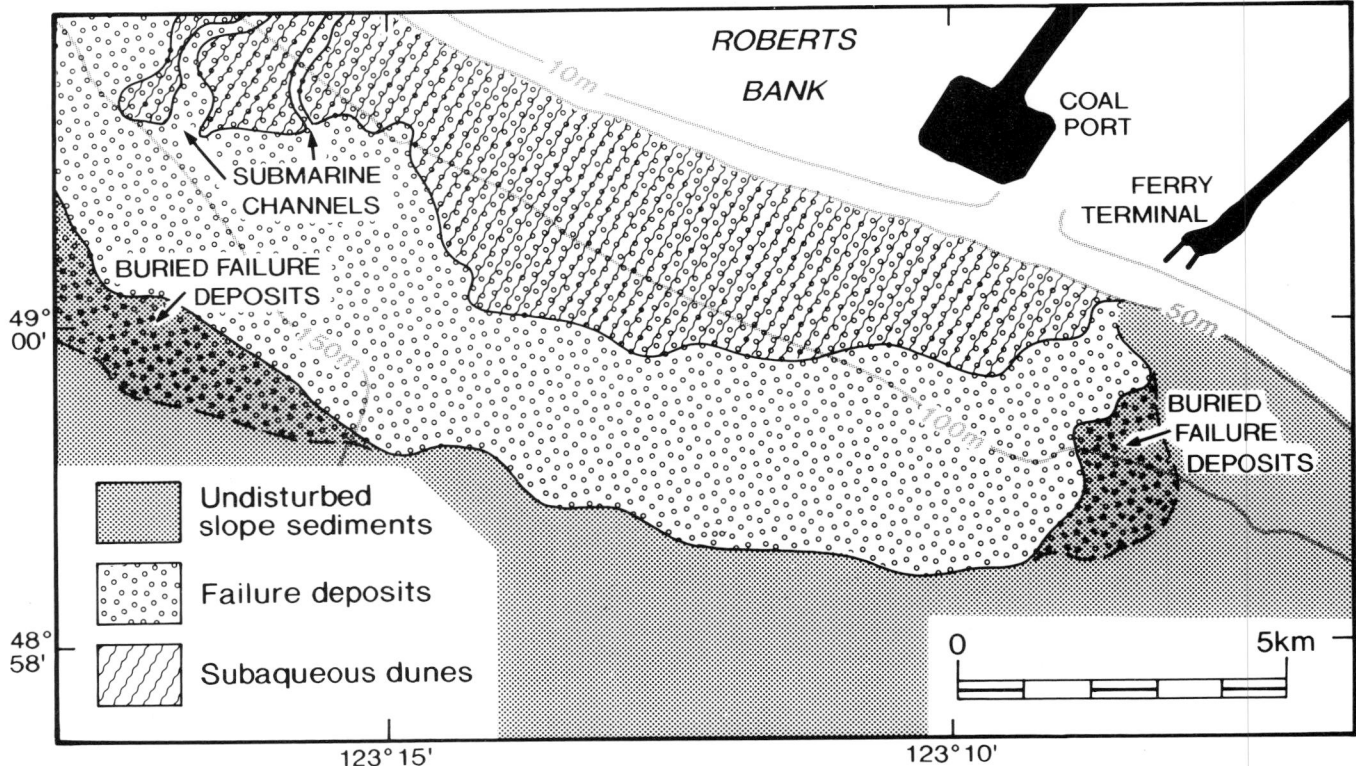

Figure 10 *Surficial geology of delta slope adjacent to Roberts Bank showing extent of area underlain by failure deposits as mapped from seismic data and extent of subaqueous dune field mapped using sonographs. See Figure 1 for location.*

these are dominantly sandy deposits, an inference borne out by bottom sampling and coring (e.g., Pharo and Barnes, 1976; Hart and Hamilton, 1992). Airgun seismic profiles penetrate more deeply (but have less resolution of detail), and show that the failure deposits are more than 75 m thick beneath some portions of the upper delta slope.

As is often the case with large submarine failure complexes (e.g., Coleman and Prior, 1988), it is not possible to positively identify any single trigger mechanism responsible for the initiation of failure. Hart and Olynyk (1994) suggested at least two episodes of large-scale retrogressive failure (involving liquefaction and subsequent flowage) of the delta slope. This conclusion suggests that the slope is affected by mass movements every few hundred years, and that the last such episode was a few hundred years ago. Such a magnitude and frequency of failure suggests that earthquake activity could have played a role in initiating failure (e.g., Clague et al., 1992; see also Chapter 30). Given the current economic importance of the area, knowledge of the susceptibility of the delta slope to seismically induced failures is vital.

Point Grey Dump Site

The Point Grey offshore dump site was established by the Ministry of Transport in 1968 seaward of the North Arm of the Fraser River in about 210 m water depth (Figs. 12, 13). This is the most frequently used offshore dump site anywhere in the Pacific region, with up to 2.5×10^6 m³ of material (mostly sand dredged from shipping channels, but also wood wastes and other types of miscellaneous debris; Packman, 1980) being dumped annually. Commercial fisheries in this area harvest at least 10 species of fish (including the bottom-dwelling sole and grey cod), and local fishermen catch 50-100 tonnes (t) of shrimp annually. Additionally, a submarine fibre optics telephone cable (part of a $16 million system to link Vancouver Island and the mainland) was laid on the sea floor close to the dump site in early 1993. Clearly, there is the potential for conflict between various activities in this area, and monitoring of the geographic range of each is essential. Dump site monitoring is an important component of the London Convention Waste Assessment Framework, and Canada has been using that accord to revise its ocean dumping regulations.

Unlike the featureless sea floor that characterizes most undisturbed portions of the Fraser delta slope, the sea floor in the vicinity of the Point Grey dump site shows abundant evidence of anthropogenic disturbance (Figs. 13A, B). Hart (1992) interpreted large (several tens of metres in diameter; Fig. 13A), circular to irregular dark patches as being produced by disposal from self-emptying (bottom-opening) barges. Linear to curvilinear successions of patches, each typically 15 m to 20 m in diameter, were interpreted as the product of repetitive spoil dumping from a moving barge (Fig. 13B). Both types of dumping operations occur at the Point Grey dump site. The dredge spoils (sandy sediments and wood wastes) are visible on the side-scan sonar records because their acoustic characteristics differ from those of the adjacent sea floor (muds). High-resolution seismic profiles collected along with the sonographs indicate that the patches represent surface features, not exposures of till or other hard substrates (see Fig. 3). Other side-scan targets (miscellaneous refuse), typically several metres in length, are also visible in this area on sonographs. Some of the larger objects (probably concrete blocks) generated craters up to 10 m in diameter upon impact with the sea floor (Fig. 13A). The sea-floor heterogeneity observable on the sonographs (e.g., Figs. 13 A, B) could also explain some of the poor reproducibility of chemical analyses between successive sampling surveys that has been noted for the Point Grey dump site (Sullivan, 1987). The dredge spoil patches are smaller than the positioning error of standard navigation systems, meaning that a contaminated patch sampled on one survey might be missed on a subsequent survey. Prior knowledge of such heterogeneity guides sampling design for biological or chemical sampling and analysis programs.

Dredge spoils were found on sonographs more than 4 km outside the dump site limit, and a well-defined trail of debris leading from Vancouver Harbour (e.g., Fig. 13B) has been mapped. Trawler marks (produced by bottom trawling) were also observed on the sonographs and, on Sturgeon Bank, dredge spoil tracks and trawler marks are found in close proximity. These results indicated that illegal dumping has occurred and, in places, these activities directly conflict with fishing activities. Repeat side-scan sonar surveys of the area would detect whether enhanced policing is needed.

DISCUSSION

The Fraser delta is a moderate-sized (by world standards; see Bhattacharya and Walker, 1992) delta that has grown in Holocene times and is now affected by urban development

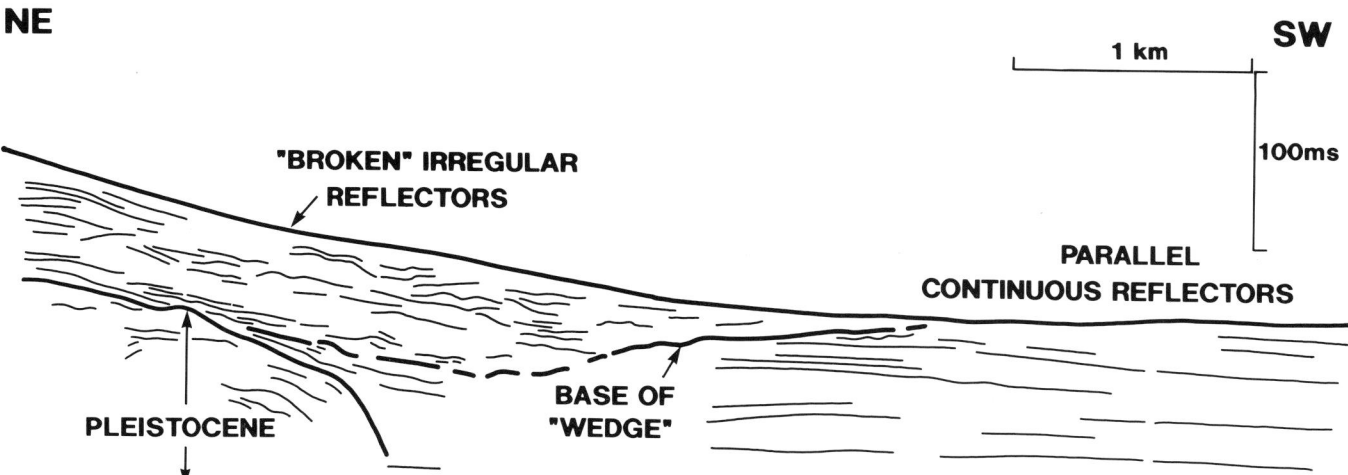

Figure 11 Line drawing interpretation of airgun seismic profile showing internal structure of delta slope adjacent to Roberts Bank, just seaward of Coal Port (Fig. 10). Failure deposits in Holocene sediments exceed 75 m thick beneath upper portion of delta slope.

and industrial activity. Comparison of this delta with modern and ancient analogs helps identify the principal factors controlling delta growth (Hart et al., 1995). High sedimentation rates, relatively high subaqueous slopes, interstitial methane gas and seismic activity combine to initiate submarine slope failures which have the potential to affect installations on the sea floor and adjacent tidal flats. The fine grain size of the sediments retards pore fluid expulsion, contributing (along with gas generation) to excess pore pressures and underconsolidation. Excess pore pressures have been associated with failures ranging in scale from the shallow rotational slides, 2-3 m thick, of the Sand Heads complex (Fig. 8; Hart et al., 1993) to the Foreslope Hills, which represent failure of the upper tens of metres of the lower delta slope (Hart, 1993), possibly as a result of earthquake activity. The Roberts Bank study identified a large failure complex in an area of signifi-cant economic importance. Geological mapping and interpretation, using a range of geophysical techniques, provide the basis for the initiation of site surveys which address the stability of the delta slope and nearby tidal flats in the event of large earthquakes. Finally, the surficial-features mapping of the Point Grey dump site has identified use conflicts and the need to enhance dump site monitoring and policing.

ACKNOWLEDGEMENTS

This paper represents a compilation of many studies conducted while the senior author was a Visiting Fellow with the Geological Survey of Canada at the Pacific Geoscience Centre, Sidney, BC. We thank Harry Olynyk (Terra Surveys Ltd.), Tark Hamilton, Kim Conway, John Luternauer, Harold Christian and David Prior. The ongoing interest and support of BC Hydro in the work is much appreciated.

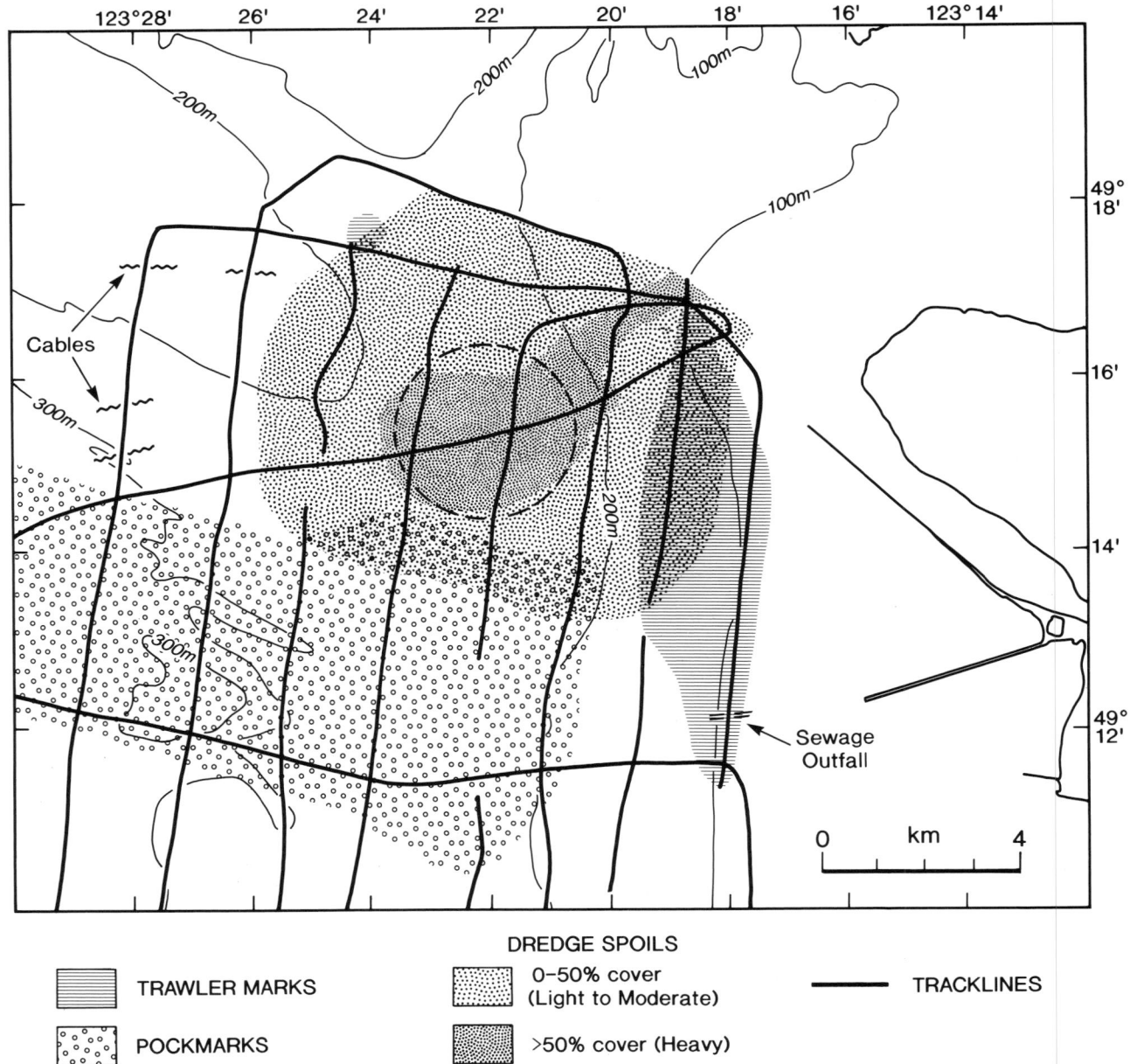

Figure 12 *Map of Point Grey dump site prepared from side-scan sonographs and seismic data. Dredge spoils detected over 4 km outside of dump site limit (dashed circle), locally in close proximity to trawler marks produced by bottom fishing. Sewage outfall from Iona treatment facility can be traced over 100 m on sonographs. See Figure 1 for location.*

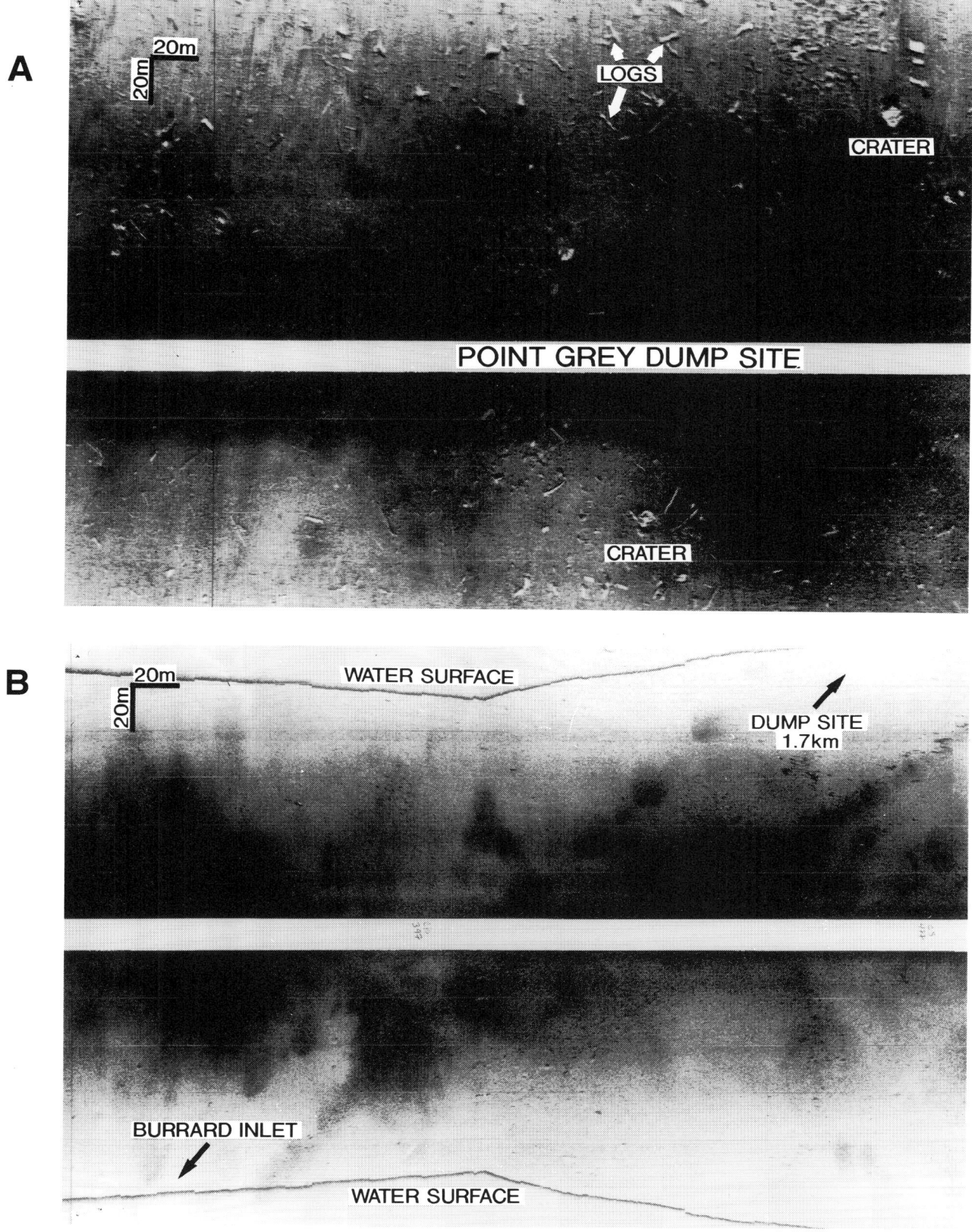

Figure 13   *Sample sonographs showing characteristics of Point Grey dump site (Figs. 1, 12). (**A**) Near continuous cover of the sea floor inside dump site limits by dark (sandy) dredge spoils, logs (from booming grounds of North Arm of Fraser River), and miscellaneous debris. Note large dark patch on right thought to be generated by disposal from self-emptying barge. (**B**) Linear trend of smaller patches thought to be generated by continued disposal from a moving barge en route from Vancouver Harbour (Burrard Inlet) to dump site. Lines marked "water surface" caused by reflection of acoustic pulse from sonar towfish (towed approximately 40 m above sea floor) on sea surface (Fig. 2).*

REFERENCES

Bhattacharya, J.P. and Walker, R.G., 1992, Deltas, in Walker, R.G. and James, N.P., eds., Facies Models. Response to Sea Level Change: Geological Association of Canada, p. 157-198.

Bouma, A.H., Stelting, C.E. and Feeley, M.H., 1983, High resolution seismic reflection profiles, in Bally, A.W., ed., Seismic Expression of Structural Styles: American Association of Petroleum Geologists, Studies in Geology Series 15, v. 1, p. 1-23.

Clague, J.J., Luternauer, J.L. and Hebda, R.J., 1983, Sedimentary environments and postglacial history of the Fraser River and lower Fraser Valley, British Columbia: Canadian Journal of Earth Sciences, v. 20, p. 1314-1320.

Clague, J.J., Naesgaard, E. and Sy, A., 1992, Liquefaction features on the Fraser delta: evidence for prehistoric earthquakes?: Canadian Journal of Earth Sciences, v. 29, p. 1734-1745.

Coleman, J.M. and Prior, D.B., 1988, Mass wasting on continental margins: Annual Review of Earth and Planetary Sciences, v. 16, p. 101-119.

Evoy, R.W., Moslow, T.F., Patterson, R.T. and Luternauer, J.L., 1993, Patterns and variability in sediment accumulation rates, Fraser River delta foreslope, British Columbia, Canada: Geo-Marine Letters, v. 13, p. 212-218.

Hamilton, T.S. and Wigen, S,O., 1987, The foreslope hills of the Fraser delta: Implications for tsunamis in Georgia Strait: International Journal of the Tsunami Society, v. 5, p. 15-33.

Hart, B.S., 1992, Side-scan sonar observations of Point Grey dump site, Strait of Georgia, British Columbia: Geological Survey of Canada, Paper 92-1A, p. 55-61.

Hart, B.S., 1993, Large scale in situ rotational failure on a low angle delta slope: the Foreslope Hills, Fraser delta, British Columbia, Canada: Geo-Marine Letters, v. 13, p. 219-226.

Hart, B.S., Barrie, J.V., Christian, H.A., Hamilton, T.S., Davis, A.M. and Law, L.K., 1993, Marine geological, geophysical and geotechnical investigations of the submarine portions of the Fraser delta: implications for submarine failures: National Research Council of Canada, Coastal Zone Engineering Program, Ottawa, ON, Canadian Coastal Conference, May 4-7, Vancouver, BC, Proceedings, p. 851-865.

Hart, B.S. and Hamilton, T.S., 1992, High-resolution acoustic mapping of shallow gas in unconsolidated sediments beneath the Strait of Georgia, British Columbia: Geo-Marine Letters, v. 13, p. 49-55.

Hart, B.S., Hamilton, T.S., Barrie, J.V., Currie, R.G. and Prior, D.B., 1992a, Marine geophysical and geological surveys of the Fraser Delta Slope and adjacent Strait of Georgia: 1991 geophysical survey tracklines and 1983-1992 core locations: Geological Survey of Canada, Open File 2543, 5 map sheets.

Hart, B.S., Hamilton, T.S., Prior, D.B. and Barrie, J.V., 1995, Seismic stratigraphy and sedimentary framework of a deep water Holocene delta: the Fraser delta, Canada, in Oti, M.N. and G. Postma, G., eds., Geology of Deltas: A.A. Balkema Publishers, Rotterdam, in press.

Hart, B.S., Horel, G., Olynyk, H.W. and Frydecky, I., 1992b, An airgun seismic survey of the Fraser delta slope: Geological Survey of Canada, Paper 92-1E, p. 33-39.

Hart, B.S., Prior, D.B., Barrie, J.V., Currie, R.A. and Luternauer, J.L., 1992c, A river mouth submarine landslide and channel complex, Fraser delta, Canada: Sedimentary Geology, v. 81, p. 73-87.

Hart, B.S., Prior, D.B., Hamilton, T.S., Barrie, J.V. and Currie, R.A., 1992d, Patterns and styles of sedimentation, erosion and failure, Fraser delta slope, British Columbia: Canadian Geotechnical Society, Geohazards '92, Proceedings, p. 365-372.

Kostaschuk, R.A., Luternauer, J.L., McKenna, G.T. and Moslow, T.F., 1992, Sediment transport in a submarine channel system: Fraser River delta, Canada: Journal of Sedimentary Petrology, v. 62, p. 273-282.

Luternauer, J.L. and Finn, W.D.L., 1983, Stability of the Fraser River delta front: Canadian Geotechnical Journal, v. 20, p. 606-613.

Luternauer, J.L. and 19 others, 1994, Fraser River delta: geology, geohazards and human impact: Geological Survey of Canada, Bulletin 481, p. 197-220.

Macdonald, R.W., Macdonald, D.M., O'Brien, M.C. and Gobeil, C., 1991, Accumulation of heavy metals (Pb, Zn, Cu, Cd), carbon and nitrogen in sediments from Strait of Georgia, B.C., Canada: Marine Chemistry, v. 34, p. 109-135.

Mathews, W.H. and Shepard, F.P., 1962, Sedimentation of the Fraser River delta: American Association of Petroleum Geologists, Bulletin, v. 46, p. 1416-1438.

McKenna, G.T., Luternauer, J.L. and Kostaschuk, R.A., 1992, Large-scale mass-wasting events on the Fraser River delta front near Sand Heads, British Columbia: Canadian Geotechnical Journal, v. 29, p. 151-156.

Milliman, J.D., 1980, Sedimentation in the Fraser River and its estuary, southwestern British Columbia, Canada: Estuarine and Coastal Marine Science, v. 10, p. 609-633.

Packman, G., 1980, An environmental assessment of the Point Grey ocean disposal area in the Strait of Georgia, British Columbia: Environmental Protection Service (Pacific), Regional Program Report 80-3, 79 p.

Pharo, C.H. and Barnes, W.C., 1976, Distribution of surficial sediments of the central and southern Strait of Georgia, British Columbia: Canadian Journal of Earth Sciences, v. 13, p. 684-696.

Prior, D.B. and Coleman, J.M. 1984, Submarine slope instability, in Brunsden, D. and Prior, D.B., eds., Slope Instability: John Wiley and Sons, Ltd., New York, p. 419-455.

Sullivan, D.L., 1987, Compilation and assessment of research, monitoring and dumping information for active dump sites on the British Columbia and Yukon coasts from 1979 to 1987: Department of Environment, Pacific Region Ocean Dumping Advisory Committee, Report 87-02, 156 p.

20. Geology and Urban Waste Management in Southern Ontario

Nicholas Eyles
Joseph I. Boyce

Environmental Earth Sciences, University of Toronto, Scarborough Campus
1265 Military Trail, Scarborough, Ontario M1C 1A4

SUMMARY

In many urban communities, waste generation commonly exceeds the capacity of management systems to handle waste by reuse, recycling, incineration and landfilling. The last is a multi-billion dollar a year industry, and involves a large effort by various public and private agencies to minimize environmental degradation of ground water by landfill leachate. Landfilling was unregulated in southern Ontario until 1971 and many sites contain toxic liquid and solid industrial wastes, but few data are available as to specific contents. Because landfill waste includes significant volumes of organic contaminants, for which drinking-water guidelines are very stringent, landfills represent a significant point source of contamination in urban areas. Fine-grained, impermeable glacial sediments afford some degree of natural protection, but have a restricted extent in southern Ontario. Future landfill sites can be fully engineered, thereby minimizing ground-water impacts, but the presence of nearly 1200 sites that lack any engineered containment system indicates the need for detailed investigation and remediation.

INTRODUCTION

Annual global production of municipal waste (about 3 km^3) is equivalent to the mean production of new rock produced each year by continental volcanism. Waste generation and disposal is one of the most serious environmental problems facing the global community, but as yet, there are few data on the relative contribution of every aspect of human activity to waste loads. This paper focusses on municipal (household or residential) waste which is normally non-hazardous, but hazardous household waste is a rapidly increasing component of the municipal waste stream. In contrast, the term urban solid waste refers to the municipal waste stream together with non-hazardous industrial, commercial and institutional waste (so-called ICI waste; Statistics Canada, 1994).

Canadian communities produced about 10.2 million tonnes of municipal waste in 1990. Average per capita production in Canada is about 1.0 kg per day, the fifth highest value in the world (Fig. 1). The highest rates of per capita generation, by province, are in Newfoundland and New Brunswick (Fig. 1). In southern Ontario, which contains 35% of the Canadian population, current municipal waste management emphasizes the burial of municipal waste in landfills, with only about 15% of the total waste stream being recycled. There are nearly 1200 active and inactive landfills sited on Pleistocene sediments in the region (Ontario Minisitry of Environment and Energy, 1991); for all but a handful there are no available data as to contents, underlying geology, or potential for groundwater contamination by leachate plumes. Most of these sites were in use prior to enactment of the Environmental Protection Act in 1971 which regulated landfilling (see below).

A detailed history of waste management in the Toronto area from 1840 to 1980 is given in Chapter 22. The purpose of this paper is to review recent developments in waste management in southern Ontario, where the distribution of glacial sediments constrains the location and design of new engineered landfills and determines the contamination potential of existing and abandoned sites.

MUNICIPAL WASTE MANAGEMENT IN THE GREATER TORONTO AREA

The Greater Toronto Area (GTA; Fig. 2) generates as much as 4 million tonnes (t) of municipal waste each year. The area's existing landfill sites are close to capacity and there is a major effort, involving government agencies, universities and the private sector, to find long-term sites that are environmentally safe. Garbage disposal is currently the focus of intense political debate. The provision for new landfill capacity in the GTA has lagged behind a steady increase in the generation of municipal waste. Annual per capita production of municipal waste in the GTA is about 600 kg which is almost 50% higher than the national average of 360 kg·a^{-1} (see above; Fig. 1). Figure 3 depicts volumes of municipal waste as currently produced every year in the five regional municipalities that comprise the GTA.

The total amount of waste landfilled in 1989 was 4.1 million tonnes. In 1989, Toronto opened a new domed stadium (SkyDome); the total amount of municipal garbage produced by the GTA in 1989 was equivalent to the volume of four SkyDomes (Fig. 4). Due to a downturn in the economy, GTA waste production has fallen recently, but this is only a temporary abatement given anticipated increases in population. Three of the four present landfill sites handling this waste (Keele Valley, which is Canada's largest landfill, Britannia Road and Brock West; Fig. 2) will be full shortly and even with planned waste diversion targets of 25% in place, the GTA has insufficient landfill capacity to meet future needs. Milton opened its own landfill in 1992 (see below). The shortfall in existing landfill capacity has been referred to as the garbage gap.

Past practice in Ontario has been to view landfilling as a cash cow to finance municipal recycling efforts. Keele Valley landfill accepted 2.25 million tonnes of municipal waste in 1990 (at a tipping fee of $150 per tonne) for a total revenue of $337.5 million. Operating costs were $27.9 million ($16.9 million as royalties to York Region, $0.5 million to the Ontario Ministry of Environment and Energy and the remainder as capital expenditure, *e.g.*, liner costs) resulting in a net income of nearly $310 million. Income was reduced to $227 million in 1991 and has fallen since in response to reduced tipping fees (currently $45 per tonne) and the diversion of waste to cheaper American dumps.

WASTE MANAGEMENT POLICIES IN ONTARIO TO 1990
Prior to the 1971, enactment of the Environmental Protection Act, unregulated dumping was widespread, and a wide variety of wastes were used to reclaim coastal marshes and infill topography (*e.g.*, Chapter 2, 18, 21). Even as late as 1977, the definition of a sanitary landfill made no mention of preventative measures designed to restrict or even assess groundwater contamination. The term sanitary referred to a daily cover of soil to prevent scavenging by animals. For most abandoned sites, there are little or no data regarding the commencement of filling, type of waste (many handled industrial and municipal solid and liquid waste) or volume, because no records were kept.

Reasonable Use Policy (1986)
Landfills constructed after 1986 have been required to satisfy Ontario Ministry of Environment (OMOE) Guidelines on Reasonable Use (Ontario Ministry of the Environment, 1986, Policy 15-08, updated 1993), which is the basis for ground quality management in Ontario. The policy sets out procedures to determine the allowable impacts on ground water in the vicinity of a landfill. The burying of waste is only acceptable if leachate is diluted sufficiently at the landfill boundary. The guidelines specify that water quality cannot be degraded by more than 50% of the difference between natural (*i.e.*, background) levels and provincial drinking-water requirements for chemical parameters listed in the Ontario Drinking Water Objectives. The allowable increase is dependent upon whether the chemical parameter is defined as health or non-health related. These guidelines are meant to ensure that landfills have a negligible impact on both ground-water quantity and quality under expected operating and failure conditions. It can be assumed that many closed sites in southern Ontario fail to satisfy OMOE Reasonable Use guidelines, since they occur on inappropriate substrates (see below).

Solid Waste Interim Steering Committee: 1989
In 1989, the solid-waste program for the GTA came under the auspices of the Solid Waste Interim Steering Committee (SWISC). Recognition of an acute shortfall in future landfill capacity prompted the five regional municipalities to set up a long-term waste management system, to be in place no later than 1996 and to be developed either by the private sector or public agencies. The main goal of the plan was to divert 50% of solid waste from landfills by reduction, reuse and recycling by the year 2000. In addition, a stand-by plan was identified to deal with the shortfall in disposal capacity between 1992 and 1996. In that plan, each region agreed to nominate one or more short-term contingency waste disposal sites to serve the GTA in that period.

The Regional Municipality of Durham nominated the P1 Contingency Landfill Site (Fig. 2) to provide interim waste capacity until long-term landfill sites were approved for use. The Regional Municipality of Peel nominated a site (V1B) near the city of Brampton, while Metropolitan Toronto nominated a controversial site (M2) on its eastern border in Scarborough, close to the Rouge Valley, in an area now designated as a park (Fig. 2). SWISC's comprehensive long-term solid-waste management programme for the GTA, planned to begin operation in 1996, was to be subject to a full Environmental Assessment Act process. This process was streamlined for interim, short-term sites (see below). Under the terms of the agreement arranged by SWISC with Durham Region, $20 million would be paid immediately and a further $21 million at the time of approval for site P1; this money would be available to fund

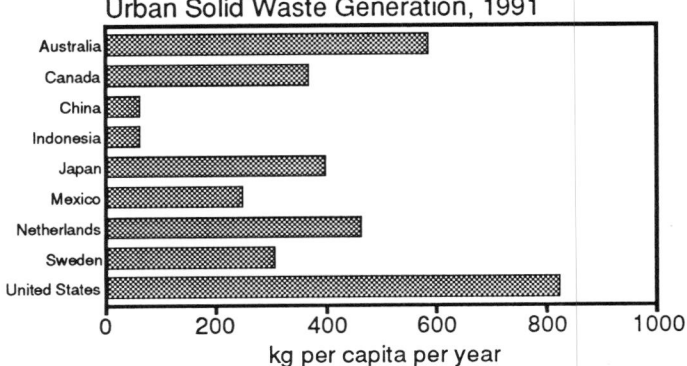

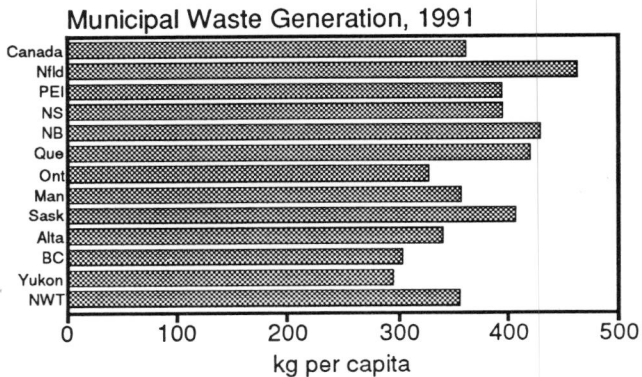

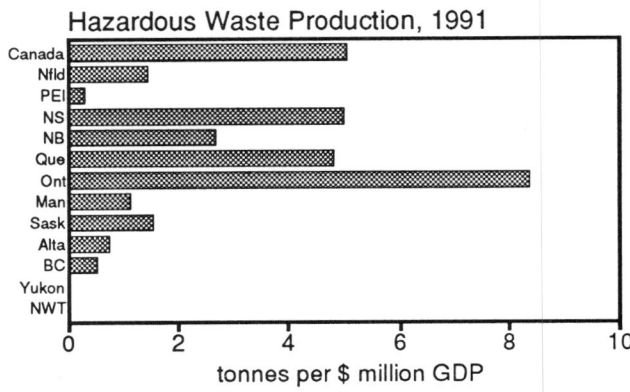

Figure 1 *Urban solid waste generation by country (**top**), by province (**middle**) and for hazardous wastes (**bottom**) for 1991 (data from Statistics Canada, 1994). Urban solid waste refers to municipal, non-hazardous industrial, commercial and institutional wastes combined.*

recycling projects in Durham. An estimate of capital expenditure on the landfill was about $51 million during a five-year period for construction, operation, closure and perpetual care; net revenue during that time would be about $210 million. Thus, there are significant financial benefits to be gained by having a landfill site approved. Commercial tipping fees at existing landfill sites increased to $150 per tonne in early 1990, representing a 700% increase since 1988.

Environmental Protection Act versus Environmental Assessment Act

The wider political dimensions of solid-waste handling and control are expressed through the legal processes, whereby landfill sites are chosen under environmental protection legislation. In some cases, legal considerations are sufficiently important to override geological considerations of the respective merits of individual landfill sites. This is not the place to review in detail the existing environmental legislation in Ontario that governs landfill practice. Reviews of legislation and compliance procedures are given by Phyper and Ibbotson (1991) and in Chapter 39.

In brief, the selection of landfill sites in Ontario is normally examined, approved and licensed under the Ontario Environmental Assessment Act (Ont. EAA), whereby the best site is located following an exhaustive screening process, taking into account potential impacts on the natural and human environment (*e.g.*, social, economic and cultural conditions; Fig. 5). The time required to complete such a process is at least five years, and would substantially exceed that of landfill exhaustion for the GTA, resulting in an intensified crisis with regard to waste disposal capacity. As a result, the municipalities of Peel and Durham applied to the Minister of the Environment in July 1990 for the V1B and P1 sites to be made exempt from the Ont. EAA. Approval was to be sought, instead, under the Ontario Environmental Protection Act (Ont. EPA), which examines the environmental impact and suitability of specific sites chosen without undergoing the comprehensive screening process demanded by the Ont. EAA (Fig. 5). Part V of the Ont. EPA governs the approval of waste management facilities. When specific criteria, designed to minimize the impact on the natural environment only, are met, the required Certificate of Approval is granted.

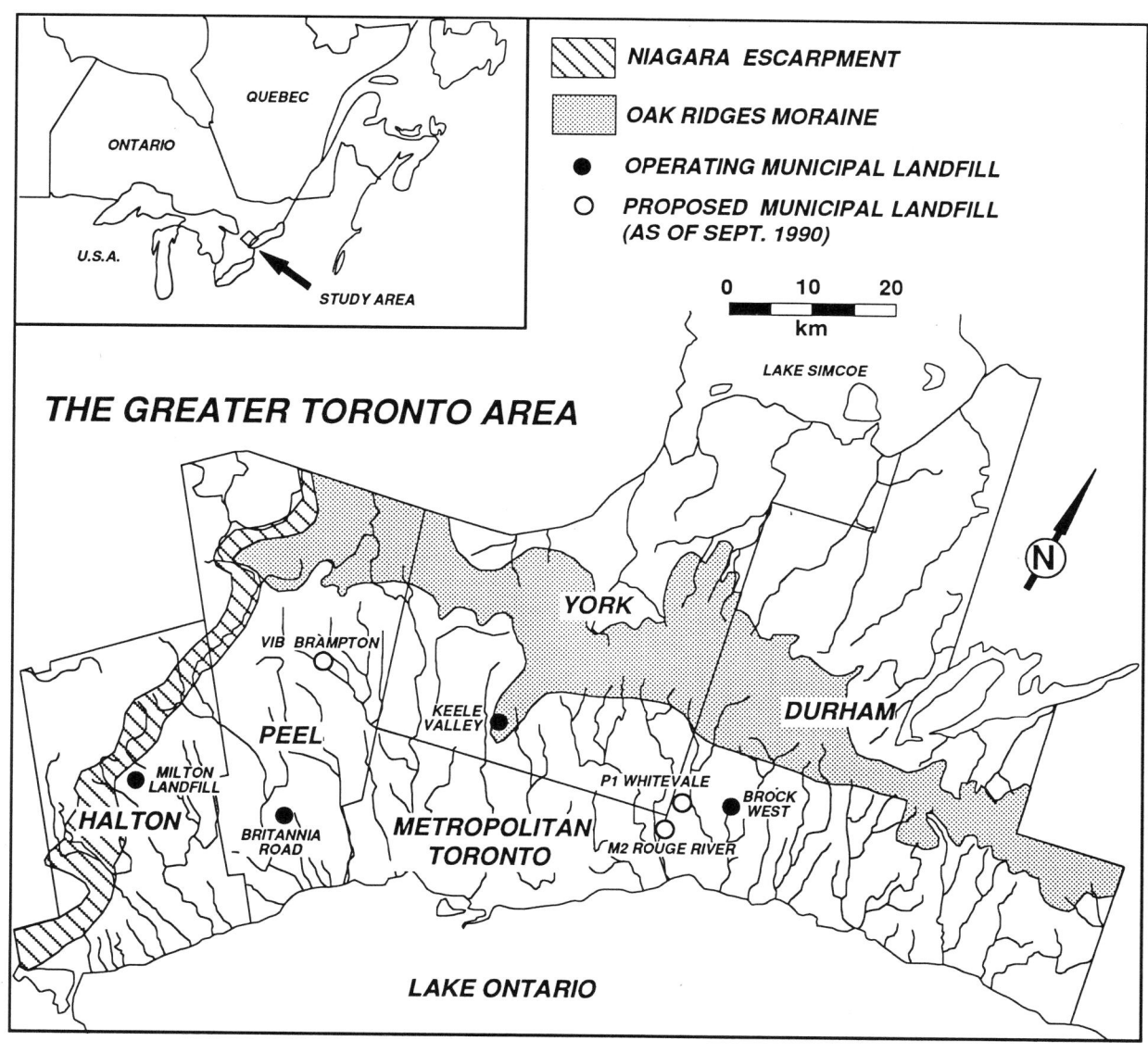

Figure 2 *Location map for Greater Toronto Area (GTA) showing currently operating municipal landfill sites and those proposed by the last Liberal Government. Keele Valley site is shown in Figure 21. Proposed sites were rejected in late 1990 and a new search initiated by the Interim Waste Authority Ltd. (Fig. 9).*

The Ont. EPA, therefore, has a much narrower definition than the Ont. EAA, and is widely known as a fast-track mechanism, compared with the lengthy and exhaustive evaluation required by Ont. EAA legislation.

This exception was granted through an order in cabinet by the then Liberal Government. A cabinet order requires no debate in the Ontario Legislature. Certain restrictions were attached, however; namely, that the sites would be closed in 1996 and that 25% of the waste stream by weight would be diverted from landfills (*i.e.*, recycled). A provincial election in September, 1990, was won by the New Democratic Party (NDP). In the election campaign, the NDP promised that, if elected, they would revoke the exemption and institute a full environmental assessment under Ont. EAA regulations for all future landfills in Ontario, including interim sites nominated by the GTA.

The Interim Waste Management Authority Limited: 1991
In November 1990, the new Environment Minister announced that the NDP would shelve the Whitevale and Brampton interim sites, arguing that they were not needed despite strong opposition from the regional chairman of the GTA. SWISC was dissolved and was replaced by a provincially established authority (the Interim Waste Authority Ltd.; IWA) with responsibility for new dump sites and technologies for recycling. One day after making this announcement, however, the same minister announced that the government was,

if necessary, prepared to use emergency powers to open the Whitevale and Brampton sites without a formal hearing, and would, furthermore, expand existing dumps (Britannia Road, Keele Valley, Brock West; Fig. 2), also without a hearing. As a result, in June 1991, a Provincial Order was issued to expand the Britannia Road site without any environmental assessment or study. At that time, a 6000-home subdivision had been planned close to the dump site on the expectation of its closure in 1992 (see Chapter 2). Brock West exceeds the original design capacity, but is still accepting waste at the time of writing. Keele Valley could accept, under emergency expansion, about 14 million tonnes, sufficient for approximately three years, given the current annual production of garbage within the GTA. Any reduction in the waste stream by recycling and reuse and compaction of existing landfill wastes would extend the capacity.

The Interim Waste Authority
Site Selection Procedure: The Seven Steps
Bill 143 was passed in the spring of 1992 to empower the Interim Waste Authority Ltd. to manage the search process, obtain environmental approvals and operate the landfills. The IWA site search process was divided into seven steps which reduced 57 candidate sites to 3 preferred sites (Fig. 6). In Step 6 (identification of the preferred site), 12 criteria were employed, encompassing consideration of such variables as impacts on agriculture and ground water to economics (Figs.

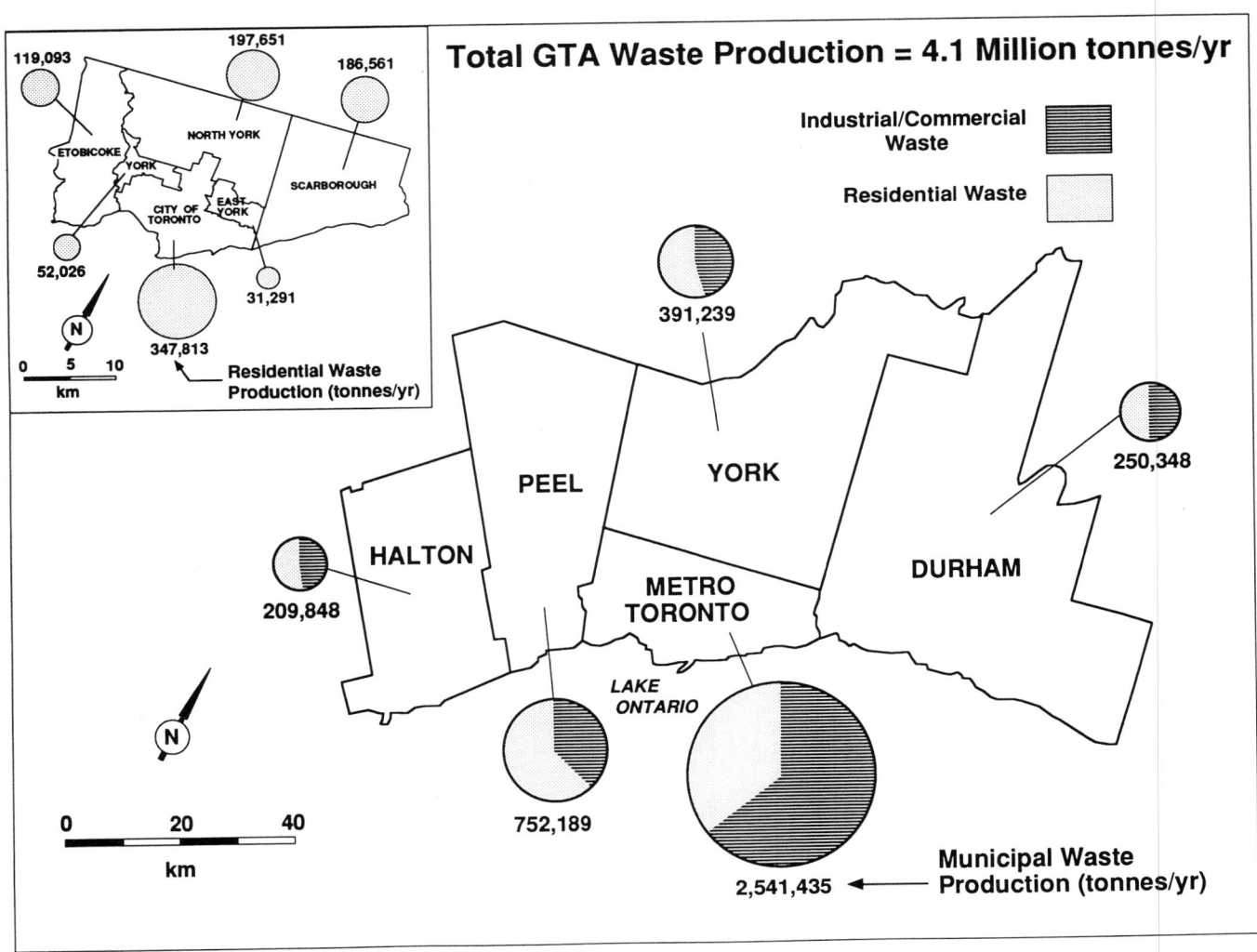

Figure 3 *Municipal waste production in 1989 for the GTA and Metropolitan Toronto (inset).*

7A, B); each site was ranked according to their scores. Each criterion was then weighted according to its relative importance, with geology and hydrogeology being given the highest weighting; several criteria were given equal weighting (Fig. 7B). An example for the Durham Region search is shown in Fig 7C. The ensuing search generated intensive studies of the subsurface glacial stratigraphy across the GTA (*e.g.,* M.M. Dillon, Ltd., 1994; Golder and Associates, 1994; Boyce *et al.,* 1995; Chapter 32). Step 7 involved detailed and lengthy investigation of the geology and hydrogeology of the preferred sites. These sites were to be reviewed, in public hearings, under the Ontario Environmental Protection Act. With the election of a new Conservative government in June 1995, the site search was abandoned.

The largest preferred site (site V4A; Fig. 8, 9) was intended to serve the combined areas of York and Metropolitan Toron-

to, with two other sites to serve Durham (site EE11) and Peel (site C34B). In the mean time, garbage production is down as 3Rs programs (reduce, reuse, recycle) take hold. A central question was, therefore, whether all three sites were, in fact, needed. Perhaps other locations could have been reconsidered that were rejected in the initial screening as being too small.

In contrast, the regional municipality of Halton exhausted its landfill capacity in 1988. Halton's new landfill at Milton (Site D; Trow Hydrology Consultants, 1986; Fig. 2), with an area of 53 ha and a capacity of 4 million tonnes, opened in 1992 after a process that cost $10 million and took 10 years. The region of Halton deliberately maintains high tipping fees ($150 per tonne) designed to promote the export of garbage to the United States and prolong the life of the Milton landfill (Buitenhuis, 1994).

Present-day situation.   With the economic recession now affecting waste production, the garbage gap has eased temporarily. Brock West, for example, may be open until 1997, given the decrease in the dimensions of the waste stream. Better-than-expected compaction at the landfill has also helped. The volume of waste dumped in Metro's landfills dropped by 31% in 1991, and the trend is continuing. The biggest decrease is in excavated fill materials, largely reflecting a slump in the construction industry. However, current recycling efforts vary widely across the GTA, and most municipalities failed to reach 1994 targets of 25% diversion. The highest proportion of recycled material was achieved in Peel Region (24%) as against 16% for the entire GTA. Recycling efforts are being spurred by rising prices for recycled newsprint ($20 to $200 in the last two years) and glass. It is probably safe to assume that all existing Metro area dumps will be expanded and will continue to be used until new IWA sites are identified. Incineration and long-distance rail haul of garbage to remote dumps in northern Ontario and the United States are now being considered.

Municipal waste exports.   Metropolitan Toronto exports up to 500,000 t of waste annually to the United States. As a

VOLUME OF SKY DOME = 1,557,435 m³

CAPACITY = 1,090,204 tonnes of garbage

= 26% of GTA annual waste production

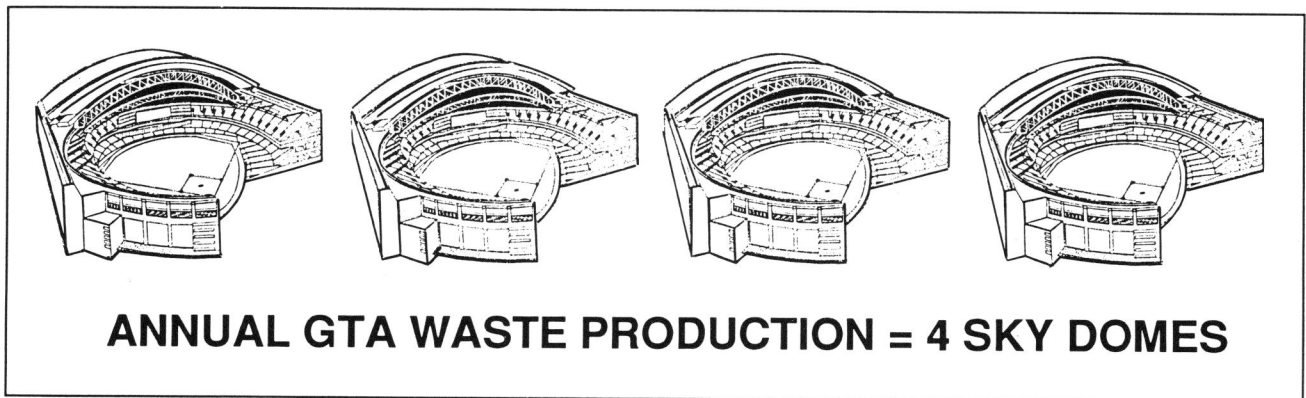

ANNUAL GTA WASTE PRODUCTION = 4 SKY DOMES

Figure 4   **(A)** *SkyDome as an interim landfill site: a cartoonist's view (courtesy of* The Toronto Star*).* **(B)** *Annual GTA municipal waste production expressed as SkyDome volumes.*

COMPARISON OF EA ACT AND EP ACT FOR WASTE MANAGEMENT UNDERTAKINGS

ENVIRONMENTAL   **V.**   **ENVIRONMENTAL**
ASSESSMENT ACT   **PROTECTION ACT**

FOCUS

- "front end" planning
- consideration/evaluation of altermatives
- detailed decision analysis (rationale) supporting the selection of the preferred alternative (undertaking) from among others considered

- regulates implementation and operation of a specific facility (specific and detailed technology at a specific site)

DEFINITION OF "ENVIRONMENT"

- "broad" definition
- proponent to consider potential effects on social, natural, cultural, economic, technical and land-use planning components and the interrelationships among them for the undertaking and its alternatives

- "narrow" definition
- proponent to consider the potential effects on the natural environment (air, land, water, plant and animal life) of the specific undertaking
- includes transportation and noise aspects

LEVEL OF DETAIL

- sufficient detail to allow approval of the preferred alternative (undertaking), *i.e.*, the preliminary design and concept
- will normally involve some site-specific assessment
- preliminary design and concept

- site-specific and detailed technical analysis of the proposal

PUBLIC/GOVERNMENT REVIEWER INVOLVEMENT

- during pre-submission consultation (PSC) (MOE Policy)
- during the Review of the EA submission
- formal public involvement occurs during minimum 30 day public review period upon completion of the Review and during optional EA Act or Consolidated Hearing (EA Act & other statutes)

- usually PSC with MOE prior to submission of Part V EP Act application
- public input on application encouraged at PSC stage
- other agencies (in addition to MOE) may be involved in reviewing the application
- formal public involvement occurs before the Hearing Board during mandatory EP Act or Consolidated Hearing (EP Act & other statutes)

DECISION ON APPROVAL

- if no hearing required, Minister makes decision on approval with Cabinet
- under EA Act and Consolidated Hearing, EA Board make decision
- the decision of the Joint Board may be appealed to Cabinet and Cabinet may vary or rescind the Board decision within 28 days

- under EP Act, EA Board makes the decision. The decision of the EA Board may be appealed to Cabinet within 30 days of the decision. Cabinet may vary or rescind the Board decision or require the EA Board to hold a new hearing

CRITERIA COVERING APPLICATION OF THE EA ACT TO WASTE MANAGEMENT ACTIVITIES

Waste Type	Landfill	Incineration	Processing Treatment (*e.g.* Sewage Treatment Plants)	Transfer Stations
Municipal	≥ 40,000 m3:EAA ≤ 40,000 m3:EPA	Banned	≥ 200 TPD:EAA < 200 TPD:EPA	≥ 300 TPD:EAA < 300 TPD:EPA
Hazardous and Liquid Industrial Waste (LIW)	EAA	EAA	≥ 200 TPD:EAA < 200 TPD:EPA	≥ 300 TPD:EAA < 300 TPD:EPA

Figure 5 *(top)* Comparison of Environmental Assessment and Environmental Protection acts as they relate to waste management undertakings in Ontario. *(bottom)* Criteria covering applications of these acts to waste management (after Ontario Ministry of Environment, 1992). Abbreviation: *TPD*, tonnes per day.

result, Metro has lost at least $60 million in disposal fee revenues to the United States, where tipping fees are much less. There is no shortage of United States dumps, which are subjected to less rigorous environmental controls and are willing to take Canadian garbage for $40 per tonne. Much of the waste is incinerated, and airborne pollutants are exported back to Ontario, where incineration is banned (see below). Even if tipping fees were greatly reduced at GTA landfills, there would be little reduction in the volume of waste being exported given that United States companies largely control the collection and haulage of such wastes. Waste is now an international commodity; competing GTA management options, as of late 1995, include long-distance rail haul to a mine in northern Ontario (500 km), to a mine in Utah (3000 km), to a landfill in Ohio (570 km) with commodities such as coal and aggregate being returned in eastbound rail cars.

Incineration.   Some countries extol the environmental virtues of incineration, but they commonly lack the geologic conditions necessary for landfills. They also ignore air quality considerations and the continuing need to landfill ash. Incineration of municipal waste is not, at present, a viable option for Ontario, and is also a touchy area in United States-Canada relations. As late as 1987, 140,000 t of garbage were incinerated each year in Ontario, but this practice was discontinued in 1988, except for a small (450 t per day) incinerator in Brampton, Hamilton (520 t per day) and London (400 t per day). In addition, sewage sludge residue is incinerated. For example, 200 t a day are incinerated at Highland Creek Sewage Treatment Plant (STP) in the city of Scarborough and 125 t day at Ashbridge's Bay STP in the city of Toronto; the ash is landfilled.

Ontario was the first jurisdiction anywhere in North America to ban the development of new solid-waste incineration facilities. In 1991, the Environment Minister issued a lawsuit against the City of Detroit to force the city to add state-of-the-art emission control systems to its incinerator. Michigan lawyers are currently seeking to prove that Ontario issued writs with unclean hands, citing that Ontario incinerators do not have the most up-to-date emission controls. Furthermore, the province exports waste to other United States incinerators that lack modern emission controls, and airborne pollutants such as dioxin are exported back to Canada by long-distance atmospheric transport, resulting in continental-scale contamination (Cohen *et al.*, 1995). The largest source of dioxin emissions in the Great Lakes region is from the incineration of medical waste (53% of total loading to atmosphere), because it contains large volumes of plastic, followed by municipal incinerators (24%). In contrast, 4.5% is derived from cement kilns that use hazardous waste as fuel, 2.8% from steel works, and 0.3% from incineration of sewage sludge (Cohen *et al.*, 1995).

Renewed discussion of incineration as an option has been generated by the plan of the provincial energy utility (Ontario Hydro) to incinerate up to 8,000 t of municipal waste a day to produce electrical power. In response, Neurath (1993) presented a quantitative assessment of the social, financial and environmental costs of different waste management options; the costs of incineration outweigh any advantages (Fig. 10). Currently, the Ontario Ministry of the Environment and Energy (OMOEE) has no legally enforceable air emission standards for incineration at source (*i.e.*, the stack). OMOEE has, instead, point-of-impingement standards (equivalent to ambient concentration in the United States) for a limited range of pollutants. These standards are meaningless for two reasons. First, the emitter need only build a taller stack and disperse the same pollutant load over a larger area. Second, the common carcinogenic pollutants produced by incinerators, such as dioxins, cadmium, chromium, lead, mercury, nickel, zinc, polychlorinated biphenyls (PCBs) and pentachlorophenols, are bioaccumulating. Ingestion *via* food over long time periods presents a greater health risk than that from direct inhalation (Webster and Connett, 1989, 1990). Neurath (1993) concluded that recycling, reuse and landfilling remain the most efficient mix of waste management options.

GEOLOGICAL SETTING OF LANDFILLS IN SOUTHERN ONTARIO

The form of the bedrock surface in southern Ontario (Fig. 11) has exerted a significant control on the thickness and preservation of overlying Pleistocene sediments. A prominent bedrock channel (Laurentian Channel; Chapter 2) contains the thickest drift cover in the area (Fig. 12). The considerable thickness of glacial drift in southern Ontario has historically prompted the disposal of waste by landfilling, and lies at the root of the current problem. Deposits of the last glaciation, such as tills (see below), are of prime interest in landfill evaluation, given that they comprise the dominant surface materials across the region and the shallow (<15 m) depths of excavation employed in landfill construction. Close to the Niagara Escarpment and on the Proterozoic Shield exposed to the northeast of the GTA, the reduced thickness of glacial drift (<10 m) necessitates that consideration be given to the hydrogeological properties of bedrock (*e.g.*, Novakowski and Lapcevic, 1988). Recent experience with proposals for landfills sited in bedrock quarries suggests that, in general, such sites are unlikely to be approved by the province in the near future given the poor understanding of ground-water flow in fractured bedrock (see Chapters 2, 39).

The map of Pleistocene geology of southern Ontario (1:1,000,000 scale), published by Chapman and Putnam (1984) with subsequent modifications by the Ontario Waste Management Corporation (1986) and the Ontario Geological Survey (Barnett *et al.*, 1991), is a valuable tool in the first stages of screening potential landfill sites. However, maps of surficial geology show only the type and distribution of surface sediments; they provide no data as to the deeper stratigraphy and commonly convey a false sense of geological homogeneity. The hydrogeological assessment of potential landfill sites is, therefore, dependent upon detailed knowledge of the distribution and geometry of sediments in the subsurface and the spatial variation in physical properties. Mapping of aquifers, water quality and ground-water yields has been conducted by the Ontario Ministry of the Environment (*e.g.*, Haefeli, 1970; Sibul *et al.*, 1977; Turner, 1977), but the coverage is neither exhaustive nor up-to-date. Although a wealth of environment-related data has been accumulated by government, private sector and other agencies in Ontario, no comprehensive management system exists for the assessment, updating and dissemination of such information.

Landfill Substrates

Figure 13 shows the distribution of near-surface (<15 m) Pleistocene sediments in southern Ontario, grouped according to dominant lithology (*i.e.*, permeability). Low-permeability glaciolacustrine silts and clays are generally considered to be favourable substrates for landfill sites, but have limited areal extent. Fine-grained, so-called matrix-dominant, and coarser (clast-dominant) tills are common surficial materials in the region. Matrix-dominant tills (Fig. 14) are characterized by high silt and clay contents (>80%) and low to moderate

INITIATION OF SEARCH PROCESS

Step 1 Primary screening (constraint analysis) to identify potential candidate areas (*e.g.* Fig.9)

GEOLOGY / HYDROGEOLOGY

Screening Criteria	Rationale	Data Sources
1. Screen out areas of ice contact stratified drift and associated hummocky topography (*e.g.* Oak Ridges Moraine: Fig. 1)	These areas have a high probability of being complex or highly variable in the subsurface. The variability or complexity of the geological regime is a measure of the variability of the hydrogeological regime. These areas were therefore considered undesirable since they would make it unlikely to be able to reliably monitor site performance; to implement effective contingency measures and to confidently assess potential impact and ensure protection for the environment.	• 1:50,000 or larger scale published geologic maps and reports • Data transferred to 1:10,000 scale mapping at the completion of this step.

Step 2 Secondary screening (constraint analysis) to identify candidate areas.
Step 3 Identification of candidate site boundaries to identify an initial long list of candidate sites (17 sites across 3 Regions)
Step 4 Comparative evaluation of initial long list of sites to identify the final long list of candidate sites.
Step 5 Comparative evaluation of final long list of candidate sites to identify a short list of candidate sites.

GEOLOGY / HYDROGEOLOGY

Evaluation of Criteria	Indicators	Rationale	Data Sources
1. Compare potential for site to provide natural geological protection from leachate.	a) Nature and relative permeability of attenuation layer b) Representative thickness of attenuation layer	Sites having higher attenuation capabilities provide greater potential for natural protection and less reliance on engineered protection of groundwater and surface water resources and users.	• Step 4 sources • Local geotechnical and hydrogeologic maps and consultants reports • Water well records • Ministry of Transportation (MTO) site investigation reports • Existing aerial photographs • Agency contacts • Road side reconnaissance
2. Compare potential for site to provide natural hydraulic protection from leachate.	a) Representative groundwater levels in the monitoring layer(s)	Sites having higher groundwater levels have a higher potential for inward groundwater flow which can provide natural hydraulic containment.	Same as above
3. Compare potential for predicting groundwater migration pathways	a) Uniformity of geological setting b) Lateral groundwater flow directions beneath the site c) Nature of monitoring layer	Sites with predictable groundwater migration pathways allow reliable monitoring of site performance, implementation of effective design and contingency measures, and a confident assessment of potential impact.	Same as above
4. Compare potential for disrupting groundwater supplies and resources.	a) Groundwater resource potential b) Distance to permitted high capacity wells.	It is desirable to minimize disruption to groundwater supplies and resources. The two components of the resource considered are potential disruption to resource quality and resource quantity.	Same as above

Step 6 Comparative evaluation of short list of candidate sites to identify the preferred site (1 for each Region; Fig. 9))

GEOLOGY / HYDROGEOLOGY

Evaluation of Criteria	Indicators	Rationale	Data Sources
• 1. Compare potential for natural protection from leachate.	a) Thickness of attenuation layer b) Nature and permeability of attenuation layer c) Vertical gradient	• Settings that afford a higher degree of natural protection are preferred	• Step 5 sources • On-site data including: - Borehole logs - Hydraulic conductivity (laboratory and field); - Water levels; - Grainsize data; - Geophysical data; - Groundwater chemistry • door to door well inventory
• 2. Compare ability to monitor groundwater and implement contingency measures	a) Number of groundwater migration pathways b) Nature of groundwater migration pathways c) Local groundwater flow directions d) Potential for interference from other sources of contamination	• Sites for which simple and reliable monitoring of site performance and simple contingency measures can be implemented are preferred.	Same as above
3. Compare potential for disrupting groundwater supplies and resources	a) Presence of groundwater resources in vicinity of site b) Number of wells in use in vicinity of site c) Proximity to existing permitted high capacity wells d) Quality of on-site groundwater resources	• It is desirable to minimize the potential for disruption to groundwater supplies and resources. Sites that are more remote from good quality, high yield groundwater resources are preferred.	Same as above

SITE C-34B (Peel Region)

SITE V4A (Metro/York Region)

SITE EE11 (Durham Region)

Step 7 Detailed geologic and hydrogeologic evaluation for the preferred site

PUBLIC HEARINGS

Figure 6 *Seven step Interim Waste Authority Ltd. site selection process (after Interim Waste Authority Ltd., 1992).*

A   SUMMARY OF TWELVE CRITERIA FOR LANDFILL SITE SELECTION

Archeology	• minimize destruction of significant archeological sites on site
Aviation	• aviation safety at airports (increase in birds)
Biology	• loss or disruption of terrestrial and aquatic systems on site, off-site, in the primary and secondary impact zones, along the access routes from hauling waste, and along the access routes affected by local road closure(s)
Design and Operations	• ease of developing waste disposal capacity
	• potential for obtaining cover and liner material
	• potential for disposal of leachate or effluent
	• ease of on-site surface water management
	• potential for existing property contamination
Economics	• displacement of on-site, business operations and public sector employers
	• disruption of off-site business operations and public sector employers within the primary and secondary impact study zones
	• disruption to business operations and public sector employers along access routes for hauling waste and roads affected by local road closure(s)
	• acquisition and haul costs related to establishing and accessing a landfill
Heritage	• minimize displacement of historic heritage or cultural features on-site and along access routes
Geology/Hydrogeology	• potential for site to provide natural geological and hydraulic protection from leachate
	• potential for predicting ground-water migration pathways
	• potential for disrupting ground-water supplies and resources
Planned Land Use	• displacement of planned land uses on-site, off-site along access routes for hauling waste and along roads affected by local road closure(s)
Social	• displacement of residents and institutional, community and recreational features located on-site, in the primary and secondary impact study zones off-site, along access routes for hauling waste and on roads affected by local road closure(s)
Surface	• impairment of surface water quality
	• potential flood hazard
Transportation	• traffic safety and operations along haul routes, and roads affected by local road closure(s)

B   MULTI-DISCIPLINARY CRITERIA GROUP RANKING AND WEIGHTING

Criteria Group	Rank	Weight
Geology/ Hydrogeology	1	22
Agriculture	2	13
Social	2	13
Transportation	4	8
Biology	5	7
Economics	5	7
Design and Operations	5	7
Surface Water	5	7
Planned Land Use	9	6
Aviation	10	5
Heritage	11	3
Archeology	12	2

C   SHORT LIST SITE NAMES

	T1	EE4	EE10	EE11	KK2
Geology/ Hydrogeology	4	11	11	1	8
Agriculture	5	2	1	2	9
Social	5	13	13	5	1
Transportation	9	13	9	6	2
Biology	5	3	4	5	5
Economics	10	3	1	1	8
Design and Operations	1	7	8	1	9
Surface Water	7	7	7	3	1
Planned Land Use	10	7	15	14	2
Aviation	1	1	1	1	1
Heritage	1	3	2	3	6
Archeology	13	5	5	2	5
Overall Site Ranking	**3**	**4**	**2**	**1**	**5**

Figure 7   (**A**) *Summary of criteria used inn landfill site selection by Interim Waste Authority Ltd. See Figure 8 for definition of primary and secondary impact zones.* (**B**) *Ranking and weighting of criteria with* (**C**), *an example from Durham Region identifying site EE11 as the preferred site (Fig. 9).*

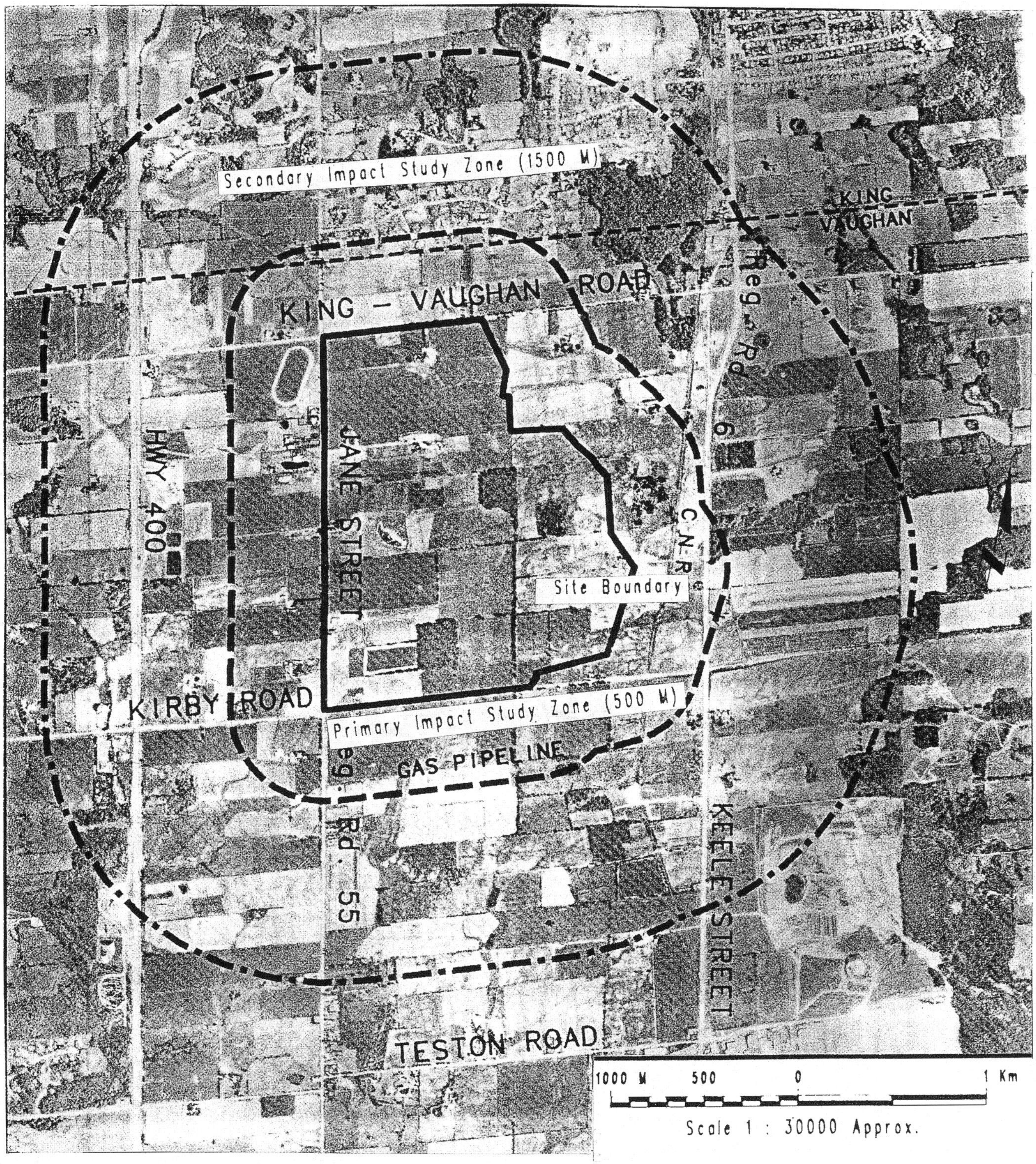

Figure 8 *Aerial photograph of Interim Waste Authority Ltd. preferred landfill site V4A in York Region (see Fig. 9 for location) showing site boundary and primary and secondary impact zones.*

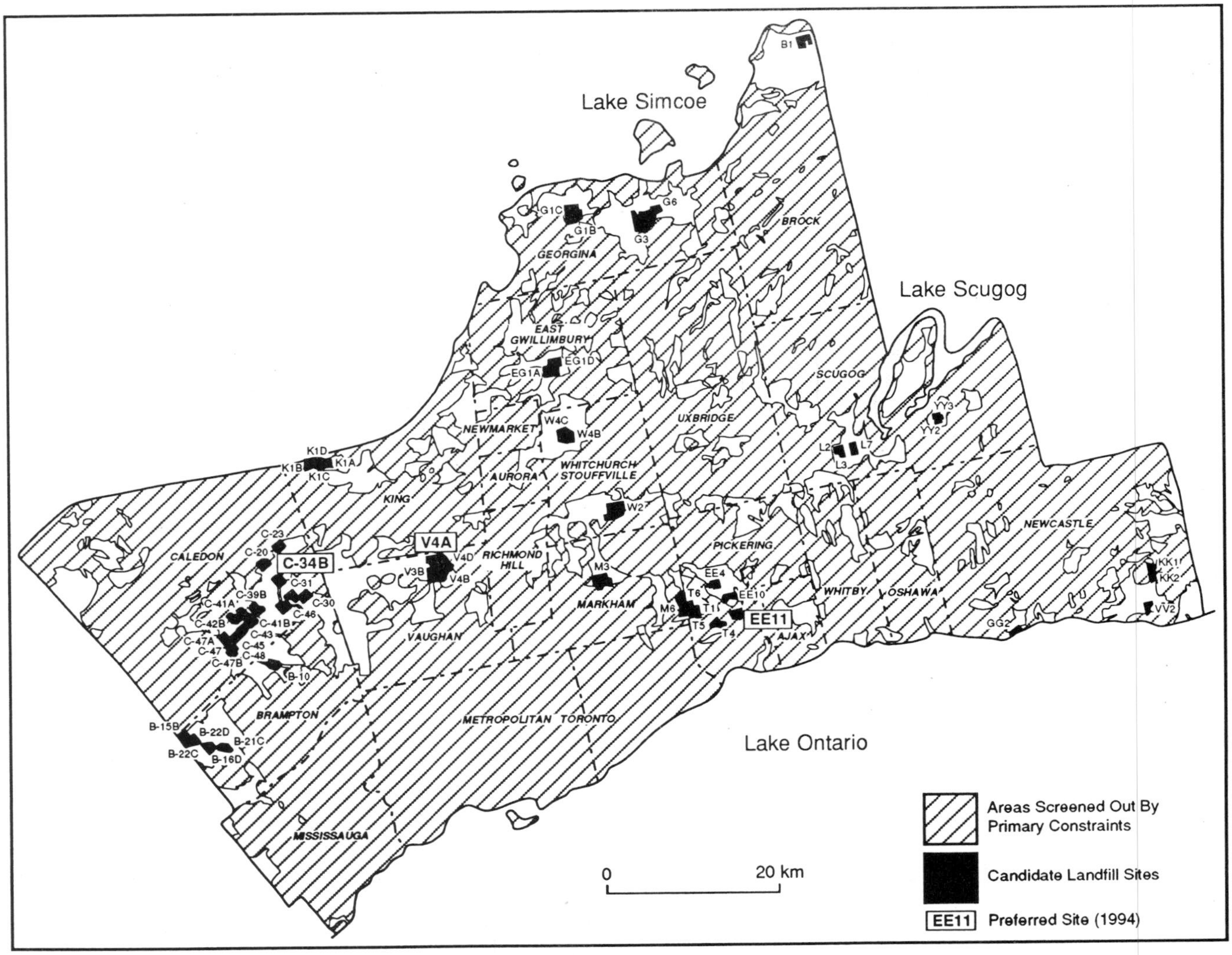

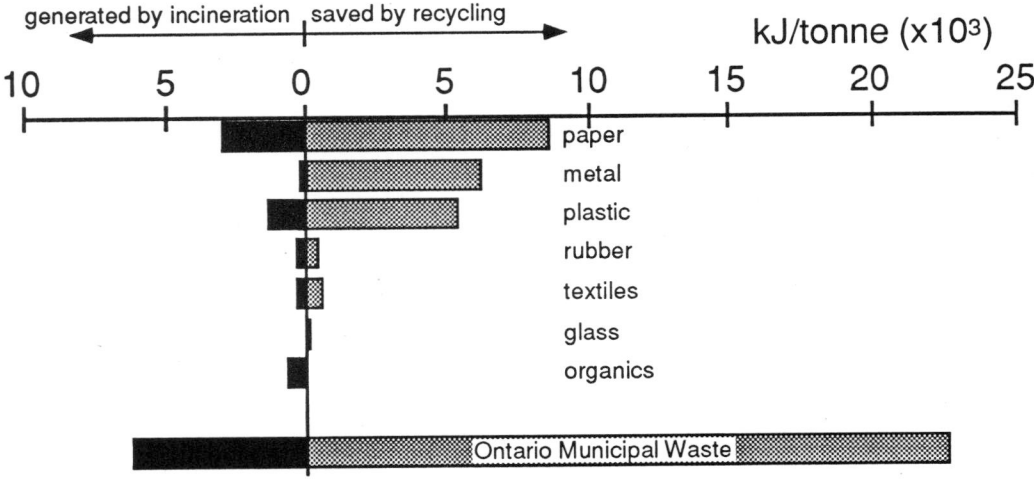

Figure 9   *Greater Toronto Area with 57 candidate landfill sites and areas screened out as being inappropriate for landfilling during search for "long term site" (to handle next 20 years of garbage) by NDP appointed Interim Waste Authority Ltd. Three preferred sites (C-34B in Peel Region; 122 ha, 40 million tonne capacity), V-4A in York (230 ha, 40 million tonne capacity) and EE11 in Durham (90 ha, 6.6 million tonne capacity) were identified from candidate sites in 1993. Halton Region was exempt from the search; it opened its own site at Milton in 1992. Not shown are over 80 potential sites identified by previous Liberal government.*

Figure 10   *Energy saved by recycling compared with incineration of same material (after Neurath, 1993). Abbreviation: **kJ**, kilojoules.*

permeability. These sediments afford some degree of natural containment and attenuation of landfill leachates as a result of their reduced permeability (see below). On the other hand, sand-rich (>50%) clast-dominant tills (Fig. 14) may not be suitable for landfill purposes, owing to their moderate to high permeabilities. Past practice has been to view matrix-dominant tills as simple stratigraphic units of low permeability, but this is a misconception arising from the use of results from permeability tests completed on small and homogenous till samples (*e.g.*, 4″ diameter drill core). In addition, better subsurface data and field assessments of permeability reveal the common presence of interbeds of coarser, more permeable sediments, fractures and other discontinuities that impart a high bulk permeability (Fig. 15). Lenticular sand and gravel horizons associated with tills locally constitute significant overburden aquifers with yields sufficient for domestic and municipal use.

Proglacial sands and gravels are widespread across the region and delineate former outwash plains, melt-water channels and moraine ridges built at the margins of ice sheet lobes. The thickness of these deposits is, in places, considerable (*e.g.*, >100 m along the Oak Ridges Moraine; Figs. 2, 12), and they constitute important areas of ground-water recharge to deeper aquifers. More than 90% of overburden wells in southern Ontario are developed in proglacial sand and gravel deposits. Although it is not out of the question to site a small attenuation-type landfill in a sandbody in southern Ontario (contaminants are rapidly diluted), in general, these substrates are unsuitable given their hydrogeologic importance as recharge areas (Chapter 10). Examples of landfills constructed within such terrains are given by Proulx and Farvolden (1989) and in Chapter 2.

Leachate Generation in a Landfill
The type and concentration of contaminants in ground water and surface water impacted by landfill wastes is dependent on many factors (Ham and Barlaz, 1993). The density of *in situ* municipal waste is usually between 0.67 and 0.80 t·m^{-3} and is generally assumed to have a hydraulic conductivity of 10^{-2} cm·s^{-1}. The composition of municipal waste is shown in Figure 16. Initially, large volumes of volatile fatty acids are produced, resulting in leachates characterized by high levels of total organic carbon and high biological and chemical oxygen demand. Thereafter, the proportion of proteins rises, and organic pollutants resulting from insecticides and solvents can be detected. Inorganic components include dissolved ions (*e.g.*, chloride), nutrients (*i.e.*, phosphorous and nitrogen compounds), and metals, principally iron. The long-term chemical evolution of leachate is described in Chapter 24.

Regional Potential for Ground-water Contamination
Landfills represent a significant point source for the release of

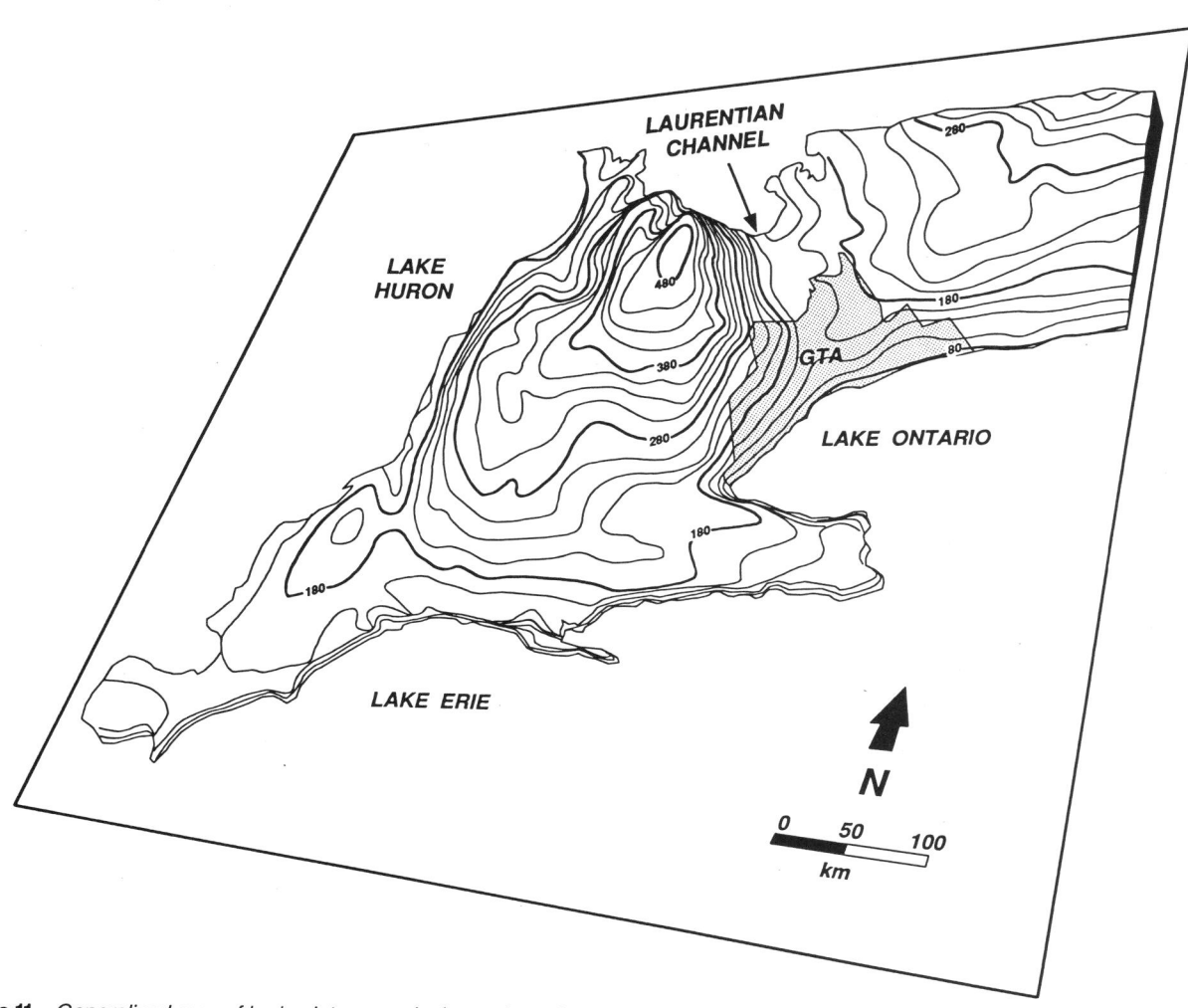

Figure 11   *Generalized map of bedrock topography in southern Ontario; contour interval 25 m.*

contaminants and there is growing concern with the impact of landfill-derived contaminants on the quality of surface and subsurface waters, especially Lake Ontario (*e.g.*, Chapter 5). Figure 13 shows the location of 1183 open and closed landfill sites in southern Ontario, superimposed on the map of surficial geology. It is likely that several hundred remain to be discovered, given the widespread former practice of unregulated landfilling (*e.g.*, Chapter 21). As much as 20% of the area of the city of Toronto is designated as landfill, much of it contaminated (Chapter 2, 27). Figure 17 gives a breakdown of the number of open and closed sites on each sediment type. The greatest number (380) are located on areas of sand and gravel. This is a reflection of the use of abandoned pits as quarries for landfilling; most sites are located in former aggregate or bedrock quarries, and are consequently associated with either sand-rich substrates or fractured bedrock in which leachate plumes are likely to develop. The extent of leachate plumes has been determined for only a restricted number of landfills in southern Ontario (*e.g.*, Cherry, 1983; Barker *et al.*, 1987; Chapter 2, 24).

Delineation of the dimensions of subsurface leachate plumes is a costly exercise. In several cases, urban development in the vicinity of landfill sites is hindered by the cost of remediation, by legal disputes as to the responsibility for bearing these costs and possible impacts on public health. Average costs for site characterization and remedial action design are about $1.5 million. Remediation costs can be as high as $12 million. Non-intrusive techniques, *i.e.*, geophysical methods, for exploring the stratigraphy of old landfills are gaining more importance given the dangers of drilling where the subsurface stratigraphy and toxicity of contents are unknown (Chapter 31). The consensus is that it is better to leave the waste in the ground than to risk further environmental damage as a result of disturbance and removal to other sites. The archeology of abandoned landfills is an emerging sub-discipline (*e.g.*, Rathje, 1991).

Controls on Leachate Migration from Landfills

The migration of leachate in the subsurface is governed by a number of hydrogeological and hydrochemical mechanisms (*e.g.*, Macfarlane *et al.*, 1983; Cherry, 1983; Barker *et al.*, 1987; Shackelford, 1993). Advection is the component of migration accomplished by the physical movement of water through the substrate. The rate is normally equal to the average groundwater velocity as determined by Darcy's Law. Thus, the grain size, porosity and presence or absence of fractures (see below) are of importance. Diffusion describes the movement of chemicals from areas of high to low concentration, and is

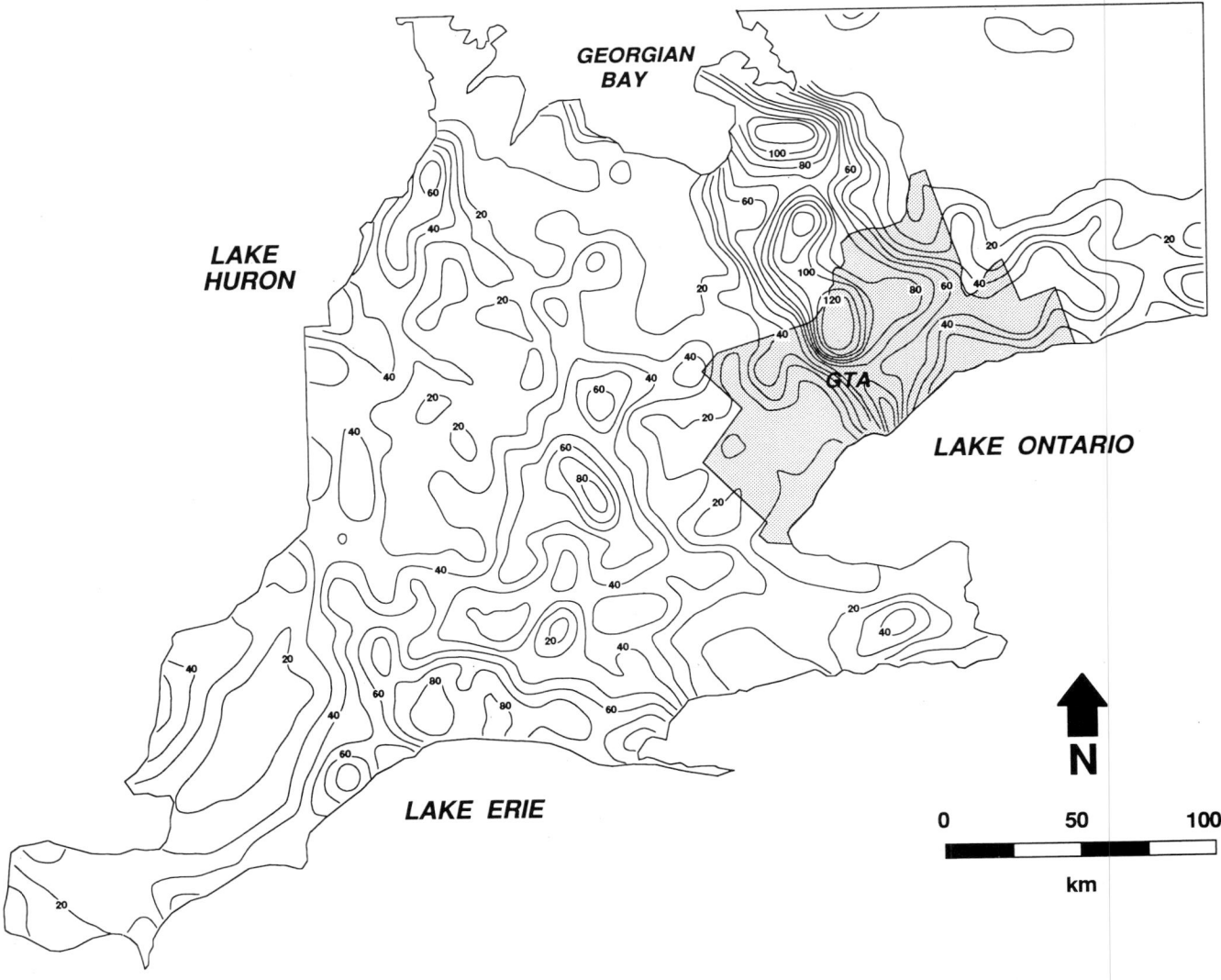

Figure 12 *Generalized map of Pleistocene glacial sediment thickness (in metres) across southern Ontario; same area as in Figure 11. Contour interval 10 m. Note thickest drift along Laurentian Channel (see Chapter 2).*

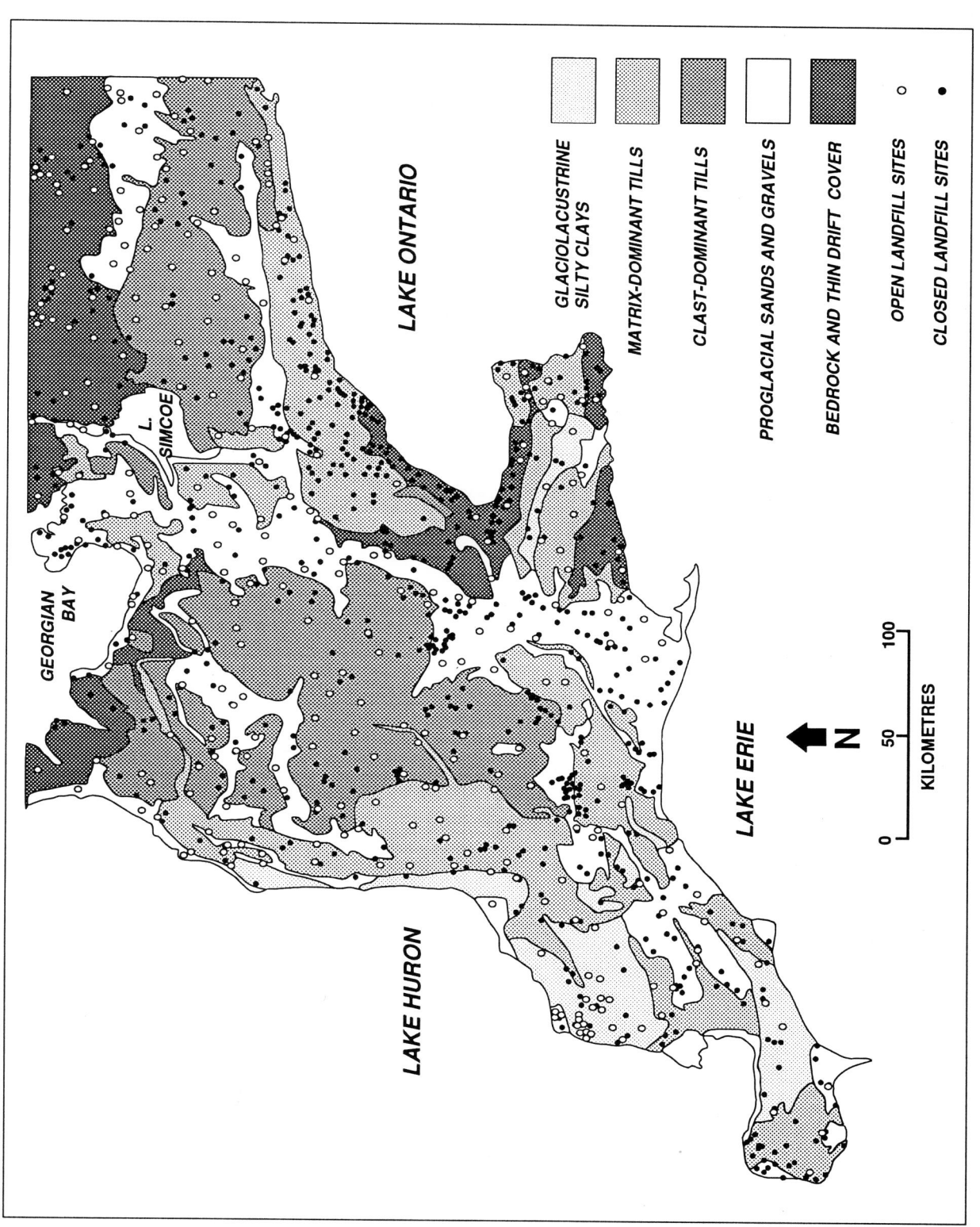

Figure 13   *Location of open (n=311) and closed (n=868) landfill sites and distribution of Pleistocene sediment types in southern Ontario (after Chapman and Putnam, 1984; Ontario Waste Management Corporation, 1986; Ontario Ministry of the Environment and Energy, 1991a; Barnett et al., 1991).*

significant in low-permeability clayey materials where advection is at a minimum. It is the dominant process when average linear ground-water velocities are less than a few centimetres per year. Mechanical dispersion of leachate occurs when particles in a sediment force ground water to deviate from a linear flow vector. Adsorption may remove leachate components by binding them to the surface of clay minerals. In addition, heavy metals are rapidly removed from solution by precipitation within carbonate-rich glacial sediments. Biodegradation of leachate will occur as a result of microbial action by bacteria present in glacial sediments. Chemical species move through the subsurface at different rates as a result of their varying ability to be retarded by the processes listed above. So-called conservative species, such as chlorides, move at the same velocity as ground water, but others are retarded and are only released to surface waterbodies after many thousands of years (see Chapter 5).

Given the requirements of the OMOE Reasonable Use criteria (Ontario Ministry of the Environment, 1986, 1994), landfills are ideally sited in areas of relatively thick, impermeable sediments where there is a high degree of natural protection afforded by attenuation of leachate within the property boundaries of the landfill (e.g., Fig. 18A). The widespread use of liners (Fig. 18B), varyingly composed of either compacted clay, till or synthetic materials, only became widespread practice in the mid-1980s (see Daniel, 1993a, b; Koerner, 1993a; see below). Under certain conditions, leachate can spill over the liner and migrate off-site in the subsurface if the liner is not overlain by a drainage layer that can be pumped. In the absence of any drainage system, leachate mounding within the landfill can result in overland transport of leachate from springs on the sideslopes of the landfill (Fig. 18B). Leachate mounding may also affect the physical stability of the landfill sideslopes and promote slumping (Oweis, 1993).

The role of fractures. The efficiency of clay liners and the rate of contaminant movement though fine-grained clayey substrates is primarily dictated by diffusion, which transports leachate from areas of high concentration to low (e.g., Desaulniers et al., 1981; Cherry, 1983; Johnson et al., 1989; Daniel, 1993a). A further problem in clay-rich substrates is

caused by the presence of fracture systems. These are of particular significance in the clay plains of southwest Ontario near Sarnia and Windsor (e.g., Soderman and Kim, 1970; D'Astous et al., 1989; Rowe, 1990; Ruland et al., 1991; McKay et al., 1993). Fracture systems are generated in Pleistocene sediments by many processes, such as deformation (glaciotectonism) below ice sheets, desiccation accompanying drawdown of the water table during Holocene dry climates, cold-climate (periglacial) processes, and neotectonic movement of underlying bedrock. Fractures in clayey glacial sediments are identified by weathering haloes, and active ground-water flow is recorded by the presence of post-1953 tritiated water and elevated ratios of $^{18}O/^{16}O$ compared to in situ pore waters in surrounding matrix. The latter, in contrast, are

Figure 14 Typical appearance of till in core sample: (**left**) massive, sandy-silt, matrix-dominant till; (**right**) pebble-rich clast-dominant silty-sand till.

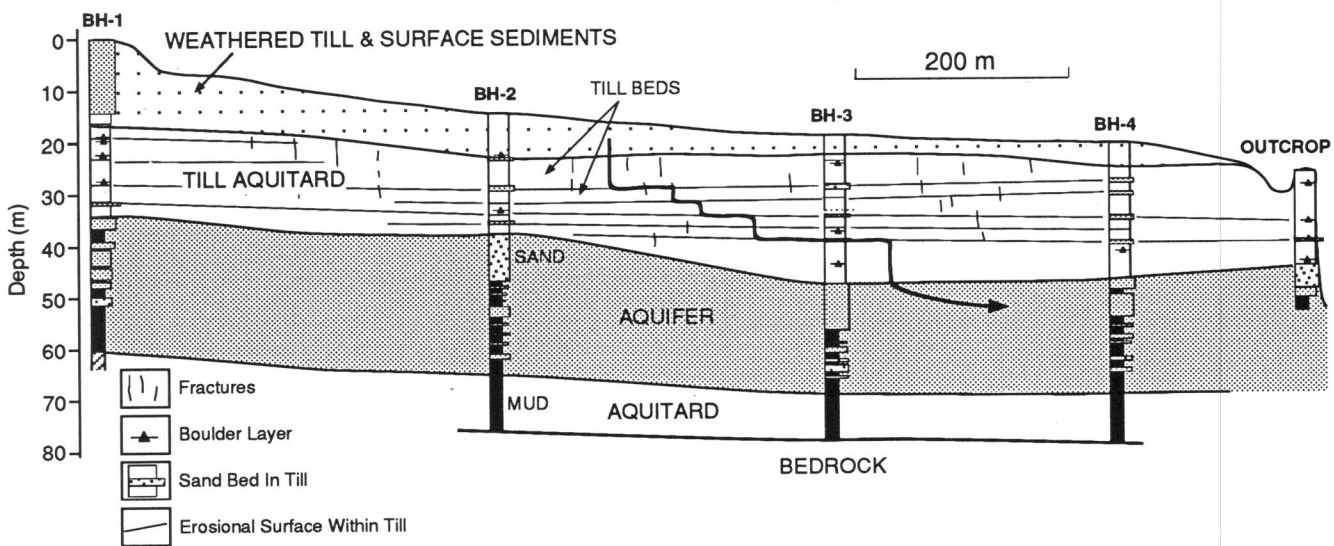

Figure 15 Schematic representation of stratigraphic complexity in thick till aquitard resting on glaciolacustrine sediments (aquifer). Till is composed of sheet-like beds, sand lenses and boulder layers. These, together with deep fractures, impart a higher bulk permeability than indicated by laboratory anaylsis of unrepresentative core samples of massive till (Fig. 14). Bold line is typical ground-water flow path to aquifer.

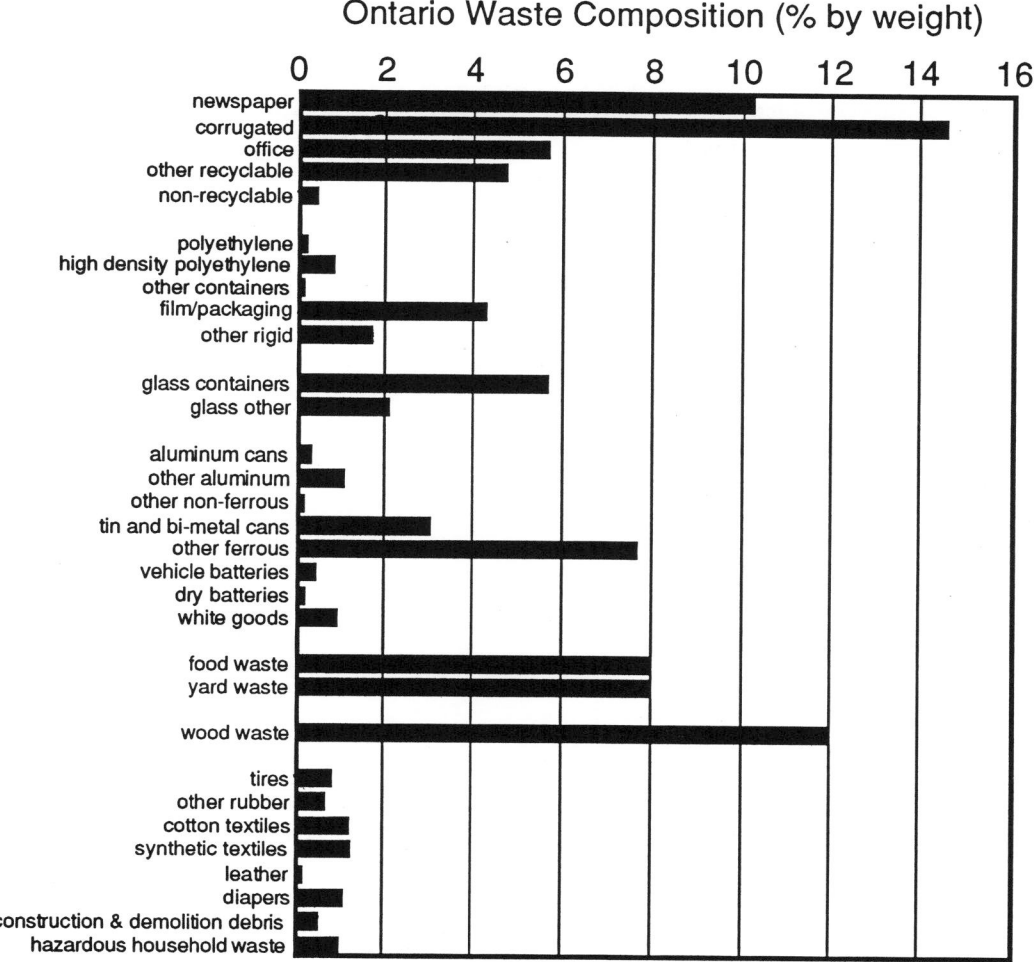

Figure 16   *Average composition of municipal waste produced by Ontario urban communities (from Bird and Hale, 1978; Ontario Ministry of the Environment and Energy, 1991b; Statistics Canada, 1994).*

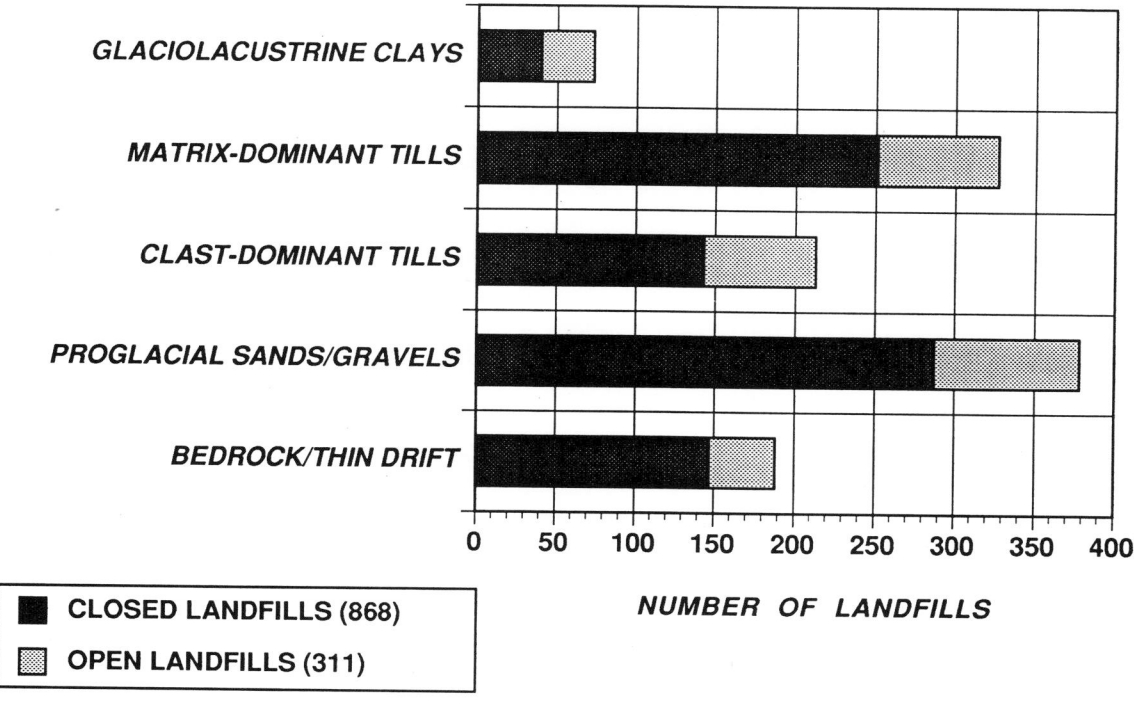

Figure 17   *Numbers of open and closed landfill sites on substrate types identified in Figure 12.*

relatively enriched in ^{16}O, typical of the original glacial melt-water source at the time of deposition. The presence of vertical fractures is not easily identified by conventional drilling, and it is common practice either to drill angled boreholes or dig large test pits when fracturing is suspected (Fig. 19).

Methane Gas Production

A landfill is a giant living organism involving biological decomposition of waste organic material to produce gas composed of methane (CH_4: ~45%) and carbon dioxide (CO_2: 55%). As with rice paddies and marshes, biological polymers such as cellulose and hemicellulose are the principal biodegradable constituents of municipal garbage, and account for over 90%

of the methane potential of such waste (Ham and Barlaz, 1993). The remainder is provided by protein (about 8%) and soluble sugars (<1%). Decomposition is by anaerobic microbes (Fig. 20A). Methane production does not commence immediately in a landfill, and may take months to years to commence; microbial activity is limited by pH (optimum between 6.8 and 7.4) and moisture content. Methane production increases with moisture content, and experiments with recycling neutralized leachate through landfills show greatly increased yields and faster decomposition of waste, resulting in higher economies for methane energy projects (see below). Future landfills will likely incorporate not only engineered leachate and gas collection systems but also systems for recycling leachate. In turn, faster decomposition of wastes

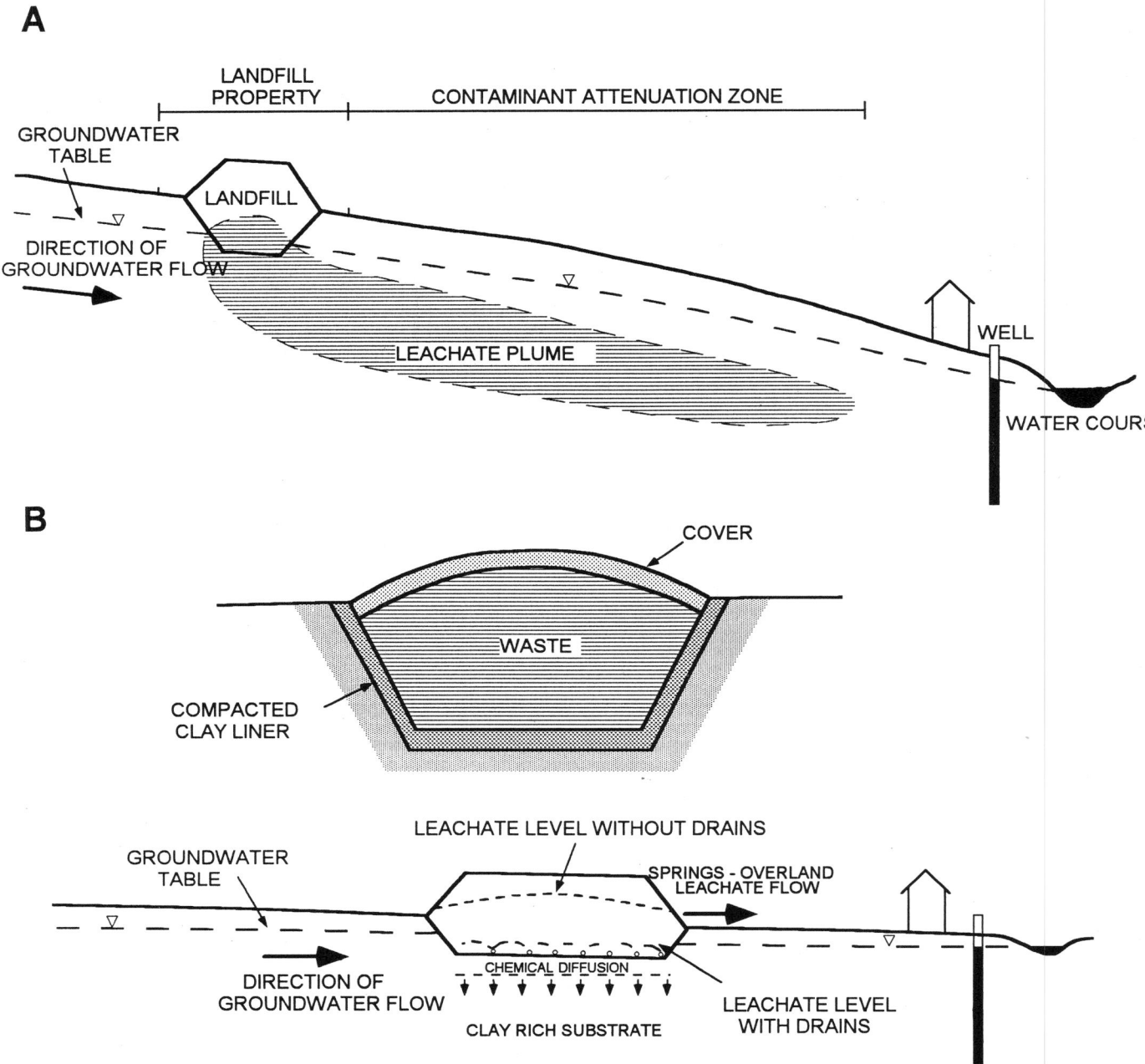

Figure 18 (**A**) *Unlined landfill, typical of most sites shown on Figure 13, with leachate plume.* (**B**) *Lined landfill showing leachate level (i) without and (ii) with drains (modified from Rowe, 1988). Note mounding of leachate within landfill without drains and the breakout of leachate springs on the landfill sideslopes.*

Figure 19 *Typical test pit designed to identify details of the near-surface (<6 m) stratigraphy particularly the presence of fractures in tills or clays. This near-surface zone is usually only poorly recovered by conventional drilling techniques because of the presence of weathered, poorly consolidated sediment and surface soil.*

A

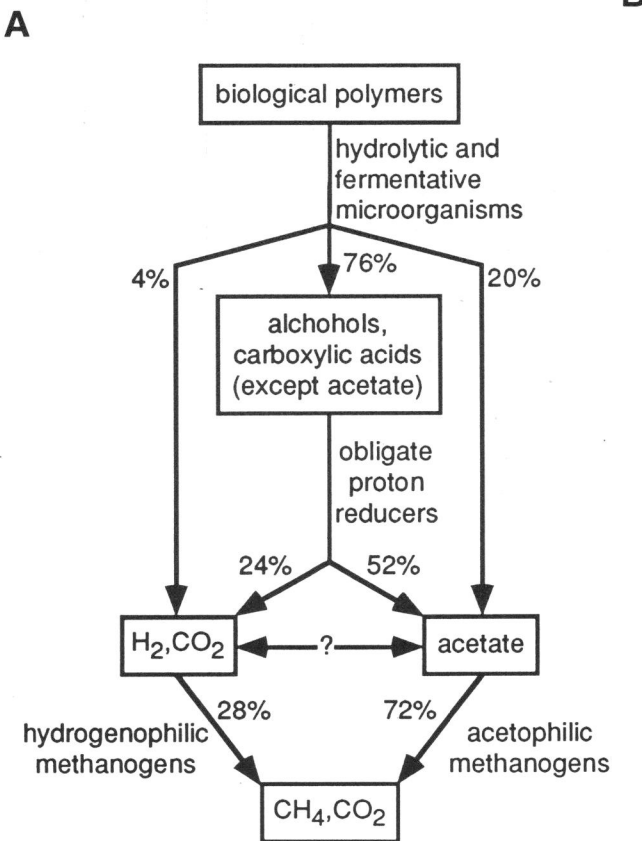

B

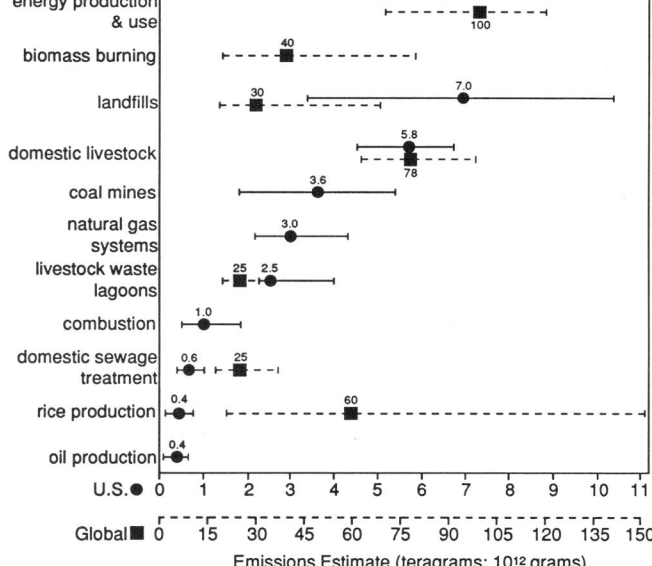

Figure 20 (**A**) *Methane production and associated evolution of chemical substrates under anaerobic conditions in municipal landfill wastes (after Ham and Barlaz, 1993). (B) Best estimates and range of United States (black circles) and global (black squares) methane emissions resulting from various human activities (data from Harriss, 1994). Note very high emission from US landfills reflecting high personal consumption patterns and disposal practices in North America; no equivalent data is available for Canadian landfills. Global landfill emissions are expected to almost double between 1990 and 2100 which is the highest expected increase for any human activity (Harriss, 1994).*

results in shortened contaminating lifespans for landfills, thereby reducing risks of contamination should (when) engineering systems fail (Koerner, 1993b).

Off-site Gas Migration

Landfill methane gas can migrate off site in surficial sediments and as a dissolved phase in ground water (Chapter 25) and pose an explosive hazard to adjacent buildings and occupants. Ontario Ministry of Environment guidelines on land use on or near landfills are given in OMOE Policy 07-07-01. This policy suggests that methane can migrate up to 500 m from the perimeter of a fill area depending on local substrates. Methane gas is lighter than air, but cannot travel through saturated sediments below the water table. Off-site migration is normally in winter when the ground surface is frozen. Urban development and installation of services on the perimeter of abandoned landfills and the attendant reduction in the water table may promote off-site gas migration. New landfills employ a gas collection system consisting of horizontal perforated pipes at different levels of the landfill; one tonne of waste has the potential to produce about 125 m^3 of gas each year, depending on the organic content of the waste. Landfill methane contains about 1% of non-methane organic compounds (NMOCs) such as vinyl chloride which can impact human health (Chapter 2; Table 1). Odour problems are usually created by the release of sulphurous gases such as mercaptans. Best management practices for landfill methane are set out by Ontario Ministry of Environment and Energy (1993a, 1993b). Landfill gas production over time can be modelled using the U.S. EPA Scholl Canyon model (United States Environmental Protection Agency, 1991, 1994).

Energy from landfill gas.   Traditionally viewed as a nuisance, landfill methane is now regarded as a cheap and effective energy source. There are currently 114 landfill-gas-to-energy projects in the United States and six in Canada. Three of the Canadian projects occur within the GTA at the Keele Valley, Brock West and Beare Road (planned) sites. At Brock West, landfill gas (average 50% CH_4, 50% CO_2 and traces of non-methane organic compounds) is used in a 23-megawatt power plant which is used in the nearby community of Ajax. Methane quality is typically low (e.g., 300-500 BTUs per cubic foot). Bacopoulos (1988) reports that the Keele Valley site has the potential to produce 803,000,000 m^3 of methane during the next 30 years, with a potential revenue generation of $158,000,000 in 1983 dollars. United States case studies are reviewed by Ham and Barlaz (1993).

Municipal waste and global climate change.   Methane is a significant contributor to global greenhouse warming. The heating co-efficient of one tonne of methane is equivalent to that of 68.6 t of carbon dioxide, although it is more rapidly degraded by reactions involving hydroxyl (–OH) radicals. Concentrations of methane in the Earth's atmosphere have increased from about 0.8 parts per million by volume (ppmv) before the Industrial Revolution to 1.6 ppmv at the present day, in parallel with the rate of increase in global population (Houghton, 1994). Some workers argue that the growth rate of atmospheric methane concentrations has decreased recently (Bekki et al., 1994), but this is likely a temporary trend.

Worldwide, landfills are estimated to contribute about 30 teragrams per year (Tg·a^{-1}) of CH_4 to the Earth's atmosphere (Bingemer and Crutzen, 1987; Peer et al., 1993; Thorneloe, 1993; Fig. 20B). North American landfills contribute about 25% of total global landfill emissions. The control of landfill-related methane, in the US at least, is considered an important aim of legislation such as the Clean Air Act of 1994

(Thorneloe, 1993). The problem is little discussed in Canada, largely because of the smaller landfill population. Methane production from landfills is not recognized as a major problem (yet) in the GTA; more emphasis has been placed on CO_2. Per capita production of CO_2 (1988 value) in the city of Toronto is about 14 t. The city was the first in the world to set a target of reducing that emission level by 20% by 2005 (Harvey, 1993). Although well meaning, this is offset by the continued uncontrolled release of methane from abandoned landfills across the region (e.g., Chapter 25).

ENGINEERED LANDFILLS

An engineered containment system consists essentially of a compacted clay liner, leachate and gas collection systems (Fig. 21) and, increasingly, an engineered drainage layer that can be naturally pressurized by ground waters in underlying deposits. In this way, ground-water flow paths are directed into the landfill, thereby ensuring no significant outward movement of contaminants. This system is termed a hydraulic trap (Rowe, 1988; Chapter 23). The basis of the system lies in ensuring that the level of leachate within the landfill is below the height of the water table in sediments surrounding the site; an inward flow of ground water results. A typical state-of-the-art engineered containment system is shown in Figure 22. The collection system is designed to prevent leachate mounding (Fig. 18B) and to collect leachate produced by infiltration of surface waters. The system consists of leachate collection pipes within a granular layer (Fig. 22) which directs leachate to a point where it can be pumped. This collection

---

Table 1   Pollutants commonly present in landfill gas with typical concentration values (after United States Environmental Protection Agency, 1991, 1994). Non-methane organics are destroyed by incineration at 800°C.

Compound	Concentration (%)	(ppm)
Methane	55	
Carbon dioxide	45	
Acrylonitrile		0.05
Benzene		0.68
Carbon tetrachloride		0.0079
Chlorodifluoromethane		8.5
Chloroform		0.039
Chloromethane		0.95
1,2–dibromoethane		0.007
m–dichlorobenzene		0.019
1,2–dichloroethane		0.11
1,1–dichloroethene		0.75
1,2–dichloroethene		0.89
Difluorodichloromethane		23.8
Dimethyldisulphide		0.02
Dimethylsulphide		3.2
Ethylbenzene		2.71
Hexane		14.9
Hydrogen sulphide		0.6
Methylene chloride		26.4
Tetracloroethylene		3.17
Toluene		12.4
1,1,1–trichloroethane		1.21
Trichloroethylene		2
Vinyl chloride		2.53
Xylenes		5.74

A

B

Figure 21   (**A**) *Keele Valley sand and gravel quarry prior to start of landfilling in 1984. Quarry is 600 m wide and up to 30 m deep.* (**B**) *Keele Valley landfill showing compacted clay liner (12 m thick with a permeability of 1×10^{-8} cm·s^{-1} or less) with leachate (red) and gas (green/white) collection systems installed. See Chapter 2 for details of site geology and hydrogeology.*

system overlies a compacted clay liner, the upper surface of which is protected by a geotextile separator layer, which, in turn, overlies a lower granular unit and second geotextile layer. The purpose of the lower granular unit is to provide a backup hydraulic control layer, whereby water could be added if there were significant drawdown of the ground-water head in the underlying sediment. Drawdown could occur if the aquifer were developed and pumped, or if anticipated climate warming were to result in reduced recharge. Monitoring of water quality in the secondary hydraulic control layer provides an early warning system should the hydraulic trap fail. If contamination is detected, then the layer can be slowly pumped and the effluent treated municipally. Ideally, the hydraulic trap is self-regulating, but provision for a hydraulic control layer ensures flexibility in the event of partial or complete failure.

The design of leachate control systems varies considerably (see Daniel, 1993b). Some designs incorporate double or triple liners and drainage layers so as to meet ground-water quality criteria at the base of the landfill (*i.e.*, the site boundary). The design for the projected IWA sites, along with the estimated service life of the various components and other engineering parameters, is shown in Figure 23. The primary leachate collection system is assumed to be operative for 50 years (see below) with secondary systems as a back-up designed to control leachate migration until contaminant levels have declined to the point where further collection is no longer needed to meet water quality criteria. As a contingency measure, clean water can be injected into the hydraulic control layers and contaminated water withdrawn, thereby reducing contaminant concentrations.

**Geologic Constraints
on the Siting of Engineered Landfills**
While the modern landfill will be fully engineered, there are several geological conditions that must be met. The long-term performance of engineered systems over hundreds of years is an open question, and this uncertainty has made landfill location more, not less, dependent on geological considerations and natural protection in the event of a total failure of the engineered system. A major consideration, therefore, is the need to understand the nature of changes that may take place to the ground-water system in the vicinity of the site, resulting from regional changes in the use of aquifers, urban development and climate change. The base contours of the landfill (typically 5-15 m below existing surface elevations) must be sufficiently low to ensure that ground water flows into the fill.

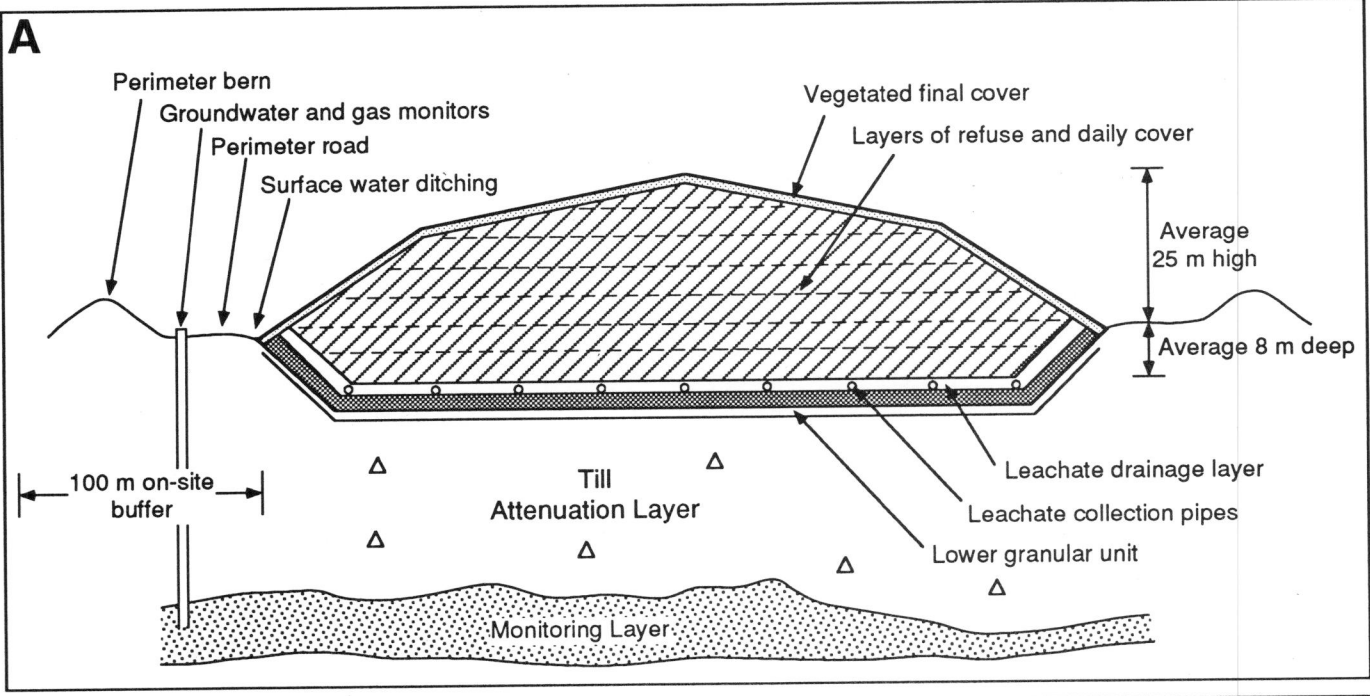

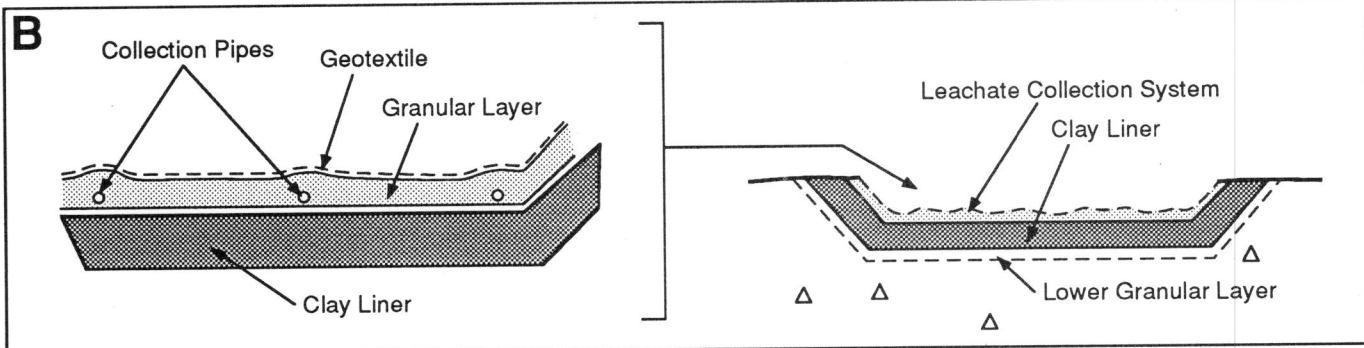

Figure 22   (A, B) *Typical design characteristics of an engineered landfill (after Maclaren Plansearch, 1990; and Interim Waste Authority Ltd., 1994). Specific designs vary and may incorporate additional liners and drainage layers as supplementary leachate control measures (Fig. 23; Chapter 23).*

Base contours should also, however, be high enough to ensure a maximum thickness of low-permeability glacial sediment under the site to form an attenuation zone. Base contours placed at too low an elevation could not only reduce the thickness of underlying sediments, but could also result in problems of liquefaction during construction of the landfill. The presence of a high-permeability stratigraphic unit below the attenuation layer is also required. This layer acts as a monitoring layer to identify and possibly control leachate that has migrated through the attenuation zone.

Long-Term Behaviour of Engineered Liner Systems

The life of an engineered containment system is usually estimated to be a minimum of about 50 years (Fig. 23), and significant deterioration of performance, including complete failure, can be expected after this time. A major problem is that there is no experience of operating an engineered landfill for long time periods. The performance of polyethylene liners at large sites may be inadequate simply because of the very large cumulative length of welded seams needed to construct the liner (Koerner, 1993b). The contaminating lifespan of a landfill is defined by the Ontario Ministry of the Environment as "the period of time during which the landfill will produce contaminants at levels that could have an unacceptable impact if they were discharged into the surrounding environment" (Ontario Ministry of the Environment, 1986). Few long-term field studies have been conducted, however, and research is commonly based in small-scale cells (lysimeters) in

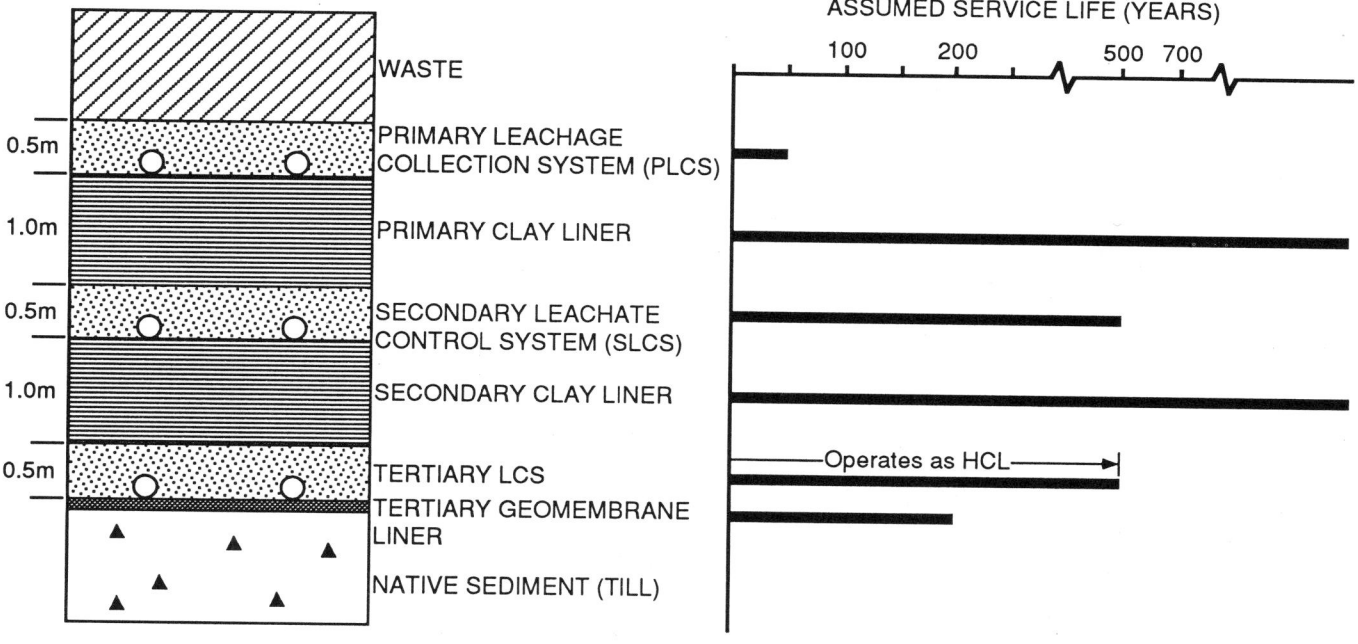

Parameter	Values
hydraulic conductivity	
clay liner	2×10^{-8} cm/s
geomembrane liner	1×10^{-12} cm/s
cover infiltration rate	0.2 m/yr
average leachate head on primary liner during operation of PLCS	0.3 m
average leachage head on secondary liner during operation of SLCS	0.3 m
effective porosity	
clay liner	0.3
drainage layer	0.3
dispersion/diffusion coefficients	
clay liner	0.016 m²/yr
geomembrane liner	6.3×10^{-5} m²/yr
drainage layer	100 m²/yr
dry soil density of clay liner	1.7 g/cm³
thickness	
clay liner	1.0 m (2 liners)
geomembrane layer	2 mm
drainage layer	0.5 m (3 layers)
service lives for the various engineered systems	See above
organic carbon content (assumed)	
clay liner	0.002
geomembrane layer	0
drainage layer	0
organic carbon/water partition coefficient for dichloromethane	350 mL/g

Parameter	Values
total mass of waste	6.6 million tonnes
apparent waste density	670 kg/m³
total porosity	0.50
field capacity (volumetric moisture content)	0.25
initial chloride concentration	2,500 mg/L
initial dichloromethane concentration	3,000 µg/L
mass percentage of chloride in waste	0.2%
mass percentage of dichloromethane in waste	0.00024%
half-life for dichloromethane	70 years

Figure 23 *Specific landfill leachate collection system design with assumed service life as adopted by Interim Waste Authority Ltd., for site EE11 in Durham Region (Fig. 9). Note incorporation of triple leachate collection systems. System and waste parameters are also shown (data from Interim Waste Authority Ltd., 1994). Abbreviation: HCL, hydraulic control layer.*

which variables are carefully monitored (see Peyton and Schroeder, 1993).

Given that there is a finite mass of contaminants in the landfill, leachate concentration levels through time take the form of an exponential first-order decay similar to radioactive decay. A half-life is the time taken to reduce the initial concentration by 50%. Therefore, it is important for the engineered containment system to operate effectively for as long as possible while contaminants are removed by the leachate collection system, biodegrade, volatilize or otherwise become immobile (see above). Contaminant transport can be modelled using a finite-layer model based on the one-dimensional advection-dispersion equation for porous media and where the hydrostratigraphy can be idealized as horizontal layers (e.g., POLLUTE; Chapter 23). Space does not permit a full discussion of the details of contaminant transport modelling, but it is common practice in Ontario to apply criteria established for reasonable use guidelines to critical contaminants, such as dichloromethane and chloride. These are present in large concentrations in leachate derived from municipal landfills, and are not easily attenuated by movement through soil. Initial source concentrations for chloride are typically ~2000 $mg \cdot L^{-1}$, with a decay half-life of 25 years. Corresponding data for dichloromethane is 6000 $mg \cdot L^{-1}$ with a half-life of 70 years (Fig. 23). In many cases, dilution of contaminants in aquifers underlying the landfill cannot be relied on to meet reasonable use criteria because the dynamics of the ground-water flow system are often not sufficiently well known. The only solution, therefore, is to increase the service life of the engineered systems. Detailed reviews of contaminant modelling and other aspects of engineered landfill systems are discussed by Shackelford (1993), Koerner (1993) and in Chapter 23.

DISCUSSION

Given the intense scutiny that projected sites will undergo in the future, new landfill sites will be fully engineered. They will only be approved where the natural geological setting provides some degree of natural attenuation combined with an ability to monitor leachate movement in the event of complete failure of the engineered containment system. Fine-grained sediments are generally favoured as landfill substrates since they afford fall-back protection, but, unfortunately, such sediments are not evenly distributed throughout southern Ontario (Fig. 13). Within the GTA, where 50% of the province's waste is generated, silt and sand-rich tills predominate and contain permeable interbeds (Fig. 15).

There is an urgent need to establish the location and detailed inventories of abandoned and existing landfill sites in the urban area of southern Ontario, paying particular emphasis to age, contents and local glacial geological and hydrogeological information in order to identify the potential for ground-water contamination. Because they contain large amounts of organic contaminants for which drinking-water guidelines are very stringent, landfills are the most significant point source of contamination in urban areas (Chapter 5).

The situation in southern Ontario, where different waste management strategies continue to be major topics of political and social debate, is typical of the garbage crisis facing most communities in North America (Taylor et al., 1991; Haight, 1992; Estrin and Swaigen, 1993). Regardless of future political developments and trends in waste management policies, properly engineered and operated landfills remain a necessary option for the foreseeable future, probably in combination with willing-host communities thereby avoiding social and political problems associated with traditional search procedures. Better communication of scientific data to the public is a key issue.

ACKNOWLEDGEMENTS

We thank many colleagues in university and industry for useful discussions during the course of landfill site selection investigations, especially Kerry Rowe, Rick Gerber and David Winfield. This paper was written with the financial support of the Natural Sciences and Engineering Research Council of Canada, the Great Lakes University Research Fund, the Interim Waste Authority Ltd., and the Ontario Ministry of the Environment and Energy.

REFERENCES

Bacopoulos, A., 1988, Operation and maintenance of the Keele Valley landfill site – 1987, Maple, Ontario: Report to Metropolitan Toronto, 87 p.

Barker, J.F., Cherry, J.A., Reinhard, M., Pankow, J.F. and Zapico, M., 1987, The occurrence and mobility of hazardous organic chemicals in ground water at several Ontario landfills: University of Waterloo, Lottery Trust Fund Project 118PL, Final Report, 116 p.

Barnett, P.J., Cowan, W.R. and Henry, A.P., 1991, Quaternary geology of Ontario, southern sheet: Ontario Geological Survey, Map 2556. Scale: 1:1,000,000.

Bekki, S., Law, K.S. and Pyle, J.A. 1994, Effect of ozone depletion on atmospheric CH_4 and CO_2 concentrations: Nature, v. 371, p. 595-597.

Bingemer, H.G. and Crutzen, P.J., 1987, The production of methane from solid wastes: Journal of Geophysical Research, v. 92, p. 2181-2187.

Bird and Hale Ltd., 1978, Municipal refuse statistics for Canadian communities over 100,000 (1976-1977), 62 p.

Boyce, J.I., Eyles, N. and Pugin, A., 1995, Seismic, outcrop and borehole geometry of late Wisconsin tills at a proposed landfill near Toronto, Canada: Canadian Journal of Earth Sciences, v. 32, p. 1331-1349.

Buitenhuis, G., 1994, Waste Management Site Newsletter, March 1994, Regional Municipality of Halton: 4 p.

Chapman, L.J. and Putnam, D.F., 1984, Physiography of Southern Ontario, Third Edition: Ontario Geological Survey, 270 p. with maps. [First Edition printed, 1951 by Ontario Research Foundation; Second Edition printed 1976 by University of Toronto Press, Toronto, ON]

Cherry, J., 1983, ed., Migration of contaminants in groundwater at a landfill; a case study: Journal of Hydrology, v. 63, p. 1-192.

Cohen, M., and 7 others, 1995, Quantitative estimation of the entry of dioxins, furans and hexachlorobenzene into the Great Lakes from airborne and waterborne sources: Centre for the Biology of Natural Systems, Queens College, Flushing, NY, 88 p.

Daniel, D.E., 1993a, Landfills and impoundments, in Daniel, D.E., ed., Geotechnical Practice for Waste Disposal: Chapman and Hall, London, UK, p. 97-112.

Daniel, D.E., 1993b, Clay liners, in Daniel, D.E., ed., Geotechnical Practice for Waste Disposal: Chapman and Hall, London, UK, p. 137-163.

D'Astous, A.Y., Ruland, W.W., Bruce, J.R.G., Cherry, J.A. and Gillam, R.W., 1989, Fracture effects in the shallow groundwater zone in weathered Sarnia-area clay: Canadian Geotechnical Journal, v. 26, p. 43-56.

Desaulniers, D.E., Cherry, J.A. and Fritz, P., 1981, Origin, age and movement of porewater in clayey Pleistocene deposits in south-central Canada, in Perry, E.C. and Montgomery, C.W., eds., Isotope Studies of Hydrologic Processes: Northern Illinois University Press, DeKalb, IL, p. 45-55.

Dillon, M.M., Ltd., 1994, Detailed assessment of the proposed site EE11 for Durham Region landfill site search: Interim Waste Authority Ltd., 436 p.

Estrin, D. and Swaigen, J., 1993, Environment on Trial: Canadian Institute for Environmental Law and Policy, Emond Montgomery Publications Ltd., Toronto, ON, 909 p.

Golder and Associates Ltd., 1994, Detailed Assessment of the proposed site C-34B, Peel Region. Appendix C – Geology/Hydrogeology: Interim Waste Authority Ltd., 215 p.

Haefeli, C.J., 1970, Regional groundwater flow between Lake Simcoe and Lake Ontario: Department of Energy, Mines and Resources, Inland Waters Branch, Technical Bulletin 23, 55 p.

Haight, M.E., 1992, ed., Municipal Solid Waste Management, Making Decisions in the Face of Uncertainty: University of Waterloo Press, Waterloo, ON, 531 p.

Ham, R.K. and Barlaz, M., 1993, Leachate and gas generation, *in* Daniel, D.E., ed., Geotechnical Practice for Waste Disposal: Chapman and Hall, London, UK, p. 113-136.

Harriss, R., 1994, Reducing urban sources of methane: An experiment in industrial ecology, *in* Socolow, R., Andrews, C., Berkhout F. and Thomas, V., eds., Industrial Ecology and Global Change, Cambridge University Press, Cambridge, UK, p. 173-182.

Harvey, L.D.D., 1993, Tackling urban CO_2 emissions in Toronto: Environment, v. 35, p. 16-23.

Houghton, J., 1994, Global Warming: The Complete Briefing: Lion Publishing Co., Elgin, IL, 192 p.

Interim Waste Authority Ltd., 1992, Comparison of the long list of candidate sites: Environmental Assessment Document II, v. 1, April 1992, 38 p.

Interim Waste Authority Ltd., 1994, Step 6 Landfill design brief for Durham Region landfill site search: Document IV, 42 p.

Johnson, R.L., Cherry, J. and Pankow, J.F., 1989, Diffusive contaminant transport in natural clay; a field example and implications for clay-lined disposal sites: Environmental Science and Technology, v. 23, p. 340-349.

Koerner, R.M., 1993a, Collection and removal systems, *in* Daniel, D.E., ed., Geotechnical Practice for Waste Disposal: Chapman and Hall, London, UK, p. 187-213.

Koerner, R.M., 1993b, Geomembrane liners, *in* Daniel, D.E., ed., Geotechnical Practice for Waste Disposal: Chapman and Hall, London, UK, p. 164-186.

Macfarlane, D.S., Cherry, J.A., Gillham, R.W. and Sudicky, E.A., 1983, Migration of contaminants in groundwater at a landfill: A case study: Journal of Hydrology, v. 63, p. 1-29.

Maclaren Plansearch Ltd., 1990, Regional Municipality of Peel Consolidated Hearings Act application – short-term contingency landfill site, Brampton site V1B: Technical Report, no. 2-3, 110 p.

McKay, L.D., Cherry, J.A. and Gillham, R.W., 1993, Field experiments in a fractured clay till 1. Hydraulic conductivity and fracture aperture: Water Resources Research, v. 29, 1149-1162.

Neurath, C., 1993, Incineration Compared to energy and waste management alternatives: A full environmental costs analysis: Pollution Probe, Toronto, ON, 42 p.

Novakowski, K.S. and Lapcevic, P.A., 1988, Regional hydrogeology of the Silurian and Ordovician sedimentary rock underlying Niagara Falls, Ontario: Journal of Hydrology, v. 104, p. 211-236.

Ontario Ministry of Health, 1986, Upper Ottawa Street landfill site study: Final Report, 24 p.

Ontario Ministry of the Environment, 1986, Incorporation of the reasonable use concept into groundwater management activities of the Ministry of the Environment: Water Resources Branch, 22 p.

Ontario Ministry of the Environment and Energy, 1991a, Waste Disposal site inventory: Waste Management Branch, 196 p.

Ontario Ministry of the Environment and Energy, 1991b, Residential waste compostion study: 150 p.

Ontario Ministry of Environment and Energy, 1992, The Ontario Environmental Assessment Act as it relates to waste management planning: 40 p.

Ontario Ministry of Environment and Energy, 1993a, Guidance manual for landfill sites receiving municipal waste, PIBS 2741: 176 p.

Ontario Ministry of Environment and Energy, 1993b, Interim Guide to estimate and assess landfill air impacts: 62 p.

Ontario Ministry of Environment and Energy, 1994, Incorporation of the Reasonal Use Policy into MOEE groundwater management activities: Ministry of the Environment and Energy, Policy 15-08, March 1993, 16 p.

Ontario Waste Management Corporation, 1986, Hydrogeological inventory for waste facilities development – southern Ontario, 137 p.

Oweis, I.S., 1993, Stability of landfills, *in* Daniel, D.E., ed., Geotechnical Practice for Waste Disposal: Chapman and Hall, London, UK, p. 244-268.

Peer, R.L., Thorneloe, S.A. and Epperson, D.L. 1993, A comparison of methods for estimating global methane emissions from landfills: Chemosphere, v. 26, p. 387-400.

Peyton, R.L. and Schroeder, P.R., 1993, Water balance for landfills, *in* Daniel, D.E., ed., Geotechnical Practice for Waste Disposal: Chapman and Hall, London, UK, p. 214-243.

Phyper, J.D. and Ibbotson, B., 1991, The Handbook of Environmental Compliance in Ontario: McGraw-Hill Ryerson, Ltd., Toronto, ON, 346 p.

Proulx, I. and Farvolden, R.N., 1989, Analysis of the contaminant plume in the Oak Ridges Aquifer: Ontario Ministry of the Environment, R.A.C. Project 261, 134 p.

Rathje, W.L., 1991, Once and future landfills: National Geographic, v. 179, p. 116-134.

Rowe, R.K., 1988, Eleventh Canadian geotechnical colloquium: Contaminant migration through groundwater – the role of modelling in the design of barriers: Canadian Geotechnical Journal, v. 25, p. 778-798.

Rowe, R.K., 1990, Contaminant migration through fractured till into an underlying aquifer: Canadian Geotechnical Journal, v. 27, p. 484-495.

Ruland, W.W., Cherry, J.A. and Feenstra, S., 1991, The depth of fractures and active groundwater flow in a clayey till plain in southwestern Ontario: Ground Water, v. 29, p. 405-417.

Shackelford, C.D., 1993, Contaminant transport, *in* Daniel, D.E., ed., Geotechnical Practice for Waste Disposal: Chapman and Hall, London, UK, p. 33-65.

Sibul, U., Wang, K.T. and Vallery, D., 1977, Groundwater resources of the Duffins Creek-Rouge River drainage basins: Ontario Ministry of the Environment, Water Resources Report 8, 109 p.

Soderman, L.G. and Kim, Y.D., 1970, Effect of groundwater levels on stress history of the St. Clair clay till deposit: Canadian Geotechnical Journal, v. 7, p. 173-187.

Statistics Canada, 1994, Human Activity and the Environment: National Accounts and Environment Division, Ottawa, ON, 300 p.

Thorneloe, S.A., 1993, Landfill gas and its influence on global climate change: United States Environmental Protection Agency, Report EPA 600/A-93/240, 17 p.

Trow Hydrology Consultants Ltd., 1986, Halton Region Landfill Technical Report Site D, Milton, Ontario: Hydrogeologic Report: 3 volumes.

Turner, N.M.E., 1977, The Oak Ridges aquifer complex: Ontario Ministry of the Environment, Water Resources Branch, Hydrogeological Map 78-2.

United States Environmental Protection Agency, 1991, Air emissions from municipal solid waste landfills – Background information for standards and guidelines: Office of Air Quality Planning and Standards, Research Triangle Park, NC, EPA-450/3-90-011a, 22 p.

United States Environmental Protection Agency, 1994, Radian Corporation, User's manual for landfill air emission estimation, v. 2, DCN 652-016-12.

Webster, T, and Connett, P., 1989, Cumulative impact of incineration on agriculture: A screening procedure for calculating population risk: Chemosphere, v. 19, p. 567-602.

Webster, T, and Connett, P., 1990, The use of bioconcentration in estimating the 2,3,7,8-TCDD content of cows' milk: Chemosphere, v. 20, p. 779-786.

21. Waste Disposal in Toronto's Past

Richard Anderson

Geography Department, York University, 4700 Keele Street, North York, Ontario M3J 1P3

SUMMARY

This paper examines waste disposal in the Toronto region between 1870 and 1980. Toronto's waste stream has fluctuated significantly over this time interval, and despite common wisdom to the contrary, there is less solid waste (by weight) per capita produced today than in the 1940s, although it has changed in bulk composition and toxicity. A particular problem is that many former dump sites, many of which accepted hazardous industrial wastes in the 1940s, go unrecorded. Up-to-date inventories of site location and contents are urgently required.

INTRODUCTION

Very little is known regarding the history of wastes produced by Canadian cities over the last 100 years. American researchers have made a significant start in compiling such information (Colten, 1994; Melosi, 1981; Rathje, 1992; Rathje and Murphy, 1992; Melosi, 1993), but Canadian studies have been much more limited (Rose, 1988; Gilbert, 1988). Inventories of old waste disposal sites are now available (*e.g.*, Ontario Ministry of Energy and the Environment, 1991), but there is a great deal of work still to be done identifying former landfills in areas that are now heavily urbanized.

Toronto, like most North American cities, has been a prodigious producer of wastes, and for at least 150 years, has been generating urban wastes at about twice the rate of European cities (Haight, 1991). Accurate records of garbage and ash tonnage have been kept since 1917. Before that, officials recorded the number of loads of waste material collected, such that it is possible to reconstruct waste tonnage back to the 1870s. The growth of the Toronto urban area is shown in Figure 1; per capita waste production is depicted in Figure 2.

The Do-It-Yourself Regime: 1840-1870

Until about 1870, Toronto, in common with many North American cities, relied on a do-it-yourself approach to waste management. Individual householders (especially women) were expected to take care of domestic wastes (Strasser, 1982). The backyard was where unwanted materials were piled on bonfires or more likely thrown onto midden heaps (Hodgetts, 1911). Excrement and some rubbish went into the backyard privy (or pit latrine), while household slops were thrown into the yard or lane. In many households, kitchen garbage fed backyard animals. Although the city had become essentially a pig-free zone by 1861, there were still thousands of chickens, geese, horses and cows living in Toronto's residential areas. Factories and commercial premises favoured on-site disposal and waste was used to infill ravines and creeks (MacIlwraith, 1991).

The Beginning of Municipal Control: 1870 to 1890

Although the habits of recycling, of salvage, and of modest consumption were the norm in the Victorian city, the do-it-yourself regime of waste management was clearly inadequate given the size and density of the urban population. Few people had the space or the means to process their own waste materials adequately. People sank pit latrines where they found the best drainage, often directly into water supplies (Hodgetts, 1911).

The city had long employed scavengers, but their job had usually been to remove putrefying animal carcasses which lay about the streets and waterfront. Occasionally, when epidemics threatened, the city organized more zealous scavenging, but the effects were temporary. In 1866, in the hope of staving off a cholera epidemic, the city began a programme of garbage collection during the summer. Each ward was provided with a cart which was supposed to troll the back lanes looking for filth. The precious cargo was then carted to the waterfront, shovelled into waiting scows, and hauled up the Don River as far as the jail. Here the prisoners unloaded the muck and spread it on the prison farm as fertilizer.

From this beginning, as an emergency measure to defeat cholera, the city's health officials pressed for the establishment of a regular programme. From 1873, the city began regular systematic collection of domestic garbage and ashes. Householders were supposed to separate the waste into garbage (which rotted and smelled bad) and ashes (which did not). Along with garbage collection, the city improved its street-cleaning services. Every spring, the city turned out crews to scrape the streets, raking up considerable quantities of mud, manure, litter and garbage. In 1890, Toronto streets generated about 200 kg per capita of sweepings and scrapings, about ten times the level for the 1980s (Gilbert, 1988).

With garbage collection in hand, health officials then turned on the pit latrine or privy, surrounding it with regulations and pressing the public to invest in flush toilets. Although the flush toilet was in widespread use by 1890, there were still thousands of pit privies in service (Hastings, 1911). In 1913, when the city finally began to force their abolition, there were still 16,000 in use. By about 1920, most of these privy vaults had been filled, with ashes and rubbish being the standard material. There are at least 25,000 of these infilled pits in Toronto's backyards.

As a part of the crusade against the privy pit, officials tightened the rules for the contractors who emptied them.

These entrepreneurs, known as night soil excavators, were licensed and subjected to regulation. They operated night soil farms in the burgeoning fringe of new suburbs, where they processed the material into fertilizer for the fruit and vegetable trade. William Berry's City Odorless Excavating Company (which had about 40% of the Toronto excrement market) obtained land in the area east of Christie Pits (Fig. 3) in the early 1880s. Here he made fertilizer out of the 2500 m³ of night soil his men collected each winter, until the city annexed the area and shut him down. Most of the contractors were also in the haulage business, and several operated sand and gravel pits.

The Challenge of Success: 1890 to 1910

The new municipal regime of waste management soon brought its own problems. Having established the practice of municipal garbage collection, the city then faced the challenge of finding dumps for all the material. At first, the city's scavengers dumped in ravines and waterbodies, but they were soon in conflict with health officials and local residents. A problem of midden heaps in back yards had been replaced by one of open municipal dumps, often quite close to new residential areas, much to the disgust of urban reformers.

By the later 1880s, officials had concluded that, although coal ash could be dumped without problem, garbage should be incinerated. After a long series of deliberations, the city built its first garbage incinerator in 1891 (Fig. 4). Known as the Eastern Crematory, it stood on newly filled land beside the Don River on Cypress Street. It was quickly followed by another, the Western Crematory, on Strachan Avenue, in 1893 (Fig. 5). This was well placed to absorb the manure of the cattle market and offal from the abattoirs. Both were low-temperature facilities, and were smoky, rather defective in design and expensive to run. The Eastern Crematory ceased operations in 1898 and subsequently collapsed, due to the settlement of its foundations. The Western facility suffered from fires, and despite renovations, broke down constantly. The city found it cheaper to dump raw garbage into Ashbridge's Bay, after constructing a special streetcar line to carry waste material.

By 1900, Toronto householders had come to identify city government as the principal manager of solid wastes. The city now had the means to deliver the household right of free regular garbage pickup. As waste volumes increased, and new subdivisions absorbed land, the dumps met increased opposition from local residents. These disputes were probably most intense in the northern sections of the city. The

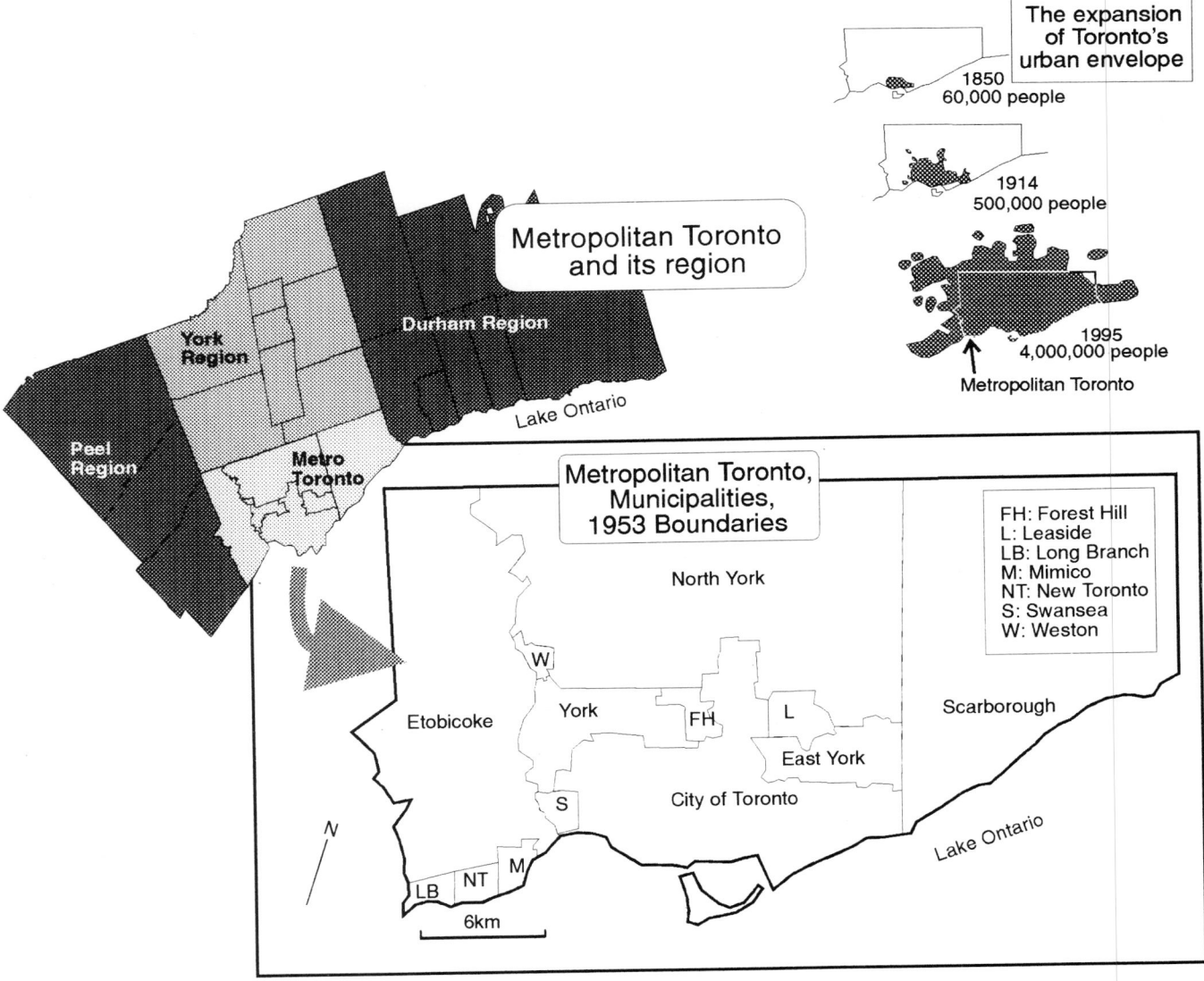

Figure 1   *Map of Toronto region, showing the municipalities of Metropolitan Toronto and historical growth of population.*

Christie Pits dump closed in 1914, for example, following pressure from local residents, who also thwarted attempts to use the site for an incinerator.

By 1910, the habit of dumping in ravines and waterbodies was rapidly obliterating several of Toronto's major creek and ravine systems. The city had filled the old course of the Don River and straightened it in the 1880s and 1890s. Large sections of Garrison and Taddle creeks (Fig. 3) had already been entombed in garbage, along with western tributaries to the Don. Low-rent housing in the Garrison Creek ravine, notably at Sully Crescent, was expropriated for dump space and demolished, as the city proceeded to fill the remaining sections of the ravine.

Edwardian Disposal: 1910 to 1917

By the first decade of the twentieth century, waste collection was much more efficient, consumption was increasing per capita, and the city was continuing to grow rapidly in population (from around 80,000 in 1880 to 500,000 in 1914). Waste generation increased as the population swelled, and was also

increasing per capita, from about 550 kg in 1890 to about 750 kg by 1910 (Fig. 2). What the officials termed ashes accounted for about 70% of the waste (Fig. 6), with the highest production in the winter. Putrescible garbage and combustible rubbish, about 30% of the total, was produced fairly constantly through the year (Jones, 1899).

By 1905, the increased waste volumes had overwhelmed Toronto's low-temperature incinerators, which had never worked reliably. Dump sites were becoming scarce inland, and every spring there was a crisis as officials struggled to dispose of the winter's accumulation of street waste. Toronto's garbage collection was dependent on the short hauling distances of horses, and many of the available dump sites were at an inconvenient carting distance.

Most residential areas, especially those on uneven ground, became pockmarked with small dumps of material. Ashes could be used to surface residential streets and lanes, and many property owners sought ashes and sweepings to fill up their lots. The city obliged, and even printed special forms on which residents could file requests. The city made be-

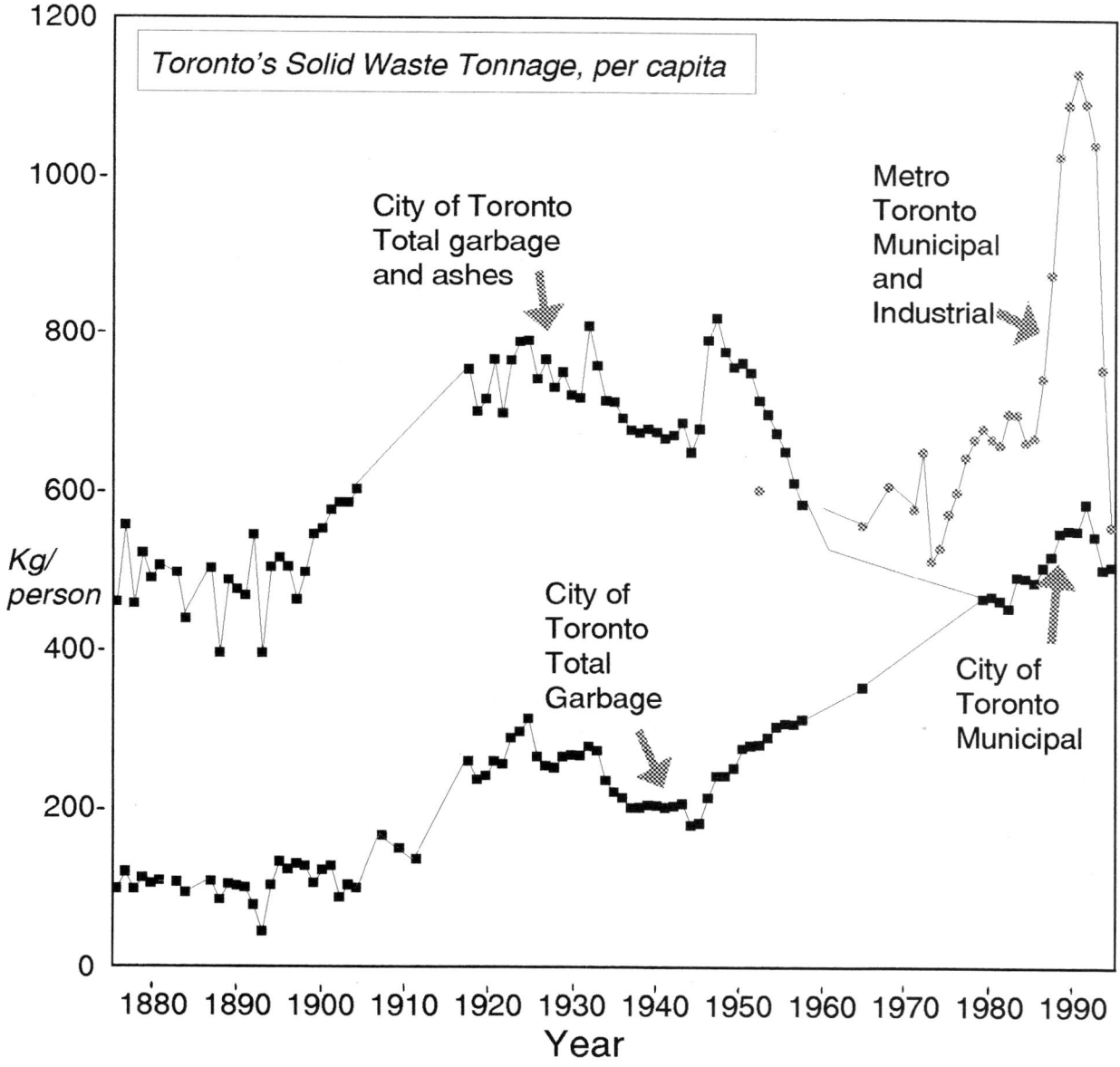

Figure 2 *Toronto's changing waste stream, 1870-1993. Figures in kilogrammes per capita, per year.*

tween 3000 and 4000 deposits of fill in Toronto's backyards between 1910 and 1955, not counting the filling-in of privy pits. The process had flourished on a substantial, but unknown, scale before then. The city preferred to use ashes and sweepings, rather than garbage, for backyard fills because of the proximity to housing.

The Renewal of Municipal Waste Services: 1917 to 1940

In 1912, under Commissioner Roland C. Harris, the City of Toronto began a major review of its waste management plans. During the second decade of the twentieth century, Harris brought in consultants, conducted detailed waste audits, and evaluated new technologies. Harris persuaded the council to sink much more capital into waste management, starting with huge investments in high-temperature incinerators. Garbage collection would be redesigned and gradually motorized. A series of transfer stations was envisaged (but not built), to gather the wastes for rail-haulage to dumps and incinerators (*The Mail & Empire*, 1913). In 1914, the city issued debentures for $1 million to build the new facilities (*The Globe*, 1913).

The first of the new incinerators, the Don Destructor, opened in 1917 on the east bank of the Don River, just north of Dundas Street. Intended to serve the eastern parts of the city, the facility was soon handling half the city's garbage. A counterpart, to replace the Western Crematory, was delayed by war until 1924-1925. Acute shortages of dump space, keenly felt before 1910, were alleviated by the landfilling projects of the Toronto Harbour Commission. There are numerous period photographs showing the disposal of solid waste along Toronto's waterfront at this time (Rose, 1988; Chapter 20). The city otherwise continued to operate ash dumps in several of the major ravines (Fig. 7). Although garbage collection and disposal remained largely labour intensive, and dependent on horse traction until the 1940s, it was beginning to develop the capability of longer hauls.

Toxicity

The occupational health literature shows that a considerable

range of toxic materials was already in common use by the 1920s. When Ontario established its Occupational Health Branch in 1918, for example, the first priority was to protect spray-painters from exposure to benzene and lead, substances which remain a significant concern today.

At the Abbott Avenue dump site in Toronto's west end, period photographs show floating drums. It is likely that these contained residues from Wiborg's printers' ink and varnish factory on Adelaide Street, an industrial waste producer which is known to have used this dump site. It also operated as an industrial sewage lagoon, also receiving oily residues and cadmium-plating wastes.

Regional Interdependence

During the 1920s, garbage processing, with incinerators, became a huge capital-intensive operation in Toronto. The new incinerators, operating by 1925, absorbed at least 88% of the city's garbage. The other 12%, which originated in industry and in the northern and extreme western regions of the city, went to the dump. The city used the usual ravine locations, dumping inside the city during the winter (when the garbage could be covered with ashes) and outside the city during the summer (Fig. 5). By the mid-1920s, neighbouring municipalities were making strenuous efforts to prohibit city garbage, and Toronto found itself in difficulties. The Harbour Commissioners, who were engaged in massive landfilling in the eastern port lands, were no longer willing to take garbage.

Because of its investment in incinerators, the city was able to trade incinerator capacity for dump space, in a series of rather one-sided deals with harassed municipalities. In return for incinerating about 5,500 tonnes (t) of Swansea's garbage between 1935 and 1940, the city was able to dump 375,000 t of ashes in the village. The arrangement continued for another 15 years, until Swansea finally ran out of dump space.

After the boom of the 1920s, in which annual waste generation reached 800 kg per capita, there was a decline which extended right through the difficult years of the Depression (Fig. 2). Ash production slackened (heating oil and domestic electricity consumption rose), but there were significant rollbacks in garbage production. The downward trend of the Thirties continued until about 1940. With World War II, there was an immediate influx of population and dramatic increases in both domestic and industrial consumption (Fig. 2).

The Ash-To-Trash Transition: 1940 to 1950

Around the time of the First World War, the average Toronto

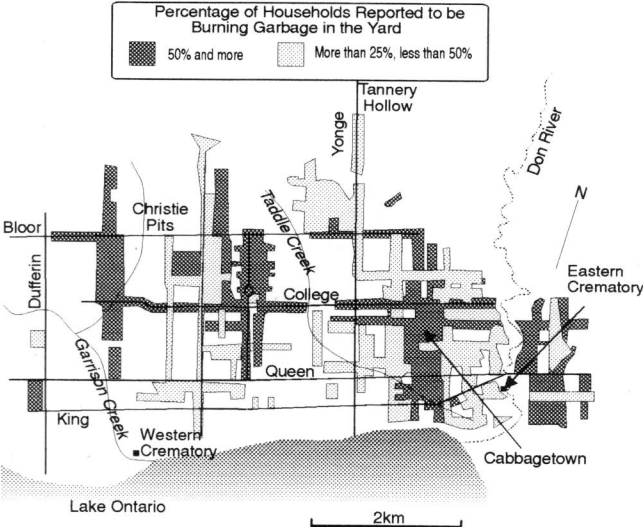

Figure 3 *Domestic garbage burning in the city of Toronto during the mid-1880s. This map uses the street-by-street sanitary surveys included in the annual reports of the Toronto Board of Health during the mid-1880s. Garbage burning was an outlawed practice by this stage, but the map demonstrates its prevalence a decade after the city began to collect garbage, especially in the poorer sections such as Cabbagetown.*

Figure 4 *The Eastern Crematory, Toronto's first garbage incinerator, built in 1891 and used until 1898.*

household burned between 5 and 7 t of coal in the typical winter. This was a pattern that continued until the Second World War. Consultants hired by the city in 1911 estimated that each person generated at least 360 kg of coal ash each year. These were quite conservative estimates since the coal-burning stoves and furnaces could absorb a fair amount of paper, wood and other combustibles, especially in winter. The result was that households generated a lot of ash, but not that much rubbish and garbage.

Dramatic shifts in household energy consumption occurred in the 1940s and early 1950s, when home heating switched from coal to oil. This began first in the newer suburbs. In this respect, Toronto was behind American cities (such as Cincinnati and New York) where the shift from coal was largely complete by the end of the 1930s (Melosi, 1981). Without coal-burning furnaces and stoves, Toronto's households had less ash to throw out, but could no longer burn combustible rubbish. The city continued to pick up the ashes from households, but this material was increasingly adulterated with other materials. It now contained so much paper and wood that its density declined dramatically, from around 430 kg·m⁻³ in the mid-1940s to around 250 kg·m⁻³ in the late 1940s.

The Waste Crisis of the 1940s and 1950s

Municipal incinerators, designed for the waste generation habits of the 1930s, were overwhelmed by the quantities of combustible waste being generated in the late 1940s. Many of

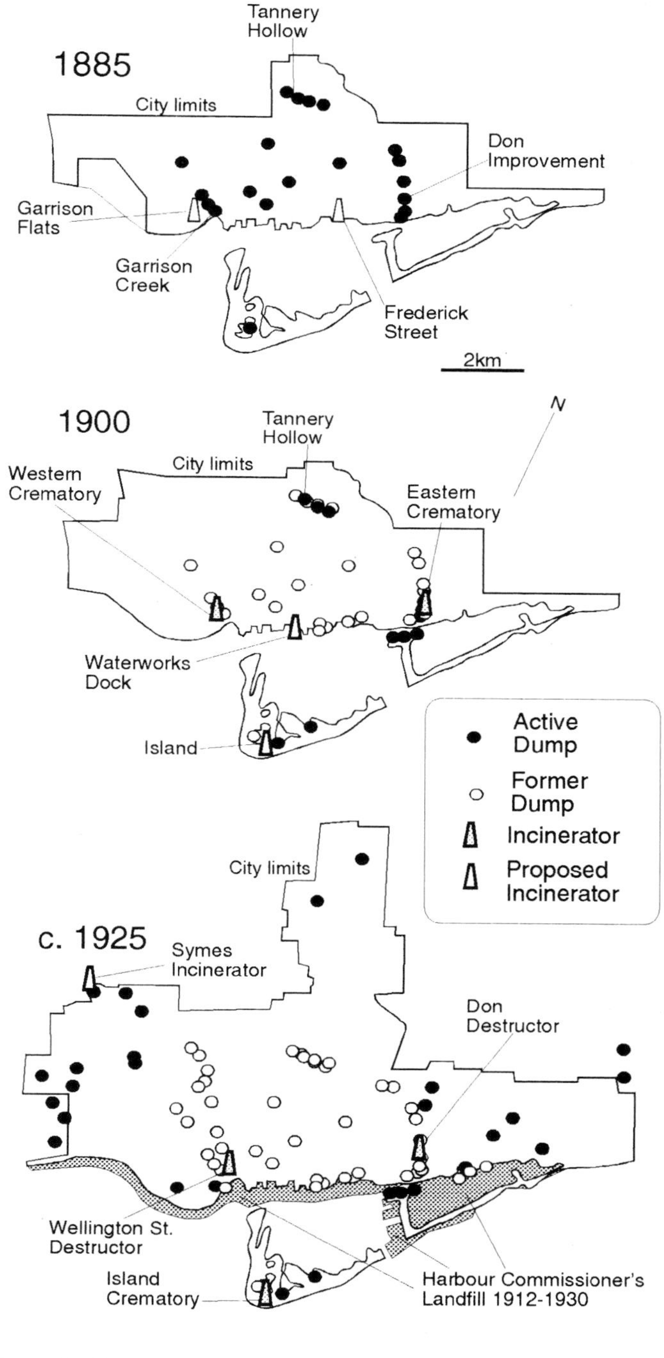

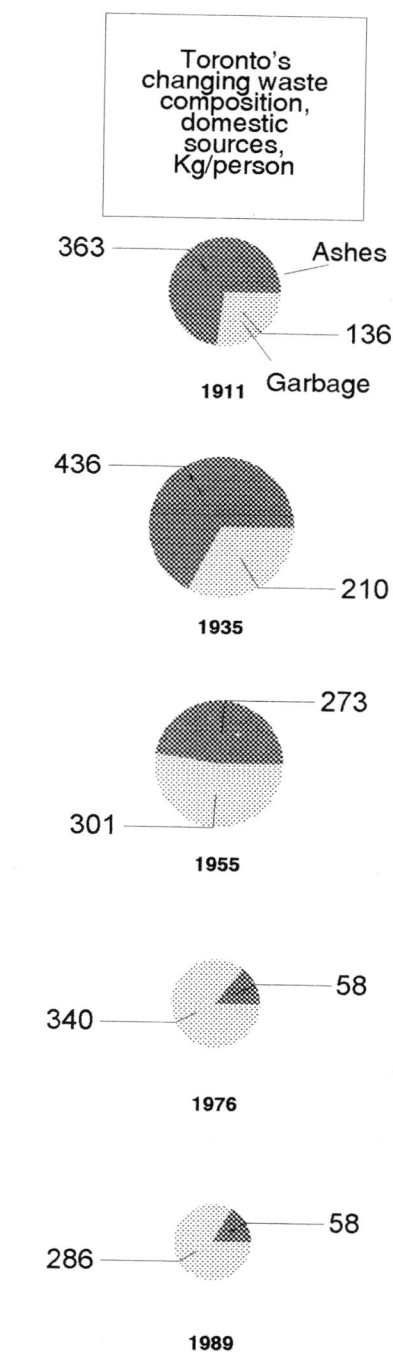

Figure 5 *Toronto's garbage disposal facilities 1880s-1920s. Generalized positions of major dumps are shown.*

Figure 6 *Changes in Toronto's waste generation, kg per person, per year between 1911 and 1989.*

the smaller municipalities were short of capacity, something which widespread municipal bankruptcy in the 1930s had not improved. There was a hurried effort to expand incinerator capacity, but, by the later 1940s, like its counterparts elsewhere in North America, the Toronto region faced a serious waste crisis (Melosi, 1981). Suburban sprawl was absorbing potential dump sites, and incinerators were beyond capacity. There were large quantities of new and toxic waste materials for which disposal facilities were limited and inappropriate.

To conserve capacity for household material, the City of Toronto first tried charging incineration fees to factories, but then had to shut its incinerators altogether to industrial waste. By 1950, most of Toronto's industrial waste was being sent to the dump. Dumps became much more prone to fire, as industrial wastes, together with the burgeoning quantities of household material, were diverted away from incinerators. Toronto's Greenwood dump, for example, showed an 8- to 10-fold increase in fires between 1947 and 1948, with more than 50 outbreaks in 1949 alone.

Many dumps practised deliberate open burning. It was standard practice for dry cleaners to burn naphtha-process waste, an activity which only ceased with the introduction of new solvent technologies. For several years, Toronto companies with flammable materials were told to take the material to the foot of Leslie Street and set fire to it. Known as the Leslie Street fire dump in the 1960s, it was used for the disposal of flammable liquids (such as petroleum-based dry-cleaning solvents). It also accepted combustible wastes from companies such as Eaton's, and from downtown commercial buildings. When Metro formulated its long-term waste disposal strategy in 1967, the need for open solvent burning remained a continuing requirement.

Investigations of the city's gas stations, dry cleaners and other small producers of toxic materials indicate that private disposal companies already had a large business in the early 1950s. Some operations, such as Disposals Services Ltd. (Crooks, 1983, 1993) acquired their own landfill space (the Maple dump in this case), while others handled unpleasant industrial wastes. A firm in North York, for example, disposed of all of the waste solvents and paints generated by CIL's Laughton Avenue factory between 1933 and 1953. Bedford Cartage, in west Toronto, accepted spent dry-cleaning solvent, while also supplying cinders for driveways. Profiting from the excess of industrial waste, some companies were salvaging and recycling materials, but many were hauling to dump sites which they preferred not to disclose. There had long been a penumbra of small disposal operations, especially in the case of toxic or flammable liquids, but, from the 1950s, their role became increasingly significant (Crooks, 1983, 1993). Municipal dumps became significant disposal points for chemical residues such as tarry ash residues from Consumers Gas, coal-laden street sweepings and sludge from dry-cleaning operations. Some municipalities sprayed the dumps with pesticides such as DDT (Fig. 8). Many of these city dumps would eventually become neighbourhood parks or sites for non-profit housing.

One municipal response to the waste crisis was to build additional garbage incinerators. These new incinerators took time to build, and did not really solve the problems faced by industry, which was beginning to create significant amounts of new and toxic materials. Solvent-laden wastes created flash-fire hazards for incinerators, an increasingly frequent hazard in the later 1940s, and a significant contributor to breakdown. Flash fires and explosions at the Mimico incinerator in the 1940s and 1950s, for example, have been traced to wastes from local factories such as Goodyear Tires, Roxalin, and CIL's Fabrikoid division (Mimico Action Committee, 1992). Even when they did not explode or catch fire, municipal incinerators usually had no air pollution controls, and could not protect downwind areas from toxic dustfall.

Away from the incinerator, the disposal of industrial wastes was haphazard, and many toxic materials were dumped, without adequate documentation, in a multitude of rural dump sites. Some of the dumped material was radioactive (Chapter 28). Battery casings and paint residues continue to be discovered during construction of residential subdivisions. These hidden legacies are a reminder of the waste disposal problems of the late 1940s.

Garbage Goes Regional: 1960 to 1980

By 1960, garbage was a regional problem. In order to accommodate industrial waste producers and harassed municipalities, Metropolitan Toronto operated a series of emergency landfills in ravines and hollows across the area (Metropolitan Toronto, 1957; Fig. 9). Because of the prevalence of war veterans in Metro Works, the programme was given the codename of Operation Overload. Overload's dumps were

Figure 7 *Greenwood dump in 1937. A vehicle has just dumped ashes and in the background is a truck from Langley's dry cleaning. This type of open dump was typical of urban dumping practice in Toronto until the 1950s (City of Toronto Archives RG8 70 533).*

Figure 8 *A Toronto street flusher sprays the McCleary Park dump with DDT in the later 1940s. This site is in the Port Industrial District (Chapter 2; City of Toronto Archives RG8 100 606).*

sanitary landfills (Metropolitan Toronto, 1964). Each day's dumping would be compacted by heavy earth-moving equipment and covered by a layer of earth. Most were sited in ravine locations, many of them close to water. Apart from combustibles, such as paper and cardboard (recycled if profitable or burned on site), there was no official policy about what could go into these dumps. They were intended to provide disposal for municipal and industrial waste.

By 1964, Metro's Operation Overload was absorbing half the region's garbage, and most of its industrial waste, on a fee basis. Its landfills had already accepted 3 million tonnes of material (Metropolitan Toronto, 1965), and what had begun as

an interim scheme had become a permanent responsibility. Following the Metro example, the province would add waste disposal to the portfolio of regional government across Ontario (Gore and Storrie, 1980; MacLaren, 1967; Proctor and Redfern, 1975). Metro's garbage strategy was plotted in the 1967 MacLaren report, which became its waste management blueprint until the 1980s. The consultants envisaged a series of suburban sanitary landfills, backed up by a programme of acquiring spare sites for future needs. The new landfill sites would be very much larger than any previous dump site (Fig. 10), offering the benefits of economies of scale. The landfills of Operation Overload had typically been in the 200,000-t

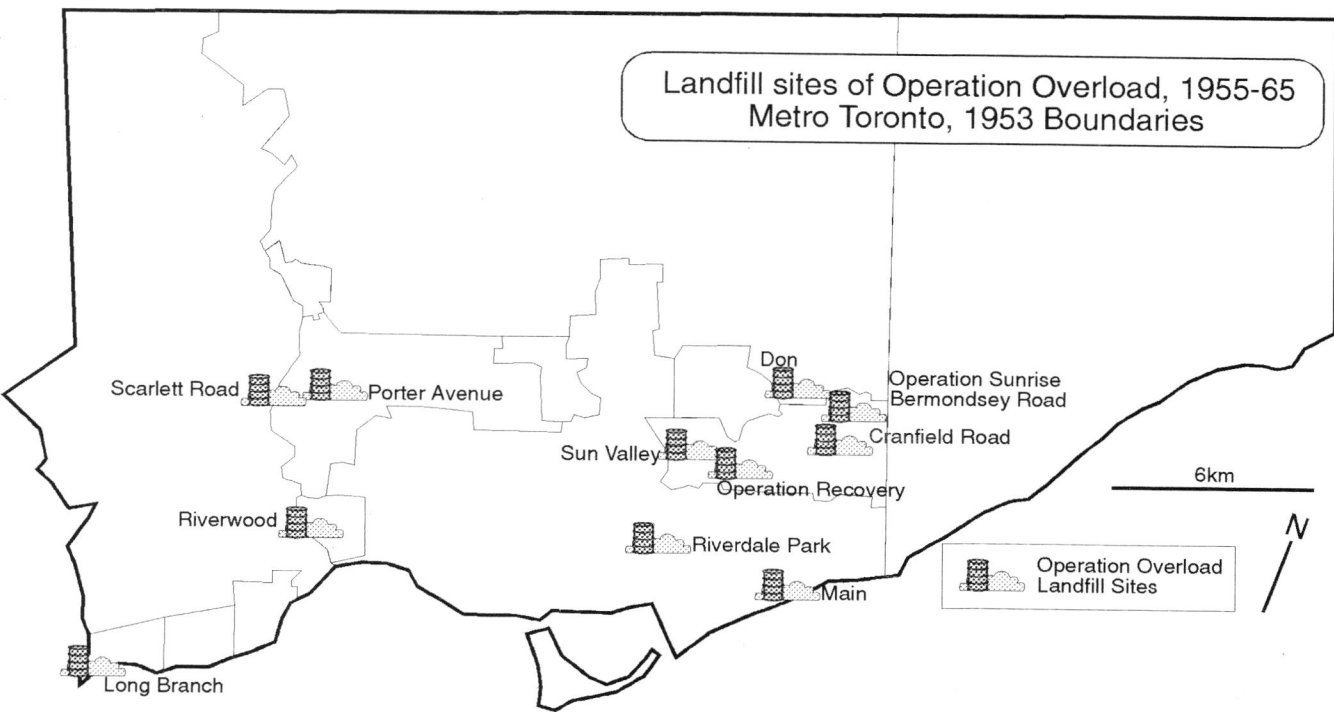

Figure 9 *Landfill sites used by Operation Overload, 1955-1965.*

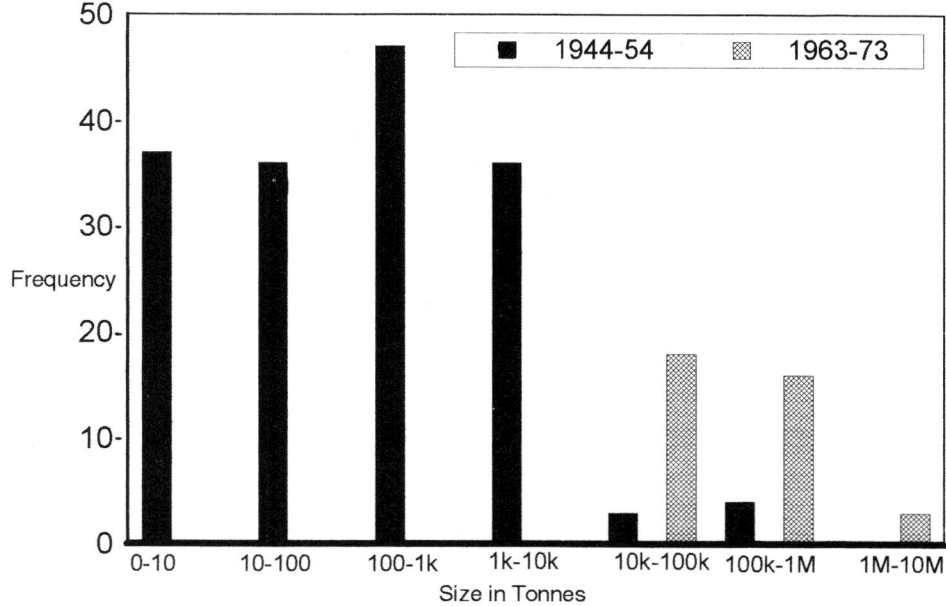

Figure 10 *Changes in the sizes of Toronto's garbage dumps 1944-1973. Size in thousands (**k**) and millions (**M**) of tonnes.*

range with the life expectancy of only a few months. The new facilities would range up to 5 or even 20 million tonnes, enough to take all of the region's solid waste for several years.

Many, if not all, the dumps operating in the Metro Toronto region prior to 1965 were planned as exercises in land improvement. They were examples of a time-honoured tradition of rehabilitating abandonned pits and quarries using waste materials (Campbell, 1955; Chapter 2). The sanitary landfill quickly became the dominant item in Metro Works capital budget (Metropolitan Toronto, 1967-1971). Despite the sense of garbage crisis at the time, the total weight of solid waste per person was actually at an historic low. At around 650 kg per capita, the solid waste from the 1960s was comparable to that of the Edwardian period by weight (Fig. 5), but was much more bulky and probably more toxic. Beginning first in Don Mills, and then in Scarborough, municipalities began to allow the use of plastic garbage bags, in response to pressure from polyethylene manufacturers. By 1966, they were in widespread use.

DISCUSSION

The environmental legacy of the 1960-1980 period in waste disposal is profound. Regional waste management at this time promoted the use of municipal and industrial wastes for the rehabilitation of former sand and gravel pits. No recognition was made of the need to prevent ground-water contamination by leachate, nor to effectively plan urban development in the vicinity of such sites (see Chapters 2, 20 and 24 for detailed discussion).

Toronto's urban landscape, like that of other Great Lake cities (*e.g.*, Chicago: Colten, 1994), has been built, in large measure, upon waste materials. It is essential to identify former dump sites and characterize their contents; the current provincial listing, for example, identifies fewer than 80 closed landfills in Metro Toronto (Ontario Ministry of Energy and the Environment, 1991), but more than 800 additional former dumps have been identified during the course of the author's research. In this regard, development of the City of Toronto Historical Land Use Database represents an important step forward in identifying contaminated sites (see Chapter 27).

REFERENCES

Anderson, R.B., 1993, Garbage disposal in the Greater Toronto Area: a preliminary historical geography: Operational Geographer, v. 11, p. 7-13.

Campbell, L.A., 1955, Problems in the utilization of municipal waste: Second Ontario Industrial Waste Conference, Ontario Water Resources Commission, p. 145-151.

Colten, C.E., 1994, Chicago's waste lands: refuse disposal and urban growth, 1840-1990: Journal of Historical Geography, v. 20, p. 124-142.

Crooks, H., 1983, Dirty Business: The Inside Story Of The New Garbage Agglomerates: Lorimer, Toronto, ON, 257 p.

Crooks, H., 1993, Giants Of Garbage: The Rise of the Global Waste Industry and the Politics of Pollution Control: Lorimer, Toronto, ON, 300 p.

Duffy, A., 1992, Metro loses $225 million to US dumps: *The Toronto Star*, Toronto, ON, October 21, 1992, p. A12.

Gilbert, R., 1988, Toronto region's looming waste management crisis: City Planning, v. 7, p. 29-33.

The Globe, 1913, Plans incinerator to cost a million: street commissioner outlines garbage disposal scheme: *The Globe*, Toronto, ON, June 24, 1913.

Gore and Storrie Ltd., 1980, Regional Municipality of York, waste management study, Final report: 266 p.

Haight, M.E., 1991, ed., Municipal Solid Waste Management: Making Decisions in the Face of Uncertainty: University of Waterloo Press, Waterloo, ON, 531 p.

Hastings, C.J., 1911, Report of the Medical Health Officer dealing with the recent investigation of slum conditions in Toronto embodying recommendations for the amelioration of the same: City of Toronto, 86 p.

Hodgetts, C.A., 1911, Report on the epidemic of typhoid fever occurring in the city of Ottawa, January 1st to March 19th, 1911: Commission of Conservation, Canada, 48 p.

Jones, J., 1899, An outline of the system of garbage collection and disposal in the city of Toronto, in Toronto, 1899, City Engineer's Report, p. 56-59.

MacIlwraith, T., 1991, Digging out and filling in: making land on the Toronto waterfront in the 1850s: Urban History Review, v. 20(1), p. 15-33.

MacLaren, J., Ltd., 1967, Report and technical discussion on refuse disposal for the Municipality of Metropolitan Toronto: 82 p.

MacLaren Plansearch Ltd., 1984, Waste Management master plan update for the Regional Municipality of Peel, Stage 1 Final report: 2 volumes.

The Mail & Empire, 1913, City will have reduction plant: *The Mail & Empire*, Toronto, ON, June 24, 1913.

Melosi, M.V., 1993, Down in the dumps: is there a garbage crisis in America?, *in* Melosi, M.V., ed., Urban Public Policy, Historical Modes and Methods: Pennsylvania State University Press, State College, PA, p. 100-27.

Metropolitan Toronto, 1957, Metropolitan Toronto works committee, report #1, January 4, 1957: Minutes of the Council of the Municipality of Metropolitan Toronto, 10 p.

Metropolitan Toronto, 1964, Metropolitan Toronto works committee, report #2, February 25, 1964: Minutes of the Council of the Municipality of Metropolitan Toronto, 6 p.

Metropolitan Toronto, 1965, Metropolitan Toronto works committee, report #3, March 9, 1965: Minutes of the Council of the Municipality of Metropolitan Toronto, 2 p.

Metropolitan Toronto, 1967-1971, Annual report of the Commissioner of Finance, Metropolitan Toronto: 15 p.

Metropolitan Toronto, 1992, Metro Toronto Works Department, Annual Report: Metropolitan Toronto, 28 p.

Mimico Action Committee, 1992, Written submission to Ontario Municipal Board in connection with the Parklawn/Lakeshore secondary plan and McGuinness Distillery development proposal: April 21, 1992, 6 p.

Munson, W., 1990, Soil contamination and port redevelopment in Toronto: Canadian Waterfront Resource Centre, Working Paper 3, 14 p.

Munson, W., 1991, The disposal of coal ash at Toronto's outer harbour: Canadian Waterfront Resource Centre, Working Paper 7, 38 p.

Ontario Ministry of Energy and the Environment, 1991, Waste disposal site inventory: Toronto, Ontario: 196 p.

Proctor and Redfern Ltd., 1975, Regional Municipality of Peel, Waste Master Plan: 112 p.

Proctor and Redfern Ltd., 1990, Towards a solid waste management master plan: a status report on Metropolitan Toronto's planning process: Metropolitan Toronto Department of Works, Solid Waste Management Division, Discussion Paper 7.1, 66 p.

Rathje, W. and Murphy, C., 1992, Rubbish! The Archaeology Of Garbage: Harper Collins, New York, 263 p.

Rose, P., 1988, Solid waste, *in* Ball, N.R., ed., Building Canada: University of Toronto Press, Toronto, ON, p. 245-261.

Strasser, S., 1982, Never Done: A History Of American Housework: Pantheon, New York, 365 p.

Toronto News, 1914, No stink pot here: *Toronto News*, Toronto, ON, June 29, 1914.

The Toronto Star, 1962, Etobicoke may have best wrapped garbage: *The Toronto Star*, Toronto, ON, May 8, 1962, p. 9.

The Toronto Telegram, 1914, West Fairbank: many rats infest dump: *The Toronto Telegram*, Toronto, ON, December 2, 1914, p. 6.

22. Hydrogeology of the Edmonton Landfill, Alberta

Laurence D. Andriashek
David G. Thomson

*Environmental Research and Engineering Department, Alberta Research Council
P.O. Box 8330, Station F, Edmonton, Alberta T6H 5X2*

Reed Jackson

Stanley Associates Engineering Ltd., Stanley Technology Centre, 10160 – 112 Street, Edmonton, Alberta T5K 2L6

SUMMARY

Detailed characterization of the hydrogeologic setting was essential to model the potential for leachate migration from the proposed City of Edmonton Waste Management Centre at the Aurum site. Conventional geologic information from boreholes limited the ability to provide a detailed interpretation of stratigraphic units made geologically complex by glacial over-riding and thrusting (glaciotectonism). Supplementary information from hydraulic tests proved to be crucial in resolving difficulties in correlation of hydrostratigraphic units as additional geologic data became available. Piezometer readings and pumping tests established that two discrete aquifers are present within an apparently single hydrostratigraphic unit. Glacial-ice thrusting and injection of a slab of displaced bedrock effectively severed the unit into two aquifers with dissimilar hydraulic properties.

INTRODUCTION

In 1987, the existing landfill site in the City of Edmonton was nearing its capacity to accept municipal waste, and a process was initiated to select a new solid-waste-only waste management site. On the basis of a number of factors, mostly socioeconomic concerns, the Aurum site was chosen as the most suitable. As part of the application to develop a sanitary landfill site, Stanley Associates Engineering Ltd. was contracted by the City of Edmonton to conduct a hydrogeological evaluation of the proposed Edmonton Waste Management Centre at the Aurum site. In the latter stages of the evaluation, the Alberta Research Council was contracted by the consultants to provide a review of the geologic and hydrogeologic interpretations of the investigations, and to carry out a numerical modelling study of potential leachate migration during the life of the site.

This paper addresses the technical aspects of the hydrogeologic investigation, focussing on the following:
1. the evolution of geologic interpretation as new information became available during the course of the investigation;
2. the limitations on geologic interpretations made from conventional borehole data; and

3. the value of hydraulic information in the interpretation of a site made geologically complex by glacial thrusting and displacement of units within the stratigraphic sequence.

Location and Background

The proposed Aurum waste management site occupies an area of about 2.5 km² in the northeast part of the City of Edmonton. The site is located on an undulating to rolling landscape which gently slopes to the North Saskatchewan River, approximately 0.5 km to the north (Fig. 1). An erosional scarp of the North Saskatchewan River forms a natural boundary along the west and northwest margin of the site. The County of Strathcona boundary limits access along the eastern and southern margins of the site. Rail-line rights of way and pipeline corridors further limit access in the southern part of the area.

Geologic and hydrogeologic information for the site was collected from boreholes drilled during three phases of the field investigation. In the preliminary phase in 1987, four rotary boreholes were drilled through the Quaternary sequence and deep (>10 m) into the underlying bedrock to establish the stratigraphic units. Piezometers were installed in some of the deeper units in these testholes. At the completion of this phase of the investigation, the site was given a poor rating as a non-engineered industrial landfill because of the presence of a thick, water-bearing sand that was encountered in each of the holes. The City of Edmonton opted to proceed with a totally engineered and lined landfill site, and a detailed hydrogeologic investigation was completed in the fall of 1988. During this second phase of the investigation, an additional 12 rotary boreholes and 27 continuous-flight, dry-auger boreholes were completed on a 200 m grid across the property. The auger boreholes were drilled through the Quaternary drift sequence and completed into the upper part of *in situ* bedrock, or what was interpreted to be *in situ* bedrock, based on cutting samples. Piezometers were installed in all major stratigraphic units within most of these boreholes. Stanley Associates Engineering Ltd. prepared a report to the City in 1989, summarizing the hydrogeologic findings and making recom-

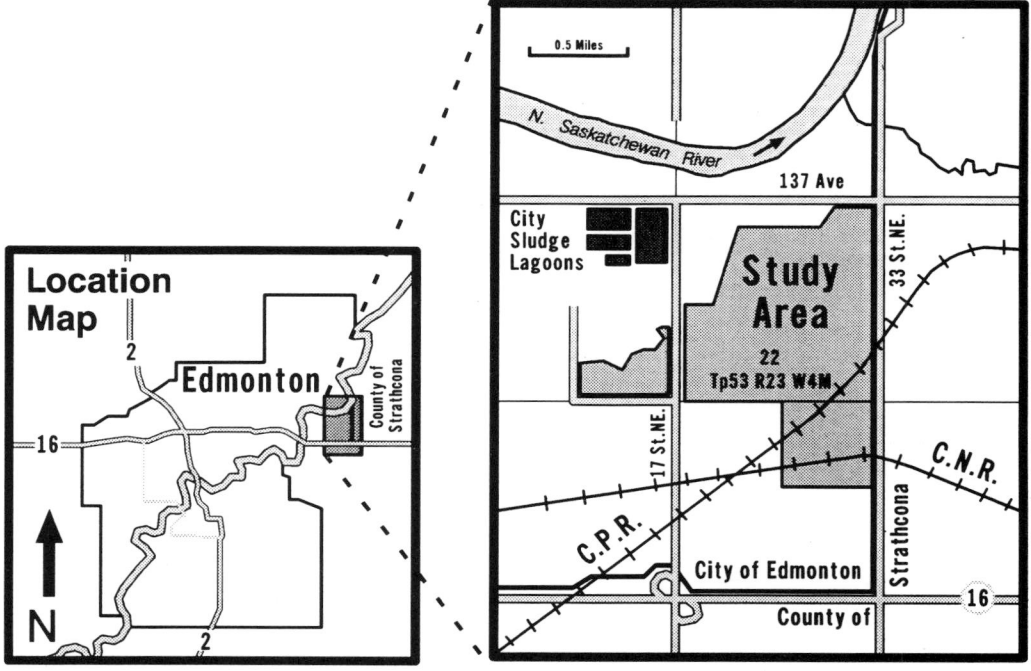

Figure 1 *Location of the proposed City of Edmonton Aurum waste management site.*

mendations to support a conceptual landfill design. The consultant was subsequently requested to further expand the subsurface investigation to support a more detailed design, and to resolve difficulties in the geologic and hydro-stratigraphic correlation. A third phase of the field investigation was initiated in 1990, at which time 24 additional auger boreholes were drilled on a 100 m spacing, primarily in the east half of the area, and two rotary holes were drilled to conduct pumping tests (Fig. 2). Borehole information from previous geologic investigations in the region provided supplementary data along the perimeter of the site.

The following section discusses the results of that investigation, focussing on the evolution of the interpretations and assumptions about the hydrogeologic setting of the site as information became available during the three phases of the investigation.

Bedrock Topography,
Buried Channels, and Drift Thickness

The Edmonton area is located on Upper Cretaceous claystone, sandstone, siltstone, ironstone and coal of the Horseshoe Canyon Formation (Irish, 1970). Most rock units within the formation are generally soft, poorly indurated, and capable of being penetrated by a dry auger. During the Late Tertiary and Early Quaternary, the bedrock surface was exposed and modified by fluvial erosion. A number of bedrock channels developed on the landscape of the region. The Beverly Channel, which is the main channel, is located north of the Aurum site (Carlson, 1967; Kathol and MacPherson, 1975; Andriashek, 1984). A tributary of the Beverly Channel, the Boag Valley (Carlson, 1967; Andriashek, 1984), trends north through the Aurum site, and dominates the bedrock topography. The valley consists of two branches. The East Channel trends north along the east property boundary, and the South Channel trends northwest along the southern boundary (Fig. 3). The talwegs of these buried channels are believed to merge southeast of the site.

Multiple glacial events during the Quaternary filled in the

bedrock channels with as much as 30 m of stratified and nonstratified sediment, masking any present-day surface expression of the valley (Fig. 4). Post-glacial fluvial erosion by the North Saskatchewan River cut down through the drift and bedrock sequence to as much as 15 m below the base of the South Channel in the northwest corner of the site. The north end of the East Channel was similarly truncated by recent erosion along Oldman Creek. Groundwater discharge in the

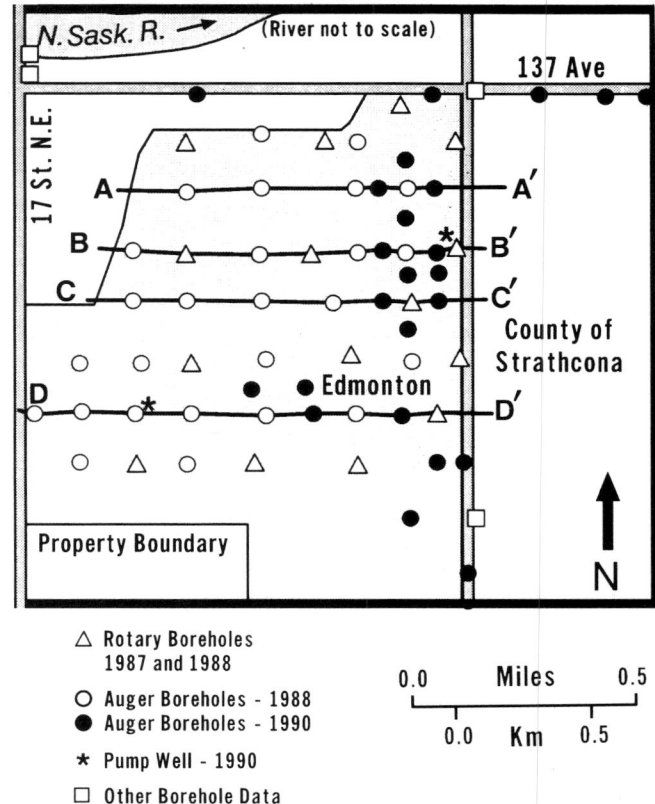

Figure 2 *Location of boreholes and geologic cross-sections.*

form of springs and seeps occurs at the outcrops of these hanging bedrock channels (Fig. 3).

SITE GEOLOGY AND HYDROSTRATIGRAPHY

The geologic interpretation of the site evolved as new information became available during the three-year investigation period. The discussion of the geology and hydrostratigraphy of the site is divided into two sections, an initial interpretation based on geologic information collected to the end of 1988 (Phases 1 and 2), and the final interpretation, which was based on additional geologic information and supplemented with the hydraulic information that was collected and compiled near the end of the third phase in 1990.

All units were described from borehole cuttings, in most cases, dry-auger flight samples. No man-made or natural exposures of the entire stratigraphic sequence are present at the site, although nearby outcrops along the banks of the North Saskatchewan River and Oldman Creek afforded examination of some of the units.

Interpretations
Following Phases 1 and 2 of the Investigation

Table 1 lists the stratigraphic units encountered within the boreholes at the Aurum site.

The lowermost surficial unit consists of a basal sand, and gravelly sand of pre-glacial fluvial origin. The unit occurs in both the east and south channels of the Boag Valley, as well as on the bedrock terrace adjacent to the channels. The sand and gravel are composed dominantly of light-coloured quartzite and dark-coloured chert rock fragments derived from the local bedrock and the Cordilleran region to the west. The unit is correlative with the salt-and-pepper coloured sand of the Empress Formation (Whitaker and Christiansen, 1972; Andriashek and Fenton, 1989), which is also referred to as the Saskatchewan Sand in the Edmonton region (Stalker, 1968). Pink-coloured granitic rock fragments, indicative of glacially transported sediment from the Canadian Shield northeast of the region, are conspicuously absent.

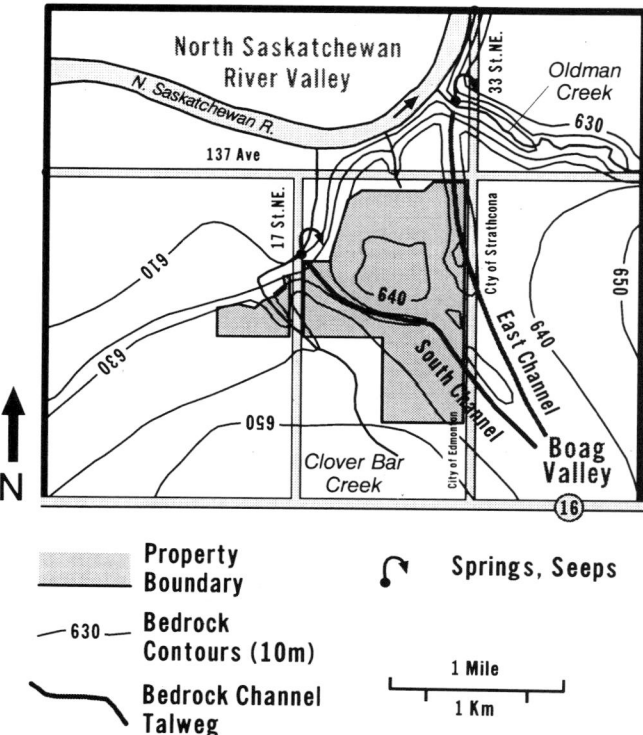

Figure 3 Regional bedrock topography showing talwegs of bedrock channels.

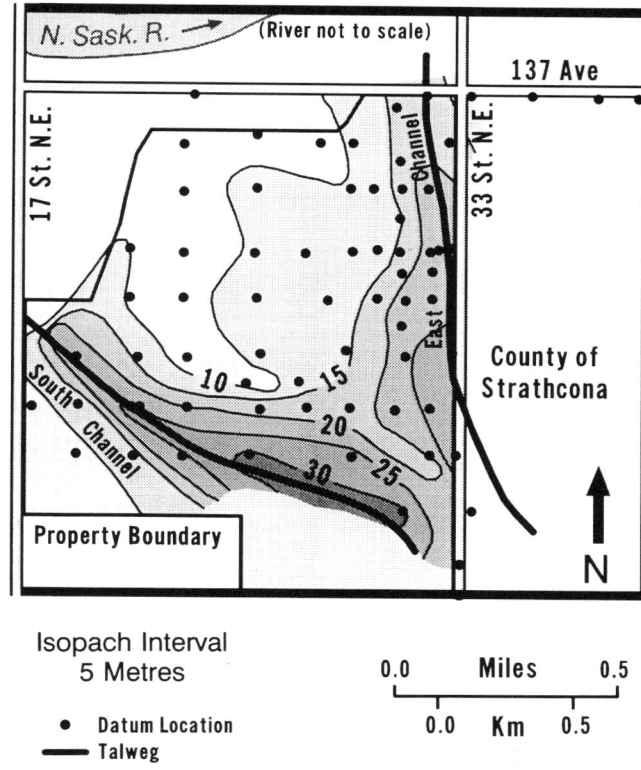

Isopach Interval
5 Metres

Figure 4 Thickness of drift overlying bedrock.

Table 1	Stratigraphy at the Aurum site, City of Edmonton.	
AGE	**UNIT**	**LITHOLOGY**
	·········· SURFICIAL ··········	
Quaternary	— Glacial Lake Edmonton	Silt and Clay
	— Upper Till	Clay Loam Till
	— Glacially Displaced Bedrock and Highly Plastic Clay Till	Claystone, Clay Till
Tertiary	Empress Formation	
	— Pre-glacial Terrace Sand	Sand, Gravel
	— Pre-glacial Channel Sand	Sand, Gravel
	·········· BEDROCK ··········	
Upper Cretaceous	Horseshoe Canyon Formation	Claystone, Siltstone, Sandstone, Coal

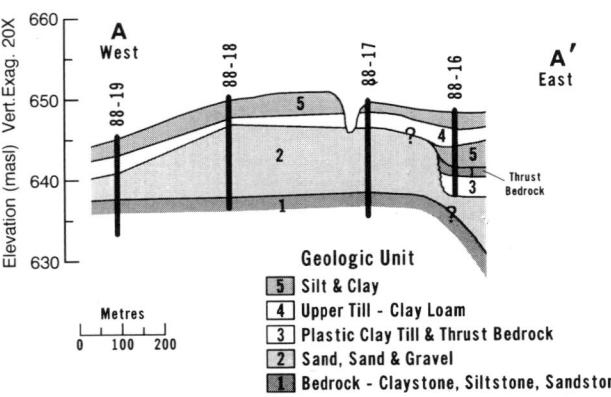

Figure 5   *Geology along cross-section A-A' (based on 1987 and 1988 borehole information).*

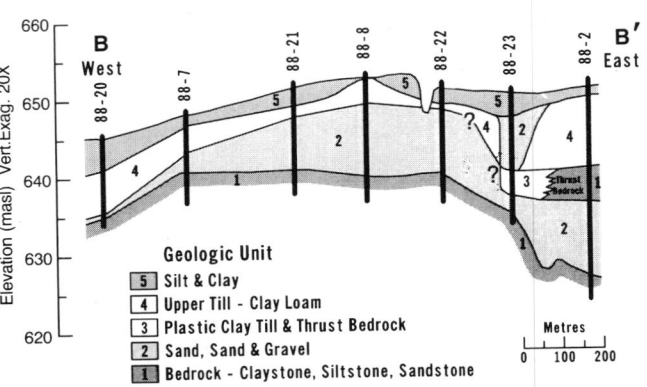

Figure 6   *Geology along cross-section B-B' (based on 1987 and 1988 borehole information).*

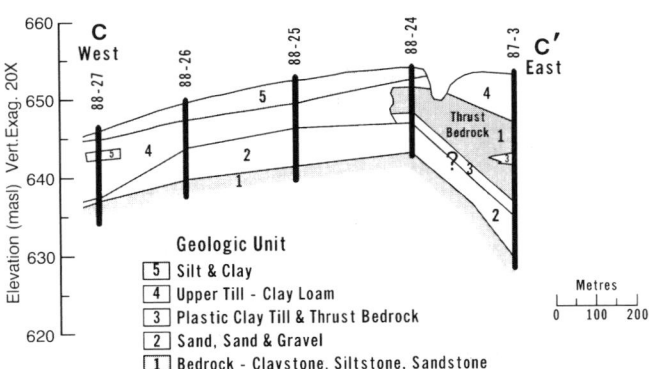

Figure 7   *Geology along cross-section C-C' (based on 1987 and 1988 borehole information).*

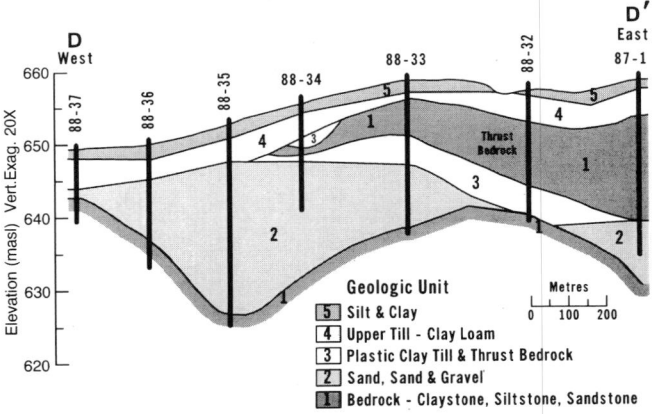

Figure 8   *Geology along cross-section D-D' (based on 1987 and 1988 borehole information).*

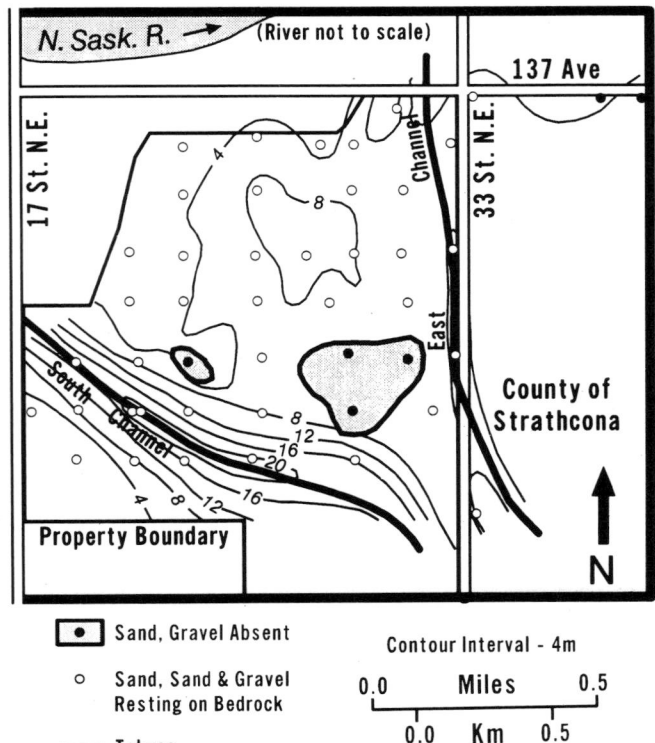

Figure 9   *Distribution and thickness of sand and gravel resting on bedrock (based on 1987 and 1988 borehole information).*

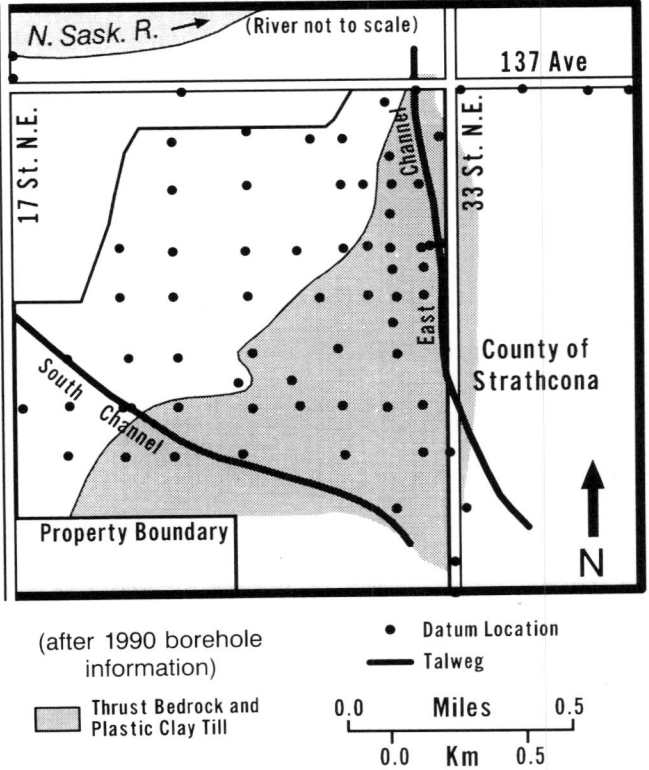

Figure 10   *Distribution of glacially thrust bedrock and plastic clay till (includes 1990 borehole information).*

Glacial sediment overlies the preglacial sand and gravelly sand over the entire study area. A mixed complex consisting of glacially thrust bedrock and a highly plastic clay till was recognized in boreholes located in the eastern part of the site. The displaced bedrock is dominantly claystone, and is indistinguishable from that of the Horseshoe Canyon Formation. The plastic clay till is composed primarily of glacially reworked claystone with numerous weak clasts of claystone and rare indurated clasts of granitic and quartzitic composition. The till is dense, has a low permeability, and is easily differentiated from the overlying regional clay loam till that lies in contact with preglacial sand in the western part of the site. Where preglacial sand is absent, the plastic clay till serves as a stratigraphic marker to differentiate glacially displaced bedrock sequences from *in situ* bedrock.

Glacial Lake Edmonton sediment (Gravenor and Bayrock, 1956; Bayrock and Hughes, 1962) forms the uppermost stratigraphic unit at the site. The sediment consists of rhythmically bedded silt and clay, with minor amounts of sand and ice-rafted till layers.

In the central and northwest part of the site, the stratigraphic sequence is interpreted to be relatively uniform, and correlations of units are straightforward. Preglacial sand and gravelly sand extends over most of the site, with deposits in terrace settings interpreted to be continuous and correlative with sand in both channels of the Boag Valley (Figs. 5, 6, 7). The sand is thickest (>20 m) along the talweg of the South Channel, but thins to less than 1 m along the edge of the escarpment that bounds the northwest corner of the site (Figs. 8, 9).

The stratigraphic sequence is more complex in the eastern and southeastern part of the site where glacially displaced bedrock and associated plastic clay till overlie preglacial sand in both channel and terrace settings (Fig. 10). The sequence of displaced bedrock and plastic clay till is as much as 10 m thick (Fig. 8), but shows great variation in thickness from borehole to borehole, making correlations based solely on stratigraphic position and elevation tenuous. This variability and abruptness in contact made it particularly difficult to establish the lower contact of the thrust bedrock with the underlying units. Of greatest hydrogeologic concern was the geometry of the contact with the sand (Figs. 5, 6, 7). Questions were raised as to whether sand was continuous beneath the thrust bedrock, or if displaced bedrock was thrust into the sand, pinching off the terrace deposits from those in the East Channel.

Preglacial sand was found to be absent only in the southeast corner of the site, where it is believed to have been eroded by over-riding glacial ice. In this area, a slab of displaced bedrock and associated basal plastic clay till was deposited on the bedrock ridge that separates the East and South channels of the Boag Valley (Figs. 8, 9).

It was assumed during the investigation stages of Phases 1 and 2 that, for the purposes of establishing the framework for the hydrogeologic model, terrace sand was correlative with sand in both channels of the Boag Valley. Further, with the exception of the southeast corner, it was believed that terrace and channel sand units were hydraulically connected and could be modelled as a single, discrete aquifer.

Interpretations Following Phase 3 of the Investigation

Uncertainty remained at the end of Phases 1 and 2 of the investigation regarding the interpretation of the thrust bedrock and plastic clay till unit along the east side of the study area. Two concerns remained unresolved. First, the displaced bedrock could not be differentiated from *in situ* bedrock in the southeast part of the site. In this area, the underlying sand and plastic clay till were absent. Second, the contact between the western edge of the thrust bedrock and the underlying sand unit could not be confirmed. This had serious implications concerning the lateral continuities of the terrace sand and channel sand.

In 1990, the third phase of the investigation was completed. Additional dry-auger boreholes were drilled on a 100 m spacing to resolve the stratigraphic complexity associated with ice thrusting, and to verify the hydraulic continuity between terrace sand, and sand in the East Channel.

The results from the 1990 field investigation are depicted in a series of east-west oriented cross-sections shown in Figures 11 through 14. They incorporate the additional geologic information, as well as the hydraulic information collected from piezometers completed in the terrace and channel sand deposits during Phases 1 to 3. The term "measured piezometric level" shown in the figures, refers to the water level measured in standpipe piezometers. The term "observed water level" refers to the level of standing water measured in open boreholes at the completion of drilling.

The additional boreholes established that the western edge of the thrust bedrock has a very abrupt lateral contact with the sand in the northeast part of the site. The geologic information was inconclusive, however, in establishing whether terrace sand was laterally continuous with channel sand, or whether displaced bedrock was thrust into *in situ* bedrock, effectively truncating the sand aquifer. For example, in cross-sections A-A' and B-B' (Figs. 11, 12) the new geologic information shows that the terrace sand and East Channel sand are correlative and laterally continuous. Information from 1990 boreholes located on cross-sections C-C' and D-D' (Figs. 13, 14), on the other hand, supports a much different interpretation. The data indicate that ice-thrust bedrock separates terrace sand from channel sand, and rests directly on *in situ* bedrock along the western margin of the East Channel. Without drilling additional boreholes on a spacing even tighter than 100 m, the configuration of the thrust bedrock-sand contact appeared to be unresolvable.

Hydraulic information proved crucial in resolving the difficulty of correlating the sand units, and in establishing the thrust bedrock-sand contact. Water-table levels from piezometers completed in the basal sand, or in the upper bedrock surface in contact with the sand, indicate that unsaturated conditions exist in both the terrace sand and sand in the South Channel (Figs. 11 to 14). Sand in the East Channel is not only saturated, but the piezometric surface is also significantly higher, with a steep hydraulic gradient separating the terrace and South Channel sand aquifer from the East Channel aquifer (Figs. 11 to 14). The steep gradient coincides with the western edge of the displaced bedrock, which supports the interpretation that glacial thrusting has either severed the connection between the East Channel sand and terrace sand along the thrust margin, or has scoured out most of the sand, thereby decreasing the transmissivity along the channel flank. A planimetric view of the piezometric surface (Fig. 15) shows that terrace sand and sand in the South Channel behave similarly, with both draining to the North Saskatchewan River escarpment to the northwest. The East Channel sand aquifer drains northward and has a steep hydraulic gradient along its western flank. The north-south orientation of this gradient is believed to approximate the subsurface configuration of the contact along which glacially displaced bedrock has been thrust through the aquifer, to rest directly on *in situ* bedrock, or where the sand has been significantly thinned by glacial scour.

Pumping test observations support the interpretation of complete hydraulic discontinuity between the two aquifer systems. Figure 16 shows the extent and effects of drawdown in the sand at the end of a week-long pumping test at test hole PW1, located in the East Channel. The well was pumped at a rate of 2.5 L·s⁻¹ to maximize drawdown as a means of delineating aquifer boundaries. The absence of drawdowns in wells located in both South Channel and terrace settings indicates that the East Channel is laterally confined by the thrust bedrock and that the aquifer is hydraulically disconnected from the terrace and South Channel aquifer. The eastern confining boundary was not established by test drilling, but is an extrapolation of the thrust bedrock slab on the bedrock surface.

CONCLUSIONS

On the basis of the geologic information from the 1990 boreholes, and the additional hydraulic information from both piezometers and pumping test, we conclude that:

1. The basal sand within the Aurum site is not continuous throughout. Sand in the terrace is separated from sand in the East Channel by a slab of ice-thrust bedrock that has severed the unit along the western margin of the channel.

2. A bedrock ridge separates the two channels in the southeast corner, with ice-thrust bedrock resting directly on *in situ* bedrock along the ridge. From geologic and hydraulic evidence, it is inferred that sand in the East Channel is not connected to sand in the South Channel.

3. Although not confirmed by test drilling, the pumping test

also confirms that the eastern boundary of the East Channel is similarly defined by thrust bedrock resting on the bedrock surface.

Based on these conclusions, the 1988 interpretation of the subsurface contact between the ice-thrust bedrock and the underlying sand (Figs. 5, 6) was revised to include the interpretations supported by the hydraulic data. In some areas where the 1990 geologic information continued to support stratigraphic correlation and lateral continuity between terrace and East Channel sand — for example, cross-section B-B′ (Fig. 12) — a somewhat more complex and convoluted geologic contact was required to account for the hydraulic observations. An example of this is illustrated in the final stratigraphic interpretation of that cross-section (Fig. 17) in which an injected tongue of displaced bedrock and plastic clay till (that has truncated the basal sand) has been introduced to the cross-section to account for the saturated and unsaturated conditions, and differential water-table elevations that exist within the aquifer.

The assumption made in 1988, that the basal sand is a single hydrostratigraphic unit resting directly on bedrock over most of the site, was subsequently revised following the conclusions based on the hydraulic information and data from the 1990 drill program. The final interpretation, depicted in Figure 18, shows sand being absent along both flanks of the East Channel, where it is believed that ice-thrust bedrock and high-plastic clay till rest directly on *in situ* bedrock. This information was used as the basis for a numerical contaminant transport model. The model had to be designed to treat the basal sand as two discrete aquifers. The terrace and

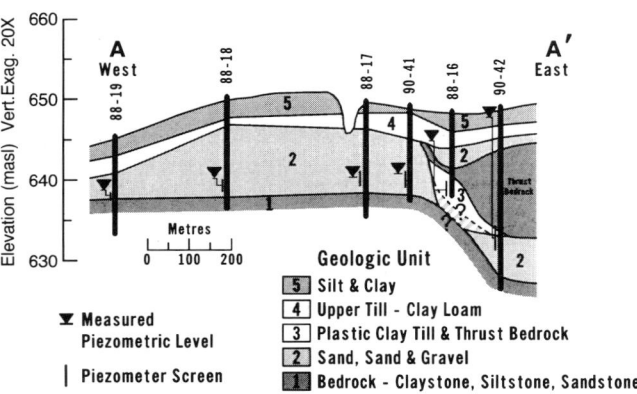

Figure 11 *Revised geologic interpretation along cross-section A-A′, including 1990 borehole information and piezometer data.*

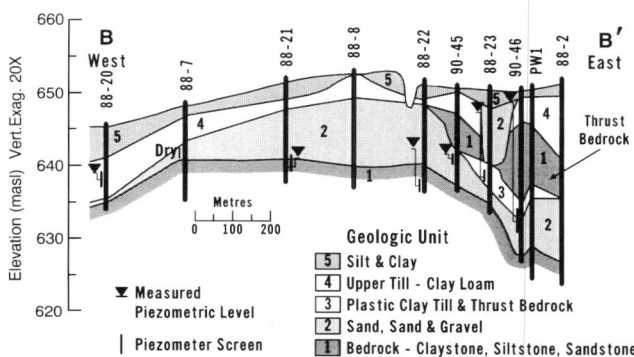

Figure 12 *Revised geologic interpretation along cross-section B-B′, including 1990 borehole information and piezometer data.*

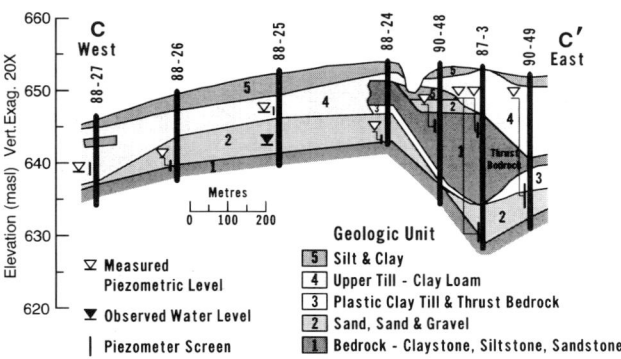

Figure 13 *Revised geologic interpretation along cross-section C-C′, including 1990 borehole information and piezometer data.*

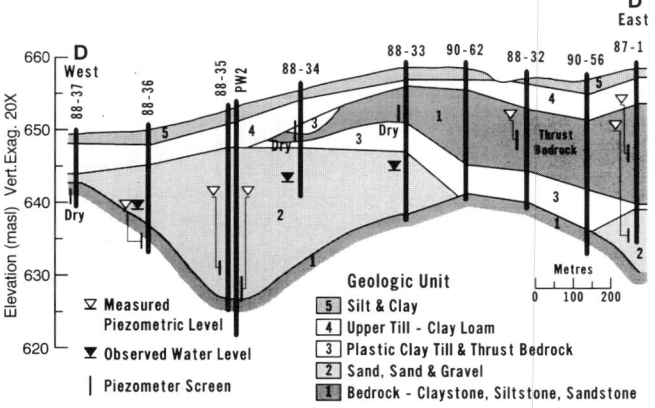

Figure 14 *Revised geologic interpretation along cross-section D-D′, including 1990 borehole information and piezometer data.*

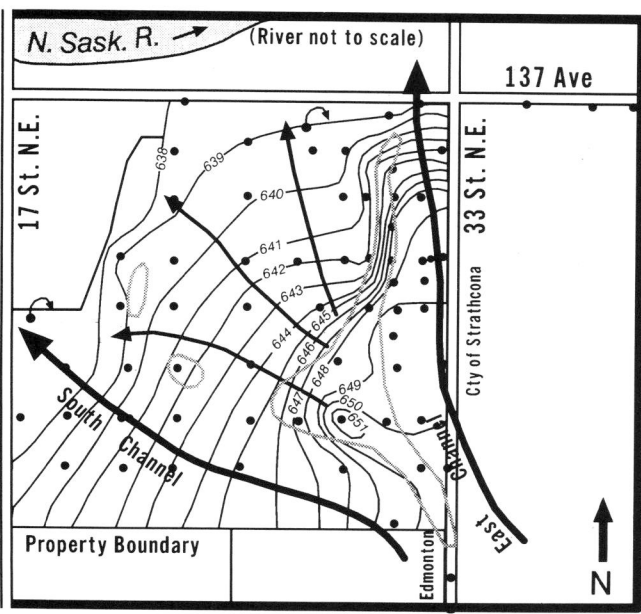

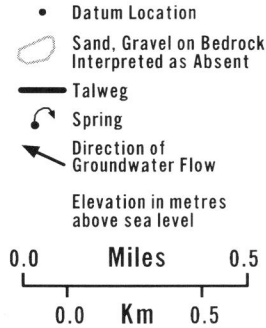

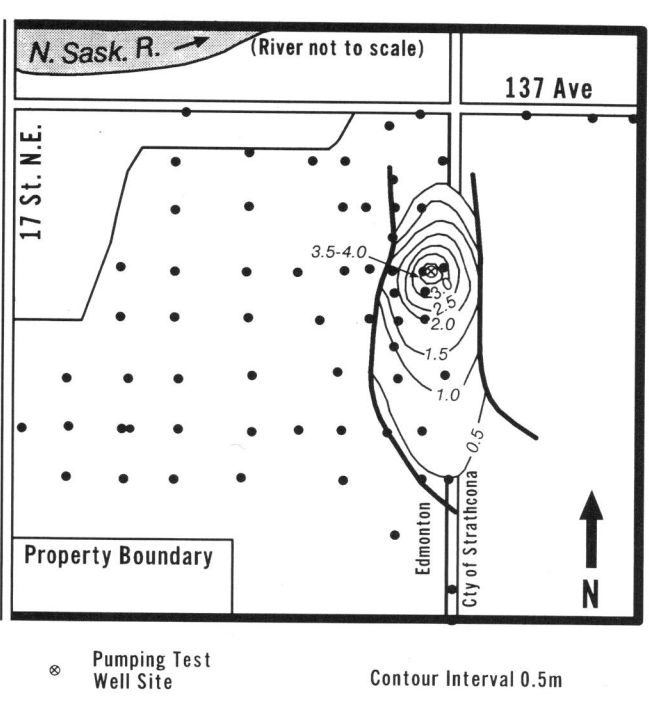

Figure 16 *Configuration of drawdown at pumping well located in East Channel sand.*

Figure 15 *(above) Piezometric surface contours showing inferred direction of groundwater flow. Note steep hydraulic gradient along western edge of the East Channel where basal sand is interpreted to be absent.*

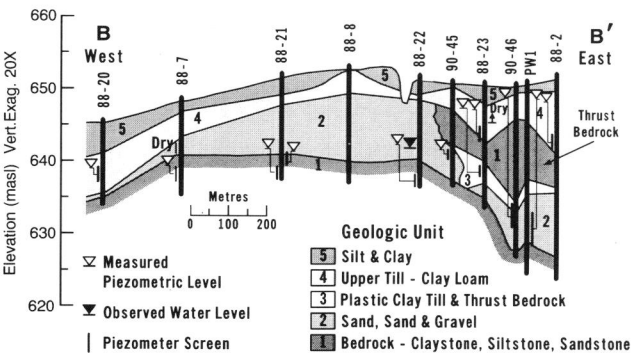

Figure 17 *Final hydrostratigraphic interpretation along cross-section B-B' based on geologic and hydraulic information. An abrupt contact between sand and thrust bedrock accounts for the saturated and unsaturated conditions in boreholes 88-23 and 90-45, respectively.*

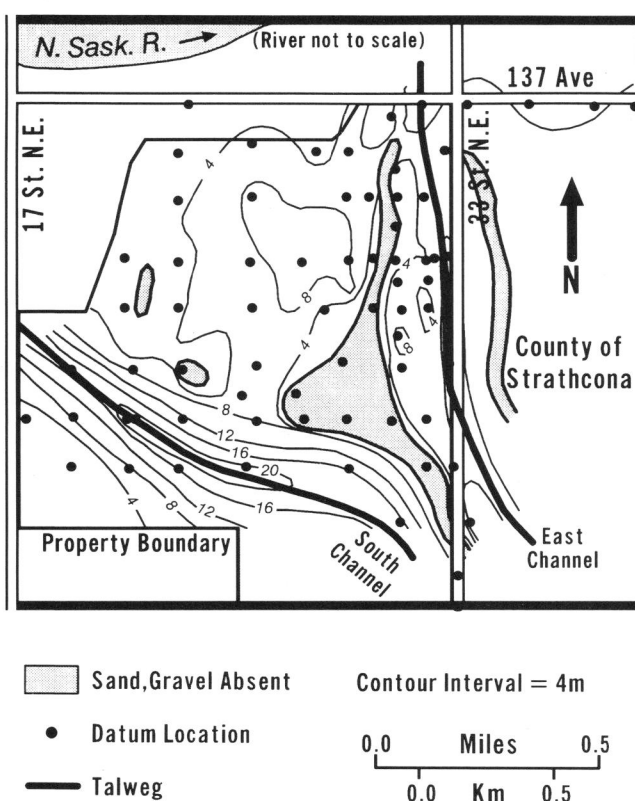

Figure 18 *Revised distribution and thickness of sand and gravel (interpretation based on 1990 borehole data, and hydraulic information).*

South Channel behaved collectively as an unconfined aquifer. The East Channel was modelled as a confined aquifer with elevated head levels.

The complex geological settings described here from Edmonton, Alberta arise principally from glacial thrusting (glaciotectonism) of bedrock and sediment. Such stratigraphic complexity is widespread in the western interior of Canada as a result of the broad extent of relatively soft and easily deformed sandstones and shales (*e.g.*, Christiansen, 1970). Consequently, the hydrogeological conditions found at Edmonton may be representative of large portions of western Canada.

REFERENCES

Andriashek, L.D., 1984, Quaternary stratigraphy of the Edmonton map area, NTS 83H: Alberta Research Council, Open File Report 1988–04, 27 p.

Andriashek, L.D. and Fenton, M.M., 1989, Quaternary stratigraphy and surficial geology of the Sand River area, 73L: Alberta Research Council, Bulletin 57, 154 p.

Bayrock, L.A. and Hughes, G.M., 1962, Surficial geology of the Edmonton district, Alberta: Alberta Research Council, Preliminary Report 62-6, 40 p.

Carlson, V.A., 1967, Bedrock topography and surficial aquifers of the Edmonton district, Alberta: Alberta Research Council, Report 66-3, 21 p.

Christiansen, E.A., ed., 1970, Physical Environment of Sasktoon, Canada: Saskatchewan Research Council and National Research Council of Canada, NRC 11378, 68 p.

Gravenor, C.P. and Bayrock, L.A., 1956, Stream-trench systems in east-central Alberta: Alberta Research Council, Preliminary Report 56-4, 11 p.

Irish, E.J.W., 1970, The Edmonton group of south-central Alberta: Canadian Petroleum Geology, Bulletin, v. 18, p. 125-156.

Kathol, C.P. and MacPherson, R.A., 1975, Urban geology of Edmonton: Alberta Research Council, Bulletin 32, 61 p.

Stalker, A.M., 1968, Identification of Saskatchewan gravels and sands: Canadian Journal of Earth Sciences, v. 5, p. 155-163.

Whitaker, S.H. and Christiansen, E.A., 1972, The Empress group in Saskatchewan: Canadian Journal of Earth Sciences, v. 9, p. 353-360.

23. Engineered Waste Disposal Facilities

R. Kerry Rowe

Geotechnical Research Centre, University of Western Ontario, London, Ontario N6A 5B9

Michael J. Fraser

AGRA Earth & Environmental, 3953 Riberdy Road, Windsor, Ontario N8W 3W5

SUMMARY
The selection and design of an engineered waste disposal facility requires consideration of the potential for protection of ground-water quality, predictability of ground-water movement, and potential for disruption of ground-water users. In the design of a waste disposal facility, engineered systems are often incorporated, and the service life of these systems must be considered when assessing their potential impact. The role of modelling in predicting the potential impacts due to the interaction between the hydrogeology and the proposed engineering is discussed, and the potential impact of different landfill designs on ground-water quality is examined.

INTRODUCTION
The selection of a suitable site and design for a waste disposal facility such as a landfill involves the interaction of many disciplines (*e.g.*, geology, geophysics, geochemistry, hydrogeology, geotechnical engineering, and landfill design) in order to characterize a particular site and then develop an appropriate engineered facility for that site. It is necessary to understand the existing site conditions and how the proposed facility will affect existing conditions both in the short term and in the long term. In this context, the potential short-term impacts may extend for up to several decades (*e.g.*, during landfill construction), while the potential long-term impacts may extend over periods of up to several centuries. This latter period of time, during which a landfill will produce contaminants at levels that could have unacceptable impact if they were discharged into the surrounding environment, is often called the contaminating lifespan of the landfill (see also Chapters 20, 24). There are a number of important factors to be considered in the selection of a suitable site, as discussed below.

Potential for Protection of Ground-water Quality
An assessment of the potential for protecting ground-water quality from degradation due to the migration of contaminated water (leachate) from a landfill may involve consideration of natural geological protection, hydraulic protection, and engineered systems.

Natural geological protection generally refers to the ability of a geological feature such as a clay till aquitard to attenuate contaminants as they migrate from the landfill through the aquitard to some potential receptor aquifer (Yanful *et al.*, 1988a, b). This potential for attenuation (*i.e.*, a reduction in concentration of contaminants) will depend on the effective thickness and bulk hydraulic conductivity of the aquitard between the base of the engineered facility and the aquifer. The effective thickness will depend on the existing thickness of the hydrogeological unit (*i.e.*, the aquitard), but also on engineering and other environmental constraints that influence the depth of excavation. Thus, neither the geology and/or hydrogeology nor the engineered design can be considered in isolation; increasing the depth of excavation may decrease other environmental impacts (*e.g.*, traffic, noise, dust, visual impacts, *etc.*) which affect nearby residents in the short term (which could be decades, as noted above), but this may be traded off against a consequent decrease in natural protection of ground-water quality in the long term. This may then need to be countered by increased engineered protection (discussed later in this paper).

The bulk hydraulic conductivity of an aquitard unit will depend on factors such as density, grain size distribution and mineralogy, and may be controlled by the presence of secondary features such as fractures (*e.g.*, D'Astous *et al.*, 1989, Herzog *et al.*, 1989). Thus, an important part of the evaluation of the potential impact on ground-water quality is an evaluation of a reasonable value (or more typically, a range of values) for the bulk hydraulic conductivity of any natural attenuation layer.

Hydraulic protection involves the use of natural ground-water levels (usually in the potential receptor aquifer) to induce a small flow into the landfill from the aquifer. Clearly, where there is ground-water flow into the landfill, there will not be an outward flow of leachate from the landfill to the aquifer. Also, the inward flow tends to reduce the outward movement of chemicals in the leachate due to the process of molecular diffusion. This concept of hydraulic protection (sometimes called a hydraulic trap) has gained popularity since the approval of the Halton Waste Management Facility (see Rowe *et al.*, 1993); however, as discussed by Rowe *et al.* (1994b) it is far simpler in concept than in implementation. In particular, it is important to consider not only the existing ground-water levels, but also the landfill base elevations, hydraulic conductivity of the aquitard and/or engineered system between the aquifer and the base of the waste, and the transmissivity of the aquifer to assess the effect of landfill construction and operation on water levels in the aquifer and the consequential potential impact on the effectiveness of the hydraulic trap.

In North America, the last decade has seen a major movement from largely uncontrolled disposal of waste in town dumps to the controlled disposal of waste in engineered landfills. The level of engineering can vary substantially, depending on the natural environment, the size of the landfill, the nature of the waste, and local regulations. As a minimum, most modern landfills have some form of engineered final cover over the waste that serves to control the infiltration of water into the waste and the consequent generation of leachate, as well as some form of engineered system for the collection of leachate. Some landfills have an engineered compacted clay liner to control the rate of migration of contaminants, others involve natural hydraulic protection combined with a backup compacted clay liner as an engineered contingency system (Rowe et al., 1993; King et al., 1993). In the United States composite liners are commonly used. These consist of a layer of plastic (typically 1-mm to 2-mm thick high-density polyethylene, HDPE) known as a geomembrane, overlying a compacted clay liner (e.g., see Koerner, 1990; Rowe et al., 1994b). A number of engineered systems will be discussed in the latter section of this paper.

Predictability of Ground-water Movement
It is important that the hydrogeology of a proposed landfill site be sufficiently well understood that it will allow reliable monitoring of the site. In addition, there needs to be some viable contingency measure that can be implemented in the event that some unexpected contamination of ground water does occur. This requirement for reasonable predictability is more restrictive with respect to what constitutes a suitable natural system than the requirement for protection of ground-water quality, since natural protection can be readily supplemented by additional engineering if needed. However, it is generally much more difficult to improve the predictability of a site using engineering methods.

Potential for Disruption of Ground-water Users
In this context, disruption of ground-water users includes both existing and potential users, particularly when there is no viable alternative water source. It may also involve potential disruption of stream baseflow due to ground-water drawdown.

The potential for disruption may be short term, either due to conventional construction drawdown or to the depressurization of an aquifer that can occur during construction of a hydraulic trap landfill. However, the potential disruption may also be long term, due to the cutoff of ground-water recharge over the area of the landfill, causing a drop in water levels, and/or due to a drop in water levels due to the operation of a hydraulic trap as discussed earlier.

A less obvious potential for disruption to ground-water users is a degradation in ground-water quality resulting from a change in water levels that induces mixing of unpotable water (e.g., saline water in fractured bedrock) with what was originally overlying fresh water. This situation creates two potential problems. First, degradation of water quality is undesirable, irrespective of whether it results directly from leachate escaping from the landfill, or indirectly due to mixing of saline or brackish ground water with overlying fresh water. Second,

this would complicate monitoring since chloride is one of the most common critical contaminants used to identify whether there has been an escape of leachate from a landfill. In this situation, it would be more difficult to identify whether an increase in chloride concentration was due to upwelling of underlying ground water or due to leachate escaping from the landfill. This problem is discussed further in Chapter 24.

Modelling
Observational techniques are used to establish existing site conditions. However, prediction of potential impacts often involves modelling which considers the interaction between the hydrogeology and the proposed engineering. The landfill design usually involves an interactive process wherein an initial design proposal is evaluated for its potential impact then revised, as necessary, to mitigate predicted impacts. For example, in the design of a hydraulic-trap landfill the engineer can control the base elevations of the landfill, and the deeper these are below water levels in the underlying receptor aquifer, the greater will be the flow into the landfill (all other things being equal), and hence the better the hydraulic trap. However, there is a tradeoff between the benefits gained due to an increased ground-water gradient into the landfill and the disadvantages of decreasing the thickness of the attenuation layer between the landfill base and the receptor aquifer. Furthermore, there is an increased potential for disruption to ground-water users due to the volume of ground water collected, with the consequent changes in local ground-water levels. Alternatively, the engineer may examine different levels of engineering (e.g., compacted clay versus composite liners, single versus double liners, etc.) when seeking to mitigate potential impacts.

Modelling will usually take the form of flow modelling and/or contaminant transport modelling. A detailed discussion on its application to engineered landfills is given by Rowe et al. (1994b). Flow modelling may range from hand calculations and simple analytical solutions (e.g., Rowe and Nadarajah, 1994) to two-dimensional cross-sectioned models (e.g., Frind and Matanga, 1985) and two-dimensional area models (e.g., Franz and Guiger, 1989). Three-dimensional modelling (e.g., Huyakorn et al., 1986) is rarely used since the data base is often not sufficiently detailed to justify the high cost of performing three-dimensional analysis relative to the improvement in understanding that can be obtained. However, there are exceptions to this observation, and in some cases three-dimensional modelling can give valuable insights (e.g., Molson and Frind, 1991, 1993; Chapter 11).

Contaminant transport models vary substantially in sophistication and ease of use. A review of a number of commonly used models is given by Pandit et al. (1993) and Franz (1993), while Panigrahi et al. (1993) described the input requirements for many of these models. Franz and Rowe (1993) discuss the application of several models for a particular landfill design situation.

The following sections illustrate how simple models can be used to quickly evaluate the potential impact of different landfill designs on ground-water quality for a hypothetical case. The migration of contaminants from the landfill into the aquifer was modelled using a finite-layer analysis model (Rowe and Booker, 1985, 1991, 1994), as implemented in the computer program POLLUTEv6 (Rowe et al., 1994a).

Since this impact is a consequence of the interaction between a particular hydrogeology and landfill design, the numerical results presented in this analysis should not be generalized beyond the level discussed in this paper.

EXAMPLE PROBLEM

In this analysis, the local hydrogeology is assumed to consist of a silty clay till overlying a gravel and sand aquifer (Fig. 1). The till is assumed to have a hydraulic conductivity of 1×10^{-8} metres per second (m·s^{-1}), a porosity of 0.4, and a diffusion coefficient of 0.02 metres per year (m^2·a^{-1}). Beneath the till is a confined aquifer consisting of gravel and sand. This aquifer is assumed to be 3 m thick, and have a porosity of 0.35. At the up-gradient edge of the landfill, the horizontal flow in the aquifer per unit width is assumed to be 30 (m^3·a^{-1})·m^{-1} (*i.e.*, a Darcy velocity of 10 m·a^{-1}). This flow will be increased at the down-gradient edge of the landfill by the downward Darcy flux originating from the landfill. The potentiometric head in the landfill is assumed to be 4 m above the top of the aquifer. The infiltration through the silty landfill cover is assumed to be 0.15 m·a^{-1}.

To quantify the impact associated with the interaction between the hydrogeology and the landfill design, the migration of chloride (a common component in municipal solid waste) was considered. The initial concentration after closure of the landfill was assumed to be 1500 mg·L^{-1}, and the mass of the chloride was assumed to represent 0.2% of the waste. In this analysis, the waste was assumed to have an average thickness of 20 m and an apparent waste density of 600 kg·m^{-3}. The landfill was assumed to be 1000 m long in the direction of ground-water flow. In assessing the impact of the landfill, the mass of contaminant was modelled as described by Rowe (1991a).

Some Landfill Design Considerations

The initial landfill design consists a 0.3 m-thick granular leachate collection system placed directly on top of the till (Fig. 1). In this and subsequent landfill designs, it is assumed that the till is excavated such that the base of the leachate collection system is 6 m above the top of the aquifer. This excavation allows for the placement of a 20-m thick waste pile.

Leachate is formed when rain water and runoff percolate through solid waste, leaching out soluble salts and bio-degraded organic products. A leachate collection system is typically a granular layer with embedded pipes, used to collect and remove the leachate at the bottom of a landfill. The primary functions of a leachate collection system are to reduce the volume of leachate in the landfill and, in particular, the pressure exerted by leachate at the base of the landfill. Removal of leachate also reduces the amount of contaminant available for transport into the hydrogeological system. By reducing the volume of leachate at the base of a landfill, the height of the leachate mound will be reduced, resulting in a lower hydraulic gradient beneath the landfill and, conse-quently, a lower Darcy velocity out of the landfill into the substrate. In this design, it is assumed that the leachate collection system is able to maintain the leachate mound at an average height of 0.3 m above the base of the landfill. The Darcy velocity beneath the landfill would then be 0.12 m·a^{-1}, which would leave 0.03 m·a^{-1} (*i.e.*, about 25%) to be collected and removed by the leachate collection system.

Due to the downward Darcy velocity and diffusion, con-taminants will migrate from the landfill through the till to the aquifer. As time passes, more and more contaminants will migrate to the aquifer at higher and higher concentrations. In this manner the mass of contaminants in the landfill is continu-ously depleted as contaminants are either removed by the leachate collection system or migrate downward. Because the mass of the contaminants is finite, the mass of contami-nants transported into the aquifer will decline. Thus, there will be an initial increase in concentration in the aquifer, followed by a decline in concentration with time, creating a peak concentration in the aquifer (Chapter 24).

Figure 2 shows the concentration of chloride in the aquifer that results from this landfill design and hydrogeology. The concentration in the aquifer reaches a peak value of about 1000 mg·L^{-1} at 45 years, and then declines. In Ontario, the Ministry of Environment and Energy Reasonable Use Policy (1994) limits the increase in the concentration of chloride in an aquifer to a maximum of 125 mg·L^{-1}, assuming that there is negligible background concentration. According to this policy the landfill design would not be acceptable.

Add a Clay Liner?

An alternative landfill design, that may result in a lower peak chloride concentration in the aquifer, would include a com-pacted clay liner beneath the primary leachate collection system (Fig. 3). This compacted clay liner would have a much lower hydraulic conductivity than the till, thus reducing the Darcy velocity beneath the landfill. In this analysis, the com-pacted clay liner is assumed to be 1 m thick, have a hydraulic conductivity of 2×10^{-10} m·s^{-1}, a porosity of 0.35, and a diffusion coefficient of 0.02 m^2·a^{-1} (Fig. 3). The resulting Darcy velocity beneath the landfill is now 0.013 m·a^{-1}, instead of the previous 0.12 m·a^{-1}. This lower Darcy velocity allows for the leachate collection system to function much more effi-ciently and collect about 91% of the leachate generated.

The concentration of chloride in the aquifer that would result from this landfill design is shown in Figure 4. This concentration reaches a maximum value of 133 mg·L^{-1} at 200 years, which is still above the maximum 125 mg·L^{-1} allowed by the Ontario Ministry of the Environment and Energy (OMOEE).

Add a Tight Cover?

A possible design change that might be considered is to add a low-permeability (tight) cover over the landfill (Fig. 5). By adding a tight cover, the amount of percolation through the waste is limited, resulting in less leachate being produced each year. This tight cover is assumed to limit the infiltration into the landfill to 0.008 m·a^{-1}, which will control the maximum Darcy velocity that can occur beneath the landfill. Thus the Darcy velocity beneath the landfill is 0.008 m·a^{-1}, and the amount of leachate that is collected by the leachate collection system is negligible.

Figure 6 shows the concentration of chloride in the aquifer for a landfill design that incorporates a tight cover. Notice that a significant amount of contaminant reaches the aquifer. These contaminants are primarily transported by the process of molecular diffusion, since the Darcy velocity is low due to the tight cover. In this design, the maximum chloride con-centration was 190 mg·L^{-1} at 500 years, which is even higher than that for the design with a permeable cover. By adding a tight cover, the peak concentration in the aquifer was delayed by 300 years, since diffusion tends to be a slower process than advection. The magnitude of the peak concentration increased since very little contaminant was removed by the leachate collection system.

Add a Geomembrane Liner?

Another possible design alternative that may reduce the amount of contaminants reaching the aquifer is to add a geomembrane on top of the compacted clay liner. This type of barrier is called a composite liner (Fig 7). The geomembrane

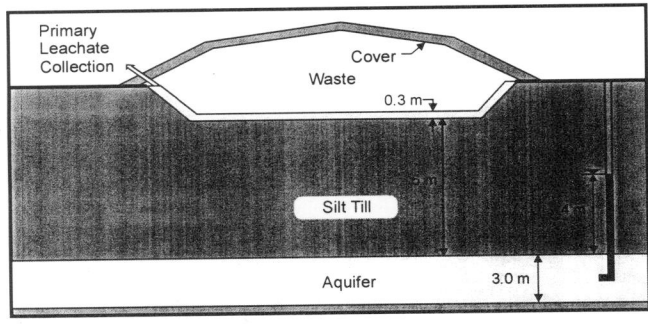

Figure 1   *Landfill design with leachate collection system only.*

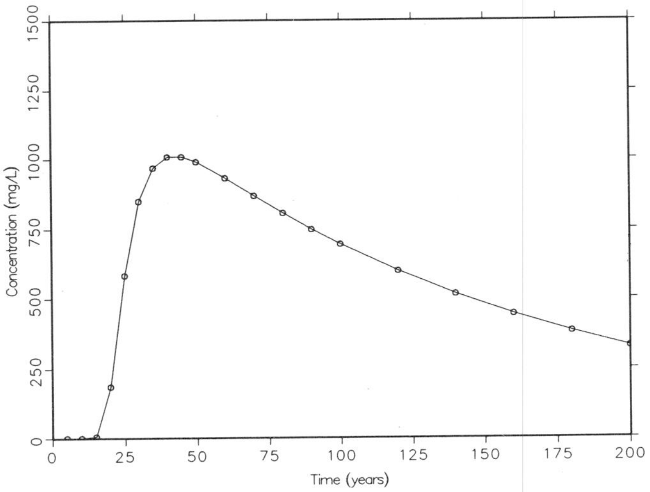

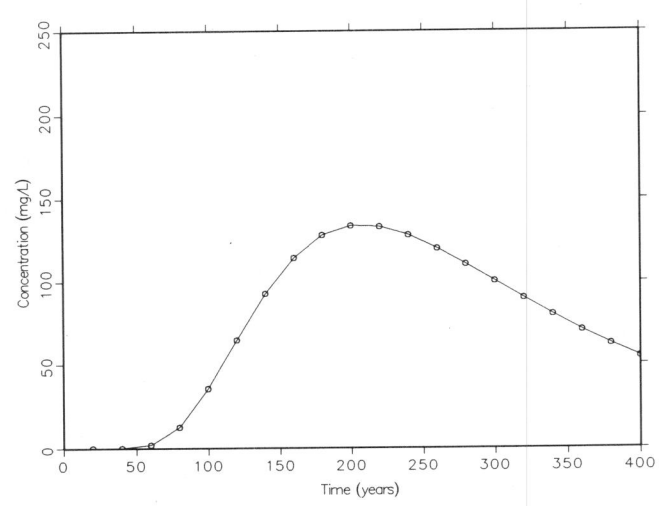

Figure 2   *Chloride concentration in aquifer for landfill design shown in Figure 1.*

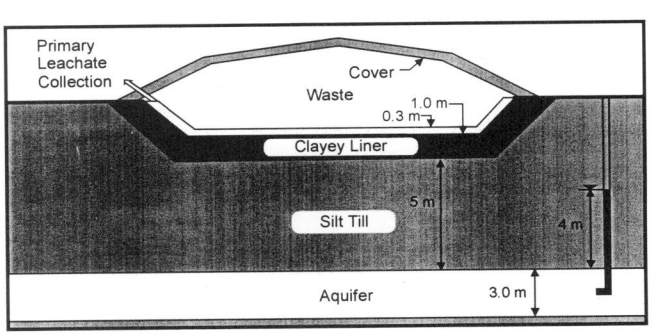

Figure 3   *Landfill with compacted clay liner.*

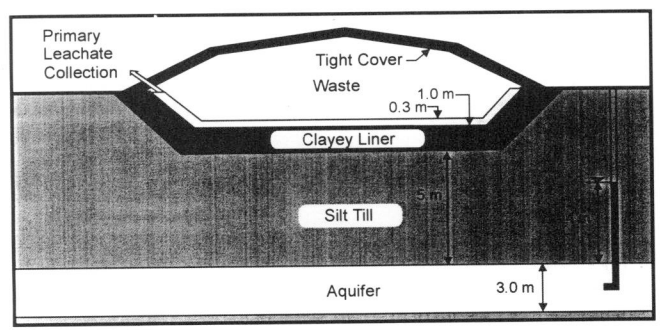

Figure 5   *Landfill with tight cover.*

Figure 4   *Chloride concentration in aquifer for clay liner.*

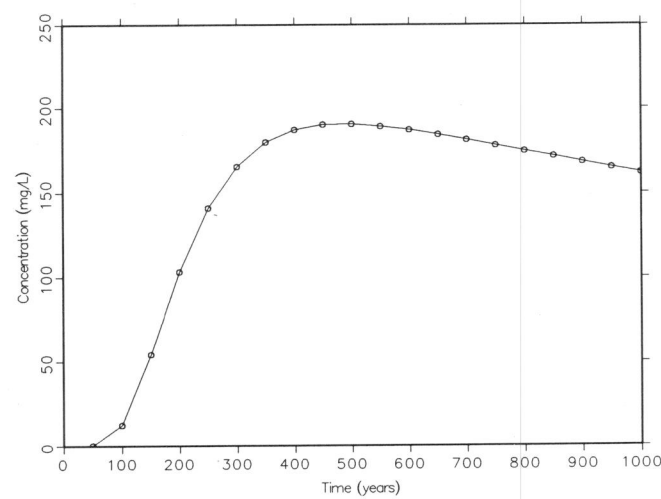

Figure 6   *Chloride concentration in aquifer for tight cover.*

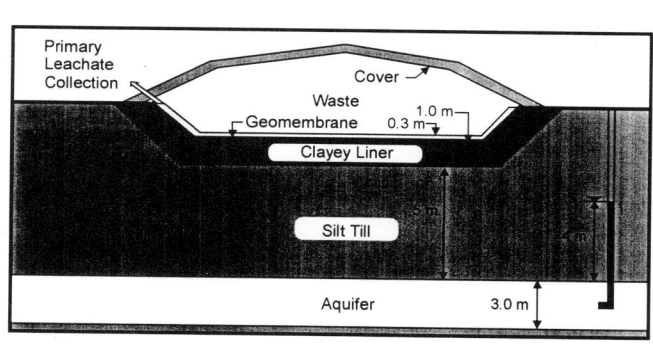

Figure 7 *Landfill design with composite liner.*

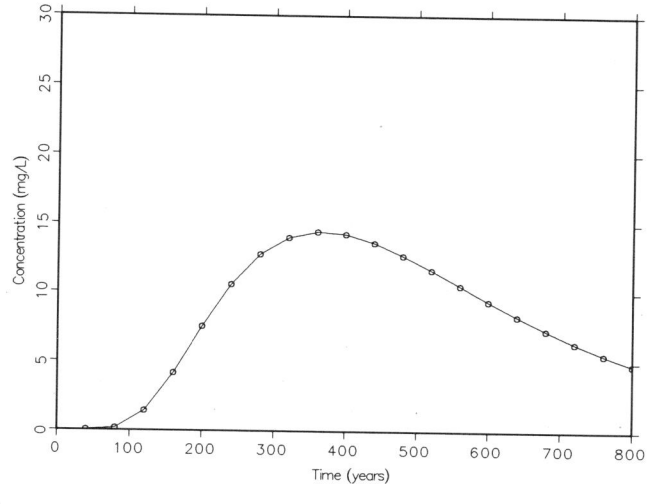

Figure 8 *Chloride concentration in aquifer for composite liner.*

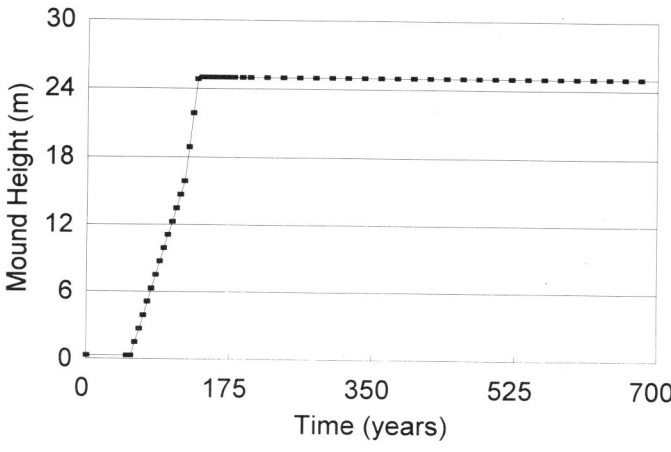

Figure 9 *Leachate mound when leachate collection system fails.*

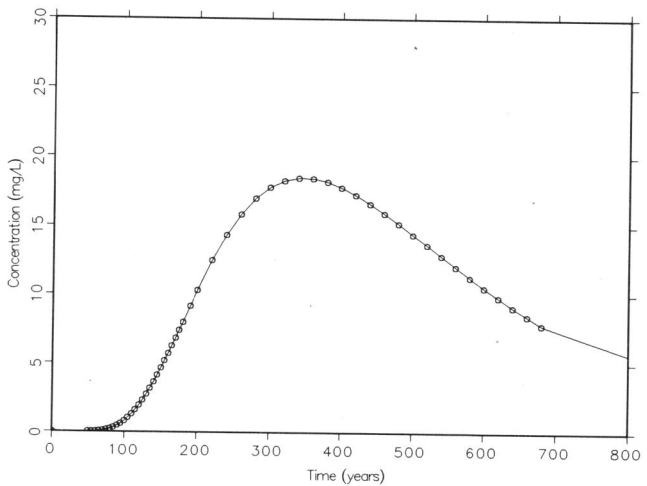

Figure 10 *Chloride concentration in aquifer for failed leachate collection system.*

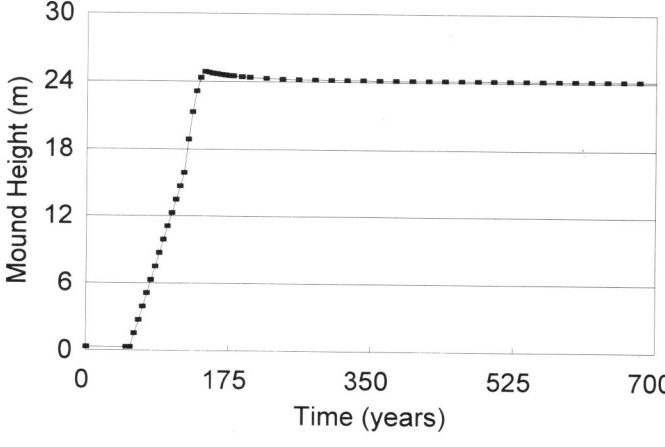

Figure 11 *Leachate mound when both leachate collection system and geomembrane fail.*

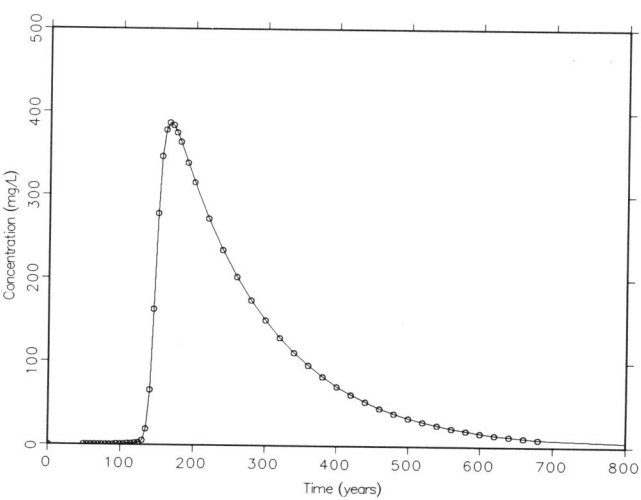

Figure 12 *Chloride concentration in aquifer for failed leachate collection system and geomembrane.*

in this design is assumed to be 1.5-mm thick high-density polyethylene (HDPE), and have a diffusion coefficient of 3×10^{-5} $m^2 \cdot a^{-1}$. During the manufacture and installation of a geomembrane, small holes or defects may be introduced into the geomembrane.

The geomembrane is assumed to have small defects of 0.1 cm^2 area with a frequency of one every acre (2.5 per ha). The effective hydraulic conductivity of the geomembrane is then 1.1×10^{-15} $m \cdot s^{-1}$, which is based upon the likely leakage through a well-constructed composite liner using information from Giroud et al. (1992). The Darcy velocity through the composite liner and silt till would be 5.3×10^{-5} $m \cdot a^{-1}$. The volume of leachate that would be collected by the leachate collection system, assuming a permeable cover, would be 0.1499 $m \cdot a^{-1}$ (i.e., essentially 100%).

The concentration of chloride in the aquifer that would result from this design, incorporating a composite liner, is shown in Figure 8. A maximum chloride concentration of 14 $mg \cdot L^{-1}$ occurs at 360 years. This maximum is well below the maximum 125 $mg \cdot L^{-1}$ specified by OMOEE.

What if the Collection System Clogs?

The time period during which an engineered leachate collection system is fully functional is defined as its service life. The service life is highly dependent on the design of the system (Rowe, 1991a, b). For example, leachate collection systems may eventually clog due to chemical and biological activity.

While the leachate collection system is functioning, the leachate mound at the base of the landfill is likely to be relatively small, in this design it is assumed to be an average of 0.3 m. If the leachate collection system fails and becomes clogged, the leachate mound will increase in height at a rate controlled, inter alia, by the infiltration through the cover and the downward Darcy velocity through the liner. The maximum height of the leachate mound is also controlled by the thickness of the waste, in this analysis assumed to average 25 m. If the leachate mound reaches this maximum height, any excess leachate generated will escape from the landfill via toe drains and seeps through the cover.

The service life of the leachate collection system in this analysis is assumed to be 50 years after closure. After this time, the leachate collection system begins to experience significant decreases in performance due to clogging, until it is no longer controlling the height of the leachate mound in the landfill (Fig. 9).

In Figure 10, the resulting chloride concentration in the aquifer is shown, assuming the leachate collection system fails. The maximum chloride concentration in the aquifer is 18 $mg \cdot L^{-1}$ at 340 years, which is only slightly more than it was when the leachate collection system did not fail. Thus, it would appear that the failure of the leachate collection system is not a major concern, assuming that the geomembrane has an infinite service life.

What if Geomembrane Degrades?

Geomembranes have a limited service life, due to degradation caused by chemical attack and other processes. This results in an increase in the effective hydraulic conductivity of the geomembrane. In this analysis, the geomembrane is assumed to have a service life of 125 years, after which it will begin to significantly degrade, until, at 150 years, it is no longer having an impact upon the Darcy velocity beneath the landfill. As the geomembrane degrades, the Darcy velocity will start to increase until it reaches a maximum value which is assumed to be that of the compacted clay beneath the geomembrane. These changes in Darcy velocity will have an effect upon the height of the leachate mound. Initially, the height of the leachate mound will be 0.3 m while both the leachate collection system and geomembrane are functioning. After the leachate collection system fails, this mound will increase to its maximum height, where it will stabilize until the geomembrane fails. When the geomembrane fails, the leachate mound may decrease in height due to the increased Darcy velocity through the landfill liner. Eventually, the leachate mound will stabilize at a new height that is controlled by the Darcy velocity through the liner and the infiltration through the cover (Fig. 11).

The calculated chloride concentration in the aquifer is shown in Figure 12 for this design, assuming finite service lives of the leachate collection system and geomembrane. Based upon these assumptions, the maximum chloride concentration in the aquifer is 387 $mg \cdot L^{-1}$ at 165 years, which would not be acceptable according to the OMOEE. At this stage, it would be necessary to further refine the design of the landfill to achieve a contaminant impact that is acceptable according to the OMOEE policy. These refinements may include (1) addition of a secondary leachate collection system and liner, (2) use of a lower permeability compacted clay liner, (3) changes in the base elevation of the landfill, and/or (4) control of the leachate mound after failure of the leachate collection system.

DISCUSSION

Irrespective of how much engineering is proposed, it is important to have an adequate understanding of site geology and hydrogeology to allow confident monitoring of the site and the development of contingency measures that could be used to mitigate any unexpected escape of leachate from the facility. The engineered design does not reduce the need for an adequate hydrogeological investigation. Most modern landfills will require some form of engineering, and the interaction between this engineering and the natural system also needs to be considered by means of flow and/or contaminant transport modelling. Consideration should also be given to the service life of the components of this system and the implications that this may have on potential impacts on ground-water quality.

REFERENCES

D'Astous, A.Y., Ruland, W.W., Bruce, J.R.G., et al., 1989, Fracture effects in the shallow groundwater zone in weathered Sarnia-area clay: Canadian Geotechnical Journal, v. 26, p. 43-56.

Franz, T., 1993, Hydrogeological computer models used in evaluating contaminant transport: Geotechnical News, v. 11, p. 50-52.

Franz, T. and Guiguer, N., 1989, FLOWPATH: Steady-state two-dimensional aquifer simulation model: Waterloo Hydrogeologic Software, Waterloo, ON, Canada.

Franz, T. and Rowe, R.K., 1993, Simulation of groundwater flow and contaminant transport at a landfill site using models: International Journal for Numerical and Analytical Methods in Geomechanics, v. 17, p. 435-455.

Frind, E.O., 1987, Modelling of contaminant transport in groundwater. An overview, in Canadian Society for Civil Engineering, Centennial Symposium on Management of Waste Contamination of Groundwater, Montreal, PQ, May, 35 p.

Frind, E.O. and Matanga, G.B., 1985, The dual formulation of flow for contaminant transport modelling 1. Review of theory and accuracy aspects: Water Resources Research, v. 21, p. 159-169.

Giroud, J.P., Badu-Tweneboah, K. and Bonaparte, R., 1992, Rate of leakage through a composite liner due to geomembrane defects: Geotextiles and Geomembranes, v. 11, p. 1-19.

Herzog, B.L., Griffin, R.A., Stohr, C.J., *et al.*, 1989, Investigation of failure mechanisms and migration of organic contaminant at Wilsonville, Illinois: Groundwater Monitoring Review, Spring 1989, p. 82-89.

Huyakorn, P.S., Springer, E.P., Guvanasen, V. and Wadsworth, T.D., 1986, A three-dimensional finite element model for simulating water flow in variably saturated porous media: Water Resources Research, v. 22, p. 1790-1808.

King, K.S., Quigley, E.M., Fernandez, F., *et al.*, 1993, Hydraulic conductivity and diffusion monitoring of the Keele Valley landfill liner, Maple, Ontario: Canadian Geotechnical Journal, v. 30, p. 124-134.

Koerner, R.M., 1990, Preservation of the environment via geosynthetic containment systems, *in* 4th International Conference on Geotextiles and Geomembranes, Proceedings, v. III, The Hague, p. 975-988.

Molson, J.W. and Frind, E.O., 1991, Regional Municipality of Durham Contingency landfill site assessment: Addendum II, 3D groundwater flow model: Institute for Groundwater Research, University of Waterloo, Waterloo, ON, Interim Waste Authority, Technical Appendix, 71 p.

Molson, J.W. and Frind, E.O., 1993, Numerical simulation of groundwater flow and contaminant transport at the Keele Valley landfill: Waterloo Centre for Groundwater Research, Waterloo, ON, 51 p.

Ontario Ministry of Energy and Environment, 1994, Incorporation of the reasonable use concept into OMOEE groundwater management activities: Policy 15-08, Ministry of the Environment and Energy, March, 1993, 16 p.

Pandit, A., Panigrahi, B.K., Peyton, L. and Sayed, S.M., 1993, Strengths and limitations of ten widely used groundwater transport models: Canadian Society of Civil Engineers – American Society of Civil Engineers, Joint National Conference on Environmental Engineering, Montreal, PQ, July, Proceedings, p. 1249-1256.

Panigrahi, B.K., Pandit, A., Hebson, C.S. and Rowe, R.K., 1993, Input parameter requirements for groundwater models: Canadian Society of Civil Engineers – American Society of Civil Engineers, Joint National Conference on Environmental Engineering, Montreal, PQ, July, v. 2, p. 1265-1272.

Rowe, R.K., 1991a, Contaminant impact assessment and the contaminating lifespan of landfills: Canadian Journal of Civil Engineering, v. 18, p. 244-253.

Rowe, R.K., 1991b, Some considerations in the design of barrier systems: Canadian Geotechnical Society, First Canadian Conference on Environmental Geotechnics, Montreal, PQ, Proceedings, p. 157-164.

Rowe, R.K. and Booker, J.R., 1985, 1-D pollutant migration in soils of finite depth: American Society of Civil Engineers, Journal of Geotechnical Engineering, v. 111, p. 479-499.

Rowe, R.K. and Booker, J.R., 1991, Pollutant migration through a liner underlain by a fractured soil: American Society of Civil Engineers, Journal of Geotechnical Engineering, v. 117, p. 1902-1919.

Rowe, R.K. and Booker, J.R., 1994, A finite layer technique for modelling complex landfill history: Geotechnical Research Centre, Research Paper, GEOT-7-94, University of Western Ontario, London, ON, 50 p.

Rowe, R.K., Booker, J.R. and Fraser, M.J., 1994a, POLLUTEv6 and POLLUTE-GUI: User's Guide: GAEA Environmental Engineering Ltd., London, ON, 304 p.

Rowe, R.K., Caers, C.J. and Chan, C., 1993, Evaluation of a compacted till liner test pad constructed over a granular subliner contingency layer: Canadian Geotechnical Journal, v. 30, p. 667-689.

Rowe, R.K. and Nadarajah, P., 1994, An analytical method for predicting the velocity field beneath landfills: Geotechnical Research Centre Report, GEOT-14-94, University of Western Ontario, London, ON, 58 p.

Rowe, R.K., Quigley, R.M. and Booker, J.R., 1994b, Clayey Barrier Systems for Waste Disposal Facilities: E & F N Spon (Chapman Hill), London, UK, 390 p.

Yanful, E.K., Nesbitt, H.W. and Quigley, R.M., 1988a, Heavy metal migration at a landfill site, Sarnia, Ontario, Canada – I: Thermodynamic assessment and chemical interpretations: Applied Geochemistry, v. 3, p. 523-533.

Yanful, E.K., Quigley, R.M. and Nesbitt, H.W., 1988b, Heavy metal migration at a landfill site, Sarnia, Ontario, Canada – II: Metal partitioning and geotechnical implications: Applied Geochemistry, v. 3, p. 623-629.

24. Leachate From Landfills Along the Niagara Escarpment

Jean Birks

Department of Geology, Queen's University, Kingston, Ontario K7L 3N7

Carolyn H. Eyles

Department of Geography, McMaster University, Hamilton, Ontario L8S 4K1

SUMMARY

This chapter describes the impacts of landfilling on ground-water quality in the vicinity of closed and open landfills along the Niagara Escarpment, near the city of Hamilton, Ontario. It discusses the relationships between leachate quality, the types and age of landfill wastes and landfill design. The chapter stresses the difficulties of characterizing diagnostic plume hydrochemistry and monitoring leachate plume behaviour in areas where ground waters are highly mineralized.

INTRODUCTION

There is considerable public concern regarding landfill-related contamination of ground waters in urban areas. Most closed landfill sites are unlined and rely on natural attenuation of leachate within underlying geologic materials (Chapters 2, 20). This approach to waste management is now widely recognized as being inappropriate; modern landfills have an engineered liner system and engineered leachate collection systems (Chapter 23).

Ground waters impacted by leachate are normally characterized by highly elevated concentrations of total dissolved solids, most notably chlorides, sulphates, total Kjeldahl nitrogen, nitrite and phenol (Table 1). However, source recognition and contaminant plume monitoring can be severely complicated in areas where the salinity of the ground water is already elevated, due to naturally saline formation waters or other sources of contamination. In such cases, reliable evaluation of landfill impact requires a detailed understanding of leachate chemistry and the processes by which it is controlled. The problem is especially evident in the Hamilton region of Ontario where numerous sources of ground-water salinity are present.

STUDY AREA

The city of Hamilton (population 300,000) lies at the western end of Lake Ontario, and has a lengthy industrial heritage, mostly as a steel-producing centre. Municipal and industrial wastes have been disposed of in landfills within the urban area. Forty-one closed landfill sites are known to exist, ten of which are monitored (Fig. 1); at least one of the monitored sites (Upper Ottawa Street) received hazardous and liquid industrial waste. All sites are unlined, and many lie within former quarries in highly fractured and permeable dolostones of the Niagara Escarpment (Fig. 1). A recent proposal to develop a landfill in a bedrock quarry near Hamilton was rejected by an environmental assessment board (Chapters 2, 39), in part because of the unacceptable risk of ground-water contamination by leachate (Environmental Assessment Board and The Joint Board, 1995).

Hamilton Harbour is severely polluted (Chapters 15, 16), and has been identified as an Area of Concern (AOC) in the Great Lakes basin by the United States-Canada International Joint Commission. The quality of ground water flowing into the harbour is of particular interest given the high estimated velocities of ground-water flow in the area, heavily industrialized catchments and the presence of numerous closed landfills. A contaminant plume emanating from the Upper Ottawa Street landfill (Fig. 1) is thought to have migrated at least as far as Lake Ontario, a distance of over 8 km (Ontario Ministry of Health, 1986).

This chapter examines three landfill sites in the Hamilton region; two closed landfill sites, the Beverly and Upper Ottawa landfill sites, and one site currently receiving municipal waste, the Glanbrook site (Fig. 1). These three sites were chosen because of the detailed hydrochemical analyses and lengthy ground-water monitoring records available for each.

Beverly Landfill

The Beverly landfill (Fig. 2) lies in a former gravel pit, and received predominantly domestic waste during its fifteen year lifespan (Table 2; Gartner Lee Associates Ltd., 1981a, b). Waste is in contact with fractured dolostone, and piezometric head measurements show a vertical hydraulic gradient consistent with the downward flow of water from the waste pile into bedrock (Gartner Lee Associates Ltd., 1981a); the velocity of ground-water flow in bedrock is about 50 m·a^{-1}.

Table 1   Minimum, maximum and average values for components of leachate and background ground water at the Beverly, Upper Ottawa Street and Glanbrook landfill sites. Many parameters show large concentration ranges in both leachate and ground water. Data from the Regional Municipality of Hamilton-Wentworth monitoring program 1980 to 1994 (Beverly and Glanbrook) and 1980 to 1986 (Upper Ottawa Street). Background water quality was obtained from wells identified as up-gradient of the landfill site (Henry well at Beverly, BH12 at Upper Ottawa Street, and BH12-II at Glanbrook). For well locations, see Figures 2, 3 and 4.

	BEVERLY LEACHATE			BEVERLY BACKGROUND			UPPER OTTAWA ST LEACHATE			UPPER OTTAWA ST BACKGROUND			GLANBROOK LEACHATE			GLANBROOK BACKGROUND		
	min	max	avg	min	max	avg	min	max	avg	min	max	avg	min	max	avg	min	max	avg
Conductivity †	190	3499	1401	265	917	479	7500	15000	11233	700	11000	2215	500	14934	6008	690	2900	1554
pH							7.6	8.3	7.9	6.8	8.2	7.3	6.1	8.2	7.1	6.7	7.8	7.3
Hardness	100	1690	762	276	444	346	1080	1300	1235	545	6900	1347	3010	11000	4885	460	2168	998
TDS	40.0	1764.0	933.0	140.0	560.0	344.7	6140.0	10108.0	8181.2									
Chloride	0	750	150	8	62	23	29	3150	1344	60	770	331	0	2175	560	1	231	50
Sulphate	0.8	147.0	75.4	1.5	34.0	18.5	100.0	5200.0	1097.5	1.02	0.04	0.03	10.0	3600.0	1023.2	36.0	1560.0	636.3
Nitrate	0.0	7.5	0.9	0.0	1.0	0.4	1.6	7.3	4.5	0.0	2.0	0.4	0.0	8.0	2.8	0.0	0.3	0.1
Alkalinity ††	98	2140	839	252	380	300				33	370	253	846	5580	3220	190	320	253
Calcium										144	620	251	116	1485	431	68	669	227
Magnesium													436	980	659	52	278	124
Potassium													30	404	140	2	18	7
Sodium										63	440	217	157	880	453	19	89	51
total Kjeldahl	3.4	129	39	0	3	1	6	1200	375	0	30	4	9	800	193	0	5	1
Free Ammonia	0.0	129.0	28.2	0.0	1.6	0.3	0.0	1200.0	445.9	0.0	1.2	0.3	12.9	309.0	124.3	0.0	1.8	0.7
Nitrite †††	0.0	90.0	6.8	0.0	10.0	0.9	1.0	2.9	2.0	0.0	6.0	1.6						
Phenol †††	0	10.2	1.0	0	20.4	2.1	1.38	331	169.9	0.0	60.1	7.1	2.4	610	140.6	0	100	9.5
BOD							119	696	402		100	30	0	2180	325	0	15	4
COD	0	283	97	0	121	24	96	4706	1772	153	2500	795	0	24000	2967	0	31	20
DOC													15	8200	1091.8	1.4	4.2	2.1
TOC													26.2	10200	1545.0	1	7	3.8
Phosphorus	1.3	1.3	1.3	0.0	0.2	0.0	2.0	6.8	3.6	0.0	0.3	0.1	0.4	2.8	1.0	0.0	0.1	0.0
Cadmium																0.02	0.02	0.020
Chromium							0.1	0.5	0.2	0	0.29	0.074	0	0.55	0.2	0.009	0.5	0.2
Copper							0.04	0.4	0.2	0	1.2	0.2	0	0.02	0.004	0.019	0.1	0.028
Iron							1.5	7.7	4.6	0.3	5.3	1.59	0.23	940	98.5	0.07	1.4	0.4
Lead													0	0.12	0.1	0.05	0.05	0.1
Manganese													0.1	18.8	2.6	0.09	0.67	0.3
Nickel							0.12	0.3	0.2	0.0	0.02	0.002	0	0.25	0.1	0.005	0.05	0.028
Zinc							0.1	0.28	0.2	0.04	1.1	0.5	0	1.9	0.3	0.010	0.02	0.014

Notes: Abbreviations: TDS, total dissolved solids; BOD, biochemical oxygen demand; COD, chemical oxygen demand; DOC, dissolved organic carbon; TOC, total organic carbon.

† conductivity expressed as $\mu mhos \cdot cm^{-1}$
†† hardness and alkalinity expressed as $mg \cdot L^{-1}$ of $CaCO_3$
††† nitrite and phenol expressed as $\mu g \cdot L^{-1}$
all other concentrations expressed as $mg \cdot L^{-1}$

Upper Ottawa Street Landfill

The Upper Ottawa Street landfill is located in a former bedrock quarry, close to the edge of the Niagara Escarpment (Fig. 3); more than 1500 households are located within 750 m of the landfill (Ontario Ministry of Health, 1986; Table 2). In its thirty years of operation, the Upper Ottawa Street landfill accepted domestic, commercial and solid and liquid industrial waste (Table 2), including up to 38 million litres of liquid industrial waste per year. Rates of ground-water flow in the fractured dolomite below the site are between 1.5 and 20 m per day (Ontario Ministry of Health, 1986). Leachate mounding occurs within the landfill, and a number of leachate seeps discharge directly into Redhill Creek (Fig. 3; Proctor and

Redfern, 1994). Widespread concern with health risks posed by the site has prompted detailed site studies (Ontario Ministry of Health, 1986).

Glanbrook Landfill

The currently operating Glanbrook landfill is underlain by a variable thickness of lacustrine silts and clays (Fig. 4) and has been receiving domestic, commercial and non-hazardous solid industrial waste since 1981 (Table 2; Proctor and Redfern, 1994). Waste is deposited on a 6- to 16-m-thick unit of glaciolacustrine silts and clays, and a toe-drain leachate collection system is in place around each of the landfilling cells.

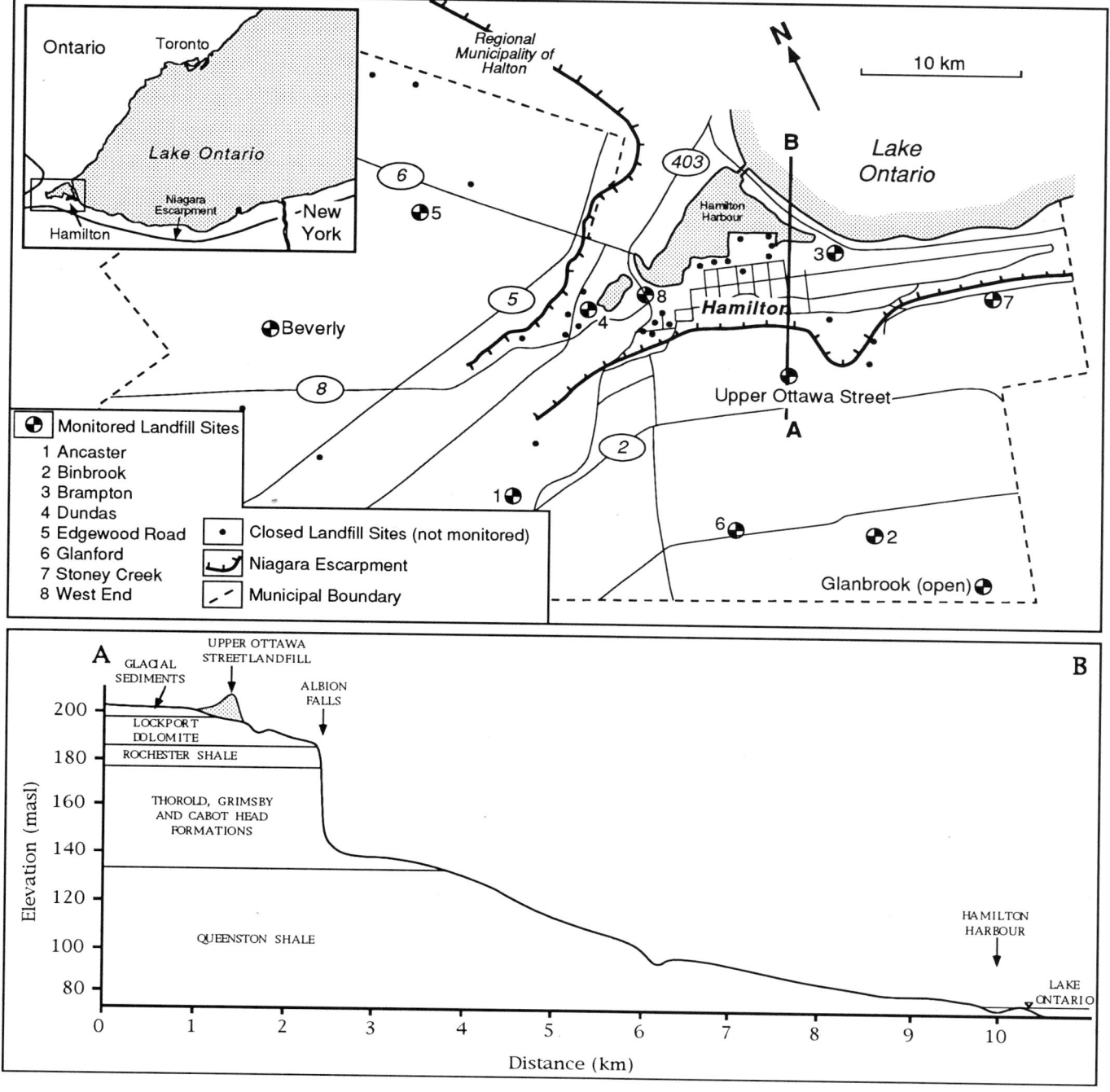

Figure 1 (*upper*) Location map showing Beverly, Upper Ottawa Street and Glanbrook landfill sites and the location of other closed landfills. (*lower*) Schematic south-north geological cross-section through the Niagara Escarpment (modified from Ontario Ministry of Health, 1986).

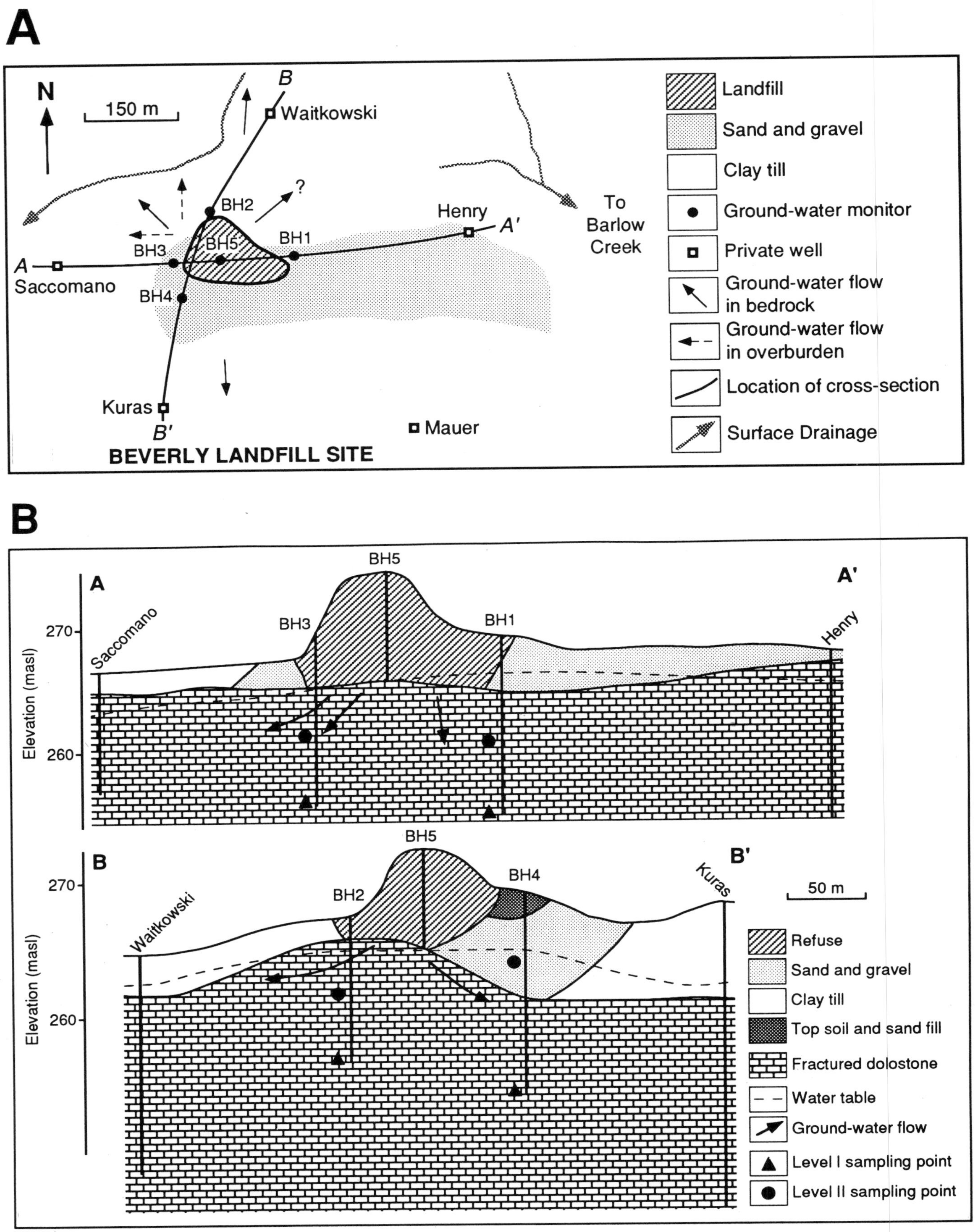

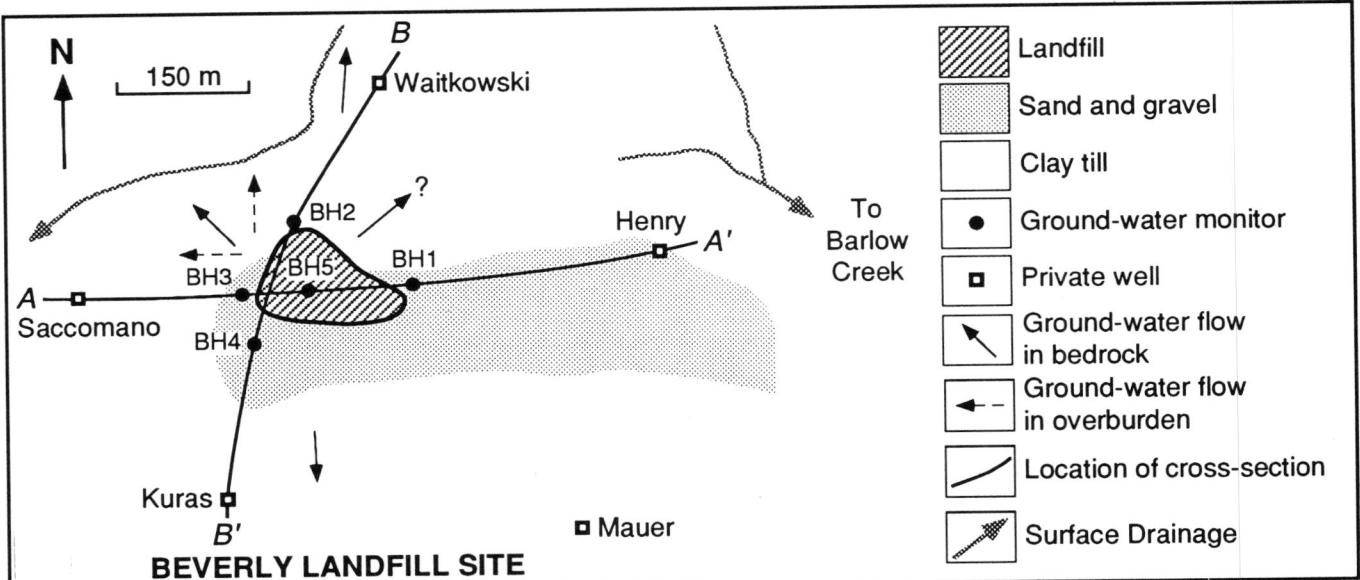

A

N

150 m

B

Waitkowski

?

BH2

Henry

To
Barlow
Creek

BH3

BH5

BH1

A

A'

Saccomano

BH4

Kuras

B'

Mauer

BEVERLY LANDFILL SITE

Landfill

Sand and gravel

Clay till

Ground-water monitor

Private well

Ground-water flow
in bedrock

Ground-water flow
in overburden

Location of cross-section

Surface Drainage

B

A   A'

BH5

270

BH3   BH1

Saccomano

Henry

Elevation (masl)

260

B   B'

BH5

270

BH4

BH2

Kuras

Waitkowski

50 m

Elevation (masl)

260

Refuse

Sand and gravel

Clay till

Top soil and sand fill

Fractured dolostone

Water table

Ground-water flow

Level I sampling point

Level II sampling point

Figure 2 *The Beverly landfill site.* (**A**) *Map and* (**B**) *schematic cross-sections of the site showing geological setting, position of monitoring wells and sampling points (levels I and II), and directions of ground-water flow. Location of cross-sections shown on A (after Gartner Lee Associates Ltd., 1981a, b with additions).*

Table 2 Physical characteristics of the Beverly, Upper Ottawa Street and Glanbrook landfill sites. For site locations, see Figure 1 (data from Gartner Lee Associates Ltd., 1979, 1981a, b; Ontario Ministry of Health, 1986).

	Site Area	Lifespan of Site	Amount of Waste	Waste Composition	Leachate Production	Site Geology	Surrounding Land Use
Beverly	1.6 ha (0.8 ha waste)	1965–1980	900 t·a⁻¹	80% domestic 10% commercial 10% agricultural	>3 million L·a⁻¹	Waste on fractured dolomite bedrock	Agricultural
Upper Ottawa Street	16 ha	1950–1980	750,000 t·a⁻¹ 0.85–38 million L·a⁻¹ liquid industrial waste	domestic, commercial, liquid industrial, solid steel industrial	>84 million L·a⁻¹	Old quarry on Niagara Escarpment; waste in contact with fractured dolomite (hydraulic conductivities 1×10^{-2} to 1×10^{-3} cm·s⁻¹)	Residential
Glanbrook	27.5 ha (100 ha proposed	1980–present		municipal, commercial, sewage sludge, incinerator ash	21 million L·a⁻¹	Waste underlain by 6–16 m of lacustrine silts and clays	Agricultural

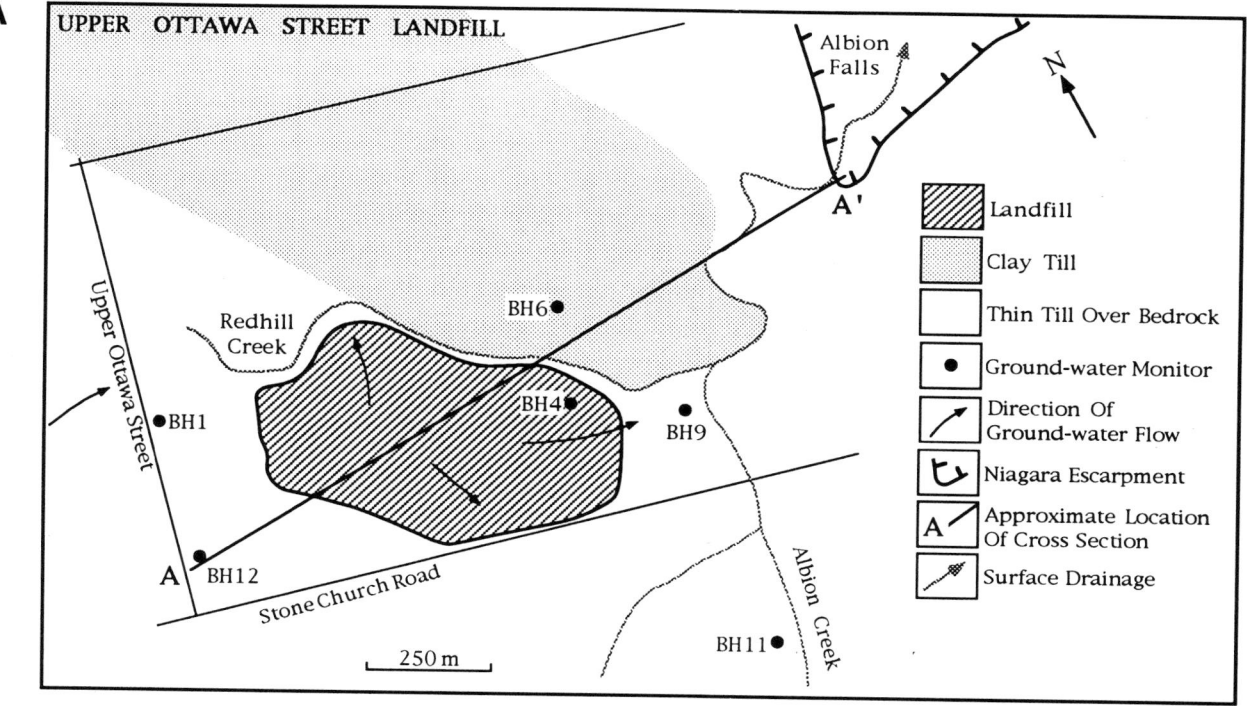

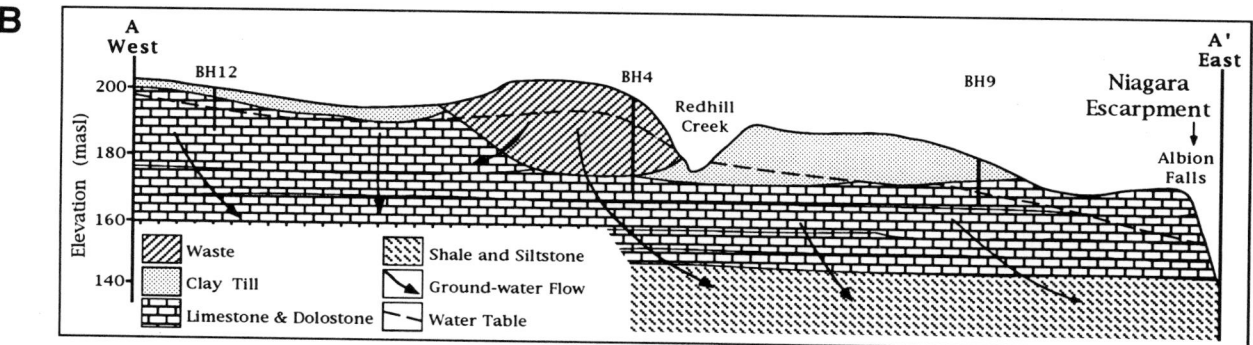

Figure 3 *The Upper Ottawa Street landfill site.* (**A**) *Map and* (**B**) *schematic cross-section of the site showing geology, position and depth of monitoring wells and directions of ground-water flow. Note proximity of the site to the Niagara Escarpment (after Ontario Ministry of Health, 1986 with additions).*

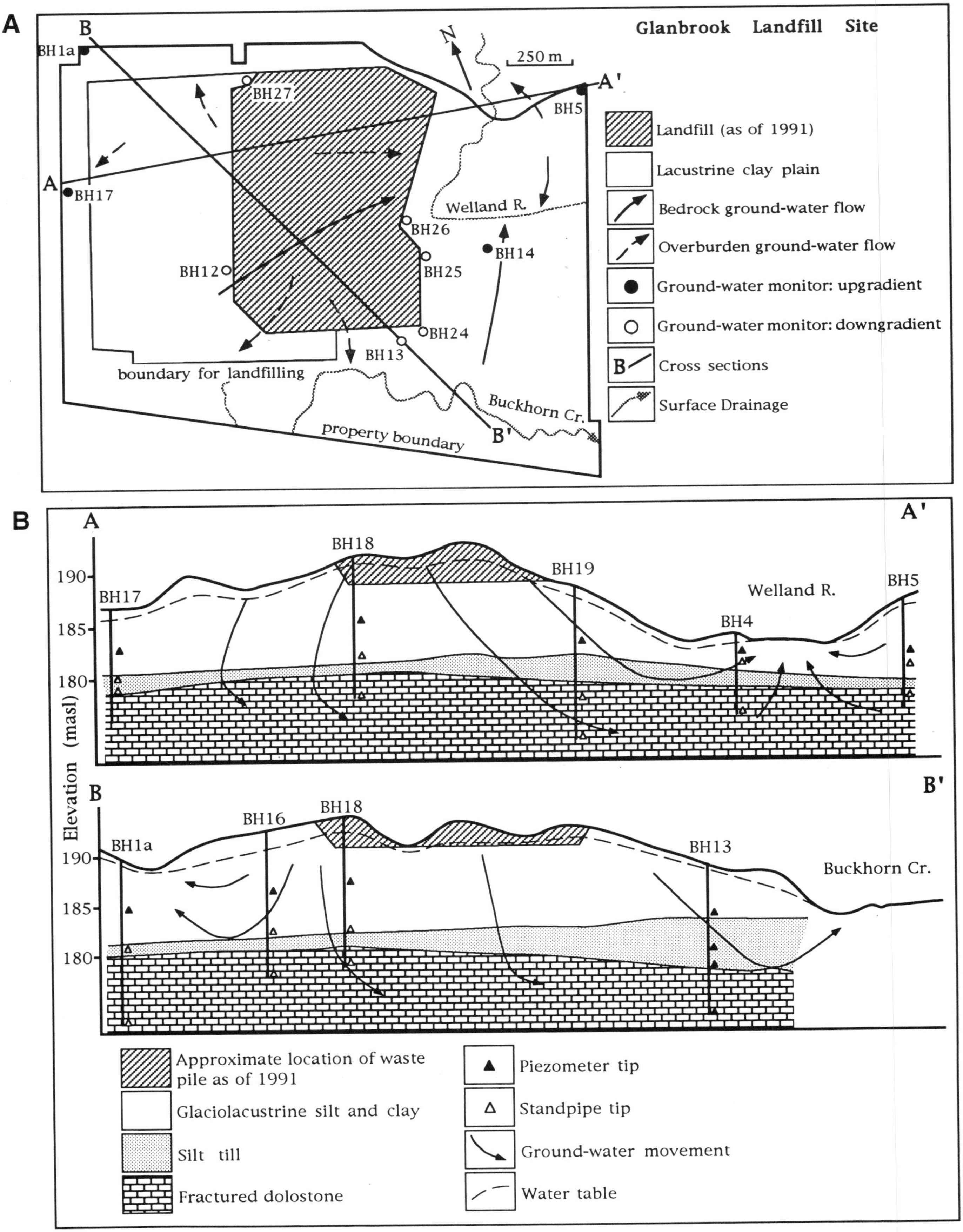

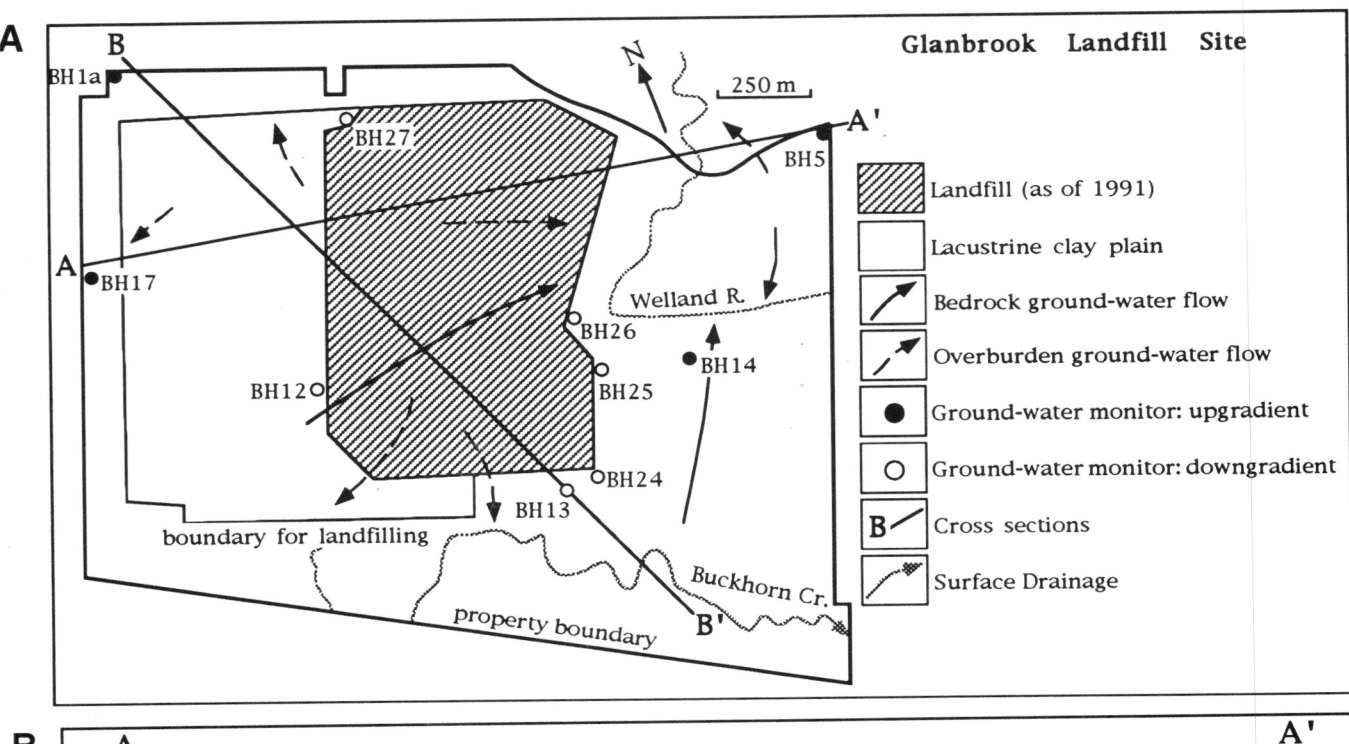

Figure 4 *The Glanbrook landfill site. (**A**) Map and (**B**) schematic cross-sections showing geology, position and depth of monitoring wells and directions of ground-water flow. This site is currently receiving waste from the Hamilton-Wentworth region and will be approximately double its present size upon closure (after Gartner Lee and Associates Ltd., 1979 with additions).*

FACTORS CONTROLLING
LEACHATE PRODUCTION AND COMPOSITION

Leachate Volume

Landfill leachate is formed as water percolates through the waste pile, extracting chemical constituents into the liquid phase (Farquhar, 1989). The quantity of leachate produced at any landfill is therefore controlled by climate, the landfill design, the nature of the waste and landfilling practices (Jones, 1991). Modern landfills control the amount of water flowing through the waste by incorporating a low-permeability material for daily cover and as a final cap (Lee and Jones, 1991). In this way, leachate generation is maintained at levels that can be handled by leachate collection systems.

The volume of leachate produced by a landfill varies considerably. Very little leachate is produced until the landfill reaches field capacity, a condition at which free drainage of water will occur. Thereafter, the amount of leachate generated is dependent on the amount of water entering the landfill. Normally, the amount of leachate will decline and stabilize after site closure and the installation of an impermeable cap (Fig. 5; Jones, 1991), but the movement of ground water into the site may keep leachate production high. For example, the Burlington landfill in southern Ontario was closed and covered with a clay cap in 1988, but currently produces large volumes of leachate. Leachate volumes collected at the site in 1992 represent recovery of almost 20% of total recorded precipitation (Henderson, Padden Environmental Inc., 1993); at least some of this leachate was generated by ground water entering the waste. Other examples of water budgets for landfills are given in Chapters 2 and 23.

Leachate Quality

Controls on Leachate Quality

Leachate quality is determined by the amount of soluble material in the waste and the rate at which water moves through the site. In addition, the degree to which the waste has been degraded by percolating water also influences leachate quality (Jones, 1991). The effects of waste composition on leachate quality are well illustrated by the Upper Ottawa Street landfill, which received considerable volumes of liquid industrial waste. Leachate from this site shows much higher average concentrations of chloride, ammonia, sulphate and phenol than leachate produced at the Beverly site (Table 1) and other sites in southern Ontario that contain municipal waste (Chapter 5; Howard *et al.*, 1996).

Daily compaction of waste, practised at most modern landfills to reduce waste volume, may reduce the internal permeability of waste, resulting in relatively low values of hydraulic conductivity (average value of 10^{-6} m·s⁻¹; Jones, 1991). As a result, low volumes of water pass through the waste in modern landfills, and the leachate generated at these sites will have higher concentrations of chemical constituents than leachate migrating away from older, closed landfills in which waste was less well compacted (Radnoff *et al.*, 1992). The reduction of water volumes passing through modern landfills slows the release of soluble constituents in waste and extends the time period over which contaminants are released (Radnoff *et al.*, 1992; Fig. 5).

The chemical characteristics of landfilled waste change with time, an aging process that results from dissolution of soluble materials and biodegradation. Consequently, leachate composition also varies with time (Table 3).

Leachate Characteristics

Leachate produced from municipal wastes has relatively high concentrations of inorganic chemicals, dissolved organics and metals (Tables 2 and 3; Cherry *et al.*, 1981; MacFarlane *et al.*, 1983; Baedecker and Apgar, 1984; Bricks, 1990; Jones, 1991). Microbial decomposition of organic materials creates reducing, anoxic and slightly acidic conditions, and results in the formation of methane (Farquhar, 1989; Jones, 1991; Barlaz and Ham, 1993; Chapters 20, 25). Higher ground-water temperatures are also observed in landfill leachate due to exothermic reactions caused by biochemical activity (MacFarlane *et al.*, 1983).

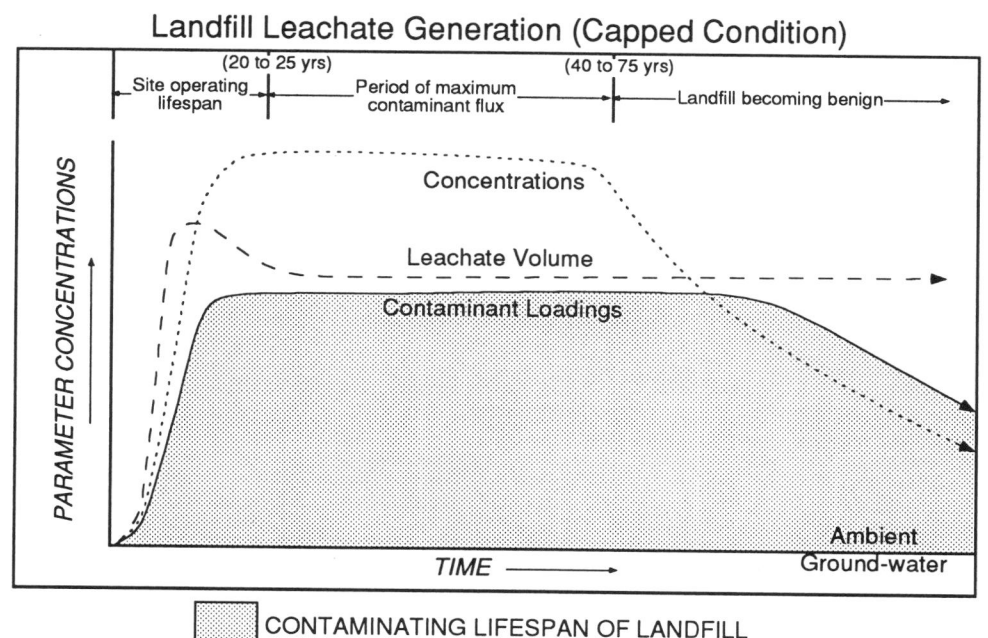

Figure 5   *Idealized leachate generation in a capped landfill (after Jones, 1991, with additions).*

Major and minor ions. Major and minor ions are released to leachate from household and industrial products, such as salts, oxides and hydroxides (Nicholson *et al.*, 1983; Jones, 1991), or from sediments used for daily cover. High concentrations of calcium and magnesium are typical (*e.g.*, Glanbrook; Table 1), and result from the enhanced dissolution of carbonates under high carbon dioxide pressure conditions within the landfill. Elevated sulphate concentrations in leachate can be the result of dissolution of gypsum from construction debris within the landfill or from the burning of waste (Nicholson *et al.*, 1983; Jones, 1991; Fritz *et al.*, 1994). Burning of waste was practised at both the Beverly and Upper Ottawa Street landfills, and may contribute to the high sulphate values measured in leachate at these sites (Table 1). Leachate released from the Glanbrook site also shows very high sulphate concentrations, but this may partly reflect the high sulphate content of ambient ground waters (Table 1).

Metal ions found in leachate originate either from the waste itself or from the daily cover material. Anoxic reducing conditions, created in the landfill site during decomposition of organic materials, increase the solubility of many metals, including iron and manganese, and cause oxide coatings to be released from soil particles (Jones, 1991). Common metals found in leachate include iron, lead, zinc, chromium, copper, manganese and nickel (Table 1).

Organic compounds. The presence of high concentrations of organic compounds in waste is reflected in the elevated biochemical oxygen demand (BOD), chemical oxygen demand (COD) and dissolved organic carbon values (DOC) of leachate (Tables 1 and 3; Jones, 1991). Phenols form as a by-product of microbial decomposition of organic wastes (Table 1). Synthetic organic compounds, such as chlorinated organic solvents (*e.g.*, trichlorethylene, tetrachloroethane, carbon tetrachloride) have their source in industrial and commercial wastes (*e.g.*, Chapter 6), and can pose significant human health problems (Tchobanoglous and Schroeder, 1985; Bredehoeft, 1992; Fetter, 1993).

Leachate is characterized by relatively high concentrations of most nitrogen species, including nitrite and the reduced nitrogen forms of ammonia and Kjeldahl nitrogen (Table 1). The source of nitrogen in the leachate is normally organic material present in the waste pile.

Temporal Variability in Leachate Composition

In a closed landfill, the amount of waste available to be leached is finite, and the concentration of contaminants within the leachate will eventually decrease, with the more soluble constituents being removed first and the organics and metals persisting (Fig. 5; Jones, 1991; Farquhar, 1989). Leachate released by any one landfill may also show compositional variability with time, as a result of changes in infiltration into the landfill produced by seasonal or annual variability in precipitation (*e.g.*, Hunter and Associates, 1987; Henderson, Padden Environmental Inc., 1993). The chemical composition of ground water entering a site can also show considerable temporal variation, and influences the final quality of the leachate. Natural variability in background water composition is well illustrated by the ranges of concentrations of analyzed parameters (maximum and minimum values) shown on Table 1.

IDENTIFICATION OF LEACHATE-IMPACTED GROUND WATERS

Monitoring Procedures

In Ontario, there are no set requirements that govern the scope or duration of ground-water quality monitoring at closed landfill sites. Instead, closure and monitoring programs are determined by the Ontario Ministry of Environment and Energy (OMOEE) on a site-by-site basis. All landfill sites currently receiving waste from 1500 people or more operate under a Certificate of Approval issued by OMOEE (Ontario Ministry of Environment and Energy, 1993) which specifies the monitoring requirements of the site during its operation, and post-closure (Table 4).

The Regional Municipality of Hamilton-Wentworth monitors eleven open and closed landfills in the region (Fig. 1; Proctor and Redfern, 1994). Water samples are collected periodically (quarterly, semi-annually or annually) from

---

Table 3 Changes in concentration of leachate parameters with age of the waste (modified from Farquhar, 1989).

Parameter (mg·L^{-1}) *	Age Category			
	0 to 5 years	5 to 10 years	10 to 20 years	> 20 years
BOD	10,000 − 25,000	1000 − 4000	50 − 100	< 50
COD	15,000 − 40,000	10,000 − 20,000	1000 − 5000	<1000
Total Kjeldahl nitrogen	1000 − 3000	400 − 600	75 − 300	< 50
Ammonia nitrogen	50 − 1500	300 − 500	50 − 200	< 30
Alkalinity	10,000 − 15,000	1000 − 6000	500 − 2000	< 500
TDS	10,000 − 25,000	5000 − 10,000	2000 − 5000	<1000
pH	5 − 6	6 − 7	7 − 7.5	7.5
Calcium	2000 − 4000	500 − 2000	300 − 500	< 500
Sodium and potassium	2000 − 4000	500 − 1500	100 − 500	< 100
Magnesium and iron	500 − 1500	500 − 1000	100 − 500	< 100
Zinc and aluminum	100 − 200	50 − 100	10 − 50	< 10
Chloride	1000 − 3000	500 − 2000	100 − 500	< 100
Sulphate	500 − 2000	200 − 1000	50 − 200	< 50
Total phosphorus	100 − 300	10 − 100		< 10

Note: * except pH

For explanation of abbreviations, see Table 1

monitoring wells containing either standpipes or multi-level piezometers, or from private water wells.

Sources of Error in Analytical Data

Errors in water sampling and handling procedures used in landfill monitoring programs can severely affect the accuracy of water chemistry data. Contamination during sample collection can be a serious problem when non-dedicated samplers are used. When water samples are analyzed for organics and metals, piezometer material and sampling equipment can be sources of contamination. The accuracy of water chemistry measurements of unstable parameters such as temperature, dissolved gasses, Eh, pH and alkalinity is also affected by delays between sample collection and analysis. Ideally, measurements of unstable parameters should be made at the wellhead and major ion analyses conducted within 48 hours of sample collection; samples should be refrigerated at a temperature of 4°C. Samples should also be filtered during collection, as suspended sediment may release metal ions during the acidification process used in subsequent chemical analyses (Gartner Lee and Associates Ltd., 1985).

To check the accuracy and precision of analytical data, a quality assurance (QA)/quality control (QC) program should be implemented using duplicate samples and/or blank samples (distilled water). At present, the Region of Hamilton-Wentworth does not have any QA/QC requirements in its ground-water monitoring program, but in the adjacent municipality of Halton (Fig. 1), approximately 15% of ground-water samples collected for landfill-monitoring purposes are used for quality control (G. Buitenhuis, pers. comm., 1994). The 1992 ground-water monitoring report for the closed Burlington landfill site in Halton Region shows that analyses for duplicate samples agreed within 10% of the original sample value for only 80% of the parameters measured, and 25% of the original sample value for 87% of the parameters (Henderson, Paddon Environmental Inc., 1993). This lack of precision is the result of inconsistent analytical processes and/or poor collection procedures.

Spatial and Temporal Variability in Water Chemistry

To help determine if ground water is impacted by leachate, concentrations of chemical parameters in the sampled waters are compared with those of naturally occurring ground waters sampled up-gradient of the landfill. Pre-landfilling water chemistry data can also be used to characterize background water quality, but these are rarely available for closed landfills.

Table 4   Monitoring program guidelines for landfill sites in Ontario (modified from Ontario Ministry of Environment and Energy, 1993).

	SMALL LANDFILL	MEDIUM LANDFILL	LARGE LANDFILL
Designed Lifetime Capacity: *Servicing Populations:*	< 40,000 m³ < 1500 people	40,000–200,000 m³ 1500–7500 people	> 200,000 m³ > 7500 people
DESCRIPTION			
1. Components of a monitoring program	Monitoring of existing adjacent wells and/or watercourses is normally adequate	(a) ground water, (b) surface water, (c) gas migration, (d) leachate, (e) liner (if installed)	(a) ground water, (b) surface water, (c) gas migration, (d) leachate, (e) liner (if installed)
2. Stages	(a) Baseline Monitoring Program, (b) Operational Monitoring Program, (c) Post Closure Monitoring Program	(a) Baseline Monitoring Program, (b) Operational Monitoring Program, (c) Post Closure Monitoring Program	(a) Baseline Monitoring Program, (b) Operational Monitoring Program, (c) Post Closure Monitoring Program
3. Monitoring Plan	Where required: (a) listing of devices to be used, (b) water quality parameters to be measured, (c) sampling and analytical procedures, (d) evaluation procedures, (e) implementation schedule	Where required: (a) listing of devices to be used, (b) water quality parameters to be measured, (c) sampling and analytical procedures, (d) evaluation procedures, (e) implementation schedule, (f) site-specific concerns for impacts on adjacent land use	Where required: (a) listing of devices to be used, (b) water quality parameters to be measured, (c) sampling and analytical procedures, (d) evaluation procedures, (e) implementation schedule, (f) site-specific concerns for impacts on adjacent land use
4. Data Records	Where required: (a) surface-water quality, (b) ground-water quality, (c) changes in ground-water levels	Where required: (a) surface-water quality, (b) ground-water quality, (c) gas migration, (d) contaminant migration rate, (e) comparison to predicted contaminant levels, (f) changes in ground-water levels	Where required: (a) surface-water quality, (b) ground-water quality, (c) gas migration, (d) contaminant migration rate, (e) comparison to predicted contaminant levels, (f) changes in ground-water levels

The OMOEE recommends that ground water should be monitored using at least one well in each aquifer hydraulically up-gradient of the landfill (to give background water quality), one well in the landfilled area and several down-gradient wells (Table 4; Ontario Ministry of Environment and Energy, 1993). Chloride is very commonly used as a leachate indicator parameter in ground-water monitoring programs because it occurs in reasonably high concentrations in leachate.

Chloride. Chloride is considered to be a conservative parameter because it undergoes very little chemical or biological change in the ground-water system, other than by the process of dilution (Baedecker and Apgar, 1984). Chloride is readily mobile, and is transported at the same velocity as the advective ground water. Elevated chloride concentrations in ground waters close to landfill sites are used to map the extent of leachate plumes (Freeze and Cherry, 1979; MacFarlane et al., 1983; Baedecker and Apgar, 1984; Jones, 1991). Chloride concentrations measured at the Beverly site show higher values in wells to the north, south and west of the site (BHs 2, 3 and 4; Figs. 2, 6), indicating possible leachate contamination in all those areas. Higher chloride concentrations in shallower rather than deeper monitors (e.g., BH 3; Fig. 6) suggest that leachate is either diluted as it migrates downward and away from the site, or is restricted to shallower ground waters. Some wells show erratic changes in chloride concentrations to levels much higher than background levels (e.g., BHs 2 and 3: Fig. 6); this may reflect variability in the concentration and amount of leachate impacting these wells or errors in the analytical data.

At the Upper Ottawa Street site, background waters show naturally high and variable chloride concentrations (Table 1), which often exceed those measured in leachate (Ontario Ministry of Health, 1986). The poor natural quality of ground water in this area makes the identification of leachate impact on the basis of elevated chloride concentrations difficult. Highly mineralized background waters are also characteristic of the Glanbrook site, where chloride concentrations show greater values in background wells than in down-gradient wells (Fig. 7). In cases such as this, alternative indicator parameters must be used to identify leachate impact.

Electrical conductivity. Electrical conductivity is a measure of the total dissolved solids in a water sample, and relatively high values are characteristic of landfill leachate (Table 1). At the Beverly site, the high electrical conductivities shown by shallow ground waters (Fig. 8) are typical of leachate-impacted waters. However, elevated electrical conductivity values are not only caused by leachate contamination, and high values may result from high natural concentrations of dissolved solids in background waters (e.g., Upper Ottawa Street and Glanbrook sites; Table 1).

An unusual trend in both chloride and conductivity values is shown by BH13 at the Glanbrook site (Figs. 7, 9). BH13 is a down-gradient well adjacent to a waste cell that has been receiving waste since the site opened. Chloride and conductivity values from this well show decreasing values after the start of landfilling in 1980, with levels not returning to their pre-landfilling values until several years later (Fig. 7, 9). This decline may be due to disruption of the ground-water flow system during excavation of the landfill. After 1991, chloride concentrations in down-gradient wells reach levels greater than pre-landfilling values and may indicate leachate impact.

Organics. Elevated concentrations of dissolved organic compounds in ground water can be used to identify leachate contamination. Reduced nitrogen species, such as Kjeldahl nitrogen (Fig. 10) and ammonia, were found in much higher concentrations in down-gradient wells than in background

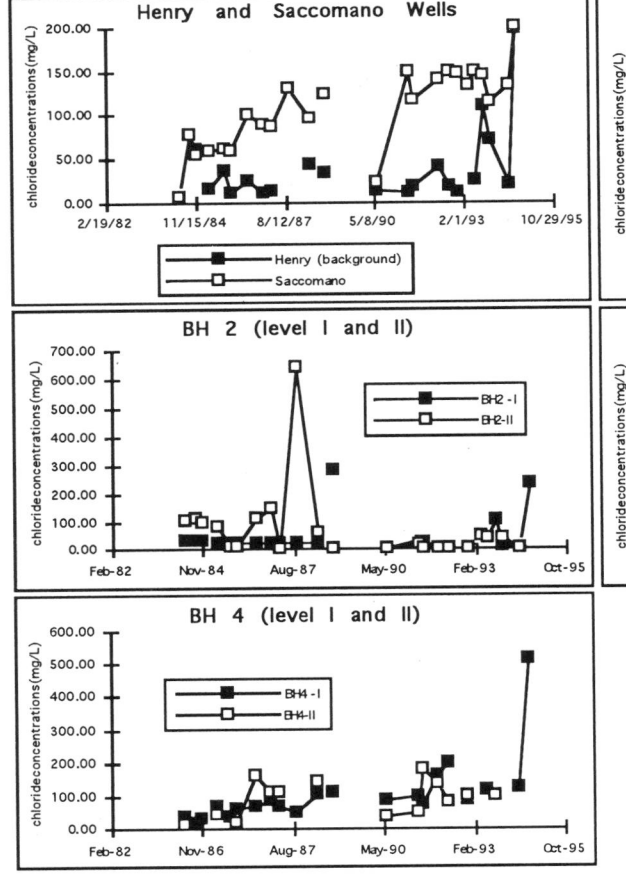

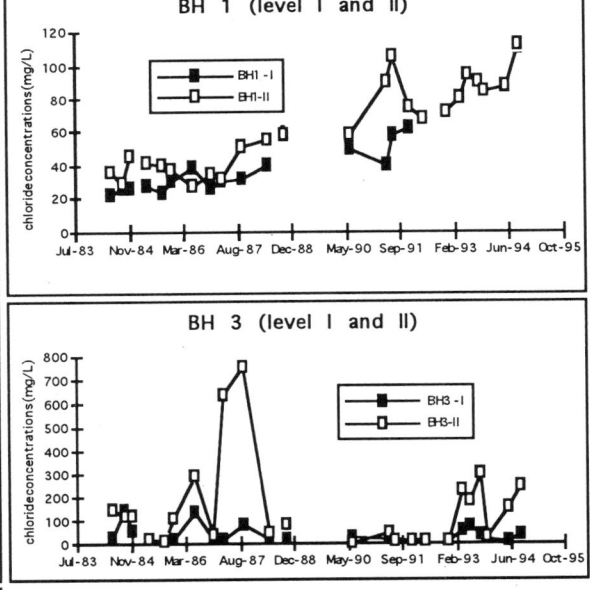

Figure 6 *Chloride concentrations in ground waters at the Beverly site (1984-94). For borehole locations, see Figure 2. The Henry well lies up-gradient of the site and is considered to represent background water. Elevated chloride concentrations in the Saccomano well, and BHs 1-4 may reflect leachate impact. Shallower monitors (level II) show higher chloride concentrations than deeper monitors (level I) which could indicate dilution of leachate with downward migration or preferential flow in the shallower parts of the aquifer.*

waters at the Beverly and Upper Ottawa Street sites. At the Upper Ottawa Street site, down-gradient concentrations of both Kjeldahl nitrogen and ammonia have decreased progressively since site closure (Fig. 10).

Most organic compounds are not uniquely associated with landfill leachate and can be derived from other sources. Elevated concentrations of organic compounds measured at the Waitkowski well (Fig. 10), which lies on a farm to the north of the Beverly site (Fig. 2), are associated with high bacterial levels (Gartner Lee Associates Ltd., 1981a) and are likely caused by agricultural activity.

Some synthetic organic compounds, such as trichloroethylene and carbon tetrachloride, can also be used as leachate indicator parameters (Fetter, 1993). However, unlike chlorides, organic compounds undergo considerable biochemical degradation during migration in ground water (Fetter, 1993), and volatile organics may be lost in the vadose zone (Freeze and Cherry, 1979). Furthermore, organic compounds usually occur in such small quantities that they are difficult to detect, especially after dilution by ground water, and may be impossible to distinguish from contaminants introduced by sampling equipment. For example, hazardous organic compounds identified in ground waters around the Upper Ottawa Street landfill site could not be confidently ascribed to leachate impact (Barker et al., 1987) as most of the compounds occur naturally in background waters or could have been introduced as contaminants from piezometer materials (Barker et al.,

1987). Hence, in a number of municipalities, organic compounds are omitted from routine ground-water analyses because of the high cost of analysis and questionable validity of the results (G. Buitenhuis, pers. comm., 1994).

Rarely can any one indicator parameter be used to confidently identify leachate impact on ground waters. In areas such as Hamilton, where highly mineralized background waters obscure the hyrochemical signature of leachate, alternative techniques involving comparison of multiple parameters on graphical plots may be required.

Graphical Representations of Major Ion Data

A number of methods are available to graphically represent the major ion chemistry of ground waters, such as Durov diagrams, anion/cation plots and major ion cross plots. These graphical techniques allow identification of anomalous or unexpected ion assemblages in the ground water which may be indicative of leachate contamination. They can also be used to investigate the nature of attenuation and mixing processes operating on the leachate as it migrates off site.

Durov Diagrams

Durov diagrams are trilinear plots designed to classify water types according to their major ion composition, and can be used to identify the processes affecting overall ground-water chemistry (Fig. 11; e.g., Lloyd and Heathcote, 1985; Howard and Beck, 1993). Durov diagrams are constructed by plotting

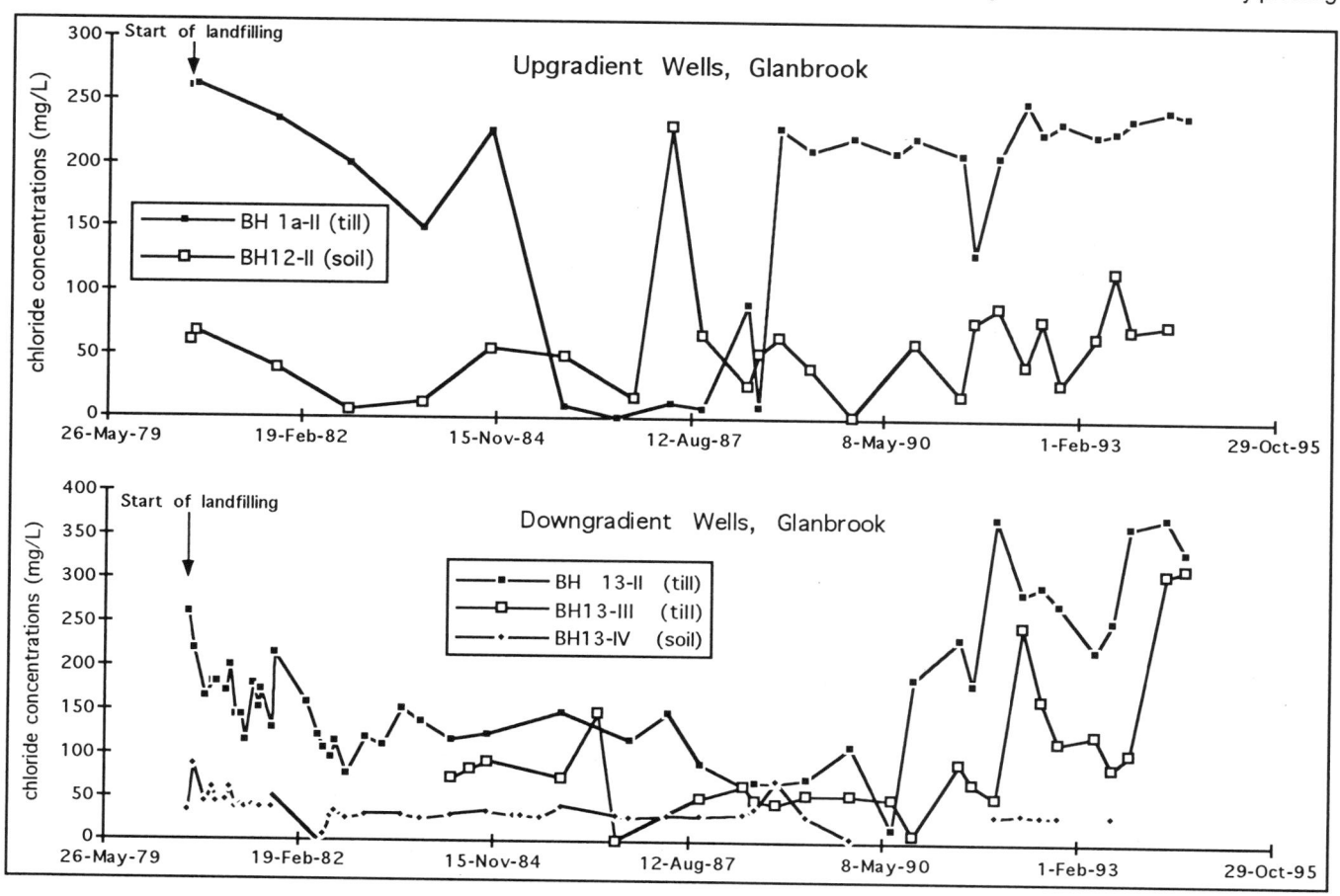

Figure 7 *Chloride concentrations for up-gradient and down-gradient wells at the Glanbrook site. Wells screened in till (BH 1a-II, BH 13-II, BH 13-III) show consistently higher chloride concentrations than wells screened in surface sediments (BH12-II, BH 13-IV). This is probably caused by relatively slow ground-water flow rates in till. Reduced chloride concentrations in down-gradient wells over the period 1980 to 1990 may be due to changes in ground-water flow paths caused by excavation of the landfill. Increased chloride concentrations to levels higher than pre-landfilling values after 1990 may reflect leachate impact in down-gradient wells.*

the relative concentrations of major cations (Ca, Mg, Na) for a water sample in the upper triangular field, and the relative concentrations of major anions (Cl, SO$_4$, HCO$_3$ and CO$_3$) in the left triangular field. Points are then projected onto the central square field for classification (Fig. 11). Hydrochemically similar waters plot as close groups in all three fields of the diagram.

Interpretation of the water sample groupings shown on a Durov diagram is based on the premise that ground water follows a predictable path of chemical evolution in natural

geologic systems, initially controlled by carbonate equilibria and then by ion exchange and dissolution processes (Baedecker and Apgar, 1984; Lloyd and Heathcote, 1985). This results in a progressive change from calcium bicarbonate waters to sodium bicarbonate waters and finally to sodium chloride waters (Fig. 11A; Freeze and Cherry, 1979; Howard and Beck, 1986).

Durov diagrams are used in this study to classify water types and identify potential leachate impact at the Upper Ottawa Street site, and to illustrate the variability of background water chemistry at the Glanbrook site. Insufficient

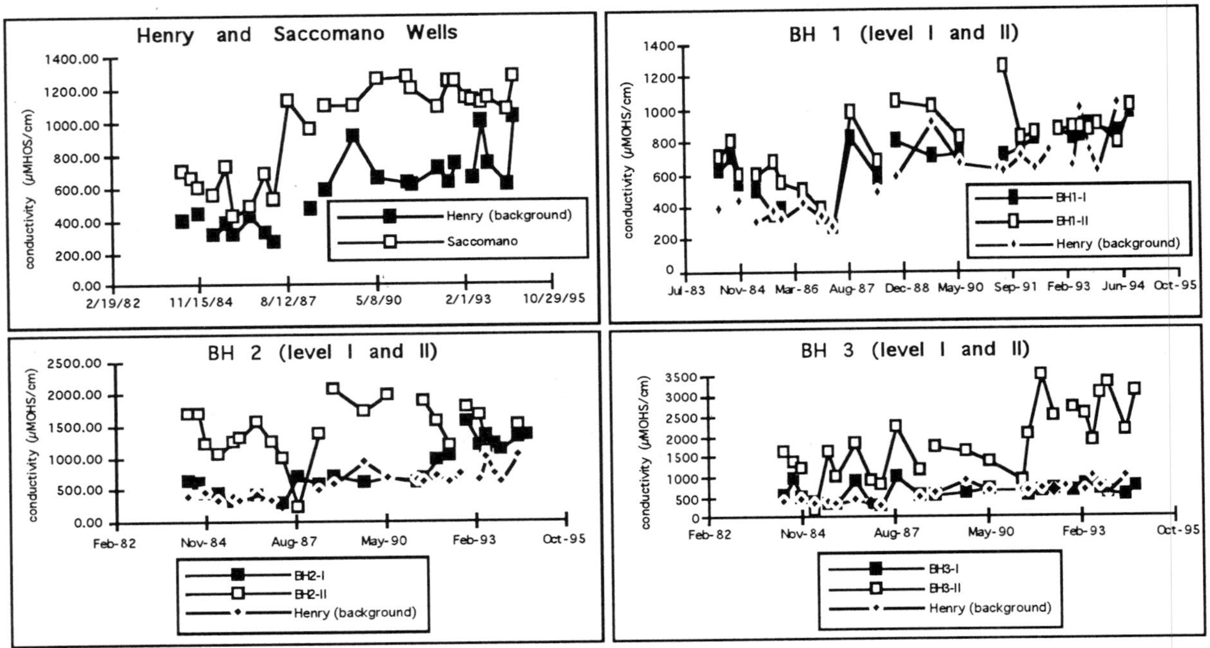

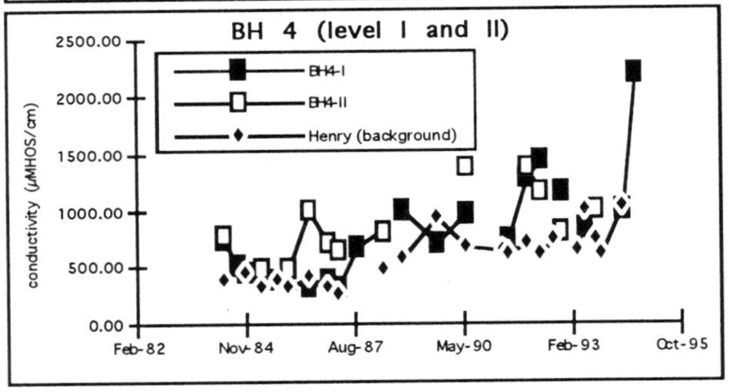

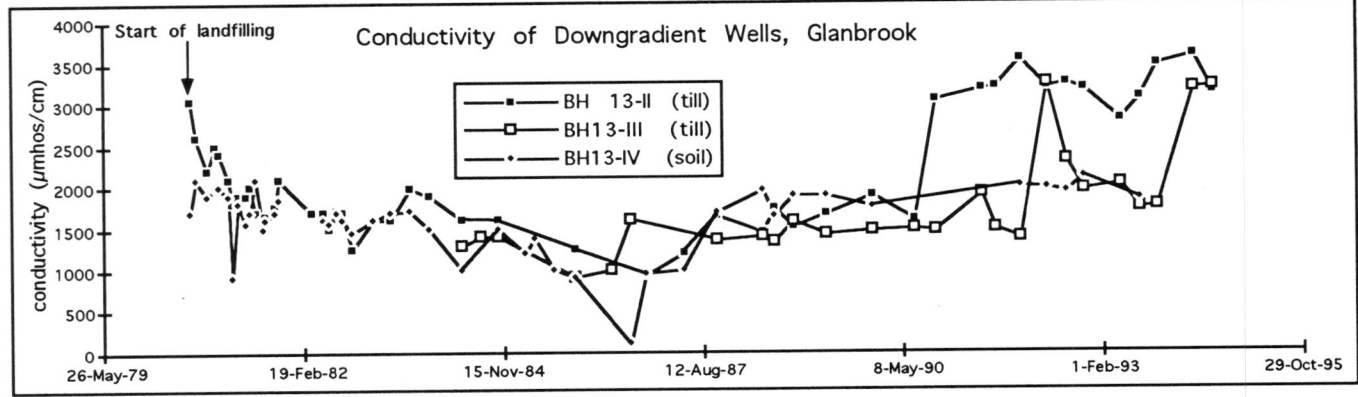

Figure 8 *Electrical conductivity values from the Beverly site 1984-1994. Conductivity values are higher for samples from shallow monitors (level II) than from deeper monitors (level I). Higher conductivity values in down-gradient wells to the north and west of the site (particularly BH 2 and BH 3) may indicate leachate contamination.*

Figure 9 *Electrical conductivity values from down-gradient wells at the Glanbrook site. Reduced conductivity values for the period 1980-1990 are similar to chloride concentration changes (Fig. 7) and may be due to changes in ground-water flow paths caused by excavation of the landfill. Increased conductivity values after 1990 may reflect re-establishment of flow systems or leachate impact.*

data were available to construct Durov diagrams for the Beverly site.

At the Upper Ottawa Street site, all ground waters sampled plot in the sodium chloride field (Fig. 11B). However, down-gradient wells (BHs 9 and 11; Fig. 11B) are enriched in calcium, possibly as a result of leachate contamination. In the anion field, two main groupings of water types are present, one with chloride as the dominant ion (BHs 11 and 12; Fig. 11B) and the other with higher amounts of sulphate (BH 9; Fig. 11B). Mixing of background waters with sulphate-rich leachate may account for the shift in dominant ions shown by BH 9 waters.

Durov diagrams constructed for ground waters at the Glanbrook site show extreme variability in background water chemistry, particularly for waters in surficial sediments (Fig. 11C). Comparison of background waters with those from down-gradient wells (Fig. 11D) suggests little impact from leachate migration at this site.

Anion/Cation Plots

Anion/cation plots provide another method of graphically characterizing water types based on the relationship between concentrations (milliequivalents per litre; meq·L⁻¹) of selected groups of anions and cations. In order to help identify waters impacted by sulphate- and chloride-rich leachate at the Upper Ottawa Street site, chloride was plotted against Ca+Mg–SO₄ (Fig. 12). Most of the up-gradient and down-gradient ground

waters plot in a similar position to leachate, except ground water from BH 11 which has a very different ionic ratio (Fig. 12). The similarity of Ca+Mg–SO₄:Cl ratios in most up-gradient and down-gradient ground waters and leachate precludes the use of these plots to identify leachate impact. However, ground waters from BH 11 are identified as having a different source to the other waters, possibly from saline formation waters.

Major Ion Cross Plots

Major ion cross plots are used to compare the actual concentration of an ion in ground water to that expected if dilution were the only factor affecting its concentration (e.g., Bricks, 1990). A theoretical dilution curve is constructed for a selected ion showing its expected concentration with respect to chloride as it becomes increasingly diluted in the groundwater system (Fig. 13). If the actual concentrations of the selected ion differ greatly from those on the theoretical curve, the ion may be undergoing attenuation by processes such as ion exchange or chemical dissolution/precipitation, or may have a source other than leachate.

Major ion cross plots showing the relationship between sodium (Na) and chloride (Cl) were constructed for BH 13, a down-gradient well at the Glanbrook site (Fig. 13). The data points are grouped and identified according to the year(s) of sample collection, and show dilution of sodium and chloride in

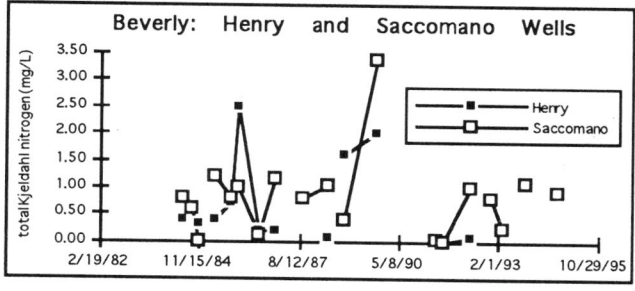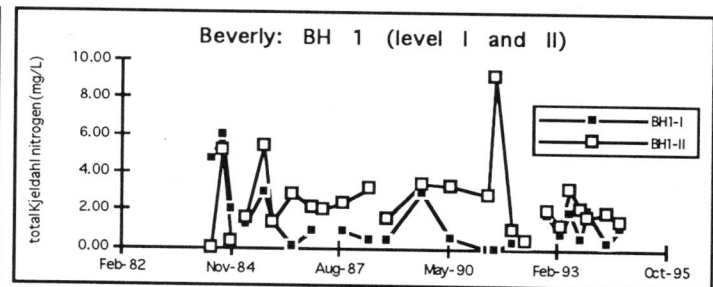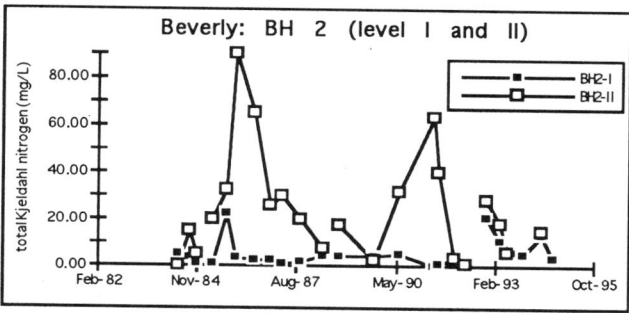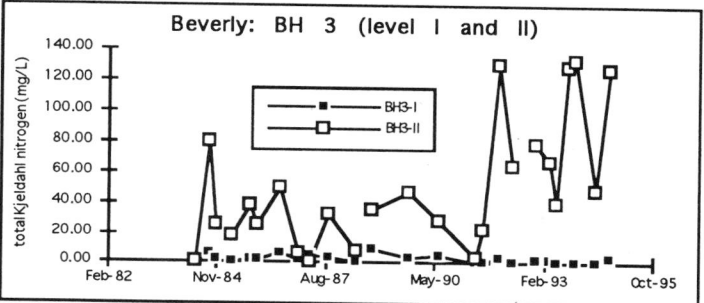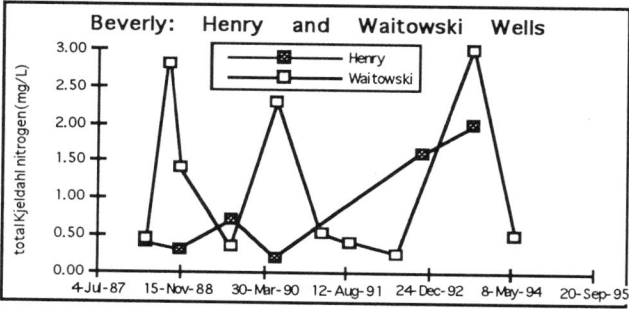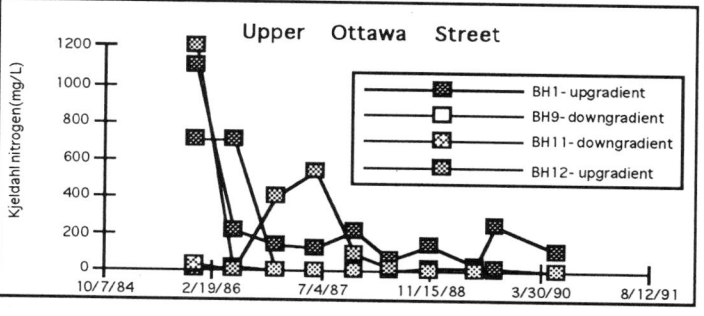

Figure 10 *Total Kjeldahl nitrogen (TKN) values for Beverly and Upper Ottawa Street. TKN values are elevated, but highly variable in shallow monitors (level II) of BHs 2 and 3, and may be due to leachate contamination. Elevated TKN values from the Waitkowski well are associated with high bacterial levels and may result from agricultural activities. Both up-gradient and down-gradient wells show considerable variability in TKN values at the Upper Ottawa Street site, making leachate impact difficult to identify.*

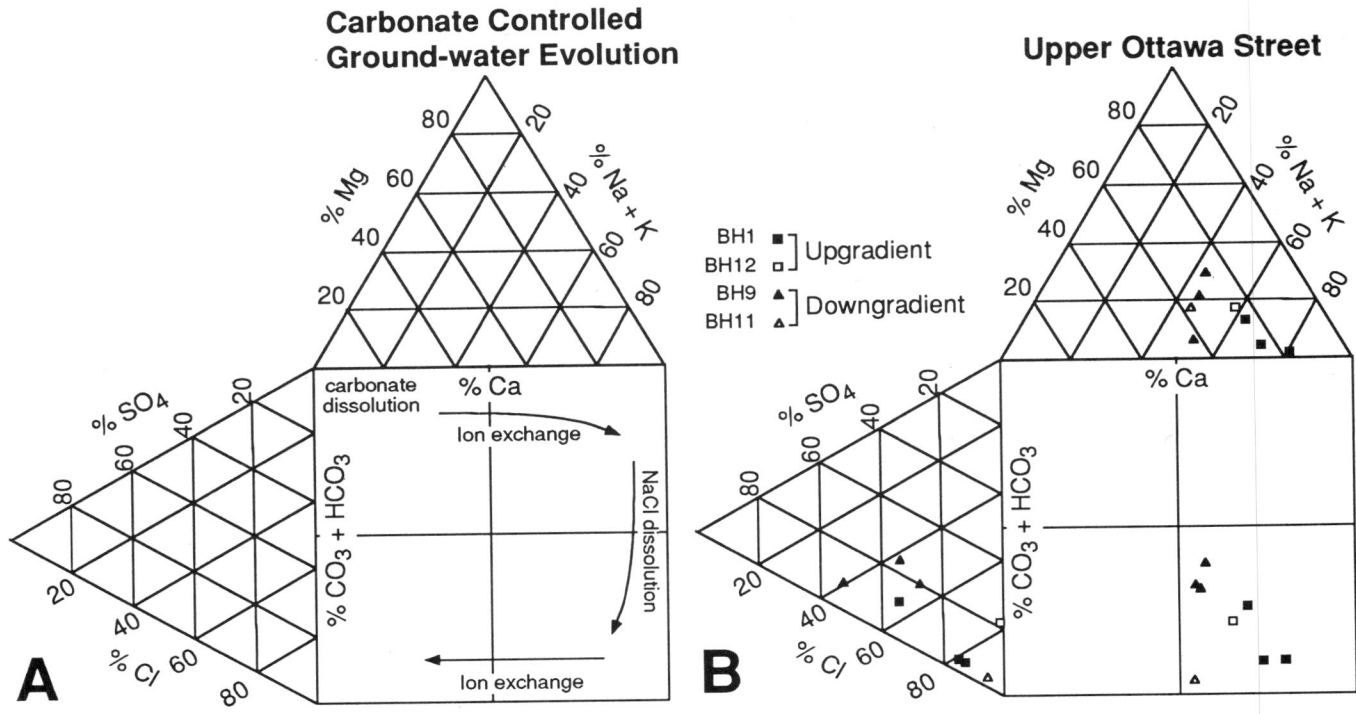

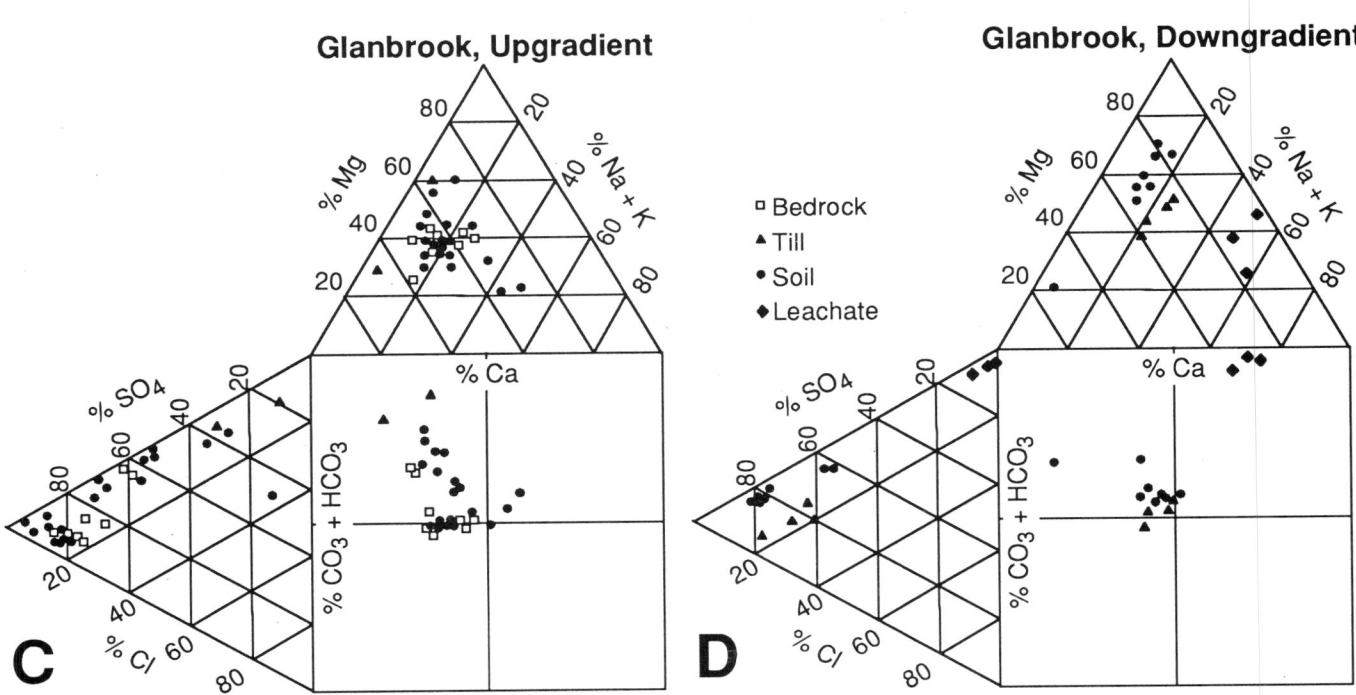

Figure 11 (A) *Durov plot showing the natural ground-water evolutionary sequence controlled by carbonate equilibria (after Hunter and Associates, 1987).* (B) *Durov plot of water samples from BHs 1,9,11 and 12 at the Upper Ottawa Street site. All water samples plot as NaCl waters. Note enrichment of Ca shown by down-gradient wells (BHs 9 and 11) and relatively high sulphate values shown by BH 9. Monitoring data from 1982-1986.* (C, D) *Durov plots from the Glanbrook landfill site (1985-1986 data). Up-gradient wells show great natural variability of water types (ranging from Ca-Mg-SO₄ waters to CaHCO₃ waters) according to the material in which the well is screened (soil, till or bedrock). Down-gradient wells also show variability of water types sampled but show no similarity to leachate plots and are unlikely to be affected by leachate. Note: only hydrochemical data with charge balance errors of less than 10% were used to construct Durov diagrams.*

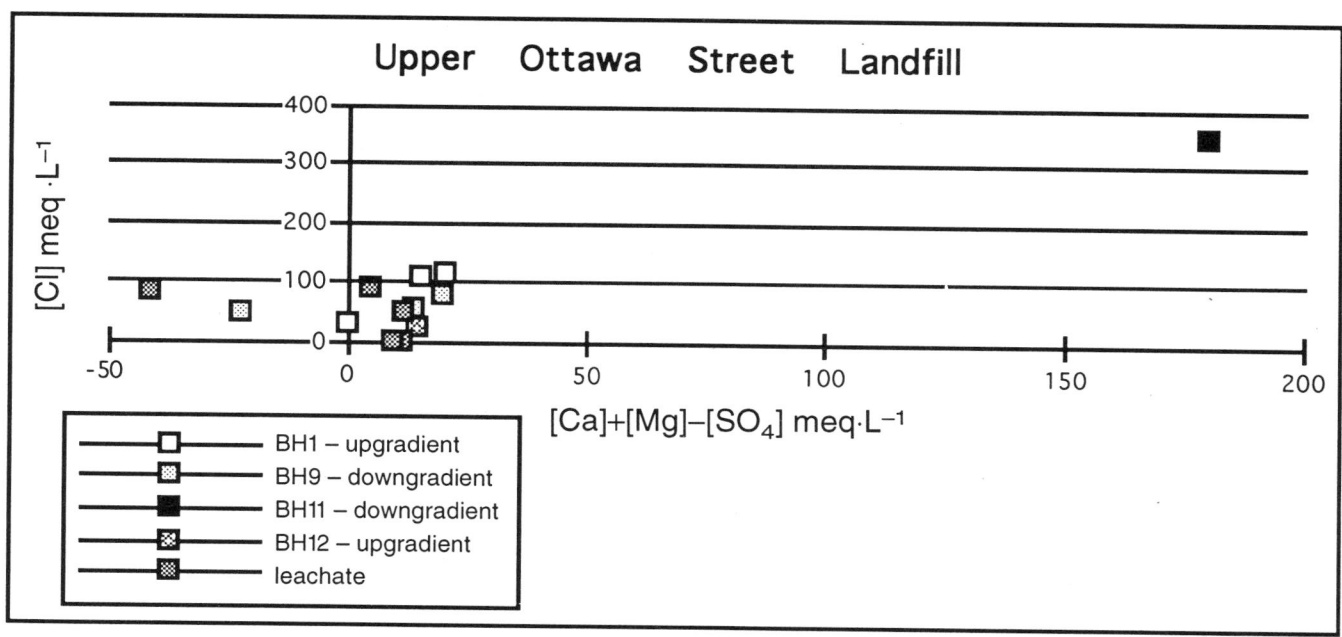

Figure 12 *Anion/cation plot showing the relationship between Cl and Ca+Mg–SO$_4$ in leachate and water samples from up-gradient and down-gradient wells at the Upper Ottawa Street landfill site. Both up-gradient and down-gradient waters plot close to leachate, except BH 11 waters which probably have a different source.*

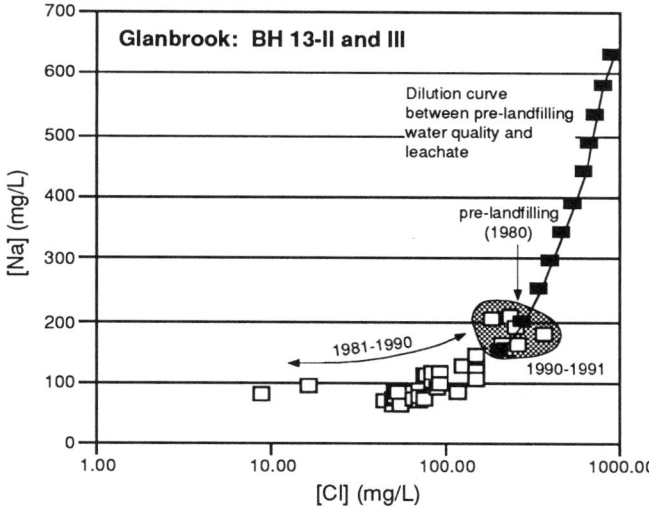

Figure 13 *Major ion cross plot of Na and Cl (dilution plot) for BH 13 (levels II and III; Fig. 4B) at the Glanbrook site. Theoretical mixing curves were constructed for leachate + fresh water and mineralized ground water + fresh water and both predict Na:Cl ratios (black squares) similar to those actually measured at BH 13 (open squares). Samples are labelled and grouped according to the year(s) of collection. Between 1980 and 1990, the ground waters became more dilute, returning to pre-landfilling values after 1990. Shaded area identifies 1990-1991 data. The Glanbrook site started receiving waste in 1981.*

ground water between 1980 and 1990, probably as a result of enhanced mixing with fresh water. This trend was also identified on chloride and conductivity plots (Figs. 7 and 9), and may relate to disturbance of the ground-water flow system by excavation and construction of the site. After 1990, the ionic ratios return to pre-landfilling values, and suggest that either the original flow systems are re-established or mixing with leachate is occurring.

Environmental Isotopes

Isotopic tracers provide an additional tool for source identification of both water and solutes in ground water. Environmental isotopes are particularly useful for the identification of leachate impact in areas with many sources of solute to the ground-water system and highly mineralized background waters.

Water molecules contain isotopes of oxygen (^{18}O and ^{16}O) which can be used to trace the source of water, and isotopes of hydrogen (tritium ^{3}H and deuterium ^{2}H) which can be used to determine the relative age of water. By analyzing a water sample for its tritium content, it can be dated as either pre-1953 (before atmospheric nuclear testing), or post-1953, which may be used to indicate a possible leachate source (*e.g.*, Barker *et al.*, 1987). However, mixing of ground waters with different ages and sources in an aquifer can make isotopic signatures extremely difficult to interpret. For example, tritium analysis of ground waters at the Upper Ottawa Street site identified two distinct groupings of waters present in the up-gradient wells. One group of waters had isotopic signatures typical of pre-1953 waters, but had concentrations of inorganic parameters and volatile aromatics even exceeding those of leachate (Ontario Ministry of Health, 1986). These poor quality waters were thought to have been in contact with bedrock containing natural hydrocarbons (*e.g.*, shale) for long periods of time. A second group of background waters appeared to be young, post-1953 waters, and had much lower concentrations of many inorganic parameters. The occurrence of these recently recharged waters as background waters at the Upper Ottawa Street site makes leachate impact impossible to identify on the basis of young isotopic signatures (Barker *et al.*, 1987).

The relationship between light and heavy isotopes of an element within a water sample can be used to determine the origin of the solutes in ground water (Fritz *et al.*, 1976; Kaplan and Mordeckai, 1986; Aravena *et al.*, 1993). Analysis of the ratio between stable isotopes of nitrogen and oxygen has allowed differentiation of ground waters containing naturally high levels of nitrate from those contaminated by nitrate originating from a septic system plume (Aravena *et al.*, 1993). Sulphur and boron isotopes in leachate can also be significantly different from isotopes derived from other sources in ground water, and have been used to identify leachate impacted waters (Davidson and Bassett, 1993; Fritz *et al.*, 1994). Stable isotope ratios can also be used to determine the composition of waste within a landfill (Fritz *et al.*, 1994).

PROBLEMS IN IDENTIFYING LEACHATE IMPACT

Difficulties involved in determining the extent of leachate impact on ground waters around landfill sites are clearly illustrated by the three Hamilton examples. At the Beverly site, poor quality bedrock waters and potential contamination by agricultural activities have made the extent of leachate impact on ground water difficult to determine. However, leachate contamination is strongly suggested by elevated chloride, nitrate, conductivity, sulphate and ammonia concentrations in the Saccomano well, which lies 170 m west of the landfill (Figs. 2, 6, 8, 10; Proctor and Redfern, 1994).

At the Upper Ottawa Street landfill, leachate impact can be identified within a radius of approximately 100 m of the landfill, but beyond this, extremely poor and variable natural background water quality obscures any leachate signature (Ontario Ministry of Health, 1986). Durov diagrams and anion/cation plots (Figs. 11, 12) suggest some alteration of groundwater chemistry in down-gradient wells, although this cannot be uniquely attributed to leachate contamination. A major problem with leachate impact assessment at this site results from the complex shape of the leachate plume. The waste lies on highly fractured bedrock, and leachate-impacted waters are most likely to migrate in narrow fracture zones. Unless monitoring wells happen to intersect these fracture zones, leachate impact cannot be detected. Rapid ground-water movement in the vicinity of the site also allows leachate to be readily diluted as it leaves the landfill, reducing the concentration of indicator parameters and increasing the problems of detection.

At the Glanbrook site, ground-water monitoring shows background waters to be highly mineralized and extremely variable in their composition (*e.g.*, Fig. 11). Leachate appears to have little impact on down-gradient water quality, although this may be difficult to ascertain given the poor natural quality of background waters. The Glanbrook landfill has a leachate collection system, designed to minimize leachate escape from the site, and is underlain by fine-grained clays which should offer a reasonable amount of natural ground-water protection through attenuation processes.

DISCUSSION

This chapter has identified the many factors controlling leachate production and quality at landfill sites, and illustrated commonly used methods of detecting leachate-impacted ground waters. Ground-water chemistry data from three landfill sites in the Hamilton region show that there are many problems involved in the evaluation of leachate impact, es-

pecially when the sites are located in former bedrock quarries. Difficulties are created by the high degree of mineralization and natural content of organic compounds in bedrock waters, and by the problems of intercepting contaminant migration pathways in highly fractured substrates. These problems may allow significant contamination of urban ground waters by landfill sites to go undetected.

Identification of leachate-impacted ground waters is particularly important for the environmental assessment of projected landfills planned for other abandoned bedrock quarries along the Niagara Escarpment (*e.g.*, Taro Aggregates Ltd., 1995). The problem of predicting and monitoring groundwater and contaminant movement in fractured bedrock is a major impediment to development of these sites (Environmental Assessment Board and the Joint Board, 1995). The characteristics of regional ground-water flow into Hamilton Harbour, and the relative contribution of landfill leachate and other industrially impacted ground water on surface-water quality is not well understood. Further investigations of leachate migration patterns in bedrock aquifers is urgently required.

REFERENCES

Aravena, R., Evans, M.L. and Cherry, J.A., 1993, Stable isotopes of oxygen and nitrogen in source identification of nitrate from septic systems: Ground Water, v. 31, p. 180-186.

Baedecker, M.J. and Apgar, M.A., 1984, Hydrochemical studies at a landfill in Delaware, *in* Bredehoeft, J., ed., Groundwater Contamination: National Academy Press, Washington, DC, p. 127-138.

Barker, J.F., Cherry, J.A., Reinhard, M., Pankow, J.F. and Zapico, M., 1987, The occurrence and mobility of hazardous organic chemicals in groundwater at several Ontario landfills, Final report: University of Waterloo Research Institute, Waterloo, ON, 116 p.

Barlaz, M.A. and Ham, R.K., 1993, Leachate and gas generation, *in* Daniel, D.E., ed., Geotechnical Practice for Waste Disposal: Chapman and Hall, London, UK, p. 113-136.

Bredehoeft, J., 1992, Much contaminated ground water can't be cleaned up: Ground Water, v. 30, p. 834-835.

Bricks, S., 1990, An examination of the major ion chemistry of the Keele Valley landfill contaminant plume: unpublished B.Sc. thesis, University of Toronto, Toronto, ON, 45 p.

Davidson, G.R. and Bassett, R.L., 1993, Application of boron isotopes for identifying contaminants such as fly ash leachate in groundwater: Environmental Science and Technology, v. 27, p. 172-176.

Egoboka, B.C.E., Cherry, J.A., Farvolden, R.N. and Frind, E.O., 1983, Migration of contaminants in groundwater at a landfill: a case study 3. Tritium as an indicator of dispersion and recharge: Journal of Hydrology, v. 63, p. 51-80.

Environmental Assessment Board and The Joint Board, 1995, Steetley-South Quarry Landfill Site, decision and reasons for decision, Office of Consolidated Hearings, Toronto, ON, 215 p.

Farquhar, G.J., 1989, Leachate: production and characterization: Canadian Journal of Civil Engineering, v. 16, p. 317-325.

Fetter, C.W., 1993, Contaminant Hydrogeology: Macmillan Publishing Company, New York, 458 p.

Freeze, R.A. and Cherry, J.A., 1979, Groundwater: Prentice-Hall, Englewood Cliffs, NJ, 604 p.

Fritz, P., Matthess, G. and Brown, R.M., 1976, Deuterium and oxygen-18 indicators of leachate movement from a sanitary landfill, *in* Interpretation of Environmental Isotope and Hydrochemical Data in Groundwater Hydrology: International Atomic Energy Agency, Vienna, Austria, p. 131-142.

Fritz, S.J., Bryan, J.D., Harvey, F.E. and Leap, D.I., 1994, A geochemical and isotopic approach to delineate landfill leachates in an RCRA study: Ground Water, v. 32, p. 743-750.

Gartner Lee Associates Ltd., 1979, Hydrogeological study – final report. Proposed sanitary landfill site for the township of Glanbrook: 102 p.

Gartner Lee Associates Ltd., 1981a, Hydrogeological study – leachate migration aspects, Report for the Hamilton-Wentworth Region: 50 p.

Gartner Lee Associates Ltd., 1981b, Operational monitoring report – Glanbrook landfill site, Report for the Hamilton-Wentworth Region: 35 p.

Gartner Lee Associates Ltd., 1985, Annual operational monitoring report – Glanbrook landfill site, Report for the Hamilton-Wentworth Region: 43 p.

Henderson, Padden Environmental Inc., 1993, 1992 Groundwater and surface water monitoring report for the closed Burlington landfill: 130 p.

Howard, K.W.F. and Beck, P.J., 1986, Hydrochemical interpretation of groundwater flow systems in Quaternary sediments of southern Ontario: Canadian Journal of Earth Sciences, v. 23, p. 938-947.

Howard, K.W.F. and Beck, P.J., 1993, Hydrogeochemical implications of groundwater contamination by road de-icing chemicals: Journal of Contaminant Hydrogeology, v. 12, p. 245-268.

Howard, K.W.F., Eyles, N. and Livingstone, S., 1996, Municipal landfilling practice and its impact on groundwater resources in and around urban Toronto: Hydrogeology Journal, v. 4, p. 64-79.

Hunter and Associates, 1987, Differentiating various sources of chlorides in domestic well waters: Ontario Ministry of Transportation and Communication, Research and Development Branch, 35 p.

Jones, M.G., 1991, Factors controlling the character of municipal landfill leachate in Ontario: Gartner Lee Associates Ltd., Internal Report, 27 p.

Kaplan, N. and Mordeckai, M., 1986, A nitrogen-isotope study of the sources of nitrate contamination in groundwater of the Pleistocene coastal plain aquifer, Israel: Water Resources Research, v. 20, p. 131-135.

Lee, G.F. and Jones, R.A., 1991, Landfills and groundwater quality: Ground Water, v. 29, p. 482-486.

Lloyd, J.W. and Heathcote, J.A., 1985, Natural Inorganic Hydrochemistry in Relation to Groundwater: Clarendon Press, Oxford, UK, 296 p.

MacFarlane, D.S., Cherry, J.A., Gillham, R.W. and Sudicky, E.A., 1983, Migration of contaminants in groundwater at a landfill: a case study: Journal of Hydrology, v. 63, p. 1-29.

Nicholson, R.V., Cherry, J.A. and Reardon, E.J., 1983, Migration of contaminants in groundwater at a landfill: a case study 6. Hydrogeochemistry: Journal of Hydrology, v. 62, p. 131-176.

Ontario Ministry of Environment and Energy, 1993, Guidance Manual for landfill sites receiving municipal waste: 156 p.

Ontario Ministry of Health, 1986, Upper Ottawa Street Landfill site study, Final report, 366 p.

Proctor and Redfern Limited, 1994, Development of a comprehensive environmental monitoring program for the open and closed landfill sites within regional juridiction: prepared for Region of Hamilton-Wentworth, 94 p.

Radnoff, D., Hollingshead, S. and Anderson, G., 1992, What legacy are we leaving with future landfill leachates?: Environmental Science and Engineering, v. 26, p. 58-60.

Taro Aggregates Ltd., 1995, Proposed East Quarry Landfill: environmental assessment, Executive Summary and Volumes I and II: 350 p.

Tchobanoglous, G. and Schroeder, E.D., 1985, Water Quality: Addison-Wesley Publishing Company Inc., Reading, MA, 768 p.

25. Isotopic and Geochemical Characterization of Landfill Methane

Steven Desrocher
Barbara Sherwood Lollar

Department of Geology, University of Toronto, 22 Russell St., Toronto, Ontario M5S 3B1

SUMMARY

The generation of methane (CH_4) from landfills can represent a significant hazard to surrounding urban communities. Microbial degradation of organic matter within glacial and glaciolacustrine sediments is a common source of methane, often in near-surface environments. Such deposits form the substrate for landfills, and may contribute to CH_4 fluxes from these sites. Additional sources include thermogenic methane from bedrock (*e.g.*, shale). Unfortunately, compositional analysis of dissolved gases in ground water underlying landfills is incapable of resolving the magnitude of substrate CH_4 and thermogenic CH_4 contribution to landfill CH_4, and thus precludes accurate assessment of the hazards posed specifically by landfill CH_4 alone. These limitations can be overcome by using an isotopic mass balance approach. This approach is demonstrated using the example of Beare Road landfill, near Toronto. Landfill-derived, thermogenic and bacterial methane sources can be clearly discriminated by reference to ^{13}C isotopic signatures and $C_1/(C_2+C_3)$ ratios. This technique allows a more realistic assessment of methane production by landfills and associated hazards.

INTRODUCTION

Methane is a gaseous constituent of many natural and anthropogenically altered ground-water systems (Schoell, 1988). Two general mechanisms exist by which natural CH_4 may be formed in such systems. First, thermal degradation of organic material may generate CH_4 of thermogenic origin, a commonly occurring process in bedrock such as shale (*e.g.*, Barker and Pollock, 1984; Sherwood Lollar *et al.*, 1994). Alternatively, CO_2 reduction or bacterial degradation of organic material under anaerobic conditions produces bacterial CH_4. This process is ubiquitous in Pleistocene glacial sediments, lacustrine sediments and wetlands.

Landfills constitute a significant anthropogenic source of bacterially generated CH_4 as a result of an abundance of organic material and prevalent anaerobic conditions (Chapter 20). The subsurface accumulation of landfill CH_4 may pose a fire and explosion hazard both on site and off site. Off-site hazards associated with CH_4 accumulation are determined by the potential for off-site migration, both as a free-gas phase in the unsaturated zone of surficial sediments and as a dissolved constituent in ground water. Off-site accumulation of both free CH_4 and CH_4 exsolved from ground water may pose a flammability hazard within areas adjacent to landfills

(Rice, 1983). Ground-water wells and building foundations represent likely locations in which the accumulation of landfill-generated CH_4 is likely to exceed the lower explosive limit for CH_4 of 5 volume percent (vol. %) concentration in air.

Compositional *versus* Isotopic Approaches to Identification of Methane Sources

Despite the potential fire and explosion hazards posed by landfill CH_4, it has frequently proven difficult to determine the precise controls on the accumulation and migration of CH_4, both within the landfill itself and off site. Much of this difficulty arises from a reliance on a strictly compositional approach that assumes that elevated CH_4 concentrations, either on or off site, are generated by the landfill. However, in situations where natural sources may be contributing to high subsurface CH_4 concentrations, simple compositional analyses are incapable of assessing the relative degree to which CH_4 from these natural sources may be contributing to elevated concentrations in the vicinity of a landfill. The advantage of isotopic analysis over a strictly compositional approach is that it permits this assessment of the number and type of CH_4 sources at a given site, as well as establishing the extent of mixing between them.

Methane Sources and Characterization

Thermogenic CH_4 is readily differentiated from biogenic CH_4 by its enrichment in the heavier ^{13}C isotope of carbon. The use of carbon isotope ratios to distinguish natural biogenic CH_4 inputs to ground water from those arising from gas storage tank leakage is discussed by Coleman and Meents (1976). Games and Hayes (1977) discuss the measurement of carbon isotope ratios in distinguishing between natural and landfill inorganic carbon fractions. Measurement of the $^{13}C/^{12}C$ isotopic ratio of CH_4 from different sources allows the extent of mixing between sources to be determined using a mass balance approach (Games and Hayes, 1977), as expressed by the equation

$$\delta_m c_m = \delta_1 c_1 + \delta_2 c_2 \qquad (1)$$

where δ_1 and δ_2 correspond to the isotopic ratios of CH_4 from sources 1 and 2, respectively, and c_1 and c_2 indicate the amounts of CH_4 contributed by each of these sources. The terms δ_m and c_m represent the isotopic ratio and the amount of carbon within the mixture of CH_4 from sources 1 and 2. By measuring the end members δ_1 and δ_2, as well as the parameters δ_m and c_m for the mixture, the ratio of the amount of CH_4

contributed by each source (c_1/c_2) can be calculated.

Assessment of the degree of CH_4 mixing by a mass balance approach also permits the determination of the probable migration pathways of CH_4 away from a landfill. By isotopically characterizing the landfill CH_4 end member, the extent and locations of off-site migration and accumulation of landfill CH_4 may be determined. Isotopic analysis may therefore provide information regarding the relative importance of specific transport pathways and, therefore, mechanisms such as advective *versus* diffusive transport or saturated zone *versus* unsaturated zone transport.

EXAMPLE

Beare Road landfill (Fig. 1) provides an excellent test site where isotopic analysis can be used to constrain the relative contribution of CH_4 from natural *versus* landfill sources. Given recent urbanization surrounding the landfill, it is important to determine the extent of off-site migration and accumulation of landfill CH_4. However, this has proven difficult to establish due to the potential existence of CH_4 sources other than the landfill itself, such as underlying Pleistocene glacial deposits containing organic matter. In addition, thermogenic CH_4 from the underlying Paleozoic bedrock may also contribute to the high methane concentrations in local ground waters.

Geological and Hydrogeological Setting

Until its closure in 1983, the Beare Road landfill (61 ha) accepted approximately 9.6 million tonnes of municipal and industrial waste (Gartner Lee Ltd., 1990), but the exact composition of the refuse is unknown (Chapter 2). Bedrock underlying the landfill consists of black shales of the Whitby Formation (Fig. 1). The upper surface of the shale commonly exhibits faults and fractures (*e.g.*, Chapter 29). Overlying Pleistocene deposits consist of the Don Beds (Fig. 1), a lacustrine unit assigned to the Sangamon interglacial, and the Scarborough Formation, a thick and regionally extensive deltaic succession (see Chapter 2). Finely laminated silts and clays comprise the lowermost 20 m of the Scarborough Formation, and are referred to as the Scarborough Clays. These clays contain a high proportion of organic matter, including thin detrital peats. Organic carbon contents range between 0.01% and 1.35% of Scarborough Formation sediments. The upper part of the deltaic succession consist of trough cross-bedded and rippled sands (the Scarborough Sands). These strata are overlain by a thick, silty clay fill (Northern till) capped by a poorly consolidated sandy till (Halton Till) and lacustrine sands (Iroquois deposits; Fig. 1)

The hydrogeology of the Beare Road landfill has never been rigorously characterized, but two ground-water flow systems are present. A shallow flow system is present within the surficial Iroquois sands (Hydrology Consultants Ltd., 1979), with a water table depth of 2 m to 6 m below ground surface (Gartner Lee Ltd., 1992). Ground water in this shallow system flows in a southerly to southwesterly direction, toward the Little Rouge River (Fig. 1; Gartner Lee Ltd., 1990; 1992). This flow direction is likely influenced by the pumping of leachate in the southwest corner of the site, in the vicinity of monitoring well GW1 (Gartner Lee Ltd., 1992). The horizontal hydraulic gradient in the shallow flow system is approximately 0.01 between wells GW15 (Fig. 1) and GW1 (Fig. 1).

A deeper flow system exists within the Scarborough Sands and underlying shale bedrock, and is characterized by ground-water flow toward the south and southwest (Gartner Lee Ltd., 1992). The horizontal gradient in this deep flow

system varies from 0.04 to 0.06, with hydraulic head decreasing toward the south and southwest (Hydrology Consultants Limited, 1979). Vertical hydraulic gradients also exist within this deeper flow system, with head values decreasing toward the surface. The magnitude of these vertical gradients is within the range of 0.3 to 0.9 (Gartner Lee Ltd., 1992). Upward hydraulic gradients are suggested by the presence of artesian conditions.

Site Instrumentation

While Beare Road landfill lacks any engineered systems for

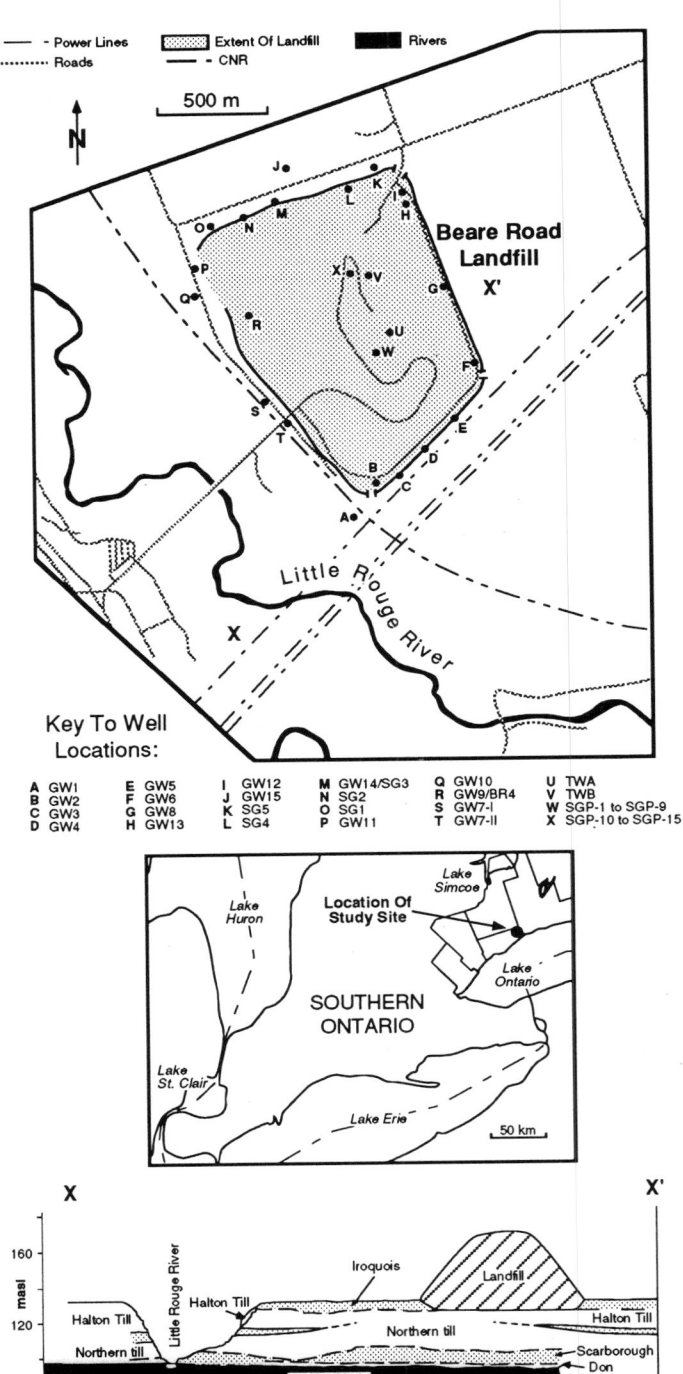

Figure 1 *Beare Road landfill, showing locations of sampled wells and geology. Perimeter ditch is approximately coincident with the solid line indicating the landfill extent.*

control of leachate and gas migration, a suite of wells surrounding the refuse mound are used for ground-water monitoring (designated GW and SG on Fig. 1). The depth of these wells ranges from 2 m to 51 m below ground surface (m bgs). In addition to these ground-water wells, two gas wells are located within the refuse mound itself (designated TWA and TWB; Fig. 1). These gas wells are surrounded by soil gas probes penetrating the landfill cover to a depth of approximately 1 to 2 m (designated SGP on Fig. 1).

A perimeter ditch, excavated to the water table, is intended to prevent the subsurface migration of landfill gas off site by venting it to the atmosphere. The assumption here is that gas migration occurs through the unsaturated zone, above the water table. Any migrating volatile plume is thought to be dispersed by mixing into the atmosphere, and no off-site migration will occur. This ignores the possibility of off-site migration of gases dissolved in ground water in the saturated zone, however. At the elevated pressures present at depths of several metres below ground surface, the increased solubility of gases in water leads to dissolved concentrations in excess of those encountered in near-surface waters. Movement of this gas-enriched ground water off site may carry with it dissolved gases which later separate from the fluid phase due to depressurization (Domenico and Schwartz, 1990).

SAMPLING AND ANALYTICAL METHODS

Sampling Locations
Sampling locations are divided into four groups based on similarity of depth and well type.

1. Deep ground-water wells extend to strata below the Northern till. Wells GW6, GW7-I, GW7-II, GW8, GW13, GW14, and GW15 were chosen from this category, based on their distribution around the landfill site (Fig. 1). Samples from GW15 are assumed to be representative of ground water uninfluenced by the landfill, as the well is situated up-gradient from the landfill site (Gartner Lee Ltd., 1992). Wells were sampled using manually operated inertial lift pumps, which are known to show high retention of dissolved volatiles relative to other pump designs (Barker and Dickhout, 1988). Dissolved gases were stripped from ground-water samples by standard headspace equilibration techniques (McAuliffe, 1971; Oremland *et al.*, 1987). Compositional analyses were typically completed within 48 hours of sample collection. Isotopic analyses were generally performed within 14 days. Water extracted from deep ground-water wells exhibited a high degree of gas ebullition due to depressurization upon reaching the surface. Gas bubbles were collected and isotopic analyses performed to establish whether fractionation

associated with gas ebullition could significantly alter the isotopic signature of ground water extracted from deep wells. Compositional analyses were performed on bubble samples in order to characterize their hydrocarbon composition.

2. Shallow ground-water wells terminate either in the Iroquois sands or in the underlying till. Of these wells, GW2, GW3, GW4, GW5, GW11, SG1, SG3, SG4, and SG5 were sampled in order to provide as complete an areal coverage of the landfill site as possible (Fig. 1). In addition, a single well in the northern part of the perimeter ditch was sampled in order to characterize shallow ground water down-gradient from the refuse mound. Sampling procedures for shallow ground-water wells were similar to those described above for deep wells, with the exception that gas ebullition was not observed.

3. Two gas wells terminating in the refuse (TWA, TWB; Fig. 1) were sampled, using a well packer, in order to indicate the composition and isotopic signature of landfill gas at its source.

4. Shallow soil gas probes (SGP; Fig. 1) terminate in the landfill cover. Soil gas probes consisted of capped 19-mm polyvinyl chloride tubing inserted into the landfill cover. Samples from these probes were obtained in order to determine whether any compositional or isotopic differences existed between gas produced within the refuse itself and gas within the overlying landfill cover.

Analytical Procedures
Gas concentration analyses were performed using a Varian 3300 gas chromatograph equipped with a flame ionization detector (FID). Isotopic analyses were performed at the Stable Isotope Laboratory of the Department of Geology, University of Toronto, on a Finnigan MAT 252 gas-source mass spectrometer interfaced to a Varian 3400 gas chromatograph. Methane and CO_2 contained in each sample were separated by the gas chromatograph at 25°C. CH_4 passing through the combustion oven was converted to CO_2 at 950°C and subsequently analyzed for stable carbon isotope ratio by the mass spectrometer. All carbon isotope ratios were expressed as ‰ difference from Pee Dee Belemnite (PDB).

RESULTS

Concentration Analyses
Dissolved CH_4, C_2H_6, and C_3H_8 concentrations in deep ground-water wells are reported in Table 1. CH_4 concentrations in all deep ground-water wells were in excess of 500 μM. The maximum CH_4 concentration for any deep well was recorded in GW15, the off-site well located up-gradient of the landfill (Fig. 1). Dissolved CH_4 concentration in ground water

Table 1   Isotopic and compositional data for deep ground-water wells. Accuracy and reproducibility for $\delta^{13}C$ measurements are ≤0.5‰. Relative error for CH_4 concentration measurements is ±0.5%. Relative error for all C_3H_8 concentration measurements is ±5%.

Well	$\delta^{13}C$ (‰ relative to PDB)	CH_4 concentration (μM)	C_2H_6 concentration (μM)	C_3H_8 concentration (μM)	$C_1/(C_2 + C_3)$
GW6–I	−65.6	562	0.139 ± 5%	0.0418	3.2×10^3 ± 4%
GW7–I	−65.5	566	0.123 ± 5%	0.0034	4.6×10^3 ± 4%
GW7–II	−64.5	508	0.069 ± 1%	0.002	7.2×10^3 ± 1%
GW8–I	−61.9	586	0.177 ± 1%	0.0023	3.3×10^3 ± 1%
GW13–I	−72.2	577	0.028 ± 5%	not detected	2.1×10^4 ± 4%
GW14–I	−69.0	549	0.032 ± 5%	not detected	1.7×10^4 ± 4%
GW15–I	−70.7	693	0.075 ± 5%	not detected	9.3×10^3 ± 4%

from this well was 693 µM, a value that exceeded the highest on-site concentration by over 100 µM. No correlation existed between methane concentration and well depth for deep ground-water wells.

Dissolved CH_4 concentrations in shallow ground-water wells were several orders of magnitude less than those observed within deep wells (Table 2). The highest CH_4 concentration recorded for a shallow well was 490 µM, detected in a shallow well within the northern part of the perimeter ditch. In general, CH_4 concentrations in shallow ground waters were <1 µM. Overall CH_4 concentrations in ground waters from shallow wells were too low for $\delta^{13}C$ isotope measurements. As with deep ground-water wells, methane concentrations in shallow wells were not strongly correlated with depth (Fig. 2).

Gas phase CH_4 concentrations in shallow gas probes ranged from 38 vol. % at SGP-4 to 64.5 vol. % at SGP-5 (Table 3). Samples obtained from the two 152-mm-diameter gas wells, TWA and TWB, exhibited CH_4 concentrations of 61 and 60 vol. %, respectively.

C_2H_6 (ethane) concentrations in deep ground-waters were <0.02 µM for all wells, ranging from 0.177 µM to 0.032 µM (Table 1). No correlation could be found between C_2H_6 concentration and well depth for deep wells. Ethane was only detected at SG3 and along the perimeter ditch (Table 2). Concentrations were within the range observed for deep ground-water wells. The occurrence of ethane in these wells could not be attributed to any difference in depth or substrate relative to the remaining shallow wells. Soil gas probes and gas wells exhibited a range of gas phase ethane concentrations (Table 3).

C_3H_8 (propane) was detected in four of the seven deep ground-water wells sampled (Table 1). All of these wells are located south of the northernmost perimeter ditch (Fig. 1). No correlation was found between concentration and depth for these four wells. Propane could not be detected in any of the

Table 2 Compositional data for shallow ground-water wells.

Well	CH_4 Concentration (µM)	C_2H_6 Concentration (µM)
BR4–I	3.65 ± 2%	Not detected
GW2–I	0.045 ± 5%	Not detected
GW3–I	0.027 ± 5%	Not detected
GW4–I	0.027 ± 5%	Not detected
GW5–I	0.036 ± 5%	Not detected
GW11–I	0.028 ± 5%	Not detected
GW17–III	0.17 ± 5%	Not detected
SG1	0.75 ± 2%	Not detected
SG3	4.30 ± 2%	0.022 ± 5%
SG4	1.26 ± 2%	Not detected
SG5	0.129 ± 2%	Not detected
Perimeter Ditch	490 ± 5%	0.013 ± 5%

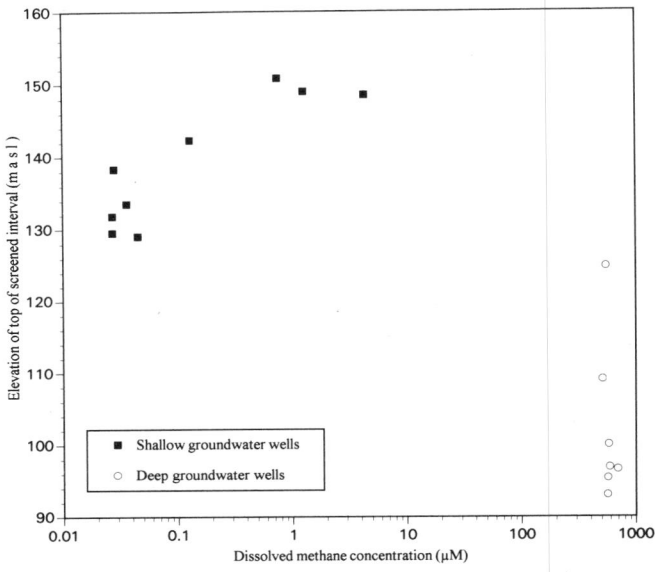

Figure 2 *Depth profile of dissolved CH_4 concentration, showing values for deep ground waters that are 2 to 4 orders of magnitude greater than concentrations in shallow ground waters. Elevation in metres above sea level (m asl).*

Table 3 Isotopic and compositional data for gas probes and gas wells. Accuracy and reproducibility for $\delta^{13}C$ measurements are ≤0.5‰ or ≤1‰ if noted (*). Relative error for CH_4 concentration measurements is ±0.8%, unless otherwise indicated. Relative error for C_2H_6 concentration measurements is ±5%. Relative error for C_3H_8 concentration measurements is ±5%.

Well	$\delta^{13}C$ (‰ relative to PDB)	CH_4 concentration (vol. %)	C_2H_6 concentration (ppmv)	C_3H_8 concentration (ppmv)	$C_1/(C_2 + C_3)$
SGP–1	−60.2	64.4	62	73	4.8×10^3 ± 5%
SGP–2	−60.4	64.4	16	7.0	2.8×10^4 ± 5%
SGP–3	−61 *	64 ± 1.5%	16	11	n/a
SGP–4	−60.1	38.0	< 2	< 2	2.15×10^5 ± 5%
SGP–5	−61.9	64.5	3.0	< 2	2.1×10^3 ± 5%
SGP–7	−59.9	64.0	286	25	1.07×10^5 ± 7%
SGP–8	−60.1	53.7	< 2	5.0	n/a
SGP–9	−60.8	64.0	< 2	< 2	5.3×10^4 ± 5%
SGP–10	−59.8	63.1	3.0	9.0	4.0×10^4 ± 5%
SGP–11	−59 *	63.4	3.0	13	4.5×10^4 ± 5%
SGP–13	−59 *	63 ± 1.5%	3.0	11	6.9×10^4 ± 5%
SGP–15	−59 *	62.5	3.0	6.0	6.8×10^4 ± 5%
TWA	−61.3	61 ± 1.5%	6.0	3.0	4.3×10^4 ± 5%
TWB	−60.2	60 ± 1.5%	3.0	3.0	

Note: n/a = not applicable (not calculated).

samples obtained from shallow ground-water wells. Among the soil gas probe samples, propane concentrations ranged from less than 2 ppm at each of SGP-4 and SGP-5 to a maximum of 73 ppm at SGP-1; the measured propane concentration in both the TWA and TWB wells was 3 ppm (Table 3).

$C_1/(C_2+C_3)$ Alkane Ratios

Figure 3 illustrates the distribution of $C_1/(C_2+C_3)$ ratios for the deep ground-water wells and soil gas probes. Given that ethane and propane could not be detected in the shallow ground-water wells, $C_1/(C_2+C_3)$ ratios could not be calculated. The ratios calculated for deep ground-water samples ranged from 3.2×10^3 to 2.1×10^4 (Table 1). Ratios for wells north of the perimeter ditch were up to 6.6 times greater than those for deep ground waters collected from wells south of the ditch.

$C_1/(C_2+C_3)$ ratios for gas probe samples varied from 2.1×10^4 at SGP-7 to 2.15×10^5 at SGP-5 (Table 3). Ratios calculated for samples collected from the TWA and TWB gas wells fell within the same range. With the exception of SGP-1 and SGP-7, ratios for the gas probes were all greater than 2.4×10^4. The reasons for this discrepancy between SGP-1 and SGP-7 and the other gas wells is not clear; contamination of the samples by ambient air may have occurred. In contrast, all deep ground-water samples have $C_1/(C_2+C_3)$ less than 2.4×10^4. There is thus a clear separation of samples into two groups on the basis of $C_1/(C_2+C_3)$ ratios, with ground waters exhibiting consistently lower ratios than gas samples.

Carbon Isotope Ratios

Samples from deep ground-water wells have different ranges of $\delta^{13}C_{CH4}$ isotopic signatures depending on their location with respect to the landfill site. Values of $\delta^{13}C$ of methane in deep ground-water samples range from $-72.2‰$ at GW13 to $-61.9‰$ at GW8 (Table 1). Wells GW13, GW14, and GW15, located north of the perimeter ditch (Fig. 1), contained the

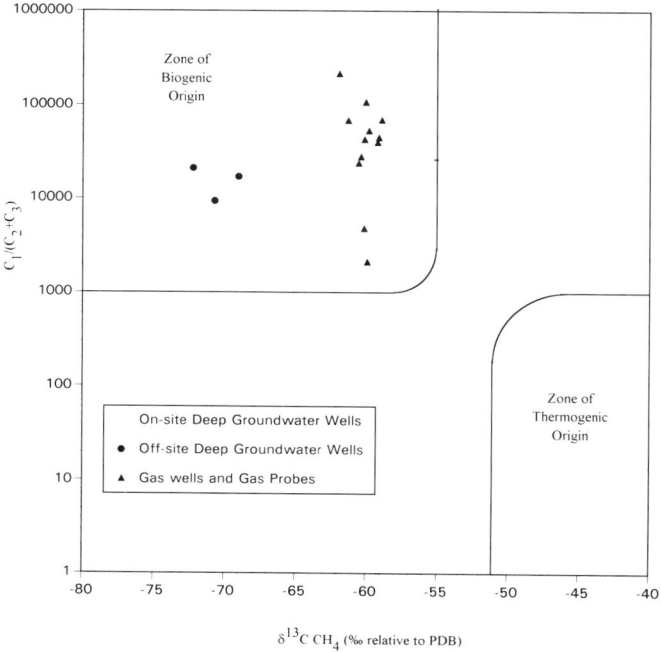

Figure 3 $\delta^{13}C_{CH4}$ versus $C_1/(C_2+C_3)$ *for on-site deep ground-water wells, off-site deep ground-water wells, and soil gas probes and gas wells, compared to the bacterial and thermogenic ranges of Bernard* et al. *(1977).*

most ^{13}C-depleted CH_4; the mean $\delta^{13}C$ for these three wells was $-71 \pm 2‰$, whereas methane in the remaining deep ground-water wells underlying the landfill site had a mean $\delta^{13}C$ of -64 ± 2 (Fig. 3). This difference in the carbon isotopic signature could not be attributed to spatial variations occurring with depth, as no correlation exists between well depth and $\delta^{13}C$. Given that the most ^{13}C-depleted ground-water samples were found within the area up-gradient of the landfill, it appears that this isotopic shift toward more ^{13}C-enriched values may simply reflect natural variation in $\delta^{13}C$ within ground waters underlying the landfill.

Samples from soil gas probes or gas wells within the landfill itself exhibit a $\delta^{13}C_{CH4}$ isotopic signature distinctly different from the ground-water wells. Methane in samples collected from the soil gas probes and gas wells exhibited $\delta^{13}C$ values approximately 5-10‰ more positive than those of dissolved CH_4 within the deep ground-water samples. Soil gas probes and gas wells showed considerably less variability in the $\delta^{13}C_{CH4}$ signature than did ground-water samples. Mean $\delta^{13}C$ for the twelve soil gas probes and two gas wells was $-60.2 \pm 0.9‰$. The close correspondence between the $\delta^{13}C$ signature of methane in the shallow gas probes and that of methane obtained from the gas wells terminating deeper within the landfill mound suggests that oxidation at shallower levels within the landfill does not alter the isotopic signature of landfill-generated CH_4.

DISCRIMINATION OF LANDFILL-DERIVED AND FORMATIONAL METHANE

Based on $C_1/(C_2+C_3)$ ratios, landfill-derived gases and formational gases (*i.e.*, those dissolved gases contained within the formations underlying the landfill) are derived from distinct methane sources. As shown in Figure 3, all but two of the gas well samples had $C_1/(C_2+C_3)$ values $>2.4 \times 10^4$. In contrast, all seven ground-water samples were characterized by $C_1/(C_2+C_3) \leq 2.1 \times 10^4$, and thus show a clear separation with respect to the gas well samples.

The distribution of carbon isotope ratios also suggests that ground-water and gas well samples represent separate gas reservoirs. All ground-water samples had $\delta^{13}C_{CH4} \leq -61.9‰$, whereas gas wells all had $\delta^{13}C_{CH4} \geq -61.9‰$. In addition to their more positive isotopic signature, gas well samples exhibited considerably less variability than did ground-water samples. Gas well $\delta^{13}C_{CH4}$ values fell between $-61.9‰$ and $-59‰$, representing a range of $<3‰$. Conversely, ground water $\delta^{13}C_{CH4}$ possessed a wider range of values, extending from $-72.2‰$ to $-61.9‰$. These results indicate two distinct sources of CH_4 in the subsurface at Beare Road landfill, an extremely isotopically depleted CH_4 in the ground-water wells derived from Pleistocene strata or Paleozoic shales (formational gas) and a more ^{13}C enriched CH_4 derived from the landfill itself.

Origin of Formational Gas

Formational gases are a CH_4 source separate from the landfill itself, and one possibility is that the formational CH_4 is generated thermogenically within the Whitby Shale, and is carried into the overlying Pleistocene strata through diffusion or advection in ground water. This possibility is unlikely, however, as $\delta^{13}C$ signatures of the dissolved CH_4 in deep ground-water samples differ markedly from that of CH_4 sourced in strata of similar age and lithology to the Whitby Shale. Sherwood Lollar et al. (1994) identified that thermogenic CH_4 from wells terminating in Ordovician carbonates and shales in southern and southwestern Ontario is characterized by $\delta^{13}C$ values ranging from −40.9‰ to −30.0‰ (Fig. 2). Such values are between 21.0‰ and 42.2‰ more ^{13}C enriched than CH_4 within deep ground waters at Beare Road. Therefore, CH_4 in the deep ground-water wells is not derived from thermogenic production in the Whitby Shale.

Bacterial activity within the Scarborough and Don formations represents a more likely origin for the formational CH_4 found in the study area. The high organic content of both geological units, and in particular the presence of large amounts of detrital peat, provides a suitable substrate for bacterial methanogenesis. A recent bacterial origin is consistent with both the $\delta^{13}C$ and $C_1/(C_2+C_3)$ signatures of this dissolved gas, both of which indicate biogenic CH_4 formation (Fig. 3). A bacterial origin for CH_4 in these deep ground waters is further supported by the similarity between the range of $\delta^{13}C_{CH4}$ observed in this study and that for bacterial CH_4 within the Alliston aquifer of southern Ontario, consisting of

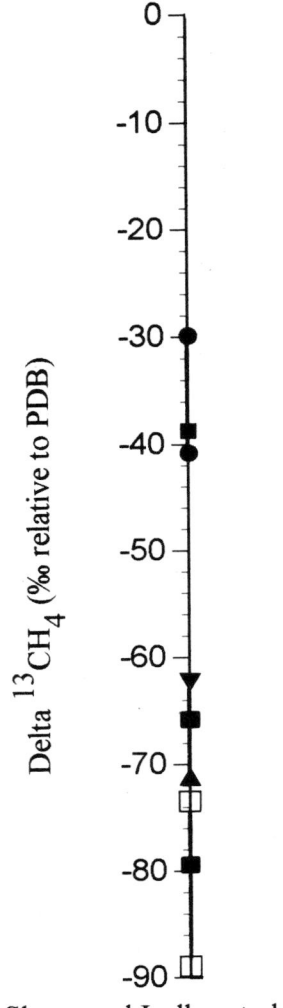

● Sherwood Lollar et al. (1994)

□ Aravena and Wassenaar (1993)

▲ This Study

Figure 4 *Range of $\delta^{13}C_{CH4}$ in deep ground-water samples from Beare Road landfill, compared to ranges reported for thermogenic methane (Sherwood Lollar et al., 1994) and for bacterially produced methane from Pleistocene strata (Aravena and Wassenaar, 1993). Solid squares indicate mean values for each range.*

organic-rich glaciolacustrine silts and sands (Aravena and Wassenaar, 1993; Fig. 4). Dissolved CH_4 generated by bacterial activity in the Scarborough and Don formations underlying Beare Road may undergo advective or diffusive transport into fractures in the underlying Whitby Shale, thus accounting for its occurrence in wells terminating in the shale. Aravena and Wassenaar (1993) invoked such a transport mechanism to account for the presence of extremely ^{13}C-depleted (−89‰ to −70‰) CH_4 in fractured Ordovician bedrock underlying the Alliston aquifer.

DISCUSSION

Methane occurring within Beare Road landfill originates from two distinct sources on the basis of its isotopic and $C_1/(C_2+C_3)$ signatures. Formational CH_4 occurring in the deep aquifers of the Pleistocene Scarborough, Don, and Whitby formations is characterized by extremely ^{13}C-depleted $\delta^{13}C$ ratios ranging from −72.2‰ to −61.9‰. This CH_4 is likely derived from *in situ* bacterial decomposition of organic carbon in the Scarborough and Don, given the similarity of the $\delta^{13}C$ signature to that of bacterial CH_4 identified in other southern Ontario aquifers. The second distinct source of CH_4 at Beare Road is bacterial decomposition within the landfill itself, which is characterized by $\delta^{13}C \geq -61.9‰$. The distinction between the two sources is further supported by compositional differences as indicated by their dissimilar ranges of $C_1/(C_2+C_3)$ ratios.

Given the similarity of the geological setting of Beare Road landfill to many other sites in glaciated terrains, it is possible that the isotopic approach highlighted in this paper may be more widely applied. Once the formational and landfill CH_4 end members have been characterized isotopically and a difference between them recognized, such data can be utilized to determine the extent and pathways of gas migration, permitting a realistic assessment of the potential hazards posed by off-site migration of landfill methane.

ACKNOWLEDGEMENTS

J.M. Lansdown provided invaluable assistance with mass spectrometric analyses. We also wish to thank N. Eyles for the discussions which led to the initiation of this study and K. Murray for providing many of the monitoring reports on Beare Road. R.G. Ferguson and E. Benda of the Metropolitan Toronto Department of Works are thanked for their permission to access monitoring wells. C. Hill and S. Taylor provided valuable field assistance. This work was supported by a Natural Sciences and Engineering Research Council Research Operating Grant to Sherwood Lollar.

REFERENCES

Aravena, R. and Wassenaar, L.I., 1993, Dissolved organic carbon and methane in a regional confined aquifer, southern Ontario, Canada: Carbon isotope evidence for associated subsurface sources: Applied Geochemistry, v. 8, p. 483-494.

Barker, J.F. and Dickhout, R., 1988, An evaluation of some systems for sampling gas-charged ground water for volatile organic analysis: Ground Water Monitoring Review, v. 8, p. 112-120.

Barker, J.F. and Pollock, S.J., 1984, The geochemistry and origin of natural gases in southern Ontario: Bulletin of Canadian Petroleum Geology, v. 32, p. 313-326.

Coleman, D.D. and Meents, W.F., 1976, Identification of leakage gas from underground natural gas storage reservoirs by isotopic analysis of methane: American Association of Petroleum Geologists, Bulletin, v. 60, p. 658.

Domenico, P.A. and Schwartz, F.W., 1990, Physical and Chemical Hydrogeology: Wiley, New York, 824 p.

Games, L.M. and Hayes, J.M., 1977, Carbon isotopic study of the fate of landfill leachate in groundwater: Journal of the Water Pollution Control Federation, v. 49, p. 668-677.

Gartner Lee Limited, 1990, Beare Road site closure report, Volume 1: A Review of monitoring data at Beare Road landfill: 48 p.

Gartner Lee Limited, 1992, Beare landfill 1992 monitoring report: 32 p.

Hydrology Consultants Limited, 1979, Hydrogeological Investigation and proposed remedial works for control of leachate at the Beare Road Sanitary landfill: 30 p.

Mackay, D. and Shiu, W.Y., 1981, A critical review of Henry's Law constants for chemicals of environmental interest: Journal of Physical and Chemical Reference Data, v. 10, p. 1175-1199.

McAuliffe, C., 1971, GC determination of solutes by multiple-phase equilibration: Chemical Technology, v. 1, p. 46-51.

Oremland, R.S., Miller, L.G. and Whiticar, M.J., 1987, Sources and flux of natural gases from Mono Lake, California: Geochimica et Cosmochimica Acta, v. 51, p. 2915-2929.

Rice, D.D. and Ladwig, L.R., 1983, Distinction between in situ biogenic gas and migrated thermogenic gas in groundwater, Denver Basin, Colorado: American Association of Petroleum Geologists, Bulletin, v. 67, p. 539-540.

Schoell, M., 1988, Genetic characterization of natural gases: American Association of Petroleum Geologists, Bulletin, v. 67, p. 2225-2238.

Sherwood Lollar, B., Weise, S.M., Frape, S.K. and Barker, J.F., 1994, Isotopic constraints on the migration of hydrocarbon and helium gases of southwestern Ontario: Bulletin of Canadian Petroleum Geology, v. 42, p. 283-295.

26. Contaminant Geochemistry of Urban Sediments and Soils

Kim A. Bolton

Environmental Earth Sciences, University of Toronto, Scarborough Campus
1265 Military Trail, Scarborough, Ontario M1C 1A4

Les J. Evans

Department of Land Resource Science, University of Guelph, Guelph, Ontario N1G 2W1

SUMMARY
Urban soils and sediments are very prone to environmental degradation through excessive inputs of toxic contaminants. These contaminants arise through both municipal and industrial sources, and include both toxic elements and organic contaminants, such as chlorinated hydrocarbons, pesticides, polychlorinated biphenyls and polycyclic aromatic hydrocarbons. Waters, soils and sediments found within urban areas generally contain elevated amounts of these contaminants. Of particular concern are toxic heavy metals, such as cadmium, mercury and lead, non-metallic elements, such as arsenic and selenium, and non-ionic hydrophobic organic compounds, such as pesticides, and benzene, toluene, ethylbenzene and xylene (referred to collectively as BTEXs).

INTRODUCTION
The term soil is used herein, in an engineering sense, to refer to a wide range of near-surface, unconsolidated materials that underlie urban areas (*e.g.*, glacial deposits, historic landfill, *etc.*). The term sediment is retained for fine-grained materials (*e.g.*, silts, clays) being deposited as a result of urban storm runoff and municipal waste-water discharge to lakes (*e.g.*, Chapter 15). Contamination of urban soils occurs primarily by disposal of wastes *via* atmospheric deposition from industrial emissions and runoff from roads and, to a lesser extent, *via* accidental spills. Sediments commonly become contaminated by discharges from sewage treatment plants (STP), storm-water runoff from roads and combined sewer overflows (CSO). Industrial sources include coal-burning power plants, non-ferrous metal smelters and iron and steel plants (Nriagu and Pacyna, 1988). An example of sources of contaminants for Hamilton Harbour, Ontario is given in Table 1.

Urban inorganic contaminants include the heavy metals, such as cadmium, mercury, lead and zinc, as well as non-

Sources of Pollutants	Suspended Solids	PAHs	Zn	Pb	Cu	Cr	CN
			(% of total loadings)				
Natural							
Lake Ontario	8.1	—	6	7.5	32.6	46.3	—
Creeks	5	0.5	0.8	—	1.6	5	—
Municipal							
Urban Runoff	8.4	11.6	10	20.5	4.1	1.3	—
Sewer Overflow	19.2	44.2	11.1	49.3	20.5	8	—
Cootes Paradise	28.5	4.5	5.1	12.9	9.8	6.3	—
Industrial							
DOFASCO	13.4	26.3	29	5	2.5	15.6	98
STELCO	5.4	1.9	24.3	1.4	7.4	4.1	—
Burlington STP	1.4	0.2	2	0.5	4.1	1.7	—
Hamilton STP	10.7	10.9	11.7	2.5	17.5	11.8	—
Total Loading (kg per day)	44,981	0.89	109	15.2	14.1	31.8	49.7

Table 1   Sources of contaminants in Hamilton Harbour (after Remedial Action Plan, 1991).

metallic elements, such as arsenic and selenium. Urban organic contaminants, particularly the non-ionic hydrophobic organic compounds, include pesticides and other agricultural chemicals; aromatic hydrocarbons derived from petroleum products, such as the polycyclic aromatic hydrocarbons (PAHs); and chlorinated hydrocarbons, such as polychlorinated biphenyls (PCBs) and industrial solvents, such as trichloroethylene.

Interim soil quality (Canadian Council of Ministers of the Environment, 1991) and draft interim sediment quality assessment values (Environment Canada, 1994) are set for some contaminants (Table 2). Soil remediation criteria have been set for three land uses: agricultural, residential/parkland and commercial/industrial. Residential/parkland remediation criteria values are given for some heavy metals and organic contaminants. Two assessment values, a lower threshold effect level (TEL) and a higher probable effect level (PEL), have been calculated for sediments. The TEL represents "the concentration below which adverse effects are expected to occur rarely" and the PEL represents the concentration "above which adverse effects are predicted to occur frequently" (Environment Canada, 1994). Both TEL and PEL values for some contaminants in fresh-water sediments are included in Table 2. Remediation criteria for sediments are lower than for soils; Phyper and Ibbotson (1991) provide a very useful overview of existing guidelines in Ontario.

The formation of insoluble precipitates controls the upper concentrations of most toxic elements in soils and sediments. These precipitates form if the chemical composition of the soil solution or water becomes supersaturated with respect to an insoluble phase. Although the application of equilibrium thermodynamic principles may suggest that a precipitate is present in a soil or sediment, kinetic constraints may often limit its formation.

Soils and sediments act as vast reservoirs for the containment of contaminants because they contain mineral and organic constituents which have surface sites that adsorb contaminant species (Evans, 1989). Contaminants can partition themselves between the various solid phases contained within soils and sediments. These phases include phyllosilicate clays, such as kaolinite and smectite; hydrous oxides of iron, aluminum and manganese, such as goethite, gibbsite and pyrolusite; and humic materials, such as organic matter and aqueous humic colloids (Bruemmer et al., 1986). While solid phases can be important for toxic element retention, only organic matter is important for non-ionic hydrophobic organic contaminants.

The surfaces of soil and sediment particles are chemically active, which allows charged ions to be adsorbed to their surfaces. This can result in both weak interactions (non-specific adsorption) and strong interactions (specific adsorption) between the element and the particle surface (Davis and Kent, 1990). The mobility of elements in soils and sediments is influenced by a number of other chemical factors that include the pH and redox potential of the aqueous phase and the type and content of the other major ions in the soil solution or water (Lumsdon and Evans, 1995). These chemical factors can affect the aqueous speciation (or the forms of occurrence) of the element in soils and sediments. The formation of aqueous metal complexes can affect the retention of elements by soils and sediments in a number of different ways. Complexant anions can compete with soil and sediment surfaces for the metal, resulting in reduced retention of the metal. Alternatively, aqueous metal complexes may be adsorbed by soil and sediment surfaces, increasing the extent of retention (Benjamin and Leckie, 1982; Lumsdon et al., 1995).

TOXIC ELEMENTS

Some of the more important toxic elements include arsenic, cadmium, chromium, copper, mercury, nickel, lead and zinc (As, Cd, Cr, Cu, Hg, Ni, Pb and Zn, respectively). Soils and sediments in proximity to steel industries are often also contaminated with cyanide (CN), the primary source of which is leachate from coke piles.

Content of Toxic Elements in Urban Soils and Sediments

The following provides a very brief overview of the principal toxic elements found in urban areas using a few well-known examples (e.g., Hamilton, Toronto, Montreal and Vancouver). More detailed treatments of contamination problems in these areas are provided elsewhere in the volume (Chapters 2, 15, 16). The purpose here is to provide background for detailed review of those factors controlling toxic elements in urban soils and sediments.

Hamilton Harbour.　　Hamilton Harbour is situated at the west end of Lake Ontario, has Canada's largest steel industries on its south shore, and has an urban population of approximately 300,000 (Chapter 16). The primary industrial

Table 2　　　Interim remediation criteria for soils (after Canadian Council of Ministers of the Environment, 1991) and sediment quality assessment values (TELs and PELs) for fresh-water sediments (after Environment Canada, 1994).

Contaminant	Soil	Sed. TEL	Sed. PEL
Element (mg·kg⁻¹)			
As	30	6	17
Cd	5	0.6	3.5
Cr	250	37	90
Cu	100	36	197
Hg	2	0.2	0.5
Ni	100	18	36
Pb	500	35	91
Zn	500	123	315
Miscellaneous Organics (µg·kg⁻¹)			
PCBs, total	50,000	34	277
Monocyclic Aromatic Hydrocarbons (µg·kg⁻¹)			
benzene	500	—	—
chlorobenzene	1000	—	—
ethylbenzene	5000	—	—
toluene	3000	—	—
xylene	5000	—	—
Polycyclic Aromatic Hydrocarbons (µg·kg⁻¹)			
phenanthrene	5000	42	515
benzo(a)anthracene	1000	32	385
benzo(a)pyrene	1000	32	782
benzo(b)fluoranthene	1000	—	—
benzo(k)fluoranthene	1000	—	—
dibenzo(a,h)anthracene	1000	—	—
naphthalene	5000	—	—
chrysene	—	57	862
fluoranthene	—	111	2355
pyrene	10,000	53	875
Pesticides (µg·kg⁻¹)			
Chlordane	—	4.5	8.9
DDT	—	7.0	4450
Dieldrin	—	2.9	67
Lindane	—	0.6	2.7

sources of contaminants are the Steel Company of Canada (STELCO) and the Dominion Foundry and Steel Company (DOFASCO). Both companies are involved in the production of iron and steel. Contaminants also enter the harbour *via* effluents from two waste-water treatment plants, Burlington STP and Hamilton STP, discharges from municipal storm sewers, and major tributaries such as Cootes Paradise. The shipping of coal, iron ore, steel scrap and petroleum products provides another potential source of contaminants.

Atmospheric deposition of contaminants emitted from industries enters Hamilton Harbour waters directly and appears to be a rather minor source (Harlow and Hodson, 1988). Automotive emissions contaminate associated roadside soils and can enter the harbour water *via* runoff from roads. Soil under the Skyway Bridge at Hamilton Harbour contains high concentrations of chromium and lead (Harlow and Hodson, 1988).

Heavy metal concentrations in nearshore Hamilton Harbour sediment samples are 3-20 times less than those in offshore sediments and higher values are also found in sediments with mean particle sizes less than 50 μm (Poulton *et al.*, 1988). This suggests that the metals are primarily associated with fine particles which presumably stay in suspension longer and are exported farther out into the harbour. Contents of total heavy metal concentrations in Hamilton Harbour suspended sediments at two depths (0.2 m and 20 m) were found to be greater in the lower suspended sediments (Poulton, 1987). Estimates of the proportion of metals existing in particulates greater than 1 μm in diameter were made by analyzing centrifuged supernatant water and comparing the data to total aqueous concentrations. Results indicated that while a significant portion (50 to 80%) of zinc was found in particulates, virtually none of the copper and nickel was bound to particulates. It is possible that copper and nickel existed as dissolved organic or inorganic complexes, or perhaps were associated with colloidal particles less than 1 μm in diameter.

Cyanide concentrations of 0.13 and 0.21 mg·L^{-1} have been measured in runoff from stored coal piles and ore piles, respectively (Harlow and Hodson, 1988). The cyanide loading for Hamilton Harbour has been reduced from 850 kg per day in 1978 to about 50 kg per day in 1989 (Remedial Action Plan, 1991), of which more than 98% is from DOFASCO. Although cyanide concentrations in Hamilton Harbour water are generally below 1 μg·L^{-1} (Harlow and Hodson, 1988), a value of 4 μg·L^{-1} was measured in 1988 (Remedial Action Plan, 1991).

Toronto Waterfront.   Former industrial areas along the Toronto waterfront lying adjacent to the Don River are extensively contaminated. Much of the area is underlain by historic landfill used to reclaim coastal wetland (see Chapter 2). Much data have been collected in the Lower Don Lands in order to characterize soil and water quality. Inorganic contaminants found in the Lower Don Land soils include heavy metals, arsenic and selenium. Maximum values of all of the elements exceed interim remediation criteria (Table 2). Soils in this region are impacted by airborne emissions from industry and automobile exhaust and most contaminants vary widely in their concentration ranges (Table 3). Secondary lead industries, including a smelter which reprocessed lead from used batteries and scrap metal and a manufacturer of new storage batteries, have been implicated in lead levels as high as 21,000 ppm in urban soils of Toronto (Linzon *et al.*, 1976). Soil samples taken from areas of the Port Industrial District of Toronto Harbour contained concentrations, per kilogram of soil, of 340 mg As, 25 mg Cd, 1260 mg Cu, 2.6 mg Hg, 480 mg Ni, 17,800 mg Pb and 6400 mg Zn (Lane, 1992). These concentrations all exceed interim remediation criteria, with As, Cu, Pb and Zn exceeding the criteria by as much as ten times.

A study examining the impact of the Don River on contamination of Toronto Harbour (Bodo, 1989) indicated that suspended-particulate metal concentrations from the Don River are quite similar to concentrations found in Toronto Harbour bottom sediments and in Metro Toronto soils (Table 4). For all elements in suspended sediments of the Don River, in Toronto Harbour bottom sediments and in surficial soils, element concentrations exceed TEL values, and lead and zinc concentrations exceed PEL values in both surface and bottom sediments (Persaud *et al.*, 1985). Element concentrations do not exceed soil remediation criteria (Table 2). Element concentrations are much lower in bottom sediments of the Don River itself. This would indicate that metals are being transported in particulate form.

St. Lawrence River.   The St. Lawrence River drains one of the largest urban industrial complexes in the world. Many of Quebec's industries and half of the provincial population of approximately 6.8 million are located on its shores (Langevin *et al.*, 1992). While many of the contaminants in the river originate from Lake Ontario, there are many urban-industrial point and non-point sources along the river. Two major petrochemical industries are located along the river, one in Cornwall, Ontario and another in Montreal. Sediments located

Table 3   Minimum and maximum concentrations of selected compounds found in soils of the Lower Don Lands (from Beak Consultants Limited and Raven Beck Environmental Ltd., 1994).

	Min. Conc'n. (mg·kg^{-1})	Max. Conc'n. (mg·kg^{-1})
Element		
Ag	n.d.	35
As	n.d.	312
Ba	1.24	2420
Be	n.d.	17.1
Cd	n.d.	98
Co	n.d.	120
Cr	n.d.	1420
Cu	n.d.	29,800
Hg	n.d.	111
Ni	n.d.	14,000
Pb	n.d.	17,800
Sb	n.d.	400
Se	n.d.	40
V	n.d.	9200
Zn	n.d.	24,600
Compound		
PCBs	n.d.	49
phenols	n.d.	31
total VOCs	n.d.	160
benzene	n.d.	10
toluene	n.d.	50
ethylbenzene	n.d.	78
xylene	n.d.	62
Total PAHs	n.d.	10,555
benzo(a)pyrene	n.d.	210
naphthalene	n.d.	7967

Note: n.d. = non-detectable concentrations

along the St. Lawrence River system were analyzed for toxic elements (Langevin *et al.*, 1992) and the following ranges per kilogram of sediment were found: 1.0-38.8 mg As, 1.0-9.2 mg Cd, 22.2-154.0 mg Cr, 8.3-164.0 mg Ni and 10.0-103.0 mg Pb. Most of the sediment element concentrations exceeded TEL values and many also exceeded PEL values. The highest sediment concentrations were found in samples taken from Lake Memphrémagog, which drains into the river from its southeast location. The Lake Memphrémagog watershed has suffered less industrialization than the river itself; however, past mining operations in the Sherbrooke area have likely led to the degradation of sediment quality.

Windsor.   Windsor is a highly industrialized city situated in the Lake Huron-Lake Erie corridor of southern Ontario. Contaminant sources for this urban area include industrial emissions from Windsor, Detroit and Michigan, as well as traffic associated with these urban areas. The concentrations of cadmium and lead in soils in the city of Windsor were compared to concentrations found in soils of the surrounding rural areas (Weis and Barclay, 1985). Cadmium and lead concentrations of city soils averaged 0.62 and 44.8 mg·kg^{-1}, respectively. While these concentrations do not exceed remediation criteria (Table 2), they were approximately 2 and 4 times larger, respectively, than concentrations found in rural agricultural soils, and were believed to be directly related to emissions from automobile exhausts (Weis and Barclay, 1985).

Vancouver.   About 1.7 million people live in the Greater Vancouver area, and this total is expected to increase by at least 600,000 over the next 20 years (Environment Canada, 1992). Between 1965 and 1985 the input of municipal effluent (all sources) to the Fraser River increased 4 times while industrial effluent increased over 400 times. Current water and sediment quality objectives and monitoring data for the Vancouver area have recently been summarized by Delaney and Turner (1994). The principal sources of sediment contamination are sewage outfalls, ocean dumping of dredge spoil and wood wastes, suphide mine tailings, waste incineration and a wide range of port activities such as marine traffic, ship building and urban runoff (Brothers, 1990; Swain and Walton, 1991; Missler, 1992). Some of these activities are described elsewhere in this volume (Chapter 19). There are four main sewage outfalls near Vancouver; daily metal loadings from individual outfalls are as high as 14.7 kg Pb, 50.5 kg Cu and 45.9 kg Zn. Remediation criteria for contaminated soil and sediment are presented in British Columbia Ministry of the Environment (1990).

Factors Controlling Element Content in Soils and Sediments

Role of Precipitates

The total concentration of toxic elements in waters is usually very low, generally less than 100 ppb and often less than 1 ppb. The maximum amount of any element in an aqueous environment is controlled by the solubility of precipitates which contain that element. These secondary minerals form if the chemical composition of the soil solution or interstitial water becomes supersaturated with respect to an insoluble phase. In soils and sediments, the most important precipitates controlling the behaviour of toxic elements are the oxides, hydroxides, carbonates, sulphides and, to a lesser extent, phosphates (Table 5).

A number of toxic elements, silver (Ag), copper, mercury, lead, tin (in the form Sn(IV)) and thallium (in the form Tl(III)), can form insoluble oxides under ambient conditions. However, the presence of these oxides in soils and sediments has yet to be unequivocally confirmed. Insoluble hydroxides of beryllium (Be), cadmium, chromium, lead, nickel and zinc have been shown to form under suitable conditions, with the hydroxides of beryllium and lead probably being important sinks for these metals in soils and sediments.

Insoluble carbonates are formed by barium (Ba), cadmium, copper, lead, nickel and zinc, with the carbonates of lead and, to a lesser extent, cadmium being important sinks for these metals. By contrast with oxides and hydroxides, the dissolution of carbonates in solutions is controlled not only by the solubility products of the individual carbonates and the pH of the water, but also by the partial pressure of carbon dioxide gas. Hence, microbial respiration in soils and sediments can influence the formation of carbonate precipitates.

Sulphide minerals are only stable in anoxic soils and sediments where conditions are sufficiently reducing to form S(-II) species. Under such conditions these precipitates control the upper concentrations of Ag, Cd, Ni and Zn and, particularly, As, Hg and Sb (antimony) in solution.

Table 4   **Element concentrations for Don River suspended particulates *(SP)*, Don River and Toronto Harbour bottom sediments *(BS)*, and Metro Toronto surficial soils (after Bodo, 1989).**

Element	Don River SP Median	Don River SP Range		Don River BS [1]	Don River BS [2]	Toronto Harbour BS [3]	Surficial Soils [4]
					(mg·kg^{-1})		
As	5.4	3.2	– 8.54	0.8	—	—	8.4
Cd	2	0.55	– 6.2	0.3	—	—	2.3
Cr	76	45	– 170	9.0	12.0	69.0	—
Cu	105	45	– 220	8.5	4.8	63.0	33.2
Hg	0.18	0.06	– 0.56	0.015	0.0045	0.23	—
Ni	30	25	– 41	6.7	6.3	29.0	34.8
Pb	200	65	– 290	13.5	33.5	220.0	292.0
Zn	370	54	– 550	33.5	50.0	300.0	154.0

Notes: [1]   Median of five to six observations at Todmorden Mills (1981-87).
            [2]   Median of four observations; 0-10 cm; Keating Channel 1980 (Persaud *et al.*, 1985).
            [3]   Median: 0-3 cm; 10 sites; Toronto Harbour 1976 (Persaud *et al.*, 1985).
            [4]   Mean concentration: 0-2.5 cm; 65 stations; Metro Toronto 1971 reported by Linzon *et al.* (1976).

The contents of toxic elements in waters associated with Hamilton Harbour, for example, are considerably higher than those in offshore Lake Ontario (Table 6). Increases range from approximately two to three times for arsenic, copper and mercury to up to 25 times for zinc. Calculations using the thermodynamic data base of Wagman *et al.* (1982) suggested that all the metals are under saturated with respect to the solid phases shown in Table 5, except for lead. Its concentration approaches that expected for solutions in equilibrium with $Pb(OH)_2$.

Role of Fine Sediment Particles and Organic Material

The surfaces of fine-grained soil and sediment mineral particles contain surface sites which are electrochemically active. The intrinsic charges on these mineral surfaces originate by two different processes of charge generation and may be either positive or negative. Permanent negative charges arise from structural imperfections within the unit structures of phyllosilicate clays, such as smectite and vermiculite, whereas pH-dependent charges are generated by proton association-dissociation reactions at the edges of phyllosilicate clays and at the surfaces of oxide minerals (Sposito, 1984). Oxide minerals include oxides such as hematite, Fe_2O_3 and birnessite, $\delta\text{-}MnO_2$; hydroxides such as gibbsite, $Al(OH)_3$; and oxyhydroxides such as goethite, $\alpha\text{-}FeO(OH)$.

The edges of clays and oxide minerals accept and donate protons and assume negative, neutral or positive charges (Davis and Kent, 1990). At the point of net zero charge (PNZC) the edges are neutral, whereas at pHs below the PNZC the sites are positively charged, and negatively charged at pHs above the PNZC. These sites can thus adsorb both cations and anions.

Many toxic anions are strongly held by mineral colloids and their retention involves the formation of covalent bonds with the surfaces of the minerals through a process of ligand exchange to form surface complexes (Hayes *et al.*, 1988). Many metallic cations, such as the toxic heavy metals, can form strong complexes with pH-dependent charged mineral surfaces found in soils and sediments, whereas others, such as Ca^{2+}, Mg^{2+} and Ba^{2+}, form weak complexes (Hayes and Leckie, 1987). Specific adsorption of cations onto mineral surfaces occurs most readily for those metals that hydrolyze in water, and there is thus a strong significant linear free-energy relationship between cation hydrolysis constants and intrinsic complexation constants (Dzombak and Morel, 1990). The extent of adsorption is dependent on pH, and increases to a maximum as the pH is raised.

The organic matter contained within soils and sediments is also an important environmental sink for the retention of toxic metals, because it contains functional groups that can act as complexant organic ligands. These organic colloids have pH-dependent negative charges which are contributed by their associated functional groups. The most important of these functional groups are the carboxylic and phenolic groups, which dissociate to release protons and generate negatively charged sites. Metallic cations, such as Pb^{2+} and Hg^{2+}, are strongly held by organic colloids, whereas Cd^{2+} is weakly held.

Role of Solution Composition

The extent of adsorption of metals onto soil and sediment particles depends not only on pH but also on the aqueous speciation of the element. Elements may exist as free ionic species, such as Zn^{2+} and Pb^{2+}, or as complexes with various inorganic and organic ligands. Complexes with inorganic ligands include hydroxo species, such as $ZnOH^+$ and $Al(OH)_4^-$; carbonate complexes, such as $CuCO_3(aq)$; and chloride complexes, such as $HgCl_2(aq)$ and $CdCl^+$. Organic complexes may be formed with relatively simple aliphatic or aromatic acids, such as citric, oxalic or gallic acids, or with more structurally complex humic substances.

Chloride complexes are particularly important in that they generally decrease metal retention by soil and sediment particles (Balistrieri and Murray, 1982). The complexes occur in significant amounts in marine sediments or in soils and fresh-water sediments which are exposed to high-chloride-containing influents. Of particular concern in urban areas is the discharge of NaCl, which is used for snow and ice removal, onto soils and into waterways (Bubeck *et al.*, 1971). Warren and Zimmerman (1994) found that increased NaCl concentrations led to decreased accumulation of cadmium, copper and zinc in the suspended sediments in the Don River, Toronto. The use of de-icing salts on roads has been suggested as a cause of the increased mobilization of lead in soils (Howard and Sova, 1993). Numerous studies have also shown decreased cadmium adsorption in soils in the presence of chloride ions (O'Connor *et al.*, 1984; Boekhold *et al.*, 1993; Lumsdon *et al.*, 1995).

Fractionation of Toxic Elements in Soils and Sediments

Toxic elements can partition themselves between the aqueous phase and the various solid phases contained within soils and sediments. Chemical fractionation schemes have been used to partition the various forms of trace elements in soils and sediments (Ducaroir *et al.*, 1990). However, studies on the use of fractionation schemes on contaminated urban soil samples are scarce in the literature. All of the schemes assume that a particular form of the element can be removed

---

Table 5   Possible toxic element precipitates in soils and sediments.

Element	Oxide	Hydroxide	Carbonate	Sulphide
As	As_2O_3			As_2S_3
Ba			$BaCO_3$	
Cd	CdO	$Cd(OH)_2$	$CdCO_3$	CdS
Co	CoO	$Co(OH)_2$		
Cr	Cr_2O_3	$Cr_2O_3 \cdot H_2O$		
Cu	CuO		$Cu_2(CO_3)(OH)_2$	CuS
Hg	HgO			HgS
Ni	NiO	$Ni(OH)_2$	$NiCO_3$	NiS
Pb	PbO	$Pb(OH)_2$	$PbCO_3$	PbS
Zn	ZnO	$Zn(OH)_2$	$ZnCO_3$	ZnS

Table 6   Contents of toxic elements in waters of Hamilton Harbour and Lake Ontario (from Harlow and Hudson, 1988).

Element ($\mu g \cdot L^{-1}$)	Hamilton Harbour	Lake Ontario
As	1	0.55
Cd	0.3	0.05
Cr	6	0.55
Cu	3.5	1.30
Hg	0.04	0.025
Pb	5	0.39
Ni	10	1.65
Zn	30	1.2

using sequentially more aggressive reagents. A detailed discussion of the benefits and the limitations associated with these schemes is given in Kersten and Förstner (1995). It should be noted that the fractions determined in all of the schemes represent operationally defined forms rather than actual soil and sediment species. Possible fractions of trace elements in natural environments include the soluble fraction, the exchangeable (or non-specifically adsorbed) fraction, the specifically adsorbed fraction, the organically bound fraction and the residual fraction. A commonly used extraction procedure is shown in Table 7.

Poulton *et al.* (1988) used a fractionation scheme similar to that of Tessier *et al.* (1979, 1980) to separate metal fractions in Hamilton Harbour sediments. The study found that most of the metals were associated with the organically bound and the residual fraction (Table 8). Large amounts of copper were associated with the organically bound fraction due to the stability of organic-copper complexes (Moore, 1990). However, the very high proportion of cadmium associated with the organically bound fraction was unexpected.

Tessier *et al.* (1980) applied a fractionation scheme to suspended sediments of the Yamaska and St. François Rivers. Both rivers are located in southeastern Quebec and are tributaries to the St. Lawrence River. Very small percentages of all the metals, except for Cd (26% and 28% for Yamaska and St. François river sediments, respectively), were found in the exchangeable form. A large percentage of Cd (49% and 31%) and smaller percentages of Cu (8% and

14%), Ni (3.9% and 3.8%), Pb (18% and 18%) and Zn (21% and 24%) were found bound to carbonates. The iron-manganese oxide bound fraction was very important for Pb (46% and 49%) and Zn (41% and 39%) and of lesser importance for Cd (12% and 13%), Cu (20% and 12%) and Ni (14% and 24%). Copper (31% and 52%) and, to a lesser extent, Pb (19% and 20%) showed a very strong affiliation for the organic fraction. Nickel (4.4% and 6.2%) and Zn (5% and 6%) had very small percentages associated with the organic fraction, and Cd bound to organic matter was undetectable.

In contrast, sediments contaminated with toxic elements in the Dalhousie and Belledune harbours in New Brunswick were found to be associated with the iron-manganese oxide bound and residual fractions (Samant *et al.*, 1990). Iron-manganese oxides were particularly important sinks for cobalt (Co), chromium and lead. It is difficult to make a comparison between this study and the previous two above as a different sequential extraction procedure was used.

ORGANIC CONTAMINANTS

Organic contaminants in soils and sediments of urban areas are related to the types of industry found within the catchment area and also depend on the extent of urban waste-water treatment in the area. Non-point sources of organic pollutants include shipping wastes and atmospheric inputs from industrial emissions, sludge incinerators and automotive emissions. A list of organic contaminants found in urban areas would be exhaustive. However, some commonly found con-

---

Table 7   The fractionation scheme of Tessier *et al.* (1979) for trace elements in soils and sediments.

Fractions	Possible forms of trace element	Possible extraction mechanisms involved	Extractants
Exchangeable	Retained by phyllosilicate clay minerals.	Displaced by mass reaction	$MgCl_2$
Carbonate bound	Adsorbed to the surfaces of carbonates.	Dissolution	NaOAc, pH 5
Oxide bound	Specifically adsorbed or isomorphously substituted into Fe, Al and Mn oxides and oxyhydroxides.	Acidic dissolution and/or complexation	$NH_2OH \cdot HCl/HOAc$
Organically bound	Chelated by organic materials.	Oxidation	H_2O_2/NH_4OAc
Residual	Isomorphously substituted into phyllosilicate clay minerals and/or primary and secondary minerals.	Acid destruction of mineral lattice structure	$HF-HClO_4$

Table 8   Mean percentages of total metal concentrations in sequential extracts of Hamilton Harbour nearshore sediments (from Poulton *et al.*, 1988).

Metal	Exchangeable	Carbonate-bound	Fe-Mn Oxide-bound (fraction [%])	Organic-bound	Residual
Cd	0.3	0.0	0.0	67	32
Cr	0.0	0.0	1.1	11	88
Cu	0.0	0.1	0.2	40	60
Ni	0.7	0.0	0.0	7.4	91.9
Pb	0.0	0.0	0.0	24	76
Zn	0.7	0.11	0.6	20	78

taminants are listed in Table 9. While the chemistry of organic pollutants is rather complex, probably the most studied group is the non-ionic hydrophobic molecules. With the exception of the chlorinated phenols and the phthalate esters, all of the compounds listed in Table 9 fall into this category.

Content of Organic Contaminants in Soils and Sediments

The Hamilton steel industry is a known source of many organic chemicals including phenols, PAHs and oil and grease in Hamilton Harbour. The harbour also receives loadings of chlorinated organic compounds from waste-water treatment plant effluents and atmospheric deposition from sewage sludge incineration (Harlow and Hodson, 1988). Non-point sources of organic contaminants include shipping discharges and atmospheric deposition.

The total PAH loading to Hamilton Harbour from point sources was 1.8 kg per day in 1986 and 0.89 kg per day in 1988, indicating that waste-water treatment has improved over the last few years (Remedial Action Plan, 1991). In 1988, major sources of PAHs included CSO (44%), DOFASCO (26%), urban runoff (12%) and the Hamilton sewage treatment plant (11%).

Metcalfe *et al.* (1990) analyzed for PCBs, organochlorine pesticides and PAHs in sediments from Hamilton Harbour and from control sites in non-polluted areas of Ontario (Table 10). Low concentrations ($\mu g \cdot kg^{-1}$ of PCBs and organochlorine pesticides) were found in all the sediment samples. While PAH concentrations varied widely, concentrations of some PAHs greatly exceeded PEL values in some samples (Table 2). PAH concentrations were generally much higher in the

harbour sediments than in the control sediments.

The presence of PAHs in suspended sediments, taken from 1 m below the water surface (surface sediments) and 1 m above the sediment-water interface (bottom sediments), has also been studied (Mayer and Nagy, 1992). Total PAH concentrations ranged from 4.41 mg to 69.35 mg per kilogram of surface sediments, and from 7.37 mg to 106.02 mg per kilogram of bottom sediments. Highest concentrations were found in samples taken from sites close to industrial areas and, with a few exceptions, concentrations were higher in the surface than in the bottom sediments, indicating an industrial point source. Mayer and Nagy (1992) also reported PAH concentrations of 26.4 mg to 89.8 mg per kilogram of benthic sediment.

Data for α-benzene hexachloride (α-BHC: an organochlorine pesticide) in harbour water show an even distribution which indicated a non-point source for this pollutant (Poulton, 1987). Lindane (γ-BHC), on the other hand, occurred at higher concentrations (as high as 59 $mg \cdot L^{-1}$) near the Hamilton STP. Concentrations of PCB in sediments varied widely within the harbour. The majority of the high concentrations were found in sediments from deep waters and along the industrial shoreline. Six PAHs were measured in harbour sediments: fluoranthene, perylene, benzo(k)fluoranthene, benzo(a)phrene, benzo(g,h,i)perylene and indene(1,2,3-c,d)pyrene. Total PAH concentrations were 2 to 3 times higher than contents found in deeper Lake Ontario sediments, and the highest concentrations were found closest to the steel mills where point source effluent concentrations were also highest. The ranges of PAHs determined by Poulton (1987) were similar to those found by Metcalfe *et al.* (1990).

The St. Lawrence River system is contaminated with many organic chemicals. Among them are chlorinated organics, PCBs and PAHs (Allan, 1988). High concentrations of PCBs have been found in bottom sediments of Lac St. François, one of the three shallow riverine lakes found along the St. Lawrence. This lake is often considered to be a temporary sink for

---

Table 9   Some organic contaminants in soils, waters and sediments.

Aromatic Hydrocarbons
naphthalene
pyrene
benzo(a)pyrene
dibenzo(a,h)anthracene
phenanthrene

Halogenated Methanes
carbon tetrachloride
trihalomethanes

Chlorinated Hydrocarbons
trichloroethylene
trichlorobenzene
PCBs
PCDDs
PCDFs

Phthalate Esters
di-n-butylphthalate
bis (2-ethylhexyl) phthalate

Chlorinated Phenols
pentachlorophenol

Pesticides
Aldrin and Dieldrin
Chlordane
DDT
Endrin
Heptachlor
Lindane

Table 10   Concentrations of polycyclic aromatic hydrocarbons in sediments collected from Hamilton Harbour compared with non-polluted control sites (from Metcalfe et al., 1990).

Compound	Control Sites	Hamilton Harbour
	---------- ($mg \cdot kg^{-1}$) ----------	
naphthalene	n.d.	2.03
acenaphthylene	n.d.	n.d. – 15.00
acenaphthene	n.d.	n.d. – 4.16
fluorene	n.d.	n.d. – 5.72
phenanthrene	n.d.–0.04	n.d. – 40.80
anthracene	n.d.	n.d. – 11.73
fluoranthene	n.d.–0.03	n.d. – 51.96
pyrene	n.d.–0.02	n.d. – 37.17
benzo(a)anthracene	n.d.–0.01	n.d. – 15.75
benzo(b)fluoranthene	n.d.	n.d. – 14.73
benzo(k)fluoranthene	n.d.	n.d. – 0.69
benzo(a)pyrene	n.d.	n.d. – 4.78
indeno(1,2,2-c,d)pyrene	n.d.	n.d. – 19.11
dibenzo(a,h)anthracene	n.d.	n.d. – 0.32
benzo(g,h,i)perylene	n.d.	n.d. – 14.24

Note: n.d. = non-detectable concentrations

PCBs. Total PCB concentrations were found to be as high as 800 µg·kg⁻¹ (which greatly exceeds the PEL) in 10% of the sediments sampled, and PAH concentrations were found to be as high as 1000 µg·kg⁻¹ (Allan, 1988). The south shore of this lake was the area of highest contamination for both PAHs and PCBs, indicating an upstream source on the United States shore of the river.

Organic contaminants in soils of the Lower Don Lands in the city of Toronto include volatile organic compounds, such as benzene, toluene and xylene, PAHs, PCBs and non-aqueous phase liquids (NAPLs), such as gasoline, lubricating oils, coal tars and creosote (Table 3). However, only naphthalene exceeded remediation criteria (Table 2). Concentrations of many of the PAHs in Port Industrial District soil samples exceeded remediation criteria. Concentrations of naphthalene, phenanthrene, pyrene, benzo(a)anthracene, benzo-(b)fluoranthene, benzo(k)fluoranthene, benzo(a)pyrene and dibenzo(a,h)anthracene in these samples were as high as 25, 24, 23.8, 13.9, 41, 14.2, 14.6 and 7.4 µg·kg⁻¹, respectively (Lane, 1992). PCB concentrations were found to be as much as five times higher in Windsor city soils than in associated rural soils (Sanderson and Weis, 1989).

Retention of Organic Contaminants in Soils and Sediments

The organic matter found in soils and sediments plays a dominant role in the sorption of non-ionic hydrophobic organic contaminants, and affects their rate of transport to surface and ground waters. Sorption by humic colloids is a hydrophobic process, the extent of which depends on both the hydrophobicity of the organic molecule and on the content of organic matter. The retention of hydrophobic organic molecules by soils and sediments increases linearly as the concentration of the solute increases (Karickhoff, 1981). This observation suggests that hydrophobic organic molecules are retained by soils and sediments by a different mechanism from that of hydrophillic molecules.

Sorption by organic matter is envisaged as a partitioning process in which the organic solute permeates, that is dissolves, into the framework of the sorbent organic matter (Chiou et al., 1979). The evidence for such a sorption process, rather than a physical adsorption to mineral particles found in soils and sediments, includes linear sorption isotherms over a considerable concentration range; an inverse relationship between aqueous solubility and the extent of sorption; absence of competitive sorption when different organic non-ionic solutes are present, and a low and exothermic heat of sorption (Rao, 1990).

This partitioning model has been used to explain a large amount of sorption data for a wide variety of natural and synthetic non-ionic organic compounds (Hance, 1969; Karickhoff et al., 1979; Brown and Flagg, 1981; Chiou et al., 1983). The model, however, is not without its critics (Mingelgrin and Gerstl, 1983).

Partition coefficients (K_d) are calculated from a linear plot of the amount of chemical retained against the concentration of the chemical remaining in the aqueous phase. The partition coefficient is then normalized to a constant organic carbon content to give a partition coefficient, K_{oc}. The partition coefficient, K_d, is usually determined by batch experiments in which increasing amounts of the organic compound are added to the soil or sediment and, after a suitable period of incubation, the amount of compound remaining in solution is determined by an appropriate analytical technique. For any particular soil or sediment, K_d can be estimated from a

knowledge of the solute's partition coefficient with organic carbon, K_{oc}, and the content of organic carbon in the soil or sediment.

To date, although there are many calculated values for K_{oc} for organic chemicals (Table 11), the data base is still rather limited. However, results from a large number of studies have shown that there is also a strong positive linear relationship between K_{oc} and the partition coefficient, K_{ow}, of the chemical between octanol and water, and a strong negative linear relationship between K_{oc} and the solubility of the chemical in water (Lyman, 1982). Values for K_{oc} can thus be estimated by reference to suitable empirical equations (Pussemier et al., 1990). These values have also been predicted for non-ionic compounds, by using a molecular topology model to calculate first-order molecular connectivity indexes (Sabjić, 1987), and extended to polar compounds by Meylan et al. (1992). Organic carbon partition coefficients have been used to model the fate of organic chemicals in various media including soils and sediments (Southwood et al., 1989; Mackay and Paterson, 1991).

It was not possible to determine the relative partitioning of organic chemicals in the soils and sediments discussed. Detailed chemical loadings, coupled with soil and sediment concentration data for individual chemicals, were not available for any of the soils or sediments.

DISCUSSION

There is a need to document, in much more detail, contaminant levels in urban soils and sediments in Canada and to ascertain the factors affecting contaminant retention. To date, much research has been focussed on the problem of polluted agricultural soils. Agricultural soils are of public concern due to the associated risk of human exposure to toxic contaminants moving through the food chain. However, contaminated soils and sediments in heavily populated urban environments also have the potential to create serious environmental problems through contamination of drinking-water supplies and direct ingestion.

Toxic Elements

More sophisticated analytical equipment has allowed for the analysis of many toxic elements in soils and sediments at the ppm and ppb level to become routine. However, more data are required on the less frequently measured and studied toxic elements such as beryllium, silver, tantalum and antimony. To evaluate the partitioning of toxic elements in soils and sedi-

Table 11   Organic carbon partition coefficients, K_{oc}, for some organic chemicals (from Lyman, 1987; and Sabjić, 1987).

	K_{oc}
Aromatic hydrocarbons	
naphthalene	1300
pyrene	68,000
dibenzo(a,h)anthracene	2,750,000
Chlorinated Hydrocarbons	
trichloroethylene	50
trichlorobenzene	590
Chlorinated Phenols	
pentachlorophenol	2900
Pesticides	
Lindane	1100
DDT	240,000

ments, the development of a universally accepted fractionation scheme is also required. Comparison of contaminant behaviour in soils and sediments with different chemical and physical composition will then be meaningful. In addition, there is a need to develop computer-based surface complexation models for toxic element retention by both soils and sediments.

Organic Contaminants

The mechanisms involved in the retention of organic contaminants in soils and sediments are well understood. However, the importance of partitioning coefficients is difficult to evaluate without more precise data on chemical loadings coupled with *in situ* soil and sediment concentrations. As well, more information on the biological degradation of organic contaminants is needed, particularly for sediments. Although half lives for many compounds have been determined in laboratory studies, few studies have been conducted under conditions directly applicable to naturally occurring soils and sediments in urban areas.

REFERENCES

Allan, R.J., 1988, Toxic chemical pollution of the St. Lawrence River (Canada) and its upper estuary: Water Science Technology, v. 20, p. 77-88.

Balistrieri, L.S. and Murray, J.W., 1982, The adsorption of Cu, Pb, Zn, and Cd on goethite from major ion sea water: Geochimica et Cosmochimica Acta, v. 46, p. 1253-1265.

Beak Consultants Limited and Raven Beck Environmental Ltd., 1994, Lower Don Lands site characterization and remedial options study: A report for the Waterfront Regeneration Trust: Beak Ref. 2896.1, 150 p.

Benjamin, M.M. and Leckie, L.O., 1982, Effects of complexation by Cl, SO_4, S_2O_3 on adsorption behaviour of Cd on oxide surfaces: Environmental Science and Technology, v. 16, p. 162-170.

Bodo, B.A., 1989, Heavy metals in water and suspended particulates from an urban basin impacting Lake Ontario: Science of the Total Environment, v. 87/88, p.329-344.

Boekhold, A.E., Temminghoff, E.J.M. and van der Zee, S.E.A.T.M., 1993, Influence of electrolyte composition and pH on the cadmium adsorption by an acidic soil: Journal of Soil Science, v. 44, p. 85-96.

British Columbia Ministry of the Environment, 1990, British Columbia standards for managing contamination at the Pacific Place site: 156 p.

Brothers, D.E., 1990, Benthic Sediment Chemistry, B.C. Ocean dumpsites: Environment Canada, Environmental Protection Program, Pacific Region, Program Report DR-90-07, 94 p.

Brown, D.S. and Flagg, E.W., 1981, Empirical prediction of organic pollutant sorption in natural sediments: Journal of Environmental Quality, v. 10, p. 382-386.

Bruemmer, G.W., Gerth, J. and Herms, U., 1986, Heavy metal species, mobility and availability in soils: Zeitschrift für Pflanzenernährung und Bodenkunde, v. 149, p. 382-398.

Bubeck, R.C., Diment, W.H., Deck, B.L., Baldwin, A.L. and Lipton, S.D., 1971, Runoff of deicing salt: effect on Irondequoit Bay, Rochester, New York: Science, v. 172, p. 1128-1132.

Canadian Council of Ministers of the Environment, 1991, Interim Canadian environmental quality criteria for contaminated sites: Canadian Council of Ministers of the Environment, Subcommittee on Environmental Quality Criteria for Contaminated Sites, CCME EPC-CS34, Winnipeg, MB, 20 p.

Chiou, C.T., Peters, L.J. and Freed, V.H., 1979, A physical concept of soil-water equilibria for non-ionic organic compounds: Science, v. 206, p. 831-832.

Chiou, C.T., Porter, P.E. and Schmedding, D.W., 1983, Partition equilibria of non-ionic organic compounds between soil organic matter and water: Environmental Science and Technology, v. 17, p. 227-231.

Davis, J.A. and Kent, D.B., 1990, Surface complexation modelling in aqueous geochemistry, *in* Hochella, M.F. and White, A.F., eds., Mineral-Water Interface Geochemistry: Mineralogical Society of America, Reviews in Mineralogy, v. 23, p. 177-260.

Delaney, T.A. and Turner, R.J.W., 1994, A preliminary directory of trace element databases available in the Vancouver area: Geological Survey of Canada, Bulletin 481, p. 299-316.

Ducaroir, J., Cabier, P., Leydecker, J-P. and Prost, R., 1990, Application of soil fractionation methods to the study of the distribution of pollutant metals: Zeitschrift für Pflanzenernährung und Bodenkunde, v. 153, p. 349-355.

Dzombak, D.A. and Morel, F.M.M., 1990, Surface Complexation Modeling: John Wiley and Sons, New York, 393 p.

Environment Canada, 1992, State of the Environment report for the lower Fraser River basin: Report 92-1, 79 p.

Environment Canada, 1994, Interim Sediment quality assessment values (draft copy): Soil and Sediment Quality Section, Guidelines Division, Ecosystem Conservation Directorate, Evaluation and Interpretation Branch, Ottawa, ON, 8 p.

Evans, L.J., 1989, Chemistry of metal retention by soils: Environmental Science and Technology, v. 23, p. 1047-1056.

Hance, R.J., 1969, An empirical relationship between chemical structure and the sorption of some herbicides by soils: Journal of Agriculture and Food Chemistry, v. 17, p. 667-668.

Harlow, H.E. and Hodson, P.V., 1988, Chemical contamination of Hamilton Harbour: A review: Fisheries and Aquatic Sciences, Canadian Technical Report 1603, 91 p.

Hayes, K.F. and Leckie, J.O., 1987, Modelling ionic strength effects on cation adsorption at hydrous oxide/solution interfaces: Journal of Colloid and Interface Science, v. 115, p. 564-572.

Hayes, K.F., Papelis, C. and Leckie, J.O., 1988, Modelling ionic strength effects on anion adsorption at hydrous oxide/solution interfaces: Journal of Colloid and Interface Science, v. 125, p. 717-726.

Howard, J.L. and Sova, J.E., 1993, Sequential extraction analysis of lead in Michigan roadside soils: mobilization in the vadose zone by deicing salts?: Journal of Soil Contamination, v. 2, p. 361-378.

Karickhoff, S.W., 1981, Semi-empirical estimation of sorption of hydrophobic pollutants on natural sediments and soils: Chemosphere, v. 10, p. 833-846.

Karickhoff, S.W., Brown, D.S. and Scott., T.A., 1979, Sorption of hydrophobic pollutants on natural sediments: Water Research, v. 13, p. 241-248.

Kersten, M. and Förstner, U., 1995, Speciation of trace metals in sediments and combustion waste, *in* Ure, A.M. and Davidson, C.M., eds., Chemical Speciation in the Environment: Chapman and Hall, London, UK, p. 234-275.

Langevin, R., Rasmussen, J.B., Sloterdijk, H. and Blaise, C., 1992, Genotoxicity in water and sediment extracts from the St. Lawrence River system using the SOS Chromotest: Water Research, v. 26, p. 419-429.

Lane, D.D., 1992, On-site remediation of contaminated soil, a case study: Demonstration of contaminated soil treatment by the Toronto Harbour Commissioners, A report by The Toronto Harbour Commissioners: 19 p.

Linzon, S.N., Chai, B.L., Temple, P.J., Pearson, R.G. and Smith, M.L., 1976, Lead contamination of urban soils and vegetation by emissions from secondary lead industries: Journal of the Air Pollution Control Association, v. 26, p. 650-654.

Lumsdon, D.G. and Evans, L.J., 1995, Predicting chemical speciation and computer simulation, *in* Ure, A.M. and Davidson, C.M., eds., Chemical Speciation in the Environment: Chapman and Hall, London, UK, p. 86-134.

Lumsdon D.G., Evans, L.J. and Bolton, K.A., 1995, The influence of pH and chloride on the retention of cadmium, lead, mercury and zinc by soils: Journal of Soil Contamination, v. 4, p. 137-150.

Lyman, W.J., 1982, Adsorption coefficients for soils and sediments, *in* Lyman, W.J. *et al.*, eds., Handbook of Chemical Property Estimation Methods: McGraw-Hill, New York, p. 4.1-4.23.

MacKay, D. and Paterson S., 1991, Evaluating the multimedia fate of organic chemicals in a level III fugacity model: Environmental Science and Technology, v. 25, p. 427-436.

Mayer, T. and Nagy, E., 1992, Polycyclic aromatic hydrocarbons in suspended particulates from Hamilton Harbour: Water Pollution Research Journal of Canada, v. 27, p. 807-831.

Metcalfe, C.D., Balch, G.C., Cairns, V.W., Fitzsimons, J.D. and Dunn, B.P., 1990, Carcinogenic and genotoxic activity of extracts from contaminated sediments in western Lake Ontario: Science of the Total Environment, v. 94, p. 125-141.

Meylan, W., Howard, P.H. and Boethling, R.S., 1992, Molecular topology/fragment contribution method for predicting soil sorption coefficients: Environmental Science and Technology, v. 26, p. 1560-1567.

Mingelgrin, U. and Gerstl, Z., 1983, Reevaluation of partitioning as a mechanism of nonionic chemical adsorption in soils: Journal of Environmental Quality, v. 12, p. 1-11.

Missler, H., 1992, A bibliography of scientific information on Fraser River basin environmental quality: Environmental Canada, Vancouver, B.C., 390 p.

Moore, J.W., 1990, Inorganic Contaminants of Surface Waters: Springer-Verlag, New York, 334 p.

Nriagu, J.O. and Pacyna, J.M., 1988, Quantitative assessment of worldwide contamination of air, water and soils by trace metals: Nature, v. 333, p. 134-139.

O'Connor, G.A., O'Connor, C., and Cline, G.R., 1984, Sorption of cadmium by calcareous soils: influence of solution composition: Soil Science Society of America, Journal, v. 48, p. 1244-1247.

Persaud, D., Lomas, T., Boyd, D. and Mathai, S., 1985, Historical development and quality of the Toronto waterfront sediment — Part I: Ontario Ministry of the Environment, Water Resources Branch, Toronto, ON, 62 p.

Phyper, J.D. and Ibbotson, B., 1991, The Handbook of Environmental Compliance in Ontario: McGraw-Hill Ryerson, Toronto, ON, 346 p.

Poulton, D.J., 1987, Trace contaminant status of Hamilton Harbour: Journal of Great Lakes Research, v. 13, p. 193-201.

Poulton, D.J., Simpson, K.J., Barton, D.R. and Lum, K.R., 1988, Trace metals and benthic invertebrates in sediments of nearshore Lake Ontario at Hamilton Harbour: Journal of Great Lakes Research, v. 14, p. 52-65.

Pussemier, L., Szabo, G. and Bulman, R.A., 1990, Prediction of the soil adsorption coefficient K_{oc} for aromatic pollutants: Chemosphere, v. 21, p. 1199-1212.

Rao, P.S.C., 1990, Sorption of organic contaminants: Water Science and Technology, v. 22, p. 1-6.

Remedial Action Plan, 1991, The Remedial Action Plan for Hamilton Harbour, Draft: Environment Canada, Burlington, ON, 240 p.

Sabjić, A., 1987, On the prediction of soil sorption coefficients of organic pollutants from molecular structure: application of molecular topology model: Environmental Science Technology, v. 21, p. 358-366.

Samant, H.S., Doe, K.G. and Vaidya, O.C., 1990, An integrated chemical and biological study of the bioavailability of metals in sediments from two contaminated harbours in New Brunswick, Canada: Science of the Total Environment, v. 96, p. 253-268.

Sanderson, M. and Weis, I.M., 1989, Concentrations of two organic contaminants in precipitation, soils and plants in the Essex region of southern Ontario: Environmental Pollution, v. 59, p. 41-54.

Southwood, J.M., Harris, R.C. and Mackay, D., 1989, Modeling the fate of chemicals in an aquatic environment: the use of computer spreadsheet and graphics software: Environmental Toxicology and Chemistry, v. 8, p. 987-996.

Sposito, G., 1984, The Surface Chemistry of Soils: John Wiley and Sons, New York, 234 p.

Swain, L.G. and Walton, D.G., 1991, Fraser River Estuary Monitoring Report on the Lower Fraser River and Boundary Bay Sediment Chemistry and Toxicity Program: British Columbia Ministry of the Environment, New Westminster, BC, 107 p.

Tessier, A., Campbell, P.G.C. and Bisson, M., 1979, Sequential extraction procedure for the speciation of particulate trace metals: Analytical Chemistry, v. 51, p. 844-851.

Tessier, A., Campbell, P.G.C. and Bisson, M., 1980, Trace metal speciation in the Yamaska and St. François Rivers (Quebec): Canadian Journal of Earth Sciences, v. 17, p. 90-105.

Wagman, D.D., Evans, W.H., Parker, V.B., Schumm, R.H., Halow, I., Bailey, S.M., Churney, K.L. and Nuttal, R.L., 1982, The NBS tables of chemical thermodynamic properties: Journal of Physical and Chemical Reference Data, v. 11, Supp. 2, p. 1-392.

Warren, L.A. and Zimmerman, A.P., 1994, The influence of temperature and NaCl on cadmium, copper and zinc partitioning among suspended particulate and dissolved phases in an urban river: Water Research, v. 28, p. 1921-1931.

Weis, I.M. and Barclay, G.F., 1985, Distribution of heavy metal and organic contaminants in plants and soils of Windsor and Essex County, Ontario: Journal of Great Lakes Research, v. 11, p. 339-346.

27. Public Health and Urban Soil Contamination

Monica Campbell

North York Public Health Department, 5100 Yonge Street, North York, Ontario M2N 5V7

Joan Campbell

School for Resource and Environmental Studies, Dalhousie University
1312 Robie Street, Halifax, Nova Scotia B3H 3E2

Stephen McKenna

Environmental Protection Office, City of Toronto Department of Public Health
100 Queen Street West, 6th Floor East Tower, Toronto, Ontario M5H 2N2

Scott MacRitchie

Environmental Consultant, 549 Euclid Street, Toronto, Ontario M6H 3G3

Miriam Diamond

Geography Department, University of Toronto, 100 St. George Street, Toronto, Ontario M5S 1A1

SUMMARY

In the past decade, soil contamination has emerged as an environmental issue of great significance for human health, social well-being, and economic viability within cities. Soil contamination can impact social well-being through constraints on housing affordability and its resultant inequities in housing opportunities and the distribution of health risk. The economic impacts of soil remediation can constrain the process of urban renewal and rehabilitation of derelict industrial lands into vibrant neighbourhoods. This chapter examines urban soil contamination from a public health perspective, using the initiatives of the Environmental Protection Office of the City of Toronto Department of Public Health as an example. In doing so, this chapter examines why soil contamination is of concern to public health, the policy and regulatory context, the role of local health units in reviewing municipal development applications, the creation of the Historical Land Use Database as a screening tool, the municipal review process, the development of a framework for assessing soil remediation options, and policy issues and future directions.

INTRODUCTION

Why Soil Contamination is Of Concern to Public Health

For many hundreds of municipalities throughout North America, the past century was a period of intensive industrial activity characterized by the use of hazardous chemicals and poor waste management practices, including the storage, leakage and deposition of wastes. Although many chemicals which have been dumped on land in the past have long since dissipated or been rendered innocuous through natural processes, other compounds, such as polycyclic aromatic hydrocarbons (PAHs), degrade very slowly (Chaudhry and Chapalamadugu, 1991) or not at all (metals such as cadmium, lead, chromium and nickel; Thornton, 1992). Consequently, lands which supported industrial and related activities in the past are likely to contain residual contaminants in their soil and ground water (Schineldecker, 1992). There is also increasing concern with accidental spillages of environmental contaminants at various stages of their manufacture, storage, use, and disposal (*e.g.*, Meharg and Osborn, 1995).

Persistent and toxic contaminants in soil and ground water are of particular concern because of the possibility of human contact (Pastenbach *et al.*, 1992). However, even toxic substances that are not persistent can be of concern if their concentration is high. The hazard and resultant adverse health effects depend on the inherent toxicity of the contaminant present, the amount of toxicant to which a person is exposed and that reaches the target site of injury in the body, and host susceptibility. Host susceptibility depends on many factors, including the genetic makeup of the individual, age, concomitant exposures to other toxicants, and overall nutritional and health status.

Contaminants in the soil that have high solubilities in water (such as phenol or benzene) may dissolve in ground water and migrate off site to adjacent properties and possibly contaminate aquifers (*e.g.*, Chapter 7). In other cases, depending on the proximity of lakes and rivers, contaminants may move with ground water to discharge into these bodies of water (*e.g.*, Chapter 5). Contamination of surface waters provides pathways of exposure to toxicants through the ingestion of drinking water and sport fish. Contaminants that are strongly hydrophobic (*i.e.*, lipid-soluble) can move easily across the biological membranes of plants and animals, and concentrate up the food chain. For instance, polychlorinated biphenyls (PCBs) are capable of bio-concentrating in fish to levels greater than fish consumption guidelines suggest, even though concentrations in lake water are barely detectable. In 1988, PCB levels in the eggs of eagles (who are at the top of the food chain) were 25 million times higher than the water of Lake Erie (Environment Canada, 1991).

Even substances that have limited water solubility (non-aqueous phase liquids or NAPLs) can migrate from the original site of deposition (Chapter 6). Off-site migration is common for tarry residues such as coal gasification by-products (*e.g.*, PAHs) or fuels from leaking underground storage tanks. Although off-site migration of certain contaminants does occur at specific sites, it is also common for other contaminants to remain relatively immobile in the soil until the soil is disturbed during clean-up and redevelopment. Examples of persistent, but relatively immobile, compounds include metals such as lead and nickel.

The presence of surficial soil contamination in the community can be an important exposure route to toxicants for young children. In the case of lead, about 64% of a child's daily intake of lead in Ontario comes from soil and the resultant dust in the home. In contrast, only about 17% of an adult's total daily intake of lead comes from soil and dust (Ontario Ministry of Environment and Energy, 1994). In the city of Toronto, empirical evidence suggests that 80% of the city's land area contains lead levels that exceed the current health-based lead guideline for lead in soil (Ontario Ministry of the Environment and Energy, 1994). Major point sources of this contamination include airborne emissions from local battery recycling plants, as well as emissions from sewage sludge and garbage incinerators. However, the most plausible explanation for the ubiquitous contamination of surficial soils throughout the city is the use of tetraethyl lead in gasoline until 1990 (Metropolitan Toronto Teaching Health Units, 1994). In Ontario, as elsewhere in North America, urban soils typically have higher concentrations of lead than rural communities in which traffic

density, and hence lead emissions from vehicles, were less (OMOEE, 1994).

The redevelopment of contaminated land can put people in direct contact with toxic substances in several ways (Shephard *et al.*, 1992; Simmons and Lewis, 1993). One route of exposure may involve gaseous diffusion of volatile contaminants, such as benzene or methane, that may be present at depth in the soil or ground water. Exposure to less volatile contaminants can occur during excavation as soil mixing promotes volatilization and wind erosion carries dust to adjacent neighbourhoods. Exposure can also occur if any remaining contaminated soil is accessible to the public through new land uses such as community gardens, day-cares, parks, schools and residential yards. In the case of urban food production, some contaminants can be taken up into the edible portion of plants (Amiro *et al.*, 1991), and aerial contamination of food crops through dust is common. Direct ingestion of contaminated soil and dust tracked into the home is common among young children under the age of four. Because toddlers have frequent hand-to-mouth activity, they may be exposed on a daily basis to soil pollutants attached to their hands and foods eaten in a contaminated setting (Hawley, 1985).

Policy and Regulatory Context

In Canada, provincial environment ministries have a central role in the assessment and remediation of contaminated sites. The Canadian Council of Ministers of the Environment (CCME) provides important technical guidance at a national level on assessment criteria for both soil and ground water (Canadian Council of Ministers of the Environment, 1991). While the CCME initiatives facilitate comprehensive and co-ordinated approaches throughout Canada regarding soil clean-up guidelines and remediation practices, regulatory authority ultimately rests with each province and territory.

In Ontario, the Ministry of the Environment and Energy (OMOEE) has recently introduced new guidelines for the clean-up of contaminated sites (Ontario Ministry of the Environment and Energy, 1994). Compared with the previous guidelines (Ontario Ministry of the Environment, 1989, 1991), the new guidelines greatly increase the number of chemicals for which soil criteria exist (new guidelines have 117 criteria *versus* 22 in the old guidelines), thereby providing a more comprehensive framework for use in the site assessment and remediation process. Given the high cost of sample collection and analysis, the guidelines do not require that each redevelopment site be tested for all 117 possible contaminants. Although the new guidelines provide a more flexible approach to site characterization (Saxe, 1994), the issue now facing officials involved in development review is to ensure that the development proponent has appropriately selected which of the 117 contaminants should be tested.

The new OMOEE guidelines permit a variety of clean-up options not considered previously. Soil clean-up options now include full clean-up to generic criteria or background levels, stratified clean-up to meet generic but differential guidelines at the surface (0 to 1.5 m) and at depth (below 1.5 m), and risk management options that meet health and environmental criteria as determined through site specific risk assessment (Ontario Ministry of the Environment and Energy, 1994; Advisory Committee on Environmental Standards, 1994; Chapter 2). While these remediation options are important in enhancing the likelihood of clean-up and reducing costs, at issue is the consistent adherence by all development proponents to sound environmental practices (discussed later) that adequately protect community health. As provincial agencies transfer greater responsibility for development review to municipal agencies, and greater accountability for the clean-up directly to the development proponent, the importance of a coherent surveillance process at the local government level increases.

In Ontario, the Commission on Planning and Development Reform (Ontario Ministry of Municipal Affairs, 1993) concluded that municipalities should be given greater control of the development process, provided that adherence to provincial policies can be ensured. This shift in responsibility extends also to soil contamination issues. Consistent with the OMOEE's 1994 clean-up guidelines, the trend in Ontario is to minimize technical scrutiny and approval responsibility at the provincial level, but to increase it at the municipal level.

The Role of Local Health Units in Municipal Development Review

As municipalities rehabilitate contaminated lands into residential neighbourhoods or public access uses, there is a need to protect the health of future residents and non-human receptors (flora and fauna) from exposure to hazardous contaminants. In Ontario, each of the province's 42 health units are mandated to meet the Health Protection and Promotion Act (1983) and the Mandatory Health Programs and Services Guidelines (Ontario Ministry of Health, 1989). Included in these requirements is the authority and responsibility to identify environmental risk areas within the each jurisdiction, and for the Medical Officer of Health to act to control or prevent health hazards. Although considerable discretionary power rests with the Medical Officer of Health in interpreting what constitutes a health hazard, several health units are now actively expanding their responsibilities to include a Healthy Environments Program. In practice, many municipalities in Canada seek the advice and recommendations of the local Medical Officer of Health concerning soil contamination issues. This local public health function is supported by the technical and policy information provided by the respective provincial ministries for environment and health.

In Ontario, appropriate site clean-up consistent with OMOEE guidelines is provided through the municipal and provincial review of development applications in accordance with the Planning Act. As such, the Ministry of Municipal Affairs circulates applications from a municipality for an Official Plan Amendment or Plan of Subdivision to relevant ministries, such as the OMOEE or the Ministry of Health. Typically, when the Ministry of Health is circulated on such applications, it refers the development application to the local area health unit for review.

Other development applications which meet the provincially approved Official Plan, but deviate from local land use planning constraints, are not usually subject to review by the Ministry of Environment and Energy or Ministry of Health. Development applications handled at the municipal level only include those that involve rezoning, site plan control or Committee of Adjustment decisions. Consequently, at the municipal level, several development approval streams exist in which scrutiny of soil and ground-water contamination issues is possible. Since responsibility for issuing development approval and building permits rests with the local planning and building departments, respectively, the role of the health unit (also referred to as Public Health Department) needs to be linked with other municipal departments routinely involved in redevelopment issues. This linkage is illustrated by examining the tools and processes developed in the city of Toronto.

City of Toronto Public Health Department

For about a decade, the Toronto Public Health Department has provided consultation services to the Planning and Development Department on development applications for which soil contamination and air quality concerns are apparent. In the 1980s, determining which development applications were associated with environmental health concerns, and hence to be referred to Public Health for further appraisal, relied on the familiarity of city planners with local historical and environmental conditions. With the establishment of the Environmental Protection Office (EPO) in the Public Health Department in 1986, this ad hoc referral process was gradually replaced with a systematic approach to screen and review development applications.

The following components were developed and put in place to ensure a systematic and consistent review process that maximized protection of community health: (1) a screening tool (i.e., Historical Land Use Database) by which municipal staff in diverse municipal departments could identify properties for which redevelopment involved potentially contaminated soil (Campbell et al., 1994); (2) a process by which planning/building staff can forward development applications to health staff for appraisal (Campbell and McKenna, 1995); (3) a consistent set of site clean-up requirements to be met by development proponents during the application process and subsequent remediation and construction phases (Campbell and McKenna, 1995); and (4) an evaluation framework for use in the review of site remediation options, including emerging soil treatment technologies (Campbell et al., 1995).

Historical Land Use Database

The City of Toronto Historical Land Use Database currently has over 22,000 records that list the addresses of properties for which previous uses are suggestive of soil contamination. First developed by the Environmental Protection Office (EPO), the database has been transferred to a Geographic Information System (GIS) by the Planning and Development Department. Activities are underway to integrate the database with the city's mainframe computer so that it is linked with the corporation's other property databases.

The most common use of the historical land use database is for city officials to screen the hundreds of applications received each year for decisions on Official Plan amendments, rezonings, site plan approvals, demolition permits and Committee of Adjustment applications. Those applications for development activity that involve properties listed in the historical land use database are then circulated to the public health department for review and comment. In this way, the use of review services by health unit staff are optimized by focussing on high-risk sites, rather than reviewing all development and building applications.

Another application of the historical land use database is to facilitate public disclosure about potential contamination. The database provides development proponents and associated legal and financial representatives with access to preliminary environmental information. Furthermore, public disclosure is facilitated during land transfer. Specific information on a given property, or copies of the database in its entirety, are not distributed to the general public; however, a property owner or prospective purchaser may request information on a specific property. All information compiled in the historical land use database is derived from publicly accessible records. When database information is disclosed, the recipient is given a disclaimer that the information presented is historical and does not necessarily reflect current conditions at the site or recent clean-up activities. The properties flagged in the database may or may not be contaminated. Confirmation of contamination must be based on detailed site characterization through soil and ground-water testing.

Components of the Database and Criteria for Inclusion

The components of the historical land use database include property address (street number and name), date (year) to which the historical information pertains, and previous activity. About 1500 of the 22,000 records involve previous industrial uses with a high likelihood of persistent and toxic residuals. For these records, additional information was collected on the names of the companies that operated on these properties, as well as the Standard Industrial Classification (SIC) code for each industrial activity (Statistics Canada, 1980). Table 1 provides a modified sample table from the EPO's hard-copy version of the historical land use database (City of Toronto Department of Public Health, 1995). Figure 1 illustrates which properties in Toronto are listed in the historical land use database, and consequently suspected of containing contaminated soil.

The purpose of including the SIC codes in the database was to link the historical land use database with the Chemical Substance List. The Chemical Substance List was compiled by public health staff to identify the types of chemicals that were associated with specific industrial processes used on the premises in the past (Campbell et al., 1994). For those historical land use database entries that have SIC codes, health unit staff can anticipate which chemicals might occur in the soil and ground water. This takes on new importance with the recent expansion in the number of chemicals covered in soil clean-up criteria in Ontario. It provides an additional surveillance tool to municipalities to ensure that site characterization is sufficiently comprehensive. For entries in the historical land use data base such as area of lakefill or area of landfill, the database can flag potentially contaminated properties; however, it can not indicate specific contaminants of concern.

In creating the historical land use database, criteria were developed for determining which properties to include in the database (Table 2). Certain criteria, such as historical industrial area, are very broad and result in the addition of thousands of records that may include properties that are not contaminated. In contrast, the priority industrial activity category is much less comprehensive in scope; however, it contains priority properties with past industrial activities known to involve persistent toxic substances. For these records, the likelihood of serious contamination is high.

Data Collection Methodology

In creating the historical land use database, health department staff collected information from sources which include historical maps, historical directories, provincial reports, and inventories.

Staff used historical maps from various archives and official plans from the Planning and Development Department to identify the maximum extent of industrial activity. In Toronto, as in most Canadian urban centres, there has been a trend toward decreasing industrialization, making it important to identify areas that supported industrial activity in the past. Early topographic maps are also useful in identifying the location of ravines that have since been filled with potentially contaminated materials, or for assessing the extent of historic lake filling (see Chapter 2).

Fire insurance plans and atlases from the late 1800s to the 1960s, such as the Goads and Underwriters series, frequently provide information on the name, address, activity and potential hazards of businesses, including industries (Hayward, 1973). In many cases, detailed site plans give clues concerning underground tanks, boilers and digesters, and chemical storage areas.

Compilation of the Toronto historical land use database also used a targeted search strategy involving trade directories. Most useful were Might's City Directories and Scott's Industrial Directories. The combined use of fire insurance plans and trade directories permitted the identification of 1500 priority industrial properties in the database. Starting with the list of industries with SIC codes, professional judgement was used to select those industries associated with persistent hazardous substances. Table 3 shows activities that were flagged as priority industrial activities and related operations. The priority industrial activities were then cross-referenced with activities identified in archival fire insurance atlases and trade directories. The information was recorded in both table and map format. The year for which the information was collected was entered in the database, resulting in a record that showed the change in company names and activities on a given property over time. Table 1 illustrates the chronological nature of the records that were created. Table 2 lists criteria for inclusion in the database and Table 3 common industrial activities for which residual soil contamination is likely.

Table 1   Sample table from City of Toronto Historical Land Use Database.

Street Number	Street Name	Year	Activity	Company Name	SIC Code
288	Anne St.	1992	Gasoline Station	Toronto Autocare Ltd.	
10	Beech Grove		Landfilling (former ravine)		
4	Colson Ave.		Lead Reduction Zone		
4	Colson Ave.	1935	Coal Tar Distillation	Ted's Tar & Chemical Co.	369
4	Colson Ave.		Lakefilling		
4	Colson Ave.		Industrial (1949 Official Plan)		
4	Colson Ave.	1951	Fuel Storage (warehouse)	Urban Service Oil Co.	511
4	Colson Ave.	1967	Lubricants Plant	Oil Canada Ltd.	361
10	Elizabeth Road		Landfilling (former ravine)		

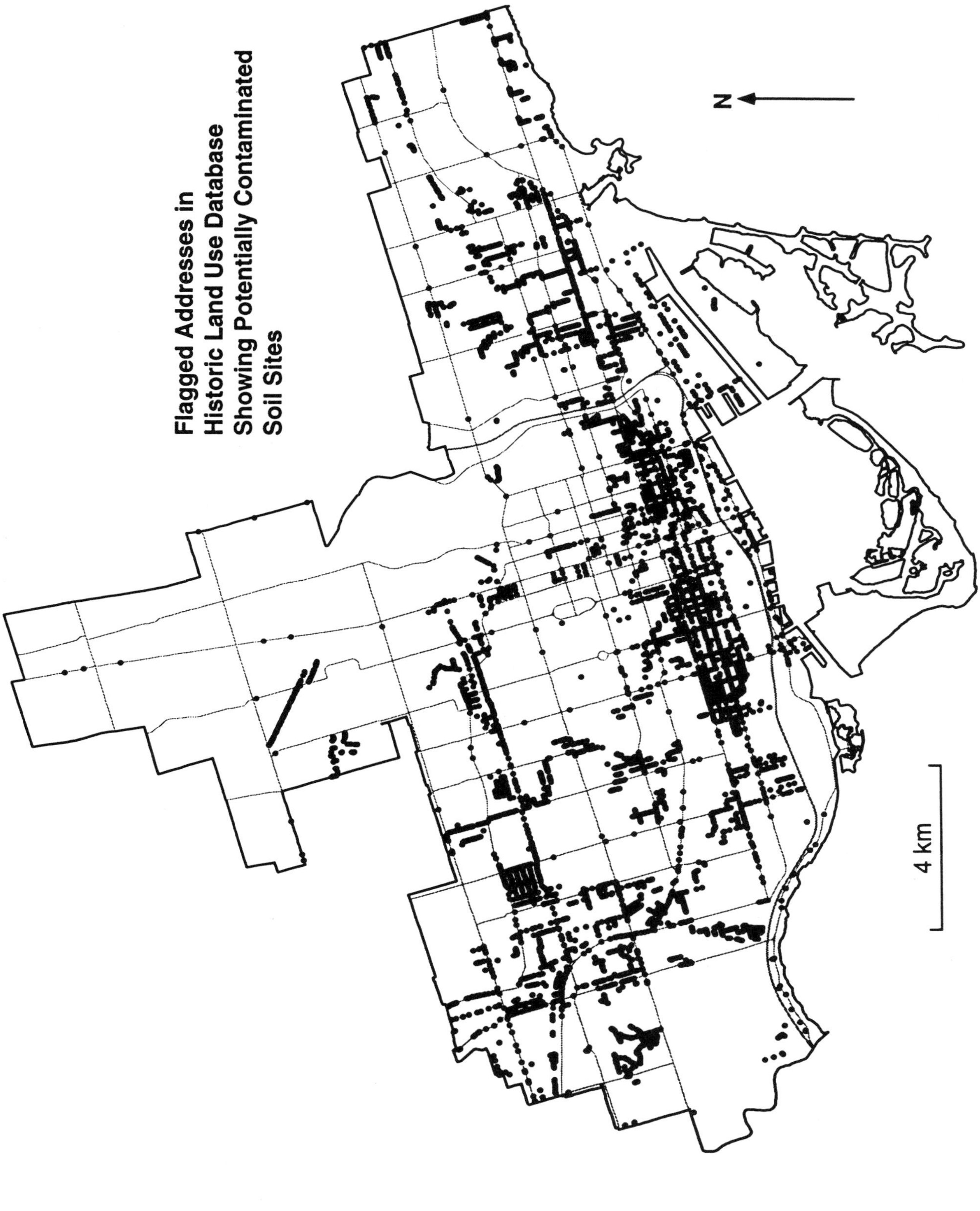

Flagged Addresses in Historic Land Use Database Showing Potentially Contaminated Soil Sites

N

4 km

Figure 1 *Areas of potential soil contamination, city of Toronto, identified to date.*

It is important to recognize that an historical land use database may contain erroneous information. This problem may be aggravated by the renaming and renumbering of some streets over time, as occurs in most municipalities. Concerns about archival materials may also extend to maps, particularly early maps (*i.e.*, early 1800s) created with rudimentary equipment. In cases where diverse information sources provide conflicting historical information, professional judgement is used to select the most reliable information source. Given these concerns, the use of records in the historical land use database should be verified and augmented by a more detailed, site-specific search conducted by the development proponent.

The Municipal Review Process

Process By Which
Development Applications with Environmental Risk
Undergo A Public Health Review

Depending on the nature of the proposed undertaking in the city of Toronto, the application is routed through one of the following processes: (1) Stream 1 – Buildings and Inspections Department processes (building permits, demolition permits); (2) Stream 2 – Planning and Development Department processes (Official Plan Amendment, rezoning, site plan control); and (3) Stream 3 – Committee of Adjustment process (minor variances from the zoning by-law, severances).

The majority of applications concerning construction and development undergo a Preliminary Zoning By-law Review by the Buildings and Inspections Department, before an application has been filed with Planning and Development (stream 2) or the Committee of Adjustment (stream 3). The Toronto historical land use database is used for screening all develop-

ment applications following streams 2 and 3. The application of the historical land use database to construction building permits (stream 1) occurs only when based on specific concerns identified by Buildings Department staff for a particular site. Application of the database to demolition permits occurs in conjunction with the results of a screening questionnaire completed by the applicant. Once the corporation flags a development application that may have some environmental health risk, the application is referred to the Department of Public Health for review and comment.

Environmental Requirements for Site Clean-Up

In 1993, the Planning and Development Department consolidated all requirements requested by the City of Toronto's various departments into a single Development Approval Manual (City of Toronto Planning and Development Department, 1993). This manual includes environmental requirements pertaining to site remediation. These requirements include (1) submission of a historical review, site audit and building audit, (2) submission of a soil and ground-water testing program and design and implementation of a soil and ground-water management plan, and (3) submission and implementation of a dust-control plan for demolition and excavation/site remediation.

Consequently, development proponents can consider such requirements during their early planning phases, well in advance of submission of a development application. A proponent can then more effectively propose to identify and remediate environmental and health risks associated with redevelopment of contaminated lands, and thereby better predict costs and plan financing arrangements.

Development proponents engage environmental consultants to provide technical reports in response to the require-

---

Table 2   Criteria for inclusion in the Historical Land Use Database.

Criteria For Inclusion	Rationale
Historical Industrial Areas	All properties that occurred within the historical industrial area (based on the earliest Official Plan available) were included. It is presumed that properties that occurred within this zoning classification supported potentially polluting activities or were sufficiently close to be contaminated through direct disposal/spills or airborne deposition of pollutants.
Priority Industrial Activities	Types of industrial and commercial activities known to involve hazardous substances were identified (see Table 3).
Lead Reduction Zones	The provincial environment ministry designated certain areas around secondary lead smelters for soil replacement due to high lead concentrations.
Historical Coal Gasification & Related Industries	Historical coal gasification and related tar processing industries have resulted in significant residual pollution that is hazardous, persistent and sometimes mobile.
Historical Waste Disposal Sites	In the past, municipal waste disposal sites received diverse wastes, including toxic household and commercial wastes. Historical landfills are also of concern because of methane gases produced by the decomposition of organic matter.
Historical and Active Gas Stations	Old underground fuel storage tanks, including those from gas stations, are associated with soil contaminated with oil and gas, as well as fuel-derived compounds such as benzene, xylene and toluene.
Areas of Lakefill	It was common in the past to deposit waste soils and construction debris at the water's edge to create new land. Much of the material deposited in the past was contaminated and exceeds current soil clean-up guidelines.
Areas of Landfill	In the past, it was common to dispose of waste soil, construction rubble and ash in ravines to level the natural terrain. Many of these fill materials were contaminated. Off-site migration of contaminants occurs by ground-water flow.

ments applied by departmental staff to their particular project. Since the city request occurs early in the development approval process, the proponent has a reasonable timeframe in which to provide the information required. The provision, through qualified environmental consultants, of the technical information, plans and recommendations ensures appropriate scientific and engineering support for the clean-up measures required. The process of review by departmental staff is designed to provide consistent application of municipal and provincial policies, practices, and guidelines. In the case of Official Plan amendments/rezonings, Public Health Department recommendations that derive from this review are included in reports to the Land Use Committee and City Council as the application proceeds.

The environmental studies and remediation plans requested must be completed to the satisfaction of the Medical Officer of Health prior to city approval of the development and/or prior to issuance of a building permit by the Department of Buildings and Inspections. Approval may be conditional on certain recommendations or conditions that must be fulfilled prior to the issuance of a building permit. Conditions arising from the environmental plans are secured through a development agreement, collateral agreement or an undertaking that the development proponent is requested to sign following City approval of the development. Adherence by the development proponent to the legal agreements is monitored by inspection staff of the Public Health Department and the Department of Buildings and Inspections, in co-operation with the city solicitor.

The environmental requirements for site remediation listed in the Development Approval Manual are applied in a phased or sequential approach, and only as justified for each application. As such, the results from the historical review, building audit and site audit inform the subsequent soil and ground-water testing. Similarly, the results of the soil and ground-water testing program need to be analyzed prior to the development and submission of a soil and ground-water management plan. Although a development proponent typically proceeds through each requirement in a sequential manner, the proponent may choose to submit a consolidated report to Public Health staff that outlines all aspects of decommissioning and site remediation in relation to the project. Some developments may require an additional monitoring and reporting system once construction activities commence. In addition, a statement of completion must be submitted in conjunction with the building permit process that includes the results of site verification testing to demonstrate that the remediation was completed in accordance with the soil and ground-water management plan.

Review of Environmental Studies
Submitted by Development Proponents

Significant progress is underway among diverse agencies in encouraging development proponents to voluntarily adopt better procedures in environmental investigations, irrespective of municipal or provincial requirements. The Canadian Standards Association (1993) has produced a set of voluntary environmental management systems and environmental auditing guidelines. The Consulting Engineers of Ontario have completed a feasibility study on generally accepted standards for environmental investigations (1993). The Ontario Ministry of Housing (1993) has prepared a handbook for environmental site assessments appropriate to housing projects on potentially contaminated lands. Through the creation of standards for environmental investigations, the overall quality and consistency of environmental consulting services offered to the development community is anticipated to increase. This would support and simplify the role of municipal agencies involved in the protection of environmental health through the development review process.

Municipalities may wish to develop a checklist based on available standards to identify aspects for environmental review of development applications. A checklist can enhance fairness and consistency among reviewers, as well as ensure the comprehensiveness of the review process. Use of a checklist, however, cannot substitute for additional thought and judgement required in dealing with complex applications, the guidelines, practices and policies from various jurisdictions, and the general public interest in terms of environment and health.

Sample questions are provided to illustrate the review of a site remediation proposal in response to environmental requirements posed by a municipality:

Historical Review, Site and Building Audit

Did the historical search go back as far as historical records were available? Were all reasonable records searched? Was sufficient detail provided concerning industrial activities and residual contaminants associated with each of these? Were surrounding land uses identified that may have contaminated the property? Was the site audit thorough in reporting storage areas for hazardous materials, underground tanks and pipes,

---

Table 3   Common industrial activities occurring in municipalities for which residual soil contamination is likely.

Battery recycling
Chemical products industries
Coal gasification
Drum & barrel reconditioning
Electrical equipment manufacture & maintenance
Electroplating
Foundries & smelters
Fuel storage depots
Furniture & fixture industries
Garages & repair shops
Gas stations
Harbour activities
Incineration
Leather industries
Lithographers
Oil refiners
Paint formulating plants
Paper and allied products
Pesticide formulators & applicators
Petroleum refining & storage
Pharmaceutical activities
Photofinishing
Power plants & substations
Printing plants
Railway depots & corridors
Scrap yards
Sewage treatment plants
Storage & warehousing
Tanneries
Textile products industries
Water treatment plants
Wood treatment plants

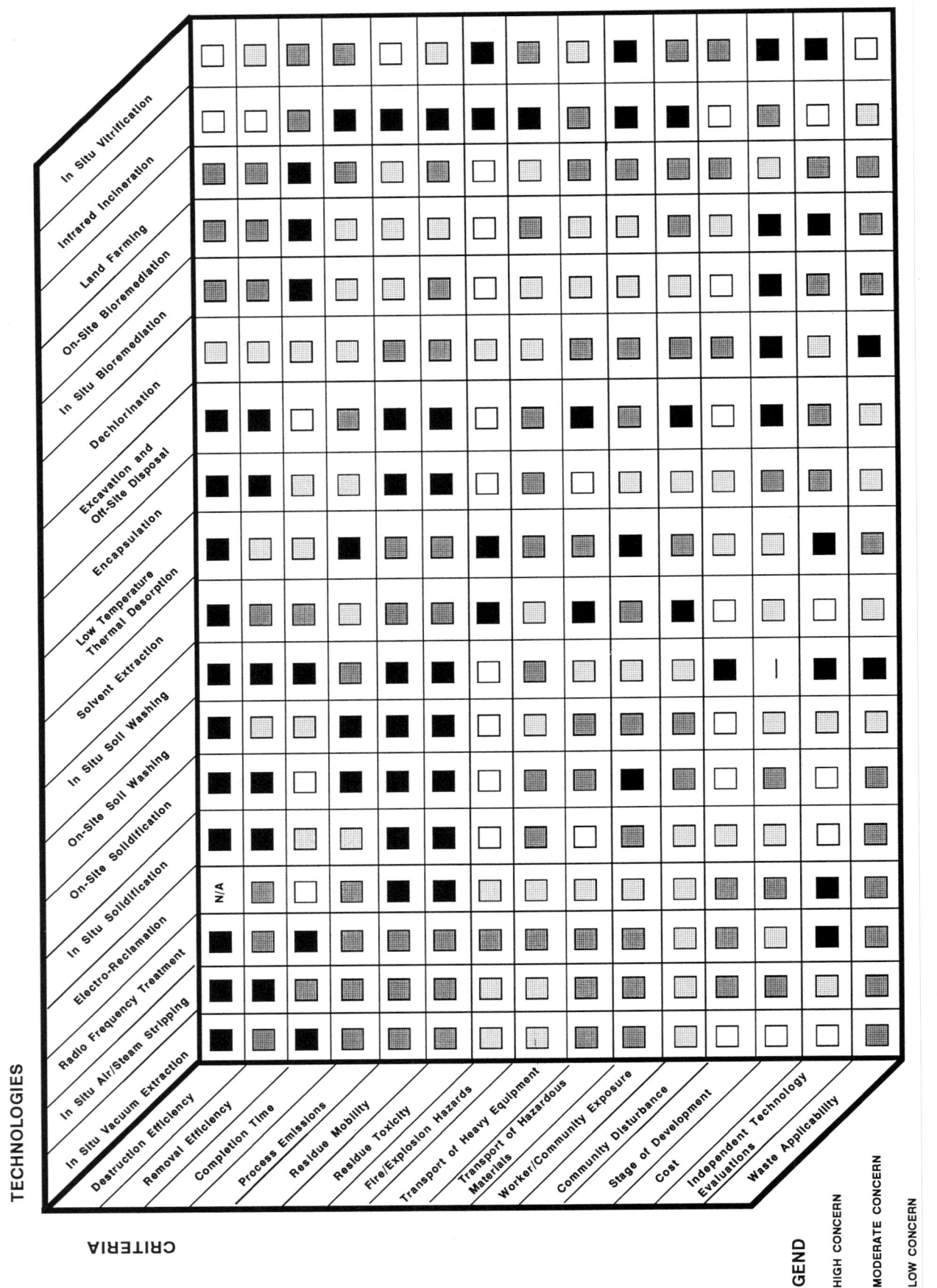

Figure 2 *Potential environmental and health concerns matrix for different soil remediation techniques described in Chapter 35.*

and vegetation dieback? Did the building audit check for asbestos, PCB-containing equipment and fixtures, and chemical storage areas? If hazardous materials occur on the premises, did the consultant identify a plan for the safe removal of these materials?

Soil and Ground-water Testing Program

Was the soil and ground-water sampling strategy justified, based on the historical review, building audit and site audit? Were the locations of boreholes appropriate, and was the borehole density adequate? Were sufficient samples taken at depth to confirm the downward extent of contamination? Did the collection, preservation and analysis of samples meet appropriate quality assurance/quality control standards? Did the report clearly summarize the sampling program and identify guideline exceedances?

Dust Control Plan

Does the plan deal with building demolition and soil excavation/remediation? Are the dust suppression measures proposed adequate? How will loading points be kept clean to prevent off-site tracking of soil? Will all trucks be swept clean of loose soil and tarped? If dust monitoring is warranted, are the monitoring measures proposed adequate?

Soil and Ground-water Management

Was the community notification/consultation procedure adequate for the remediation proposed? Is there evidence in the report that the consultant considered on-site remediation and off-site reuse/recycling/treatment of soil? If a risk assessment approach was used to support the management plan, does the risk assessment adhere to the provincial guidelines for assessment of human health and ecological risk? Is there something unique about this site that requires particular measures regarding site clean-up and/or management? How will excess soil be disposed? How will contaminated ground water be treated/disposed? Will a qualified environmental consultant oversee the clean-up? Will verification sampling be conducted upon completion of the clean-up?

A Framework For Assessing Soil Remediation Options

To date, the most common method of dealing with contaminated soil in North America continues to be to excavate and transport it to other locations for landfilling. While this solution removes the health risk to future occupants of the rehabilitated site, the transfer of contaminated soil to landfills elsewhere does not minimize its volume or reduce its toxicity. In a broader geographical horizon, the contamination is simply shifted in location and becomes the long-term concern of the receiving area. From an environmental perspective, a pollution prevention approach to site clean-up is preferred. In this approach, the contaminants associated with soil are separated, treated or destroyed, and the site is restored with clean soil. This approach minimizes the transfer of contaminants to other communities and other parts of the ecosystem.

In recent years in Toronto, several soil clean-up initiatives based on innovative emerging technologies have been proposed by private and public sector proponents. Consequently, the Environmental Protection Office (EPO) investigated the potential environmental and health concerns associated with 18 generic soil remediation technologies (City of Toronto Department of Public Health, 1991). In addition, a framework was established to permit the systematic review of new proposals for site remediation within the city. While the city supports and encourages pilot testing of innovative site

remediation technologies that are consistent with pollution prevention principles, many of these technologies are new and not commercially proven. The risks associated with new technologies must be adequately addressed in terms of community exposure and environmental concerns. Community concerns should form an important part of the decision-making process to determine the suitability of a particular technology for a given site.

Site Remediation Technologies

In preparing the site remediation framework, types of technologies were first assessed. A full review of remediation technologies and the advantages and disadvantages of each is given in Chapter 35. The next stage in preparing the assessment framework was to identify 15 criteria with which to evaluate the different technologies. These criteria form the framework upon which public health staff assess proposals subject to municipal development review. Evaluation criteria include the following:

Destruction efficiency. The destruction efficiency is the percentage of contaminants destroyed through the application of technology. Technologies with a high destruction efficiency tend to provide a permanent, long-term solution that does not require future monitoring of the contaminated site or disposal facility.

Removal efficiency. The removal efficiency is the percentage of the total contamination extracted from the soil as a result of the treatment process. Technologies with a high removal efficiency either separate contaminants from soil, or concentrate them in a smaller fraction of soil, thereby reducing contamination in the original soil.

Completion time. Completion time can influence costs as well as increase worker and community exposure to hazards associated with the process. To evaluate the completion time of a technology, a daily rate, expressed in tonnes of soil treated per day, was calculated for each technology. Completion time required can be an important consideration for community acceptance.

Process emissions. This criterion summarizes the ability of the technology to minimize residual waste streams. Judgement is necessary when evaluating technologies that produce relatively small volumes of concentrated waste *versus* large volumes of dilute waste.

Residue mobility. Residue mobility is a measure of the tendency of remaining contaminants to move in the environment through desorption from soil, movement in ground or surface water, and/or release to the air.

Residue toxicity. Many remediation processes remove or destroy the less toxic, more soluble and more bioavailable organic chemicals, while leaving the strongly sorbed, hydrophobic and toxic organics.

Fire/explosion hazards. The use of flammable or explosive materials, as well as processes involving high temperatures, are important risk factors to consider.

Transportation of heavy equipment. The amount and size of heavy equipment transported on or off site contributes to ancillary chemical and dust emissions, to risk for site workers, and to the likelihood of vehicular accidents and general disruption to the community. This is of greatest concern in high-density areas.

Transportation of hazardous materials. The amount and nature of hazardous materials transported on or off site affects the probability of community and worker exposure to such materials.

Worker/community exposure. The potential exposure to

various waste streams including air, water and solids are assessed with respect to community and worker exposure.

Community disturbance.   Noise, time, odour, dust and visual impact are factors used to evaluate disturbance to the community. Contributing to this is the visual impact of operations.

Stage of development.   Technology is characterized as conceptual, bench scale, pilot scale and commercial scale operations. Emerging, rather than proven, technologies usually require a greater level of technical scrutiny, because few data may be available on removal or destruction efficiencies, waste emissions, completion time and cost (Chapter 35). Cost and independent evaluations of the technology are additional factors.

Assessment of Remediation Technologies

The assessment criteria noted above were then used to evaluate the 18 generic remediation technologies. Each criterion was ranked according to four levels of concern (none, low, moderate and high), depending on the severity of potential impacts or disadvantages.

Figure 2 shows the matrix on potential environmental and health concerns that resulted from the technology assessment process. The rating system used to create the matrix was not weighted, because the intent was to highlight potential environmental health concerns for a given technology, rather than to evaluate or disqualify the technology from consideration at a specific site.

DISCUSSION

Clean-up Costs

The high cost of remediation is a significant impediment to both private and public landowners planning to undertake site clean-up. Clean-up costs impact the affordability and accessibility of housing in the urban core of many Canadian cities (Canada Mortgage and Housing Corporation, 1993). As urban planners try to implement policies to curtail urban sprawl and foster urban renewal, historical soil contamination in the city core threatens to promote further development on the relatively pristine suburban periphery (Bailes, 1994).

Role of Government

Soil contamination will remain undiminished in a community unless clean-up is triggered by government for specific properties shown to be leaking contamination off site, or if properties are being redeveloped and fall under the development review process. For properties undergoing redevelopment, the technical review of site remediation is shifting from provincial experts to municipal reviewers or third-party assessors retained by development proponents. Issues of accountability and expertise remain to be resolved.

Municipal officials who receive innovative soil clean-up proposals in their community need to develop the expertise to assess the environmental and health concerns associated with these technologies. The assessment framework developed for Toronto is applicable to other municipalities facing similar soil clean-up initiatives.

Healthy Environments and Public Input

Although the revised guidelines provide greater clarity on protecting human health, issues such as ecological risk and potential impacts on the biophysical environment remain elusive and difficult to assess. Limits on contaminant concentrations that are protective for humans may not protect sensitive ecological receptors, as demonstrated by many metals which are more toxic to aquatic organisms than to humans (Means and Cooper, 1994), or organochlorines which biomagnify and threaten top predators. At issue is the extent and process for involving the public in site remediation decisions that affect them and their local environment.

REFERENCES

Advisory Committee on Environmental Standards (ACES), 1994, Proposed guideline for the clean-up of contaminated sites in Ontario: Recommendations to the Minister of the Environment and Energy: Toronto, ON, 90 p.

Amiro, B.D., Zhuang, Y. and Sheppard, S.C., 1991, Relative importance of atmospheric and root uptake pathways for $^{14}CO_2$ transfer from contaminated soil to plants: Health Physics, v. 61, p. 825-829.

Bails, J.D., 1994, Cleanup standards and urban sprawl: a policy dilemma: Health and Environment Digest, v. 8, p. 21-22.

British Standards Institution, 1989, Code of practice for the identification of potentially contaminated land and its investigation: London, UK, 92 p.

Campbell, M. and McKenna, S., 1995, Utilizing the municipal development review process to ensure proper site remediation of contaminated sites: Environmental Health Review, v. 38, p. 88-93.

Campbell, M., McKenna, S. and Campbell, J., 1994, Historical land use database: a municipal tool to screen potentially contaminated properties: Environmental Health Review, v. 38, p. 32-38.

Campbell, M., MacRitchie, S., Diamond, M. and McKenna, S., 1995, A framework for assessing environmental and health concerns associated with site remediation technologies: Environmental Health Review, v. 39, p. 32-37, 56.

Canada Mortgage and Housing Corporation (CMHC), 1993, The relationship between urban soil contamination and housing in Canada: Prepared for CMHC by Gardner Church & Associates, Ottawa, ON, 50 p.

Canadian Council of Ministers of the Environment (CCME), 1991, Interim CCME environmental quality criteria for contaminated sites: Prepared by the Canadian Council of Ministers of the Environment (CCME) Subcommittee on Environmental Quality Criteria for Contaminated Sites, Ottawa, ON, 20 p.

Canadian Standards Association, 1993, Phase I Environmental Site Assessment: Ottawa, ON, 49 p.

Chaudhry, G.R. and Chapalamadugu, S., 1991, Bioremediation of halogenated organic compounds: Microbiological Reviews, v. 55, p. 59-79.

City of Toronto Department of Public Health, 1991, Identification of potential environmental and health concerns of soil remediation technologies: Environmental Protection Office, Toronto, ON, 240 p.

City of Toronto Department of Public Health, 1995, Historical land use inventory issue Number 3: Environmental Protection Office, Toronto, ON, 86 p.

City of Toronto Planning and Development Department, 1993, Development approval manual: Toronto, ON, 290 p.

Consulting Engineers of Ontario, 1993, Generally accepted standards for environmental investigations: Toronto, ON, 81 p.

Environment Canada, 1991, Toxic chemicals in the Great Lakes and associated effects. Synopsis: Department of Fisheries and Oceans, and Health and Welfare Canada, Ottawa, ON, 65 p.

Hawley, J.K., 1985, Assessment of health risk from exposure to contaminated soil: Risk Analysis, v. 5, p. 289-302.

Hayward, R.J., 1973, Insurance plans and land use atlases: sources for urban historical research: Urban History Review, v. 1, p. 2-9.

Means, B. and Cooper, D., 1994, EPA looks at superfund reform: Health and Environment Digest, v. 8, p. 19-20.

Meharg, A.A. and Osborn, D., 1995, Dioxins released from chemical accidents: Nature, v. 375, p. 353-354.

Metropolitan Toronto Teaching Health Units and South Riverdale Community Health Centre, 1994, Why barns are red: Health risks from lead and their prevention: A resource manual to promote public awareness: Toronto, ON, 81 p.

Ontario Ministry of Health, 1989, Mandatory health programs and services guidelines: Queen's Printer for Ontario, Toronto, ON, 41 p.

Ontario Ministry of Housing, 1993, Non-profit housing environmental site assessment review handbook — November 1993: Toronto, ON, 52 p.

Ontario Ministry of the Environment (OMOE), 1989, Guidelines for the decommissioning and clean-up of sites in Ontario: Queen's Printer for Ontario, Toronto, ON, 28 p.

Ontario Ministry of the Environment (OMOE), 1991, Soil clean-up guidelines for decommissioning of industrial lands: background and rationale for development: Prepared by the Phytotoxicology Section, Queen's Printer for Ontario, Toronto, ON, 33 p.

Ontario Ministry of the Environment and Energy (OMOEE), 1994, Scientific criteria document for multimedia environmental standards development — Lead: Standards Development Branch, Toronto, ON, 310 p.

Ontario Ministry of the Environment and Energy (OMOEE), 1994, Proposed guideline for the clean-up of contaminated sites in Ontario: Toronto, ON, 74 p.

Ontario Ministry of Municipal Affairs, 1993, A new approach to land use planning: Consultation paper: Queen's Printer for Ontario, Toronto, ON, 35 p.

Pastenbach, D.J., Jernigan, J.D., Bass, R., Kalmes, R. and Scott, P., 1992, A proposed approach to regulating contaminated soil: identify safe concentrations for seven of the most frequently encountered exposure scenarios: Regulatory Toxicology and Pharmacology, v. 16, p. 21-56.

Saxe, D., 1994, The new guideline for the cleanup of contaminated sites in Ontario: Hazardous Materials Management, v. 6, p. 60-61.

Schineldecker, C.L., 1992, Handbook of Environmental Contaminants: A Guide for Site Assessment: Lewis Publishers, Chelsea, MI, 371 p.

Shephard, S.C., Gaudet, C., Sheppard, M.I., Cureton, P.M. and Wong, M.P., 1992, The development of assessment and remediation guidelines for contaminated soils — a review of the science: Canadian Journal of Soil Science, v. 72, p. 359-394.

Simmons, S.A. and Lewis, W.K., 1993. Health and Safety, *in* Cairney, T., ed., Contaminated Land: Problems and Solutions: Chapman and Hall, London, UK, 351 p.

Statistics Canada, 1980, Standard Industrial Classification: Standards Division, Statistics Canada, Minister of Supply and Services Canada, Ottawa, ON, 350 p.

Thornton, I., 1992, Sources and pathways of cadmium in the environment: International Agency for Research on Cancer Scientific Publications, v. 118, p. 149-62.

28. Radon in Urban Environments

Imshun Je

Environmental Earth Sciences, University of Toronto, Scarborough Campus
1265 Military Trail, Scarborough, Ontario M1C 1A4

SUMMARY

Radon is an inert, naturally occurring radioactive gas produced by the decay of uranium and thorium in rocks and soils. It has a half-life of 3.825 days and breaks down into two polonium daughters (^{218}Po and ^{214}Po). These radioactive species are solids, and become attached to particulates and aerosols in the air which can then be inhaled into the lungs. It has been estimated that about 10% of all lung cancers in the United States may be due to radon exposure.

The distribution and concentration of radon in the environment is primarily limited by its short half-life, therefore, the transport processes that bring radon from source rocks to the surface need to be well characterized. The principal natural geologic sources are uranium-rich granites, gneisses and black shales; such strata underlie many urban areas in midcontinent North America. Locally, glacial deposits and soils containing clasts derived from these strata represent a significant radon source.

The purpose of this paper is to review the origin of radon, what is currently understood of the associated health risks to urban communities, and the problems associated with the historic unregulated dumping of radioactive wastes at several sites in Ontario, Canada.

RECENT IDENTIFICATION OF RADON AS AN ENVIRONMENTAL HAZARD

Radon is the single most important source of radioactivity that people are routinely exposed to, but, as a radiation hazard, has received comparatively very little attention. Natural background radiation levels are approximately 3 mSv·a^{-1} (CNA, 1991; Clarke and Southwood, 1989). A sievert (Sv) is a measure of the actual biological damage to be expected when the body is irradiated. A dose from a single medical X-ray is equivalent to 10^{-5} Sv. A dose of 0.1 to 7 Sv will cause chronic radiation sickness; greater than 7 Sv is fatal (Upton, 1982). For the average person, natural background radiation constitutes over 70% of the total typical exposure to radiation, which includes X-rays, cosmic rays and microwaves (Fig. 1), of which two-thirds is from indoor radon. The health risks associated with residential radon have only recently been recognized and are still under much investigation, and only tentative guidelines have been proposed for acceptable levels of radon concentrations. It has long been known (since the sixteenth century in fact) that there is a much higher frequency of lung-related deaths among uranium miners than the rest of the general population. After the discovery of radioactivity in the early twentieth century, Margaret Uhlig proposed, in 1921, that radium emanations might be the cause of the lung cancers (Cothern and Smith, 1987). Since then, the association between radon and lung cancer has been well established for miners, but it was not until 1984 that radon was recognized as a hazard affecting communities at large.

On December 2, 1984, a Philadelphia nuclear power plant worker named Stanley Watras set off the radiation-detecting alarms as he made his way *into* work that morning (Brown *et al.*, 1992; Pearce, 1987). He continued to set off the alarms every morning for a week, until it was finally discovered that

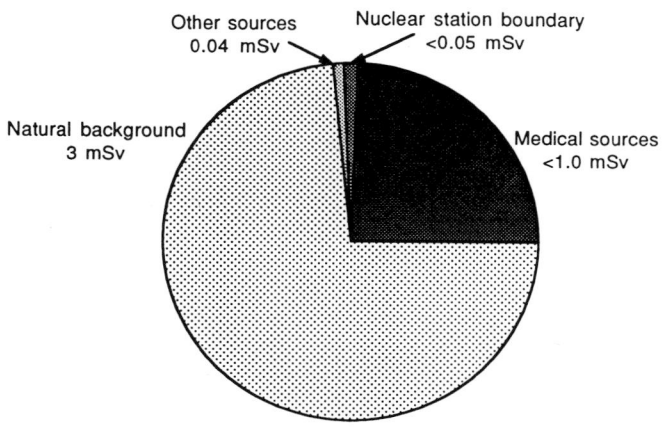

Source	Amount of Radiation (mSv)(1987)
Radon from earth materials (98% from indoor exposure)	2.0
Terrestrial γ-rays	0.28
Ingested natural radionuclides	0.39
Medical exposures	0.53
Cosmic rays	0.27
Fallout (including Chernobyl)	0.0006
Miscellaneous sources	0.05-0.13
Occupational exposure	0.009
Radioactive effluent discharges	0.0005
Total	**3.6**

Figure 1 *Sources of radiation and average annual radiation dose in Canada (from Clarke and Southwood, 1989; and Canadian Nuclear Association, 1991).*

the source of the radioactivity was in his own home. It was found that the Watras house was built on an excavated vein of uranium that forms part of a uraniferous geological formation, called the Reading Prong, that trends in a northeast-southwest direction stretching from Pennsylvania through to New York State. Breathing the air in the Watras' house posed the same risk of contracting lung cancer as smoking 135 packs of cigarettes per day (Pearce, 1987). Although the amount of radon radiation found in this instance represents the most extreme case, more than one-third of North America has potentially high geological radon production (Gundersen et al., 1992; Tilsley, 1992; see below).

PRINCIPLES OF RADIOACTIVITY

A radioactive atom results from the natural tendency for a nucleus to move from a higher to lower energy state in which the protons and neutrons are more tightly bound (Cothern, 1990; Draganic, 1993; Wilkening, 1990). The nucleus achieves this lower energy state by the emission of radiation, of which there are three types:

1. Alpha (α) emission.   An α-particle is a positively charged helium nucleus. These are relatively heavy particles that can only travel a few centimetres in the air before combining with free electrons to become neutral helium atoms. In the case of radon, this process occurs in following form:

$$^{222}Rn = {}^{218}Po + {}^4He^{2+} \qquad (1)$$

These particles can easily be stopped by a sheet of paper or by human skin. However, they can be very biologically damaging when consumed or inhaled into the human body.

2. Beta (β) emission.   A β-particle is simply a fast-moving electron. By its expulsion from the nucleus, an atom gains one more proton and becomes a different element.

$$n = \beta^- + p \qquad (2)$$

This type of radiation can penetrate 1 cm of water and human flesh, but can be stopped by a thin sheet of aluminum several millimetres thick. In air, it can travel less than a metre before it combines with positive ions and forms neutral atoms.

3. Gamma (γ) emission.   A γ-ray is part of the electromagnetic spectrum, consisting of high-energy photons. It is highly penetrating, with a greater amount of penetrability than X-rays, but can be stopped by a 1-m thick block of concrete or by appropriate thicknesses of other materials. Its emission creates no change in the element.

$$n = \gamma + n \qquad (3)$$

Inherent instability of the nucleus seems to occur among elements with 82 to 92 protons (Pb to U) and mass numbers (neutrons plus protons) between 204 to 238 (Wilkening, 1990), although there are radioactive elements with smaller numbers of protons, such as tritium (^3H) and radiocarbon (^{14}C).

As radioactive decay tends to proceed from lower to higher stability, the proton number decreases. In nature, there are three major decay series, two that begin with uranium (isotopes ^{238}U and ^{235}U) and the third with thorium (^{232}Th), all of which terminate with lead (Fig. 2). The stability of radioactive elements is normally measured by their half-life, which is the time required for half of the nuclei in a sample of a particular nuclear species to decay. Common units of radioactivity measurements are listed in Table 1.

Properties of Radon

Radon is a naturally occurring radioactive gas produced within the Earth's crust from the decay of uranium and thorium. It is the element 86 of the Periodic Table and is one of the noble gases along with helium, neon, argon, krypton, and xenon. Radon has no colour or smell and is almost chemically inert, although it is known to form compounds such as fluorides and clathrates. Clathrates form when radon atoms become incorporated into the crystal lattices of certain hydrogen compounds. Radon has a density of 9.73 g·L^{-1}, which makes it the heaviest gas under standard conditions (Wilkening, 1990). It also has the highest melting point, boiling point, critical temperature, and critical pressure of any gas (see Table 2). It is soluble in cold water, and its solubility decreases with increasing temperature.

The three most common isotopes of radon are ^{219}Rn, ^{220}Rn and ^{222}Rn, all of which are radioactive. The isotope ^{219}Rn has a very short half-life of only 3.96 seconds, and is formed through the decay of ^{235}U. This form of uranium has a low relative abundance of 0.7% of all naturally occurring uranium (Cothern, 1990; Fauré, 1986; Wilkening, 1990). With such a low abundance and short half-life, ^{219}Rn is not considered to be environmentally significant.

Radon-220 also has a short half-life (56 seconds), and is formed through the decay of ^{232}Th, which is more abundant in nature than uranium, having a global average concentration of 11 ppm in the crust, compared with that of uranium which is only 2 to 3 ppm (Wilkening, 1990). The isotope ^{220}Rn (also referred to as thoron) may constitute a significant fraction of natural radon gas emissions in some environments. Martell (1985) and Schery et al. (1989) have identified thoron's significance to overall radioactivity in soil gas.

The most important isotope of radon as an environmental hazard is ^{222}Rn, which has a half-life of 3.82 days. It breaks down through a succession of decay products into ^{210}Pb, which is easily removed from the atmosphere by precipitation. During the decay process, two radioactive polonium daughters (^{218}Po and ^{214}Po) are produced, and become attached to particulates and aerosols in the air which can then be inhaled into the lungs and cause damage to surrounding tissues by the emission of alpha (α) particles. Although ^{222}Rn is also a relatively short-lived radioactive isotope, it is able to be transported by soil gas diffusion, atmospheric convection or ground-water flow and become widely distributed away from its source. This form of radon is one of the progeny of the ^{238}U decay series. An important parent radionuclide is ^{226}Ra because of its mobility under reducing conditions. Of total uranium, ^{238}U is by far the most abundant naturally occurring uranium isotope (99.3%).

Since radon is soluble in water, it may travel long distances from ground-water sources and surface waters. It can enter into households from wells and be released into kitchens and bathrooms (Tilsley, 1992). It is more soluble in air than in water, and will readily pass into the gas phase when in contact with air, especially if water is agitated mechanically, such as in a bathroom shower.

Radioactivity of radon is commonly measured in picocuries (pCi). Ambient air has an average radon level of about 0.2 pCi per litre of air; ambient soil has levels between 20 pCi·L^{-1} to >100,000 pCi·L^{-1} and that associated with radon dissolved in ground waters can be as high as 3,000,000 pCi·L^{-1} (Otton, 1992; Fig. 3). Indoor air has radon levels that averages about 2 pCi·L^{-1}, and can be as high as 5000 pCi·L^{-1} in uranium mines, but as high as 7000 pCi·L^{-1} in homes (Ennemoser et al., 1993; see below).

GEOLOGIC CONTROLS ON RADON

The concentration of naturally occurring radon in soils is primarily controlled by two factors: (1) source rock, and (2) transport processes from source rocks into soil pore spaces. Even with a low relative abundance of the source material, rapid transport processes, such as the movement of ground water, can cause indoor radon accumulations to exceed the proposed action levels. This is a very important consideration where radon potential studies are carried out relying solely on geological and gamma-ray surveys. Several studies (*e.g.*, de Jong *et al.*, 1993; Hand and Banikowski, 1988; Kodosky, 1994) have clearly shown that such surveys can grossly underestimate actual soil gas concentrations of radon. Detailed soil-sampling surveys are required to determine the abundance of radon.

Source Rocks

The most important natural sources of radon are uranium-rich rocks and soils. There are two principal isotopes of uranium, ^{238}U and ^{235}U, with relative abundances of approximately 99.3% and 0.7%, respectively (Dyck, 1978). A third isotope, ^{234}U, is a decay product within the ^{238}U decay series, but it is only present in 0.0054% of occurrences (Fauré, 1986). The half-life of ^{238}U is about 4.5×10^9 years, which is nearly equal to the estimated age of planet Earth.

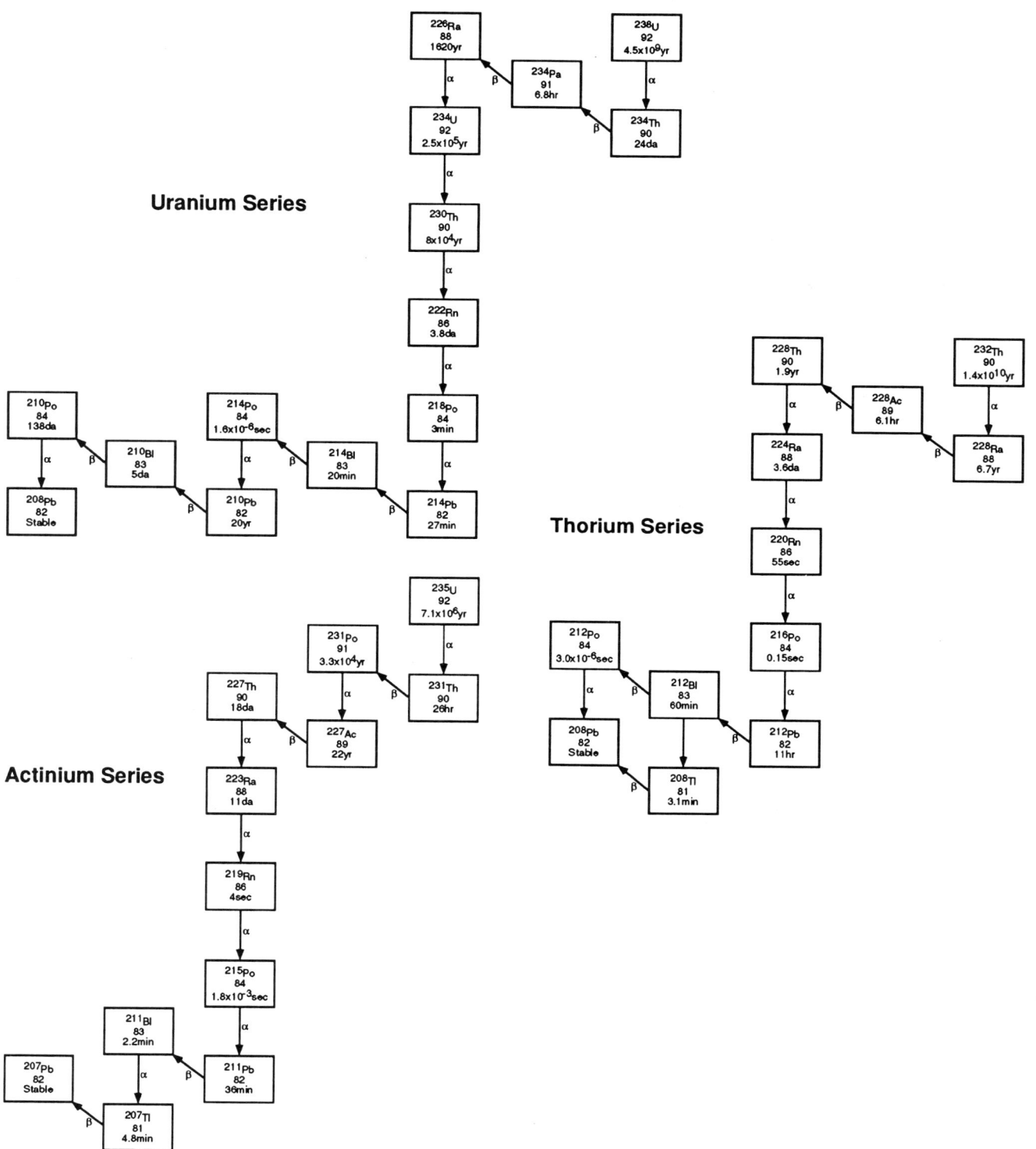

Figure 2   *Decay series for uranium, thorium and actinium showing generation of radon isotopes (highlighted; from Schmalz, 1990).*

Uranium concentrations display a very wide range, depending on rock type (Table 3). In general, sedimentary rocks that contain organic material are usually moderately enriched in metals, including uranium (Bell, 1978; Dyck, 1978; Tilsley, 1992). Organic-rich black shales and rocks containing phos-

phate show the highest values, some approaching 1000 ppm. Additionally, glacially eroded debris derived from uranium-rich source rocks has been transported long distances and spread over large areas in tills and outwash deposits (Gundersen et al., 1992; Tilsley, 1992).

Uranium usually occurs as oxides such as uraninite and carnotite, with phosphates and monazite sands being of importance in some cases. In general, ^{226}Ra is in equilibrium with ^{238}U. Thorium has a crustal abundance of 11 ppm, and is found in monazite sands in India, Brazil, the former Soviet Union and areas in North Carolina and Virginia in the United States. Monazite sands contain approximately 10% thorium, providing high external dose rates for local residents.

Table 3 gives the normal ranges of ^{238}U and ^{232}Th concentrations that can be expected in nature. Although ^{232}Th exceeds ^{238}U in the continental crust by a factor of 3.8, the activity of ^{232}Th exceeds that of ^{238}U by about 20%. Since the ^{232}Th decay series can only yield the ^{220}Rn isotope, which has a half-life of only 55 seconds, ^{238}U is the primary radioactive source for indoor radon, and is the main focus of the following discussion.

Uranium in Igneous and Metamorphic Rocks

Uranium, as well as thorium, has a tendency to be enriched in the more volatile phases in molten or partially melted rocks as they cool. Rocks with low melting points, such as granites, have a higher uranium content (10-40 ppm) than in higher temperature rocks, such as diorites and basalt (0.5-5 ppm). Generally, the higher the silica content of igneous rocks, the higher the uranium content. During metamorphism, rocks become more depleted in uranium as heating proceeds and the uranium becomes remobilized and concentrated into the

---

Table 1   Units of radioactivity (Cothern and Smith, 1987).

1 Curie (Ci)	= 3.7×10^{10} radioactive decays per second
1 picoCurie (pCi)	= 0.037 radioactive decays per second
1 becquerel (Bq)	= 1 radioactive decay per second
1 Bq	= 27 pCi
1 pCi·L^{-1}	= 37 Bq·m^{-3}
1 Working Level	= 100 pCi·L^{-1}
(WL)	= 1.3×10^5 MeV
1 gray (Gy)	= 1 joule per kilogram
	= 1.0×10^4 erg per gram = 100 rad
1 sievert (Sv)	= 100 rem = 100 rad

Table 2   Physical properties of radon (^{222}Rn; after Cothern, 1987; Schmalz, 1990; and Wilkening, 1990).

Density at 1 atm and 0°C	9.73	g·L^{-1}
Boiling point	–62	°C
Melting point	–71	°C
Critical temperature	104	°C
Critical pressure	62	atm
Diffusion coefficient in free air	0.1	cm^2·s^{-1}
Viscosity at 1 atm and 20°C	229.0	micropoise
Solubility in water at 20°C and 1 atm partial pressure	230	cm^3·kg^{-1} water
Solubility in various liquids at 1 atm and 18°C:		
Glycerine	0.21	cm^3·kg^{-1} liquid
Ethyl alcohol	7.4	cm^3·kg^{-1} liquid
Petroleum (liquid paraffin)	9.2	cm^3·kg^{-1} liquid
Toluene	13.2	cm^3·kg^{-1} liquid
Carbon disulphide	23.1	cm^3·kg^{-1} liquid
Olive oil	29.0	cm^3·kg^{-1} liquid

Table 3   Typical concentrations of uranium, uranium-238 and thorium-232 in rocks and soils (after Wilkening, 1980).

Rock Type	^{238}U (ppm)	^{238}U (Bq·kg^{-1})	^{232}Th (ppm)	^{238}U (Bq·kg^{-1})
Igneous				
Basalt	0.5–1	7–10	3–4	10–15
Granite	3	40	17	70
Sedimentary				
Shale, Sandstone	3.7	40	12	50
Carbonate rocks	2	25	2	8
Black shale	8			
Continental upper crust (average)	2.8	36	10.7	44
Soils	1.8	66	9	37

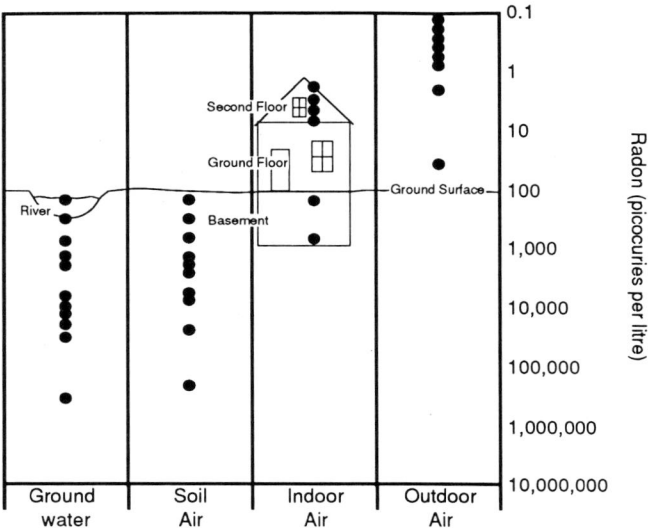

Figure 3 *Variation in radon concentration in ambient air, soil, ground water and residences (after Otton, 1992).*

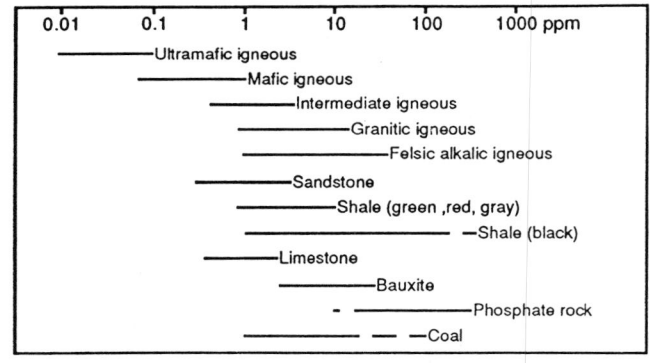

escaping gases and liquids; pegmatite veins become enriched in uranium and thorium by the same processes (Keppler and Wyllie, 1990).

Uranium in Sedimentary Rocks

Under certain conditions, sedimentary rocks can become a significant reservoir for the adsorption of uranium. The presence of clay minerals and organic matter enhances the adsorption of uranium within sediments once uranium has been weathered from its host rock (Flexser et al., 1993). Coprecipitation with iron oxide and formation of secondary minerals in pore spaces or fractures also commonly scavenge uranium and radium from solution. Thus, the surface of sedimentary grains can be relatively enriched in radium. Flexser et al. (1993), de Jong et al. (1993), Keller et al. (1992), Loureiro et al. (1990), and McCallum (1992) confirm this apparent relationship between grain size and radium content, with increasing content of clay being associated with higher radium content.

Black shales. Black, organic-rich shales are associated with high uranium concentrations, and were first reported for lower Paleozoic shales of the Baltic region in the 1890s (Bell, 1978). A shale is considered organic-rich if it has an organic carbon content greater than 2% (Bell, 1978). The uranium content of black shales is usually around 8 ppm (compared with average crustal abundance of 2-3 ppm), and is linked with the organic fraction of the black shales (Bell, 1978; Otton, 1992). During deposition, uranium is sorbed onto organic

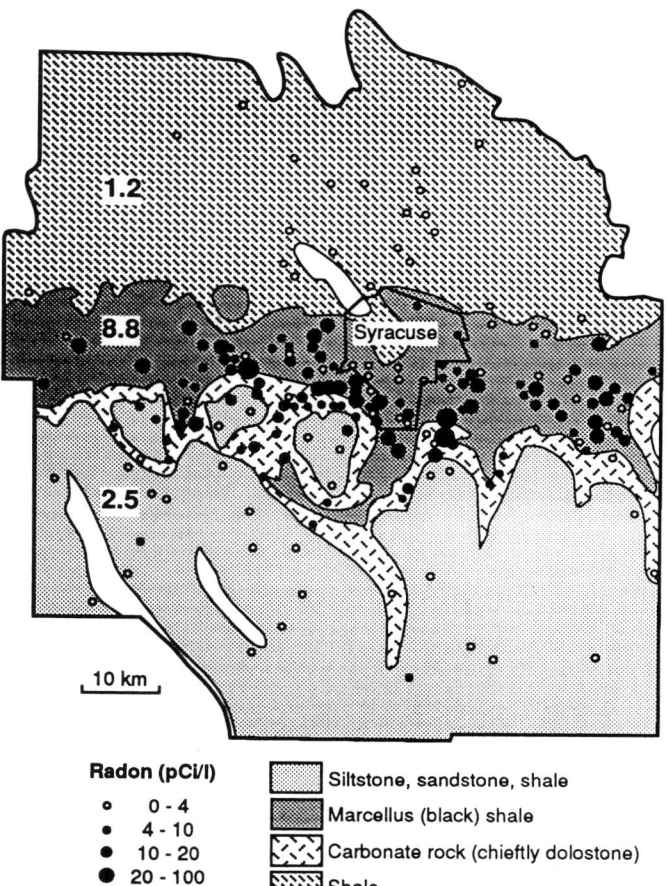

Radon (pCi/l)

○	0 - 4
•	4 - 10
●	10 - 20
●	20 - 100
⬤	>100

Siltstone, sandstone, shale

Marcellus (black) shale

Carbonate rock (chiefly dolostone)

Shale

Figure 4 *Radon activity (pCi·L⁻¹) in basements of single family dwellings in Onondaga County, near Syracuse, New York State. Large numbers are mean values for geologic units. Note restriction of high values to Marcellus (black) Shale (from Hand and Banikowski, 1988).*

compounds, and is directly precipitated as colloidal uraninite, both at the sediment interface and at shallow depths within the sediments under the influence of hydrogen sulphide (Harrel et al., 1991). Other metals also become preferentially adsorbed in black shales due to the high organic content. In anoxic waters, typical of deep marine basins in which black shales accumulate, reducing conditions dominate, and cause hexavalent uranium to be converted into a tetravalent form which is insoluble and is readily adsorbed on organic matter and becomes incorporated into uranium-organic complexes (Gates and Gundersen, 1989; Hand and Banikowski, 1988). Much of the organic material is composed of terrestrial cellulose and lignin and humic derivatives, which are very effective scavengers of uranium from sea water.

A case study in Syracuse, New York (Hand and Banikowski, 1988) found elevated levels of radon associated with Paleozoic sedimentary strata (Fig. 4). The Marcellus (black) Shale of Silurian age is enriched in uranium, and is a source rock for uranium found in associated limestone and dolostone strata and transported by hydrothermal ground waters. The redistribution of uranium in the shales has occurred over several tens of millions of years, and gives rise to what has been termed a hot belt of radon 12 km wide. The geological setting of the Syracuse area is typical of much of mid-continent North America, where shales are an important component of large intracratonic basins such as the Appalachian, Michigan and Illinois basins (see Leighton et al., 1990).

Harrell et al. (1991) investigated the potential radon hazards associated with Devonian shales in Ohio, and found similar results to those reported by Hand and Banikowski (1988). The average uranium concentrations found in continental crustal materials are typically around 2-3 ppm, and this results in radon radioactivity levels on the order of 1.0-1.5 pCi·L⁻¹ inside houses. The Devonian shale in Ohio has a high uranium concentration (10-40 ppm), resulting in a mean radon concentration of 1165 pCi·kg⁻¹ within the shale. Measured radon levels in homes within the study area correlated very well with the radon levels in the rocks, with indoor values ranging between 2 pCi·L⁻¹ to 20 pCi·L⁻¹ with a maximum of 94 pCi·L⁻¹.

In southern Ontario, the northern part of the Greater Toronto Area is underlain by the Middle Ordovician Collingwood Member of the Lindsay Formation, which is a bituminous oil shale (Churcher et al., 1991). Background uranium concentrations of 6 ppm are reported by Bell (1978), but this figure is based on few samples and is not easily related to likely radon levels in soils, given the ability of transport processes to concentrate radon in near-surface sediments (e.g., Kodosky, 1994). Of greater importance, is the radon levels in glacial sediments and soils.

Glacial deposits and soils. Glacial deposits such as tills, outwash and glaciolacustrine sediments cover a very large area of mid-continent North America, and contain granite and gneiss clasts derived from the Canadian Shield. The role of glacial deposits in controlling radon levels is not well known. In some instances, the glacial overburden acts as a barrier to upward migration of radon, especially if it is thick or of low permeability. However, some locations with greater thicknesses of overburden also display high levels of radon, suggesting that uranium-bearing clasts and boulders can make a significant contribution to overall radon levels (Harrel et al., 1991).

A study of Saskatchewan soils by de Jong et al. (1993) demonstrated a clear correlation between radon gas levels and the clay content of soils. Soils derived from lacustrine sediments showed 50% greater radon levels than tills. Radio-activity levels ranged from 3 to 8 Bq·kg⁻¹ in tills, with a mean of 6.0 Bq·kg⁻¹, and from 9 to 13 Bq·kg⁻¹ in the lacustrine soils, with a mean of 10.6 Bq·kg⁻¹. Levels of ^{40}K, ^{238}U and ^{232}Th decreased as soil texture became coarser.

The importance of clay content in determining ^{226}Ra concentrations in soils is supported by data from Ontario. Mc-Callum (1992) determined ^{226}Ra concentrations in sandy soils associated with coniferous forests in northern Ontario and in soils typical of mixed forests in southern Ontario. The mean ^{226}Ra concentration of Ontario soils ranges from 0.035 Bq·g⁻¹ to 0.053 Bq·g⁻¹, with the highest concentrations in fine-textured soils of southern Ontario. This study was completed to determine natural background variation in ^{226}Ra in soils, in order to establish ALARA (as low as reasonably achievable) values for remedial action projects involving the clean-up of sites contaminated by ^{226}Ra. Upper limits of normal (ULN) concentrations were also of interest, again as a basis to establish the degree and extent of contamination of sites where low-level radioactive waste is present. The ULN values reported by McCallum (1992) for Ontario soils range from 0.063 Bq·g⁻¹ to 0.113 Bq·g⁻¹ with an average of 0.083 Bq·g⁻¹. Because of the considerable range in ULN values, it is difficult to select any one value as a threshold for uniquely identifying soils as being radioactive at contaminated sites. This is because native soils having a higher ULN would be required to be classified as contaminated. The study of McCallum identifies the need for further study of background variation, and particularly the nature of ^{226}Ra profiles with depth in soils.

Tilsley et al. (1993) report results of a regional soil radon gas sampling program in southern Ontario, and found a distinct correlation of elevated radon levels in soils with areas of known oil and gas deposits within underlying Paleozoic strata. It was suggested that the transport of uranium and radium to surface environments was by volatile hydrocarbons escaping upward along faults and fractures. Surface release of gas near faults is also reported by Noor et al. (1992; see Chapter 34). Regional bedrock faults and fracture systems are known to have controlled the development of subsurface hydrocarbon reservoirs in southern Ontario (Sanford et al., 1985; Sherwood Lollar et al., 1994) and guided the cutting of a channelled topography on the bedrock surface (Chapter 2), suggesting that faults and fractures extend from depth to surface. Overlying Pleistocene deposits are also fractured (Wills et al., 1992). The same structures may localize the release of radon in urban environments and control the spatial variation in soil radon gas.

A survey of 632 public schools in Metropolitan Toronto identified two schools with basement radon levels equal to or above the proposed action level of 2 pCi·L⁻¹; three schools had levels between 1 pCi·L⁻¹ to 2 pCi·L⁻¹ (Becker and Moridi, 1992). Controlling factors were not investigated. Given that the largest urban centre in southern Ontario (the Greater Toronto Area) lies above a network of bedrock channels and faults, research aimed at identifying the relationship between underlying geological structures and surface radon emanation should be actively pursued. It has already become common practice in California to predict earthquake events by closely monitoring fluctuations in radon levels in groundwater wells close to known faults (Pipken, 1994). Bedrock topography mapping may be a useful tool in identifying priority areas for soil radon gas surveys in southern Ontario, given the close relationship between such topography and geological structure and seismic activity (Chapters 2, 29). In general, there are few data regarding the systematic variation in radon across urban communities in Canada (see below), and such data are urgently required. The same conclusions apply to other elements such as lead.

Transport Processes from Source to Surface

Given the short half-life of radon, radon levels in surface environments are controlled by transport distances from source to surface. For example, where it takes a period equivalent to six half-lives (about 23 days) to pass through near-surface fractured rock and soil, about 99% of the gas will decay to non-mobile solids before reaching the surface (Harrell et al., 1991). Even in dry, porous, high-permeability materials, radon will be diminished 100-fold (only 1% will survive) after a diffusion distance of only 6 m. The three main transport mechanisms that bring radon from its source in soils or rocks to the surface environments are (1) transport from the solid phase to either gas or liquid in pore spaces by alpha recoil, (2) transport of radon relative to the gas or liquid by molecular diffusion, and (3) transport of radon within soil gas or ground water by advection.

Alpha Recoil

Alpha recoil is a microscopic-scale process that transfers radon from the solid grains of the source rock into the pore spaces. During alpha decay of radium to radon, the energy released causes the alpha particle to be fired out from the nucleus, and the recoil affects the newly formed radon progeny, which is fired in the opposite direction to that of the alpha particle (Fig. 5). Energies involved in this process are 1000 to 10,000 times typical chemical bond energies (Michel, 1987). For radon, the typical recoil range is about 20 to 70 nm, and several possible recoil paths can be identified. In the case of a radium atom located at a depth within the grain greater than the recoil range, the radon progeny remains embedded in the grain (A: Fig. 5). Other atoms may escape from the original grain, but become trapped in the adjacent grain (B: Fig. 5). Other atoms escape from the original grain and are fired into

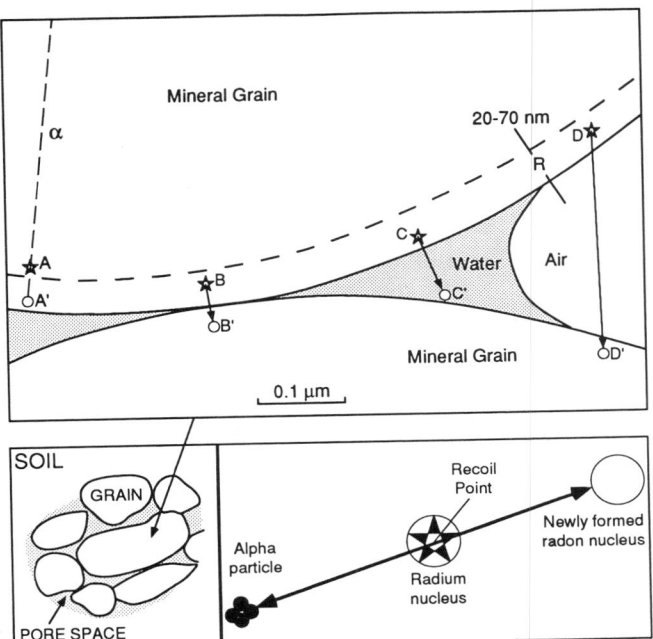

Figure 5 *(top) Radon emanation process; see text for details (from Michel, 1987). (bottom) Decay of radium atom to produce radon and an alpha particle. Radon is fired into mineral grains or pore spaces (after Otton, 1992).*

pore water, which stops the atom by absorbing any remaining recoil energy (C: Fig. 5). It is now free to diffuse through the pores. Finally, atoms can also be fired into pore air where little energy is absorbed. These may travel across pore spaces and become embedded in adjacent grains (D: Fig. 5).

The proportion of radon atoms that end their recoil path in a pore space, relative to the total number of recoils, is termed the emanation coefficient and the emanated atoms are called the direct-recoil fraction. The presence of water in pore spaces significantly increases the direct-recoil fraction, since water slows the recoiling atom travel better than air. Fleischer (1988) found that radon emanation is maximized by a thin film of water covering the solid grains, which has the effect of not only stopping atoms from travelling into adjacent grains, but also leaching any radon atoms that are embedded in the surface of adjacent grains (Rose *et al.*, 1990).

The recoil range, R, (Fig. 5) is a critical factor in determining how much radon escapes from the solid grain. In grains much larger than the recoil range, few recoil atoms will be able to escape by direct recoil. If the grain is 1 µm in diameter, about 5% of the atoms may recoil out of the grain, whereas for millimetre-size grains, only 0.005% leave the grain by direct recoil. Generally, in coarsely crystalline rocks or coarse-textured soils with a homogeneous distribution of radium, the emanation coefficient is less than 0.01. In fine-grained rocks and soils, the normal range is between 0.15 to 0.35 (Michel, 1987) . Radon emanation measured from various geologic materials is much greater than that calculated assuming direct recoil only, suggesting other processes of radon loss from mineral grains that, as yet, are poorly understood (Michel, 1987). Once a radon atom has been fired into a pore space, it can be transported by diffusion and advection.

Diffusion

Diffusion is the migration of the gas relative to the host medium; radon obeys the standard laws of diffusion, so the flux is proportional to the concentration gradient. Diffusion coefficients for radon within various media are given in Table 4. The major factors governing radon diffusion in a rock or sediment involve properties such as grain size, distribution of radium within the solid, porosity, the amount of interstitial fluids and the presence of fractures and their degree of interconnectedness. The relative contributions of each factor vary with varying sediments and rock types.

Diffusive transport of radon in soils is limited because of the short half-life of radon. Migration by diffusion ranges from about 5 m in gravel to about 2 cm in saturated mud or clay, and distances greater than 1 m are probably unusual (Michel, 1987; Rose *et al.*, 1990; Tilsley *et al.*, 1993). In unfractured crystalline rocks, the diffusion coefficients for radon are ex-

tremely small, and much of the radon will decay before moving any appreciable distance. The presence of water in the pore spaces also tends to decrease radon migration, because diffusion in water is about three orders of magnitude smaller than in air.

Advection

Advection either by ground-water movement or pressure-driven air flow can transport radon over significantly greater distances than by diffusion alone. Since radon is basically chemically inert, its transport in ground-water systems is often controlled by the velocity of ground-water flow itself. The limiting factor in the transport length of radon in ground water is its half-life. In 30 days, the radon content of ground water will be less than 1% of its original activity (Michel, 1987).

Transport of radon in air caused by pressure gradients is enhanced at times of low atmospheric pressure, high winds, or significant temperature gradients between the soil air and the atmosphere (Kokotti *et al.*, 1992; Loureiro *et al.*, 1990; Nazaroff *et al.*, 1987). The pressure-driven flow of radon is directly dependent on the permeability *(k)* of soils where the overall mean grain size is larger than silt or fine sand, and where $k > 1.0 \times 10^{-12}.m^{-2}$; below this value transport by diffusion dominates (Loureiro *et al.*, 1990).

In most cases, the transport distance of radon by diffusion and advective processes combined is less than 3 m and, thus, the source of radon is usually nearby within the soil or surficial sediment, and not in the deeper underlying bedrock (Michel, 1987). However, Harrell *et al.* (1991) demonstrated that nearly all the radon measured in residential homes built above black shales in Ohio had been transported distances of well over 30 m.

Radon Entry Into Residences

As a result of diffusive and advective transport, radon is able to enter residences *via* many possible pathways and entry points (Fig. 6). Due to its relative density, there tends to be a pronounced vertical gradient in radon concentration in ambient air. A child breathing at a height of 0.5 m is exposed, on the average, to about 16% higher radon levels than an adult breathing at 1.5 m (Michel, 1987). Kodosky (1994) found that basements tend to have radon levels 2 to 3 times higher than upper-level rooms. Hence, the general emphasis on basements in most radon studies.

The most important entry pathways are cracks and holes in the foundation, and uncapped sumps. These were responsible for 23% of radon seepage into the residences studied by Kodosky (1994) in southeast Michigan. Radon advected in ground water accounted for 11% of the indoor concentrations in homes that relied on private wells as their water source (see below). House depressurization and escape of heated air are particularly significant as they allow radon to be drawn into the houses through basement fractures. Negative-pressure conditions in houses, caused by central heating in winter months, enhance the release of radon from near-surface strata in contact with basement walls. Nazaroff *et al.* (1987) found that an indoor-outdoor pressure difference of between 25-50 pascals (Pa) substantially increased the rate of transport of radon through soil air and into houses; observed transport velocities exceed 1 m per hour. The entry rate of radon by pressure-driven flow can exceed that due to diffusion by an order of magnitude (Nazaroff *et al.*, 1987). Reduced indoor air pressures tend to occur in newer energy-efficient houses due to their airtight structure, and it has been suggested that these types of houses are at greater risk than older, better ventilated structures.

Table 4   Radon diffusion coefficients for various media (after Michel, 1987).

Medium	Diffusion Coefficient ($cm^2 \cdot s^{-1}$)	Diffusion Length (m)
Air	10^{-2}	2.4
Water	10^{-5}	
Sand	3×10^{-2}	1.5
Argillite	8×10^{-2}	
Concrete	2×10^{-5}	0.04–0.26
Mineral crystals	10^{-9}–10^{-20}	

Kokotti *et al.* (1992) found that the radon entry rate exponentially increases with increases in the outdoor-indoor pressure difference. The pressure difference provides the driving force behind convective flow and the pressure difference tends to increase with increasing air-exchange rate. This is a cause for concern, as it has been assumed in other studies that a higher ventilation rate in houses will dilute indoor radon concentrations (Hamilton, 1993; Kodosky, 1994). Kokotti *et al.* (1992) conclude that a high ventilation rate does not guarantee lowered indoor radon levels.

In residences in Umhausen, Austria, the summer median for basement areas is lower than in the winter by a factor of about 10 (Ennemoser *et al.*, 1993). About 40% of the basements register radon levels above 10,000 Bq·m^{-3} in winter, but only 7% reach those levels in summer. This clearly illustrates the importance of year-round monitoring for radon levels with particular attention to winter peak levels when central-heating systems are operative.

Natural radon levels in homes can approach those found in uranium mines. In uranium mines with only natural ventilation, typical radiation levels are in a range between 75,000 to 190,000 Bq·m^{-3} (2,000 pCi·L^{-1} to 5,000 pCi·L^{-1}; Wilkening, 1990). Forced air ventilation of mines in the late 1940s reduced these levels to about 2,000 Bq·m^{-3} (50 pCi·L^{-1}). In Umhausen, Austria, Ennemoser *et al.* (1993) identified extremely high indoor radon concentrations (up to 274,000 Bq·m^{-3}). The town is built on alluvial fan sediments having an extremely high permeability; the emanation factor is enhanced by the presence of fractures in underlying granite gneiss bedrock. These soil characteristics, coupled with the slightly enriched uranium content of the granite gneiss, are responsible for the elevated indoor radon levels. The medians measured were 3750 Bq·m^{-3} in winter and 361 Bq·m^{-3} in summer in the basements, and 1180 Bq·m^{-3} and 210 Bq·m^{-3} on the ground floors, respectively. Unusually high radon concentrations in Umhausen coincide with a statistically significant increase in lung cancer mortality.

The difference between median concentrations measured in winter and summer by Ennemoser *et al.* (1993) can not only be ascribed to central-heating effects (see above), but also to the great difference of soil permeability in summer and winter. In winter, the soil surface is frozen and radon becomes sucked into the houses by a low indoor air pressure. During summer, most of the soil emanates radon directly into the atmosphere, and little is able to penetrate into basements. No association was found between radon concentrations and building type; houses with earth basement floors showed the same radon levels as those with concrete basement floors. It is likely that the broad findings established by Ennemoser *et al.* (1993) can be applied to Canadian urban settings, given similar climatic conditions and heating practices.

HEALTH RISKS FROM RADON EXPOSURE

Most data appertaining to health risks is derived from United States studies; to date very little work has been conducted in Canada. Nero *et al.* (1986) made quantitative measurements in randomly selected buildings and homes throughout the United States, and concluded that approximately a million homes are exposed to radon levels that exceed the recommended action level of 2 working level months (WLM) per year (see Table 1). About 2% of the 60 million single-family residences in the United States are expected to have levels exceeding 8 pCi·L^{-1}. and it has been estimated that between 7,000 to 30,000 lung cancer deaths are attributed to radon each year; this is equal to approximately 10% of all lung

cancers in the United States (National Research Council, 1994). Cohen (1995) and Alavanja *et al.* (1994) concluded that no simple relationship between radon levels and lung cancer rates can be identified, and suggested additional confounding factors. As yet, relatively minor efforts have been put forward by governments to ensure public safety. With regard to radiation exposure in general, prolonged exposure at low doses is associated with more risk than shorter exposure at higher doses, although it has yet to be determined whether this relationship holds for the low radon levels found in residential areas. Exposures in childhood carry no greater risk than exposures at older ages and, not surprisingly, smokers are at much greater risk than non-smokers (NRC, 1994; Pershagen *et al.*, 1994).

As a noble gas, radon is chemically inert and does not form compounds. The major part of inhaled radon is, therefore, exhaled again. However, short-lived radioactive isotopes such as polonium accumulate in the respiratory system. Between 20% and 50% of inhaled radon daughters are retained in the respiratory tract, where alpha-emitting progeny expose the lung to radioactivity (Ennemoser *et al.*, 1993). Typically, most damage occurs to the surface walls of the bronchii leading to the lungs, because the mucous layer on the surface of the bronchial tubes is not heavy enough to absorb alpha particles and damage to the underlying basal cells results.

In the United States, the National Research Council (1994) warn that they cannot state with any degree of certainty the level of radiation below which there is no health risk. There have been protests raised in the United States that the costs of remediating all homes that have radon levels above the EPA guideline of 4 pCi·L^{-1} would amount to more than $50 billion (Horgan, 1994). The agency argues that some 15% of

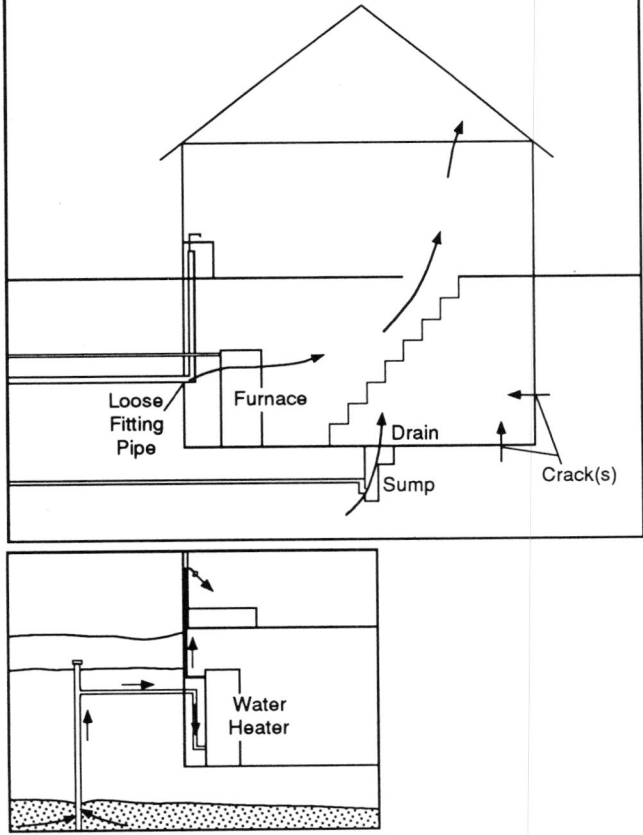

Figure 6 *Radon entry into residences (after Otton, 1992).*

lung cancer deaths caused by radon exposure could be avoided by reducing radon levels greater than 4 pCi·L⁻¹ (estimated to occur in 5 million of the 60 million homes in the US; Horgan, 1994). In addition, the International Commission on Radiological Protection of the United States has found that the risk of developing cancer from exposure to low levels of radiation from X-rays and gamma rays is three to four times as high as previously thought (Vaughan, 1990).

Proper remediation in residential basements need not be costly or complicated. Houses with earthen basement floors can be outfitted with several sheets of high- density poly-ethylene plastic laid loosely on the floor, a plastic pipe beneath the sheets to collect the radon gas and a mechanical fan in the attic that exhausts the gas outside, above the roofline. This technique successfully reduced indoor radon levels in houses on Mackinac Island, Michigan from between 12.9 pCi·L⁻¹ to 82.3 pCi·L⁻¹ to below 4 pCi·L⁻¹, and remained effective three years after its initiation (Hamilton, 1993). In other types of basements, entry points can be sealed by various readily available, inexpensive methods and in areas of very high radon levels, ventilated fans can also be installed. Typical costs range between $500 to $1500, depending on the size of the house (Tilsley, 1992). The remediation of homes with high radon levels does not pose a problem. In absence of any systematic coverage in Canadian communities, the major problem is in determining which areas are at risk.

Radon In Drinking Water

Radon dissolved in drinking water is an important secondary source of radon introduced into residences, particularly where water is derived from ground water. Consumption of radon in water is not as important as inhalation of radon that has escaped from the water source. Dissolved radon can be released into indoor air, however, if water is agitated by the use of common devices such as showers and faucets. In general, water containing a radon activity of 10,000 pCi·L⁻¹ will release 1 pCi·L⁻¹ into indoor air (Otton, 1992). It is estimated that 137 people die every year in the United States from exposure to radon in water, although drinking water only represents 1% of total exposure from all sources of environmental radon (Swistock et al., 1993). Extreme cases have been reported in Maine where radon from private wells drilled in granite bedrock contributes nearly 100% of indoor air levels, which were as high as 2000 pCi·L⁻¹ (74 Bq·L⁻¹) prior to remediation (Lowry et al., 1987). The United States Environmental Protection Agency (U.S. EPA) standard for radon in drinking water has been set at 300 pCi·L⁻¹ (11.1 Bq·L⁻¹; Valentine and Stearns, 1994). However, even at this level, there is still a cancer risk of 1 in 10,000 (Swistock et al., 1993).

Waterborne radon levels in rivers and lakes tend to be negligible (2.2 pCi per 100 kg; NRCC, 1983), since exchange with the atmosphere is rapid enough to allow significant amounts of radon to escape from the water phase. Thus, areas reliant on surface-water supplies are at a minimal risk. Large cities also generally do not encounter any problems with radon in drinking water, since processing in municipal treatment facilities aerates the water sufficiently to allow much of the existing radon to be released into the air. Furthermore, the residence time in the supply reservoirs is usually long enough that most of the remaining radon will have decayed to insignificant levels before ever entering residential outlets (Otton, 1992). However, in areas where ground water is used as the main source of water, dissolved radon can pose a hazard. The closed systems and short transit times of small public waterworks and private domestic wells

are not able to remove radon from drinking water, or allow it enough time to decay.

Systematic surveying of radon in Canadian drinking water has not been completed, but available data from pilot surveys suggest that relatively high levels of radon in Canadian drinking-water supplies are not uncommon. Beck and Brown (1987) identified radon levels above the U.S. EPA guideline in municipal and domestic wells of southern Ontario, with values up to 600 pCi·L⁻¹ (1650 Bq·L⁻¹). However, no provincial objective for radon in drinking water yet exists. The most widely tested and available method for remediating drinking water containing high levels of radon is through the use of granular activated carbon (GAC). This type of treatment system is able to achieve a steady-state removal greater than 99.95% over several years of use, even with radon levels of 37,000,000 Bq·m⁻³ in water supplies (Lowry et al., 1987). The disadvantage of this system is that, over time, the GAC itself will give off radiation as the accumulated radon decays, reaching levels as high as 3 to 5 times greater than background (Shapiro and Sorg, 1988). The simplest method is to increase indoor ventilation in rooms of heavy water use (such as in bathrooms), since inhalation of airborne radon is more hazardous than ingesting it in dissolved form (Otton, 1992).

RADON MEASURING TECHNIQUES

There are several methods available for the detection and measurement of indoor levels of radon. These include active and passive alpha track detectors, charcoal canisters, Electret Passive Environmental Radon Monitors (EPERMs), Continuous Radon Monitoring (CRMs) and quick sampling. There are certain advantages and disadvantages to each method. Usually the limiting factor is the cost of the equipment and the length of the study period. EPERMs and CRMs are the most expensive (between US $2500 to $10,000), and charcoal canisters the least at US $10 to $25 per unit (Anonymous, 1986).

Radiation from radon and radon daughters is recorded by alpha track detectors as tiny pits in a small piece of plastic. The number of pits is counted using a microscope or an automated counting system, and gives an estimate of the concentration of radioactive atoms (Anonymous, 1986). The accuracy of the results is increased if multiple detectors are used in the same area. Active detectors use a small pump to force air through a filter that traps the radon daughters, and only require a test period of 7 days, as opposed to a minimum of 30 days required by the use of passive detectors (Becker, 1991).

Charcoal canisters use activated charcoal to trap radon over a period of several days. Air diffuses through a screen on the canister, and radon will adsorb onto the charcoal and begin to decay. These devices do not generally provide an accurate average reading, due to the day-to-day fluctuations in radon levels and the short half-life of radon. Using this method requires that the canisters be quickly returned for analysis before much of the collected radon decays.

EPERMs have a plastic chamber that contains an electrostatically charged disc, or scintillation cell. The disc becomes discharged when it becomes exposed to radon in air. The concentration of radon is recorded as the difference in surface voltage on the disc, measured before and after exposure. This device can be used for short-term (2 to 7 days) and long-term (2 to 52 weeks) study periods (Becker, 1991). Similar devices, Continuous Radon Monitors (CRMs), use an air pump, and can be programmed to run continuously to record hourly radon concentrations for up to 24 hours (Anonymous, 1986).

Quick sampling, or grab sampling, pumps air through a filter or a measurement cell lined with zinc sulphide phosphor for about 5 minutes. Using a filter, the total alpha activity is counted and that number is converted to disintegrations. With the zinc sulphide cell, a photomultiplier tube counts the light pulses (scintillations) produced by alpha decays in the sample that react to the inner coating. The number of pulses is proportional to the radon concentration in the cell. This method gives a rough estimate of radon level and its probable source within a house. It is used only as a quick and general indication of the radon risk involved. It cannot give accurate measurements of average radon levels, since these levels can show large fluctuations on a daily and weekly basis, depending on ambient air pressure gradients, weather, temperature, etc. (Gundersen and Wanty, 1993).

Soil gas measurements can be made either by burying a passive monitor device, such as a charcoal canister, or by analyzing samples in an alpha-scintillometer. A new method was developed by the United States Geological Survey involving the latter technique (Gates and Gundersen, 1992; Otton, 1992). A hollow carbon steel probe is driven 1 m into the soil, and soil gas is drawn into the probe through small holes at the bottom tip and into a 20-mL syringe at the top. The gas samples are analyzed at the end of the day through an alpha-scintillometer. An older method involves digging holes, up to 2 m deep and 10 cm in diameter, and directly circulating samples through an alpha- scintillometer under high vacuum pressure for 5 to 15 minutes. The newer method is preferred, since it involves minimal disturbance of the soil and is less time consuming.

RADON RISK FROM UNREGULATED DUMPING OF RADIOACTIVE WASTES

The above discussion has focussed on naturally occurring radon, but, locally, some urban areas are at risk from unregulated historic dumping of low-level radioactive wastes. In the 1940s and 1950s, radioactive products were in widespread industrial use without any understanding of health risks associated with the manufacture, use and disposal of these products. Much radioactive material was simply dumped in outlying rural areas and forgotten. In the meantime, urbanization has encroached into these areas, resulting in unexpected discoveries of waste sites in residential areas. There is an estimated 1.2 million cubic metres of historic low-level waste (LLW) in Canada which will remain hazardous over the next 500 years (Auditor General of Canada, 1995). Most of this waste was produced in Port Hope, Ontario; other sites are located in Scarborough, Ontario (see below), in Surrey, British Columbia and in various areas in Alberta and the Northwest Territories.

Much of historic LLW has been recovered, and is now stored at temporary facilities, but several sites, principally in the Port Hope area and at several abandoned uranium mines in northern Ontario, have not been remediated. Unremediated sites at Port Hope are not thought to pose an immediate risk to the public, but could become a hazard if physical containment is disrupted or if proper waste management practices are not followed (see Chapter 2). Tailings at uranium mines that ceased operations prior to 1976, in the Elliot Lake area of northern Ontario and the Bancroft area of central Ontario, are not subject to current regulations, and have not been inspected or monitored since their abandonment (Auditor General, 1995). It was not until 1976 that the Atomic Energy Control Board (AECB) introduced a licensing system to ensure that uranium-mining companies comply with provincial

health and safety regulations. Since the impact on the environment remains unknown, the communities living near these tailings sites may be exposed to hazardous levels of contamination, of which radon is of primary concern. The AECB has begun discussions with known owners of the sites to include them under current regulatory control.

Example: Malvern, Scarborough, Ontario

Abnormally high indoor radon gas concentrations were detected in 1983 in two residential subdivisions in the southwest Malvern area of Scarborough, Ontario, at McClure Crescent-Burrows Hall Boulevard and McLevin Avenue (Fig. 7). Contamination results from the dumping of waste products of a small World War II radium incineration and processing plant which produced radium-luminous paint and nighttime markers. At that time, use of radium-226 (^{226}Ra) was unregulated, and wastes were dumped on nearby farms (McCallum, 1992). Above-background levels of radon in residential basements are the product of insoluble radium salts such as $RaSO_4$ (Haque, 1982). The health hazard associated with radon gas present is still under investigation, and only tentative guidelines have been proposed for acceptable levels of radon gas concentrations in basements. A level of 2 pCi·L^{-1} has been set as a tentative guideline in Canada (Haque, 1982). Typical background levels of Ontario soils are 0.1 Bq·g^{-1} or 2-3 pCi·g^{-1} (McCallum, 1992). The measured 98th percentile concentration for radium in Toronto area soils is 0.073 Bq·g^{-1} (2 pCi·g^{-1}).

At McLevin Avenue, remedial action has been taken to reduce radon concentrations seeping through basement walls, and approximately 2500 m^3 soil has been excavated and sorted (Acres International Limited, 1993). Radioactive markers were shipped to the Low-Level Radioactive Waste Management Office warehouse at the Atomic Energy of Canada Limited (AECL) Chalk River Laboratories for temporary storage. Some of the remaining mildly contaminated soil is temporarily stored in a single mound on the site that is surrounded by fencing. The nearest residences are 250 m away to the north. Radiation levels rapidly decrease to background levels within about 10 m of the mound (Acres International Ltd., 1993). There still remains subsurface contamination in 3867 m^3 of soil, in a band 60 m wide under the McLevin Avenue road allowance, which narrows to 15 m wide under the pathway to the Malvern Town Centre. It is considered to pose no health threat to pedestrians in the area and no residences are nearby. Further excavation and clean-up of the site is still required.

By 1983, 39 homes had been identified at McClure Crescent and Burrows Hall Boulevard as having contaminated soil that required removal (Fig. 8). In 1989, 15 homes had been identified with levels of interior radon and radon daughter levels exceeding the established criterion of 2 pCi·L^{-1}. It has been estimated that 7200 m^3 of contaminated soil requires excavation, but no soil has yet been excavated. On average, ^{226}Ra concentration levels in the contaminated soils had a value of 1.4 Bq·g^{-1} and a maximum value of up to 12 Bq·g^{-1} (320 pCi·g^{-1}). The exposure associated with living inside a house on uncontaminated soil all year round with natural background levels is equal to about 20 X-rays. The Malvern exposure is thought to be double that, equal to about 40 X-rays (Acres International Ltd., 1993). A site investigation is currently underway at the location for a semi-permanent storage facility for additional excavated and treated contaminated soils (see Chapter 32) until a permanent facility can be developed in perhaps 5 to 10 years. The proposed site is located immediately south of Passmore Avenue between Tapscott Road and Neilson Road (Fig. 7).

Example: Port Hope, Ontario

Port Hope was a major centre for radium processing from 1933 to 1952, which resulted in severe contamination of the inner harbour area (Hart *et al.*, 1986). In 1983, the International Joint Commission found that uranium concentrations in harbour waters exceeded the maximum acceptable concentration for drinking water (20 mg·L⁻¹; International Joint Commission, 1983). Surveys carried out in 1981 and 1982 showed that radium activity levels in the harbour were near or below the detection limit of 37 mBq·L⁻¹ (1.0 pCi·L⁻¹), and a 1986 survey showed that sediments within the inner harbour had radium levels around 1.56 Bq·g⁻¹ (42.12 pCi·g⁻¹; Hart *et al.*, 1986).

Several investigations have identified deformities in the larvae of midge flies (chironomids) in Port Hope Harbour. Hart *et al.* (1986) identified bioaccumulation factors (ratios of the amount of a trace substance incorporated into body tissue to the amount in the organism's environment) for radionuclides and heavy metals in 40 benthic invertebrate taxa (including chironomids). Radium-226 had the second highest bioaccumulation factor (1.94) after ²²⁸Th (9.59). Warwick (1991) studied deformities in chironomid larvae across several aquatic basins in western Canada, along the St. Lawrence Seaway and including Port Hope Harbour. Port Hope Harbour has the highest frequency of deformities (12.28%) of any of the other sites except Lac St. Louis, along the St. Lawrence Seaway, where the index chironomid species has completely disappeared due to the high levels of contamination. Warwick *et al.* (1987) estimated the radiation dose to chironomids in Port Hope Harbour at 1 mGy per day. This dose is significant enough to suggest that radionuclides in the harbour are the

cause of the observed deformities. However, because of high concentrations of heavy metals that also occur in the harbour, it is difficult to isolate the effects of the radionuclides alone. Apart from chironomids, similar radiation doses have been found not to be significantly detrimental to other invertebrates or small mammals (Warwick *et al.*, 1987). There is a clear need to determine what levels of radionuclides pose a hazard to different types of aquatic organisms before any one organism can be employed as a reliable bio-indicator of the degree of radioactive contamination in a given environment.

Example: Elliot Lake, Ontario

Elliot Lake was, until recently, the pre-eminent uranium-producing region of Canada; production has since shifted to Saskatchewan. Large-scale mining operations in this area have existed since 1954 (Dubrovsky *et al.*, 1984), and resulted in the dumping of 120 million metric tonnes of uranium tailings over an area of 600 ha (Dave *et al.*, 1985; Wren *et al.*, 1987; Payne, 1995). Uranium that could not be extracted was left as solid residues, and discarded along with the rest of the tailings to eventually become a source of radionuclide contamination through decay into radium, polonium and lead. Tailing waters have a pH between 6 to 9, but seepage and runoff from abandoned tailings have pH values of 2 to 3 (Silver *et al.*, 1985); progressive acidification of the tailings promotes the leaching of radionuclides into ground and surface waters (Moffett and Tellier, 1978). Acidification results from the chemical and biological oxidation of pyrite within the waste material (see Chapter 14); typical levels of radium in Elliot Lake tailings are about 336 pCi·g⁻¹.

Predatory mammals (*e.g.*, otter, raccoon, mink) that feed upon benthic aquatic organisms near tailing sites accumulate radium. Detectable levels of ²²⁶Ra have been found in otters, with values ranging from 0.2 pCi·g⁻¹ to 12.6 pCi·g⁻¹ (Wren *et*

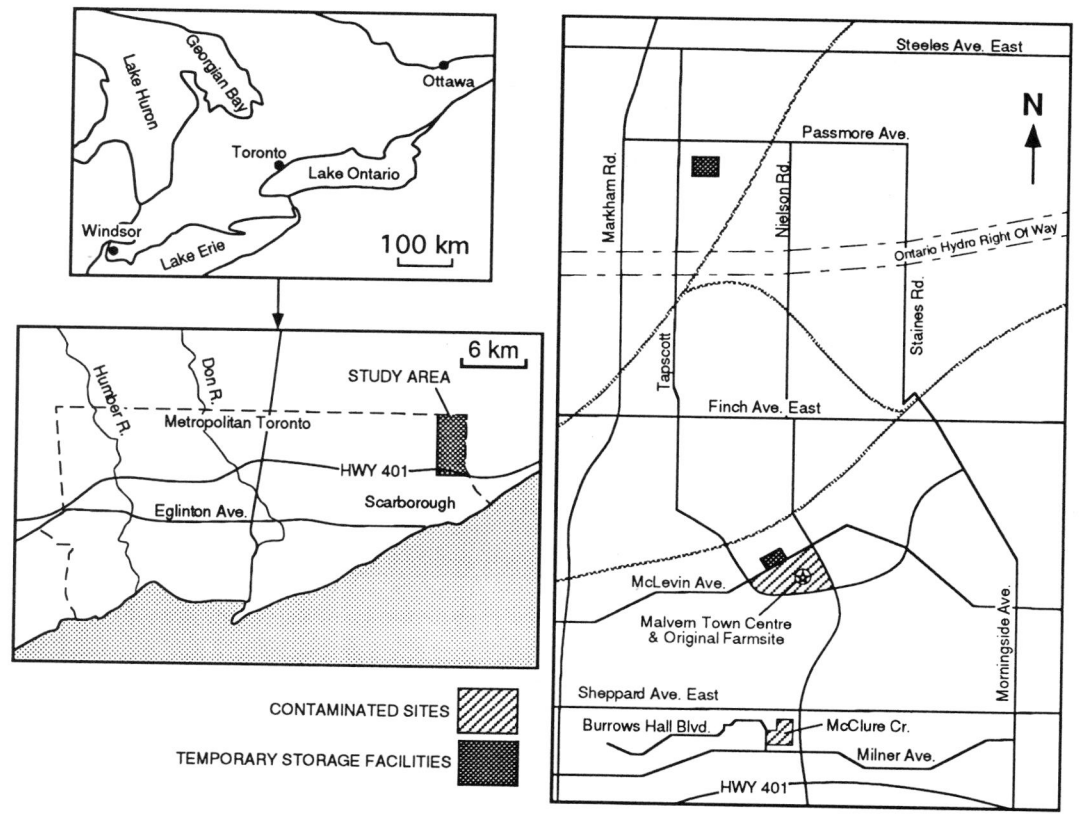

Figure 7   *Contaminated sites in Malvern Area of Scarborough, Ontario (after Acres International, 1993).*

al., 1987). A control otter from a remote area did not show detectable levels of radium. The source of radium is likely fish and clams. River and lake waters sampled near the uranium tailings contained 118.1 mBq·L⁻¹ (3.19 pCi·L⁻¹) of dissolved radium, compared to a control site near Sudbury, Ontario, which had a value of 12.1 mBq·L⁻¹ (0.33 pCi·L⁻¹).

Clulow et al. (1992) found elevated radium levels in ruffed grouse near tailing sites in the Elliot Lake watershed. Mean values were found to be 28.5 mBq·g⁻¹ (0.77 pCi·g⁻¹). The leaves of aspen and fungal material upon which they feed had mean radium levels of 52.7 mBq·g⁻¹ (1.42 pCi·g⁻¹) and 215.4 mBq·g⁻¹ (5.82 pCi·g⁻¹), respectively, which does not suggest strong bioaccumulation of radium in grouse. Overall, the consumption of grouse from Elliot Lake does not constitute a health problem. Above-normal levels of radium have been found on the surface of blueberries from tailings spill sites, but it is highly unlikely that a person would consume a large enough quantity to exceed the recommended dose limit. Samples collected within 500 m of tailing piles had radium levels between 20 mBq·g⁻¹ to 290 mBq·g⁻¹ (0.54 pCi·g⁻¹ to 7.83 pCi·g⁻¹), while natural background levels were 2 Bq·g⁻¹ to 6 Bq·g⁻¹ (0.054 pCi·g⁻¹ to 0.162 pCi·g⁻¹). To exceed the dose limit, a person would have to consume 47 kg of blueberries per

year. Wind dispersal of dust is primarily responsible for transporting and depositing radium onto the blueberries.

The Auditor General of Canada (1995) warns that radon gas is being released from uranium tailings, and may be a hazard to those living close to the sites or if the tailings are used in construction. Radioactive waste will continue to pose a hazard for tens of thousands of years. Impacts on the environment and public safety of tailings sites abandoned prior to 1976 are not known, leaving the possibility that people are exposed to unacceptable levels of radioactivity and chemical toxicity.

DISCUSSION

The occurrence of radon in urban environments appears to be a common, but poorly understood, phenomenon. Work in the United States and Europe suggests radon can be a significant hazard; in the absence of data in Canadian settings, there is no room for complacency. With improvements in understanding of the geologic controls on radon, its importance as a potential hazard is increasingly apparent. Recent work clearly shows that natural background radiation from radium in natural soils may in fact be sufficiently high to warrant classification of natural materials as radioactive. This stresses the

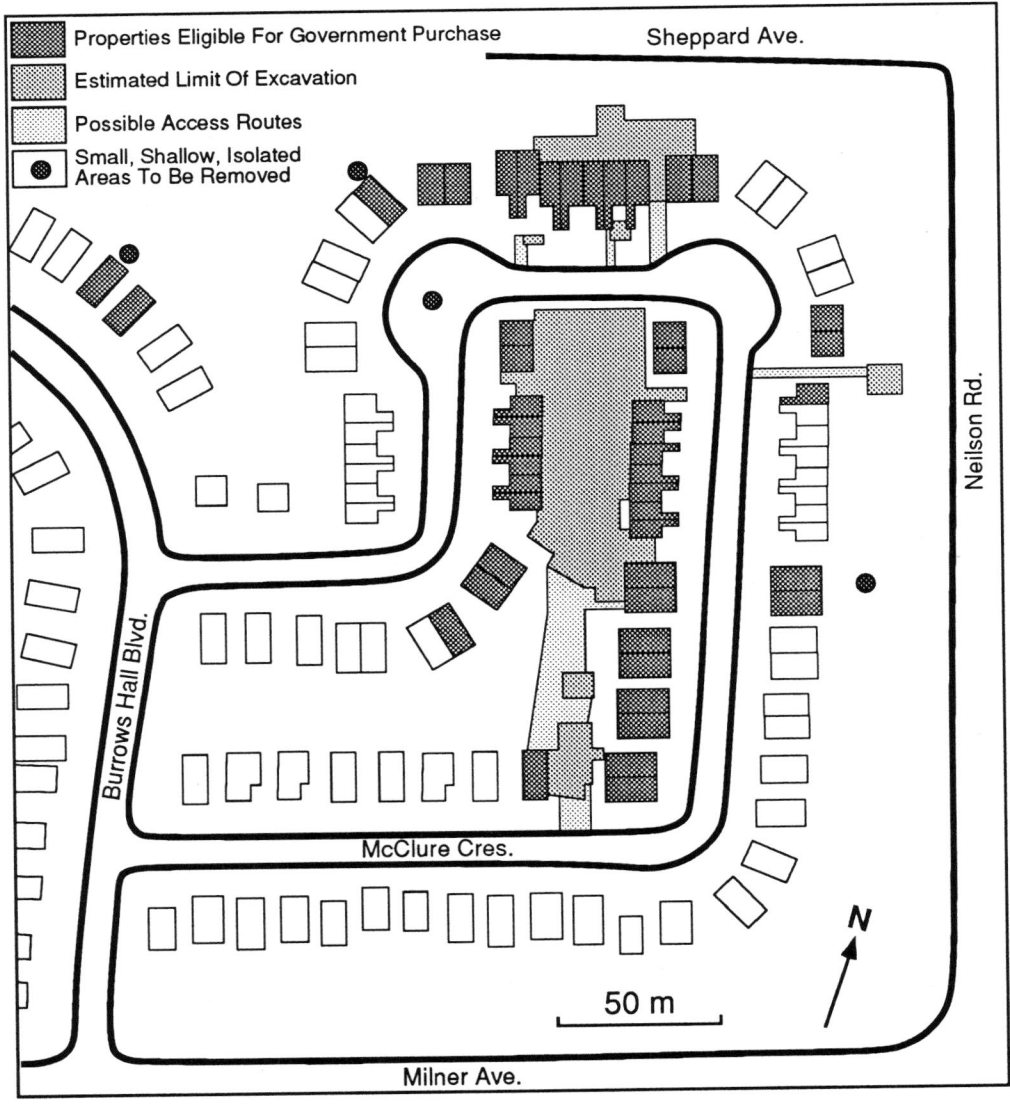

Figure 8 Contaminated sites in residential area of McClure Crescent (after Acres International, 1993; see Fig. 7 for location).

need for an accelerated program of radon surveys in urban areas, combined with systematic mapping of geologic structures such as faults and fractures. A detailed inventory of historic, unregulated radioactive waste is also urgently required.

ACKNOWLEDGEMENTS

I am very grateful to J.E. Tilsley for assistance and discussions. This research was funded by Natural Sciences and Engineering Research Council of Canada grant to N. Eyles.

REFERENCES

Acres International Limited (AIL), 1993, Revised environmental screening of the Malvern remedial project, draft for public review: Acres International Limited, Report M93-05, 157 p.

Alavanja, M.C., Brownson, R.C., Lubin, J.H., Berger, E., Chang, J.C. and Boice, J.D., Jr., 1994, Residential radon exposure and lung cancer among non-smoking women: National Cancer Institute, Journal, v. 80, p. 1829-1837.

Anonymous, 1986, Standard procedures for radon measurement developed by EPA: Journal of Environmental Health, v. 49, p. 163-165.

Auditor General of Canada, 1995, Federal radioactive waste management, *in* Report of the Auditor General of Canada to the House of Commons, volume 1: Minister of Supply and Services Canada, 34 p.

Beck, P.J. and Brown, D.R., 1987, Hydrogeologic controls on the occurrence of radionuclides in groundwater of southern Ontario, *in* Graves, B., ed., Radon in Ground Water: Lewis Publishers, Somerset, NJ, p. 449-473.

Becker, E., 1991, Radon in the home: Canadian Institute for Radiation Safety, Toronto, ON, 2 p.

Becker, E. and Moridi, R., 1992, Radon survey in metropolitan Toronto schools, *in* Proceedings, IRPA8: International Radiological Protection Association, Montreal, PQ, v. II, p. 1582-1585.

Bell, R.T., 1978, Uranium in black shales - a review, *in* Kimberley, M.M., ed., Uranium Deposits, Their Mineralogy and Origin: University of Toronto Press, Toronto, ON, p. 307- 329.

Brown, C.E., Mose, D.G., Mushrush, G.W. and Chrosniak, C.E., 1992, Statistical analysis of the radon-222 potential of rocks in Virginia, U.S.A.: Environmental Geology and Water Sciences, v. 19, p. 193-203.

Canadian Nuclear Association (CNA), 1991, Nuclear facts: Seeking to generate a better understanding: Canadian Nuclear Association, 4 p.

Churcher, P.L., Johnson, M.D., Telford, P.G. and Barker, J.F., 1991, Stratigraphy and oil shale resource potential of the Upper Ordovician Collingwood Member, Lindsay Formation, southwestern Ontario: Ontario Geological Society, Open File Report 5817, 98 p.

Clarke, R.H. and Southwood, T.R.E., 1989, Risks from ionizing radiation: Nature, v. 338, p. 197-198.

Clulow, F.V., Lim, T.P., Dave, N.K. and Avadhanula, R., 1992, Radium-226 levels and concentration ratios between water, vegetation, and tissues of ruffed grouse (*Bonasa umbellus*) from a watershed with uranium tailings near Elliot Lake, Canada: Environmental Pollution, v. 77, p. 39-50.

Cohen, B.L., 1995, Test of the linear No Threshold Theory of radiation carcinogenesis for inhaled radon decay products: Health Physics, v. 68, p. 157-174.

Cothern, C.R., 1987, Properties, *in* Cothern, C.R. and Smith, J.E., eds., Environmental Radon: Plenum Press, New York, p. 1-29.

Cothern, C.R., 1990, Radon properties, *in* Majumdar, S.D., Schmalz, R.F. and Miller, E.W., eds., Environmental Radon – Occurrence, Control and Health Hazards: Pennsylvania Academy of Science, Phillipsburg, NJ, p. 54-63.

Cothern, C.R. and Smith, J.E. eds., 1987, Environmental Radon: Plenum Press, New York, 363 p.

Dave, N.K., Lim, T.P. and Cloutier, N.R., 1985, Ra-226 concentrations in blueberries *Vaccinium angustifolium* Ait near an inactive uranium tailings site in Elliot Lake, Ontario, Canada: Environmental Pollution, series B, v. 10, p. 301-314.

de Jong, E., Acton, D.F. and Kozak, L.M., 1993, Naturally occurring gamma-emitting isotopes, radon release and properties of parent materials of Saskatchewan soils: Canadian Journal of Soil Science, v. 74, p. 47-53.

Draganic, I.G., 1993, Radiation and Radioactivity on Earth and Beyond: CRC Press, Boca Raton, FL, 349 p.

Dubrovsky, N.M., Morin, K.A., Cherry, J.A. and Smyth, D.J.A., 1984, Uranium tailings acidification and subsurface contaminant migration in a sand aquifer: Water Pollution Research Journal of Canada, v. 19, p. 55-89.

Dyck, W., 1978, The mobility and concentration of uranium and its decay products in temperate surficial environments, *in* Kimberley, M.M., ed., Uranium Deposits, Their Mineralogy and Origin: University of Toronto Press, Toronto, ON, p. 57-100.

Ennemoser, O., Ambach, W., Brunner, P., Schneider, P., Oberaigner, W., Purtscheller F. and Stingl, V., 1993, Unusually high radon concentrations: Atmospheric Environment, v. 27A, p. 2169-2172.

Eyles, N., Boyce, J. and Mohajer, A., 1993, Bedrock topography in the western Lake Ontario region: Evidence of reactivated basement structures?: Géographie physique et Quaternaire, v. 47, p. 269-283.

Fauré, G., 1986, Principles of Isotope Geology: John Wiley and Sons, New York, 589 p.

Fleischer, R.L., 1988, Alpha-recoil damage: Relation to isotopic disequilibrium and leaching of radionuclides: Geochimica et Cosmochimica Acta, v. 52, p. 1459-1466.

Flexser, S., Wollenberg, H.A. and Smith, A.R., 1993, Distribution of radon sources and effects on radon emanation in granitic soil at Ben Lomond, California: Environmental Geology, v. 22, p. 162-177.

Gates, A.E. and Gundersen, L.C.S., 1989, Role of ductile shearing in the concentration of radon in the Brookneal zone, Virginia: Geology, v. 17, p. 391-394.

Gates, A.E. and Gundersen, L.C.S., 1992, eds., Geological Controls on Radon: Geological Society of America, Special Paper 271, 88 p.

Gundersen, L.C.S., Schumann, R.R., Otton, J.K., Dubiel, R.F., Owen, D.E. and Dickinson, K.A., 1992, Geology of radon in the United States, *in* Gates, A.E. and Gundersen, L.C.S., eds., Geological Controls on Radon: Geological Society of America, Special Paper 271, p. 1-16.

Gundersen, L.C.S. and Wanty, R.B., 1993, eds., Field Studies of Radon in Rocks, Soils, and Water: C.K. Smoley, Boca Raton, FL, 334 p.

Hamilton, M.A., 1993, Radon reduction: a three year follow-up: Journal of Environmental Health, v. 56, p. 19-21.

Hand, B.M. and Banikowski, J.E., 1988, Radon in Onondaga County, New York: paleohydrogeology and redistribution of uranium in Paleozoic sedimentary rocks: Geology, v. 16, p. 775-778.

Haque, K.E., 1982, Radium extraction from the Scarborough contaminated soil: CANMET Mineral Sciences Laboratories, Report MRP/MSL 82-8, 6 p.

Harrel, J.A., Belsito, M.E. and Kumar, A., 1991, Radon hazards associated with outcrops of Ohio Shale in Ohio: Environmental Geology, v. 18, p. 17-26.

Hart, D.R., McKee, P.M. and Burt, A.J., 1986, Benthic community and sediment quality assessment of Port Hope Harbour, Lake Ontario: Journal of Great Lakes Research, v. 12, p. 206-220.

Henshaw, D.L., Eatough, J.P. and Richardson, R.B., 1990, Radon as a causative factor in induction of myeloid leukemia and other cancers: Lancet, v. 335, P. 1008-1012.

Horgan, J., 1994, Radon's risks: Scientific American, v. 271, p. 14-16.

Humphreys, C.L., 1987, Factors controlling uranium and radium isotopic distributions in ground waters of the West Central Florida phosphate district, *in* Graves, B., ed., Radon in Ground Water: Lewis Publishers, Somerset, NJ, p. 171-189.

International Joint Commission (IJC) 1983, Report on Great Lakes water quality: International Joint Commission, Windsor, ON, 30 p.

Keller, G., Schneiders, H., Schutz, M., Siehl, A. and Stamm, R., 1992, Indoor radon correlated with soil and subsoil radon potential — a case study: Environmental Geology, v. 19, p. 113-119.

Kepler, H. and Wyllie, P.J., 1990, Role of fluids in transport and fractionation of uranium and thorium in magmatic processes: Nature, v. 348, p. 531-533.

Kodosky, L.G., 1994, An evaluation of residential air radon concentrations and related variables in southeast Michigan, USA: Environmental Geology, v. 23, p. 65-72.

Kokotti, H., Kalliokoski, P. and Jantunen, M., 1992, Dependency of radon entry on pressure difference: Atmospheric Environment, v. 26A, p. 2247-2250.

Leighton, M.W., Kolata, D.R., Oltz, D.F. and Eidel, J.J., 1990, eds., Interior Cratonic Basins: American Association of Petroleum Geologists, Memoir 51, 819 p.

Loureiro, C.O., Abriola, S.M., Martin, J.E. and Sextro, R.G., 1990, Three-dimensional simulation of radon transport into houses with basements under negative pressure: Environmental Science and Technology, v. 24, p. 1338-1348.

Lowry, J.D., Hoxie, D.C. and Moreau, E., 1987, Extreme levels of ^{222}Rn and U in a private water supply, in Graves, B., ed., Radon in Ground Water: Lewis Publishers, Somerset, NJ, p. 363-375.

Martell, E.A., 1985, Enhanced ion production in convective storms by transpired radon isotopes and their decay products: Journal of Geophysical Research, v. 90, p. 5909- 5916.

McCallum, B.A., 1992, A gamma spectroscopy analysis of the distribution of ^{226}Ra in Ontario soils, a preliminary study: Atomic Energy Canada Limited Research, 17 p.

Michel, J., 1987, Sources, in Cothern, C.R. and Smith, J.E., eds., 1987, Environmental Radon: Plenum Press, New York, p. 81-130.

Moffett, D. and Tellier, M., 1978, Radiological investigations of an abandoned uranium tailings area: Journal of Environmental Quality, v. 7, p. 310-314.

National Research Council (NRC), 1994, Health effects of exposure to radon: Time for reassessment?: National Academy Press, Washington, DC, 94 p.

National Research Council Canada (NRCC), 1983, Radioactivity in the Canadian aquatic environment: National Research Council Canada, Ottawa, ON, 292 p.

Nazaroff, W.W., Lewis, S.R., Doyle, S.M., Moed, B.A. and Nero, A.V., 1987, Experiments on pollutant transport from soil into residential basements under pressure- driven airflow: Environmental Science and Technology, v. 21, p. 459-466.

Nero, A.V., Schwehr, M.B., Nazaroff, W.W. and Revzan, K.L., 1986, Distribution of airborne radon-222 concentrations in U.S. homes: Science, v. 234, p. 992-997.

Noor, I., Novakowski, K.S. and Egden, J., 1992, Soil-gas surveys as a tool for delineating pathways of natural hydrocarbon loading in southern Ontario: Implications for shallow ground water: International Association of Hydrogeologists, Hamilton, ON, meeting, Proceedings, p. 709-723.

Otton, J.K., 1992, The geology of radon: United States Geological Survey, Reston, VA, 28 p.

Paynes, R.A., 1995, Decommissioning of Elliot Lake mining properties, in Hynes, T.P. and Blanchettee, M.C., eds., Sudbury '95: Mining and the Environment, Conference Proceedings, CANMET, Ottawa, ON, p. 1199-1210.

Pearce, F., 1987. A deadly gas under the floorboards: New Scientist, v. 113, p. 33-35.

Pershagen, G., Akerblom, G., Axelson, O., Clavensjo, B., Damber, L., Desai, G., Enflo, A., Lagarde, F., Melander, H., Svartengren, M. and Swedjemark, G.A., 1994, Residential radon exposure and lung cancer in Sweden: New England Journal of Medicine, v. 330, p. 159-164.

Pipken, B.W., 1994, Geology and the Environment: West Publishing Company, New York, 476 p.

Rose, A.W., Washington, J.W. and Greeman, D.J., 1990, Geology and geochemistry of radon occurrence, in Majumdar, S.D., Schmalz, R.F. and Miller, E.W., eds., Environmental Radon: Occurrence, Control and Health Hazards: Pennsylvania Academy of Science, Easton, PA, p. 64-77.

Sanford, B.V., Thompson, F.J. and McFall, G., 1985, Plate tectonics - A possible controlling mechanism in the development of hydrocarbon traps in southwestern Ontario: Bulletin of Canadian Petroleum Geology, v. 33, p. 52-71.

Schery, S.D., Whittlestone, S., Hart, K.P. and Hill, S.E., 1989, The flux of radon and thoron from Australian soils: Journal of Geophysical Research, v. 94, p. 8567-8575.

Schmalz, R.F., 1990. Geology and occurrence of radon precursors, in Majumdar, S.D., Schmalz, R.F. and Miller, E.W., eds., Environmental Radon: Occurrence, Control and Health Hazards: Pennsylvania Academy of Science, Easton, PA, p. 78-89.

Shapiro, P.S. and Sorg, T.J., 1988, Reduction of radon from household water supplies: Radiation Protection Dosimetry, v. 24, no. 1/4, p. 523-525.

Sherwood Lollar, B., Weise, S.M., Frape, S.K. and Barker, J.F., 1994, Isotopic constraints on the migration of hydrocarbon and helium gases of southwestern Ontario: Bulletin of Canadian Petroleum Geology, v. 42, p. 283-295.

Silver, M., Ritcey, G.M. and Cauley, M.P., 1985, A lysimeter comparison of the effects of uranium tailings deposition methods on the release of environmental contaminants: Hydrometallurgy, v. 15, p. 159-172.

Swistock, B.R., Sharpe, W.E. and Robillard, P.D., 1993, A survey of lead, nitrate and radon contamination of private individual water systems in Pennsylvania: Journal of Environmental Health, v. 55, p. 6-12.

Tilsley, J.E., 1992, Radon: sources, hazards and control: Geoscience Canada, v. 19, p. 163-166.

Tilsley, J.E., Veldhuyzen, H. and Nicholls, R.R., 1993, Soil radon gas study of southern Ontario: Ontario Geological Survey, Open File Report 5847, 148 p.

Upton, A.C., 1982, The biological effects of low-level ionizing radiation: Scientific American, v. 246, p. 41-49.

Valentine, R.L. and Stearns, S.W., 1994, Radon release from water distribution system deposits: Environmental Science and Technology, v. 28, p. 534-537.

Vaughan, C., 1990, Hiroshima study shows higher risks of low-level radiation: New Scientist, v. 1698, p. 28.

Warwick, W.F., 1991, Indexing deformities in ligulae and antennae of Procladius larvae (Diptera: Chironomidae): application to contaminant-stressed environments: Canadian Journal of Fisheries and Aquatic Sciences, v. 48, p. 1151-1165.

Warwick, W.F., Fitchko, J., McKee, P.M., Hart, D.R. and Burt, A.J., 1987, The incidence of deformities in Chironomus spp. from Port Hope Harbour, Lake Ontario: Journal of Great Lakes Research, v. 13, p. 88-92.

Wilkening, M., 1990. Radon in the Environment: Elsevier, New York, 137 p.

Wills, J., Howell, L., McKay, L., Parker, B. and Walter, A., 1992, Smithville C.W.M.L. site: Characterization of overburden fractures and implications for DNAPL transport: International Association of Hydrogeologists, Hamilton, ON, meeting, Proceedings, p. 501-515.

Wren, C.D., Cloutier, N.R., Lim, T.P. and Dave, N.K., 1987, Ra-226 concentrations in otter, Lutra canadensis, trapped near uranium tailings at Elliot Lake, Ontario: Bulletin of Environmental Toxicology, v. 38, p. 209-212.

29. Earthquake Hazard in the Greater Toronto Area

Alex A. Mohajer

Environmental Earth Sciences, University of Toronto, Scarborough Campus
1265 Military Trail Road, Scarborough, Ontario M1C 1A4

SUMMARY

Accelerated urbanization in recent decades has heightened concerns for the protection of communities against natural environmental hazards such as earthquakes. A history of past earthquakes in a region is traditionally used as a guide to assessing the future potential seismic hazard. In areas of sparse or low seismic activity, with a short history of less than three centuries of settlement (*e.g.*, southern Ontario), a lack of destructive earthquakes in the past does not necessarily rule out the possibility of future large magnitude events.

In southern Ontario, prominent crustal basement structures, delineated on the basis of geophysical data, have continued to be reactivated to the present time. Statistical extrapolation of current small magnitude earthquakes indicates that there is a potential for the Greater Toronto Area to experience moderate to large earthquakes. The probabilities of earthquakes of $M = 5$, $= 6$ and $= 7$, in the next 50 years, are about 57%, 6% and less than 1%, respectively. These risk levels are acceptable for ordinary buildings that are designed in accordance with the provisions of the National Building Code. However, the expected ground motion at lower risk levels exceeds those previously considered, and necessitates a re-evaluation of the seismic hazard impact on nuclear and other public facilities in the area.

INTRODUCTION

Urban seismic risk is growing worldwide, and is becoming a greater problem, for metropolitan areas of developing countries as well as for unprepared cities in eastern North America (Tucker *et al.*, 1994). The media provides constant reminders of the global occurrence of natural hazards, such as floods and earthquakes. On average, about 20 large magnitude earthquakes ($M > 7$) occur somewhere on this planet every year. So far, most have occurred in remote and unpopulated areas, with limited consequences. Nonetheless, destructive earthquakes have been responsible for many disasters in modern urban areas in recent years. They include the 1976 earthquake which produced 655,000 casualties in Tanshang, China, the 1985 event which killed 5000 and caused more than $10 billion damage in Mexico City, and the 1994 Northridge, California, tremblor that resulted in 62 casualties and $30 billion worth of damage. Lastly, the massive 1995 earthquake near Kobe, Japan killed at least 5092, injured about 27,000, and resulted in an estimated loss in excess of $150 billion. These cases illustrate the vulnerability of present-day communities to earthquakes. Because of the rapidly growing cities in many earthquake-prone areas of the world, realistic seismic hazard assessments and methods to mitigate the adverse effects of an earthquake must be addressed by all levels of government. This comprises not only assessing the locations and magnitudes of previous earthquakes, but identifying, and carrying out detailed investigations of, regionally extensive faults, to determine whether or not they have been repeatedly active throughout their history, and to ascertain the age of last movement. Investigations such as these apply, in particular, to the impact of seismicity on facilities such as dams and nuclear power plants.

EARTHQUAKES IN EASTERN NORTH AMERICA

Canada and the United States, east of the Rocky Mountains, have similar crustal properties, and make up Eastern North America (ENA), which is known as one of the more stable continental regions of the world (Johnston, 1989; Johnston *et al.*, 1994). The rate of crustal deformation and seismic activity interpretation in ENA is lower than along the plate boundary areas of western North America and elsewhere in the world. Nonetheless, there are more than 12 large magnitude earthquakes documented in ENA, three of which, with $M > 8.2$, occurred near New Madrid, Missouri, in 1811 and 1812, and are among the largest ever reported. Despite events such as this, there is still a public perception that destructive earthquakes cannot happen in eastern Canada, but this is a misconception stemming principally from the short, documented seismological history. Regardless, a few damaging earthquakes have been centred in eastern Canada, and include the Charlevoix, Grand Banks, Timiskaming, Saguenay and Lac Turquoise earthquakes, all of which were greater than $M = 6$ (Adams *et al.*, 1995). Several others with $M > 5$, including the $M = 5.6\text{-}6.0$ earthquake near Montreal in 1732 (Leblanc, 1981) and Cornwall, in 1945, have also been documented.

The history of earthquakes in a region is usually used as a guide to assessing future potential seismic hazard. In areas of low seismic activity with a short history of settlement, such as southern Ontario, a lack of destructive earthquakes in the past does not necessarily rule out the possibility of future big events. Earthquakes signify the presence of a fault plane or zone on which strain energy is released under the ambient crustal stress. It is, therefore, necessary to search for paleoseismic evidence, to delineate exposed or subsurface faults, and to investigate the history of the faults in order to assess their potential for producing future large-magnitude earthquakes.

In the Greater Toronto Area (GTA), the most populated centre in Canada, major deep-seated fault zones are expressed as aeromagnetic and gravity lineaments (Wallach and Mohajer, 1990). Knowledge of these faults, and their potential for reactivation under the present-day stress field, is limited; however, Mohajer et al. (1992) identified faulted Paleozoic and Quaternary strata within the Georgian Bay linear zone (GBLZ), near its intersection with the Niagara-Pickering linear zone (NPLZ; Fig. 1). Eyles et al. (1993) showed a spatial

correlation between the GBLZ and a linear depression of the bedrock surface, buried under Quaternary and recent sediments, and Mohajer (1993) identified the seismically active Toronto-Hamilton seismic zone (THSZ) The geology and age of these structures is described in Chapter 2; the relationship between seismicity and detectable soil gas anomalies is briefly reviewed in Chapter 34.

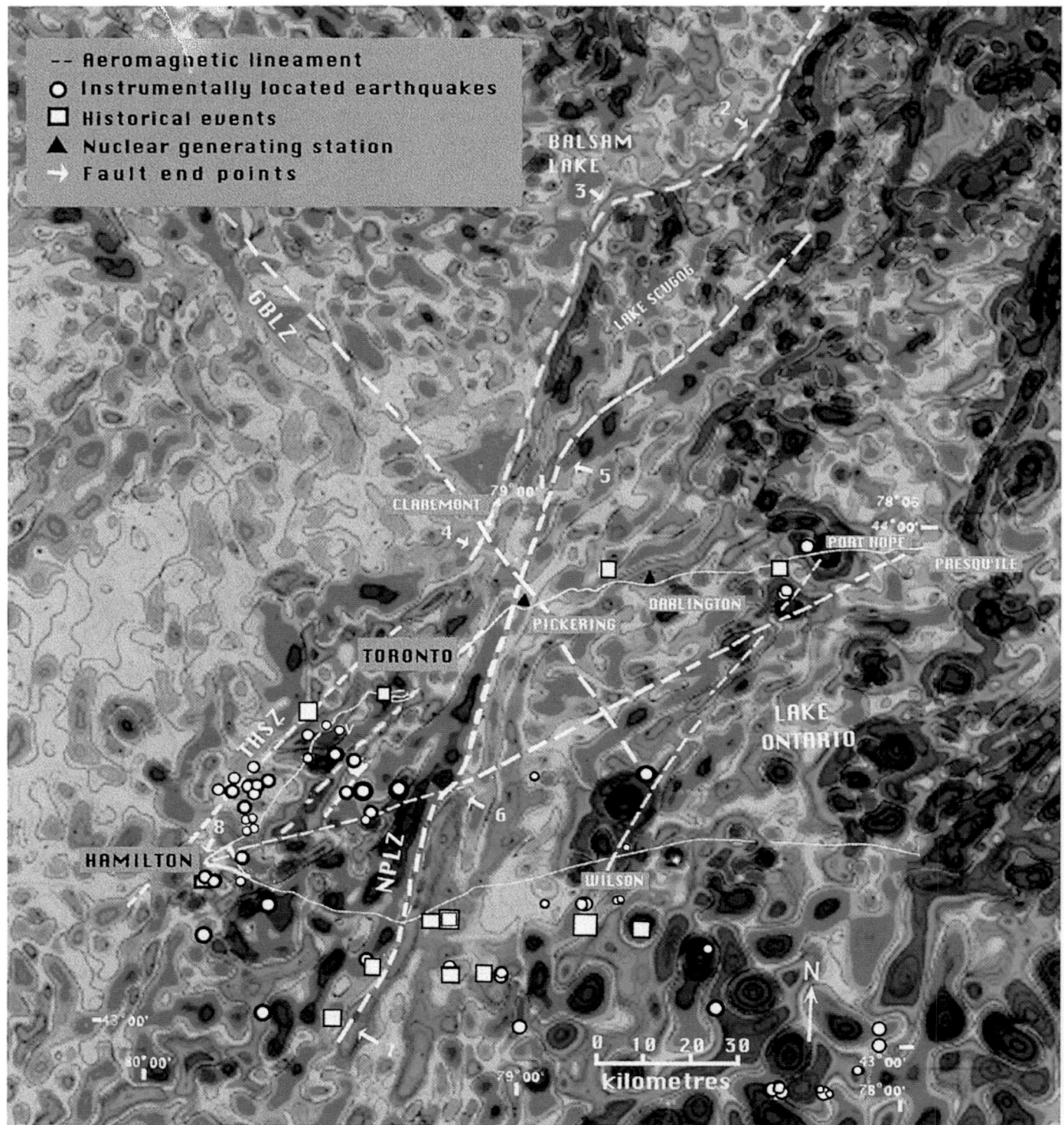

Figure 1 *Seismicity and aeromagnetic lineaments in the western Lake Ontario region. Earthquake locations are from the Geological Survey of Canada's national earthquake catalogues (Canadian Earthquake Epicentre File) and Mohajer (1987) recomputation of earthquake locations. Abbreviations: **GBLZ**, Georgian Bay linear zone; **NPLZ**, Niagara-Pickering linear zone; **THSZ**, Toronto-Hamilton seismic zone.*

Marine surveys across some of the above lineaments where they cross Lake Ontario have been conducted using ship-borne side-scan sonar, seismic reflection, and echo sounding (Thomas *et al.*, 1993; Lewis *et al.*, 1995). Offshore surveys reveal numerous indications of deformation of the bedrock and overlying young lake bottom sediments, including offset and discontinuous reflectors and pop-ups (bedrock pressure ridges; Chapter 2). Some of the pop-ups and bedrock fractures were confirmed by direct submersible observations south of Toronto Island (Fig. 2). Lewis *et al.* (1995) suggested that the widespread discontinuities at the base of Holocene lake sediments may result from rapid gas/water venting under the influence of seismic shaking. This may provide a record of paleo-earthquake shaking events, several between 11 ka and 13 ka and one at approximately 6-8 ka.

The objectives of this paper are to present information relevant to the potential occurrence of the large magnitude earthquakes in the GTA and to characterize local seismic source zones, in an attempt to reduce the uncertainties regarding the current earthquake risk assessment in the area. The conclusions are presented in a way that is appropriate for subsequent use in probabilistic and deterministic ground-motion analyses.

Recent Faulting in the Greater Toronto Area

Identification of the paleoseismicity of an area is dependant on examination of the geological record of ground shaking and fault rupture. Fault movements recur episodically throughout time, in response to the build-up of stress. Thus, in order to show that a fault is inactive, it must be shown either that the stress no longer exists, or that the fault has not moved for a period of time since the onset of the current stress field, as long as, perhaps, the last tens of millions of years. Failing either, it is reasonable to assume that all faults in an area where seismicity has been documented have the potential to be reactivated (Muir Wood and Mallard, 1992).

An exposure along the Rouge River valley (section 3; Fig. 3) reveals a series of normal (extensional) faults in both interglacial sediments and the underlying bedrock (Fig. 4). Interglacial sediments are represented by the Don and Scarborough formations, with respective approximate ages of 125 ka and 70 ka, which establish the minimum age of faulting.

Figure 2 *Bedrock pop-up along the NPLZ on the floor of Lake Ontario, south of Toronto Island. A manned submersible (on loan from Harbour Branch Oceanographic Institute, Fort Pierce, FL) was used for the Lake Ontario observations. An image (approximately 4 m × 3 m) is used as scale.*

The maximum vertical displacement observed was 1.2 m at section 1 (Figs. 3, 4), which was interpreted by Mohajer *et al.* (1992) as a possible consequence of crustal tectonics and the reactivation of bedrock joints. Those faults were also examined, briefly, by Adams *et al.* (1993a, b), who indicated that they formed during a single episode of faulting accompanying glaciotectonism (ice push) during the last glaciation. Field work by J.L. Wallach (pers. comm.) revealed that there were at least two, and possibly three, periods of normal faulting, none of which were likely to have formed by glacial deformation, because such activity is a predominantly compressional phenomenon (Croot, 1987; Aber, 1992).

Normal faults (Fig. 5), were also recognized along the Rouge River valley, upstream from the aforementioned location (section 13, Fig. 3). An apparent throw of nearly 4 m is indicated on the main fault (fault D, Fig. 5) by Hibbert *et al.* (1993), who identified repeated episodes of fault displacement. A maximum age of about 40 ka for the faults at section 13 and a minimum age can also be estimated because a Late Wisconsin till is only present on the downthrown block of fault D, which indicates that the most recent movement along fault D is younger than 13 ka. These faults can therefore be classified as "active" (see Appendix A).

The close correspondence between the trends of the faults at section 13 and bedrock fractures and faults throughout the lower Rouge drainage basin (Figs. 3, 4, 5) strongly suggests that faulting was controlled by joints in the underlying bedrock.

High-resolution seismic data were collected along four variably oriented shallow seismic reflection profiles within the Rouge River valley (Mohajer *et al.*, 1995). The profiles, totalling more than 1.5 km in length, were selected, where possible, to run across the projected strike of the faults mapped in the Pleistocene deposits and bedrock outcrops exposed along the Rouge River.

Profile A reveals discontinuous bedrock reflectors, one of which is offset by about 12 milliseconds (ms), indicating a maximum vertical displacement of the bedrock surface of 15 m (Fig. 6). That displacement occurs precisely where profile A intersects the projection of the main fault (fault D; Fig. 3), exposed at section 13, meaning that fault D is recorded in the profile. The maximum displacement of the Quaternary strata along fault D, at section 13, is approximately 4 m near the surface, which indicates an upward decrease in displacement along the fault. This suggests a tectonic origin for the faults. In addition to the displaced reflectors seen along profile A, deeper reflections were also seen. They occur within the bedrock, with a two-way travel time of between 100 ms and 130 ms, and show an anomalously low-angle tilting of strata which warrants further investigations.

Systematic examination of the Quaternary stratigraphic record of the Toronto area, paying particular attention to sedimentary structures that identify paleoseismic activity and ground shaking, is currently in progress (Chapter 2).

Local Earthquake Monitoring

After an extensive background noise survey of ambient vibration caused by human activity, such as farming and quarry blasts, a five-station seismic network was deployed. The area covered is located between the Pickering and Darlington nuclear power plants on Lake Ontario to the south, and Lake Scugog to the north (Fig. 7). The detection threshold obtained for two of the stations allows recording of local events of $M_L \geq$ 0.5, a magnitude range which is usually not detected by regional seismic networks. The other stations were able to detect earthquakes as small as $M_L \geq 1.5$. An analysis of several thousand triggered signals resulted in identifying about 120 naturally occurring, small magnitude events (*e.g.*, Fig. 8).

The preliminary results indicate that there are distinct clusters of microearthquakes, with those in each cluster showing identical wave forms and spectra, indicating a common origin (Fig. 9). The closely spaced clusters were deduced from the fixed distance arrivals at one station, based on the difference in arrival times of S and P waves (S-P time plot) *versus* number of events (*e.g.*, Fig. 10), which suggests that most of them occur within the eastern block of the Central Metasedimentary Belt boundary zone (CMBBZ). Prior to the establishment of this network, an average of 25 to 30 local events per year were being missed, because they are below

the detection threshold of the regional and national networks.

The linear slope of the magnitude-frequency relationship for the microearthquakes detected in this study (Fig. 11) is aligned with, and matches, that for the larger events which occurred in the last two centuries in the western Lake Ontario region (Fig. 11). Therefore, it is justified to extrapolate the linear curve in Figure 12 to estimate the periodicity of magnitude 6 and 7 earthquakes likely to affect the GTA (Fig. 12).

Seismicity in the Western Lake Ontario Region
In the western Lake Ontario region, data on more than 58 small to moderate magnitude earthquakes were compiled and used for the seismotectonic interpretation. An update of the data file, in addition to a recomputation of some of the

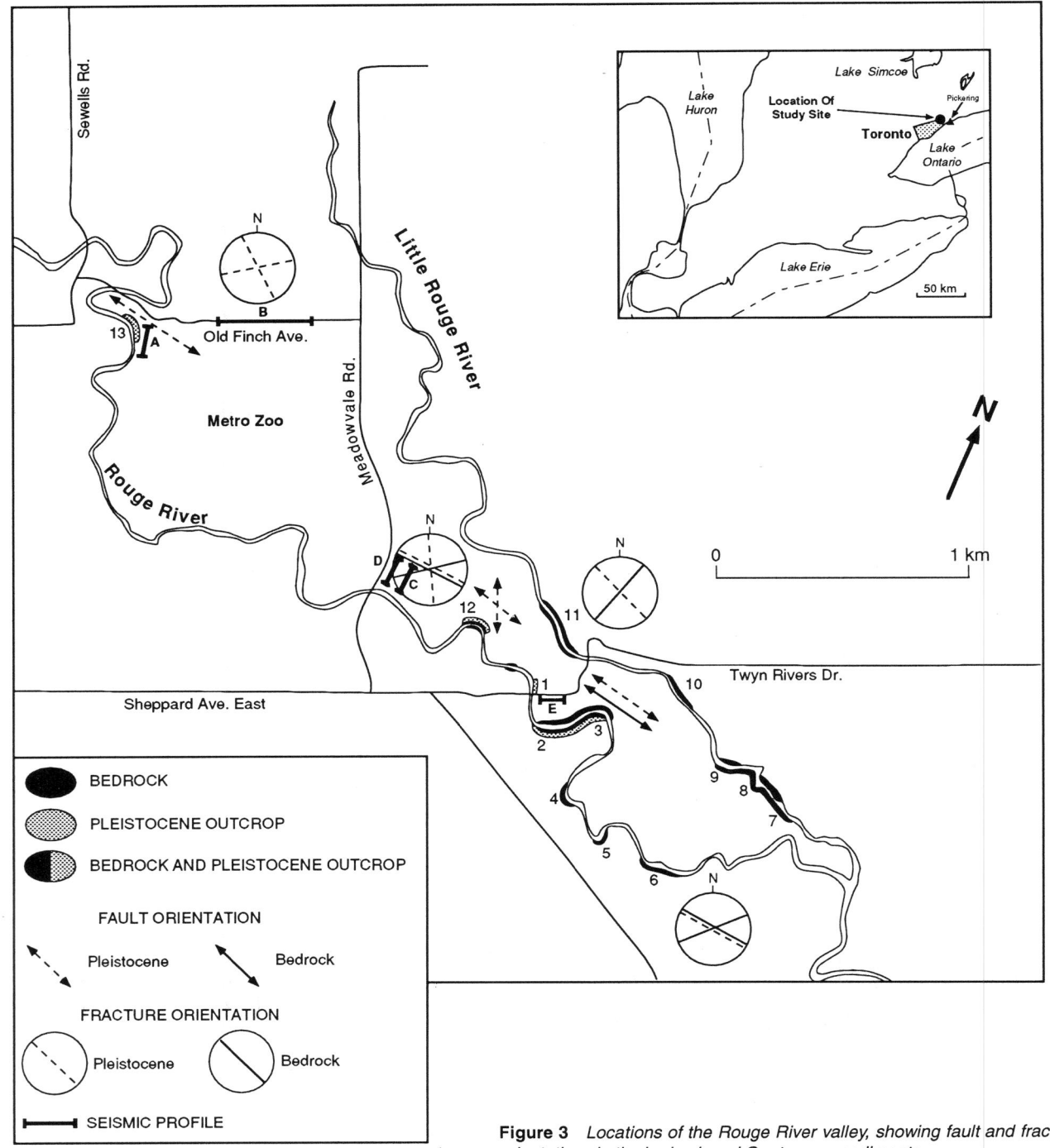

Figure 3 *Locations of the Rouge River valley, showing fault and fracture orientations in the bedrock and Quaternary sediments.*

locations where ever possible, was carried out by Mohajer (1987). More than 80% of the known events are apparently confined to a 20-km wide zone located between Toronto and Hamilton (THSZ), which is also bounded by magnetic and gravity lineaments (Mohajer, 1993). Nonetheless, the remaining earthquakes in the western Lake Ontario region are, constrained to the NPLZ to the east and the Niagara Peninsula to the south.

The depth distribution of the existing earthquakes shows a relatively shallow concentration, ranging from 1 to 20 km (Mohajer, 1993). Nevertheless, the better located group confirms that more than half of the local earthquakes are generated at depths in excess of 2 km, and are, therefore, not simply surficial stress relief phenomena. Wallach *et al.* (1993) showed that pop-ups, which are surficial stress relief phenomena, are kinematically congruent with the focal mecha-

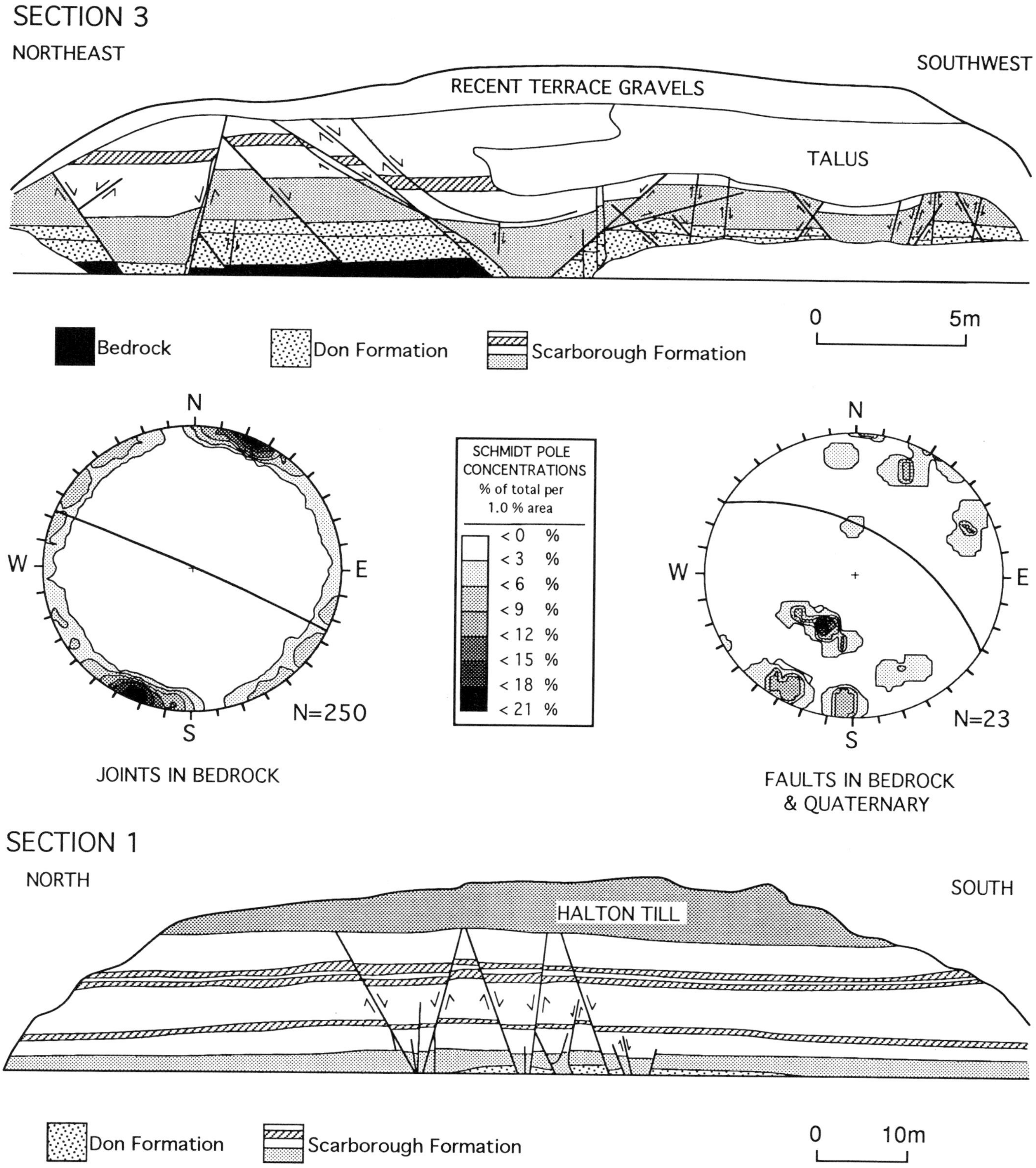

SECTION 3

NORTHEAST   SOUTHWEST

RECENT TERRACE GRAVELS

TALUS

0   5m

 Bedrock   Don Formation   Scarborough Formation

N

W   E

S

N=250

JOINTS IN BEDROCK

SCHMIDT POLE
CONCENTRATIONS
% of total per
1.0 % area

< 0	%
< 3	%
< 6	%
< 9	%
< 12	%
< 15	%
< 18	%
< 21	%

N

W   E

S

N=23

FAULTS IN BEDROCK
& QUATERNARY

SECTION 1

NORTH   SOUTH

HALTON TILL

Don Formation   Scarborough Formation   0   10m

Figure 4 *Quaternary and bedrock faulting at sections 1 and 3, in the Rouge River valley (refer to Fig. 3).*

nisms of deeper earthquakes, and compatible with the current ambient stress field. They concluded, therefore, that pop-ups on the surface may well be indicators of areas subject to future large magnitude earthquakes. The same comments apply to shallow earthquakes.

Fault Segmentation and Earthquake Potentials

Relating earthquake magnitude and rupture parameters, such as fault dimensions or displacements, is a significant input for any realistic hazard assessment. This is particularly important for areas in which the documented seismological record is short, or only low-magnitude seismicity is occurring, because it is the only means of estimating the maximum magnitude (M_x) of a future earthquake. Empirical relations have been developed for plate boundary areas with visible surface faulting (*e.g.*, Slemmons, 1977). In intraplate environments, where active faults are not readily visible on the ground surface, indirect observations, such as the extent of an aftershock zone, and seismicity pattern recognition have been utilized (Wells and Coppersmith, 1994).

One of the better established relations in observational seismology is that between earthquake magnitude and fault rupture parameters, such as length and amount of displacement. It is now common knowledge that ruptures do not usually propagate throughout the entire fault length during a single event, but tend to stop at sharp bends or intersections with other faults, often called tectonic-knots. This has provided a basis for the estimation of the largest potential earthquake that could occur in a particular area (Slemmons, 1977; Wells and Coppersmith, 1994).

In order to estimate the M_x in southern Ontario, it is necessary to define the segment lengths of the potentially capable faults. A lack of observable surface faulting makes this task rather difficult for the GTA, nonetheless, there is sufficient geological information to identify structural intersections, or sharp changes in fault orientation along strike (Fig. 1), thereby permitting the lengths of individual structural

segments to be determined. For the GTA, the segment lengths of the CMBBZ and GBLZ were measured, and their associated M_x were estimated using the suggested empirical relations of Wells and Coppersmith (1994), which are summarized in Table 1. There, it is shown that, deterministically, an earthquake of $M \gg 7$ can be considered credible for the GTA.

The 15-m offset at the interface of the bedrock and Quaternary deposits, along fault D at section 13 (Fig. 6), may also provide some insight into the return period of large magnitude earthquakes (*e.g.*, $M < 7.5$) in the GTA. If each large magnitude earthquake is associated with a 2-m-high displacement, a worldwide average suggested by Wells and Coppersmith (1994), at least seven such earthquakes would have occurred along fault D since deposition of the Don Formation, the oldest interglacial unit known along the Rouge River. The oldest possible age for the Don Formation is 125 ka (Mohajer *et al.*, 1992), which suggests a recurrence rate of about 18,000 years, corresponding to an annual occurrence of about 5.6×10^{-5}, for each large local earthquake, due solely to movement along fault D. This does not take into account displacements along other faults, elsewhere in the GTA, which may also produce substantial earthquakes and thereby increase the annual probability of occurrence of large magnitude earthquakes to 10^{-4}, or even larger. This low-probability estimate does not apply to the building code requirements for ordinary structures, but has considerable implications for the hazard assessment and design parameters of public buildings and critical facilities such as dams and nuclear power plants.

Deterministic Hazard Estimate for the Greater Toronto Area

Areas of sparse seismicity with a relatively short history of earthquakes lack a sufficient data base for statistical (deterministic) manipulations. The deterministic approach is, therefore, adapted to estimate the peak ground acceleration that could be produced along each segment of two of the principal linear zones that pass within the GTA, the CMBBZ and the

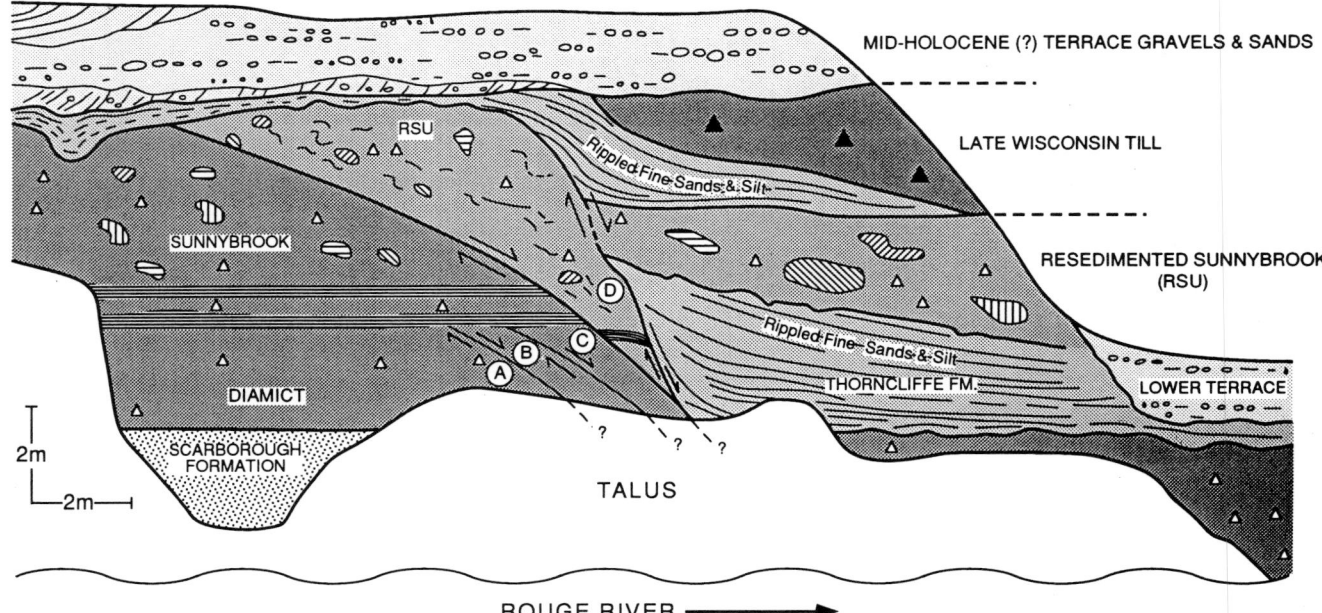

NW   **SE**

Figure 5   *Sketch of normal faulting observed at section 13, in the Rouge River valley (refer to Fig. 3).*

METRO ZOO WEST
PROFILE A

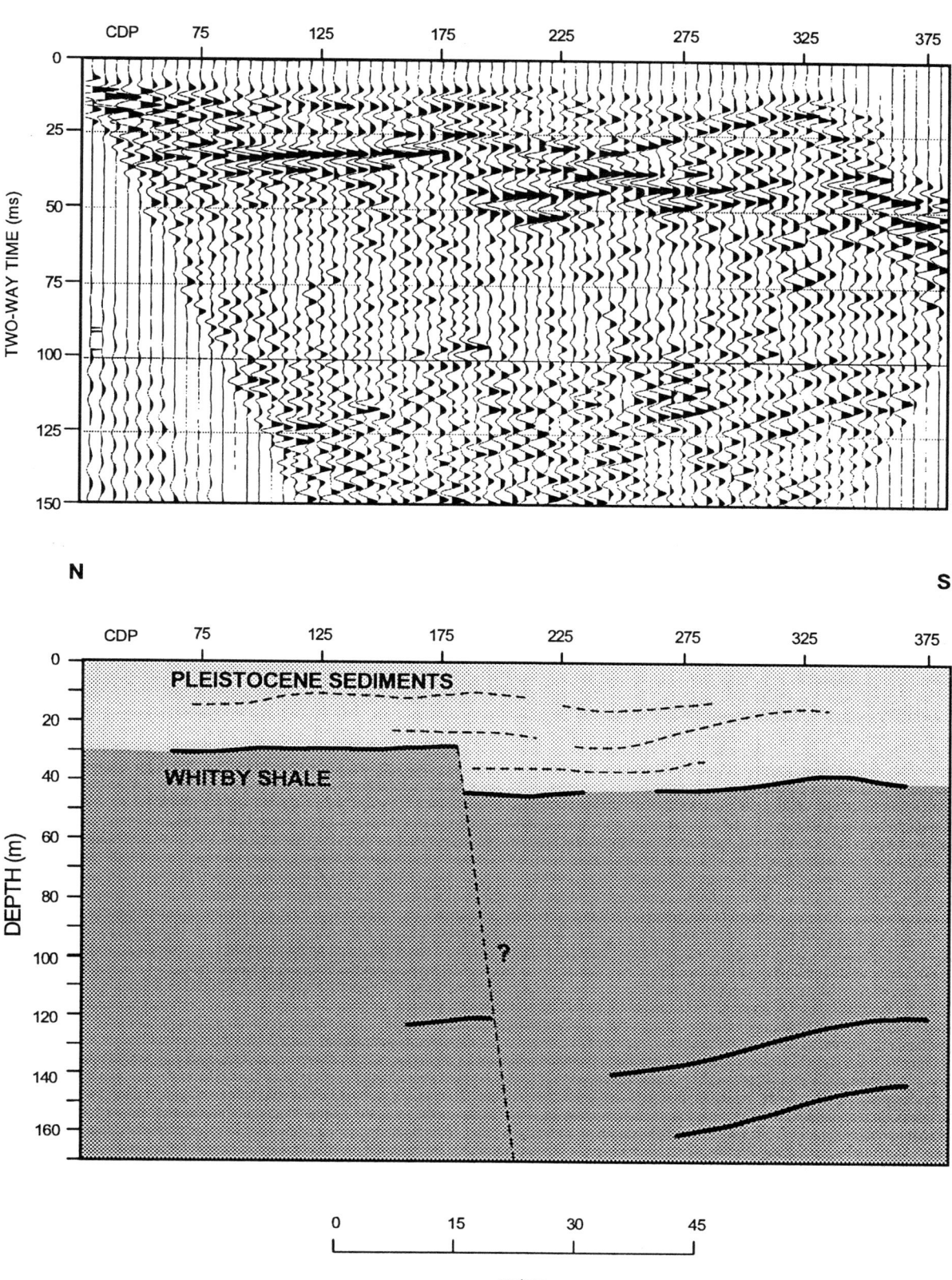

Figure 6 *Seismic reflection profile across the fault exposure near section 13 (profile A, Fig. 3), showing 15-m offset of the bedrock-Quaternary sediment interface.*

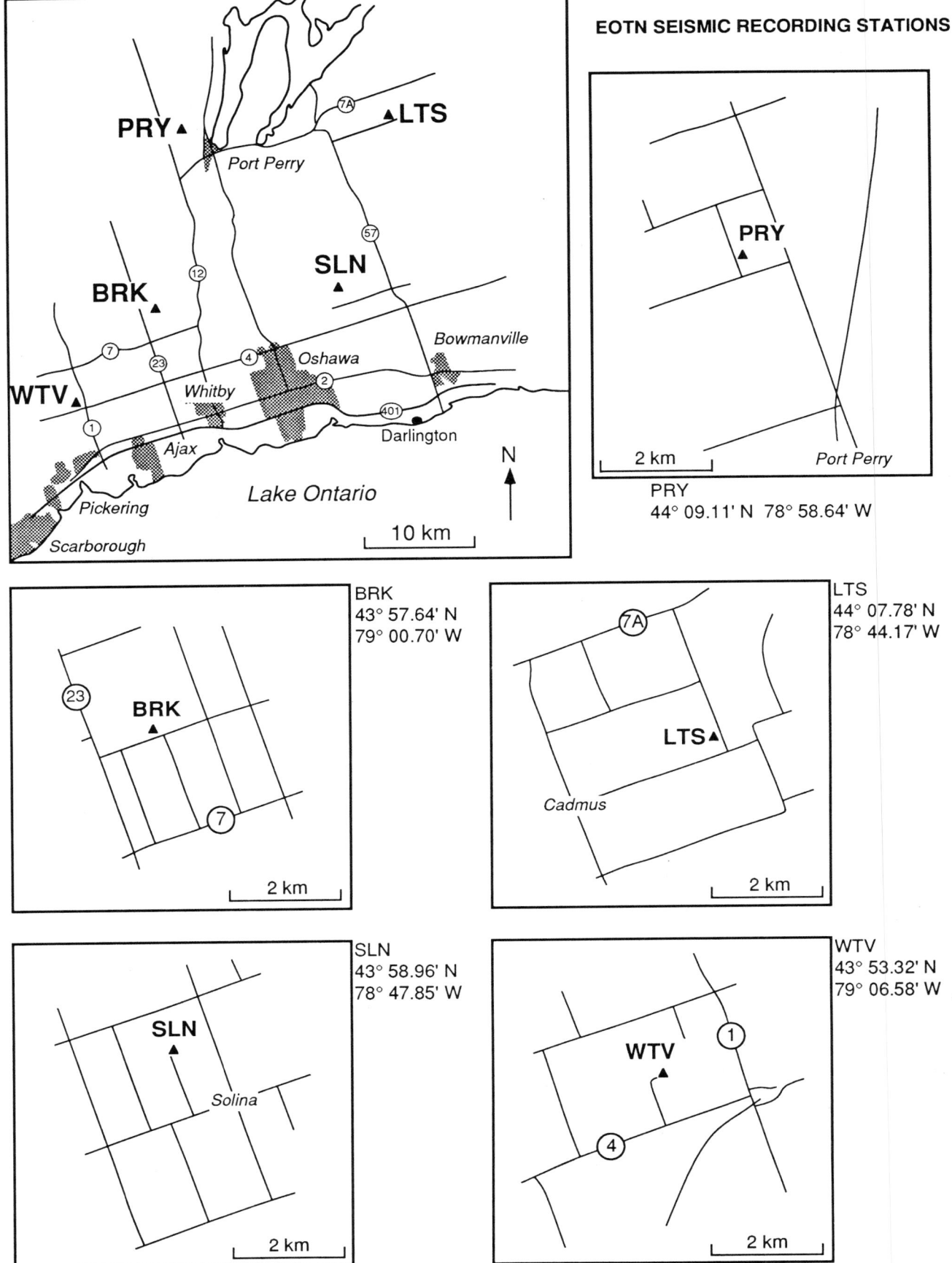

Figure 7 *Seismographic station locations in the east of Toronto network (EOTN). Abbreviations:* **BRK**, *Brooklin;* **LTS**, *Lotus;* **PRY**, *Port Perry;* **SLN**, *Solina;* **WTV**, *Whitevale.*

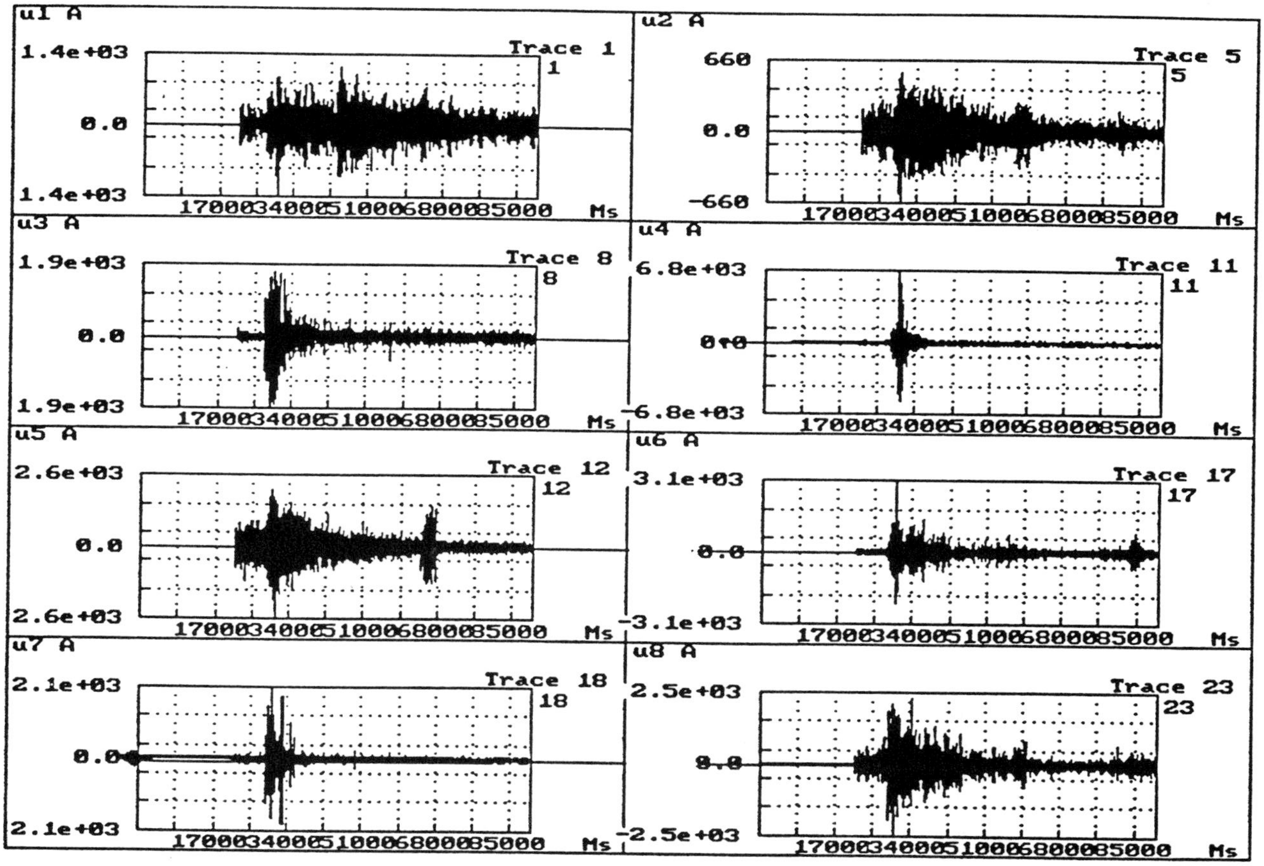

Figure 8   *Typical examples of the wave form time-history of the recorded signals in the EOTN, which are related to local sources of stress release.*

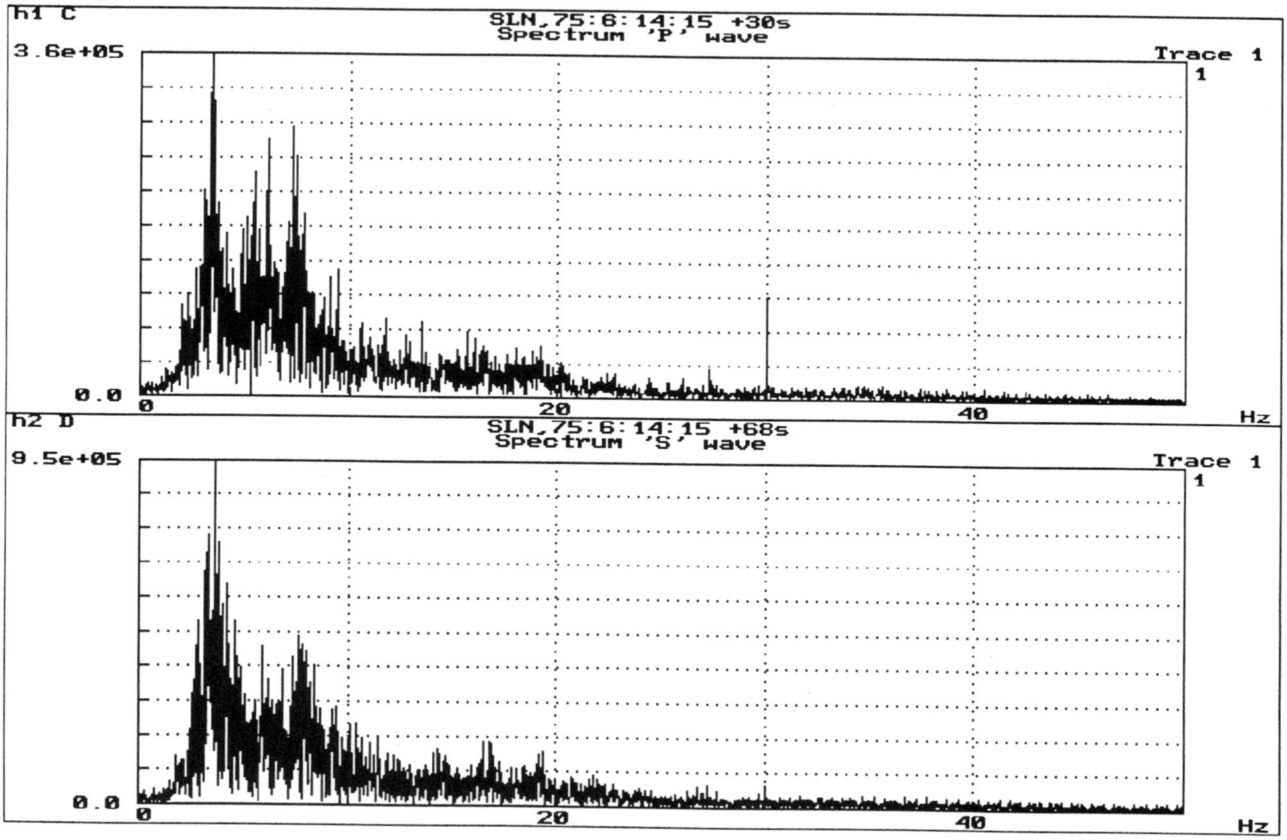

Figure 9   *Identical ground response spectra for typical local events recorded in a single station of the EOTN.*

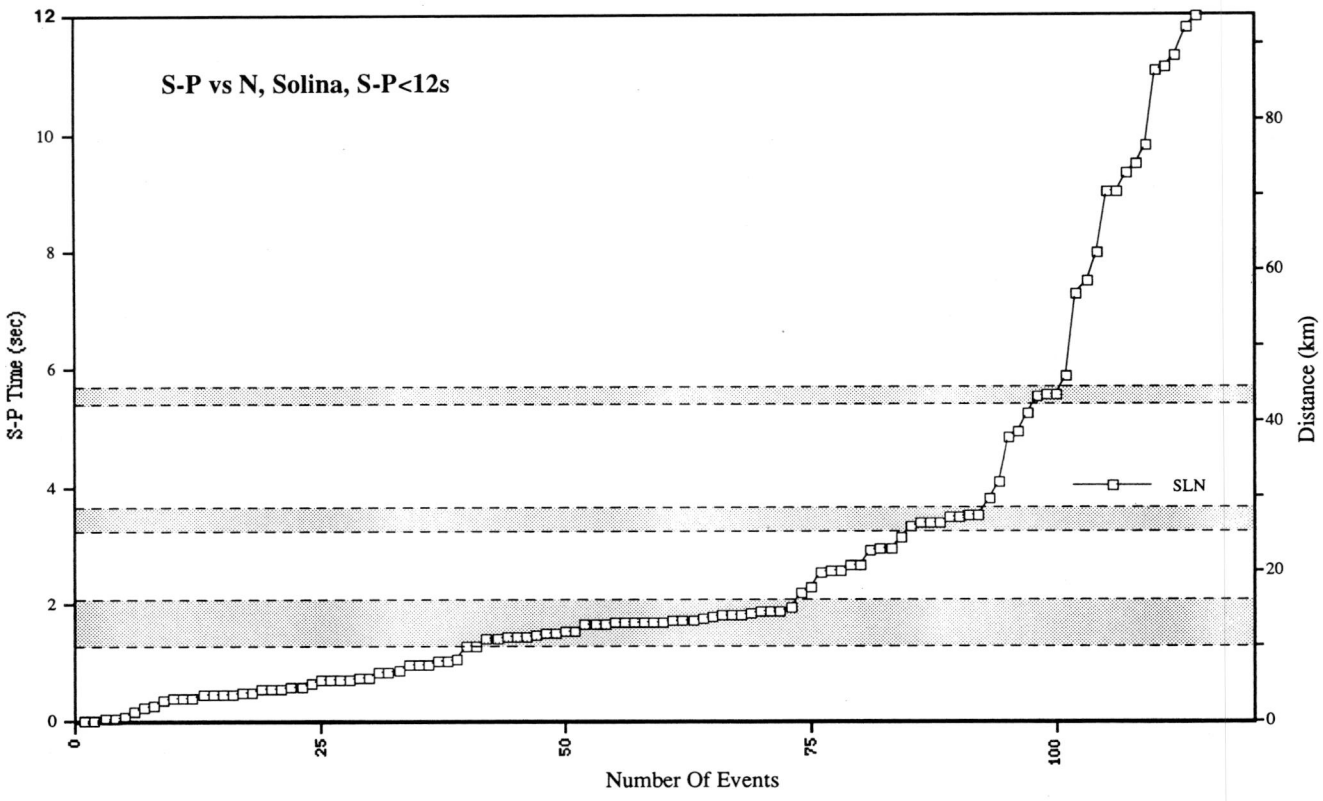

Figure 10 *Cumulative plot of the microearthquakes recorded at Solina station of the EOTN, which shows clustering of the events from specific distances.*

could be produced along each segment of two of the principal linear zones that pass within the GTA, the CMBBZ and the GBLZ (Table 1). Using the attenuation relation of Atkinson and Boore (1995), the least robust of the attenuation relations assessed by Leblanc and Klimkiewicz (1994), and assuming a 30-km focal depth, slightly greater than the 29-km depth of the $M = 6.5$ Saguenay earthquake, a peak ground acceleration of 300 cm·s^{-2} (>>30% *g*), can be estimated for a $M = 7$ earthquake. The corresponding ground velocity will be in the order 7.5 cm·s^{-1}. Typical response spectra, based on the above deterministic input parameters, was developed for 5% damping factor, and compared with the design response spectra for some of the existing structures in the area, which are anchored to the 0.05 *g* recommended ground acceleration by the recent Building Codes (Fig. 13). These values may be too conservative to be used for ordinary buildings, however they may serve as a minimum requirement for design or seismic margin studies of critical public facilities in the GTA.

Probabilistic Hazard Estimate
For the Greater Toronto Area

The probabilistic method for seismic hazard assessment has been widely used since the mid-1970s because it provides a basis to compare the results with those of other natural and man-made hazards and engineering design decisions (McGuire, 1977). In addition, this approach facilitates a consistent treatment of uncertainties by testing the sensitivity of the results to various input parameters. This method may not be readily applicable to local sources in the western Lake

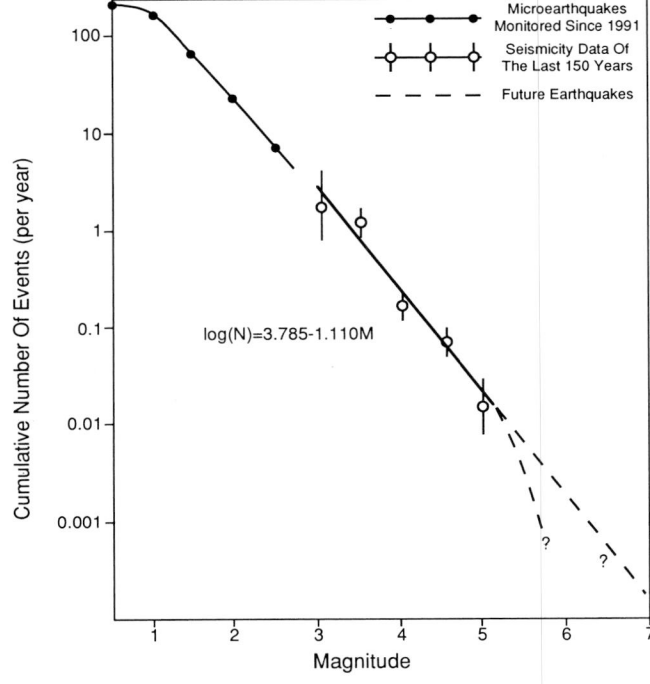

Figure 11 *Magnitude-recurrence relations developed for the GTA, based on the local seismic monitoring east of Toronto compared to the historical data files (Canadian Earthquake Epicentre File, Geological Survey of Canada) in the western Lake Ontario region.*

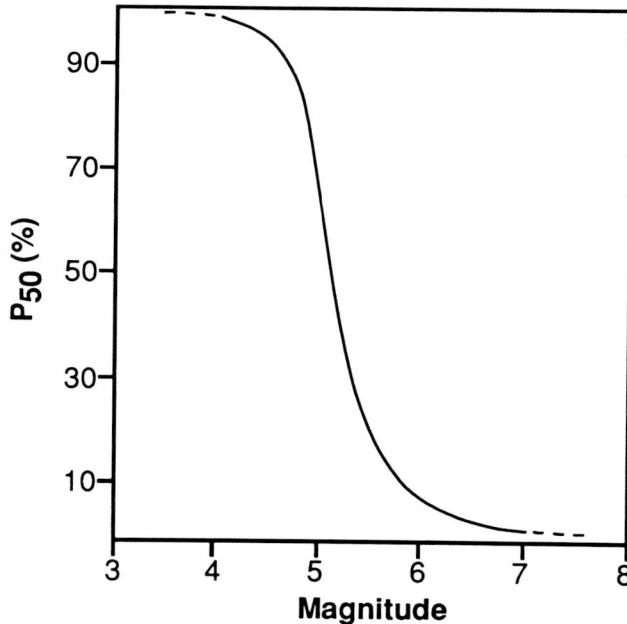

M	Annual Risk	Probability in 50 years (%)
4.0	1.9×10^{-1}	99
4.5	5.9×10^{-2}	95
5.0	1.7×10^{-2}	57
5.5	4.7×10^{-3}	21
6.0	1.3×10^{-3}	6
6.5	3.7×10^{-4}	1.8
7.0	1.0×10^{-4}	0.5

Figure 12 *(top) Probabilistic risk assessment for the GTA, based on the projection of a random distribution (Poisson process) of the past earthquakes. (bottom) Results are presented in terms of annual risk and probability, in 50 years, for various magnitude ranges.*

Ontario region, because there is a limited number of earthquakes above $M = 4.5$ to 5. Therefore, the probabilistic hazard assessment in this region is usually influenced by distant sources, with a history of larger magnitude earthquakes, and may result in underestimating the hazard from local sources in the western Lake Ontario region. To eliminate this shortcoming, a simple probability function is used by assuming a statistically random distribution (a commonly used Poisson process) for the reported small- to moderate-magnitude earthquakes to estimate the likelihood of the future earthquake activities in the GTA. All the shocks recorded or reported within a 60 km radius around Toronto have been used to estimate the annual risk of various magnitudes.

More than 80% of the earthquakes in the western Lake Ontario region are constrained to the THSZ (Mohajer, 1993). Nonetheless, all the nearly 60 reported events used for statistical computations are confined to an area (1800 km²) bounded by the Toronto-Hamilton gravity lineament on the west, NPLZ on the east, and the Niagara Peninsula on the south. The probability of each earthquake magnitude within the above area, in the next 50 years, has been calculated, and shows that for earthquakes of $M = 5$, $= 6$ and $= 7$, the probability is about 57%, 6% and less than 1%, respectively (Fig. 12). These risk levels are within the provisions of the National Building Code of Canada (1985, 1990) for ordinary buildings, but the expected ground motion at lower risk levels exceeds the design basis for critical facilities such as the nuclear facilities at Pickering and Darlington. In this regard, it should be noted that the Ontario reactors are underdesigned compared with reactors in the northeastern United States in a similar seismogenic zone. The United States reactors are designed to withstand an earthquake with a ground motion equivalent to 15% of the force of acceleration due to gravity *(g)* (Design Basis Earthquake; DBE). The DBE for Pickering A and B is 4% and 5%, respectively, that for Darlington is 8%. In January 1986, an earthquake ($M \approx 5$) occurred 17 km from the Perry, Ohio, plant and resulted in ground motions of between 18% and 20% *g*. The plant was closed briefly, but not damaged.

Table 1   Proposed fault segmentation and earthquake potential for the western Lake Ontario region.

Fault Zone	Segment Number	Location (end points)	Orientation	Length (km)	M_x
CMBBZ	1.1	Pembroke-Kamaniskeg Lake	NE	70	7.12
	1.2	Dorset (OGS map)	E-W	95	7.32
	(1.2)	(Babtist)(aeromag)	E-W	80	7.20
	1.3	Balsam Lake	NNE	75	7.17
	1.4	Claremont	NNE	80	7.20
NPLZ	2.1	Pickering-Dundas	NNE	45	6.84
	2.2	Lake Erie shore	NE	70	7.12
THSZ	3.1	Toronto-Hamilton	NE	65	7.08
GBLZ	4.1	Georgian Bay-Oshawa (uncertain)	NW	>130	7.5
WPLZ	5.1	Wilson-Dundas	NE	45	6.84
	5.2	Port Hope	NE	40	6.80
DFZ	6.2	Hamilton-NPLZ	ENE	60	7.05
	6.3	Port Hope	ENE	65	7.08
	6.4	Pres'quile	ENE	50	7.00

Abbreviations: CMBBZ, Central Metasedimentary Belt boundary zone; DFZ, Dundas fault zone; GBLZ, Georgian Bay linear zone; NPLZ, Niagara-Pickering linear zone; OGS, Ontario Geological Survey; THSZ, Toronto-Hamilton seismic zone; WPLZ, Wilson-Dundas–Port Hope linear zone

DISCUSSION

Intraplate earthquakes pose a poorly recognized risk in the heavily urbanized western Lake Ontario region, where critical infrastructures such as nuclear power plants are ill prepared for their effects. Earthquake awareness is generally low in the region, and, therefore, there is considerable resistance to recommendations to review earthquake hazards. Seismic design requirements were first introduced in the National Building Code of Canada in 1953, but local by-laws delayed its adoption until 1967. Nonetheless, this code does not cover facilities such as dams and nuclear power plants in sufficient depth. Such facilities normally require site-specific seismic risk assessment and analysis for their evaluation and design.

Microzonation maps, delineating seismic sources and substrata having a potentially detrimental influence during earthquakes, would assist in mitigating earthquake damage. Environmental geologists and earthquake engineers have a moral obligation to promote public safety by providing reliable data, and informing and educating the public regarding hazard assessment and earthquake loss reduction.

ACKNOWLEDGEMENTS

This work was supported by the Atomic Energy Control Board. Special thanks are due to J.L. Wallach for generously devoting time for field discussions and for the review of this manuscript. J. Bowlby, N. Eyles and G. West are acknowledged for helpful discussions and their valuable comments on the original text. Shallow seismic reflection data were collected and processed in collaboration with J. Boyce, B. Koseoglu and H. Siakoohi. Assistance in the field was provided by D. Popovitch, M. Doughty and S. Salvatori.

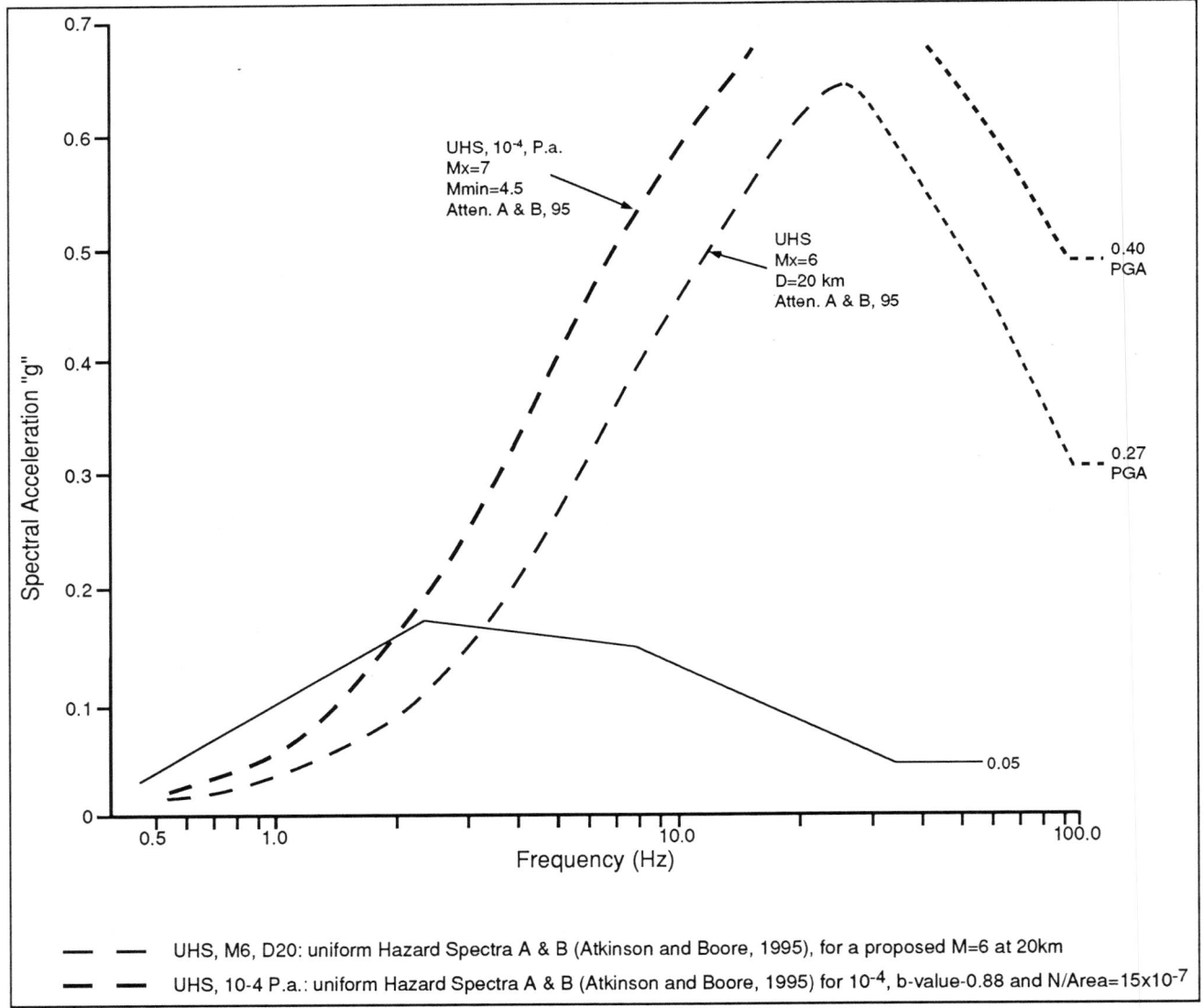

Figure 13 *Typical uniform hazard spectra (UHS) for the GTA based on deterministically estimated $M_x = 7$, which also corresponds to an annual risk of about 10^{-4} (Fig. 14), suggested for design review of the critical public facilities, compared to a M = 6 spectra, based on the attenuation relationships of Atkinson and Boore (1995) at 5% damping. A standard design response spectra anchored to 0.05 g ground acceleration, used for some of the existing facilities in the GTA, is also shown for comparison.*

REFERENCES

Aber, J.S., 1992, Glaciotectonic structures and landforms: Academic Press, San Diego, CA, Encyclopedia of Earth System Science, v. 2, p. 361-378.

Adams, J., Dredge, L., Fenton, C., Grant, D.R. and Shilts, W.W., 1993a, Late Quaternary faulting in the Rouge River Valley, southern Ontario: Seismotectonics or glaciotectonics?: Geological Survey of Canada, Open File 2652, 58 p.

Adams, J., Dredge, L., Fenton, C., Grant, D.R. and Shilts, W.W., 1993b, Comment on "Neotectonic faulting in Metropolitan Toronto: implications for earthquake hazard assessment in the Lake Ontario region, by Mohajer *et al.*, 1992": Geology, v. 21, p. 863.

Adams, J., Basham, P.W. and Halchuk, S., 1995, Northeastern North America
earthquake potential – new challenges for seismic hazard mapping: Geological Survey of Canada, Paper 95-1D, p. 91-99.

Atkinson, G. and Boore, D., 1995, Ground-motion relations for eastern North America: Seismological Society of America, Bulletin, v. 85, p. 17-30.

Croot, D.G., 1987, ed., Glaciotectonic Forms and Processes: A.A. Balkema, Rotterdam, The Netherlands, 212 p.

Eyles, N., Boyce, J. and Mohajer, A.A., 1993, The bedrock surface of the western Lake Ontario region: Evidence of reactivated basement structures?: Géographie physique et Quaternaire, v. 47, p. 269-283.

Hibbert, J., Mohajer, A.A. and Eyles, N., 1994, Faulting in unconsolidated sediments and bedrock east of Toronto: Atomic Energy Control Board, Report 2.263.2, 50 p.

Johnston, A.C., 1989, The seismicity of "Stable Continental Interactions", *in* Gregersen, S. and Basham, P.W., eds., Earthquakes at North Atlantic Passive Margins: Neotectonics and Postglacial Rebound: Kluwer Academic Publishers, Dordrecht, The Netherlands, p. 299-327.

Johnston, A.C., Coppersmith, K.J., Kanter, L.R. and Cornell, C.A., 1994, The earthquakes of stable continental regions: Assessment of large earthquake potential, *in* Schneider, J.F., ed., EPRI report TR-102261: Electrical Power Research Institute, Palo Alto, CA, 180 p.

Leblanc, G., 1981, A closer look at the September 16, 1732, Montreal earthquake: Canadian Journal of Earth Sciences, v. 18, p. 539-550.

Leblanc, G.A. and Klimkiewicz, G.C., 1994, Seismological issues: history and examples of earthquake hazard assessment for Canadian nuclear generating stations: Geological Survey of Canada, Open File 2929, 75 p.

Lewis, C.F.M., Cameron, G.D.M., King, E.L., Todd, B.J. and Blasco, S.M., 1995, Structural contour, isopach and feature maps of Quaternary sediments in western Lake Ontario: Atomic Energy Control Board, Project 2.243.1, INFO-0555, 80 p.

McGuire, R., 1977, Seismic design spectra and mapping procedures using hazard analysis based directly on oscillator response: International Journal of Earthquake Engineering Structural Dynamics, v. 5, p. 211-234.

Mohajer, A.A., 1987, Relocation of earthquakes and delineation of seismic trends in southern Ontario: prepared for Atomic Energy Control Board/Ontario Hydro, Multi-agency Group for Neotectonics in Eastern Canada, Contribution 87-01, 50 p.

Mohajer, A.A., 1993, Seismicity and seismotectonics of the western Lake Ontario region: Géographie physique et Quaternaire, v. 47, p. 353-362.

Mohajer, A.A., Boyce, J. and Eyles, N., 1995, Shallow seismic reflection of bedrock and Quaternary strata in the Rouge River valley, Toronto: Atomic Energy Control Board, Project 2.263.3, Report INFO-0557, p 26.

Mohajer, A.A., Eyles, N. and Rogojina, C., 1992, Neotectonic faulting in Metropolitan Toronto: Implications for earthquake hazard assessment in the Lake Ontario region: Geology, v. 20, p. 1003-1006.

Muir Wood, R. and Mallard, D.J., 1992, When is a fault "extinct"?: Geological Society [London], Journal, v. 149, p. 251-254.

National Building Code of Canada, 1985, Associate committee on the National Building Code, National Research Council, Ottawa, ON, 106 p.

National Building Code of Canada, 1990, Associate Committee on the National Building Code, National Research Council, Ottawa, ON, 181 p.

Slemmons, D.B., 1977, Faults and earthquake magnitude. State-of-the-art for assessing earthquake hazards in the United States: Office Chief of Engineers, United States Army, Washington, DC, Report 6, Miscellaneous Paper S-73-1, 129 p.

Thomas, R.L., Wallach, J.L., McMillan, R.K., Bowlby, J.R., Frape, S., Keyes, D. and Mohajer, A.A., 1993, Recent deformation in the bottom sediments of western and southeastern Lake Ontario and its association with major structures and seismicity: Géographie physique et Quaternaire, v. 47, p 325-335.

Tucker, B.E., Erdik, M. and Hwany, C.N., 1994, Issues in Urban Earthquake Risk: Kluwer Academic Publishers, Noswell, MA, 352 p.

Wallach, J.L. and Mohajer, A.A., 1990, Integrated geoscientific data relevant to assessing seismic hazard in the vicinity of Darlington and Pickering nuclear power plants: Canadian Geotechnical Society, Conference, October 1990, Quebec City, PQ, Proceedings, p. 679-686.

Wallach, J.L., Mohajer, A.A., McFall, G.H., Bowlby, J.R., Pearce, M. and McKay, D.A., 1993, Pop-ups as geological indicators of earthquake-prone areas in intraplate eastern North America: Quaternary Proceedings, v. 3, p. 67-83.

Wells, D.L. and Coppersmith, K.J., 1994, New empirical relationships among magnitude, rupture length, rupture width, rupture area and surface displacement: Seismological Society of America, Bulletin, v. 84, p. 974-1002.

30. Earthquake Hazard in the Greater Vancouver Area [1]

John J. Clague

Geological Survey of Canada, 100 West Pender St., Vancouver, British Columbia V6B 1R8

SUMMARY

There have been nine moderate to large (M = 6-7) earthquakes within 300 km of Vancouver since the late 1800s, but none of these has occurred close enough to the city to damage it. However, the geologic record suggests that large earthquakes have rocked the Vancouver area on several occasions during the prehistoric period. Some of these earthquakes occurred at the boundary between the North America and Juan de Fuca plates, were probably larger than magnitude (M) = 8, and may have strongly shaken several tens of thousands of square kilometres of the Pacific Northwest. Others were local, with sources either at relatively shallow depth in the North America plate or deeper, within the Juan de Fuca plate. Large prehistoric earthquakes in the Pacific Northwest have deformed the crust, caused some types of sediment to liquefy, and triggered tsunamis and landslides. They provide clues as to what can be expected from comparable earthquakes in the future; considerable damage in and around Vancouver is likely to result from liquefaction, ground motion amplification, and landslides. Great subduction earthquakes in the Pacific Northwest occur, on average, once every 500-700 years. The intervals between them, however, range from a few hundred years, or less, to more than 1000 years, making it impossible to accurately predict the time of the next plate-boundary event. Perhaps even more worrisome, in terms of hazard, are M = 6-7 intraplate earthquakes in the North America and Juan de Fuca plates. These are much smaller than subduction earthquakes, but are about 50 times more frequent. A local earthquake with an epicentre near Vancouver could cause losses comparable to those of the 1989 Loma Prieta earthquake in the San Francisco area (M = 7.2, US$10 billion) or the 1994 Los Angeles earthquake (M = 6.7, US$15 billion).

INTRODUCTION

Vancouver, British Columbia (BC) is the third largest city in Canada and lies on the Pacific "Ring of Fire", an area of intense crustal deformation, high seismicity, and active volcanism. It is situated at the leading edge of the North America lithospheric plate, just to the east of the Juan de Fuca plate which is subducting beneath the continent and causing much of the deformation in the region (Fig. 1). Hemmed in by the sea on the west and mountains to the north and east, Vancouver and its satellite communities are vulnerable to damage and isolation during a large earthquake. Highways and rail lines, which are the economic arteries of the city, pass through

mountain valleys where they would be blocked by seismically triggered landslides. Key facilities, including the Vancouver International Airport, a ferry terminal, gas and hydroelectric lines, and critical bridges, might be damaged by earthquake-induced liquefaction, ground settlement, or slope failure (see also Chapter 19). Heightened awareness of seismic hazards, stemming in part from earthquake disasters in the San Francisco area in 1989, Los Angeles in 1994, and Kobe, Japan, in 1995, has led to new concerns about local safety and emergency preparedness.

There have been nine moderate to large (M = 6-7) earthquakes within 300 km of Vancouver during the historical period (Fig. 2). This suggests that the earthquake hazard in the region is relatively high. The historical record, however, is short (*ca.* 150 years) and may understate the actual hazard: an earthquake larger than any in the last 150 years could strike southwestern British Columbia in the future, seriously damaging Vancouver or other cities in the region.

The historical record of seismicity can be extended, albeit incompletely, through the use of geological information. Notably, some coastal sediments near Vancouver and elsewhere in the Pacific Northwest are archives of paleoseismic information. Geologists have studied these sediments to document large earthquakes that occurred hundreds to thousands of years ago. This, in combination with new information from geodetic and other geophysical studies, has led to an improved understanding of seismic hazards in southwestern British Columbia. Of particular importance, it is now known that the region has experienced not only strong local earthquakes, but also great (M = 8+) plate-boundary events. Geological studies also have provided new insights into how the ground responds to seismic shaking, and have shown that earthquake damage is strongly influenced by local soil conditions.

The objectives of this paper are (1) to outline and discuss the earthquake hazard at Vancouver, and (2) to show how geological information contributes to the assessment of this hazard.

TECTONIC SETTING

The present tectonic regime of southwestern British Columbia is controlled mainly by the motions of the Pacific, North America, Juan de Fuca, and Explorer plates (Fig. 1; Riddihough, 1977). The Pacific, North America and Explorer plates intersect in a triple junction off northern Vancouver Island (Riddihough *et al.*, 1980). South of the triple junction, a system of spreading ridges and short transform faults forms the boundary between the Pacific plate on the west and the Juan de Fuca and Explorer plates on the east (Barr and Chase,

[1] Geological Survey of Canada Contribution No. 51194

1974; Riddihough *et al.*, 1983).

The boundary between the North America and Juan de Fuca plates is thought to be a zone of convergence and subduction, referred to as the Cascadia subduction zone. It is well established that subduction has occurred along the coasts of British Columbia, Washington, and Oregon during the last few million years, but there has been some debate as to whether or not it is continuing at present. Riddihough and Hyndman (1976) and Rogers (1988) reviewed relevant geo-

logical and geophysical data bearing on this problem and concluded that subduction is continuing south of 50°N latitude, although at a low rate in southwestern British Columbia. This region shares most of the classic features of other subduction zones: (1) a zone of earthquakes at depths of 35-80 km, distinct from crustal earthquakes, inclined downward to the east from the coast (Crosson, 1983; Rogers, 1983; Cockerham, 1984; Taber and Smith, 1985); (2) an active chain of andesitic volcanoes extending from northern California to

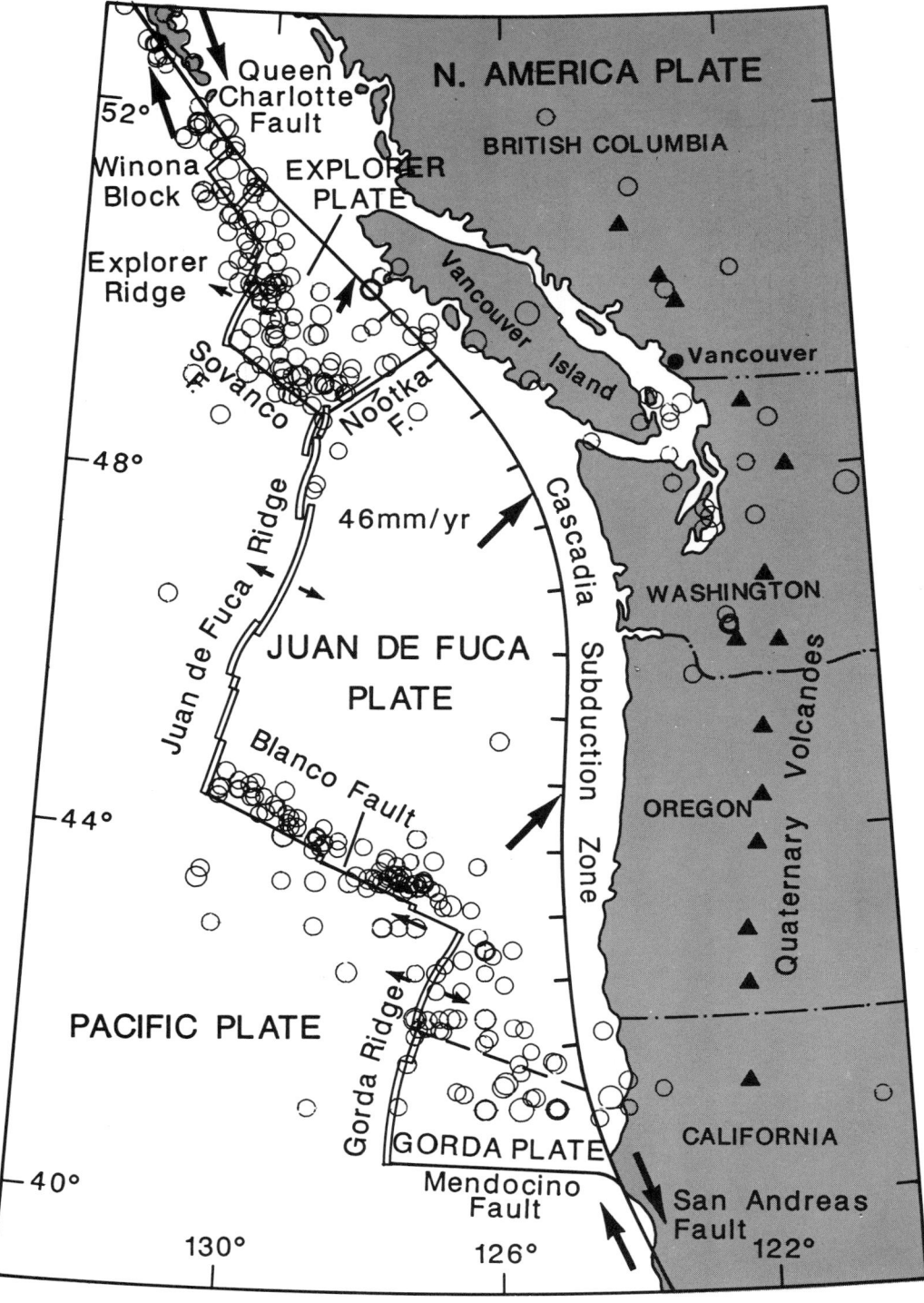

Figure 1 *Plate tectonic regime of the Pacific Northwest. Circles are epicentres of earthquakes of magnitude 5 and larger (data from Canadian Earthquake Epicentre File of the Geological Survey of Canada). All earthquakes prior to 1990 (United States) and 1992 (Canada) are shown. Arrows indicate directions of plate motion (modified from Dragert et al., 1994).*

southern British Columbia (McBirney and White, 1982); (3) a zone of low heat flux oceanward of the volcanic chain (Riddihough and Hyndman, 1976; Blackwell *et al.*, 1982; Hyndman, 1984); (4) a pattern of free-air and Bouguer gravity anomalies consistent with subduction, *i.e.*, an elongated gravity low along the toe of the continental slope and a gravity high inland over the coastal mountains (Riddihough, 1979); and (5) a wedge of deformed Tertiary and Quaternary sediments at the continental margin where the converging plates meet. The most obvious signs of contemporary subduction are current vertical and horizontal deformation (Ando and Balazs, 1979; Savage *et al.*, 1981, 1991; Riddihough, 1982; Adams, 1984; Dragert, 1985; Dragert and Lisowski, 1990; Savage and Lisowski, 1991; Dragert *et al.*, 1994), and a level of seismicity on offshore fracture zones that is consistent with the rate of motion deduced over the past few million years from an analysis of sea-floor magnetic anomalies (Hyndman and Weichert, 1983).

HISTORICAL SEISMICITY

The southwest corner of British Columbia has more than 200 earthquakes each year and is Canada's most seismically active region (Fig. 1). Almost all of these earthquakes are too

A

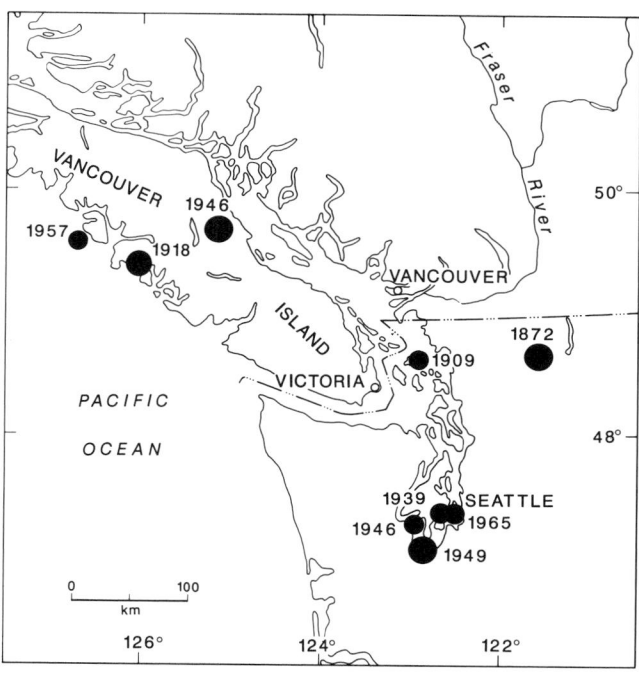

B

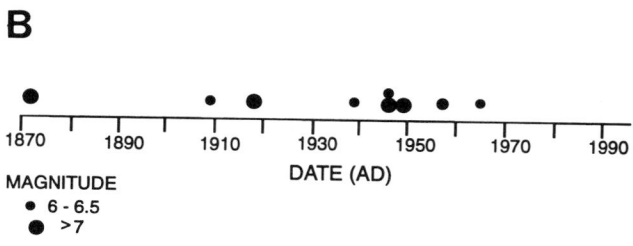

Figure 2 (**A**) *Locations and* (**B**) *dates of large* (M ≥ 6), *historical earthquakes in southern British Columbia and Washington. Smaller dots,* M = 6-6.9 *earthquakes; larger dots,* M ≥ 7 *earthquakes.*

small to be felt, but one capable of major structural damage has occurred, on average, once every 10 years during this century (Table 1).

The most recent large earthquake in southwestern British Columbia, in 1946, was a $M = 7.2$ event centred northwest of Courtenay on Vancouver Island (Fig. 3; Rogers and Hasegawa, 1978). Although widely felt throughout British Columbia, damage was limited because the area of strongest shaking was sparsely populated. In towns nearest the epicentre, chimneys fell, windows broke, and walls cracked. Roads and water and utility lines were also damaged. The earthquake triggered hundreds of small landslides (Mathews, 1979), and there were numerous instances of liquefaction (Rogers, 1980). Were a comparable earthquake to occur today near Vancouver, damage would likely be in the billions of dollars (Munich Reinsurance Company of Canada, 1992).

Damaging earthquakes in southwestern British Columbia may originate within the North America plate (crustal earthquakes), within the subducting Juan de Fuca plate (intraplate quakes), at the boundary between the North America and Juan de Fuca plate (subduction or plate-boundary quakes), and possibly at the boundary between the North America and Explorer plates (Shedlock and Weaver, 1991; Rogers, 1994). All historical earthquakes in the region are of the first two types. The Juan de Fuca-North America plate boundary is presently aseismic, but there is abundant geophysical evidence that it is locked, accumulating strain, and thus capable of great earthquakes (Heaton and Kanamori, 1984; Heaton and Hartzell, 1986; Savage *et al.*, 1991; Hyndman and Wang, 1993; Dragert *et al.*, 1994; Nelson *et al.*, 1995). There also is considerable geological evidence, summarized below, that such earthquakes have occurred many times during the Holocene, most recently about 300 years ago. Although the foci of these earthquakes are more than 200 km distant from Vancouver, the amount of energy released is so large that the city might suffer considerable damage.

EVIDENCE FOR PAST LARGE EARTHQUAKES

Infrequent large earthquakes along the Cascadia subduction zone have caused sudden land-level changes, tsunamis, and strong ground shaking. Buried marsh or forest soils record sudden subsidence at more than a dozen estuaries between northern California and central Vancouver Island (Atwater, 1987, 1992; Darienzo and Peterson, 1990; Clarke and Carver, 1992; Nelson, 1992; Clague and Bobrowsky, 1994a; Darienzo *et al.*, 1994). Sediment types and fossil foraminifera, diatoms, and vascular plants show that the burial of soils at many of the estuaries resulted from at least 0.5 m of sudden subsidence (*e.g.*, Guilbault *et al.*, 1995). The subsidence allowed tides to deposit mud on land that was previously at or above high-tide level (Fig. 4A). Radiocarbon ages on fossil plants suggest that hundreds of kilometres of coast may have subsided during some earthquakes (Atwater *et al.*, 1991; Carver *et al.*, 1992; Nelson and Atwater, 1993; Nelson *et al.*, 1994). The pattern of subsidence during the last earthquake about 300 years ago is thought to be analogous to the widespread subsidence of great historical earthquakes along subduction zones in south-central Alaska and southern Chile (Nelson *et al.*, 1995).

Sheets of sand deposited by tsunamis mantle some of the buried soils in British Columbia (Fig. 5; Clague and Bobrowsky, 1994b), Washington (Atwater, 1987, 1992; Atwater and Yamaguchi, 1991), and Oregon (Darienzo and Peterson, 1990). Stems and leaves of fossil plants rooted in the uppermost buried soil are covered by, or extend into, the overlying sand, implying that the sand was deposited no more than a

Table 1   Moderate and large (M = 5 to 7) earthquakes felt at Vancouver (from Milne *et al.*, 1978; Rogers, 1983, 1994; and the earthquake data file of the Geological Survey of Canada).

Date	Epicentre	Magnitude (M)	Comment
29 Oct. 1864	49°N, 123°W	~5.5	Felt strongly in Vancouver area
14 Dec. 1872	48°N, 120°W	~7.4	Large earthquake felt through most of settled British Columbia
11 Jan. 1909	48.7°N, 122.8°W	~6.0	Felt strongly at Vancouver
6 Dec. 1918	49.4°N, 126.2°W	7.2	Felt at Vancouver, damage on west coast of Vancouver Island
24 Jan. 1920	48.6°N, 123.0°W	~5.5	Felt strongly at Vancouver
17 Sept. 1926	50°N, 123°W	~5.5	Felt by many at Vancouver
13 Nov. 1939	47.4°N, 122.6°W	6.2	Large earthquake in southern Puget Lowland
15 Feb. 1946	47.3°N, 122.9°W	6.4	Large earthquake in southern Puget Lowland
23 June 1946	49.8°N, 125.3°W	7.2	Much damage on Vancouver Island; felt strongly at Vancouver
13 April 1949	47.2°N, 122.7°W	7.1	Much damage in Seattle and Tacoma; felt at Vancouver
16 Dec. 1957	49.6°N, 127.0°W	6.3	Large earthquake on western Vancouver Island
29 April 1965	47.4°N, 122.3°W	6.5	Much damage in Seattle; felt by most at Vancouver
16 May 1976	48.8°N, 123.4°W	5.4	Felt by most at Vancouver

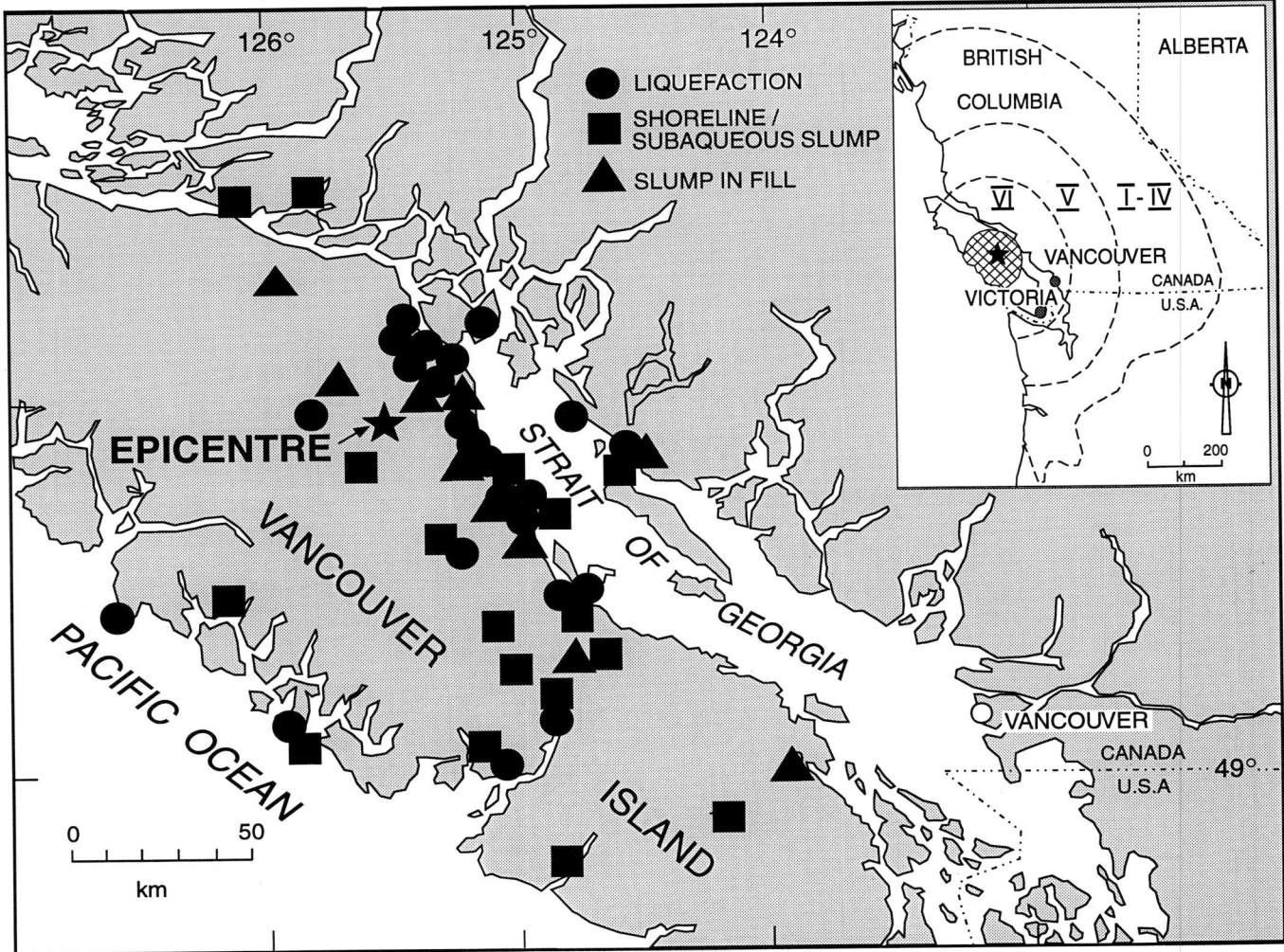

Figure 3   *Known sites of soil failure during the 1946 Vancouver Island earthquake. The inset is an isoseismal (Modified Mercalli Intensity) map; liquefaction occurred in areas of intensity VI and greater (modified from Rogers, 1980).*

few years after the soil subsided (Fig. 4B; Atwater *et al.*, 1991; Atwater, 1992). This strongly suggests that the tsunami and subsidence were caused by the same phenomenon, namely an earthquake.

Liquefaction features, which provide evidence for ground shaking during the last great subduction earthquake, have been found in banks of the lower Columbia River up to 60 km inland from the coast. The river banks expose sand that flowed upward along fractures and was expelled onto a subsided surface about 300 years ago (Fig. 4C; Obermeier *et al.*, 1993). Shaking from earthquakes on the Cascadia subduction zone also has been inferred from Holocene turbidites off the Pacific coast of Washington and Oregon (Adams, 1990).

Geological evidence for large prehistoric earthquakes has been found close to Vancouver. Studies of sediments beneath tidal marshes and bogs south of Vancouver and in the vicinity of Victoria on southern Vancouver Island have shown that sea level changed suddenly about 3600 and 1900 years ago (Fig.

6; Mathewes and Clague, 1994). Slight emergence at sites near Vancouver and submergence on southern Vancouver Island during the older event are consistent with the deformation expected from an earthquake on the Cascadia subduction zone. The younger event is characterized by submergence throughout the region, and may record a subduction earthquake or a very large crustal or intraplate earthquake.

The younger event is particularly well recorded in several exposures along Serpentine River south of Vancouver, where a woody fresh-water peat is abruptly overlain by intertidal mud (Fig. 6; Mathewes and Clague, 1994). At one of the Serpentine River sites (SR3, Fig. 6), submergence can be directly linked to injection and venting of mud. Here, dykes of silty mud containing upward displaced peat clasts cut the lower part of the exposed section. The dykes broaden upward and are continuous with an eroded mound of erupted mud that mantles the peat and is draped by intertidal mud. The stratigraphic relations show that injection and venting occurred at the same time as submergence, and suggest that both phenomena are the result of a large earthquake.

Finally, there is abundant evidence on the Fraser River delta, just south of Vancouver, of earthquake-induced liquefaction in the recent past (Clague *et al.*, 1992; Naesgaard *et al.*, 1992). Sand dykes and sills are common in near-surface sediments over much of the delta plain. At a few sites, fluidized sand was expelled onto a former subaerial or intertidal surface, producing sand blows (Fig. 7). The source of the liquefied sediment is a shallow sheet of distributary channel sands, deposited by the Fraser River as it built its delta into the Strait of Georgia during the Holocene. Liquefaction and the upward movement of sand and water deformed some of the intruded sediments, causing the delta plain to subside locally. All observed liquefaction features on the Fraser delta are less than 3800 years old, and at least some are less than 2500 years old. Offshore data are briefly reviewed in Chapter 19.

EFFECTS OF A LARGE EARTHQUAKE

Ground Motion Amplification

The ground surface response to seismic shaking of thick or soft sediments differs markedly from that of bedrock. Potential adverse effects of seismic shaking on such sediments include ground motion amplification and liquefaction.

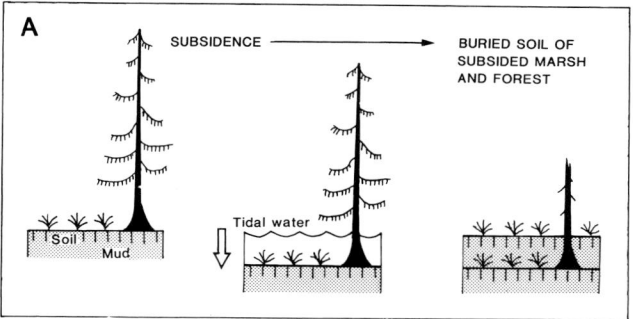

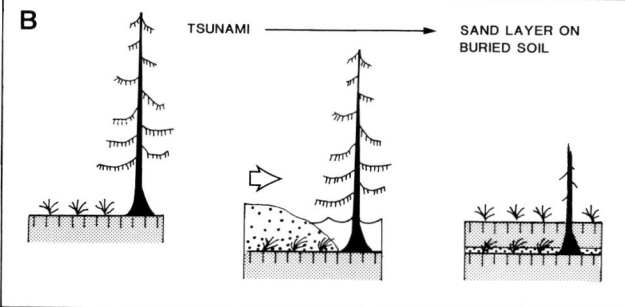

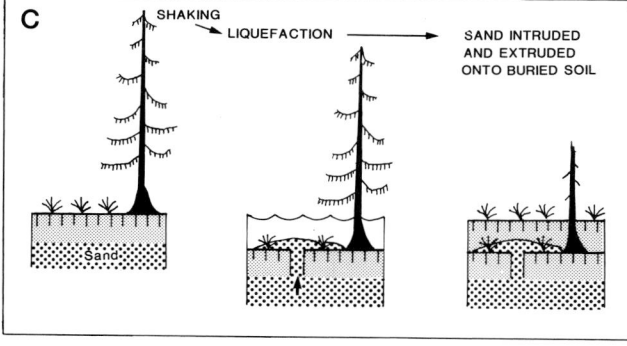

Figure 4 *Inferred origin of the main coastal features cited as evidence for prehistoric earthquakes in southwestern British Columbia and Washington. (**A**) Soil is buried by tidal mud after earthquake-induced subsidence lowers the land into the intertidal zone. (**B**) A sheet of sand is deposited on a subsided surface by a tsunami shortly after an earthquake. (**C**) Liquefied sand is erupted through and onto a subsided surface as a result of shaking during an earthquake (modified from Atwater* et al.*, 1995).*

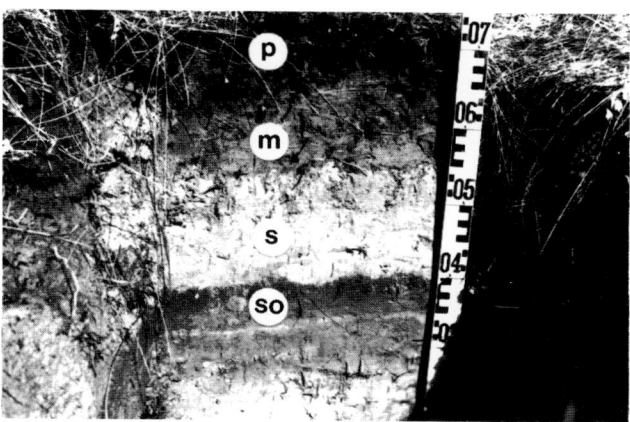

Figure 5 *Massive sand bed (**s**), interpreted to be a tsunami deposit, sharply overlying an eroded peaty soil (**so**) in a pit at a tidal marsh near Tofino, British Columbia. The sand is abruptly overlain by intertidal mud (**m**) which grades upward into peat of the present-day marsh (**p**) (from Clague and Bobrowsky, 1994b).*

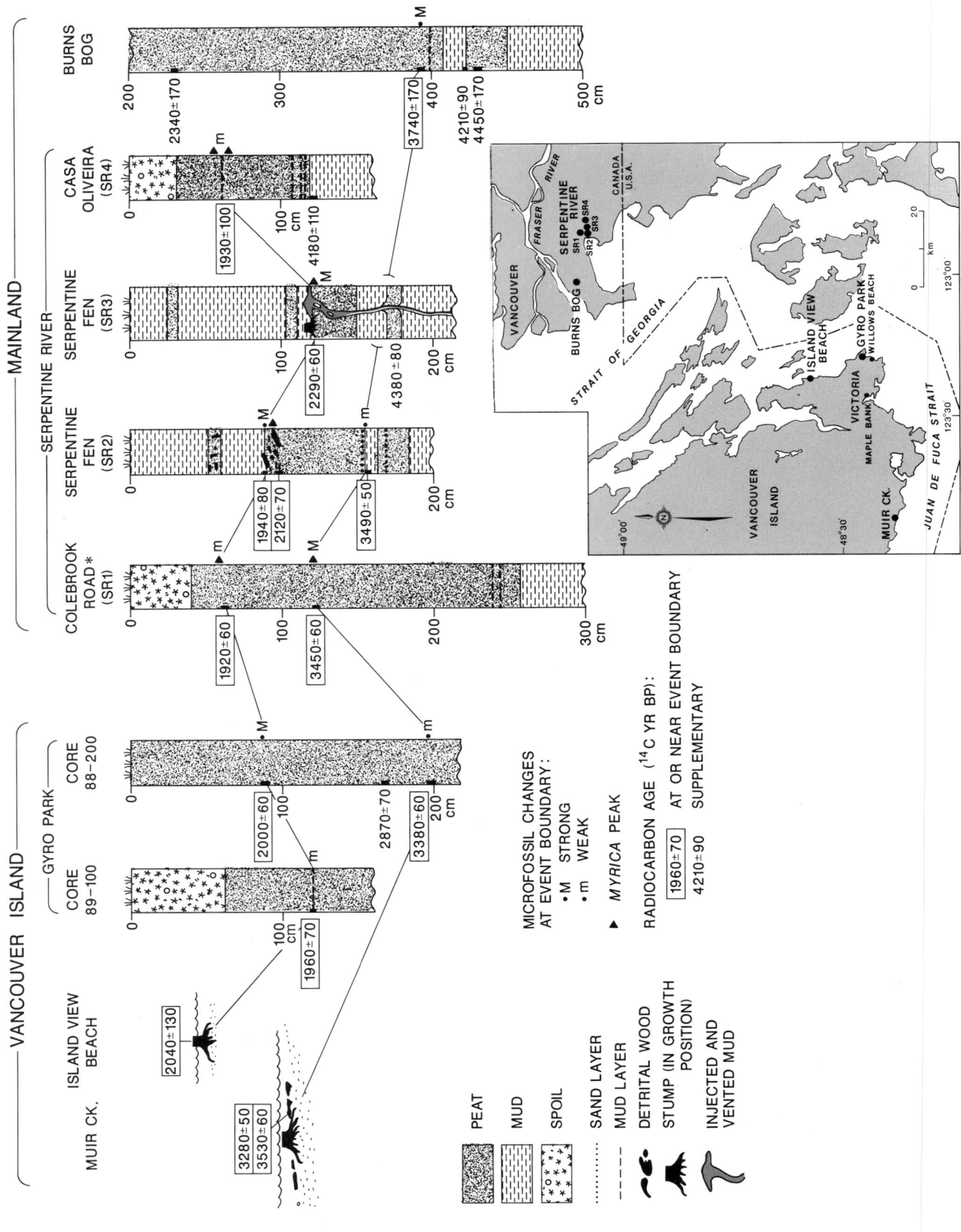

Figure 6 *Stratigraphy at study sites on the British Columbia mainland near Vancouver and on southern Vancouver Island, showing correlations of two inferred earthquakes about 3600 and 1900 years ago (3400 and 2000 14C years BP). Correlations are based on radiocarbon ages in boxes and microfossil changes (modified from Mathewes and Clague, 1994).*

Thick Holocene sediments prone to ground motion amplification are common in the Vancouver metropolitan area. The delta plain of the Fraser River, for example, is underlain by up to 235 m of loose, water-saturated, clay, silt, and sand of Holocene age. This area is of particular concern because it supports a large population (*ca.* 200,000 people) as well as critical transportation, energy, and communication lines. Accordingly, considerable effort has been made to determine how the delta might respond to shaking during a large earthquake.

One-dimensional computer models of ground motion amplification have been applied to the Fraser delta (Finn and Nichols, 1988). The models require the shear-wave velocity structure of the sediment pile from the ground surface down to, and including, bedrock. Holocene sediments of the Fraser delta generally have lower shear-wave velocities than underlying Pleistocene deposits, hence the depth of the buried Pleistocene surface is an important parameter in modelling ground motion amplification. In some areas, the Pleistocene surface has been delineated by means of drilling, high-resolution seismic reflection surveys, and shear-wave refraction surveys. This work has shown that the top of the Pleistocene succession is an irregular surface, ranging in depth from 8 m to more than 200 m (Clague *et al.*, 1991). The local presence of shallow Pleistocene sediments significantly alters and possibly lowers the ground motion that would occur during an earthquake.

A simple, first-order technique for estimating ground motion amplification in unconsolidated sediments using shear-wave velocities has been used effectively by the United States Geological Survey (Fumal and Tinsley, 1985; Joyner and Fumal, 1985) and has been applied to sites on the Fraser delta (Luternauer *et al.*, 1994). The technique involves comparison of thickness-weighted, average, shear-wave velocities of sediments to the shear-wave velocity of bedrock, assumed to be 1500 m·s^{-1}. In the case of the Fraser delta, the thick sequence of low-velocity Holocene sediments has a marked effect on the thickness-weighted average velocity. As a result, ground motion amplification is higher than for similar types of deposits in California that have been analyzed in the same way (Hunter *et al.*, 1993).

Liquefaction

Liquefaction occurs when stresses propagating through saturated, non-cohesive sediments cause shear and volumetric straining of the sediment structure. This results in an increase in pore-water pressure and a decrease in grain-to-grain contact forces. As the grain-to-grain contact force approaches zero, grains become suspended in the pore water. The sediment then loses most of its strength, behaves essentially as a fluid, and is said to have liquefied. The liquefied sediment may flow upward along fractures to the surface where it accumulates as mounds, or blows. This may be accompanied by lateral spreading, ground subsidence, and a loss of sediment-bearing capacity.

Liquefaction was common in the epicentral region of the 1946 Vancouver Island earthquake (M = 7.2), the strongest seismic event of the historical period in southern British Columbia. Most of the occurrences were along the east coast of Vancouver Island, but other areas were also affected (Fig. 3; Hodgson, 1946; Rogers, 1980). The greatest damage of the earthquake was to a fish cannery 100 km south-southeast of the epicentre. Pilings supporting the cannery settled in liquefied sediment during 30 seconds of strong shaking, necessitating $100,000 in repairs (1946 dollars).

Liquefaction accompanied other historical earthquakes in the region, including a M = 7 quake at Olympia, Washington, in 1949, and a M = 6.5 quake at Seattle, Washington, in 1965. One of the lessons of these events is that considerable damage will result from liquefaction during any future large earthquake near Vancouver or any other population centre in southwestern British Columbia.

Water-saturated, silty and sandy sediments and non-compacted fill are the materials that are most prone to liquefaction. Areas of highest hazard in and around Vancouver include the Fraser delta and floodplain, smaller deltas, landfill sites, and shorelines (Fig. 8; Watts *et al.*, 1992).

Geological engineers have conducted extensive tests of Fraser delta sediments to determine how they would respond to seismic shaking. The susceptibility of the sediments to liquefaction has been estimated by empirical correlations with standard penetration and cone penetration test data and, more recently, from shear wave velocity data (Finn *et al.*, 1989; Hunter *et al.*, 1991, 1992; Robertson *et al.*, 1992a, b). The studies indicate that liquefiable sediments are widespread at shallow depth beneath much of the Fraser delta plain.

Landslides

Ground shaking during earthquakes may trigger landslides that damage or destroy buildings and bridges, bury roads and rail lines, and kill and injure people. The 1976 Guatemala earthquake (M = 7.5), for example, generated more than 10,000 landslides, which claimed hundreds of lives over an area of approximately 16,000 km^2, and severely disrupted road and rail traffic (Harp *et al.*, 1981).

Rock and soil falls and small slides may be triggered by

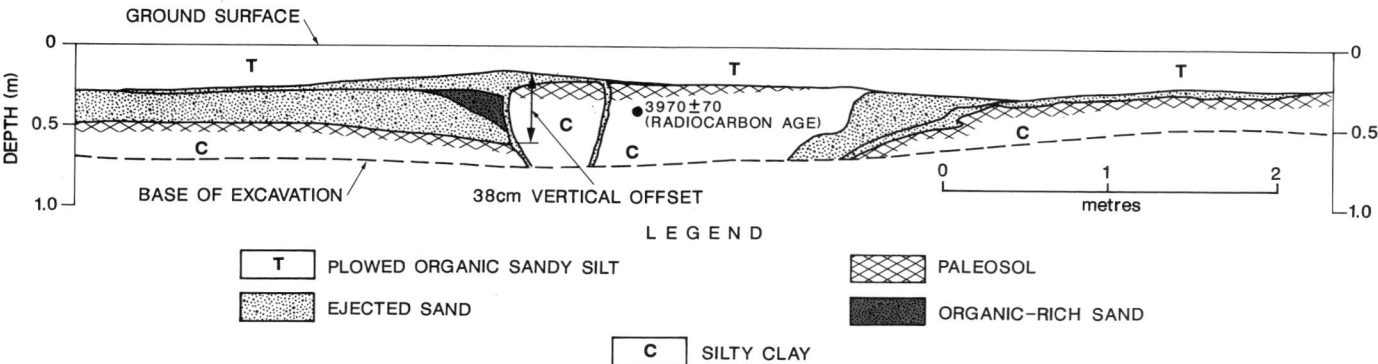

Figure 7 *Profile of sand dykes and vented sand exposed in the shallow wall of an excavation on the Fraser River delta just south of Vancouver. The vented sand directly overlies a brown paleosol and displays stratification indicative of emplacement during either multiple events or multiple pulses within a single event. Note the vertical displacement of the paleosol and the subjacent mud along the two main feeder dykes (Clague et al., 1992).*

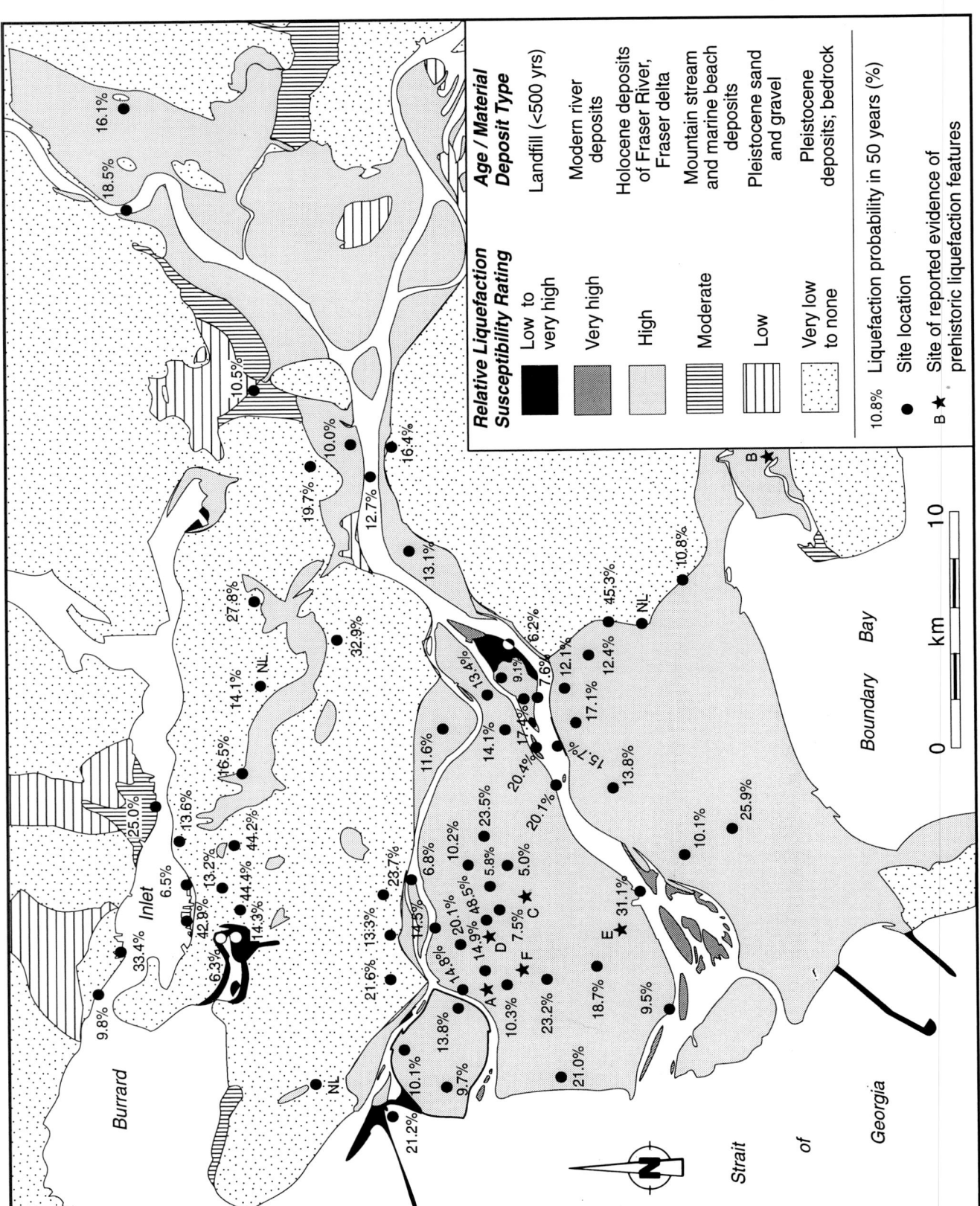

Figure 8 *Liquefaction hazard map of the greater Vancouver area, simplified from a map prepared by B.C. Hydro and Klohn Leonoff Limited (1992). The map is generalized and regional in scope and cannot be used to determine whether or not sediments will liquefy at any specific site; site-specific geotechnical investigations are needed.*

relatively small (M = 4-5) earthquakes (Keefer, 1984). In contrast, more coherent, deeper seated slides and slumps require stronger shaking, and large rock and soil avalanches and lateral spreads generally occur only during large earthquakes. There is also a relation between earthquake magnitude and the areal extent of landsliding (Keefer, 1984): the affected area increases from 0 for a M = 3-4 quake to 500,000 km^2 for a M = 9 event.

The 1946 Vancouver Island earthquake triggered more than 300 landslides over an area of about 20,000 km^2 (Mathews, 1979), giving some indication as to what might happen around Vancouver in the event of a future large earthquake there. Most of the slope failures in 1946 were rockfalls and small rock and soil slides on steep slopes. Similar slope failures would occur in the mountains north and east of Vancouver during a large earthquake. The main highways, rail lines, and energy transmission lines pass through valleys and canyons bordered by steep, potentially unstable slopes, and probably would be blocked by landslides (Fig. 9). A major earthquake might cause numerous blockages over a large area, disrupting economic activity and restricting access to Vancouver. In this context, even small landslides, which are the vast majority of those triggered by earthquakes, can be severely disruptive. The pro-

blem would be worse if an earthquake were to occur after a lengthy period of rain, when soils were saturated and pore-water pressures high.

Landslides would not be as severe a problem within the city of Vancouver and its suburbs because relief in these areas generally is low. There are, however, some steep slopes in Pleistocene sediments that might slump or slide during an earthquake (Eisbacher and Clague, 1981). In addition, liquefaction could trigger destructive lateral spreads on the Fraser River delta and flood plain.

Tsunamis

Tsunamis are ocean waves generated by underwater disturbances in the Earth's crust. They are triggered by earthquakes and, less commonly, by submarine landslides and volcanic eruptions. The waves are imperceptible on the open ocean, where they have amplitudes of less than 1 m and move at velocities up to 1000 km·h^{-1}, but may pile up to heights of 30 m or more in shallow water and wreak havoc on coastal areas.

The most destructive historical tsunami in British Columbia was triggered by the great Alaska earthquake of March 27, 1964. A series of waves radiated from the epicentre near the head of Prince William Sound and, within a few hours,

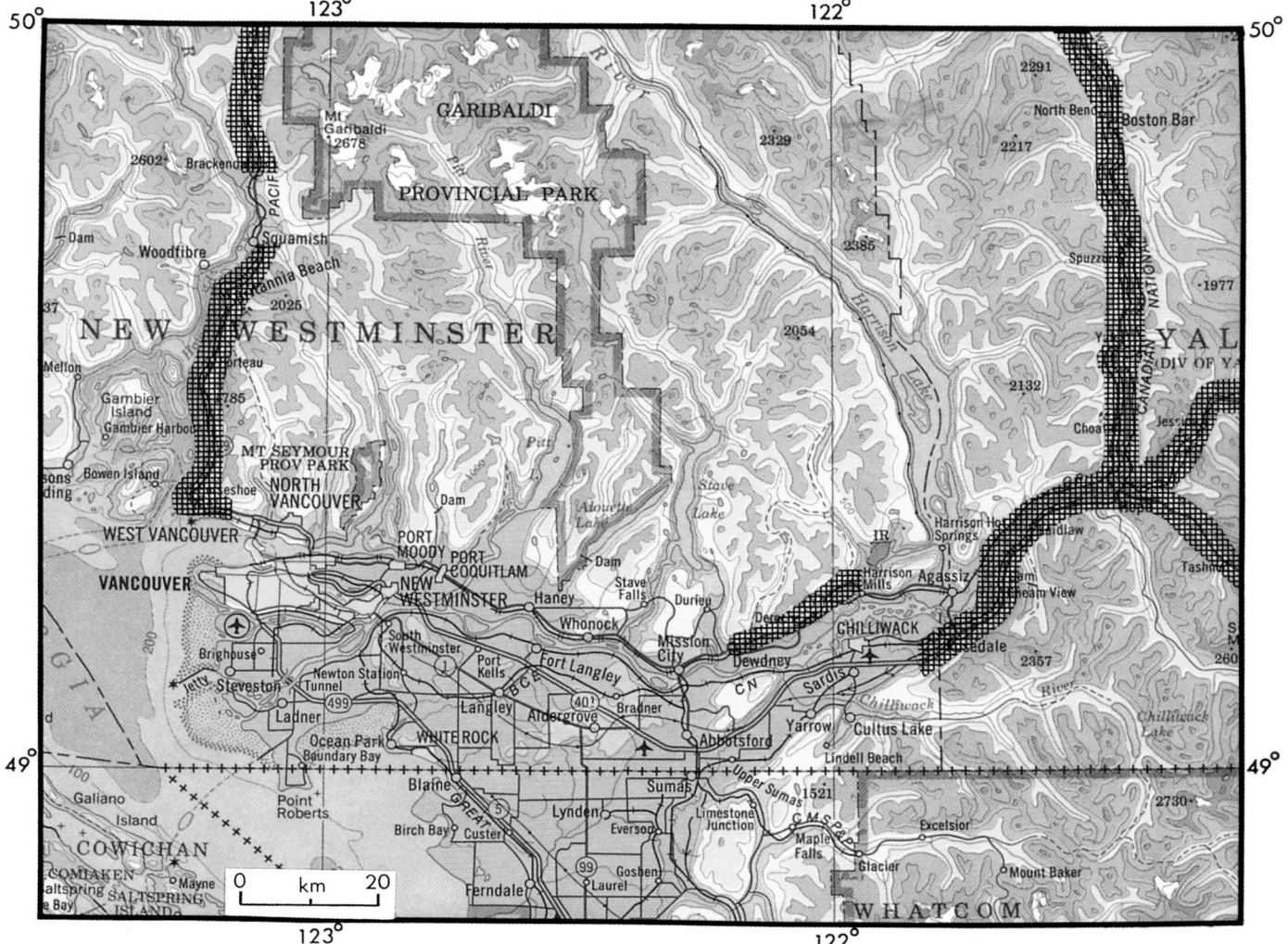

Figure 9 *Map of southwestern British Columbia showing sections (cross-hatched pattern) of major transportation and energy lifelines into Vancouver that are vulnerable to damage or blockage by earthquake-triggered landslides. Offshore portions of the Fraser Delta are also at risk (Chapter 19).*

reached the outer coast of British Columbia, damaging several communities on Vancouver Island (Wigen and White, 1964; Murty and Boilard, 1970). Although there was no loss of life in British Columbia, most of the 130 deaths attributable to the earthquake were caused by tsunamis (Hansen *et al.*, 1966).

Estimates have been made of maximum wave heights and velocities on the British Columbia coast for hypothetical tsunamis from different sources in the Pacific Ocean. Dunbar *et al.* (1989, 1991) developed computer models to generate and propagate tsunamis from several regions in the North Pacific Ocean. They verified model predictions against water level measurements made at tide gauge stations along the British Columbia coast after the 1964 Alaska earthquake. The model results show the areas most vulnerable to far-travelled tsunamis are the central mainland coast and the west coasts of Vancouver Island and the Queen Charlotte Islands. In contrast, there would be very little wave run-up along the shores of the Strait of Georgia.

A large tsunami would be produced by an earthquake on the Cascadia subduction zone. A numerical model has been used to simulate three tsunamis, involving rupture of different sections of the plate boundary (Ng *et al.*, 1990, 1991). As might be expected, the west coast of Vancouver Island would be the most severely affected area in British Columbia. Maximum wave amplitudes for a simulated $M = 8.5$ earthquake that ruptures the northern section of the subduction zone are approximately the same as the magnitude of the sea-floor displacement (5 m). Amplification, up to a factor of three, is predicted for some inlets on western Vancouver Island. Much energy would be lost as the waves passed through narrow passages connecting Juan de Fuca Strait to Puget Sound and the Strait of Georgia, and tsunami amplitudes would be reduced to 1 m or less by the time the waves reached Seattle and Vancouver (Fig. 10). A tsunami would, of course, be superimposed on the tides, which in south-coastal British Columbia have a range of up to 5 m. A positive sea-level displacement at low tide would have much less impact than one at high tide. However, a tsunami could persist through a tidal cycle, thus extreme water levels would likely combine tsunami and tidal effects (Ng *et al.*, 1991).

It would take little time for a Cascadia tsunami to reach the British Columbia coast. The first wave would strike western Vancouver Island in minutes to, at most, one hour (Hebenstreit and Murty, 1989; Murty, 1992). Travel times to Victoria and Vancouver might be as little as 1.5 and 3 hours, respectively. In contrast, the 1964 Alaska tsunami arrived on the west coast of Vancouver Island about 5 hours after the earthquake.

A potentially damaging tsunami could also be triggered by a large earthquake within the North America or Juan de Fuca plate. Locally generated waves could damage areas that would not be affected by tsunamis of distant earthquakes. For example, a shallow crustal earthquake near Seattle about 1000 years ago triggered a large tsunami in Puget Sound (Atwater and Moore, 1992). Wave runups along the shores of Puget Sound during this event may have been comparable to the runups of large, far-travelled tsunamis on the exposed, outer coast.

Simulations have been made of tsunamis triggered by hypothetical earthquakes in the North America plate (Murty and Hebenstreit, 1989; Murty, 1992). Earthquakes used in the simulations were similar in magnitude and motion to the 1946 Vancouver Island earthquake (Rogers and Hasegawa, 1978), but were positioned in Juan de Fuca Strait near Victoria, in Puget Sound near Seattle, and in the Strait of Georgia near Vancouver. In all cases, the tsunamis were small (largest

waves = 0.5-1 m) and of local effect. Little energy passed through the Gulf Islands between Juan de Fuca Strait and the Strait of Georgia. It should be noted, however, that a local shallow earthquake of $M = 8$, much larger than the 1946 event, could produce waves 3 m or more in height along the shores of the Strait of Georgia (Murty and Hebenstreit, 1989). Such a tsunami probably would flood low-lying areas and might severely damage shoreline development. No geological evidence has yet been found around the Strait of Georgia for such a tsunami.

HAZARD EVALUATION

Local Earthquakes
On average, there has been one $M = 6$-7 earthquake in

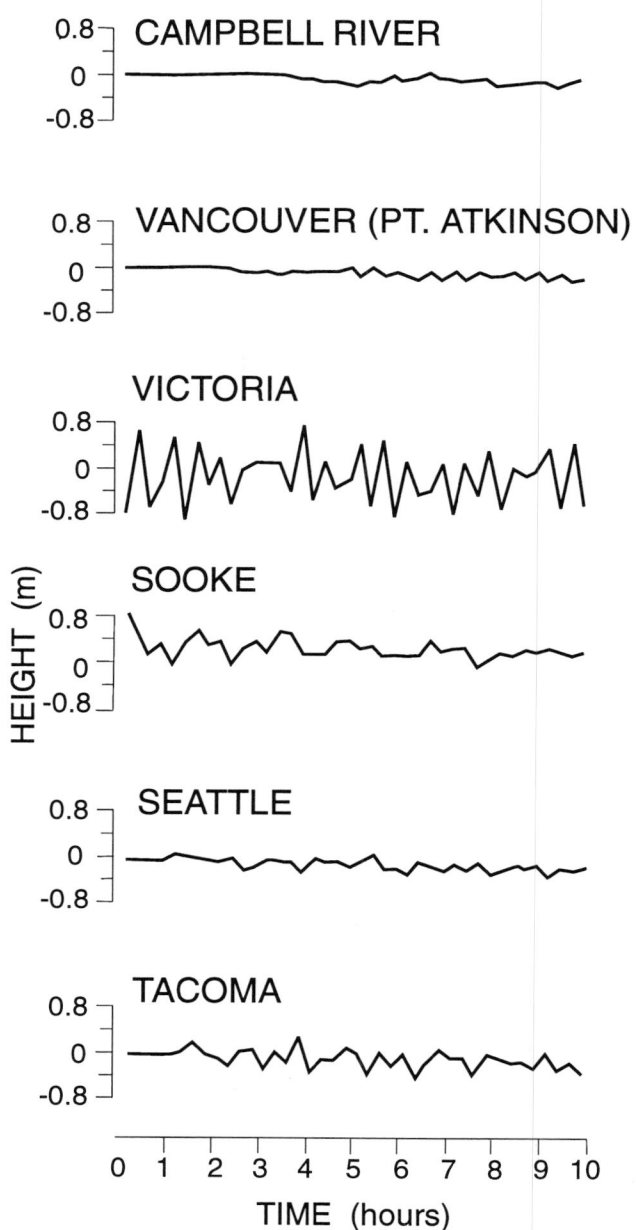

Figure 10 *Time series of sea-level displacements generated by a hypothetical* $M = 7$ *crustal earthquake similar to the 1946 Vancouver Island earthquake, but centred in Juan de Fuca Strait near Victoria. The tsunami is small and localized (modified from Murty and Hebenstreit, 1989).*

southwestern British Columbia and northwestern Washington every 15 years since the late 1800s. Probability analysis indicates that the next large earthquake is likely to happen within 20 years, and is almost certain to occur within 50 years. It is not possible, however, to predict the time or location of this earthquake. Efforts to do so are thwarted by an inadequate understanding of the relationship between historical seismicity and geological structure.

The instrumented record does suggest, however, that large crustal and intraplate earthquakes do not occur randomly throughout the region (Fig. 2A). The three largest earthquakes in southwestern British Columbia in the twentieth century have been relatively shallow events beneath central Vancouver Island; their epicentres lie along the extrapolated trend of the transform fault that separates the Explorer and Juan de Fuca plates. In contrast, most earthquakes of similar size in northwestern Washington have been deep events, with foci in the Juan de Fuca plate; four of these, including the most recent in 1965, were centred in the southern Puget Lowland. There has been only one $M = 6$ earthquake in the northern Puget Lowland (1909) and none beneath the Strait of Georgia (however, a $M = 5.4$ earthquake occurred deep under Pender Island in 1976).

Similarly, large historical earthquakes in the region have not been evenly spaced in time (Fig. 2B, Table 1). Five occurred between 1946 and 1957, but there have been none since 1965. The 30-year period since the 1965 Seattle earthquake is the longest separating two $M = 6$-7 events during this century. Based on this meagre evidence, it can be argued that the region is overdue for another large earthquake.

Modified Mercalli Intensity plots of historical earthquakes in the Pacific Northwest (Fig. 3) show that a $M = 7$ event can produce significant damage within 50-100 km of the epicentre; in contrast, the radius of widespread damage for a $M = 6$ event is only 20-50 km. The damage that might result from a future $M = 6$-7 earthquake would, of course, depend on the location of the epicentre. Obviously, a $M = 7$ earthquake beneath the southern Strait of Georgia, within 50 km of Vancouver, would be much more damaging that one on central Vancouver Island, some 200 km away. A recent study of the economic impact of a hypothetical $M = 6.5$ crustal earthquake with a focus 10 km beneath Vancouver provides a worst-case scenario of the damage that could be expected from a large earthquake in the region. The study concluded that the total economic loss from such an earthquake would be between $14 and $32 billion (1992 dollars; Munich Reinsurance Company of Canada, 1992).

Ground motion and secondary effects, notably landslides, liquefaction, and fire, would cause much of the damage to Vancouver during a large local earthquake. The economic infrastructure of the affected area might be disrupted for weeks or months due to closure of critical roads, rail lines, bridges, airports, and communication facilities. Most wood-framed houses and newer multi-story buildings would withstand the shaking, but old masonry buildings, as well as structures on poor ground, might be severely damaged. On the positive side, it is unlikely that such an earthquake would trigger a large tsunami, and any co-seismic subsidence or uplift probably would be localized. Only an exceptionally large, shallow, crustal earthquake, such as occurred at Seattle about 1000 years ago, would produce surface rupture, significant uplift and/or subsidence, and a destructive tsunami.

Subduction Earthquakes

A subduction earthquake might also damage Vancouver.

Ground motion during a future earthquake along the Cascadia subduction zone would be constrained by the position of the plate boundary, which is known from micro-earthquake seismicity (Rogers *et al.*, 1990) and seismic reflection and seismic refraction surveys (Drew and Clowes, 1990; Hyndman *et al.*, 1990). In addition, information from other subduction zones can be used to estimate rupture behaviour, and thus surface ground motion. The most important factor in estimating hazard is the location and extent of the locked portion of the plate boundary, which is where earthquakes can nucleate. Geophysical modelling suggests that the zone is narrow and shallow off Vancouver Island, because the subducting plate is relatively hot and steep and is mantled by thick sediments; it is thought to lie beneath the continental slope and outer shelf, some 200 km from Vancouver (Hyndman and Wang, 1993). In contrast, the locked zone may be wider and extend closer to the coast off Washington, because the subducting plate is slightly cooler and dips more gently there.

Although there is little doubt that subduction earthquakes occur in the region, there is still uncertainty as to how much of the Cascadia subduction zone ruptures during each event. It is possible that the entire 1000 km length of the subduction zone has ruptured during an earthquake, analogous to the situation in Chile in 1960. If this were to happen in the future, the resulting earthquake would exceed magnitude 9 (Rogers, 1988) and would severely damage Vancouver.

An alternative scenario is one in which a section of the subduction zone, 100 km to several hundred kilometres in length, fails, producing a $M = 8$ earthquake. This could happen if the subduction zone is segmented and strain is accumulating at different rates in different areas. Several indirect lines of evidence suggest that this scenario may be more likely than one involving rupture of the entire plate boundary. The orientation of the subduction zone changes from northwest-southeast off Vancouver Island to north-south off Washington, coinciding with a change in the rate and direction of convergence. In addition, the distribution of foci of intraplate earthquakes indicates that the subducting Juan de Fuca plate arches upward beneath southern and central Puget Sound (Weaver and Baker, 1988). Beneath southwestern Washington, the plate dips to the east, whereas in northwestern Washington and southwestern British Columbia, it dips northeast. Finally, semi-independent microplates at the north and south ends of the Juan de Fuca plate complicate the plate tectonic regime and may promote segmentation.

The issues of segmentation and the maximum magnitude of subduction earthquakes have not yet been resolved; however, radiocarbon and tree-ring dating (Yamaguchi *et al.*, 1989; Atwater *et al.*, 1991; Carver *et al.*, 1992; Nelson *et al.*, 1995) do not rule out the possibility that the last earthquake, about 300 years ago, ruptured the entire subduction zone and thus was a $M = 9+$ event. The alternative is that a series of lesser ($M = 8$) earthquakes, involving different sections of the subduction zone, occurred over a period of, at most, a few decades. Earthquakes of even smaller size ($M < 8$), that rupture less than 100 km of the plate boundary, are much less likely. The 40-50 mm·a^{-1} convergence of the Juan de Fuca and North America plates requires that such small ruptures occur more frequently (Dragert *et al.*, 1994), perhaps once every hundred years or less, and none has been reported in the past 150 years.

These issues are critical in evaluating the threat posed by subduction earthquakes. A $M = 9$ earthquake releases ap-

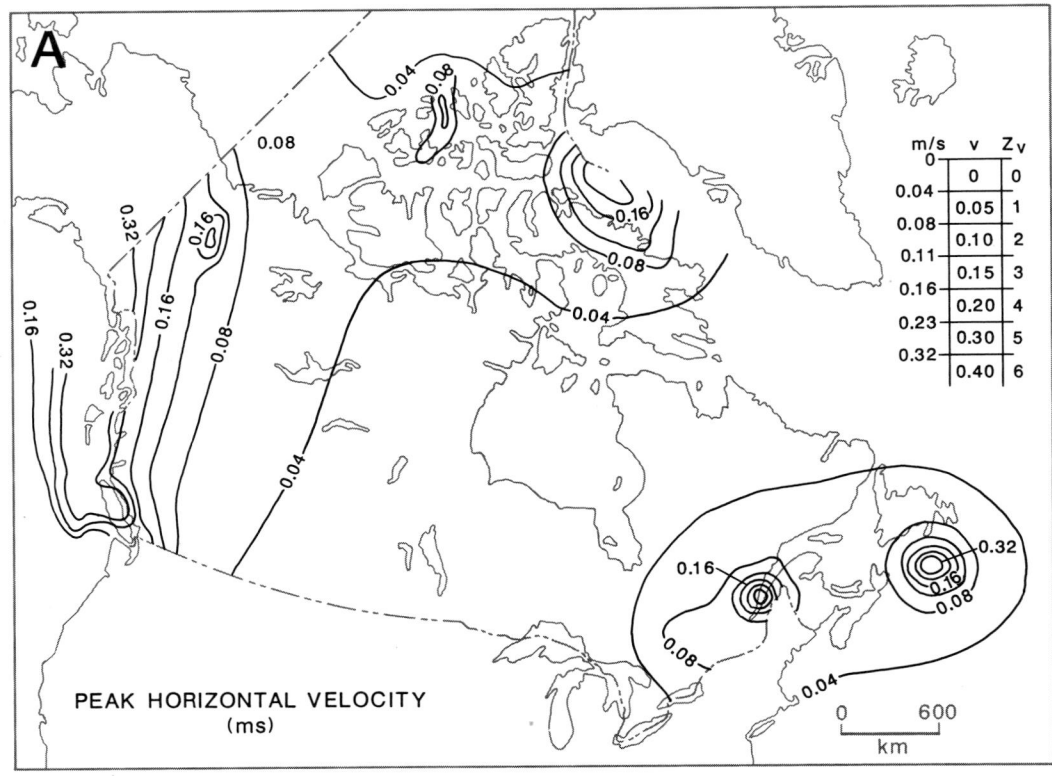

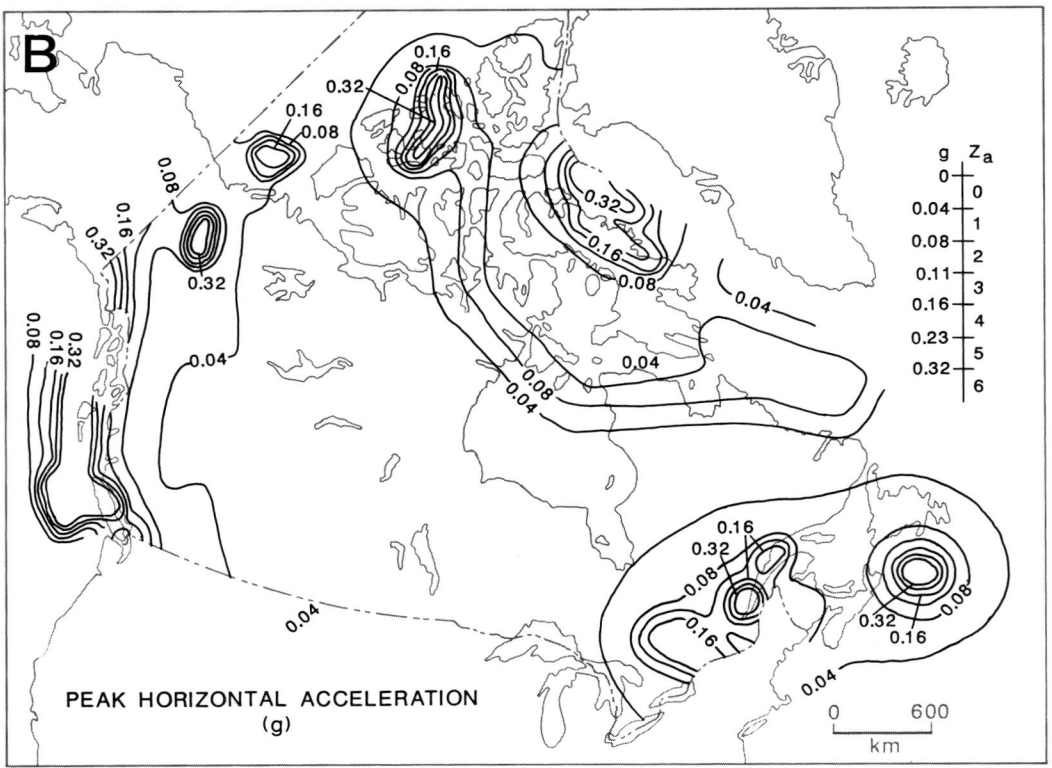

Figure 11 *Current seismic ground motion zoning maps of Canada. (A) Peak horizontal velocity zoning map; contour intervals for the seven velocity zones and zonal velocities (Zᵥ) used in National Building Code of Canada applications are shown in the legend. (B) Peak horizontal acceleration map; contour intervals for the seven acceleration zones (Zₐ) are shown in the legend. The maps are based on information available in the early 1980s, and do not take into account possible great subduction earthquakes off the west coast of British Columbia, nor moderate- to large-magnitude intraplate earthquakes in mid-continent areas such as the western Lake Ontario region (see Chapter 29; modified from Basham, 1985).*

proximately 30 times the energy of a $M = 8$ event, and thus affects a much larger area. Should the entire Cascadia subduction zone release during a $M = 9$ event, all cities in western Oregon, western Washington, and southwestern British Columbia would likely be damaged. The intensity of shaking in many areas might be no greater than that of a nearby, large, crustal earthquake, but strong shaking would last much longer, probably 1-3 minutes as opposed to 30 seconds or less during a local quake. In addition, the seismic waves of a subduction earthquake would have relatively long periods, and might be more damaging to taller buildings than the high-frequency waves of local earthquakes.

In contrast, a $M = 8$ subduction earthquake would affect a smaller total area. The shaking might not last as long, but would still be severe along the coast adjacent to the rupture. Although a $M = 8$ earthquake centred on the southernmost part of the Cascadia subduction zone might have little effect on Vancouver, a succession of such earthquakes, rupturing different parts of the subduction zone over a short period of time (days to perhaps years), might produce as much damage to the entire coastal Pacific Northwest as a single $M = 9$ event. A particularly disturbing prospect is that a $M = 8$ earthquake would not fully release the accumulated strain along part of the plate boundary, and might lead to another earthquake in the same area a short time later, destroying already damaged and weakened buildings. The fear of a second or third great earthquake might impede reconstruction and the resumption of normal economic activity and life in cities affected by the first earthquake, leading to what Gary Carver has referred to as a "decade of terror" (Carver *et al.*, 1992).

Fortunately, subduction earthquakes are rare, occurring, on average, once every 500-700 years in this region. Intervals between successive earthquakes, however, have ranged from a few hundred years, or less, to more than 1000 years, making it difficult to estimate the time of the next event. In detail, recurrence intervals are poorly known, because the number and ages of the earthquakes are uncertain (Atwater *et al.*, 1995). These uncertainties result in a broad range of probabilities (from a few percent to perhaps 30%) that a great earthquake will occur somewhere along the Cascadia subduction zone in the next 50 years.

There are, on average, about 50 $M = 6-7$ earthquakes for each $M = 8-9$ subduction event; consequently, it is likely that the next earthquake to damage Vancouver will be one centred in the North America or Juan de Fuca plate, rather than at the plate boundary. In view of the fact that the $M = 6.7$ Los Angeles earthquake in 1994 caused more than US$15 billion damage, we would be well advised to be at least as concerned about $M = 6-7$ earthquakes as larger, much rarer, plate-boundary events.

Preparing for Earthquakes
An accurate assessment of seismic hazards requires knowledge of the causes and characteristics of significant earthquakes, as well as the nature of the ground motion produced by these earthquakes. Geological studies, such as those mentioned in this paper, contribute to seismic hazard assessment by providing information on earthquake sources, magnitude, recurrence, and effects. This information is utilized in the periodic revision of seismic zoning maps for the National Building Code of Canada (NBCC; Fig. 11), which provides guidelines for structural design.

The current (1985) edition of the NBCC is based on knowledge that was available in the early 1980s. Since then, there have been significant advances in our understanding of earth-

quakes in southwestern British Columbia, perhaps the most important of which is the realization that great earthquakes occur along the Cascadia subduction zone. This improved understanding has, in turn, led to better estimates of potential seismic ground motions, which are required for the next generation of seismic zoning maps for the year-2000 NBCC. The new maps will include, for the first time, a deterministic treatment of Cascadia subduction zone earthquakes.

In the last decade, public awareness of the earthquake hazard at Vancouver has increased considerably, due largely to media coverage of recent geological and geophysical findings and destructive earthquakes in Mexico (1985), San Francisco (1989), and Los Angeles (1994). During the same period, emergency preparedness has improved, earthquake education has become more common in Vancouver's schools, and some critical facilities, including bridges, schools, and hospitals, have been seismically upgraded or replaced. For example, British Columbia Hydro and Power Authority has upgraded several of its hydroelectric dams in southern British Columbia, and the hospital in the downtown core of Vancouver has been replaced. At present, however, there is no co-ordinated, adequately funded plan to reduce structural damage and loss of life due to a strong earthquake in southwestern British Columbia. The devastation wrought by the 1995 Kobe earthquake, in a country widely acknowledged to have the highest level of earthquake preparedness in the world, provides a reminder of the need for such a plan.

ACKNOWLEDGEMENTS
Tonia Williams drafted the figures. Chris Spindler provided a plot of the earthquake epicentres shown in Figure 1. Reviews by Nick Eyles, Jim Hunter, A.A. Mohajer, and an unidentified referee improved the paper and are much appreciated.

REFERENCES

Adams, J., 1984, Active deformation of the Pacific Northwest continental margin: Tectonics, v. 3, p. 449-472.

Adams, J., 1990, Paleoseismicity of the Cascadia subduction zone: Evidence from turbidites off the Oregon-Washington margin: Tectonics, v. 9, p. 569-583.

Ando, M. and Balazs, E.I., 1979, Geodetic evidence for aseismic subduction of the Juan de Fuca plate: Journal of Geophysical Research, v. 84, p. 3023-3028.

Atwater, B.F., 1987, Evidence for great Holocene earthquakes along the outer coast of Washington state: Science, v. 236, p. 942-944.

Atwater, B.F., 1992, Geologic evidence for earthquakes during the past 2000 years along the Copalis River, southern coastal Washington: Journal of Geophysical Research, v. 97, p. 1901-1919.

Atwater, B.F. and Moore, A.L., 1992, A tsunami about 1000 years ago in Puget Sound, Washington: Science, v. 258, p. 1614-1617.

Atwater, B.F., Nelson, A.R., Clague, J.J., Carver, G.A., Yamaguchi, D.K., Bobrowsky, P.T., Bourgeois, J., Darienzo, M.E., Grant, W.C., Hemphill-Haley, E., Kelsey, H.M., Jacoby, G.C., Nishenko, S.P., Palmer, S.P., Peterson, C.D. and Reinhart, M.A., 1995, Summary of coastal geologic evidence for past great earthquakes at the Cascadia subduction zone: Earthquake Spectra, in press.

Atwater, B.F., Stuiver, M. and Yamaguchi, D.K., 1991, Radiocarbon test of earthquake magnitude at the Cascadia subduction zone: Nature, v. 353, p. 156-158.

Atwater, B.F. and Yamaguchi, D.K., 1991, Sudden, probably coseismic submergence of Holocene trees and grass in coastal Washington state: Geology, v. 16, p. 706-709.

Barr, S.M. and Chase, R.L., 1974, Geology of the northern end of Juan de Fuca Ridge and sea-floor spreading: Canadian Journal of Earth Sciences, v. 11, p. 1384-1406.

Basham, P.W., Weichert, D.H., Anglin, F.M. and Berry, M.J., 1985, New probabilistic strong ground motion maps of Canada: Seismological Society of America, Bulletin, v. 75, p. 563-595.

B.C. Hydro and Klohn Leonoff Limited, 1992, Liquefaction hazard assessment for the Lower Mainland region: British Columbia Hydro and Power Authority, Report H2474a.

Blackwell, D.D., Bowen, R.G., Hull, D.A., Riccio, J. and Steele, J.L., 1982, Heat flow, arc volcanism and subduction in northern Oregon: Journal of Geophysical Research, v. 87, p. 8735-8754.

Carver, G.A., Stuiver, M. and Atwater, B.F., 1992, Radiocarbon ages of earthquake-killed trees at Humboldt Bay, California [abstract]: EOS, v. 73, p. 398.

Clague, J.J. and Bobrowsky, P.T., 1994a, Evidence for a large earthquake and tsunami 100-400 years ago on western Vancouver Island, British Columbia: Quaternary Research, v. 41, p. 176-184.

Clague, J.J. and Bobrowsky, P.T., 1994b, Tsunami deposits beneath tidal marshes on Vancouver Island, British Columbia: Geological Society of America, Bulletin, v. 106, p. 1293-1303.

Clague, J.J., Luternauer, J.L., Pullan, S.E. and Hunter, J.A., 1991, Postglacial deltaic sediments, southern Fraser River delta, British Columbia: Canadian Journal of Earth Sciences, v. 28, p. 1386-1393.

Clague, J.J., Naesgaard, E. and Sy, A., 1992, Liquefaction features on the Fraser delta: Evidence for prehistoric earthquakes?: Canadian Journal of Earth Sciences, v. 29, p. 1734-1745.

Clarke, S.H., Jr. and Carver, G.A., 1992, Late Holocene tectonics and paleoseismicity, southern Cascadia subduction zone: Science, v. 255, p. 188-192.

Cockerham, R.S., 1984, Evidence for a 180-km-long subducted slab beneath northern California: Seismological Society of America, Bulletin, v. 74, p. 569-576.

Crosson, R.S., 1983, Review of seismicity in the Puget Sound region from 1970 through 1978: a brief summary, in Yount, J.C. and Crosson, R.C., eds., Earthquake Hazards of the Puget Sound Region, Washington State: United States Geological Survey, Open-file 83-19, p. 6-18.

Darienzo, M.E. and Peterson, C.D., 1990, Episodic tectonic subsidence of late Holocene salt marshes, northern Oregon, central Cascadia margin: Tectonics, v. 9, p. 1-22.

Darienzo, M.E., Peterson, C.D. and Clough, C., 1994, Stratigraphic evidence for great subduction-zone earthquakes at four estuaries in northern Oregon: Journal of Coastal Research, v. 10, p. 850-876.

Dragert, H., 1985, A summary of recent geodetic measurements of surface deformation on central Vancouver Island: Royal Society of New Zealand, Bulletin, v. 24, p. 29-37.

Dragert, H., Hyndman, R.D., Rogers, G.C. and Wang, K., 1994, Current deformation and the width of the seismogenic zone of the northern Cascadia subduction thrust: Journal of Geophysical Research, v. 99, p. 653-668.

Dragert, H. and Lisowski, M., 1990, Crustal deformation measurements on Vancouver Island, British Columbia: 1976 to 1988, in Vyskocil, D.R., Reigber, C. and Cross, P.A., eds., Proceedings of the 125th Anniversary Meeting of the International Association of Geodesy: Springer-Verlag, New York, p. 241-249.

Drew, J.J. and Clowes, R.M., 1990, A re-interpretation of the seismic structure across the active subduction zone of western Canada, in Green, A.G., ed., Studies of Laterally Heterogeneous Structures Using Seismic Refraction and Reflection Data: Geological Survey of Canada, Paper 89-13, p. 115-132.

Dunbar, D., LeBlond, P.H. and Murty, T.S., 1989, Maximum tsunami amplitudes and associated currents on the coast of British Columbia: Science of Tsunami Hazards, v. 7, p. 3-44.

Dunbar, D., LeBlond, P.H. and Murty, T.S., 1991, Evaluation of tsunami amplitudes for the Pacific coast of Canada: Progress in Oceanography, v. 26, p. 115-177.

Eisbacher, G.H. and Clague, J.J., 1981, Urban landslides in the vicinity of Vancouvver, British Columbia, with special reference to the December 1979 rainstorm: Canadian Geotechnical Journal, v. 18, p. 205-216.

Finn, W.D.L. and Nichols, A.M., 1988, Seismic response of long-period sites: Lessons from the September 19, 1985, Mexican earthquake: Canadian Geotechnical Journal, v. 25, p. 128-137

Finn, W.D.L., Woeller, D.J., Davies, M.P., Luternauer, J.L., Hunter, J.A. and Pullan, S.E., 1989, New approaches for assessing liquefaction potential of the Fraser River delta, British Columbia: Geological Survey of Canada, Paper 89-1E, p. 221-231.

Fumal, T.E. and Tinsley, J.C., 1985, Mapping shear-wave velocities of near-surface geologic materials, in Ziony, J.I., ed., Evaluating Earthquake Hazards in the Los Angeles Region — An Earth-Science Perspective: United States Geological Survey, Professional Paper 1360, p. 127-150.

Guilbault, J.-P., Clague, J.J. and Lapointe, M., 1995, Amount of subsidence during a late Holocene earthquake — evidence from fossil tidal marsh foraminifera at Vancouver Island, west coast of Canada: Palaeogeography, Palaeoclimatology, Palaeoecology, in press.

Hansen, W.R., Eckel, E.B., Schaem, W.E., Lyle, R.E., George, W. and Chance, G., 1966, The Alaska earthquake March 27, 1964: Field investigations and reconstruction effort: United States Geological Survey, Professional Paper 541, 111 p.

Harp, E.L., Wilson, R.C. and Wieczorek, G.F., 1981, Landslides from the February 4, 1976, Guatemala earthquake: United States Geological Survey, Professional Paper 1204-A, 35 p.

Heaton, T.H. and Hartzell, S.H., 1986, Source characteristics of hypothetical subduction earthquakes in the northwestern United States: Seismological Society of America Bulletin, v. 76, p. 675-703.

Heaton, T.H. and Kanamori, H., 1984, Seismic potential associated with subduction in the northwestern United States: Seismological Society of America, Bulletin, v. 74, p. 933-941.

Hebenstreit, G.T. and Murty, T.S., 1989, Tsunami amplitudes from local earthquakes in the Pacific Northwest region of North America; Part 1: The outer coast: Marine Geodesy, v. 13, p. 101-146.

Hodgson, E.A., 1946, British Columbia earthquake, June 23, 1946: Royal Astronomical Society of Canada, Journal, v. 40, p. 285-319.

Hunter, J.A., Luternauer, J.L., Neave, K.G., Pullan, S.E., Good, R.L., Burns, R.A. and Douma, M., 1992, Shallow shear-wave velocity-depth data in the Fraser River delta from surface refraction measurements, 1989, 1990, 1991: Geological Survey of Canada, Open File 2504, 271 p.

Hunter, J.A., Woeller, D.J., Addo, K.O., Luternauer, J.L. and Pullan, S.E., 1993, Application of shear wave seismic techniques to earthquake hazard mapping in the Fraser River delta, British Columbia [abstract]: Society of Exploration Geophysicists, 63rd Annual Meeting, Washington, DC.

Hunter, J.L., Woeller, D.J. and Luternauer, J.L., 1991, Comparison of surface, borehole, and seismic cone penetrometer methods of determining the shallow shear-wave velocity structure in the Fraser River delta, British Columbia: Geological Survey of Canada, Paper 91-1A, p. 23-26.

Hyndman, R.D., 1984, Juan de Fuca Plate map, JFP 10: Geothermal heat flux: Department of Energy, Mines and Resources, Pacific Geoscience Centre, Sidney, BC.

Hyndman, R.D. and Wang, K., 1993, Thermal constraints on the zone of major thrust earthquake failure: the Cascadia subduction zone: Journal of Geophysical Research, v. 98, p. 2039-2060.

Hyndman, R.D. and Weichert, D.H., 1983, Seismicity and rates of relative motion on the plate boundaries of western North America: Geophysical Journal of the Royal Astronomical Society, v. 58, p. 667-683.

Hyndman, R.D., Yorath, C.J., Clowes, R.M. and Davis, E.E., 1990, The northern Cascadia subduction zone at Vancouver Island: Seismic structure and tectonic history: Canadian Journal of Earth Sciences, v. 27, p. 313-329.

Joyner, W.B. and Fumal, T.E., 1985, Predictive mapping of earthquake ground motion, in Ziony, J.I., ed., Evaluating Earthquake Hazards in the Los Angeles Region — An Earth-Science Perspective: United States Geological Survey, Professional Paper 1360, p. 203-220 .

Keefer, D.K., 1984, Landslides caused by earthquakes: Geological Society of America Bulletin, v. 95, p. 406-421.

Luternauer, J.L. and 19 others, 1994, Fraser River delta: Geology, geohazards and human impact, *in* Monger, J.W.H., ed., Geology and Geological Hazards of the Vancouver Region, Southwestern British Columbia: Geological Survey of Canada, Bulletin 418, p. 197-220.

Mathewes, R.W. and Clague, J.J., 1994, Detection of large prehistoric earthquakes in the Pacific Northwest by microfossil analysis; Science, v. 264, p. 688-691.

Mathews, W.H., 1979, Landslides of central Vancouver Island and the 1946 earthquake: Seismological Society of America, Bulletin, v. 69, p. 445-450.

McBirney, A.R. and White, C.M., 1982, The Cascade Province, *in* Thorp, R.S., ed., Andesites: John Wiley and Sons, Chichester, MA, p. 115-135.

Milne, W.G., Rogers, G.C., Riddihough, R.P., McMechan, G.A. and Hyndman, R.D., 1978, Seismicity of western Canada: Canadian Journal of Earth Sciences, v. 15, p. 1170-1193.

Munich Reinsurance Company of Canada, 1992, Earthquake: Economic impact study: Toronto, ON, 99 p.

Murty, T.S., 1992, Tsunami threat to the British Columbia coast, *in* Geotechnique and Natural Hazards: BiTech Publishers, Vancouver, BC, p. 81-89.

Murty, T.S. and Boilard, L., 1970, The tsunami in Alberni Inlet caused by the Alaska earthquake of March 1964, *in* Mansfield, W.M., ed., Tsunamis in the Pacific Ocean: International Symposium on Tsunamis and Tsunami Research, Honolulu, 1969: East-West Center Press, Honolulu, HI, Proceedings, p. 165-187.

Murty, T.S. and Hebenstreit, G.T., 1989, Tsunami amplitudes from local earthquakes in the Pacific Northwest region of North America; Part 2: Strait of Georgia, Juan de Fuca Strait, and Puget Sound: Marine Geodesy, v. 13, p. 189-209.

Naesgaard, E., Sy, A. and Clague, J.J., 1992, Liquefaction sand dykes at Kwantlen College, Richmond, *in* Geotechnique and Natural Hazards: BiTech Publishers, Vancouver, BC, p. 159-166.

Nelson, A.R., 1992, Holocene tidal-marsh stratigraphy in south-central Oregon — Evidence for localized sudden submergence in the Cascadia subduction zone, *in* Fletcher, C.P. and Wehmiller, J.F., eds., Quaternary Coasts of the United States: SEPM (Society for Sedimentary Geology), Special Publication 48, p. 287-301.

Nelson, A.R. and 11 others, 1995, Radiocarbon evidence for extensive plate boundary rupture about 300 years ago at the Cascadia subduction zone: Nature, v. 378, p. 371-374.

Ng, M.K.-F., LeBlond, P.H. and Murty, T.S., 1990, Numerical simulation of tsunami amplitudes on the coast of British Columbia due to local earthquakes: Science of Tsunami Hazards, v. 8, p. 97-127.

Ng, M.K.-F., LeBlond, P.H. and Murty, T.S., 1991, Simulation of tsunamis from great earthquakes on the Cascadia subduction zone: Science, v. 250, p. 1248-1251.

Obermeier, S.F., Atwater, B.F., Benson, B.E., Peterson, C.D., Moses, L.J., Pringle, P.T. and Palmer, S.P., 1993, Liquefaction about 300 years ago along tidal reaches of the Columbia River, Oregon and Washington [abstract]: EOS, v. 74, p. 198-199.

Riddihough, R.P., 1977, A model for recent plate interactions off Canada's west coast: Canadian Journal of Earth Sciences, v. 14, p. 384-396.

Riddihough, R.P., 1979, Gravity and structure of an active margin — British Columbia and Washington: Canadian Journal of Earth Sciences, v. 16, p. 350-363.

Riddihough, R.P., 1982, Contemporary vertical movements and tectonics on Canada's west coast: a discussion: Tectonophysics, v. 86, p. 319-341.

Riddihough, R.P., Beck, M.E., Chase, R.L., Davis, E.E., Hyndman, R.D., Johnson, S.H. and Rogers, G.C., 1983, Geodynamics of the Juan de Fuca Plate, *in* Cabre, R., ed., Geodynamics of the Eastern Pacific Region, Caribbean and Scotia Arcs: American Geophysical Union, Geodynamics Series, v. 9, p. 5-21.

Riddihough, R.P., Currie, R.G. and Hyndman, R.D., 1980, The Dellwood Knolls and their role in triple junction tectonics off northern Vancouver Island: Canadian Journal of Earth Sciences, v. 17, p. 577-593.

Riddihough, R.P. and Hyndman, R.D., 1976, Canada's active western margin — the case for subduction: Geoscience Canada, v. 3, p. 269-278.

Robertson, P.K., Woeller, D.J. and Finn, W.D.L., 1992a, Seismic cone penetration test for evaluating liquefaction potential under cyclic loading: Canadian Geotechnical Journal, v. 29, p. 686-695.

Robertson, P.K., Woeller, D.J., Kokan, M., Hunter, J. and Luternauer, J.L., 1992b, Seismic techniques to evaluate liquefaction potential: 45th Canadian Geotechnical Conference, Toronto, ON, Proceedings, p. 5-1–5-9.

Rogers, G.C., 1980, A documentation of soil failure during the British Columbia earthquake of 23 June 1946: Canadian Geotechnical Journal, v. 17, p. 122-127.

Rogers, G.C., 1983, Seismotectonics of British Columbia: Ph.D. thesis, University of British Columbia, Vancouver, BC, 227 p.

Rogers, G.C., 1988, An assessment of the megathrust earthquake potential of the Cascadia subduction zone: Canadian Journal of Earth Sciences, v. 25, p. 844-852.

Rogers, G.C., 1994, Earthquakes in the Vancouver area, *in* Monger, J.W.H., ed., Geology and Geological Hazards of the Vancouver Region, Southwestern British Columbia; Geological Survey of Canada, Bulletin 418, p. 221-229.

Rogers, G.C. and Hasegawa, H.S., 1978, A second look at the British Columbia earthquake of 23 June 1946: Seismological Society of America, Bulletin, v. 68, p. 653-676.

Rogers, G.C., Spindler, C. and Hyndman, R.D., 1990, Seismicity along the Vancouver Island LITHOPROBE corridor, *in* Cook, F.A., ed., Southern Cordillera Transect Workshop II: University of British Columbia, Vancouver, BC, LITHOPROBE Secretariat, Report No. 15, p. 166-169.

Savage, J.C. and Lisowski, M., 1991, Strain measurements and the potential for a great subduction zone earthquake off the coast of Washington: Science, v. 252, p. 101-103.

Savage, J.C., Lisowski, M. and Prescott, W.H., 1981, Geodetic strain measurements in Washington: Journal of Geophysical Research, v. 86, p. 4929-4940.

Savage, J.C., Lisowski, M. and Prescott, W.H., 1991, Strain accumulation in western Washington: Journal of Geophysical Research, v. 96, p. 14,493-14,507.

Shedlock, K.M. and Weaver, C.S., 1991, Program for earthquake hazards assessment in the Pacific Northwest: United States Geological Survey, Circular 1067, 29 p.

Taber, J.J. and Smith, S.W., 1985, Seismicity and focal mechanisms associated with the subduction of the Juan de Fuca Plate beneath the Olympic Peninsula, Washington: Seismological Society of America, Bulletin, v. 75, p. 237-250.

Watts, B.D., Seyers, W.C. and Stewart, R.A., 1992, Liquefaction susceptibility of Greater Vancouver area soils, *in* Geotechnique and Natural Hazards: BiTech Publishers, Vancouver, BC, p. 145-157.

Weaver, C.S. and Baker, G.E., 1988, Geometry of the Juan de Fuca plate beneath Washington and northern Oregon from seismicity: Seismological Society of America, Bulletin, v. 78, p. 264-275.

Wigen, S.O. and White, W.R., 1964, Tsunami of March 27-29, 1964, west coast of Canada: Canada Department of Mines and Technical Surveys, Ottawa, ON, 6 p.

Yamaguchi, D.K., Woodhouse, C.A. and Reid, M.S., 1989, Tree-ring evidence for synchronous rapid submergence of the southwestern Washington coast about 300 years ago [abstract]: EOS, v. 70, p. 1332.

31. Geophysical Techniques in Urban Areas

John E. Scaife

multiVIEW Geoservices Inc., 1091 Brevik Place, Mississauga, Ontario L4W 3R7

SUMMARY

Just as a doctor uses an X-ray as a non-intrusive diagnostic tool to identify internal problems within the human body, site investigations in urban environments can utilize a variety of geophysical tools to review subsurface site conditions without disturbing the ground. The features which can be delineated depend on the site conditions, the nature (composition) of the objects and the size of objects which must be found. A series of case histories are included which describe the application of three different geophysical technologies in urban environments.

INTRODUCTION

Environmental site investigations in urban areas commonly involve the exploration of former industrial sites to determine the extent of contamination potential environmental damage. Investigators must effectively characterize the subsurface conditions to evaluate all potential sources of environmental hazards and the spatial extent of any subsurface contamination, and to determine whether this contamination is migrating, and, if so, where.

A standard approach for subsurface investigations includes an information search to reveal past site history, followed by an intrusive investigation, which typically consists of boreholes for soil and ground-water samples and test pits for confirming the suspected location of remnant underground features such as tanks, drums, pipes and building foundations.

The use of surficial geophysical techniques is gaining acceptance as a tool to assist in the investigation and characterization of subsurface site conditions (Scaife, 1990, 1993). The various geophysical methods can define the location of buried metallic objects (*i.e.*, underground storage tanks, drums, pipelines, *etc.*), areas of unusual soil conditions (*i.e.*, natural soils *versus* fill materials *versus* contaminated soils), and potential pathways for the migration of contaminants (*i.e.*, linear occurrences of coarse-grained soils and/or bedrock channels). The application of these techniques can permit intelligent planning of an effective borehole and/or excavation program to ensure that all features of interest are adequately investigated.

CASE HISTORIES

Three case histories are presented to illustrate the successful application of various geophysical techniques for environmental site investigations in urban environments. These are drawn from actual contracted work.

Delineating a Former Petroleum Service Station

An environmental site investigation must define all potential sources of subsurface contamination. One common source of soil contamination is corroded or leaking underground storage tanks (Chapters 4, 7, 34, 35). The position of underground storage tanks at a service station is normally recorded on the fire insurance plans or the as-built drawings. However, when urban commercial properties change hands, upgrading and modification of facilities, and land-use changes are often not well recorded.

This first case history succinctly illustrates the lack of continuity in documentation. A property owner, who was considering selling his small commercial plot, was requested to complete an environmental audit of the site by a prospective buyer. The audit indicated that the property had been previously used as a service station, but all buildings were demolished and the entire site had been covered with asphalt prior to purchase by the existing owner. Plans revealed three underground storage tanks used by the service station, but no record existed as to whether they were ever removed. The existence of underground storage tanks and potential for hydrocarbon contamination was of immediate concern to both the current owner and the prospective buyer. As a result, a geophysical site investigation was initiated.

Since the objective of the geophysical survey was to delineate any underground storage tanks made of steel, a magnetic survey was selected as the technique of choice, since this technology will map ferrous objects (*i.e.*, iron or steel). The site was bounded by fences, a major thoroughfare and a building, all of which are sources of interference for magnetic investigations. Since the dimensions of the site were only 20 m by 20 m, these urban constrictions dictated the use of a highly sensitive magnetic instrument with a relatively small instrument footprint, or lateral range of exploration.

This investigation was completed with an FM-18 vertical gradient fluxgate magnetometer, manufactured by Geoscan Research Limited, which was initially designed for archeological applications. By traversing with this instrument, archeologists can locate items such as clay-fired kilns and footpaths, based on the localized re-orientation of ferromagnetic minerals. These minerals, if freed from their original orientation through melting or physical pounding, will re-orient themselves in the direction of the Earth's magnetic field at the time of their re-solidification or re-deposition. The use of magnetic measurements for a variety of subsurface explorations is well documented (Breiner, 1973; see also Chapters 2, 16, 29).

The vertical-gradient fluxgate magnetometer combines the benefits of vertical-gradient methodology with a narrower frequency range of measurement. This permits the acquisition of high-quality magnetic data in close proximity to above-ground cultural features. Simply put, this instrument has a very high sensitivity and yet has a very small instrument footprint. For this investigation, the magnetic data were acquired at a station interval of 0.5 m, along survey lines spaced 1 m apart. The entire site was surveyed within four hours and resulted in the acquisition of 980 magnetic readings.

Figure 1 shows the magnetic data presented in colour contour map format. The background values are represented by the yellow-orange and green colours, ranging from ±1000 nanoTesla per metre ($nT \cdot m^{-1}$). The position of large ferrous objects is identified by purple areas having values in excess of 4000 $nT \cdot m^{-1}$.

The geophysical interpretation of these data is also presented on Figure 1, with the three large purple areas near the centre of the site inferred to represent the location of underground storage tanks. The position of the demolished service station building is inferred from the group of isolated ferrous targets which likely correspond to the distribution of rubble from the former building.

The subsurface information produced by this geophysical investigation is quite clear, even though the site was covered in asphalt. This geophysical investigation not only confirmed the existence of underground storage tanks at this site, but defined their position and orientation, and was completed without disturbing the normal course of business and without exposing the owners, occupants or passers-by to any environmental risk.

Delineating Abandoned Municipal Landfills

Prior to the current environmental regulations regarding the engineering, installation and monitoring of landfills, garbage was a convenient way to fill quarries and borrow pits. Many are now being used as public parks and fields for local public schools (Chapter 21). Others underlie residential areas.

Since permitting requirements for quarries rarely required an accurate survey of the extent of the quarry upon their closure, and the convenience of filling pits obviated the need for recording its position, the precise position and extent of the municipal garbage is not known.

A city required an environmental site investigation of abandoned landfills to determine their effect on the local environment, including the health and safety of school-children using the school yards and residents. Information was required for mapping any leachate plumes, thickness of the clay cap and other engineering considerations in light of modern-day standards. Before these data could be gathered, the city had to first determine the actual spatial limits of the landfills. A geophysical investigation of several properties was commissioned.

A technique was required which would distinguish between landfill refuse and natural soil conditions and yet permit broad areal coverage. The landfills could likely be distinguished, based on fluctuating soil conductivity values, due to the heterogeneous nature of the municipal landfill materials in contrast to the relatively uniform soil conductivity values from the surrounding natural soils. The sites were all school yards or parks, such that interference from cultural features would likely be limited to a few fences, underground utilities and playground facilities.

This investigation was completed using an EM31 terrain conductivity meter manufactured by Geonics Limited. This instrument measures the bulk conductivity of the soil within an exploration footprint of approximately 5 m, and its primary measurement is the apparent conductivity of the surrounding environment. The magnitude of the signal detected by the receiver is approximately proportional to the ground conductivity when the unit is over uniform ground, however, the term apparent is applied to this measurement because the environment where the measurements are made is seldom uniform. The apparent conductivity measurement is given in units of milliSiemen per metre ($mS \cdot m^{-1}$). By traversing with this instrument, and recording soil conductivity values at regular intervals, a distribution map of soil apparent conductivity can be created (McNeill, 1980). For this investigation, electromagnetic data were acquired, at a station interval of 2 m, along survey lines spaced 5 m apart. One large site was surveyed within 6 hours, and resulted in the acquisition of 2226 electromagnetic readings.

Figure 2 shows the apparent conductivity data presented in colour contour map format acquired from one sites. The background values are shown in blue and green, ranging from 10 to 20 $mS \cdot m^{-1}$. The position of the conductive landfill is easily recognizable in the mass of fluctuating, generally elevated, conductivity values. Localized, highly variable measurements acquired in the vicinity of the playground apparatus due to their metallic construction, can be seen along the east edge of the site. Gaps, or blank areas, in the data are due to other school yard apparatus and baseball backstops. This subsurface information was derived in one day, under a grass-covered school yard without disturbing the soils or the school activities and with no environmental risk to school-children or passers-by.

Mapping Subsurface Voids under a Warehouse Floor

The environmental inventory of existing industrial sites commonly includes documentation of installed subsurface features such as underground storage tanks, utilities, steam lines, product lines and other conduits. These features are ideally installed at their location as defined on the as-built construction drawings, as are other engineered items such as type and compaction requirements for backfilled soils. Unfortunately, the position of these features and adherence to specified engineering constraints are not always as requested.

Such is the case of a warehouse which was constructed as a 20-cm reinforced concrete slab on grade. The slab was poured onto a crushed gravel base overlying tamped backfill. This backfill was required to have all organic materials removed and to be installed in specified lift thicknesses, which were to be tamped to compaction criteria as determined from laboratory conditions.

The warehouse was used to stockpile finished products stored in rows of metal racks standing on the concrete floor, with the racks mounted roughly 8 m high. The horizontal spacing between the racks was roughly 4 m, enough to permit pallet trucks and plant personnel relatively easy access. In a short period of time, the concrete floor, reinforced with steel rebar, started to show stress cracks radiating from the footings of the racking system. The industrial facility commissioned a very limited borehole program, in the vicinity of the cracking, to evaluate whether the concrete was poorly cured or whether the cracks were due to other circumstances. The borehole program indicated that there were no problems with the quality of the concrete, but revealed subsurface voids under the concrete floor in the vicinity of the stress cracks.

Since the network of cracks covered a large area of the

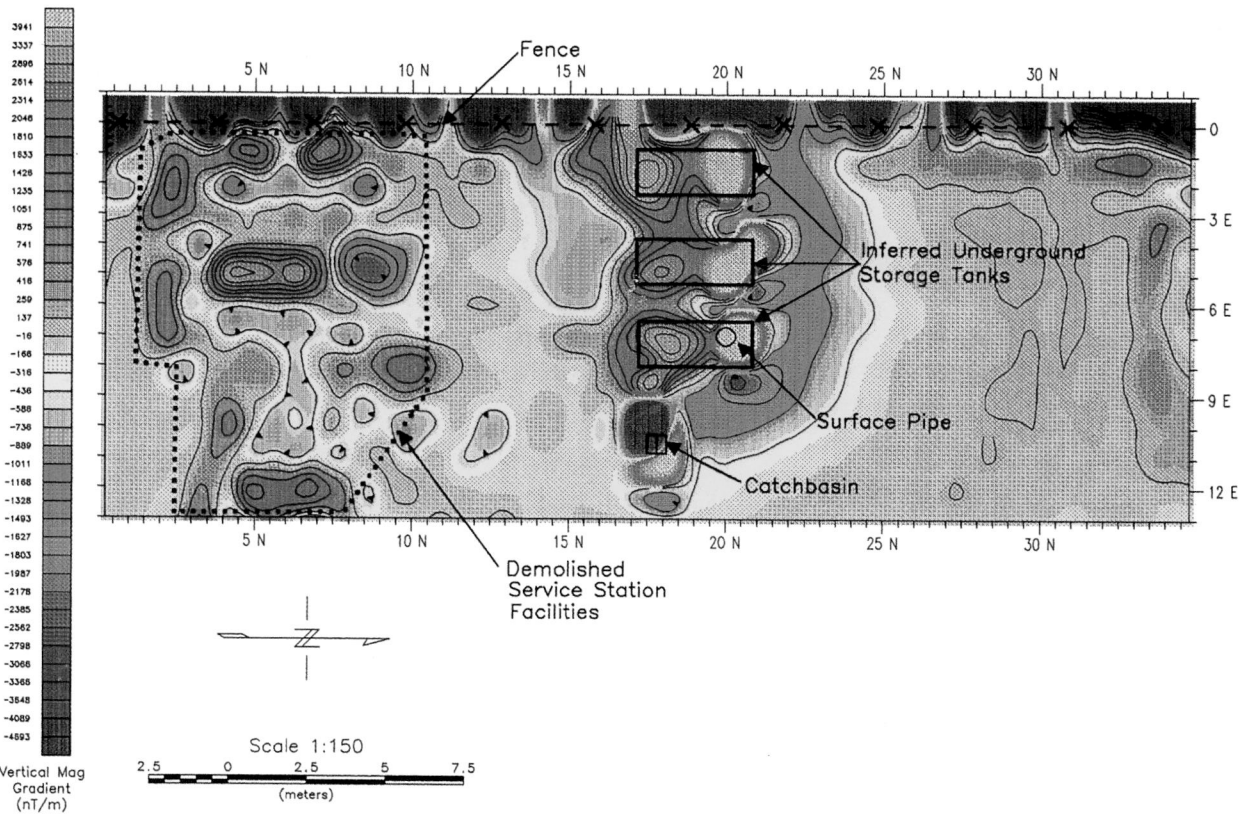

Figure 1   *Colour contour map of vertical-gradient fluxgate magnetometer data showing buried storage tanks at a former service station.*

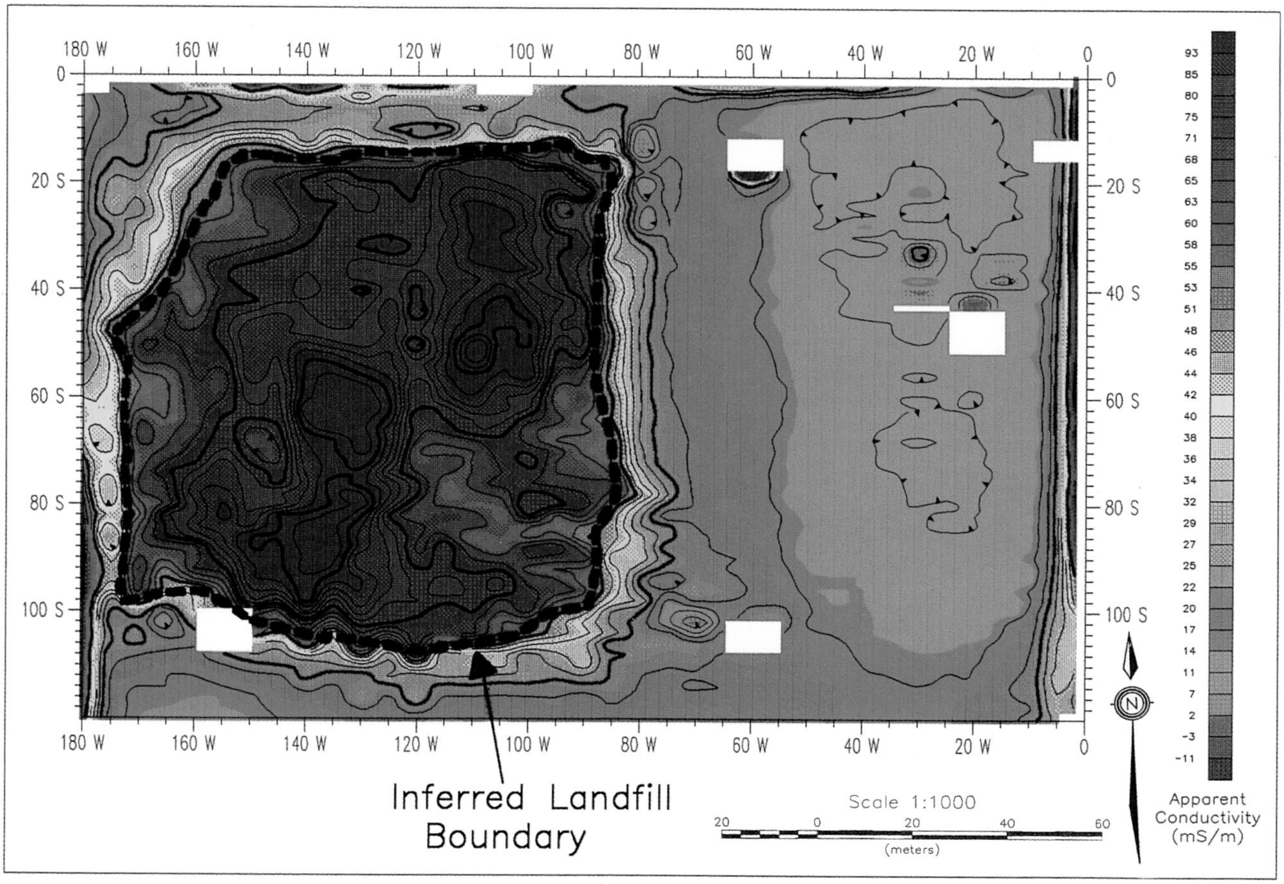

Figure 2   *Colour contour map of electromagnetic apparent conductivity data showing inferred extent of waste within former landfill site, now used as a school yard. Blank areas are due to school yard apparatus.*

warehouse facility, an investigative borehole program would not only be prohibitively expensive, but would seriously disrupt on-going business activities of the warehouse. The industrial facility commissioned a geophysical investigation to determine the extent of subsurface voids under the concrete floor.

The objective of this investigation was to map air-filled cavities, or areas of poor-compaction conditions, requiring a geophysical technique which could delineate these contrasts and yet function in the culturally cluttered environment of the warehouse. The cultural conditions of the warehouse environment included numerous overhead power lines, the metallic racking system and finished products. No conduits were suspected either within or below the concrete floor. It was anticipated that the voids could be mapped based on their localized change in electrical properties with respect to the well-compacted soil areas.

This investigation was completed using a pulseEKKO 1000 high-frequency ground-penetrating radar system manufactured by Sensors and Software Inc. Radar is similar in principle to seismic and sonar techniques (see Pilon, 1992). The radar produces a short duration pulse of high-frequency (10 MHz to 1000 MHz) electromagnetic energy which is transmitted into the ground. The reflected signals are detected and amplified at the receiver. The received signals are digitized and stored on disk on a field computer for post-survey processing.

Radar penetration into geological material is controlled by the electrical conductivity of the material. As the electrical conductivity increases, the radar (pulsed electromagnetic) signal is dissipated as heat. Geological horizons that exhibit different conductivities from those above can be delineated by the radar due to their contrast in electrical properties. Similarly, man-made objects and subsurface voids or cavities can be detected, since these objects will have markedly different electrical properties than the host material.

For this investigation, radar antennas operating at central frequencies of 450 MHz and 900 MHz were mobilized to the site. A series of test data was acquired in the vicinity of the known voids to evaluate the effectiveness of the radar system in mapping these targets. A series of initial tests is commonly completed prior to starting a production survey to ensure the

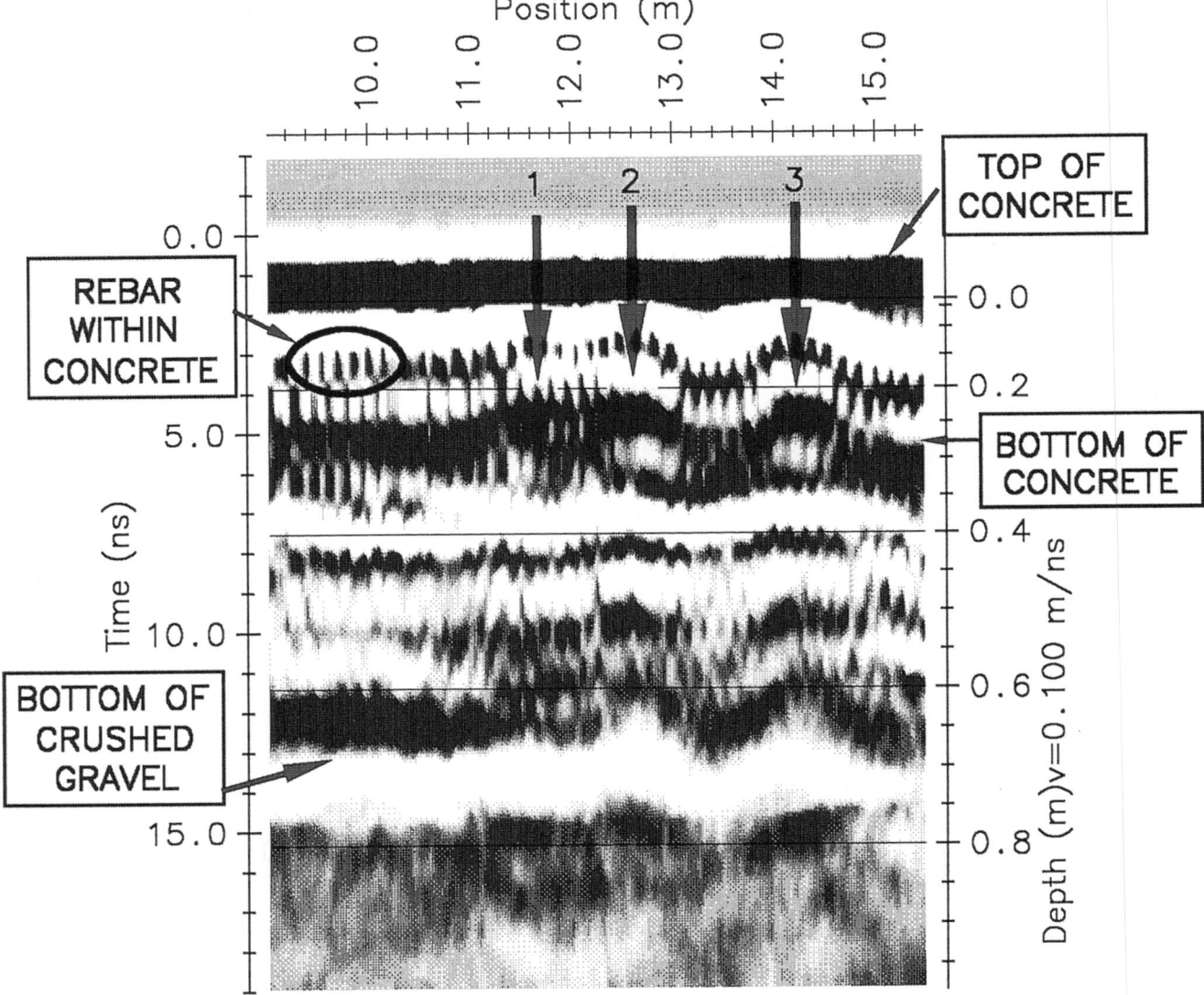

Figure 3 *Processed ground penetrating radar data showing location of sub-concrete voids numbered 1, 2 and 3. The upward bulging of the reflectors at the sites (pull-ups) indicates air-filled cavities and faster radar propagation velocities.*

radar data are recorded with the optimum acquisition parameters (Annan and Cosway, 1992). The on-site interpretation of these preliminary data provided satisfactory results, and an investigation program was outlined by an industrial facility representative. This program consisted of four survey lines, roughly 45 m in length, with the radar data collected at approximately 5 cm intervals along each survey line. This investigation was completed in roughly six hours and resulted in the acquisition of 3956 data points.

Figure 3 shows an example of the processed 450 MHz radar data from one of the survey lines. These data show the top of the concrete, the rebar within the concrete, the bottom of the concrete and the bottom of the crushed gravel or top of the compacted backfill. The elevation of the rebar can be seen to vary within the concrete, which indicates either a variation in the local composition of the concrete (*i.e.*, change in aggregate mixture or different curing processes) or actual deviations in the layout of the rebar installed by the contractor.

Of greater interest, the apparent depth to the bottom of crushed gravel/top of backfill seems to vary significantly along this short section of line. This indicates that either the crushed gravel layer locally varies in thickness by a factor of roughly 20%, or pockets of less compact material are located between the bottom of the concrete and the top of the fill in these areas. Radar waves propagate at the speed of light in air and at velocities roughly one-third of that in soils and rocks. If the radar waves experience an increase in velocity due to a change in the media through which they are passing, it will take less time for the energy to propagate down to the geological horizons and return to the surface. The apparent pull-up of the top of backfill was interpreted not to be a result of thinner gravels, but indicative of air-filled cavities directly below the concrete, resulting in localized faster propagation velocities. This localized velocity increase can be identified by apparent shallower stratigraphy, hence, the position of a stratigraphic pull-up may infer the position of a sub-concrete void.

This initial interpretation was presented to the owners of the industrial facility who commissioned a limited borehole program to verify and calibrate the results of the radar survey. This borehole program identified open cavities of 1.25, 2.5 and 1.25 cm below the concrete at the locations marked as 1, 2 and 3 on the radar profile (Fig. 3). These findings confirmed the interpretation, and the remainder of the radar data were reviewed using this scenario as a model.

A total of 103 inferred voids were identified along the 145 line-metres of radar data acquired. The subsurface voids caused by differential settlement of the backfill were obviously a significant problem at this facility. This geophysical investigation was completed within one day, revealing information under a concrete floor without disturbing the floor and having minimal impact on the day-to-day business activities of the warehouse facility.

DISCUSSION

Geophysical technology can assist in revealing subsurface objects, structures and geological conditions without disturbing the ground surface. Another very commonly employed technique, that of seismic reflection surveying, is described in Chapter 32. Geophysical measurements can be of tremendous benefit to environmental geologists planning an intrusive investigation program, since their program can now be designed in a more cost-effective manner through the use of real data describing the actual subsurface conditions at any given site.

Each of the described case histories illustrates the application of only one geophysical technology to assist with outlining the problem. Frequently, the simultaneous application of several different technologies can reveal additional features, or assist in confirming objects; *e.g.*, a magnetometer and electromagnetic conductivity survey would both delineate the position of buried steel underground storage tanks.

There are other cases where the geophysical technologies discussed in this paper will not be applicable. For example, detection of small-diameter plastic pipes at depths of greater than 3 m cannot be achieved realistically under most circumstances. Additionally, simple delineation of electrically resistive materials, such as pockets of hydrocarbons at a former industrial site where the local soils and fill materials have been extensively reworked, is virtually impossible given the present-day technological limitations. In this case, drilling of closely spaced boreholes is the only alternative (Chapter 7), combined with soil gas surveys (Chapter 34).

Application of geophysical techniques in urban environments requires additional care, since urbanized areas are congested with cultural noise sources which tend to mask the contrast between properties and, hence, reduce the odds of detection. Urban industrialized sites commonly contain a large number of existing ferrous metal objects (*i.e.*, fences, buildings, utilities, parked vehicles, *etc.*) which would influence magnetic and electromagnetic readings. The investigative survey must be designed with these constraints in mind to assure the effective application of the technology.

Successful implementation of geophysical techniques, including survey planning and interpretation, requires an experienced practitioner who is skilled in the use of these exploration technologies in an urban environment. As additional experience is gained through the emerging use of these geophysical tools, a greater understanding will result in more widespread usage, which will ultimately improve the instrumentation, utilization and acceptance of these non-intrusive technologies for application in the urban environment.

REFERENCES

Annan, A.P. and Cosway, S.W., 1992, Ground penetrating radar survey design: Society of Engineering, Mining and Environmental Geophysicists, Golden, CO, Symposium of Applied Geophysics to Engineering and Environmental Problems, April 26-29, 1992, Oakbrook, IL, Proceedings, p. 329-351.

Breiner, S., 1973, Applications manual for portable magnetometers: GeoMetrics, Sunnyvale, CA, 58 p.

McNeill, J.D., 1980, Electromagnetic terrain conductivity measurement at low induction numbers: Geonics Limited, Technical note TN-6, 15 p.

Pilon, J.A., 1992, Ground Penetrating Radar: Geological Survey of Canada, Paper 90-4, 237 p.

Scaife, J.E., 1993, A practical introduction to surficial geophysical techniques for environmental site investigations, *in* Remediation of Contaminated Sites, Regulation and State-of-the-Art Technologies, May 27-28, 1993, Vancouver, BC, Proceedings: In-Sight Press, Toronto, ON, 23 p.

Scaife, J.E., 1990, Using geophysical techniques in environmental site assessments: Municipal & Industrial Water & Pollution Control, CXXVIII, 4 (August, 1990). p. 4-5.

32. Shallow Seismic Reflection Profiling of Waste Disposal Sites

Joseph I. Boyce

Earth and Environmental Sciences, University of Toronto, Scarbororough Campus, 1265 Military Trail, Scarborough, Ontario M1C 1A4

Berkant B. Koseoglu

Geophysics Division, Department of Physics, University of Toronto, Toronto, Ontario M5S 1A7

SUMMARY

Conventional approaches to environmental site characterization usually involve drilling of relatively dense networks of continuously sampled and geophysically logged boreholes. In glaciated terrains, problems are often encountered in determining the lateral continuity and three-dimensional subsurface geometry of sediments between borehole locations. An improved and potentially more cost-effective approach combines drilling and borehole sampling with high-resolution shallow seismic reflection profiling. This approach is illustrated with case studies from candidate landfill sites. Seismic data provide critical information regarding the lateral continuity and geometry of sediments and pathways for ground water and contaminants.

INTRODUCTION

Detailed subsurface geological information is an important requirement for hydrogeological evaluation of waste disposal sites and other environmental geological investigations carried out in urbanized areas. In the Toronto urban area of southern Ontario (Fig. 1), applied investigations are dependent on a detailed understanding of the subsurface stratigraphy and geometry of thick Pleistocene sediments (up to 200 m) which overlie bedrock across the region. Conventional site characterization typically involves intensive field programmes of drilling and core sampling, downhole geophysical logging, well monitoring, and hydrochemical analysis. However, even with closely spaced networks of continuously sampled boreholes, significant uncertainties can arise in determining the lateral continuity and true subsurface geometry of glacial sediments between drilling locations.

Stratigraphic resolution can be significantly improved in applied investigations by integrating conventional borehole information with continuous subsurface-imaging techniques such as shallow seismic reflection profiling and ground-penetrating radar (GPR; Chapter 31). Shallow seismic reflection methods are now employed routinely in Pleistocene geological studies (Pugin and Rossetti, 1992; Genau *et al.*, 1994;

Koseoglu, 1995; Boyce *et al.*, 1995) and in regional ground-water investigations (Hunter *et al.*, 1987; Pullan *et al.*, 1994). With the advent over the last decade of less expensive, more powerful seismographs and PC-based processing software (*e.g.*, Somanas *et al.*, 1987), shallow seismic reflection surveys have become efficient and cost-effective for private sector use.

This paper presents an overview of shallow seismic reflection methods and illustrates their application in case studies from three candidate landfill sites and a low-level radioactive soil remediation site in the Greater Toronto Area (GTA; Fig. 1). Searches for new municipal and hazardous waste sites are being carried out in many other parts of Canada and the northern United States where thick glacial sediments are widespread (*e.g.*, Curry *et al.*, 1994). The seismic reflection methods and results reported here, therefore, have a wider relevance for landfilling and applied ground-water investigations in other urbanized areas.

SEISMIC REFLECTION METHODS

The seismic reflection method has been used in the petroleum industry for more than 60 years for evaluating the deep subsurface structure and oil and gas potential of sedimentary basins (Telford *et al.*, 1990). Artificial seismic sources (most commonly explosives or mobile vibrators) are employed to introduce elastic waves (principally compressional or p-waves) into the subsurface (Fig. 2). In a manner similar to sound echoing in air, reflections are produced in the subsurface, where p-waves impinge upon boundaries between geological layers with contrasting densities and/or seismic velocities. The reflectivity of a formation boundary is a function of the change in acoustic impedance which is given by the product of the bulk density and velocity of a formation. Reflected waves arriving at the surface are detected with a linear array of geophones (Fig. 2) or other seismometers (*e.g.*, accelerometers) which output a voltage signal proportional to the vertical velocity of the ground motion. The signals from the receivers are amplified, converted to digital words, and recorded on a multi-channel seismograph. The ground motion recorded by each receiver is displayed as an individual trace on a seismic field record which is a plot of time *versus* signal amplitude. The travel time-amplitude data are then manipulated in a number of processing stages which culminate with the production of a seismic reflection section. The latter outwardly resembles a geological cross-section (with the time axis corresponding qualitatively with depth). The important difference is that the changes in the vertical position of reflecting horizons may be due to changes in either velocity, reflector depth, or both. In order to generate a true geological section, showing the location and structural attitude of reflectors, time-domain seismic sections need to be converted to depth using the average seismic velocities and travel times of reflections.

Shallow Seismic Reflection Methods

During the past two decades, seismic reflection methods have been adapted for exploration of the shallow subsurface (≈5 m to 200 m depth) in engineering and ground-water

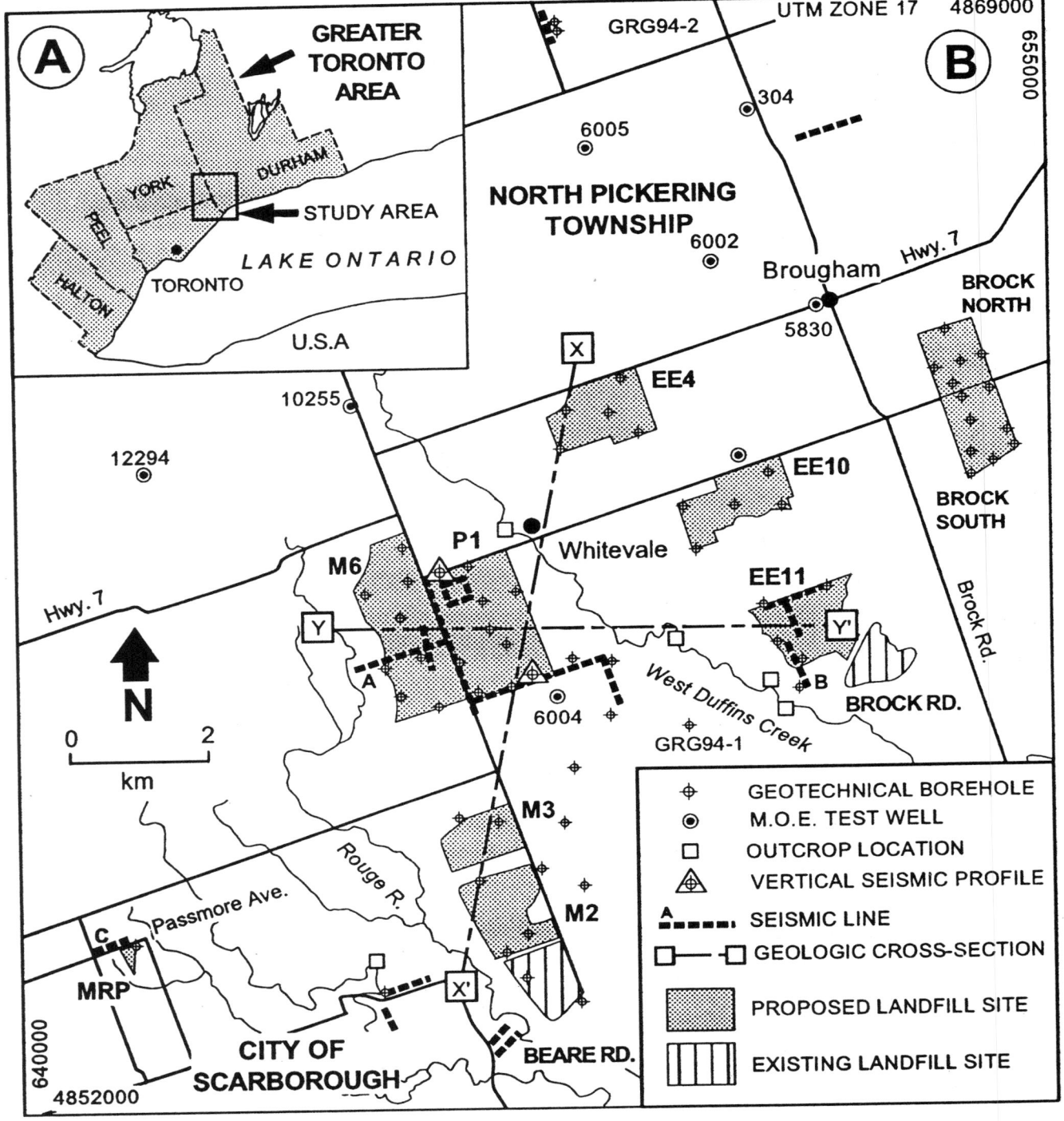

Figure 1   (A) *Location of study area.* (B) *Location of proposed and existing landfill sites, geologic cross-sections and seismic reflection profiles.*

investigations (Schepers, 1975; Hunter *et al.*, 1984, 1987; Miller and Steeples, 1994). Resolution of layering down to about 1 m in thickness is now routinely achieved in shallow seismic work although, under near-ideal substrate conditions, beds as thin as 0.5 m have been imaged (*e.g.*, Jongerius and Helbig, 1988). Shallow seismic methods have been used to locate faults, sinkholes and void spaces (Steeples *et al.*, 1986; Branham and Steeples, 1988; Treadway *et al.*, 1988).

Among the most important requirements for shallow reflection work are the use of high-frequency energy sources and low-cut filtering of data during field acquisition (Knapp and Steeples, 1986a, b; Hunter *et al.*, 1987). In contrast to petroleum-scale surveys which employ source frequencies in the 10-90 Hz range, frequencies of several hundred Hz are required to obtain high resolution at shallow depths of investigation. This stems from the fact that the vertical resolution, or thinnest bed that can be successfully imaged, is dependent upon the wavelength of the seismic impulse (wavelength = velocity/frequency). In theoretical terms, reflecting boundaries must be a minimum thickness of a quarter wavelength apart in order to be resolved; below this limit, destructive interference between reflections from the top and bottom of a layer results in greatly reduced reflection amplitude (Sheriff, 1985). Field filtering involves the use of pre-emphasis low-cut filters on the recording instrument to counter the selective attenuation of high-frequency signals which occurs in the subsurface (Knapp and Steeples, 1986b). In general, fine-grained, water-saturated sediments are more conducive to transmission of high frequencies and collection of high-resolution seismic data than dry, coarse-grained overburden materials (Hunter *et al.*, 1987). Besides low-cut filtering, high-frequency geophones with high natural resonant frequencies (between 50 Hz and 100 Hz) can be used to reduce some of the unwanted low-frequency components. The instrumentation and optimal field parameters for high-resolution seismic work are discussed by Knapp and Steeples (1986a, b) and Hunter *et al.* (1987). Generally, steps must be taken to prevent low-frequency signal components from dominating the recording.

The necessity for preserving high-frequency components of the seismic record has led to much development and testing of source types, in particular, downhole rifles and shot-

guns which are now routinely employed in shallow reflection work (*e.g.*, Miller *et al.*, 1986; Seeber and Steeples, 1986; Pullan and MacAulay, 1987; Miller *et al.*, 1992). Two acquisition strategies are employed routinely in shallow seismic reflection work, the common-depth point method and the optimum-offset method, both of which are reviewed here.

Optimum-Offset Method

The optimum-offset method was developed by the Geological Survey of Canada as a method that could be implemented with a minimum of equipment and computing facilities (Hunter *et al.*, 1984). This method involves collection of end-to-end single-fold spreads, using an optimized source-receiver offset that allows reflections to be recorded with minimum interference from groundroll and shallow refraction events. The optimum-offset range is determined by performing a series of walk-away noise tests with increasing source-receiver separations. A single offset is then selected within this window, and the survey proceeds by shooting one trace at a time, using the same offset for all traces. Alternately, multi-channel records are collected and the traces corresponding to the selected offset are assembled later during processing. In the optimum-offset method, each subsurface reflection point is sampled once, providing single-fold, or 100%, coverage. Case studies demonstrating the successful application and some potential pitfalls of the optimum-offset method are discussed by Hunter *et al.* (1987) and Slaine *et al.* (1990).

Common-Depth Point Method

The common-depth point method (CDP) is a continuous-profiling technique in which subsurface reflection points are sampled repeatedly over a range of source-receiver separations (Mayne, 1962; Knapp and Steeples, 1986a, b; Steeples and Miller, 1990). The strategy employs roll-along shooting in which the shot point and geophones are advanced along the survey line at increments equal to the geophone spacing or some multiple of this distance (Fig. 3). The resulting overlap between spreads leads to a redundancy of data collected for each reflection mid-point in the subsurface (Fig. 3). These mid-points are variously referred to as common-depth points (CDP) or common mid-points (CMP) and have a spacing which is one-half the surface shot station separation. Figure 4 illustrates the CDP concept for a case where a reflection point is sampled from six different source-receiver separations.

Continuous profiling in CDP surveys is usually facilitated with a roll-along switch which is used to increment a number of live recording channels through successive spreads. For example, during roll-along shooting with a spread of 24 geophones and a 12-channel recording instrument, geophones 1-12 would record shot 1, geophones 2-13 shot 2, geophones 3-14 shot 3, and so on. The fold, or redundancy, of subsurface coverage is determined by the number of active channels and the shot and receiver intervals; for the 12-channel setup described, six-fold or 600% coverage (*e.g.*, Fig. 4) would be obtained where shots are performed at every geophone location, and three-fold data where the shot spacing was set at twice the geophone interval. Exploration seismographs available at the present are commonly equipped with 24, 48 or 96 channels and 12-, 24- or 48-fold coverages are now employed routinely in shallow seismic reflection work.

CDP surveys are usually acquired using either end-on or a split-spread field layout (Fig. 3). For a fixed number of channels, the end-on layout has the advantage of providing a wider range of source-receiver offsets for a given geophone spacing when compared to the split-spread design. This additional

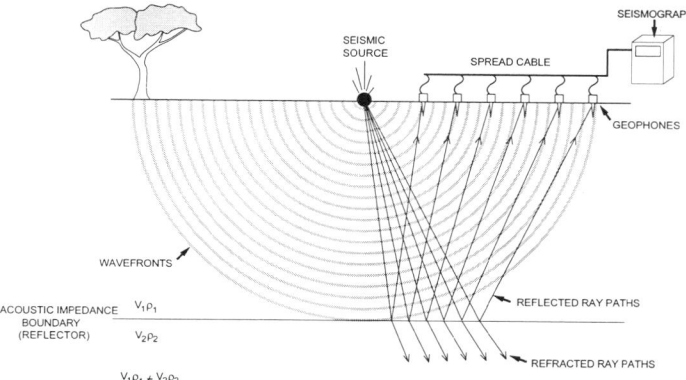

Figure 2   *Basic principle of seismic reflection method. Elastic waves (principally p-waves) are propagated into the subsurface using an artificial seismic energy source (e.g., explosives) and are reflected and refracted at acoustical discontinuities between layers with contrasting mass density and/or seismic velocity (acoustic impedance = V_p). The arrival times and amplitudes of reflected waves are detected at the surface by an array of receivers and recorded by a seismograph. The depth and seismic velocity above the reflecting interfaces are obtained from analysis of the recorded travel time-amplitude series.*

offset information is often critical for accurate determination of moveout velocities of reflectors (see below) and, in most instances, permits a larger number of traces to be recorded without interference from groundroll energy (*e.g.*, Fig. 5A).

Figure 5A shows two raw field records, or shot gathers, collected with shotgun and hammer sources. A number of events can be identified on the field records including reflections, shallow refractions, surface waves and a ground-coupled airwave (air blast from source). Reflection events are identified by their hyperbolic curvature, which is a result of the increasing ray path length as the source-receiver separation is increased (*e.g.*, Figs. 3, 4). The degree of curvature of the hyperbola is also determined by average seismic velocity above a reflector as well as its depth and dip angle. The time difference between a reflection arrival time at a geophone remote from the source compared to a detector at the source is referred to as the normal moveout (NMO). The normal moveout of a reflector is used to estimate the average seismic velocity above the reflecting horizon, and must be removed prior to stacking of CDP data. Refractions, in contrast, are recognized as linear arrival events which are characteristically lower frequency than reflections and show a number of cycles (Fig. 5A). The first arrivals on the records in

Figure 5 represent shallow refractions from the water table and base of the surface low-velocity layer (weathering layer).

The chief advantage of the CDP method over single-fold techniques (*e.g.*, optimum-offset method) is the significant increase in signal-to-noise ratio that potentially can be achieved through the stacking (summation) of redundantly sampled reflection points. Before data can be stacked, a number of processing operations must be performed on CDP data which are summarized briefly here. More in-depth discussions of processing techniques are given by Yilmaz (1987) and others (*e.g.*, Robinson and Treitel, 1980; Sheriff and Geldart, 1982). The main objective of CDP processing is to enhance reflection events at the cost of all other signals and to produce a synthetic zero-offset seismic section in which each trace simulates a vertical-incident ray path between the source and the reflector (Fig. 4). The initial stages of processing carried out on raw shot records involve manual editing of traces, application of static corrections, and muting of refractions and noise events (*e.g.*, airwave) that do not contribute usefully to the reflection stack (Fig. 6A, B). Static corrections involve removal of time shifts which result from differences in the elevations of source and receivers (elevation correction) and the changes in thickness and velocity of near-surface low-velocity layers (weathering correction). Static corrections are calculated from elevation information and analysis of the first-arrival times of refractions from the shallow low-velocity layer (refraction statics). Various types of filtering are applied to remove unwanted frequency components of the seismic signal including surface waves, airwaves and random background noise (*e.g.*, Fig. 5A) which can damage reflection coherency.

The next stage of processing involves sorting of shot gathers into common-depth point gathers (Fig. 6C) and application of normal moveout (NMO) corrections. The latter involves applying time and offset-dependent time shifts to each trace in a CDP gather which corrects for the non-vertical incidence of the reflected ray paths (*e.g.*, Fig. 4). This effectively simulates placing all shot and receivers pairs at a zero-offset position directly above the reflection point, and results in flattening of the reflection curve (Fig. 6D). The required NMO corrections are determined from an analysis of the moveout, or stacking velocities, of individual reflectors on some typical CDP gathers. This is done by fitting curves to the

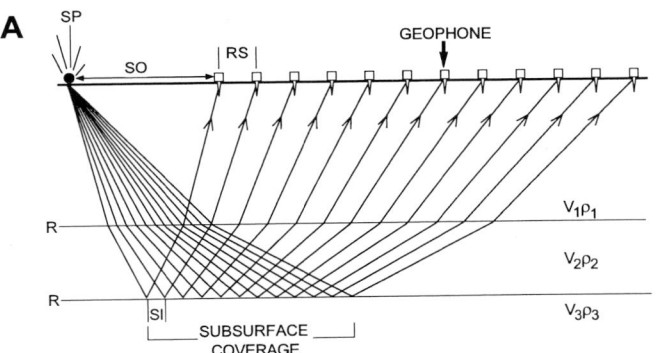

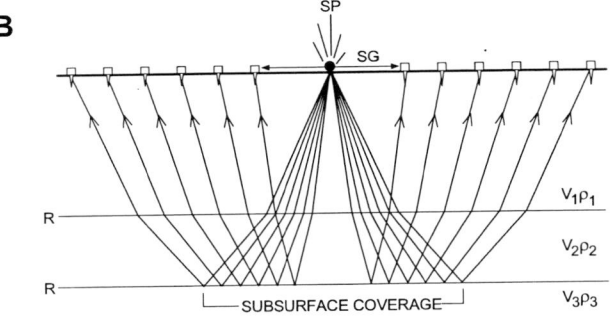

Figure 3   *Two field layouts commonly used in acquisition of CDP seismic reflection data with idealized reflection raypaths shown for a simple horizontally layered case. (**A**) End-on geometry. Shot point (**SP**) is positioned at set source offset (**SO**) and fired into line of receivers which are moved incrementally along the line using a roll-box or by repositioning of receiver spread. The reflection sampling interval (**SI**) is one-half the receiver spacing (**RS**). Continuous roll-along shooting using the 12-channel arrangement shown and with a shot interval equal to the receiver spacing would result in a maximum 6-fold (600%) subsurface coverage. (**B**) Split-spread geometry. Shot point is positioned mid-spread and a shot gap (**SG**) is used to separate near-shot receivers from energy source. Spread is rolled-along in same manner as in end-on layout to provide continuous subsurface coverage.*

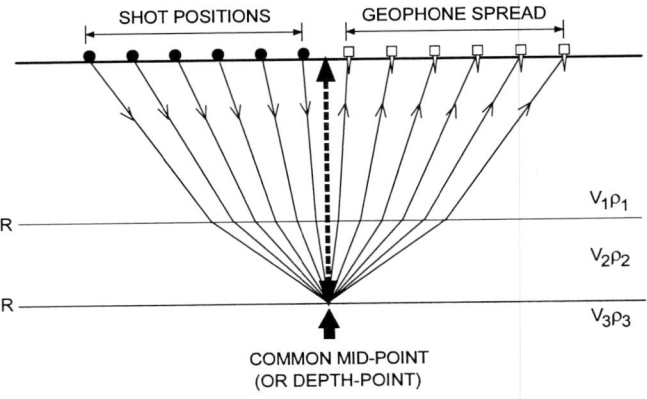

Figure 4   *The common-depth point (**CDP**) concept. Reflector is repeatedly sampled with source-receiver pairs centred over a common-depth point (or common mid-point) in the subsurface. The example shown is for the case of a 12-channel seismograph with shots fired at all geophone locations (e.g., Fig. 3) which results in a maximum 6-fold subsurface coverage. Bold dashed line shows the vertically incident ray path simulated by normal moveout (**NMO**) correction of CDP data during processing.*

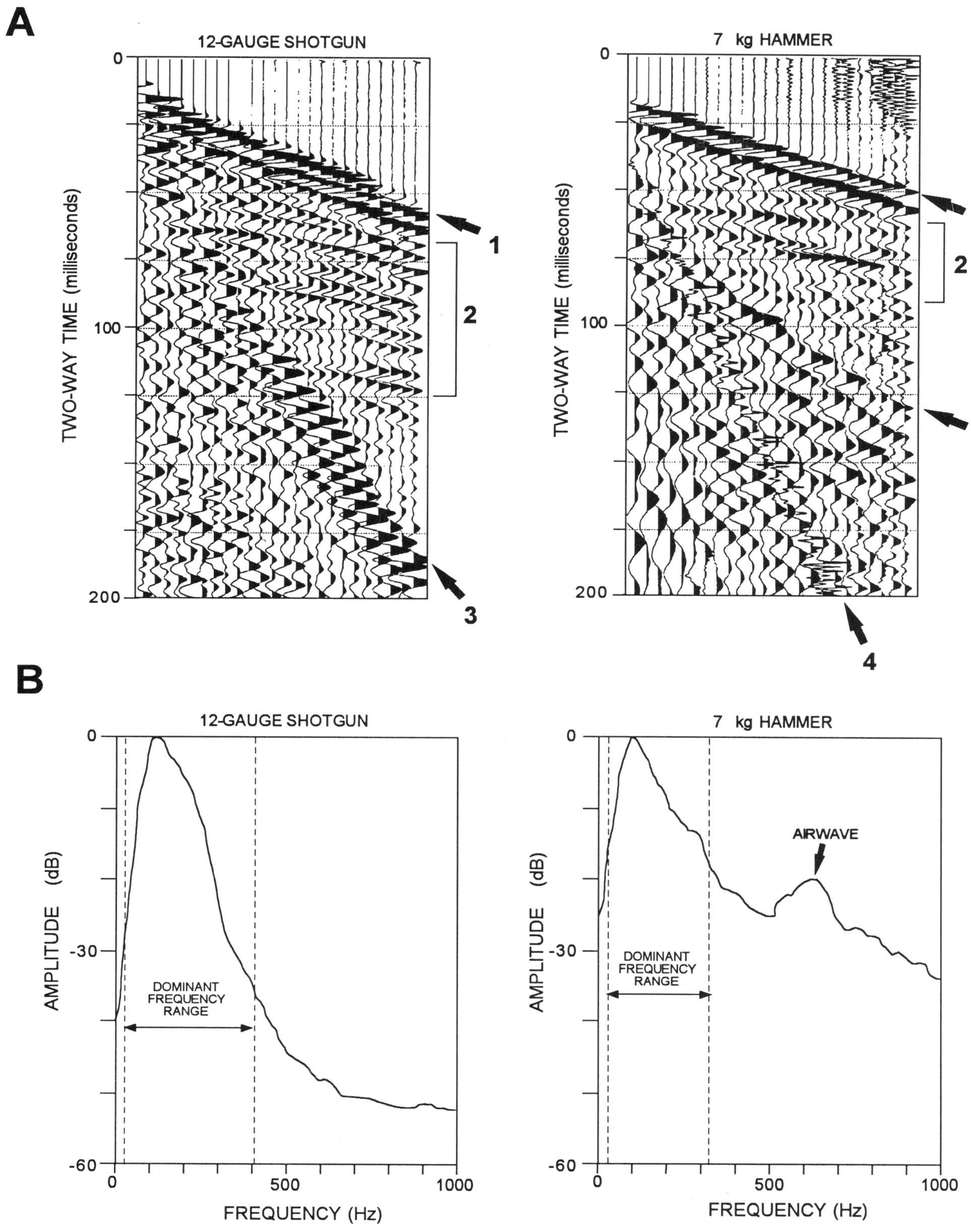

Figure 5   (**A**) *24-channel shot gathers acquired with shotgun (**left**; single shot) and hammer source (**right**; 5 stacks) with refraction (**1**), reflection (**2**) and groundroll (**3**) arrivals identified. Ground-coupled airwave (**4**) records sound energy of hammer impact carried through the air.* (**B**) *Frequency amplitude spectra for field records shown in* (A) *above.*

reflection hyperbola or by using constant-velocity stacking and semblance analysis techniques (see Yilmaz, 1987). The velocity analysis is commonly augmented with velocity information obtained from vertical seismic profiles (VSP) collected in boreholes adjacent to the seismic line (see Chapter 33). Once the correct stacking velocities have been determined for selected CDPs, the NMO corrections are applied universally to all CDP gathers in preparation for stacking. Stacking involves summing of the moved-out traces in each CDP gather to produce a single composite trace (Fig. 6E). The combining of traces results in cancellation of out-of-phase noise energy and reinforcement of in-phase reflection events. From a theoretical standpoint, the improvement in signal-to-noise ratio achieved by stacking increases proportionally with the square root of the CDP fold. As final steps, various types of filtering and gain-balancing operations are performed to enhance the appearance of reflection events and to equalize the trace amplitudes on the final stacked section (Fig. 6F).

EXAMPLES
In this section, examples illustrating the application of CDP seismic reflection profiling are presented from work completed at three proposed landfills and other test sites (Fig. 1).

Whitevale Candidate Sites
The Whitevale P1, M6 and EE11 sites (Fig. 1B) were investi-

gated as possible locations for a landfill to accept 6 million tonnes of municipal waste. Site investigations involved intensive programmes of drilling, core sampling, hydrochemical analysis, and hydrogeological field monitoring (M.M. Dillon Ltd., 1990; Fenco MacLaren Ltd., 1994). These studies demonstrated the presence of unsuspected hydraulic windows through a thick (up to 50 m), compact till unit below the sites (Northern till; Figs. 7, 8) which had previously been considered to be a highly impermeable aquitard capable of restricting the movement of ground water and landfill leachates to underlying aquifers (Boyce et al., 1995; Gerber and Howard, in press). Seismic reflection investigations were carried out in an attempt to better resolve the internal stratigraphy and nature of potential ground-water pathways in the till.

Seismic Data Acquisition and Processing
Seismic data were acquired using two 24-channel exploration seismographs and 50 Hz geophones. Profiles were collected end-on (e.g., Fig. 3A) with 3-9 m source offsets and a geophone and shot spacing of 3 m. Continuous 12- or 24-fold coverage was obtained using a roll-along switch or by repositioning of receiver spreads (see Koseoglu, 1995). During initial trials of the CDP method at sites P1 and M6, a 7-kg hammer and aluminum plate were used as the source, with five blows stacked per record. The hammer source was found to be highly repeatable, and provided good high-frequency response on the compacted road beds around the sites. The

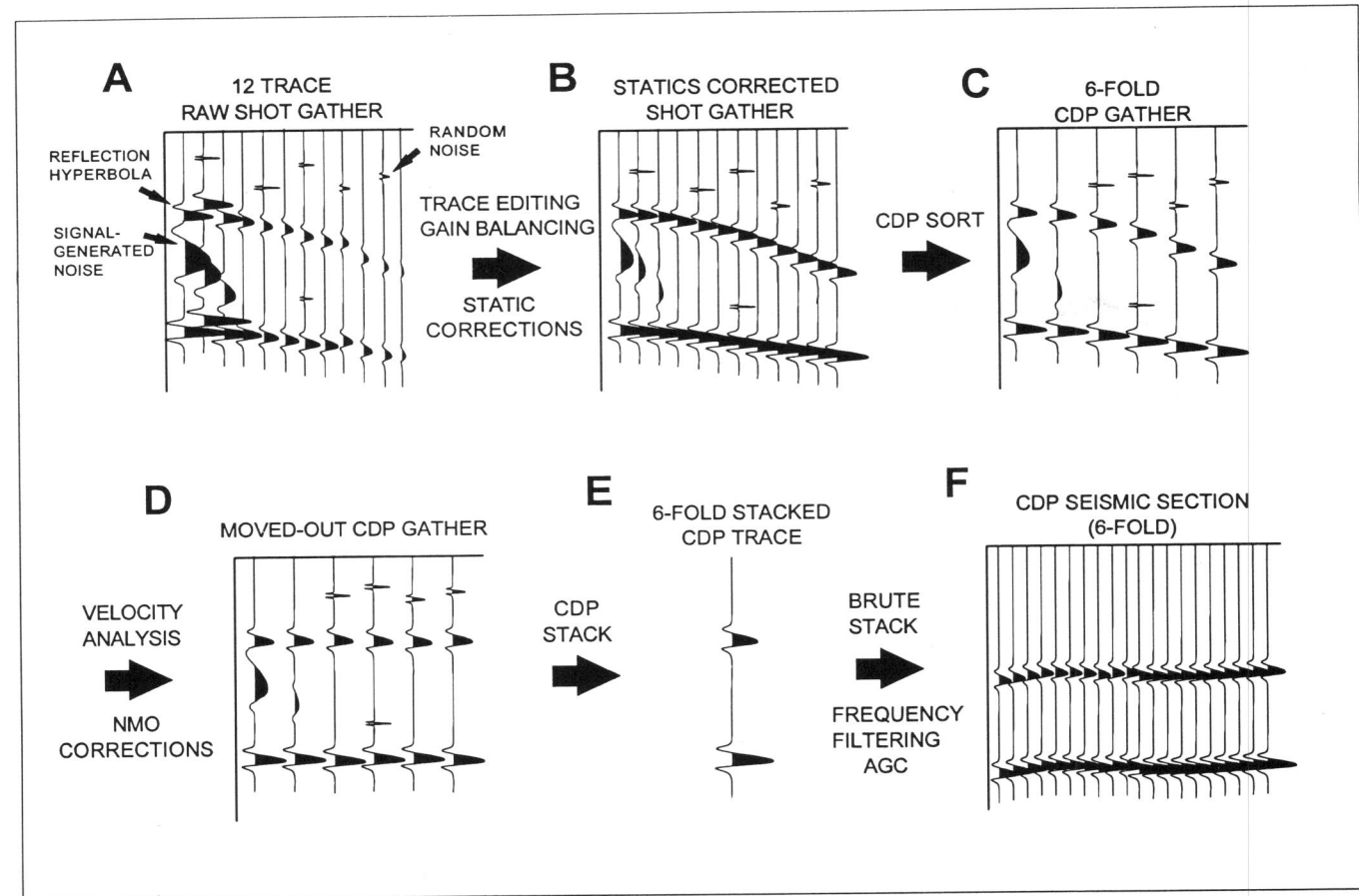

Figure 6 *Principal steps involved in the processing of CDP seismic reflection data: (A) raw shot gathers are edited, (B) corrected for elevation and weathering-layer statics, and (C) are sorted into CDP gathers containing all traces associated with a single reflection point. A velocity analysis is performed and dynamic corrections are made to remove the normal moveouts (NMO) of reflectors prior to (D) stacking. (E) Traces in each CDP gather are then stacked, resulting in re-inforcement of reflections and attenuation of other out-of-phase signals. (F) Filtering and gain-balancing operations (automatic gain control; AGC) are applied to enhance the appearance of the final stacked section.*

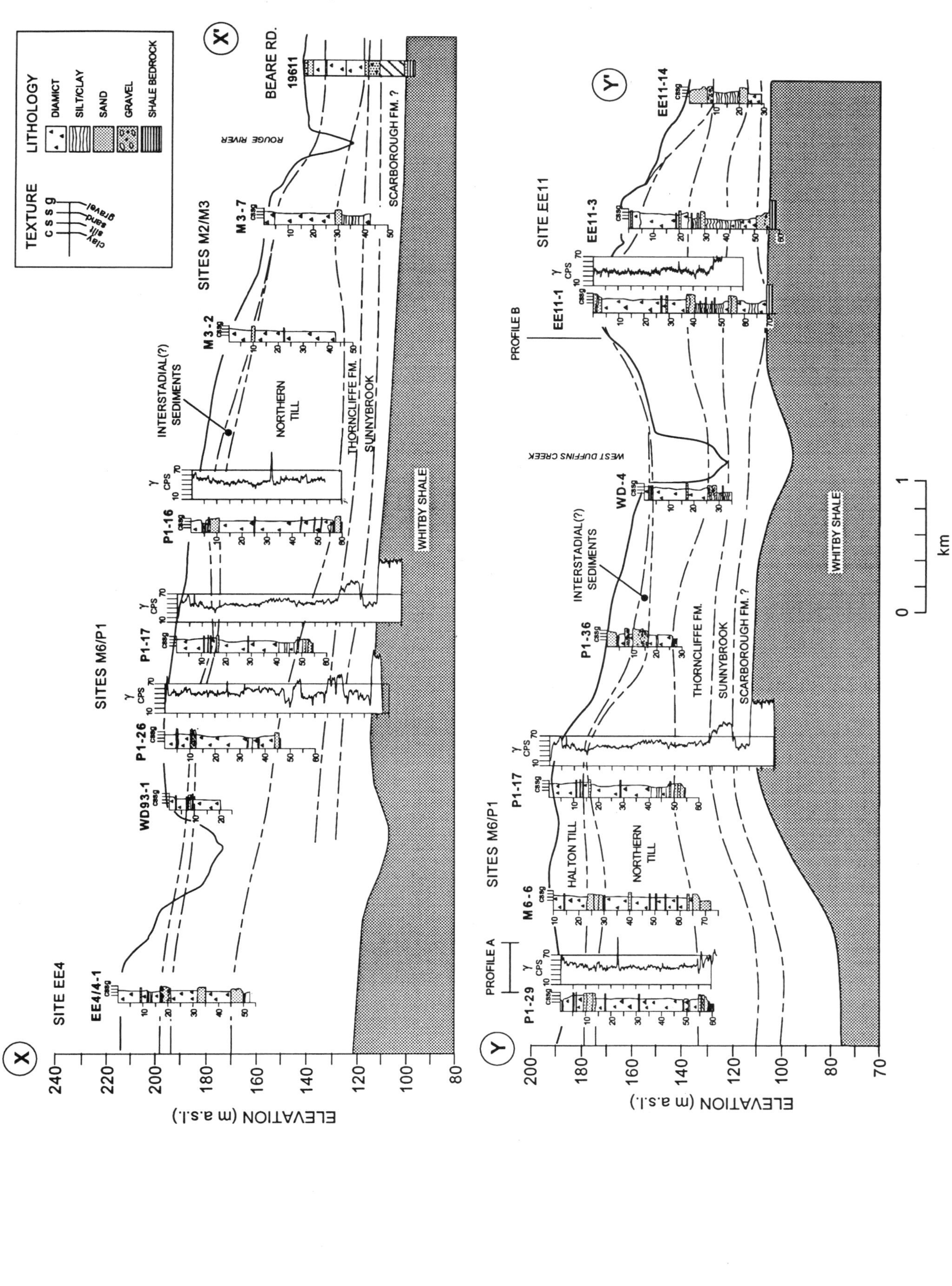

Figure 7 *Geological cross-sections showing details of Pleistocene stratigraphy and bedrock surface below Whitevale candidate sites (locations shown in Fig. 1B). Borehole lithologic and geophysical data (natural gamma; units in counts per second, CPS) also shown. Note presence of thin sands and gravels within Northern till aquitard. Vertical exaggeration is 18:1.*

excellent seismic response is in part due to the presence of a shallow water table within 1 m or so of the surface. The hammer source was found to be generally unsuitable for surveying in other off-road areas, due to problems with high-amplitude surface waves and rapid attenuation of high frequencies by a loose surface-weathering layer. In these areas, a 12-gauge Buffalo gun (Pullan and MacAulay, 1987) was found to produce a significant improvement in high-frequency content of the data. The gun was detonated in 1 m deep, water-filled shot holes, with one shot registered for each 24-channel record. Although production rates are somewhat slower than with hammer, the downhole shotgun has been found to provide more consistent frequency response under varying substrate conditions, and is the preferred source for most applications (e.g., Hunter et al., 1987). Band-pass filtering was employed during data acquisition (A/D input filters 140-1000 Hz) to attenuate low-frequency signals and prevent saturation of the records by ground roll energy.

Seismic data were processed on Unix and PC platforms using Inverse Theory Applications (ITA) and Vista (Seismic Image Software Ltd.) software packages. The processing, in general terms, followed that described in previous sections. A detailed discussion of the processing methods and parameters employed is given by Koseoglu (1995).

Sites P1 and M6.   A 350 m segment of a 1.2 km east-west profile collected at sites P1 and M6 is shown in Figure 9 along with an interpreted depth-converted section (lower panel). A number of laterally continuous reflection events are recognized on the seismic profile (A through F: Fig. 9) which can be correlated with sequence boundaries within Pleistocene sediments and the bedrock interface (Fig. 7). Figure 8 summarizes the lithologic characteristics and range of interval velocities for the seismostratigraphic units defined by these reflection events.

The first coherent event observed on the section is a gently undulating reflection (A: Fig. 9) occurring at a two-way time of 18-20 ms, which is correlated with the top of the Northern till, identified in boreholes P1-29 and M6-6, at depth of 14 and 20 metres below ground surface (m bgs), respec-

tively. The unit above this reflector consists of a sandy surficial till and underlying sands and gravels (Halton Till complex; Fig. 7) which either lack marked acoustical contrasts or are too shallow to be resolved on seismic sections. The velocity for this uppermost interval is in the range of 1800-2000 m·s^{-1} (Fig. 8), and is consistent with the low level of consolidation and sandy character of the Halton Till complex. Interval velocities within the Northern till are considerably higher ($\approx$2200-2800 m·s^{-1}), as a result of the high degree of compaction (bulk density $\approx$ 2.3 g·cm^{-3}, porosity $\approx$12-14%) and silt-rich texture of this unit. A number of semi-continuous, low-amplitude reflectors are also recognized within the Northern till (Fig. 9). These internal reflectors are correlated with thin (<1.5 m thick) sand and gravel beds. The presence of sand and gravel beds in the Northern till was observed in some core samples (e.g., P1-29, M6-6; Fig. 7), but was more generally indicated during drilling operations by zones of increased drilling rates and poor core recovery (M.M. Dillon Ltd., 1990).

The base of the Northern till is marked by a high-amplitude reflection (B: Fig. 9) at a two-way time of 45-50 ms, which marks the contact with underlying unconsolidated sands and silts (Thorncliffe Formation). The interval velocities in the Thorncliffe Formation are low in comparison to the overlying till (Fig. 8), and result in a marked acoustic impedance contrast at this boundary. The lower part of the Thorncliffe Formation passes transitionally into a dense pebbly clay (Sunnybrook Till; Fig. 7) which is identified on geophysical logs by a zone of high gamma counts (>80 cps) occurring at depth of 65-80 m bgs. The upper and lower contacts of the pebbly clay are correlated with reflection events at two-way times of about 70 ms and 85 ms, respectively (C, D: Fig. 9).

Bedrock is identified by a high-amplitude reflector occurring between 100-105 ms (E: Fig. 9) which marks a shift to much higher velocities (>3000 m·s^{-1}; Fig. 8). This reflector indicates a relatively flat to gently undulating bedrock surface at a depth of about 115 m bgs. A number of deeper internal bedrock reflections are also recorded, including a discontinuous event at 130 ms (F; Fig. 9). These likely result from the alternation of shales with thin siltstone beds known to be

BOUNDING REFLECTORS	SEISMOSTRATIGRAPHIC UNIT	LITHOLOGY	HYDROGEOLOGIC SIGNIFICANCE	INTERVAL VELOCITY (m/s)
A	HALTON TILL COMPLEX	VARIABLY CONSOLIDATED SANDY TILL AND UNDERLYING INTERSTADIAL SANDS AND GRAVELS	LOCAL AQUIFER	~ 1500 - 2200
A-B	NORTHERN TILL	OVERCONSOLIDATED SANDY SILT TILL WITH THIN (< 1.5 m) SAND AND GRAVEL INTERBEDS AND SAND LENSES	REGIONAL AQUITARD LANDFILL SUBSTRATE	~ 2200 - 2800
B-C	THORNCLIFFE FM.	STRATIFIED LACUSTRINE SILT AND SAND AND LAMINATED SILTY CLAYS	REGIONAL AQUIFER	~ 1600 - 1900
C-D	SUNNYBROOK TILL	WELL-CONSOLIDATED PEBBLY CLAY INTERCALATED WITH LAMINATED SILTY CLAYS AND SANDY GRAVELS	REGIONAL AQUITARD	~ 2200 - 2400
D-E	SCARBOROUGH FM. AND OLDER SEDIMENTS(?)	LAMINATED LACUSTRINE SILTY CLAYS, SANDS AND UNDERLYING COMPACT SANDY DIAMICT UNITS (TILL)	DEEP REGIONAL AQUIFER	~ 2000 - 2800
E	WHITBY SHALE	BLACK SHALE WITH THIN SILTSTONE BEDS	FRACTURED BEDROCK AQUIFER	~ 3000 - 3200

Figure 8   *Seismostratigraphic units identified in the study region with a summary of their lithologic characteristics, hydrogeological significance and approximate range of interval velocities and bounding reflectors (**A-E**) depicted in subsequent figures.*

present in the Whitby Formation.

Site EE11. Figure 10 shows a 450-m segment of a seismic profile acquired at the southwest corner of the EE11 candidate landfill site (Fig. 1B). This shows a number of reflective horizons that can be directly correlated with seismostratigraphic units identified at sites P1 and M6 (*e.g.*, Fig. 9). The upper surface of the Northern till (Fig. 7) lies at a shallow depth (<12 m), and is not resolved on seismic profiles. The base of this unit is correlated with a gently undulating reflector occurring at two-way time of about 25 ms (B: Fig. 10), which records a shift to much lower velocities (1700-900 $m \cdot s^{-1}$) in the underlying Thorncliffe Formation. A single low-amplitude reflector recorded above this event at about 15 ms correlates with a thin sand and gravel bed identified in the Northern till at a depth of 14 m bgs in borehole EE11-2.

The Thorncliffe Formation at this site shows a number of low-amplitude internal reflection events which record the alternation of sand beds with silty clay units (*e.g.*, EE11-2; Fig. 10). The base of the Thorncliffe unit is marked by a clearly defined reflector at 50 ms, which correlates well with the contact with the Sunnybrook at a depth of 46-50 m bgs. This unit is of hydrogeological importance at site EE11 as an aquitard separating aquifers in the Thorncliffe Formation from a deeper bedrock aquifer system (Fig. 8).

Malvern Radioactive-Soil Treatment Site, City of Scarborough

Work at this Passmore Avenue site (Fig. 1B) is designed to evaluate the feasibility of conducting seismic reflection surveys in urban and industrial areas where surface conditions (*e.g.*, fill, paved surfaces) and cultural noise from roadways, and overhead and underground services (*e.g.*, gas lines, anode-protected pipes, power lines) could potentially limit the use of seismic methods.

The site was investigated in 1993-1994 for emplacement of a temporary radioactive-soil treatment facility (Acres Ltd., 1994; Chapter 28). The site was approved in early 1995, and is currently under use for radium-contaminated soil being removed from a nearby housing project. At this site, only shallow (<25 m) borehole information is available. Interpretation of the deeper seismic stratigraphy is based on comparison of interval velocities and reflection characteristics with seismic data at Whitevale (Figs. 9, 10) and with the regional stratigraphic framework (Fig. 7).

A 300-m segment of a test profile is shown in Figure 11. The data were acquired by shooting in 1-m deep shot holes in a water-filled roadside ditch which provided excellent coupling and penetration of high frequencies. Borehole data show that the near-surface stratigraphy consists of a thin (<5 m) surface

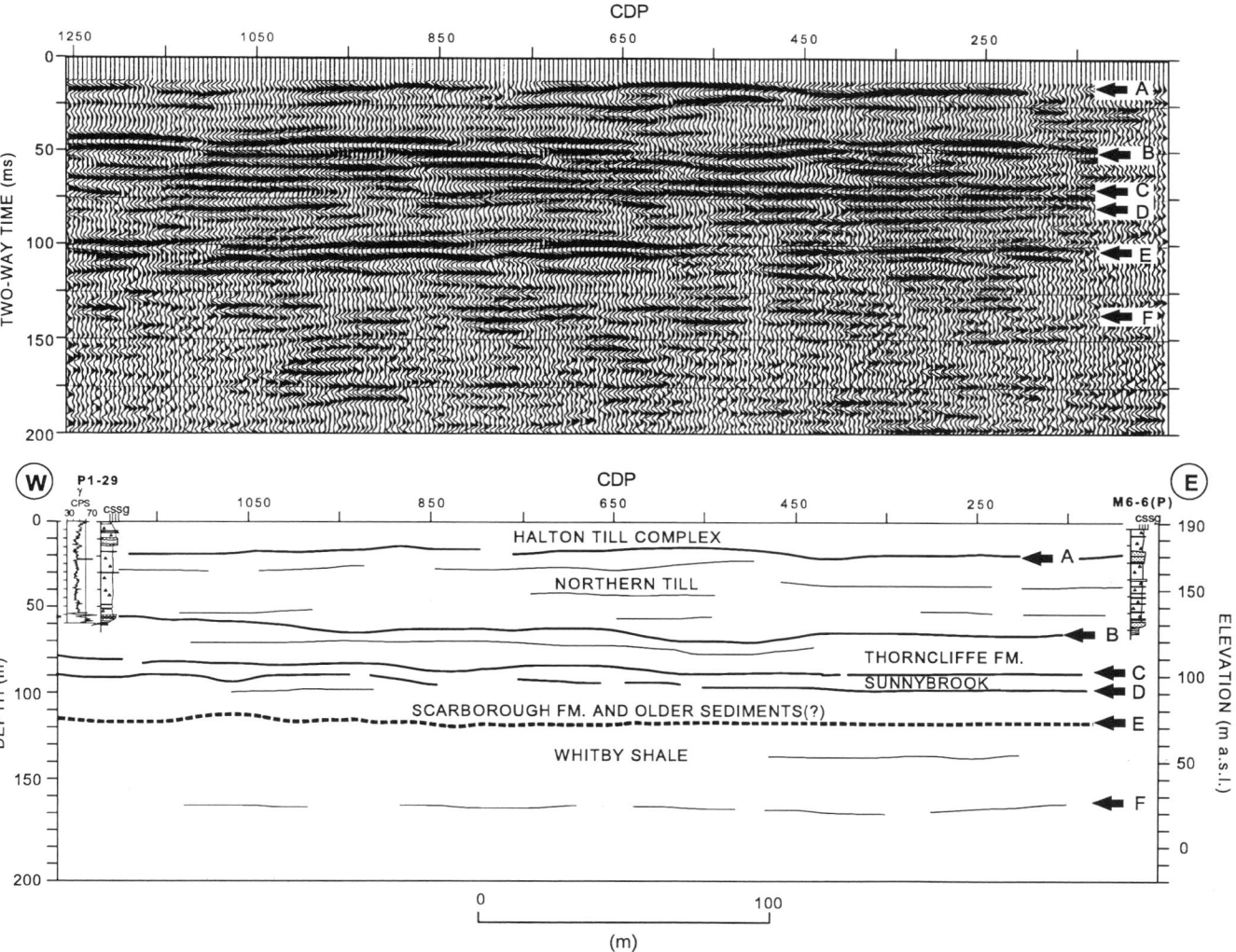

Figure 9 *24-fold seismic reflection profile (A, Fig. 1B) acquired across site M6 and adjacent to site P1. Interpreted section below shows correlations between reflection events (labelled A to F; Fig. 8) and lithologic and natural gamma logs from adjacent boreholes. Horizontal and vertical scales on interpreted panel are 1:1.*

till unit (Halton Till) overlying up to 19 m of sand and silt. The first coherent reflection event at 25-27 ms (A: Fig. 11) is characterized by very low stacking velocities ($\approx$1530-1650 m·s^{-1}) which are consistent with the presence of a thick sand unit. The underlying interval, bounded by reflectors B and C (Fig. 11), shows a shift to much higher interval velocities (>2200 m·s^{-1}) which are consistent with the presence of the compact Northern till. A number of gently undulating internal reflection events are recognized in this interval, and are most likely associated with the presence of sorted sediment lenses in the till.

Below the Northern till, the low velocity and thickness ($\approx$25 m) of the interval between reflectors B and C (Fig. 11) suggests a possible correlation with the Thorncliffe Formation. The underlying interval bounded by reflectors C and D shows a marked shift to higher interval velocities, and likely represents the Sunnybrook based on its stratigraphic position. The bedrock interface below the Passmore site is correlated with a relatively strong reflection event occurring at about 115 ms (at about 80 m asl).

Data quality was not degraded significantly by the cultural noise levels at the Passmore site. The use of a 60-Hz notch filter on the recording instrument was found to be effective in suppressing noise from overhead lines and an anode-protected gas pipe which runs along the length of Passmore

Avenue, about 3 m north of the seismic line. The use of a 140-Hz low-cut filter and careful timing of shots between the passing of vehicles was found to be effective in limiting traffic noise at this site.

DISCUSSION

In conventional drilling approaches, major stratigraphic boundaries and, in some cases, individual lithofacies can be correlated between drilling locations with an acceptable degree of confidence (e.g., Fig. 7). Delineation of the geometry of bounding surfaces, however, can be problematic, particularly where bed contacts are non-planar or where rapid changes in the thickness of units occur as a result of erosion or non-deposition (e.g., Fig. 7; EE11-14). Problems also frequently arise in correlating relatively thin, (<1 m) lensate beds within otherwise dense tills, and in locating their terminations, or pinch-outs, in the subsurface. When borehole data are integrated with the continuous two-dimensional subsurface coverage provided by seismic reflection profiling, resolution of the continuity and geometry of stratigraphic boundaries is significantly enhanced (Figs. 9, 10, 11).

The vertical resolution achieved at the Whitevale sites, based on the dominant reflection frequencies (150-400 Hz) and velocities within the Pleistocene interval (1500-2800; Fig. 8), is estimated at 1 m to 5 m using the quarter-wavelength

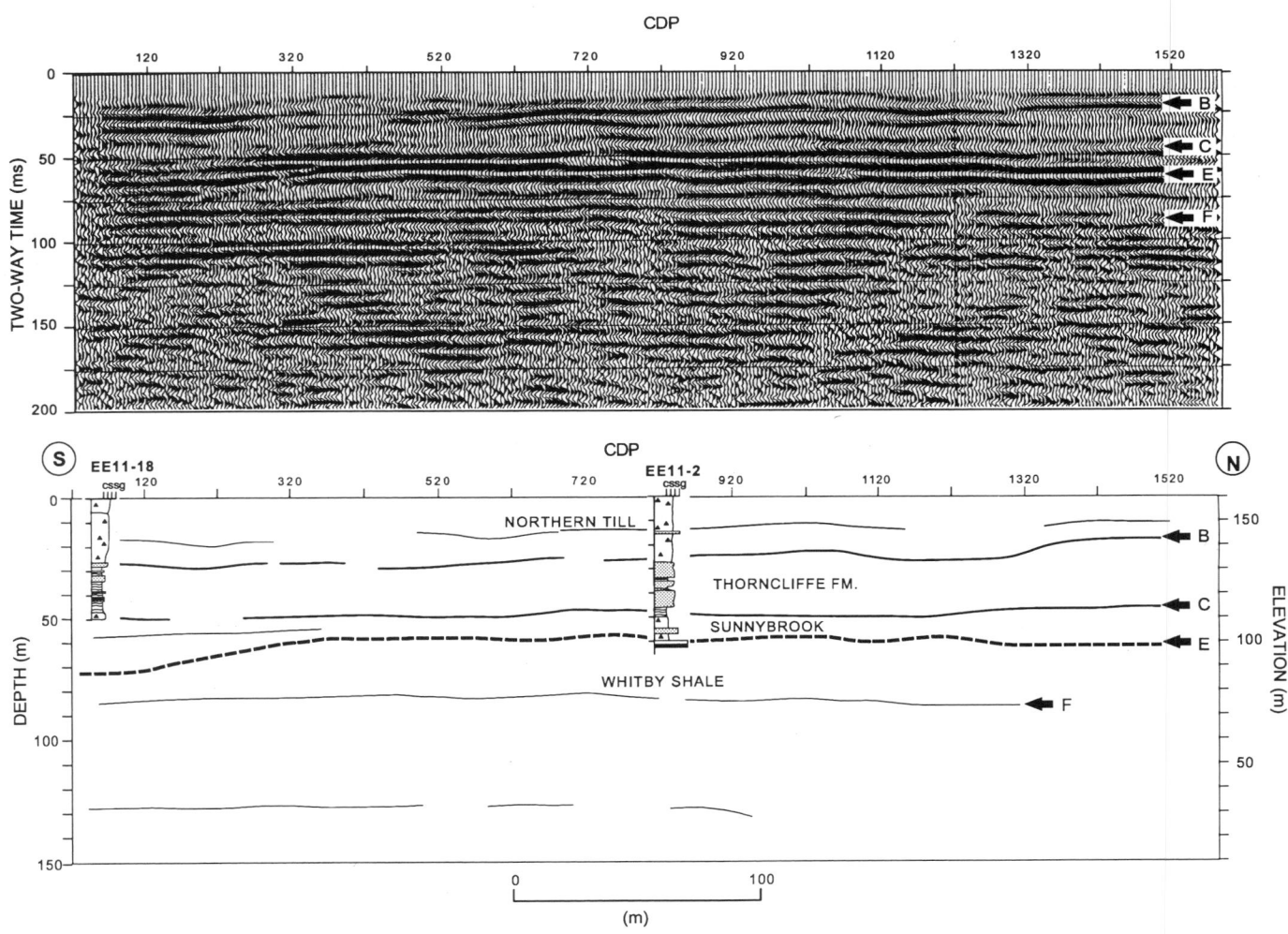

Figure 10 *12-fold seismic reflection profile (B, Fig. 1B) collected on site EE11 with interpreted section. Horizontal and vertical scales on interpreted panel are 1:1.*

criteria. In general, the larger value is associated with later reflection events on seismic sections (*e.g.*, bedrock, E: Fig. 9) due to the attenuation of high frequencies and overall increase in velocity with depth.

It should be noted that the field layout described was designed to resolve the Pleistocene sediments and bedrock surface at depths of up to 110 m bgs. This geometry, however, produced very wide angle reflections and loss of resolution in the near surface (≈15 m bgs). Resolution of the uppermost interval would require use of much smaller geophone spacings (*e.g.*, <1 m) and source-receiver offsets. The resulting reduction in the spread length would considerably increase the time and cost of field acquisition, due to the larger number of shot points, but is technically feasible where information from the very near subsurface is critical. From a practical stand point, the shallow limit of the seismic method is probably about 5 m for most applications, although reflections from as shallow as 2 m have been obtained (*e.g.*, Birkelo *et al.*, 1987).

Seismic profiles from P1 and M6 demonstrate clearly defined planar to gently undulating reflection events within the Northern till, which can be traced continuously over distances of up to 110 m (Fig. 9). This geometry is consistent with the characteristics of laterally extensive sand and gravel beds, up to 1 m in thickness, identified in the Northern till in outcrop and in core (Fig. 7), which can be correlated with internal reflectors in the till (Boyce *et al.*, 1995). The continuity of these horizons, as suggested by outcrop and seismic data together with their coarse-grained texture and thickness, suggests their importance as ground-water pathways in the Northern till.

In addition to being able to identify small-scale stratigraphic variability, the use of shallow seismic reflection surveys has the potential to reduce the costs of applied investigations by reducing the total number of boreholes required to characterize a site. Landfill investigations at sites P1 and EE11 in the Whitevale area were conducted at a total cost of more than $2 million; drilling costs alone amounted to some 15% of the total expenditure for field studies. During investigations at site EE11, seismic studies were implemented during the last stages of field investigations to confirm the results of

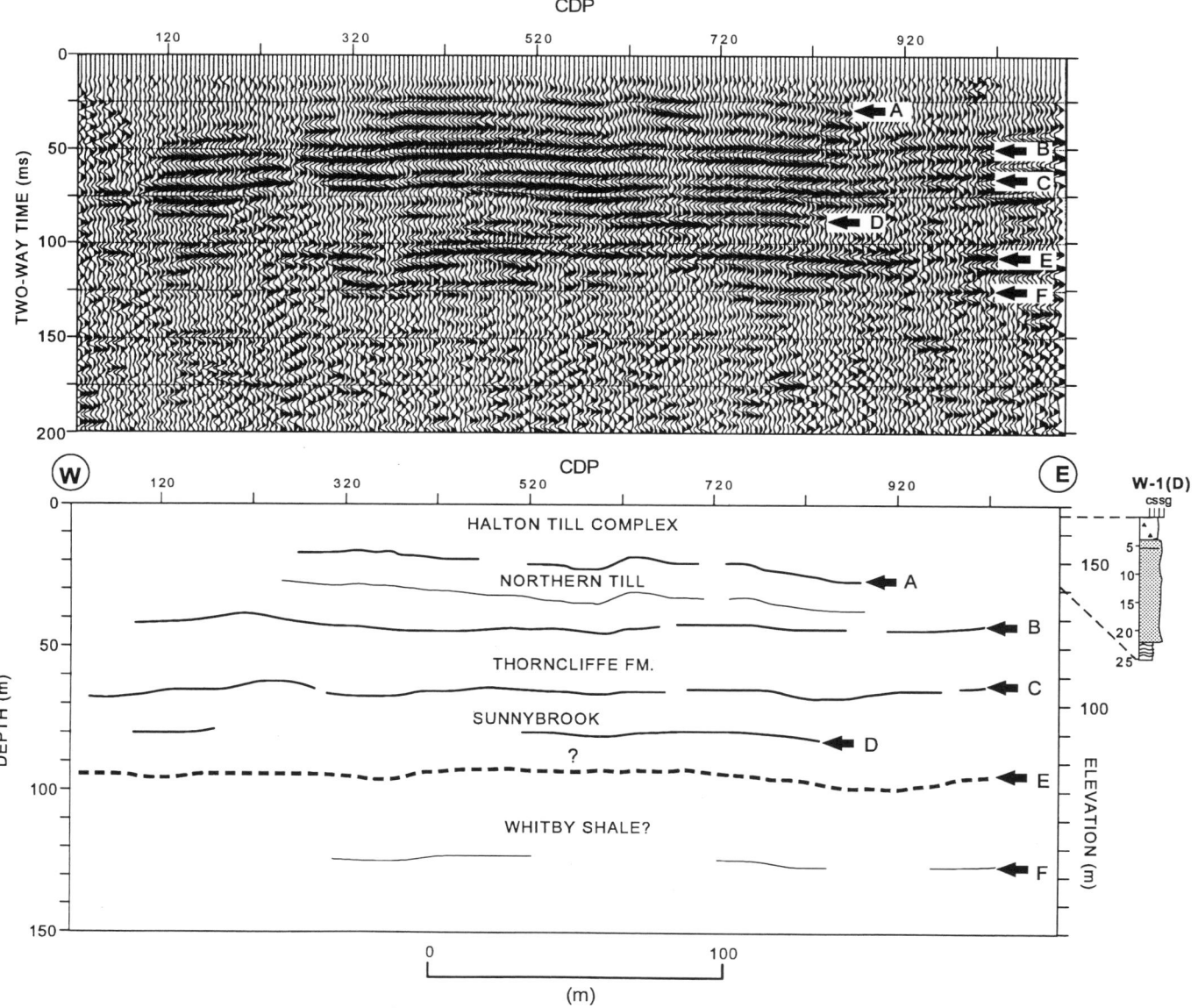

Figure 11   *12-fold seismic reflection profile (C, Fig. 1B) collected on Passmore Road adjacent to Malvern Remedial Project (MRP) radioactive-soil treatment site with provisional interpretation. Horizontal and vertical scales on interpreted panel 1:1.*

drilling operations. A better approach, based on this experience, would be to integrate seismic studies at a much earlier stage and would involve drilling of number of strategically placed test boreholes tied by shallow seismic reflection lines. In this way, the placement of subsequent drill holes could be more effectively selected to target the specific data requirements of the study.

ACKNOWLEDGEMENTS

This research was supported through Natural Sciences and Engineering Research Council of Canada grants to N. Eyles and G.F. West. The Interim Waste Authority Ltd. and the Ministry of Government Services are thanked for allowing access to properties. We are particularly grateful for technical advice provided by Gordon West, Hamid Siakoohi and André Pugin, and for discussions during the course of landfill investigations with Nick Eyles and David Winfield. Mike Doughty, David Popovich, and Romel Sargezei are thanked for assistance in the field.

REFERENCES

Acres, Ltd., 1994, Geotechnical and hydrogeological investigations for Passmore Avenue candidate site: Report submitted to Malvern Remedial Project, Scarborough, March, 1994, 63 p.

Birkelo, B.A., Steeples, D.W., Miller, R.D. and Sophocleous, M.A., 1987, Seismic-reflection study of a shallow aquifer during a pumping test: Ground Water, v. 25, p. 703-709.

Boyce, J.I., Eyles, N. and Pugin, A., 1995, Seismic reflection, borehole and outcrop geometry of Late Wisconsin tills at a proposed landfill site near Toronto, Ontario: Canadian Journal of Earth Sciences, v. 32, p. 1331-1349.

Branham, K.L. and Steeples, D.W., 1988, Cavity detection using high-resolution seismic reflection methods: Mining Engineering, v. 40, p. 115-119.

Curry, B.B., Troost, K.R. and Berg, R.C., 1994, Quaternary geology of the Martinsville alternative site, Clark County, Illinois – A proposed low-level radioactive waste disposal site: Illinois State Geological Survey, Circular 556, 83 p.

Dillon, M.M., Ltd., 1990, Regional Municipality of Durham, P1 contingency landfill site assessment, technical support volume B (Part I) – hydrogeology: Report submitted to Durham Region, Pickering, 220 p.

Fenco MacLaren Inc., 1994, IWA landfill site search Metro/York Region – step 6 hydrogeological report site M6, Background report to EA Document IV, Appendix G: Interim Waste Authority Ltd., Toronto, 180 p.

Genau, R.B., Madsen, J.A., McGeary, S. and Wehmiller, J.F., 1994, Seismic-reflection identification of Susquehanna River paleochannels on the mid-Atlantic coastal plain: Quaternary Research, v. 42, p. 166-175.

Gerber, R. and Howard, K.W.F., in press, Evidence for recent groundwater flow through Late Wisconsin till near Toronto, Ontario: Canadian Geotechnical Journal, in press.

Hunter, J.A., Pullan, S.E., Gagne, R.E. and Good, R.L., 1984, Shallow seismic reflection mapping of the overburden-bedrock interface with the engineering seismograph – some simple techniques: Geophysics, v. 49, p. 1381-1385.

Hunter, J.A., Pullan, S.E., Burns, R.A., Gagne, R.M. and Good, R.L., 1987, Applications of a shallow seismic reflection method to groundwater and engineering studies: Ontario Geological Survey, Special Volume 3, p. 704-715.

Jongerius, P. and Helbig, K., 1988, Onshore high-resolution seismic reflection profiling applied to sedimentology: Geophysics, v. 53, p. 1276-1283.

Knapp, R.W. and Steeples, D.W., 1986a, High-resolution common-depth point seismic reflection profiling: Instrumentation: Geophysics, v. 51, p. 276-282.

Knapp, R.W. and Steeples, D.W., 1986b, High-resolution common-depth point seismic reflection profiling: Field acquisition parameter design: Geophysics, v. 51, p. 283-294.

Koseoglu, B.B., 1995, Seismic reflection surveys around a possible municipal waste disposal site on stratified Pleistocene deposits near Pickering, Ontario: unpublished M.Sc. thesis, University of Toronto, Toronto, ON, 30 p.

Mayne, W.H., 1962, Common-reflection-point data stacking techniques: Geophysics, v. 27, p. 927-38.

Miller, R.D., Pullan, S.E., Steeples, D.W. and Hunter, J.A., 1992, Field comparison of shallow seismic sources near Chino, California: Geophysics, v. 57, p. 693-709.

Miller, R.D., Pullan, S.E., Waldner, J.S. and Haeni, F.P., 1986, Field comparison of shallow seismic sources: Geophysics, v. 51, p. 2067-2092.

Miller, R.D. and Steeples, D.W., 1994, Applications of shallow high-resolution seismic reflection to various environmental problems: Journal of Applied Geophysics, v. 31, p. 65-72.

Pugin, A. and Rossetti, S., 1992, Acquisition of land-based high resolution seismic profiles in glacial basins, two case studies in the Alpine foreland of Switzerland: Eclogae geologicae Helvetae, v. 85, p. 491-502.

Pullan, S.E. and Hunter, J.A., 1985, Seismic model studies of the overburden-bedrock reflection: Geophysics, v. 50, p. 1684-1688.

Pullan, S.E. and MacAulay, H.A., 1987, An in-hole shotgun source for engineering seismic surveys: Geophysics, v. 52, p. 985-996.

Pullan, S.E., Pugin, A., Dyke, L.D., Hunter, J.A., Pilon, J.A., Todd, B.J., Allen, V.S. and Barnett, P.J., 1994, Shallow geophysics in a hydrogeological investigation of the Oak Ridges Moraine, Ontario: Symposium on the Application of Geophysics to Engineering and Environmental Problems, Boston, MA, March 27-31, 1994, Proceedings, p. 143-161.

Robinson, E.A. and Treitel, S., 1980, Geophysical Signal Analysis: Prentice-Hall, Inc., Englewood Cliffs, NJ, 466 p.

Schepers, R., 1975, A seismic reflection method for solving engineering problems: Journal of Geophysics, v. 41, p. 367-384.

Seeber, M.D. and Steeples, D., 1986, Seismic data obtained using a .50-calibre machine gun as a high-resolution seismic source: American Association of Petroleum Geologists, Bulletin, v. 70, p. 970-976.

Sheriff, R.E., 1985, Aspects of seismic resolution, in Berg, O.R., and Woolverton, D.G., eds., Seismic Stratigraphy II: American Association of Petroleum Geologists, Memoir 39, p. 1-10.

Sheriff, R.E. and Geldart, L.P., 1982, Exploration Seismology, Volume 1: Cambridge University Press, Cambridge, MA, 245 p.

Slaine, D.D., Pehme, P.E., Hunter, J.A., Pullan, S.E. and Greenhouse, J.P., 1990, Mapping overburden stratigraphy at a hazardous waste facility using shallow seismic reflection methods, in Ward, S.H., ed., Geotechnical and Environmental Geophysics: Society of Exploration Geophysicists, v. 2, p. 273-280.

Somanas, D., Bennett, B. and Chung, Y., 1987, In-field seismic CDP processing with a micro-computer: The Leading Edge, v. 6, p. 24-26.

Steeples, D.W., Knapp, R.W. and McElwee, C.D., 1986, Seismic reflection investigations of sinkholes beneath interstate Highway 70 in Kansas: Geophysics, v. 51, p. 295-301.

Steeples, D.W. and Miller, R.D., 1990, Seismic reflection methods applied to engineering, environmental and groundwater problems, in Ward, S.H., ed., Investigations in Geophysics No. 5: Society of Exploration Geophysicists, v. 1, p. 1-30.

Telford, W.M., Geldart, L.P. and Sheriff, R.E., 1990, Applied Geophysics: Cambridge University Press, Cambridge, MA, 770 p.

Treadway, J.A., Steeples, D.W. and Miller, R.D., 1988, Shallow seismic study of a fault scarp near Borah Peak, Idaho: Journal of Geophysical Research, v. 93, p. 6325-6337.

Yilmaz, O., 1987, Seismic data processing, in Doherty, S.M., ed., Investigations in Geophysics 2: Society of Exploration Geophysicists, 526 p.

33. Urban Geophysics in the Kitchener-Waterloo Region, Ontario

George W. Schneider
David C. Nobes
Michael A. Lockhard
John P. Greenhouse

Waterloo Centre for Groundwater Research and Quaternary Sciences Institute
Department of Earth Sciences, University of Waterloo, Waterloo, Ontario N2L 3G1

SUMMARY

During the past decade, the geophysical methods of well logging, and borehole and surface seismic reflection have been adapted and applied to the delineation of overburden aquifers in the Regional Municipality of Waterloo of southern Ontario. We present here an overview of the methodologies that have been developed for this purpose at the University of Waterloo. Some examples are presented, and future research directions for this area of geophysics are discussed.

INTRODUCTION

The Regional Municipality of Waterloo (RMW), situated 100 km southwest of Toronto, includes the cities of Waterloo, Kitchener and Cambridge. Its population of 387,000 is spread over an area of 1330 km². Like many communities, the RMW relies largely on overburden aquifers for its residential and industrial water requirements. Sustainable growth depends on securing and expanding this important resource. This growth could potentially degrade the quality of water if not managed properly.

To manage and protect the groundwater, a detailed understanding of the aquifer/aquitard system is required. Owing to the complexity of these overburden deposits, the lateral resolution required for this degree of understanding cannot be achieved cost-effectively with drilling alone. Geophysics has long been used successfully as a tool to image stratigraphy and assess reserves in the oil and gas industry. Significant progress has been made to adapt these methods to water resource investigations (Keys, 1989; Hunter *et al.*, 1989). Here we describe the progress of the geophysics group at the University of Waterloo in addressing the problems we encounter in the RMW, recognizing our dependence on developments within the larger research community.

GEOLOGICAL SETTING

Underlying the glacial sediments that make up the overburden in the RMW are the Silurian and Devonian shales and limestones of the Michigan Basin. These are the Bois Blanc, Bass Island, Salina and Guelph formations (Fig. 1).

Several layers of glacial sediments lie unconformably on these shales and limestones. They are less than 100,000 years in age and in some places exceed 100 m in thickness. Exposed at the surface are the Wentworth, Port Stanley, Tavistock and Maryhill tills as well as the Waterloo, Paris and Galt moraines. At depth, drilling encounters the Catfish Creek Till, as well as at least two pre-Catfish Creek tills (Karrow *et al.*, 1990b; Chapter 2).

The exact extent and nature of these deposits is subject to

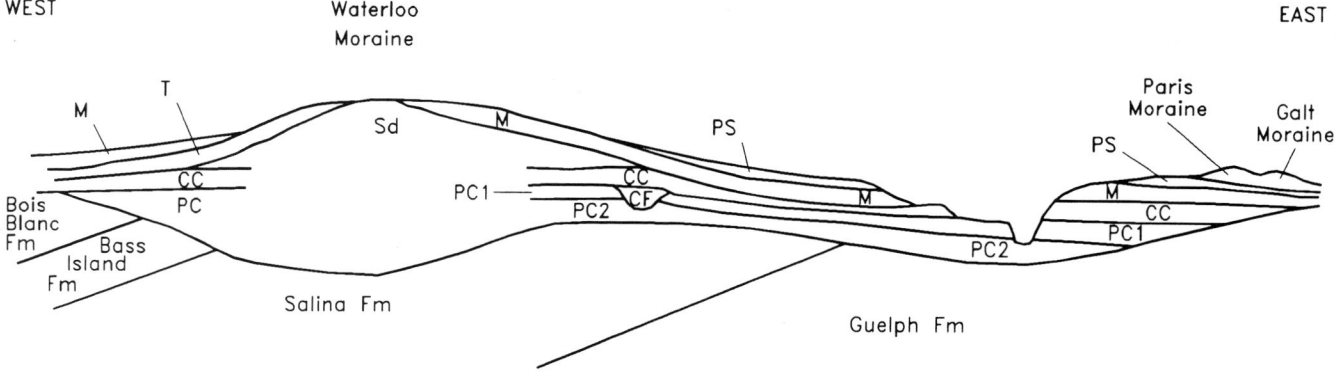

Figure 1 *Cross-section of the RMW showing sequence of deposits (from Karrow et al., 1990b). Symbols:* **CC**, *Catfish Creek Till;* **CF**, *Stream Channel Fill;* **M**, *Maryhill Till;* **PC1,PC2**, *Pre-Catfish Creek Tills;* **PS**, *Port Stanley Till;* **Sd**, *Sand;* **T**, *Tavistock Till;* **W**, *Wentworth Till.*

continual refinement as information from each new borehole is acquired.

WELL LOGGING

Since our acquisition of a well-logging system in the early 1970s, the University of Waterloo has been routinely conducting the geophysical logging of boreholes as they are drilled in the RMW. A complete suite of geophysical logs (Fig. 2) typically consists of natural gamma, density (gamma-gamma), epithermal neutron, short/long normal resistivity, and conductivity responses. Since the early 1980s, all data have been collected in digital form, most recently by the computer interface hardware and the software system LOGGITT. A total of 15 cased boreholes are now available, throughout the RMW, for teaching and experimentation with geophysical logging. Many of these have continuous core for comparison with the geophysical data. These boreholes use a standard 7.5 cm schedule 40 PVC casing (Greenhouse and Pehme, 1986; Karrow et al., 1990a).

Assessing Physical Properties
of Quaternary Deposits: Calibration Pits

Pehme (1984) first undertook a methodical investigation of borehole geophysics as a cost-effective method of identifying Quaternary deposits in the RMW. As part of this study, he constructed a series of calibration pits (Fig. 3) to standardize tool detectors and to calibrate log responses for physical properties. From the data collected in these pits, empirical mathematical relationships were developed to convert gamma-gamma, neutron and natural gamma counts to bulk density, water content, and clay content in the standard cased RMW boreholes.

Automated Identification of Quaternary Deposits
Using Cross-Plots and Principal Component Analysis

Using geophysical logs and continuous core data at four boreholes, Pehme (1984) constructed gamma-neutron and gamma-density log cross-plots to describe the geophysical responses of the common lithologies and stratigraphic units encountered. Figure 4 shows the gamma-neutron cross-plot distributions for core materials described as gravel, sand, silt and clay. Identification of the four lithological groups can be improved to some extent by including other logs, but considerable overlap still exists. Leask (1985), Kassenaar and Dusseault (1987) and Kassenaar (1989) describe attempts to develop objective lithology and stratigraphy recognition algorithms from the calibrated logs. Leask used cross-plots, while Kassenaar and Dusseault used principal component analysis. Although successful in some of their objectives, these methods have not yet proved practical.

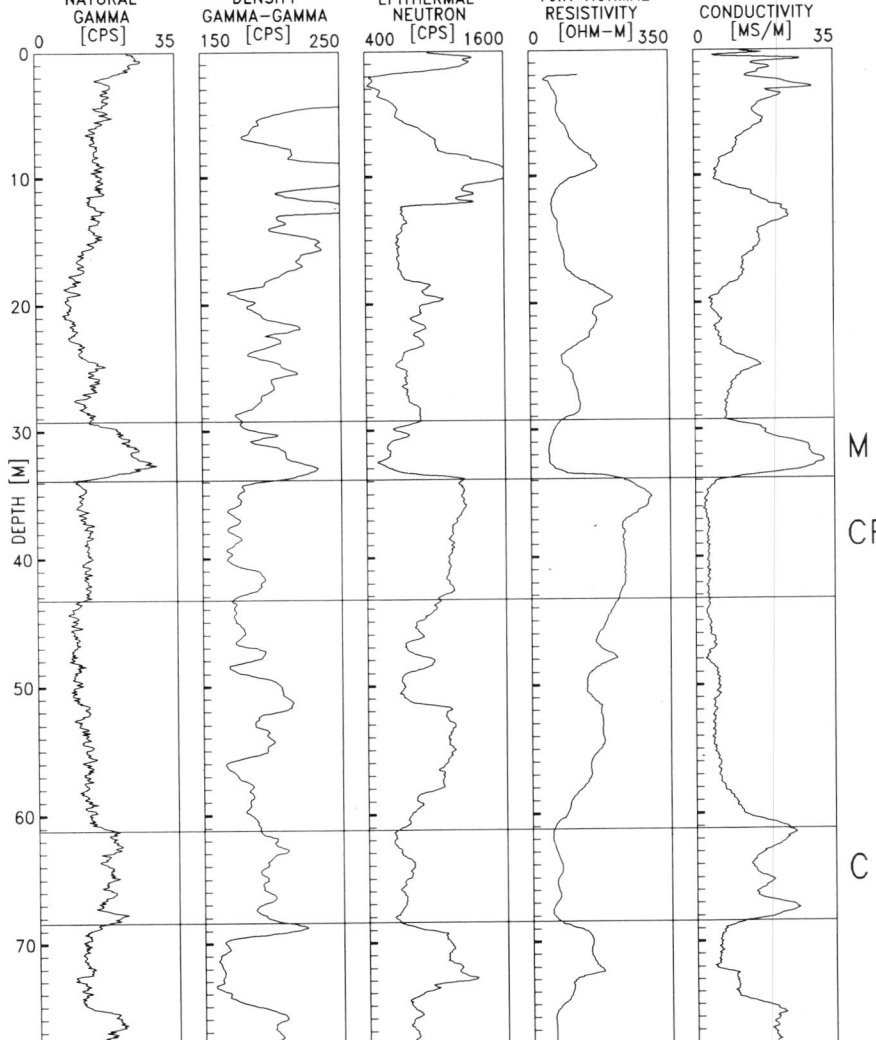

Figure 2 *Geophysical log suite at the Mannheim 1-87 borehole. The Maryhill (**M**), Catfish Creek (**CF**) and Canning (**C**) tills are prominent horizons recognizable on the logs.*

Geophysical Log Catalogue

As understanding of log responses and their relationships to sediment texture and composition grew, so did the volume of logs available for interpretation. It soon became apparent that all of these logs, as well as others provided from local consultants, needed to be brought together into some kind of database if they were to be maintained for posterity and used collectively to unravel the RMW's stratigraphy. In 1988, greatly assisted by Dirk Kassenaar's development of the log analysis software VIEWLOG, the first Well Log Catalogue for the RMW was compiled by Schneider and Greenhouse (1989), containing data from 32 boreholes. The third edition was recently compiled by Schneider (1993), in a format suitable for inclusion in a regional database of geological information. It contains geophysical and geological data for 291 boreholes drilled and logged geophysically between 1959 and 1992 across the entire RMW.

The first catalogue provided several benefits. For example, a direct result of this compilation of all the data in one place was the recognition of a new regional marker horizon, the Canning Till. It is a silty clay till encountered below the Catfish Creek Till in the southern part of the RMW. Another benefit was the concept of electrofacies.

Electrofacies

Many of the logs, for example the gamma-neutron pair in Figure 2, suggest that there are a number of roughly repeatable sequences, or "electrofacies", throughout the RMW. Electrofacies are a common descriptor in sedimentary basin analysis. Doveton (1986) describes electrofacies as "collective associations of log responses which appear to typify certain zones and differentiate them from others". Eyles *et al.* (1985) have used similar concepts in dealing with Quaternary deposits in the Lake Ontario Basin.

Figure 5 shows our electrofacies subdivision of the section at Mannheim 1-87 (see also Fig. 2), which serves as a type-section for the southern part of the RMW. There is still debate as to whether these subdivisions are regional features, and whether they are, in fact, the manifestation of cyclic depositional processes such as the repeated advance and retreat of ice.

If validated, the electrofacies do imply a regional order that had not been previously recognized. In Figure 5, the sequences are compared to the generalized hydrostratigraphy. Of the sequences, II and III are most frequently encountered and are of principal importance to the hydrogeology. They suggest that the main aquifer system in the RMW is divided into two distinct zones, each with a coarser and/or mixed-grainsize top and a finer grained bottom. The top of each zone is generally characterized by higher neutron response (lower porosity) and lower gamma (lower clay content) response than the base of the zone.

At the surface, the Maryhill Drift is frequently encountered, which is glaciofluvial in origin and quite variable in composition. Either this drift or local till units comprise the top of sequence I. The bottom of sequence I is the clay-rich Maryhill Till, the top of II is the Catfish Creek Till, a silty gravel unit which is commonly cemented in places. The bottom of II, the Catfish Creek Drift, has a low neutron response, but no high gamma response, suggesting that it is a fine silt rather than a clay. The top of III has been described by Ross (1986) as a sand till with similarities to the Catfish Creek Till. It has thus been designated as a pre-Catfish Till or PC1. The bottom of III is identified with the Canning Till. Unit IV is presumed to be a second pre-Catfish Creek Till, designated PC2.

The geophysical logs are an essential part of a regional stratigraphic correlation. They help to establish a reference section for the RMW against which new data can be judged. For example, occasionally we drill into materials that, based on their geophysical logs, clearly have no equivalent in the majority of local boreholes. These anomalous materials are probably infill within channels that have been eroded through the standard units.

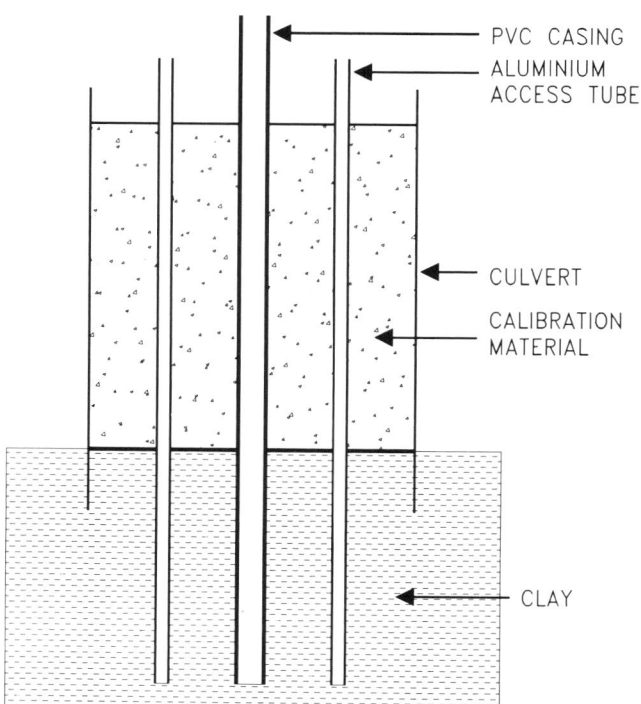

Figure 3 *Standard calibration pits for determination of physical properties. Pits were constructed from 90 cm diameter steel culverts and filled with various earth materials. The standard PVC casing and access tubes penetrate through into the Wentworth clay till below.*

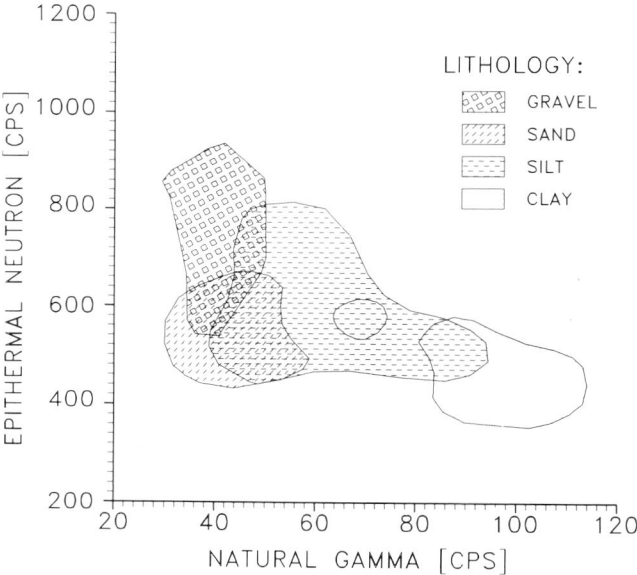

Figure 4 *Natural gamma versus neutron cross-plot showing lithology distributions, based on four boreholes in the RMW which were both geophysically logged and cored (adapted from Pehme, 1984).*

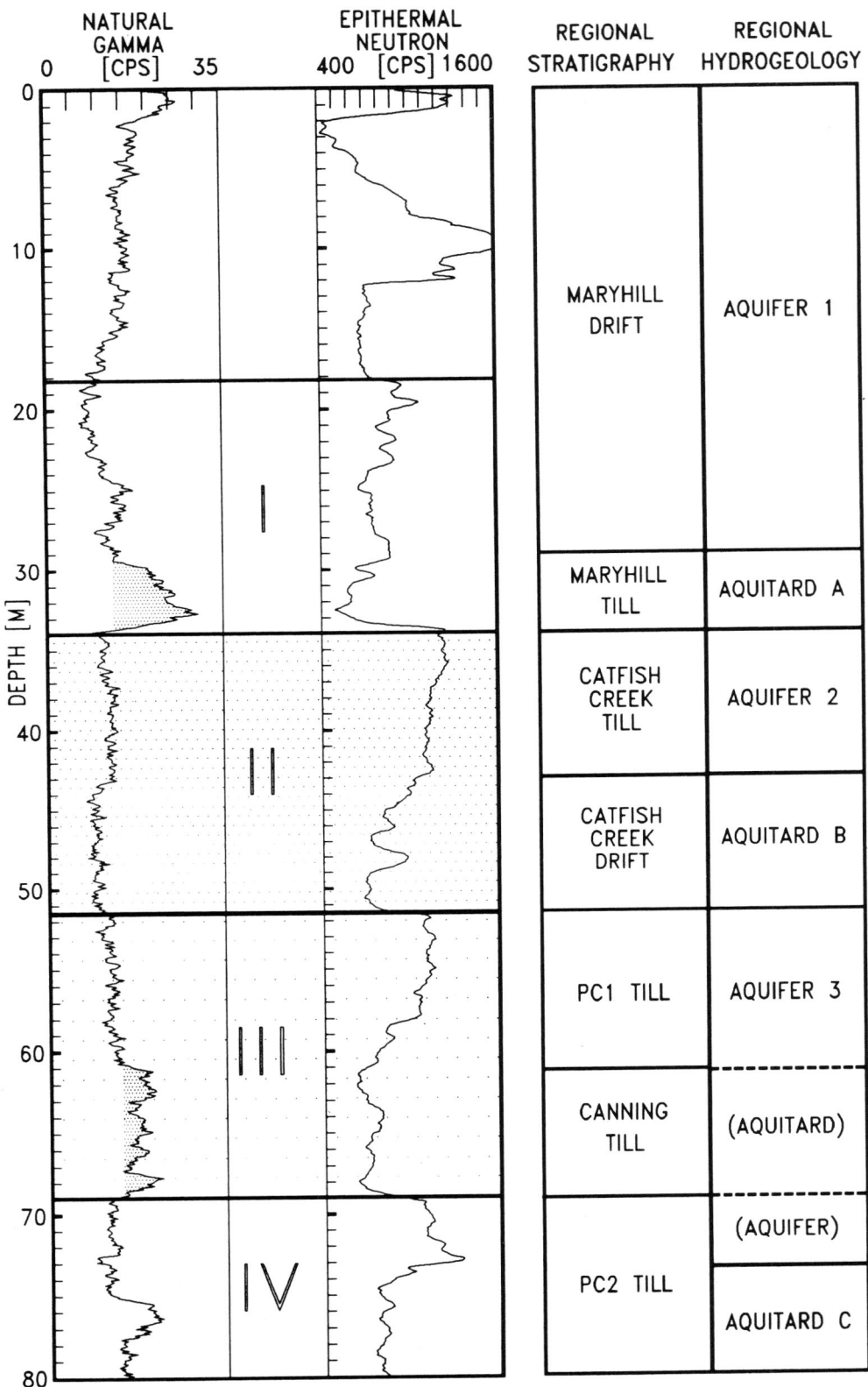

Figure 5 *Electrofacies classification at the Mannheim 1-87 borehole. Electrofacies are correlated to the aquifer classification system used by area hydrogeologists. Sequences I to III are characterized by downward decreasing neutron response and a fine grained (clay or silt) basal unit. See text for comments on the electrofacies/aquifer correlation.*

SEISMIC REFLECTION

The well logs provide detailed vertical resolution of the stratigraphy in the RMW. An inventory of the aquifer system also requires that the lateral variations from borehole to borehole be mapped. A modified seismic reflection method for these urban environments is being developed for this purpose.

Seismic reflection has proved successful in the imaging of Quaternary deposits (*e.g.*, Hunter *et al.*, 1984; Chapter 32). In an urbanized environment, there are a number of problems that must be overcome.

Clear access routes between boreholes must be secured in order to conduct seismic reflection surveys. Vibrational noise from cultural features such as factories and traffic is superimposed on the seismograms, and must be dealt with. Energy loss due to near-surface fill, coarse-grained sediments, and an often thick unsaturated zone must be addressed. Scattered arrivals from buried cultural features such as sewer pipes must be recognized. Induced electrical signals from power lines must be filtered out. Finally, reflection events must be correlated to known stratigraphy from borehole data.

Figure 6 shows a common offset reflection record obtained along a railroad right-of-way through the city of Waterloo. The poor quality of the image — due primarily to the low average frequency of the recorded signals — is typical of what is obtained with conventional techniques in this area (Lockhard, 1992). Chapter 32 summarizes recent advances in seismic profiling techniques made at the University of Toronto. Considerable progress has been made with enhancing the resolution of the technique and in using seismic profiling in heavily urbanized areas characterized by cultural noise.

A New Seismic Energy Source

Currently under development is an alternative seismic energy source, similar in certain respects to the familiar "vibroseis" source of the oil industry and the borehole seismic source described by Wong *et al.* (1983). The source is a vibrating transducer consisting of a coil moving in a permanent magnetic field. The transducer is coupled to the ground by an auger attached to the moving coil. The coil is fed a "maximum length sequence" signal, a random but repeatable sequence of 16,000 pulses delivered over 8 seconds. The impulse response of the earth is calculated by cross correlation of the recorded geophone response with the generated input signal to the transducer. The attractiveness of this source is its ability to stack signals and its relatively broad band frequency output. Initial tests of this source have been very encouraging.

VERTICAL SEISMIC PROFILING

Vertical seismic profiling (VSP) of selected boreholes in the RMW has been used to correlate seismic reflection arrivals to geophysical electrofacies and core stratigraphy (Meleski, 1988; Greenhouse *et al.*, 1990). VSP data are collected by placing an array of hydrophones in a borehole and detonating a shotgun blast at a point on the surface, typically 20 m from the borehole. The resulting seismogram can be used to determine seismic velocity as a function of depth and to identify the origin of seismic reflections observed at the surface.

A good example at the Bauer-87 borehole is shown in Figure 7. Neutron and gamma logs, and the interpreted electrofacies I through IV, are shown for comparison. The first or direct arrivals at the downhole geophone form a gently curving event, moving downward from left to right, that defines the seismic P-wave velocity. Reflections are evident as events moving upward from left to right and originating where the direct arrival intersects the reflector depth. Reflected arrivals that reach the near-surface geophone at 81 milliseconds (ms) and 63 ms are particularly strong and represent reflections from the bedrock and the top of sequence IV, respectively. Weaker reflected events reach the surface at approximately 38 ms and perhaps 34 ms. A weak reflected event may perhaps also originate on the II/III boundary. Downward moving events arriving later are probably multiples, and appear to originate from the intersection of upward moving reflections with stratigraphic boundaries. The low angle upward moving event originating at the bedrock is thought to be a tube wave, a slowly propagating mode which is confined to the borehole casing.

Not all the VSPs we record are of such good quality. The problem of separating up-going and down-going events on VSP records was investigated by Bloomer (1989) using singular valued decomposition, who found that the technique was poorly suited to this application. This separation is currently performed by frequency-wave number (F-K) filtering, since the up-going and down-going events occupy different spaces in the F-K domain. Down-going events are identified and suppressed in F-K space and then data are inverse trans-

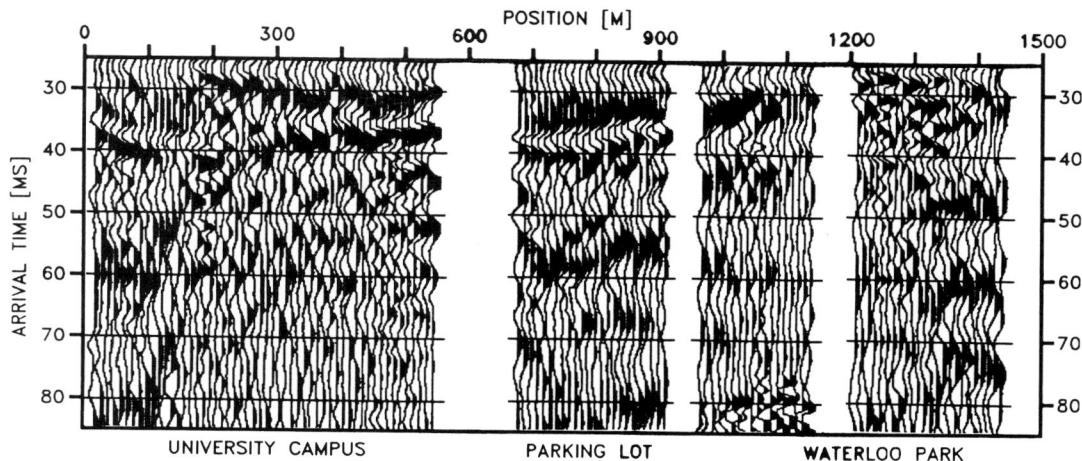

Figure 6 *This 30-metre common offset reflection record, obtained along railroad right-of-way near the University of Waterloo campus, demonstrates the typical poor resolution obtained using conventional field recording techniques. Cultural noise, nearsurface fill material and a relatively deep water table all contribute to the poor results obtained. Gaps are present where streets were crossed during the survey.*

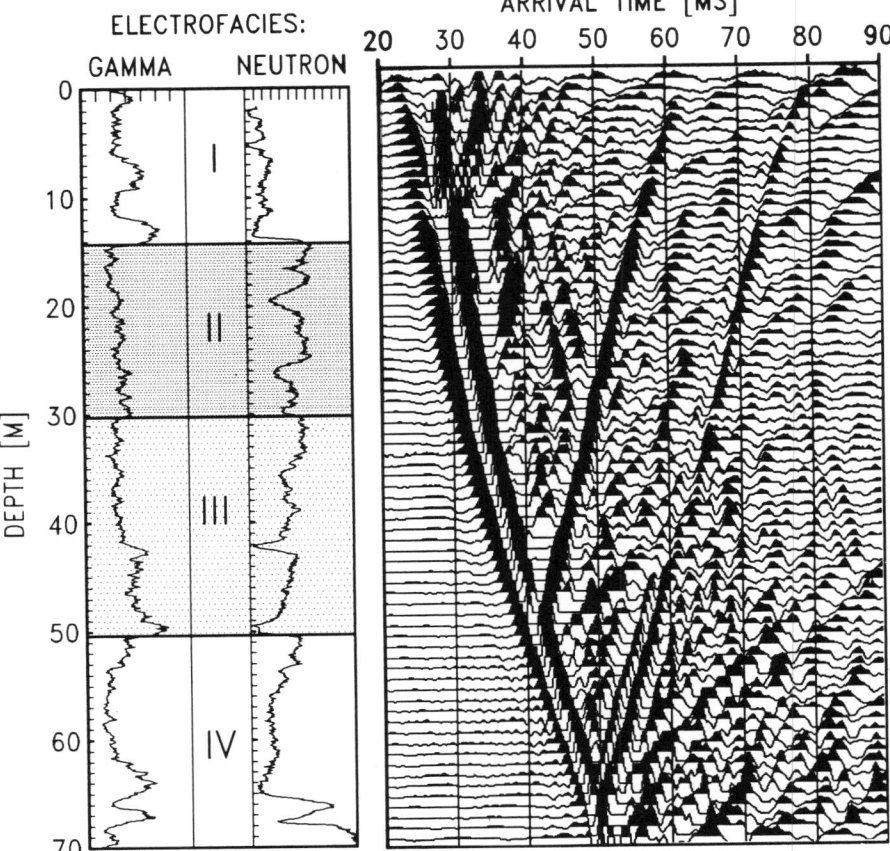

Figure 7 *A vertical seismic profile record at the Bauer-87 borehole. Seismic energy is seen to reflect upward from the bedrock (base of sequence IV), as well as from the III–IV, II–III (less obvious) and I–II boundaries. Note the multiples generated from the underside of some of these boundaries, and what appear to be upward travelling tube waves generated at the bedrock interface.*

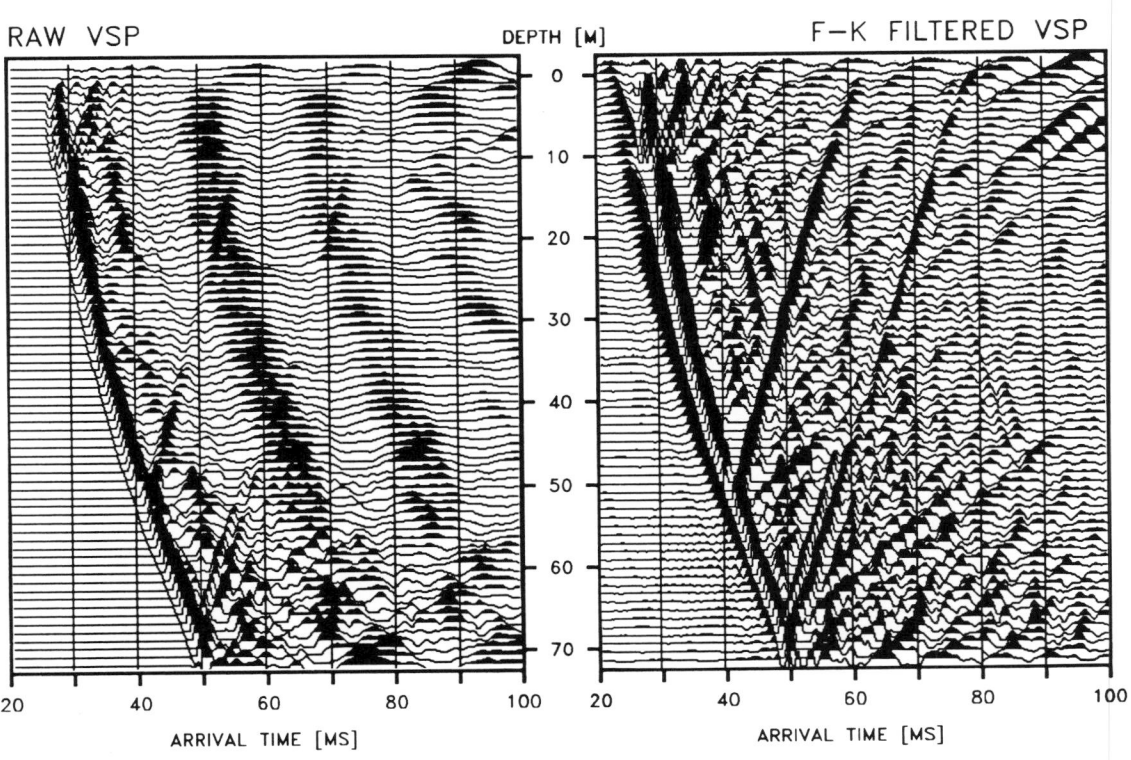

Figure 8 *F-K filtering used for VSP processing. This technique not only enhances high frequencies in the record, but also enhances the upgoing events while supressing the downgoing events. This example was recorded near the Bauer-87 borehole.*

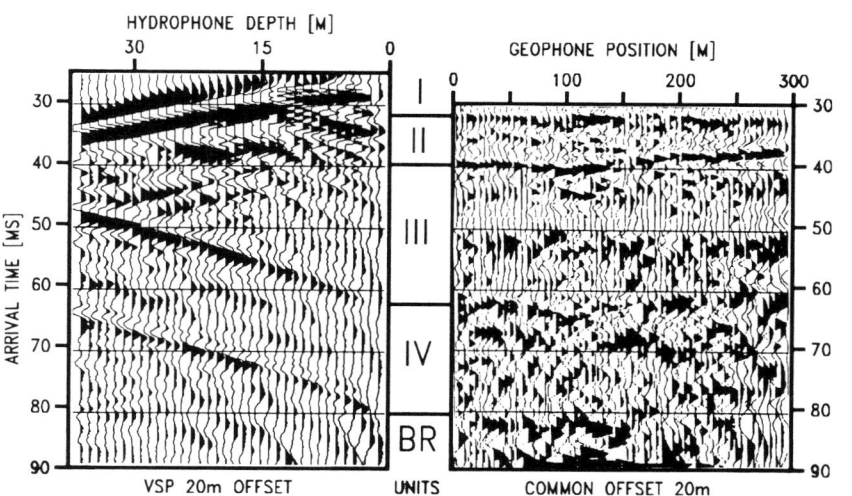

Figure 9 *VSP correlation to surface common offset records. The VSP record of Figure 7 is compared to a CO record. Both were shot with a 20 metre shot to geophone/borehole offset. The right-most trace on the (rotated) VSP record has the same shot/detector geometry as the left-most trace on the surface recording. Seismic reflection events are also correlated to the boundaries between electrofacies sequences seen on the geophysical log data.*

formed back to time-distance space. An example is shown in Figure 8, in which the raw data on the left have been processed to enhance the upward moving events at higher frequencies. The result is superior to band-pass filtered records and reveals more upward moving events (probably tube waves) on the right of the record. Multiples of the first arrivals are a problem on records obtained at particular boreholes. We are currently examining predictive deconvolution as a technique for removing multiples on records that are dominated by these multiples.

Correlation of Vertical Seismic Profiling to Common Offset

Clearly the wealth of subsurface information in these borehole records is extremely valuable when it comes to interpreting the surface seismic data. In particular, note that the near surface geophone in a VSP survey will mimic the record of a common offset surface survey having the same shot-to-geophone offset (20 m in the case of Fig. 7). The uppermost trace of the VSP record can, therefore, be compared directly with the surface data recorded adjacent to the borehole. It is particularly instructive to observe the deterioration of the reflected signals as they approach the surface in Figures 7 and 8, and therefore how useful the VSP records can be in identifying them on surface records.

In Figure 9, the VSP record of Figure 7 has been turned on its side so that it can be compared directly with a 20 m common offset record shot whose first geophone lies immediately beside the borehole. The common offset record quality is not good, but we can begin to define the origins of the various events on the basis of their continuation with depth on the VSP record. Four events are identifed and can be correlated to the boundaries between electrofacies units. The arrival times for the events I–II and II–III on the VSP and CO records do not correspond exactly, presumably because seasonal variations in the water table change the near-surface P-wave velocity. At depth, the arrival times for III–IV and IV–BR correspond quite well. There is also evidence of an event separating the top and bottom of III; it is particularly noticeable on the common offset record.

CONCLUSIONS

By integrating the geophysical methods of well logging, vertical seismic profiling, and seismic reflection, overburden stratigraphy is being resolved in detail that could not have otherwise been achieved.

Well logs have provided much important information about overburden stratigraphy in the RMW. They are indicators of lithology, stratigraphy and physical properties. The concept of Quaternary electrofacies can serve as a correlational tool across the RMW. When fully developed, we believe that these geophysical techniques can effectively serve communities in their efforts to secure and manage their overburden groundwater supplies.

REFERENCES

Bloomer, S.F., 1989, Application of singular valued decomposition filtering to vertical seismic profiles from the Kitchener-Waterloo Region: Internal Report of the Geophysics Laboratory, University of Waterloo, Waterloo, ON, 23 p.

Doveton, J.H., 1986, Log Analysis of Subsurface Geology: Concepts and Computer Methods: John Wiley and Sons, New York, 273 p.

Eyles, N., Clark, B.M., Kaye, B.G., Howard, K.W.F. and Eyles, C.H., 1985, The application of basin analysis techniques to glaciated terrains: an example from the Lake Ontario Basin, Canada: Geoscience Canada, v. 12, p. 22-32.

Greenhouse, J.P. and Pehme, P.E., 1986, An approach to determining stratigraphic and physical properties of overburden using borehole geophysics, *in* Killeen, P.G., ed., Borehole Geophysics for Mining and Geotechnical Applications: Geological Survey of Canada, Paper 85-27, p. 7-11.

Greenhouse, J.P., Nobes, D.C. and Schneider, G.W., 1990, Groundwater beneath the city: a geophysical study, *in* Proceedings of the Fourth National Outdoor Action Conference on Aquifer Restoration, Ground Water Monitoring and Geophysical Methods: National Water Well Association, Las Vegas, NV, May 17-19, p. 1179-1191.

Greenhouse, J.P., Nobes, D.C., Schneider, G.W. and Lockhard, M.A., 1991, Modification of the shallow reflection seismic method for geophysical studies in southern Ontario, OGRF Grant #386: Ontario Geological Survey, Miscellaneous Paper 156, p. 121-129.

Hunter, J.A., Pullan, S., Burns, R., Gagne, R. and Good, R., 1984, Shallow reflection seismics: some simple techniques: Geophysics, v. 49, p. 1381-1385.

Hunter, J.A., Pullan, S.E., Burns, R.A., Gagne, R.M. and Good, R.L., 1989, Applications of a shallow seismic reflection method to groundwater and engineering studies, *in* Garland, G.D., ed., Exploration '87: Third Decennial International Conference on Geophysical and Geochemical Exploration for Minerals and Groundwater, Proceedings: Ontario Geological Survey, Special Volume 3, p. 704-715.

Karrow, P.F., Greenhouse, J.P. and Dusseault, M.B., 1990a, Subsurface Quaternary stratigraphy using borehole geophysics: Open File Report 5734, Ontario Geological Survey, 249 p.

Karrow, P.F., Warner, B.G., Ellis, C.J. and MacDonald, J.D., 1990b, What's beneath our feet in the Waterloo Region: Quaternary Sciences Institute, University of Waterloo, Waterloo, ON, Publication 1, 8 p.

Kassenaar, J.D.C., 1989, Automated classification of geophysical well logs: unpublished M.Sc. thesis, University of Waterloo, Waterloo, ON, 112 p.

Kassenaar, J.D.C. and Dusseault, M.B., 1987, Expert system analysis of log data: Second International Symposium on Borehole Geophysics for Minerals, Geotechnical and Groundwater Applications, Denver, CO, October 6-8, Proceedings, p. 199-232.

Keys, W.S., 1989, Borehole Geophysics Applied to Groundwater Investigations: National Water Well Association, Dublin, OH, 313 p.

Leask, N.E., 1985, An expert system to interpret lithology and stratigraphy from geophysical logs in the Waterloo Region: unpublished B.Sc. thesis, University of Waterloo, Waterloo, ON, 37 p.

Lockhard, M.A., 1992, Modifications to the shallow seismic reflection technique for application in southern Ontario: unpublished M.Sc. thesis, University of Waterloo, Waterloo, ON.

Meleski, M.J., 1988, Vertical seismic profiling and its application to boreholes in the Kitchener-Waterloo Region: unpublished B.Sc. thesis, University of Waterloo, Waterloo, ON, 73 p.

Pehme, P.E., 1984, Identification of Quaternary deposits with borehole geophysics in the Waterloo Region: unpublished M.Sc. thesis, University of Waterloo, Waterloo, ON, 124 p.

Ross, L.C., 1986, A Quaternary stratigraphic cross-section through the Waterloo Region: unpublished M.Sc. thesis, University of Waterloo, Waterloo, ON, 56 p.

Schneider, G.W., 1993, Geophysical Log Catalogue of the Waterloo Region, volumes 1-5: Quaternary Sciences Institute, University of Waterloo, Waterloo, ON, Publication 9, 699 p.

Schneider, G.W. and Greenhouse, J.P., 1989, Geophysical well logs for Waterloo and the surrounding areas, 1976-1988: Internal Report of the Geophysics Laboratory, University of Waterloo, Waterloo, ON, 196 p.

Wong, J., Hurley, P. and West, G.F., 1983, Cross-hole seismology and seismic imaging in crystalline rocks: Geophysical Research Letters, v. 10, p. 686-689.

34. Site Assessment Using Soil Gas Surveys

Iqbal Noor

Global Environmental Management Solutions, 4635 Crosswinds Drive, Mississauga, Ontario L5V 1G6

SUMMARY
Soil gas surveys are used to identify the presence of hydrocarbons in urban soils, and can also differentiate natural from anthropogenic hydrocarbon sources. The composition and concentrations of light hydrocarbons, such as methane, ethane, propane and butane, together with volatile organic compounds, particularly aromatic hydrocarbons such as benzene, toluene, ethylbenzene and xylene, is a particularly useful discriminant of natural and anthropogenic hydrocarbons. Concentrations of hydrocarbon vapours are used to identify anthropogenic source areas such as buried tanks, to map contaminant plumes, and to identify shallow ground-water aquifers impacted by natural hydrocarbon-bearing strata.

INTRODUCTION
A soil gas survey is a rapid and cost-effective method of collecting preliminary information regarding the presence and concentration of light petroleum hydrocarbons and other volatile organic compounds (VOCs) in soil. The VOCs in soil can occur through natural geological processes or may result from chemical spills and leaking underground gasoline tanks. A soil gas survey is conducted to detect concentrations of volatile hydrocarbon vapours and track their movement in the shallow subsurface. A high concentration of soil hydrocarbons generally indicates soil and ground-water contamination. Concentrations of soil hydrocarbon vapours are used to identify source areas, such as buried tanks, and map subsurface contaminant plumes. For an excellent introduction to hydrocarbons in ground water see Eslinger *et al.* (1994).

In many areas, faults and fractures intersect hydrocarbon-bearing geological formations. These fractures may provide preferred pathways for natural gas seepage (Noor *et al.*, 1992a). This natural gas accumulates in shallow ground-water aquifers, causing severe degradation of the quality of drinking water. In these areas, soil gas surveys can be used as a tool for delineating pathways of natural hydrocarbon loading (Noor *et al.*, 1992b).

Volatile hydrocarbons from leaking underground storage tanks (USTs) or other spill-related sources generally form a small pool floating on the ground-water table (Chapters 7, 35). As the ground water moves, some of the volatile hydrocarbon vapours are released into the unsaturated zone. Soil gas surveys are used to detect the presence and concentration of these VOCs. Once the presence of a contaminating substance, such as gasoline or diesel fuel, is established in soil or shallow ground water, the extent of the presence may often be delineated using the magnitude and composition of light hydrocarbons (C_1-C_4) in soil gas (Pirkle *et al.*, 1991).

Soil gas survey is a rapid method of mapping the size, shape and areal extent of contaminant plumes. It is cost effective and saves exorbitant costs of drilling monitoring wells. It also causes very little operational interference. In regional studies, soil gas surveys can be used to delineate migration pathways, and hence shallow aquifers susceptible to natural hydrocarbon loading. Thus, soil gas surveys are useful for the long-term management of ground-water resources.

There are two types of soil gas surveys, passive and active systems (Richers *et al.*, 1986). In the passive-system gas survey, adsorbent material, such as carbon cubes or rods, are implanted in the ground to adsorb hydrocarbon vapours. After several days, the adsorbent material is removed for analysis. In active soil gas surveys, soil gas probes are inserted directly into the contaminated soil, and a gas sample is withdrawn often for instant analysis (Noor and Novakowski, 1992). This paper deals mainly with the active soil gas method.

DIFFERENTIATING NATURAL FROM ANTHROPOGENIC HYDROCARBON CONTAMINATION
Soil gas surveys of light petroleum hydrocarbons can be used to differentiate the presence of natural hydrocarbons, generally methane, from anthropogenic hydrocarbon contamination. These hydrocarbons are members of the alkanes group where the general formula is C_nH_{2n+2}.

Uncontaminated soils may contain some biogenic methane (CH_4; produced by microbial alteration of organic matter) or thermogenic methane (produced as a result of thermal alteration of organic matter over geological time). Biogenic methane is also produced in old landfills (Chapters 2, 20, 25). The presence and concentration of methane (and any other hydrocarbon associated with it) in soil is a function of the soil type, the presence of source beds, and the subsurface geology, including faults and fractures which may serve as migration pathways. In all these cases, however, higher molecular weight hydrocarbons beyond propane (C_3; C_3H_8) are very rarely present in soil gas.

Thermogenic methane in near-surface environments migrates from deep geologic sources within a sedimentary basin. During the migration process, hydrocarbons above ethane (C_2), such as propane (C_3) and butane (C_4), are generally lost. Methane is always found in considerable excess, ranging from 60% to 99+% of the total C_1-C_4; the remaining hydrocarbons decrease rapidly with increasing number of carbons, *i.e.*, $C_1 > C_2 > C_3 > C_4$ (Pirkle *et al.*, 1991). The natural occurrence of butane rarely exceeds 2% of total and often less than 0.5% (Pirkle *et al.*, 1991).

Gasoline, on the other hand, is a complex mixture of

hydrocarbons with composition generally in the range C_5-C_{10} (Chapter 7). However, minor amounts of butanes are present as a result of distillation separation to gasoline from crude oil. The concentration of butane, although minor in gasoline, is large relative to its concentration in ambient soil gas. When contaminated with gasoline vapours, soil gas often contains in excess of 50% butanes in the total C_1-C_4 hydrocarbons (Pirkle et al., 1991). Thus, anomalous butane concentration in soil relative to other light hydrocarbons is an indication of anthropogenic hydrocarbon or gasoline contamination.

In cases where shallow ground water is also contaminated, additional evidence of gasoline contamination is commonly provided by the presence of aromatic hydrocarbons, benzene, toluene, ethylbenzene and xylene (referred to collectively as BTEX) in ground-water samples. Gasolines of different brands have different ratios of BTEX constituents. Though also occurring naturally, BTEX is present in high concentration in gasoline, and its presence in ground water is typical of an advancing contaminant plume moving away from a gas station.

Ground waters impacted by natural gas seepage have BTEX ratios different from those of gasoline-contaminated ground water. Multivariate plots of water-soluble BTEX concentrations in ground water are sometimes used to differentiate between natural and anthropogenic hydrocarbon sources (National Water Well Association, 1987). Lesage and Lapcevic (1990) used these plots to differentiate between anthropogenic and natural hydrocarbons in a deep injection well near Sarnia, southwestern Ontario, where natural hydrocarbons seep into ground water from petroleum-bearing geological formations. Anthropogenic hydrocarbons come from waste industrial fluids disposed of in the well. Their plots show that BTEX (in ground water) from both natural and anthropogenic sources can be clearly differentiated.

SAMPLING METHODS

Sampling Plan

The sampling plan is an important part of the study design. Sampling for a soil gas survey is preceded by a careful examination of the physical aspects of a site. These aspects include depth to ground water, soil texture and heterogeneity, soil moisture conditions, and local capping by asphalt. These conditions are taken into account when developing the sampling plan and in determining the choice of sampling equipment for a soil gas survey (Tillman et al., 1989). Figure 1 is an example of a sampling plan down-gradient from a gasoline station. The gasoline station is a potential source of hydrocarbon contamination. Sampling locations are equally spaced on the grid. The direction of ground-water flow is the axis of the grid. The points are placed at regular intervals along lines which intersect the plume axis perpendicularly.

Sampling Techniques

Gas samples are obtained in the field using a soil gas probe or directly from auger holes, monitoring well and piezometers. In the probe method, gas samples are obtained from 1 to 2 m depth with a commercially available stainless steel probe (Fig. 2). Soil gas enters the probe through a stainless steel screen located about 5 cm above the tip of the probe. Inside the stainless steel screen is a gas-sampling chamber, which is connected to a port at the top of the column through a small-diameter teflon tube. A gas sample is withdrawn from the sample chamber and run through a gas chromatograph for sample analysis.

The auger hole method is used in field situations where the ground is paved by thick concrete or where soil consists of impermeable clay. This method is also ideal for large sites. Sampling stations are equally spaced at 25 to 30 m, and the drilling is done by a small mobile rig. This method involves the drilling of up to 6-m deep and 10-cm outer diameter holes in which the probe is lowered 3 to 4 m immediately after drilling. Each hole is sealed with a cover through which the gas port was fitted. Covering the hole reduces the dilution effect that might otherwise result from free air circulation in the open hole. A portable gas chromatograph, set in survey mode, is attached to the sampling port, and the total concentration of hydrocarbon gas is directly measured. At sampling stations with total hydrocarbon concentrations of more than a 100 ppm, gas samples are taken in air-tight syringes for compositional analysis of hydrocarbon compounds (see "Analytical Procedures").

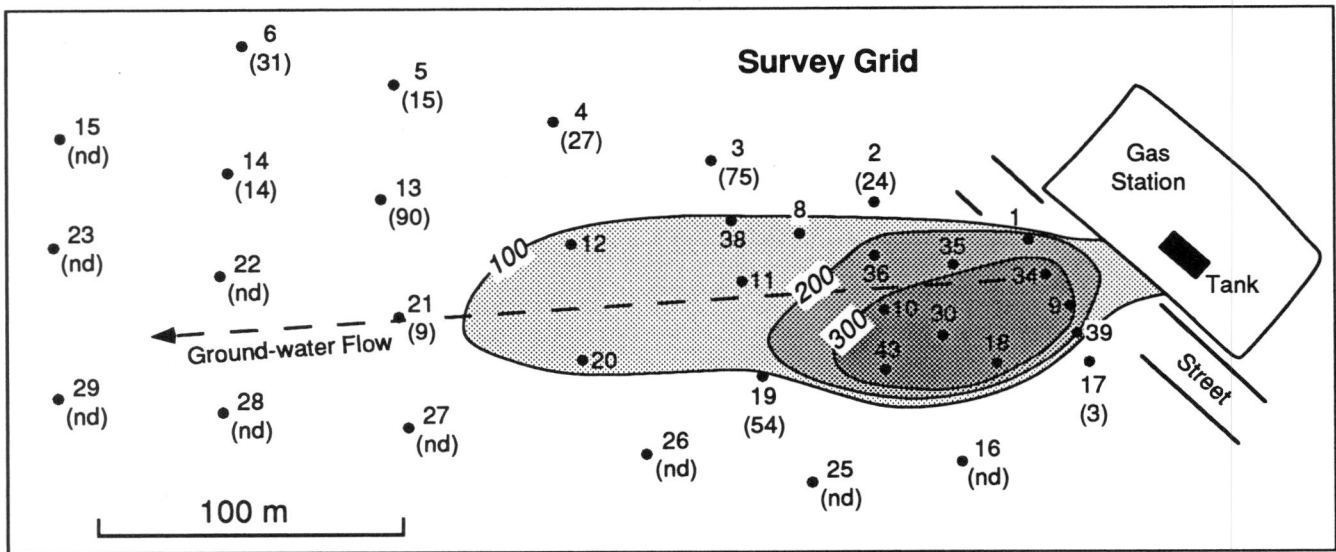

Figure 1 Soil gas survey grid showing sampling plan (numbered points) down-gradient from a leaking underground storage tank containing gasoline (after KVA Analytical Systems, 1990). See also Figure 4. Numbers in parentheses and contours are concentrations of volatile organic compounds in mg·kg⁻¹ and identify a contaminant plume. Abbreviation: **nd**, not determined.

In some sites, monitoring wells are already installed for ground-water studies. In these sites, gas samples can be obtained directly from wells using the same procedure as used in the auger hole method. The latter method is most effective for soil gas surveys in low-permeability tills, where the soil gas probe is used with little or no success. However, the probe is useful in silty and sandy tills. It is also useful in measuring methane concentrations in the vicinity of sanitary landfills, which generally contain an abundance of methane. The failure of the gas probe in tight clay tills is primarily due to the lack of a small quantity of oxygen, required by the flame ionization detector (FID) of the oxygen vapour analyzer (OVA) to remain functional during gas analysis.

ANALYTICAL PROCEDURES

An airtight syringe is used to withdraw a small volume of gas from the glass sample container which is injected into a portable gas chromatograph. One of the units, which is commonly employed in these surveys, is the Century™ OVA 128 Organic Vapour Analyzer, developed by Foxboro of Massachusetts. The OVA 128 contains a Flame Ionization Detector (FID) for measuring volatile organic vapours, and can be operated in two modes, a survey mode, and a gas chromatography (GC) mode.

In the survey mode, the OVA gives only total hydrocarbons within a gas sample. The instrument is calibrated with methane or some other standard of certain ppm concentration (generally with a 95-ppm methane standard supplied by the manufacturer). Thus, total organic vapour concentration in a sample is expressed in ppm methane equivalent units and read on a readout assembly.

In the GC mode, a fixed volume of sample is injected into the analytical column, which is filled with packing material. When the injected sample passes through the column, its components demonstrate varying affinity with the stationary phase. This varying affinity with the column material causes gas components to move through it at different rates and, thus, elute at different times. After elution, each component exits the column, and enters the ionization detector where it is ionized and proportional output voltage is recorded on an integrator and shown as a chromatogram or peaks on an output device. The chart of the chromatogram is used to measure the retention time for each component.

Soil gas components are identified, based on comparison of their retention times with the retention time of known reference standards run under the same conditions. For compositional analysis of hydrocarbon gas, the gas chromatograph is calibrated for methane, ethane, propane and butane with a commercial gas mixture. A portable computer and any of the commercially available software (*e.g.*, Peak Simple II™ software developed by SRI Instruments of Torrance, California) can be used as an integrator and a printer as an output device.

DATA ANALYSIS AND PRESENTATION

Soil gas samples are analyzed on site to obtain: (1) the composition of soil gas and concentration of its individual components (by analyzing a gas sample in the GC mode of the OVA 128); and (2) the concentration of the total VOCs obtained in the survey mode of the organic vapour analyzer.

Pictograms

Pictograms are generally used to show the soil gas concentration anomalies along a linear transect. Figure 3 shows an example of a typical pictogram, based on a soil gas survey conducted over a natural gas storage reservoir in southwestern Ontario. Soil gas readings are taken at predetermined fixed intervals. The location of the data points are plotted on the X-axis, and soil gas concentrations are plotted along the Y-axis. Pictograms are useful in delineating hydrocarbon impacted areas. For instance, in Figure 3B, the area between stations 100 to 140 is impacted by natural gas seepage along fractures associated with a large subsurface fault in bedrock.

Depth profiles

Depth profiles are constructed to show the vertical distribution of VOCs in the unsaturated or vadose zone. Horizontal fractures, bedding planes, and sand and silt layers generally provide preferred pathways of gas migration. Vertical profiles are useful to delineate these pathways in a single borehole as shown in Figure 4. Depth-profiling is done during the construction of a borehole, where the auger is withdrawn at fixed intervals of depth and the gas probe is lowered into the hole to take the gas reading. Gas concentrations are plotted on a graph, with sampling depths and hydrocarbon concentrations as axes.

Isopleth maps

An isopleth map is used to show VOC concentrations across a survey site. It gives a plan view of a contaminant plume. The isopleth map is constructed based on the sampling grid as

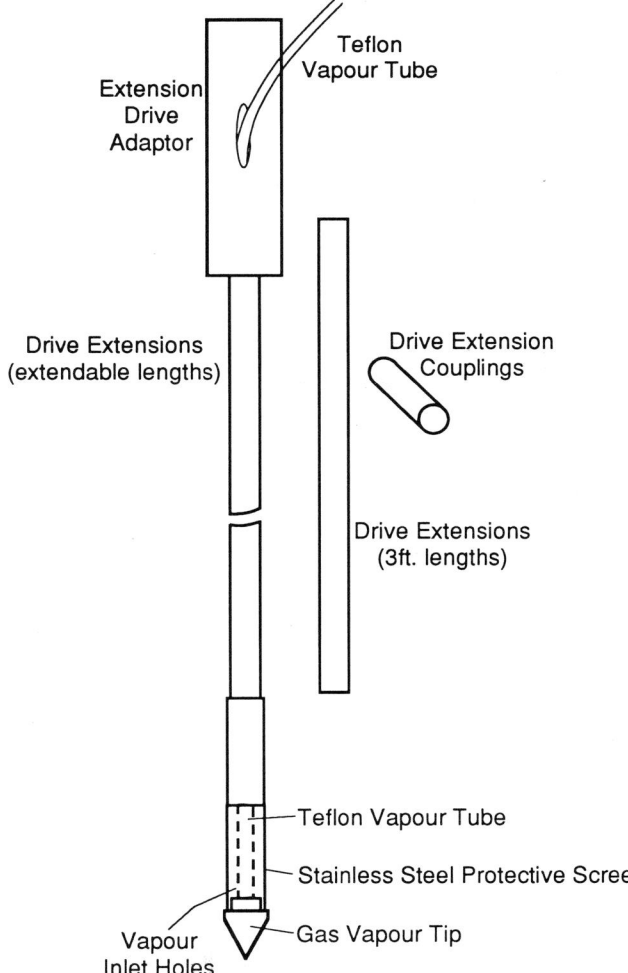

Figure 2   *Schematic diagram of a soil gas probe assembly.*

Teflon
Vapour Tube

Extension
Drive
Adaptor

Drive Extensions
(extendable lengths)

Drive Extension
Couplings

Drive Extensions
(3ft. lengths)

Teflon Vapour Tube

Stainless Steel Protective Screen

Gas Vapour Tip

Vapour
Inlet Holes

shown in Figure 1, which illustrates sample locations, scale, ground-water flow direction and VOC values. It is important to point out that only the VOC values for the same depth at all locations are used in the construction of the map. For sites with a large number of data points, an isopleth map of soil gas

concentration is plotted using any commercially available software (e.g., SURFER™ developed by Golden Software Inc., Golden, Colorado). To see anomalies in data, mean and standard deviations are calculated for the whole population.

APPLICATION IN REGIONAL STUDIES

In many sedimentary basins, hydrocarbon-bearing subsurface formations are intersected by faults and large fractures (e.g., Sanford et al., 1985). These structures can serve as conduits along which natural gas may seep into the shallow subsurface. Fractures associated with such a fault zone provide detectable soil gas anomalies. These anomalies could be used as a tool by which to delineate natural gas seepage. Thus, soil gas survey, in combination with geophysi-

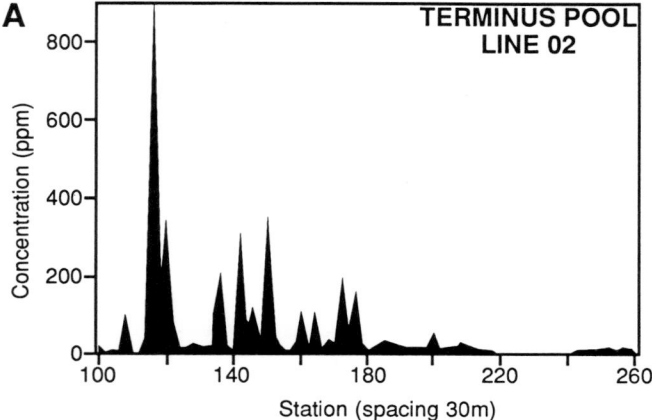

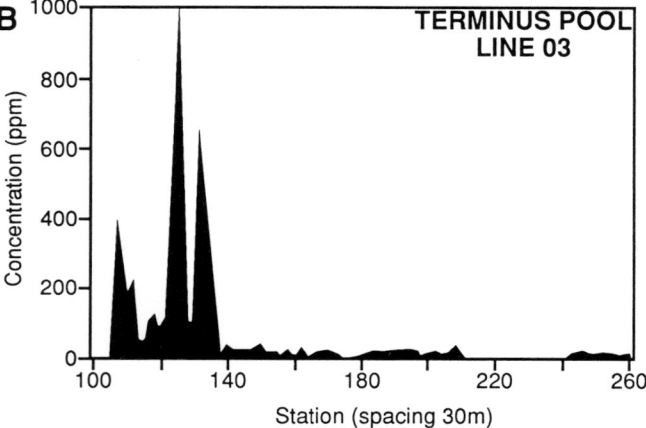

Figure 3 (**A, B**) *Methane concentration along two east-west transects in the Terminus field of southwestern Ontario, near Lambton. The highest concentrations, in the east, coincide with the presence of subsurface fractures in bedrock which control the upward escape of gas (see also Sanford* et al., *1985).*

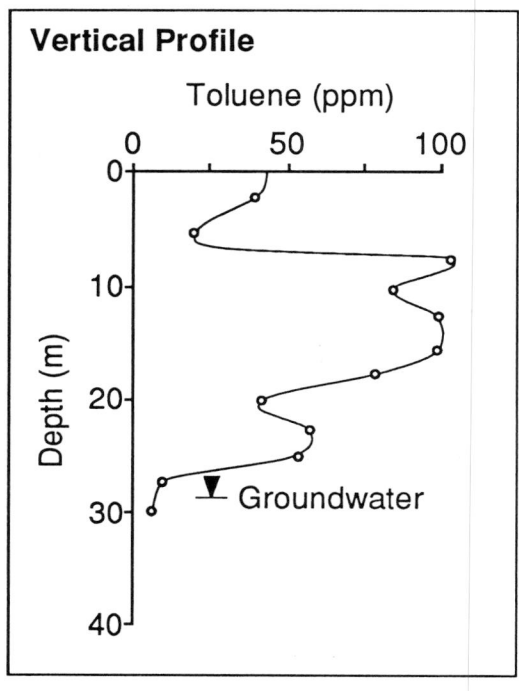

Figure 4 *A vertical profile showing the distribution of VOCs (toluene) in the vadose (unsaturated) zone down-gradient of a leaking underground storage tank containing gasoline (Fig. 1).*

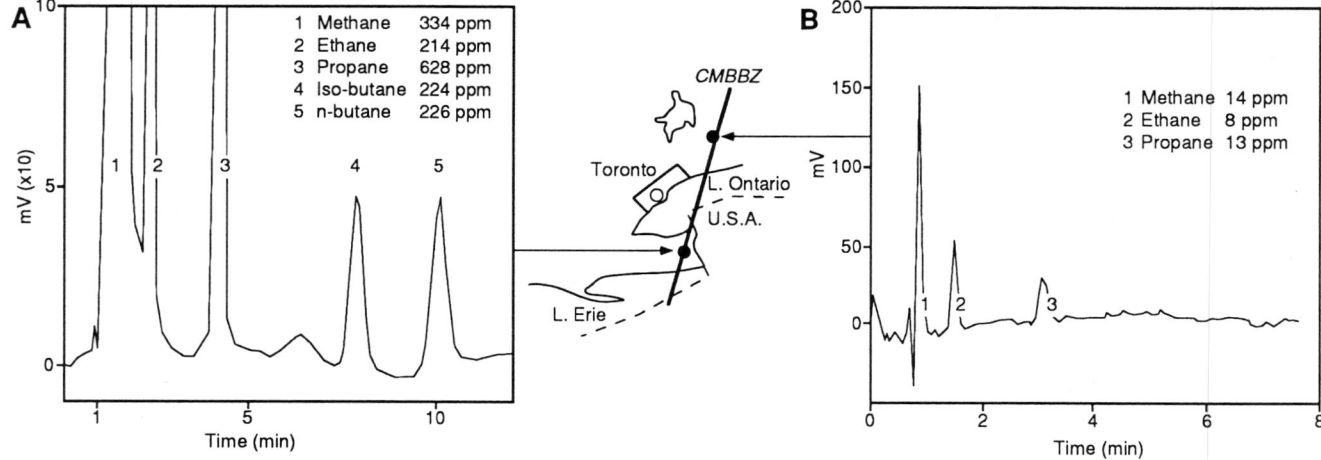

Figure 5 *Chromatograms of naturally occuring hydrocarbon gas at two locations along the Central Metasedimentary Belt boundary zone, an active fault zone in southern Ontario (see Chapters 2, 29). (**A**) Typical composition of natural hydrocarbon gas collected from five instrumented boreholes in the Niagara Falls area. (**B**) Soil gas composition showing C_1-C_3 hydrocarbons near the town of Valentia in the Lake Scucog area. Many faults show anomalies of high hydrocarbon concentrations and serve as regional pathways for natural gas migration from the subsurface (see Chapter 29).*

cal techniques, such as electromagnetic (EM) and seismic surveys, is used as an oil and gas exploration tool.

Since large fractures associated with fault zones can provide pathways for natural gas loading into shallow aquifers, soil gas surveys could also be used to delineate and map areas where shallow ground water might be impacted by natural gas from deeper sources. This application is particulary important in southwestern Ontario and Alberta.

Regional soil gas surveys are preceded by a suitable geophysical survey to obtain structural details of the bedrock fractures associated with a fault zone. This is useful in narrowing down the survey area. In western New York and southern Ontario, soil gas surveys were conducted across three large NE trending faults, namely the Clarendon-Linden fault zone, Niagara-Pickering linear zone and Dawn Fault by Noor *et al.* (1992a, b). These surveys show anomalies of high hydrocarbon concentrations.

Figure 3 shows the results of two soil gas survey transects across the Dawn Fault that crosses the Terminus Pool, southern Lambton County, Ontario. High concentrations (> 100 ppm) of soil hydrocarbons are associated with subsurface fractures in the fault zone. Comparable results, but with lower concentration than in Terminus Pool, were also noted in the soil gas surveys across the Central Metasedimentary Belt boundary zone (CMBBZ). The CMBBZ is a regional fault that occurs underneath the Lake Scugog area and extends across western Lake Ontario into Niagara Falls (Chapters 2, 29). Figure 5 shows chromatograms of gas samples from two locations (soil gas in the Lake Scugog and gas sample from an instrumented borehole in Niagara Falls) along this fault zone. Both samples contain C_1-C_3 hydrocarbons. Gas samples from Niagara Falls also contains iso- and n-butane. Soil gas where ethane and propane occur with methane is regarded as thermogenic, and comes from a deeper basinal source. The implication is that the fault zone is active, and may be serving as a regional pathway for natural gas migration to the ground surface.

ACKNOWLEDGEMENTS

I am indebted to Kent Novakowski, Chief, Groundwater Remediation Project, National Water Research Institute, for providing an opportunity to develop soil gas surveying techniques and applying these in various fields in southern Ontario. His generous support and encouragement to undertake these studies is much appreciated. I thank Mike Doughty for drafting the figures.

REFERENCES

Eslinger, E., Oko, U., Smith, J.A. and Holliday, G.H., 1994, Introduction to Environmental Hydrogeology: Society for Sedimentary Geology, Short Course Notes 32, 126 p.

KVA Analytical Systems, 1990, Soil gas equipment and brief field techniques: KVA Analytical Systems, Division of K-V Associates Inc., P.O. Box 574, Falmouth, MA 02541, 50 p.

Lesage, S. and Lapcevic, P.A., 1990, Differentiation of the origin of BTEX in ground water, using Multivariate plots: Ground Water Monitoring Review, Spring Issue, p. 102-105.

National Water Well Association, 1987, Introduction to groundwater geochemistry — analytical and sampling techniques: National Water Well Association Field Course, July 7-9, Boston, MA, p. 60-65.

Noor, I. and Novakowski, K.S., 1992, Soil gas surveys of natural hydrocarbons in southern Ontario: Environment Canada, National Water Research Institute, Contribution 92-121, 53 p.

Noor, I., Novakowski, K.S. and Egden J., 1992a, Structural control on natural gas seepage in southern Ontario: Ontario Petroleum Institute, London, ON, 31st Annual Conference, October 28-30, Niagara Falls, ON, Technical Paper 2, 24 p.

Noor, I., Novakowski, K.S. and Egden J., 1992b, Soil gas surveys as a tool for delineating pathways of natural hydrocarbon loading in southern Ontario: Implications for shallow groundwater, *in* Modern Trends in Hydrogeology: 1992 Conference of the Canadian National Chapter, International Association of Hydrogeologists, Hamilton, ON, p. 709-723.

Pirkle, R.J., Price, V. and Looney, B.B., 1991, Soil gas survey methods for gasoline and diesel fuel contamination [abstract]: 4th Chemical Congress of North America and 202nd American Chemical Society Meeting, Abstracts with Program, v. 31(2), p. 188-189.

Richers, D.M., Jones, V.T., Mathews, M.D., Maciolek, J., Pircle, R.J. and Sidle, W.C., 1986, The 1983 Landsat soil gas geochemical survey of Patrick Draw Area, Sweetwater County, Wyoming: American Association of Petroleum Geologists, Bulletin, v. 70, p. 869-887.

Sanford, B.V., Thompson, F.J. and McFall, G., 1985, Plate tectonics as a possible controlling mechanism in the development of hydrocarbon traps in southwestern Ontario: Bulletin of Canadian Petroleum Geology, v. 33, p. 52-71.

Tillman, N., Ranlet, K, and Meyer, T.J., 1989, Soil gas surveys: Part II: Procedures: Pollution Engineering, August, p. 79-83.

35. Ground-water and Soil Remediation Techniques

Paul Beck

Raven Beck Environmental Ltd., 7500 Woodbine Avenue, Suite 201, Markham, Ontario L3R 1A8

SUMMARY

Urban areas typically contain numerous sites underlain by soils or ground waters which are contaminated to levels that exceed clean-up guidelines and are hazardous to public health. Contamination most commonly results from the disposal, careless use and spillage of chemicals, or the historic importation of contaminated fill onto properties undergoing redevelopment. Contaminants of concern in soil and ground water include: inorganic chemicals such as heavy metals, radioactive metals, salt and inorganic pesticides, and a range of organic chemicals included within petroleum fuels, coal tar products, polychlorinated biphenyl oils, chlorinated solvents, and pesticides. Dealing with contaminated sites is a major problem affecting all urban areas and a wide range of different remedial technologies are available. This chapter reviews the more commonly used methods for ground-water and soil remediation, paying particular regard to efficiency and applicability of specific treatments to different site conditions.

INTRODUCTION

The built landscapes of urban areas are widely characterized by contaminated substrates and ground waters. A recurring theme throughout this volume has been the need to identify sources of chemical contamination and characterize the extent of contamination in the subsurface. In general, the impact of contaminants on the environment depends on the physical characteristics of the contaminated site, the toxicity and mobility of the contaminant, and the geometry of the contaminant source.

The need for clean-up of soil and ground water is frequently triggered either by complaints from adjacent property owners, who are affected by off-site migration of contaminants, or by the sale and/or redevelopment of contaminated properties. Site investigations prior to property transactions are now routine. A wide range of clean-up options is available, and great care is needed in choosing the most appropriate technology.

The recent development of remedial technologies for the treatment of soil and ground water has been greatly influenced by the environmental threat increasingly presented by organic chemicals known as non-aqueous phase liquids (NAPLs). NAPLs are organic liquids which have a low solubility in water, but their solubility is often significantly greater than drinking-water clean-up criteria (*e.g.*, Chapters 6, 7). NAPLs can be less dense than water (LNAPLs) including such liquids as fuels and light oils, or more dense than water (DNAPLs) including chlorinated solvents and coal tars. Soil and ground-water contamination involving NAPLs often re-

sults from the leakage of underground storage tanks or spillage at surface during handling and transfer operations.

The release of a large volume of LNAPL from an underground storage tank (Fig. 1) results in downward movement of LNAPL through the unsaturated zone toward the water table under the influence of gravity and capillary action, leaving residually saturated LNAPL in the unsaturated zone. The capillary fringe and the water table provide a barrier to further downward migration of the LNAPL. LNAPL accumulates as free product spreading horizontally on top of the water table. A plume of dissolved phase LNAPL develops in ground water and moves in the direction of ground-water flow.

Leakage of DNAPL from an underground storage tank is depicted in Figure 2. DNAPL moves under the influence of gravity through the unsaturated zone to the water table. Because it has a density greater than water, DNAPL migrates across and below the water table until it reaches a low-permeability base or horizon, where it will accumulate and migrate laterally. A plume of dissolved phase DNAPL migrates in the direction of ground-water flow. A detailed discussion of the subsurface controls affecting DNAPL migration is provided in Chapter 6.

The distribution of contaminants resulting from surface or near-surface spills can be extremely complex. As a result, the methods and strategies used to remediate contaminated soil and ground water vary according to the nature and distribution of contaminants. An important consideration is that contamination commonly involves mixtures of different contaminants including metals and organics in soil, *versus* metals alone; or where the same contaminant such as gasoline occurs in both the soil above the water table and in ground water. Consequently, at complex sites, it is unlikely that a

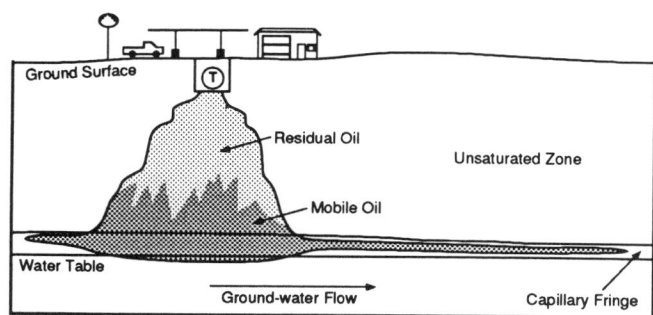

Figure 1 *Migration of spilled LNAPL in subsurface resulting from sudden large volume loss from underground storage tank (T). Also see Chapter 7.*

single treatment train will be suitable for clean-up; several different processes may be required.

Classification of Treatment Technologies

Treatment technologies can be broadly classified as either (1) isolation methods, (2) separation methods, or (3) destructive methods.

Isolation is a form of contaminant treatment/management that physically isolates and contains the contaminant from the environment; the purpose of isolation is to decrease further spread of the contaminant within the environment, or to reduce exposure to human or environmental receptors. Barrier walls are an example of an isolation method.

Separation methods involve physical and/or chemical processes which separate contaminants from the matrix, leaving a clean matrix and a contaminant concentrate which requires further treatment. Soil-washing and filtration processes are examples of separation methods.

Destructive methods apply only to organic compounds, and involve chemical or biological processes which cause a change in the chemical structure of the contaminant to transform the compound into non-toxic by-products. Examples of destructive technologies include bioremediation and incineration. However, both of these technologies can, in some cases, produce by-products that are more toxic than the original contaminant. For example, incomplete biodegradation of trichloroethylene can produce the highly toxic intermediate vinyl chloride. Incineration of polychlorinated biphenyls (PCBs) at too low a temperature can produce 2,3,7,8–tetrachlorodibenzo-p-dioxin (TCDD).

Treatment technologies can also be classified according to the dominant treatment process as thermal, physical, chemical or biological. Some technologies can be operated on site without removing soil or ground water. These methods are termed *in situ* methods. Other technologies require excavation of soil or pumping of ground water and treatment on surface or off site. These methods are termed *ex situ* methods.

A list of treatment technologies that will be discussed later in the paper is provided in Table 1.

Status of Treatment Technologies

The field of ground-water and soil treatment technologies is a dynamic one, constantly undergoing change as existing technologies are improved and new ones emerge. Because the

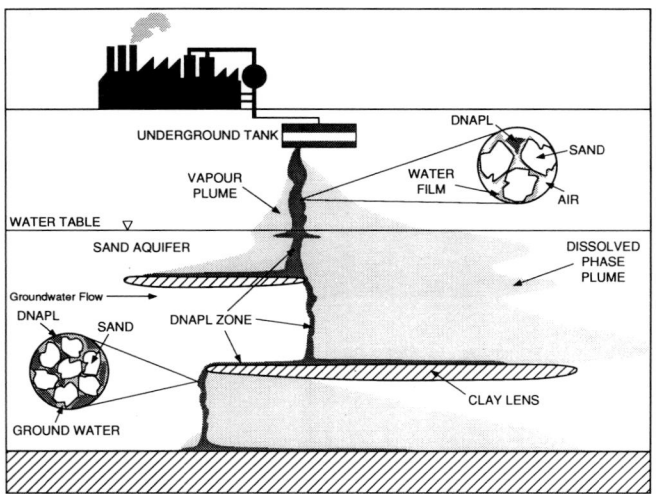

Figure 2 *Migration of leaking DNAPL in the subsurface. Also see Chapter 6.*

field is relatively new, and because of the high costs associated with treatment and monitoring, there are very few technologies that have undergone rigorous testing under a variety of conditions for a variety of contaminants. Consequently, there may be no history of performance for a vendor technology under a given set of conditions. It is therefore necessary, prior to the application of a technology for full-scale clean-up, that some pilot testing or field testing over a small area of the site, or the testing of a large (several kilograms to several tonnes) sample size taken from the site, be conducted first to establish appropriate operating parameters of the technology, and to provide a reasonable estimate of full-scale clean-up costs. Emerging technologies which show promise at the laboratory scale may not be capable of completing full-scale clean-up.

Several major government-funded programs in the United States and Canada have provided support for the demonstration of soil/sediment and ground-water treatment technologies. These include: (1) US EPA SITE Program (evaluation of innovative technologies for the clean-up of SUPERFUND sites in the United States); (2) Environment Canada and Provinces DESRT Program (evaluation of technologies to clean up contaminated sites in Canada); (3) Environment Canada and Provinces COSTTEP (evaluation of contaminated sediment treatment technologies in Canada); and (4) OMOEE ETP Program (promotion and development of environmental technologies in Ontario).

The United States Environmental Protection Agency (US EPA) SITE Program was established to support the demonstration of innovative technologies that provide alternatives to existing established technologies and landfilling. Innovative technologies are those technologies for which cost and performance data is limited, while established technologies are those which have a history of use, and consequently performance and costs are well documented. The US Environmental Protection Agency considers incineration, solidification/stabilization, and soil aeration as established technologies. The distribution for the application of 666 technologies to SUPERFUND sites through fiscal year 1993 is shown in Figure 3. Clean-up plans for 56% of the sites involved established technologies, the majority of which involved incineration and solidification/stabilization. The remaining 44% of site clean-up plans involved innovative technologies such as soil vapour extraction, bioremediation and thermal desorption. Soil vapour extraction accounts for more than two-thirds of soil treatment, with most of the remaining treatment being completed by *in situ* soil flushing, thermal desorption, *in situ* bioremediation, and soil washing.

The Development and Demonstration of Site Remediation Technology (DESRT) Program was established under the National Contaminated Sites Remediation Program through the federal-provincial-territorial Canadian Council of Ministers of the Environment. Thirty-four demonstration projects were funded. Thirty of the 34 involved soil treatment technologies, of which 50% involved bioremediation. A summary of DESRT Program demonstration projects is shown graphically in Figure 4.

The Contaminated Sediment Treatment Technology Program (COSTTEP) was initiated by Environment Canada under the Great Lakes Action Plan to assess remedial technologies at bench, pilot and full scale in support of efforts to clean up 17 Areas of Concern around the Great Lakes (see Chapter 15). Four of the eighteen projects were conducted at pilot scale, with the remainder at bench scale.

The Ontario Ministry of Environment and Energy's Environmental Technologies Program (ETP) was implemented to assist Ontario-based technology vendors in the development and commercialization of environmental technologies, with particular emphasis on technologies which had the potential to be marketed internationally (OMOEE, 1994).

TREATMENT TECHNOLOGIES

As discussed previously, some contaminant situations may require multiple or integrated treatment trains, but, for the purposes of this discussion, we will describe individual technologies under separate headings for soil and ground water.

Soil Remediation Technologies

Incineration

Incineration is a well-developed *ex situ* technology that is suitable for destruction of organic compounds. Complete combustion is controlled by the incineration temperature, the residence time of the contaminant in the burner, the availability of oxygen, and the physical and chemical characteristics of the waste streams. lincinerator designs include rotary kilns, fluidized-bed and circulating-bed combustors, plasma arc, and infra-red. They can be stationary facilities or track-

---

Table 1   Summary of treatment technologies/management methods described in text.

SOIL

	TREATMENT TECHNOLOGY	TYPE	MAJOR PROCESS	CONTAMINANTS AFFECTED
1.	Incineration	destructive	thermal	organics
2.	Low Temperature Thermal Desorption	separation	thermal	volatile, semi-volatile organics
3.	Solidification/Stabilization	isolation	physical	inorganics, (organics?)
4.	Soil Washing	separation	physical	inorganics, organics
5.	Solvent Extraction	separation	chemical	inorganics, organics
6.	Reduction-oxidation	destructive	chemical	organic
7.	Bioremediation	destructive, immobilization	biological, physical	organic metals
8.	Soil Vapour Extraction	separation	physical	volatile organics
9.	Excavation/Landfilling	separation/isolation	physical	inorganics, organics
10.	Steam Extraction	separation	physical	inorganics
11.	Hydraulic Fracturing	N/A	physical	improve permeability
12.	Naturalization	isolation, destructive	physical/biological	organics

GROUNDWATER

	TREATMENT TECHNOLOGY	TYPE	MAJOR PROCESS	CONTAMINANTS AFFECTED
1.	Hydraulic Fracturing	n/a	physical	improve permeability
2.	Free Product Recovery	separation	physical	NAPLs
3.	Groundwater Pump and Treat			
	• granular activated carbon	separation	physical/chemical	organics, inorganics
	• filtration	separation	physical	organics, inorganics
	• air stripping	separation	physical	volatile organics
	• advanced oxidation	destructive	physical	organics
	• bioreactors	destructive	biological	organics
4.	Chemical Treatment	separation	physical, chemical	inorganics, organics
5.	Chemical Washing	separation	chemical	organics, inorganics
6.	Bioremediation	destructive	biological	organics
7.	Air Sparging	separation/destructive	physical	organics
8.	Reactive Walls	destructive	chemical	organics
9.	Manufactured Wetlands	separation/destructive	physical/chemical	inorganics, organics
10.	Isolation	isolation	physical	inorganics, organics

mounted. In a rotary kiln, waste is conveyed to a rotating chamber and incinerated at temperatures ranging from 800° to 1500°C. A secondary combustion chamber may be required to ensure combustion of gases generated in the primary kiln. Incinerators should operate under high combustion efficiencies to prevent the formation of products of incomplete combustion (PICs) which are highly toxic. A schematic for a rotary kiln is shown in Figure 5.

Low Temperature Thermal Desorption

Thermal desorption is an *ex situ* separation technology that applies heat to soil to separate volatile and semi-volatile organic compounds from the soil matrix. The heated contaminant airstream is directed to a second process unit where the contaminants are destroyed by thermal or catalytic oxidation to carbon dioxide and water, or are separated by adsorption onto activated carbon.

The technology can be fixed or transportable. It is particularly suited to the treatment of gasoline and petroleum fuels. Treatment costs for dry sandy soils are competitive with landfilling. Increased soil moisture and clay content will increase treatment costs. The technology is not currently approved in Ontario for the treatment of chlorinated organics, and cannot be used to treat contaminated soils containing hazardous levels of metals because of the possibility of volatilizing certain metals. A schematic of the low-temperature thermal desorption process is provided in Figure 6.

Solidification/Stabilization

Solidification/stabilization technologies are applied *ex situ* or *in situ*, and involve the addition and mixing of reagents to the contaminated soil. The reagents serve to bind the contaminants through adsorption processes and solidify the material to form a stable, solid mass which is mechanically competent, and resistant to the effects of freeze/thaw and weathering.

The low permeability of the processed material minimizes the release and distribution of contaminants into the environment by such processes as leaching, volatilization or wind. Portland cement, lime diatomaceous earth, and fly-ash are common reagents in the solidification/stabilization process.

Solidification/stabilization is suitable for treatment of heavy metals, but while some vendors claim effectiveness in

treating organic wastes, there is very little evidence to support these claims. It is also not clear whether solidified/stabilized material will withstand tens and hundreds of years of weathering. The addition of reagents could increase the initial volume of material by 10-30%. Heat generated by the addition of reagents may result in significant emissions of volatile chemicals from the contaminated soil. In some jurisdictions, solidified materials must be landfilled, which adds to the treatment costs. High clay and organic content in contaminated materials will result in a structurally weaker solid mass.

Other forms of stabilization include asphalt batching and encapsulation. Asphalt batching incorporates contaminated material into asphalt materials for reuse in road surfacing materials. Contaminated materials used for this process are restricted by clay content and the amount of any fuel or solvent contamination, since these materials will lower the strength of the final asphalt product. Encapsulation involves the mixing of plastic resins, which surround and encapsulate the contaminant, producing a solid material that has very low permeability.

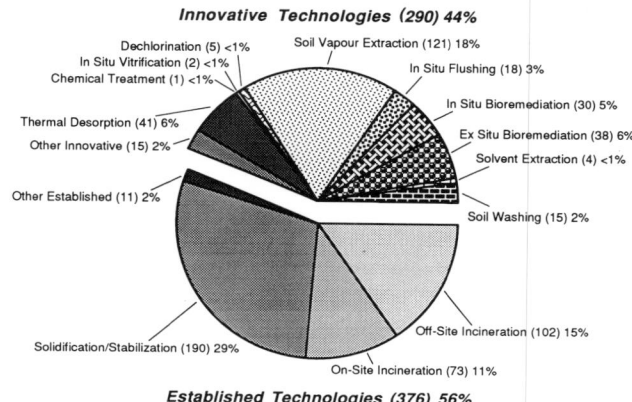

Figure 3 *United States Superfund remedial actions: summary of alternative treatment technologies from 1982 to 1993. Percentage values identify frequency of use over that time period relative to total number of remedial site cleanups. Numbers in parentheses identify number of times that technology was used; more than one technology is commonly used at any one site.*

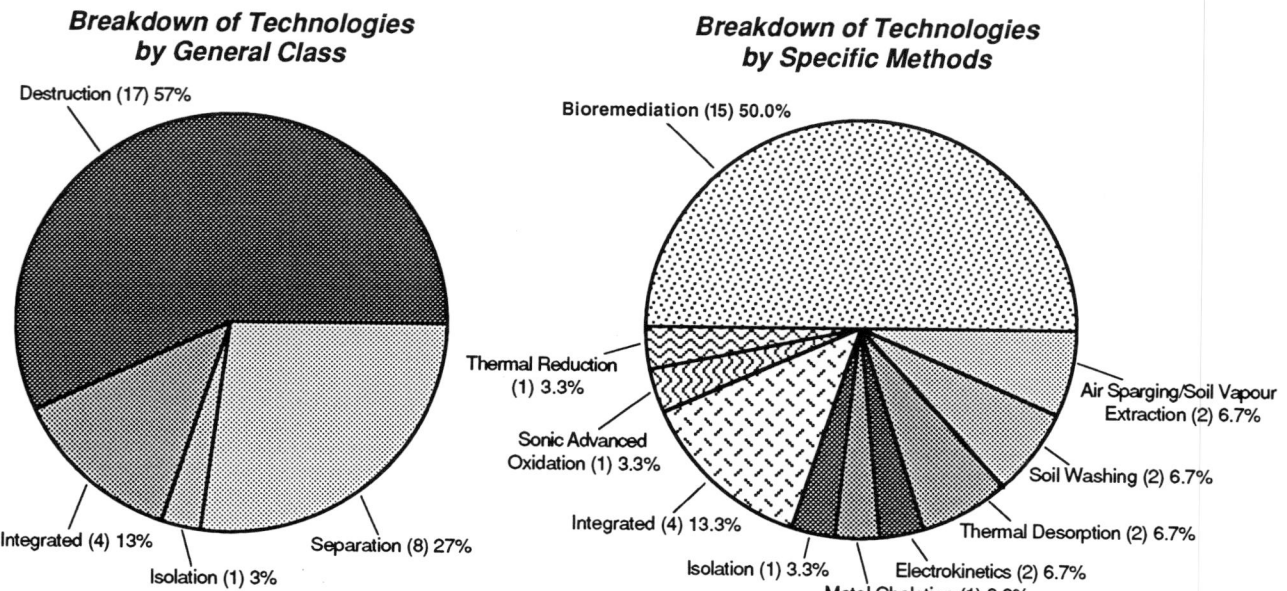

Figure 4 *Summary of treatment technologies funded by DESRT Program in Canada.*

Soil Washing

Soil washing is a well-developed *ex situ* technology for the treatment of contaminated soil. The process technology has a long history of use in Canada in the mining and metal processing industry, but has only recently been applied to large-scale clean-up of contaminated soils. Soil washing is a separation technology that is suitable for the treatment of inorganic and organic contaminants.

Contaminated soil is first pretreated by screening out large objects. The soil is then directed to a vessel containing water which agitates the soil, and through the process of attrition scrubbing, the mechanical shearing action of the soil particles against one another removes adsorbed contaminants. The soil is separated by density and size. The coarse particle fractions are cleaned quickly and can be returned to the site as clean fill. The finer grained fractions may require additional treatment. Variations on this process include the use of high-pressure washing to remove adsorbed contaminants. This process works best on coarse-grained materials such as sands and gravels. With increasing clay content, the washing process becomes more complex, resulting in higher treatment costs. Chemical additives such as acids and surfactants can be added to facilitate the removal of specific contaminants. Chelating agents can be added to metal-contaminated soils to recover specific metals which can be sold for reuse.

Soil-washing facilities can be established as fixed or mobile facilities. Treatment costs are competitive with alternative treatment for hazardous waste, but are likely too expensive at the present time for treatment of non-hazardous waste. The process separates the contaminants from the coarser fraction of soil. The final product includes a cleaned soil and a concentrated contaminant sludge which requires further treatment and/or disposal.

Hydraulic Fracturing

Fine-grained materials pose a problem for clean-up because their low permeability decreases advective transport and increases the time required to remove contaminants from the soil. Hydraulic fracturing is an *in situ* technology that has been used in the oil industry for secondary recovery of crude oil, and in the water well drilling industry to improve the yield of water wells. It is now being applied to contaminated soils to enhance permeability for the removal of volatiles by soil vapour extraction and to promote bioremediation. The mechanics of hydraulic fracturing in rock are well understood. Hydraulic fracturing in unconsolidated materials, for the purpose of enhancing permeability, is a more recent area of research.

As hydraulic pressure is applied, the generation of fractures occurs at a critical pressure. As hydraulic pressure increases, fractures are propagated and as propagation continues, there is a rapid decline in pressure as the fractures develop. The procedure is shown in Figure 7. The application of hydraulic-fracturing technology is described by US EPA (1993b) for enhancing permeability of soil for the purpose of soil vapour extraction. The following results were observed: (1) fractured wells showed increases in vapour flow rates of fifteen times those of unfractured wells; (2) contaminant mass recovery rates were seven to fourteen times greater in fractured wells than non-fractured wells; and (3) the radius of fracture propagation was estimated to be 8 m.

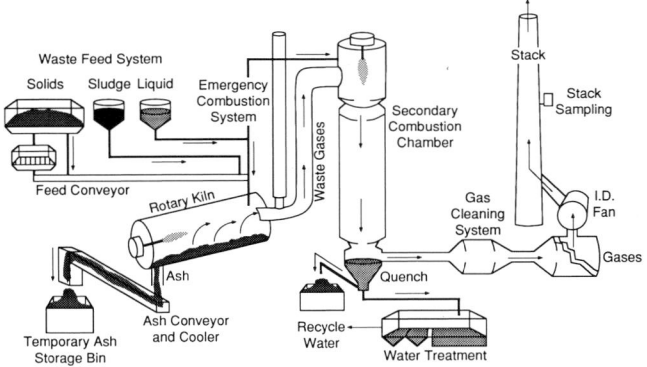

Figure 5 *Typical rotary kiln incinerator.*

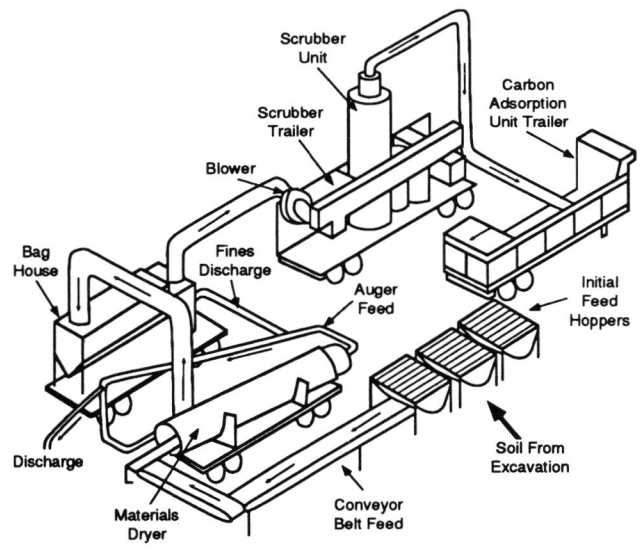

Figure 6 *Low-temperature thermal aeration.*

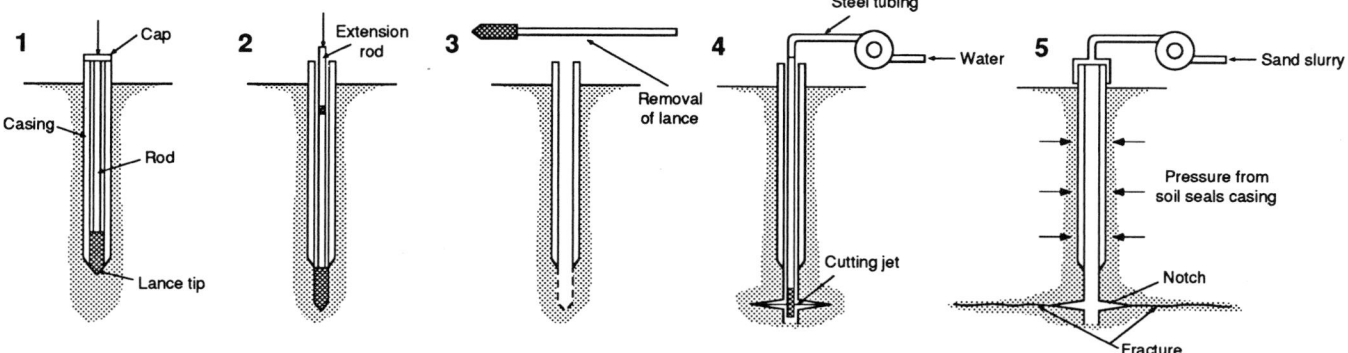

Figure 7 *Method of hydraulic fracturing.*

Steam Injection

The injection of hot water and steam *in situ* has been used extensively in the oil industry as a technique for secondary oil recovery. The application of this technology to the treatment of sites contaminated with volatile and semi-volatile organic compounds is in the pilot scale and field scale stage of development. Steam is generated from an on-site steam boiler, and is injected into the zone of organic contamination by means of injection wells. A steam front advances through this zone, decreasing viscosities and interfacial tensions between the water and the organic compound and resulting in improved contaminant mobilization. Volatile components of organic contaminants are distilled from the zone of contamination and subsequently condensed into the steam front, producing a front of mobile organic liquids. Extraction wells are used to pump organic liquids and ground water to the surface for treatment and to recover steam and vapourized contaminants. Ground water is directed through an oil/water separator prior to treatment, and is discharged to a storm or sanitary sewer. The vapour phase contaminants can be treated using thermal or catalytic oxidation, or can be condensed for recycling or treatment.

A schematic of a steam-enhanced recovery system is provided in Figure 8. The technology has been demonstrated at sites contaminated with volatile fuels, chlorinated solvents, and semi-volatile fuels and oils (Hughes Environmental Systems Inc., 1992; Udell Technologies, Inc., 1992). The use of steam injection in the recovery of DNAPL compounds requires considerable care in design and operation to prevent uncontrolled mobilization of DNAPL.

Naturalization

Certain vegetation types have been found to be effective in enhancing the biodegradation of pesticides, chlorinated solvents, light-fraction aromatic and aliphatic petroleum hydrocarbons and polycyclic aromatic hydrocarbons in their root zones, and the uptake of metal contaminants. The root zone, or rhizosphere, typically shows increased microbial activity of one or more orders of magnitude compared to soil outside of the root zone. The composition of microbial communities is dependent on several factors including root type, plant species, plant age, soil type, and the type of contaminant that is present in the soil. The vegetation itself influences conditions in the rhizosphere that are significant in promoting the development of biomass. These conditions include oxygen and carbon dioxide concentration, osmotic and redox potentials, moisture content, and pH.

Plant root systems provide the supporting structure on which microbial populations can develop. Plant species that develop fibrous root systems provide a larger surface area than tap root systems, and can support a larger microbial population. As well, root cells secrete nutrients that are necessary to maintain microbial respiration. Research into the assessment of specific plant species to degrade or uptake specific contaminants is ongoing.

The use of naturalization is a long-term strategy, and should only be considered in areas that do not require immediate remediation. Restricted access to areas undergoing remediation by this method may be necessary. Long-term monitoring of soil, and possibly ground-water, quality would be necessary to determine whether naturalization processes are effective.

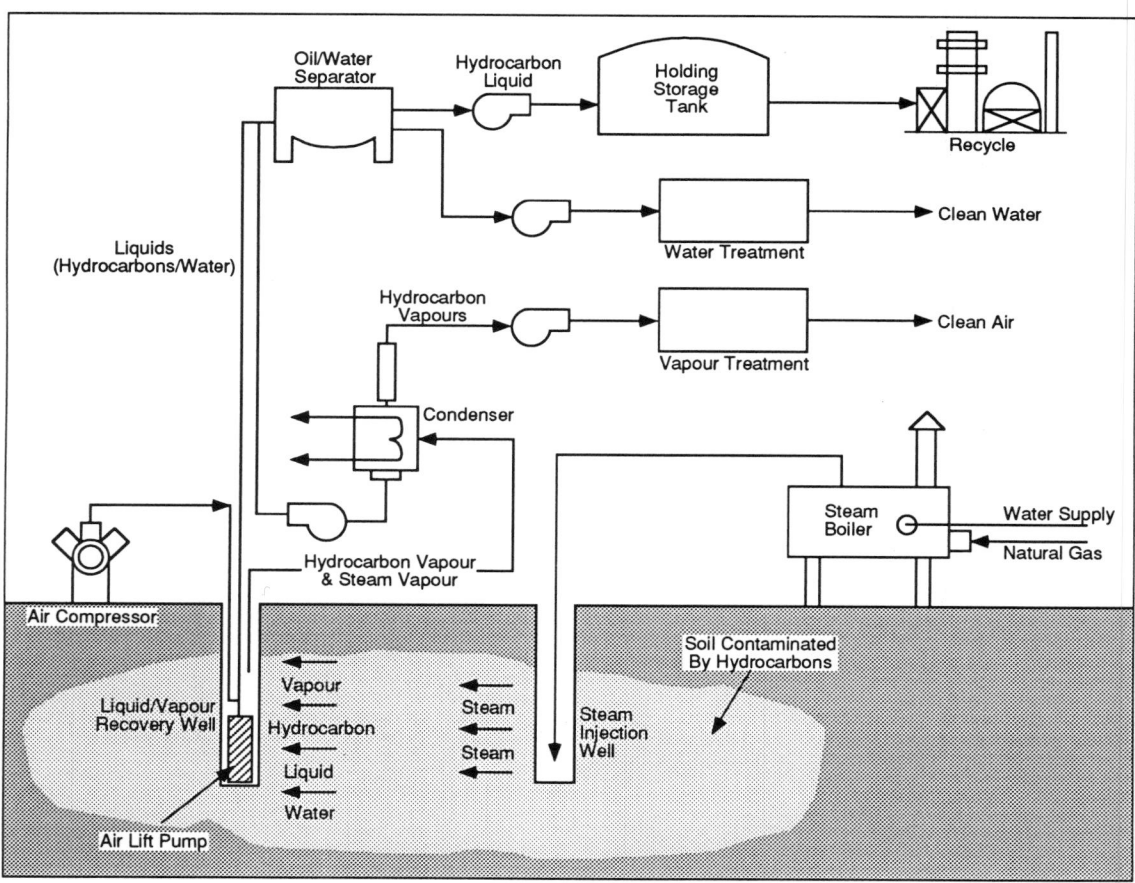

Figure 8 *Steam-enhanced recovery process.*

Excavation/Landfilling

Excavation and landfilling of soil is currently the most popular method of dealing with non-hazardous contaminated soil in Canada, given the relatively low cost for disposal. Very few treatment technologies are able to compete with the current low tipping fees at landfills. As the cost of disposal rises, alternative methods of treatment become economically viable. In some jurisdictions in the United States, soils containing certain levels of hazardous materials can no longer be landfilled, and require alternative methods of disposal. This has favoured the development of alternative technologies.

Disposal of contaminated soil at a landfill is perceived by some to be a band-aid solution, since the environmental issues associated with contaminated soils are not resolved, but moved from one jurisdiction to another. As well, in some jurisdictions, owners of contaminated soil who dispose of their soil at landfills may be subject to future liability, should there be any pollution claims against the landfill as a result of contamination of water supplies from landfill leachate.

Soil Vapour Extraction

Soil vapour extraction is an *in situ* technique that is used to remove volatile contaminants from the unsaturated zone. It is commonly used to remediate gasoline spills and is useful for remediation around structures where excavation could result in damage to the structure. A typical soil vacuum extraction system includes an array of wells installed in the areas of contamination in the unsaturated zone. The wells are connected to headers, which are attached to a knockout drum and vacuum pump. When a vacuum is applied to the wells, the lower pressure at the well screen causes air flow to occur in the soil from areas of higher pressure (atmospheric) located away from the wells. The air stream that is extracted at the well includes volatile contaminants and soil moisture. A knockout drum is attached in front of the vacuum pump to remove moisture from the airstream.

Soil vapour extraction is a separation technology. The

separated contaminants in the airstream require further treatment to prevent release of contaminants to the atmosphere. The most common treatment methods include activated carbon, thermal and catalytic oxidation, and biofilters. The use of activated carbon would require further processing or treatment when the carbon becomes saturated with the contaminant. Further treatment could involve landfill disposal or regeneration of the carbon.

Soil vapour extraction is most effective with volatile contaminants which have vapour pressures of greater than 0.5mm Hg and dimensionless Henry's Law constants of greater than 0.01 (United States Environmental Protection Agency, 1991), as shown on the nomograph in Figure 9. Finer grained soils such as silts and clay silts would require permeability enhancement using methods such as hydraulic fracturing described above.

Depending on the depth of contamination, air flow can be controlled by sealing the surface to prevent the short-circuiting of air flow. If the vacuum extraction wells are located near the water table, the lower pressure in the wells may cause the water table to upwell in the vicinity of the wells, reducing the efficiency of the process. In such cases, dewatering and possibly ground-water treatment may need to be considered.

Prior to full-scale clean-up, a pilot air extraction test should be conducted to provide site-specific design parameters including radius of influence of extraction wells, contaminant concentration in air, vacuum pressures and changes in water level. This data will be used to design the full-scale system for such design parameters as number and location of extraction wells, vacuum pump size, estimated mass removal rate, length of time of treatment. A schematic for soil vapour extraction is provided in Figure 10.

Solvent Extraction

Solvent extraction is an *ex situ* chemical separation technology that uses solvents to solubilize contaminants and remove them from the soil matrix. The process is similar to soil washing (see above). Soil is screened for size and conveyed to a reaction vessel where the solvent is added. The waste stream from this process includes a cleaned soil stream and a mixed stream containing the solvent, contaminant and fines. Further separation of the mixed stream is conducted to redirect the fines back into the system for further treatment if necessary, separate the solvent for reuse, and collect the contaminant for disposal or resale. Solvent extraction technologies can be complex systems because they involve two different processes: the separation of the contaminant from the soil matrix and the separation of the contaminant from the solvents. As well, treatment of any process water may be required to meet discharge criteria.

Thermal Reduction

Thermal reduction is a process that uses a high-temperature, highly reducing hydrogen environment to destroy chlorinated organic compounds including PCBs, chlorinated solvents, and dioxins and furans. The process is not considered to be an incineration process because oxygen is not involved in the reaction.

The main component of the process is the reactor module. The waste soil is preheated and injected into the reactor in the presence of hydrogen. The mixture is heated to 850°C as it falls through the reactor. Particulates up to 5 mm in diameter that are not entrained in the gas mixture, impact the walls of the reactor and volatilize organic matter adsorbed onto the particulate. The particulate falls through the reactor into a

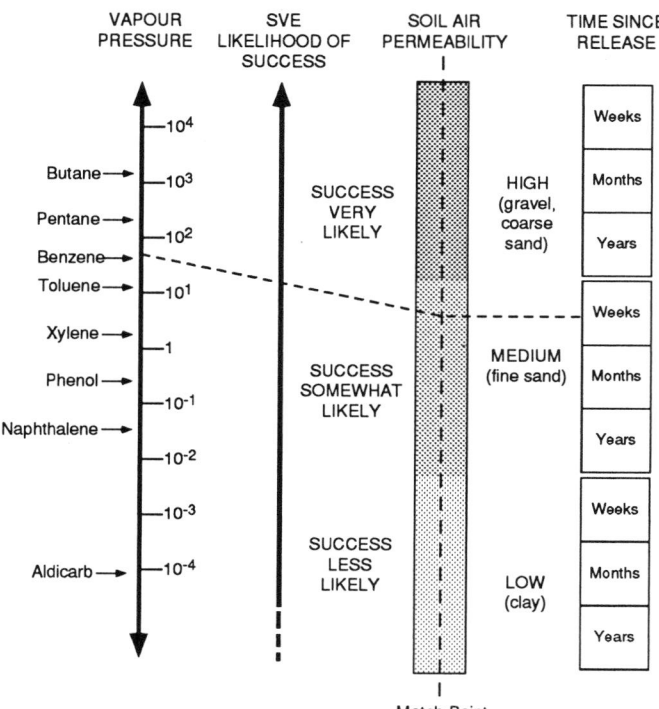

Figure 9   *SVE (soil vapour extraction) applicability (see Fig. 10).*

quench tank and is removed (United States Environmental Protection Agency, 1994b).

PCBs are destroyed in the process by the following reactions:

$$PCB + 5H_2 \rightarrow benzene + 4HCl \qquad (1)$$

$$Benzene + 9H_2 \rightarrow methane \qquad (2)$$

The process has undergone a number of field demonstrations and is commercially available. The system is mobile and can be transported to the site on two tractor trailers. Following the reaction, gases are cooled with scrubber water which removes hydrogen chloride. The process produces a sludge and decant water waste streams. The final gas stream consists of hydrogen, methane and ethylene, most of which is recirculated back into the reactor or is used to preheat the incoming waste stream.

Reduction-Oxidation Reactions

Reduction-oxidation reaction technology is a chemical method that destroys organic contaminants through the addition of reagents that break down the chemical structure of the contaminant. Reduction reaction technology is used in the destruction of chlorinated compounds such as PCBs. An example of a reduction reaction is where an alkali metal reacts with glycol to produce an alkoxide. The alkoxide substitutes for chlorine on the chlorinated compound to form an ether and an alkali metal salt. The process involves the pre-screening of contaminated soil for size. The reagents are pre-mixed and added to the soil in a closed reaction vessel to form a slurry. The slurry is heated for some period of time, depending on the nature of the contaminant and matrix. Following the reaction, unreacted reagent is removed. The cleaned soil is washed with water. Volatile emissions and wash water are treated prior to disposal. This process has been demonstrated at field scale at several PCB contaminated sites in the United States (United States Environmental Protection Agency, 1989).

Bioremediation

Bioremediation is a destructive technology that has had widespread application in the clean-up of soil and ground water contaminated with light hydrocarbons such as fuels and light oils. Research and field studies are being conducted to determine its application to the treatment of more recalcitrant organic compounds such as polycyclic aromatic hydrocarbons (PAHs), and chlorinated aliphatic and chlorinated aromatic compounds.

Bioremediation is a process that can occur over a wide range of chemical conditions in the presence of free oxygen (aerobic conditions) or without free oxygen (anaerobic conditions). Biodegrading microbes include bacteria and fungi which are naturally occurring in soil. Bioremediation technology involves the stimulation of growth of indigenous microbes to break down contaminants into less harmful transformation products and, ultimately, to carbon dioxide. Some vendors add proprietary genetically engineered microbes to stimulate biodegradation, but it is believed that these are generally short-lived, and the advantage of their addition is a matter of debate. The requirements for successful in situ bioremediation include (1) appropriate microbial conditions, and (2) appropriate hydraulic conditions. Growth of the microbial biomass requires a source of carbon (the contaminant), an electron acceptor (e.g., oxygen, nitrate, sulphate, carbon dioxide), nutrients (phosphorus and nitrogen) and subsurface conditions friendly to the microbes (e.g., temperature and pH).

Bioremediation is a destructive method that is suitable for clean-up of organic compounds, particularly the light- to mid-range fraction of petroleum hydrocarbons and the lighter fraction PAHs (Chapter 7). Several methods have been developed using bioremediation technology either in situ or ex situ and include bioventing and various types of bioreactors.

Bioventing is an in situ, aerobic bioremediation method that involves the movement of air through the subsurface, directly by means of blowers or indirectly by means of vacuum pumps. The circulation of air containing oxygen facilitates in situ bioremediation, and would be suitable for those areas containing petroleum hydrocarbons in the soil zone above the water table. A schematic showing bioventing with nutrient and moisture addition is provided in Figure 11.

Slurry bioreactors are an ex situ bioremediation technology that requires excavation of contaminated soil and transfer into a reactor vessel. The vessel could include some form of enclosed tank or an open lagoon. Contaminated soil is added to water to form a slurry that is 10-20 weight percent of solids to maintain particle suspension (Brox and Hanify, 1989). Nutrients and electron acceptors can be added in appropriate concentrations to match contaminant concentrations in the slurry and optimize biodegradation. While slurry bioreactors are restricted by the size of the vessel as to the volume of material that can be treated, residence time can be considerably shorter than other types of reactors because of the ability to control conditions that increase rates of biodegradation.

Prepared-bed reactors are an ex situ bioremediation method that involves piling contaminated soil into prepared reactor beds (also termed engineered bioremediation cells, see Chapter 36 for a case example). They operate at a slower rate than slurry bioreactors, but are capable of treating larger volumes of soil. Space requirements for prepared-bed reactors are greater than for slurry bioreactors. In situ bed reactors involve in situ deep tilling of soil, with the addition of organic soil amendments which facilitate the bioremediation of the heavier and more recalcitrant petroleum hydrocarbons and PAHs.

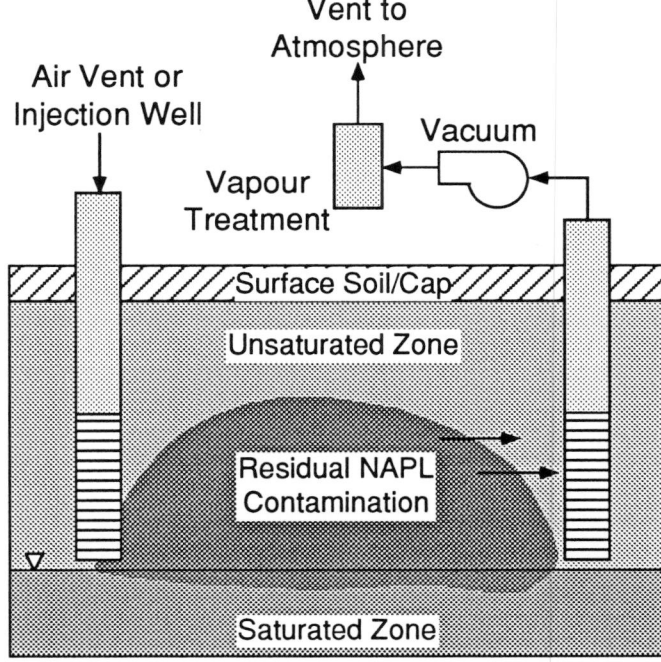

Figure 10 *Soil vapour extraction.*

Electrokinetics

Electrokinetics involves the application of a low-voltage direct current across a saturated soil mass, which results in the transport of ionic species toward the current electrodes (Acar and Alshawabkeh, 1993). Electrokinetics is a separation technology that is suitable for the treatment of metal contaminants in soil that occur below the water table. Metal ions migrate toward the cathode, where fluids containing the metal contaminants are removed from the subsurface by pumping, and the fluid is treated on surface for metal recovery using standard treatment methods. Application of electrokinetics to soil and ground-water clean-up is still at the field or pilot scale stage in North America. It has been used in site clean-up in the Netherlands since 1987 (Lageman, 1993).

Ground-water Remediation Technologies

In the United States, more than three-quarters of the SUPER-FUND sites use pump-and-treat systems to remediate con-

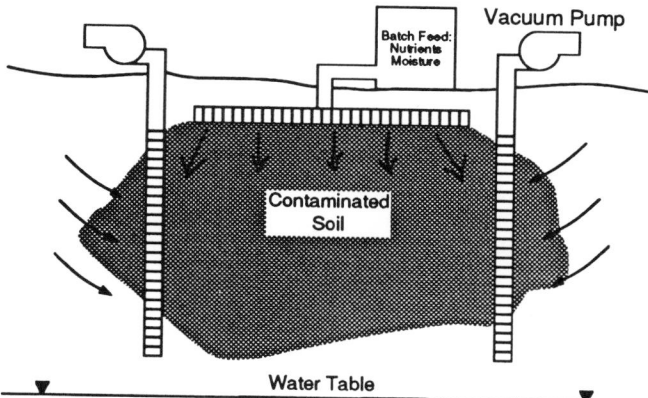

Figure 11   *Bioventing (after National Research Council, 1994).*

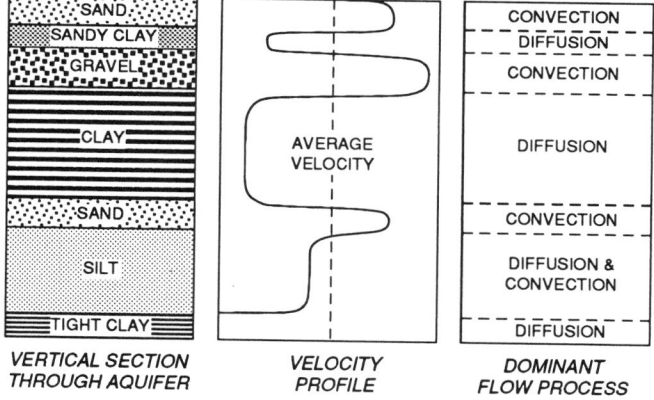

Figure 12   *Influence of geological stratification on ground-water flow.*

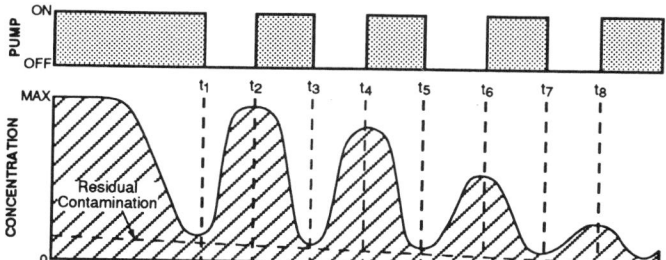

Figure 13   *Variation in contaminant concentration with pulsed pumping (after Mercer et al., 1990).*

taminated ground water. Pump-and-treat involves the pumping of contaminated ground water to the surface through a series of extraction wells where it is treated and discharged.

Early on in the application of pump-and-treat, it was thought that by pumping a volume of water equal to the pore space of the aquifer, a significant volume of contaminant would be removed. Due to heterogeneities within a stratigraphic sequence, contaminant migration may be dominated by convection in permeable zones or by diffusion in low-permeable materials as illustrated in Figure 12. Depending on site conditions and the age of the contamination, pump-and-treat type treatment is often characterized by decreasing concentrations of contaminants with pumping time. The contaminants may reach a level that is below ground-water clean-up guidelines, at which time the pumps are shut off. After some period of no pumping, the contaminant concentration starts to increase. This observed increase in concentration with time is attributed to contaminant desorption and diffusion from fine-gained materials adjacent to or within the aquifer. The pump-and-treat process must then be started again, until such time as the contaminant reduces to a level that is below ground-water clean-up guidelines. This cycle of pulsed pumping is illustrated in Figure 13.

Experience in the United States has shown that pump-and-treat is severely limited in its ability to restore contaminated aquifers, because contaminants remain within the soil matrix (see Chapter 6 for discussion).

Hydraulic Fracturing

Hydraulic fracturing (see above) can be applied to the saturated zone to improve hydraulic conductivity and improve the efficacy of pumping for pump-and-treat and *in situ* methods of ground-water treatment.

Free-Product Recovery

Petroleum hydrocarbon fuels such as gasoline and diesel fuel (LNAPLs, Fig. 1) are common contaminants in soil and ground water. Recovery can be completed using pumps or interceptor trenches. Figure 14 depicts a recovery well. A water pump located near the bottom of the well is used to lower the water level in the well and create a cone of depression around the well. This cone of depression induces free product to flow into the void space previously occupied by the water, and allows a greater thickness of product to accumulate. A separate product pump is used to pump product from the well to the surface, where it is stored for disposal or re-refining.

Alternative methods of product recovery, which may be suitable for shallow ground-water conditions, involve the use of inceptor trenches which are dug below the water table. Product flows into the trench under natural hydraulic up-gradients, and is removed by a floating skimmer. The down-gradient side of the trench is covered with an impermeable barrier, such as plastic film, to contain the product in the trench until it can be pumped out. A schematic of an interceptor trench system is shown in Figure 15. Chapter 7 provides case examples of the use of recovery wells and interceptor trenches to remove petroleum products from the shallow subsurface.

Pump-and-Treat

The following five common treatment technologies are reviewed: granular activated carbon, air stripping, filtration processes, advanced oxidation, and bioreactors.

Granular activated carbon.   Granular activated carbon (GAC) has been used to remove organic contaminants from drinking water since the 1940s when its use in treating municipal water supplies became widespread. GAC is manufactured from a variety of source materials which include coal, wood, peat and coconut shell. The source material is first crushed to a powdered form, then roasted to produce charcoal. The charcoal is then activated by roasting again in a low-oxygen environment, in the presence of steam, which creates numerous pits and grooves in the charcoal particles. This action significantly increases the surface area of the particles (Lehr, 1991). GAC works by adsorbing organic contaminants onto its surface, where the forces of attraction between the carbon surface and the organic contaminant are greater than the forces that keep the organic contaminant in solution. Organic compounds that readily absorb onto GAC are those with a relatively low water solubility and a large molecular weight.

Physical properties of GAC that are used to determine its effectiveness for treatment include surface area, density, mesh size, abrasion resistance, and ash content (Delthorn and Mazzoni, 1986). The larger the surface area of the GAC, the more efficient the removal process will be, provided that the pore size in the GAC is suitable for the size of the contaminant molecule. Mesh size describes the size of individual GAC particles. A smaller mesh results in a larger pressure drop across the carbon bed, resulting in longer retention time within the bed and greater adsorption of the contaminant. Abrasion resistance reflects the ability of the carbon to resist breakage during handling, and is an important consideration where back-washing of the carbon bed is required. Ash content reflects the purity of the carbon. Impurities in the carbon may be released during back-washing and may affect water quality.

In addition to the treatment of liquid phase contaminants in ground water, GAC can be used for the treatment of gas phase contaminants which would be generated during air stripping or soil vapour extraction. GAC can be purchased in bulk or in pre-packaged systems ranging in size from 45-gallon drums to large trailers. Once all the adsorption sites on the carbon are full, the carbon has no further capacity to adsorb the contaminant, and breakthrough of the contaminant can occur. Therefore, it is common practice to establish GAC treatment

in series, so that contaminant breakthrough can be contained. Monitoring of the influent and effluent stream is a key component of GAC treatment.

Air stripping.   Air stripping is a well-developed technology for the separation and removal of volatile organic compounds from contaminated ground water. The ability of air strippers to remove contaminants from ground water is a function of the contaminant's ability to partition from water into air. This partitioning can be quantified by the Henry's Law constant for the contaminant, which is derived from the ratio of the contaminant's vapour pressure to its solubility in water (Chapter 7). Contaminants with a Constructedlow solubility in water and a high vapour pressure have large constants and will tend to partition readily from water to air. These contaminants are good candidates for air stripping.

Contaminants with Henry's Law constants of greater than about 5×10^{-3} atm m$^3 \cdot$mol^{-1} can be considered as strippable. This would include common contaminants such as tetrachloroethene (perchloroethylene), trichloroethene, benzene, xylene and toluene. Air-stripping systems can consist of packed towers, diffusers, trays or spray systems. Each is designed to increase the contact time between the influent contaminated ground-water stream and a counter-current flow of air which removes the contaminant from the water. The stripping process generates a contaminated air stream which requires further treatment, which commonly includes granular activated carbon, oxidation processes, or bioremediation.

Filtration processes.   Filtration involves physical/chemical processes that remove or separate the contaminant from the liquid, leaving a contaminant concentrate that requires further treatment or disposal. Filter membranes can include manufactured membranes such as cellulose acetate or silicone rubber, or formed-in-place membranes consisting of layers of inorganic and polymeric coatings on rigid, porous plates or tubes. Filtration technologies are described in US EPA (1992a, b, 1993c, d). Filter membranes and formed-in-place membranes are porous materials that allow ions and compounds less than a critical size to pass through, while retaining larger sized species or compounds. Filters can be used to treat ground-water waste streams containing heavy metals, as well as organic compounds ranging from large molecular-weight organics such as polyaromatic hydrocarbons to lower molecular-weight organics such as toluene and trichloroethylene.

Dissolved metals require pre-treatment prior to filtration in order to increase the diameter of the contaminant and prevent it from passing through the membrane. Pre-treatment includes the addition of reagents to form larger metal precipitates or metal polymer complexes which can then be effectively filtered.

A filter operating in cross-flow mode, as shown in Figure

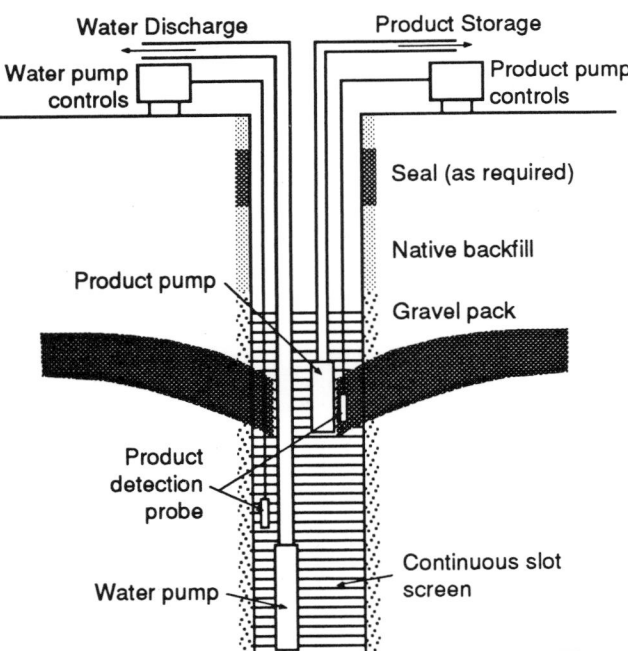

Figure 14 *Recovery of free product using two-pump system (after Domenico and Schwartz, 1990). Also see Chapter 7.*

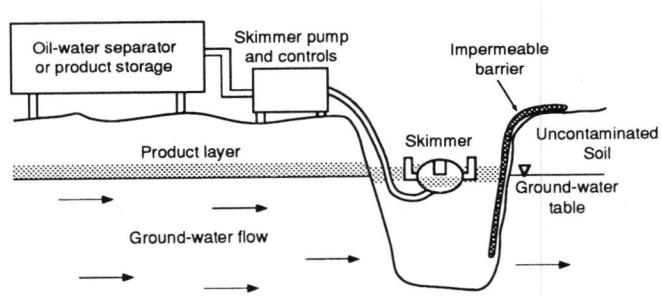

Figure 15 *Recovery of free product using an interceptor trench (after Domenico and Schwartz, 1990).*

16, involves a series of hollow tubes through which the contaminated ground water is pumped. The filter selectively prevents the contaminant from passing through the filter, while allowing the permeate to flow through the sides of an inner tube into an outer tube. The permeate is collected and further treated as necessary prior to discharge. The contaminant is concentrated in the inner tube and collected for recycling or disposal.

Filtration technology at bench and pilot scale indicated greater than 95% removal rates for mercury, cadmium and lead (United States Environmental Protection Agency, 1992a). Treatment of metals may require pre-treatment to remove elevated levels of iron, sulphate and calcium in the feed. Removal efficiencies of 95 to 99% were reported for toluene, trichloroethylene and PAHs (United States Environmental Protection Agency, 1992b, 1993d).

Advanced oxidation. Advanced oxidation processes using photo-oxidation are used in the treatment of organic compounds and offer advantages over filtration and other separation technologies because photo-oxidation results in the destruction of the contaminant. Advanced oxidation involves the formation of a hydroxyl radical (OH^{2-}), a powerful oxidizing agent that reacts extremely rapidly with almost all organic compounds, at rates which are typically one million to one billion times faster than reactions with molecular ozone alone.

The hydroxyl radical can be generated from the reaction between ozone or hydrogen peroxide and ultraviolet light (United States Environmental Protection Agency, 1993e, f), cavitation (United States Environmental Protection Agency, 1993g), or electron beam (United States Environmental Protection Agency, 1992c, 1993h, i). Once generated, the hydroxyl radical reacts rapidly to oxidize organic compounds in solution to carbon dioxide and water. Where chlorinated organic compounds are involved, chloride is also produced. In addition to oxidation of organic compounds by the hydroxyl radical, organic compounds also undergo photochemical destruction. The ultra violet source produces light at a wavelength of less than 3×10^{-9} m (300 nm). Many of the problematic organic contaminants, such as the chlorinated solvents and many of the intermediate species generated during the photochemical/oxidation reactions, react strongly to light with wavelengths at or below the range of 250-300 nm (United States Environmental Protection Agency, 1993f).

Advanced oxidation using photo oxidation is suitable for the treatment of most organic contaminants including petroleum hydrocarbons, PAHs, chlorinated solvents, pesticides, PCBs and chlorinated phenolics. Commercial skid-mounted reactors are custom built to the nature of the contaminant and the contaminant through-put. The reactor consists of one or more vertical cylinders equipped with a UV lamp and a quartz

sleeve. The influent-contaminated ground water is pumped through a line where hydrogen peroxide or ozone oxidants are added and mixed prior to entering the reactor. The ground water is treated in the reactor, and the effluent is sampled to ensure compliance with discharge criteria. A schematic showing an advanced oxidation reactor is provided in Figure 17.

Bioreactors. Bioremediation has previously been discussed with respect to the clean-up of contaminated soil, and is discussed later with respect to *in situ* methods of ground-water treatment. Bioreactors used to treat ground water (*ex situ*) that has been pumped from a contaminated area utilize a reaction vessel to optimize biodegradation of the contaminant inside the vessel. Design criteria for bioreactors include the type and concentration of contaminant, ground-water flow rate, clean-up criteria for contaminated ground water, and additional constituents in the ground water such as heavy metals that may inhibit microbial activity.

Chemical Flushing

Flushing of chemicals through the contaminated porous medium to enhance the solubility of contaminants has been applied at sites contaminated with NAPLs, as well as metals. The use of surfactants to mobilize petroleum hydrocarbons has been used in the petroleum industry to enhance oil recovery. The purpose of the surfactant is to (1) increase the solubilization of the contaminant, (2) increase desorption of the contaminant from the aquifer matrix, and (3) lower the interfacial tension of the contaminant with water. The surfactant is added to water in concentrations typically ranging from 0.5% to 5% by weight of the solution. In water, the surfactant forms colloidal clusters of molecules, called micelles, which consist of a hydrophilic sheath surrounding a hydrophobic core. At the interface between the NAPL and water, the micelles become re-oriented, so that the hydrophobic group is in contact with the NAPL while the hydrophilic group remains in contact with the water. The non-polar hydrophobic group allows non-polar NAPLs to solubilize into the interior of the micelle. The process is illustrated in Figure 18.

The formation of micelles leads to lowering of the interfacial tension (IFT) between the NAPL and water, thereby reducing the capillary forces on the NAPL and increasing the mobility of NAPL. While the lowering of IFT is an important property of the surfactant, maximum reduction in IFT may not be desirable because of the possibility of the uncontrolled remobilization of the NAPLs, particularly those that have a higher density than water (DNAPLs). Surfactants can be selected that provide significant reduction in IFT without causing vertical migration.

Experiments on the biodegradability of surfactants have

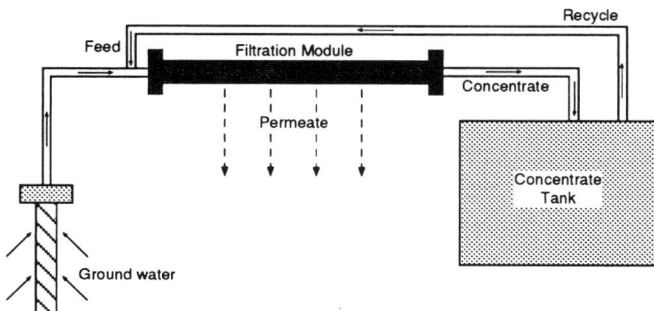

Figure 16 *Schematic diagram of filtration process (after United States Environmental Protection Agency, 1992c).*

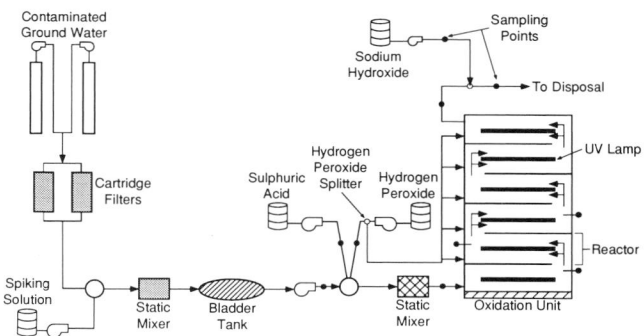

Figure 17 *Schematic of photo-oxidation treatment system (after United States Environmental Protection Agency, 1993e).*

shown that surfactants can be selected to biodegrade under aerobic and anaerobic conditions. As well, there is evidence that surfactants can co-metabolize recalcitrant chlorinated organic compounds by acting as primary substrates for microbial activity (Jackson et al., 1992).

Field tests using surfactants on a controlled spill of perchloroethylene (PCE) indicated that, while pump-and-treat without surfactants showed virtually no effect on PCE removal, the addition of surfactants to enhance pump-and-treat resulted in the removal of free-phase PCE from the field study without any apparent vertical remobilization of the PCE (Jackson et al., 1992). Figure 19 shows the decrease in PCE concentrations versus extracted pore volumes.

The lowering of the interfacial tension decreases the capillary pressure of contaminant within the aquifer matrix and increases contaminant mobility. The increase in mobilization, solubilization and desorption results in an increase in the concentration of the contaminant in the ground water of up to several orders of magnitude, thereby significantly reducing the pumping time required to flush the contaminants from the aquifer.

Bioremediation

Ex situ bioremediation of pumped ground water has been discussed previously. In situ bioremediation, as with any other in situ process, requires careful characterization of the physical and chemical characteristics of the site and an understanding of the system hydraulics. As well, laboratory bench scale testing of the site microbiology should be undertaken to identify specific degrading microbes and conditions that would optimize contaminant degradation. Following bench scale testing, the system should be assessed to determine

whether conditions for optimum contaminant degradation can be developed at the site.

The design of an aerobic system to treat a contaminated ground-water plume involves extraction wells at the toe of the plume, a nutrient and oxygen supply at surface, and infiltration wells within the zone of contamination up-gradient from the extraction wells. Pumped water containing organic contaminants removed from ahead of the treatment front are treated prior to re-injection. A schematic showing an in situ bioremediation system is provided in Figure 20.

The electron acceptor under aerobic conditions is free oxygen (O_2) ,which can be delivered to the system by in situ air sparging (see following discussion), circulation of oxygenated water and the addition of hydrogen peroxide. Air sparging can raise dissolved oxygen concentrations to 8-12 $mg \cdot L^{-1}$. The addition of hydrogen peroxide can raise dissolved oxygen concentrations to 40 $mg \cdot L^{-1}$, but is subject to instability which can reduce its effectiveness. Under anaerobic conditions, electron acceptors can include nitrate, sulphate and carbon dioxide. Geological conditions (e.g., permeability, heterogeneities) and electron acceptors are often the limiting factors to successful in situ bioremediation of ground water.

Air Sparging

Air sparging is used to treat dissolved and residual volatile organic compounds below the water table. The technology can effectively treat light petroleum hydrocarbon contaminants by volatilization and bioremediation (Marley and Droste, 1995).

Air sparging involves forcing air below the water table, through the zone of contamination, using a surface-mounted blower and an array of vertically or horizontally emplaced sparge points (Fig. 21). Vertically emplaced sparge points are typically constructed of materials similar to conventional ground-water monitor wells. They are completed in the saturated zone to a depth below the zone of contamination.

Sparge wells are connected to a header leading to a blower which is used to force air through screened portions of the sparge points. Air bubbles are forced into the formation and rise upward to the zone of contamination. Air bubbles strip VOCs that are residually saturated in the aquifer and dissolved in ground water. Contaminants that are amenable

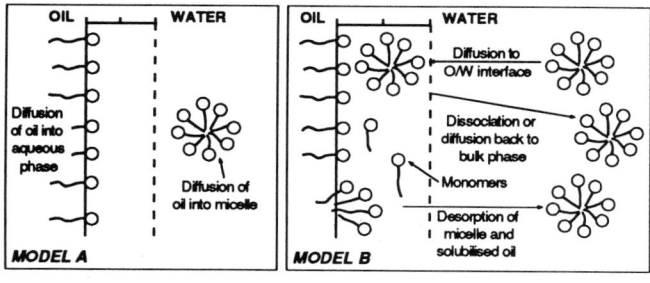

Figure 18   *Solubilization of oil by surfactant micelles (after Jackson et al., 1992).*

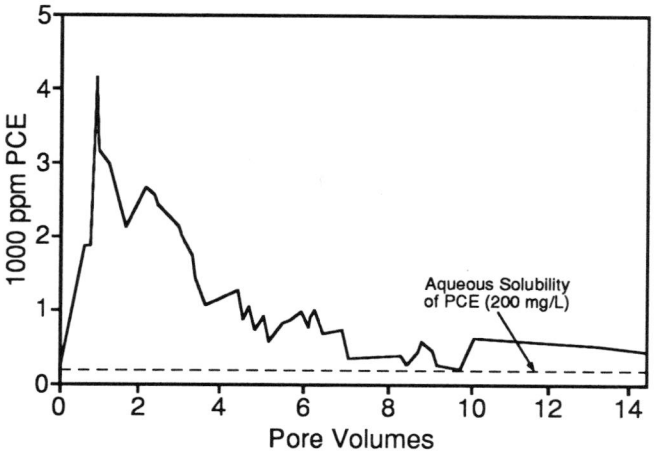

Figure 19   *Effluent concentration of PCE (from Jackson et al., 1992).*

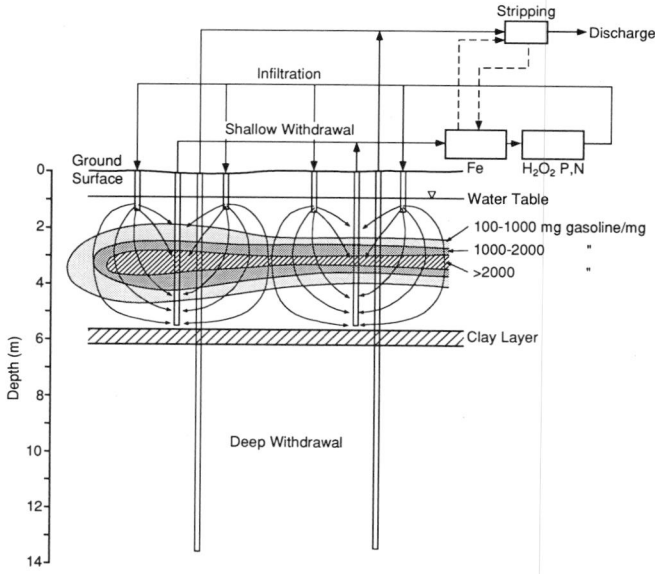

Figure 20   *In situ bioremediation treatment system (after Stapps, 1990). Ex situ techniques requiring excavation of contaminated soil are discussed in Chapter 36.*

to stripping include components of gasoline and chlorinated solvents.

The rising air/contaminant vapour stream results in increasing concentrations of VOCs in the unsaturated zone above the water table. In order to prevent uncontrolled migration of the contaminant in the unsaturated zone and uncontrolled emissions to the atmosphere, air sparging should be coupled with a soil vapour extraction system. Soil vapour extraction involves the installation of arrays of vertical or horizontal extraction wells in the unsaturated zone which are attached to a vacuum pump. Air containing VOCs is pumped from the unsaturated zone through a knockout pot to remove moisture, and directed to a treatment system to remove VOCs, using one of several technologies including activated carbon, catalytic oxidation or biofilters prior to discharge to the atmosphere. In addition to the *in situ* stripping of volatiles from below the water table, air sparging can also enhance *in situ* bioremediation of VOCs that are conducive to aerobic biodegradation (Fig. 22). A schematic showing the use of horizontal wells is shown in Figure 23.

The impact of air sparging on the concentration of VOCs recovered during soil vapour extraction is provided in Figure 24. The figure shows that when the air-sparging system is turned on, there is an immediate increase in the concentration of VOCs in soil gas.

Reactive Walls

Reactive walls are an *in situ* technology in which a permeable wall containing a suitable reactive material is placed downgradient from a plume of contaminated ground water. As the ground water flows through the wall under natural gradients, the specific contaminant in the ground water reacts either directly or indirectly with the reactive material in the wall. A reactive wall is shown schematically in Figure 25. This technology can also be applied above ground where *in situ* applications are not feasible. In above-ground applications, the reactive material is held within a container and contaminated ground water is pumped through the reactive material.

Research at the University of Waterloo identified abiotic reductive dehalogenation reactions involving elemental iron as an electron source to degrade halogenated methane, ethanes and ethenes to non-toxic reduced species (O'Hannesin and Gillham, 1992). A laboratory study showed significant degradation of TCE after 50 hours contact time with a reactive metal as shown in Figure 26.

Greater control of system hydrodynamics and plume control can be obtained through the use of funnel-and-gate technology, where impermeable walls are used to funnel

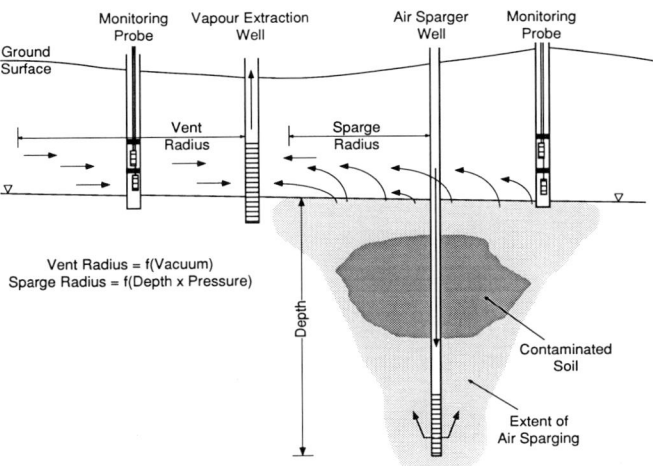

Figure 21 *Typical air-sparging system.*

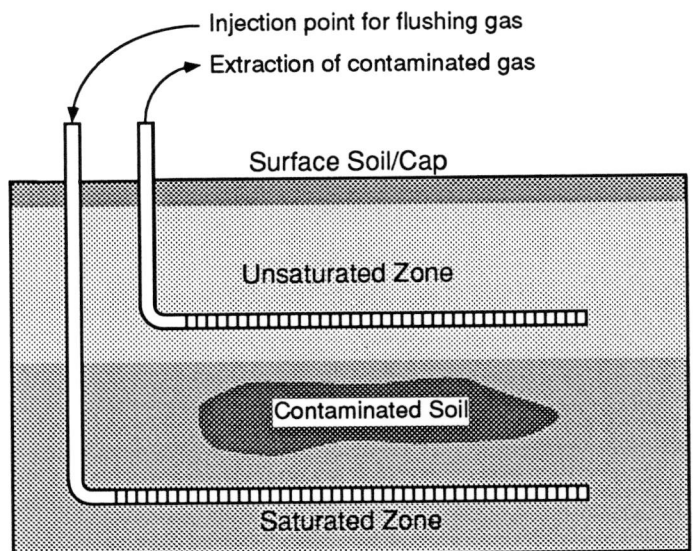

Figure 23 *Air sparging using horizontal wells (after National Research Council, 1994).*

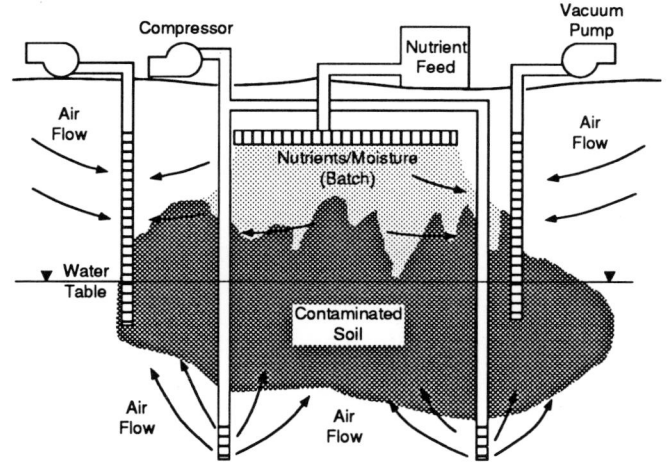

Figure 22 *Air-sparging system with nutrient feed and vapour extraction (after National Research Council, 1993).*

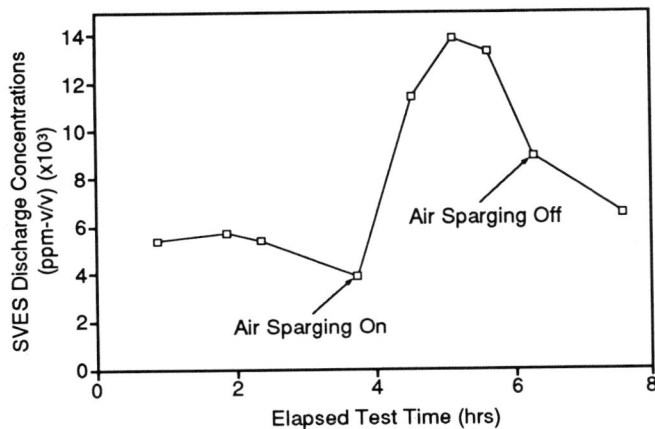

Figure 24 *Impact of air sparging on concentrations of volatile organic compounds in soil vapour (after Marley et al., 1992).*

contaminated ground water through permeable gates in the wall that contain reactive material. A funnel-and-gate configuration is illustrated in Figure 27. Recent research in reactive wall technology involves the use of a proprietary mixture containing magnesium oxides that slowly releases oxygen when contacted with water. This process has potential application to enhance *in situ* aerobic bioremediation of petroleum hydrocarbons such as gasoline (Thomas and Mackay, 1994).

Constructed Wetlands
These are passive treatment systems that mimic natural conditions and have been applied to the treatment of acid mine drainage problems, municipal wastes and urban storm runoff. A more detailed review is provided in Chapter 12.

Isolation
Isolation is a method of managing contaminated soil and/or ground water at a site by containing the contamination within physical barriers. Physical barriers could include low-permeability caps and walls.

The use of capping material over contaminated areas reduces the exposure of human and environmental receptors to the contaminant, and/or reduces the infiltration of precipitation through the contaminated zone. The reduction in infiltration can result in lower contaminant loadings from the unsaturated zone to ground water. Capping materials can include natural materials such as clay, or paving materials such as asphalt or concrete. Redevelopment of such sites can incorporate design and landscape features such as parking lots, roadways, structural envelopes, berms or raised gardens over contaminated areas.

Where contaminant migration off-site must be controlled, low-permeability barrier walls constructed of grouted materials, clay, plastic membranes, or sheet piling can be installed. Barrier walls extend below the zone of contamination, and are tied into a low-permeability zone to facilitate hydraulic con-

tainment. Hydraulic containment can be further enhanced by maintaining inward hydraulic gradients into the contained area through ground-water pumping. Pumped ground water may require treatment prior to discharge. The appropriate isolation method will depend on the mobility and toxicity of the contaminant and the intended use of the property.

Contaminated sites that are managed by isolation methods should be registered on title, and with regulatory agencies and local government building departments, so that existing and future owners and users are aware of the type, extent and location of the contaminants, in the event that redevelopment occurs or subsurface maintenance is required. As well, sites should be subject to long-term monitoring to ensure that containment features operate as designed, and contingency plans should be implemented in the event of containment failure. Isolation is often considered at those sites where clean-up is either technically or economically not feasible.

DISCUSSION
Soil and ground-water treatment technologies are continually evolving in an attempt to provide cost-effective solutions to contaminant clean-up problems. The remediation of soil and ground water to regulatory clean-up criteria has severe long-term economic implications relating to treatment costs and the abandonment and/or under use of contaminated land in

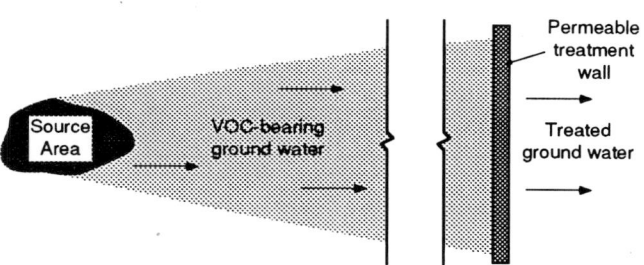

Figure 25 *Schematic diagram of a reactive wall.*

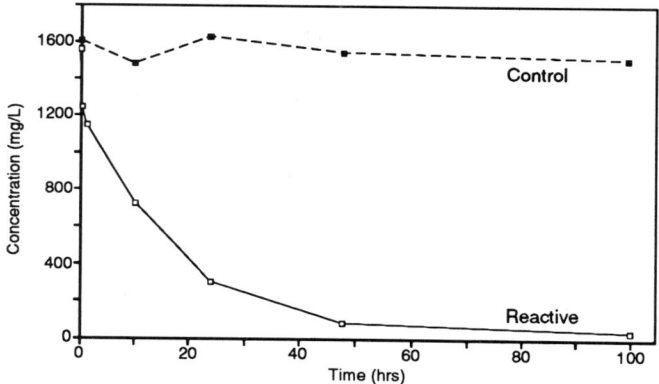

Figure 26 *Degradation of TCE with reactive metal contact time.*

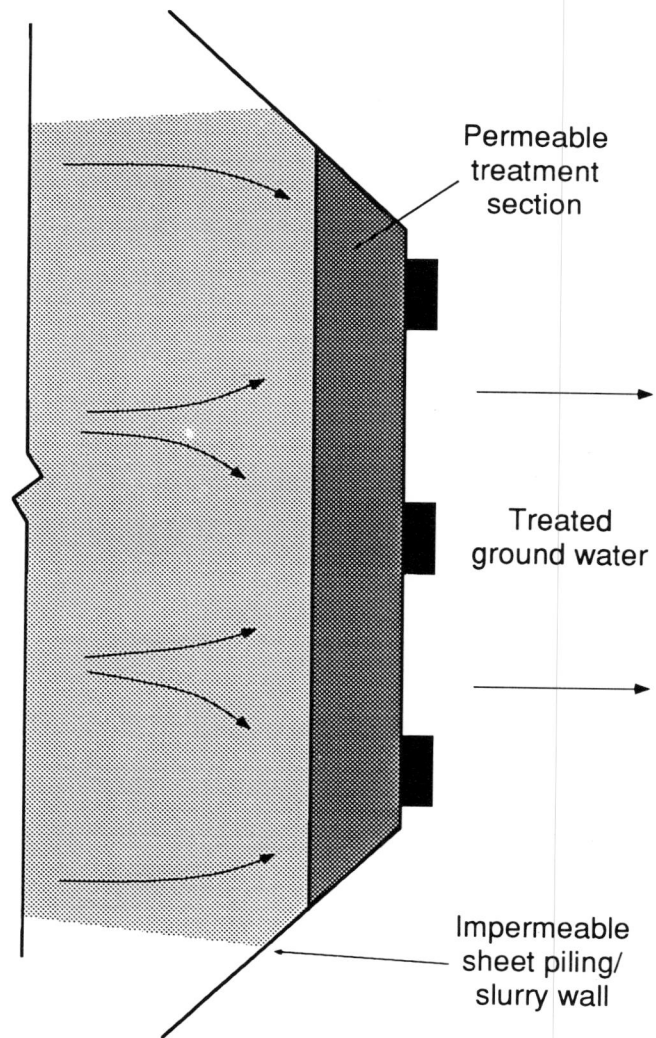

Figure 27 *Schematic of funnel-and-gate technology.*

prime urban settings (see Chapter 27 for discussion). These conditions, along with the landfill ban on hazardous chemicals in some jurisdictions, have led to an increase in the use of alternative technologies. However, because of the relatively short time frame within which alternative technologies have developed, there is a general lack of good quality assessment information relating to the efficiency and applicability of specific technologies to a given contaminant problem. Part of the problem stems from the high cost of conducting such assessments, which typically includes third-party review of a given technology. Government-sponsored demonstration programs have assisted in the evaluation of a number of treatment technologies.

In some situations, particularly in the application of *in situ* methods, the limiting factor to successful treatment is less an issue of technical performance of the technology, and more an issue of the understanding of subsurface conditions and the ability to control the treatment process in the subsurface. For example, heterogeneities in aquifer permeability may be sufficient to prevent ground-water removal from low-permeability zones within the area of contamination, resulting in inefficient and lengthy pumping. Heterogeneities also hinder the distribution of additives which are part of some *in situ* treatment methods (*e.g.*, surfactants, electron acceptors), consequently, contaminated fine-grained materials within and adjacent to an aquifer may be unaffected by pumping strategies or *in situ* treatment. In some situations, inadequate site characterization information and incomplete technology treatability studies can result in inappropriate or improper application of a technology.

Problems associated with soil and ground-water clean-up can include (1) inadequate site characterization; (2) insufficient vendor information indicating results of previous applications and indicating limitations of the technology; (3) incomplete treatment at the site; (4) inadequate monitoring to confirm clean-up; (5) inability of the clean-up program to meet clean-up criteria; and (6) difficulty in obtaining regulatory approvals. Ideally a vendor should be able to show application of his/her technology to the client's situation through (7) documentation of technology demonstration involving third party review; (8) documentation of technology through peer reviewed literature; (9) references of clients that used the technology; and (10) ability to conduct bench and/or pilot scale treatability studies of the technology using an industry accepted protocol that accounts for mass balance of the contaminant.

ACKNOWLEDGEMENTS

Figures were draughted by Mike Doughty.

REFERENCES

Acar, Y.B. and Alshawabkeh, A.N., 1993, Principles of electrokinetic remediation: Environmental Science and Technology, v. 27, p. 2638-2647.

Anderson, T.A., Guthrie, E.A. and Walton, B.T., 1993, Bioremediation in the rhizosphere: Environmental Science and Technology, v. 27, p. 2630-2636.

Brox, G.H. and Hanify, D.E., 1989, A new solid/liquid contact bio-slurry reactor making bio-remediation more cost-competitive, *in* SUPERFUND '89, Proceedings of the 10th National Conference, November 27-29, 1989, Washington, DC, p. 371-3.

Delthorn, R.T. and Mazzoni, A.F., 1986, Activated carbon, what it is, how it works: Water Technology, v. 9, p. 6-29.

Domenico, P.A. and Schwartz, F.W., 1990, Physical and Chemical Hydrogeology: John Wiley & Sons, New York, 824 p.

Hughes Environmental Systems Inc., 1992, Steam-enhanced recovery process, *in* SUPERFUND Innovative Technology Evaluation Program Technology Profiles, Fifth Edition: EPA/540/R-92/077, Nov. 1992, p. 106-7.

Jackson, R.F., Fountain, J.C. and Wunderlich, R.W., 1992, Chemically enhanced solubilization of DNAPL: Advancing the state of pump and treat remediation by injection and withdrawal of biodegradable surfactants: International Association of Hydrogeologists, Hamilton, Ontario, May 11-13, 1992, Proceedings, p. 104-116.

Lageman, R., 1993, Electroreclamation: Environmental Science and Technology, v. 27, p. 2648-2650.

Lehr, J.H., 1991, Granular Activated Carbon (GAC): Everyone knows of it, few understand it: Ground Water Monitoring Review, Fall 1991, p. 5-8.

Marley, M.C. and Droste, E.X., 1995, Sparging strips trapped contaminants: International Ground Water Technology, v. 1, p. 18-23.

Marley, M.C., Hazebrouck, D. and Walsh, M.T., 1992, The application of *in situ* air sparging as an innovative soils and groundwater remediation technology: Ground Water Monitoring Review, Spring, p. 137-144.

Mercer, J.W., Skipp, D.C. and Giffin, D., 1990, Basics of Pump-and-Treat Ground-Water Remediation Technology: U.S. EPA/600/-8-90/003, 31 p.

National Research Council, 1993, In Situ Bioremediation — When Does It Work?: National Academy Press, Washington, DC, 207 p.

National Research Council, 1994, Alternatives for Ground Water Cleanup: National Academy Press, Washington, DC, 315 p.

O'Hannesin, S.F. and Gillham, R.W., 1992, A permeable reaction wall for *in situ* degradation of halogenated organic compounds [abstract]: 45th Canadian Geotechnical Society Conference, Toronto, Ontario, October 25-28, 1992, conference volume, p. 16.

Ontario Ministry of Environment and Energy (OMOEE), 1994, Ontario's Environmental Technologies Program: Partnerships for a Cleaner Environment, Current Projects, 33 p.

Raven Beck Environmental Ltd., 1995, Report on demonstration projects for soil and ground water management at Ataratiri: prepared for the Waterfront Regeneration Trust, 42 p.

Staps, J.J.M., 1990, International evaluation of *in situ* biorestoration of contaminated soil and groundwater: National Institute of Public Health and Environmental Protection, Bilthoven, The Netherlands, 192 p.

Thomas, D.G. and Mackay, D.M., 1994, Oxygen enhancement of groundwater using a solid oxygen releasing compound [abstract]: Air and Waste Management Association Annual Conference, Burlington, Ontario, May 1-3, 1994, Proceedings, p. 12.

Udell Technologies, Inc., 1992, In situ steam enhanced extraction, *in* SUPERFUND Innovative Technology Evaluation Program Technology Profiles, Fifth Edition: EPA/540/R-92/077, Nov. 1992, p. 184-185.

United States Environmental Protection Agency, 1989, Chemical on-site treatment utilizing KPEG process at Wide Beach, New York, in Demonstration of Remedial Action Technologies for Contaminated Land and Groundwater: NATO/CCMS Third International Conference, Montreal, PQ, p. 7-36.

United States Environmental Protection Agency, 1991, Soil Vapour Extraction Technology Reference Handbook: EPA/540/2-91/003, 316 p.

United States Environmental Protection Agency, 1992a, Chemical treatment and ultrafiltration: Emerging Technology Bulletin, SITE Program, EPA/540/F-92/002.

United States Environmental Protection Agency, 1992b, Membrane filtration: Demonstration Bulletin, SITE Program, EPA/540/MR-92/014, 136 p.

United States Environmental Protection Agency, 1992c, Electron beam treatment for removal of trichloroethylene and tetrachloroethylene from streams and sludge: Emerging Technology Bulletin, SITE Program, EPA/540/F-92/009, 51 p.

United States Environmental Protection Agency, 1993a, SUPERFUND Innovative Technology Evaluation Program, Technology Profiles, Sixth Edition: EPA/540/R-93/526, 231 p.

United States Environmental Protection Agency, 1993b, Hydraulic fracturing technology: applications analysis and technology evaluation report: EPA/540/R-93/505, 90 p.

United States Environmental Protection Agency, 1993c, A cross-flow pervaporation system for removal of VOCs from contaminated wastewater: Emerging Technology Bulletin, SITE Program, EPA/540/F-93/503, 15 p.

United States Environmental Protection Agency, 1993d, Microfiltration technology: Demonstration Bulletin, SITE Program, EPA/540/MR-93/513, 21 p.

United States Environmental Protection Agency, 1993e, Peroxidation Systems, Inc. perox-pure chemical oxidation technology: Technology Demonstration Summary, SITE Program, EPA/540/SR-93/501, 6 p.

United States Environmental Protection Agency, 1993f, Destruction of organic contaminants in air using advanced ultraviolet flashlamps: Emerging Technology Summary, SITE Program, EPA/540/MR-93/520, 10 p.

United States Environmental Protection Agency, 1993g, CAV-OX Ultraviolet Oxidation Process Magnum Water Technology: Demonstration Bulletin, SITE Program, EPA/540/MR-93/520, 18 p.

United States Environmental Protection Agency, 1993h, Electron beam treatment for the removal of benzene and toluene from aqueous streams and sludge: Emerging Technology Bulletin, SITE Program, EPA/540/F-93/502, 22 p.

United States Environmental Protection Agency, 1993i, Removal of phenol from aqueous solutions using high energy electron beam irradiation: Emerging Technology Bulletin, SITE Program, EPA/540/F-93/509, 123 p.

United States Environmental Protection Agency, 1994a, Innovative treatment technologies: Annual Status Report, Sixth Edition: EPA/542/R-94/005, 37 p.

United States Environmental Protection Agency, 1994b, Eco Logic International Gas Phase chemical reduction process and thermal desorption unit, Middleground Landfill, Bay City, MI: SITE Technology Demonstration Summary, EPA/540/5R-93/522, 8 p.

36. On-site *Ex Situ* Bioremediation of Petroleum-Contaminated Soils

Paul J. Hubley
Andrew W. Panko

Arcturus Environmental Limited, Niagara Falls, Ontario L2E 6S5

Doug W. Boocock

Petro-Canada Products, North York, Ontario M2N 6L6

SUMMARY

Landfilling of petroleum-contaminated soils has historically been an easy and inexpensive solution to site remediation. However, with the shortage of landfill space, increasing landfill costs, and problems of liability a low-cost alternative to landfilling is required in many urban areas.

This paper presents a case study for an ongoing *ex situ* bioremediation project, involving excavation and placement of contaminated soil, bacteria and nutrients in an above-ground engineered bioremediation cell at a petroleum distribution facility. The case study represents a good example of an urban environmental geology investigation due to the interrelation of many physical and chemical, hydrogeological, and historical considerations. In addition, the proximity of the site to a sensitive watershed and residences has resulted in the involvement of numerous regulatory agencies, each with particular interests.

INTRODUCTION

Ex situ bioremediation refers to the excavation of contaminated soils and their treatment on surface or off-site (Chapter 35). The term bioremediation describes the use of bacteria to degrade organic compounds. On-site *ex situ* bioremediation of approximately 6000 tonnes (t) of diesel-fuel-contaminated soil is currently ongoing at a former petroleum distribution facility located in a southern Ontario city. *Ex situ* bioremediation was identified as the most cost-effective remediation strategy, based on a site-specific remedial option feasibility study.

Implementation of the remedial strategy required co-operation from a variety of provincial, regional, and local regulatory agencies and special interest groups, and most importantly, local residents and neighbouring business people. *Ex situ* bioremediation using indigenous (native) micro-organisms was considered favourable by regulatory agencies due to the low potential impact to the surrounding neighbourhood and local watershed and the lack of landfilling required.

Implementation of *ex situ* bioremediation involved three main activities: first, a site assessment was performed to assess the chemical and physical characteristics of the contaminants, ground-water flow regime, stratigraphy, and numerous other factors; second, a detailed remedial strategy was developed with regard to the desired remedial objectives, and necessary permits were obtained; and third, the remedial strategy was implemented and monitored.

BACKGROUND

The site, including an active gas bar/car wash and a former bulk storage area, covers approximately 0.8 ha along the banks of a tributary of the Grand River, in south-central Ontario. The banks of the river had been asphalt-capped, with up to 1 m of asphalt, to stabilize the banks prior to placement of imported fill required to support commercial buildings on the river bank.

The former bulk plant had been operated for several decades by a large Canadian petroleum company, but had been decommissioned for some time. In the bulk plant area, several grades of heating oils and diesel fuel had been stored in a total of seven underground, and two above-ground, fuel storage tanks, for later distribution. In addition, three underground storage tanks, containing gasoline for use at the retail gas bar, had been recently removed from the bulk plant portion of the property.

The surrounding land use is primarily commercial and residential; industrial activities are located within view upstream. The riverine environment, including mammals, fish, and waterfowl, is under the jurisdiction of the Grand River Conservation Authority; recreational fishing is common.

Stratigraphy and Hydrogeology

The upper part of the site stratigraphy consists of fill materials including highly permeable sand and gravel, brick and concrete blocks, placed to a depth of between 4 to 6 m below current grade. Below the fill is generally a thin organic clayey silt with occasional sand lenses, then a coarse sand and gravel outwash deposit to an indeterminate depth of at least

6.5 m below grade. The average hydraulic gradient across the site is 0.08. The average hydraulic conductivity of the sand and gravel fill is in the order of 1.5×10^{-5} cm·s^{-1}, based on recovery tests in on-site wells. Ground water from the area of contamination generally flows above the native material directly toward the Grand River which, in turn, empties into Lake Erie.

Environmental Site Assessment

An environmental site assessment and delineation study was conducted, and this confirmed that free-phase (liquid) hydrocarbons were present in the subsurface near former underground and above-ground storage tanks. The liquid hydrocarbons were suspected to be a mixture predominantly consisting of diesel fuel, with minor amounts of other fuels. Ground water and soil in a large (750 m²) area surrounding the former storage tanks had been impacted by free-phase product. This is a typical situation commonly encountered in the vicinity of underground storage tanks (Riser-Roberts, 1992; Flatham *et al.*, 1994; Chapters 7, 35).

Soil and ground-water sampling was conducted as part of the environmental site assessment to determine the lateral and vertical extent of soil and ground-water contamination. Samples were analyzed for benzene, toluene, ethylbenzene, m/p-xylenes and o-xylene (referred to collectively as BTEX), total petroleum hydrocarbons (TPH) and lead. One sample was submitted for Ontario Ministry of Environment and Energy (OMOEE) Regulation 347 Schedule 4, Leachate Quality Criteria testing. Results of soil and ground-water sampling defined the aerial extent of soil requiring remediation (Fig. 1), representing approximately 6000 t of petroleum hydrocarbon-contaminated soil.

Gas chromatography analysis of contaminated soil identified that nearly all (>90%) of the petroleum hydrocarbons were between carbon number C_8 and C_{22}, and therefore considered semi-volatile. The main contaminants were middle distillates, which contain little to no BTEX and lead. Therefore, the zone of contamination was based on TPH exceedance of provincial guidelines. Similarly, remediation was driven by TPH exceedances. Based on site hydrogeological characteristics, the location and extent of con-

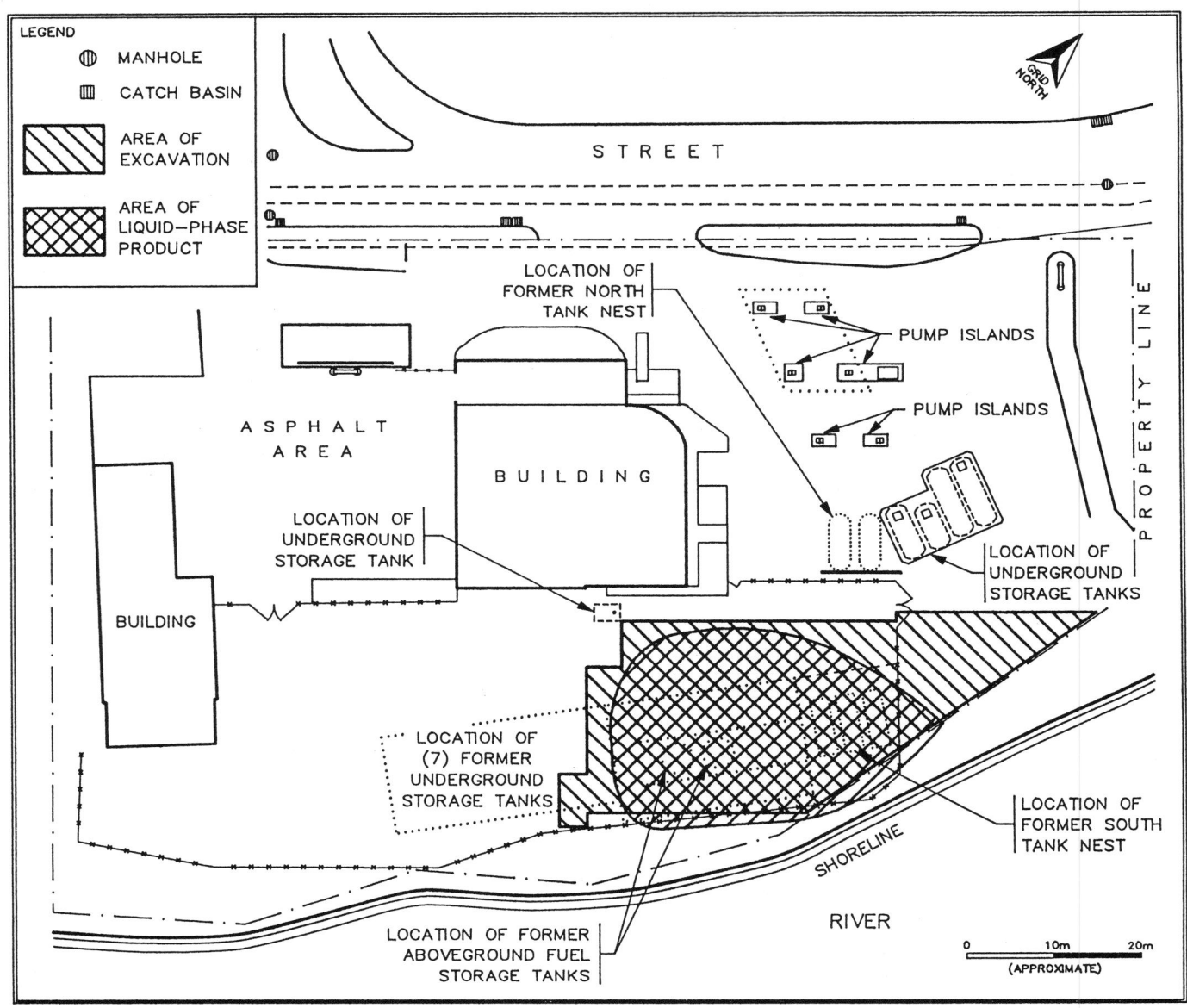

Figure 1   *Extent of liquid phase product in the vicinity of former petroleum storage tanks.*

tamination, and the nature of contamination (petroleum hydrocarbons), a significant potential existed for impact to the adjacent river.

REMEDIAL STRATEGY

Remedial Objectives

Remedial objectives were based on Ministry of Consumer and Commercial Relations Fuels Safety Branch (FSB) Gasoline Handling Act Standard GH12 Guidelines for the Remediation of Petroleum Contamination at Operating Commercial Sites in Ontario (May 1992). Using the decision-tree approach provided in the guidelines, a site sensitivity analysis

was used to set remedial objectives as level II (Fig. 2). FSB Standard GH12 Guidelines are currently part of OMOEE Interim Guidelines for the Assessment and Management of Petroleum Contaminated Sites in Ontario (August, 1993).

Remedial Option Feasibility Study

Based on the results of the environmental site assessment, and concerns regarding potential for further impact to the riverine ecology, a variety of remedial strategies was considered as part of a remedial option feasibility study (Table 1). Many available remedial strategies were not cost-effective and/or were inadequate for complete remediation of the 6000 t of petroleum-contaminated soil. Some available methods

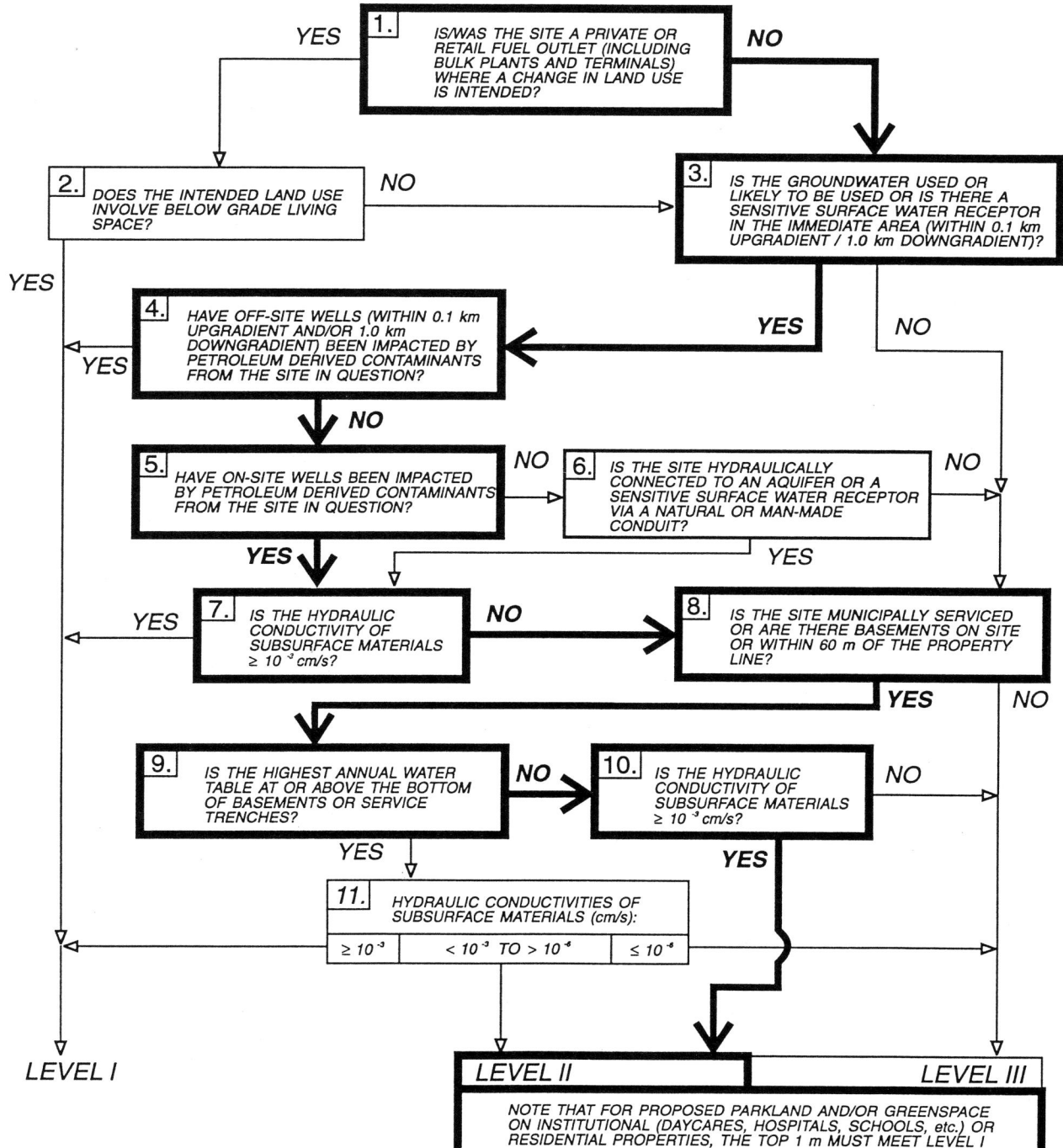

Figure 2   *Site sensitivity analysis.*

using chemical oxidants, surfactants, or nutrient solutions were considered too risky due to the potential impact to the adjacent river. Technologies requiring the release of chemicals *in situ* may be cost prohibitive, due solely to the time and effort required to obtain necessary permits and ensure some control on the release of the chemicals. Other invasive (removing source soils for treatment above-grade) technologies proved cost-prohibitive, or relied on stabilization rather than destruction of contaminants.

Of particular note in Table 1 are the numerous advantages of landfilling over other available remedial options. Clearly, landfilling in the province of Ontario has remained as the preferred remedial option, largely due to the potential for a rapid turnaround of real estate, high potential for complete site remediation, and the ease of obtaining required permits to dump at the landfill. However, with the available volume in landfills diminishing, and costs rising, landfilling is no longer the remedial strategy of choice. The question of liability for environmental impacts arising from landfilling also arises (see Chapter 35).

On-site, but *ex situ*, bioremediation exhibited the most advantages for remediation of the 6000 t of middle distillate contaminated soil at this site. Unlike other remedial options considered in the feasibility study, *ex situ* bioremediation represented a viable, cost-effective alternative to landfilling, and has been well documented as an effective method of soil remediation for soils containing fresh and weathered distilled products of oil.

Permits. Implementation of the remedial action plan required the combined efforts of a diverse group of regulating agencies, including the provincial Ministry of Environment and Energy and Ministry of Consumer and Commercial Relations Fuels Safety Branch, regional and city works departments, local fire department, and the local conservation authority. On-site *ex situ* bioremediation was generally well received by regulators, as it meant immediate source removal and provided the opportunity for complete breakdown of contaminants to inert compounds. In addition, the strategy supports the OMOEE Reduce, Reuse, and Recycle ("3Rs") directives, by direct reuse of all soil on-site.

Table 1   Site-specific remedial options study. See Chapter 35 for a detailed review of each option.

Option	Cost Effective	Complete Remediation Likely	Reasonable Treatment Time	Minimal Regulatory Requirements	Resources Immediately Available	Comments
In situ Bioremediation introduction of nutrients, oxygen, and biological organisms	no	no	no	no	yes	• not possible with the volume of free product *in situ* • risk of nutrient loading to the river
Ex situ Bioremediation excavation and manipulation of contaminated soil at surface in a lined containment unit or bioreactor	yes	yes	yes	yes	yes	• rapid treatment; suitable for these shallow source soils (~ 6000 t) • very effective on source soils
Landfill disposal of non-hazardous materials in an approved waste disposal site	no	yes	yes	yes	yes	• considerable landfill space would be used
LTTD on-site treatment of contaminated soil using low temperature thermal desorption	no	yes	yes	no	yes	• not cost-effective to treat 6000 t of source soils • significant regulatory approvals
Air Sparging *in situ* recovery/treatment of volatile contaminants from groundwater by injecting air into the water table	no	no	no	no	yes	• not feasible with the volume of free product present in the subsurface
Active/Passive Bioventing *in situ* oxygenation of subsurface to induce bioremediation	no	no	no	no	yes	• not feasible with the volume of free product present in the subsurface
Incineration *ex situ* destruction by high temperature incineration	no	yes	yes	no	no	• not cost effective to treat 6000 t of soil • significant regulatory approvals
Soil Vapour Extraction *in situ* recovery/treatment of volatiles from the subsurface	no	no	no	no	yes	• not feasible with the volume of free product present in the subsurface

REMEDIAL DESIGN

Microbiology And Bioremediation

Petroleum hydrocarbons in the middle distillate (diesel fuel) range are readily degraded in the presence of oxygen by aerobic micro-organisms. Microbially mediated breakdown of chemicals is referred to as biodegradation. Biodegradation of diesel-type compounds in soil is likely mainly due to the action of soil bacteria and fungus. Typical microbially mediated reactions for breakdown of diesel-type compounds are:

$$C_8H_8 + 10O_2 \rightarrow 8CO_2 + 4H_2O \qquad (1)$$
xylene

$$C_{10}H_{22} + 15\tfrac{1}{2}O_2 \rightarrow 10CO_2 + 11H_2O \qquad (2)$$
decane

Although fresh and weathered petroleum products have been demonstrated in numerous laboratory and bench scale studies as readily degradable (Riser-Roberts, 1992), the effectiveness of a particular bioremediation effort is highly site-specific. Physicochemical properties of the soil, method of remediation *(in situ* or *ex situ)*, remedial end-point(s), and microbial considerations (indigenous or inoculated populations, nutrient availability, *etc.*) are important to the success of a particular bioremediation effort. Physical parameters conducive to effective bioremediation of diesel fuel usually include pH between 6 and 8, moisture between 50% and 80% of field capacity, temperature between approximately 10°C and 30°C, and carbon:nitrogen:phosphorus ratios of approximately 100:10:4.

Sorption of diesel fuel to soil particulates, which is of primary concern with weathered contaminants in fine-grained soils, provides both inhibitory and enhancing effects with respect to biodegradation. Inhibitory effects include the unavailability of hydrocarbons for biodegradation, by limiting the concentration of hydrocarbons in the aqueous phase, and thereby limiting bioavailability of the hydrocarbons to indigenous micro-organisms. Enhancing effects are mainly due to the reduction of low molecular weight (carbon numbers C_5 to C_{10}) alkanes and cyclic alkanes, which are toxic to micro-organisms (Leahy and Colwell, 1990).

Option	Cost Effective	Complete Remediation Likely	Reasonable Treatment Time	Minimal Regulatory Requirements	Resources Immediately Available	Comments
Recovery Wells (Pump and Treat) recovery and treatment of groundwater and light non-aqueous phase liquids, to burn, export, refine, or reuse	no	no	no	yes	yes	• ineffective due to adjacent river influence • hydraulic connection to river will prevent the development of required cone of depression
Soil Flushing removal of organic and/or inorganic contaminants from soil for recovery and treatment using a treatment liquid *in situ* or *ex situ*	no	yes	no	no	yes	• unacceptable risk of chemical contamination to the nearby river
Chemical Degradation introduction of reagents to immobilize or reduce contaminant toxicity *in situ* or *ex situ*	no	no	no	no	no	• unacceptable risk of chemical contamination to the nearby river
Impermeable Barriers temporary physical barriers placed *in situ* to isolate contaminated groundwater	no	no	no	no	no	• not a treatment, only containment • contaminants would remain in subsurface
***In situ* Vitrification** contaminated soils are converted into a chemically inert stable glass and crystalline product through heating	no	no	no	no	no	• not a proven or accepted technology • affected area is too large • contaminants would remain in the subsurface
Solidification/Stabilization contaminated soil is mixed into a bituminous cement or insoluble matrix *in situ* or *ex situ*	no	no	no	no	no	• not a treatment, only containment • contaminants would remain in subsurface
Impermeable Barriers temporary physical barriers placed *in situ* to isolate contaminated areas and assist control of contaminated groundwater	no	no	no	no	no	• not a treatment, only containment • contaminants would remain in subsurface

Table 1   *cont'd.*

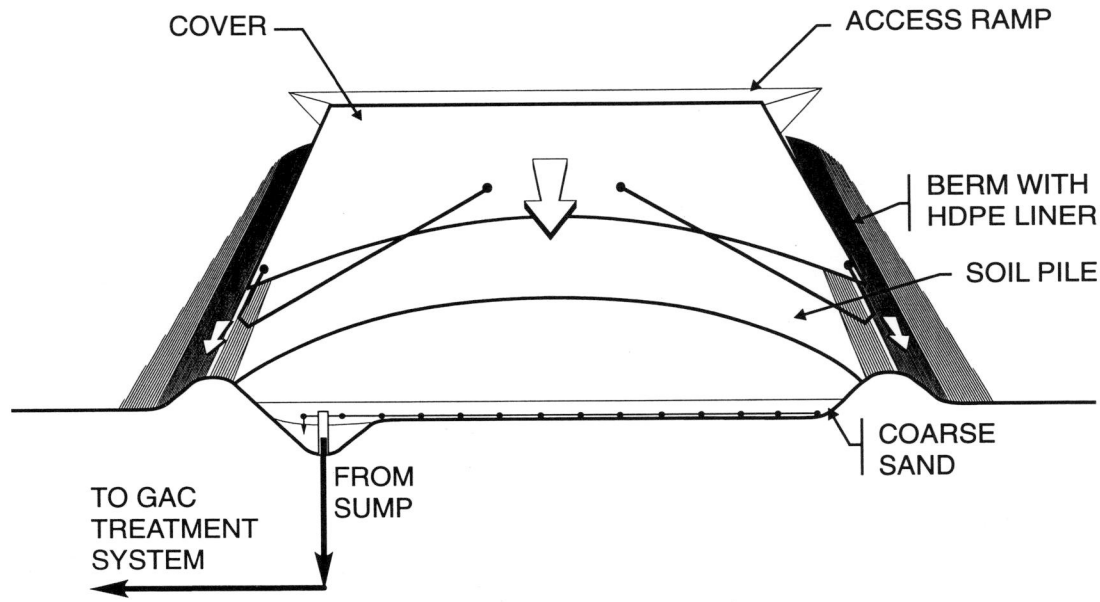

Figure 3 *Schematic of engineered bioremediation cell (EBC).*

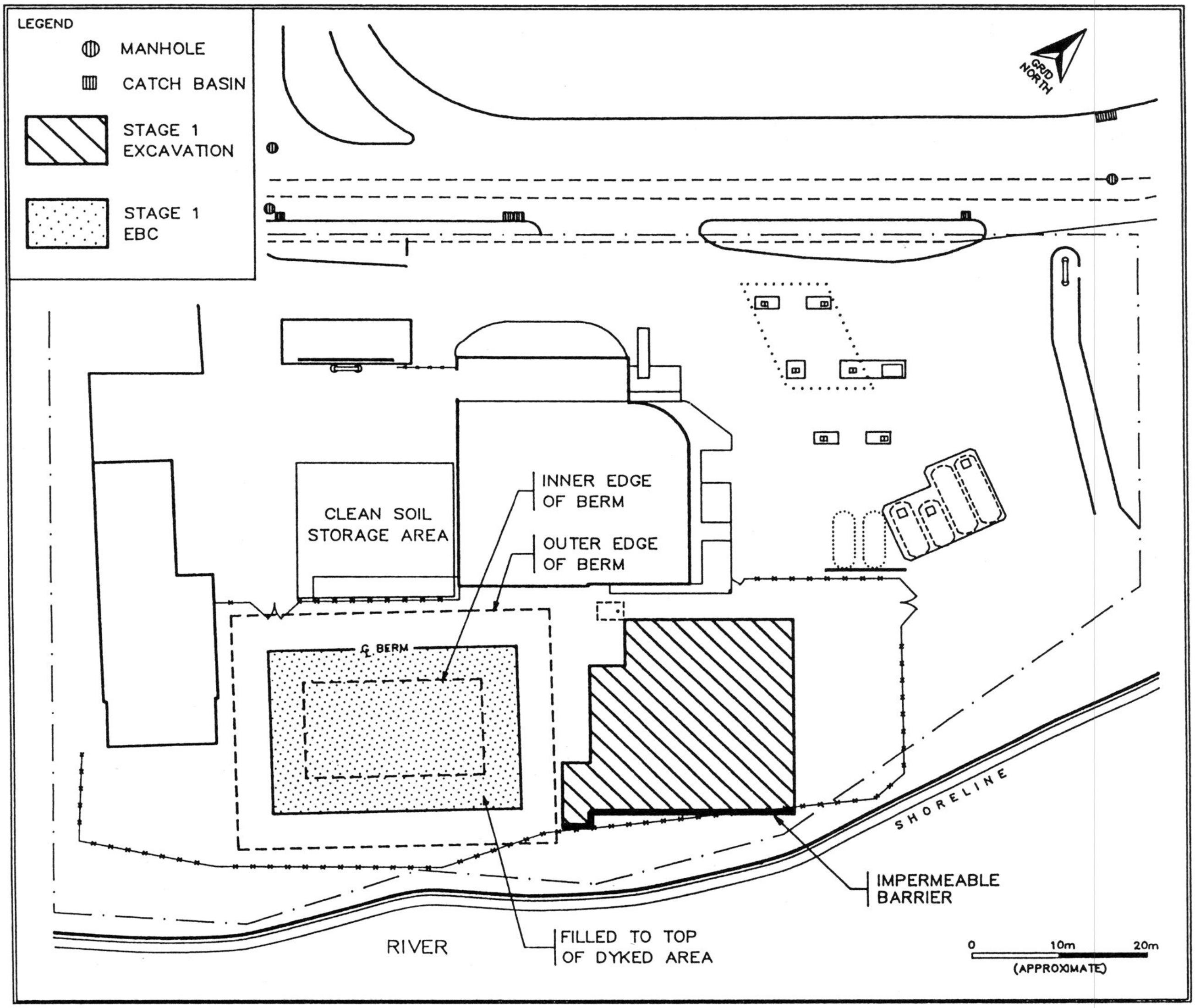

COVER

ACCESS RAMP

BERM WITH HDPE LINER

SOIL PILE

COARSE SAND

TO GAC TREATMENT SYSTEM

FROM SUMP

LEGEND

MANHOLE

CATCH BASIN

STAGE 1 EXCAVATION

STAGE 1 EBC

GRID NORTH

CLEAN SOIL STORAGE AREA

INNER EDGE OF BERM

OUTER EDGE OF BERM

C BERM

FILLED TO TOP OF DYKED AREA

IMPERMEABLE BARRIER

SHORELINE

RIVER

0   10m   20m
(APPROXIMATE)

Figure 4 *First stage EBC and excavation.*

Preliminary plate-counting techniques were used at this site to assess the total bacteria and hydrocarbon (diesel)-utilizing bacteria in contaminated soil. Based on the results of the microbial assessment, it was concluded that a viable microbial population existed in the soil, and that the native microbial population would be able to degrade site contaminants. Augmentation with endogenous (cultured) bacteria would not be necessary. Nutrient additions were necessary to achieve the desired C:N:P ratio of 100:10:4. Nutrients included inorganic (nitrogen, potassium and phosphorus mix) and organic (urea nitrogen) supplements. The top 1 m of soil in the pile was regularly tilled to allow adequate air (oxygen) to the micro-organisms. Moisture was adjusted when necessary by sparging.

Engineered Bioremediation Cell

Ex situ bioremediation of petroleum-contaminated soil was achieved within an engineered bioremediation cell (EBC) shown in Figure 3. The EBC was lined with high-density polyethylene (HDPE) plastic to prevent leaching of contaminants below the liner and escape of vapours from the pile. The EBC was constructed using basal drainage piping graded to a sump surrounded by a gravel collection gallery. Contami-

nated water was removed from the sump and pumped through a treatment system, including an oil/water separator and containers with granular activated carbon (GAC). Treated water was then discharged to the sanitary sewer.

Placement of the Engineered Bioremediation Cell

As often the case in urban areas, available space for soil storage and vehicle movement was limited, and therefore placement of the EBC required overlap with the area of excavation. Therefore the excavation of the 6000 t of contaminated soil was conducted concurrently with construction of the EBC in two stages. The first stage (Fig. 4) involved approximately 60% of the total construction and excavation. Prior to soil excavation, an impermeable HDPE liner was placed parallel to the river to a depth below the ambient water level, in order to sever the hydraulic connection between the river and a small amount of inaccessible soil containing residual petroleum hydrocarbons.

In the second stage, the remaining section of contaminated soil was excavated, and the EBC was completed (Fig. 5). At this time, the impermeable barrier between the excavation and river was extended. Nutrient supplements were added upon soil loading into the EBC. Clean soil excavated

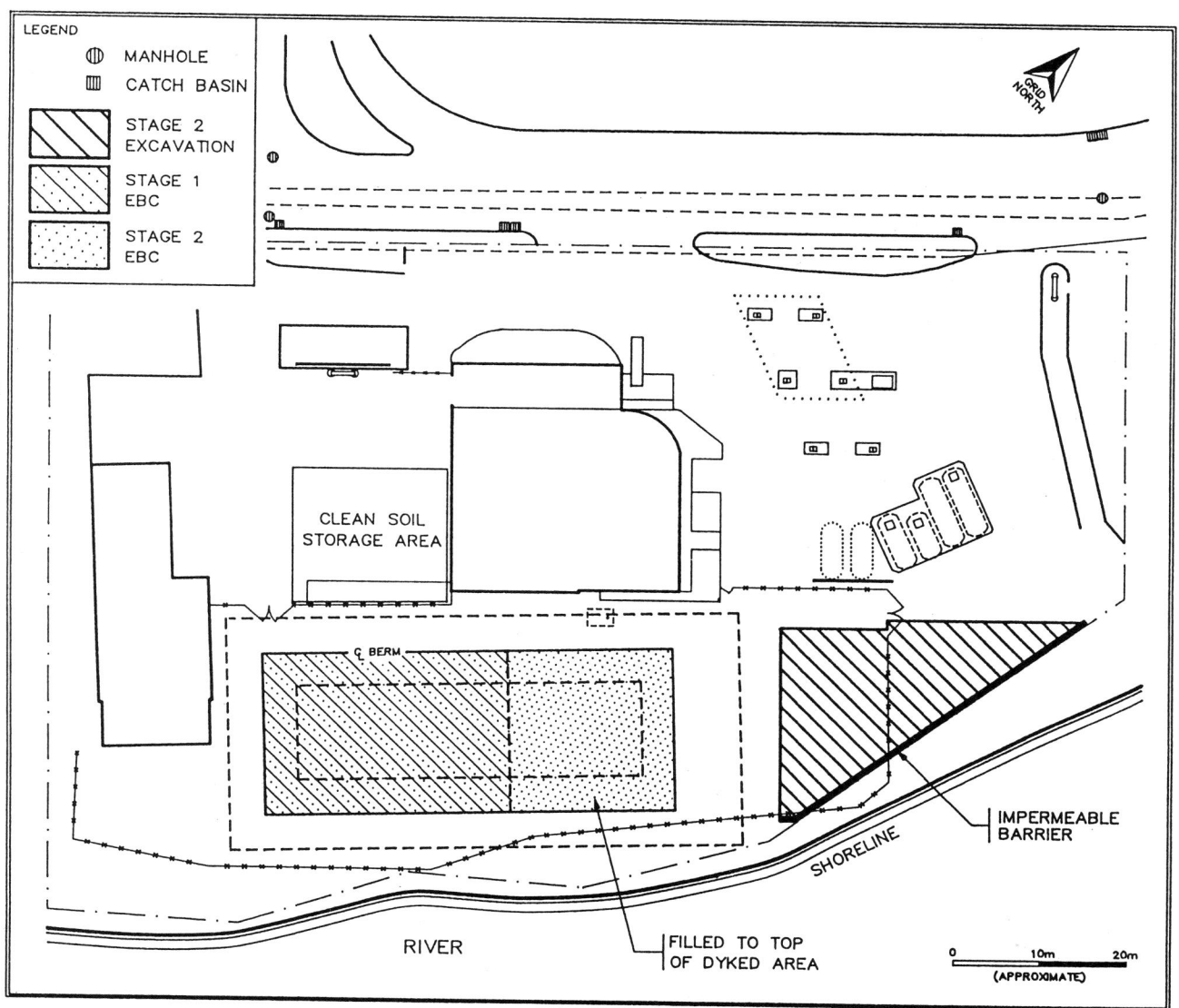

Figure 5   *Second stage EBC and excavation.*

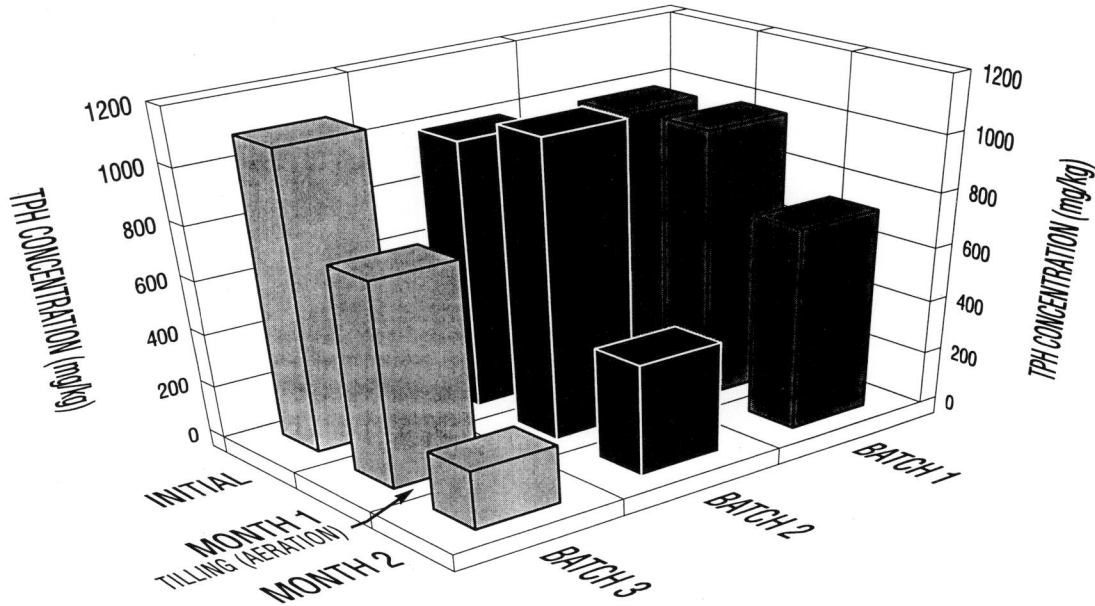

Figure 6 *Total petroleum hydrocarbons (TPH) concentration with time.*

during EBC construction was used to partially backfill the excavated area.

Removal/placement of treated soil.   Each month, treated soil was removed in lifts of between 0.5 m and 1.0 m, and placed in the former excavation as backfill. At this time, the next layer was tilled to supply necessary oxygen, and nutrients and moisture were added as required. The site is expected to be restored to grade following eight warm weather months.

Preliminary Results

Composite soil samples, representing approximately 12 sub-samples per 1000 t soil, were collected during remedial activities, and preliminary soil treatment results have been compiled.

The mean TPH concentration in soil is presented in Figure 6 for three batches (lifts of approximately 1000 t each) of soil. As shown in Figure 6, the mean TPH concentration in each batch was tracked for a period of two months, by sampling initially after one month, and then re-sampling approximately one month after soil tilling. As discussed, nutrients were added upon initial placement of soil in the EBC. As shown in Figure 6, an apparent lag time, represented by a general lack of TPH reduction by month 1, indicated that nutrients alone may not have been sufficient for significant biodegradation of petroleum hydrocarbons. However, significant biodegradation of soil contaminants was observed by month 2, following aeration by tilling. The increased TPH loss following tilling may be attributed to the need for sufficient available oxygen for aerobic micro-organisms to break down petroleum hydrocarbons efficiently. Air sampling, conducted during tilling, confirmed that no significant volatilization has occurred as a direct result of tilling activities, and therefore supports biodegradation as responsible for observed TPH loss.

At the time of the remedial option study, average landfill costs were approximately $75 per tonne soil. Preliminary costs using on-site *ex situ* bioremediation are approximately $41 per tonne, representing a cost savings averaging approximately 45%.

DISCUSSION

Ex situ bioremediation of petroleum-contaminated soils is a cost-effective and sensible alternative to landfilling. Following a review of available treatment technologies, on-site *ex situ* bioremediation was the most cost-effective solution for remediation of 6000 t of contaminated soil to meet the desired objectives. Using simple aeration and nutrient additions, the native micro-organism population is sufficient to break down diesel-type contaminants. *Ex situ* bioremediation may be implemented on urban sites exhibiting space limitations, by using a staged approach to excavation and EBC construction. Process odours can be effectively contained so as not to effect neighbouring residents and businesses.

REFERENCES

Leahy, J.G. and Colwell, R.R., 1990, Microbial degradation of hydro-carbons in the environment: Microbiological Reviews, v. 54, p. 305-315.

Flatham, P.E., Jerger, D.E. and Exner, J.H., 1994, Bioremediation: Field Experience: Lewis Publishers, Boca Raton, FL, 548 p.

Ontario Ministry of Environment and Energy, August 1993, Interim guidelines for the assessment and management of petroleum contaminated sites in Ontario: 24 p.

Riser-Roberts, E., 1992, Bioremediation of Petroleum Contaminated Sites: CRC Press, Boca Raton, FL, 197 p.

37. Geoscientific Information Systems for Environmental Geology

Nicholas Eyles
Michael Doughty

Environmental Earth Sciences, University of Toronto, Scarborough Campus
1265 Military Trail, Scarborough, Ontario M1C 1A4

Derek Mack-Mumford

Works and Environment Department, City of Scarborough
300 Consilium Place, Suite 1000, Scarborough, Ontario M1H 3G2

SUMMARY
Information is the Achilles' heel of environmental geology. In most urban areas, there is an inadequate understanding of regional geological and hydrogeological conditions despite the existence of a wealth of subsurface data. This wealth, unfortunately, is in hard-copy format only, usually of variable quality, and widely dispersed across government agencies, the private sector and universities. Where sufficient digitized data have been collated, the construction of three-dimensional images of the subsurface geology is now possible using Geoscientific Information Systems (GSIS). This paper reviews commercially available GSIS systems and provides examples of urban areas that have adopted GSIS techniques; particular emphasis is given to the city of Scarborough in southern Ontario. The collection and pooling of multifaceted information into environmental geological data banks is an urgent priority in urban areas with regard to ground water and contaminant transport modelling, land use planning, site remediation design, litigation, public access to information and effective use of public monies. Submission of data to a municipal electronic data bank, using a standard digitized format, should be a mandatory requirement for all development projects; much can be learned from practices followed by the oil industry.

INTRODUCTION
Rapid land use changes in urban areas and the release of contaminants from a wide range of sources pose a threat to air, ground and surface waters, natural habitats and, ultimately, public health (Frönzle, 1993). As yet, however, contaminant pathways through urban environments, from sources to receptors, are not well known. A particular problem is identifying subsurface transport pathways through complex near-surface (<100 m depth) sediments and rocks. Even in areas where much hard-copy stratigraphic data has been

collected, subsurface geological information is commonly scattered across many agencies, and not readily available or in a usable format. There is, as a result, an insufficient understanding and visualization of local or regional geological and hydrogeological variability. This is an obstacle to ground-water resource modelling and protection, identification and tracking of contaminant sources in public health and litigation cases, the design of remedial work at contaminated sites, and dissemination of information to the public.

The purpose of this paper is to briefly introduce the more commonly used computer techniques designed to handle and portray subsurface geological data. Examples are given of urban areas in Canada and the United States that are using such techniques for integrating environmental geology information into the municipal planning process. Adoption of such systems represents a significant saving in direct costs (data are readily available and do not have to be recollected) and indirect costs (environmentally sensitive planning results in reduced need for expensive after-the-event remedial action).

The Problem
Common questions that are routinely asked at the beginning of any environmental geology investigation are: "What geological/hydrogeological information is available for this area? Where can I find it and what format is it in?". Assembling existing data can be difficult and costly; many studies are based on data that are outdated. In most urban areas, subsurface geological data is often dispersed across many different agencies (consultants, government agencies, universities, *etc.*), mostly in hard-copy format, and not collated into a single data base. Much public money is expended in collecting such data, which, at the conclusion of any one project, is then deeply interred in reports, theses and filing cabinets, or simply discarded. Later, large sums of public money are again spent to locate, disinter and translate such data into a usable format

or, often unknowingly, collect the same information.

Filling Data Gaps

In many urban areas, the rate of development and the pace at which planning decisions are made have outstripped the ability to gather data and quantitatively model the environmental impact of existing and future development (the data gaps referred to in Chapter 2). Other sub-disciplines of geology have identified and closed these gaps, most notably in the oil industry. In the mid-1980s, the vast increase in computing power available to model hydrocarbon reservoirs greatly exceeded the availability of quantitative input data. This need was to foster a virtual explosion of effort to quantify stratigraphic information such as permeability, bed thickness, bed geometry and other properties (what geologists call geological heterogeneity) from microscopic to regional scales. This, most importantly, led to much better communication between geologists and users (in this case, production engineers) because the practical and economic significance of detailed geological information was realized in the development of oil fields. Multinational oil companies designed many three-dimensional geological modelling software packages (*e.g.*, Exxon's GEOSET: Krum and Johnson, 1993; Shell's MON-ARCH 3-D: van Vark *et al.*, 1994; SESIMIRA: Gundeso and Egelund, 1990; MOHERES: Tyler *et al.*, 1994; STRATA-MODEL: Bryant *et al.*, 1990; Lasseter, 1990; and many others in development: see Martin, 1994 for an excellent review). Such techniques are now in routine use for stratigraphic modelling and reservoir characterization in the oil and mining industries (*e.g.*, Cross, 1990; Agterberg and Bonham-Carter, 1990; Franseen *et al.*, 1991; Miall and Tyler, 1991; Bryant and Flint, 1993; Martin, 1994; see below) and are finding increasing application for what are essentially similar geological problems and management needs in urban settings.

Urban databases

In Canada, there have been few attempts to establish centralized urban environmental geology databases. Prominent examples are those developed for such cities as Ottawa (Belanger, 1974; Belanger and St. Onge, 1974), Hamilton (Morin, 1975), Saskatoon (Christiansen, 1970) and Edmonton (Kathol and McPherson, 1975). The present-day utility of such databases is very limited, because they are dependent on the then current computing technology and graphics capability; most existing data bases are essentially archival systems only capable of producing simple contour maps. Computer hardware and software for environmental applications have evolved very rapidly in the last few years, and it is now possible to depict three-dimensional visualizations (models, pictures) of subsurface conditions.

Geographic *versus* Geoscientific Information Systems

Geologists must possess the ability to conceptualize (*i.e.*, think) in multiple dimensions. A conceptual model of geological conditions in any one area however, is of limited use unless it can be reproduced (*i.e.*, drawn on a piece of paper) and shown to others for comment, analysis, testing and modification. Two computer-based systems have been developed for portraying geological data and these are briefly reviewed below.

Geographic Information Systems

Geographic information systems (GIS) combines the technology of database management systems (DBMS) and high-performance computer graphics to describe the Earth's sur-face through geographically referenced data. These systems utilize two types of structure, raster or vector. The raster type is based on grids of cells where any image is defined by reference to row and column number; the resolution (*i.e.*, cell size) varies from system to system. In contrast, vector types use XY co-ordinates akin to an A to Z street map where any one place has unique co-ordinates (*e.g.*, A3). Lines, polygons and dots can be portrayed (Figs. 1, 2A, B). Regardless of the structure used, information is retrieved from databases by a data base engine and then displayed in relation to other objects using the modelling engine (Oloufa *et al.*, 1992). Different data sets (*e.g.*, roads, topography, geology, vegetation, *etc.*) are stored as individual layers, but can be viewed and printed together as a single, composite map (Fig. 1). GIS thus allows the capture and integration of both spatial and aspatial features of the Earth's surface into a single logical data model. Statistical and quantitative operations can be conducted on the data sets.

GIS is essentially designed to portray two-dimensional (2-D) surfaces (*e.g.*, road systems, land use, forest conditions, rock, sediment and soil types, *etc.*; Lo and Shipman, 1990; Adams *et al.*, 1993; Lortie *et al.*, 1995; Fig. 2A) with the use of contoured surfaces (gridded elevation matrices or triangular meshes) or isometric views to mimic three-dimensional (3-D) analysis. In these cases, the elevation (or "z" value) consists of a single value. This is referred to as 2.5-D or quasi-3-D (Turner, 1992a). In contrast, geological applications require representation of multiple surfaces and layers at any one site, such as in a layer cake. The tops and bottoms of multiple stratigraphic units cannot be represented in a 2-D (2.5-D, quasi-3-D) GIS.

Geoscientific Information Systems

A three-dimensional GIS is necessary for areas of complex or multi-layered geology/hydrogeology (Fig. 2A). A fully three-dimensional GIS can accept and depict many surfaces at any one location and the term Geoscientific Information System

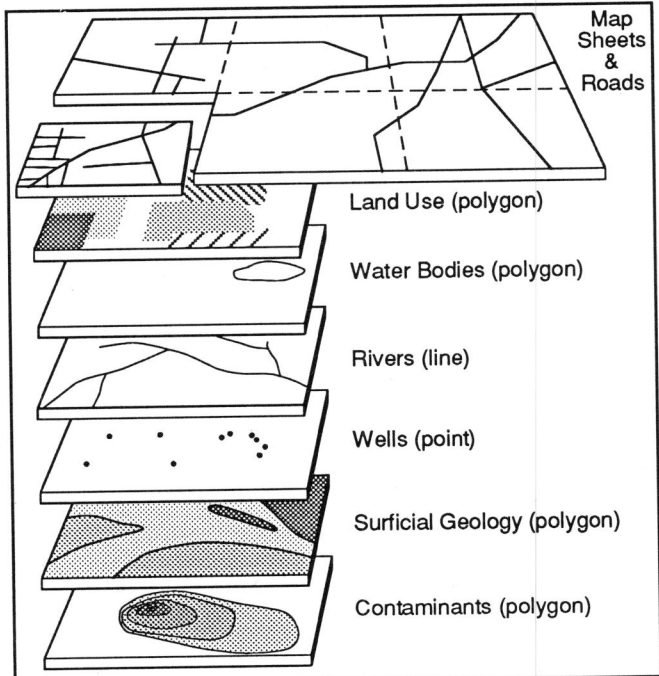

Figure 1 *Layers showing points, polygons and lines from which a composite GIS map image is produced (after Lortie et al., 1995).*

A

Two Dimensional (GIS) (typically, all displays are static)	Three-Dimensional (GSIS) (displays can be interactively manipulated)
Base Maps Contour Maps Perspective Views Block Diagrams Cross Sections Fence Diagrams	3-D Base Maps Rectangular Solids (Block Diagrams)[1] Model Clipped By Bounding Polygon[2] Model Clipped By 2-D Surfaces[3] Model Sliced Along X, Y and Z Axes[4] Isovalue Shell Peels[5] Chair Mode Display[6] Chair Mode Display With Isovalue Shells in Chair Void[6] Multiple Zone Displays[7] 3-D Well Paths With Log Curves 3-D Flow Vectors

1 - Rectangular Solids (Block Diagrams).

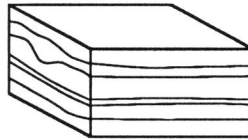

2 - Display boundary set along user-specified polygon (allowing lateral truncation of display at precise non-rectangular boundaries).

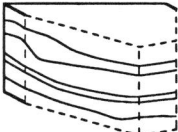

3 - Display built within volume between top and bottom truncating surfaces (and thus allowing multiple zone displays).

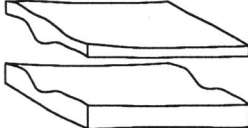

4 - Display may be cut along any X, Y and Z axis to allow viewing of a specific part of the total volume.

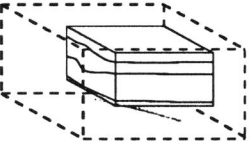

5 - User can specify volumes with a specific value (or range) to either be removed or displayed (e.g. contaminant plume concentrations; refer to Figure 7B).

6 - Chair mode display formed by removing a section of the model perpendicular to the X, Y and Z axes. Chair void is bounded by vertical walls perpendicular to the X and Y axes and a horizontal floor perpendicular to the Z axis. Isovalue surfaces may be extruded into the 'chair void' (A). (Removing a piece from a block to create what resembles a "chair".)

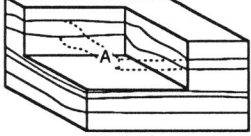

7 - Zones (determined by 2-D surface clipping) within display may contain an independent model of an input property (P) uninfluenced by adjacent zones.

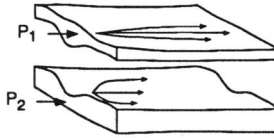

(GSIS) is used to differentiate these geologically oriented systems from the standard 2-D GIS (Turner, 1992a; Belcher and Paradis, 1992; Fig. 2A). One of the largest Geoscientific Information Systems in current use is being employed to evaluate the complex geological and hydrogeological conditions at Yucca Mountain, Nevada, in the United States, in preparation for its possible use as a permanent disposal site for high-level nuclear waste (Turner *et al.*, 1991). Complex projects involving landfill design, petroleum reservoir characterization, and assessment of regional, local and site-specific ground-water flow and contaminant modelling can all benefit from the use of GSIS (*e.g.*, Turner, 1992a). Examples of many other applications are provided by Houlding (1995).

GSISs can integrate various 3-D and 2-D data such as geophysical and geological information pertaining to the geometry and composition of strata and the nature and properties of associated fluids (see below). Such systems are also capable of interactive 3-D graphics as well as the simultaneous display of plan views and cross-sections/well logs. Fundamentally, they allow the rapid 3-D visualization of subsurface logging data collected from boreholes or outcrops. (*e.g.*, Turner *et al.*, 1990; Adams, 1994). Of course, this ability to picture the subsurface is dependent upon having sufficient data. Kelk (1992) has outlined potential problems with GSIS applications, most notably the general sparseness of 3-D data for many areas (see also Harbaugh and Martinez, 1992).

Commercially available GSIS systems (specifically for environmental geological applications) are marketed by Intergraph Canada (MGE application suite), ESRI (Arc/INFO) and Core Technologies Development Corporation (EVS). The Intergraph system is described below with regard to data base management by the city of Scarborough, Ontario. Arc/INFO is a vector-based GIS designed to be a georelational data model. This consists of (1) a topologic data model (based on United States Geological Survey digital line graph cartographic standards) to represent locational data in a 2-D co-

B

System	Additional Data Structures	Platform/Environment
Vector-based GIS		
ESRI – Arc/INFO		Unix, DOS, Windows
ESRI – ArcView 2	Display Images	All (except DOS)
ATLAS/GIS		DOS, Windows
EPS/PAMAP	Raster	Unix, Windows
INTERGRAPH	Limited Raster	All (except DOS)
MAPINFO	Display Images	Windows
SPANS	Limited Raster	OS/2, Windows
AUTOCAD Overlays		
ESRI – ArcCAD	Vector (only)	DOS, Windows
Raster-based GIS		
EPP7	Limited Vector	DOS
ERDAS	Limited Vector	Unix
GRASS	Vector	Unix
ESRI – GRID		Unix
IDRISI	Limited Vector	DOS, Windows
PCI (EASI PACE)	Some Vector	OS/2, Windows

Figure 2 (**A**) *Essential features of Geographic Information Systems contrasted with Geoscientific Information Systems (after Belcher and Paradis, 1992) with additions.* (**B**) *Summary of commercially available GIS systems.*

ordinate system; and (2) a conventional relational data base model to represent thematic information. The system is, thus, a hybrid of two existing data models (see Oloufa *et al.*, 1992).

EVS (Environmental Visualization System) is designed for 3-D geological applications and is particularly useful for the management of data and visualization of data relating to plumes of chemical contaminants in the subsurface (see below). The software allows for input of ASCII data and CAD drawings. Modules allow the construction of 3-D fence diagrams, modelling of 3-D structures, volume and mass calculations, display of sampling points and chemistry distributions and the overlaying of aerial photographs.

EXAMPLES

City of Scarborough, Ontario

The city of Scarborough is one of the fastest growing urban communities in Canada. Ground- and surface-water resources are increasingly threatened by many point and non-point contaminant sources represented by industrial, residential, municipal and agricultural activities. There have also been dramatic shifts in land use resulting from rapid urbanization and the new economy requiring remediation of former industrial sites and waste disposal sites (landfills), including those where low-level radioactive waste was dumped (Chapters 2, 28). In revising its Official Plan, the city is faced with the need to make environmentally sensitive planning decisions differentiating between the potential environmental impact of land use X or land use Y or some combination thereof.

The city of Scarborough covers 18,800 ha on the north shore of Lake Ontario. It has a current population in excess of 500,000 which is expected to double in the next 30 years. About 50% of the city area is residential, 20% is industrial and commercial, with the remainder being public parkland, school site, institutional, utility corridor or open space. There are some 300,000 households and 128,000 land parcels. Scarborough is the only municipality in Metropolitan Toronto still developing new subdivisions from rural land and thus straddles the urban fringe where land use change is particularly rapid.

A collaborative venture between the city, the University of Toronto and Intergraph Canada was initiated in 1994 to establish a central environmental database for use by the city's Works and Environment Department. Significant progress has already been made by the city in providing over-the-counter PC-based computer maps to the public (User-Defined Mapping Application; Sussman, 1994). The provision of subsurface geological data is a further objective.

Hardware

The GSIS under joint development by the city of Scarorough and the University of Toronto is based on an Intergraph TD-1 workstation which is a 80486DX 66-Mhz computer with 32 Mb (70ms) of RAM memory and 256 kb external cache. It includes: a 20″ non-interlaced multisync monitor (.28 dpi) with a resolution of 1280 × 1024; a VESA-VL accelerated graphics system (OpenGL 3D) with a S3 928 chipset and 2 Mb of VRAM; storage areas include a 540 Mb hard disk (with a 8.5 ms seek time, 13.3 ms access time and 2.5-4.6 Mb/s transfer rate) and a 600 Mb CD-ROM (double speed); 2 fast SCSI-2 channels (allowing an addition of 4 mass storage devices) and 3 EISA slots; onboard Ethernet (10BaseT and AUI); flash BIOS (for upgrading hardware); zero-wait secondary cache (zero-wait on every cycle) with high performance synchronous burst SRAM memory, write-back cache logic and

address pipelining (allowing two address accesses at once and reducing the number of calls to slower main memory, this increases CPU performance 30-60%). The TD-1 runs under Windows NT (version 3.1) a 32-bit pre-emptive multi-tasking operating system. This allows the use of native Windows NT, Windows, OS/2 1.x (text based), POSIX and MS-DOS applications. Windows NT features high system availability and quick recovery through protected subsystems, memory protection and hardware isolation.

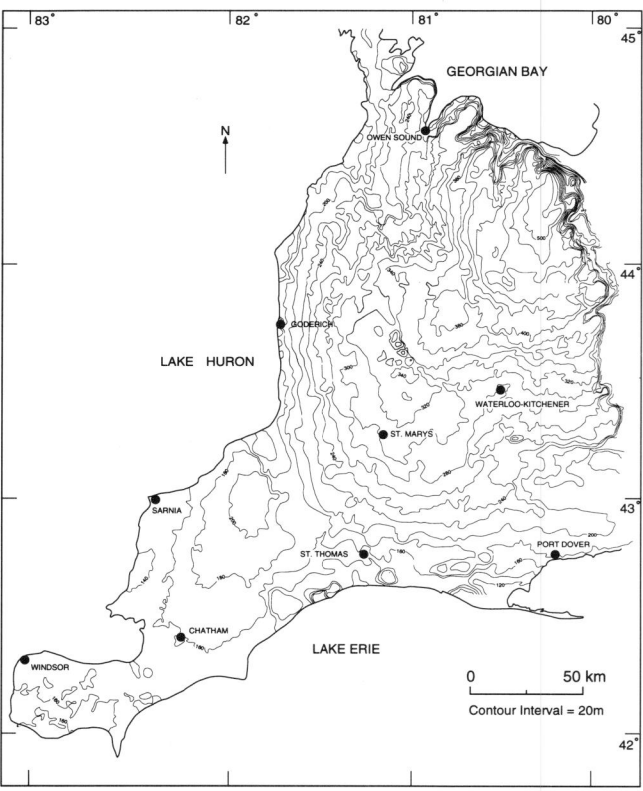

Figure 3 (**A**) *Geographic coverage of 95,000 Ontario Ministry of Environment and Energy (OMOEE) water wells in southern Ontario. Wells are in Pleistocene glacial sediments and penetrate to underlying bedrock. Grid is Universal Transverse Mercator (UTM). (**B**) Bedrock topography map of southwestern Ontario generated from regional OMOEE waterwell database. Elevations in metres above sea level (m asl). See Chapter 2 for bedrock topography of southeastern Ontario.*

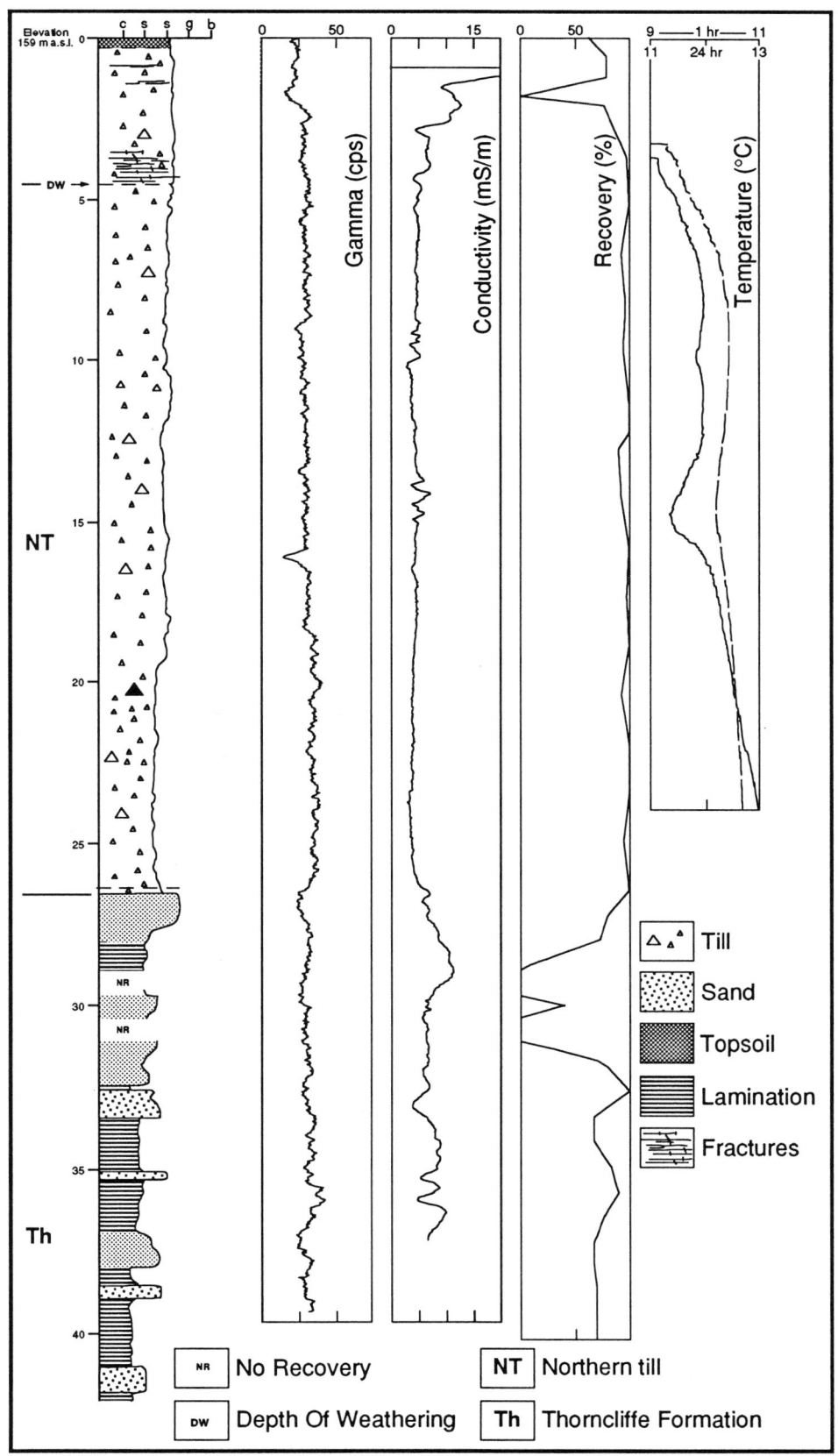

Figure 4 *Typical borehole log through till and underlying glaciolacustrine sediments showing (**from left**) sedimentologic graphic log, and hydrogeologic and geophysical data such as gamma counts, conductivity and temperature logs.*

Software

The principal software components are Intergraph's Micro-Station, Relational Interface System (RIS), Insitu and Oracle Relational Data Base Management System. MicroStation is the core graphics software program for Intergraph applications. It is a general purpose CAD software package (2-D and 3-D) which provides links to various relational databases including Informix, Oracle and RIS, providing Structured Query Language (SQL) support. Linkages are made between database information and visual graphic elements in a design file (for reporting and display functions). This has the advantage over standard CAD programs that if the data base information changes, annotations within a file are updated automatically. Also, as data is contained elsewhere, the size of design files remain manageable. Being file based rather than memory based provides data protection because information is written directly to disk on-the-fly through a write-through memory cache to maintain speed without noticeable delay time being experienced while working. Employing NURBS curves (2nd to 16th order) allows easy modification of model geometry without losing precision. Microstation supports a wide variety of input/output standards (DXF, IGES,

CGM, etc.) giving the ability to share design files between software packages running various operating systems.

Oracle is an SQL database management system based on the relational database model. Database queries from Micro-Station/Insitu are routed through RIS and handled by Oracle. RIS provides links with standard relational databases as well as third-party software for statistical analysis. RIS is used as a middleman between Oracle and Insitu.

Insitu is used to efficiently manage and interpret large quantities of field and laboratory data. With well data as a base, it combines field investigations, laboratory analyses, details of boreholes, site planning, cross-section generation and subsurface modelling into one package. Insitu uses Oracle and RIS to maintain a project data base (providing powerful query, report and data management capabilities for both individual and grouped boreholes and other data sets) and MicroStation to provide interactive 2-D and 3-D graphics. Drawings generated include boring logs, site plans, fence diagrams and contour maps.

Intergraph's MGE (Modular GIS Environment) applications allow data to be input, captured, converted and validated from diverse sources and combined into an MGE database. RIS

Borehole Attribute - Identifier Level

*ID	ID	City Well			MOE Well		MTM		Elevation	Owner	Completion Date			Contractor Code
Bore #	Data Type	Project #	Contract #	Boring #	County Code	Seq. No.	Easting	Northing			Day	Month	Year	
7	2	10	10	2	2	5	11	12	5	20	2	2	4	4

Borehole Attribute - Stratigraphy / Hydrogeology

*ID	Formation (20 Total)			Materials				Substrata (5 Total)					Samples (20 Total)				N		Grain Size			Water (5 Total)		Water Comment
Bore #	Formation #	Depth To Bottom (m)	Colour	1st	2nd	3rd	4th	Top	Thickness	Colour	1st	2nd	Top	Sample Length	Type	Moisture Content	Lab	Site	Sand	Silt	Clay	Depth Found	Kind	
7	2	4*	1	2	2	2	2	4*	3*	1	2	2	4*	2*	1	2	3*	3	2	2	2	4*	1	19

Borehole Attribute - Casing Information

*ID	Casing (6 Total)			Screen (2 Total)				Plug (3 Total)	
Bore #	I. Dia	Material	Depth	Slot No.	Diameter	Length	Depth To	Depth	Thickness
7	2	1	4*	3	5	2*	4*	4*	3*

Borehole Attribute - Pumping Test Results

*ID	Test Period	Pumping			Levels		Levels During Pumping Or Recovery (R)				Flow Rate	Clear/Cloudy	Recommended		Specific Capacity	Use		Drill Method
Bore #		Rate	Duration		Static	After Pump	15 min	30 min	45 min	60 min			Setting	Rate		1st	2nd	
			Hrs.	Mins.														
7	1	4	2	2	3*	3*	3*	3*	3*	3*	4	1	3	4	4	1	1	1

* Indicates number expressed in tenth's

Figure 5 *Attribute table for characterizing information from individual boreholes included in the City of Scarborough environmental data base.*

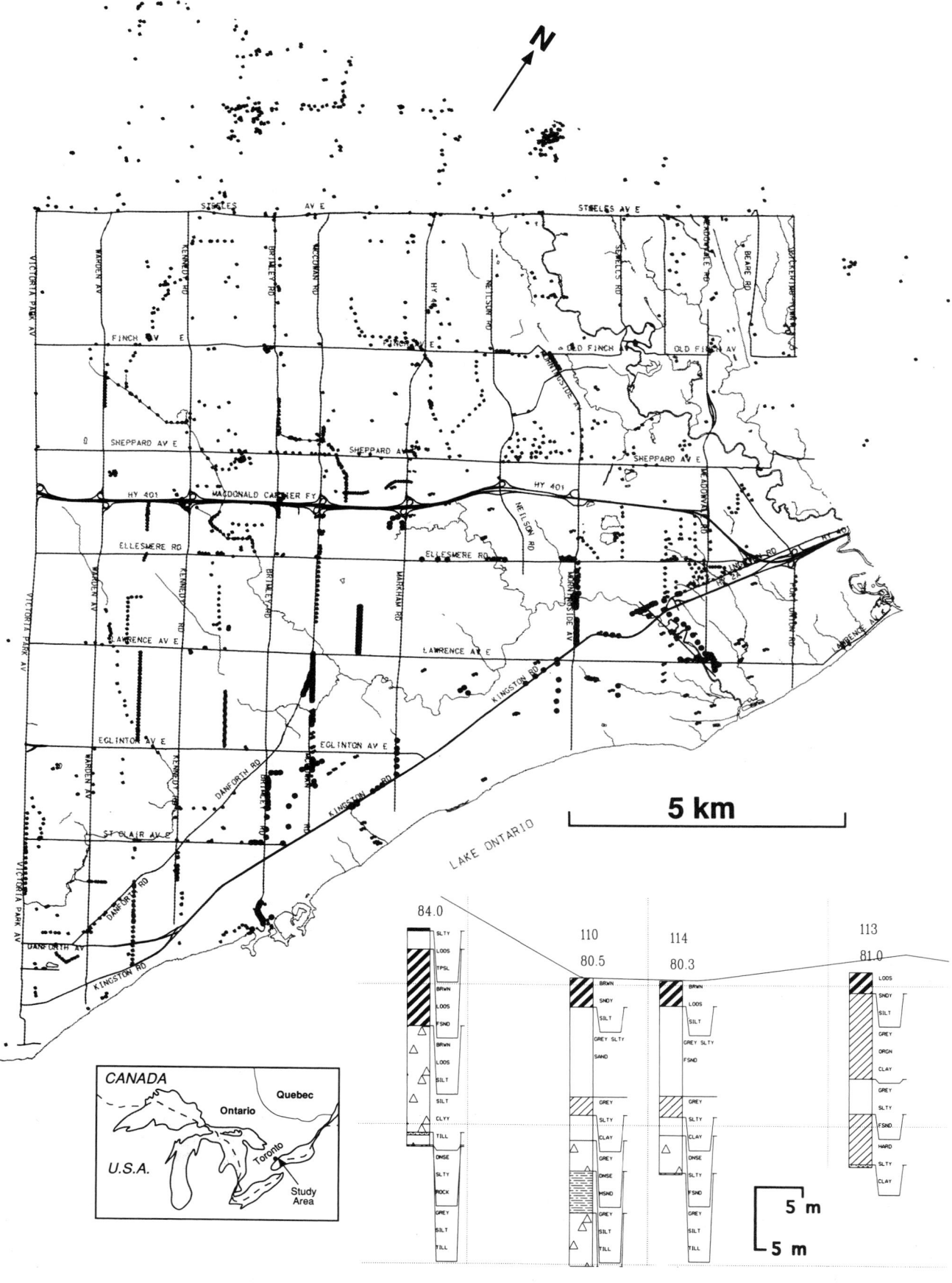

Figure 6 *Map showing borehole locations in eastern part of City of Scarborough with stratigraphic cross-section generated using Insitu software.*

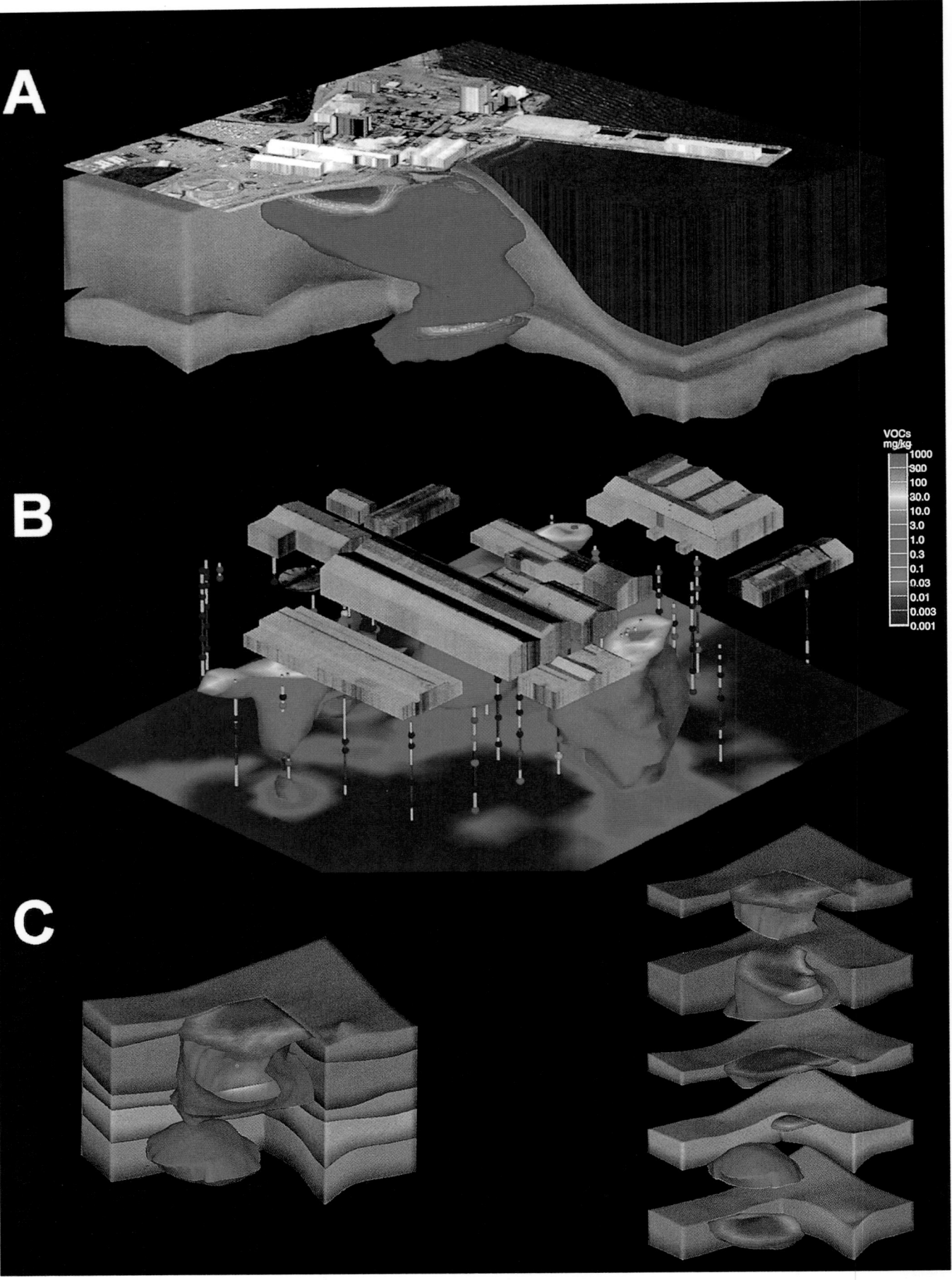

(see above) enables multiple MGE workstations, such as in many municipal planning and engineering departments, to communicate simultaneously with an MGE database server over a network, without concern as to where the data actually resides, thereby minimizing data redundancy between applications.

Input Data

Geological data sources are varied and include digitized surficial geology and soil maps of the city, outcrop information, Ontario Ministry of the Environment and Energy (OMOEE) water well records (to 1985; Fig. 3), hard-copy borehole logs from construction and research projects (dating from the late 1950s to present day; *e.g.*, Fig. 4) and digital road network and river maps. Other data include water chemistry, hydraulic conductivity, permeability and other hydrogeological attributes, seismic data, aerial photographs, various remote sensing (RS) images and a digital terrain model (DTM) of surface topography. The city database is very large because for each borehole log a wide range of attributes is commonly available (Fig. 5). The distribution of boreholes in the eastern portion of the city is shown in Figure 6.

Data can be directly input through individual workstations, alphanumeric or PC-networked terminals, or bulk loaded using ASCII data generated through a standard text editor. All incoming data is checked (location, stratigraphy, *etc.*) and potential errors identified before incorporation into the project database.

Output

The OMOEE water well data file for southern Ontario consists of nearly 200,000 data points, of which about 95,000 (Fig. 3A) penetrate to bedrock, therby enabling regional and local maps of bedrock topography to be printed (Fig. 3B). For the City of Scarborough data base, cross-sections (generated by selecting boreholes in plan view) of subsurface strata can be portrayed between any two or more geographic locations (Fig. 6). Boundaries between different stratigraphic units (*i.e.*, unconformities) are marked and used to develop distinct surfaces and topologically triangulated networks (TIN) for final model analysis. These are stored as named unconformity surfaces and used to generate volumes, contours, isopachs or 3-D models. These can be integrated with geochemical data to produce 3-D images of subsurface contaminant plumes associated with industrial sites and leaking fuel storage tanks (*e.g.*, Fig. 7).

The use of a standard database allows integration with Intergraph's Modular GIS Environment (MGE) and Environmental Resource Management Applications (ERMA). Thus, Insitu data can be directly imported into MGE Voxel Analyst for visualizing 3-D slices of the developed model. Industry-accepted ground-water modelling packages (*e.g.*, MODFLOW and MODPATH; Chapters 9, 11) can be applied as well as smooth integration with GIS related applications such as

engineering, construction and automated mapping (AM) and facilities management (FM). Interactive graphics facilitate the analysis of subsurface data. Three-dimensional data sets can be integrated with terrain surfaces using map projection systems for geographic referencing. Where data are irregular, missing values may be estimated using metric, multi-quadratic, Shepard's, thin-plate spline and volume spline methods.

OTHER EXAMPLES

Other North American cities have attempted to incorporate a wide variety of geological and environmental information into a single data base, with links into a true GSIS system for querying and geometric modelling of subsurface geological and hydrogeological conditions. These cities have found that project costs significantly decrease because information obtained in previous investigations can be readily reused. The following provides a few representative examples of incorporation of urban geological databases using a GIS or GSIS system.

Washington, District of Columbia.   Batten (1989) reports the development of a 3-D model of the Monumental Core, Washington, DC, to illustrate the potentials of such a model for use in the urban planning process. This project is a cooperative effort between the United States Geologic Survey (USGS), the National Capital Planning Commission (NCPC) and the United States Bureau of Census (BOC) to develop tools for urban planning applications. The database includes both vector and raster data derived from aerial photographs, building construction documents, local planning documents, topographic data (digitized), USGS Digital Line Graph (DLG) data, BOC statistical and geographic data (as part of the TIGER [Topologically Geographic Encoding and Reference system] database) and NCPC data sets. The intial study allowed for the analysis of existing and proposed new buildings, network analysis for urban transportation and simulation modelling. This includes the ability to update cartographic and attribute databases subsequent to design changes, model transportation patterns before and after a new structure has been built, and make the planning review process more efficient and reliable. The techniques developed during the project also applied to 3-D geological, ground-water and geophysical studies and data visualization. Most significantly, combining GIS and CAD technologies has the effect of depicting environmental and geological data in a fashion that is much more readily understandable to the public, thereby facilitating faster and more effective public review of development proposals.

Bridgetown, Nova Scotia.   Using an ESRI Arc/INFO GIS, Pietsch and Akhavi (1989) developed a hydrogeological data base for Bridgetown, Annapolis Valley, Nova Scotia. Information within the study incorporated both digitized and remote sensing (RS) data, and included land use/land cover classifications, geologic lineaments, surficial geology, geomorphology, hydrology, topography, soils, drainage systems, water

---

Figure 7 *(facing page)* (**A**) *Industrial facility along lake shoreline showing water, sand and gravel as three layers exploded by a short distance. Water depth is 10 m. Plume shows extent and degree of contamination by volatile organic compounds (mg·kg⁻¹) in sediments. Buildings were modelled using AutoCad and imported into Environmental Visualization Systems 2.0, and texture mapped using an air photograph (courtesy of Reed Copsey, Environmental Visualization Systems, Core Technologies Development Corporation).* (**B**) *Volatile organic compound contamination below an urban industrial site. Borings and samples are represented by tubes and spheres. Deepest borehole is about 5 m. Three-dimensional subsurface plumes outlined by 3 mg·kg⁻¹ value. Concentrations at the underlying water table are displayed on a flat coloured surface. Three-dimensional CAD models of buildings are derived from an air photograph.* (**C**) *Use of exploded layers to show vertical extent of contaminated sediment below an urban industrial site and within each geologic layer. Total depth shown is 10 m; uppermost layer is historic landfill.*

well and bedrock depths, soil permeability and road networks. A Landsat Thematic Mapper image was chosen as a base map. The results of the study allowed for the siting of areas of possible industrial subdivision expansion including the availability of ground-water supplies, runoff sites, potential recharge areas, infiltration and the potential for ground-water contamination/pollution at such sites. The lack of *in situ* hydrogeological data, necessary to verify the results of the investigation, was also noted. Note that this is only a 2.5-D or quasi-3-D study (as most information dealt with contoured surfaces only, not actual 3-D objects or points).

Honolulu, Hawaii.   Oloufa *et al.* (1992) describe a PC Arc/INFO GIS system, used to incorporate geotechnical report data (boreholes) within a GIS system to construct a 3-D representation of soil strata below Honolulu. This was done on a 80386-class machine equipped with a Nth three-dimensional engine graphics card. They found that borehole data, in the form of boring logs, can be displayed entirely within Arc/INFO using its drawing and annotation capabilities and its SML macro language. As Arc/INFO is based on 2-D representation, it could not internally create a 3-D body representing substrata from boreholes. This was also the case for another study (Camp and Brown, 1993) using Arc/INFO on a Sun SPARC Station 2 to store, analyze and manage geological and hydraulic data from well logs in Shelby County, Tennessee. In this case, a TIN was created for each stratigraphic layer with each 2-D data set incorporated into MODFLOW for ground-water modelling. Quasi-3-D was obtained through control point elevations (of strata) which may be contoured for terrain analysis. For 3-D representation, they suggested the use of HOOPS (an object oriented graphics library callable from within FORTRAN) to actually visualize and examine substrata in 3-D after extracting the information using Arc/INFO. The data can be pulled by borehole or street layer. The need for a knowledge-based system that regulates topology and layer extents or allows user intervention guided by past experience (or a combination of the two) to interpret raw borehole data was stressed by Oloufa *et al.* (1992).

In a similar study for the island of Oahu, Hawaii, Oloufa *et al.* (1994), used MapInfo, a GIS for Windows, and Knowledge Pro, a visual programming language (used to develop a graphical user interface and facilitate input, output and display the boring logs), to incorporate bore logs into a database. Data entry was accomplished in a typical spreadsheet manner, and data were then accessed through SQL statements from within the GIS (with links into dBase files). This system creates in plan and cross-sectional views, but does not create a 3-D representation of subsurface strata.

Dane County, Wisconsin.   Adams (1994) describes a system (termed GeoGIS) for managing subsurface data and supporting site characterization by integrating several public-domain and commercial packages. These include: GRASS 4.0 (public domain raster and vector GIS, developed by the United States Army Corps of Engineers), RIM (public domain relational database management package developed at the University of Washington, used to store and retreive maps), INGRES and INGRES Windows4GL (commercial RDBMS with the ability to create custom applications and interfaces), GKS graphics kernel system (2-D graphics toolbox used to plot 1-, 2- and 3-D borelogs), GSLIB (a geostatistical software library developed at Stanford University) and Xgen (a GUI builder developed by the United States Army Construction Engineering Lab). This system was used to successfully compile and model 1275 borehole logs.

DISCUSSION

The need for data banks to visualize, assess, manage and remediate environmental geological problems in urban areas has focussed attention and research on a number of issues. These are briefly reviewed here.

It is a common experience that large volumes of data, collected at considerable expense, are routinely discarded by consulting companies at the end of many projects; there is no economic incentive to store and maintain them in a form that is widely accessible. Submission of data to an electronic data bank, using a standard digitized format, should be made a mandatory requirement for all publicly funded projects and development proposals. Proprietary restrictions can be overcome by limiting access for an initial time period as is routinely applied in the oil industry.

In 1994, the Shell Oil Co. donated its entire Midland, Texas, core repository, one of the largest United States drill core collections, to the University of Texas at Austin, with the intention that it become part of a National Geoscience Data Repository. A similar facility (Core Research Centre), financed by the Province of Alberta, is located in Calgary and a smaller scale facility exists in London, Ontario (Petroleum Research Centre), operated by the Ontario Ministry of Natural Resources. The Calgary facility has about 100,000 wells on record represented by core and/or drill cuttings. As yet, there is no provision for the centralized storage and retrieval of sediment or rock core in Ontario; core is routinely discarded at the conclusion of a project.

A major concern is one of identifying which agency or organization will take a leadership role in establishing a regional environmental geological data base and making information available. So far in Ontario, no government agency has come forward with a proposal to collate such data, review the quality of data, standardize data classifications and facilitate data exchanges. Such efforts should, perhaps, be initiated at the municipal level because of the shift in responsibility for environmental planning from the federal and provincial governments to local government (see Chapters 2, 27).

Start-up costs for the various commercially available GSIS systems identified above vary considerably (from a few thousand to several tens of thousands of dollars), but it needs to be emphasized that the cost of any one system is typically much less than the cost of acquiring and capturing environmental geological data. The expense of collecting new or existing data, and the time spent reformatting and inputting such information, can be very considerable. The collection and entry of data into a database is a key process, since it cannot be done without ongoing scientific assessment of data reliability and significance in order to construct a realistic model of the subsurface environment. Good models are fundamental to the correct and cost-effective design of remedial measures and land use plans. It is also now being realized that the success or failure of development proposals and litigation cases is dependant on effective visualization of subsurface data to convey complex geological and hydrogeological information to the public, planners and legislators.

A concern with regard to GSIS databases, especially with regard to litigation cases, is that the quality, and hence reliability, of much urban geological and environmental information is very variable. Documentation as to the origin of the data and an indication of its quality should be included within the data set itself (so-called metadata; Carter, 1992; Adams, 1994). This allows users to make appropriate decisions about the reliability and use of the information within the database.

Data need to be precisely defined by a geographic referencing system determined through geodetic co-ordinates or through the use of GPS technology (Adams, 1994). These aspects often loom large in litigation and remediation cases because the precise source, volume and off-site pathways of contaminants is the central issue in question.

ACKNOWLEDGEMENTS

We are grateful to Andrew Mackie of Intergraph Canada, Mississuaga, Ontario, for ongoing support and Reed Copsey of Core Technologies Development Corporation of Port Angeles, Washington, for colour images used in this study. Research on which this paper is based is jointly funded by grants to the senior author from the Natural Science and Engineering Research Council of Canada and the city of Scarborough.

REFERENCES

Adams, T.M., Yang, A.Y.S and Wiegand, N., 1993, Spatial data models for managing subsurface data: Journal of Computing in Civil Engineering, v. 7, p. 260-277.

Adams, T.M., 1994, GIS-based subsurface data management: Microcomputers in Civil Engineering, v. 9, p. 305-313.

Agterberg, E.P. and Bonham-Carter, G.F., 1990, eds., Statistical Applications in the Earth Sciences: Geological Survey of Canada, Paper 89-9, 588 p.

Antenucci, J., Brown, K., Croswell, P. and Kevany, M., 1991, Geographic Information Systems: A Guide to the Technology: Van Nostrand Reinhold, New York, 301 p.

Batten, L.G., 1989, National capital urban planning projects: development of a three-dimensional GIS model: Auto-carte 9, Proceedings of Symposium, Baltimore, MD, p. 336-340.

Belanger, J.R., 1974, Urban geology automated information system (U.G.A.I.S): Geological Survey of Canada, Paper 74-60, p. 95-98.

Belanger, J.R. and St.Onge, D.A., 1974, Environmental geology prototype study; Ottawa-Hull area: Geological Survey of Canada, Paper 74-1A, p. 215.

Belcher, R.C. and Paradis, A., 1992, A mapping approach to three-dimensional modelling, in Turner, K.A., ed., Three-Dimensional Modeling with Geoscientific Information Systems: Kluwer Academic Publishers, Amsterdam, The Netherlands, p. 107-122.

Bryant, I.D. and Flint, S.S., 1993, Quantitative clastic reservoir geological modelling; problems and perspectives, in Flint, S.S. and Bryant, I.D., eds.: International Association of Sedimentologists, Special Publication 12, p. 3-20.

Bryant, I.D., Paarkedam, A.H.M., Davies, P. and Rudding, M.C., 1990, Integrated reservoir characterization of cycle III, Brent Group, U.K. North Sea, in The Integration of Geology, Geophysics and Reservoir Management: American Association of Petroleum Geologists, 1st Archie Conference, Houston, TX, Proceedings, p. 405-411.

Camp, C.V. and Brown, M.C., 1993, GIS procedure for developing three-dimensional subsurface profile: Journal of Computing in Civil Engineering, v. 7, p. 296-309.

Carter, J.R., 1992, Perspectives on sharing data in geographic information systems: Photogrammetric Engineering & Remote Sensing, v. 58, p. 1557-1560.

Christiansen, E.A., 1970, ed., Physical Environment of Saskatoon, Canada: Saskatchewan Research Council and National Research Council of Canada, NRC 11378, 68 p.

Cross, T.A., ed., 1990, Quantitative Dynamic Stratigraphy: Prentice Hall, Englewood Cliffs, NJ, 625 p.

Franseen, E.K., Watney, W.L., St.C Kendall, C. and Ross, W., 1991, Sedimentary modelling: Computer simulations and methods for improved parameter definition: Kansas Geological Survey, Bulletin 233, 524 p.

Frönzle, O., 1993, Contaminants in Terrestrial Environments, Springer Verlag, Berlin, 454 p.

Gundeso, R. and Egelund, O., 1990, SESIMIRA — A new geologic tool for 3D modelling of heterogenous reservoirs, in Buller, A.T., ed., North Sea Oil and Gas Reservoirs: Graham and Trotman, London, UK, p. 363-372.

Harbaugh, J.W. and Martinez, P.A., 1992, Two major problems in representing geological well data and seismic data in pretroleum-bearing regions via 3-D geographic information systems, in Turner, K.A., ed., Three-Dimensional Modeling with Geoscientific Information Systems: Kluwer Academic Publishers, Amsterdam, The Netherlands, p. 291-302.

Houlding, S., 1995, 3D Geoscience Modeling: Computer Techniques for Geological Characterization: Springer-Verlag, Berlin, 309 p.

Kathol, C.P. and McPherson, R.A., 1975, Urban geology of Edmonton: Alberta Research Council, Bulletin 32, 61 p.

Kelk, B., 1992, 3-D modelling with geoscientific information systems: the problem, in Turner, K.A., ed., Three-Dimensional Modeling with Geoscientific Information Systems: Kluwer Academic Publishers, Amsterdam, The Netherlands, p. 29-37.

Krum, G.L. and Johnson, C.R., 1993, A 3D modelling approach for providing a complex reservoir description for reservoir simulation, in Flint, S.S. and Bryant, I.D., eds.: International Association of Sedimentologists, Special Publication 12, p. 253-258.

Lasseter, T.J., 1990, A 3-D interactive modelling system for an integrated reservoir interpretation, in The Integration of Geology, Geophysics and Reservoir Management: American Association of Petroleum Geologists, 1st Archie Conference, Houston, TX, Proceedings, p. 182-193.

Lo, C.P. and Shipman, R.L., 1990, A GIS approach to land-use change dynamics detection: Photogrammetric Engineering & Remote Sensing, v. 56, p. 1483-1491.

Lortie, B., van Huyssteen, E. and Landriault, Y., 1995, Practical applications of GIS to mine environmental problems, in Hynes, T.P. and Blanchette, M.C., eds., Rehabilitation Methods: Sudbury '95. Mining and the Environment: CANMET, Ottawa, ON, v. III, p. 1151-1160.

Martin, J.H., 1994, The Geological Basis of Reservoir Simulation: Principles, Problems and Practical Advice: Fifth International Forum on Reservoir Simulation, Muscat, Oman: J.H. Martin and Associates, 150 Croxted Road, Dulwich, London, UK, 42 p.

Miall, A.D. and Tyler, N., 1991, eds., The Three Dimensional Facies Architecture of Terrigenous Sediments and Its Implications for Hydrocarbon Discovery and Recovery: Society of Economic Paleontologists and Mineralogists, Concepts in Sedimentology and Palaeontology, v. 3, 309 p.

Morin, F., 1975, Environmental geology-Hamilton urban area, Ontario: Geological Survey of Canada, Paper 75-1A, p. 369-370.

Oloufa, A.A., Papacostas, C.S. and Espino, R., 1992, Construction applications of relational data bases in three-dimensional GIS: Journal of Computing in Civil Engineering, v. 6, p. 73-84.

Oloufa, A.A., Eltahan, A.A. and Papacostas, C.S., 1994, Integrated GIS for construction site investigation: Journal of Construction Engineering and Management, v. 120, p. 211-222.

Pietsch, R.W. and Akhavi, M.S., 1989, Integration of remote sensing and GIS for a hydrogeological assessment of the Bridgetown, Nova Scotia area: Seventh Thematic Conference on Remote Sensing for Exploration Geology; Methods, Integration, Solutions, p. 649-663.

Statistics Canada, 1994, Databases for Environmental Analysis: Ottawa, ON, Report 11-527E, 257 p.

Sussman, R., 1994, User-defined mapping application, City of Scarborough: Geographic Information System, Group Report: 6 p.

Turner, K.A., 1992a, ed., Three-Dimensional Modeling with Geoscientific Information Systems: Kluwer Academic Publishers, Amsterdam, The Netherlands, 443 p.

Turner, K.A., 1992b, Applications of three-dimensional geoscientific mapping and modeling systems to hydrogeological studies, in Turner, K.A., ed., Three-Dimensional Modeling with Geoscientific Information Systems: Kluwer Academic Publishers, Amsterdam, The Netherlands, p. 327-364.

Turner, K.A., Downey, J.S. and Kolm, K.E., 1990, Potential applications of three-dimensional geoscientific information systems (GSIS) for regional groundwater flow-system modeling, Yucca Mountain, Nevada: EOS, v. 71, p. 1316.

Turner, K.A., Ervin, E.M. and Downey, J.S., 1991, Evaluation of geographic information systems for three-dimensional ground-water modeling, Yucca Mountain, Nevada, in Stow, S.H., ed., Second Annual International Conference on High Level Radioactive Waste Management, Las Vegas, NV, April 28-May 3, 1991: United States Geological Survey, Denver, CO, p. 520-528.

Tyler, K., Henriquez, A., Macdonald, A., Svanes, T., Holden, L. and Hektoen, A.L., 1994, MOHERES — A collection of stochastic models for describing reservoir heterogeneities in clastic reservoirs, in North Sea Oil and Gas Reservoirs III, Norwegian Institute of Technology: Kluwer Academic Press, Amsterdam, The Netherlands, p. 213-221.

van Vark, W., Paarkedam, A.H.M., Brint, J.F., van Lieshout, J.B. and George, P.M., 1994, Construction and validation of a numerical model of a reservoir consisting of meandering channels: Society of Petroleum Engineers, Reservoir Engineering, v. 9, p. 9-14.

38. Geographic Information Systems and Remote Sensing Techniques in Environmental Assessment

Frank M. Kenny

Provincial Remote Sensing Office, Ontario Ministry of Natural Resources
90 Sheppard Avenue E., 4th Floor, North York, Ontario M2N 3A1

SUMMARY

Digital map products and spatial inventories are becoming increasingly available from geological surveys, agricultural, natural resource, environmental, energy, transportation and forestry departments. As well, there are now multitudes of specialized digital airborne and satellite image products available. This wide availability of geographically referenced data and the advances in spatial data analysis software are providing geoscientists with new tools and new ways of viewing traditionally used data.

Through several examples, this paper will demonstrate how remote sensing and GIS technologies can contribute to environmental assessment of an urban fringe area. Nowhere is the need for spatial inventories and mapping greater than in such areas, where pre-existing information becomes rapidly outdated. A 260-km² site, north of Metropolitan Toronto was chosen as a study area. A spatial data base was constructed which included imagery from three different satellite sensors, a Digital Terrain Model (DTM), a digital drainage network, and a digital copy of the Ontario Geological Survey's Quaternary geology map.

INTRODUCTION

Urban planning is critically dependent upon map data and analysis of spatial information. Currently, an evolution is happening in spatial data analysis and manipulation that is having a great impact on environmental geology. Many spatial analysis functions, once treated separately and analyzed by separate software packages, can now be found in single spatial analysis software packages. It is now common to find functions of Image Analysis Systems (IAS), vector Geographic Information Systems (GIS), raster modelling Geographic Information Systems, Computer Aided Design (CAD) systems, geophysical processing software, and 3-dimensional modelling software available in one software package. As well, all the information contained in and generated by this software can be linked *via* a central Relational Data Base Management System (RDBMS), providing further analytical capabilities. Introductions to these different techniques can be found in Sabins (1986), Richards (1986), Lillesand and Kiefer (1987), Aronoff (1989), and Antenucci *et al.* (1991).

STUDY AREA

The urban fringe area chosen for this study is on the southern flank of the Oak Ridges Moraine (ORM), 20 km north of Metropolitan Toronto (Fig. 1). There is increased awareness and appreciation of the environmental importance of the moraine (Chapter 9).

The study area is 20 km × 13 km in size and contains the towns of Claremont, Stouffville and Goodwood (Fig. 2). Elevation varies from 140 m asl in the southeast corner, to 410 m asl at the crest of the moraine in the northeast corner. The lower relief (the southern half) of the study area is dominated by the Halton Till (Chapters 2, 9). Also present in the south-central portion of the study area are large areas of glaciolacustrine silt and clay deposits (Fig. 3). The higher relief (the northern portion) of the study area is on the ORM, and is dominated by coarse-grained glaciofluvial and ice-contact sand and gravel deposits (Fig. 3).

Land cover in the study area varies from predominately agricultural in the southern relatively flat areas, to natural and reforested tracts in the hummocky areas on the moraine. Scattered throughout the area are numerous wetlands and kettle lakes. A north-south drainage divide is present in the northern portion of the study area. The study area also contains numerous cold-water springs that feed the head waters of several streams and tributaries, including Duffins Creek, Reesor Creek and Little Rouge Creek, all flowing to the south, and Black Creek, Pefferlaw Brook and Uxbridge Brook, all flowing to the north (see Chapter 10). Along many of these streams and tributaries the slopes are steep and remain in a naturally forested state. There are several large aggregate mining operations in the glaciofluvial and ice-contact deposits along the crest of the moraine.

EXAMPLES

The four examples presented in this paper were selected to provide an overview of GIS and remote-sensing capabilities to aid environmental investigations. These examples will demonstrate (1) semi-automated feature extraction from multi-spectral satellite imagery, (2) image enhancement and integration techniques for visual interpretation, (3) Digital Terrain Models for extracting terrain parameters, and (4) modelling using a raster-based Geographic Information System. The first three examples will each generate additional thematic information, and the fourth example will demonstrate how this interpreted (or derived) information can be used as input for a model to further characterize the area.

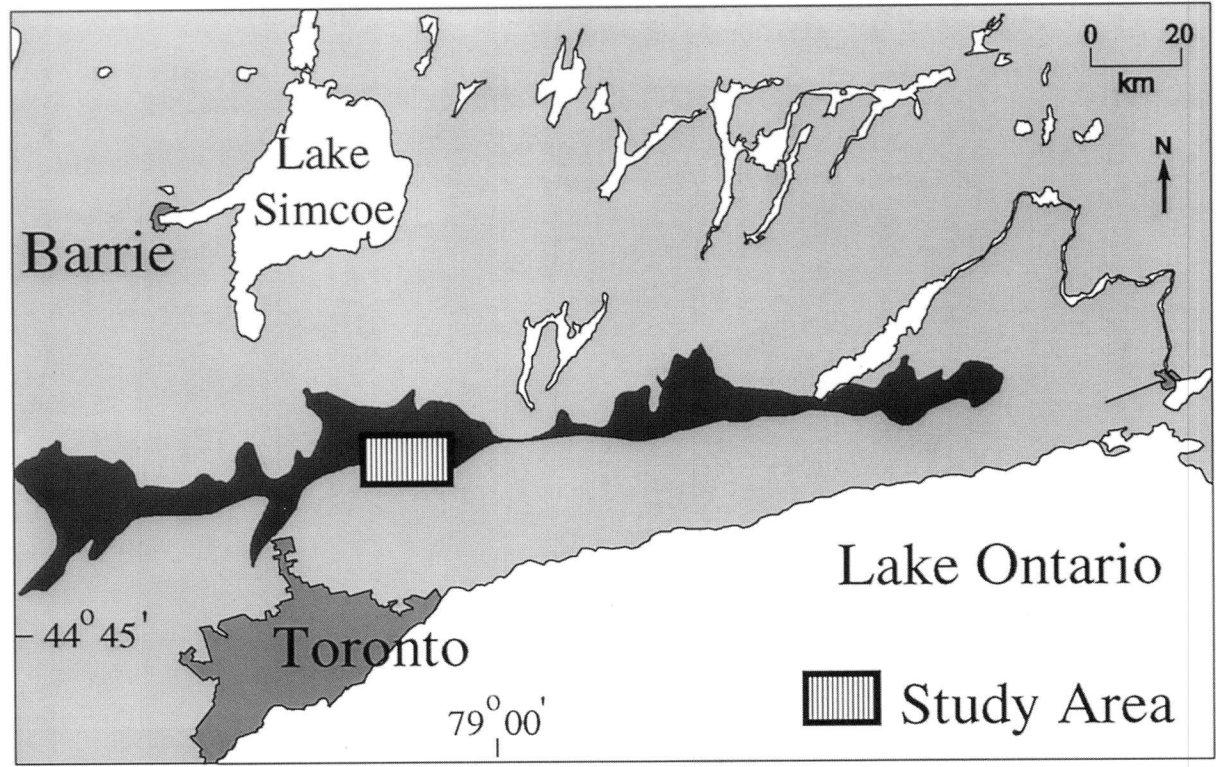

Figure 1 *The Oak Ridges Moraine and location of study area.*

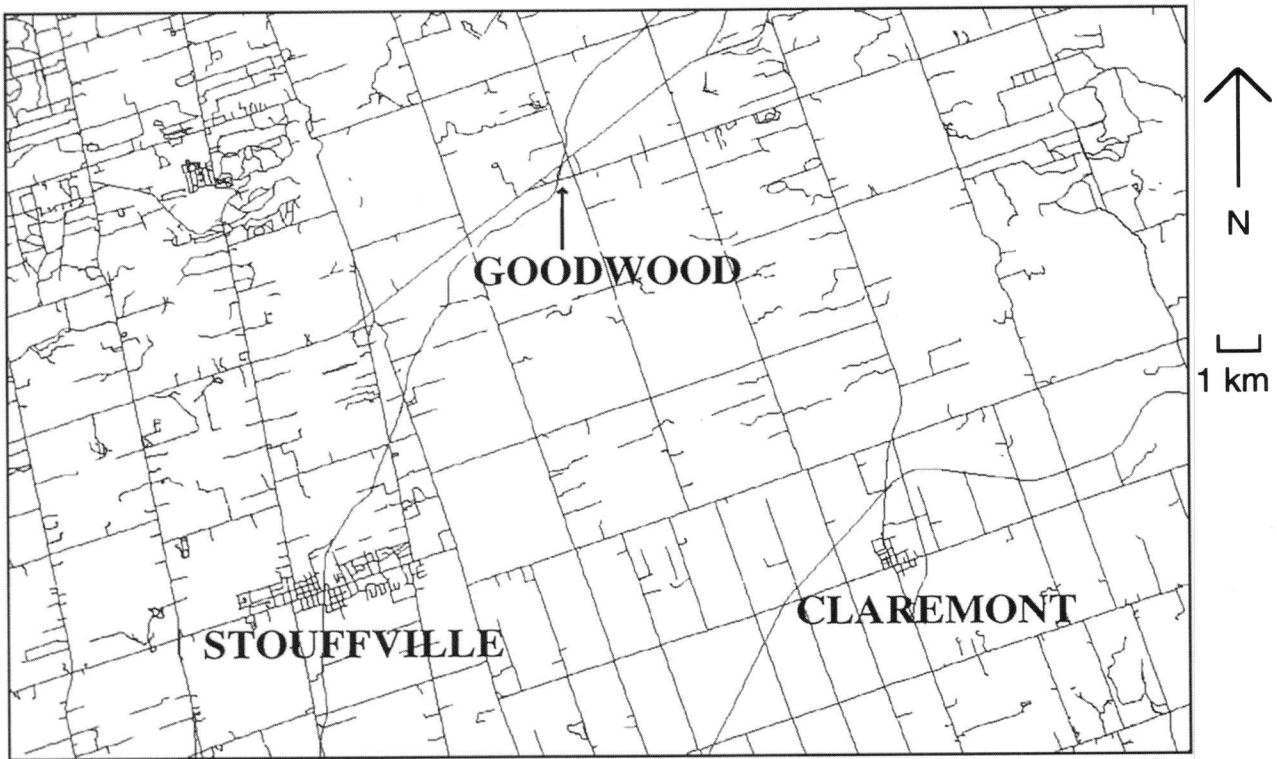

Figure 2 *Location of study area. Geographic extent: NW corner 635000E, 4880000N; SE corner 655000E, 4867000N. Note this is the geographic extent for all subsequent images and maps.*

LANDSAT
TM
April 29, 1985

↓

Supervised
Maximum
likelihood
Classification

↓

Manually
Edit
Themes

↓

Post
Classification
Smoothing

↓

Geo-reference
and Resample
to UTM grid

↓

Final 10 m.
Classification

Digital Trans.
Map
1:10,000

↓

Raster
Conversion
10 m.

↓

Perform Bit-Map
Morphological
"Closing"

↓

"Burn" Urban
Class into
Classification

←

Figure 5   *Flow chart of classification methodology employed.*

Semi-Automated Feature Extraction
From Multi-Spectral Satellite Imagery

One application where optical satellite imagery is routinely and operationally used is land cover mapping. Two of the most commonly used sources of multi-spectral satellite imagery are the United States National Aeronautics and Space Administration's (NASA's) Landsat series and the French SPOT series of satellites. The imagery used for generating the land cover map for this test site was a Landsat Thematic Mapper (TM) image recorded April 27, 1985 (Fig. 4A). Landsat TM is a 28-m-resolution instrument, that records imagery on seven spectral bands: three in the visible, one in the near infra-red, two in mid infra-red and one in the thermal infra-red portion of the electromagnetic spectrum.

Of the numerous mathematical approaches of image classification or spectral pattern recognition, the method used in this study is the supervised Gaussian Maximum Likelihood (GML) method. The first step in this method is to train the classifier, *i.e.*, to gather spectral statistics for areas of known cover type. This is a user-interactive task, where the user delineates areas of known composition on screen. To ensure an accurate training of the system, a user typically uses a combination of field surveys, aerial photography, and/or a

Table 1   Tabular summation of land cover units for the study area.

Cover Type	Area (hectares)	Area (%)
Agricultural – crops	8195	31.52
Agricultural – bare fields	6464	24.86
Urban/Infrastructure	2785	10.71
Deciduous Forest	2489	9.57
Mixed Forest	1922	7.39
Pasture/Reforested	1836	7.06
Wetlands	919	3.53
Gravel Pits/Construction	599	2.40
Water Bodies	150	0.57

OBM TRANSPORTATION MAP

URBAN/INFRASTRUCTURE CLASS

Figure 6   *A portion of the OBM vector transportation map and derived raster Urban/Infrastructure class.*

knowledge of the area in this training stage. Once representative statistics are acquired for each land cover type, a complete classification for the area can be generated. The spectral signature of each pixel in the image is statistically compared to each of the known signatures and assigned to the cover type of the highest probability. For a full description of the GML technique and other multi-spectral classification techniques, see Lillesand and Kiefer (1987) and Sabins (1986). A flow chart of the processing methodology used for this classification is presented in Figure 5.

From the Landsat image ten land cover classes were derived: (1) agricultural fields bare of crops, (2) agricultural fields with emergent crops, (3) agricultural lands, pasture or idle, (4) waterbodies, (5) wetlands, (6) deciduous forest, (7) coniferous forest, (8) mixed forest, (9) gravel pits/construction sites, and (10) urban and infrastructure (*e.g.*, roads). The result of this classification is not, however, registered to a ground co-ordinate system, and it is necessary therefore to geometrically correct this classification so that it can be used in subsequent analysis. The referencing process involves selecting identifiable points on the original image, registering these points to ground co-ordinates, and then warping, or rubber sheeting, this image or map to a ground co-ordinate system. For a full description of geometric correction methods, see Lillesand and Kiefer (1987) and Sabins (1986).

Two problems were encountered with this classification that required rectification: gravel pits and quarries could not be spectrally distinguished from roads, and it was not possible to accurately classify the urban areas and portions of the transportation network. The first problem was overcome by using aerial photography as a reference and manually editing the pits and quarries to a separate class. This was not a large problem, as there were less than 20 pits and/or quarries in the test site. The second problem is a common land cover classification problem, and is the result of the high variability in land cover types in an urban setting. Urban areas can be composed of many cover types including roads, lawns, gardens,

buildings, trees, swimming pools, *etc.* Most pixels from a 28-m instrument such as Landsat TM are the result of more than one cover type, and therefore produce mixed signatures, which can not be classified using this approach. This problem was resolved by using a digital map of the transportation network (primary and secondary roads, rail lines, *etc.*). This vector transportation map was brought to a 10-m raster representation and converted to a binary bit-map where a morphological closing operation was performed on it (Fig. 6). This process involved performing nine bitmap dilations, followed by eight bitmap erosions. This operation had the combined effect of providing a 15-m buffer around the transportation corridors while at the same time closing, or making homogeneous, those areas where the transportation network is very dense, such as suburbs and industrial areas. A 15-m buffer around most roads provided allowances for the road surface, the road shoulder, ditches and fence lines, all of which produced mis-classified pixels in the classified image. This new infrastructure/urban class was then burned, or superimposed, onto the satellite classification. Presented in Figure 4B is a generalized version of the final classification, and shown in Table 1 is a tabular summation of the land cover units in hectares. Archives of Landsat imagery go back to 1972, making it also possible to monitor and quantify land cover changes over time.

Image Enhancement and Integration for Visual Interpretation

There are many airborne and satellite sensors in operation that record digital imagery in various portions of the electromagnetic spectrum, including the visible, colour infra-red, thermal infra-red and radar frequencies. Imagery from each portion of the spectrum can contain unique geoscientific information. This data is usually in digital format, which allows it to be referenced, digitally enhanced and integrated with other image products for visual interpretation.

Some of the most common functions include contrast

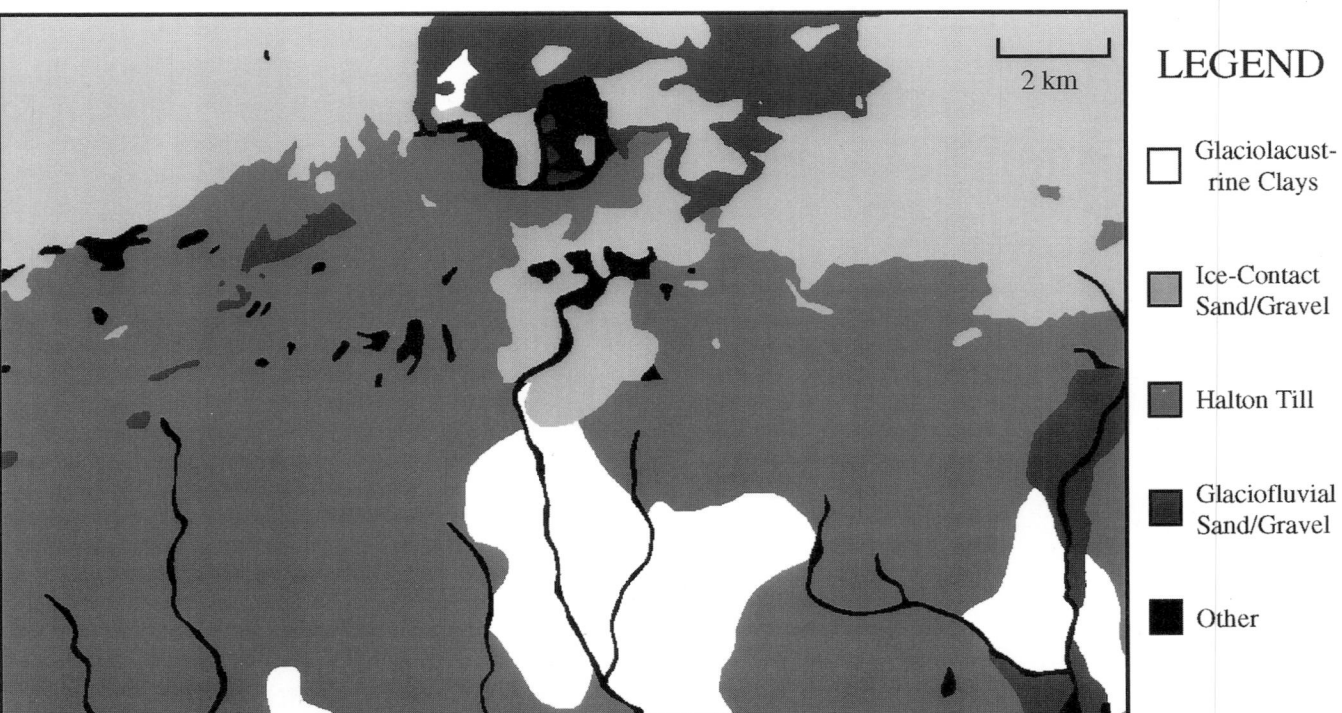

LEGEND

☐ Glaciolacustrine Clays

▨ Ice-Contact Sand/Gravel

▨ Halton Till

▧ Glaciofluvial Sand/Gravel

■ Other

Figure 3   *Generalized Quaternary geology of the study area (after Gwyn and DiLabio, 1973; and Hewitt, 1969).*

stretching, spatial filtering, textural filtering, principal component analysis, band ratioing, composite generation and colour space enhancements. As well, it is possible to generate merged images by integrating imagery from different sensors. The objective of combining imagery from different sensors is to exploit the advantages of each sensor, and create a more informative and easier to interpret image for visual interpretation. For a full discussion on image enhancement techniques for remotely sensed imagery, see Richards (1986) or Sabins (1986). For a comparison of multi-sensor image integration techniques for geological applications, see Harris *et al.* (1994).

In this example, imagery from two different satellites sen-

Figure 4 (**A**) *Decimated, geometrically corrected Landsat TM false colour composite of the study area. Composed of band 4 (near infra-red), band 3 (red) and band 2 (green).*

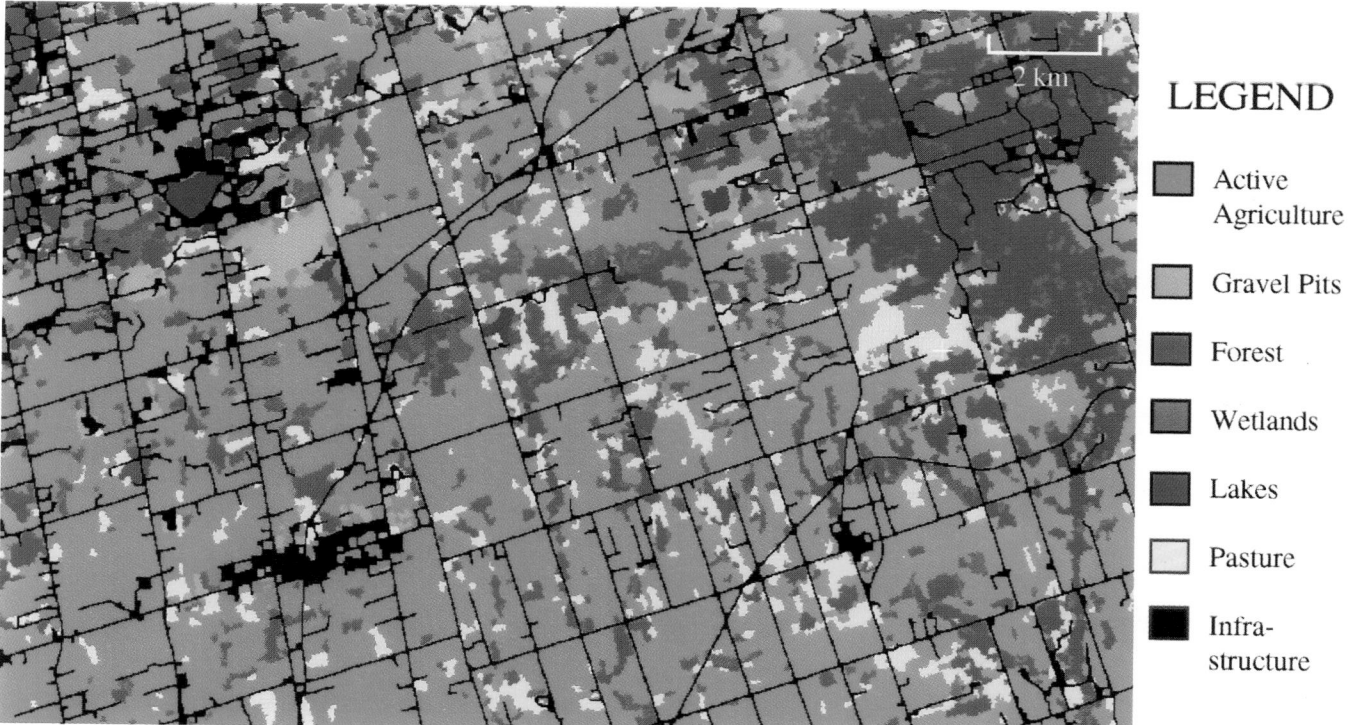

LEGEND

- Active Agriculture
- Gravel Pits
- Forest
- Wetlands
- Lakes
- Pasture
- Infrastructure

Figure 4 (**B**) *Land cover classification of the study area.*

sors (SPOT and ERS-1) and the elevation information obtained from digital topographic maps were enhanced and digitally merged for the purpose of extracting Quaternary geology information. Each data set was referenced, filtered, visually enhanced and then merged to a single image to provide a product that was more informative and easier to interpret than the individual data sets. A flow chart of the image-processing and integration methodology employed in this example is shown in Figure 7. A full description of the image-processing methodology employed for this study is beyond the scope of this paper, but can be found in Kenny *et al.* (1994). A brief description of the processing methodology employed and results obtained is given here.

The ERS-1 scene selected was an early spring 1992 image. The ERS-1 is a 30-m instrument operating in the radar portion of the electromagnetic spectrum. Radar images are very complex and can be resultant from many ground variables including land cover, ground cover, topographic relief, surface roughness, surficial material texture and soil moisture content. Each of these parameters can have a direct or indirect relationship with the surficial materials or landform morphology, making radar imagery a valuable source of Quaternary geology information. To reduce the speckled appearance, which is typical of radar imagery, a median filter was applied. This filter has the combined effect of producing a smoother image while at the same time preserving edges and apparent contacts (Fig. 8).

A SPOT Panchromatic scene of acquisition date, June 4, 1985, was selected for this study. SPOT Panchromatic is a single band at 10-m spatial resolution that spans the green, red and near infra-red portions of the spectrum, similar to black and white infra-red photography. The choice of the image acquisition date was to correspond to a near vegetation-free surface, allowing for observation of the bare soil surface in the agricultural areas. In this scene, contrasts in

soil moisture conditions, representative of different material types and textures, can be observed. These same observations would not have been possible within weeks of this acquisition date, as the agricultural crops in this area would have fully emerged. To enhance visually the spatial characteristics of this image, a sharpening filter (high pass) was applied (Fig. 9).

The digital terrain data for the test site was contained in 12 separate 1:10,000 5-m contour Ontario Base Maps (OBM). For the purpose of this study, this contour information was brought to a single vector data base. Through several GIS

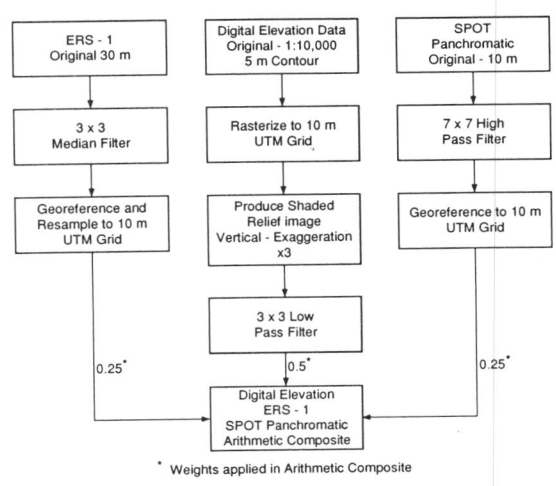

Figure 7 *Flow chart of image enhancement and integration methodology.*

Figure 8 *Decimated, geometrically corrected ERS-1 satellite image of the study area.*

processing steps, this vector data base was interpolated and rasterized at a 10-m grid cell to provide a Digital Terrain Model (DTM; Fig. 10). With the elevation data in raster format, it was then transferred to an Image Analysis System (IAS) where image enhancement techniques could be applied. The most

effective terrain enhancement method was found to be a shaded relief representation of the surface. This technique is usually associated with airborne geophysical data, but is equally effective on DTMs. Of the many views evaluated, the most informative for Quaternary terrain information was

Figure 9   *Decimated, geometrically corrected SPOT Panchromatic image of the study area.*

Figure 10   *Decimated, Digital Terrain Model of the study area.*

found to be at a sun azimuth of 53°, a sun elevation of 26°, and a vertical exaggeration of 3 (Fig. 11). To further enhance this image, a smoothing filter (low pass) was applied. It was found that the application of this filter gave the terrain surface a more realistic appearance.

The processing steps described in the previous section have preconditioned the images, such that when they are produced in composite these same features can be interpreted with much greater ease.

Various arithmetic combinations of the three images were evaluated for their usefulness in surficial and terrain mapping. The most useful composite was found to be a weighted

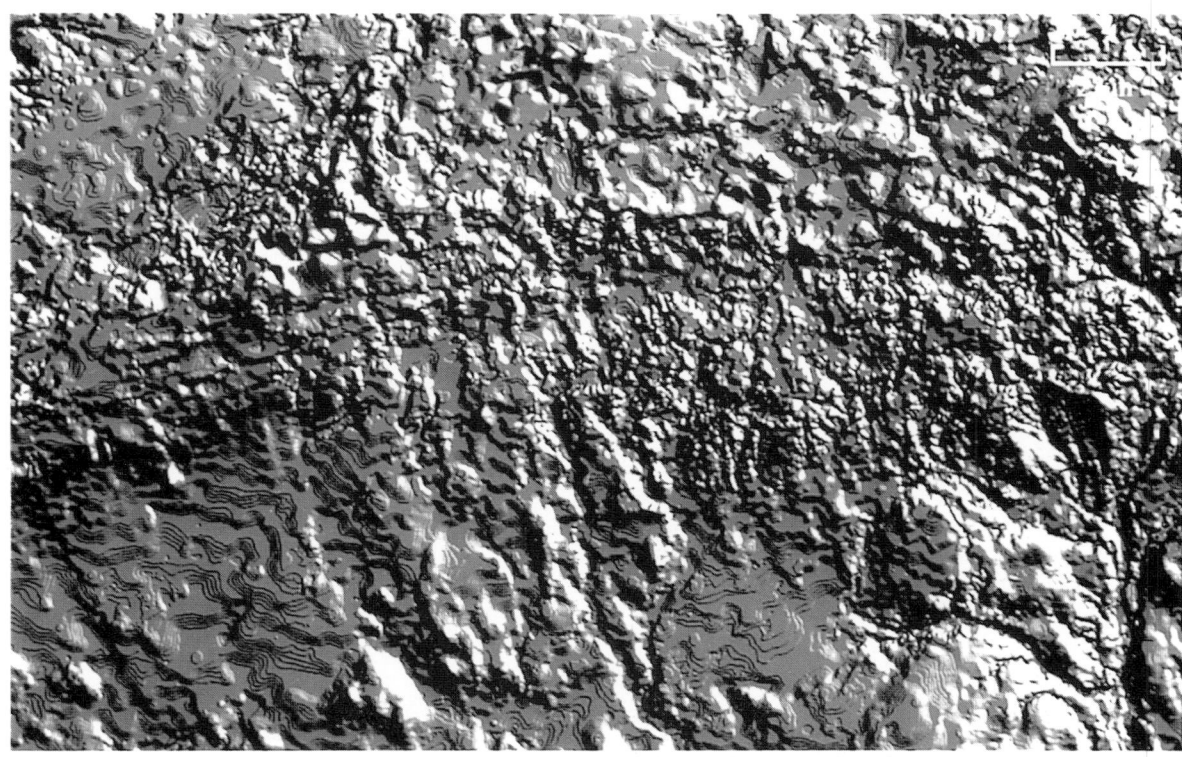

Figure 11   *Shaded relief enhancement of Digital Terrain Model.*

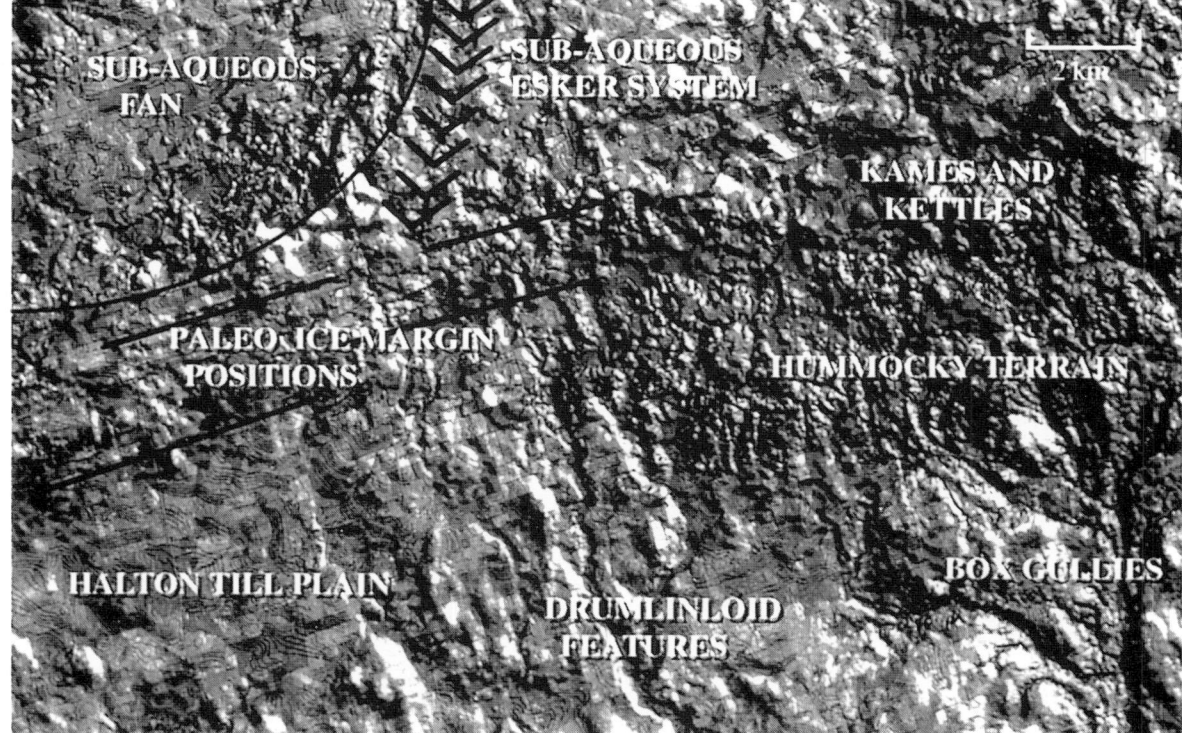

Figure 12   *Interpreted, decimated, arithmetic composite image of ERS-1, SPOT and DTM.*

average of the three images (Fig. 12). The weights applied to each image were 0.25 for both the SPOT Panchromatic and ERS-1 images, and 0.5 for the digital elevation data. The benefits of viewing the imagery synergistically can be seen in the final composite (Fig. 12). Some of the recognizable features include drumlins and drumlinoid features, a till plain, an area of hummocky disintegration moraine, a subaqueous fan, numerous kames and kettles, a subaqueous esker complex, and several paleo ice-marginal positions. This map and others are currently being evaluated and integrated into an Ontario Geological Survey-Geological Survey of Canada mapping program on the Oak Ridges Moraine. These products, will be used to assist in the revision of the current geology map (Fig. 3).

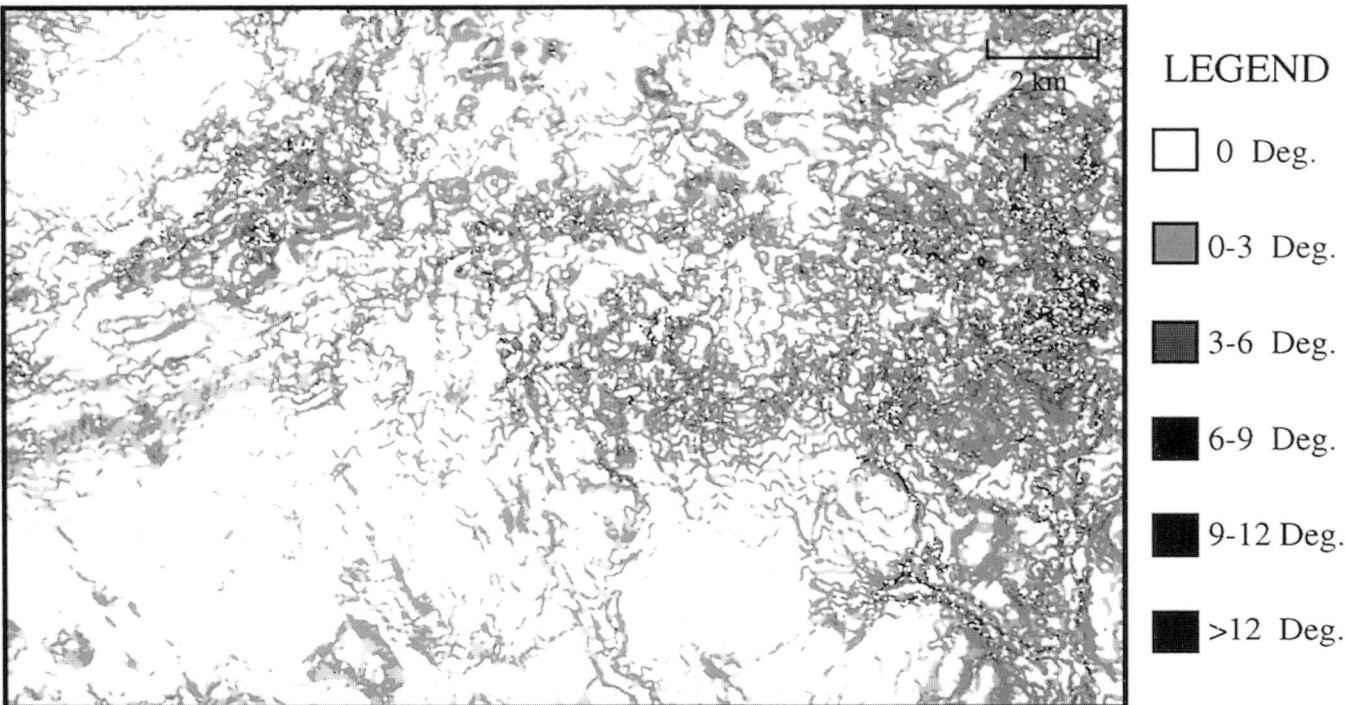

LEGEND

☐ 0 Deg.

▨ 0-3 Deg.

▨ 3-6 Deg.

◼ 6-9 Deg.

◼ 9-12 Deg.

◼ >12 Deg.

Figure 13 *Slope map of study area.*

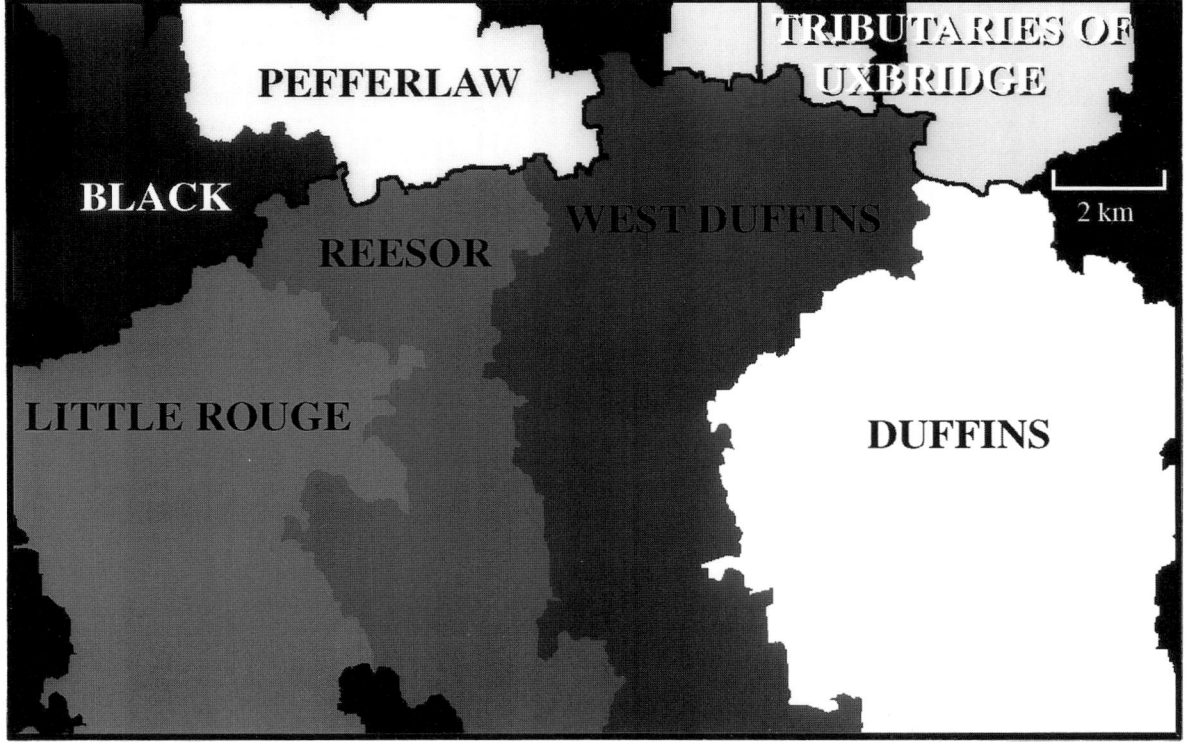

PEFFERLAW

TRIBUTARIES OF UXBRIDGE

BLACK

WEST DUFFINS

REESOR

LITTLE ROUGE

DUFFINS

Figure 14 *Watersheds of the study area (see also Chapter 10).*

Digital Terrain Models for Extracting Terrain Features

In the previous example, we saw how a DTM can be used in a qualitative manner, where it can be treated as an image product, and enhanced, integrated and interpreted with other image products. In this example, we will examine how the same DTM (Fig. 10) can be used quantitatively in a raster GIS environment to extract the terrain parameter of slope and to delineate individual watersheds. By knowing the elevation of every pixel in an image, it is a simple calculation to generate the slope of the terrain at each pixel in the image. Figure 13 is a thematic representation of slope derived from the DTM in three-degree increments for the study area. Similar operations can also provide slope aspect and slope length.

A further application using a DTM is the delineation of watersheds or sub-watersheds. Of the many algorithms that exist for watershed calculation, one of the most refined, and the method used in this study, is that developed by the United States Geological Survey (Jensen and Dominique, 1988). This program uses the raw DTM as input and, through a series of processing steps, generates an expected drainage network for the given terrain. This program can then calculate the individual drainage basins for user-specified or system-selected tributaries. Figure 14 is a sample output from this program. Seven major streams were manually seeded at the image edges: Duffins Creek, Reesor Creek, Little Rough Creek, Black Creek, Uxbridge Brook, West Duffins Creek and Pefferlaw Brook. In this image, it is interesting to note not only the delineation of the individual basins, but also a north-south divide where the northern-flowing basins (Pefferlaw, Black and Uxbridge) contact the southern-flowing basins (Reesor, Little Rough, and Duffins). As many environmental geoscience projects are completed on a watershed basis, the ability to clearly define the basin can be essential (*e.g.*, Chapter 10).

Raster Geographic Information System Modelling

This example will use the information generated from the previous examples to examine the relationship between the ERS-1 radar image (Fig. 8) and the mapped Quaternary geology (Fig. 3). As stated previously, radar backscatter can be a function of many ground variables, including topographic relief, landcover, groundcover, surface roughness and soil moisture. Each of these parameters can have a direct or indirect relationship with Quaternary materials or landform morphology, but to understand the contribution of each is difficult. In this example, we will attempt to isolate those areas where one of the parameters, soil moisture, is believed to be the dominant backscatter parameter, and then use these areas to examine the relationship between radar backscatter and Quaternary sediments. The only areas where soil moisture is the dominate radar backscatter parameter in the spring of the year are thought to be in the agricultural, flat areas.

As seen previously, there are a full range of glacially derived surficial materials present in the test site. They range from very fine-grained, low-permeability glaciolacustrine clays and tills to very coarse-grained, high-permeability, glaciofluvial and ice-contact sediments (Fig. 3). An advantage of using a spring ERS-1 scene (May 8, 1992), is that the ground has already thawed, and the winter accumulation of snow and ice has entered the ground-water flow system. Contrasts in surface soil moisture, at this time of year, are at or near maximum, and can often be related to the textural properties of the surficial materials. Another advantage of using spring imagery is that many of the agricultural areas are bare of crops and crop residues, which permits direct obser-vation of the surficial materials without the influence of ground or land cover.

By using a raster-based Geographic Information System, the agricultural flat areas were delineated for each of the four major geological material types present in the study area (Fig. 15). The information necessary to delineate these areas was obtained from the geology map (Fig. 3), the land cover map (Fig. 4B) and the slope map (Fig. 13). These four areas were then used as masks, under which ERS-1 statistics were calculated for the four major geological materials (Fig. 16).

The various surficial materials have distinct, near gaussian ERS-1 signatures (Fig. 16). The well-drained, drier, coarser grained materials, such as the glaciofluvial and ice-contact deposits, as expected, have a lower radar response. The signatures of these two sediments are almost overlapping. This was to be expected as the difference between these two mapped units is based on their origins, and they, in fact, have a similar composition and similar hydrogeologic properties. The finer grained materials, such as the Halton Till and glaciolacustrine deposits, exhibit a much stronger radar signal. These low-permeability materials are still quite wet, and this is evident in the radar signatures. One would expect these materials to have similar hydrogeological responses. This is seen in the signatures. The radar curves are very similar in shape, but the glaciolacustrine clay has a slightly brighter shift. This difference can be attributed to the glacio-lacustrine clays having a slightly lower permeability and, therefore, being wetter than the Halton Till. These early results demonstrate the potential use of satellite radar imag-ery for ground-water studies, in addition to its use as a geology mapping tool.

DISCUSSION

This paper demonstrates only a few of the spatial data analy-sis capabilities that are now possible with current digital data bases and new spatial data analysis software products. There are several other technological trends in the geomatics sector occurring concurrently that will also have a significant impact on the environmental geosciences.

Advances in raster scanners are presenting new ways of viewing traditionally used spatial data. Digital raster scanners have developed to the point where they can fully capture the information content of photographs or photographic nega-tives. Systems have been developed that enable an interpre-ter to view, interpret and annotate aerial photographs stereo-scopically on screen. Software also exists that can geo-metrically correct aerial photographs and annotated inter-pretations for camera lens and terrain distortions. These developments bring aerial photographs and aerial photo-graph interpretation into the digital realm, where a full suite of image analysis and GIS functions can be applied to the imagery.

The availability of spatially referenced data is rapidly in-creasing. Where once the public would purchase hard-copy maps, government agencies are now distributing their data in both hard-copy and digital format. By the late 1990s, there will be a dramatic increase in the quality and quantity of spe-cialized satellite imagery, and this imagery will be easily accessible in a more competitive market. There are now advanced satellite sensors being developed in the visible, reflected infra-red, thermal infra-red and microwave spectral regions. Some of these sensors will have a spatial resolution of down to one metre. One of the largest beneficiaries of this imagery will be the environmental sciences. NASA's recently announced Smallsat satellites are expected to be launched before 1997. One of these satellites will provide a panchromatic channel at 3-m spatial resolution and be capable of stereo imaging. The other Smallsat will be a hyperspectral imaging satellite recording 30-m resolution images on 384 spectral bands ranging from 0.4 μm to 2.5 μm (visible to reflected infrared). To promote the use of this data, NASA plans to offer this imagery free of charge for the first year after launch. Three

private American companies, Lockheed, WorldView and Orbital Sciences, have also announced plans to initiate their own satellite programs, and expect to be marketing specialized satellite imagery within the next few years.

A joint American-Japanese program is now developing an advanced resource satellite thermal sensor, ASTER (Advanced Spaceborne Thermal Emission and Reflection Radi-

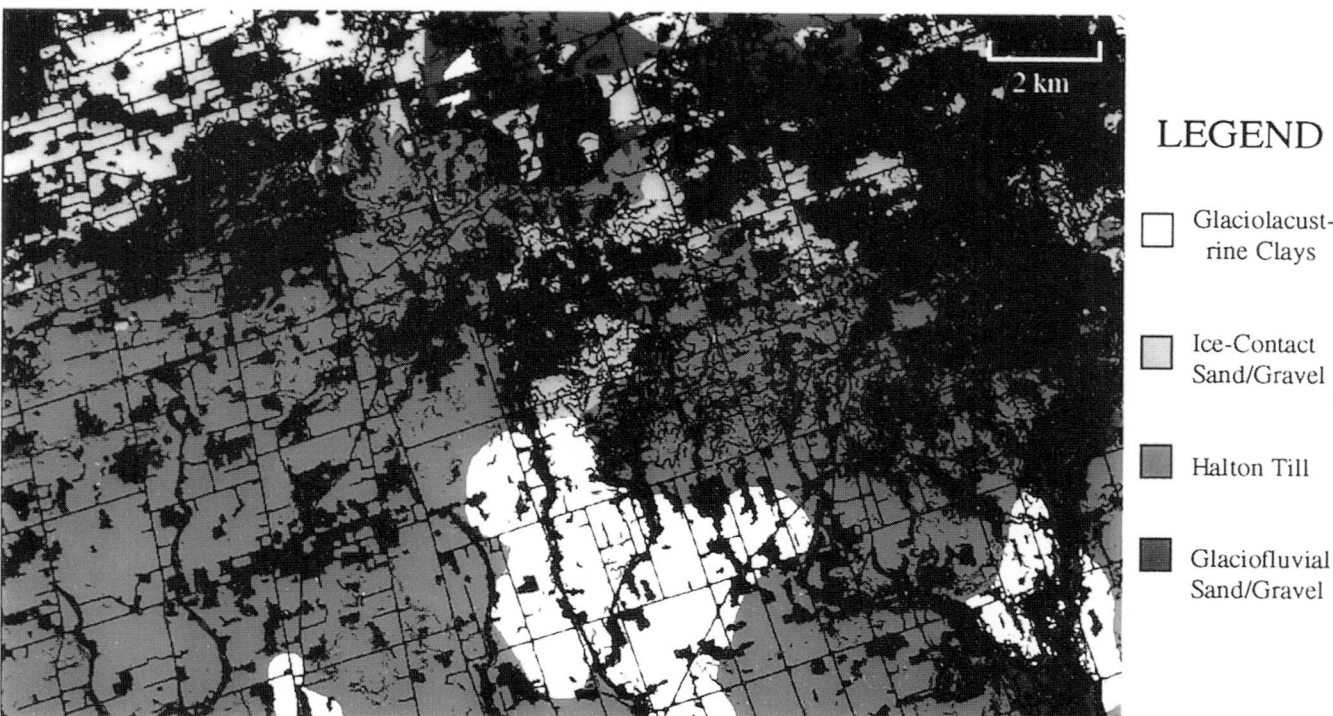

Figure 15   *Quaternary geology masks for the study area.*

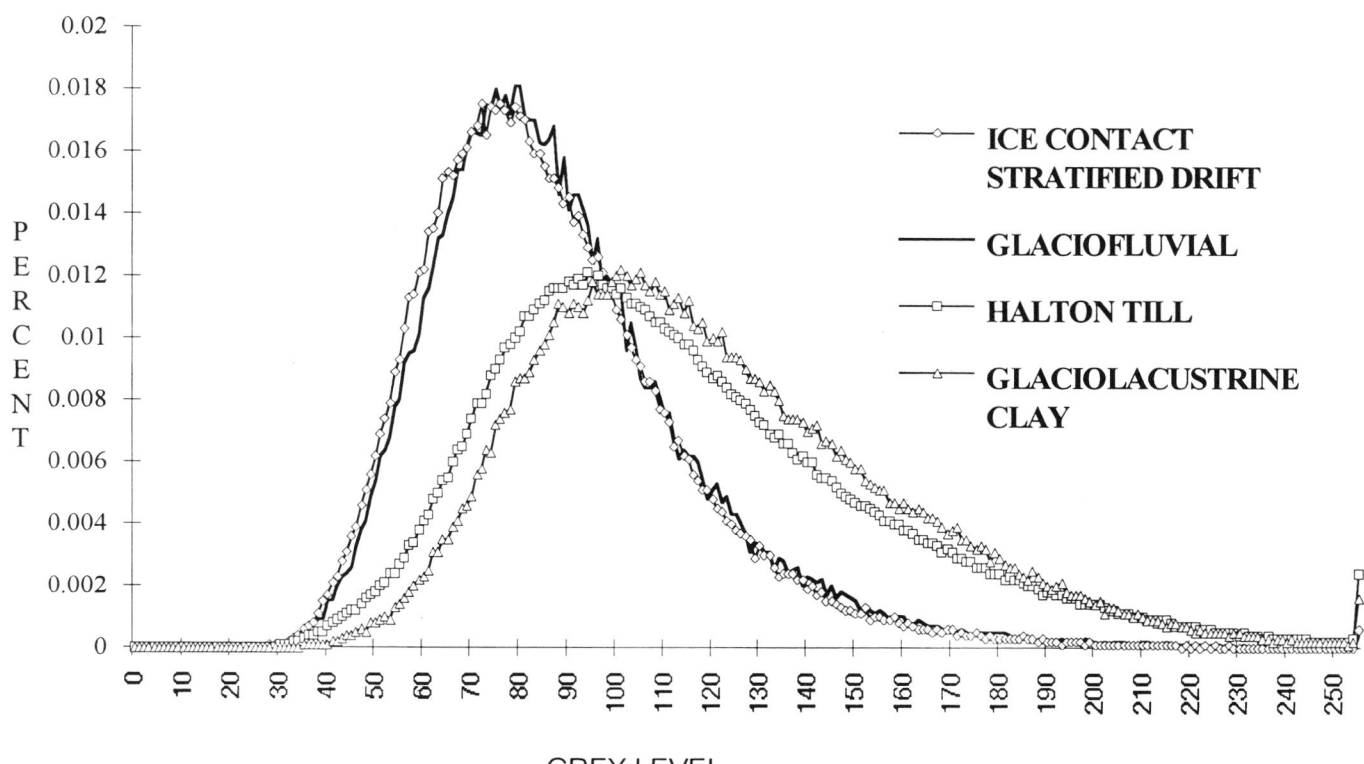

Figure 16   *ERS-1 radar signatures under masks used to construct Figure 15.*

ometer). The multi-spectral thermal infra-red imagery from ASTER will provide a thermal picture of the Earth we have not seen before. Similarly, the Canadian RADARSAT satellite, launched in 1995, will provide the most advanced radar imagery of the planet yet available. In addition to these new programs, imagery from Russian and Indian satellites is now being marketed internationally.

ACKNOWLEDGEMENTS

I would like to acknowledge RADARSAT International for providing the ERS-1 scene for this study and the Geological Survey of Canada for providing digital copies of the Quaternary geology maps for this area. I would also like to acknowledge Trevor Bain of Geomatics International for reviewing an early draft of this manuscript.

REFERENCES

Antenucci, J., Brown, K., Croswell, P. and Kevany, M., 1991, Geographic Information Systems: A Guide to the Technology: Van Nostrand Reinhold, New York, 301 p.

Aronoff, S., 1989, Geographic Information Systems: A Management Perspective: WDL Publications, Ottawa, ON, 294 p.

Gwyn, Q.H.J. and DiLabio, R.N.W., 1973, Quaternary geology of the Newmarket area, southern Ontario: Ontario Division of Mines, Map 836, scale 1:50,000.

Harris, J.R., Bowie, C., Rencz, A.N. and Graham, D., 1994, Computer enhancement techniques for the integration of remotely sensed, geophysical, and thematic data for the geosciences: Canadian Journal of Remote Sensing, v. 20, p. 210-221.

Hewitt, D.F., 1969, Industrial mineral resources of the Markham-Newmarket area: Ontario Department of Mines, Industrial Mineral Report 24, 41 p.

Jensen, S.K. and Dominque, J.O., 1988, Extracting topographic structure from digital elevation data for Geographic Information System analysis: Photogrammetric Engineering and Remote Sensing, v. 54, p. 1593-1600.

Kenny, F.M., Singhroy, V.H. and Barnett, P.J., 1994, Integration of SPOT panchromatic, ERS-1, and digital elevation data for terrain and surficial mapping: Tenth Thematic Conference on Geological Remote Sensing, San Antonio, Texas, 9-12 May 1994, Proceedings, p. 503-516.

Lillesand, T.M. and Kiefer, R.W., 1987, Remote Sensing and Image Interpretation: John Wiley and Sons, New York, 721 p.

Richards, J.A., 1986, Remote Sensing Digital Image Analysis: An Introduction: Springer-Verlag, Berlin, 281 p.

Sabins, F.F., 1986, Remote Sensing Principles and Interpretation: W.H. Freeman and Company, New York, 449 p.

39. Environmental Law And Assessment

Carolyn H. Eyles

Department of Geography, McMaster University, Hamilton Ontario L8S 4K1

SUMMARY
Growing concern among the public about the deterioration of environmental quality and its influence on human health has resulted in a dramatic increase in the number and scope of government regulations covering environmental issues at both a national and provincial level. In addition, the realization that environmental protection is less costly and more effective through prevention rather than remediation has led to the widespread introduction of environmental assessment as a planning tool. It is the purpose of this chapter to summarize the fundamental elements of Canadian and provincial environmental law and environmental assessment procedures, especially those aspects that affect urban geological issues. Case studies are described reviewing the environmental assessment process. With increased emphasis on regulatory control of environmental problems, environmental geologists, government employees and the public must be familiar with the fundamentals of environmental law and assessment procedures.

INTRODUCTION
This chapter will begin with an overview of the Canadian legal system and how it operates. It will also describe the fundamentals of environmental law and outline significant environmental regulations. The second part will focus on environmental assessment procedures at both the federal and provincial level. In the area of provincial environmental assessment, the focus will be on the Ontario system for a number of reasons. The Ontario Environmental Assessment Act (Ont. EAA), proclaimed in 1976, has been used to evaluate the environmental effects of government undertakings over a 20-year time period. This is much longer than any other provincial or federal assessment procedure has been in operation. The Ont. EAA is generally recognized as a relatively stringent piece of legislation and has been effective in the environmental planning of most government activities, although a criticism made often is that it does not adequately address developments proposed by the private sector.

FUNDAMENTALS OF CANADIAN LAW
In Canada, there are three levels of government, federal, provincial and municipal, and each plays a role in environmental legislation. This chapter will concentrate on environmental legislation enacted by the federal and provincial governments.

Federal and provincial governments consist of a cabinet of elected ministers, with either a prime minister or premier, and various executive committees consisting of elected representatives (Fig. 1). Cabinet members are mostly assigned a portfolio, allowing them to direct a certain department or agency (Fry, 1991). Both federal and provincial parliaments have a Department of the Environment, headed by the Minister of the Environment. All cabinet members are elected officials and hence are accountable to the public. Executive bodies at all government levels are assisted by thousands of civil servants, who are not elected and thus are not accountable to the public.

Canadian society is governed by two main bodies of law, common law and statute law. Common law refers to the body of rules and legal principles that has evolved through decisions made by judges over the past centuries as a result of the settling of disputes between people (Estrin and Swaigen, 1993). Common law embodies the key rights and principles about many things, including environmental quality. Statute law refers to new laws, called statutes or acts, which are made by federal parliament or provincial legislatures. Statutes are public laws intended to regulate conduct in the public interest and set out the rights and responsibilities of people in general terms; they often need to be expanded through a series of more detailed rules and procedures, such as regulations (Estrin and Swaigen, 1993).

Statutes or Acts
In order for a statute or act to become law, it must pass through a series of procedures (Fig. 2). A bill is presented to the federal parliament or provincial legislature, and must undergo first, second and third readings and receive Royal Assent before it can be passed. Upon receiving Royal Assent, it becomes an act. Many acts must then be proclaimed in force by the Governor General (or federal cabinet) or provincial equivalents before the law can take effect.

The length of time between passing and proclamation of an act can be considerable; the Canadian Environmental Assessment Act (CEAA) received royal assent as Bill C-13 in 1992, but was not proclaimed until January 1995. Ontario's "spills bill" (part X of the Environmental Protection Act) took six years and a change of government to be proclaimed (Estrin and Swaigen, 1993). Not all parts of an act have to be proclaimed at once. For example, the section in Ontario's Environmental Assessment Act establishing an Environmental Assessment Board was proclaimed in April 1976, but the section giving the board real power was not proclaimed until October of that year.

Despite problems and delays implementing new environmental laws, acts are intended to work in the best interest of the public, as they are subject to scrutiny by elected members of Parliament or the provincial legislatures who can criticize and suggest amendments prior to the act being passed. In addition, environmental statutes or acts cannot be amended unless the proposed change is brought before parliament or the legislature; this is seen as a safeguard for both the public and the environment.

Regulations

Regulations are usually found at the back of the concerned act. The regulations section of an act provides the government with a list of powers for regulation making; acts create offences, whereas regulations specify various ways in which the offence can occur.

Regulations are prepared by civil servants, often in closed-door consultation with industry, rarely in consultation with the public (Estrin and Swaigen, 1993). Regulations are approved by Cabinet and do not have to pass through parliament or the legislature where members can criticize them or propose amendments. Critics argue that regulations reflect only the views of the party in power, and that governments are strongly influenced by powerful members of the industries they are supposed to be regulating (regulator is becoming the regulated). Hence, the resulting standards are seen to give little protection to the environment.

Government Policies (Guidelines, Criteria, Objectives)

Many of the rules that industries, government bodies and the public are expected to comply with are not found in statutes or regulations that are legally binding, but in government policies which may be guidelines, criteria or objectives (Estrin and Swaigen, 1993). Further rules may be made by specialized boards or tribunals (*e.g.*, Ontario Municipal Board). Guidelines, objectives and criteria are private documents that do not need to receive consent from any elected official; they are set largely without public input or scrutiny.

Policies govern how agencies will interpret, administer or enforce the laws (Estrin and Swaigen, 1993), and are often followed when decisions are made regarding the issue of licences or approval of developments. Policies provide the government with quasi-legal documents that allow fast action and flexibility in emergency situations when passing of an act would take too long. They also provide a means for government to work out a site-by-site method of regulation that may be appropriate for specific environmental concerns.

However, government and its agencies are not bound to follow policies and can easily change them or ignore them under pressure. There is no requirement for politicians or bureaucrats to tell the public exactly what their policies are. Enforcement of policies, such as environmental standards or objectives, is usually by gentlemen's agreement, and is based on co-operation between government officials and the operators of a company. In some cases, this may be formalized into a program of approval under an environmental protection act. If the company does not keep up its end of the bargain and there is enough public pressure, the government will prosecute. However, this is not a safeguard for effective regulation, as public intervention is more often than not required to initiate government action.

Guidelines can be used by government to put pressure on those engaged in environmentally unsound practices. Guidelines provide advice on acceptable levels of pollutants, and may be used as levers in cases where certificates of approval are required. Ministry guidelines are often issued when serious concerns are raised about the effects of a pollutant, but insufficient scientific data are available to establish a specific objective (see below). Most of Ontario's water quality guide-

GOVERNMENT OF CANADA

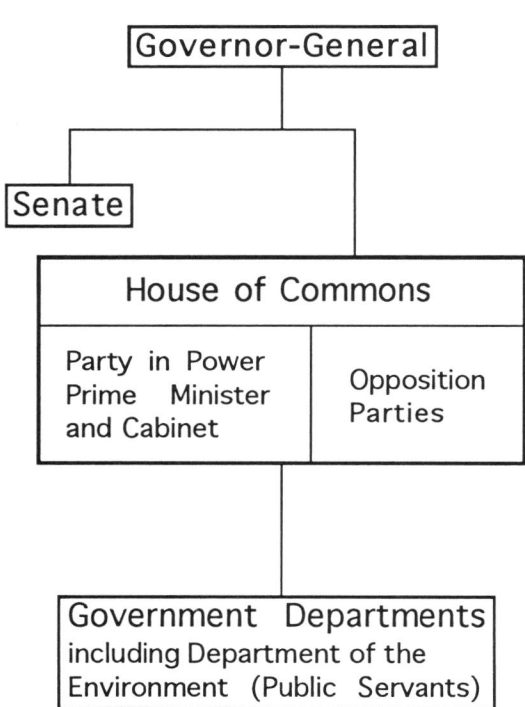

PROVINCIAL GOVERNMENTS

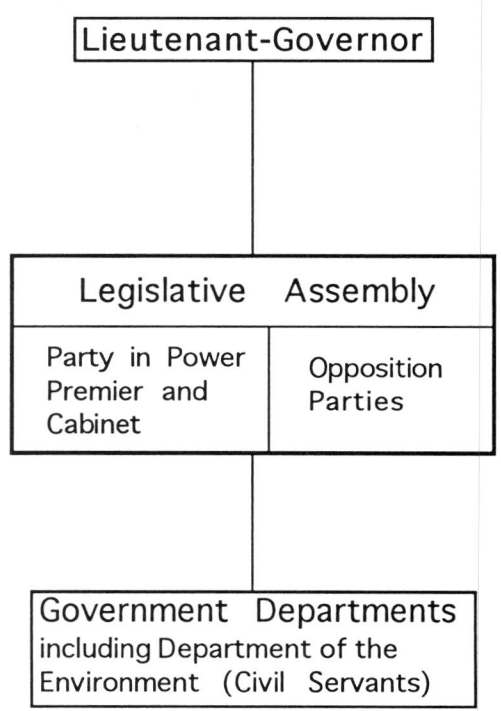

Figure 1   *Structure of the Canadian federal and provincial governments.*

lines and objectives are published in the Ontario Ministry of Environment and Energy (OMOEE)'s policy manual (Ontario Ministry of the Environment, 1984), and new effluent standards are set out in the Municipal-Industrial Strategy for Abatement (MISA) program. OMOEE also publishes "Guidelines for the Resolution of Groundwater Interference Problems" (policy no. 15-10), and "Guidelines for the Protection and Management of Aquatic Sediment Quality in Ontario", while the Ontario Energy Board sets "Environmental Guidelines for Locat-

ing, Constructing, and Operating Hydrocarbon Pipelines in Ontario".

Criteria are similar to guidelines and may be used, for example, to set acceptable limits for contaminants in water used for different purposes such as drinking, irrigation and industry. Criteria are not legally enforceable, but can be used in decision making over new projects that have potential to cause pollution or the issuing of orders for existing plants to abate their emissions.

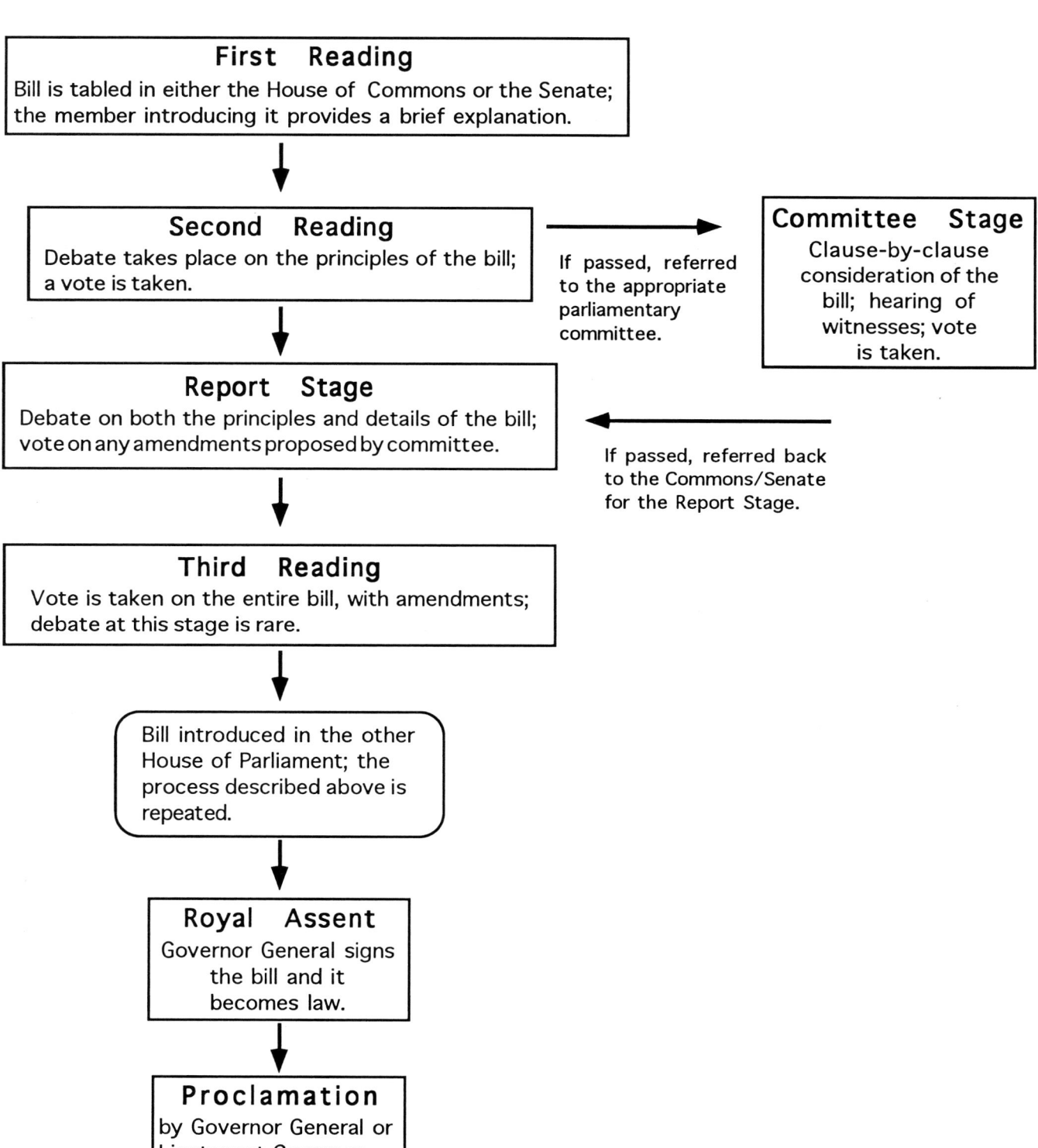

First Reading
Bill is tabled in either the House of Commons or the Senate; the member introducing it provides a brief explanation.

Second Reading
Debate takes place on the principles of the bill; a vote is taken.

If passed, referred to the appropriate parliamentary committee.

Committee Stage
Clause-by-clause consideration of the bill; hearing of witnesses; vote is taken.

Report Stage
Debate on both the principles and details of the bill; vote on any amendments proposed by committee.

If passed, referred back to the Commons/Senate for the Report Stage.

Third Reading
Vote is taken on the entire bill, with amendments; debate at this stage is rare.

Bill introduced in the other House of Parliament; the process described above is repeated.

Royal Assent
Governor General signs the bill and it becomes law.

Proclamation
by Governor General or Lieutenant Governor

Figure 2 *From bill to statute (after Brooks, 1993).*

Table 1   Summary of Canadian environmental legislation at the federal and provincial (Ontario) levels.

	Federal	Ontario
Discharge to air, water or soil	Canadian Environmental Protection Act (c.16, 4th Supp.)	Environmental Protection Act (E-19), Transboundary Pollution Reciprocal Access Act (T-18, no am.), Conservation Authorities Act (C-27, no am.)
	Air: Energy Supplies Emergency Act (E-9) *Water:* Arctic Waters Pollution Prevention Act (A-12), Canada Shipping Act (S-9), Canada Water Act (C-11), Fisheries Act (F-14), Carriage of Goods by Water Act (1993, c.21, no am.), Coastal Fisheries Protection Act (C-33), Navigable Waters Protection Act (N-22), Northwest Territories Water Act (1992, c.39, no am.), Yukon Waters Act (1992, c.40, no am.) *Soil:* Fertilizers Act (F-10), Pest Control Products Act (P-9)	*Air:* Mining Act (M-14, no am.) *Water:* Ontario Water Resources Act (O-40), Water Transfer Control Act (W-4), Fisheries Loans Act (F-19, no am.), Beds of Navigable Waters Act (B-4, no am.) *Soil:* Public Lands Act (P-43)
Waste disposal and transportation	Hazardous Products Act (H-3), Excise Act (E-14), Oil and Gas Production and Conservation Act (O-7)*, Transportation of Dangerous Goods Act (1992, c.34), Canadian Environmental Protection Act (c.16, 4th Supp.)	Environmental Protection Act (E-19), Ontario Waste Management Corporation Act (O-39, no am.), Conveyancing and Law of Property Act (C-34), Lakes and Rivers Improvement Act (L-3, no am.), Ontario Water Resources Act (O-40), Dangerous Goods Transportation Act (D-1, no am.), Waste Management Act (1992, c.1, no am.), Health Protection and Promotion Act (H-7)
Special substances and controlled products	Canadian Environmental Protection Act (c.16, 4th Supp.), Hazardous Products Act (H-3), Hazardous Materials Information Review Act (c.24, 3rd Supp.), Explosives Act (E-17), Radiation Emitting Devices Act (R-1), Pest Control Products Act (P-9)	Pesticides Act (P-11, no am.), Forest Tree Pest Control Act (F-25, no am.), Dangerous Goods Transportation Act (D-1, no am.), Gasoline Handling Act (G-4), Environmental Protection Act (E-19), Energy Act (E-16), Occupational Health and Safety Act (O-1)
Energy	Atomic Energy Control Act (A-16, no am.), Oil and Gas Production and Conservation Act (O-7)*, National Energy Board Act (R.S.C. 1959, c.46), Oil Substitution and Conservation Act (O-8), Energy Efficiency Act (1992, c.36, no am.), Canadian Laws Offshore Application Act (1990, c.44)	Energy Efficiency Act (E-17, no am.), Petroleum Resources Act (P-12, no am.), Public Utilities Act (P-52)
Spills	Transportation of Dangerous Goods Act (1992, c.34), National Energy Board Act (R.S.C. 1959, c.46), Canada Shipping Act (S-9), Canadian Environmental Protection Act (c.16, 4th Supp.)	Environmental Protection Act (E-19), Gasoline Handling Act (G-4)
Fire protection	Criminal Code (C-46), Explosives Act (E-17), Forestry Development and Research Act (F-30), Indian Act (I-5), Canada Shipping Act (S-9), National Parks Act (R.S.C. 1952, c.189)	Forest Fires Prevention Act (F-24, no am.), Fire Marshals Act (F-17), Hotel Fire Safety Act (H-16, no am.), Municipal Act (M-45), Theatres Act (T-6)
Other emergencies	An Act to Authorize Taking of Special Temporary Measures to Ensure the Safety and Security During National Emergencies and to Amend Other Acts in Consequences Thereof (c.22, 4th Supp.)	Power Corporation Act (P-18), Highway Traffic Act (H-8), Shoreline Property Assistance Act (S-10, no am.), Energy Act (E-16), Pesticides Act (P-11, no am.), Ontario Water Resources Act (O-40)
Environmental Assessment Process	Canadian Environmental Assessment Act (1992, c.37), Fisheries Act (F-14)	Environmental Assessment Act (E-18), Environmental Protection Act (E-19), Consolidated Hearings Act (C-29, no am.), Ontario Water Resources Act (O-40), Intervenor Funding Project Act (I-13)**, Waste Management Act (1992, c.1, no am.), Class Proceedings Act (1992, c.6, no am.)
Other	National Parks Act (R.S.C. 1952, c.189)	Provincial Parks Act (P-34), Conservation Authorities Act (C-27), Aggregate Resources Act (A-8), Endangered Species Act (E-15, no am.), Niagara Escarpment Planning and Development Act (N-2), Niagara Parks Act (N-3, no am.)

Notes: *Renamed the Canada Oil and Gas Operations Act
        ** As stated within the Act, the Intervenor Funding Project Act was repealed on April 1, 1996
• Unless a date is given whereby the act can be found in the Statutes of Canada or the Statutes of Ontario of that year, chapter references (c.) are from the Revised Statutes of Canada, 1985 for federal acts and from the Revised Statutes of Ontario, 1990. Supplements (supp.) are to the Revised Statutes volumes.   • Unless otherwise stated (no am.), all the above acts have a number of amendments since the referenced date. The Statute Citator indexes should thus be consulted.

Objectives are used to set maximum desired levels of contamination, and usually serve as goals for the future. For example, OMOE's "green book" outlines Ontario Drinking-Water Objectives (Ontario Ministry of the Environment, 1983b), which establish the maximum acceptable concentrations of substances considered harmful to human health or aesthetically unacceptable in drinking water. The objectives do not state how much of a substance may be discharged by a polluter into a waterbody, but set out maximum limits of a pollutant desirable in a water supply (Estrin and Swaigen, 1993). A number of problems are recognized in the process of setting meaningful objectives. The toxicological effect of many substances is poorly understood, particularly over the long term and on the full range of organisms that exist in natural environments. In addition, the combined effects of more than one pollutant in any environment are not taken into account when objectives or guidelines are established.

OVERVIEW OF ENVIRONMENTAL LEGISLATION

Federal, provincial and municipal powers are all involved in environmental legislation in Canada, and it is often difficult to determine which level of government is responsible for which environmental problem. The federal government has authority to make laws concerning matters that were considered national in scope at the time of Confederation, such as trade and commerce, navigation and shipping, sea coasts and inland fisheries, and undertakings that are interprovincial or international in nature (shipping, telephone, telegraphs and railways) or works that are declared to be for the general advantage of Canada (*e.g.*, grain elevators in the Prairies and nuclear power plants; Estrin and Swaigen, 1993; Table 1).

Provincial governments have authority over matters of local or private nature, such as the control of natural resources (except uranium which is under federal control), the management and sale of public lands and timber belonging to the province, and property and civil rights within the province (*e.g.*, Table 1). However, there are many areas of overlap. Both governments can legislate direct taxation, or make laws concerning immigration or agriculture.

Federal Legislation

Environment Canada is the federal agency primarily responsible for environmental matters, and administers statutes such as the Canadian Environmental Protection Act (CEPA; Table 1). CEPA replaces a number of separate pieces of environmental legislation, such as the Environmental Contaminants Act, the Clean Air Act, the Ocean Dumping Control Act and parts of the Canada Water Act and the Department of the Environment Act. CEPA is intended to protect human health and the environment from risks related to chemicals and toxic substances in all stages of their life cycles (cradle-to-grave philosophy), and regulates aspects of waste handling, waste disposal, spills and air emissions that come under federal jurisdiction. The federal government also administers the Transportation of Dangerous Goods Act (Table 1), which provides a comprehensive set of regulations designed to enhance safety and promote uniformity in the transportation of dangerous goods by land, sea, air and rail. Canada's Fisheries Act provides for the conservation and protection of fish and the waters frequented by fish by controlling potentially harmful emissions into federal jurisdiction waters such as the Great Lakes and the Pacific and Atlantic coasts. The recently proclaimed Canadian Environmental Assessment Act (CEAA) replaces the environmental assessment process set out in the Environmental Assessment Review Process

Guidelines Order (EARP Guidelines) and is designed to guide federal-government planning. CEAA is administered by the Canadian Environmental Assessment Agency (CEA Agency) which replaces the Federal Environmental Assessment Review Office (FEARO).

Provincial Legislation

In Ontario, the two central agencies charged with responsibility for the protection of the environment are the Ontario Ministry of Environment and Energy (OMOEE) and the Ontario Ministry of Natural Resources (OMNR). The OMNR is responsible for regulating Ontario's resource extraction industries, commercial hunting, fishing and logging and administering all Crown lands. OMNR administers the Provincial Parks Act, the Conservation Authorities Act, the Public Lands Act, the Aggregate Resources Act, the Endangered Species Act, the Lakes and Rivers Improvement Act and the Beds of Navigable Waters Act (Table 1).

Despite the ability of the OMNR to take a proactive stance on environmental issues, it has two conflicting mandates, conservation of the environment and regulation of commercial and recreational exploitation of natural resources. This is often recognized as a problem in the effective administration of OMNR environmental legislation.

The OMOEE is responsible for administering the bulk of environmental legislation in Ontario and has power to approve, license or impose conditions on any facility that may cause pollution. The Ontario Environmental Protection Act (Ont. EPA; Table 1) is designed to provide for the protection and conservation of the natural environment. It deals with air and water emissions and waste management, and addresses both preventative and remedial issues. The Ont. EPA regulates discharge of contaminants or pollutants into the natural environment, and places liability upon corporations, including all directors and officers and past, present and future owners, landlords, or trustees, for the source of contamination (deep pockets approach). Specific regulations attached to the Ont. EPA control air quality and atmospheric emissions (Regulation 308/346), waste emissions (Regulation 309/347, Part V of Ont. EPA, 3Rs Regulations), water quality and liquid discharges (Regulation 695/88, Municipal and Industrial Strategy for Abatement [MISA]), spills (Part X of Ont. EPA, Regulation 360), Certificates of Approval, contaminated properties, orders, offences and penalties.

OMOEE also administers the Ontario Water Resources Act (OWRA), which regulates surface and ground-water quality and the discharge of contaminants into water, the Gasoline Handling Act, the Occupational Health and Safety Act, the Pesticides Act, and the Dangerous Goods Transportation Act (DGTA; Table 1). Environmental planning acts such as the Environmental Assessment Act, the Consolidated Hearings Act, the Ontario Waste Management Corporation Act, the Waste Management Act, and the Niagara Escarpment Planning and Development Act (Table 1) also come under the jurisdiction of OMOEE.

Compliance with Environmental Regulations

Environmental acts and regulations create offences for disobeying certificates, orders, licences, terms or conditions of approval. Penalties on conviction of environmental offences vary, but in most cases are substantial. In Ontario, violations of the EPA that involve discharge of contaminants or hazardous waste, carry fines of up to $25,000 per day for individuals ($200,000 per day for corporations) and up to one year in jail. Maximum fines allowed under the Nova Scotia Environment Act (1994) have been increased to $1,000,000, and terms of imprisonment can be up to two years (Nova Scotia Department of the Environment, 1993). The Nova Scotia Environment Act also stipulates that a convicted person can be ordered to repay any profits earned from an offence. This is designed to deter those who decide that the profits gained from an illegal activity outweigh the penalties imposed (Nova Scotia Department of the Environment, 1993).

There is a recent trend toward charging all individuals responsible for an environmental offence made by a corporation, which may include officers, directors, and employees, in order to instil a message of individual accountability for the corporate action. In Ontario, the chairman and two directors of Bata Industries Limited were charged with illegal disposal of waste under the OWRA and Ont. EPA; the company was convicted and fined $90,000, and the directors were each fined $6,000. Total costs incurred, including legal fees, probably approached $1,000,000.

Individuals charged with environmental offences can use the defence of due diligence, which requires that the accused show that all reasonable steps were taken to prevent the commission of the offence. The defence of due diligence can be used for both provincial and federal offences (CEPA, Ont. EPA, Nova Scotia EA). In using due diligence as a defence, it is the responsibility of the defendent to show that the care taken was consistent with common standards used in the business in question (Phyper and Ibbotson, 1991). In Ontario, this means that the company must show that they attempted to comply with environmental laws through appropriate pollution prevention programs, health and safety systems, and that remedial and contingency plans are in place for spills (if appropriate), and ongoing environmental and health and safety audits are carried out. The company must also demonstrate that training programs are in place and that directors or officers review environmental compliance reports and promptly address environmental concerns brought to their attention.

Legislation of Environmental Rights

As the public becomes more concerned about the quality of the environment in which they live, there is more interest in the concept of environmental rights and the question of whether individuals have a fundamental right to a healthy environment. In Canada, there is no constitutional recognition of a public right to a clean environment (Saxe, 1990). However, the Canadian Charter of Rights and Freedoms, which is part of the Canadian Constitution, states its fundamental purpose as the constitutional protection of basic rights and freedoms in Canadian society. The rights given to each Canadian citizen through this charter allow any person to challenge the government if their basic human rights are being threatened by government action. In the environmental field, a citizen or public interest group could issue a Charter challenge to the Ministry of the Environment for failure to enforce its own standards regarding air quality under section 7 "Life, liberty and security of the person". However, the Charter applies only to government action and cannot be used between private litigants. Failure to incorporate any form of environmental rights or protection into the Canadian constitution is seen by many as a major problem. Canada is certainly behind many states in the United States and several other nations in terms of constitutional rights to environmental protection. The concept of a human right to a healthy environment has been incorporated into environmental bills of rights in Michigan, Minnesota and Pennsylvania, and is guaranteed by statute in

some nations; only Ontario, the Yukon and Northwest Territories have environmental bills of rights in Canada. The constitution of the state of Pennsylvania (article 1, section 27) states that:

> "The people have a right to clean air, pure water and to the preservation of the natural, scenic, historic and aesthetic values of the environment. Pennsylvania's public natural resources are the common property of all the people, including generations yet to come. As trustee of these resources, the Commonwealth shall conserve and maintain them for the benefit of all the people."

In Ontario, the recently proclaimed Bill 26, the Environmental Bill of Rights, grants the public much greater opportunity to participate in, and scrutinize, environmental decision making. However, the bill does not include an unqualified right to a healthy environment, although it does recognize the legitimacy of such a right (Estrin and Swaigen, 1993).

Any two residents of Ontario can now apply to the environment commissioner for an investigation if they believe any environmental laws or regulations have been contravened. The Ontario Environmental Bill of Rights also provides the public with earlier and better access to information related to Certificates of Approval and proposed legislation.

It is now recognized that one of the most effective ways to protect the environment is to prevent the development of projects or proposals that are likely to have negative environmental impacts through a process of legislated planning and assessment. Many recent developments in the field of environmental law have been in the area of environmental assessment legislation, a process which aims to minimize future environmental harm by regulating environmental planning procedures.

ENVIRONMENTAL ASSESSMENT LEGISLATION

Environmental Assessment

Environmental assessment (or environmental impact assessment) is primarily a planning aid, designed to anticipate the effects that any program, plan, project or other undertaking may have on the environment. The aim of environmental assessment (EA) is to identify and assess the direct and indirect costs of an undertaking prior to its initiation. The EA process is thus designed to make planners, officials and the public aware of any environmental problems an undertaking may cause before a firm decision has been made to go ahead with it.

The term environmental assessment refers to both the process of identifying and evaluating the effects on the environment of a project and its alternatives, and to the document which describes the carrying out of the proposed undertaking and its assessment. All environmental assessments require more or less the same procedure, and include some initial determination of the likely impact of the proposed development and alternatives to the development, a detailed assessment of the preferred alternative, government and public review, and a decision to proceed or not to proceed.

The Need for Environmental Assessment

There is now a substantial need for the EA process in urban environmental planning. Traditionally, the decision-making process for major projects was based on cost-benefit or economic analysis with little consideration given to social and environmental costs. As a result, promoters of projects stressed the benefits of increased employment, transportation and communication, but would not mention hidden social and environmental costs that are much more difficult to quantify. These costs are usually paid for by the public once the plan or project is in operation and when any remedial action is very costly. Environmental assessment is a method of identifying and assessing public costs of a project before government approval for a project is given. This shifts the emphasis of environmental protection toward pollution prevention and away from remedial action, which has limited effect and is extremely costly. An additional benefit is that the public can identify who bears the cost of any adverse effects before the project is approved. This should reduce the amount and cost of environmental clean-ups paid for from the public purse.

There is much concern at present in the insurance industry over the number of orphaned polluted sites in Canada (48 known) for which the government will not take responsibility. The federal government is proposing that insurance companies should pay for clean-up costs, but the insurance industry is arguing that it should not be responsible for environmental damage done in the past when environmental standards were different (*The Financial Post*, June 8, 1995). It is hoped that effective environmental assessment of potential impacts of future developments will put an end to this kind of problem.

Environmental Assessment and Sustainable Development

Environmental assessment procedures are consistent with the principle of sustainable development which aims to meet "the needs of the present without compromising the ability of future generations to meet their own needs" (United Nations World Commission on Environment and Development, 1987). Sustainable development requires that both physical and human environmental effects of any project be evaluated before that project is undertaken. Environmental assessment is thus considered to be one of the keys to acheiving sustainable development (Jacobs and Sadler, 1991). Legislation implementing environmental assessment procedures in Ontario (Environmental Assessment Act) has been constantly criticized and, in 1989, leaked documents (Project X) revealed that the provincial treasurer tried to dismantle the Ontario EA process saying that environmental assessment was not consistent with sustainable development. The reasoning behind this statement was that, in trying to conserve the environment, the Ont. EAA was not encouraging development, and hence was not compatible with sustainable development (Estrin and Swaigen, 1993).

Environmental Assessment Legislation in the United States and Canada

The need for a formalized process of environmental assessment has been the driving force behind environmental assessment legislation in Canada and much of the United States. The United States National Environmental Policy Act (NEPA) became law in 1970 and requires that federal agencies take into account environmental consequences of major legislative proposals or other agency actions before decisions are made (Kubasek and Silverman, 1994). NEPA requires that an Environmental Impact Statement (EIS) is prepared for every major federal proposal having a significant impact on the quality of the human environment. The Canadian Environmental Assessment Act (CEAA), proclaimed in January 1995, also requires that environmental assessments be carried out for federal projects, but does not apply to policies or programs implemented by the federal government.

In Ontario, the Environmental Assessment Act (1975) establishes a relatively rigorous environmental planning process designed for "the betterment of the people of the whole, or any part of, Ontario by providing for the protection, conservation, and wise management of the environment".

Environmental assessment legislation in other Canadian provinces is not as well established as in Ontario, although all provinces and territories now have some form of impact assessment included in their environmental laws (Table 2).

In June 1995, British Columbia proclaimed its Environmental Assessment Act, which replaces the current Mine Development Assessment Process, Energy Project Review Process and Major Project Review Process. The new BC EA legislation is seen to be a great improvement over existing assessment processes as it includes a broad and comprehensive definition of environmental effects, and provides a single, consistent process to be used in the assessment of a range of major industrial, energy, mining, waste, water management, aquaculture, food processing, transportation and tourism projects (British Columbia EA Act — Quickfacts, 1994).

In February 1995, Nova Scotia proclaimed its Environment Act, which replaces the Nova Scotia Environmental Assessment Act (NSEAA), but retains many provisions made in the NSEAA (Nova Scotia Department of the Environment, 1993). Several other provinces are in the process of amending and updating their EA legislation (*e.g.*, Saskatchewan, Table 2).

Each of the provincial EA processes and the federal EA process differ from one another in terms of the scope of their application, the nature of environmental impacts considered, the actual assessment procedure to be followed, the extent and timing of public involvement, and the decision-making and implementation process. Two environmental assessment processes will be discussed in detail below, the recently introduced federal EA process and the long-established Ontario EA process.

FEDERAL ENVIRONMENTAL ASSESSMENT

Canadian Environmental Assessment Act

The newly proclaimed Canadian Environmental Assessment Act (CEAA) governs the federal assessment process and applies to projects proposed by the federal government, projects carried out on federal lands, or projects funded by a federal authority. The CEAA replaces the discretionary Environmental Assessment and Review Process (EARP) which was administered by the Federal Environmental Assessment Review Office (FEARO). The CEAA has four stated objectives: (1) to ensure that the environmental effects of a project receive careful consideration before the responsible authority (RA) takes action; (2) to encourage RAs to take actions that promote sustainable development; (3) to ensure that projects to be carried out in Canada or on federal lands do not cause significant adverse environmental effects outside jurisdictions in which the projects are carried out; and (4) to ensure that there is an opportunity for public participation in the EA process (CEAA Training Compendium, 1994).

The CEAA also establishes the Canadian Environmental Assessment Agency which replaces the Federal Environmental Assessment and Review Office (FEARO). The new agency operates independently of any other government agency, but reports directly to the Minister of the Environment. Regulations attached to the CEAA include a Law List which designates a number of federal permitting, licensing and approval processes under the Act. Other regulations attached to the CEAA are the Inclusion List Regulations, Exclusion List Reg-

Table 2   Environmental assessment legislation in Canadian provinces (data from Couch, 1988; Statistics Canada, 1994). The environmental legislation of Quebec, New Brunswick and the Northwest Territories makes no specific reference to environmental assessment (EA). Although there is no specific EA legislation in Alberta, Manitoba, Nova Scotia and the Yukon, the EA process is outlined and regulated under umbrella-type environmental legislation. Other provinces, such as British Columbia, Saskatchewan, Ontario, and Newfoundland and Labrador, have specific acts dealing with EA.

Province	Legislation	Comments
British Columbia	Environmental Assessment Act (1995)	• Royal Assent July 8, 1994 • Proclaimed June 10, 1995 • Previously, 45 different statutes made reference to EA
Alberta	Environmental Protection and Enhancement Act (S.A. 1992, c.E-13.3)	• EA covered in Part II section s37-s57 • Mandatory and Exempted Activities (Alberta Regulations 111-93 & 112-93)
Saskatchewan	Environmental Assessment Act (S.S. 1979–80, c.E-10.1)	• Amended in 1983, 1988, 1989 • White paper on EA reform released 1994 and awaiting comments: no changes to date
Manitoba	Environment Act (S.M. 1987–88, c.26 as am.) Planning Act (R.S.M. 1987, c.P-80 as am.)	• Environmental approval process under the Planning Act
Ontario	Environmental Assessment Act (R.S.O. 1990, c.E-18 as am.); Environmental Protection Act (R.S.O. 1990, c.E-19 as am.); Consolidated Hearings Act (R.S.O. 1990, c.C-29); Ontario Water Resources Act (R.S.O. 1990, c. O-40 as am.); Intervenor Funding Project Act (R.S.O. 1990, c.I-13 as am.); Waste Management Act (1992, c.1); Class Proceedings Act (1992, c.6)	• Consolidated Hearings Act: joint board reviews undertaking which could be reviewed under 12 other statutes (e.g., EAA, EPA, Ont. Water Resources Act, Municipal Act and Planning Act)
Quebec – Southern	Environmental Quality Act (R.S.Q. 1977, c.Q-2 as am.)	• Amended 1978
Quebec – Cree and Inuit territories	James Bay and Northern Quebec Agreement (1975)	• Sections c.22 & c.23 • Also in EQA, c.11 • Five advisory committees (Cree: Evaluating and Review Committee; Inuit: Kativik Environment Quality Commission, James Bay Advisory Committee on the Environment, Kativik Environmental Advisory Committee)
Quebec – Naskapi Territory	North Eastern Quebec Agreement (1978)	
New Brunswick	Clean Environment Act (R.S.N.B. 1973, c.6 as am.) and Environmental Impact Assessment Regulation	
Nova Scotia	Environment Act (1995) Environmental Protection Act (R.S.N.S. 1989, c.150); Water Act (R.S.N.S. 1989, c.500)	• Effective February 1995
Prince Edward Island	Environmental Protection Act (R.S.P.E.I. 1988, E-9 as am.) Planning Act (R.S.P.E.I. 1988, c.P-8 as am.) Electric Power and Telephone Act (R.S.P.E.I. 1988, c.E-4 as am.)	• Appeals re: impact can be made to the Land Use Commission under the Planning Act • EIA must be submitted by a public utility to Public Utility Commission for construction or extension when cost ≥ $5000
Newfoundland and Labrador	Environmental Assessment Act (R.S.N. 1990, c.E-14) Environmental Assessment Regulations (1984)	• Individual assessment committees are appointed to review EIS or E. Preview Report. Separate EA Board is appointed if EIS reveals concerns.
Yukon & Northwest Territories	Environment Act (S.Y. 1991, c.5 as am.) Area Development Act (R.S.N.W.T. 1988, c.A-8)	

ulations and Comprehensive Study List (see below).

The following discussion of the new federal environmental assessment procedure is summarized from the Canadian Environmental Assessment Agency document "The Canadian Environmental Assessment Act, Training Compendium" (1994).

Projects Subject to
the Canadian Environmental Assessment Act

The CEAA defines a project as "a physical work that a proponent proposes to construct, operate, modify, decommission, abandon or otherwise carry out". This restricts the type of project subject to the CEAA to those involving some form of physical change, not plans or transactions, and does not include federal policy and program decisions (Northey, 1995). An environmental assessment is required by the CEAA if a federal authority either proposes the project, grants money or any form of financial assistance to the project, transfers the control of land to the project, or issues any licences or permits required for the project. The CEAA can apply to both public and private sector projects. Certain federal agencies, such as harbour commissioners, Crown corporations and band councils under the Indian Act, have been excluded from the CEAA process, although they can be made subject to the process by regulations. Other projects that may be excluded are those carried out in response to an emergency or that are on an exclusion list. The exclusion list takes the form of a regulation and incorporates projects that have minor anticipated environmental significance or have minimal input from federal agencies or where an EA would interfere with national security.

Assessment Procedure

If a project is not on an exclusion list or otherwise exempted from assessment by a regulation, then some form of environmental assessment will be carried out. This may either be a comprehensive study including examination of all aspects of construction, operation, modification, and abandonment of the project, or a more cursory screening or class screening. The responsibility of carrying out the assessment is given to a responsible authority (RA) which may be the proponent agency itself or an agency asked to provide support or approval in the form of funding, land, or issuing a permit or licence.

The scope of any environmental assessment under the CEAA is limited to consideration of effects on the natural environment, defined in terms of air, land and water; impacts upon socio-economic and cultural conditions are only considered where the effects are the result of changes to the natural environment (Estrin and Swaigen, 1993). A project is subject to different degrees of scrutiny depending upon the predicted severity of environmental impacts. The most rigorous assessment is a comprehensive study which should include discussions of the purpose of a project, its environmental effects, the significance of these effects, proposed mitigation measures, monitoring programs and alternative methods. Projects included on the Comprehensive Study List include major projects for national parks, water management projects, development of oil, gas and natural resources, pulp and paper mills, defence works, navigable waterways development and waste management projects (Estrin and Swaigen, 1993).

Projects not included on the Comprehensive Study List are subject to a screening process in which the basic environmental impacts of a project are identified. If the project is considered to be routine or repetitive in nature, with insignificant environmental effects, it may be included in a class screening. Class screenings may be used for projects such as dredging, highway maintenance, rail and tie replacement or rebuilding of facilities on the same site. It is anticipated that the majority of federal projects (95% or more), will be assessed through a screening or class screening.

The proponent of a project subject to the CEAA must first determine the scope of the project by identifying those components of the development that require assessment. Secondly, the nature of environmental effects to be addressed in the assessment and considered in the decision-making process must be determined. The EA process is considered to be largely self directed, and it is the responsibility of the proponent (or responsible authority) to determine the scope of the EA and factors to be considered, to manage the EA process and to ensure that the EA report is prepared.

Key Steps of
the Self-Directed Environmental Assessment

A self-directed federal EA should consist of eight major steps (Fig. 3):

Step 1.   Scoping

The RA (Responsible Authority, or proponent) establishes the boundaries of the screening or comprehensive study, and focuses the analysis on directly relevant issues and concerns.

Step 2.   Assess the Environmental Effects

Three tasks are involved in this process:

Description of the project.   This includes information on location, physical layout and design, construction plans and scheduling, standard control procedures and mitigation measures, operating procedures and decommissioning plans, and details about funding, permits or licences requested. In the case of industrial projects, it is necessary to specify the quantity and quality of all emissions, as well as any pollution control devices to be installed.

Description of existing environment.   This should identify the most important environmental elements of the region being examined and explain the selection of the boundaries of the study area. Only elements of the biophysical environment need to be considered.

Identification of project-environment interactions.
This includes a discussion of how, where and when a project's activities interact with and affect environmental components. These interactions can often be identified using map overlays and matrix tables, although they are not always readily apparent.

The RA must also ensure that the screening or comprehensive study consider the full range of environmental effects as defined in the CEAA. This may include some socio-economic or cultural effects that are directly related to the environmental changes caused by the project, such as effects on human health, effects on socio-economic conditions, and effects on physical and cultural heritage.

Step 3.   Mitigate the Environmental Effects

This step involves identification of technically and economically feasible measures that will mitigate (eliminate, reduce or control) a project's likely environmental effects. Mitigation measures may include replacement, restoration, and compensation.

Step 4.   Determine
the Significant Environmental Effects

This involves the determination of whether or not a project is likely to cause significant environmental effects after mitigation measures have been taken into account. Assessment of the significance of effects should be based on environmental standards, guidelines and objectives, or risk assessment measures, although public input is also important at this stage.

Step 5. Prepare the Environmental Assessment Report

The report should describe the results of the EA and explain how the assessment arrived at its conclusion. It should also describe any proposed mitigation measures and outline follow-up actions the RA believes are necessary.

Step 6. Review of the Environmental Assessment Report

Once the self-directed EA is completed, the report is submitted for review. If only a screening was carried out, the RA must ensure that all relevant federal departments have an opportunity to comment on the report, but public involvement is discretionary. A comprehensive study report must be reviewed by the Canadian Environmental Assessment Agency and made available to the public.

Step 7. Decision by the Responsible Authority and the Minister

When the review is completed, a decision is made on whether the RA can provide federal support for the project (i.e., proceed as proponent, grant funds, licences, or the land required). At the screening level, this decision can be made by the RA, but the minister determines the next step in the process for a comprehensive study.

In the case of screening, one of three decisions may be made by the RA: (1) to provide federal funds to the project if it is not likely to produce significant adverse environmental effects; (2) not to provide federal funds for a project if it is likely to cause significant adverse environmental effects that cannot be justified; (3) to request that the minister refer the project to a public review if there are uncertainties as to the significance of the effects or justification of the effects, or if there is sufficient public concern.

Step 8. Post-Decision Activity

The RA has a number of obligations following completion of the screening report or comprehensive study report and its decision. These obligations are (1) to provide public notice

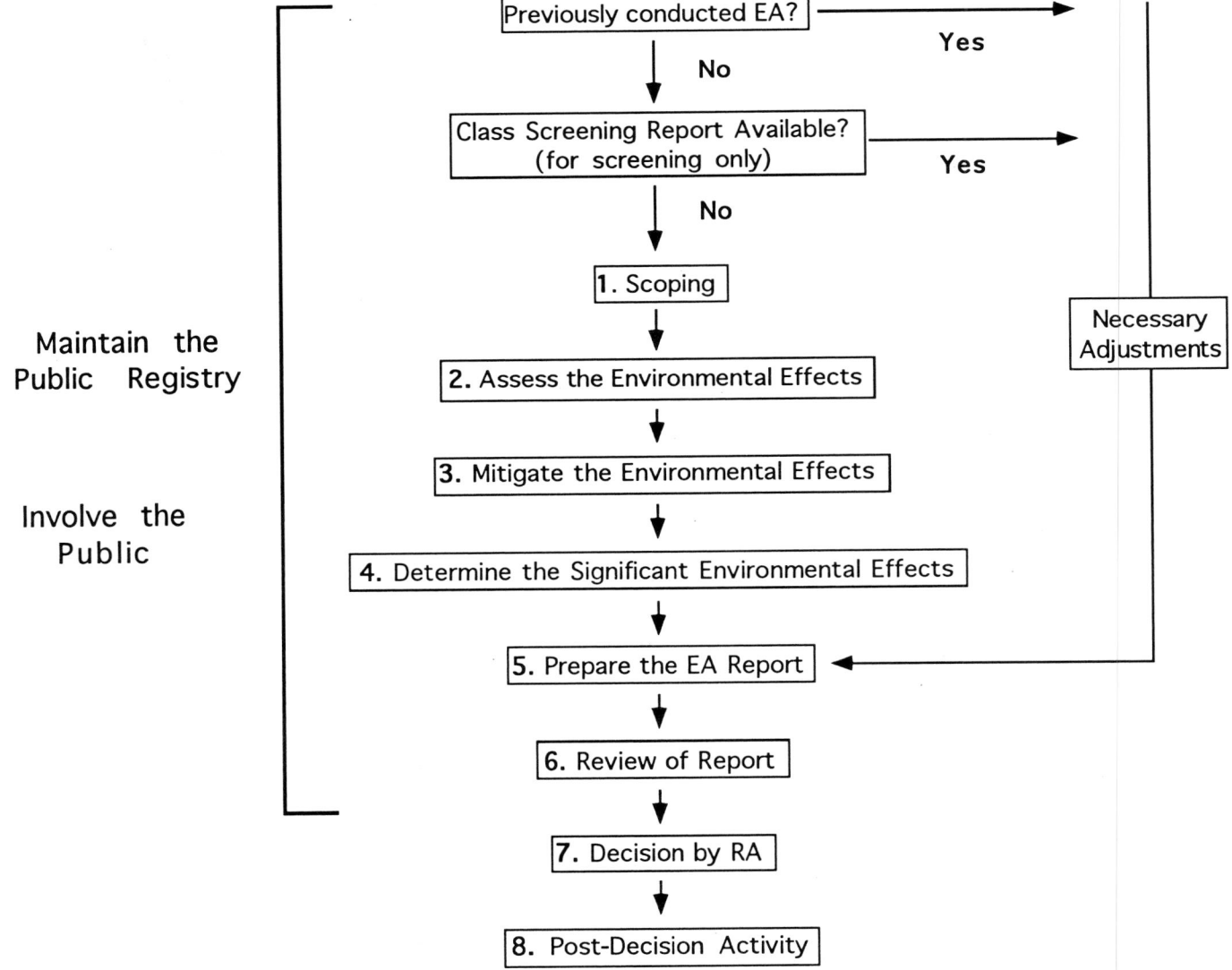

Figure 3　*Key steps of the self-directed federal environmental assessment process (modified from Canadian Environmental Assessment Act Training Compendium, 1994).*

about the course of action; (2) to decide whether a follow-up program to determine the accuracy of the assessment and effectiveness of mitigation measures is appropriate; and (3) to ensure the implementation of appropriate mitigation measures.

Public Participation

The CEAA allows the public an opportunity to comment on a proposed project at the comprehensive study or screening stage of the assessment and also to participate in the mediation process or a hearing. Public access to information on which the EA is based is provided through a public registry. A participant-funding program is authorized in the CEAA to allow the public to participate in mediations and assessments by review panels, but not at any earlier stage of the EA process.

If the responsible authority identifies that significant environmental impacts could result from a project, or if there is sufficient public concern, the project may be referred to a mediator or panel. The mediation process is designed to facilitate voluntary resolution of disputes between parties, and may be most effective when decisions regarding the methods of implementing a project are at issue. If necessary, the Minister of the Environment may establish an unbiased panel to hold a public hearing, but as yet it is uncertain what form the hearings will take. The panels appointed under CEAA have the power to subpoena witnesses, which suggests that hearings will be more formal than the public reviews held under EARP, but it is not certain whether sworn testimony will be required and cross-examination of witnesses and experts allowed (Estrin and Swaigen, 1993).

Decision-Making Process

If a mediation or review panel report has been prepared, it must go to cabinet for a decision (Northey, 1995). The project will only be granted approval if the project is unlikely to cause significant adverse environmental impacts or if the impacts can be justified in the circumstances. The responsible authority must design a follow-up program to determine the accuracy of the assessment and the effectiveness of mitigation measures implemented.

Federal agencies have a considerable amount of discretion whether to implement the findings of an environmental assessment, as the recommendations of a CEAA panel are not legally enforceable and the decisions made are advisory only. Projects may be approved without terms and conditions, and follow-up programs have restricted application. One of the greatest criticisms of the CEAA is that it provides insufficient opportunity for independent assessment of projects, as the responsible authorities have substantial control over the terms of reference of the study, whether or not to provide intervenor funding, and whether to apply the findings and recommendations of the assessment (Estrin and Swaigen, 1993). The quality of public participation in the CEAA process has also been questioned.

ONTARIO ENVIRONMENTAL ASSESSMENT

Ontario Environmental Assessment Act

The Environmental Assessment Act (1975; Ont. EAA) is the chief statute governing environmental assessments in Ontario (Ontario Ministry of Environment and Energy, 1994b). The Ont. EAA is regarded as perhaps the most sophisticated environmental legislation in Canada. The Ont. EAA "provides for the protection, conservation and wise management of Ontario's environment. The act is intended to promote good environmental planning and to ensure the public has the opportunity to comment on projects that may ultimately affect them".

There are three basic requirements of the Ont. EAA which are central to achieving its stated purpose:

(1) All aspects of the potentially affected environment must be considered, natural, social, economic, cultural and technical components (section 1 [c]).

(2) An evaluation of the advantages and disadvantages of the undertaking, the alternative methods of carrying out the undertaking and the alternatives to the undertaking, including the do-nothing alternative (section 5 [3]), must be made. This feature sets the Ont. EAA apart from other environmental statutes in Ontario (such as the Ont. EPA).

(3) Government ministries and agencies and the public should have an opportunity to provide comment to the Minister of the Environment and Energy before a decision is made on acceptance or approval of the undertaking.

The Ont. EAA defines the environment as air, land or water, plant and animal life, including human life; the social, economic, and cultural conditions that influence the life of man or a community; any building, structure, machine, or other device or thing made by humans; any solid, liquid, gas, odour, heat, sound, vibration, or radiation resulting directly or indirectly from human activities; or any part or combination of the foregoing and the interrelationships between any two or more of them, in or of Ontario (section 1). This definition is much broader than that provided in other environmental statutes, such as the CEAA or Ontario's Environmental Protection Act which restrict their definition to the natural environment.

Projects Subject to the Ontario Environmental Assessment Act

The Ontario Environmental Assessment Act applies to all undertakings (projects, plans, programs, *etc.*) of the Ontario government and its departments, all public bodies (*e.g.*, Ontario Hydro, universities, conservation authorities) and municipalities. Private-sector undertakings are not routinely subject to the Ont. EAA, but can be designated by regulation if the potential adverse environmental impacts of the proposed undertaking are considered to be significant. Designated private-sector activities in Ontario include schedule C undertakings for new municipal works such as sewage or water supply systems, road services to residents, and waste management activities which involve servicing a community of greater than 1500 persons (or the construction of landfills greater than 40,000 m³ capacity). Twenty municipal and private-sector proposals have been designated under the Ont. EAA since it was proclaimed in 1976 (Table 3). Members of the public, groups or government agencies may request designation of an individual project, and the Minister of the Environment and Energy, subject to cabinet approval, decides whether designation is warranted.

Exemptions

A major problem in setting up the Ont. EAA was the question of whether all projects should be subject to the Ont. EAA unless exempted, or whether only designated projects would be subjected. The first option was selected as environmentalists feared that many harmful projects would escape assessment. Certain classes of projects carried out by government agencies are exempted from the Ont. EAA, such as fish and wildlife habitat improvement schemes, flood proofing, and reforestation projects proposed by the Metro Toronto Regional Conservation Authority (MTRCA). Some individual projects may also be exempted from the EA process if it is felt that it is not in the public's interest to carry out an EA. A proponent can make an exemption request to the Minister of the Environment on the following grounds:

Minimal environmental impact anticipated.   If the project is not anticipated to have significant impacts, it may be exempted. It is suggested that proponents ask the following kinds of questions when screening projects for degree of environmental impact (Ontario Ministry of the Environment, 1983).

Might the proposed undertaking: (1) conflict with the environmental goals, objectives, plans, standards, criteria or guidelines adopted by the province or the community where the project is to be located? (2) have an effect on any unique, rare or endangered species, habitat or physical feature of the environment? (3) have effects on an area of ten acres or greater? (4) necessitate the irreversible commitment of any significant amount of non-renewable resources? (5) pre-empt the use, or potential use, of a significant natural resource for any other purpose? (6) result in a substantial detrimental effect on air or water quality, or on ambient noise levels for adjoining areas? (7) cause substantial interference with the

movement of any resident or migratory fish or wildlife species? (8) adversely affect human health? (9) substantially change the social structure or demographic characteristics of the surrounding neighbourhood or community?

If the answers to these questions are "no", an application for exemption may be warranted; if any are "yes" or "maybe", then an EA should be prepared.

An emergency.   Exemption is appropriate where emergency action is required to prevent damage, injury or interference to people or property. For example, emergency repair or replacement of structures such as dams or sewage treatment plants would be exempted from assessment.

Overwhelming public interest.   Exemption may be granted where the implemenation of a particular undertaking is a matter of such overwhelming urgency that the public interest would not be served by applying the Ont. EAA to it. This was used as grounds for exemption from the EA process for the Etobicoke Motel Strip Waterfront Park Project proposed by the MTRCA (Exemption Order MTRCA-A-2). In this instance, as the project was essentially designed to enhance current environmental conditions and stabilize the shoreline, it was decided that the people of the city of Etobicoke and Metro Toronto would not have their best interests served by delaying

---

Table 3   Private and municipal projects designated under the Ontario Environmental Assessment Act (Ont. EAA) 1976–1995.

Number	Proponent	Project Name/Description	Designation Date (Status)
1	Private	• Great Lakes Paper Forest Resource Corporation (Reed Limited)	October 1976 (Dormant)
2	Private	• International Nickel Company of Canada (INCO): Spanish River Hydroelectric Proposal	June 1977 (Withdrawn)
3	Private	• Onakowana Limited: Lignite Development Proposal	March 1978 (Withdrawn)
4	Municipal	• City of Detroit: Fighting Island Sludge Management Proposal	1981 (Dormant)
5	Municipal	• Regional Municipality of Hamilton-Wentworth: Redhill Creek Expressway	June 1980 (EA approved)
6	Municipal	• Victoria Hospital Corporation: Energy from Waste Plant	November 1983 (EA approved)
7	Private	• Petrosun/SNC: EFW Plant	April 1987 (EA approved, 1988)
8	Private	• Trintek Systems Inc.: EFW Plant	December 1987
9	Private	• KAM1 Hydroelectric Project	May 1988
10	Private	• RSI Reclamation Systems Inc.: Acton Quarry Landfill	January 1989 (Hearing)
11	Private	• Unitec Landfill Project	November 1989 (EA approved, 1993)
12	Private	• Steetley Landfill	May 1989 (EA approval denied by Board, March 1995)
13	Private	• Browning Ferris Industries (BFI): Biomedical	October 1989 (Withdrawing EA)
14	Private	• Consolidated Professor Mines: Shoal Lake	August 1989
15	Private	• Laidlaw/Tricil: Storrington Township Landfill	October 1989 (EA approved, 1993)
16	Private	• LASCO Landfill	January 1991 (EA approved, 1994)
17	Private	• Laidlaw: Warwick Landfill	October 1991 (Dormant)
18	Private	• St. Lawrence Cement RDF	July 1990
19	Private	• Laidlaw Rotary Kiln	EA Withdrawn
20	Private	• Taro Aggregates Ltd.: East Quarry Landfill	EA Under Review

the project with a full EA. Similarly, Ontario Provincial Parks are currently operating under an interim exemption order (MNR-59/2) while a Class EA for provincial park management is prepared. The exemption order states that it would not be in the best interests of the public, or the environmental and recreational features within the parks, to prepare separate environmental assessments for proposed undertakings. The order notes that delays in implementing park management projects could cause considerable damage to the natural environments that the parks aim to protect.

Most exemption orders are issued with an expiry date and are subject to a number of terms and conditions that the proponent must adhere to. An exemption order may be extended by the minister with cabinet approval.

Types of Environmental Assessment in Ontario

For projects that are subject to the Ont. EAA, either individual assessments or class assessments may be carried out.

Individual Assessment

An individual assessment is the most comprehensive form of assessment and may be carried out on an undertaking which is either a single project or consists of several distinct, but associated, activities. A proposed landfill site of greater than 40,000 m³ would be subject to an individual EA (Ontario Ministry of Environment and Energy, 1993a), as would proposed major highway construction. In some instances, two separate, but geographically connected, highway projects could be dealt with as similar projects in one proposal. Similarly, a combination of projects involved in the construction of a flood control dam, downstream channelization and bank stabilization works could be viewed as dissimilar projects within the same proposal. Individual EAs include detailed descriptions of the undertaking itself, the environmental effects of the undertaking, alternatives to the project, evaluation criteria and methodologies used in the assessment of environmental effects and mitigation measures proposed (see below).

Class Assessment

Class environmental assessments are used for the many provincial and municipal projects which occur frequently and have a predictable range of relatively minor effects on the environment. Class EAs are viewed as a generic form of assessment for an entire class of project. For example, many highway-widening and freeway-upgrading projects are included in the Class EA for Municipal Road Projects; sewage line installations may fall under the Class EA for Municipal Water and Waste Water Projects. Class EAs are generally used when a group of undertakings is identified that has a common process of planning, design and construction, a predictable range of effects, occurs frequently, and consists of projects that are relatively small in scale (*e.g.*, Association of Conservation Authorities of Ontario, 1993).

Class EAs have to comply with all requirements of the Ont. EA Act and must be submitted, reviewed, accepted and approved in a manner similar to the individual assessments. However, they follow a series of planning requirements that are streamlined and standardized for use solely on that particular group of projects. The Class EA is subject to government review and public comments, and may also require an Environmental Assessment Board hearing. Hearings for the Class EA Timber Management Plan for the OMNR took over four years to complete, and resulted in great improvements to the forestry management plan.

Once a Class EA has been approved, the proponent plans, constructs and operates the project according to procedures set out in the class assessment document. All individual projects are then exempted from the individual assessment process as long as approved procedures are followed. However, provisions exist for the public to bump-up the class assessment to an individual assessment if more significant environmental effects are anticipated or if sufficient public concern exists. Requests for bump-ups are made in writing to the Minister of Environment and Energy who has the authority to demand an individual assessment be made.

Example.   Erosion control measures, designed by the MTRCA to protect parts of the Lake Ontario shoreline along Scarborough Bluffs east of Metro Toronto, were assessed within the Class EA for Remedial Flood Control and Erosion developed by the Association of Conservation Authorities of Ontario (Association of Conservation Authorities of Ontario, 1993). These erosion control measures form the Sylvan Avenue Shoreline Regeneration Project (see Chapter 18). There were two requests for this project to be bumped-up to an individual EA on the grounds that the project should not be started until an integrated shoreline management plan had been completed. These requests were rejected by the Minister of Environment and Energy in November 1994, as it was felt that many of the issues raised had already been publicly addressed, and that by delaying the construction of the erosion control measures the public interest would not be well served.

What Must an Environmental Assessment Contain?

An undertaking subject to the Ont. EAA must have EA documentation prepared by the proponent and submitted to the Minister of Environment and Energy. Section 5 (3) of the Environmental Assessment Act is the starting point for a description of the legal requirements of an Environmental Assessment document. Section 5 (3) states:

"5(3) An environmental assessment ... shall consist of:
(a) a description of the purpose of the undertaking;
(b) a description of and a statement of the rationale for
(i) the undertaking
(ii) the alternative methods of carrying out the undertaking, and
(iii) the alternatives to the undertaking;
(c) a description of
(i) the environment that will be affected or that might reasonably be expected to be affected, directly or indirectly,
(ii) the effects that will be caused or that might reasonably be expected to be caused to the environment, and
(iii) the actions necessary or that may reasonably be expected to be necessary to prevent, change, mitigate or remedy the effects upon or the effects that might reasonably be expected upon the environment,
by the undertaking, the alternative methods of carrying out the undertaking and the alternatives to the undertaking; and
(d) an evaluation of the advantages and disadvantages to the environment of the undertaking, the alternative methods of carrying out the undertaking and the alternatives to the undertaking. (R.S.O. 1990, c.E-18)."

Regulation 334 R.R.O. 1990 has created additional information required under subsection 5(3) of the Act:

"2(1) An environmental assessment submitted to the Minister shall contain, in addition to the information required under subsection 5(3) of the Act,
(a) a brief summary of the environmental assessment

organized in accordance with the matters set out in subsection 5(3) of the Act;

> (b) a list of studies and reports which are under the control of the proponent and which were done in connection with the undertaking or matters related to the undertaking;
> (c) a list of studies and reports done in connection with the undertaking or matters related to the undertaking of which the proponent is aware and that are not under the control of the proponent."

All environmental assessments conducted in Ontario must contain documentation sufficient to fulfill these requirements. EAs can be rejected by a hearing panel if insufficient or inappropriate documentation is provided.

THE ONTARIO ENVIRONMENTAL ASSESSMENT PROCESS

The Ontario environmental assessment process may be divided into four phases: a planning phase, a review phase, a decision phase, and if necessary, a hearing phase (Northey, 1995).

The Planning Phase

The planning phase of the EA is extremely important, and proponents are strongly advised to follow recommendations set out in a number of OMOEE policy and guideline documents covering pre-submission consultation processes (Ontario Ministry of the Environment, 1987), and the preparation of EA proposals (Ontario Ministry of the Environment, 1992). The planning phase involves all stages of preparation of the EA and final submission of the EA document to the OMOEE (Fig. 4). In order to successfully plan and undertake an EA, it is recommended that a number of key features should be addressed. These are (1) consult with affected parties, (2) consider all reasonable alternatives, (3) consider all aspects of the environment, (4) systematically evaluate net environmental effects, and (5) provide clear, complete documentation.

Consultation with Affected Parties

To make the planning process a co-operative venture with affected parties, early consultation is essential. Pre-submission consultation is viewed as a mechanism to ensure that concerns of all affected parties can be identified and addressed before irreversible decisions or commitments are made. Early consultation carries with it a number of benefits including improved understanding of environmental concerns, identification and resolution of issues before the EA is submitted, and the promotion of mutually acceptable, environmentally sound solutions.

The Ontario Ministry of the Environment's document "Guidelines on Pre-Submission Consultation in the EA Process" (1987) encourages proponents to consult with members of the review team, the EA branch, the OMOEE technical staff, provincial and federal agencies and other potentially affected or concerned parties throughout the planning stage. The Environmental Assessment Branch of OMOEE assigns an adviser to each undertaking to guide the proponent as to the kind and degree of information and consultation required (Estrin and Swaigen, 1993). Proponents are encouraged to include members of the public, often in the form of liason committees or community study groups (*e.g.*, Taro Aggregates Ltd., 1995), in the earliest stages of the process to promote a high degree of integrity in the process. However, there are still discrepancies as to what adequate public participation consists of. Private proponents with limited

experience of public involvement have found it difficult to administer public consultation programs and have been criticized for turning the process into a public relations exercise (Estrin and Swaigen, 1993). Members of the public have also criticized the process by saying that consultation has only been introduced once key decisions have already been made and that their input is purely cosmetic. These problems are recognized by the EA board who have been particularly critical of inadequate and inappropriate public consultation processes in recent decisions they have published. The summary of the decision for the Steetley South Quarry landfill site proposal states

> "The Board also found that the proponent failed to include the community from the earliest stages of the project in a meaningful consultation process that would have allowed input into the planning process and provided an opportunity to the public to influence the proponent's decision-making process." (Environmental Assessment Board and The Joint Board, 1995, Summary).

Consider Reasonable Alternatives

One of the most distinctive features of the Ontario EA process is the requirement for the proponent to identify and consider alternatives to the undertaking and alternative methods of carrying out the undertaking. This procedure is meant to ensure that the undertaking is environmentally preferable to all other feasible alternatives (the preferred alternative). The proponent of a project must thus look at the ramifications of the project from perspectives other than those of economics or convenience. This approach aims to give adverse environmental effects the same consideration as financial costs when establishing preference (Northey, 1995).

EA planning must consider alternatives to the undertaking, which fulfill the purpose of the undertaking in functionally different ways, and alternative methods of implementing a particular type of alternative. The do-nothing alternative must also be considered as this addresses the question of need for the undertaking.

Example. Proponents of landfill undertakings are expected to investigate alternatives to their proposal such as incineration, waste export, reduction, reuse and recycling. The Ontario Waste Management Act (1992, c.1) restricted the range of alternatives to be considered by the now defunct Interim Waste Authority (IWA) in their search for waste management facilities to serve the GTA, by eliminating incineration and waste export as options. The IWA proposed to use an OMOEE document "Greater Toronto Area 3Rs Analysis" (Ontario Ministry of Environment and Energy, 1993c) as its input on alternative methods for the EA. Changes to the Ont. EAA proposed by the current Conservative government may allow the minister to exempt aspects of a project from environmental review. This may mean that proponents of waste management projects will no longer have to demonstrate consideration of alternatives.

Proposed alternative methods of carrying out the undertaking are different ways of doing the same activity and involve consideration of the questions: How? What? and Where?. The first two questions address the alternate technologies available to carry out the undertaking and the different designs that may be used. For a waste management undertaking, alternative methods may include discussion of centralized *versus* local composting facilities, landfill sites with different liner designs or linked systems such as a landfill connected to a composting facility. The third question requires consideration of different locations for the undertaking and generally involves various levels of constraint mapping and site suitability analysis. For example, there are two parts to the process of identifying potential landfill site locations. First, areas where landfilling would be unsuitable are identified on the basis of established screening criteria. This is known as constraint mapping. Second, remaining unconstrained areas are refined into sites and evaluated. Comparative criteria are used to explore tradeoffs among sites and to identify the preferred location.

This process of successive screening of alternatives allows the proponent to narrow down the possibilities and select a methodology and site that offers the least adverse environmental impacts. The proponent should, however, clearly explain the way in which the initial set of alternatives was determined, including the criteria and assumptions that were used. Also, it is important to clearly explain alternatives that were rejected and the unreasonable and negative features that resulted in the decision to reject.

It should be noted that two recent negative EA decisions published for waste management facilities (Environmental Assessment Board and The Joint Board, 1994, 1995) criticized the proponents for inadequate consideration of alternative methods and alternatives to the undertaking. Although there may be some flexibility in the implementation of Ont. EAA requirements regarding the range of alternatives to be considered by private sector proponents, extreme care must be taken to supply adequate, reasoned and documented justification for the assessment given.

Consider All Aspects of the Environment

Given the very broad and comprehensive definition of environment given in the Ont. EAA, the EA planning process must consider effects of an undertaking not only on the natural or biophysical environment, but also on "the social, economic and cultural conditions that influence the life of man or a community" and their interrelationships. This means that a large number of criteria have to be examined in any EA, which for waste management undertakings may range from geology and hydrogeology to transport, economics, culture and heritage (see Chapter 20). Proponents should ensure that government ministries and agencies and the public are consulted for their comments on the relevance of proposed criteria.

Reasons for selecting the criteria used in the EA should be clearly defined and explained in the EA document (see

Chapter 20). This allows all affected parties to understand the judgements made when selecting criteria. Where possible, the same set of criteria should be used for the entire evaluation process. However, as the level of detail in which alternatives are evaluated normally increases as the proponent

---

PURPOSE OF THE STUDY
-problem or opportunity

DESIGN OF PLANNING PROCESS
-selection of evaluation method(s)
-public involvement program

DEVELOPMENT OF CRITERIA AND ASSUMPTIONS FOR:
-determining study area
-establishing initial set of alternatives
-initial screening
-evaluation of alternatives

GENERATION AND EVALUATION OF ALTERNATIVES
-data collection
  -environment affected
  -environmental effects
  -mitigation/enhancement
  -advantages/disadvantages
-initial screening
-phased evaluation of reasonable alternatives
-description of rationale at each decision point

DETAILED DESCRIPTION OF UNDERTAKING
-environment affected
-environmental effects
-mitigation/enhancement
-advantages/disadvantages
-purpose of the undertaking
-rationale
-implementation strategy
-monitoring program

ENVIRONMENTAL ASSESSMENTS
-EA submitted to the Minister of Environment, Government Review, acceptance and approval (possible hearing).

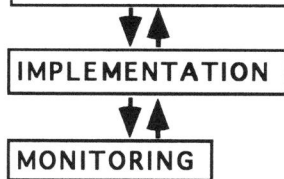

IMPLEMENTATION

MONITORING

Figure 4 *The planning framework for an Ontario EA (modified from Ontario Ministry of the Environment, 1989). Input from affected parties should occur at all stages of the process.*

procedes through the planning process, changes in the criteria used may also be made.

Systematically Evaluate Net Environmental Effects

Once comprehensive investigations of potentially impacted criteria have been carried out, proponents must explicitly evaluate alternatives, in light of their advantages and disadvantages, through a net effects analysis. Net Environmental Effects are defined as the remaining environmental effects, both positive and negative, after reasonable ways to minimize negative effects (mitigation) and increase positive effects (enhancement) have been considered and incorporated as appropriate. They are used to compare the advantages and disadvantages of alternatives. For example, a proponent might decide that a landscaped buffer zone would be provided for all alternative landfill sites to reduce the negative visual effects of the proposed facility. Where a variety of mitigation or enhancement measures are available, the relative merits of each should be considered. In some cases, the consideration of mitigation could change the alternative which is preferred.

The EA planning framework is based on a phased sequence of decision making, with points in the planning process where alternatives are evaluated on their net environmental effects and either rejected or carried forward for further study. In the early stages of assessment or screening, alternatives may be eliminated on the basis of a less detailed evaluation of their net environmental effects than in later stages of the process.

Because of the broad definition of the environment used in the Ont. EAA, one alternative is rarely preferred to all others on the basis of net environmental effects. Relative advantages in one area may be offset by disadvantages in another. A consistent basis for trade-offs is therefore important so that a solid case can be made for the selection of the preferred alternative. Often this is the most difficult stage of the evaluation and should be documented as clearly as possible.

Evaluation Methods.   Although the evaluation of alternatives is a fundamental requirement of environmental assessments in Ontario, the EAA does not specify the types of evaluation methods that should be used. The relative net environmental effects of alternatives may thus be evaluated in different ways either within a single project or between projects. However, it is important that a proponent establish one or more methods for predicting and evaluating the net environmental effects of an undertaking. Evaluation methods are defined as formal procedures for establishing an order of preference among alternatives (Ontario Ministry of the Environment, 1990a) and are important in making the assumptions and judgements used in the decision-making process both systematic and explicit. Proponents are advised to select evaluation methods that clearly identify the relative differences and key impacts of various alternatives in order to identify the trade-offs used in selecting the preferred alternative.

There are four major requirements of any evaluation process. First, criteria must be selected on which to base the decision. These criteria are the features or considerations used to compare alternatives and will vary from project to project. Examples of criteria used in recent landfill site selection processes are illustrated in Chapter 20 and include geology, hydrogeology, transportation, surface water, biology, archaeology, social, economic and cultural factors. Determination of the criteria to be used in an assessment is a critical matter and should involve public input at all stages. Criteria used in the Interim Waste Authority search for landfill sites in the GTA were revised several times, and included comments and concerns of the public wherever possible (Interim Waste Authority, 1992a, b). Some criteria used in assessments act as screening criteria and set minimum conditions or requirements that must be fulfilled by alternatives (see fig. 6 in Chapter 20). Screening criteria are usually applied at the earliest stages of the assessment, to eliminate unacceptable alternatives and to reduce the number of alternatives considered to a reasonable number.

The second stage of an evaluation process involves the prediction of impacts on each criterion by each alternative. In any evaluation exercise, all criteria identified are not of equal importance, and the relative weight of each criterion needs to be established (see fig. 7 in Chapter 20). Weights should be set in accordance with the willingness to trade off the criterion, and can take the form of single values that are multiplied by rates to give scores, or more sophisticated equations that allow for incremental changes (Ontario Ministry of the Environment, 1990a).

Once criteria for the evaluation and weightings for the criteria have been determined, then the impacts predicted for each criterion and each alternative can be compared. The fourth and final stage in evaluation is the application of an evaluation method which combines the weights and rates for each of the criteria to identify the preferred alternative. The OMOE document entitled "Evaluation Methods in Environmental Assessment" (Ontario Ministry of the Environment, 1990a) identifies six categories of evaluation methods appropriate for EA work.

1. Ad hoc methods in which descriptions of alternatives and their relative strengths and weaknesses are given in narrative terms.

2. Checklists compare and evaluate alternatives against a specified set of criteria with no compensatory rules or trade-offs. Checklists may take the form of unordered or ordered lists of criteria, or may involve satisficing (see Appendix B. Glossary), in which only alternatives that satisfy certain specified conditions are considered. Constraint mapping, or elimination of unsuitable geographic locations, may also be used to identify unconstrained areas suitable for a particular undertaking.

3. Matrix methods use matrices for comparison and evaluation of alternatives. Descriptive matrices rely on professional judgement to order preferences. Mathematical matrices, such as simple additive matrices, use mathematical operations to order preferences and allow trade-offs between attributes to be made. The Simple Additive Weighting Method (SAWM) attempts to classify each alternative using a single score intended to represent the attractiveness of the project (Ontario Ministry of the Environment, 1990a). Numerical raw data (*e.g.*, hectares, kilometres, dollars) have to be standardized into dimensionless units by establishing a value for each alternative relative to the alternative which is most impacted. Once data are standardized to a common scale, mathematical analyses can be performed. Specifically, the results can be multiplied by criterion weights and the weighted scores can be summed to give the final weighted score for a given criterion or criteria group.

4. Economic methods apply economic principles and methods to evaluate alternatives in monetary terms. Economic evaluation methods include cost-benefit analysis, cost-effectiveness analysis, cost-minimization analysis and planning balance sheets. Economic evaluation methods are not generally used in EAs in Ontario, but Canadian federal government policy requires that major new regulations in areas relating to health, safety or fairness are subject to socio-economic impact analysis.

5. Pair-wise comparisons involve the sequential comparison of alternatives in pairs, in order to organize preferences using mathematical techniques. It is argued that people are more able to consider two things at a time than many things. Various forms of concordance and discordance analysis may be used in pair-wise evaluations, including Saaty's analytical heirarchy procedure (AHP) which was used by Ontario Hydro in its southwestern Ontario transmission line study (Hoglund and Buck, 1987).

The Concordance Method is useful where a combination of different data types are being examined. The method compares alternative sites in pairs for each criterion or indicator, identifying the better of the two sites. The site which is the best of the pair receives points equivalent to the weight of the criterion or indicator under which the two sites are being compared. If the sites are equally good for that criterion, the points are divided equally between the sites. This pair-wise comparison is carried out for all sites for each criterion; results are placed in a Concordance Matrix. Points attributed to each site are summed across rows and divided by the sum of the weights for the criteria. Scores can then be used to identify differences in levels of impacts between sites and to rank the sites.

6. Mathematical programming methods use mathematical techniques to search for the alternative that best meets a specific objective. These methods identify the optimum solution which best satisfies predefined goals and objectives while staying within specified constraints. Specific methods are linear programming, dynamic programming, and goal programming.

Despite the wide range of evaluation methods available, EAs in Ontario have tended to use a only a limited number of relatively simple techniques such as *ad hoc* methods, satisficing (including constraint mapping) and simple additive methods. This is partly due to the fact that very formal and complex evaluation methods have high levels of traceability and accountability and can expose vulnerabilities in the proposal that a proponent may find difficult to explain. In addition, many of the personnel involved in EA evaluation in Ontario may be unfamiliar with all of the available techniques. In assessing the significance of evaluation methods, it must be remembered that evaluation is only one part of the EA process, and any evaluation method is of little use if the impacts of an undertaking are poorly predicted.

Provide Clear, Complete Documentation

It is the proponent's responsibility to provide clear and complete documentation of the EA process in a form that is concise, logical and accessible. The language and format used should be understandable to all potentially interested parties and should satisfy both expert and lay readers. In particular, all stages in the decision-making process should be clearly documented to fulfill the Ont. EAA requirements for traceability. Reasons for decisions made should be clearly explained, and negative as well as positive aspects of the proposal discussed. The EA document must include technical support for the EA application and any explanation and information needed for non-specialists to understand the planning process that was followed and the proposed undertaking. Detailed technical analysis and data should be placed in appendices and support documents. A flowchart representing the proponent's planning and decision-making process is recommended (Ontario Minstry of the Environment, 1989).

Document Format.   The following list gives the eight headings under which the planning process and undertaking are to be described in the EA: (1) Description of the Undertaking; (2) Description of the Purpose of the Undertaking; (3) Description of Alternatives to the Undertaking and Alternative Methods of Carrying out the Undertaking; (4) Description of the Environment Affected; (5) Description of the Environmental Effects; (6) Description of Mitigation/Enhancement Measures; (7) Evaluation of Advantages and Disadvantages; and (8) Description and Statement of Rationale for the Undertaking and Alternatives.

The description of the undertaking should be more detailed than the description of the alternatives and should specify clearly and comprehensively what the proponent is seeking approval for. The description might include descriptions and discussion of some or all of the following: location, dimensions, construction, operation, maintenance, decommissioning, industrial processes, nature and sources of fuels and raw materials, by-products, emissions, effluents, support services required, products and services supplied, scale of employment, schedule for production and operation.

The statement of rationale represents a proponent's summary explanation of the decisions that were made throughout the process of selecting the undertaking from the alternatives considered.

Regulations in the Ont. EAA also require an executive summary be included in the EA which must contain a brief summary, a list of studies and reports under the control of the proponent, a list of studies and reports related to the undertaking but not under the proponent's control, and two unbound, well-marked and legible maps of the proposed undertaking. Proponents should make the EA available at community libraries and to members of the public. The EA branch of OMOEE will provide copies to the Ministry of Government Services and to local District and Regional Offices of the OMOEE.

The Review Phase

After all the research is complete, the EA is submitted to the Minister of Environment and Energy who arranges for a review to be prepared by ministry staff and other relevant agencies. The review is co-ordinated by the Environmental Assessment Branch of OMOEE. The main purpose of the review is to provide a broad evaluation of the strengths and weaknesses of the EA and to determine the extent to which the requirements set out in subsection 5(3) of the Ont. EAA are met (Ontario Ministry of the Environment, 1989). If ministry staff participating in the review are not satisfied that the EA complies with the requirements of the Ont. EAA, they can suggest amendments or order the proponent to do further studies. If the Environmental Assessment Branch feel that issues are not covered in sufficient detail or quality to satisfy the review agencies, they may make a finding that the EA does not meet the requirements of section 5(3) of the Ont. EAA.

Two questions are asked when evaluating the extent to which an EA meets the Ont. EAA's requirements: (1) Are all required components of subsection 5(3) of the Ont. EAA present? and (2) Is the technical quality and level of detail of the information satisfactory and was an appropriate range of alternatives considered? (Ontario Ministry of the Environment, 1989).

The EA Branch evaluates and consolidates the comments of the various reviewers into a document referred to as the "Review". The review is released for comment to the public, municipalities, government ministries and agencies and the proponent before a decision on the acceptability of the EA is made. During a minimum 30-day period after public notice of the government review, anyone may submit written comments to the Minister of the Environment and Energy, and the minister may decide that a public hearing is required (Fig. 5). The proponent or any other affected person can request a public hearing if reasons are given in writing. The minister must comply with this request unless "in his absolute discre-

tion he considers that the requirement is frivolous or vexatious or that the hearing is unnecessary or may cause undue delay" (Environmental Assessment Act, R.S.O. 1990, c.E-18).

The Decision Phase

Two decisions are made after the review of the EA has been published and the minimum 30-day public review period has ended: (1) whether to accept the EA as a basis for making a

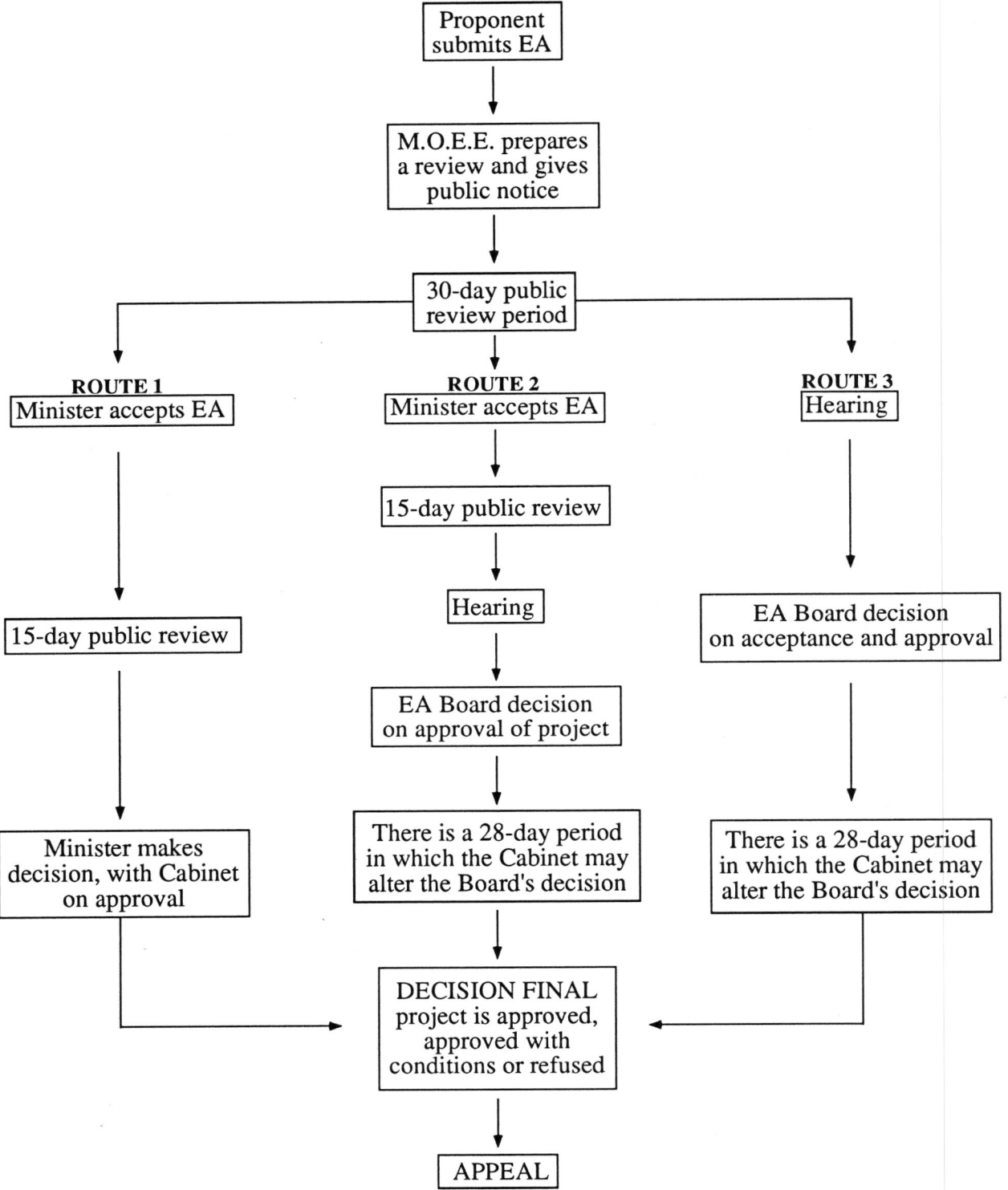

Figure 5 *The decision-making process for Ontario EAs (after Ontario Ministry of the Environment brochure, "A Proponents Guide to Environmental Assessment").*

decision on the undertaking, and (2) whether to approve the undertaking. Not all EAs are subject to the same decision route following submission (Fig. 5). If there are no serious objections raised to the EA and no hearings are called, then the decisions on acceptability of the EA and approval of the undertaking are made by the minister (route 1, Fig. 5). However, if a public hearing is called to decide on the acceptability of the EA and approval of the undertaking, the public hearing board will make the decisions (route 3, Fig. 5). The third decision route lies between the previous two and may involve a decision on EA acceptability made by the minister, and reference of the decision on approval of the undertaking to a hearing (route 2, Fig. 5).

Acceptability of the Environmental Assessment

The minister must decide whether the EA is acceptable as submitted, is acceptable with certain amendments, requires further research or changes/additions by the proponent to be acceptable, or should have its acceptability decided by a hearing board rather than the minister (Ontario Ministry of the Environment, 1989). However, the minister cannot reject an assessment document if it continues to be inadequate, only order the proponent to carry out further studies (Estrin and Swaigen, 1993). The proponent or any affected party may also request, with reasons given, a public hearing on the acceptability of the EA.

Approval of the Undertaking

The second decision to be made in the EA process is the approval of the undertaking. The undertaking is either given approval to proceed, refused approval to proceed, or given approval to proceed with certain terms and conditions specified. When the decision on approval is made by the minister together with cabinet, it is final. If the decision is made by a hearing board acting under the Ont. EAA, the decision is final only if the minister does not intervene within 28 days (Fig. 5). In this situation, the minister can modify the decision in any way, by reversing it, substituting a different decision, instructing the board to reconsider its decision, or ordering a new hearing to be held (Ontario Ministry of the Environment, 1989).

If the hearing was held by a Joint Board under the Consolidated Hearings Act (1981), any member of the public may also appeal the decision to Cabinet within 28 days. Cabinet may modify the decision in any way.

The Hearing Phase

A public hearing on EA acceptance or project approval may be called by either the proponent or the minister, or may be requested by any member of the public. The hearing is usually held before the Environmental Assessment Board (EAB), but if approvals for the undertaking are required under other statutes (such as the Environmental Protection Act, or the Planning Act) the proponent can request a consolidated hearing by a Joint Board under the Consolidated Hearings Act. The Consolidated Hearings Act creates a joint board consisting of members of the EAB and the Ontario Municipal Board which can consider matters dealt with under 12 statutes listed in the act and can issue necessary approvals. This is seen as a streamlining mechanism designed to reduce the time and cost involved in multiple hearings. Joint hearings under the Consolidated Hearings Act are now more common than hearings under the Ont. EAA alone. All EA hearings for waste disposal sites have been carried out under joint boards. Members of the EAB are independent and not employed by any ministry, although they are appointed by cabinet.

Public Participation in the Hearing Process

A preliminary hearing is usually held prior to the main EA hearing in order to identify interested parties and establish the position of these parties in support or opposition to the proposed undertaking. During the preliminary hearing, interested parties may also be grouped together into "umbrella" groups if similar ideas or concerns are expressed.

Participation in an EA hearing is a long and costly process and requires the employment of lawyers and expert witnesses to represent the interests of ordinary people. In order to allow effective public input into the hearings process, the Intervenor Funding Project Act was passed to provide funding to public interest groups participating in hearings under the joint board, the EAB and the Ontario Energy Board. This act was only intended as a pilot project and expired in 1991, although it has been extended into 1996. Intervenors can qualify for funds to help with hearing costs under the Intervenor Funding Project Act if they can demonstrate to a funding panel at an Intervenor Funding Hearing that the issues they intend to address are in the public's interest, that they have need for funds, that they have a clear plan for the use of funds and that their intervention will help the board and contribute to the hearing. Eligibility criteria for intervenor funding are set out in section 7 of the Ont. EAA. Intervenor funding is normally provided by the proponent (Robb, 1994). Steetley Quarry Products Inc. paid out over $310,000 to one group of intervenors in the EA hearing for their proposed landfill site (Hal Miettinen, pers. comm., 1995); hearing costs covered by Steetley reached $400,000 per month over a period of almost a year. This system of participant funding can make it extremely difficult for small companies or municipalities to propose undertakings that require Ont. EAA approval.

Many criticisms have been made of the funding systems available to allow the public to effectively participate in the EA process. Although the Intervenor Funding Project Act has helped a great deal with public participation at the hearing level, there are no provisions for funds to allow public involvement early in the assessment process. In 1992, as part of the EA process for three new landfill sites in the GTA, the Interim Waste Authority (IWA) granted a form of participant funding to three technical committees made up of members of the public. This funding was intended for the committees to employ independent experts to comment on technical studies carried out by the IWA, and is seen as a new approach to participant funding in the EA process.

The Hearing Process

The procedures followed by an Ont. EAA hearing are similar to those followed by waste management hearings under the Ontario Environmental Protection Act. The proponent is generally the first to speak and is represented by a lawyer who will call a number of expert witnesses to explain the planning process, and details of the proposed undertaking, mitigation measures and any other relevant information (Estrin and Swaigen, 1993). Lawyers for other recognized parties, whether they support or oppose the application, will then have the opportunity to cross-examine each witness brought forward by the proponent. Following the proponent's presentation, the board will invite lawyers representing other parties to make presentations. These parties may include ministries of the provincial government, municipal officials, conservation authorities and special interest groups. Individuals attending the hearing are also given an opportunity to make representations after all witnesses called by counsel have been heard (Estrin and Swaigen, 1993).

After the hearing, the board will adjourn to consider its decision, which is generally announced several months later together with a published document summarizing the reasons for the decision. The board may accept or amend an EA, and it may approve, reject or approve with terms and conditions the proposed undertaking. The board decision is final unless the minister (with cabinet approval) intervenes within 28 days of receipt of the decision, or an appeal is made by any party to the hearing.

The decision made by the board may be either appealed to cabinet on the basis of the content of the decision, or may be appealed to divisional court on a point of law. Many EA decisions released by a hearing board are appealed to cabinet; in these cases, cabinet has either upheld the board decision or has suggested that some changes be made to the proposed undertaking.

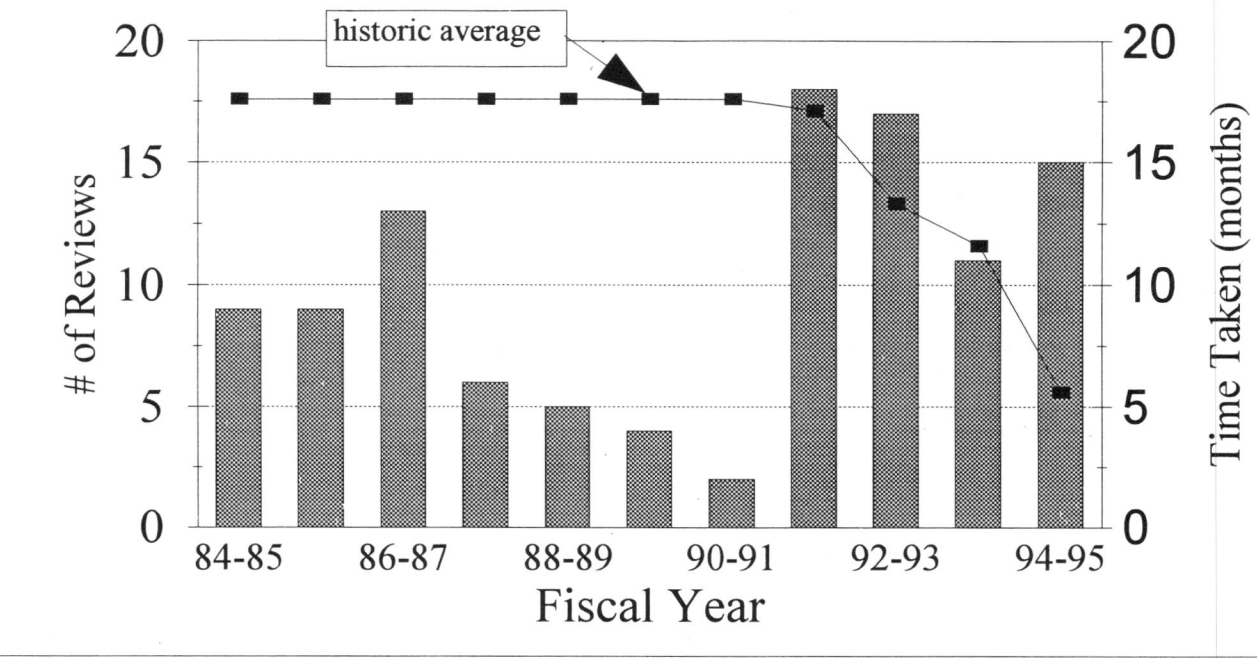

Figure 6 *Number of EAs reviewed and time taken for review, between 1984 and 1995 (data from Ontario Ministry of Environment and Energy, 1994a).*

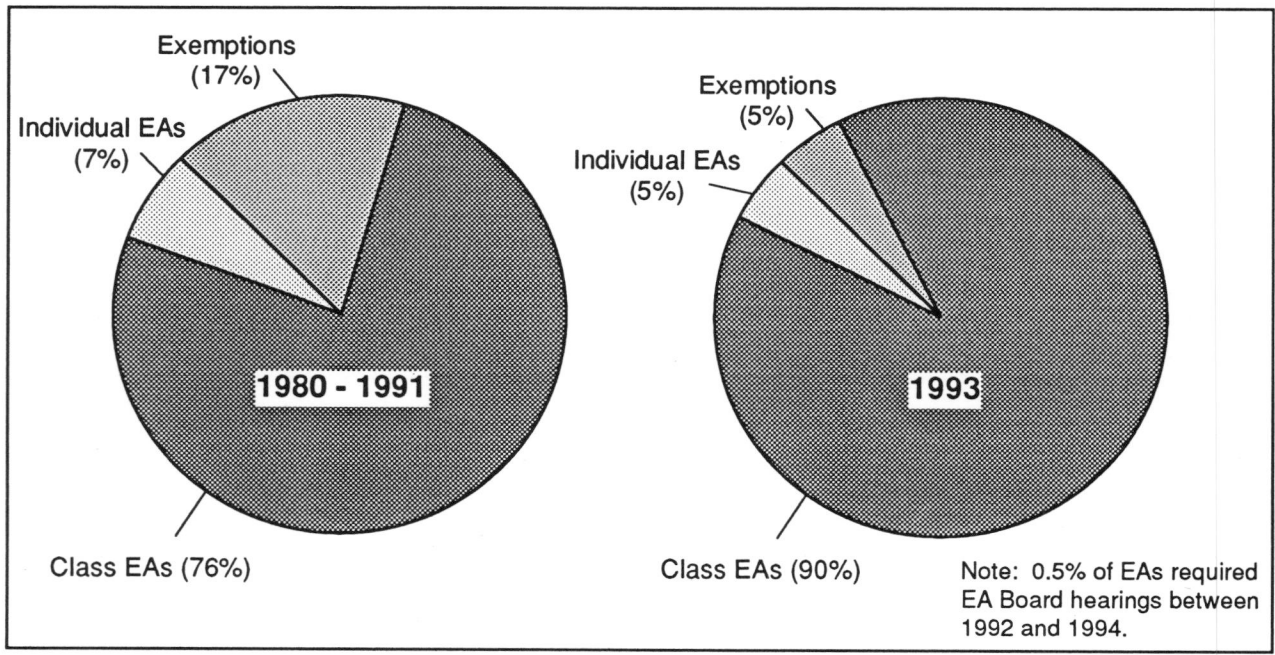

Figure 7 *Types of EAs planned and approved between 1980-91 and in 1993. Note increased emphasis on Class EAs and reduction in number of exemptions granted in 1993 (data from Ontario Ministry of Environment and Energy, 1994a).*

ENVIRONMENTAL ASSESSMENT REFORM IN ONTARIO

The Ontario environmental assessment process has received much criticism for becoming more costly, more time consuming and more complex each year (Robb, 1994). The major criticisms of the process are in the areas of length of review time, excessive costs involved, adversarial nature of the hearings process, scope of application of the Ont. EAA and the inadequacy of the Class EA process.

Review Time

At present, there are no legislated time limits on any phases of the review and decision-making process (Estrin and Swaigen, 1993) and reviews have taken up to 21 months to complete (Environmental Assessment Board and The Joint Board, 1995). The OMOEE is now committed to reducing the review period by two-thirds and intends to do this by establishing deadlines for document reviews, a standard review format and encouraging concurrent public and government reviews of certain EA documents (Robb, 1994). Recent streamlining of the EA review process has resulted in an improvement of average review time from an historic average of 17.6 months to 11.6 months in 1993-1994 (Fig. 6: Ontario Ministry of Environment and Energy, 1994a). The number of EA reviews completed annually ranges from 2 in 1990-1991 to 18 in 1991-1992; it is estimated that a total of 15 EA reviews will be completed in 1994-1995 (Fig. 6). Government reviews for the IWA landfill sites in the Greater Toronto Area were completed in only 60 days because of the perceived urgency in having the sites approved (Ontario Ministry of Environment and Energy, 1994a).

Cost

Even when the government review procedure is streamlined, the Ontario EA process as a whole is very costly. In the Greater Toronto Area of southern Ontario, the IWA siting process for three landfill sites cost over $80 million prior to the dissolution of the IWA in July 1995. The Steetley South Quarry landfill proposal cost over $11 million to take through an EA hearing (Russ Hughes, pers. comm., 1995), while the hazardous waste treatment facility proposed by the Ontario Waste Management Corporation involved over $80 million in hearings costs.

The EA process relating to waste facility siting is recognized to be extremely complex and costly, particularly for small communities. In an attempt to reduce this problem, the OMOEE has prepared a series of guidance documents to help proponents, review agencies and the public through the process without having to rely too heavily on consultants. These documents are termed Environmental Assessment Plans (EAPs) and set out work plans and procedures to be followed by proponents. It is thought that by following such EAPs, the EA process could be greatly streamlined and, in cases where an effective public consultation program is implemented, may eliminate the need for a hearing.

Hearings Process

A major problem with the EA hearings process is that it is viewed as being overly adversarial and formal (Ontario Ministry of the Environment, 1990b), with many costs in terms of time, funds, delays and frustration. In particular, the current hearings process is seen as intimidating to public participants. The EAB recognizes these problems and has suggested that a mixture of adversarial and investigative practices should be applied in the hearings process (Robb, 1994). Improvements to the hearing process recommended by the EAB include direct access by the public to the hearing (a more user-friendly setting), allowing direct probing of issues by the board, and effective use of expert staff by the board to help interpret specialized and technical information (Robb, 1994). In addition, an attempt is being made to reduce the number of hearings required by mediation and resolution of contentious issues between the proponent and concerned parties (Ontario Ministry of Environment and Energy, 1994a). The number of EAs approved by hearing boards has reduced from 1 in 3 in the 1985-1990 period, to 1 in 9 in 1991-1994. Between January 1992 and September 1994, 29 individual EA approvals were granted, 26 of which did not require a formal hearing (Ontario Ministry of Environment and Energy, 1994a). The three hearings that were held during this period represent only 0.5% of the total number of EAs submitted (Fig. 7).

Scope of Application

The scope of application of the Ont. EAA to proposed projects has also come under criticism. The Ont. EAA has very restricted application to undertakings carried out by the private sector, and is seen to be inconsistently applied to government plans and programs. In 1993, the Ontario government decided not to amend the Ont. EAA to apply to the private sector (Ontario Ministry of Environment and Energy, 1993b), but preferred an option to designate certain private sector undertakings through regulations. A commitment was made at this time to develop regulations that would apply the Ont. EAA comprehensively to private sector waste undertakings. Of the 18 private sector projects that were designated under the Ont. EAA during the 1976-1992 period, 13 were waste management related (Ontario Ministry of the Environment, 1992b). In addition, the government recently passed a regulation making privately constructed municipal roads, sewage and water projects subject to class EAs.

Class Environmental Assessments

Recent emphasis on the development of Class EAs has also been criticized. Class EAs are seen as providing inadequate assessment of all individual undertakings included in the generalized class. This has resulted in the identification of sub-class groups, designated on the basis of severity of anticipated environmental effects, to better accommodate all parties included within an EA class. The increased emphasis on Class EAs has produced a decrease in the number of applications for exemption orders. This is due to agencies such as OMNR and Ontario Hydro incorporating formerly exempt projects into Class EA documents (Ontario Ministry of Environment and Energy, 1994a). The number of exemptions granted in 1993 was only 5% of the total number of EAs submitted. This contrasts with 17% of exemptions granted in the period between 1980 and 1991 (Fig. 7). The ten exemptions made in 1993 were granted for emergency work, for environmentally beneficial projects or to allow comprehensive EAs to be concluded (Ontario Ministry of Environment and Energy, 1994a).

Any member of the public concerned about the appropriateness of a Class EA to a project can request a bump-up. During the period 1985-1993, 162 requests for bump-ups out of a total of 3400 projects, were received. Of these, only four were actually bumped-up by the minister. In the other cases, additional conditions were imposed on the proponent or revisions to the planned undertaking were made.

JOINT FEDERAL-PROVINCIAL ENVIRONMENTAL ASSESSMENTS

Some projects may require authorization by both provincial or territorial governments and the federal government. The CEAA gives the Minister of the Environment power to enter into agreements or arrangements with any jurisdiction for the purposes of environmental assessment of projects where both parties have authorization responsibilities (Canadian Environmental Assessment Act Training Compendium, 1994). This allows for co-operation of both federal and provincial bodies on EAs, including screening measures, mediation, joint panels and cost-sharing. Provincial and territorial governments are also committed to improve harmonization between the CEAA and their own assessment procedures.

In 1992, the Canadian Council of Ministers of the Environment (CCME) approved the "Framework for Environmental Assessment Harmonization" which aims to provide a basis for bilateral agreements. The framework sets out each government's jurisdictional responsibilities in the EA process, acknowledges the need for clear and consistent rules that eliminate unnecessary duplication, affirms the need for a single-window approach to EA, and establishes the mechanism to allow for intergovernmental co-operation at all steps

of the federal EA process. The recently released "Environmental Management Framework Agreement" includes proposals to transfer the responsibility for national environmental leadership to the CCME, and administration and enforcement of key federal environmental laws to the provinces (*The Toronto Star*, June 20, 1995). This is criticized by environmental law groups as evidence that the federal government is willing to abandon its responsibilities for the protection of the environment to provinces that do not necessarily have the commitment or resources to implement progressive environmental standards. Many environmental lawyers see this as a move toward diminished environmental protection in Canada, and feel that federal and provincial governments should continue to share responsibility for the protection of Canada's environment (Canadian Institute for Environmental Law and Policy, 1995).

The first bilateral agreement set up under the harmonization framework is the Canada-Alberta Agreement for Environmental Assessment Co-operation. This agreement includes provisions for early notification of projects of shared interest,

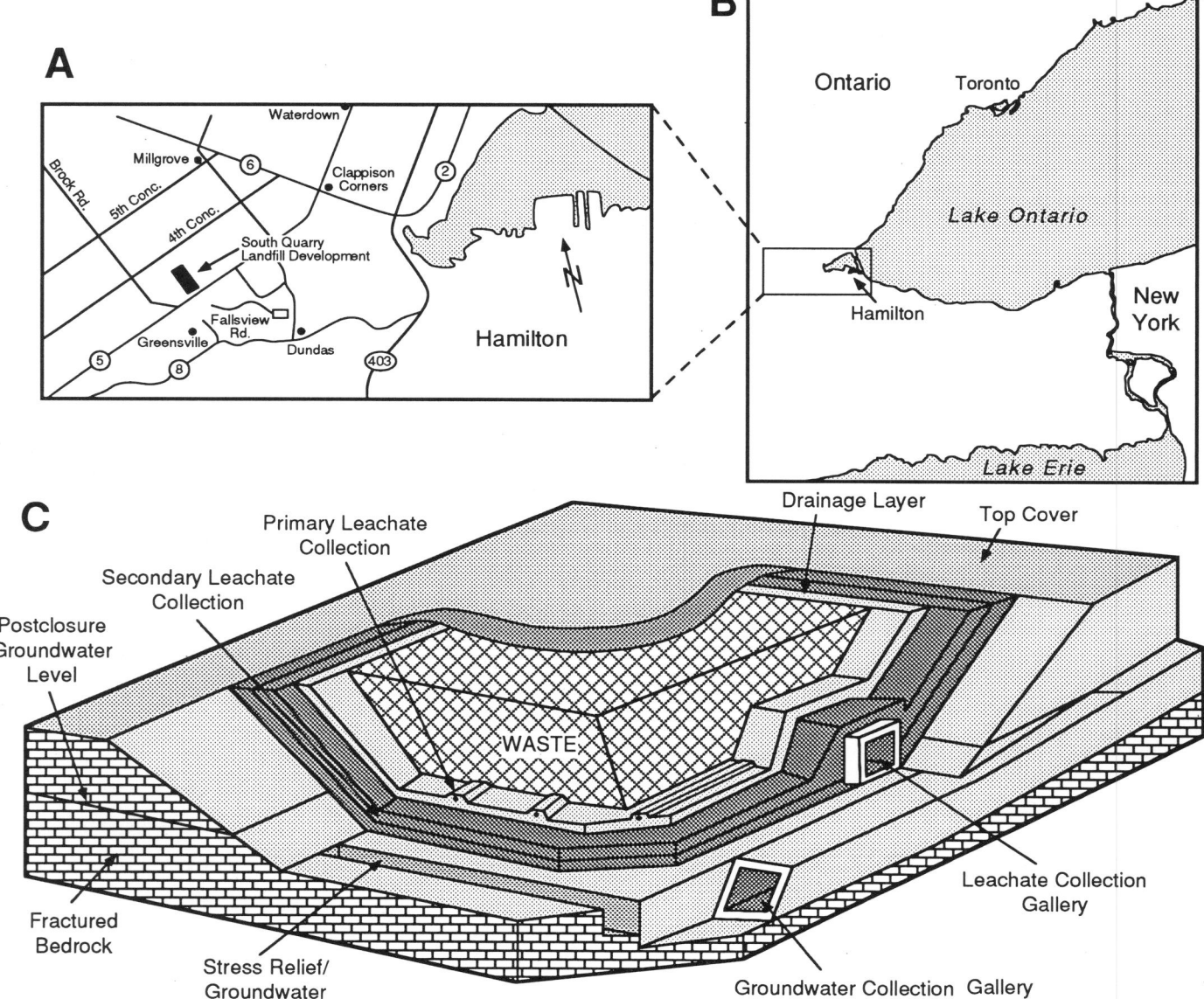

Figure 8 *Steetley South Quarry landfill development (modified from Steetley Quarry Products Ltd., 1990).* (**A, B**) *Location of proposed development.* (**C**) *Schematic cross-section of proposed design for the landfill. Note primary and secondary leachate collection systems and stress release slots excavated in bedrock.*

co-ordination of decision making by both parties with mutually agreeable time frames, and guidelines for the establishment of joint review panels consistent with federal and provincial legal and operational requirements. Provisions in British Columbia's Environmental Assessment Act also allow for a British Columbia-Canada agreement for the joint administration of the environmental assessment process (British Columbia Environmental Assessment Act — Quickfacts, 1994).

There have been no joint federal-provincial EAs in Ontario as yet, but an agreement was reached between the federal and provincial governments on the joint assessment of an Ontario Hydro project described as the Proposed Little Jackfish Hydroelectric Development (Northey, 1995). The Ont. EAA applied to the project, as Ontario Hydro is a designated public body under the act, and a permit was required under the federal Navigable Waters Protection Act. Hence, environmental assessment hearings were required for both federal and provincial purposes and it was planned to hold a joint hearing to cover both requirements. However, the harmonization agreement drawn up between the federal and Ontario governments for hearings for the Little Jackfish Hydroelectric project was not needed as the proponent withdrew the project from the EA process (Ontario Ministry of Environment and Energy, 1994a). Provisions made in the agreement will, however, be incorporated into the planned Canada-Ontario EA Harmonization Agreement.

PROVINCIAL AND FEDERAL-PROVINCIAL ENVIRONMENTAL ASSESSMENTS IN PRACTICE: TWO CASE STUDIES

Case Study 1: Steetley South Quarry landfill site, Hamilton, Ontario

Introduction
In 1986, Steetley Quarry Products Inc. (now a subsidiary of Redland PLC) commenced a feasibility study to rehabilitate their South Quarry, located near Hamilton, in southern Ontario, by landfilling with solid non-hazardous waste (Fig. 8). The geology of the site is discussed in Chapter 2. The accessible reserves of dolostone in the South Quarry had almost been depleted, and it was estimated that the resulting landfill would have capacity for 26 million tonnes of waste. The final land use was proposed to be a 27-hole championship golf course. The proposed landfill was designed as a hydraulic trap system, with a double composite liner and two separate leachate collection systems to prevent contaminant leakage into the underlying fractured dolomite bedrock (Fig. 8C). The Steetley South Quarry is located in a scenic and attractive area, well endowed with natural resources and fertile farmland. Municipal water service is not available in the area and local communities rely on ground-water aquifers for drinking-water supplies (Environmental Assessment Board and The Joint Board, 1995).

Environmental Assessment Process
Steetley Quarry Products Inc. first advised the public of their landfilling initiatives early in 1988, after completion of market studies and preliminary hydrogeological studies of the South Quarry site. In June of 1988, Steetley was advised that the project would probably be designated by the Minister of the Environment as subject to environmental assessment; designation was finally legislated in May 1989. A draft EA was completed and distributed to government agencies in September 1989 and the final EA was submitted in November

1990. The government review of the EA document, co-ordinated by the OMOEE, was completed in August 1992, and preliminary hearings began in December 1992. The hearing of evidence on the merits of the Steetley EA commenced in May 1992, and consumed approximately 140 hearing days over a period of 12 months. The hearing concluded in June 1994, and the Board's decision was released in March 1995. In total, the process (from first inception to the board's decision) lasted almost 9 years and cost Steetley approximately $11 million.

EA Decision
The Joint Board decision (Environmental Assessment Board and The Joint Board, 1995) rejected the EA document submitted by Steetley, and did not give approval to the proposed undertaking. In its summary of the decision, the Board described its reasons for rejecting the EA document. It states:

> "that the rationale for the undertaking was not adequately established, that the assumed need was unrealistic, that alternatives to the undertaking and alternative methods of carrying out the undertaking were inadequately addressed, and that the environmental assessment process was not "rational, traceable and consistent" " (Environmental Assessment Board and The Joint Board, 1995, summary).

The Board also refused approval of the undertaking proposed by Steetley and concluded that:

> "the location of a 26 million tonne landfill within the fractured bedrock of the South Quarry would pose an unacceptable risk to local ground-water and surface-water resources. In particular, it found that impacts to the natural environment were not satisfactorily addressed, that off-site impacts could not be discounted, that the proponent did not reliably demonstrate that hydraulic containment could be maintained in the landfill site throughout the potentially century-long contaminating life span of the site, and that the issue of treatment and disposal of leachate required further study. According to the Board, landfilling operations could dramatically change the character of the area surrounding the quarry, and residents within surrounding communities would suffer continuing and further social impacts."
> (Environmental Assessment Board and The Joint Board, 1995, summary).

Observations of the Ontario Environmental Assessment Board
A number of important points are raised by the EA Board in their decision document that relate both to the manner in which the EA process was carried out and inconsistencies between the nature and location of the proposed undertaking and OMOEE policies. In response to criticisms that the EA process was "lengthy, costly and difficult for all involved" (Environmental Assessment Board and The Joint Board, 1995, p. 214), the Board identified several flaws in the processing of the Steetley EA. First, the Steetley EA was not assigned an EA co-ordinator, as recommended in the OMOEE guidelines (Ontario Ministry of the Environment, 1989). As a result, OMOEE policies, staff input and information requests were not addressed thoroughly, resulting in additional length and cost of the hearings process (Environmental Assessment Board and The Joint Board, 1995). Government review of the EA document took 21 months, an excessively long time.

The board also recognized that a number of OMOEE policies were not adequately addressed prior to the hearing. These policies concerned pre-submission consultation processes, potential adverse effects on local residents, preference for sites with a high degree of natural protection, and water management policies (Environmental Assessment Board and The Joint Board, 1995).

The criticisms of the Steetley EA process are aimed both at Steetley (the proponent) and at the OMOEE for failing to follow their own guidelines and policies. Opponents of the undertaking questioned how the OMOEE EA Branch could conclude that the Steetley EA had fulfilled the basics of the Ont. EAA and OMOEE policies given their obvious concern with issues regarding liner design and contaminating lifespan of the proposed site. The board concluded that the government review "tended to gloss over serious flaws in the Steetley EA process and inconsistencies between the undertaking and OMOEE policies" (Environmental Assessment Board and The Joint Board, 1995, p. 215). Had the OMOEE rejected the adequacy of the Steetley EA document at the review stage, much time and money would have been saved at the hearings.

However, criticisms are also made of the Board and its interpretation of the requirements of the Ont. EAA. Some participants in the Steetley EA process feel that the OMOEE and Steetley did fulfill the requirements of the Ont. EAA to the best of their understanding, and the board rejected the EA and the undertaking on the basis of a different interpretation of the requirements. These criticisms illustrate the difficulties involved in applying the Ontario EA process to a private undertaking, and the ambiguity of some of the requirements set out in the Ont. EAA.

Case Study 2:
Halifax Harbour Clean-up Project

Introduction
Halifax Harbour, located on the eastern coast of Nova Scotia, is bounded by two important port cities, Dartmouth and Halifax, which have a combined population of 225,000 (Fig. 9). Over the past 30 years, many citizens and environmental groups have expressed concerns over the practice of dumping raw sewage into the harbour. Provincial, federal and municipal governments have all been involved in a number of studies, proposals and recommendations for a waste-water treatment system for the growing urban region around Halifax Harbour (Halifax Harbour Cleanup Project, 1993). The various geologic and anthropogenic factors relating to water quality in Halifax Harbour are summarized in Chapter 17.

The Existing Sewage Collection System
The existing sewage collection system for the Halifax and Dartmouth region consists of a combined sewer system comprised of collection pipes carrying both sanitary sewage and storm runoff. There are 39 municipal outfalls releasing untreated sewage into the Halifax Harbour and over 60 industrial, commercial and institutional outfalls. The newer municipalities of Eastern Passage and Mill Cove have separate sewer systems and primary and secondary level treatment

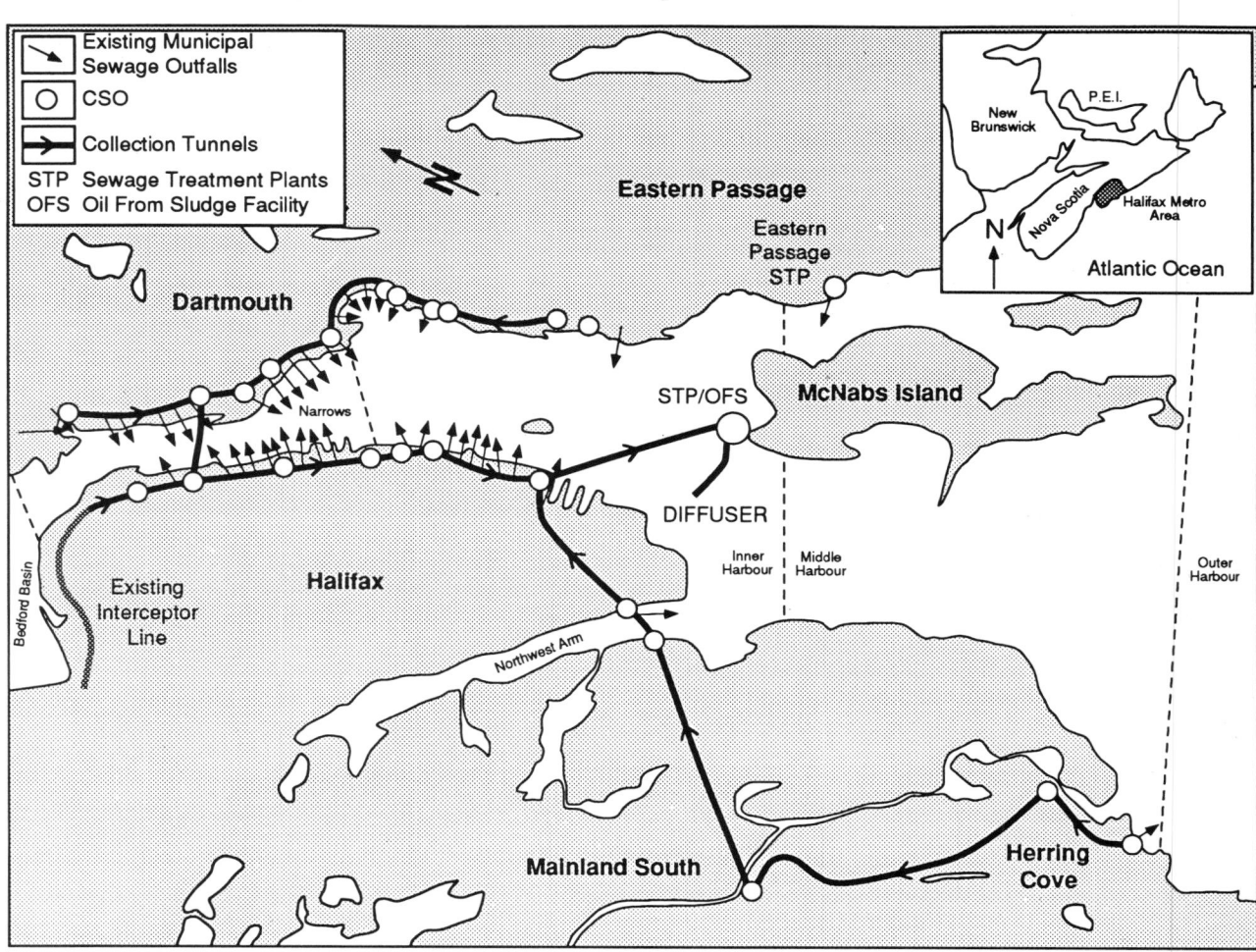

Figure 9 *Proposed sewage collection system and Sewage Treatment Plant/Oil From Sludge (STP/OFS) facility for Halifax Harbour (modified from Halifax Harbour Cleanup Project, 1993). 39 municipal outfalls (small arrows) and 60 industrial, commercial and institutional outfalls (not marked), currently discharge raw sewage into the harbour (Chapter 17). The proposed collection system would allow processing of sewage at a centralized STP/OFS facility. Proposed combined sewer overflows (CSOs) would allow untreated effluent to enter the harbour during storms.*

plants. However, both of these plants are operating at full capacity, and untreated sewage and runoff is often released into the harbour during storms (Fournier, 1990). It has been noted that despite the two new sewage treatment plants, 80% of the area's raw sewage remains untreated, and 135 million litres of untreated sewage enter the harbour every day. Sewage contamination of the harbour has many implications, including ecological and environmental degradation, human health risks, loss of the harbour's aesthetic value, and negative impacts on the harbour's recreational and commercial uses.

The Proposed Project

The Halifax Harbour Cleanup Project aims to rectify environmental problems caused by the release of raw sewage into the harbour through the construction of a waste-water management system and implementation of an environmental management plan. The proposed water management system consists of three components: (1) a collection system of interceptor sewers, tunnels, pumping stations and combined sewer overflows (CSOs) to intercept waste water from 39 existing outfalls around the harbour; (2) a single regional Sewage Treatment Plant/Oil From Sludge (STP/OFS) facility providing primary treatment, disinfection and sludge conversion, to be located on an artificial island constructed at the north end of McNabs Island (Fig. 9); and (3) a diffuser to discharge treated sewage to the harbour (Halifax Harbour Cleanup Project, 1993).

This system is designed to deal with municipal outfalls only; industrial, institutional and commercial outfalls will not be included in the project. The construction of an extensive collection network along the shores of the inner harbour, the Narrows and on the mainland will allow most of the sewage generated in the region to be directed to one regional treatment facility offering primary level treatment. However, as the network will remain a combined sanitary sewage and storm-water system, some untreated effluent will overflow into the harbour during storms. It is estimated that 75% of the yearly produced raw sewage and storm water from the Halifax-Dartmouth region would be processed by the proposed sewage treatment facility. The total cost of the clean-up project was estimated to be over $385 million (Halifax Harbour Cleanup Project, 1993).

The preferred site for the STP/OFS treatment facility, on an artificial island immediately to the north of McNabs Island, was selected after consideration of 16 potential sites. This site was selected by Halifax Harbour Cleanup Inc. (HHCI) due to its isolation from most residential communities and its proximity to a suitable outfall site in the inner harbour (Halifax Harbour Cleanup Project, 1993).

Environmental Assessment Process

The problem of contamination of Halifax Harbour by raw sewage and its impacts on environmental quality was first recognized in 1924. However, it was not until 1988, under a joint Federal-Provincial Agreement, that the regional sewage treatment facility was proposed, with costs to be shared by the federal, provincial and municipal governments (Fournier, 1990). The task of evaluating the potential impacts of this proposal was assigned to the Environmental Control Council of Nova Scotia, who recommended a comprehensive study of the harbour area be undertaken. In 1989, the province and municipalities agreed to establish the Halifax Harbour Cleanup Inc. (HHCI) and the Halifax Harbour Task Force to design and assess the impacts of the proposed treatment facility on the surrounding harbour area. In reviewing the feasibility of

the project, the Halifax Harbour Task Force focussed on a variety of issues, including (1) desired level of water quality, (2) multiple or single treatment plant, (3) extent of the area covered by the regional management plan, (4) control of toxic contaminants, (5) primary or secondary treatment, (6) selection of outfall and plant location, (7) containment or disposal of contaminants, (8) impact of storm overflows and sewage treatment on water quality and contaminants in existing sediments, (9) public participation, and (10) sewage treatment and overall harbour management (Fournier, 1990).

In 1990, the Task Force made 18 recommendations on harbour management (Fournier, 1990). One of the recommendations was to carry out a full federal-provincial environmental assessment of the proposed regional treatment system in order to fulfill the requirements of the Nova Scotia Environmental Act and the federal Environmental Assessment Review Process. A joint federal-provincial Environmental Assessment Panel (EAP) was established in November 1990. The Panel consisted of four members and was assisted by two technical specialists in the fields of environmental science and physical oceanography. The mandate of the EAP was to conduct a public environmental review of the proposed Halifax Harbour Cleanup Project, in order to advise both federal and provincial government ministers and agencies of its environmental acceptability. The panel were given the task of evaluating alternative methodologies for the waste-water treatment facility, alternative sites for the treatment plant and outfall, and alternative levels of treatment.

Environmental Assessment Panel Recommendations

The environmental assessment panel (EAP) published a report in July 1993 which strongly supports the proposed clean-up project and outlines a number of recommendations for implementing the project and improving its design (Halifax Harbour Cleanup Project, 1993). One of the recommendations made is that the preferred site selected by the proponent (HHCI) for the location of the STP/OFS facility and outfall diffuser may not be the most suitable site available, and that a site at the Dartmouth Ocean Terminal/Georges Island should be investigated further.

Although the EAP review of the Halifax Harbour Cleanup Project was favourable, there has been no action on the project to date and raw sewage continues to be discharged into the harbour (Chapter 17). Costs involved in the prelimary investigations and assessment process are estimated to be around $25 million (R. Cunningham, pers. comm., 1995).

IMPLICATIONS

This chapter summarizes the main elements of environmental law and environmental assessment processes that are likely to affect urban geological undertakings at both provincial (dominantly Ontario) and federal levels. A number of themes emerge from this which are worthy of note.

Compliance with Environmental Regulation

The effectiveness of laws and regulations designed to protect the environment is largely determined by the severity of the penalties imposed for non-compliance. Disincentives for non-compliance with environmental regulations should be severe (*i.e.*, high fines and penalties) and rigorously enforced by government agencies.

Data Collection/Analysis and the Environmental Assessment Process

The value of any EA lies in its ability to be able to accurately predict the types and severity of impacts that a particular development or undertaking may have on the environment. Hence, a substantial amount of basic scientific data are required to determine baseline conditions in the environment prior to development, to identify particularly sensitive elements of the environment and to predict the full range of likely impacts. In many instances, such baseline datasets do not exist, even in large urbanized regions such as the GTA, and new data have to be collected specifically for the assessment. Unfortunately, most EAs require data to be collected over a relatively short time span (often less than a year) and any mid- to long-term temporal changes in environmental conditions may not be adequately addressed. Similarly, as data for separate criteria groups are collected independently, by experts or consultants in specialized fields, little attention is paid to the many and complex interrelationships between individual criteria and their compound impacts over time. Future environmental assessments should involve more temporal and multi-disciplinary quantitative modelling of these complex ecosystems.

Devolution of
Environmental Assessment Responsibility
Recent concern by many environmental lawyers is that the environmental assessment process is being progressively devolved to more local levels of government where the commitment to environmental protection may not be consistent or strong. Proposed devolution of federal environmental responsibilities to the provinces could result in the reduction of overall environmental standards. Implementation of many provincial undertakings that require EAs is the responsibility of local or municipal governments who do not have their own in-house experts to carry out the assessments. This may result in much weaker EAs being produced and less valid conclusions being drawn.

Public Education and Communication
The environmental assessment process can only be truly effective when the public are productively involved at all stages. To accomplish this, greater emphasis needs to be placed on educating and informing citizens about the aims, requirements, procedures and limitations of environmental assessment and how they can contribute to the process.

ACKNOWLEDGEMENTS
I would like to thank Gina Remy and Emmanuelle Arnaud for their enormous help in the preparation of this manuscript. Emmanuelle Arnaud and Mike Doughty are thanked for drafting the figures. Russ Hughes and John Bullen advised me on many aspects of the Ontario EA process; Nigel Cowie, Jill McColl, Verne Perry, Gordon Fader and Jay Rosenblatt are thanked for information they provided. However, any errors or omissions are entirely the responsibility of the author. This work was supported in part by an NSERC individual operating grant and an NSERC summer studentship to Arnaud.

REFERENCES

Association of Conservation Authorities of Ontario, 1993, Class Environmental Assessment for remedial flood and erosion control projects: 65 p.

British Columbia EA Act — Quickfacts, 1994, Province of British Columbia: 35 p.

Brooks, S., 1993, Canadian Democracy: An Introduction: McLelland and Stewart Inc., Toronto, ON, 412 p.

Canadian Institute for Environmental Law and Policy, 1995, Environmental harmony or environmental discord?: Position paper and press release, April 25, 1995, 8 p.

Canadian Environmental Assessment Act Training Compendium, 1994, Canadian Environmental Assessment Agency, Hull, PQ, 300 p.

Couch, W.J. ed., 1988, Environmental Assessment in Canada: 1988 Summary of Current Practice: Canadian Council of Resource and Environment Ministers, Federal Environmental Assessment Review Office, 45 p.

Environmental Assessment Act, R.S.O. 1990, c.E-18

Environmental Protection Act, R.S.O. 1990, c.E-19

Environmental Assessment Board and The Joint Board (Consolidated Hearings Act), 1994, Ontario Waste Management Corporation: Decision CH-87-02, 12 Chapters.

Environmental Assessment Board and The Joint Board (Consolidated Hearings Act), 1995, Steetley - South Quarry landfill site, decision and reasons for decision: Office of Consolidated Hearings, Toronto, ON, 215 p.

Estrin, D. and Swaigen, J., 1993, Environment on Trial, Third edition: Emond Montgomery, Toronto, ON, 909 p.

The Financial Post, 8 June, 1995, Passing the cleanup buck.

Fournier, R.O., 1990, Halifax Harbour Task Force Final Report: Halifax, NS, 84 p.

Fry, E.H., 1991, The Canadian Political System: Association for Canadian Studies in the United States, Washington, DC, 50 p.

Halifax Harbour Cleanup Project, 1993, Report of the Federal-Provincial Environmental Assessment Review Panel for the Halifax-Dartmouth Metropolitan Wastewater Management System: 79 p.

Hoglund, G.M. and Buck, A. 1987, Southwestern Ontario Transmission III: decision-making techniques in complex route and system: 4th Symposium on Environmental Concerns in Rights-of-Way management, Selection Studies Proceedings, Indianapolis, IN, 15 p.

Interim Waste Authority Ltd., 1992a, Revised approach and criteria, Supplemental criteria for Durham Region landfill site search: Environmental Assessment document II, part I, v. I, 235 p.

Interim Waste Authority, 1992b, Identification of the preferred site for Peel Region landfill site search: Environmental Assessment document IV, v. 1, 324 p.

Jacobs, P. and Sadler, B., 1991, Sustainable development and environmental assessment: perspectives on planning for a common future: Canadian Environmental Assessment Research Council, 100 p.

Kubasek, N.K. and Silverman, G.S., 1994, Environmental Law: Prentice Hall, Englewood Cliff, NJ, 294 p.

Northey, R., 1995, Environmental Assessment in Ontario (from a federal and provincial standpoint): 1995 Institute of Continuing Legal Education, Canadian Bar Association, Toronto, ON, 28 p.

Nova Scotia Department of the Environment, 1993, The Nova Scotia Environment Act: Towards a healthy environment and a sustainable future, a summary: 18 p.

Ontario Ministry of the Environment, 1983a, Project screening and application for exemption orders under section 29 of the Environmental Assessment Act: Environmental Assessment Branch, 15 p.

Ontario Ministry of the Environment, 1983b, Ontario Drinking Water Objectives: 22 p.

Ontario Ministry of the Environment, 1984, Water management: goals, policies, objectives and implementation procedures of the Ministry of the Environment: 39 p.

Ontario Ministry of the Environment, 1987, Guidelines and policy on pre-submission consultation in the EA process: 35 p.

Ontario Ministry of the Environment, 1989, Interim guidelines on environmental assessment planning and approvals, Environmental Assessment Branch: 45 p.

Ontario Ministry of the Environment, 1990a, Evaluation methods in environmental assessment: Environmental Assessment Board, 140 p.

Ontario Ministry of the Environment, 1990b, Toward improving the Environmental Assessment program in Ontario: Environmental Assessment Task Force Report, 19 p.

Ontario Ministry of the Environment, 1992a, Guideline for preparing environmental assessment proposals: Environmental Assessment Branch, 12 p.

Ontario Ministry of the Environment, 1992b, EA Update: Environmental Assessment Branch, 16 p.

Ontario Ministry of Environment and Energy, 1993a, Guidance manual for landfill sites receiving municipal waste: 207 p.

Ontario Ministry of Environment and Energy, 1993b, Environmental Assessment Reform: a report on improvements in program administration: 24 p.

Ontario Minstry of Energy and Environment, 1993c, Greater Toronto Area 3R's Analysis: 9 volumes.

Ontario Ministry of Environment and Energy, 1994a, Environmental Assessment Branch Annual Report: 19 p.

Ontario Ministry of Environment and Energy, 1994b, A plain language guide to the Environmental Assessment Act: Environmental Assessment Branch, 24 p.

Phyper, J-D. and Ibbotson, B., 1991, The Handbook of Environmental Compliance In Ontario: McGraw-Hill Ryerson, Toronto, ON, 346 p.

Robb, J., 1994, Ontario's Environmental Assessment Process: unpublished document, Environment Assessment Board Vice-Chair, 18 p.

Saxe, D., 1990, Environmental Offences: Corporate responsibility and executive liability: Canada Law Book Inc., Aurora, ON, 254 p.

Statistics Canada, 1994, Human Activity and the Environment: Minister of Industry, Science and Technology, 300 p.

Steetley Quarry Products, 1990, Environmental Assessment — South Quarry Landfill Development, Final report, 350 p.

Taro Aggregates Ltd., 1995, Proposed East Quarry Landfill Environmental Assessment: Executive Summary and Volumes I and II, 315 p.

The Toronto Star, June 20, 1995, Environmental harmony hits a sour note.

United Nations World Commission on Environment and Development, 1987, Our Common Future: Oxford University Press, New York, 43 p.

APPENDIX A

A GEOLOGIC TIME SCALE

APPENDIX A   A Geologic Time Scale

Era	Period	Epoch	Age		Major Environmental Events
			0		World population about 5.7 billion (5.7×10^9)
			50		Beginning of the Atomic Age (December 2, 1942)
		Holocene	3	ka	Beginning of the Iron Age
			6	ka	Agricultural Revolution: urbanization begins
			10	**ka**	
			12	ka	Intense deglaciation; giant floods down the Mississippi Valley
	Quaternary		20	ka	Maximum of the last ice age
			125	ka	Temperature maximum of the late interglacial: appearance of *Homo sapiens sapiens* and *Homo sapiens neanderthalensis*. World population about 10,000
		Pleistocene			
			250	ka	Disappearance of *Homo erectus*; appearance of *Homo sapiens "praesapiens"*
			1.5	Ma	Disappearance of *Paranthropus*; appearance of *Homo erectus*
			1.64	**Ma**	Appearance of *Hyalinea baltica*
			2.5	Ma	Beginning of extensive northern hemisphere glaciations
Cenozoic			3.0	Ma	Appearance of *Australopithecus africanus*
		Pliocene	3.2	Ma	Closing of Central American isthmus
			3.6	Ma	Appearance of *Australopithecus afarensis*
	Neogene		**5.2**	**Ma**	Opening of Gibraltar passage
			6.2	Ma	Isolation of the Mediterranean; salt deposits on its floor
		Miocene	14	Ma	Arrival of Antarctic sheet at ocean; *Ramapithecus*
			23.3	**Ma**	Alpine orogeny
		Oligocene	30	Ma	African primate: *Aegyptopithecus*
			35.4	**Ma**	Sudden expansion of the Antarctic ice sheet; separation of Australia from Antarctica; appearance of artiodactyls, perissodactyls and apes
	Paleogene	Eocene			
			56.5	**Ma**	Appearance of globorotalids; radiation of placental mammals; first primates; flowering plants spread
		Paleocene			
			65	**Ma**	Giant asteroidal impact; extinction of cycadeoidales, globotruncanids, ammonoids, belemnoids, ichthyosaurs, plesiosaurs, dinosaurs
	Cretaceous				First angiosperms and marsupials; opening of the South Atlantic Ocean; the White Cliffs of Dover
			85	Ma	Laramide orogeny in western North America; thrusting of Rocky Mountains
Mesozoic			145.6	Ma	Opening of the North Atlantic; first coccoliths and planktic foraminifera; first birds
	Jurassic				
			175	Ma	Columbian orogeny in western Canada
			208	**Ma**	Palisades sill, New York; Carrara marble; appearance of dinosaurs, lizards, turtles; first mammals. Pangea begins to break up.
	Triassic				
			245	**Ma**	Appalachian/Hercynian-Variscan orogeny; New Red Sandstone in Europe; extinction of tetracorals, cystoids, placoderms
	Permian				
			290	**Ma**	Widespread formation of coal; cyclothems; first reptiles, winged insects and glaciation in the southern hemisphere
Paleozoic					
	Carboniferous		300	Ma	Alleghenian orogeny in eastern North America
			350	Ma	Continents assembled in one large megacontinent: Pangea ("All the Lands")

APPENDIX A A Geologic Time Scale

... continued

Era	Period	Epoch	Age		Major Environmental Events
Paleozoic, cont'd.	Devonian	———————	**362.5**	**Ma**	Old Red Sandstone; Queenston-Juniata red beds; first sharks; first amphibia
	Silurian	———————	408.5	Ma	Acadian-Caledonian orogeny
					Lockport dolostone; first bony fishes; first trees
	Ordovician	———————	**439**	**Ma**	
			450	Ma	Taconic orogeny in eastern North America; glaciation in north Africa
					First corals; first vertebrates (jawless fishes)
	Cambrian	———————	**510**	**Ma**	Appearance of trilobites, brachiopods, echinoderms and shelled molluscs, Rodinia breaks up: continents disperse
	Ediacaran	———————	540	Ma	Appearance of *Archaeocyatha*
					Ediacaran fauna — widespread glaciation
Proterozoic		———————	**590**	**Ma**	Appearance of metazoa
			650	Ma	Continents assembled into one megacontinent: Rodinia
			1.7	Ga	Increasing O_2 in atmosphere; appearance of eucaryota
			2.2	Ga	Glaciation (Gowganda Tillite)
Archean			**2.7**	**Ga**	Oldest stromatolites
			3.5	Ga	Earliest bacteria (heterotrophs)
			3.8	Ga	Oldest glaciation (Pangola sequence)
Hadean			**4.0**	**Ga**	Oldest terrestrial rocks; formation of planet Earth completed
			4.44	Ga	Oldest lunar rocks
			4.6	Ga	Age of the meteorites
			4.7	Ga	Formation of the solar system
Gamowian			**4.7**	**Ga**	Nucleogenesis in the region of the solar system
					Formation and evolution of stars, quasars and galaxies; general increase in the relative abundances of the heavier elements (continuing to the present)
			16.5	Ga	**300,000–800,000 years:** electrons captured by nuclei; formation of H and He atoms and of H_2 molecules; the universe becomes transparent
					3.8 minutes: stabilization of ^2H, ^3He and ^4He nuclei; relative abundances: 74% H + 26% He (by mass), or 92% H + 8% He (by number of atoms)
					10 seconds: stabilization of electrons
					10^{-5} **to** 10^{-3} **s:** stabilization of protons and neutrons
					10^{-8} **to** 10^{-5} **s:** quark-antiquark annihilation
					10^{-10} **to** 10^{-8} **s:** stabilization of quarks
					10^{-10} **s:** electromagnetic and weak forces separate
					10^{-33} **to** 10^{-32} **s:** inflation
					10^{-35} **s:** strong force and electroweak force separate
					10^{-43} **s:** gravity separates
Planckian		———————	**16.5**	**Ga**	**0 to** 5.390×10^{-44} **s:** appearance of space, time, energy and superforce
			16.5	**Ga**	

After
Emiliani, C., 1992, Planet Earth: Cosmology, Geology and the Evolution of Life and Environment: Cambridge University Press, Cambridge, MA, 710 p.

APPENDIX B
GLOSSARY OF TERMS

Note: Use of *italics* indicates cross reference to another entry.

A

"A" Horizon
Uppermost horizon of a *soil*, usually containing significant organic material and dark coloured.

Absorption
The weak physical attraction of one material for another, whereby one material incorporates relatively large proportions of the second material, and then reversibly gives up the second material when only small amounts of energy are applied. Typically refers to incorporation of liquids in solids, or gases in liquids.

Acceleration
Parameter used for categorizing seismic motion during earthquakes. Expressed in $cm \cdot s^{-2}$ or as a per cent of the force of gravity acting on a unit of mass ($g = 980.6 \, cm \cdot s^{-2}$). See also *design basis seismic ground motion*.

Acid Mine Drainage (AMD)
Ground or surface waters that move through sulphur-rich ore or mining wastes (tailings) and become acidified (see *pH*). The converse is referred to as *basic mine drainage* (BMD).

Activated Sludge
Sludge particles produced in raw or settled waste water (primary *effluent*) by the growth of organisms, including bacteria in aeration tanks in the presence of dissolved oxygen. The term "activated" comes from the fact that the particles contain bacteria, fungi and protozoa.

Active Face (Working Face)
The portion of a landfill site where waste is currently being deposited, spread and/or compacted prior to the placement of cover material.

Active Fault
Defined, in the State of California, as a fault that has moved in the last 10,000 years. The US Nuclear Regulatory Commission, on the other hand, defines a fault a "capable" if it has moved during the last 50,000 years or on two or more occasions in the last 500,000 years. See also *seismic source zone characterization* and *seismic hazard assessment*.

Active Layer
Refers to zone of shallow depth (about 1 m) that thaws during summer in *permafrost* terrains.

Acoustic Impedance
Product of the seismic velocity and bulk density of a sediment or rock.

Acoustic Masking
Term describing the effect of methane gas on the geophysical properties of sediment where gasified sediment commonly becomes "transparent," such that its origin and stratigraphy cannot be readily identified. Particularly problematic with *seismic reflections surveys*.

Actual Evapotranspiration
Amount of water that evaporates from soils and is transpired by vegetative covers. This water is not available to recharge underlying ground water. See also *potential evapotranspiration*.

Adsorption
The capture of molecular or ionic species at liquid or solid surfaces by strong, attractive free energy forces. By comparison with adsorption, relatively large amounts of energy are required to free the attracted species (desorption). The process involves cation exchange with clay minerals, organic material or iron oxides and is pH dependent; at low pH, metals are mobile, but, as pH rises, metals are captured by adsorption. pH control may provide a means of remediating ground and surface waters, thereby limiting the movement of contaminants especially metals.

Advection
The process by which natural chemical constituents, or introduced contaminants, in ground water are transported by the bulk motion of the flowing ground water.

Adverse Environmental Impact
Any direct or indirect undesirable effect on the environment (see *impact*) resulting from an emission or discharge, which is caused or likely to be caused by human activity.

Aerobic Bacteria
Bacteria which will live and reproduce only in an environment containing oxygen which is available for their respiration (breathing), namely atmospheric oxygen or oxygen dissolved in water. Oxgyen combined chemically, such as in water molecules (H_2O), cannot be used for respiration by aerobic bacteria. Contrast with *anaerobic*. See also *Eh*.

Aerobic Process
Biological process that occurs in the presence of oxygen.

Aeromagnetic data
Data acquired from an airplane identifying the changing magnetic properties of buried rocks. Very useful in mapping old, deeply buried rocks and their structures such as faults. The term aeromagnetic lineament is a generic term used for a well-defined feature on an aeromagnetic map where the geologic origin or significance is not well understood.

Aggregate
Crushed rock, or sand and gravel, used for construction purposes.

AGST

Above-ground storage tank.

Air Sparging

Forcing of air below the water table to strip volatile organic compounds from ground water.

Air Stripping

Refers to use of various techniques involving the flow of air and used to remove common organic contaminants such as perchloroethylene, benzene and toluene from ground water.

Alluvium

Sediments deposited by streams and rivers. Contrast with *colluvium*.

Alpha Recoil

Process by which a radon atom is moved away from decaying radium atom when an alpha (α-) particle is fired out from the nucleus. Principal means by which radon is transferred from solid rock to pore spaces and thence to the environment. See *radon*.

Alternatives

Term used in *environmental assessment* to refer to different methodologies, locations, technologies, *etc*. These choices, or alternatives, are screened or narrowed during a selection process (see *satisficing*) to arrive at a preferred alternative. Selection of a preferred alternative results in a plan, proposal or a program of activity (an *undertaking*).

AMD

See *acid mine drainage*.

Anaerobic

Biological process (decomposition or treatment) that occurs in the absence of oxygen, *e.g.*, decomposition of landfill waste with consequent generation of methane gas. Also pertains to organisms that can live in the absence of free oxygen. Contrast with *aerobic*.

Anchorturbation

Term used to refer to disturbance of sea- or lake-floor sediments by dragging of ships' anchors. Buried contaminants may be resuspended during such disturbance and be transported in the water column to pollute other sites; in Great Lakes harbours this can be a problem in the vicinity of drinking water intakes.

ANSI

In Ontario, the abbreviation for an Area of Natural and Scientific Interest; an area of land and water containing natural landscapes or features that have values related to protection, natural heritage, appreciation, scientific study or education.

Anticline

Used to describe folded rocks that are convex upward (in the form of an arch). Anticlines may also be defined as folds with older rocks toward the centre of curvature. Contrast with *syncline*.

Approved Landfill Design and Operations Plan

A Design and Operations Plan that has received formal approval by the appropriate regulatory authority (*e.g.*, the *OMOEE* in Ontario), through the issuance of the Certificate of Approval. A plan is submitted as part of the application for the certificate, and once approved, the plan is a supporting document to the Certificate of Approval.

Aquifer

A subsurface body of sediment (or rock) that contains sufficient saturated permeable material to conduct ground water and yield economically significant quantities of potable water to wells and springs. Divided into *confined aquifer* and *unconfined aquifer* types.

Aquiclude

A subsurface body of sediment (or rock) that is impermeable and incapable of transmitting or storing water in significant quantities.

Aquitard

Generally used to describe less permeable beds in a stratigraphic sequence. These beds may be permeable enough to store water in quantities that are significant in the study of regional ground-water flow, but their permeability is not sufficient to allow the completion of production wells within them.

Area of Concern (AOC)

Coastal body of water in the Great Lakes system designated as such by the Canada-United States International Joint Commission because of severely degraded water-sediment quality. There are 43 designated Areas of Concern in the Great Lakes basin. Remedial Action Plans (*RAPs*) are in progress in these areas and are designed to ultimately remove the AOC designation.

Argillaceous

Descriptive term for a rock containing appreciable clay material. The term is applied to all rocks or substances composed of clay, or having a notable proportion of clay in their composition, such as slate, shale, argillaceous dolostone, *etc*.

Artesian Condition

Refers to ground water under sufficient *hydraulic head* to rise above the aquifer containing it, *i.e.*, where the static water level would be above the ground surface in a well. Typical of *confined aquifers*. See also *piezometric surface* and *potentiometric surface*.

Artificial Recharge

See *exfiltration*.

Asthenosphere

Outermost part of the Earth's mantle and about 100 km thick, where rocks are close to melting. The overlying layer is called the lithosphere and is broken up into tectonic plates that move over the asthenosphere (see *Plate tectonics*).

Attenuation

Natural process by which the concentrations of landfill-generated *leachates* and other contaminants are reduced to safe levels by moving through sediment or bedrock and being subject to *diffusion, adsorption, dispersion, biodegradation* and *dilution*.

Atterberg Limits

Describes the critical water contents of a sediment separating its mechanical behaviour as a liquid or plastic (the liquid limit) and a plastic and semi-solid state (the plastic limit).

Average Linear Ground-water Velocity (v)

Refers to the volumetric flux (Q) of ground water which passes through a specified area of an aquifer. The equation for average linear ground-water velocity is

$$v = Q/(A \times n),$$

where A is the cross-sectional area and n is the porosity. The usual units are metres per second (m·s^{-1}) or metres per year (m·a^{-1}).

B

"B" Horizon

Soil horizon below the *"A" horizon*, usually brownish in colour and enriched in *clay* and iron oxides washed down from above. See *elluviation* and *illuviation*.

B-Spline

A spline using piecewise polynomial approximation passing through a set of points (poles). See *spline*.

Background Level

Concentration of potential *contaminants* present in the natural environment prior to establishment, start-up and operation of a landfill facility or other environmental *undertaking*.

Backscatter

In radar, the portion of the microwave energy scattered by the terrain surface directly back to the antenna. The term is also used where sea- or lake-floor sediments scatter an acoustic (sound source); gravelly sediments result in high backscatter, flat muddy bottoms give rise to low backscatter (see *sidescan sonar*).

Bailer

Bucket-like device used to remove water from a well.

Bailing

Removal of water from a well with a *bailer*.

Baseflow

The water contribution to streams from ground water following infiltration of precipitation (rain, snow) into the soil. The flow to which a stream will recede after a storm when surface runoff drops to zero.

Base Contour

The contours of the base of a landfill.

Basement

Geologic term used for old, commonly Proterozoic and older, rocks (Appendix A), found at depths of tens to thousands of metres below younger sedimentary rocks in mid-continent North America. The Canadian Shield is where basement is exposed at surface; the same rocks become progressively more deeply buried below younger rocks away from the exposed shield.

Bathymetry

Pertaining to water depth, *e.g.,* a bathymetric chart of a lake or other water body showing bathymetric contours (isobaths).

Becquerel

A unit used to identify level of radioactive decay (radioactivity) in a substance. For example, a wheelbarrow of soil from a garden contains 1000 Becquerels (Bq) of naturally occurring radium. A 30 year old person has about 4000 Bq of potassium-40 in his/her body. Water from Lake Ontario, used for municipal drinking water supplies, contains on average 40 Becquerels per litre ($Bq \cdot L^{-1}$) of tritium as a result of spills from nuclear plants. See also *Sievert*.

Bed

Defined as a layer of bedrock or sediment at least 1 cm thick. A bed has lateral continuity and is defined by distinct lower and upper surfaces (bounding surfaces) and having a composition, texture or structure different from beds above or below. Units thinner than 1 cm are defined as *laminae*.

Bedrock

Any solid rock exposed at the surface or overlain by *soil* or other unconsolidated sediment (see *overburden*).

Bentonite Clay

Clay of the montmorillonite (smectite) group, commonly mixed with water to produce a heavy slurry used to support the sidewalls of wells or trenches. See *slurry wall*.

Berm

A ridge of earth constructed around landfills to provide visual screening and attenuation of noise and dust impacts; also applies to protective walls to reduce coastal erosion of cliffs.

Bioaccumulation

Refers to the progressive accumulation of trace metals, toxins and other contaminants in organisms at successively higher trophic levels of the food chain (synonymous with *biomagnification*). The bioaccumulation factor is the ratio of concentration of the particular substance being investigated in the tissues of an organism to background environmental concentrations.

Bioavailability

The susceptibility of a contaminant or nutrient element to being incorporated within the tissues of an organism.

Biochemical Oxygen Demand (BOD)

Commonly used measure of water quality referring to amount of oxygen consumed by bacterial decomposition of organic wastes. In decomposition, organic matter serves as food for the bacteria and energy results from the oxidation. Urban activity results in high contributions of organic material (*e.g.,* sewage) to waters and the uptake of oxygen by bacteria resulting in reduced oxygen levels. This creates an environment inimicable to fish and other organisms. Measured as milligrams per litre ($mg \cdot L^{-1}$) of oxygen consumed over 5 days at 20°C.

Biodegradation

Decomposition of organic compounds (such as the aromatic compounds, benzene, toluene, and xylene) which dissolve in water and can be broken down under oxidizing conditions in the unsaturated zone above the groundwater table by the natural or induced activity of microorganisms to form carbon dioxide. Other contaminants (such as the solvents perchloroethylene and trichloroethene) are more easily broken down under reducing conditions and are rapidly broken down below the water table. See also *bioremediation*.

Biologic uptake

Ingestion by biota and incorporation within tissue or skeletal mass.

Biomagnification

The process whereby the body burden of some persistent contaminants (mainly organic) in aquatic animals increases with higher trophic levels in the food chain. This is caused by preferential accumulation and concentration of these contaminants in fatty tissue.

Biopile

Used in *bioremediation* where excavated contaminated soil is mixed with nutrients and bacteria and placed in piles (also called reactor beds).

Bioremediation

A process, usually employing some engineered controls, utilizing micro-organisms to break down chemicals to more innocuous products (see also *biodegradation*). Remediation may occur *in situ* (in place) or *ex situ*, where contaminated soils are removed and treated in a nearby disposal facility.

Bit Map

Binary *raster image* of *pixel* values 0 or 1.

BMD

Basic mine drainage water. Contrast with *acid mine drainage water*.

BOD

See *biochemical oxygen demand*.

Body Wave

Term used in study of earthquakes (see *seismology*) for waves that pass through the interior of planet Earth. *P* (for Primary) waves have a velocity of up to 7 kilometres per second (km·s⁻¹), *S* waves (for Secondary) have velocities up to 3.5 km·s⁻¹ and cannot travel through fluids (*i.e.*, the Earth's molten outer core zone). Body waves are used to map the changing physical properties of rocks within the deep interior of the planet and to locate the origin of earthquakes (see *focus* and *epicentre*).

Boomer

A sound source for a seismic reflection system that provides sub-bottom information on sediments beneath bodies of water such as lakes and seas. A boomer system generates a wide spectrum of sound frequencies by the flexure of a metal plate produced by electromechanical means.

Borehole

A hole bored or drilled into the ground to determine sediment or rock characteristics, collect representative samples of each soil or rock encountered, and permit the installation of a water well or observation well.

Bouguer correction

Correction made to measurement of local gravitational field allowing for the gravitational attraction of rocks between the measurement station and sea level. Gravity surveys provide important data regarding the density of underlying rocks and earth materials and, therefore, are used to map subsurface variation in geologic conditions (see *free-air correction*).

Boulder Clay

Obsolete term for *till*.

Boulder Pavement

Concentration of boulders commonly resulting from removal (winnowing) of finer particles from *till*.

BP

Years before present, commonly calculated prior to A.D. 1950.

BTEX

Acronym for the aromatic hydrocarbons: benzene, toluene, ethylbenzene and xylene. These are present in high concentrations in gasoline and are frequent *contaminants* in urban ground waters. See also *LNAPL*.

Buffer Area

Area of land surrounding the fill area of a landfill, but limited in extent to the landfill property boundary, assigned to provide space for remedial measures, contaminant control measures, and for the reduction or elimination of adverse environmental impact caused by migrating contaminants.

Built Landscape

Refers to man-made features of urban areas such as *historic landfill* and *infrastructure* such as roads, buildings, pieplines, *etc.*

Bulk Permeability

The ability of a sediment or rock to transmit water as measured in the field. This includes the effects of fractures and other discontinuities which are not present when permeability is measured in small laboratory samples. See *permeability*.

Bulk Properties

Refers to the physical properties of *strata* (rock, sediment) measured *in situ* in the field. Such properties (*e.g.*, density, shear strength, hydraulic conductivity, *etc.*) can vary significantly from results detered *ex situ* in the laboratory on small samples. See *representative sample*.

Buried Valley

A channel or depression in an ancient landscape or bedrock surface now buried by younger deposits. Buried valleys comprise part of a paleotopographic surface.

C

"C" Horizon

Unweathered bedrock or sediment below the *"B" horizon* of a soil.

Calcareous

Containing predominant amounts of limestone or other calcium minerals.

Calcite

A mineral formed of calcium carbonate ($CaCO_3$).

Capable Fault

See *active fault*.

Capillary Zone

Also called the capillary fringe or tension-saturated zone; the zone directly above the water table in which the pore spaces are fully saturated, but pressure is less than atmospheric.

Carbon-13

A naturally occurring stable isotope of carbon. Stable means it is not radioactive. The average terrestrial abundance of carbon-13 (^{13}C) is approximately 1.11% of the total carbon.

Carbon-14

This is a radioactive isotope of carbon with a half-life of approximately 5730 years. The steady-state concentration of carbon-14 (^{14}C) in the atmosphere is about one carbon-14 atom in 10 atoms of ordinary carbon. While any biologic organism is alive, it contains carbon-14 in steady-state concentrations. After death, the amount of radioactive carbon decreases by radioactive decay; precise counting of the level of radioactivity in dead tissue, bone, cloth, wood, ground water and ice provides a means of determining age. This is a useful technique for measuring ages back to about 45 *ka*.

Cell

A space or contained area within an active landfill identified and prepared for receiving waste during any stage of landfilling, and subsequently compacted, enclosed by soil or other cover material.

Certificate of Approval

The permit issued by the appropriate regulatory agency for the use, operation, establishment, alteration, enlargement, or extension of a landfill site. It is issued to the owner of the site with terms and conditions of compliance stated therein.

CDF

Confined disposal facility where contaminated sediments dredged from harbours and waterways are placed.

Chemical Flushing

Flushing of ground water contaminated with hydrocarbons with surfactants (*i.e.*, detergents).

Chironomid

Larval stage of an insect (non-biting midge) commonly used as a bioindicator of the state of health (*i.e.*, degree of contamination) of surface waters.

Chlorination

The application of chlorine to water or waste water, generally for the purpose of disinfection, but frequently for accomplishing other biological or chemical results.

Clay

Fine-grained cohesive soil with particle size no greater than 0.002 mm (see *particle size classification*).

Closing

In remote-sensing studies, a mathematical morphological dilation followed by erosion.

Closure Plan

Document detailing the process of closing a landfill site at the end of its active life. This document also includes post-closure requirements, such as long-term monitoring.

Coagulation

The use of chemicals that cause very fine particles to clump together into larger particles. This makes it easier to separate solids from liquids by settling, skimming, draining or filtering.

Cohesion

Strength of a soil or rock not related to interparticle friction. Describes a sediment that has strength when air-dried, *e.g.*, clay. Non-cohesive sediments (*e.g.*, sand, gravel) have no cohesive strength when air dried. Cohesion is derived from physical attraction of clay particles. See *internal friction* and *shear strength*.

Cohesive Shore

Erosional shoreline composed predominantly of glacial and glaciolacustrine silts and clays often characterized by high erosion rates. Fundamentally different from non-cohesive shorelines with plentiful supply of coarse-grained beach material.

Cold-water Stream

A stream which is capable of supporting salmonids (*e.g.*, trout and salmon species); maximum summer water temperatures are normally less than 21°C. Contrast with *warm-water stream*.

Colloids

Very small, finely divided solids (particles that do not dissolve) that remain dispersed in a liquid for a long time due to their small size and electrical charge. A particle size less than 0.00024 mm.

Colour Composite Image

Colour image prepared by projecting individual black and white multi-spectral images, each through a different colour filter. When the projected images are superposed, a colour composite image is produced.

Colluvium

Geologic term referring to generally fine-grained sediment (clay to sand) that accumulates at the base of gentle slopes and which is transported downslope by water. Contrast with *alluvium* and *talus*.

Combined Sewer

Urban water transport system wherein the conduits for sanitary sewage and for storm-water collection are shared, with the result that in cases of intense rainfall, the volume delivered to the sewage treatment plant (*STP*) exceeds its capability and is discharged untreated directly into nearby surface bodies such as rivers, lakes or seas.

Combined Sewer Outfall (CSO)

Point of discharge into a lake, sea or river for a *combined sewer* designed to carry both sanitary waste waters and storm- or surface-water runoff.

Comet Marks

Refers to sediment on sea floor which has accumulated in the lee of an obstacle and which can provide information as to dominant current directions (see also *obstacle--induced sediment drift*).

Common Depth Point (CDP)

A common reflecting point in the subsurface sampled over a range of source-receiver offset distances.

Common Depth Point (CDP) Gather

Seismic record containing all the traces which have a common reflecting point in the subsurface.

Common Depth Point (CDP) Method

A seismic reflection profiling method which employs summation of subsurface reflection points (*common depth points*) to achieve an increase in signal to noise ratio.

Common Depth Point (CDP) Stacking

The summing of traces associated with a *common depth point* following application of static and normal movement corrections.

Compaction Test

A laboratory procedure to determine the optimum water content at which a soil can be compacted so as to yield the maximum density (dry unit weight). The method involves placing a soil sample at a known water content in a mold of given dimensions, subjecting it to a compactive effort of controlled magnitude, and determining the resulting unit weight. The procedure is repeated for various water contents sufficient to establish a relation between water content and unit weight. The maximum dry density for a given compactive effort will usually produce a sample whose saturated strength is near maximum.

Comparative Evaluation

A process used to identify the advantages and disadvantages to the environment resulting from alternative environmental undertakings (*e.g.*, prospective landfill sites) after others have been eliminated from further consideration. Elimination is the result of a screening evaluation process to decide which is the preferred site.

Composting

A controlled method of decomposing organic matter by the natural activity of micro-organisms to yield a humus-like product, usually for soil enrichment purposes. A method for reducing the volume of organic wastes going to *landfills* or incinerators.

Conductivity

A measure of the capacity of a sediment or rock to conduct an electric current which is related to the total concentration of ionized substances in water and the temperature at which the measurement was made. A diagram showing the variation of conductivity with depth in a borehole is called a conductivity log (see *downhole logging*). Also refers to the ability of soils or other materials to transmit water, *etc.* (see *hydraulic conductivity*).

Cone of Depression

Distance from a ground-water well where changes in the ground-water surface (*i.e.*, water levels) can be detected as a result of pumping from that well. Theoretically, in an homogenous aquifer, the cone of depression will be circular, but, in practice, is often irregular. Synonymous with *zone of influence*.

Confined Aquifer

Where an *aquifer* is overlain by an *aquitard* or *aquiclude* and the pressure of ground water is sufficient that the water level in a well will rise above the base of the overlying *confining bed*. Where pressure is great, water may rise above the ground surface as a flowing well (termed *artesian conditions*). See also *piezometric surface* and *unconfined aquifer*.

Confined Disposal Facility (CDF)

Shoreline enclosure constructed to receive and isolate dredged materials too contaminated for open-water disposal.

Confining Bed (Layer)

A body of impermeable or distinctly less permeable material stratigraphically adjacent to one or more aquifers (synonymous with *aquitard*).

Confluence

The location where two or more streams join together as a single river. The opposite to diffluence where one or more distributary streams flow away from a river.

Conservation Area

Watershed area requiring protection and conservation, through the administration of the Conservation Authorities Act by the Ontario Ministry of Natural Resources and the local Conservation Authority.

Conservative

The property of a stable element, where any change in its concentration in the environment is related primarily to dilution rather than to chemical transformation or radioactive decay. Commonly used to describe contaminants (such as chloride from a landfill) that move at the same velocity as ground water and are not retarded by chemical transformations. See *attenuation, contaminant attenuation zone, Reasonable Use Policy* and *retardation*).

Consolidated Hearings Act, 1981

This Province of Ontario Act is intended to streamline the approvals process for undertakings which may require a hearing under more than one statute. For example, hearings required under the Environmental Assessment Act (*Ont. EAA*), the Environmental Protection Act (*Ont. EPA*), and the Planning and Expropriations Acts, may be consolidated into one hearing under the Consolidated Hearings Act (CH Act). This Act can apply to projects which are not subject to the *EA Act*. For a complete listing of the Acts to which the CH Act applies, refer to the Schedule of the CH Act or any of its amendments or regulations which may be set from time to time. (See *hearing, Joint Board*).

Consolidation

Term used to describe the process by which loose (*i.e.*, unconsolidated) sediment is subject to compressive forces under the weight of overlying sediment and undergoes a decrease in volume and an increase in density. See also *overconsolidated*. Final result over millions of years and deep burial is lithification to form a *sedimentary* rock.

Consolidated-Undrained Test (quick test)

A soil test in which essentially complete consolidation under the vertical load (in a direct shear test) or under the confining pressure (in a triaxial test) is followed by a shear at constant water content. See also *shear strength*.

Constructed Wetland

Refers to artificial wetland built to mimic natural functions of wetlands in treating and restoring contaminated surface waters.

Contaminant

A compound, element or physical parameter resulting from human activity, or found at elevated concentrations. Those that have, or may have, a harmful effect on public health or the environment are termed *pollutants*.

Contaminant Audit

Accounting process used in assessment or modelling of ground-water quality whereby all possible sources of contamination in a study area are identified and the resulting population of contaminants listed. Identification of total mass of contaminants resident in any one area can be compared with local drinking-water guidelines and the potential for ground-water contamination can be established.

Contaminating Lifespan

The period of time in which a *landfill* will continue to produce contaminants at levels that could have unacceptable impacts if discharged to surrounding environments. Typically, fifty to a hundred years in duration. Engineered *liners* are designed to contain the bulk of leachate within the landfill during this phase (see *contingency plan*).

Contaminant Migration Path

Route by which a contaminant will move in ground waters from a landfill site or other source, into adjacent properties or the natural environment. See *plume*.

Contamination Attenuation Zone

The zone beneath the surface, located beyond the landfill site boundary, where contaminants will be naturally attenuated to predetermined levels. Also see *attenuation* and *Reasonable Use Policy*.

Contingency Plan

A documented plan detailing a co-ordinated course of action to be followed to control and remedy unforeseen occurrences that could threaten the environment and public health, *e.g.*, failure of an engineered liner system at the base of a *landfill*.

Cover Material

Usually soil material used to cover deposited waste at a landfill site. Its use may be for daily, interim or final cover. Final covers are intended to control the amount of precipitation infiltrating into the landfill and give a final landscaped appearance.

Crust

The outermost shell of planet Earth; thickest (up to 70 km) under the continents and thin (<6 km) under the oceans. Dominated by granitic rocks under the continents and basalt under the oceans.

Cultural Noise

Existing man-made infrastructure (*e.g.*, buried pipes, power lines, *etc.*) that can adversely affect the quality of data recorded by a geophysical instrument being used to investigate the subsurface geology. The source of the noise can be either a physical object which has similar properties to the target being sought or random emissions at similar frequencies to the recording range of the geophysical instrument.

Culvert

A drain carrying water under a road, railway, *etc.*

D

Darcy's Law

Equation relating the velocity of water movement in the saturated zone where velocity is equivalent to *hydraulic conductivity* (= *permeability*) multiplied by *hydraulic gradient*. Because the volume of flow is determined by the velocity of ground-water flow and the cross-sectional area, Darcy's law can be used to calculate the volume of water moving through an aquifer.

DDT

Abbreviation for dichlorodiphenyltrichloroethane, a toxic chlorinated organic compound.

Decomposition

Processes that convert unstable materials into more stable forms by chemical or biological action. When organic matter decays under *anaerobic* conditions (putrefaction), undesirable odours are produced. *Aerobic processes* in common use for waste water (*e.g.*, sewage) treatment produce much less objectional odours.

Deformation Till

Poorly sorted glacial sediment containing a wide range of particle sizes, produced by subglacial deformation of pre-existing sediments overrun by an ice sheet (see *lodgement till*). It is likely that most till deposits are the product of some combination of deformation and lodgement. Regardless, most tills contain sand and gravel beds deposited by subglacial melt waters and these will control the overall permeability of the till deposit. See also *drift* and *diamict*.

Delta

Commonly triangular-shaped tract of land, usually extending beyond the general trend of the coastline, deposited at the mouths of rivers where they enter lakes or seas.

Density Current

Refers, in lakes and seas, to downslope movement of water and sediment as a turbulent mixuture (turbidity current). Movement is maintained by higher density of current compared to ambient water and may be triggered by slope failure.

Designation

In Ontario, the Lieutenant Governor in Council may make a regulation under section 40 which makes the EA Act apply to any proposal, plan, program, major commercial business, enterprise or activity. See *exemption*.

Design Basis Seismic Ground Motion

Worst-case seismic ground motion used in design of facilities so as to ensure adequate protection against earthquake-induced damage during lifetime of facility. Design Basis Earthquake (DBE) describes the ground motion that a facility is thought to be capable of withstanding; expressed as % of acceleration due to gravity (*g*).

Design Capacity

The maximum amount of waste that is planned to be disposed of at a landfill site.

Detection Limit

Concentration under which a parameter cannot be quantitatively measured.

Deuterium

This is a naturally occurring stable isotope of hydrogen. Stable means it is not radioactive. Deuterium (^2H) comprises approximately 0.016% of the hydrogen in sea water.

Diamict

Generic, non-specific term used for any poorly sorted admixture of clay, silt, sand, gravel and/or boulders regardless of origin (contrast with *till*).

Diffusion

The random movement of particles caused by thermal agitation that results in the uniform mixing of gases, miscible liquids or suspended solids. It is the process whereby ionic or molecular constituents move under the influence of their kinetic activity in the direction of their concentration gradient. Diffusion can occur in the absence of any bulk hydraulic movement of the solution. If the solution is flowing, diffusion is a mechanism, along with mechanical dispersion, that causes mixing of ionic or molecular constituents. See *attenuation*.

Digester

A tank in which sludge is placed to allow decomposition by micro-organisms. Digestion may occur under *anaerobic* (more common) or *aerobic* conditions.

Dike

Artificial embankment contructed on low-lying areas to prevent flooding.

Dilation

In remote sensing, a mathematical morphological function; in a binary image, a feature is dilated to all adjacent *pixels*.

Dilution

Reduction in concentration of a dissolved substance in ground and surface waters as a result of various *attenuation* processes.

Dimensionless numbers

Numbers that do not have measurement units, such as length, mass or time.

Direct Shear Test

A shear test in which sediment or rock under an applied normal load is stressed to failure by moving one section of the sample or sample container (shear box) relative to the other section. Measures the *shear strength* of a sediment or rock.

Disinfection

Process designed to kill most micro-organisms in waste water, including essentially all pathogenic (disease-causing) bacteria. There are several ways to disinfect, with chlorine being most frequently used in municipal drinking-water and waste-water treatment plants.

Disintegration Moraine

(synonymous with dead-ice moraine, hummocky moraine, stagnation moraine). Refers to undulating glacial topography consisting of enclosed depressions (*kettle holes*) commonly filled with water and conical hills (*kames*) produced by melt of buried glacial ice.

Dispersion

An important *attenuation* mechanism. During flow through porous media, substances dissolved in the ground water (known as *solutes*), tend to spread out from the path they would be expected to follow based only on the ground-water flow path and velocity. This spreading phenomenon is called hydrodynamic dispersion. It causes a reduction in concentration of the contaminant (*dilution*) and occurs because of mechanical mixing during flow through the pore spaces and because of molecular *diffusion*.

Discharge

In ground-water studies, discharge refers to the flow of water from the saturated zone across the *water table*, together with the associated flow toward the water table within the saturated zone. See *discharge area*.

Discharge Area

That part of a drainage basin where ground-water flow is directed to the ground surface as evidenced by seeps, springs or high ground-water tables.

DNAPL

Dense Non-aqueous Phase Liquid (such as chlorinated hydrocarbons; synthetic compounds, such as 1,1,1–trichloroethane, carbon tetrachloride and tetrachloroethene) that is heavier than water and which sinks through ground water to impermeable layers at depth where it pools and migrates through the site stratigraphy. Remediation is difficult as drilling may accelerate dispersal. Dissolved plumes commonly move away from the column through which DNAPL moved to depth. Contrast with *LNAPL*.

Dolomite

A hexagonal rhombohedral mineral of composition $CaMg(CO_3)_2$, commonly with some Fe replacing Mg.

Dolostone

A term proposed for the sedimentary rock traditionally known as dolomite, in order to avoid confusion with the mineral of the same name. Dolostone is defined as a carbonate sedimentary rock composed of at least 50% fragmental, concretionary or precipitated dolomite of organic or inorganic origin.

Domestic Waste

Non-hazardous wastes defined as domestic (synonymous with *municipal waste*) in Ontario Regulation 309. Contrast with *ICI waste*.

Downhole Logging

Refers to the use of various geophysical tools (see *gamma log, gamma-gamma log, vertical seismic profiling, resistivity log, conductivity log, neutron log*) which, when lowered or raised in a horehole, identify geophysical responses of the surrounding sediment or rock (*i.e.*, electrofacies). A diagram showing variation in *electrofacies* with depth is called a geophysical log.

Drawdown

The lowering of the water level in a well as a result of pumping. The water table surrounding the well is drawn down in the form of a *cone of depression*.

Drift

A general term applied to all unconsolidated (*i.e.*, loose) sediment (*clay, silt, sand*, gravel, boulders) transported by a glacier and deposited directly by or from the ice, or by running water emanating from a glacier. *Till* is an important component of drift. Drift is synonymous with *overburden*.

Drinking-water Objectives

These address drinking-water quality as against ambient water quality (see *Water Quality Guidelines*). OMOEE has three objectives: Maximum Acceptable Concentrations (MAC) which sets limits to substances with known health effects, Interim Maximum Acceptable Concentration (IMAC) describing limits for toxic substances for which no MAC value has been established, and Maximum Desirable Concentration (MDC) which sets upper concentrations for substances that are aesthetically objectionable.

Drumlin

Smoothly rounded elongate ridge commonly built of till and commonly cored by older strata. Results from deposition and/or erosion below an ice sheet; the long axis of the drumlin is parallel to ice flow. Commonly has blunt nose that points up-glacier and a smoother slope tapering down-glacier much like an upturned row boat.

Dry Unit Weight (soil)

The unit weight of solids per unit of total volume of soil mass.

DTM

Digital Terrain Model. A three-dimensional surface, representing an area of the Earth's surface, represented in x,y and z values. Can be created in either vector or raster domains in a *Geographic Information System*. See also *vector coverage* and *raster image*.

Due Diligence

Legal obligation acting on individuals, government organizations and business to identify and meet all applicable environmental regulations. Ignorance is no defence under the law and many environmental geologists are employed to provide specific advice on the current regulatory framework and likely future changes, and to demonstrate how such regulations are being complied with.

Durov Diagram

A trilinear plot that classifies ground- or surface-water types according to their major ion composition and used to identify processes that affect overall water chemistry.

Dyke (geologic)

Vertical or near vertical, tabular body of igneous rock intruded into older strata. In American usage, it is spelled dike. Term also used for vertical intrusions composed of sand or other sediment, created by ground shaking during an earthquake and liquefaction.

E

EA

Environmental Assessment in the province of Ontario (see *environmental assessment*).

EAA or EA Act

Environmental Assessment Act, Revised Statutes of Ontario, 1989, Chapter E-18 (Ont. EAA, R.S.O. 1989, c.E-18). One of the primary acts of legislation intended to protect, conserve and wisely manage Ontario's environment through regulating, planning and development.

EA Advisory Committee

A committee established by the Ontario Minister of the Environment to provide advice to the government through the minister on requests for exemption or designation of undertakings under the *EA Act*.

EA Update

Periodic publication of the Environmental Assessment Branch of the Ministry of the Environment to inform the public about the application of the *EA Act* or *EA* issues.

Ebullition

Term used to describe the bubbling of dissolved gas from a liquid due to depressurization.

Echosounder

A device for measuring the depth of water below a boat, based on the time it takes for sound pulses transmitted from the boat to travel to the sea floor and return. Repeated pulses can be used to build a profile of the relief on the sea or lake floor.

Ecosystem

Complexly interrelated systems of plants, animals, and micro-organisms, together with non-living components of their environment, and humans.

Ecosystem Integrity

An ecosystem is considered to have integrity when its inherent potential is realized, its condition is stable, its capacity for self-repair when disturbed is undiminished, and minimal external management is needed.

Effluent

Any raw or partially treated liquid and associated material discharged into a surface watercourse or discharged on land as a means of final disposal.

Eh

Identifies potential for reduction-oxidation (redox potential) of elements, *i.e.*, the ability of the chemical environment to supply or remove electrons. Eh compliments *pH* which measures the ability to supply or remove protons (H^+). Negative values of Eh indicate reducing conditions, positive values indicate oxidizing conditions. Plots of Eh against pH provide useful information regarding the stability and mobility of mineral constituents in ground and surface waters, soils, *etc*. Most elements are oxidized in near-surface soil layers; reducing conditions occur in underlying water-saturated oxygen-poor layers where *anaerobic* bacteria are common.

Electrofacies

Geophysical responses, typically measured by *downhole logging* which characterize certain subsurface strata (rock, sediment) and which can be used to differentiate such strata from others (see *facies*).

Electrokinetics

Refers to use of low-voltage, direct current to remove metals from contaminated sediments.

Electromagnetic Spectrum

Continuous sequence of electromagnetic energy arranged according to wavelength or frequency.

Electromagnetics

A survey method in which the electromagnetic and/or magnetic fields associated with artificially generated subsurface currents are measured. Instruments are typically designed to determine variations in electrical conductivity with depth or position.

Elluviation

Downward movement and loss of soluble or suspended material in a soil usually from the *"A" horizon* to *"B" horizon*. Contrast with *illuviation*.

Emanation

Refers herein to the release of *radon* atoms by *alpha recoil* from within the solid grains of a host rock or sediment; the ratio of radon atoms escaping into pore spaces compared with the total number of recoils is the emanation coefficient. The direct recoil fraction refers to the radon atoms that escape into *pore spaces*.

Energy From Waste (EFW)

The process of converting used or waste material into fuel or any form of energy.

Engineered Landfill

A landfill constructed using engineered *leachate collection systems*, compacted clay, soil liners and/or artificial *liners* (see *polyethylene*).

Environment

The sum of all the features and conditions surrounding an organism that may influence it. In Law, the definition of "environment" in the Ontario *Environmental Assessment Act*, includes not only the **physical** and **natural** environment, but also includes man and social, economic, and cultural factors, and their interrelationships.

Clause 1(c) of the EA Act defines "environment" as follows:

(i) air, land or water;

(ii) plant and animal life, including man;

(iii) the social, economic and cultural conditions that influence the life of man or a community;

(iv) any building, structure, machine or other device or thing made by man;

(v) any solid liquid, gas, odour, heat, sound, vibration or radiation resulting directly or indirectly from the activities of man, or

(vi) any part or combination of the foregoing and the interrelationships between any two or more of them, in or of Ontario.

Environmental Assessment

Environmental assessment (EA) is the identification and evaluation of the effects of an undertaking and its alternatives on the environment. The term EA refers both to the process of identifying and evaluating the alternatives, and to the document which describes the carrying out of that process resulting in the selection of the proposed undertaking. The contents of an EA are described in subsection 5(3) of the *EA Act* of Ontario.

Environmental Assessment Board

In Ontario, the Environmental Assessment Board is a quasi-judicial body which has the authority to conduct hearings when they are required by the Minister under the EA Act. The Board has the authority to decide on questions of acceptance or amendment and acceptance of the EA, and approval of the undertaking. See *hearing, Joint Board*.

Environmental Geology

The term *geology* identifies the scientific study of planet Earth, its internal structure and the varied processes that operate on the surface and within it. *Environment* refers to the sum of all or any of the features and conditions that surround and influence an organism. Because planet Earth provides the fundamental physical environment in which organisms live, virtually the entire discipline of geology can be regarded as *Environmental Geology*. However, this term is increasingly being reserved for the application of geological sciences to understanding environmental problems arising from the **mutual interaction of humans, ecosystems and the earth**. Soil and water contamination, hazardous geological processes, environmental problems caused by urbanization, mining and abstraction activity, management of land use planning and decision making all fall within the scope of environmental geology. An increasingly important component of this emerging field is education and dissemination of knowledge aimed at ameliorating the impact of humans on the planet and *vice versa*.

Environmental geology is characterized by its broad scope and close interdisciplinary relations with chemistry and biology and, increasingly, planning, the law and public health.

Environmentally Significant Area (ESA)

An area of land identified using a variety of evaluation criteria and designated to protect the biological significance of the landscape.

EPA

Environmental Protection Act, Revised Statutes of Ontario, 1990, Chapter E-19 (Ont. EPA, R.S.O., 1990, c.E-19). EPA is the primary provincial legislation governing the protection of the natural environment of the province.

In the United States, EPA identifies the Environmental Protection Agency (referred to in the text as US EPA).

Epicentre

Point on Earth's surface directly above the origin of an earthquake (see *focus*). Point where greatest damage can be expected to occur.

Epidemiology

The study of the relationship between human health and exposure to a wide range of environmental contaminants and biological organisms in air, water and soil.

Erosion

In remote sensing, a mathematical morphological function. In a binary image, a feature is eroded from adjacent pixels.

Erosion

In geology, refers to physical or chemical removal of weathered rock or sediment and results in topographic smoothing and/or dissection of landscapes. See *weathering*.

Erratic

A "foreign" boulder transported long distances by a glacier or ice sheet, *e.g.*, granite boulders from the Canadian Shield transported into areas of southern Canada and the northern United States where the shield is deeply buried by younger rocks and nowhere exposed.

ERS-1

European Resource Satellite 1; a 30-m resolution C-band active radar satellite.

Esker

Elongate ridge of sand and gravel deposited by melt-water rivers under, in, or on ice sheets.

Etiology

Study of the causes of disease.

Evaporite

A non-clastic sedimentary rock composed primarily of minerals produced from a saline solution as a result of extensive or total evaporation of the solvent. Examples include gypsum, anhydrite, rock salt, primary dolomite, and various nitrates and borates.

Evapotranspiration

The combined loss of water from soil and plant surfaces by direct evaporation and transpiration. Evaporation is defined as the physical transformation of a liquid to a gas at any temperature below its boiling point. Transpiration is defined as the process by which water absorbed by plants, usually through the roots, is evaporated into the atmosphere from the plant surface. See also *actual* and *potential evapotranspiration*.

Exemption

An exemption removes the need for a proponent to carry out specific requirements of the Ontario *EA Act*. It may exempt a *proponent* or an *undertaking* entirely from the Act and may be granted with or without terms and conditions. All provincial and municipal undertakings and those of various bodies which have been designated public bodies by regulation for the purposes of the EA Act, as well as certain private sector undertakings which have been designated by regulation, are subject to the EA Act, unless exempted by Order of the Minister with the approval of the Lieutenant Governor in Council (section 29) or regulation (section 39). Exemption Orders under Section 29 of the EA Act are filed and published as regulations.

Exfiltration

Where liquid wastes and liquid-carried contaminants either unintentionally or intentionally leak out of a sewer pipe system, storm-water pond or other utility and into ground water.

Export

In waste management programmes refers to the transport of waste outside the jurisdiction within which it was generated.

Ex situ

Refers to testing, remediation, or treatment techniques that are used on sediments that are excavated and moved.

F

Facies

A rock or sediment, regardless of origin and age, having characteristics that distinguish it from overlying, underlying or adjacent units, *e.g.*, colour, particle size, fossil or mineral content. The term is usually prefixed by the general rock or sediment type in question *e.g.*, *sedimentary* facies, *metamorphic* facies, *glacial* sedimentary facies, *etc. Electrofacies* is used to identify rocks or sediments having different geophysical properties.

Failure

Engineering term for downslope movement of sediment or rock, *i.e.*, slope failure.

Failure Complex

Term used to describe the hummocky topography and disturbed near-surface stratigraphy of areas that have experienced downslope slumping. Can be on the sea floor or on land. The identification of such complexes is important in order to minimize future damage to urban infrastructure during any future earth movements, especially during earthquakes and associated ground shaking.

False Colour

In remote sensing, the use of one colour to represent another; for example, the use of red emulsion to represent infra-red light in colour infra-red film.

Fault

A surface in geologic strata along which movement has occurred. Normal faults are the result of extensional movements where strata are pulled apart. Thrust faults (otherwise known as reverse faults) are where displacement is a result of compressional movement where strata are pushed together. Faults range in scale from those that affect an local outcrop or worksite to major regional features that are on a continental scale (*e.g.*, San Andreas Fault). See also *lineament* and *suture*.

FDD

Freezing degree days; climatic term which identifies, for any site, the total number of days that the air temperature is below freezing multiplied by the number of degrees below freezing, in any one year.

Field Capacity

The volume of water held by a soil, sediment or rock against the pull of gravity.

Fill Area

The part of a landfill site designed and designated for the disposal of waste.

Filter, digital

Mathematical procedure for modifying numerical data.

Filtration Process

Refers to use of filter membranes to treat contaminated ground water containing organic compounds and metals.

Final Contours

The topography of a landfill site once it has reached the limit of the approved capacity provided for by the Certificate of Approval.

Final Cover

Soil material or soil in combination with synthetic membranes, overlain by vegetation in a planned landscape, placed over a waste cell that has reached the end of its active life. See also *cover material*.

Flocculation

The gathering together of small suspended particles in water to form larger particles.

Floodplain

An area, usually low lands, adjoining a watercourse which has been, or which may be, covered by flood water.

Flux

Rate of flow, for example of landfill gas or ground water, through a given cross-sectional area.

Focus

Term used by seismologists for the point within the Earth's interior where a particular earthquake is produced. Earthquakes can occur at depths of as much as 700 km in deep *subduction zones*, such as occurs along the western boundary of North America (see also *epicentre*).

Formation

The fundamental unit in stratigraphic subdivision and correlation of rocks and sediments. Thickness is not a determining factor and a formation may be defined on the basis of multiple bedding units (see *bed*) having distinct characteristics and being capable of being mapped across the region. Formations can be combined into *groups* or subdivided into *members*. A formation is usually from tens to hundreds of metres in thickness.

Fracture Porosity

Porosity which is due to fractures. This is a particular type of *secondary porosity* common in rocks. See *secondary porosity*.

Free-air correction

Correction made to measurements of gravitational field made during a gravity survey to allow for difference in elevation above sea level (*i.e.*, distance from centre of the Earth). See *Bouguer correction*.

Free Product

Any substance (usually a hydrocarbon) present as a nonaqueous phase liquid (*i.e.*, not dissolved in water). See *LNAPL*.

Free Product Recovery

Removal of hydrocarbon fuels that are pooled on the water table.

Frequency (geophysics)

Cycles per second (units = Hertz; Hz).

G

Ga

Abbreviation for Giga-annum used by geologists to refer to the age of a geologic event in billions of years (*e.g.*, 1,000,000,000 or 10^9 years). The first letter is always capitalized (unlike the prefix kilo-annum or *ka*). In contrast, the duration of a geologic event is always expressed in millions of years as m.y. See also *Ma*.

Gamma Log

Refers to technique of measuring downhole variation in natural gamma radiation emitted from sediments and rocks penetrated by a drill hole. Used to distinguish fine-grained clayey units containing potassium and that have high gamma-ray counts from coarser materials such as sands or gravels. See also *facies* and *downhole logging*.

Gamma-Gamma Log

Refers to technique of measuring downhole variation in the backscattering of gamma rays emitted from a source lowered or raised in the borehole; the degree of scattering is a function of porosity of the surrounding strata and is used to determine density. See *downhole logging*.

Gas Collection System

An engineered system to contain and collect migrating landfill gas for safe dissipation, energy recovery or incineration.

Gas Chromotography

A process for separating gases or vapours from one another.

Gas Extraction Well

A constructed well, within or outside waste disposal areas, intended to draw in landfill gas for collection. Gas extraction wells are part of a landfill gas collection system.

GDD

Growing degree-days; climatic term which refers, for any one site, to the number of days that the air temperature is above a specified temperature (*e.g.*, 10°C) multiplied by the number of degrees above that level, in any one year.

Geographic Information System (GIS)

Any manual or computer based set of procedures used to store and manipulate geographically referenced data. Contrast with *Geoscientific Information System (GSIS)*.

Geology

The scientific study of the nature, origin and development of planet Earth. See also *Environmental Geology*.

Geologic Timescale

See Appendix A.

Geomatics

A term encompassing the disciplines of surveying, mapping, remote sensing, cartography, photogrammetry and the use of *Geographic Information Systems (GIS)* and *Geoscientific Information Systems (GSIS)*.

Geomembrane

A highly impermeable synthetic membrane made of plastic or rubber-based material and designed to prevent passage of liquids or vapours. Used in the construction of engineered landfills to prevent or reduce leakage of contaminants from base of landfill (see *polyethylene*).

Geometric Correction

Image-processing procedure that corrects spatial distortions in an image.

Geophone

A mechanical and/or electronic device (essentially a microphone) for detecting *seismic waves* placed in the ground and which produces a voltage proportional to the displacement, velocity, or *acceleration* of ground motion, within a limited frequency range. See also *hydrophone*. Used in *seismic reflection surveys*.

Geophysics

Study of planet Earth and earth materials by quantitative physical methods. Methods employ sound waves, radio waves and magnetic fields to map subsurface features. Increasingly employed in urban environmental geology projects where drilling is too expensive and where a broad coverage of subsurface conditions is needed. Drilling may also disturb subsurface contaminants and accelerate their spread into the natural environment. Geophysical techniques are rapid and non-intrusive. The term "borehole geophysics" refers to the application of geophysical techniques that employ instruments placed in a borehole.

Geoscientific Information System (GSIS)

Any manual or computer-based set of procedures used to store and manipulate three-dimensionally referenced data (*e.g.*, subsurface geological data). Contrast with *GIS*.

Geotechnical

The application of scientific methods and engineering principles to the acquisition, interpretation, and use of knowledge of geological materials of the Earth's crust for the solution of engineering problems.

Geotextile

A permeable textile made of synthetic fibres, constructed into woven or non-woven fabrics, commonly referred to as filter fabric or erosion control cloth. For this application, geotextiles will be placed between various layers in the landfill liner design to perform as either a separation layer or to act as a filter medium.

GIS

See *Geographic Information System*.

Glacial

Pertaining to the action of glaciers and ice sheets (contrast with *periglacial*).

Glacial Lake

A lake fed predominantly by glacial melt water or a lake in a depression in part closed by glacial ice. Includes lakes lying against or on a glacier. See also *glaciolacustrine*.

Glacial Maximum

Time of maximum extent of ice sheets, *e.g.*, about 18 ka for the last glaciation of mid-continent North America by the *Laurentide Ice Sheet*.

Glaciofluvial

Pertaining to melt-water streams flowing from glaciers and to their deposits. See also *outwash*.

Glacio-isostatic

Refers to vertical movement of Earth's surface caused by loading below a large ice sheet. Can be downward (depression) or upward (rebound). Glacio-isostatic movement continues for many thousands of years after final demise of ice sheet; rate decreases exponentially until movement is completed.

Glaciolacustrine

Pertaining to or characterized by both glacial and lacustrine (lake) conditions. Sediment deposited in lakes affected by glacier ice, or deposited by melt water flowing directly from glaciers into a lake.

Glaciomarine

Pertaining to or characterized by both glacial and marine conditions. Sediment deposited in seas and oceans affected by glacier ice, or deposited by melt water flowing directly from glaciers into the marine environment.

Glaciotectonic

Refers to small- (cm) to large- (km) scale deformation structures in rock or sediment caused by overriding of a glacier or ice sheet.

Global Positioning System (GPS)

Typically, a hand-held instrument using satellites to determine location, elevation, *etc.* of sites on the Earth's surface.

Gneiss

(Pronounced "nice"); a very coarse-grained metamorphic rock characterized by crude layering (foliation). Typical of very old rocks produced by heat and pressure at great depths in the Earth's crust and now commonly exposed at surface (*e.g.*, the Canadian Shield) as a result of erosion and uplift. Also found at depth (as *basement*) below younger cover rocks. Though often complexly folded and difficult to decipher, such rocks record a long history of *plate tectonic* processes.

GPS   See *Global Positioning System*.

Granular Activated Carbon (GAC)

Refers to use of charcoal to remove organic contaminants from ground water.

Groundroll

Low-frequency surface wave noise generated by a seismic source (explosive, shotgun, hammer, *etc.*) which can be removed by processing of seismic data.

Ground Truth

Term used for the collection of field data and information to verify geophysical or remotely sensed data such as *seismic reflection profiling* or satellite imagery.

Ground Water

Subsurface water occurring below the *water table* in the saturated zone where all pores are filled with water under hydrostatic pressure. This contrast with subsurface water above the water table, which is referred to as soil moisture. Ground water accounts for about 4% of the total global water balance. About 2% is in ice caps and glaciers and almost all the rest is in the oceans and seas.

Ground-water Divide

The boundary between two or more adjacent ground-water basins represented by a high in the water table.

Ground-water Flux

See *specific discharge*.

Group

Unit in stratigraphic studies composed of two or more *formations*. Usually of considerable thickness (several hundreds of metres) and regional extent.

Grout

A cement slurry of high water content, fluid enough to be poured or injected into spaces and thereby fill or seal them.

Grout Curtain

Refers to use of closely spaced boreholes or trenches which are filled with concrete or bentonite to contain contaminated ground water within a site.

Groyne

Man-made structure extending perpendicular to shoreline of sea or lake designed to trap littoral sediment and reduce coastal erosion.

GSIS

See *Geoscientific Information System*.

GTA

Greater Toronto Area (see Chapter 2 for maps).

H

ha

Hectare (see Appendix C)

Hardness (Water)

A property of water causing formation of an insoluble residue when the water is used with soap, and forming a scale in vessels in which water has been allowed to evaporate. It is primarily due to the presence of ions of calcium and magnesium, but also to ions of other alkali metals, other metals (*e.g.*, iron) and even hydrogen. Hardness of water is generally expressed as parts per million (*ppm*); also as milligrams per litre (mg·L^{-1}).

Hardpan

Informal term used by drillers for resistant units in sediments, usually *till*.

Hazardous Waste

Waste that poses a present or potential danger to human beings or other organisms because it is toxic, flammable, radioactive, explosive, or has some other property that produces substantial risk to biological organisms and public health.

HDPE

See *polyethylene*.

Head

The difference in elevation between intake and discharge points for a liquid. In environmental geology, most commonly used in connection with the movement of ground water from one elevation to another. Ground water flows through an aquifer when the water levels within it are at different elevations or heads. The difference in elevation is sometimes called head loss (see *hydraulic gradient*).

Hearing

In Ontario, a quasi-judiciary public hearing on an *EA* may be conducted by the *Environmental Assessment Board* or a *Joint Board* on whether to accept or amend and accept an EA, and whether to approve the undertakings with or without terms and conditions. (See *Consolidated Hearings Act*, 1981).

Henry's Law Constant

Describes the tendency of an organic compound dissolved in water to volatilize into a gas.

Historic Landfilling

Refers to practices of using waste materials to infill topographic lows (valleys, wetlands) to create new land for development. Such activity was particularly prevalent during the mid-19th to mid-20th centuries, but is now restricted by environmental regulations. Historic landfill is an important component of the *built landscape* of urban areas.

Holocene

Most recent and continuing epoch of Earth history that started approximately 10,000 years ago (see Appendix A). Broadly refers to time interval that has elapsed since the disappearance of the last Northern hemispheric continental ice sheets (*e.g.*, *Laurentide Ice Sheet* in Canada).

Hydraulic Conductivity

The rate of ground-water flow expressed in units of metres per day or centimetres per second ($cm \cdot s^{-1}$) through a cross-section of an aquifer under a specific *hydraulic gradient*. The ability of soil or rock to transmit water; the higher the hydraulic conductivity, the greater the ability to transmit water (see *transmissivity*). It is a high value for sands and gravels and low for silts and clays.

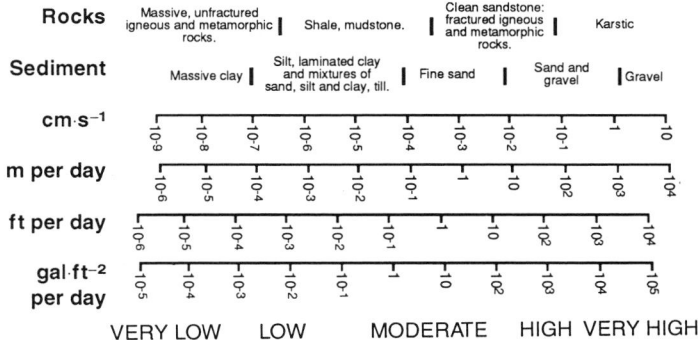

Hydraulic Control Layer

A saturated, permeable (usually coarse stone) layer at the base of a landfill that can be pumped to control hydraulic gradients across a clay liner. Can be used to induce an inward *hydraulic gradient* and create a *hydraulic trap*.

Hydraulic Fracturing

In situ technology using hydraulic pressure to fracture rock or fine-grained sediment to develop secondary permeability and enhanced ground-water flow.

Hydraulic Gradient

The rate of change of *hydraulic head* per unit of distance away from any given well and in a given direction. The head of underground water divided by the distance of travel between two points. If the change in head is 10 m for two points 100 m apart, the hydraulic gradient is 0.1. When head and distance of flow are the same, the hydraulic gradient is 1.0.

Hydraulic Head

The height (usually expressed in terms of elevation above sea level) of the free surface of a body of water (*water table* or *static water level*). The difference in elevation of water levels in two wells is called head loss.

Hydraulic Trap

Landfill design where ground-water flow is into the landfill and which restricts the outward migration of contaminants.

Hydrocarbon

Organic compound composed mostly of carbon and hydrogen. Divided into "aliphatic" (*e.g.*, paraffin, naphthene, alkene, alkyne) and "aromatic" (*e.g.*, benzene) types.

Hydrogeology

The scientific study of the occurrence, movement and chemistry of ground water in relation to the geological environment.

Hydrograph

A graph showing the discharge of a river *versus* time. In natural streams, the hydrograph for flood events shows a slow rise and fall; urban rivers, in contrast, show distinct peaks where flow rises and falls very rapidly due to a lack of infiltration through hardened urban surfaces (*e.g.*, asphalt).

Hydrophone

Essentially, a microphone-like device akin to a *geophone* towed behind a ship to record seismic waves reflecting from subsurface sediment/rock layers, during a *seismic reflection survey*.

Hydrostratigraphic Unit

Part or all of a subsurface body of rock or sediment having considerable lateral extent and either differing in its ability to transmit water or having waters of differing characteristics.

I

Ice-contact

Refers to sediments deposited close to an ice sheet margin, typically ice-contact sands and gravels.

Ice Wedge Cast

Fossil ice wedge. An ice wedge forms by contraction and deep cracking of deeply frozen ground (see *periglacial*) and the freezing of water in the crack to form a downward tapering carrot-shaped wedge of ice. With time and repeated cracking, the ice wedge thickens and may ultimately form an ice body several metres wide and many metres deep. Upon climate warming, the ice decays and is slowly replaced by sediment and debris forming a v-shaped cast of the original ice wedge.

Ichnofossil

Term used to describe the fossilized trace of an organism in sediment and/or rock created by locomotion, burrowing, boring, feeding, *etc.* See also *oligochaete*.

ICI Waste

Non-hazardous industrial, commercial and institutional waste.

Illuviation

Accumulation of soluble or suspended material in a soil as a result of downward transport from upper part of soil (see *elluviation*).

Impermeable Fill

Soil material that is placed as filling material that is sufficiently cohesive and fine grained to impede and restrict the flow of water through it.

Impact

The predicted effect on public health, safety and the natural environment by the introduction of a proposed environmental undertaking. The impact is positive or negative and is identified by an investigation called an impact assessment.

Incineration

Controlled burning of contaminated soil or solid waste and gases in the presence of oxygen for the purpose of waste destruction and/or achieving volume, weight reduction or to change waste characteristics. Waste is incinerated at temperatures between 800°C and 1500°C.

Industrial Waste

Any waste that is the direct or indirect by-product of the manufacturing of a product or the performance of a service by industry or business.

Inert Fill

Earth or rock fill that contains no *putrescible* materials or soluble or decomposable substances.

Infiltration

The percolation of water from precipitation into soil or rock through pores or fractures and which eventually reaches the ground-water table (see *recharge*). Note that the term is also used for the seepage of ground water into a sewer system, including service connections (see also *inflow*). Seepage frequently occurs through defective or cracked pipes, pipe joints, connections or manhole walls.

Infiltration Rate

The maximum rate at which soil can absorb water.

Inflow

Water entering a sewer system from sources other than regular connections. This includes flow from yard drains, foundation drains and around manhole covers. Inflow differs from infiltration (see above) in that it is a direct discharge into the sewer rather than a leak in the sewer itself. See also *exfiltration*.

Influent

Waste water or other liquid, raw or partially treated, flowing into a reservoir, basin, treatment process, or treatment plant. Contrast with *effluent*.

Infrastructure

Physical structures that form the foundation for urban development. Includes sewage and water works, waste management systems, power-generating facilities, communications and transportation corridors and facilities, oil and gas pipelines and any associated facilities.

Inorganic Waste

Waste material, such as sand, salt, iron, calcium, and other mineral materials, which are only slightly affected by the action of organisms. Inorganic wastes are chemical substances of mineral origin, whereas organic wastes are chemical substances usually of animal or plant origin.

In situ

Testing or treatment done on-site, in the field, of sediments or rocks in their original state. Contrast with *ex situ*.

Instrument Footprint

An area within which an instrument measures or explores. A microscope has a very tiny "footprint", whereas a telescope has a very large one.

Interflow

Lateral movement of water within the unsaturated zone above the water table.

Interglacial

Variably employed either for time period or sediments deposited between two major glaciations. Interglacials are of several thousands of years in duration (commonly less than 20,000 years). Deposits of the last (Sangamon) interglacial in mid-continent North America lie between tills of the Illinoian and Wisconsin glaciations (see also *Interstadial*, with which it is commonly confused).

Interim Waste Authority (IWA)

Ontario provincial authority (1990-1995) charged with finding long-term landfills for municipal waste from the Greater Toronto Area.

Interlobate

The area between two ice sheet lobes.

Internal friction (shear resistance)

The portion of the strength of a sediment or rock resulting from friction between constituent particles. It is usally considered to be due to the interlocking of the soil or rock grains and the resistance to sliding between the grains. Sands and gravels are able to generate considerable internal friction, whereas silts and clays cannot. Contrast with *cohesion*.

Interstadial

Short-lived, typically a few hundred years long, episode of climatic warming during a glaciation in which the ice sheet margin may retreat and expose land surfaces. Commonly recorded in stratigraphic sections by peaty deposits, bones, periglacial structures and deposits of rivers and lakes. Not to be confused with *interglacial*, which is a longer time period of enhanced climatic warmth and associated deposits separating major glaciations and their deposits.

Intraplate

Meaning within a tectonic plate. Used, for example, to refer to mid-continent earthquakes in North America, *i.e.*, those occurring away from the active margins of the North American plate (*e.g.*, California, Alaska, British Columbia). The most famous intraplate earthquakes occurred at New Madrid, Missouri, 1811-1812 as a result of movement along a deeply buried *rift*. Intraplate earthquakes are are source of concern in eastern North America, traditionally thought to be "stable". They occur infrequently and an absence of damaging earthquakes since European colonization is no guarantee of future safety. See also *plate tectonics*.

Isochrone

Lines on a map showing time taken by contaminants to move away from source areas in ground-water flow system. See *time of travel*.

Isolation

Method of managing contaminated sediment and/or ground water by containing contamination within physical barriers (see *slurry wall*). Used where cost of remediation is excessive or where suitable technology is not available.

Isopach map

Map showing variations in thickness of a rock or sediment layer across an area of interest.

Isotope

Different forms of an element having a different atomic mass. An element is defined by the number of protons in its nucleus, but the number of neutrons varies, giving rise to isotopes of that element. A carbon atom has six protons, but may have six neutrons (^{12}C) or eight neutrons (^{14}C). Two isotopes of oxygen (^{16}O, ^{18}O) are employed in studies of past climate; the heavier isotope (^{18}O) is present in oxygen at a concentration of about 1 in 500 as compared with ^{16}O. When water evaporates, remaining water becomes progressively enriched in the heavier isotope; the growth of continental ice sheets leaves ocean water enriched in ^{18}O as the lighter isotope is concentrated in snow that falls on the ice cap. Information regarding the past concentration of ^{18}O in ocean water over millions of years is stored in the corals and other fossil organisms and provides important information on the past extent of ice sheets on planet Earth. See also *carbon-14*.

Iteration

Reconsideration of decisions made in the past in the light of new data; often difficult to put into practice with large *environmental assessments* taking many years to complete.

J

Joint

Vertical or near-vertical fracture in bedrock, or consolidated sediments, across which no vertical or lateral movement has occurred (contrast with *fault*). Open joints are where the original fracture has been widened. Measurement of joint orientation frequently reveals a systematic regional trend (regional joint system). Joints are commonly propagated upward from bedrock into overlying sediments (most commonly till or clays) and control ground-water movement and mechanical strength characteristics.

Joint Board

In Ontario, Joint Board is a quasi-judicial body established under the authority of the *Consolidated Hearings Act* (1981). It is comprised of one or more members of either or both the Environmental Assessment Board and the Ontario Municipal Board.

K

ka

Shortened version of kilo annum (thousand years), *i.e.,* 10^3 years, used to denote age of a geologic event. Unlike the prefixes Mega (M) and Giga (G), the initial letter (k) is not capitalized.

Kame

Isolated hill composed of outwash sands and gravels dumped along, on top of, or under the margin of a glacier or ice sheet. See also *disintegration moraine*.

Karst

Dissolution of carbonate bedrock (limestone, dolostone) to form underground cave systems, conduits and voids. Gives rise to unpredictable ground-water flow that is difficult to model and collapse of overlying strata to form *sinkholes*.

Kettle Hole

Enclosed, steep-sided crater-like depression formed by the melt of glacier ice buried under rapidly deposited glacial outwash sediments which collapse to form a kettle hole. Commonly filled with water (kettle lake).

L

Lamina

The thinnest recognizable unit layer of original deposition in a sediment or sedimentary rock, differing from other layers in colour, composition, or particle size; specifically such a sedimentary layer less than 1 cm in thickness. Plural: laminae.

Landfilling

Refers to the disposal of a wide range of waste materials by deposition, under controlled conditions, on land; includes compaction of waste into a *cell* and covering the waste with *cover material* at regular intervals. The term is also used to describe use of waste materials to reclaim low-lying land or to infill irregular topography (see *historic landfilling*). Many cities are built on landfill and much is contaminated resulting in expensive *remedial action*, and substrates that are prone to *liquefaction* during earthquakes.

Landfill Gas

Gas (primarily *methane* and carbon dioxide) generated by the decomposition of organic waste materials. Also contains *NMOCs*.

Landfill Site

A parcel of land where solid waste is disposed of on land, under controlled conditions, for the purposes of waste management.

Landsat

NASA series of resource satellites in operation from 1972 to the present. Either 80 mor 30 m spatial resolution and with up to 7 spectral bands in visible, colour infra-red and thermal infra-red spectral regions.

Laurentide Ice Sheet

Specific name (and capitalized throughout) given to last continental ice sheet that formed over central and northern North America during the interval from approximately 80 ka to approximately 6 ka (*i.e.,* the interval lasted approximately 74 *k.y.*). The maximum thickness was at least 2 km in its inner portions forming a broad dome with two or more outflow centres. Extended from the Rocky Mountains in the west to offshore Newfoundland in the east and from Ohio in the south to the High Arctic. A similar ice sheet has formed and disappeared many times in the past as a result of repeated glacial/interglacial cycles cause by systematic variation in Earth's orbit (see *Milankovitch Theory* and *isotope*). Traditionally, only four such cycles were recognized, but it is now known that there have been many tens of such cycles over the last 3 m.y. Ice sheets have fundamentally altered the topography of glaciated North America and left a thick cover of complex glacial strata. The depositional record in any one locality in North America is biased toward the last one or two cycles because older deposits are removed by erosion below later ice sheets. See *drift, till, glacio-isostatic, glaciotectonic, glacio-lacustrine, glaciomarine*.

Leachate

The liquid formed in the waste disposal area of a landfill when water (typically from infiltrating precipitation) comes into contact with refuse and picks up or leaches out contaminants. Contains dissolved, suspended and/or microbial contaminants derived from solid waste.

Leachate Breakout

Location where leachate comes to the ground surface as a seep or spring along a natural slope or the side slope of a landfill mound. Results from *leachate mounding* in landfill.

Leachate Collection System

Pipes or other media (*e.g.,* layers of gravel) which collect leachate under or adjacent to a landfill and transport the leachate to a pick-up point. Serves as a barrier to prevent or reduce the amount of leachate that reaches and mixes with ground water.

Leachate Compatibility Test

A test conducted on materials to determine if the material is suitable for use in an environment where the material will be in contact with leachate. Commonly conducted on materials to be used as a liner at the base of a landfill (see *liner*).

Leachate Indicator Parameters

Parameters and/or constituents that have been found to be characteristic of a specific landfill leachate and, hence, can be used to monitor the movement of leachate in ground water.

Leachate Monitoring System

A system of strategically placed wells or other measuring devices for scrutinizing and assessing qualitatively the movement of leachate and its effect on adjacent ground- and surface-water resources.

Leachate Mounding

Situation where leachate builds uip in a landfill mound, most commonly because leachate collection system is clogged. Results in *leachate breakouts* from sideslopes of landfill mound.

Leachate Plume

Ground water, beyond the waste fill limits, that has been impacted by the leachate from a landfill site; commonly takes the physical form similar to that of a plume of smoke drifting downwind from a fire.

Limestone

A bedded sedimentary deposit consisting chiefly of calcium carbonate ($CaCO_3$) in the form of calcite. Limestones can be of organic, chemical or detrital origin and are the most important and widely distributed carbonate rocks. See also *dolostone*.

Limit of Filling

The outermost limit at which waste has been disposed of, or approved or proposed for disposal, at a landfill.

Lineament

A term used to describe any linear and relatively narrow geologic feature apparent on a map and where the origin is unclear. Lineaments may be metres to many hundreds of kilometres in length and from less than a metre to several tens of kilometres wide. Commonly identified using various geophysical techniques which are able to identify structures buried at depth and not amenable to direct observation and analysis. See *aeromagnetic data* and *geophysics*. Most lineaments are faults; a major problem is determining whether such structures are currently active or dormant and capable of being reactivated and producing earthquakes. See also *suture*.

Liner

A constructed, continuous layer of reworked natural soil (usually *clay*), or synthetic materials (see *polyethylene*) placed beneath and on the sides of a landfill, or waste cell that restricts the downward or lateral migration of leachate or landfill gas.

Liquefaction

Refers to the ability of certain sediments to become liquid-like when shaken such as occurs in an earthquake or during construction activity. Such sediments lose their ability to support buildings and roads resulting in collapse and damage. Landfilled areas in cities are prone to liquefaction. See *landfill* and *quick clay*.

LNAPL

Light non-aqeous phase liquid (such as gasoline and other petroleum products) which cannot move through ground water but accumulates as "free product" on top of the ground-water table. Many components are immiscible in water (benzene, toluene, ethylbenzene and xylene: referred to collectively as BTEX) and move as complex contaminant plumes away from the original site of contamination. Remediation involves pump-and-treat, isolation and various bioremediation techniques.

Lodgement Till

Poorly sorted, dense and heavily *over-consolidated sediment* having a concrete-like appearance and consisting of admixtures of boulders to clay size materials. Deposited below a moving ice sheet or glacier by being "smeared" against underlying substrate. Commonly described as *hardpan* by drillers. See also *deformation till*.

Loess

Term used for deposits of wind-blown silt derived from glacial outwash rivers. Forms extensive and thick deposits in mid-continent North America where it formed marginal to the *Laurentide Ice Sheet*. Deposited in a very dry state, it is prone to collapse when wetted during construction activity. See *compaction test*.

Log

A continuous record as a function of depth, usually graphic and plotted to scale on a narrow paper strip, of observations made on the rocks and fluids of the geologic section penetrated by a borehole or observed in outcrop.

Loss on Ignition

Refers to loss of weight of a dried sediment sample after combustion; commonly used as a measure of organic content in a sediment.

Low-level Radioactive Waste

Equipment and waste materials from nuclear operations having concentrations of radiation in the range of 1 microcurie per cubic metre ($1\ \mu Ci \cdot m^{-3}$).

Low-temperature Thermal Desorption

Method for remediating sediments contaminated with volatile organic compounds (*e.g.*, gasoline) using heat and thermal or catalytic oxidation.

LTTD

Low-temperature thermal desorption.

Lysimeter

Generally, large containers of soil set in the field to represent prevailing soil and climatic conditions used for the purpose of obtaining data on *in situ* conditions. May be installed at the base of, or within, a landfill to monitor ground-water movement.

M

Ma

Abbreviated form of Mega-annum meaning a million years (*e.g.*, 1,000,000 or 10^6 years) to denote the age of a geologic event. The initial letter of Ma is always capitalized, unlike *ka* (kilo-annum). Note: the duration of a geological event is always expressed in millions of years (m.y), *e.g.*, the Cretaceous period started at approximately 145 Ma and ended at approximately 65 Ma, thus lasting approximately 80 m.y. (see Appendix A).

Magnetics

A geophysical survey method with which the natural magnetic fields of the Earth, minerals or ferrous objects are measured. Typically designed to determine magnetic variations at various positions over an area of interest with the objective of locating concentrations of magnetic minerals or objects (*e.g.*, buried storage tanks).

Magnetic Susceptibility

Measure of the ease at which a material, such as a sediment sample, can be magnetized. It is a function of the amount of magnetic material present, grain size and mineralogy of the sediment. Volume susceptibility, κ, is the ratio of the volume magnetization (M) induced in a material by an applied field (H), *i.e.*, M/H and is therefore dimensionless. Susceptibility is commonly measured by the change in current or voltage in an electrical circuit, where the changing inductance of a measurement coil is recorded when a sediment sample is placed within it. Variations in magnetic susceptibility in geologic materials provide important clues as to origins and can also be used to identify and map sediments contaminated by industrial contaminants which are usually associated with the release of fine particles of magnetic minerals.

Magnitude (M)

Used to describe the size of an earthquake. The most widely used is the Richter scale defined as the amplitude of seismic waves (see *P wave* and *S wave*) at a distance (100 km) from the *epicentre* of the earthquake. It is a logarithmic scale from 1 to 10 where each unit increase corresponds to a ten-fold increase in the amplitude of the signal. The energy released by an earthquake can be related to signal amplitude; each unit increase in Richter scale corresponds to a thirty-fold increase in energy, *i.e.*, the difference in energy released by a Richter $M = 4$ and $M = 7$ earthquakes is $30 \times 30 \times 30 = 27,000$ times. Earthquakes with $M < 3$ are recorded only by a *seismograph*; a $M = 3$ earthquake is felt by some people. Damage to buildings occurs at $M = 5$ with serious damage at greater magnitudes. The world distribution of earthquakes shows pronounced clustering along boundaries of the planet's tectonic plates; some occur within plates (intraplate earthquakes). See *plate tectonics* and *intraplate*.

Magnitude/Frequency (occurrence) Relationship

Simply put, large events (*e.g.*, big earthquakes) occur less commonly than small ones and, further, that there often exists a distinct quantifiable relationship between size and frequency. Widely used in analysis of earthquake risk (see *seismic hazard asessment*) to estimate probablility of occurrence of different-sized earthquakes.

m asl

Abbreviation used to describe elevation in metres above sea level.

m bgs

Abbreviation of metres below ground surface.

Member

Stratigraphic unit subordinate to a *formation* and consisting of one or more *beds*.

Mercalli scale

A scale of earthquake intensity describing the degree of ground shaking and extent of damage, from I (detectable only by instruments) to XII (complete destruction). The energy released by an earthquake (*i.e.*, "size" of an earthquake) is given by the Richter scale (see *Magnitude*).

Metadata

Data describing properties of data such as source, where collected and by whom, with what purpose, accuracy, limitations, structure, *etc.*

Metamorphic rock

Rock whose original constituents or structure have been altered by burial pressures and/or heat within the Earth's crust. Increasing pressure and temperature result in a systematic change from low-grade to high-grade metamorphic rocks (*i.e.*, slate-phyllite-schist-gneiss).

Methane

A colourless, odourless gas (CH_4) that is lighter than air and which is produced by the decomposition of organic matter in an oxygen-poor environment. Methane is the main component of natural gas and is the main combustible component of a landfill gas. It is flammable in air at concentrations between 5 and 15 per cent by volume. See also *landfill gas*.

Middle Distillate

Refers to a particular group of hydrocarbons (*i.e.*, diesel oil, fuel oil No.2) produced by the distillation of crude oil at moderate ($160°$-$360°C$) temperatures.

Milankovitch Theory

Named after Serbian astronomer Milutin Milankovitch (1879-1958) who argued that major climatic changes (glacial/interglacial cycles) and formation of continental ice sheets (*e.g.*, *Laurentide Ice Sheet*) on planet Earth result from systematic, predictable variations in Earth's orbit. Recent work using oxygen isotopes has demonstrated the essential validity of the Milankovitch variables as a so-called climatic "pacemaker". The theory can be used for hindcasting and forecasting long-term natural climate changes.

MISA

Municipal Industrial Strategy for Abatement. A program developed by the *OMOEE* for the regulation of the quality of effluent and discharges, and intended to protect human health and the environment by enforcement of rigorous quality-control standards.

Mitigation Measures

Refers to methods that can be undertaken to lessen or modify potential negative impacts of an environmental undertaking. Steps which can be taken to reduce the potential negative impact which an alternative or alternative method might have upon the environment as defined in the Ontario *Environmental Assessment Act*.

Monitoring

Regular or spontaneous procedures used to methodically inspect and collect data on the performance of a landfill site relating to environmental quality (*i.e.*, air, leachate, gas, ground or surface water, unsaturated soils, *etc.*). Program may continue for several decades (or centuries) following site closure to ensure the facility complies with environmental regulations.

Monitoring Well

A water well used for the purpose of monitoring groundwater conditions.

Monitor Recovery

The change in water level in a *piezometer* or monitoring well toward static conditions as a function of time. This information can be analyzed to assess the *hydraulic conductivity* of the screened interval.

Moraine

Ridge-like topographic high underlain by glacial sediments deposited along the margin of an ice sheet or glacier.

MTSS

Metropolitan Toronto Sewer System

Multibeam Bathymetry

Device used to measure the depth (bathymetry) of a body of water employing multiple *echosounders* or ship-mounted fan-shaped arrays of acoustic beams. Water depth information is collected in a wide swath. Images or maps of the sea floor are produced from such information.

Multi-channel Seismograph

An instrument for amplifying, recording and manipulating seismic waveforms detected by an array of *seismometers* or *hydrophones*.

Multi-level Piezometer

A *piezometer* that allows ground-water properties to be measured and samples to be taken at multiple elevations in any one well. See also *piezometer nest*.

Multiple Liners

A system of layers of reworked natural soil or artificial materials or a combination of both, placed beneath or on the sides of a landfill or waste cell, that restrict the downward or lateral movement of solid waste, leachate or landfill gas. See *liner*.

Municipal Waste

Under Ontario Regulation 347, municipal waste refers to any non-hazardous waste, whether or not it is owned, controlled or managed by a municipality, except, hazardous waste, liquid industrial waste, or gaseous waste, and solid fuel. In a general sense, municipal waste refers to materials discarded by individuals in the course of their daily activities at home (*Domestic Waste*) and by business as a result of normal operating activities and not including industrial or hazardous wastes (see *ICI Waste*).

Municipal Well

Commonly a large-diameter well used for high-volume abstraction of ground water to service an urban area. Several wells may be drilled close together in a well field.

Mutagenic

Capable of producing genetic mutations in organisms.

N

Natural Attenuation

Refers to assumption that contaminants will be dispersed and diluted within sediments and their associated groundwater systems. Landfills have been widely constructed on tills and reliant on natural attenuation of leachate. Such substrates are now recognized as providing inadequate protection of ground-water resources. See *leachate collection system, liner, engineered landfill.*

NEC

Abbreviation for Niagara Escarpment Commission.

Neotectonic

Definitions vary, but usually applied to geologically recent (*neo*) tectonic activity within the last 5 million years (*i.e.*, after the Miocene Epoch: Appendix A).

NEP

Abbreviation for Niagara Escarpment Plan.

Neutron Log

Refers to technique of measuring downhole variation in the intensity of radiation produced when surrounding strata are bombarded by neutrons from a source lowered or raised in a borehole; indicates presence of fluids and degree of saturation.

Nitrification

A two-stage biological treatment process by which ammonia is converted first to nitrite and then to nitrate.

NMOCs

Abbreviation for Non-methane Organic Compounds, such as vinyl chloride, ethylbenzene, *etc.*, associated with decomposition of waste in a landfill and present in *landfill gas*.

Non-point Source

Areal source of *contaminants* that may enter ground water, *e.g.*, road de-icing chemicals, pesticides, *etc.* Contrast with *point source*.

Normal Move

The increase in two-way travel time of a reflection event as a result of increasing distance from the source or reflector dip in a *seismic reflection survey*.

Normal Moveout Correction

Correction applied to simulate a vertically incident ray path between shot and receiver (also known as dynamic correction) in *seismic reflection surveys*.

Nuisance Impact or Effect

A deleterious impact upon the environment which is generally considered less severe than a negative effect.

O

OBM

Ontario Base Map referring to a provincial government digital mapping program at a scale of 1:10,000 in southern Ontario and 1:20,000 in the north.

Obduction

The collision and overthrusting of one of the Earth's tectonic plates by another where *subduction* and related volcanic activity does not occur.

Observation Well

See *monitoring well.*

Odour Control

Minimizing or eliminating the nuisance and undesirable impact of objectionable or unpleasant odours arising out of waste disposal operations.

ODWO

Abbreviation for Ontario Drinking Water Objectives.

Official Plan

Title given to planning document used by municipal governments to identify current land use in an area or to guide future land use. See *OP*.

Offset

The horizontal distance between the shot point and a detector in a *seismic reflection survey.*

Oligochaete

Burrowing worm commonly present in sediments. Responsible for mixing of sediments and downward or upward transport of contaminants. See *ichnofossil.*

OMB

Abbreviation for the Ontario Municipal Board.

OMNR

Abbreviation for the Ministry of Natural Resources of Ontario.

OMOE

Ontario Ministry of the Environment (presently Ontario Ministry of Environment and Energy).

OMOEE

Ontario Ministry of Environment and Energy.

OMTO

Abbreviation used for the Ministry of Transportation of Ontario.

OP

Abbreviation for *Official Plan*, such as published by urban municipalities to identify current land use and future developments.

OPA

Abbreviation for Official Plan Amendment.

Ore

Any naturally occurring geologic material from which a mineral (usually metals, *e.g.*, copper, nickel, iron, *etc.*) can be extracted economically.

Organic Constituents

Constituent/compounds in leachate and ground water that contain carbon combined with hydrogen along with other elements (*e.g.*, phenol, benzene, toluene, trichloroethylene, *etc.*).

Organic Waste

Waste material which comes mainly from animal or plant sources. Organic waste generally can be consumed by bacteria and other small organisms. Inorganic wastes are chemical substances of mineral origin.

ORM

Oak Ridges Moraine; large area of upland glacial morainic topography in southern Ontario on the northern outskirts of Greater Toronto Area. It forms the principal recharge area for *GTA* rivers and hosts a ground-water resource of provincial significance.

ORMGTA

That portion of the Oak Ridges Moraine lying within the Greater Toronto Area.

Orphan Site

Former industrial, mining or other site where ownership cannot be traced and where costs of remediation must be borne by government.

Orogeny

Refers to large-scale deformation processes within the Earth's crust arising from the collision of tectonic plates. Usually, but not always, associated with formation of mountainous terrain (orogenic belts).

Outcrop

Refers to sites, such as cliffs or quarries, where rocks or sediments can be clearly observed and documented free of any *overburden*, soil or sediment cover.

Outwash

Catchall term used for sandy and gravelly deposits of glacial melt-water rivers. See also *glaciofluvial*.

Overburden

Sediments resting on bedrock: synonymous in mid-continent North America with glacial *drift*.

Overconsolidated

Term used to describe *consolidated* sediment or rock that formerly has experienced high compressive stresses under the weight of overlying strata (or an ice sheet), but is now exposed at surface under low stresses. Such materials are prone to breakage (see *joint*) and an increase in moisture content resulting in a loss of strength and enhanced ability to transmit water. See *porosity, secondary porosity and permeability*.

Oxidation

Oxidation is the addition of oxygen, removal of hydrogen, or the removal of electrons from an element, or compound. In waste-water treatment, organic matter is oxidized to more stable substances.

P

P wave   See *body wave*.

Packer Test

Hydraulic testing performed in boreholes whereby a pair of inflatable seals (packers) are used to isolate the interval of the borehole to be tested. Typical tests include *slug tests* and *hydraulic head* measurements.

PAH

Polycyclic aromatic hydrocarbon, *e.g.*, pyrene, fluorene, anthracene and napthalene.

Paleoseismology

The study of prehistoric earthquakes involving detailed study of the geological record for evidence of ground shaking, faulting and liquefaction.

Particle Size Classification System

Note: A sediment that contains a wide range of particle sizes is poorly sorted; one where the grains are about the same size is well sorted. A common source of confusion is that in the geotechnical literature, the term "well graded" refers to a poorly sorted sediment, with a wide range of particles present.

Particle Size Classification System, *cont'd.*

Name	Qualifying Term	Particle Size (mm)
Boulders		>256
Cobbles		60-256
Pebbles		4-60
Granules		2-4
Sand	Coarse	0.5-2.0
	Medium	0.25-0.50
	Fine	0.06-0.25
Silt		0.004-0.06
Clay		<0.002

Pathogenic Organism

Bacteria, viruses or cysts which can cause disease (typhoid, cholera, dysentery). There are many types of bacteria which do not cause disease and which are not pathogenic. Many beneficial bacteria are used in waste-water treatment processes to clean up organic wastes (see *bioremediation*).

PCB

Polychlorinated biphenyl.

Peneplain

Featureless topographic surface of low relief cut across bedrock and usually of considerable regional extent. The exposed and buried surface of the Canadian Shield is a peneplain of continental scale.

Perched Ground Water

Water contained in an isolated saturated zone lying above and separated from the underlying regional *water table*. Arises because of the presence of a layer of low permeability which retards downward movement of water. Common in glacial sediments where there is rapid vertical and lateral changes in sediment type.

Periglacial

Pertaining to processes, sedimentary deposits and associated structures that form under cold climates either in the absence of, or peripheral to glaciers and ice sheets. Sediments are exposed to deep freeezing, contraction cracking and other deformations produced by severe cold and mechanical disturbance arising from the growth of subsurface ice masses (see *permafrost*). Typical of much of the present-day Canadian sub-Arctic and Arctic.

Permafrost

Term used for perennially frozen ground where summer thawing is limited to shallow depth (usually less than one metre; the *active layer*). Permafrozen rocks or sediments are often characterized by the growth of subsurface ice masses (see *ice wedge cast*). Freezing can extend to depths of several hundred metres or more. Much permafrost in Canada is relict and results more from the severe climate conditions during the last glaciation. See also *talik*.

Permeability

The capacity of a porous medium to transmit a liquid or gas subjected to an energy gradient (in the case of water, *hydraulic gradient*). Permeability is the result of the specific *in situ* properties of the sediment or rock such as grain size, grain sphericity, roundness and packing, presence or absence of fractures, *etc*. Pores have to be interconnected to create a permeable material. Materials of high *porosity* may not be capable of transmitting fluids (*i.e.*, are impermeable) if pores are not connected. Conversely, materials of low porosity may be permeable if pores are connected. See also *porosity*.

Petroliferous

Containing or yielding petroleum, which is defined as material occurring naturally in the earth composed predominantly of mixtures of chemical compounds of carbon and hydrogen with or without other non-metallic elements such as sulphur, oxygen, nitrogen, *etc.* and existing in gaseous, liquid or solid phase.

Pesticide

General term used for chemicals used as herbicides, insecticides, fungicides, nematicides and rodenticides. About 120 are in use in Ontario, the most prevalent being triazine herbicides, which are a widely reported contaminant in ground waters in rural areas.

pH

Acidity indicator; a pH of 4 and below is acidic, above 7 is alkaline. Identifies concentration of H^+ ions in a chemical solution (*e.g., water*). The mobility of many *contaminants* is dependent upon pH of ground and surface waters.

Photo-oxidation

Use of hydroxyl radical (OH^-) as an oxidizing agent to oxidize organic compounds in contaminated ground water.

Piezometer

A small-diameter tube set in the ground which allows for the determination of the elevation of the ground-water table (see *static water level*). It is usually constructed of plastic with a slotted screen at the bottom that permits inflow of water, but not geologic materials; it is sealed above the screened section and is open to the atmosphere at the top.

Piezometer Nest

Refers to a number of piezometers completed at various depths in closely spaced boreholes at one location. Typically used where hydrogeological information is required from different geological strata at depth. See also *multi-level piezometer*.

Piezometric Surface

Older term, now disfavoured, for *potentiometric surface*.

Piston Corer

A type of coring system for collecting subsurface samples of sea or lake bed sediments. It consists of a long narrow tube, fitted with a plastic liner, with a heavy weight on top. The system is dropped to the sea or lake bed and penetrates the bottom sediments. A movable piston inside the tube assists in retaining a complete core.

Pixel

Picture **el**ement; the smallest element in a raster image, usually square.

Plate Tectonics

A theory of how planet Earth works as a large machine. The Earth's lithosphere (crust and outer mantle) is broken into large tectonic plates which are produced by volcanism at mid-ocean ridges, interact with one another at their boundaries (producing earthquakes) and are destroyed by being pushed and dragged down subduction zones where one plate overrides another. This process is relentless, has obtained for much of Earth history and has conditioned the biological and climatic evolution of planet Earth. Plate movement is driven by internal heat produced by radioactive decay and convection in the deep mantle; the rate of plate motion will eventually slow and the plate tectonic activity will cease.

Pleistocene

Last 1.6 million years of Earth history, characterized by major global climate changes and episodic growth of ice sheets (see Appendix A). Terminated 10,000 years ago with the commencement of the *Holocene* epoch.

Plume

Identifies three-dimensional form of contaminated ground water moving away from a point source of contamination (*e.g.*, a landfill), or non-point source (*e.g.*, road salt applied to a road). See also *leachate plume*. Analogous to a plume of smoke drifting away from a fire.

Pockmark

Sea- or lake-floor pit or crater caused by escape of gas or fluids from buried strata and subsequent collapse of overlying sediments.

Point Source

Localized source of contaminants that may enter ground water, *e.g.*, underground tanks of a gas station or dry-cleaning store.

Pollutant

Refers to any *contaminant*, in addition to viruses or bacteria, harmful to biological organisms and which has a negative impact on *ecosystem integrity*.

Pollution

Contamination of the environment (ecosystem) such that individual organisms are no longer capable of fully serving their original functions.

Polyethylene

A partially crystalline lightweight thermoplastic that is resistant to chemicals and moisture. It is available in high-density (HDPE) or low-density (LDPE) forms. Commonly used as a *liner* below landfills.

Pop-up

Geologic term for upward buckling of near-surface layers of rock in response to horizontal compressive stresses resulting from movement of the planet's tectonic plates. Buckling occurs rapidly and commonly produces a small magnitude earthquake.

Porosity

The ratio of the volume of voids to the total volume of a sediment sample, *i.e.*, the fraction of a unit volume of sediment that is void space. Usually separated into *primary* and *secondary porosity*. Primary porosity is that of the parent unaltered rock or sediment; secondary porosity results from weathering, disturbance, jointing, dissolution of cements and results in an increase in porosity. Where pores are interconnected, the material is said to be permeable and can transmit and store fluids and gases (see *permeability*).

Pore Space

Refers to open pores and voids within a sediment and which may contain water or gas at various pressures. See *porosity*.

Pore-water Pressure

Stress transmitted through the pore water.

Potential Evapotranspiration

Amount of water that would evaporate from soil surface and be transpired from vegetative covers if water were available in unlimited supply. See *actual evapotranspiration*.

Potentiometric Contours

Lines connecting points of equal potentiometric pressure on a map (see next entry).

Potentiometric Surface

An imaginary surface to which water in a *confined aquifer* would rise by hydrostatic pressure, which is defined as the pressure exerted by the water at any given point in a body of water at rest. The hydrostatic pressure of ground water is generally due to the weight of water present at higher levels in the same zone of saturation. When the potentiometric surface is above the topographic ground surface, *artesian* conditions obtain. The *water table* is a potentiometric surface for an *unconfined aquifer*.

ppm

Parts per million; equivalent to one thimbleful of a contaminant in five bathtubs of water.

ppb

Parts per billion; equivalent to one thimbleful of a contaminant in two olympic-sized swimming pools.

ppt

Parts per trillion; equivalent to one thimbleful of a contaminant in 2000 olympic-sized swimming pools (or the contents of about 1600 one-litre milk cartons in Lake Ontario).

Precipitate

To separate (as substance) out in solid form from a solution, as by the use of a reagent. The substance is precipitated.

Preliminary Treatment

The removal of metal, rocks, rags, sand and similar materials which may hinder the operation of a waste-water treatment plant. Preliminary treatment is accomplished by using equipment such as racks, screens and grit removal systems.

Pre-submission Consultation

In Ontario, this is the process of consultation among the proponent, the review co-ordinator, various government ministries and the public, before the formal submission of an *Environmental Assessment*. It is intended to perform the following functions:

a) identification of most important concerns;

b) location of other relevant reports or policies;

c) development of awareness of all required approvals, licenses or permits;

d) provision of advice to proponents on preparing the EA document;

e) sharing of information which the review agencies possess;

f) discussion of requests for further information and analysis; and

g) establishment of a project timetable.

Primary Porosity

Essentially the amount of intergranular pore space present in a rock or sediment. If pore spaces are connected, the material is permeable, but materials of high porosity may be poorly permeable if pore spaces are not interconnected. *Secondary porosity* results from fracturing or dissolution of the material.

Primary Treatment

A waste-water treatment process that takes place in a rectangular or circular tank and allows those substances in waste water that readily settle or float to be separated from the water being treated.

Principal Component Image

Digitally processed image produced by a transformation that recognizes maximum variance in multispectral images.

Priority Contaminants

List of contaminants identified by government agencies as having the most severe effects on water quality and aquatic biota.

Private/Domestic Well

A water well used for private household or farm water supplies. Contrast with *municipal well*.

Progradation

The building out of a beach or delta as a result of sediment deposition or a fall in lake or sea level. See *regression*.

Proponent

A proponent is the person, agency or government ministry who carries out or proposes to carry out, an environmental *undertaking*, or is the owner or person having charge, management or control of an undertaking.

Public Hearing

A quasi-judicial process, whereby the public or any affected parties have the opportunity to voice concerns or otherwise address studies and the planning process carried out by the proponent.

Public Record

Section 30 of the Ontario *Environment Assessment Act* states that a record which is available for public inspection shall be kept of every environmental undertaking. This record is commonly referred to as the public record. It is an ongoing record of each EA and consists of the EA, the review of the EA, any written submissions by the public, or any decisions by either the Environmental Assessment Board or the *Joint Board*, or the minister, together with written reasons for them and any notices and/or orders made by the minister relating to the EA Act. The record may also contain materials from *pre-submission consultation* and information required to be submitted pursuant to an EA for an approved undertaking or pursuant to any conditions of approval.

In Ontario, complete public records are maintained by the Environmental Assessment Branch, 250 Davisville Avenue, Toronto. Regional offices of the ministry also have copies of public record documents which pertain to that region. Public records with equivalent information are also maintained for exemptions at the EA Branch.

Pump-and-Treat

Use of pumping wells to remove contaminated ground water thereby allowing treatment.

Pump Test

A procedure whereby ground water is pumped from a *borehole* and ground-water levels in the pumping wells and nearby observation wells are monitored. The information obtained from the pump test can be analyzed to assess the *transmissivity, storativity* and *hydraulic conductivity* of the *aquifer*.

Purge Well

A drilled well equipped with pumps to allow pumping down of the *water table* or *leachate* level to promote flow toward the well. Typically, a well placed close to the site boundary of a landfill and used to prevent or control the off-site migration of a *leachate plume*.

Putrescible Wastes

Wastes which rot or decay.

Quaternary

The youngest period of the Cenozoic Era (see Appendix A). The Quaternary is subdivided into the *Pleistocene* and *Recent* (or *Holocene*) epochs. It comprises all geologic time and deposits from the end of the Tertiary (approximately 1.6 million years ago) up to, and including, the present.

Quick Clay

Clay sediment of *glaciomarine* origin noted for its ability to liquefy (*i.e.*, change from a solid to liquid state) when shaken by earthquakes or disturbed during construction. Prevalent along St. Lawrence River Valley, in coastal Alaska and in western Norway. All these areas have experienced *glacio-isostatic rebound* and the uplift of glaciomarine clays above sea level. Properties result from presence of clay-sized, but non-cohesive minerals and high water content. See *cohesion, liquefaction*.

R

Radiation

Refers to alpha (α-), beta (β-), gamma (γ-) or X-rays, also neutrons, high-energy electrons, protons and other atomic particles, but not radio or sound waves, visible, infra-red or ultraviolet light.

Radioactivity

Refers to ability of some elements, such as uranium, for the nucleii of atoms to decay spontaneously and emit alpha, beta and/or gamma rays.

Radon

Inert, naturally occurring radioactive gas produced by decay of uranium (U) and thorium (Th). Radon (Rn) has a half life of 3.8 days and decays into radioactive polonium (Po) which is associated with lung cancer if inhaled.

RAP

Remedial Action Plan; a plan of work involving government agencies, the private and educational sectors, and the public designed to restore and remediate areas of environmental degradation. See *Area of Concern, remedial action.*

Raster Image

An image based on rows and columns of *pixels.*

Ratio Image

An image prepared by processing digital multi-spectral data as follows: for each *pixel*, the value of any one spectral band is divided by that of another. The resultant digital values are displayed as an image.

Reactive Wall

Refers to a deep trench filled with reactive material placed down-gradient of a plume of contaminated ground water.

Reasonable Use Policy

A policy developed by the *OMOEE* (Policy 15-08) to stipulate limits to the level of ground-water quality impairment that may be permitted to occur at site property boundaries, to allow the reasonable use of adjacent properties or land without adversely affecting public health and the environment. Policy differs for health- and non-health-related chemical parameters. Policy designed to prevent any contaminant originating from a landfill site from causing an increase, on or under adjacent properties, in the maximum allowable concentration (C_m) for that contaminant within a period of 1000 years.

$$C_m = C_b + x(C_r - C_b),$$

where C_r is the *drinking-water objective* for the contaminant in question, C_b is the background concentration of the contaminant prior to landfilling, and x is a constant (= 0.5 for non-health-related contaminants and 0.25 for health-related contaminants).

Recent

Synonymous with *Holocene, i.e.,* the latest and continuing epoch of Earth history that started approximately 10,000 years ago (Appendix A). Broadly equivalent to time since disappearance of last continental ice sheet, *i.e.,* the present interglacial. Term is always capitalized.

Receiving Water

A stream, river, lake or ocean into which treated or untreated waste water (*effluent*) is discharged.

Recharge

The downward movement of surface precipitation (*i.e.,* rain, snow melt) to the water table and underlying saturated zone (see also *interflow*). Essentially the process by which surface water becomes ground water. See also *infiltration.*

Recharge Area

The portion of the drainage basin in which the net flow of ground water is directed downward, away from the water table. See also *zone of capture, wellhead protection.*

Recycling

Important component of "3Rs" program (Reduce, Reuse and Recycle). Refers to sorting, collecting or processing waste materials that can be used as a substitute for the raw materials in a processor activity for the production of (the same or other) goods. For example, the Blue Box system, in-plant scrap handling, or raw material recovery systems. Recycling is also the marketing of products made from recycled or recyclable materials.

Recycling Facility or Plant

A facility where recycling of used or waste material is carried out.

Redox

Means reduction-oxidation, where one atom or molecule receives an electron (and is reduced) and where another gives up an electron (and is oxidized).

Reduction (of waste or component of 3Rs program)

Those actions, practices or processes which result in the production or generation of less waste (see *recycling*).

Reduction-Oxidation Reactions

Using chemical reagents to destroy organic contaminants such as *PCB*s.

REE

Rare Earth Element. Refers to oxides of a series of fifteen metallic elements from lanthanum (La) to lutetium (Lu), and of three other elements: yttrium (Y), thorium (Th) and scandium (Sc).

Reflecting Horizon

A subsurface geologic layer or structure capable of reflecting seismic waves and thus being recorded by a *seismograph.*

Regression

Refers to offshore movement of shoreline over time in response to lowered lake or sea level. Associated with *progradation* of shoreline deposits such as deltas or beaches. Opposite of *transgression.*

Rehabilitation

The treatment of land from which aggregate has been excavated so that the use or condition of the land is restored to its former use or condition or is changed to another use or condition that is or will be compatible with the use of adjoining land.

Remedial Action

Corrective action taken to clean up or remedy a spill, an uncontrolled discharge of a contaminant, or a breach in a facility or its operations, in order to minimize the consequent threat to public health and the environment.

Remote Sensing

Collection and interpretation of information about an object without being in physical contact with that object, *e.g., geophysics.*

Representative Sample

A small portion of soil, water, *etc.,* which can be subjected to testing and analysis, that is expected to yield results that will reliably represent the identical characteristics of the *in situ* material or of a larger body of material. Note that the *bulk properties* of a rock or sediment can be completely different from that of a small sample collected for laboratory analysis. Laboratory data must be interpreted with care.

Resistivity Log

Diagram showing downhole variation in ability of sediments or rocks penetrated by a drill hole to conduct an electrical current; fine-grained, clayey materials are less resistant that coarser grained materials. See also *electrofacies*.

Resource Recovery

Salvaging of valuable resources from waste material. Resource methods include recycling of used products to provide material for manufacturing and the conversion of waste to energy (*EFW* or Energy From Waste).

Retardation

A term used for processes that act to remove contaminants and other solutes in ground water by chemical transformations. As a result contaminants will move more slowly than advecting ground water. See *advection, contaminant migration path, plume, conservative, attenuation*.

Reuse (component of 3Rs program)

The use of an item again in its original form, for a similar purpose as originally intended, or to fulfil a different function (see *recycling*).

Revetment

A protective wall built to protect and stabilize eroding coastlines or slopes. Also known as a berm.

Richter scale

See *Magnitude* and *Mercalli scale*.

Rift

Geologic term for a narrow trough bounded by normal *faults* and resulting from regional extensional stresses where the Earth's crust is pulled apart and broken (rifted). Rifting is the initial process by which continents may split, intially forming narrow seaways and eventually, large oceans. Not all rifts develop into oceans; these so-called failed rifts (or aulacogens) become filled with younger sediments and become deeply buried. They are of interest to environmental geologists because associated faults may be reactivated many hundreds of millions of years after rifting stopped to produce *intraplate* earthquakes where the regional tectonic setting suggests apparent stability. In Canada, the St. Lawrence River valley marks the site of a former rift which may extend into Lake Ontario. The Ottawa River valley is another such rift.

Riparian Vegetation

Streamside vegetation which provides temperature control (shading), habitat diversity, bank stability, food and shelter to aquatic organisms and their habitats.

Rotational Landslide

Describes landslides in cohesive fine-grained sediments involving the downslope movement of large blocks of coherent sediment along curved failure surfaces.

ROV

Remotely operated vehicle.

Runoff

The part of precipitation (rain water, snow melt) that flows overland to rivers, lakes or oceans and does not infiltrate the surface material (soil or rock).

S

S wave

See *body wave*.

Sand

See *particle size classification*.

Sanitary Sewer

A sewer intended to carry waste water from homes, businesses, and industries. Storm-water runoff should be collected and transported in a separate system of pipes. See also *combined sewer*.

Sanitary Landfill

As originally applied (*circa* 1970s) a landfill where a daily layer of clean fill (*i.e.*, soil) is used to minimize odours and to control pests.

Satisficing

Term used in *environmental assessment* for initial process of "narrowing down" or screening of *alternatives* (*i.e.*, methodologies, locations, technologies, *etc.*) using strict criteria. Examples are where a landfill has to be sited on an area of low permeability and the site geology has to satisfy a specified minimum criteria such as a certain value of permeability, where an airport has to be sited a minimum distance from an urban area, or where a particular cleanup technology has to achieve a maximum allowable concentration of contaminants in soil or water. A site is either acceptable or unacceptable on the basis of the chosen criteria.

Saturated

Where the interstices or pores of a material are filled with a liquid, usually water. It applies whether the liquid is under greater than or less than atmospheric pressure, so long as all connected interstices are full.

Saturated Zone

The subsurface zone in rock or sediment soil where all voids are filled with water, *i.e.*, in general, conditions below the ground-water table. See also *unsaturated zone*.

Sea-floor Permafrost

Refers to those areas in the far north where sea floor is underlain by permanently frozen sediment or rock (see *permafrost*). Usually the result of rising sea level and flooding of deeply frozen coastlines.

Secondary Porosity

An increase in porosity due to such phenomena as weathering, mechanical disturbance, dissolution of cements holding grains or particles together and/or fracturing. See *primary porosity*.

Secondary Treatment

A waste-water treatment process used to convert dissolved or suspended materials into a form more readily separated from the water being treated. Usually the process follows primary treatment by *sedimentation*. The process commonly is a type of biological treatment process followed by secondary clarifiers that allow the solids to settle out of the water being treated.

Sediment

Geological term for non-cemented (unlithified) materials (*e.g.*, gravels, sand, clays, *etc.*). In areas that have experienced glaciation, the terms *drift* or *overburden* are commonly used. Contrast with use of the term *soil* by engineers. The two terms are used interchangeably in this volume, except where specified.

Sedimentary

Broad range of rock types deposited by chemical precipitation or sedimentation and cementation of sedimentary particles such as mineral grains at, or close to, the Earth's surface.

Sedimentation (geologic)

Naturally occurring geological process whereby mineral or rock particles are transported and deposited in rivers, lakes, seas, by wind, glaciers, *etc.* to ultimately produce *sedimentary* rocks.

Sedimentation (waste-water treatment)

The deposition of fine-grained soil in an undesirable location, as a result of scouring, erosion and transportation of earth materials by urban runoff from streets, *etc.*

Sedimentation Pond

An impoundment within a natural topographic depression, man-made excavation, or dike arrangement that is used to control and minimize sedimentation off-site that would cause an adverse environmental effect.

Sediment Focussing

Transport of sediment and its adsorbed contaminant burden toward the deepest parts of a basin by repeated cycles of resuspension and gravity settling.

Seiche

Refers to sudden and short-lived episode of high water and accompanying flooding, on the downwind side of large lakes, caused by strong winds piling up water (to heights of several metres) above normal seasonal levels. French term used originally to describe such episodes on Lake Geneva in Switzerland; common process along the shores of the Great Lakes. In June 1954, a seiche 2.5 m high killed seven people in Chicago. An equivalent term is "wind setup."

Seismic Hazard Assessment

Procedure carried out to determine the risk to society and infrastructure (*e.g.*, roads, power plants, *etc.*) posed by earthquake activity. One of two approaches is usually followed. The "deterministic" approach tries to recognize the most severe "worst-case" earthquake that may affect an area (see *seismotectonic province*) and to design infrastructure accordingly. The "probabilistic" approach identifies a range of earthquake magnitudes and identifies the probability of occurrence of each (see *magnitude/ frequency relationship*) allowing policy makers and regulators to define acceptable risks.

Seismic Reflection Survey

Employing the return of a seismic wave, generated at the ground surface (or close to it) by either an explosion (shotgun, dynamite, *etc.*) or a sledgehammer striking a metal plate, from a subsurface reflecting horizon back to the ground surface. Returning waves are recorded on a geophysical instrument known as an exploration *seismograph* to map (profile) subsurface structures and geologic layers. Can be used for shallow investigations (<100 m) or deep (tens of km). Natural seismic waves produced by large earthquakes (and man-made waves produced by nuclear explosions) are used to map the deep structure and composition of planet Earth.

Seismic Source Zone Characterization

This is a fundamental first step in *seismic hazard assessments* where geological structures, such as faults, are mapped and their past activity identified (see *active fault*). This procedure is used to identify *seismotectonic provinces*.

Seismic Velocity

The rate of propagation of an elastic wave, usually measured in $m \cdot s^{-1}$; typically up to 2000 $m \cdot s^{-1}$ for materials such as dense till, and 5000 $m \cdot s^{-1}$ for bedrock.

Seismic Wave

Energy released naturally during an earthquake or during man-made explosion (see *seismic reflection survey*). Earthquake seismic energy is divided into *body waves* that pass through the interior of the Earth, and surface waves that travel through the surface.

Seismograph

Recording instrument and associated *geophones* that magnify and record seismic waves produced either naturally by an earthquake or by man (explosions) in order to determine subsurface geologic materials and structures. In the latter case it is referred to as an exploration seismograph. See *seismic reflection survey*. A seismogram is the record made by a seismograph used to record natural earthquakes.

Seismology

The scientific discipline relating to the study of earthquakes.

Seismotectonic Province

Term used in seismic risk assessment for a geographic region of similar geological and seismological characteristics and thus assumed to have uniform potential for earthquakes.

Semi-orphan Site

Former industrial, mining or other site where partial ownership can be traced.

Semi-volatile

Compounds that exhibit moderate vapour pressures and are capable of being easily dispersed through soil or air as a result of evaporation.

Sensitivity Analysis

An analysis that tests the results of varying the values, data or assumptions used in a calculation of the environmental impact of any *undertaking* or *alternative* within an expected range of uncertainty.

Sensor

A device that receives electromagnetic radiation and converts it into a signal that can be recorded and displayed as either numerical data or as an image.

Septic

This condition is produced by *anaerobic bacteria*. If severe, the waste water turns black, gives off a foul odour, contains little or no dissolved oxygen and creates a heavy oxygen demand.

Septic Tank

Underground tank used in areas not served by municipal sewers for residential waste-water treatment. Solids are trapped in a tank where they undergo *biodegradation* by *anaerobic bacteria*. Waste water (*effluent*) spills over into tile beds where it is treated by *aerobic* biodegradation and *adsorption*. Commonly contaminates shallow ground waters; present practice is to ensure that the lot size is sufficiently large to reduce any impact at lot boundary.

Settlement

In waste management, the subsidence or compaction of a landfill or waste cell under its own weight. Also used in a geological sense for post-depositional compaction of sediment.

Shear Strength

The internal resistance offered by a sediment or rock to applied shear stresses. It is measured by the maximum shear stress that can be sustained without failure (breakage). Two extreme conditions are recognized: drained conditions, under which the changes in stress are applied so slowly with respect to the ability of the soil to drain that no excess pore pressures develop; and undrained conditions, under which the stresses are changed so rapidly with respect to the ability of the soil to drain that no dissipation of *pore-water pressure* takes place. See *undrained test*. Shear strength can be measured using many field and laboratory devices (see *triaxial compression test, and vane test*).

Shot
Explosive or other energy source used to introduce elastic waves into the subsurface.

Shot Gather
Seismic record containing the traces associated with a common shot point.

Sidescan Sonar
A system that produces images of the sea or lake floor using fan-shaped acoustic beams transmitted from a boat. Sediment texture and bed topography can be assessed using this technique. The images so produced (called sidescan sonograms) resemble air photographs taken from a plane at low sun angles.

Sievert (Sv)
Measure of radiation dose or biological effect arising from presence of radioactive materials in various substances. For example, eating one carrot (which contains potassium-40 from soil) gives a radiation dose of about 0.1 mSv.

Silt
See *particle size classification*.

Sinkhole
Enclosed depression created by collapse of near-surface strata into underlying voids or caverns, such as those created by *karst* processes in limestone.

Site Capacity
The maximum amount of waste that is planned to be disposed (design capacity) or that has been disposed of at a landfill site.

Site Characterization
Term referring to process where the geology and hydrogeology of a site is identified using a wide range of techniques such as boreholes, outcrop study, geophysics.

Site Closure
The planned and approved cessation or termination of landfilling activities at a landfill site upon reaching its site capacity.

Site Life
Normally refers to the period of time a landfill site is actively receiving wastes for land disposal, up to, and including, site closure, when it reaches the limits of its approved capacity.

Sludge
Refers to any solid, semi-solid, liquid, waste generated from municipal, commercial or industrial waste-water treatment plants.

Slug Test
A method of determining *in situ* hydraulic conductivity by causing an instantaneous change in the water level by adding or removing a known volume of water in a well and measuring recovery time.

Slurry Bioreactor
The placing of excavated contaminated soil into an open lagoon with nutrients and bacteria to optimize *biodegradation*.

Slurry Wall
A narrow (<1 m), deep (many metres) trench filled with bentonite clay slurry and commonly intended to prevent contaminated ground water from moving out from a contaminated site. See also *reactive wall, isolation*.

Soil
Geotechnical engineers use this term for any material resting on bedrock and regardless of its thickness, *i.e.*, materials for which geologists employ the terms *sediment, overburden, drift*, or, in some areas, *landfill*. Geologists and others also use the term soil specifically for near-surface weathered materials, usually no more than about 1 m thick, that can support plant growth (synonymous with topsoil). The terms sediment and soil (engineering sense) are used interchangeably in this volume, except where specified.

Soil Vapour Extraction
Use of air pumped from a well to enhance removal of organic contaminants in sediment.

Soil Washing
Use of water to remove inorganic and organic contaminants from sediment.

Solidification/Stabilization
Refers to addition of reagents such as cement, lime and fly-ash to contaminated sediments to form a stable solid mass. Commonly used to treat soils contaminated with heavy metals.

Solute
Dissolved ions and molecules in ground water or surface water.

Solvent Extraction
Use of solvents to solubilize and remove contaminants from sediments.

Sparging
To flush air through contaminated soil lying below the water table. Promotes oxygenation and bacterial growth (see *bioremediation*) and mobility of hydrocarbons that can then be captured and removed as a result of volatilization.

Spatial Resolution
In remote sensing, the smallest area being sensed, usually represented by one *pixel*.

Specific Discharge (v)
The quantity of ground water flowing through a porous media per unit area perpendicular to the flow direction. The equation for specific discharge is $v = Q/A$, where Q is the volumetric flux and A is the area perpendicular to flow. The usual units are metres per second ($m \cdot s^{-1}$) or metres per year ($m \cdot a^{-1}$).

Specific gravity
The ratio of the weight in air of a given volume of solids at a stated temperature to the weight in air of an equal volume of distilled water at a stated temperature.

Specific Yield
Refers to volume of water that can be drained from an *aquifer* under the influence of gravity. Some water will remain trapped by surface tension forces in pore spaces and is termed specific retention.

Spectral Reflectance
Reflectance of electromagnetic energy at specified wavelength intervals.

Spectral Resolution
The range of the electromagnetic spectrum that is being sensed.

Spline
A type of mathematical function used to approximate curves and surfaces.

Split-Spoon Sample
Sampling spoons for exploratory borings commonly consist of a pipe with an inside diameter of about 35 mm and an outside diameter of about 51 mm. The pipe is split lengthwise. Consequently, the sampler is called a split spoon. While the sample is being taken, the two halves of the spoon are held together at the ends by short pieces of threaded pipe. One piece serves to couple the spoon to the drill pipe. The other, which has been sharpened, serves as the cutting edge while the spoon is driven into the soil.

SPOT

Système Probatorie d'Observation de la Terre. French series of resource satellites operating from 1986 to present. Either 10-m or 20-m spatial resolution operating in the visible and near infra-red spectral regions.

Spread

A layout of *geophones* used in a *seismic reflection survey* to record data from a single shot.

Stabilize

To convert to a form that resists change. Organic material is stabilized by bacteria which convert the material to gases and other relatively inert substances. Stabilized organic material generally will not give off obnoxious odours.

Stadial

Climatic term for short-lived episode, during a glacial period, of severe cold and typically a few hundred years in duration. Ice sheets thicken and ice margins advance during a stadial. Stadials alternate with *interstadials* when climate warms and ice sheet thins and retreats. Final disappearance of an ice sheet marks the beginning of *interglacial* conditions.

Static Corrections

Removal of time shifts in seismic records produced by surface topography and low velocity near surface layers (weathering layer).

Static (Water) Level

Water level in a well that is not being affected or disturbed by withdrawal of ground water (see *piezometer*). Synonymous with the *water table*.

Steam Injection

Refers to injection of steam into subsurface to allow removal of organic contaminants.

Storativity

This term (sometimes called storage coefficient) is used to refer to the volume of water contained in a unit volume of an aquifer. In an *unconfined aquifer*, storativity is equivalent to *specific yield*.

Storm Sewer

A separate sewer that carries runoff from storms, surface drainage, street wash and snow melt, but does not include domestic sewage and industrial wastes. See also *combined sewer*.

Storm Water

Runoff that occurs as a direct result of a storm event or thaw. Storm water in urban areas is usually contaminated with pollutants washed from streets and other surfaces.

Storm-water Detention

Control of storm water by the construction of impoundments or structures for the purpose of regulating storm-water flows during high-intensity rainfall events, that would otherwise transport excessive amounts of sediment, cause soil erosion or cause flooding. Also used to control contaminants present in runoff.

Strata

General term used for multiple *stratigraphic units* composed of rocks or sediments regardless of origin, but most commonly applied to layered or bedded *sedimentary* rocks and sediments. Widely used as a catchall term for rocks and sediments below the ground surface (*e.g.*, subsurface strata) or exposed in outcrop.

Stratigraphic Unit

A single body of rock or sediment having lateral continuity and thickness and recognized as different from surrounding material on the basis of colour, fossil content, age, and many other properties or attributes (see *group, formation, member* and *bed*). See also *hydrostratigraphic unit*.

Stratigraphy

The study of the three-dimensional arrangement of geologic layers (see *strata*) in the subsurface. Also refers to the branch of geology which deals with the formation, composition, sequence and correlation of rocks and sediments in order to understand Earth history. A good understanding of stratigraphy is fundamental to hydrogeological investigation and remediation of contaminated sites.

STP

Sewage treatment plant.

Subglacial

Refers to processes or deposits (*e.g.*, *till*) that occur below the base of a glacier or ice sheet.

Subduction

Refers to the process whereby lithospheric plates are destroyed by undergoing deep underthrusting and melting below another lithospheric plate. Associated with volcanism and large magnitude earthquakes. Contrast with *obduction*. See *plate tectonics*.

Substrate

Term used to denote near-surface layers of a site, usually to depths of a few tens of metres (*i.e.*, shallow substrate).

Supercity

Term used for city containing more that 10% of national population.

Supernatant

Liquid removed from settled sludge. Supernatant commonly refers to the liquid between the sludge on the bottom and the scum on the surface of an anaerobic digester. This liquid is usually returned to the influent of the treatment plant or to the primary clarifier.

Surfactant

Material akin to detergent flushed through contaminated soils to mobilize hydrocarbons or metals. See *chemical flushing*.

Surge

Refers to very rapid advance of an ice sheet or glacier margin.

Suspended Solids

Solids that either float on the surface of, or are in suspension in, water, waste water or other liquids, and which are largely removable by laboratory filtering.

Sustainable

Use of land or a resource without the loss or reduction of *ecosystem integrity*.

Suture

Term used by geologists for regional tectonic boundary resulting from collision of tectonic plates (see *terrane, orogeny, plate tectonics, lineament*). Sutures usually consist of thrust faults and may be buried below younger strata. In mid-continent North America, buried sutures can be reactivated to produce intraplate earthquakes.

Syncline

A fold in which the strata dip inward from both sides toward the axis (in the form of a basin). Synclines may also be defined as folds with younger rocks toward the centre of curvature, providing the geological history has not been unusually complex. Contrast with *anticline*.

2.5-D or Quasi–2-D
A surface defined by XYZ co-ordinates, with one Z value for any XY co-ordinate to produce the appearance of three dimensions.

Tailings
Waste residue produced by the extraction of economic minerals from *ore*.

Talik
Unfrozen zone at depth with permanently frozen rock or sediment. Contains unfrozen water and may host locally important *aquifers*.

Talus
Loose, coarse rock fragments that are released from a steep slope and which accumulate by falling, rolling or sliding as at the foot of the slope. Synonymous with scree. Contrast with *colluvium*.

Teratogenic
Capable of producing deformities in organisms.

Terrain (physiographic)
Physiographic term used for describing areas of the Earth's surface having distinct topographic characteristics, such as relief or landforms, *e.g.*, mountain terrain, lowland terrain, *etc*. Glaciated terrain, for example, is distinguished by the presence of *till plains, drumlins, eskers, etc*. Commonly confused with *terrane*; the terms are not synonymous.

Terrane (geologic)
Geologic term used for a piece or block of the Earth's crust (tens to thousands of kilometres in extent, and extending to depths of several tens of kilometres) having distinctive geologic characteristics (*e.g.*, age, fossil content, rock types, structures, history) that distinguish it from adjacent terranes. Many continents have grown over geologic time by the addition and amalgamation (accretion) of terranes that have been transported by *plate tectonic* processes from great distances. Commonly confused with *terrain*; the terms are not synonymous.

Terrestrial Ecology
The study of relationships between organisms living on land (not water) and their environment.

Test Pit
Large hole dug to enable observation of near surface geologic conditions. Maximum practical depth is about 6 m.

Thermal Reduction
Use of a high-temperature, highly reducing hydrogen environment to destroy chlorinated organic compounds. The process is not considered *incineration* because oxygen is not involved.

Till
Specific term used for any poorly sorted admixture (see *diamict*) of *clay, silt, sand* and gravel, often with large boulders deposited directly by glacier ice. See *deformation till, lodgement till*.

Till Plain
Flat topography underlain by *till*; commonly drumlinized forming undulatory topography. Very common in glaciated terrains.

Time of Travel (TOT)
Time required for a contaminant to reach a ground-water well from a contaminant source, *e.g.*, a gas station or a municipal landfill. Can be depicted as *isochrones* (lines of equal travel time) on a map.

TIN
Topologically triangulated network.

Thalweg
The longitudinal profile of a river. German for "valley way".

TL Dating
Refers to use of mineral property called thermoluminescence (TL) to age-date sediments. Minerals that compose sediments emit thermoluminescence when subject to increasing temperature, but the TL signal loses its intensity when exposed to light (bleaching). When the mineral becomes buried during sediment deposition, the latent TL signal regenerates in response to ambient ionizing radiation. The present-day TL signal of sediments when measured in the laboratory reflects the time elapsed since burial (*i.e.*, the age of the sediment). See also *Carbon-14*.

Topography
Relating to surface form and the physical features of the land. See also *terrain*.

Topology
Mathematical relationship that determines spatial properties and relationships between features in the vector domain (*e.g.*, points, arcs and polygons).

Topsoil
Geological term for the uppermost layer of soil (see *soil*) containing appreciable organic materials in mineral soils, and typically having adequate fertility to support plant growth. Commonly shows distinct layers (see *"A", "B", "C" horizons*).

Total Petroleum Hydrocarbons
Common laboratory term for the entire group of petroleum hydrocarbons in a media (*i.e.*, soil or sediment) and usually reported as a concentration per unit of media.

Transfer Facility or Station
A facility where solid wastes are brought by smaller refuse collection vehicles and transferred to larger trucks to be hauled to a disposal site, processing facility or resource recovery facility.

Transgression
Refers to inland migration of shoreline over time, caused by rising lake or sea level. Opposite of *regression*.

Transport Pathways
The route taken by a particle or contaminant from source to sink.

Transmissivity
The rate at which water is transmitted through a unit width of aquifer under a unit hydraulic gradient. It is the *hydraulic conductivity* multiplied by the full thickness of saturated *aquifer*.

Triaxial Compression Test
A test in which a cylindrical specimen of sediment encased in an impervious membrane is subjected to a confining pressure and then loaded axially to failure. Measures the *shear strength* of rocks or sediments.

Tritium
This is a radioactive *isotope* of hydrogen (^3H) with a half-life of approximately 12.3 years. Only released into the natural environment by nuclear bomb tests (starting in 1942) and from nuclear energy plants; its presence in ground waters indicates a recent source. Before bomb testing, the concentration of tritium in rain water was 10^{-18}. A concentration of 1 atom of tritium per 10^{18} of ^1H is referred to as a tritium unit (T.U.).

TSS
Total suspended solids in water.

Tsunami

Sea wave produced by sudden movement of the sea floor during earthquake activity. These waves move at high velocity (up to 1000 km·h⁻¹) away from the source zone and, although of long wavelength (up to 200 km) and low height (less than 1 m), pile up in the shallow waters of coastal embayments to create devastating waves up to 30 m high. Particularly common around the Pacific Ocean. Derived from the Japanese term for "harbour wave"; commonly and erroneously called a "tidal wave."

Tuples

A relational table with two elements and n records.

Turbid

Typically applied to waters having a cloudy or muddy appearance as a result of suspended sediment.

Unconfined Aquifer

Also known as a water table aquifer. That part of an aquifer that is not overlain (confined) by an impermeable bed (see *aquitard, aquiclude*) and which receives direct *recharge* from infiltrating precipitation. The top of an unconfined aquifer (*i.e.*, the water table) commonly shows strong seasonal fluctuations in level.

Unconfined Compression Test

A special condition of a *triaxial compression test* in which no confining pressure is applied.

Unconformity

A distinct surface, often showing signs of weathering, separating different geologic strata and marking a gap in the geologic record characterized by weathering, erosion and non-deposition. May record many millions of years of erosion and non-deposition. See also *peneplain*.

Unconsolidated

Refers to loose sediment that is easily disaggregated. Contrast with *overconsolidated*.

Undertaking

In Ontario, an undertaking is an enterprise, activity or proposal, plan or program in respect of an enterprise or activity which a proponent initiates or proposes to initiate. The status of a given undertaking may be determined by consulting section 3 and clause 1 of the *Environmental Assessment Act*, and the regulations, exemption orders and approvals issued under the EA Act.

Undrained Test (quick test)

A soil test in which the water content of the test specimen remains practically unchanged during the application of the confining pressure and the additional axial (or shearing) force. See *shear strength*.

Unsaturated Zone (also Vadose Zone)

The zone in a porous soil or sediment, where the voids are not completely water-filled, but contain some air-filled voids. Limited above by the land surface and below by the water table.

Urban Area

Generally defined as a community having a population greater than 1000 persons and having a density of greater than 400 persons per square kilometre.

Urban Shadow

Term used to refer to land use pressures that urban areas exert on outlying rural regions such as caused by the need of the urban area for water aggregate and other resources, recreation areas, transport, waste disposal sites, *etc.*

UST

Underground storage tank.

UTM

Abbreviation for Universal Transverse Mercator, which is a type of map projection used to represent a spherical round Earth on a flat plane, *i.e.*, a map.

Vane Test

A test to measure the shear strength of a fine-grained cohesive soils and other soft deposits. A rod with four flat radial blades, or vanes, projecting at 90-degree intervals is forced into the soil and rotated; the torque required to rotate the rod is a measure of the material's *shear strength*.

Vector

In public health studies, a disease carrier and transmitter, usually an insect or rodent.

Vector Coverage

In remote sensing, a map theme of geographically stored attributes, points, lines and polygons.

Vertical Seismic Profiling

Involves placing *geophones* in a borehole to measure downhole variation in *seismic velocity* of surrounding strata; commonly to ground truth *seismic reflection profiles*.

Vibrocorer

A type of sediment-coring system consisting of a vibrating device which transmits energy to the core tube which assists in penetration of the coring tube into sediments (see also *piston corer*). Can be used on land or under water in wet, fine-grained materials.

Volatile

A volatile substance is one that is capable of being evaporated or changed to a vapour at a relatively low temperature.

Volatile Organic Compound (VOC)

Organic compounds which will readily volatilize (convert from liquid to gas phase) at conditions normally found in natural environments.

Voxel

Term used for **vo**lume **el**ement (see *pixel*); in other words, a 3-D pixel.

Vug

A small cavity in a vein or rock, usually lined with crystals of a different mineral composition from the enclosing rock. Common in *limestones* and *dolostones* and, where interconnected, capable of storing and transmitting large volumes of water and, at depth, oil and gas.

W

Warm-water Stream

A stream which is not capable of supporting salmonids (*e.g.*, trout and salmon species) since maximum summer water temperatures exceed 21°C. Contrast with *coldwater stream*.

Waste

Under the *Environmental Protection Act* of Ontario, waste includes ashes, garbage, refuse, *domestic waste, industrial waste*, or *municipal waste* and other used products as are designated in the regulations. In a general sense, waste may be commonly described as material intended for disposal or recycling.

Waste Disposal Site (see also *Landfill*)

Any land or land covered by water upon, into, in, or through which, or building or structure in which, waste is deposited or processed and any machinery or equipment or operation required for the treatment or disposal of waste.

Waste Management System

All facilities, equipment and operations for the complete management of *waste*, including the collection, handling, transportation, storage, processing and disposal thereof, and may include one or more waste disposal sites.

Water Balance

Amounts of water to various components in a system so that the amount of water entering the system equals the amount of water contained within and discharged out of a system. Typically applied to studies of regional ground-water systems aimed at identifying long-term sustainability of resource.

Water Quality Guidelines

Used by provincial, territorial and federal agencies to assess water-quality problems, manage competing uses of water, and protect public water supplies. Updates are issued perodically by the Water Quality Branch of Environment Canada. Guidelines are not legally enforceable, but include numerical criteria (Water Quality Objectives) representing a desirable level of water quality with regard to conventional parameters, metals, turbidity, pesticides, fertilizers, radioactive material and organic compounds.

Water Table

The surface of the *saturated zone* where hydrostatic pressure is in equilibrium with atmospheric pressure (see also *unsaturated zone*). Essentially the upper surface of an *unconfined aquifer* (or so-called "water table aquifer"). The water table is found by boring wells and measuring water levels.

Water Table Map

A map showing lines of equal elevation of the *water table* for any one area.

Weathering

A geological term used to refer to the destructive processes by which Earth materials on exposure to atmospheric agents are changed in colour, texture, composition, firmness, or form with little or no transport of the loosened or altered material; specifically, the physical disintegration and chemical decomposition of rock that produces sediments for transportation by surface water, ice, wind, *etc.* See *erosion*.

Weathered

A term used in chemistry to describe a compound, or mix of compounds, that has undergone changes (*i.e.*, leaching, volatilization) due to environmental factors.

Well Casing

The pipe that is used to construct a well.

Wellhead Protection

Refers to recharge zone around a pumping well that supplies ground water to that well or well field and the measures (regulation, legislation) taken out to prevent contamination of *recharge areas* by regulating land use.

YDSS

York-Durham Sewer System.

Zone of Attenuation (ZOA)

Term applicable to contaminants moving from a contaminant source (*e.g.*, a landfill or gas station) located within the *zone of capture* (ZOC) of a ground-water well or well field. Some contaminants are retarded (see *attenuation*) and so will be immobilized or reduced to acceptable levels, before reaching the well. The ZOA is defined by the use of *time of travel* (TOT) data. The zone of attenuation is thus smaller than the *zone of capture* for any one well or well field.

Zone of Capture (ZOC)

That part of an *aquifer* that recharges a well or well field.

Zone of Influence (ZOI)

See *cone of depression*.

APPENDIX C

UNITS OF MEASUREMENT

Prefixes for Multiples and Submultiples

With large or very small quantities, a set of prefixes is used with SI units.

Giga	1,000,000,000	= 10^9
Mega	1,000,000	= 10^6
kilo	1,000	= 10^3
hecto	100	= 10^2
Deka	10	= 10^1
deci	0.1	= 10^{-1}
centi	0.01	= 10^{-2}
milli	0.001	= 10^{-3}
micro	0.000001	= 10^{-6}
nano	0.000000001	= 10^{-9}
pico	0.000000000001	= 10^{-12}

UNITS OF MEASUREMENT

METRIC MEASURE	NON-METRIC MEASURE	CONVERSIONS

LENGTH

METRIC MEASURE	NON-METRIC MEASURE	CONVERSIONS
1 kilometre (km) = 1000 metres (m)	1 mile (mi) = 5280 feet (ft)	1 kilometre (km) = 0.6214 mile (mi)
1 metre (m) = 100 centimetres (cm)	= 1760 yards (yd)	1 metre (m) = 1.094 yards (yd)
1 centimetre (cm) = 10 millimetres (mm)	1 yard (yd) = 3 feet (ft)	= 3.281 feet (ft)
1 millimetre (mm) = 1000 micrometres (μm) (formerly called microns)	1 foot (ft) = 36 inches (in) = 12 inches (in)	1 centimetre (cm) = 0.3937 inch (in) 1 millimetre (mm) = 0.0394 inch (in)
1 micrometre (μm) = 0.001 millimetre (mm)	1 fathom (fath) = 6 feet (ft)	1 mile (mi) = 1.609 kilometre (km)
1 angstrom (Å) = 10⁻⁸ centimetres		1 yard (yd) = 0.9144 metre (m)
		1 foot (ft) = 0.3048 metre (m)
		1 inch (in) = 2.54 centimetres (cm) = 25.4 millimetres (mm)
		1 fathom (fath) = 1.8288 metres (m)

AREA

METRIC MEASURE	NON-METRIC MEASURE	CONVERSIONS
1 square kilometre (km²) = 1,000,000 square metres (m²) = 100 hectares (ha)	1 square mile (mi²) = 640 acres (ac) = 4840 square yards (yd²)	1 square kilometre (km²) = 0.386 square mile (mi²)
1 square metre (m²) = 10,000 square centimetres (cm²)	1 acre (ac) 1 square foot (ft²) = 144 square inches (in²)	1 hectare (ha) = 2.471 acres (ac) 1 square metre (m²) = 1.196 square yards (yd²) = 10.764 square feet (ft²)
1 hectare (ha) = 10,000 square metres (m²)		1 square centimetre (cm²) = 0.155 square inch (in²)
		1 square mile (mi²) = 2.59 square kilometres (km²)
		1 acre (ac) = 0.4047 hectare (ha)
		1 square yard (yd²) = 0.836 square metre (m²)
		1 square foot (ft²) = 0.0929 square metre (m²)
		1 square inch (in²) = 6.4516 square centimetres (cm²)

UNITS OF MEASUREMENT, *cont'd.*

METRIC MEASURE	NON-METRIC MEASURE	CONVERSIONS
VOLUME		
1 cubic metre (m³) = 1,000,000 cubic centimetres (cm³)	1 cubic yard (yd³) = 27 cubic feet (ft³)	1 cubic kilometre (km³) = 0.24 cubic miles (mi³)
1 litre (L) = 1000 millilitres (mL) = 0.001 cubic metre (m³)	1 cubic foot (ft³) = 1728 cubic inches (in³)	1 cubic metre (m³) = 264.2 gallons (US) (gal) = 35.314 cubic feet (ft³)
1 centilitre (cL) = 10 millilitres (mL)	1 barrel (oil) (bbl) = 42 gallons (US) (gal) = 34.9722 gallons (imperial)	1 litre (L) = 1.057 quarts (US) (qt) = 33.815 ounces (US fluid) (fl. oz.)
1 millilitre (mL) = 1 cubic centimetre (cm²)		1 cubic centimetre (cm³) (formerly cc) = 0.0610 cubic inch (in³)
		1 cubic mile (mi³) = 4.168 cubic kilometres (km³)
		1 acre-foot (ac-ft) = 1233.46 cubic metres (m³)
		1 cubic yard (yd³) = 0.7646 cubic metre (m³)
		1 cubic foot (ft³) = 0.0283 cubic metre (m³)
		1 cubic inch (in³) = 16.39 cubic centimetres (cm³)
		1 gallon (gal) = 3.784 litres (L)
		1 cubic metre (oil) (m³) = 6.28981 barrels (oil) (bbl)
		1 barrel (oil) (bbl) = 158.98284 litres (L)
Flow Rates		
1 cubic foot per second (ft³·s⁻¹) = 448.8 gallons per minute = 0.0283 cubic metres per second (m³·s⁻¹)		
1 cubic foot per minute = 4.72×10^{-4} m³·s⁻¹ = 7.4805 gallons per minute		
1 gallon per minute = 6.31×10^{-5} m³·s⁻¹		
1 cubic metre per second (m³·s⁻¹) = 2118.9 cubic feet per minute		
MASS		
1000 kilogram (kg) = 1 metric tonne (t)	1 short ton (sh.t.) = 2000 pounds (lb)	1 metric tonne (t) = 2205 pounds (avrdp.) (lb)
1 kilogram (kg) = 1000 grams (g)	1 long ton (l.t.) = 2240 pounds (lb)	1 kilogram (kg) = 2.205 pounds (avrdp.) (lb)
	1 pound = 16 ounces (avoirdupois) (oz) = 7000 grains (gr)	1 gram (g) = 0.03527 ounce (avrdp.) (oz) = 0.03215 ounce (Troy) (Tr.oz) = 15,432 grains (gr)
	1 ounce (avoirdupois; avrdp.) (oz) = 737.5 grains (gr)	1 pound (lb) = 0.4536 kilogram (kg)
	1 pound (Troy) (Tr.lb) = 12 ounces (Troy) (Tr.oz)	1 ounce (avrdp.) (oz) = 28.35 grams (g)
	1 ounce (Troy) (Tr.oz) = 20 pennyweight (dwt)	1 ounces (avrdp.) (oz) = 1.097 ounces (Troy) (Tr.oz)

UNITS OF MEASUREMENT, *cont'd.*

PRESSURE

1 pascal (Pa)	= 1 newton per square metre $(N \cdot m^{-2})$
	= 0.0209 $lb \cdot ft^{-2}$
1 $kg \cdot cm^{-2}$	= 0.96784 atmosphere (atm)
	= 14.2233 $lb \cdot in^{-2}$
	= 0.98067 bar
1 bar	= 0.98692 atmosphere (atm)
	= 10^5 pascals (Pa)
	= 10^2 kilopascals (kPa)
	= 14.7 $lb \cdot in^{-2}$
	= 1.02 $kg \cdot cm^{-2}$
1 $lb \cdot in^{-2}$	= 6.895 $kN \cdot m^{-2}$
20.90 $lb \cdot ft^{-2}$	= 1 $kN \cdot m^{-2}$
	= 0.145 $lb \cdot in^{-2}$

FORCE

1 N	= 0.2248 lb
1 kN	= 224.8 lb
1 lb	= 4.448 N
1 dyne	= 1×10^{-5} N
	= 2.25×10^{-6} lb
	= 1 $(g\text{-}cm) \cdot s^{-2}$

TEMPERATURE

To change from Celsius (C) to Fahrenheit (F):

$$°F = (°C \times 1.8) + 32°$$

To change from Fahrenheit (F) to Celsius (C):

$$°C = \frac{(°F - 32°)}{1.8}$$

ENERGY and POWER

Energy

1 joule (J)	= 1 newton-metre (N-m)
	= 2.390×10^{-1} calorie (cal)
	= 9.47×10^{-4} British thermal units (BTU)
	= 2.78×10^{-7} kilowatt-hour (kWh)
1 calorie (cal)	= 4.184 joule (J)
	= 3.968×10^{-3} BTU
	= 1.16×10^{-6} kWh
1 British thermal unit	= 1055.87 J
	= 252.19 cal
	= 2.928×10^{-4} kWh
1 kilowatt-hour (kWh)	= 3.6×10^6 J
	= 8.60×10^5 cal
	= 3.41×10^3 BTU

Power (energy per unit time)

1 watt (W)	= 1 joule per second $(J \cdot s^{-1})$
	= 3.4129 BTU per hour
	= 1.341×10^{-3} horsepower (hp)
	= 14.34 calories per minute
1 horsepower (hp)	= 7.46×10^2 watts (W)

VELOCITY and ACCELERATION

1 $m \cdot s^{-1}$	= 3.28 feet per second
1 $cm \cdot s^{-1}$	= 1.97 feet per minute
1 foot per second	= 0.305 $m \cdot s^{-1}$
1 mile per hour	= 1.61 kilometres per hour
1 $m \cdot s^{-2}$	= 3.28 feet per square second
1 foot per square second	= 0.3048 $m \cdot s^{-2}$

INDEX

Note: Page numbers in *italics* indicates figures and/or tables.